Evolution

Third Edition

Evolution
Making Sense of Life

Douglas J. Emlen   Carl Zimmer

Third Edition

Evolution

Making Sense of Life

Douglas J. Emlen | Carl Zimmer

macmillan learning

Austin • Boston • New York • Plymouth

Vice President, STEM: Daryl Fox
Editorial Program Director: Andrew Dunaway
Senior Program Manager: Jennifer Edwards
Director of Development: Lisa Samols
Senior Development Editor: Debbie Hardin
Visual Development Editor: Emiko-Rose Paul
Editorial Assistants: Casey Blanchard, Justin Jones
Marketing Manager: Will Moore
Director of Content: Jennifer Driscoll Hollis
Senior Media Editor: Mary Tibbets
Media Editor: Doug Newman
Director of Digital Production: Keri deManigold
Senior Media Project Manager: Chris Efstratiou
Director, Content Management Enhancement: Tracey Kuehn
Senior Managing Editor: Lisa Kinne
Senior Workflow Project Manager: Paul Rohloff
Senior Content Project Manager: Edward Dionne
Project Manager: Kathi Townes
Director of Design, Content Management: Diana Blume
Design Services Manager: Natasha Wolfe
Interior Design: Patrice Sheridan
Art Manager: Matthew McAdams
Illustrations: Emiko-Rose Paul
Cover Design Manager: John Callahan
Permissions Manager: Jennifer MacMillan
Photo Researcher: Krystyna Borgen, Lumina Datamatics, Inc.
Permissions Associate: Michael McCarty
Production Supervisor: Jose Olivera
Composition: Lumina Datamatics, Inc.
Printing and Binding: LSC Communications
Front Cover and Title Page Images: photos_martYmage / iStock / Getty Images
Back Cover Image: Art Wolfe / Science Source
Banner Photo: Wila_Image / Shutterstock

ISBN-13: 978-1-319-07986-4
ISBN-10: 1-319-07986-5

Library of Congress Control Number: 2019937273

Printed in the United States of America
1 2 3 4 5 6   24 23 22 21 20 19

Macmillan Learning
One New York Plaza
Suite 4600
New York, NY 10004-1562
www.macmillanlearning.com

In 1946, William Freeman founded W. H. Freeman and Company and published Linus Pauling's *General Chemistry*, which revolutionized the chemistry curriculum and established the prototype for a Freeman text. W. H. Freeman quickly became a publishing house where leading researchers can make significant contributions to mathematics and science. In 1996, W. H. Freeman joined Macmillan and we have since proudly continued the legacy of providing revolutionary, quality educational tools for teaching and learning in STEM.

To Kerry, Cory, and Nicole, the center of my beautiful world. — D. J. E.

*To Grace, who awakens me to the life around us, and to
Charlotte and Veronica, who have grown up amid my books. — C. W. Z.*

Brief Contents

Contents

Wolfgang Poelzer/Getty Images

Steve Brusatte

U.S. Fish and Wildlife Service

Mark Kostich / Getty Images

10 Natural Selection: Empirical Studies in the Wild 352

11 Sex: Causes and Consequences 390

Ray Hennessy/Shutterstock

Vladone/iStock/Getty Images

15 Intimate Partnerships: How Species Adapt to Each Other 548

16 Brains and Behavior 588

Paul Freed / Animals Animals—Earth Scenes

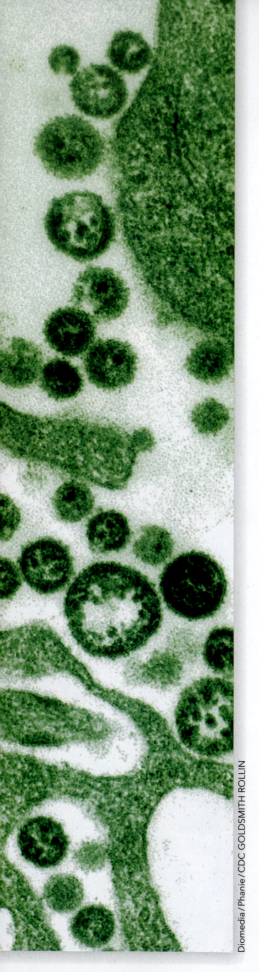

Diomedia/Phanie/CDC GOLDSMITH ROLLIN

Preface

About 40,000 years ago, someone entered a cave on the island of Borneo and drew a portrait of a thick-bodied, spindly legged animal. In 2015, a team of researchers climbed jungle-covered mountains to the same cave and discovered the image. When the researchers dated it, they realized that it was the oldest known example of figurative art ever made by humans—a profoundly important milestone in the history of our species. It indicates that by 40,000 years ago our ancestors had gained the ability to think symbolically—a capacity not found in other species (Aubert et al. 2018).

Researchers are uncovering such extraordinary chapters in the history of life at an astonishing pace. This ongoing work made creating the Third Edition of *Evolution: Making Sense of Life* an especially enjoyable experience for us. News of the Bornean cave art came to light too late to appear in the second edition of this textbook, published in 2016. But we've added that study to the Third Edition, along with dozens of others on a range of evolutionary topics that have come out over the past few years.

Although we've infused the Third Edition of *Evolution: Making Sense of Life* with many changes, we have not wavered from our original mission: to create a textbook that would capture the imagination of biology majors and inspire an abiding curiosity about evolution.

First and foremost, we have strived to create an engaging narrative. Narratives can draw students into scientific subjects and help them see how scientists actually do their research (Hillis 2007; Avraamidou and Osborne 2009). Throughout *Evolution*, we present stories about evolutionary biologists and how they've advanced our understanding of the history of life. We recount how these scientists set out to test hypotheses, how they developed experiments or went in search of fossils, and how they interpreted their results.

In telling these stories, we demonstrate the integrative nature of evolution. We show how scientists weave together different scientific methods and concepts to gain new insights about the history of life and the evolutionary process. Some stories will resonate with students because they are adventures—traveling to the Arctic in search of our earliest tetrapod relatives, for example. Other stories, of researchers fighting emerging diseases, highlight the vital importance of evolution to our global health.

Another strategy we use is to organize the chapters so that students become familiar, step by step, with the fundamental aspects of evolution. We then weave these lessons together to help students understand increasingly complex concepts. Rather than encountering ideas once and then never meeting them again, students revisit key examples of evolution several times, gaining a deeper understanding with each new discussion. When introducing the mechanisms of inheritance in Chapter 5, for example, we guide students from mutations to phenotypes, and we also discuss interactions between genes and the environment and phenotypic plasticity. By joining these concepts together in the minds of students, we help them better understand the links between genetic variation and phenotypic variation when we return to them later in the book.

We reinforce this understanding with a series of three chapters that delve into the details of how populations evolve, beginning with what we know from the action of mechanisms like selection and drift as they act on alleles (Chapter 6) and then turning to how these mechanisms operate at the level of phenotypes (Chapter 7). We then devote an entire chapter (Chapter 10) to field studies of natural selection on phenotypes in the wild. We illustrate these concepts with vivid examples—some from classic papers and others from more recent studies. In Chapter 7, for example, we introduce readers to the work of Hopi Hoekstra, of Harvard University, on the genetic basis of coat color in oldfield mice. We return to Hoekstra's research in the following chapter to show how she uses her insights into mouse genetics to learn how natural selection has produced strikingly different coat colors over just a few thousand years. We chose Hoekstra's work and the other examples with great care, looking for case studies that effectively convey the essential links between genotype and phenotype we have built over these chapters.

Another theme of *Evolution* is the importance of thinking in terms of phylogenetics, what many instructors call "tree thinking," throughout evolutionary biology. We introduce phylogeny in the first chapter and then incorporate it into almost every chapter in the book. We introduce phylogenetic methods to students in two parts, incrementally building their understanding of this difficult yet vital aspect of evolutionary biology. In Chapter 4, we begin with the most fundamental phylogenetic concepts. Having introduced the fossil record in the previous chapter, we demonstrate how scientists can use morphological data from fossils and extant taxa to reconstruct branching patterns. We also introduce students to the essential logic of reading trees as well as using trees to test explicit hypotheses about the deep past. In the process, we address the most common misconceptions students typically have about tree thinking.

After introducing students to molecular evolution, we revisit phylogeny in Chapter 8. There we show how molecular information is being used to build trees and to test hypotheses by using trees. By this stage, the basic concepts of building and using trees should be so familiar to students that we can begin discussing complex methods such as maximum likelihood and Bayesian inference. The genomes of organisms contain rich troves of information about the past, and scientists are now mining this information to learn about far more than phylogenetic relationships among species. Students will learn about the ways scientists explore the histories of individual genes and why, sometimes, the phylogeny of a gene may be very different from the phylogeny of the species that carry it. Students will also learn about the process of coalescence and how estimates of the coalescence times of alleles are being used by biologists to study historical changes in population size, as well as identify genes evolving in response to selection.

In Chapter 9, we combine many of these concepts to address the evolution of new adaptations. We examine the evolution of developmental biology—"evo-devo"—and give modern examples to illustrate how genetic variation can act to alter organismal phenotypes in biologically relevant ways. Using phylogenies, we illustrate how these adaptations evolve through stepwise tinkering rather than mysterious leaps.

Another important theme of *Evolution* is the origin—and current crisis—of biodiversity. We introduce the factors that drive the diversification of life, including sexual selection (Chapter 11) and coevolution (Chapter 15), and dedicate a chapter to how this diversification leads to speciation (Chapter 13). We explore macroevolution, showing how scientists reconstruct the ebbs and flows of diversity over hundreds of millions of years (Chapter 14). Here we look at patterns of diversity across time as well as biogeographic patterns across space. We demonstrate how such paleontological studies are now informing conservation biology by showing how ecosystems have collapsed in the past.

We also weave the theme of behavioral evolution throughout the book. In chapters on sexual selection (Chapter 11) and parental care and life history (Chapter 12), we illuminate numerous ways in which behaviors raise or lower reproductive success. We demonstrate how behavior plays a pivotal role in speciation (Chapter 13) as well as in the coevolution of ecological partners (Chapter 15). We then return to these themes in a chapter on the evolution of behavior (Chapter 16), where we examine memory, learning, and the costs and benefits of social behavior. This chapter opens the way for a chapter on human evolution, which focuses on the origin of uniquely human behavior such as language and complex toolmaking (Chapter 17).

By the time students reach Chapter 17, they have already encountered our own species many times in *Evolution*. Students are naturally curious about *Homo sapiens*, and we take advantage of their interest to illustrate many different concepts in evolution. We discuss human examples of population bottlenecks (the descendants of the crew of the *Bounty*) and of inbreeding (the fall of the Hapsburg dynasty in Spain). In Chapter 17, we bring together many different strands of evidence—from genomics to psychology, neuroscience, anthropology, and other disciplines—to present a synthesis of the latest advances in the study of human evolution.

We then end the book, in Chapter 18, with a look at medicine. We show students how medicine is, in fact, a form of applied evolution. We explore the origin of new diseases, the dynamics of antibiotic resistance, and the evolution of virulence in pathogens.

We also examine the vulnerabilities of the human body through an evolutionary perspective, considering disorders ranging from cancer to allergies. In addition, we show how scientists use insights from evolution to investigate new avenues of treatment.

Along with the structure of the book and the narrative style, we have also used a wealth of illustrations to draw students into the story of evolution and to convey complex concepts. To integrate multiple lines of evidence, for example, we use innovative "evograms." First proposed by Kevin Padian, of the University of California at Berkeley (Padian 2008), evograms present not just the phylogeny of species but also homologous traits in each lineage, some of the major transitions in their diversification, and a timeline showing the ages of relevant fossils. *Evolution* is also replete with striking images of organisms, both alive and extinct, to convey to students the rich tapestry of life that evolutionary biology helps to explain.

Our narrative approach can inspire students without sacrificing intellectual heft. While we explain most concepts verbally in the main text of the chapters, we also go into mathematical detail in stand-alone boxes. To balance our focus on the stories of individual scientists, we provide topical overviews. We also introduce and define the essential scientific terms that students will need to explore the literature on their own. We help students keep track of their own progress with key concepts at the end of each section. At the end of each chapter, we provide study questions and an extensive list of references for students who want to explore further.

Updates for the Third Edition

Throughout the Third Edition, we have made a number of improvements to text and art. We have updated chapters with new research, we have rearranged some material for better flow, and we have also strengthened our discussions of several concepts. Here is a select list of changes:

Chapter 1:
- New research on the H7N9 virus
- Updated discussion on cetacean brains

Chapter 2:
- Streamlined illustrations throughout

Chapter 3:
- New worked example dating Earth
- Heavily revised section on early Earth—primordial disk formation, early bombardments, crust formation, plate tectonics

Chapter 4:
- New chapter-opening story on *Zhenyuanlong suni*
- Updated discussion on divergence of common ancestors of living mammals

Chapter 5:
- Revised illustration to clarify protein structure

Chapter 6:
- New discussion on the debate among scientists regarding the importance of natural selection versus genetic drift

Chapter 7:
- Extensive new section on physical linkage, linkage disequilibrium, and inversions maintaining linkage disequilibrium

Chapter 8:
- Rearranged Chapters 8 through 10 so that the discussion of the history of genes and adaptation falls immediately after the discussion of quantitative genetics
- New section on gene trees
- New section on coalescence
- New discussion on human adaptation to high elevation
- New discussion on how the cancer-linked gene *BRCA1* has experienced strong natural selection

Chapter 9:
- New discussion on gene control regions
- New example of bacteria surviving harsh conditions by producing spores
- New discussion on adaptations in bacteria and archaea, including the role of horizontal gene transfer
- Updated discussion of the evolution of the nervous system through changes in developmental regulation

Chapter 10:
- Updated discussion of how climate change affects snowshoe hares in the Pacific Northwest
- Updated coverage of the evolution of resistance to glyphosate in weeds

Chapter 11:
- Streamlined illustrations throughout

Chapter 12:
- Updated discussion of menopause, including new research on whales

Chapter 13:
- Revised discussion on Lord Howe Island palm trees as an example of sympatric speciation
- New section on hybridization and the origin of a new species
- Updated discussion of microbial speciation, including the role of horizontal gene transfer

Chapter 14:
- Entire chapter extensively rewritten for greater clarity
- Revised discussion of key innovations with new examples from flowers and insects
- New discussion on punctuated equilibria
- New discussion on Cambrian explosion
- New discussion on recent studies focusing on volcanoes as a cause of extinction
- Updated IUCN data on extinctions and extinction rate

Chapter 15:
- Updated examples throughout

Chapter 16:
- New discussion about the debate over the phylogeny of sponges and ctenophores, and its implications for the evolution of the animal nervous system
- New discussion of ecological and social intelligence

Chapter 17:
- Heavily rewritten chapter
- New phylogeny of primates and new hominin taxa chart
- New material on recently discovered fossils of early *Homo*
- New discussion tracing first toolmakers back at least 3.3 million years ago
- Revised discussion of the history of *Homo sapiens*, pushing back human species by about 100,000 years
- New discussion of ancient DNA, describing new studies on the origin and inter-breeding of Neanderthals, Denisovans, and modern humans
- New focus on recent research on the evolution of the human brain
- New discussion on the evolutionary history of oxytocin and vasopressin

Chapter 18:
- Heavily rewritten chapter
- New chapter-opener about Harvard University biologist Pardis Sabeti and her pioneering work on the evolution of diseases like Lassa fever
- Revised discussion of evolution of pathogens within individual hosts, featuring new research on HIV and cystic fibrosis
- Revised discussion on bacterial resistance to antibiotics

- Revised discussion of the evolution of influenza, with new research on the origins of the 1918 pandemic
- New discussion on long-term coevolution between viruses and virus-evading host proteins
- New discussion on antagonistic pleiotropy and how it relates to diseases in older humans
- New discussion on how diet choices humans make are informed by our evolutionary past
- Introduction of the Old Friends hypothesis to replace the hygiene hypothesis
- Revised discussion on cancer evolving resistance to chemotherapy
- New showcase of applications of evolutionary medicine, including vaccines, disease forecasting, and evolution-proof antibiotics

Additional Resources

For this edition, we are adding exciting new videos that augment the book coverage of topics many students find difficult. We include seven whiteboard videos that walk students through the most difficult quantitative reasoning in evolution, including Hardy-Weinberg, mathematical models of drift, quantitative inheritance, and the breeder's equation—all written and presented by Bronwyn Bleakley, associate professor of biology at Stonehill College and former National Science Foundation international research fellow.

We also provide animated tutorial videos for this edition that further explain topics that students sometimes struggle with, including reading phylogenetic trees, natural selection and population size, mechanisms of speciation, and genetic drift.

This Third Edition of *Evolution: Making Sense of Life* is accompanied by a media and supplements package that facilitates student learning and enhances the teaching experience. Fully loaded with our interactive e-book and all student and instructor resources, SaplingPlus has a variety of assessments, videos, and animations aligned with each chapter of the text.

Created and supported by the authors and other educators, SaplingPlus's instructional online homework drives student success and saves educators time. Every homework problem contains hints, answer-specific feedback, and solutions to ensure that students find the help they need.

In addition, *Evolution* is accompanied by a supplemental chapter on creationism, religion, and evolution, by Louise S. Mead, PhD, the Education Director of the BEACON Center for the Study of Evolution in Action at Michigan State University. Professors wishing to spend time addressing these important issues will find much useful information there. The booklet is available as a free PDF. Please contact your sales representative for access to this chapter.

• • •

The Third Edition of *Evolution* reflects our shared vision for what modern textbooks can be: exciting, relevant, concept oriented, and gorgeously illustrated. We want it to be a reading adventure that grabs students' imagination and shows them exactly why it is that evolution makes such brilliant sense of life.

Douglas J. Emlen
Carl Zimmer

References

Aubert, M., P. Setiawan, A. A. Oktaviana, A. Brumm, P. H. Sulistyarto, et al. 2018. Palaeolithic Cave Art in Borneo. *Nature* 564: 254–7.

Avraamidou, L., and J. Osborne. 2009. The Role of Narrative in Communicating Science. *International Journal of Science Education* 31 (12): 1683–1707.

Hillis, D. M. 2007. Making Evolution Relevant and Exciting to Biology Students. *Evolution* 61 (6): 1261–4.

Padian, K. 2008. Trickle-Down Evolution: An Approach to Getting Major Evolutionary Adaptive Changes into Textbooks and Curricula. *Integrative and Comparative Biology* 48: 175–88.

Steve Brusatte

The Latest Research in Evolutionary Biology

Douglas Emlen and Carl Zimmer have painstakingly reviewed the latest research in evolution and incorporated these updates in the text to ensure students see evolution for the dynamic and exciting discipline that it is.

Updates include the following:

- Formation of Earth and early fossil record of life
- Revised approach to coalescence
- Promiscuous proteins
- Looking at menopause by studying whales
- Revisiting the Cambrian explosion

- New fossils of hominins
- Evolutionary history of oxytocin and vasopressin
- Genome sequencing in HIV patients
- Antagonistic pleiotropy and Alzheimer's Disease + Human Evolution

CHAPTER 4

The Tree of Life

Elegant design and spectacular photos and illustrations throughout engage students—making the reading experience accessible and fun.

How Biologists Use Phylogeny to Reconstruct the Deep Past

Learning Objectives

- Construct a simple phylogeny and identify the different components.
- Explain how different lines of evidence can lead to different conclusions about species' taxonomical relationships.
- Discuss how scientists can determine the timing of branching events.
- Explain how phylogenies can be used to develop hypotheses about the evolution of tetrapods.
- Explain how the bones of the middle ear can be used to trace the evolution of mammals.
- Discuss exaptations and explain their relevance in the evolution of birds.
- Name three examples of convergent evolution.

I n November 2014, Steven Brusatte stepped into Beijing's bustling central railway station. Even though he spoke virtually no Mandarin, he managed to find a train headed for the city of Jinzhou. For the next four hours, Brusatte and a colleague, Junchang Lü, from the Chinese Academy of Geological Sciences, traveled north, almost as far as the border with North Korea. They had a date to see a dinosaur.

Brusatte, who teaches at the University of Edinburgh, has dug up dinosaur fossils in many parts of the world (**Figure 4.1**). He has studied those fossils to address many questions about dinosaurs—how they first evolved some 240 million years ago, how they rose to dominate terrestrial ecosystems, and how they were decimated in a mass extinction 66 million years ago. But there are still thousands of species of dinosaurs with us today (look into the branches of any tree and you may see some), and Brusatte and Lü hoped that the fossil they were traveling to see would shed light on how dinosaurs managed to endure.

Zhenyuanlong was a ground-running dinosaur with feathers that lived about 125 million years ago. The phylogeny of such dinosaurs reveals that they were the ancient relatives of birds and helps us understand how the unique body plan of birds gradually evolved.

Steve Brusatte

99

Built-in pedagogy makes rigorous science

Figure 6.17 By sequencing the genomes of bacteria sampled every 500 generations, Lenski and his colleagues were able to reconstruct a complete picture of the evolution of each of their experimental populations, documenting the appearance of every new mutation and the fates of each of these new alleles. Here we show changes in relative frequency over time for all of the alleles that arose in one of their experimental populations (gray lines). A: Some of these alleles proved to be beneficial, eventually becoming fixed in the population by selection (colored lines). B: Other alleles arose and persisted for a while, but were eventually lost either because their effects were deleterious to the bacteria or simply due to the action of genetic drift. Studies like this provide an unusually comprehensive view of the mechanisms of evolution in action. (Data from Good et al. 2017)

It's not immediately obvious how the mutation benefits the bacteria, but Lenski and his colleagues have some clues. In bacteria with the *BoxG1* mutation, less GlmS and GlmU is expressed. It's possible that the bacteria divert resources from building thick membranes to other functions, speeding up their reproduction.

The *BoxG1* mutation is just one of a growing collection of beneficial mutations that Lenski and his colleagues have identified in their long-term evolution experiment. These mutations arose sequentially in the bacterial lines, building on the increased fitness of previous mutations.

Lenski's team has also documented other alleles that arose, persisted for a period of time, and eventually disappeared. Figure 6.17 shows the fates of new beneficial and deleterious alleles for one of the 12 experimental populations, providing an unusually clear picture of 60,000 generations of evolution.

Large-scale comparisons of these mutations are revealing lessons about how beneficial mutations interact. Some mutations, for example, are beneficial only when they follow certain other mutations. That's because their effects on the bacteria interact with alleles at other loci in a process known as **epistasis**.

Finally, Lenski's experiment also makes it possible to observe how predictable evolution is at the molecular level. Only one line of *E. coli* evolved the *BoxG1* mutation. But other mutations arose independently in several different lines, and the scientists found three genes that mutated in all 12 lines. Although evolution moved in the same overall direction in their experiment—a rapid increase in fitness followed by a tapering off—the mutations that drove this change were not the same. We'll revisit the contingency and convergence of adaptations in Chapter 9.

Epistasis occurs when the effects of an allele at one genetic locus are modified by alleles at one or more other loci.

Key Concepts
- Selection occurs whenever genotypes differ in their relative fitnesses.
- The outcome of selection depends on the frequency of an allele as well as its effects on fitness.
- Both drift and selection are potent mechanisms of evolution, and their relative importance depends critically on population size. When populations are very small, the

200 CHAPTER 6 / THE WAYS OF CHANGE

Key Terms and their definitions appear in the margins just at the time students need them.

Key Concepts appear at the end of each section, encouraging students to test their understanding as they go.

New to This Edition!

Interpret the Data, a quantitative assessment at the end of each chapter designed to help students increase their scientific literacy

INTERPRET THE DATA Answer can be found at the end of the book.

11. Which of the following statements about the graph that follows is *not true*?

Evolution of beak size

a. The drought of 1977 caused a dramatic increase in beak size.
b. Mean beak size remained constant for medium ground finches from 1975 to 1985.
c. By the early 2000s, mean beak sizes were close to pre-drought sizes.
d. Mean beak size in ground finches fluctuated from 1990 to 2000.

accessible and engaging

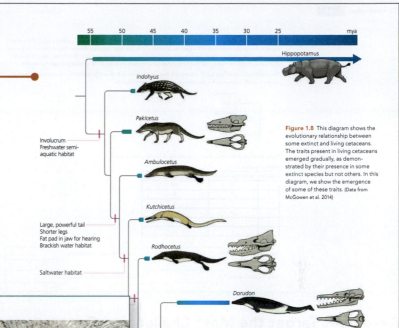

Figure 1.8 This diagram shows the evolutionary relationship between some extinct and living cetaceans. The traits present in living cetaceans emerged gradually, as demonstrated by their presence in some extinct species but not others. In this diagram, we show the emergence of some of these traits. (Data from McGowen et al. 2014)

Labels on diagram: Hippopotamus; Indohyus; Pakicetus — Involucrum, Freshwater semi-aquatic habitat; Ambulocetus; Kutchicetus — Large, powerful tail, Shorter legs, Fat pad in jaw for hearing, Brackish water habitat; Rodhocetus — Saltwater habitat; Dorudon; Basilosaurus; Odontocetes; Mysticetes

1.1 Whales: Mammals Gone to Sea 9

BOX 4.3

The Myth of the "Primitive"

For millennia, naturalists organized living things into "lower" and "higher" forms, envisioning life arrayed along a ladder-like scale. Darwin's metaphor of a tree of life is fundamentally opposed to this view of biology. Everything alive today, from bacteria to jellyfish to humans, is the product of a 3.7-billion-year-old lineage. There is nothing intrinsically superior about animals, let alone humans. Indeed, if we want to judge by numbers alone, the most successful organisms are viruses, which have an estimated population size of 10^{31} (Zimmer 2011).

When we look at phylogenies, we must be very careful not to slip into ladderlike thinking. Consider, for example, the mammal tree in Figure 4.27. We've arranged the phylogeny to make it easy for you to follow the series of steps by which the modern mammal body—including the middle ear—evolved. But we could just as easily have swung the nodes around in a different way, so that the tree seemed to lead to platypuses instead of eutherians like ourselves.

Because platypuses have so few close relatives, their position off by themselves on the side of the mammal tree may make them seem like they must be "primitive." It is certainly true that platypuses have some traits that are ancestral to those found in therian mammals. Platypuses lay eggs, for example, whereas marsupial and placental mammal embryos develop in the uterus, attached to a placenta. Platypuses produce milk from mammary glands, but they lack teats (Box Figure 4.3.1).

Still, these ancestral traits are only part of platypus biology. Platypuses have a unique bill, which they sweep through water to detect electrical signals from their prey. This bill evolved only after the ancestors of platypuses and the ancestors of the eutherian clade split from each other. Platypuses also have webbing on their feet and produce venom. All of these traits are likely derived in the platypus

Box Figure 4.3.1 Platypuses are mammals that lay eggs. That does not mean they are "primitive," however. (age fotostock/Superstock, Inc.)

lineage (Box Figure 4.3.2). These are just the tip of the iceberg when it comes to derived traits in platypuses. In 2008, an international team of scientists sequenced the platypus genome and found that it contained thousands of derived segments of DNA, including protein-coding genes that likely were involved in adaptations to their particular ecological niche (Warren et al. 2008).

We can certainly gain some clues to the nature of the common ancestor of platypuses and other mammals by comparing their biology and mapping the acquisition of traits on their phylogeny. But we cannot treat living platypuses as a stand-in for that ancestor. Indeed, when it comes to electroreceptors and other derived platypus traits, we humans are the "primitive" mammals (Omland et al. 2008).

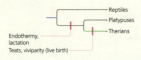
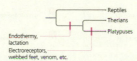

Left tree: Reptiles; Platypuses; Therians — Endothermy, lactation; Teats, viviparity (live birth)

Right tree: Reptiles; Therians; Platypuses — Endothermy, lactation; Electroreceptors, webbed feet, venom, etc.

Box Figure 4.3.2 For some traits, platypuses retain the ancestral state relative to therian mammals. But platypuses also have derived traits not found in therian mammals. We could build a similar tree using a "platypus-centric" set of characters and give the impression that our own species has not evolved much.

4.6 Homology as a Window into Evolutionary History 129

With Sapling Learning, every problem counts

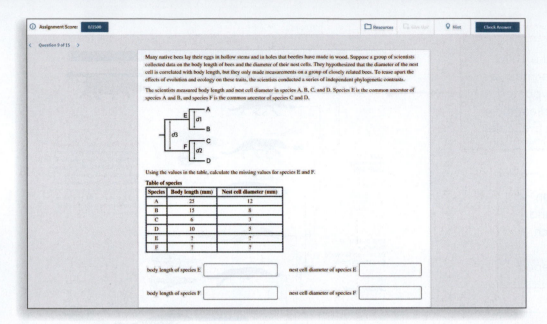

Resources Target the Most Challenging Concepts and Skills in the Course

Evolution: Making Sense of Life, Third Edition, is accompanied by a media and supplements package that facilitates student learning and enhances the teaching experience. Fully loaded with our interactive e-book and all student and instructor resources, SaplingPlus is organized around a set of pre-built units available for each chapter in *Evolution: Making Sense of Life*, Third Edition.

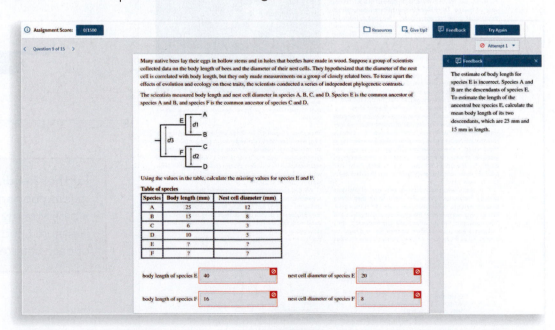

Created and supported by educators, SaplingPlus's instructional online homework drives student success and saves educators time. Every homework problem contains hints, answer-specific feedback, and solutions to ensure that students find the help they need.

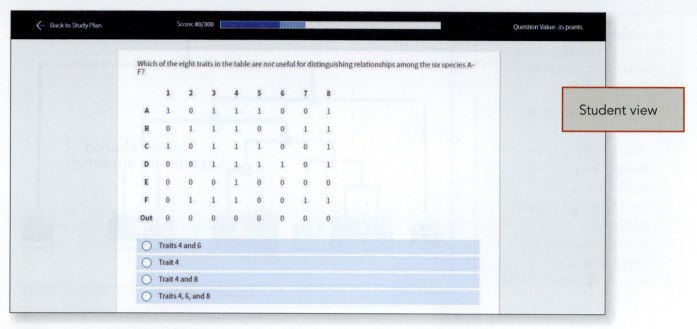

LearningCurve
macmillan learning

Put "testing to learn" into action. Based on educational research, LearningCurve really works: Game-like quizzing motivates students and adapts to their needs based on their performance. It is the perfect tool to get students to engage before class and review after class. Additional reporting tools and metrics help teachers get a handle on what their class knows and doesn't know. Available in SaplingPlus and Achieve Read & Practice.

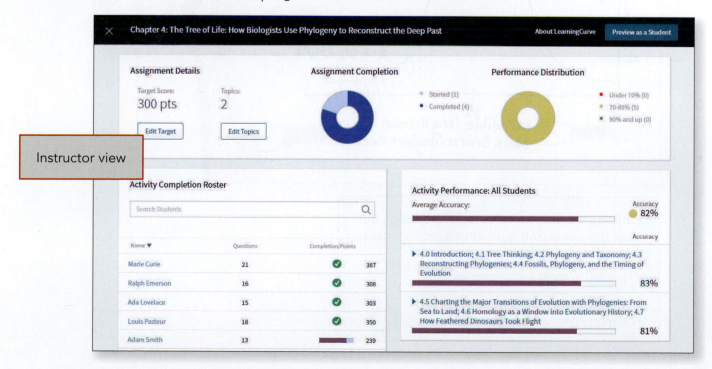

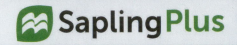

Concept tutorial videos reinforce teaching of the most difficult concepts in evolution, including

- Reading phylogenetic trees
- Hardy-Weinberg theorem
- Mechanisms of speciation
- Heritability and selection coefficients
- Natural selection and population size

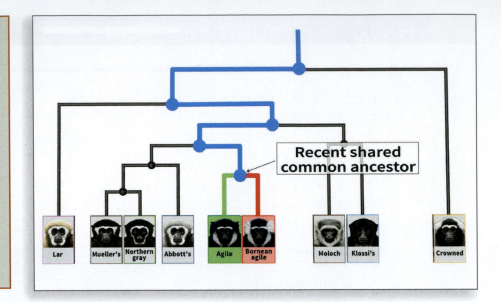

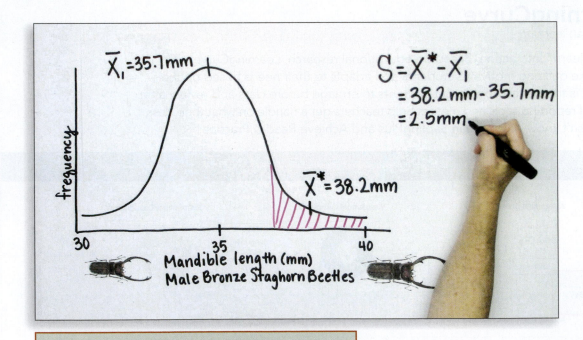

Whiteboard videos walk students through the most difficult quantitative coverage, including

- Mathematical models of drift
- Breeder's equation
- Quantitative inheritance
- Parent−offspring regression

Instructor Resources

The Hub of Active Learning

For the Third Edition, we've developed a suite of active learning resources to help instructors make their classes more engaging. Each chapter has a variety of activities based on the content in the text. Resources include handouts for students, instructor guides for each activity, and PowerPoint slideshows.

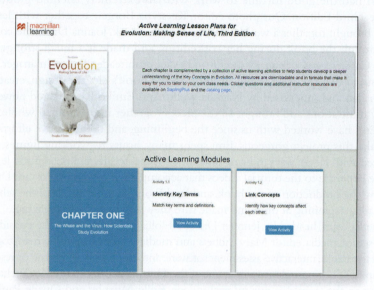

Test Bank

The Test Bank provides a wide range of questions appropriate for assessing student comprehension, interpretation, analysis, and synthesis skills. Each question is tagged for difficulty, level on Bloom's Taxonomy, book sections and number, as well as learning objectives.

Lecture Outlines for PowerPoint

Adjunct professors and instructors who are new to the discipline will appreciate these detailed companion lectures, perfect for walking students through the key ideas in each chapter. These rich, pre-built lectures make it easy for instructors to transition to the book.

Clicker Questions

Clicker questions allow instructors to jump-start discussions, illuminate important points, and promote better conceptual understanding during lectures.

i-clicker.

Get more out of your classroom with iClicker. iClicker's innovative solutions make it easier to track attendance, increase participation, facilitate quizzes, and measure performance.

Acknowledgments

We are grateful to the many people who helped us with content creation for *Evolution*, beginning with Debbie Hardin, our senior development editor. Debbie made the revision process not just possible but enjoyable. In addition to managing all of the many deadlines and keeping us constantly on track, Debbie solicited and analyzed feedback from external reviewers and made substantial improvements to the text and organization throughout. She also coordinated the production of online video resources developed for this new edition. Her help was absolutely invaluable. Our deep thanks also goes to our inexhaustible program manager, Jennifer Edwards. Jen has been our constant champion, advocating for our vision and moving aside any stumbling blocks along the way. Jen helped assemble the team that ultimately brought this Third Edition to fruition and was a valuable leader throughout the revision process. We also thank Andy Dunaway, who has been a steadfast supporter of the project and of its editorial team members.

We also had the good fortune to work with an extremely talented production team on the Third Edition. Our photo editor, Krystyna Borgen, together with Jennifer MacMillan, brought together a wealth of images for the book. Joanna Dinsmore combed carefully through the book during her copyediting and made it a much stronger volume. Karen Carriere was an equally careful and exact proofreader. Project manager Kathi Townes ensured that the page layout lived up to the book's elegant design, created by Patrice Sheridan and supervised by Natasha Wolfe. And Emiko Paul created powerfully explanatory illustrations and brought all the pieces together into a unified whole. Both Kathi and Emi have worked with us since the beginning, and their creative efforts continue to make this project gorgeous. Emi spent many hours re-working figures, as we experimented with clever ways to bring difficult concepts to life, and in addition lent her considerable expertise to the tutorial videos that accompany this edition. Edward Dionne, who oversaw the production of the book, and Paul Rohloff, who managed the schedule, kept all the plates spinning at once and made sure they landed without breaking.

Our media team, headed by Jennifer Driscoll Hollis and comprising the considerable energies of senior media editor Mary Tibbets and media editor Doug Newman, together coordinated test banks, interactive assessment software, and the invaluable instructor resources that make up the exciting online component of the Third Edition. Cheryl McCutchan lent her expertise to the scripts of our tutorial videos. Kelly Murphy and Quade Paul were innovative contributors to the subsequent storyboards and videos as well. We also thank the amazing Bronwyn Bleakley for writing and performing in the whiteboard videos, and Tyler DeWitt for producing them. Lee Hershey helped curate our active learning exercises.

To all of our team at Macmillan Learning, we extend our deepest thanks.

To ensure that our textbook continues to serve as an accurate introduction to the vast scope of evolutionary biology, we consulted closely with professors to check the accuracy of the material and to make sure it was helpful to students. Our deep appreciation goes to all the people who assisted us in reviewing the Third Edition:

Book Reviewers:

Eric Baack, *Luther College*

Stacy Bennetts, *Augusta University*

Randall Bernot, *Ball State University*

Mark W. Bland, *University of Central Arkansas*

Bronwyn H. Bleakley, *Stonehill College*

Stephen C. Burnett, *Clayton State University*

Douglas Causey, *University of Alaska, Anchorage*

Norman A. Douglas, *University of Florida*

Anthony Friscia, *University of California, Los Angeles*

Sharon L. Gillies, *University of the Fraser Valley*

Joel Gramling, *The Citadel*

David Gray, *California State University, Northridge*

Mandy J. Guinn, *United Tribes Technical College*

Charles T. Hanifin, *Utah State University, Uintah Basin Campus*

Michelle H. Hersh, *Sarah Lawrence College*

David Hoekman, *Southern Nazarene University*

Colin Hughes, *Florida Atlantic University*

Christopher T. Ivey, *California State University, Chico*

Edgar Lehr, *Illinois Wesleyan University*

Jeffrey Marcus, *University of Manitoba*

Luana S. Maroja, *Williams College*

Joel McGlothlin, *Virginia Polytechnic Institute and State University*

Jeffrey Meldrum, *Idaho State University*

J. Jeffrey Morris, *University of Alabama at Birmingham*

Barbara Musolf, *Clayton State University*

Douglas Norman, *University of Florida*

Christopher O'Connor, *Maryville University*

Robin E. Owen, *Mount Royal University*

Tami M. Panhuis, *Ohio Wesleyan University*

Tom Ray, *University of Oklahoma*

Veronica Riha, *Madonna University*

Michael Russello, *University of British Columbia*

Amanda Solem, *Hastings College*

Chrissy Spencer, *Georgia Institute of Technology*

Andrew D. Stewart, *Canisius College*

Judy Stone, *Colby College*

Jared Strasburg, *University of Minnesota, Duluth*

Jeffrey Thomas, *Queens University of Charlotte*

Dustin Wilgers, *McPherson College*

Pamela Yeh, *University of California, Los Angeles*

Rebecca Zufall, *University of Houston*

Media Reviewers:

Susan Bailey, *Clarkson University*

Bonnie Bain, *Dixie State University*

Peter Bednekoff, *Eastern Michigan University*

Mark W. Bland, *University of Central Arkansas*

Bronwyn H. Bleakley, *Stonehill College*

Mirjana Milosevic Brockett, *Georgia Institute of Technology*

Edmund Brodie, *University of Virginia*

Mark A. Buchheim, *University of Tulsa*

Stephen Burnett, *Clayton State University*

Bruce J. Cochrane, *Miami University*

Mrinal Das, *Grant MacEwan University*

Norman A. Douglas, *University of Florida*

M. Caitlin Fisher-Reid, *Bridgewater State University*

Amy E. Faivre, *Cedar Crest College*

David H. A. Fitch, *New York University*

Kasey Fowler-Finn, *Saint Louis University*

Sharon L. Gillies, *University of the Fraser Valley*

Jennifer Gleason, *University of Kansas*

David Gray, *California State University, Northridge*

William Hahn, *George Mason University*

Mark Hammer, *Wayne State College*

Charles T. Hanifin, *Utah State University, Uintah Basin Campus*

Eric A. Hoffman, *University of Central Florida*

Colin Hughes, *Florida Atlantic University*

Susan M. Jenks, *The Sage Colleges*

Matthew Julius, *St. Cloud State University*

Stephen Kajiura, *Florida Atlantic University*

Kristopher B. Karsten, *California Lutheran University*

William Kristan, *California State University, San Marcos*

Sarah Maveety, *Brevard College*

Joel McGlothlin, *Virginia Polytechnic Institute and State University*

Jeffrey K. McKee, *The Ohio State University*

Louise S. Mead, *Michigan State University*

Dr. Jeffrey Meldrum, *Idaho State University*

Paulina Mena, *Central College*

J. Jeffrey Morris, *University of Alabama at Birmingham*

Martha Muñoz, *Virginia Polytechnic Institute and State University*

Christopher O'Connor, *Maryville University*

Robin E. Owen, *Mount Royal University*

Leslee A. Parr, *San Jose State University*

Raymond Pierotti, *University of Kansas*

Rebecca A. Pyles, *East Tennessee State University*

Veronica Riha, *Madonna University*

Eric Routman, *San Francisco State University*

Michael Russello, *University of British Columbia*

Jonathan Shaw, *Duke University*

Stephen A. Smith, *University of Michigan*

Christina D. Steel, *Old Dominion University*

Andrew D. Stewart, *Canisius College*

Bradley J. Swanson, *Central Michigan University*

Martin Tracey, *Florida International University*

E. Natasha Vanderhoff, *Jacksonville University*

Michael Whiting, *Brigham Young University*

Dustin Wilgers, *McPherson College*

Ken Yasukawa, *Beloit College*

Olga Zhaxybayeva, *Dartmouth College*

Rebecca Zufall, *University of Houston*

Many thanks also go to the reviewers of the first two editions.

Second Edition:

Windsor Aguirre, *DePaul University*

Lisa Belden, *Virginia Tech University*

Stewart Berlocher, *University of Illinois at Urbana–Champaign*

Annalisa Berta, *San Diego State University*

Gregory Bole, *University of British Columbia*

Jeffrey L. Boore, *University of California, Berkeley*

Brent Burt, *Stephen F. Austin State University*

Nancy Buschhaus, *University of Tennessee at Martin*

Douglas Causey, *University of Alaska Anchorage*

Robert Cox, *University of Virginia*

Maheshi Dassanayake, *Louisiana State University*

Robert Dowler, *Angelo State University*

Abby Drake, *Skidmore College*

Devin Drown, *University of Alaska Fairbanks*

David Fastovsky, *University of Rhode Island*

Charles Fenster, *University of Maryland*

Caitlin Fisher-Reid, *Bridgewater State University*

David Fitch, *New York University*

Jennifer Foote, *Algoma University*

Anthony Frankino, *University of Houston*

Barbara Frase, *Bradley University*

Nicole Gerlach, *University of Florida*

Jeff Good, *University of Montana*

Charles Goodnight, *University of Vermont*

Neil Greenspan, *Case Western Reserve University*

T. Ryan Gregory, *University of Guelph*

David Hale, *United States Air Force Academy*

Benjamin Harrison, *University of Alaska Anchorage*

Sher Hendrickson-Lambert, *Shepherd University*

David Hoferer, *Judson University*

Luke Holbrook, *Rowan University*

Elizabeth Jockusch, *University of Connecticut*

Charles Knight, *Cal Poly San Luis Obispo*

Patrick Krug, *California State University*

Simon Lailvaux, *University of New Orleans*

David Lampe, *Duquesne University*

Hayley Lanier, *University of Wyoming at Casper*

Kari Lavalli, *Boston University*

Amy Lawton-Rauh, *Clemson University*

Brian Lazzarro, *Cornell University*

Matthew Lehnert, *Kent State University Stark*

Kevin Livingstone, *Trinity University*

John Logsdon, *University of Iowa*

Patrick Lorch, *Kent State University*

J. P. Masley, *University of Oklahoma*

Lauren Mathews, *Worcester Polytechnic Institute*

Rodney Mauricio, *University of Georgia*

Joel McGlothlin, *Virginia Tech University*

Steve Mech, *Albright College*

Matthew Miller, *Villanova University*

Nathan Morehouse, *University of Pittsburgh*

James Morris, *Brandeis University*

Brian Morton, *Barnard College*

Barbara Musolf, *Clayton State University*

Mohamed Noor, *Duke University*

Steve O'Kane, *University of Northern Iowa*

Cassia Oliveira, *Lyon College*

Brian O'Meara, *University of Tennessee*

Daniel Pavuk, *Bowling Green State University*

Rob Phillips, *California Institute of Technology*

Marcelo Pires, *Saddleback College*

Patricia Princehouse, *Case Western Reserve University*

Sean Rice, *Texas Tech University*

Christina Richards, *University of South Florida*

Ajna Rivera, *University of the Pacific*

Sean Rogers, *University of Calgary*

Antonis Rokas, *Vanderbilt University*

Cameron Siler, *University of Oklahoma*

Sally Sommers Smith, *Boston University*

Chrissy Spencer, *Georgia Tech*

Joshua Springer, *Purdue University*

Christina Steel, *Old Dominion University*

Judy Stone, *Colby College*

Thomas Turner, *University of California, Santa Barbara*

Steve Vamosi, *University of Calgary*

Matthew White, *Ohio University*

Christopher Wills, *University of California, San Diego*

Peter Wimberger, *University of Puget Sound*

Christopher Witt, *University of New Mexico*

Lorne Wolfe, *Georgia Southern University*

Danielle Zacherl, *California State University, Fullerton*

Robert Zink, *University of Minnesota*

First Edition:

John Alcock, *Arizona State University*

Diane Angell, *St. Olaf College*

Robert Angus, *University of Alabama at Birmingham*

Dan Ardia, *Franklin & Marshall College*

Steven N. Austad, *University of Texas Health Science Center, San Antonio*

Christopher C. Austin, *Louisiana State University*

Lisa Belden, *Virginia Tech University*

Antoine D. Bercovici, *Lund University*

Neil Blackstone, *Northern Illinois University*

Bronwyn H. Bleakley, *Stonehill College*

Gregory Bole, *University of British Columbia*

Jeremy Bono, *University of Colorado, Colorado Springs*

Helen C. Boswell, *Southern Utah University*

Sarah Boyer, *Macalester College*

Christopher G. Brown, *Shorter University*

Stephen C. Burnett, *Clayton State University*

Diane L. Byers, *Illinois State University*

Bryan C. Carstens, *Louisiana State University*

Ashley Carter, *Long Beach State University*

Marty Condon, *Cornell College*

Jennifer M. Dechaine, *Central Washington University*

Jacobus de Roode, *Emory University*

Eric Dewar, *Suffolk University*

Siobain Duffy, *Rutgers, The State University of New Jersey*

Jacob Egge, *Pacific Lutheran College*

Stephen T. Emlen, *Cornell University*

William J. Etges, *University of Arkansas*

Paul Ewald, *University of Louisville*

Charles B. Fenster, *University of Maryland*

Anthony Frankino, *University of Houston*

Jessica Garb, *University of Massachusetts Lowell*

George Gilchrist, *The College of William and Mary*

Matt Gilg, *University of North Florida*

Jennifer Gleason, *University of Kansas*

Kenneth Gobalet, *California State University, Bakersfield*

Jeffrey Good, *University of Montana*

David A. Gray, *California State University, Northridge*

Katherine R. Greenwald, *Eastern Michigan University*

Shala J. Hankison, *Ohio Wesleyan University*

Luke Harmon, *University of Idaho*

Scott Henry Harrison, *North Carolina A&T State University*

Hopi Hoekstra, *Harvard University*

Brett Holland, *California State University, Sacramento*

Christopher T. Ivey, *California State University, Chico*

Rebecca Jabbour, *St. Mary's College (CA)*

Jeff A. Johnson, *University of North Texas*

Michele Johnson, *Trinity University*

Nicole King, *University of California, Berkeley*

Jacob Koella, *Imperial College London*

Robert A. Krebs, *Cleveland State University*

Patrick J. Krug, *California State University, Los Angeles*

Lori LaPlante, *Saint Anselm College*

John Logsdon, *Iowa State University*

Jonathan B. Losos, *Harvard University*

Katy Lustofin, *Marietta College*

Tim Maret, *Shippensburg University*

Luana Maroja, *Williams College*

Andy Martin, *University of Colorado, Boulder*

Gregory C. Mayer, *University of Wisconsin–Parkside*

Andrew C. McCall, *Denison University*

John McCormack, *Louisiana State University*

Louise Mead, *Michigan State University*

Jeff Meldrum, *Idaho State University*

Tamra Mendelson, *University of Maryland, Baltimore County*

James Morris, *Brandeis University*

Barbara Musolf, *Clayton State University*

Cynthia G. Norton, *St. Catherine University*

Patrik Nosil, *University of Colorado, Boulder*

Karen Ober, *College of Holy Cross*

Kevin Omland, *University of Maryland, Baltimore County*

Chris Organ, *University of Utah*

Kevin Padian, *University of California, Berkeley*

Leslee Parr, *San Jose State University*

Adrian Paterson, *Lincoln University*

Manus Patten, *Georgetown University*

Kathryn E. Perez, *University of Wisconsin–La Crosse*

Richard Phillips, *Wittenberg University*

David Pindel, *Corning Community College*

Gordon Plague, *Fordham University*

Colin Purrington, *Swarthmore College*

Stan Rachootin, *Mt. Holyoke College*

David Reznick, *University of California, Riverside*

Leslie Rissler, *University of Alabama*

Fred Rogers, *Franklin Pierce University*

Sean M. Rogers, *University of Calgary*

Eric Routman, *San Francisco State University*

Scott Sampson, *Utah Museum of Natural History*

Jeffrey D. Silberman, *University of Arkansas*

Dawn Simon, *University of Nebraska Kearny*

Erik Skully, *Towson University*

Scott Solomon, *Rice University*

Theresa Spradling, *University of Northern Iowa*

Nancy Staub, *Gonzaga University*

Judy Stone, *Colby College*

Lena Struwe, *Rutgers, The State University of New Jersey*

Bradley J. Swanson, *Central Michigan University*

Ian Tattersall, *American Museum of Natural History*

Hans Thewissen, *Northeast Ohio Medical University*

Chad Thompson, *Westchester Community College*

John Thompson, *University of California, Santa Cruz*

Robert Thomson, *University of Hawaii at Manoa*

Joe Thornton, *University of Oregon*

Stephen G. Tilley, *Smith College*

Steven M. Vamosi, *University of Calgary*

Sara Via, *University of Maryland*

Donald Waller, *University of Wisconsin–Madison*

Joseph Walsh, *University of Connecticut Medical School*

Colleen Webb, *Colorado State University*

Alexander Werth, *Hampden-Sydney College*

Chris Whit, *University of New Mexico*

Lisa Whitenack, *Allegheny College*

Michael Whitlock, *University of British Columbia*

Justen Whittall, *Santa Clara University*

Jeannette Whitton, *University of British Columbia*

Barry Williams, *Michigan State University*

Paul Wilson, *California State University, Northridge*

Peter Wimberger, *University of Puget Sound*

Wade B. Worthen, *Furman University*

Stephen Wright, *University of Toronto*

Sam Zeveloff, *Weber State University*

Kirk Zigler, *University of the South*

Rebecca Zufall, *University of Houston*

About the Authors

Douglas J. Emlen is a professor at the University of Montana. He is a fellow of the American Academy of Arts and Sciences and a recipient of the U.S. Presidential Early Career Award for Scientists and Engineers, multiple research awards from the National Science Foundation including its five-year CAREER award, and the E. O. Wilson Naturalist Award from the American Society of Naturalists. In 2014 he was awarded UM's Distinguished Teaching Award and in 2015 the Carnegie/CASE Professor of the Year Award for the state of Montana. His 2014 book *Animal Weapons: The Evolution of Battle* won the Phi Beta Kappa Award in Science, and he recently starred in documentaries about his work on BBC (Nature's Wildest Weapons) and NOVA (Extreme Animal Weapons).

Mark Finkenstaedt

Carl Zimmer is one of the country's leading science writers. A columnist for the *New York Times*, he is the author of 13 books, including *She Has Her Mother's Laugh: The Powers, Perversions, and Potentials of Heredity*, which *The Guardian* named the best science book of 2018. Zimmer is professor adjunct at Yale University, where he teaches science writing. Among his many honors, Zimmer has won the Stephen Jay Gould Prize, awarded by the Society for the Study of Evolution, and the National Association of Biology Teachers Distinguished Service Award.

Mistina Hanscom

CHAPTER 1

The Whale and the Virus

How Scientists Study Evolution

Learning Objectives

- Define biological evolution.
- Write three questions biological evolution can potentially address.
- Using evidence from fossil whales, explain how lineages change through time.
- List the characteristics of viruses that make them difficult to control.
- Describe three different kinds of evidence that scientists use for understanding evolution.
- Define natural selection and explain how it is important for evolution.

The blue whale, the biggest animal on Earth, can reach more than 30 meters long and weigh up to 100,000 kilograms—equal to the weight of a town of more than a thousand people. To grow to this staggering size, blue whales search for swarms of krill and other small animals. They drop open their lower jaws, and enough water to fill two school buses rushes in. Swinging their mouths shut, the whales ram their tongues forward, forcing the water back out through rows of bristle-covered plates called baleen. The baleen traps the krill and other animals; once all the water is gone, the whales swallow their meal. In a single lunge, scientists calculate, a blue whale can gather half a million calories (Goldbogen et al. 2010). You'd need to eat a thousand hamburgers to get the same amount of energy—and you'd need to eat them all at once.

Blue whales are remarkable not just for their size or for their massive gulps. Although they have fishlike bodies well suited for swimming, they cannot survive an hour underwater. They must periodically rise to the ocean's surface to breathe, opening a blowhole to draw air into their lungs. Whales lack the gills that sharks and other fishes use to draw oxygen from water. Instead of laying eggs, like most species of fish do, whales give birth to live young. The calves must

Ambulocetus was a 49-million-year-old relative of today's whales and dolphins. It evolved from terrestrial mammals and still had limbs.

travel with their mothers for years, drinking milk instead of searching for their own food.

At the other end of life's spectrum are viruses. They are minuscule, typically measuring about 100 nanometers across. That's about a thousand times smaller than the width of a human hair, and more than 10 billion times smaller than a blue whale. Unlike animals, viruses can be exquisitely simple. We have about 20,000 protein-coding genes, for example. The influenza virus has only 13 (Jagger et al. 2012).

Viruses may be small and simple, but that doesn't mean they're not extremely important. As they spread through their hosts, they cause a wide range of diseases, some of which are devastating. In 2016 alone, for example, the human immunodeficiency virus (HIV) killed 1 million people (UNAIDS 2017). Viruses infect almost every species of animal, plant, fungus, protozoan, and bacterium on Earth. They can even infect blue whales. In a single drop of seawater, there may be millions of viruses. In the world's ocean, there are about 10^{31} of them—equivalent in weight to 75 million blue whales (Simmonds et al. 2017).

As different as blue whales and viruses may be, a single explanation can account for them both—along with all the other species they share the planet with: they are all the products of evolution (**Figure 1.1**).

In 1973, the biologist Theodosius Dobzhansky wrote one of the most eloquent accounts for evolution's place in the study of life. He titled his essay, "Nothing in Biology Makes Sense Except in the Light of Evolution" (Dobzhansky 1973). "Seen in the light of evolution, biology is, perhaps, intellectually the most satisfying and inspiring science," he wrote. "Without that light it becomes a pile of sundry facts—some of them interesting or curious but making no meaningful picture as a whole."

By understanding evolution, Dobzhansky explained, we can understand why the natural world is the way that it is. We can understand the similarities among different species, as well as the differences. We can understand why some species are present in some parts of the world and not others. We can understand the adaptations of living things, as well as their weaknesses.

The insights that we get from studying evolution can lead to practical applications. By studying evolution, we can find new ways to fight the viruses and bacteria that make us sick. We can understand how insects become resistant to the pesticides that farmers apply to their fields. We are altering the environment on a planetary scale by introducing invasive species to new habitats, spreading pollution, and altering the climate. Globally, we are witnessing a wave of extinctions the likes of which Earth may not have seen for tens of millions of years. By studying evolution's history, we can learn how today's extinctions compare with those of the past, begin to make predictions about which current populations will vanish, and devise strategies to slow their decline.

But evolution is also useful in an even more profound way: it helps us find answers to some of the biggest questions we ask about ourselves. How did we get here? How did we acquire our powers of reasoning and language? To fully address these questions, we must first appreciate how and why populations change over time. We must understand the basic principles of **biological evolution**.

Figure 1.1 Paleontologist Hans Thewissen discovered the first fossil of *Ambulocetus* in 1993. (Photo by J. G. M. Thewissen, NEOMED)

Biological evolution is any change in the inherited traits of a population that occurs from one generation to the next (that is, over a time period longer than the lifetime of an individual in the population).

This book is an introduction to evolutionary biology—the study of both the processes by which life evolves and the patterns these processes have generated over the past 4 billion years. It is also about how scientists study evolution. When Charles Darwin studied evolution in the mid-1800s, the most sophisticated tool he could use was a crude light microscope. Today, scientists study evolution by using high-throughput sequencing technologies to analyze DNA. They probe the molecules of ancient rocks to determine the age of **fossils**. They use powerful computers to apply new statistical equations to the diversity of life. They observe evolution unfolding in their laboratories. And they synthesize these different lines of evidence into a unified understanding of how life has evolved.

As an introduction to this book, and to evolutionary biology in general, let's turn back to those remarkable extremes, whales and viruses.

A **fossil** is preserved evidence of life from a past geological age, including impressions and mineralized remains of organisms embedded in rocks.

- By understanding evolution, we can understand why the natural world is the way it is. ●

Key Concept

1.1 Whales: Mammals Gone to Sea

Whales, dolphins, and porpoises are collectively known as cetaceans (that's because they're members of the order Cetacea; **Figure 1.2**). There are 92 species of living cetaceans, and all of them share a number of traits (Perrin 2017). They have fishlike bodies—sculpted with the same sleek curves you can find on tuna and sharks—that allow them to use relatively little energy to shoot through the water. Their tails narrow at the end and then expand into horizontally flattened flukes. Cetaceans lift and lower their flukes to generate thrust, much like sharks and tuna generate thrust by moving their tails from side to side.

It's clear that cetaceans have a superficial similarity with fishes, but they also possess many traits found only in mammals. Cetacean embryos develop in the uterus, forming a placenta to extract nutrients from their mothers. They are born

Figure 1.2 Living cetaceans all share a number of traits, such as blowholes and horizontal tail flukes. Within this group of species, there are two different kinds of cetaceans. Some species have filter-like growths called baleen in their mouths that they use to sieve small animals from seawater (left). Other species, such as dolphins (right), have peg-shaped teeth that they use to grab larger prey, such as seals and fishes. (Left: James Michael Dorsey/Shutterstock; right: Styve Reineck/Shutterstock)

Figure 1.3 Whales give birth to live young, which survive on milk from their mothers for months. In other words, they retain mammalian reproductive traits after having returned to the ocean more than 40 million years ago. (Westend61/Getty Images)

alive and then drink milk produced by their mothers until they're old enough to eat solid food (**Figure 1.3**). Whales and dolphins even have tiny bones embedded in their flesh just where the hips would be on land mammals.

Evolution as a Gradual Process

In his 1859 book *The Origin of Species*, Darwin proposed a straightforward explanation for this puzzling pattern of similarities and differences. Cetaceans descended from mammals that lived on land, and their **lineage** gradually evolved into marine mammals through a process he dubbed **natural selection**. (See Chapter 2 for a detailed exploration of Darwin's general theory.) The ancestors of modern whales lost their hindlimbs, and their front legs became shaped like flippers. Yet whales retain some traits from their mammalian ancestors, such as lungs and mammary glands. The mammary glands of whales and land mammals are examples of **homology**—structural characters that are shared because they are inherited from a common ancestor.

Darwin proposed that evolution was a gradual process. If that were to be true, then intermediate species of cetaceans must have existed in the past that had bodies specialized for life on land. In Darwin's time, paleontologists were only starting to find fossils that could illuminate the past, and they knew of no such intermediate cetaceans. But starting in the late 1800s, researchers found a growing number of these fossils. They knew that these fossils were cetaceans because they had a set of distinctive traits in their skeletons that are found today only in living cetaceans.

Figure 1.4 illustrates an example, the 40-million-year-old *Dorudon atrox*. Fossils of *Dorudon* revealed that its flippers and long vertebral column are very much like those found in living whales. But when biologists want to classify a species of mammal, they focus much of their attention on anatomical details of the skull. The teeth, ear bones, and other parts of the skull are especially helpful at revealing evolutionary links between species.

Mammals all hear with a series of tiny bones that vibrate in response to sound. These bones sit in an air-filled cavity that's enclosed in a hollow shell known as the ectotympanic. Living cetaceans have a distinctive ectotympanic. The inner wall of the ectotympanic, called the involucrum, forms a thick lip of dense bone. The involucrum of *Dorudon* has this same dense, thick form (Thewissen et al. 2009).

Although *Dorudon* shares some features with living cetaceans, it also was different in some important respects. Living cetaceans have two basic kinds of mouth anatomy, reflecting two different ways of feeding. Blue whales belong to the group known as baleen whales, or mysticetes. They use baleen to filter small animals from the water (see Figure 1.2, left panel). The other group is known as odontocetes, or toothed whales (right panel). This group includes sperm whales, dolphins, and killer whales. They have uniform peg-shaped teeth, which they use to hunt for prey.

Lineage refers to a chain of ancestors and their descendants. A lineage may be the successive generations of organisms in a single population, the members of an entire species during an interval of geological time, or a group of related species descending from a common ancestor.

Natural selection is a mechanism that can lead to evolution, whereby differential survival and reproduction of individuals cause some genetic types to replace (outcompete) others.

A **homologous** characteristic is similar in two or more species because it is inherited from a common ancestor.

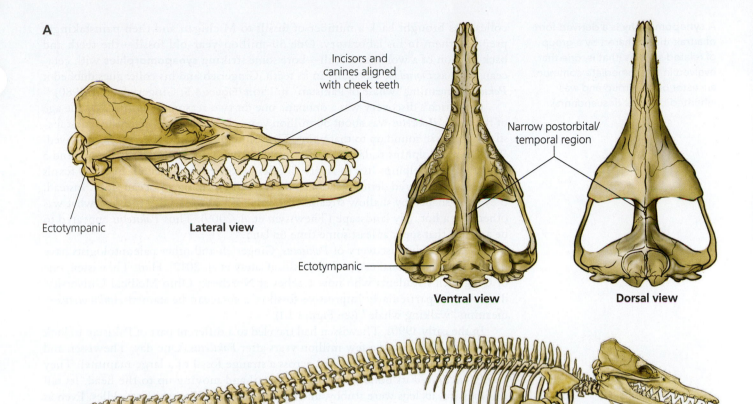

A

Incisors and canines aligned with cheek teeth

Ectotympanic

Lateral view

Narrow postorbital/temporal region

Ectotympanic

Ventral view

Dorsal view

Figure 1.4 A: Early clues to whale evolution come from the teeth of whale fossils. *Dorudon*, a 40-million-year-old whale, has different types of teeth, some of which have complicated surfaces. They bear a strong resemblance to the teeth of some extinct land mammals. When paleontologists found older cetacean skulls, they used traits such as the ones shown here to determine that they were related to *Dorudon* and living cetaceans. B: An artist's reconstruction of *Dorudon*. (*The Walking Whales: From Land to Water in Eight Million Years*, by J. G. M. "Hans" Thewissen, © 2014 by the Regents of the University of California. Published by the University of California Press)

B

Dorudon had teeth as well. But its teeth had different shapes—pointed incisors in the front and shearing molars in the back. The cheek teeth had complex patterns of cusps and facets. *Dorudon*'s teeth resembled those of land mammals more than they did those of living odontocetes. This difference suggested that the evolution of peg-like teeth and baleen occurred long after cetaceans had become aquatic.

It wasn't until the late 1900s that whale evolution started to come into sharp focus. In 1979 Philip Gingerich, a paleontologist from the University of Michigan, traveled to Pakistan to investigate a geological formation—rocks that formed from 56 million to 34 million years ago (mya), from the Eocene period—with the intent of documenting the mammals existing in that region during that time. He and his

A **synapomorphy** is a derived form of a trait that is shared by a group of related species (that is, one that evolved in the immediate common ancestor of the group and was inherited by all its descendants).

colleagues brought back a number of fossils to Michigan and then painstakingly prepared them in his laboratory. One 50-million-year-old fossil—the teeth and back portion of a wolf-sized skull—bore some striking **synapomorphies** with cetaceans such as *Dorudon*, especially in its teeth. Gingerich and his colleagues dubbed it *Pakicetus*, meaning "whale of Pakistan" in Latin (**Figure 1.5**; Gingerich et al. 1980).

Gingerich's discovery was a dramatic one for two reasons. The first was the age of the fossil: *Pakicetus* was about 50 million years old, making it by far the oldest fossil of a cetacean found up to that point. The second reason was where *Pakicetus* lived. Whales and dolphins today live only in the water—most live in the ocean, and a few species of dolphins live in rivers. Studies on the rocks in which *Pakicetus* fossils have been discovered demonstrate that they did not form on an ocean floor. Instead, they were shaped by shallow streams that only flowed seasonally through what was otherwise a hot, dry landscape (Thewissen et al. 2009). Thus *Pakicetus* appeared to be a whale that spent at least some time on land.

Ever since the discovery of *Pakicetus*, Gingerich and other paleontologists have been finding additional early whale fossils (Gatesy et al. 2012). Hans Thewissen, one of Gingerich's students who now teaches at Northeast Ohio Medical University, discovered a particularly impressive fossil of a cetacean he named *Ambulocetus*—meaning "walking whale" (see Figure 1.1).

In the early 1990s, Thewissen had traveled to a different part of Pakistan to look for mammals that lived a few million years after *Pakicetus*. One day, Thewissen and his Pakistani colleagues happened across a strange fossil of a large mammal. They slowly excavated its bones, starting at the tail and moving up to the head. Its tail was massive, its legs were stubby, and its rear feet were shaped like paddles. Even as Thewissen was digging up the fossil, he could see that its head was long like an alligator's, but it had teeth like those of fossil whales.

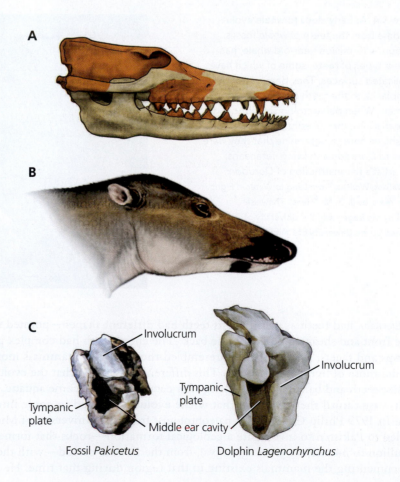

Figure 1.5 A: The 50-million-year-old *Pakicetus* was the first terrestrial cetacean ever discovered. This artist's reconstruction is based on several different fossils. Fossil material is colored orange; the tan areas are reconstructed based on skulls of related mammals. B: An artist's reconstruction of a living *Pakicetus*. C: Although *Pakicetus* may not look like a whale, synapomorphies reveal its relationship to more familiar cetaceans. Here, we show the right ectotympanic bones of *Pakicetus* and a living dolphin species. The ectotympanic bone surrounds the middle ear cavity. The inner wall of the ectotympanic, known as the involucrum, is thick and dense in cetaceans—and in fossils such as *Pakicetus*. But this trait is found in no other living or fossil mammals. Other synapomorphies include features of the teeth and skull indicated on the skull of *Dorudon* in Figure 1.4A. (Uhen 2010; A, B: photo by J. G. M. Thewissen, NEOMED)

Thewissen brought the fossil (**Figure 1.6A**) to the United States, where he continued preparing and analyzing it. He discovered more cetacean synapomorphies, such as a thickened involucrum. He concluded that the fossil was, in fact, a whale that could walk (**Figure 1.6B**; Thewissen et al. 1997; Zimmer 1998).

In the years since Thewissen's discovery, paleontologists have found many fossils of different species of whales with legs. Thewissen himself, for example, has found additional bones of the older cetacean ancestor *Pakicetus*. Fragments of its skull revealed it to have a cetacean-like involucrum and also showed that its eyes sat on top of its head. Fossil bones from other parts of its skeleton led Thewissen to reconstruct *Pakicetus* as a wolf-sized mammal with slender legs and a pointed snout. Meanwhile, in Pakistan, Gingerich found the fossils of a whale that was later named *Rodhocetus*, whose short limbs resembled those of a seal.

As the evidence that cetaceans evolved from land mammals grew, scientists wondered which group of land mammals in particular they had evolved from. One way to address a question like this is to compare the DNA of living species. As we'll see in Chapter 8, scientists can use genetic information to draw evolutionary trees showing the relationships among species. In the 1990s, several research groups began comparing small snippets of DNA from cetaceans and other mammals. They concluded that cetaceans were most closely related to a group of mammals called artiodactyls, which includes cows, goats, camels, and hippopotamuses. As they compared more DNA from these species, the cetacean–artiodactyl link only grew stronger. In fact, the scientists concluded that cetaceans were most closely related to hippopotamuses (Nikaido et al. 1999).

Paleontologists were surprised by this result. In the first fossils of whales they had discovered, they found no compelling **morphological** evidence linking them to artiodactyls. But the paleontologists changed their minds when they began to find the *limbs* of fossil whales. One trait thought to be unique to artiodactyls can be found on an ankle bone called the astragalus (plural, astragali). The ends of the astragalus are pulley-shaped in artiodactyls. Obviously, living cetaceans don't have an astragalus because they don't have hind legs. But when paleontologists finally discovered the ankle region of ancient cetaceans with limbs, they discovered that the animals had double-pulley astragali (**Figure 1.7**). The presence of this trait in artiodactyls and cetaceans suggests that the two groups share a more recent common ancestor than either group does to other mammals. It may seem incredible that whales and cows are like cousins, and yet we can see how two lines of evidence independently lead us to that same conclusion.

Figure 1.6 A: The 49-million-year-old skeleton of *Ambulocetus* has an involucrum, as well as other traits that are found today only in whales. It was the first fossil whale to be discovered with legs. B: An artist's reconstruction of *Ambulocetus* shows how it may have used its stout, powerful limbs to swim underwater in pursuit of prey. (A: Photo by J. G. M. Thewissen, NEOMED; B: Carl Buell)

Morphology refers to the form and structure of organisms.

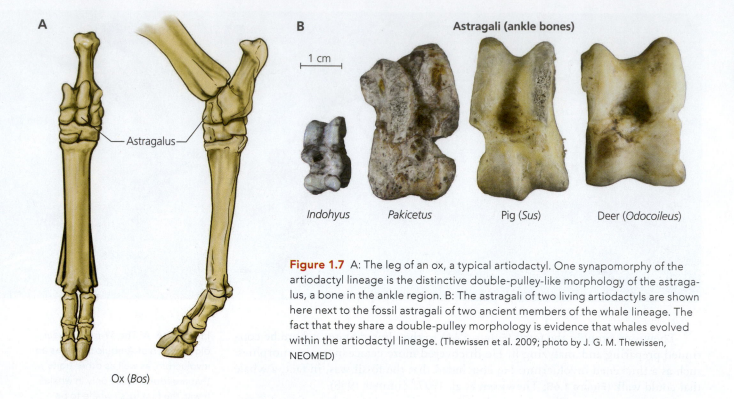

A

Astragalus

Ox (*Bos*)

B Astragali (ankle bones)

1 cm

Indohyus   Pakicetus   Pig (*Sus*)   Deer (*Odocoileus*)

Figure 1.7 A: The leg of an ox, a typical artiodactyl. One synapomorphy of the artiodactyl lineage is the distinctive double-pulley-like morphology of the astragalus, a bone in the ankle region. B: The astragali of two living artiodactyls are shown here next to the fossil astragali of two ancient members of the whale lineage. The fact that they share a double-pulley morphology is evidence that whales evolved within the artiodactyl lineage. (Thewissen et al. 2009; photo by J. G. M. Thewissen, NEOMED)

Phylogeny is a visual representation of the evolutionary history of populations, genes, or species.

Visualizing Evolutionary History

Darwin maintained that species evolved like a branching tree in which new lineages split off from old ones. To reconstruct this evolutionary branching pattern, known as a **phylogeny**, scientists can analyze morphology (Chapter 4) and genes (Chapter 8). The phylogeny of cetaceans and their relatives is illustrated in **Figure 1.8**. We will include a number of similar figures in this book and use them to explore the patterns of evolution. Figure 1.8, for example, allows us to synthesize a great deal of evidence to get an overall sense of the complex evolutionary history of cetaceans that transformed them from ordinary-looking mammals into fishlike creatures.

Indohyus and *Pakicetus,* which belong to the two deepest branches on this tree, have an unusually dense type of bone in their limbs. This feature is also found in hippopotamuses, which you'll remember are the closest living relatives of cetaceans. Hippos and other living mammals with this sort of dense bone are able to stay underwater and walk on the bottom of rivers and lakes. It's therefore possible that the common ancestor of hippopotamuses and cetaceans had already spent time at the bottom of bodies of freshwater.

Other lineages became even more specialized for life in water. *Ambulocetus,* which lived about 49 million years ago along the coastline of what is now Pakistan, had short legs and massive feet. It probably swam like an otter, kicking its large feet and bending its tail. *Rodhocetus,* with its seal-like limbs, probably could only drag itself around on land.

As cetaceans lived more and more in the ocean, legs became obsolete—or even costly. Later cetaceans evolved flat flippers that helped them steer through the water. Other parts of their terrestrial anatomy also adapted to life in the water. They continued to breathe air into their lungs, but their nostrils gradually shifted up to the top of their skulls, allowing them to efficiently take in air each time they rose to the surface of the water.

Fossils show that the loss of hind legs took millions of years. By 40 million years ago, fully aquatic species such as *Dorudon* had evolved. But *Dorudon,* despite being 6 meters long, still had hind legs—complete with ankles and toes—that were smaller than those of a human child.

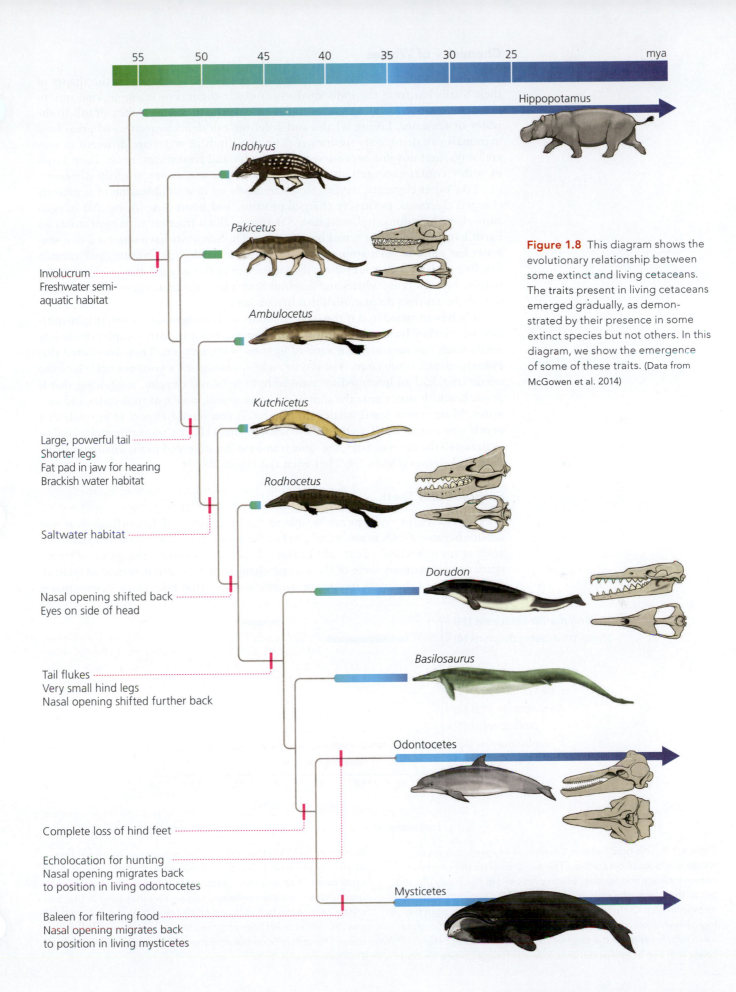

55   50   45   40   35   30   25   mya

Hippopotamus

Indohyus

Pakicetus

Involucrum
Freshwater semi-
aquatic habitat

Ambulocetus

Kutchicetus

Large, powerful tail
Shorter legs
Fat pad in jaw for hearing
Brackish water habitat

Rodhocetus

Saltwater habitat

Dorudon

Nasal opening shifted back
Eyes on side of head

Basilosaurus

Tail flukes
Very small hind legs
Nasal opening shifted further back

Odontocetes

Complete loss of hind feet

Echolocation for hunting
Nasal opening migrates back
to position in living odontocetes

Mysticetes

Baleen for filtering food
Nasal opening migrates back
to position in living mysticetes

Figure 1.8 This diagram shows the evolutionary relationship between some extinct and living cetaceans. The traits present in living cetaceans emerged gradually, as demonstrated by their presence in some extinct species but not others. In this diagram, we show the emergence of some of these traits. (Data from McGowen et al. 2014)

Chemistry of Whales

Scientists can get clues about the origins of cetaceans not just from the shapes of their fossils but from the individual atoms inside them. For example, one way to tell if ancestors were terrestrial or aquatic is to look at whether they drank freshwater or seawater. Living whales and dolphins can drink seawater, whereas land mammals can drink only freshwater. The two kinds of water are different in several ways, and not just because seawater is salty and freshwater is not. Both kinds of water contain oxygen atoms, but the oxygen atoms are slightly different.

Like other elements, oxygen atoms are made up of a combination of negatively charged electrons, positively charged protons, and neutral neutrons. All oxygen atoms have 8 protons, and most have 8 neutrons. But a fraction of oxygen atoms on Earth have extra neutrons, making them heavier. Scientists have observed that seawater has more oxygen atoms with 10 neutrons than does freshwater, and animals that live on land and at sea reflect this difference in the oxygen atoms incorporated in their bones. Living whales and dolphins have a larger percentage of heavy oxygen in their bones than do mammals that live on land.

Thewissen wondered if the oxygen atoms in ancient whale fossils might indicate where they lived. So he and his colleagues ground up tiny samples of ancient whale teeth and measured the ratio of light to heavy oxygen. They discovered that *Pakicetus* drank freshwater. *Ambulocetus*, which belongs to a younger branch of the whale tree, had an intermediate ratio of light to heavy oxygen, suggesting that it drank brackish water near the shore—in other words, a mix of freshwater and seawater. More recent fossil whales had the ratios you would expect in animals that drank seawater alone. Together, these isotope ratios chart a transition from land to estuaries to the open ocean—the same transition documented in the changing shape of their skeletons (**Figure 1.9**; Thewissen and Bajpai 2001).

Embryos and Teeth

It's likely scientists will never be able to read the genes of 40-million-year-old whales because DNA is too fragile to last for more than a few hundred thousand years at most (Orlando et al. 2013). But it is possible to study the genes of living whales to learn about some of the genetic changes that occurred as they adapted to water. Some of the most important genetic changes that take place during major

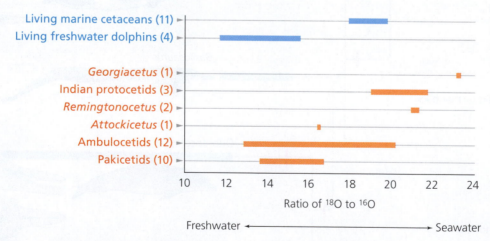

Figure 1.9 This graph shows the ratio of oxygen isotopes in modern and fossil cetaceans. (The numbers after the cetacean names indicate the number of samples used in this study.) To the left of the graph, the ratio of oxygen-18 (^{18}O, "heavy oxygen") to oxygen-16 (^{16}O, "normal oxygen") is low, and to the right it is high. Freshwater has a relatively low ratio of ^{18}O to ^{16}O. Seawater has a higher ratio. As the top two rows indicate, living cetaceans that live in freshwater and seawater have a corresponding proportion of ^{18}O in their bones. The isotopic ratio of early fossil whales suggests that they drank freshwater. *Ambulocetus* and other forms that were more adapted to living in water had ratios spanning the two environments, suggesting they lived in brackish waters or traveled between the open ocean and rivers. Cetacean species that evolved later acquired the isotopic signature of seawater. The shift reflects their permanent habitation in the ocean. (Data from Roe et al. 1998)

evolutionary transitions involve the timing and pattern of gene activity in embryos (Chapter 9).

When legs begin to develop in the embryos of humans or other land vertebrates, a distinctive set of genes becomes active. Thewissen and a team of embryologists discovered that these leg-building genes also become active in dolphin embryos. The genes help build tiny buds of tissue, but these buds stop growing after a few weeks and then die back (**Figure 1.10**). Thewissen's discovery revealed that the evolutionary loss of hind legs in the ancestors of dolphins and whales took place when a chance mutation occurred in a regulatory gene, turning off the expression of a developmental pathway. As a result, their hindlimbs began to form but then stopped growing (Thewissen et al. 2006). To better understand the origin of whales, scientists can integrate these insights from developmental experiments with the fossil record, which shows the reduction of the legs and pelvis (**Figure 1.11**).

The earliest ancestors of whales are long extinct. The two lineages alive today, the mysticetes and odontocetes, evolved from a common ancestor that lived about 40 million years ago. Odontocetes evolved muscles and special organs that could be used to produce high-pitched sounds in their nasal passages, allowing them to hear the echoes that bounced off animals and objects around them in the water (Geisler et al. 2014). Today, dolphins and other toothed whales use these echoes to hunt for their prey. The other living lineage, the baleen whales (mysticetes), lost their teeth and evolved baleen in their mouths that allowed them to filter prey from huge volumes of water.

Scientists are now beginning to find important new clues to the origins of both lineages. Fossils from about 25 million years ago, for example, show that the toothed ancestors of baleen whales probably grew small patches of baleen from their upper jaw. Only later did their teeth disappear. Baleen whales still carry genes for building teeth, but all of these genes have been disabled by mutations, just like the genes that were used to build the hind legs (McGowen et al. 2014.)

Cetacean Brains

Biologists have long been impressed with the size and complexity of the brains of living cetaceans. Aside from humans, dolphins have the biggest brains, in proportion to their bodies, of any animal (**Figure 1.12**; Marino 2007). The structure of the cetacean brain gives these animals a sophisticated level of behavior. Scientists who

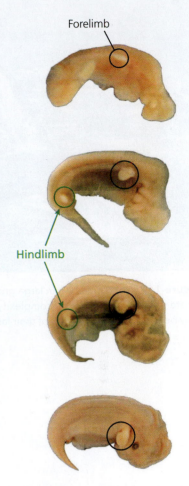

Weeks 4–9 of embryonic development

Forelimb

Hindlimb

Figure 1.10 A developing dolphin grows forelimbs, which will eventually become flippers. It also sprouts hindlimb buds, but the genes that maintain their growth eventually become inactive, and the buds are absorbed back into the embryo. Through a change in the timing by which genes turn on and off, cetaceans have evolved dramatically different bodies from their terrestrial ancestors. (Thewissen et al. 2009; photo by J. G. M. Thewissen, NEOMED)

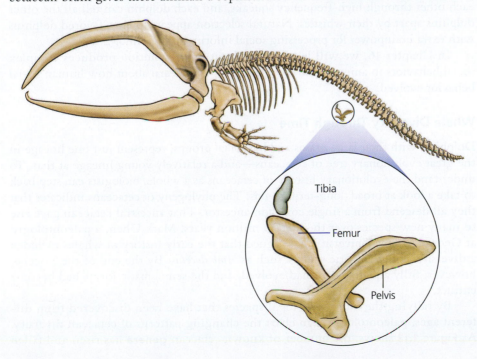

Tibia

Femur

Pelvis

Figure 1.11 Early cetaceans had a relatively large pelvis with wide surfaces adapted for the attachment of leg muscles. In a living whale, the pelvis is still present, but only as a tiny vestige embedded in the whale's body.

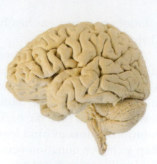

Human
(*Homo sapiens*)

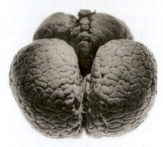

Bottlenose dolphin
(*Tursiops truncatus*)

Figure 1.12 A: Dolphins live in large groups and can communicate with each other. B: The complexity of dolphin social life may be linked to the evolution of their large brains. Dolphins are second only to humans in brain size relative to their body size. (A: David M. Schrader/Getty Images; B: Jesada Sabai/Shutterstock; Naruaki Onishi/Getty Images)

study dolphins have found, for example, that they can solve remarkably complicated puzzles to get a reward.

Evolutionary biologists can illuminate the history of the evolution of cetacean brains in a variety of ways. For example, they can look at the size and shape of brain cavities in early cetacean fossils. This analysis shows that the earliest whales had relatively small brains. Only after they adapted to the water did whale brains begin to change dramatically.

Studies of living cetaceans can shed light on the selective pressures that drove these changes. Dolphins appear to have evolved their especially big brains to solve a particular kind of natural puzzle: figuring out how to thrive in a large social group. Dozens of dolphins live together, forming alliances, competing for mates, and building relationships that persist throughout their long lives. They communicate with each other through high-frequency squeaks, and each dolphin can tell all the other dolphins apart by their whistles. Natural selection appears to have favored dolphins with extra brainpower for processing social information (Connor 2007).

In Chapter 16, we will learn more about how evolution produces complex social behaviors in animals, and in Chapter 17, we'll learn about how human social behavior evolved.

Whale Diversity Through Time

Dolphins, with their large brains and big social groups, represent just one lineage in the great evolutionary tree of cetaceans—and a relatively young lineage at that. To understand the evolutionary history of cetaceans as a whole, biologists can step back to take a look at broad, long-term trends. The phylogeny of cetaceans indicates that they all descend from a single common ancestor. That ancestral cetacean gave rise to many new species over the past 50 million years. Mark Uhen, a paleontologist at George Mason University, has argued that the early history of whales included a diversity of semiaquatic species such as *Ambulocetus*. By the end of the Eocene, however, fully aquatic forms had evolved, and the semiaquatic forms had become extinct.

By tallying the number of fossil species that have been discovered from different ages, paleontologists can chart the changing patterns of cetacean diversity. As **Figure 1.13** shows, the number of known cetacean **genera** has risen and fallen

Genus (plural, **genera**) is a taxonomic group that includes species. (*Canis*, for example, is the genus that includes dogs, wolves, and coyotes.)

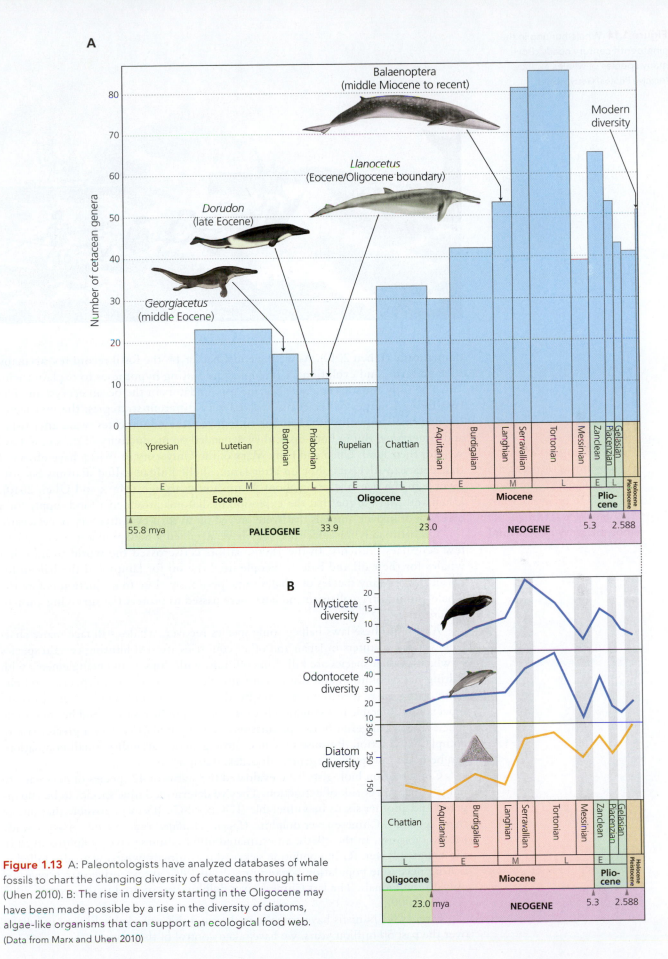

Figure 1.13 A: Paleontologists have analyzed databases of whale fossils to chart the changing diversity of cetaceans through time (Uhen 2010). B: The rise in diversity starting in the Oligocene may have been made possible by a rise in the diversity of diatoms, algae-like organisms that can support an ecological food web. (Data from Marx and Uhen 2010)

Figure 1.14 Whale hunting in the nineteenth century nearly drove many species of whales extinct. (Archive Photos/Getty Images)

dramatically (Uhen 2010). As we'll see in Chapter 14, the fossil record reveals many such patterns, and evolutionary biologists are testing hypotheses to explain them.

The fully aquatic cetaceans became top predators in the ocean ecosystem. The two living lineages of cetaceans evolved very different strategies: the mysticetes trapped small animals with their baleen, whereas the odontocetes swam after fishes and larger prey. But in both cases, they depended on productivity of the ocean food web for their food. Uhen and Felix Marx of the University of Bristol have observed that cetacean diversity shot up after large, shelled algae called diatoms became diverse in the ocean starting about 20 million years ago (Marx and Uhen 2010). Uhen and Marx argue that the explosion of diatoms provided a food supply for a diversity of marine animals that were preyed on in turn by a diversity of cetaceans.

Unfortunately for cetaceans, they have attracted a new predator over the past few centuries: humans. In the 1800s, sailors crisscrossed the world to hunt big whales for their oil and baleen (people used the oil for lamps and the baleen for corset stays). Many species of whales came perilously close to extinction before the whale-oil industry collapsed and laws were passed to protect the surviving animals (**Figure 1.14**).

Although these laws helped some species recover, whales still face many challenges today. Hunters in Japan and other countries are still hunting certain species of whales. Other species are killed in collisions with ships or in entanglements with fishing nets. Pollution and diseases are also threatening cetacean species, and climate change may threaten cetaceans by altering their food supply. Cetaceans are especially vulnerable to extinction because they reproduce slowly and because some species are at dangerously low populations. Small populations face a greater risk of complete extinction. Because they have little genetic variability, small populations can be more susceptible to genetic disorders (Chapter 6).

Conservation biologists have evaluated the status of 42 species of cetaceans to judge which are at risk of extinction. They've determined nine species to be endangered and another six to be vulnerable (IUCN-SSC). It's even possible that one of these species—Chinese river dolphins (*Lipotes vexillifer*; **Figure 1.15**)—have already become extinct. In the 1950s, an estimated 6000 Chinese river dolphins lived in the Yangtze River. Rising pollution sickened the animals, which were also killed in fishing nets. The population began to crash; in 1997, an extensive search revealed only 13 individuals. The last Chinese river dolphin was spotted in the river in 2007 (Turvey 2009).

Even as we humans have come to understand the remarkable history of whales over the past 50 million years, we have taken control of their fate.

Key Concepts

- Whale and fish lineages, evolving independently, converged on body forms that are superficially similar.

- *Ambulocetus* is a fossil whale with legs. This animal had traits that were intermediate between those of modern whales and their terrestrial ancestors.

- Scientists use different lines of evidence to study evolution. The chemistry of fossil whales documents a transition from land to estuaries to the open ocean—the same transition documented in the changing shape of their skeletons.

- As further evidence of their evolution from terrestrial mammalian ancestors, whales began to develop hindlimbs. ●

1.2 Viruses: The Deadly Escape Artists

In March 2013, two elderly men in the Chinese city of Shanghai died of the flu. Though tragic, the deaths were not terribly unusual—influenza viruses kill about half a million people worldwide every year. The type of flu, however, was unusual. It was a kind that had never been recorded infecting humans before, and it worried public health experts (**Figure 1.16**). When new kinds of influenza emerge, they sometimes become devastating global outbreaks.

Figure 1.16 Medical staff in a Chinese hospital treat a patient infected with a deadly form of bird flu in 2017. The strain of influenza virus, called H7N9, had never been seen before. Evolutionary biologists helped solve the mystery of its emergence. (STR/AFP/Getty Images)

This concern is not simply hysteria fueled by science fiction. Influenza viruses have caused these outbreaks before, killing hundreds of millions of people, and experts agree they will do it again. The question is, when? The world's health depends on the future evolution of influenza viruses. In fact, the threat of another influenza pandemic is so significant that the World Health Organization maintains an extensive network of scientists actively monitoring the progression of this virus's evolution. Whereas paleontologists track the evolution of whales with fossils that date back tens of millions of years, influenza researchers work on much shorter timescales, observing evolutionary change over months or even days.

Anatomy of a Virus

To understand how deadly outbreaks of influenza occur, we must start with the influenza virus itself. As shown in **Figure 1.17**, an influenza virus is basically a membrane and protein shell encasing strands of RNA. To replicate, the virus must invade a host cell. Influenza viruses have proteins called hemagglutinins on their surfaces that allow the viruses to bind to the receptors naturally present on epithelial cells in the host's respiratory tract. After binding to the receptors, the viruses trigger the cells to open a passageway into the cell, through which the viruses enter. Once inside, viruses recruit our cells' genetic machinery to mass-produce copies of themselves. Each infected cell can make thousands of new copies of the viral genome and proteins, which are then packaged into new protein shells that bud out from the cell membrane. The viruses then use a protein called neuraminidase (see Figure 1.17) to cut themselves free from the cell.

The new viruses are imperfect copies of the originals. As we'll see in Chapter 5, all living things reproduce by making copies of their genes. But in each generation they make a few mistakes, known as **mutations**. As a result, ancestors and their descendants do not have precisely identical genes (**Figure 1.18**). Some mutations will affect the organism's ability to replicate. In the case of viruses, for example, a mutation may leave a virus unable to escape its host cell and infect new cells.

But other mutations can benefit flu viruses, for example, by letting them evade the immune system. When we get sick with the flu, our immune system responds by

A **mutation** is any change to the genomic sequence of an organism.

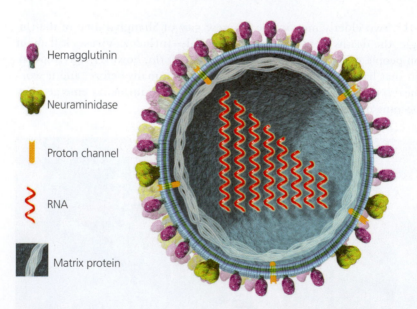

Hemagglutinin

Neuraminidase

Proton channel

RNA

Matrix protein

Figure 1.17 The influenza virus contains only eight segments of RNA that can make 13 proteins. It is surrounded by a shell of proteins and a lipid bilayer. Proteins on its surface allow it to enter and exit host cells.

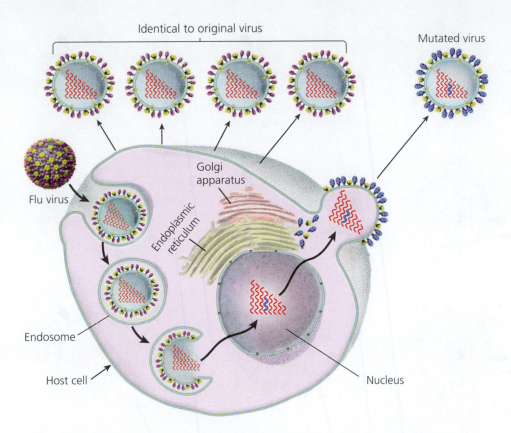

Identical to original virus

Mutated virus

Flu virus

Golgi apparatus

Endoplasmic reticulum

Endosome

Host cell

Nucleus

Figure 1.18 Flu viruses enter host cells, at which point their genes enter the cell nucleus. The cell makes new copies of the flu genes as well as the proteins in which they are encased. The parts of the virus are assembled at the cell membrane, and mature viruses are extruded from the cell to infect other cells. The replication of genes is not always precise. When a flu virus invades a host cell, some of its offspring may end up carrying mutations. (Not to scale. Viruses are thousands of times smaller than cells.)

making antibodies that can bind precisely to proteins on the surface of the viruses. The effect can be like gluing a bit of metal to a key so that it can't fit into a lock: the viruses can no longer invade cells. If the immune system can make these antibodies quickly enough, it can wipe out an infection. And if the same type of influenza virus causes a new infection later on, the antibodies can wipe it out before we even feel sick.

Mutations to genes for surface proteins can change their structure. Some of those mutations can make it harder for antibodies to bind tightly to viruses, so they can still invade cells. Viruses with these beneficial mutations reproduce more than other viruses; as a result, their descendants come to dominate the virus population (**Figure 1.19**). This is precisely how natural selection operates, and it helps create a diversity of new strains that change over time (**Figure 1.20**; Drake and Holland 1999).

As we'll discuss in more detail in Chapter 18, all strains of human influenza virus ultimately have descended from strains that infected other animals, most commonly birds. In birds, the virus infects cells in the intestines and is spread in bird droppings. From time to time, bird flu "crosses over" and begins to replicate in humans.

Viral Reassortment

This transition is aided by the ability of viruses to swap genes, in a process known as **viral reassortment**. When two viruses infect a cell at once, their genes sometimes are shuffled together as they become packaged into protein shells. Reassortment

Viral reassortment occurs when genetic material from different strains is mixed into new combinations within a single cell.

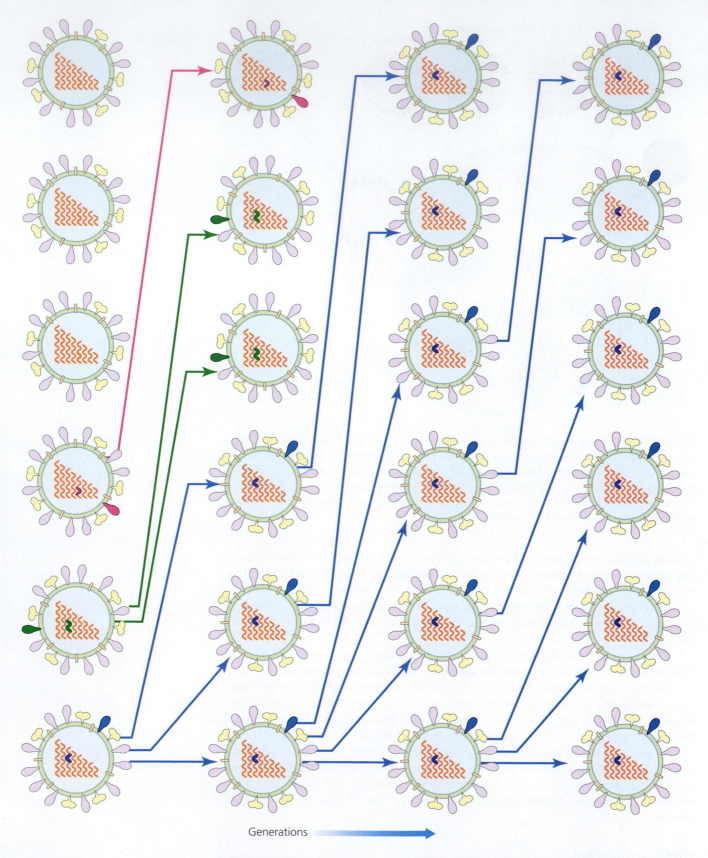

Figure 1.19 Some mutations are harmful to flu viruses, whereas others have no effect. A few may be beneficial in one way or another, such as helping a virus evade the immune system. Viruses with these mutations will reproduce more, making the mutation more common in the virus population. This is one way that natural selection can change the genetic makeup of a population.

Generations

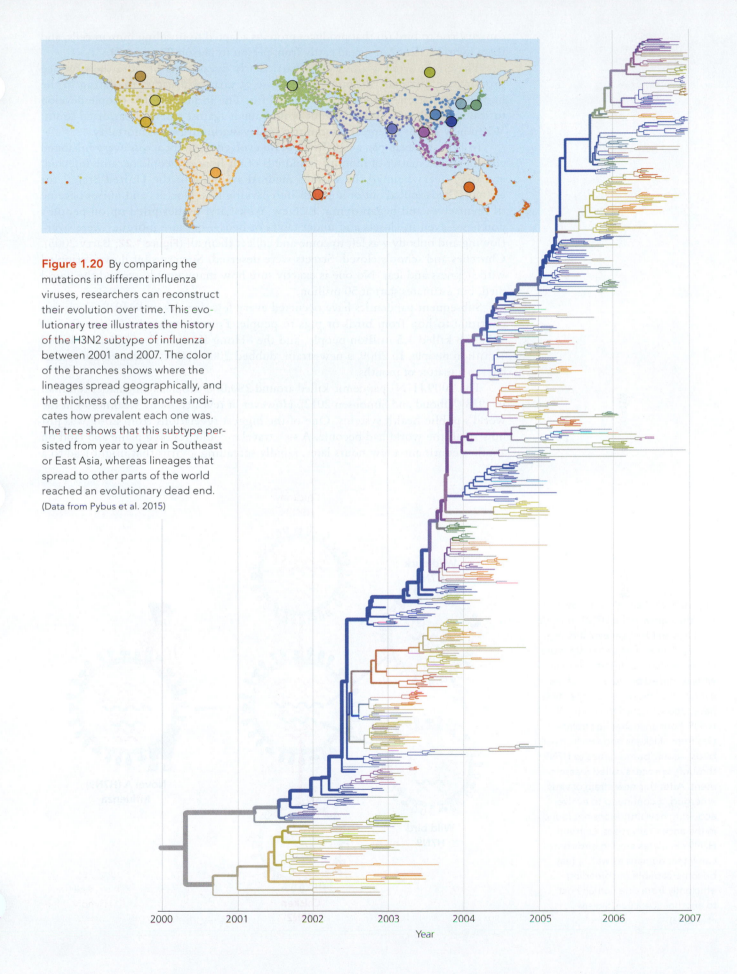

Figure 1.20 By comparing the mutations in different influenza viruses, researchers can reconstruct their evolution over time. This evolutionary tree illustrates the history of the H3N2 subtype of influenza between 2001 and 2007. The color of the branches shows where the lineages spread geographically, and the thickness of the branches indicates how prevalent each one was. The tree shows that this subtype persisted from year to year in Southeast or East Asia, whereas lineages that spread to other parts of the world reached an evolutionary dead end. (Data from Pybus et al. 2015)

can give rise to bird flu that produces proteins adapted to invading human cells, and these hybrid strains can spread easily from person to person (**Figure 1.21**).

These reassorted strains can be particularly devastating because their surface proteins are distinctly different from those of viruses already circulating in the human population, the so-called seasonal flu. The immunity that people develop to seasonal influenza strains cannot protect them against a newly reassorted strain. When new strains emerge, they can trigger devastating waves of mortality.

In 1918, for example, a new strain of flu evolved. It swept across every inhabited continent in a matter of months. Hundreds of millions of people became infected, and they overwhelmed the world's medical systems. In the United States, for example, beds spilled out of hospitals into parking lots. Doctors and nurses became ill themselves and began dying. In New York City, bodies piled up on people's porches—even in closets and under tables or beds—because morgues were overflowing and nobody was left to come and collect them all (**Figure 1.22**; Barry 2005). Churches and schools closed. Streets were deserted. Society shut down, crippled with sickness and fear. No one is exactly sure how many people around the world died, but estimates start at 50 million.

Subsequent pandemics have occurred every few decades since 1918, as viruses continue to hop from birds or pigs to people. For example, the "Asian flu" of 1957–58 killed 1.5 million people, and the "Hong Kong flu" of 1968–69 killed 1 million people. In 2009, a new strain, dubbed 2009 H1N1, spread to 200 countries in a matter of months.

The 2009 H1N1 pandemic killed around 280,000 people, a far smaller toll than in 1918 (Viboud and Simonsen 2012). However, it revealed many weaknesses in the world's public health systems. One of the biggest revelations from 2009 was just how small the world had become. A sick traveler could get on a plane and land on another continent a few hours later, rapidly spreading a virus that might lead to a

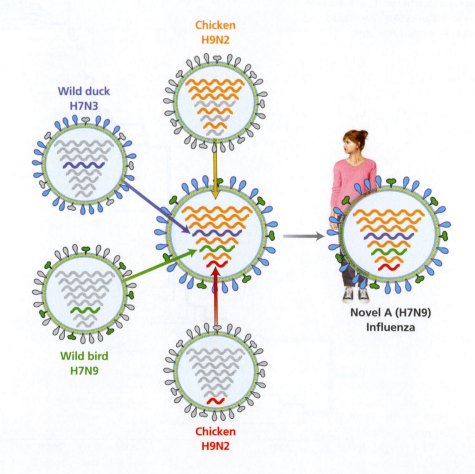

Figure 1.21 Scientists discovered that the genes in the H7N9 strain do not have an identical evolutionary history. Some of its genes are more closely related to different bird flu viruses. Based on the study of the evolution of these genes, scientists have reconstructed the origin of H7N9. Four ancestral flu strains—two from chickens and two from wild birds—contributed genes to H7N9 through a process called reassortment. After this new strain of virus emerged, it continued to evolve, acquiring new mutations not found in the ancestral viruses. Currently, H7N9 circulates among birds but can infect humans as well. It may become capable of spreading efficiently from one human host to another. (Data from Morens et al. 2013)

Figure 1.22 Studying the evolution of flu is essential to understanding why an outbreak in 1918 was able to kill at least 50 million people—and to preventing such an outbreak from happening again. (U.S. Army/Science Source)

pandemic. Another unsettling revelation was that before crossing into humans, the virus strain probably circulated among pigs for years—without ever being detected.

Experiences with recent pandemics like H1N1 have brought an added urgency to detecting new flu strains with the potential to infect humans. Scientists now monitor populations of domestic chickens and other poultry, as well as wild birds, for bird flu viruses. It's particularly worrisome when one of these bird flu strains infects a person. However, such an infection doesn't mean that a new pandemic has begun. In most cases, bird flu viruses are incapable of being passed from one person to another, which keeps them from spreading. But the potential exists for the virus to evolve that capability.

Detecting New Viruses

When two men died of the flu in Shanghai in 2013, doctors examined the virus that had killed them. The flu had a type of hemagglutinin called H7 and a type of neuraminidase called N9, a combination that had never been seen before in either humans or birds. The virus turned up in chickens and ducks across much of China, and sometimes people who handled poultry got sick. The virus proved more deadly than seasonal influenza. In 2017, for example, 759 people contracted H7N9 influenza, and of these, 281 died. Fortunately, scientists could find only a few cases in which H7N9 spread from one person to another. It remained a bird flu for the time being.

Genetic studies on H7N9 revealed that it emerged through reassortment (see Figure 1.21). Viruses that infected wild birds were picked up by ducks, which then spread them to chickens. There, these wild viruses reassorted with another virus that was already circulating in chickens. The new H7N9 virus then swept rapidly through chickens and ducks in many regions of China. Along the way, it evolved modifications to its surface proteins.

In 2017, Yoshihiro Kawaoka of the University of Wisconsin and his colleagues ran an experiment on H7N9 to learn about its risk to humans. They studied ferrets, which are very similar to people in how they get sick with the flu. Kawaoka and his colleagues found that H7N9 sprayed into the noses of ferrets readily made them sick. More worryingly, the virus could also spread through the air from one ferret to the next. And the most common antivirals used to fight influenza were not able to stop the infections (Imai et al. 2017).

For now, public health workers are doing what they can to slow the spread of H7N9. They regularly order the slaughter of infected flocks of poultry to keep the

Figure 1.23 The emergence of H7N9 has spurred new measures to control the spread of bird flu to humans. Blocking its transmission can prevent the evolution of a strain that can spread easily between humans. (VCG/Getty Images)

virus from infecting more birds and more people, for example (**Figure 1.23**). They are also monitoring animals and people for new strains of H7N9 that have evolved new adaptations. Kawaoka's research shows that H7N9 might need very few additional mutations to become the next pandemic.

Key Concepts

- Natural selection favors new variants of influenza virus that can escape detection (or destruction) by the immune system.

- Reassortment of the influenza virus is terrifying because it could blend dangerous elements of pig or bird flu with the infectious potential of human flu, instantaneously generating a strain that is both deadly and infectious.

- By monitoring bird flu, scientists can identify the evolution of new flu strains as they occur. ●

1.3 Evolution: A Tapestry of Concepts

These two case studies, of whales and viruses, illustrate that evolution is a complex phenomenon that plays out over both short and long timescales. In this book, we'll investigate the key concepts in evolutionary theory and observe how mechanisms of evolution combine to produce the diversity of life.

In Chapter 2, we'll examine how the history of biology led up to Darwin's theory of evolution through natural selection. Darwin argued that similarities among living species are the result of descent with modification from common ancestors. To explain how that modification took place, Darwin noted that variation occurred in every generation. Some of this variation caused individuals to reproduce and pass down their traits more successfully than others. Over time, natural selection and other processes would drive large-scale changes, even giving rise to new species.

In the decades since Darwin first published his theory, the evidence for evolution has increased exponentially. In Chapter 3, we investigate the fossil record as it stands today, chronicling more than 3.7 billion years of life on Earth. Scientists can analyze these fossils to reconstruct phylogenies. In Chapter 4, we'll explore how such phylogenies are built and what they can tell us about the history of life.

Chapter 5 begins our investigation of the molecular mechanisms that allow the long-term changes of evolution to occur. We examine how DNA encodes RNA and proteins and how mutations and other processes give rise to genetic variation. In Chapters 6 and 7, we look at the rise and fall of the frequency of gene variants in populations. Many mutations eventually disappear, whereas others spread widely. Sometimes these changes are the result of chance, a process called **genetic drift**; at other times they are more predictable, resulting from natural selection. Mutations spread if they cause their bearers to perform well—to thrive and be especially successful at reproducing. Mutations may decrease in frequency or be lost completely if their bearers perform poorly.

The discovery of DNA and mutations revolutionized the study of evolution. As we'll see in Chapter 8, your genes and the genes of all living things store a huge amount of historical information about evolution. As we saw with whales and flu viruses in this chapter, genes provide a separate line of evidence about the phylogeny of species. They can also allow scientists to reconstruct ancient episodes of natural selection, pinpointing the molecular alterations that gave some individuals a reproductive advantage.

Phylogenies based on fossils and DNA shed light on how new adaptations have evolved over the history of life. In Chapter 9, we examine the origins of adaptations. We focus in particular on how mutations affect the development of animals and other multicellular organisms. Relatively simple changes to developmental genes can significantly affect an organism's **phenotype**.

The effect of a mutation depends on more than just the mutation itself. It may be influenced by other genes that an organism carries. The environment in which an organism lives can also have a huge effect. As a result, the same mutation to the same gene may be devastating in one individual and harmless in another. Depending on the particular circumstances, natural selection may favor a mutation or drive it to oblivion. Charles Darwin believed that natural selection occurred so slowly that he couldn't hope to observe it directly. But in Chapter 10, we survey a number of cases in which scientists have documented natural selection taking place in our own lifetime.

Some complex adaptations, like the eye, are essential for an organism's survival. Other adaptations, such as antlers or bright tail feathers, serve another vital function: attracting or defending mates in a process known as sexual selection. In Chapter 11, we look at how profoundly animals are influenced by sex and sexual selection. In Chapter 12, we turn our attention to what happens after sex—namely, how evolution shapes the ways in which parents rear their young.

Natural selection and sexual selection alike can play important roles in the origin of new species, the subject of Chapter 13. And over millions of years, the origin of new species leads to complex patterns of diversity through both time and space. In Chapter 14, we look at the evolution of biological diversity and examine the current drop in diversity due to widespread extinctions. Chapter 15 is dedicated to coevolution, the process by which different lineages respond to each other's adaptations and create ecological partnerships.

The phenotype of an organism is more than just the anatomy we can see. It's also an organism's behavior. In Chapter 16, we look at how behavior evolves. We consider studies of the genetics of behavior and explore how the mating systems of species help shape their social behaviors. Behavior is an enormous part of the story of human evolution, the subject of Chapter 17, in which we look closely at the evidence of human evolution to be found in the fossil record, in the human genome, and in psychology experiments that probe the mind.

Throughout this book, we pay special attention to the practical side of evolutionary biology. We discuss, for example, how insects and weeds evolve resistance to the chemicals farmers use. One of the most exciting areas of applied evolutionary biology is the subject of Chapter 18: medicine. Doctors seek to treat our bodies, which are the product of some 3.7 billion years of evolution. They would do well to understand the role of evolution in forming our bodies and minds.

Genetic drift is evolution arising from random changes in the genetic composition of a population from one generation to the next.

A **phenotype** is a measurable aspect of organisms, such as morphology (structure), physiology, and behavior. Genes interact with other genes and with the environment during the development of the phenotype.

TO SUM UP . . .

- Biological evolution is a process by which populations of organisms change over time.

- Mutations are changes in the genome of an organism. They can be neutral, detrimental, or beneficial.

- Whales evolved from land mammals about 50 million years ago. Evolution has shaped many aspects of their biology, from development to behavior.

- Natural selection occurs when heritable characteristics cause some individuals to survive and reproduce more successfully than others.

- Natural selection is a mechanism of evolution that can cause the genetic composition of a population to shift from generation to generation.

- The rapid evolution of the influenza virus makes it difficult to fight.

- Evolutionary biologists use many different lines of evidence to test hypotheses about evolution.

- Evolutionary theory explains the patterns of life observable in the natural world and the processes by which that life has evolved.

MULTIPLE CHOICE QUESTIONS Answers can be found at the end of the book.

1. Define biological evolution.

   a. Any process by which populations of organisms change over time.
   b. Any change in the inherited traits of a population that occurs from one generation to the next.
   c. Change within a lineage due to natural selection and other mechanisms.
   d. All of the above.

2. What type of evidence has been used to infer that whales evolved from mammals that used to live on land (rather than from fishes, for example)?

   a. Whales share traits with other mammals, like needing to breathe air and giving birth to live young that feed on milk.
   b. Whales still have developmental genes for traits they no longer have, like hind legs.
   c. Fossils of early whales display combinations of traits that reveal a gradual transition from terrestrial to aquatic life.
   d. All of the above.

3. Why do baleen whales still have genes for building teeth?

   a. These genes are now used to make baleen.
   b. Their ancestors had teeth, and they inherited these genes from them, even though the genes no longer function.
   c. Their descendants might need teeth, so evolution keeps the genes around.
   d. Evolution can't take away genes, only add new ones.

4. What makes hemagglutinin important in the evolution of influenza viruses?

   a. It allows a virus to attack the red blood cells of its host.
   b. It is the basic building block of virus cell walls.
   c. It allows the virus to bind to the cells of its host.
   d. All of the above.

5. Which of the following is a *true* statement?

   a. The ancestors of whales needed more food than could be found on land, so they evolved features that allowed them to survive in the water.
   b. Whales and humans share a common ancestor.
   c. Mutations always cause the improvement of a trait.
   d. Viruses mutate because they want to have the best-adapted hemagglutinins throughout their evolution.

6. Which of the following is *not* a place that scientists look for evidence of evolution?

   a. The fossil remains of extinct animals.
   b. Comparison of homologous traits in various species.
   c. Change during an individual organism's lifetime.
   d. Change across generations in wild populations.

7. Which of these statements about phenotypes is *true*?

   a. Individuals that adjust their phenotypes in response to their environment cannot be favored by natural selection.
   b. Natural selection does not act on phenotypes.
   c. An individual's behavior is not part of its phenotype.
   d. An individual's phenotype is a result of its genes and the environment in which it develops.

8. Why do scientists overwhelmingly accept the theory of evolution?

   a. Because the theory has overwhelming evidentiary support.

   b. Because the theory explains and predicts independent lines of evidence.

   c. Because scientists have tested and retested its predictions.

   d. All of the above.

9. Which of the following best describes natural selection?

   a. Natural selection is a mechanism that allows individuals to pass on to their offspring those traits that are favorable.

   b. Natural selection occurs when heritable characteristics cause some individuals to survive and reproduce more successfully than others.

   c. Natural selection is a mechanism that allows for greater reproductive success in all species within a given environment.

   d. All of the above.

10. Why is influenza a worldwide threat?

   a. Because influenza has the potential to kill millions of people from a single outbreak.

   b. Because mutations in the influenza virus can allow the virus to evade the human immune system.

   c. Because viral reassortment can result in the flu crossing over from another species to humans.

   d. All of the above.

INTERPRET THE DATA Answer can be found at the end of the book.

11. In the figure below, what do the arrows above the host cell represent?

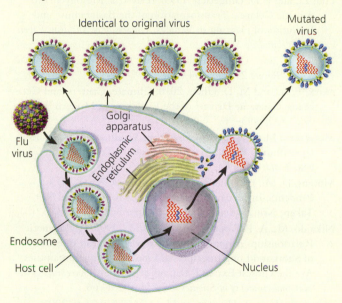

Identical to original virus · Mutated virus · Golgi apparatus · Endoplasmic reticulum · Flu virus · Endosome · Host cell · Nucleus

   a. Transfer of surface proteins from one virus to the next.

   b. Movement of viruses from one host to the next.

   c. Reassortment of virus genes from one time point to the next.

   d. Copies/daughter viruses produced by each virus from one generation to the next.

SHORT ANSWER QUESTIONS Answers can be found at the end of the book.

12. Recall the quotation from Theodosius Dobzhansky: "Nothing in biology makes sense except in the light of evolution." Give one example of how evolutionary principles help us "make sense" of biological observations.

13. What kinds of evidence do evolutionary biologists use to test hypotheses about how different species are related to each other?

14. What are some characteristics that distinguish whales from sharks and tunas?

15. How do mutations become more or less common in a population over the course of generations?

16. What role does viral reassortment play in flu pandemics?

17. Explain how phylogeny helps us understand the evolutionary history of populations, genes, and species.

ADDITIONAL READING

Gatesy, J., J. H. Geisler, J. Chang, C. Buell, A. Berta, et al. 2012. A Phylogenetic Blueprint for a Modern Whale. *Molecular Phylogenetics and Evolution* 66:479–506.

Mayr, E. 2001. *What Evolution Is*. New York: Basic Books.

Morens, D. M., J. K. Taubenberger, and A. S. Fauci. 2013. Pandemic Influenza Viruses—Hoping for the Road Not Taken. *New England Journal of Medicine* 368:2345–48.

Pyenson, N. D. 2017. The Ecological Rise of Whales Chronicled by the Fossil Record. *Current Biology* 27:R558–64.

Thewissen, J. G. M. 2014. *The Walking Whales: From Land to Water in Eight Million Years*. Berkeley, CA: University of California Press.

Uhen, M. 2010. The Origin(s) of Whales. *Annual Review of Earth and Planetary Sciences* 38:189–219.

University of California Museum of Paleontology. 2011. Misconceptions About Evolution. http://evolution.berkeley.edu/evolibrary/misconceptions_faq.php (accessed August 17, 2018).

Zimmer, C. 1998. *At the Water's Edge: Macroevolution and the Transformation of Life*. New York: Free Press.

PRIMARY LITERATURE CITED IN CHAPTER 1

Barry, J. 2005. *The Great Influenza: The Story of the Greatest Pandemic in History*. New York: Penguin Books.

Connor, R. C. 2007. Dolphin Social Intelligence: Complex Alliance Relationships in Bottlenose Dolphins and a Consideration of Selective Environments for Extreme Brain Size Evolution in Mammals. *Philosophical Transactions of the Royal Society B: Biological Sciences* 362 (1480): 587–602.

Cooper, L. N., J. G. M. Thewissen, and S. T. Hussain. 2009. New Early–Middle Eocene Archaeocetes (Pakicetidae and Remingtonocetidae, Cetacea) from the Kuldana Formation of Northern Pakistan. *Journal of Vertebrate Paleontology* 29:1289–99.

Dobzhansky, T. 1973. Nothing in Biology Makes Sense Except in the Light of Evolution. *American Biology Teacher* 35:125–29.

Drake, J. W., and J. J. Holland. 1999. Mutation Rates Among RNA Viruses. *Proceedings of the National Academy of Sciences USA* 96 (24): 13910–13.

Gatesy, J., J. H. Geisler, J. Chang, C. Buell, A. Berta, et al. 2012. A Phylogenetic Blueprint for a Modern Whale. *Molecular Phylogenetics and Evolution* 66:479–506.

Geisler, J. H., M. W. Colbert, and J. L. Carew. 2014. A New Fossil Species Supports an Early Origin for Toothed Whale Echolocation. *Nature* 508:383–86.

Gingerich, P. D., K. D. Rose, and D. W. Krause. 1980. Early Cenozoic Mammalian Faunas of the Clark's Fork Basin–Polecat Bench Area, Northwestern Wyoming. *University of Michigan Papers on Paleontology* 24:51–68.

Goldbogen, J. A., J. Potvin, and R. E. Shadwick. 2010. Skull and Buccal Cavity Allometry Increase Mass-Specific Engulfment Capacity in Fin Whales. *Proceedings of the Royal Society B: Biological Sciences* 277:861–68.

Imai, M., T. Watanabe, M. Kiso, N. Nakajima, S. Yamayoshi, et al. 2017. A Highly Pathogenic Avian H7N9 Influenza Virus Isolated from a Human Is Lethal in Some Ferrets Infected via Respiratory Droplets. *Cell Host & Microbe* 22:1–12.

IUCN-SSC. Status of the World's Cetaceans. http://www.iucn-csg.org/index.php/status-of-the-worlds-cetaceans/ (accessed November 11, 2017).

Jagger, B. W., H. M. Wise, J. C. Kash, K. A. Walters, N. M. Wills, et al. 2012. An Overlapping Protein-Coding Region in Influenza A Virus Segment 3 Modulates the Host Response. *Science* 337 (6091): 199–204.

Luo, Z., and P. D. Gingerich. 1999. Terrestrial Mesonychia to Aquatic Cetacea: Transformation of the Basicranium and Evolution of Hearing in Whales. *University of Michigan Papers on Paleontology* 31:1–98.

Marino, L. 2007. Cetacean Brains: How Aquatic Are They? *Anatomical Record* 290 (6): 694–700.

Marx, F. G., and M. D. Uhen. 2010. Climate, Critters, and Cetaceans: Cenozoic Drivers of the Evolution of Modern Whales. *Science* 327:993–96.

McGowen, M. R., J. Gatesy, and D. E. Wildman. 2014. Molecular Evolution Tracks Macroevolutionary Transitions in Cetacea. *Trends in Ecology and Evolution* 6:336–46.

Morens, D. M., J. K. Taubenberger, and A. S. Fauci. 2013. Pandemic Influenza Viruses—Hoping for the Road Not Taken. *New England Journal of Medicine* 368:2345–48.

Nikaido, M., A. P. Rooney, and N. Okada. 1999. Phylogenetic Relationships Among Cetartiodactyls Based on Insertions of Short and Long Interspersed Elements: Hippopotamuses Are the Closest Extant Relatives of Whales. *Proceedings of the National Academy of Sciences USA* 96:10261–66.

Nummela, S., S. T. Hussain, and J. G. M. Thewissen. 2006. Cranial Anatomy of Pakicetidae (Cetacea, Mammalia). *Journal of Vertebrate Paleontology* 26 (3): 746–59.

Orlando, L., A. Ginolhac, G. Zhang, D. Froese, A. Albrechtsen, et al. 2013. Recalibrating Equus Evolution Using the Genome Sequence of an Early Middle Pleistocene Horse. *Nature* 499:74–78.

Perrin, W. F. 2017. World Cetacea Database. http://www.marinespecies.org/cetacea (accessed November 11, 2017).

Pybus, O. G., A. J. Tatem, and P. Lemey. 2015. Virus Evolution and Transmission in an Ever More Connected World. *Proceedings of the Royal Society B: Biological Sciences* 282:20142878.

Roe, L. J., J. G. M. Thewissen, J. Quade, J. R. O'Neill, S. Bajpai, et al. 1998. Isotopic Approaches to Understanding the Terrestrial-to-Marine Transition of the Earliest

Cetaceans. In J. G. M. Thewissen (ed.), *The Emergence of Whales: Evolutionary Patterns in the Origin of Cetacea* (pp. 399–422). New York: Plenum Publishing.

Simmonds, P., M. J. Adams, M. Benkö, M. Breitbart, J. R. Brister, et al. 2017. Consensus Statement: Virus Taxonomy in the Age of Metagenomics. *Nature Reviews Microbiology* 15:161–68.

Thewissen, J. G. M., and S. Bajpai. 2001. Whale Origins as a Poster Child for Macroevolution. *Bioscience* 15:1037–49.

Thewissen, J. G. M., M. J. Cohn, L. S. Stevens, S. Bajpai, J. Heyning, et al. 2006. Developmental Basis for Hind-Limb Loss in Dolphins and Origin of the Cetacean Bodyplan. *Proceedings of the National Academy of Sciences USA* 103:8414–18.

Thewissen, J. G. M., L. Cooper, J. George, and S. Bajpai. 2009. From Land to Water: The Origin of Whales, Dolphins, and Porpoises. *Evolution: Education and Outreach* 2 (2): 272–88.

Thewissen, J. G. M., S. I. Madar, E. Ganz, S. T. Hussain, M. Arif, et al. 1997. Fossil Yak (Bos Grunniens: Artiodactyla, Mammalia) from the Himalayas of Pakistan. *Kirtlandia (Cleveland)* 50:11–16.

Turvey, S. T. 2009. *Witness to Extinction: How We Failed to Save the Yangtze River Dolphin*. Oxford: Oxford University Press.

Uhen, M. 2010. The Origin(s) of Whales. *Annual Review of Earth and Planetary Sciences* 38:189–219.

UNAIDS. 2017. *UNAIDS Data 2017*. Joint United Nations Programme on HIV/AIDS (UNAIDS).

Viboud, C., and L. Simonsen. 2012. Global Mortality of 2009 Pandemic Influenza A H1N1. *The Lancet Infectious Diseases* 12 (9): 651–53.

Zimmer, C. 1998. *At the Water's Edge: Macroevolution and the Transformation of Life*. New York: Free Press.

CHAPTER 2

From Natural Philosophy to Darwin

A Brief History of Evolutionary Ideas

Learning Objectives

- Identify early naturalists and discuss their contributions to evolutionary theory.
- Explain the role the fossil record played in the development of the concept of evolution.
- Discuss how Darwin's observations of nature led to the inferences he developed regarding natural selection.
- Give three examples of homologies.

In the Pacific Ocean, seven hundred miles west of Ecuador, lies an isolated cluster of volcanic islands known as the Galápagos. On these strange outcrops are strange kinds of life. Large seabirds with bright blue feet tend their nests. Scaly iguanas leap into the ocean to eat seaweed and then wade back out to bask on the rocks. Giant tortoises chew peacefully on cactuses. The finches of the islands are so tame that they will let you hold them in your hand.

Every year, dozens of scientists come from across the world to the Galápagos Islands to study these species, many of which exist nowhere else in the world. The islands are like a laboratory of evolution where scientists can study an isolated example of how life has changed over millions of years. The journey to the remote islands is not easy—but a journey to the Galápagos Islands today is not nearly as arduous as the journeys by steamer that scientists took a hundred years ago. And *those* steamer trips were much faster than the voyage of a British surveying ship that sailed to the Galápagos Islands in 1835. On board the HMS *Beagle* was a young British naturalist named Charles Darwin (1809–82) (**Figure 2.1**).

Darwin had been traveling aboard the *Beagle* for almost four years, during which time he had studied the marine life of the Atlantic, trekked in the jungles of Brazil, and climbed

The Galápagos Islands in the Pacific are home to several species of tortoises found nowhere else on Earth. They descend from a common ancestor that arrived on the islands millions of years ago and then diversified into different forms.

Figure 2.1 Charles Darwin (1809–82) at age 31. Five years before this portrait was painted, in 1835, he visited the Galápagos Islands. The experience helped shape his theory of evolution. (Iberfoto/Superstock)

A **fossil** is preserved evidence of life from a past geological age, including impressions and mineralized remains of organisms embedded in rocks.

the Andes. But even after all that, Darwin was astonished by the Galápagos Islands. "The natural history of this archipelago is very remarkable: it seems to be a little world within itself," he wrote later in his book *The Voyage of the Beagle*. Darwin spent five weeks on the islands, clambering over jagged volcanic rocks and gathering plants and animals. The experiences he had during his voyage around the globe would later lead him to a scientific revolution.

Darwin was born in 1809, at a time when most people—including the world's leading naturalists—thought the world was only thousands of years old, not billions. They generally believed that species had been specially created, either at the beginning of the world or from time to time over Earth's history. But after Darwin returned from his voyage around the world in 1836, his experiences in places like the Galápagos Islands caused him to question those beliefs. He opened a notebook and began jotting down ideas for a new theory of life, one in which life evolved.

In this chapter, we explore the historical foundations of evolutionary biology. We'll begin not with Darwin but with the philosophers in ancient Greece who lived 2500 years earlier. It is to them that we attribute the first overarching explanations for life and its diversity. They conceived of life as forming a fixed hierarchy from lower to higher forms, a view that continued to dominate Western thought during the Renaissance some two thousand years later. In the 1600s, naturalists developed more sophisticated ways to classify species according to their similarities. As we'll see, these early naturalists also recognized that **fossils** are mineralized remains of living things, and as a result geologists then began to chart Earth's history by studying layers of rock. By the early 1800s, a few naturalists were beginning to argue that life had changed over vast stretches of time. Darwin was just starting out in his career as a scientist as these debates were unfolding. In this chapter, we'll see how he masterfully synthesized the work of previous researchers with his own observations to produce a theory of evolution that remains the foundation of biology today. ●

2.1 Conceptions of Nature Before Darwin

Darwin first began to learn about nature as a teenager in the 1820s. The concepts he was taught had emerged over previous centuries, as naturalists pondered two questions: what were the patterns in nature's diversity, and how did those patterns come to be?

The Great Chain of Being

Understanding the diversity of life had long been a practical necessity. People needed names for different kinds of plants and animals, for example, so that they could pass on their wisdom about which kinds were safe to eat or useful as medicines. For thousands of years, people had been well aware that some kinds of animals and plants were similar to other kinds. Cats and cows and humans all nourished their young with milk, for example.

Figure 2.2 Medieval European scholars envisioned nature as a "Great Chain of Being," from lower to higher forms. This illustration accompanies a book by the philosopher and theologian Ramon Llull published in 1303. Here, nature's hierarchy is illustrated as a series of steps, labeled as stone, flame, plant, animal, man, heaven, angel, God. (Album / Kurwenal / Prisma / Album / Superstock, Inc.)

But beyond those practical needs, early Greek philosophers also searched for an all-encompassing explanation for the patterns of diversity they observed. Many of them saw species as being arranged on a scale from lower to higher forms—a concept that eventually came to be known as the Great Chain of Being (Lovejoy 1936; Wilkins 2009). According to this view, plants were the lowest forms of life, having only the capacity to grow. Animals occupied a higher position on the chain because, in addition to growth, they also had the ability to move. Humans were situated even higher, thanks to their powers of reason (**Figure 2.2**). Whereas the Great Chain of Being had its origins in ancient Greek philosophy, European scholars in the Middle Ages and the Renaissance adapted it to a Christian view of the cosmos—as a divine plan established by God at creation.

The Great Chain of Being might have been an attractive theological concept, but it wasn't very useful for classifying life's diversity at a finer scale. A marigold and a lily are both lower forms of life according to this view; however, it's obvious that they are not identical to each other. In the 1600s, naturalists dissatisfied with the Great Chain of Being began to develop a systematic way to classify species. They came up with rules for naming species and schemes for sorting species into different groups. The Swedish naturalist Carolus Linnaeus (1707–78) brought this method to its pinnacle in the mid-1700s (**Figure 2.3**).

Figure 2.3 Carolus Linnaeus (1707–78) invented a system to classify biological entities into groups based on their similarities. (Hirarchivum Press / Alamy Stock Photo)

Linnaean Classification

Linnaeus organized all living things known at the time into a nested hierarchy of groups, or **taxa** (Koerner 1999). Humans belong to the mammal class, for example; and within that class, the primate order; and within that order, the family Hominidae; and within that family, the genus *Homo*; and within that genus, the

A **taxon** (plural, **taxa**) refers to groups of organisms that a taxonomist judges to be cohesive units, such as species or orders.

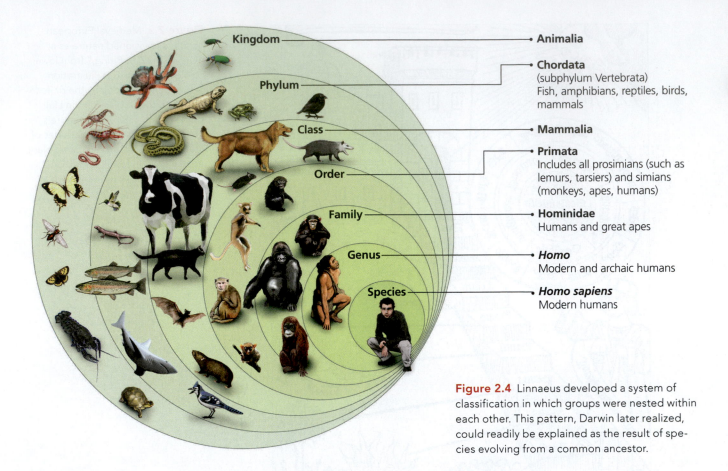

Kingdom	**Animalia**
Phylum	**Chordata** (subphylum Vertebrata) Fish, amphibians, reptiles, birds, mammals
Class	**Mammalia**
Order	**Primata** Includes all prosimians (such as lemurs, tarsiers) and simians (monkeys, apes, humans)
Family	**Hominidae** Humans and great apes
Genus	***Homo*** Modern and archaic humans
Species	***Homo sapiens*** Modern humans

Figure 2.4 Linnaeus developed a system of classification in which groups were nested within each other. This pattern, Darwin later realized, could readily be explained as the result of species evolving from a common ancestor.

Taxonomy is the science of describing, naming, and classifying species of living or fossil organisms.

species *Homo sapiens* (**Figure 2.4**). Linnaeus could assign every species to a particular genus, family, or order according to the traits it shared with other species. His system of **taxonomy** was so useful that biologists still use it today.

Linnaeus believed that the pattern of his system reflected a divine plan. "There are as many species as the Infinite Being produced diverse forms in the beginning," he wrote. In some cases, Linnaeus thought that species had later changed. He believed that two species of plants could sometimes interbreed, producing a new hybrid species. For the most part, though, Linnaeus assumed that the overall patterns of life's diversity had not changed since the biblical creation of the world.

The Fossil Record

While Linnaeus studied the diversity of life in its present form, other naturalists were looking back over life's history. They discovered that the remains of animals and plants were sometimes preserved and transformed into stone. One of the first naturalists to realize this was Nicolaus Steno (1638–86), a seventeenth-century Dutch anatomist and bishop in the Catholic Church. In 1666, some fishermen brought him a giant shark they had caught. As Steno studied the shark's teeth, it occurred to him that they looked just like triangular rocks that were known at the time as tongue stones. Steno proposed that tongue stones had started out as teeth in living sharks. He came to realize that after the sharks died, their teeth gradually were transformed into stone (**Figure 2.5**; Cutler 2003).

But if fossils really were the remains of once-living things, Steno would have to explain how it was that stones shaped like seashells had come to be found on top of certain mountains. How could animals that lived in the ocean end up so far from home? Steno argued that originally a sea must have covered the mountains. Shelled animals died and fell to the ocean floor, where they were covered over in sediment. As sediments accumulated, they turned to rock. The layers of rocks exposed on the

TABULA I.
LAMIAE PISCIS CAPVT.

EIVSDEM LAMIAE DENTES.

Figure 2.5 Nicolaus Steno (1638–86) recognized that triangular rocks known as tongue stones were in fact fossils of teeth from sharks. (Left: Sarin Images/GRANGER—All rights reserved.; right: www.Lowcountrygeologic.com)

sides of mountains, Steno recognized, had been laid down in succession with the oldest layers at the bottom and the youngest ones at the top. (These layers came to be known as strata, and their study was named **stratigraphy**.)

Steno was still a traditional believer in a biblical Earth that was just a few thousand years old. Nevertheless, he was able to introduce a radically new idea: life and the planet that supported it had a history filled with change, and Earth itself kept a record of that history.

Natural Theology

Steno's deep knowledge of anatomy allowed him to recognize the similarities between tongue stones and sharks' teeth. Such expertise was new in Europe, fostered by the rediscovery of anatomical research carried out in ancient Greece. Building on that classic work, seventeenth-century European anatomists studied the organs of humans and other species to determine their functions. They came to appreciate the similarities as well as the differences between species. The English physician William Harvey (1578–1657), for example, compared the hearts of humans, snakes, and fishes. Their similarities helped him discover that they all share the same function: to pump blood in a circular course through the body.

Anatomists in the 1600s saw their research as further evidence of the divine order of nature reflected in the Great Chain of Being. The anatomy of each species was exquisitely adapted to its particular way of life, and its structure also reflected its position in the Great Chain of Being. For example, the English physician Thomas Willis (1621–75), a disciple of Harvey, carried out the first large-scale survey of animal brains in the mid-1600s (Zimmer 2004). He observed that lobsters and other invertebrates—considered to be low on the Great Chain of Being—had simple, small brains. Mammals such as cows and sheep had much larger brains with more intricate structures. Humans, Willis recognized, had much larger and more complex brains than other animals did.

The seventeenth century also saw the rise of physics and engineering, which influenced how anatomists interpreted the functions of the bodies they studied. They treated bodies as natural machines, made up of pumps, engines, and channels. The mechanical function of living things came to be seen as evidence of God's design. Robert Hooke (1635–1703), an assistant to Willis who went on to become

Stratigraphy is the study of layering in rock (stratification).

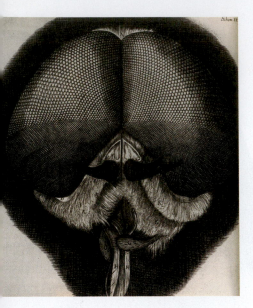

Figure 2.6 An engraving of a fly's eyes, from Robert Hooke's 1665 book *Micrographia*. Hooke saw the microscopic complexity in nature as evidence that it had been designed. (Courtesy of the U.S. National Library of Medicine)

a physicist, argued in 1665 that God had furnished each plant and animal "with all kinds of contrivances necessary for its own existence and propagation . . . as a Clock-maker might make a Set of Chimes to be a part of a Clock" (**Figure 2.6**; quoted in Ayala 2013). This argument, which became more popular in the eighteenth century, came to be known as natural theology.

The most influential formulation of natural theology was published in 1802 by the English minister William Paley (1743–1805). In *Natural Theology, or Evidences of the Existence and Attributes of the Deity Collected from the Appearances of Nature* (Paley 1802), Paley asked his readers to picture themselves walking across a heath. If they tripped on a rock, they would consider that rock nothing more than a simple, natural part of its surroundings. But what if they stumbled across a watch lying on the ground? They would draw a very different conclusion about its origin. It could not have formed by accident. The organization of its parts, which together carry out the function of telling time, was clearly evidence of design.

Anatomy provided the same evidence for design, Paley argued. Just as a watch implies a watchmaker, the intricate design of life is evidence of a Divine Creator. In *Natural Theology*, Paley cataloged many examples of exquisitely complex organs and noted their similarity to human-made machines. The human eye and the telescope both took advantage of the same laws of physics to focus light through lenses, for example.

Two decades after the minister's death, Darwin would occupy Paley's rooms at Cambridge. There, Darwin would read *Natural Theology* and be deeply impressed by Paley's arguments. But, as we'll see later in this chapter, Darwin would go on to find a more powerful explanation for the anatomical complexity of life: it emerged through evolution by natural selection.

Key Concepts

- The Great Chain of Being, which ranked life on a scale, was a dominant concept of nature for nearly two thousand years.

- Carolus Linnaeus is considered the founder of modern taxonomy because his system for grouping organisms into a nested hierarchy is still in use today (although many of the groupings he proposed are not).

- Stratigraphy is the study of layering in rock (stratification). Nicolaus Steno, the founder of geology, pioneered the use of stratigraphy as a method for reconstructing the past.

- William Paley proposed that the mechanical complexity of animal organs provided evidence for the existence of a Divine Creator. ●

2.2 Evolution Before Darwin

The concept that life changes over the course of vast stretches of time—what came to be known as evolution—was already being vigorously debated before Darwin was born. One of the earliest evolutionary thinkers was the eighteenth-century French nobleman Georges-Louis Leclerc, Comte de Buffon (1707–88) (**Figure 2.7**). Buffon was the director of the King's Garden in Paris, and he owned a huge estate in Burgundy, where he harvested timber for the French navy and carried out research on the strength of different kinds of wood. Buffon also spent years writing an encyclopedia in which he intended to include everything known about the natural world (Roger 1997).

Like other thinkers in the mid-1700s, Buffon recognized that the new sciences of physics and chemistry offered a radically different way of thinking about the universe. It had become clear that the world was made up of minuscule particles, which we now call atoms and molecules. These particles reacted with each other according to certain laws, and when they came together into larger objects, the objects obeyed certain laws as well. They were attracted to one another by gravity, for example, and either attracted or repelled by electric charge. Following these laws, the particles moved about, and the complexity of the universe emerged spontaneously as a result.

Buffon proposed that Earth had formed according to the laws of physics. A comet struck the Sun, he argued, breaking off debris that formed into a planet. The scorching Earth cooled down and hardened, and oceans formed and dry land emerged. The entire process took more than 70,000 years, Buffon calculated—a span of time too vast for most people in Buffon's day to imagine.

The fact that living things were made from the same kinds of particles found in rocks and water struck Buffon as profoundly important. He argued that each species had a supply of organic particles that somehow transformed an egg or a seed into its adult form. He envisioned that these organic particles had first come together in the hot oceans of early Earth. Animals and plants sprang into existence in the process, and, as the planet cooled, they retreated to the warm tropics. Those migrations could explain the stunning discovery in the mid-1700s of fossil elephants in Siberia and North America, far from the tropics where elephants live today.

When life first emerged, Buffon proposed, it was already divided into a number of distinct types. Each "internal mould," as he called it, organized the organic particles that made up any individual creature. But life could also be transformed. As a species moved to a new habitat, its organic particles changed, and its mould changed as well. Buffon was proposing something quite close to a central tenet of modern evolutionary theory: populations change over time.

Figure 2.7 Georges Buffon (1707–88) was one of the earliest naturalists to argue that life had changed over time. (Pictorial Press Ltd/Alamy Stock Photo)

Fossils and Extinctions

Steno's realization that fossils were the remains of living things helped open up a new science that came to be known as **paleontology**. Some of the most startling discoveries were fossils of species that no longer lived at the sites where they were unearthed—such as the Siberian elephants (Rudwick 1985). When the French naturalist Georges Cuvier (1769–1832) compared the elephant fossils to the skeletons of living elephants from Africa and India, he discovered that some of the fossils were distinct from living elephants in crucial ways, such as the shapes of their teeth. These fossil animals, which he called mammoths and mastodons, were species that no longer existed. They had, in other words, become extinct (**Figure 2.8**; Rudwick 1997).

Paleontology is the study of prehistoric life.

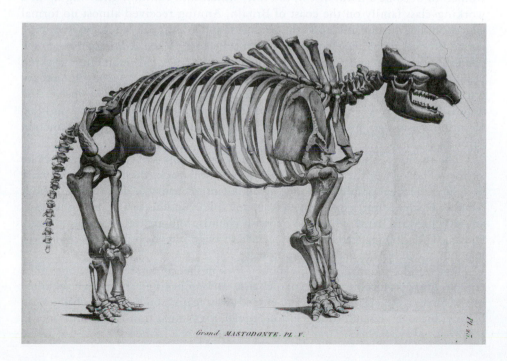

Grand MASTODONTE. Pl. V.

Figure 2.8 In the late 1700s, paleontologists recognized that some fossils belonged to species that no longer existed, such as this mastodon, a relative of elephants that became extinct 11,000 years ago. (Courtesy of the Library & Archives of the Academy of Natural Sciences of Drexel University, Philadelphia, PA)

Figure 2.9 A: Mary Anning discovered fossils of marine reptiles and other strange animals. B: The fossils supplied compelling evidence that species do indeed become extinct. (A: The Natural History Museum/Alamy Stock Photo; B: The Natural History Museum, London/Science Source)

Extinction refers to the permanent loss of a species. It is marked by the death or failure to breed of the last individual.

Extinction was a difficult concept to incorporate into the explanations that European naturalists had developed over the previous centuries. If species became extinct, they would leave gaps in the Great Chain of Being. The concept of extinction thus met with a great deal of resistance. But other naturalists built on Cuvier's research and found other fossils of species that clearly no longer existed. Some of the most spectacular of those early fossils were discovered by a British naturalist named Mary Anning (1799–1847; **Figure 2.9A**).

Although many evolutionary biologists today are women, it was rare for a woman to become a naturalist in the early nineteenth century. Growing up in a working-class family on the coast of Britain, Anning received almost no formal education, but during childhood she became familiar with paleontology by helping her family find fossils that they then sold to collectors. When Anning grew up, she opened a fossil shop and filled it with the bones she collected—sometimes risking death from drowning in high tides or being buried in landslides. Among her discoveries were spectacular reptiles unlike any living today—giant sharklike species and winged flying forms (**Figure 2.9B**). The discovery of these species, which vanished 65 million years ago, provided a powerful argument for the concept of extinction (Emling 2011).

The reality of extinction raised the question of how, exactly, species became extinct. The answer turned out to be hidden in the rocks themselves—or, more precisely, their geography. During the eighteenth century, a debate raged about whether Earth's features had formed from volcanic eruptions or floods. An important step forward came when the Scottish chemist and amateur geologist James Hutton (1726–97) realized that rocks formed through imperceptibly slow changes—many of which we can see around us today.

Rain erodes mountains, while molten rock pushes up to create new ones. The eroded sediments form into layers of rock that can later be lifted above sea level, tilted by the force of the uprising rock, and eroded away again. Some of these changes can be tiny; but over enough time, Hutton argued, they could transform landscapes in

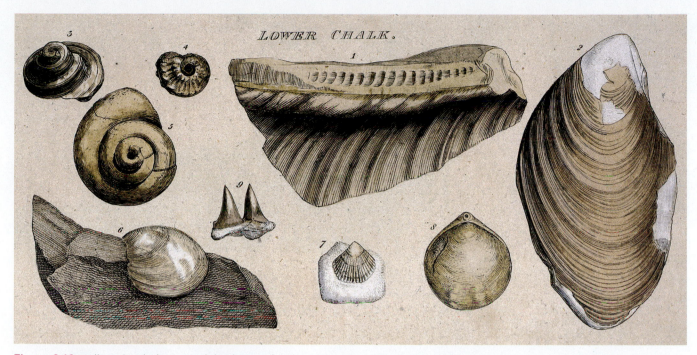

Figure 2.10 William Smith discovered that layers of rocks contain distinctive groups of fossils. (The Natural History Museum/Alamy Stock Photo)

dramatic ways. Earth must therefore be vastly old—Hutton envisioned Earth as a sort of perpetual-motion machine passing through regular cycles of destruction and rebuilding that made the planet suitable for humankind.

Hutton's vision of a slowly transformed Earth came to be accepted by most geologists in the 1800s. They looked closely at layers of exposed rock and began to determine how they were formed by volcanoes and deposits of sediment. And they began to figure out the order in which those layers had formed. Some of the most important clues to the geological history of Earth came from fossils. William Smith (1769–1839), a British canal surveyor, came to this realization as he inspected rocks around England to decide where to dig canals (**Figure 2.10**). He noticed that the same kinds of fossils tended to appear in older rocks, but different ones appeared in younger layers. Smith could find the same sets of fossils in rocks separated by hundreds of miles (**Figure 2.11**; Winchester 2001).

By the early 1800s, then, geologists came to agree that the surface of the planet had been sculpted gradually over vast spans of time. Smith realized that each type of animal had lived across a wide geographical range for a certain period of time, and so the rocks formed during that time preserved their fossils. As those animals became extinct and new ones emerged, younger rocks contained their own sets of fossils. By marking the places where he found certain fossils, Smith was able to organize strata into a geological history, from oldest to youngest.

Figure 2.11 William Smith learned how to recognize the same layers of rocks in different parts of England by looking at the fossils they contained. In this map, Smith used color to identify rocks from the same layer. (The Trustees of the Natural History Museum, London)

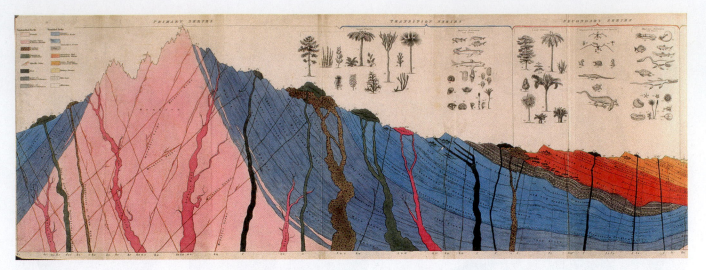

Figure 2.12 Early geologists organized surface features of Earth into strata based on where certain fossils were found. This early engraving of a mountain depicts strata and the associated fossils. (The Trustees of the Natural History Museum, London)

Other researchers, including Cuvier, later used the same method to map the geology of other parts of the world (**Figure 2.12**). They discovered that formations of rock exposed in one country could often be found in others. They began giving names to the sequences of these far-flung rock formations (see the geological chart at the front of this book). Many fossil species were restricted to just a few layers of rock. Larger groups of species spanned more geological history, but they had their own beginnings and endings as well. In the early 1800s, for example, fossil hunters discovered the bones of gigantic reptiles; some of them had lived on land and some in the sea. These fossils came only from rocks dating back to the Mesozoic era (252 to 66 million years ago) and disappeared abruptly at its end.

Why species emerged and disappeared over the history of life was a subject of fierce debate. Cuvier, for example, rejected Buffon's earlier suggestion that life had evolved. He believed that life's history was punctuated by revolutions that had wiped out many species and brought many new ones to take their place. But one of his colleagues at the National Museum of Natural History in Paris was about to make a new case for evolution.

Lamarck's Theory of Evolution

Figure 2.13 Jean-Baptiste Lamarck (1744–1829) argued that complex species had evolved from simple ones. (Hulton Archive/Getty Images)

Cuvier's colleague, Jean-Baptiste Pierre Antoine de Monet, Chevalier de Lamarck (1744–1829), was an expert on plants and invertebrates (**Figure 2.13**). He was struck by the anatomical similarities between some of the species he studied. He was also impressed by the fossil record, which at the time was becoming detailed enough to reveal a dynamic history of life (Burkhardt 1977). Lamarck eventually concluded that the diversity of life he saw around him was the product of evolution.

Although Lamarck was one of the first naturalists to offer a detailed theory of evolution, his theory differed from the modern conception of evolution in some important ways. Lamarck argued that life was driven inexorably from simplicity to complexity and that humans and other large species descended from microbes. In this respect, Lamarck continued to rely on the old concept of the Great Chain of Being, in which lower forms of life became higher. To explain why microbes still existed, Lamarck argued that primitive life was being spontaneously generated all the time. Today's bacteria are just the newest arrivals.

Lamarck also believed that animals and plants could adapt to their environment. If an animal began to use an organ more than its ancestors had, the organ would

BOX 2.1

Lamarck's View of Inheritance

The inheritance of acquired characteristics is an idea dating at least as far back as Aristotle and Hippocrates, although today it's associated most strongly with Lamarck. Although it has had various incarnations, the basic idea is that the physiological and physical changes acquired over an organism's lifetime can be transmitted to its offspring. Lamarck extended this idea to its logical conclusion: over vast stretches of time, the inheritance of acquired characteristics could lead to the emergence of new forms adapted to changing environments.

The concept was popular because it was intuitive. People could see with their own eyes how the phenotypes of their fellow humans changed over their lifetimes. A blacksmith hammering horseshoes for years developed a very different physique than a physician who pored over books and potions. (Today, we call this type of within-individual change phenotypic plasticity; see Chapters 5 and 7). What's more, the blacksmith's physique seemed well fitted to his way of life: requiring strength, he became strong. When naturalists looked at other species, they noted a similarly adaptive fit between the forms of organisms and their habitats. It was natural to connect these two types of patterns under a common mechanism, as Lamarck did. Today, this linking is called Lamarckism.

Biologists rejected Lamarckism as a mechanism of biological evolution in the early 1900s. With the discovery of genetic mechanisms of inheritance, it became clear that phenotypes were built anew each generation from genes inherited from the parents. The information contained in the DNA of germline cells (eggs and sperm) was transmitted to offspring; the rest of the parental phenotypes were not. Thus there seemed to be no way for traits acquired during the lifetime of an individual to be inherited.

For many decades, this was the end of the story. But recent advances in genetics and medicine have revealed a new twist to the story of inheritance. Sometimes information that lies *outside* of genes can be transmitted across generations. Molecules that adhere to DNA control whether particular genes are turned on or off. The experiences that parents have over their lifetimes can alter this scaffolding and thus influence how genes are expressed—a phenomenon known as epigenetics (see Chapter 12). And some of these epigenetic changes are reproduced in the cells of their offspring. In this way, acquired phenotypic states of parents can occasionally be transmitted to offspring.

Is it right to call this Lamarckian inheritance? It depends on what we mean by the term. The examples that emerge from epigenetics show little evidence of being adaptive. In fact, in many cases they can be maladaptive. And in many cases, an epigenetic "mark" seems to be passed down for only a few generations before disappearing. Nor have there been compelling examples in which entire populations have changed through inherited epigenetic changes. Thus Lamarckian inheritance does not seem to give rise to Lamarckian evolution (Zimmer 2018).

---

change during its lifetime. Lamarck argued that if a giraffe repeatedly stretched its neck for leaves high up on trees, for example, it would cause a "nervous fluid" to flow into its neck, making it grow longer. Lamarck claimed that these changes could be passed down from an animal to its offspring. A giraffe could inherit a longer neck; if the offspring continued stretching for leaves, it would pass on an even longer neck to its descendants. (See **Box 2.1** for more on Lamarck's view on inheritance.)

Lamarck gained attention across Europe for his shocking notion that species were not fixed. Cuvier, who did not believe that life evolved, led the scientific attack against Lamarck's theory. Lamarck envisioned life evolving up a seamless scale of nature from simple to complex. But Cuvier, who had undertaken an ambitious comparative study of major groups of animals, argued that they were divided by huge gulfs with no intermediates to join them. His ideas rejected, Lamarck died in poverty and obscurity in 1829 (Appel 1987).

However, just eight years after Lamarck's death, a young British naturalist, newly returned from a voyage around the world, quietly embraced the notion that life had evolved. And three decades after Lamarck's death, that naturalist—Darwin—would publish *The Origin of Species* and change the science of biology forever.

- Early naturalists contributed to evolutionary theory long before Darwin. Georges-Louis Buffon proposed that new varieties of a species could arise in response to new habitats. However, he did not believe that *species* could arise this way.

- Some of the first compelling evidence for extinction came from research conducted by Georges Cuvier, a pioneer in comparative anatomy and paleontology and an ardent anti-evolutionist.

- James Hutton envisioned a world with a deep history shaped by gradual transformations of landscapes through imperceptibly slow changes.

- The first geological map of fossils and rock layers was developed by William Smith, an English geologist and land surveyor.

- Jean-Baptiste Lamarck was an early proponent of evolution as a process that obeyed natural laws. ●

2.3 The Unofficial Naturalist

By the time Darwin was born in 1809, Lamarck was already famous (and infamous) for arguing that life had changed over a long history. Fifty years later, when Darwin finally presented his own theory of evolution, he could not be so easily dismissed. He had assembled a towering edifice of evidence and argument for evolution.

The story of Darwin's life makes his breakthrough all the more remarkable. He was born into comfortable wealth, thanks to the fortune his mother's family made manufacturing china and pottery. Darwin's father, a physician, expected Charles and his brother, Erasmus, to follow him into medicine, and he sent them to Edinburgh for training. There Charles also learned about geology, chemistry, and natural history, and he soon realized that he would much rather spend his life studying nature than practicing medicine. It was common in Darwin's day for well-to-do young men interested in nature to train in theology and become clergymen, using their spare time to pursue their investigations. Darwin started down that path, leaving Edinburgh to study theology at the University of Cambridge.

But Darwin grew restless. He reveled in a journey to Wales, where he was able to study geological formations. He devoured books about the travels of great naturalists to distant tropical countries. And then, at age 22, Darwin got a chance to go on a voyage of his own.

In 1831, Darwin was invited to join the company of a small British navy ship, HMS *Beagle*, on its voyage around the world (**Figure 2.14**). The captain, Robert Fitzroy, feared that the long journey might drive him to suicide. So Fitzroy began to search for a gentleman who might act as an unofficial naturalist for the voyage and whose companionship might keep him from succumbing to depression. He eventually settled on the 22-year-old Darwin, who was thrilled by the opportunity to explore the world. He would not return home for five years.

The *Beagle* traveled from England to South America. During his voyage, Darwin gathered fossils of extinct mammals. He trapped birds and collected barnacles. He observed the ecological complexity of the jungles of Brazil. Darwin also learned a great deal about geology in South America. He recognized the layers of rock that had gradually formed and were then reworked into mountains and valleys. He experienced an earthquake in Chile, and he observed that the shoreline had been lifted a few feet as a result. When Darwin set out on the *Beagle*, one of the books he brought with him was the first volume of *The Principles of Geology* (1830–33), which the Scottish lawyer and scholar Charles Lyell (1797–1875) had just published. Lyell made the provocative argument that Earth's landscapes had been created not by gigantic catastrophes but by a series of many small changes (a school of thought that grew out of Hutton's work and came to be called **uniformitarianism**). During the

Uniformitarianism is the idea that the natural laws observable around us now are also responsible for events in the past. One part of this view, for example, is the idea that Earth has been shaped by the cumulative action of gradual processes like sediment deposition and erosion.

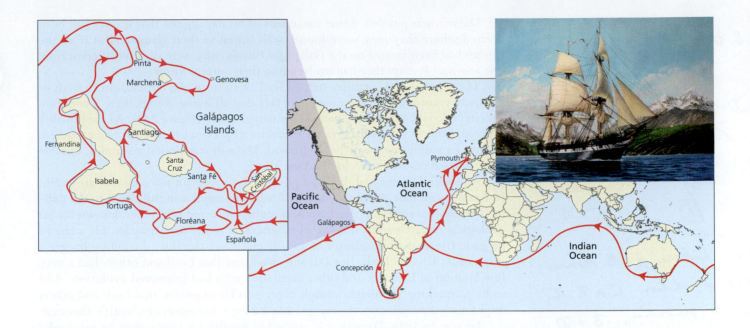

earthquake in Chile, Darwin saw firsthand one of these small changes taking place. And through his travels, Darwin became a passionate "Lyellian."

Darwin did not realize the full importance of his observations until he returned to England in 1836. His experiences on the Galápagos Islands proved to be especially inspirational. There he observed tortoises found nowhere else on Earth. But even on these tiny islands, there were different types of tortoises—some with domed shells and some with saddle-shaped ones.

Darwin also collected a number of Galápagos birds that had dramatically different beaks. Some had massive beaks, good for crushing seeds; others had slender, needlelike beaks for feeding on cactus plants. Darwin assumed he had found species of blackbirds, wrens, and finches. But when he gave his preserved birds to a London ornithologist named John Gould, Gould made a surprising discovery: the birds were all finches. Despite their radically different beaks, they shared a number of telltale traits found only in finches (**Figure 2.15**).

Figure 2.14 Charles Darwin spent five years aboard HMS *Beagle*, traveling the world and gathering clues that he would later use to develop his ideas about evolution. (ullstein – Olaf Rahardt / The Image Works)

Figure 2.15 Darwin was surprised to discover that although some birds from the Galápagos Islands had dramatically different beaks, they were all finches. (Row 1 left to right: Michael Stubblefield / Getty Images, Michael Stubblefield / Getty Images, iStock / Getty Images; row 2 left to right: iStock / Getty Images, NHPA / Superstock, Mary Plage / Getty Images)

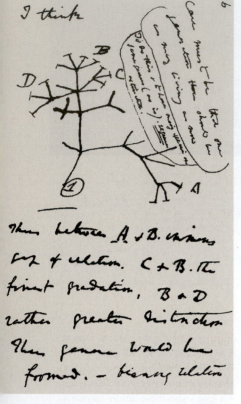

Figure 2.16 After returning to England, Darwin began to develop his ideas about evolution in a series of notebooks. He drew this tree in 1837 to illustrate how different lineages evolve from a common ancestor.

Darwin was puzzled. Some naturalists of his day argued that species had been created where they now were found, well suited to their climate. But if all the finches had been created on the Galápagos Islands, why were they so different from one another? Perhaps they had evolved into their current forms.

The finches and tortoises helped lead Darwin to the conclusion that all of life had evolved. Only evolution—the fact that all living things share a common ancestry—could explain the patterns in nature today (**Figure 2.16**). As for how life evolved, Darwin rejected the vague mechanisms previous naturalists had proposed, such as Lamarck's nervous fluids. Instead, Darwin envisioned a simpler, testable process based on variation and selection.

Although Darwin had the basic pieces of his theory in place by the late 1830s, it would be another 20 years before he finally presented it in full detail to the public. During that time, Darwin became a highly respected researcher, known for his geological research and a massive monograph he wrote on barnacles. He came to be good friends with England's most distinguished naturalists, including his great inspiration, Charles Lyell. But Darwin also knew that Lyell and others had a very low opinion of Lamarck and other naturalists who had promoted evolution. And so he painstakingly worked through every possible objection that Lyell and others might raise about his own theory (see **Box 2.2** for a discussion of scientific theories).

Finally, in 1858, Darwin was spurred to publish his ideas when he received a letter from Indonesia. The letter was from another English naturalist named Alfred Russel Wallace (1823–1913). Wallace, 14 years Darwin's junior, had patterned his own life after Darwin's famous travels. He had spent years in the jungles of South America and the Malay Archipelago, gathering plants and animals that he sold to museums and wealthy collectors in Europe. Wallace also kept careful records of the diversity of life he saw, and as he reflected on his observations, he concluded that life had indeed evolved. He even described a mechanism for evolution very much like Darwin's idea of natural selection. Wallace wrote to Darwin to share his new ideas, and he asked Darwin to present them to the Linnean Society, one of England's leading scientific organizations.

If Wallace were to publish first, Darwin knew, his own years of work could be cast into shadow. Darwin also knew that he had worked out his own argument in far more detail than Wallace. On the advice of Lyell and others, Darwin decided to turn the matter over to the Linnean Society. In July 1858, letters from both Wallace and Darwin were read at a meeting of the Linnean Society and later published in the society's scientific journal.

Strangely, though, neither the letters nor the article made much of an impression. It was not until Darwin wrote a book about his theory and published it in 1859 that the world sat up and took notice.

On the Origin of Species by Means of Natural Selection, or the Preservation of Favoured Races in the Struggle for Life was an immediate sensation, both in scientific circles and among the public at large. Some scientists immediately embraced Darwin's argument, engaging in fierce public debates with those who rejected it. Darwin himself, however, did not personally enter the fray. He went on working quietly and patiently at his rural home. There he continued to carry out experiments to investigate his theory, studying everything from orchids to earthworms. Darwin went on to write more books about evolution and other aspects of nature—including human nature, which was the subject of his 1871 book *The Descent of Man, and Selection in Relation to Sex.*

When Darwin first argued for evolution, he shocked many readers. But, over the years, much of the public came to accept a good deal of what he had to say. When *The Descent of Man* came out 12 years after *The Origin of Species*, it generated far less controversy. As the botanist Joseph Hooker, Darwin's friend, wrote in a letter, "I dined out three days last week, and at every table heard evolution talked about as accepted fact, and the descent of man with calmness."

As for scientists, some were skeptical of Darwin's proposal regarding how evolution had occurred, but few disagreed that life had indeed evolved. Perhaps most

BOX 2.2

Theories in Science

Darwin developed a theory of evolution. The word "theory" in scientific context is often misunderstood. Many people use the word to mean a hunch, a vague guess based on little evidence. When they hear scientists speak of "evolutionary theory," they assume that it's mere speculation, far less certain than a fact.

But that's not what scientists mean when they speak of a theory. A **theory** is an overarching set of mechanisms or principles that explain a major aspect of the natural world. Many theories develop from tentative explanations or ideas called hypotheses. The word *hypothesis* may suggest an idea that is closer to a guess, but a **hypothesis** is actually grounded in evidence. Testing and verification of a hypothesis lead to greater understanding and explanation. Ultimately, a theory makes sense of what would otherwise seem like an arbitrary, mysterious collection of data. And a theory is supported by independent lines of evidence.

Modern science is dominated by theories, for example, the theory of gravitation (**Box Figure 2.2.1**), the theory of plate tectonics, the germ theory of disease, and the theory of evolution. Each of these theories came about when scientists surveyed research from experiments and observations and proposed an explanation that accounted for them in a consistent way. Scientists can use theories to generate additional hypotheses, which they can test with new observations and experiments. The better a scientific theory holds up to this sort of scrutiny, the more likely it is to become widely accepted. If subsequent observations do not support a theory, it will be revised in light of the new discovery.

Modern evolutionary theory still embraces many of Darwin's central insights, such as the mechanism of natural selection and the concept of the tree of life. But evolutionary theory has matured, and it now takes a form that Darwin might not entirely recognize. That change has come about as one generation of scientists after another has examined the theory and tested aspects of it in specific ways, such as measuring natural selection in our species or running experiments to see how sexual selection influences mating success.

In addition, modern scientists have been able to examine evidence, such as fossils and DNA, that Darwin couldn't see. Today, evolutionary theory is as well supported as any of the other leading theories of modern science, but even so, scientists are still learning about evolution. New fossils are discovered every year. The DNA of humans and other species is yielding profound insights about how evolution works— insights that scientists are only now beginning to understand.

Box Figure 2.2.1 Sir Isaac Newton (1642–1726) was instrumental in developing a universal theory of gravitation. (iStock/Getty Images)

The great expanses scientists have yet to explore do not diminish the importance of the theory of evolution, however. As with other theories, scientists value evolutionary thinking for what it has helped them to understand so far. A good theory is like a powerful flashlight helping scientists make their way in the dark. It's ironic that those who would reject evolution often say it is "just" a theory, implying that it's inferior to facts. Facts are important—they are well-confirmed, objective observations—but they are not superior to theories. Nor are the laws that describe relationships in science. For scientists, a good theory organizes facts and laws, changing them from a loose collection of details into a meaningful, well-supported picture of the past and a valuable tool that shapes important tests for the future.

A **theory** is an overarching set of mechanisms or principles that explain major aspects of the natural world. Theories are supported by many different kinds of evidence and experimental results.

A **hypothesis** is a tentative explanation for an observation, phenomenon, or scientific problem that can be tested by further investigation or experimentation.

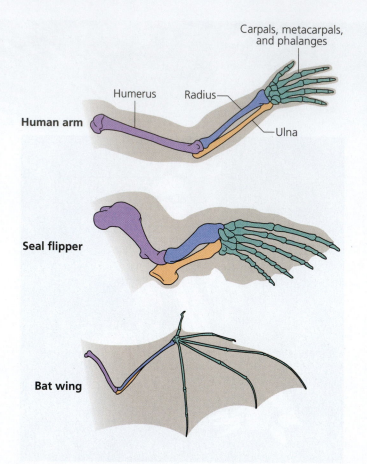

Carpals, metacarpals, and phalanges

Human arm

Humerus   Radius

Ulna

Seal flipper

Bat wing

Figure 2.17 Humans, seals, and bats have seemingly different limbs, which they use for different functions. But the numbers and arrangement of bones in one species correspond to those in the others. Darwin argued that this similarity was a sign of common ancestry.

Homology refers to the similarity of characteristics resulting from shared ancestry.

A **homologous characteristic** is similar in two or more species because it is inherited from a common ancestor.

Descent with modification refers to the passing of traits from parents to offspring. Darwin recognized that, over time, this process could account for gradual change in species' traits and homology.

important, Darwin had established evolution as a subject that could be studied scientifically: by running experiments, by comparing species, and by thinking of processes that could explain the patterns of nature. When Darwin died in 1882, he was buried in Westminster Abbey, in the company of kings and queens, great writers and prime ministers, and other great scientists, including Isaac Newton.

Common Descent

One of Darwin's great achievements was to show that humans and all other species on Earth are related, like cousins in a family tree. Closely related species belonged to close branches on the tree of life. For evidence of these relationships, Darwin pointed to the patterns of nature that had puzzled naturalists for so long.

In the mid-1800s, anatomists became keenly aware that the diversity of life had many common themes. Consider a seal's flippers, a bat's wings, and your arms (**Figure 2.17**). The seal uses its flippers to swim through the ocean; the bat uses its wings to fly; and people use their arms to cook, sew, write, perform surgery, and drive cars. These appendages serve very different functions, and yet they have a deep similarity. The bones, for example, are arranged in the same way. A long bone (the humerus) extends from the shoulder. On its far end, it meets two thin, parallel bones (the radius and ulna), and the joint between them allows the arm or wing to bend at an elbow. At the end of the radius and ulna is a cluster of wrist bones. The same set of bones can be found in each species' wrist. Extending from the wrist are five digits. Of course, any given bone in one species is somewhat different from the corresponding bone in the other species. A seal's humerus is short and stout, for example, whereas a bat's looks more like a chopstick. But those differences don't obscure the arrangement that all of those limbs share. Naturalists call this similarity **homology**, and characteristics that display homology are known as **homologous characteristics**.

What accounts for this combination of differences and similarities? Some anatomists in the mid-1800s argued that each species was created according to an archetype—a fundamental plan to which some variations could be added. Darwin preferred a simpler, less transcendental explanation: seals, bats, and humans all shared a common ancestor that had limbs with wrists and digits. That ancestor gave rise to many lineages. In each of them the limbs evolved, yet the underlying legacy of our common ancestor survived.

Darwin's case for **descent with modification** was strengthened by the fact that many homologies are found together in the same groups of species. Bats, humans, and seals don't just share limbs, for example. They also have hair, and the females of each species secrete milk to nurture their young. Linnaeus had used these traits to classify humans, bats, and seals as members of the same category: all three species are mammals. Darwin argued that the very fact that we can classify species in this way is consistent with the notion that they evolved from a common ancestor. Although new traits can evolve in different lineages (we can't fly like bats, for example), each species descends from an ancestral mammal. And that mammal, Darwin argued, shared an even older ancestry with other animals. For example, we humans share many characteristics with fishes. We have eyes with the same arrangement of lenses, retinas, and nerves. We have skulls, livers, and many other organs in common.

Of course, we are different in some important ways. Just about all vertebrates on land have lungs. So do vertebrates that have gone back to the ocean, such as whales

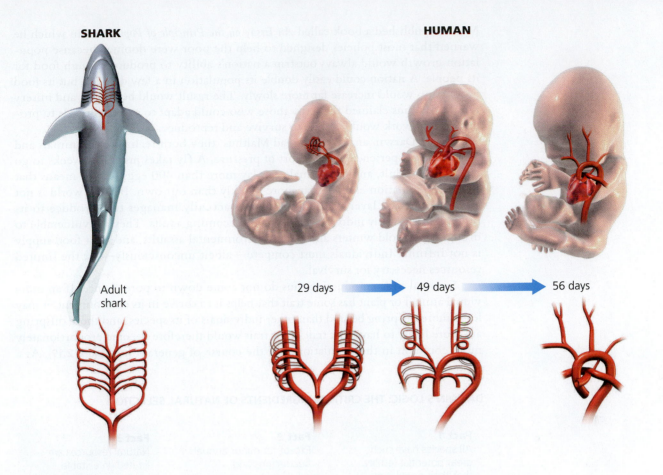

SHARK　　　　　　　　　　　　　　　　　　　　**HUMAN**

Adult
shark

29 days　　→　　49 days　　→　　56 days

and seals (see Chapter 1). Some fishes have lunglike structures for breathing. But all fishes have gills, which let them draw in dissolved oxygen from water.

　　Darwin argued that these differences might not actually be as profound as they first appear. In some cases, homologies are clear only when animals are still embryos, not when they are adults. While fishes and land vertebrates are still embryos, for example, they all develop the same set of arches near their heads. In fishes, those arches go on to become gills. In land vertebrates like us, they go on to form a number of different structures in the head and neck, including the lower jaw. A human embryo initially develops blood vessels in the same pattern seen in fish gills. But later, the blood vessels are modified (**Figure 2.18**). Darwin argued that those arches are homologies inherited from a common ancestor. In our ancestors, the arches that once supported gills evolved to take on a new function in adulthood.

Figure 2.18 Fishes have a series of branching blood vessels to absorb oxygen in their gills. Human embryos (at 29 days) grow blood vessels in the same arrangement, but later the vessels change to allow us to absorb oxygen through our lungs.

Natural Selection

Darwin argued that patterns in biology—the homologies, fossil record, and so on—could be explained by the inheritance of these features from common ancestors: in other words, evolution. He also argued for a new understanding of a mechanism that drove much of that change. To account for evolution, Darwin's predecessors usually proposed mysterious, long-term drives. Lamarck, for example, claimed that the history of life followed a trend toward "higher" forms. Many German biologists in the early 1800s argued that life evolved much as an embryo develops in the womb, from simple to complex. What made the new ideas of Darwin and Wallace so important was that they depended on processes that were not just natural but observable. One of the most important of these processes is **natural selection.**

　　Darwin and Wallace both found inspiration for the idea of natural selection in the writings of an English clergyman named Thomas Malthus (1766–1834). In 1798,

Natural selection is a mechanism that can lead to adaptive evolution, whereby differences in the phenotypes of individuals cause some of them to survive and reproduce more effectively than others.

Malthus published a book called *An Essay on the Principle of Population*, in which he warned that most policies designed to help the poor were doomed because population growth would always outstrip a nation's ability to produce enough food for its people. A nation could easily double its population in a few decades, but its food production would increase far more slowly. The result would be famine and misery for all. Malthus claimed that only those who could adapt to society's needs to produce useful work would be able to survive and reproduce.

When Darwin and Wallace read Malthus, they both realized that animals and plants must experience just this sort of pressure. A fly takes just a few weeks to go from egg to adult, and each female can lay more than 400 eggs, which means that the fly's population explodes far more quickly than our own. But the world is not buried in a thick layer of flies. No species actually manages to reproduce to its full potential. Many individuals die before becoming adults. They are vulnerable to droughts and cold winters and other environmental assaults, and their food supply is not infinite. Individuals must compete—albeit unconsciously—for the limited resources necessary for survival.

Survival and reproduction thus do not come down to pure chance. If an individual animal or plant has some trait that helps it to thrive in its environment, it may leave more offspring behind than other individuals of its species, and those offspring are more likely to have that trait. That trait would therefore become proportionately more common in the population over the course of generations (**Figure 2.19**). As a

DARWIN'S LOGIC: THE CRITICAL INGREDIENTS OF NATURAL SELECTION

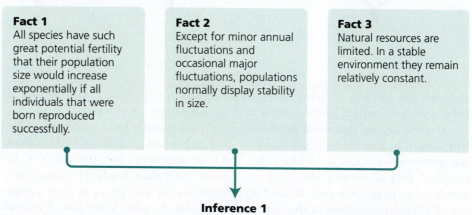

Fact 1
All species have such great potential fertility that their population size would increase exponentially if all individuals that were born reproduced successfully.

Fact 2
Except for minor annual fluctuations and occasional major fluctuations, populations normally display stability in size.

Fact 3
Natural resources are limited. In a stable environment they remain relatively constant.

Inference 1
Because more individuals are produced than can be supported by the available resources and population size remains stable, it means that there must be a fierce struggle for existence among the individuals of a population, resulting in the survival of only a part, often a very small part, of the progeny of each generation.

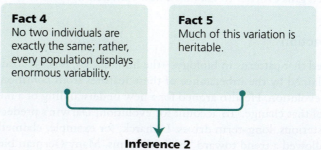

Fact 4
No two individuals are exactly the same; rather, every population displays enormous variability.

Fact 5
Much of this variation is heritable.

Inference 2
Survival in the struggle for existence is not random but depends in part on the inherited traits of the surviving individuals. This unequal survival constitutes the process of natural selection.

Inference 3
Over the generations this process of natural selection will lead to a continuing gradual change of populations, that is, to evolution and to the production of new species.

Figure 2.19 Darwin's theory of natural selection consisted of three inferences based on five facts derived from population ecology and principles of inheritance. (Data from Mayr 1982)

case in point, Darwin noted that grouse have feathers that closely match the brown color of the heath where they live. Birds with colors that didn't closely match the heath would stand out, becoming easy targets for predators. Natural selection thus favored the birds that blended more effectively into their background.

Darwin saw an analogy to natural selection in the way people bred animals and plants. Pigeon breeders had produced a remarkable range of birds—pigeons with ruffles around the neck, for example, or brilliant white feathers, or thick plumage running down their legs. The breeders achieved this diversity by selecting a few birds from each generation with the traits they desired. Over many generations, this selective breeding caused the traits to become exaggerated.

Darwin recognized similarities between breeding, or **artificially selecting** a trait, and natural selection. Pigeon breeders artificially select certain individual animals to reproduce. Natural selection, on the other hand, takes place because some individuals happen to be better suited than others to surviving and reproducing in a particular environment. Given enough time, Darwin and Wallace argued, natural selection could produce new **adaptations** ranging from wings to eyes.

Artificial selection is the selective breeding of animals and plants to encourage the occurrence of desirable traits. Individuals with preferred characteristics are mated or cross-pollinated with other individuals having similar traits.

An **adaptation** is an inherited aspect of an individual that allows it to outcompete other members of the same population that lack the trait (or that have a different version of the trait). Adaptations are traits that have evolved through the mechanism of natural selection.

2.4 Darwin's Ideas in the Twenty-First Century

In this chapter, we have taken a journey through two thousand years of biological research, ending with Darwin's publication of *The Origin of Species* in 1859. This chapter surveys the foundation of modern biology, but it does not mark the end of evolutionary biology's growth as a science.

Darwin himself went on to publish more books on different aspects of evolution. In his 1871 book, *The Descent of Man and Selection in Relation to Sex*, for example, Darwin investigated how humans had evolved, a subject we'll delve into in great detail in Chapter 17. *The Descent of Man* was also where Darwin first fully explored how mating behavior could shape evolution, through a process he called sexual selection (**Figure 2.20**). As we'll see in Chapter 11, sexual selection is pervasive in the natural world. It shapes not just visible weapons and ornaments but also invisible aspects of biology such as how fast embryos grow in the uterus.

Heredity was a crucial element of Darwin's theory, and yet no explanation existed in Darwin's day for the molecular biology of inherited traits (Zimmer 2018). As we'll see in Chapter 5, the Czech monk Gregor Mendel made the first key discoveries in **genetics** in the 1860s, but his work went overlooked for decades. It wasn't until the early 1900s that scientists realized that entities called **genes** controlled inherited traits. And it wasn't until the 1950s that scientists figured out how

Heredity is the transmission of characteristics from parent to offspring.

Genetics is the study of heredity, or how characteristics of organisms are transmitted from one generation to the next.

A **gene** is a segment of DNA whose nucleotide sequence codes for proteins, codes for RNA, or regulates expression of other genes.

Figure 2.20 Darwin recognized that some traits evolve not because they help organisms survive but because they help organisms mate more often. In some species, for example, male beetles grow huge mandibles that they use in battles with other males as they compete over access to females. (Igor Siwanowicz)

DNA, or deoxyribonucleic acid, is a long, double-stranded molecule containing genetic information for development, life, and reproduction.

Genetic drift is evolution arising from random changes in the genetic composition of a population from one generation to the next.

genetic information is stored as **DNA**. The birth of molecular biology transformed evolutionary biology. In the mid-1900s, a number of scientists joined genetics, paleontology, and ecology into a new vision of evolution, known as the modern evolutionary synthesis (Provine 1971).

As we'll see in Chapters 6 through 8, scientists can now uncover the precise changes in genes that underlie evolutionary events. They can distinguish between genetic changes due to natural or sexual selection and changes arising simply due to chance (a process called **genetic drift**). And yet, after all these discoveries, Darwin's theory of natural selection remains a powerful explanation for the diversity of life. Researchers have demonstrated that natural selection is a real, powerful force in nature.

Evolutionary biology in the twenty-first century is a remarkably rich body of knowledge and sophisticated explanatory theory, extending well beyond the foundation laid by Darwin in *The Origin of Species*. Just as Isaac Newton opened a door for generations of physicists, Darwin opened a door for biologists (**Figure 2.21**).

Figure 2.21 It's been more than 150 years since Charles Darwin first offered his argument for evolution. His insights are now the foundation of modern biology. (Huntington Library/Superstock, Inc.)

TO SUM UP . . .

- In the seventeenth and eighteenth centuries, naturalists devised systems for classifying life, and they recognized fossils as the remains of living things.

- Georges Buffon proposed that Earth was very old and that life had gradually changed during its history.

- Georges Cuvier helped establish that many fossils were the remains of extinct species.

- The geological record revealed a succession of different species that lived on Earth.

- Jean-Baptiste Lamarck developed an early theory of evolution, based in part on the idea that acquired traits are passed down through a mechanism of heredity.

- Charles Darwin and Alfred Russel Wallace independently developed a theory of evolution by means of natural selection.

- Homology is the similarity of characteristics in different species resulting from their inheritance of these characteristics from a common ancestor.

- Natural selection—a mechanism by which differential reproduction of individuals causes some genetic types to replace (outcompete) others—can lead to evolutionary change.

MULTIPLE CHOICE QUESTIONS Answers can be found at the end of the book.

1. Which of these statements is a concept found in Georges Buffon's ideas about evolution but not the way we understand evolution now?
   a. Populations can change over time.
   b. Life can be divided into a number of distinct types that are not related to each other.
   c. Living things are made of the same particles found in rocks and water.
   d. Life took more than a few thousand years to evolve.

2. What would Jean–Baptiste Lamarck and Charles Darwin have agreed on?
   a. One generation can pass on its traits to the next.
   b. Individual animals and plants can adapt to their environment.
   c. Life is driven from simplicity to complexity.
   d. Both a and b.

3. What is a correct definition of homology?
   a. Common traits due to shared inheritance from a common ancestor.
   b. Common function of traits due to similar usage.
   c. Structure of limbs that are common among all mammals.
   d. Traits in separate lineages that converged on a similar shape.

4. What set Darwin and Wallace's concept of natural selection apart from earlier ideas of evolution?
   a. Their concept explained why organisms were related to each other.
   b. Their concept depended on a process that is observable.
   c. Their concept depended on the inheritance of characteristics from one generation to the next.
   d. Their concept suggested that change was very gradual.

5. Darwin is generally given more credit than Wallace for discovering natural selection as a mechanism of evolution because
   a. Wallace already knew of Darwin's work, so he didn't arrive at the idea independently.
   b. Darwin was part of the academic establishment (a fellow of the Royal Society), whereas Wallace was not.
   c. Darwin's paper explained natural selection more effectively than Wallace's.
   d. Darwin's journals established that he had arrived at the idea many years earlier than Wallace.

6. Which of the following is *not* a reason Darwin's concept of descent with modification fits so well with Linnaeus's system of taxonomy classification?
   a. Linnaeus formulated his classification system to account for species changing gradually over time.

b. Linnaeus's taxonomic groups (genus, family, order, class, phylum, kingdom) nest within each other in much the same way that branches converge on bigger branches and then trunks on Darwin's "tree of life."

c. Linnaeus's groups are defined based on shared characteristics, which fits well with Darwin's idea that descendent species inherit homologous traits from their most recent common ancestor.

d. Descent with modification can explain the patterns Linnaeus observed and the hierarchical classification system he formulated to describe them.

7. Lamarck's idea that organisms inherit their parents' acquired characteristics has experienced a recent resurgence in popularity because

a. modern genomic studies cast doubt on traditional concepts of gene function and inheritance.

b. phenotypic plasticity—changes in traits acquired during the lifetime of an organism in response to environmental conditions—is now a thriving field of research in biology.

c. it has become clear that some acquired traits, such as physiological responses to stress, can be transmitted epigenetically to offspring (at least for a generation or two).

d. phenotypic plasticity is now considered a form of biological evolution.

8. Why is Linnaeus considered the founder of modern taxonomy?

a. Because his system of grouping organisms into a nested hierarchy is still in use today.

b. Because the groupings of organisms within his original hierarchy are still supported today.

c. Because he was the first to use stratigraphy as a method for reconstructing the past.

d. Because he proposed that the mechanical complexity of animal systems was evidence of a Divine Creator.

9. Homologous characteristics are those that

a. are similar characteristics shared by individuals of the same species.

b. are similar in two or more species because they are inherited from a common ancestor.

c. serve similar functions but are dissimilar.

d. All of the above.

10. Why is embryology important in determining evolutionary relationships among species?

a. Embryos give insight into analogous structures.

b. Humans and fishes share characteristics that are first evident in embryos.

c. Some homologies are evident only when animals are embryos.

d. The inheritance of features from common ancestors is only evident in embryos.

INTERPRET THE DATA Answer can be found at the end of the book.

11. Modern taxonomy closely resembles the original scheme developed by Linnaeus. One aspect that is characteristic of both Linnaeus's and modern taxonomic classification is the hierarchical grouping of taxonomic categories (genera are nested within families, and families are nested within orders, for example). Taxonomic groups are defined by features shared by included species. Chordates, for example, all have a hollow nerve cord called the notochord and a circulatory system that uses vessels to transport blood. Mammals all share the capacity to produce milk (mammary glands). Mammals also have hair and a particular combination of three inner ear bones. Using the figure as a guide, which of the following statements is false?

a. The three inner ear bones in humans are homologous with the inner ear bones of bats.

b. Snakes should have a notochord and circulatory system but not three inner ear bones or mammary glands.

c. Orangutans should have mammary glands and hair but not a notochord.

d. Both gorillas and cats should have mammary glands, hair, and circulatory systems.

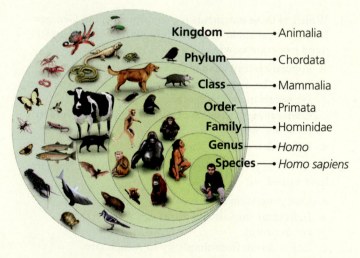

SHORT ANSWER QUESTIONS Answers can be found at the end of the book.

12. Which of Carolus Linnaeus's many contributions to biology are most valuable to scientists studying evolution? Which of Linnaeus's concepts have been shown to be incorrect?

13. How did the researcher Georges Cuvier combine the geological discoveries and theories of James Hutton and William Smith and his own observations to decide that geological formations from very different geographic locations were from the same time period?

14. Define stratigraphy and explain how this field of study helped scientists understand evolution.

15. How did the work of Thomas Malthus influence Darwin and Wallace in creating their theories of natural selection?

16. How did pigeon breeding influence Darwin's conception of natural selection?

17. Why was Lamarck's view of inheritance seemingly intuitive? What observations would be needed to support his theory?

ADDITIONAL READING

Bowler, P. J. 2003. *Evolution: The History of an Idea.* 3rd ed. Berkeley: University of California Press.

Browne, E. J. 1995. *Charles Darwin: A Biography.* New York: Knopf.

———. 2006. *Darwin's Origin of Species: A Biography.* New York: Atlantic Monthly Press.

Darwin, C. 1859. *On the Origin of Species by Means of Natural Selection, or the Preservation of Favoured Races in the Struggle for Life.* London: John Murray.

Gregory, T. R. 2008. Evolution as Fact, Theory, and Path. *Evolution Education Outreach* 1:46–52.

Magner, L. N. 2002. *A History of the Life Sciences.* 3rd ed. New York: M. Dekker.

Mayr, E. 1982. *The Growth of Biological Thought: Diversity, Evolution and Inheritance.* Cambridge, MA: Harvard University Press.

Padian, K. 2008. Darwin's Enduring Legacy. *Nature* 451: 632–34.

Young, D. 2007. *The Discovery of Evolution.* 2nd ed. Cambridge: Cambridge University Press, in association with Natural History Museum, London.

PRIMARY LITERATURE CITED IN CHAPTER 2

Appel, T. A. 1987. *The Cuvier-Geoffroy Debate: French Biology in the Decades Before Darwin, Monographs on the History and Philosophy of Biology.* New York: Oxford University Press.

Ayala, F. 2013. Evolution and Religion. In J. B. Losos et al. (eds.) *The Princeton Guide to Evolution.* Princeton, NJ: Princeton University Press.

Burkhardt, R. W. 1977. *The Spirit of System: Lamarck and Evolutionary Biology.* Cambridge, MA: Harvard University Press.

Cutler, A. 2003. *The Seashell on the Mountaintop: A Story of Science, Sainthood, and the Humble Genius Who Discovered a New History of the Earth.* New York: Dutton.

Emling, S. 2011. *The Fossil Hunter: Dinosaurs, Evolution, and the Woman Whose Discoveries Changed the World.* New York: Palgrave Macmillan.

Koerner, L. 1999. *Linnaeus: Nature and Nation.* Cambridge, MA: Harvard University Press.

Lovejoy, A. O. 1936. *The Great Chain of Being: A Study of the History of an Idea.* Cambridge, MA: Harvard University Press.

Mayr, E. 1982. *The Growth of Biological Thought: Diversity, Evolution and Inheritance.* Cambridge, MA: Harvard University Press.

Paley, W. 1802. *Natural Theology, or Evidences of the Existence and Attributes of the Deity Collected from the Appearances of Nature.* London: R. Faulder.

Provine, W. B. 1971. *The Origins of Theoretical Population Genetics.* Chicago: University of Chicago Press.

Roger, J. 1997. Buffon: A Life in Natural History. In L. P. Williams (ed.), *Cornell History of Science Series.* Ithaca, NY: Cornell University Press.

Rudwick, M. J. S. 1985. *The Meaning of Fossils: Episodes in the History of Palaeontology.* Chicago: University of Chicago Press.

———. 1997. *Georges Cuvier, Fossil Bones, and Geological Catastrophes: New Translations & Interpretations of the Primary Texts.* Chicago: University of Chicago Press.

Wilkins, J. S. 2009. *Species: A History of the Idea.* Oakland: University of California Press.

Winchester, S. 2001. *The Map That Changed the World: William Smith and the Birth of Modern Geology.* New York: HarperCollins.

Zimmer, C. 2004. *Soul Made Flesh: The Discovery of the Brain and How It Changed the World.* New York: Free Press.

———. 2018. *She Has Her Mother's Laugh: The Power, Perversion, and Potential of Heredity.* New York: Dutton.

CHAPTER 3

What the Rocks Say

How Geology and Paleontology Reveal the History of Life

Learning Objectives

- Describe how radioactive elements are used to determine the age of rocks.
- Use the strontium-rubidium system to determine the age of a rock.
- Explain how the fossil record relates to patterns of fossilization.
- Explain how behaviors observed today can be used to understand plants and animals of the past.
- Explain how isotopes and biomarkers can be used to examine past habitats.
- Discuss why different lines of evidence are important in examining Earth's history.
- Describe the earliest forms of life on Earth.
- Describe the origins of multicellular life.
- Evaluate the contributions Ediacaran and trilobite fossils have made to our understanding of animal evolution.
- Define tetrapods, and discuss their significance to human evolution.
- Give one potential explanation for the patterns of diversity observed in animal and plant species alive today.

Abigail Allwood searches for clues to the evolution of life in one of the most remote, inhospitable places on Earth. Allwood (Figure 3.1), a geologist at NASA's Jet Propulsion Laboratory, travels with her colleagues deep into the outback of Australia, where there are plenty of lizards and cockatoos but virtually no people. Water is scarce among the bare outcrops and hills, and the days can be scalding hot. The name of one of the geological formations where Allwood works is a grim joke: North Pole.

Allwood hikes along the exposed rocks, taking photographs and sometimes hammering off pieces to take home to study further. To the inexperienced eye, nothing in the rocks looks like

Stromatolites are layered mats of bacteria. They are relatively rare today.

it was ever alive. The most notable thing about the rocks is their fine layers, which curve and sag into strange shapes. In some rocks, the layers look like upside-down ice cream cones. In others, they look like egg cartons.

It may be hard to believe, but these rocks are some of the oldest evidence of life on Earth (Allwood et al. 2006). They formed 3.43 billion years ago by vast mats of bacteria that stretched across the floor of a shallow sea.

By studying these rocks, Allwood is learning about the early evolution of life on Earth. NASA supports her search for fossils because it may also provide clues about how life evolved on other planets, such as Mars.

Allwood belongs to an army of scientists thousands strong who traverse the globe in search of traces of the deep history of life. In this chapter, we'll meet some of these researchers and learn how they study evolution from fossils and other geological clues. We will start this exploration by revisiting the great geological debates of the nineteenth century that led to the realization that Earth—and life along with it—is immensely old. Next, we'll examine the techniques that scientists use to estimate how old rocks and fossils are. As we'll see, fossils can tell us more than just their age. Scientists can gain clues to how these extinct species lived—what they looked like, how they behaved, and how they fit into their ecosystems.

Figure 3.1 Geologist Abigail Allwood studies fossil stromatolites to get clues about the earlier period of life on Earth. (Abigail Allwood)

After exploring the methods that scientists use to reconstruct the past, we will embark on a journey through time. We'll start our trip at the dawn of life by looking at the ancient fossils that Allwood and others are digging up. We'll then move onward, observing how the world's ecosystems gradually changed and how the components of today's natural world slowly emerged.

This chapter serves as scaffolding for the rest of the book. Once we have become acquainted with the fossil record, we can then turn to the mechanisms of evolution that gave rise to its patterns—the emergence of new species, for example (Chapter 13), or the extinction of major lineages (Chapter 14), or the emergence of ecological partners (Chapter 15). And perhaps most important, we will see in this chapter just how vast the history of life is—and just how recently we have become a part of it. ●

3.1 Debating the Age of Earth

Charles Darwin is now best known as an evolutionary biologist, but he first came to fame as a geologist. On his journey aboard the HMS *Beagle*, he made careful observations of mountains, islands, and other geological formations. He recognized that these formations were the result of gradual changes spread over vast amounts of time. Given how slowly these changes took place, Darwin joined many other nineteenth-century geologists in rejecting the widely accepted belief that the world was only a few thousand years old.

Figure 3.2 Top: An early geological map of the Weald. Bottom: The rolling hills today. (Top: Image courtesy of History of Science Collections, University of Oklahoma Libraries; bottom: Loop Images/Slawek Staszczuk/Getty Images)

Darwin and his fellow geologists at the time could not figure out the exact age of a particular fossil or rock. But they could get some clues about how long geological formations needed to form, for example, by studying how long it took for sediments to accumulate on riverbanks and in coastal waters. While working on *The Origin of Species,* Darwin applied this method to the Weald, a stretch of ridges and valleys in southeastern England (**Figure 3.2**). He concluded that it formed through gradual erosion over the course of 300 million years.

If it had taken hundreds of millions of years for a relatively small geological formation to reach its current state, Darwin surmised, then Earth itself must be billions of years old. On such an ancient planet, Darwin argued, a gradual process of evolution would have had enough time to produce the diversity of life we see today.

But some scientists disputed the idea that Earth is ancient. The most prominent of those skeptics was the eminent physicist William Thomson (Lord Kelvin; 1824–1907). Kelvin, who pioneered research on the flow of heat through solids,

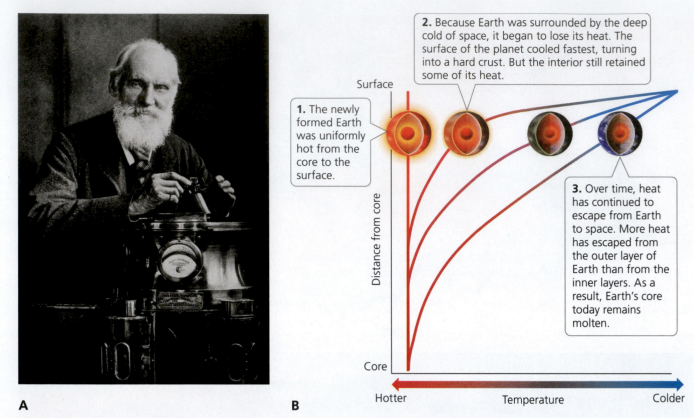

1. The newly formed Earth was uniformly hot from the core to the surface.

2. Because Earth was surrounded by the deep cold of space, it began to lose its heat. The surface of the planet cooled fastest, turning into a hard crust. But the interior still retained some of its heat.

3. Over time, heat has continued to escape from Earth to space. More heat has escaped from the outer layer of Earth than from the inner layers. As a result, Earth's core today remains molten.

Surface

Distance from core

Core

Hotter — Temperature — Colder

A   **B**

Figure 3.3 A: Lord Kelvin (William Thomson, 1824–1907) was a leading nineteenth-century physicist. He argued that Earth was less than 20 million years old. B: His calculation was based on a careful measurement of Earth's rate of heat loss. The early Earth (red line on the far left) was hot throughout. Over time, the planet shifted to the blue curve. Kelvin assumed, incorrectly, that Earth had had no additional source of heat since it was formed. (A: Science and Society/Superstock, Inc)

used this work to argue that Earth could not possibly be hundreds of millions of years old (**Figure 3.3**).

When Earth formed, it was a ball of molten rock. Kelvin argued that the cold temperature of space caused the surface of the planet to cool to a solidified crust. The interior of Earth, however, remained hot. The heat remaining in the interior of the planet steadily flowed to the surface, where it escaped to space. Because a hot rock cools at a steady rate, Kelvin reasoned, the current temperature of rocks could be used to estimate how long they had been cooling.

Kelvin reckoned that rocks on the planet's surface would not give a reliable estimate because they were heated by the Sun every day and cooled every night. The rocks deep underground in mine shafts, on the other hand, stayed at the same temperature year-round. Based on the warmth of those mine rocks, Kelvin calculated that Earth could be only 20 million years old at most.

Darwin was deeply concerned by Kelvin's calculations, although he didn't engage publicly with the great physicist. Other geologists challenged Kelvin to explain the geological record on Earth, but none of them could explain why Earth's rocks were so warm. Only much later did it become clear why Kelvin was wrong.

The source of Kelvin's error lay in his assumptions. To calculate Earth's heat flow, he had assumed that the planet was a rigid sphere. In the twentieth century, geophysicists would discover that the planet's interior is dynamic. Hot rock rises through the mantle, cools, and then sinks back down again. This movement drives the motion of tectonic plates across the surface of Earth. It also makes the upper layers of Earth warmer than in Kelvin's model (England et al. 2007).

Key Concept • Nineteenth-century scientists debated the age of Earth. Early estimates were based on flawed assumptions about the structure of the planet's interior. ●

3.2 A Curious Lack of Radioactivity

It was not until the early 1900s, shortly before Kelvin's death, that indisputable evidence of an ancient Earth came to light. A new generation of physicists discovered the structure of atoms, and their findings led to a reliable, precise way to estimate the ages of rocks. To most everyone's surprise, it turned out that many rocks contain, in effect, an atomic clock—one that has been quietly ticking away for millions—or billions—of years.

Radioactive Decay

This new kind of timekeeping was based on the discovery that atoms were in either a permanently stable state or an unstable state. In both cases, the atom's state depended on its combination of subatomic particles.

All atoms are made of positively charged protons, negatively charged electrons, and neutral neutrons. The number of protons in an atom determines which element it belongs to, but the number of neutrons in atoms of the same element can vary. Take carbon: all carbon atoms have 6 protons, but 98.93% of all carbon atoms on Earth have 6 neutrons, 1.07% have 7, and one in a trillion carbon atoms have 8. These isotopes of carbon are named, respectively, carbon-12, carbon-13, and carbon-14. (Isotopes are often represented with the total number of neutrons and protons as a superscript preceding their elemental symbol, such as ^{14}C.)

The number of neutrons in an isotope determines whether it is stable or unstable. Carbon-12 is stable, for example, because it has an equal number of protons and neutrons. An atom of carbon-12 will remain carbon-12 tomorrow and a billion years from now. Carbon-14, on the other hand, has two extra neutrons. This configuration means that carbon-14 will spontaneously decay, becoming a more stable configuration of 7 protons and 7 neutrons, known as nitrogen-14.

In any time interval, each unstable isotope has a fixed probability of decaying (Lanphere 2001). We can calculate the number of unstable atoms, N, that remain from an original supply, N_0, with the equation

$$N = N_0 e^{-\lambda t}$$

where λ is the probability of an atom decaying in a given time interval, t.

Unstable—or radioactive—isotopes have different probabilities of decaying. Some are highly likely to decay; others are much less so. These probabilities determine how quickly a group of radioactive isotopes will decay into a stable one.

Scientists typically measure this decay rate in terms of how long it takes for half of a given sample of atoms to decay. This measure is known as an isotope's half-life. Carbon-14 has a half-life of 5730 years, which means that half of the carbon-14 in a sample will become nitrogen-14 in 5730 years. After 11,460 years—two half-lives—a quarter of the original carbon-14 will be left. Some isotopes have a half-life measuring just a fraction of a second, whereas others have half-lives of billions of years (Table 3.1).

When scientists discovered radioactivity, they became interested in the origin of radioactive isotopes on Earth and their rates of decay. Most of the atoms on Earth—both radioactive isotopes and stable ones—formed in ancient stars that exploded billions of years ago. This stellar debris condensed into a disk-shaped cloud. The gas and dust at the center of the cloud collapsed in on itself and became our Sun. The rest of the disk then condensed into the planets and other bodies of our solar system.

Although most radioactive isotopes were present on Earth when it formed, there are some exceptions. For example, high-energy particles that come from space into the atmosphere, known as cosmic rays, sometimes collide with nitrogen atoms, converting them to the radioactive isotope carbon-14. Aside from the few isotopes

Table 3.1 Some Naturally Occurring Radioactive Isotopes and Their Half-Lives

Radioactive Isotope (parent)	Product (daughter)	Half-Life (years)
Samarium-147	Neodymium-143	106 billion
Rubidium-87	Strontium-87	48.8 billion
Rhenium-187	Osmium-187	42 billion
Lutetium-176	Hafnium-176	38 billion
Thorium-232	Lead-208	14 billion
Uranium-238	Thorium-234	4.5 billion
Potassium-40	Argon-40	1.25 billion
Uranium-235	Thorium-231	703.8 million
Carbon-14	Nitrogen-14	5730

that can be replenished in this way, Earth's natural supply of radioactive isotopes has been steadily dwindling ever since the planet formed.

This shrinking supply gives geologists a powerful tool for learning about the history of Earth. Consider, for example, the distribution of half-lives of isotopes in Earth's rocks. All the isotopes that geologists have found have half-lives of more than 80 million years. Physicists have experimentally generated isotopes with far shorter half-lives. But those short-lived isotopes don't exist in rocks—although their decay products do.

These short-lived isotopes were therefore present on early Earth, but they've since decayed. Their absence tells us that Earth is ancient. If Kelvin had been right that Earth was less than 20 million years old, we would expect to find many non-decayed, short-lived isotopes in rocks. The pattern we do find is consistent with a much older age for Earth. Our planet has existed for so long that all of its short-lived isotopes have decayed to levels too low for us to detect (Miller 1999).

Geologists can use radioactive isotopes not only to determine that Earth is very old, but also to determine precisely *how* old Earth is. They can determine how long ago rocks formed, too. To do so, they measure the concentrations of certain radioactive isotopes and stable isotopes in a sample. The steady decay rate of radioactive isotopes allows scientists to estimate how long the decay has been occurring in a rock—and thus to estimate the rock's age.

Radiometric dating is a technique that allows geologists to estimate the precise ages at which one geological formation ends and another begins.

Box 3.1 provides a detailed explanation of how this method, known as **radiometric dating**, works. The discovery of radiometric dating profoundly transformed geology. Before, geologists could say only that Earth was very old, but now they can estimate its age to be 4.568 billion years old. In the previous chapter, we saw how geologists discovered geological formations and mapped them across the planet. Radiometric dating allowed geologists to estimate the precise ages when one geological formation ended and another began. **Figure 3.4** shows a stratigraphic chart with the latest estimates for the ages of different geological periods.

Key Concepts

- Many elements have both stable and unstable (radioactive) isotopes.
- Unstable isotopes have a fixed probability of decay.
- Isotopes with high decay probabilities decay rapidly, and those with low probabilities decay slowly. ●

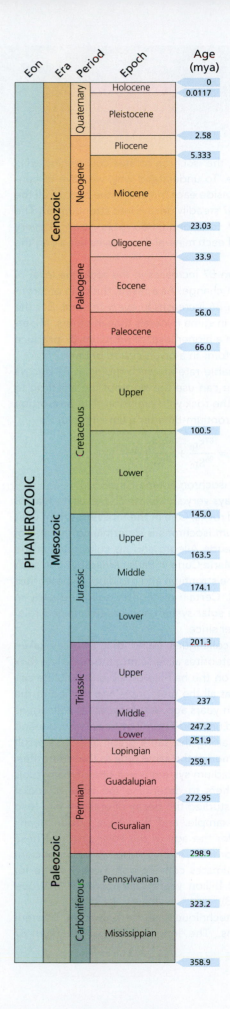

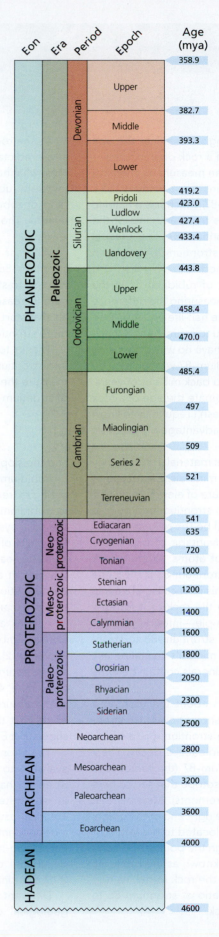

Figure 3.4 Geologists in the nineteenth century identified rock layers around the world and mapped them in chronological order, associating them with time blocks delineated by geological and biological events. Radiometric dating has allowed geologists to precisely date the boundaries between these geological periods.

BOX 3.1

Clocks in Rocks

In this box, we'll discuss how geologists estimate the age of rocks. Let's suppose that you find a rock on a mountainside and take it to a lab where you can measure its chemical composition. You find that it contains trace amounts of a radioactive isotope called rubidium-87, which has a half-life of 48.8 billion years. It also contains trace amounts of the decay product of rubidium-87, strontium-87. The gradual transformation of rubidium-87 into strontium-87 has been taking place ever since the rock formed.

If you knew the original amount of rubidium-87 in the rock, you could calculate its age by measuring the difference between the levels then and now. The half-life equation (see earlier in this chapter) would allow you to estimate how much time has passed. Unfortunately, you have no way of knowing what the rock's original level of rubidium-87 was, unless you own a time machine that can take you back millions of years.

Fortunately, we have a way to estimate the age of a rock based on its current isotopes that doesn't require us to know its original state. This method takes advantage of two basic facts about how rocks form.

First, rocks are not uniform slabs of material. Instead, they are mosaics composed of different minerals. Each type of mineral incorporates a different mixture of elements when it forms. Some minerals may be rich in strontium and poor in rubidium, whereas other minerals will be the reverse.

The second fact we can take advantage of is that the rock formed from a single source of atoms. When a rock forms, the ratio of strontium-87 to strontium-86 is the same in every mineral (**Box Figure 3.1.1**). This is because elements like strontium diffuse freely through the rock as it forms—while magma is still molten, for example—infusing all parts of it with the same starting ratio of strontium-86 to strontium-87. Once the rock cools and its minerals crystalize, any strontium or other elements these minerals contain become locked in place.

We can visualize these two facts about the rock on a graph (**Box Figure 3.1.1C**). The y axis shows how much strontium-87 the rock contains relative to strontium-86, a stable isotope that is not the product of radioactive decay. The x axis shows the proportion of rubidium-87 (the radioactive isotope) to strontium-86 (the stable isotope that serves here as a reference for our other measurements). Minerals that are rich in rubidium are located to the right end of the graph; minerals that are rubidium-poor are located to the left.

We can draw a straight line through the points representing the different minerals. This line, known as the **isochron**, starts out with a slope of zero when the rock forms. Its horizontal slope reflects the identical ratio of strontium-86 to strontium-87 in all minerals in the rock.

As the rock ages, the isochron rotates counterclockwise at a predictable rate. To understand why, we have to consider what happens inside each mineral in the rock. In all the minerals, rubidium-87 steadily decays into strontium-87. The ratio of rubidium-87 to strontium-86 goes down in every mineral. The value of each mineral thus moves left along the x axis.

Whereas strontium-87 increases in the rock, the level of strontium-86 doesn't change. As a result, the ratio of strontium-87 to strontium-86 increases in every mineral. But the ratio increases more in some minerals than in others. Minerals that start out with a higher level of rubidium-87 produce more strontium-87 atoms in a given period of time.

Thanks to the reliable rate at which rubidium-87 decays into strontium-87, we can use the slope of the isochron to measure the age of the rock with precision. The proportion of strontium-87 to strontium-86 after a time interval t is

$$\frac{^{87}\text{Sr}_t}{^{86}\text{Sr}_t} = \frac{^{87}\text{Sr}_0}{^{86}\text{Sr}_0} + \frac{^{87}\text{Rb}_t}{^{86}\text{Sr}_t}[e^{\lambda t} - 1],$$

and the slope of the isochron at time t is $e^{\lambda t} - 1$.

Rubidium-87 decays very slowly, which makes it ideal for dating extremely old rocks. One of the most important uses for rubidium-strontium isochrons is determining the age of the solar system. Claude Allegre and his colleagues at the University of Pierre Marie Curie in Paris carried out one such study by analyzing the strontium and rubidium in meteorites (Minster et al. 1982). Certain meteorites solidified early in the formation of the solar system and have remained relatively unchanged ever since.

Allegre and his colleagues found that samples taken from a number of meteorites all fell along the same line (**Box Figure 3.1.2**). Based on the half-life of rubidium-87, the scientists estimated that all the meteorites had formed at the same time, 4.5 billion years ago. Allegre and his colleagues also measured rubidium and strontium in Earth rocks and found that they fell along the same graph line. This result shows that Earth formed at the same time as the meteorites.

The strontium-rubidium system is just one of many tools available to researchers to determine the ages of rocks. Scientists have also studied the decay of uranium into lead in meteorites, for example, and they have gotten nearly identical estimates for the age of the solar system. These independent tests provide strong support for an ancient Earth. The latest estimates converge on a precise age for our planet of 4.568 billion years (Dalrymple 1991; Wood 2011; Mattinson 2013).

Different dating techniques can be better for different geological questions. The long half-life of rubidium-87

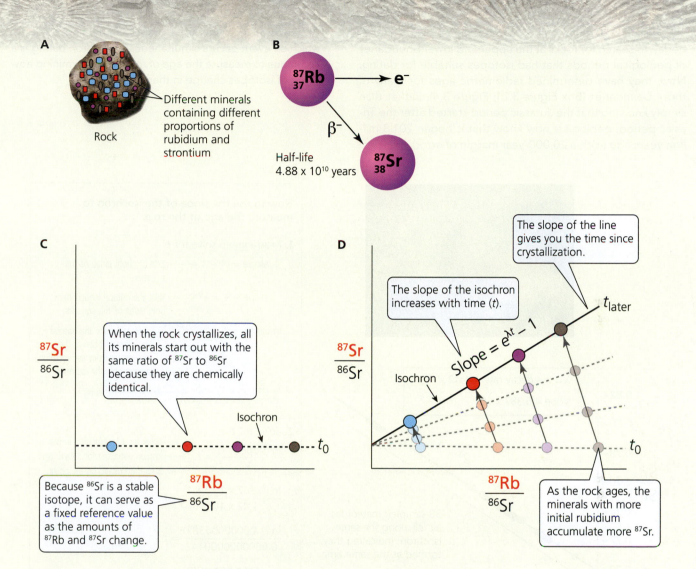

A: Rock — Different minerals containing different proportions of rubidium and strontium

B: $^{87}_{37}\text{Rb} \rightarrow e^-$

β^-

Half-life 4.88×10^{10} years

$^{87}_{38}\text{Sr}$

C: $\frac{^{87}\text{Sr}}{^{86}\text{Sr}}$

When the rock crystallizes, all its minerals start out with the same ratio of ^{87}Sr to ^{86}Sr because they are chemically identical.

Isochron

t_0

$\frac{^{87}\text{Rb}}{^{86}\text{Sr}}$

Because ^{86}Sr is a stable isotope, it can serve as a fixed reference value as the amounts of ^{87}Rb and ^{87}Sr change.

D: The slope of the line gives you the time since crystallization.

The slope of the isochron increases with time (*t*).

t_{later}

Slope $= e^{\lambda t} - 1$

Isochron

$\frac{^{87}\text{Sr}}{^{86}\text{Sr}}$

t_0

$\frac{^{87}\text{Rb}}{^{86}\text{Sr}}$

As the rock ages, the minerals with more initial rubidium accumulate more ^{87}Sr.

Box Figure 3.1.1 A: Rocks contain various minerals, each having various trace amounts of rubidium (Rb) and strontium (Sr). B: Rubidium-87 decays to strontium-87. After a rock crystallizes, this radioactive decay increases the strontium-87 concentration. C: When a rock first forms, the ratio of strontium-87 to strontium-86 is the same throughout the rock, regardless of the proportion of rubidium to strontium. The values form a straight horizontal line, called an isochron. D: As the strontium-87 decays, the ratios change, increasing the slope of the isochron. The slope of the isochron can be used to calculate the elapsed time (*t*) since the rock formed. (Data from Lanphere 2001)

makes it good for measuring very old rocks. In younger rocks, however, so little decay has occurred that it's hard to measure the isochrons precisely using this method. Potassium-40, by contrast, breaks down into argon-40 more rapidly—its half-life is only 1.25 billion years. As a result, this dating tool provides a more accurate clock for dating the age of younger rocks.

These different dating tools have enabled geologists to bring a precision to the geological record that did not exist in Darwin's day. In the nineteenth century, geologists mapped layers of rocks and named the periods of time when the layers formed. The Jurassic period, for example, was named after the Jura Mountains in Switzerland, where rocks of that

(continued)

Box 3.1 Clocks in Rocks (continued)

age were first identified. In the twentieth century, scientists began to search for rocks at the upper and lower boundaries of geological periods that had isotopes suitable for dating. Now, they have determined radiometric ages for many of these boundaries (Box Figure 3.1.1; Figure 3.4). Rather than simply knowing that the Jurassic period started after the Triassic period, geologists now know that it began 201.3 million years ago (with a 20,000-year margin of error).

An **isochron** is a line on a graph, connecting points at which an event occurs simultaneously. Isochron dating is a technique to measure the age of rocks by determining how ratios of isotopes change in them over time.

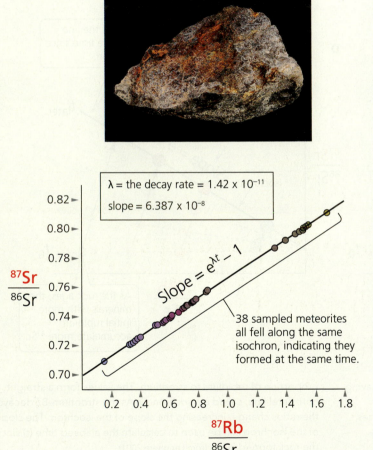

How to use the slope of the isochron to measure the age of the rock:

1. Rearrange to solve for *t*

$slope = e^{\lambda t} - 1$ ←— add 1 to both sides of the
$+ 1 \qquad + 1$ ⠀ equation

$slope + 1 = e^{\lambda t}$ ←— take the natural logarithm of both sides of the equation

$\ln(slope + 1) = \ln e^{\lambda t}$ ←— when you take the natural logarithm of base e, the ln e drops from the equation, leaving only the exponent

$\dfrac{\ln(slope + 1)}{\lambda} = \dfrac{\lambda t}{\lambda}$ ←— divide both sides by λ

2. Solve for *t*

$\dfrac{\ln(slope + 1)}{\lambda} = t$ ←— solve for t by plugging in the known variables for slope and λ as determined by the graph

$\dfrac{\ln(6.387 \times 10^{-8} + 1)}{1.42 \times 10^{-11}} = t$

$\dfrac{\ln(1.00000006387)}{0.0000000000142} = t$

$t =$ **4497.887 million years**

or

$t =$ **4.498 billion years**

$\lambda =$ the decay rate $= 1.42 \times 10^{-11}$

slope $= 6.387 \times 10^{-8}$

$\dfrac{^{87}Sr}{^{86}Sr}$

Slope $= e^{\lambda t} - 1$

38 sampled meteorites all fell along the same isochron, indicating they formed at the same time.

$\dfrac{^{87}Rb}{^{86}Sr}$

Box Figure 3.1.2 Isochrons are typically estimated using different minerals within a single rock, but they can also be used to compare the ages of multiple rocks. Some meteorites (see inset) have remained virtually unchanged since the formation of the solar system. In one study, 38 different meteorites all fell along the same isochron, indicating that they all formed at the same time—the time of the formation of our solar system. Researchers used the slope of the isochron to calculate that the meteorites formed 4.498 billion years ago. (Data from Minster et al. 1982; inset: Susan E. Degginger/Alamy Stock Photo)

3.3 A Vast Museum

In Chapter 2, we saw how Darwin used evolution to explain the fossil record. He argued that the pattern of remains preserved in rocks chronicled the history of species as they emerged, adapted, and become extinct. But critics in Darwin's time pointed out that the fossil record provided an incomplete chain of fossils for all the transitions that Darwin's theory implied.

Darwin had an answer for why those fossil chains had not been discovered. "I believe the answer mainly lies in the record being incomparably less perfect than is generally supposed," Darwin wrote in *The Origin of Species*. "The crust of the Earth is a vast museum; but the natural collections have been imperfectly made, and only at long intervals of time."

Over the past 150 years, scientists have confirmed Darwin's conclusion that the fossil record is far from complete. To understand why most living things don't turn to stone and a few do, researchers have studied the process of fossilization. They've observed how dead animals and other organisms decay over time, and they have replicated some of the chemistry that turns living tissues into rock.

Most organisms don't fossilize. One reason is simple—because other organisms eat them before this can happen. When an elephant dies, for example, scavengers such as hyenas or vultures typically strip the muscles and organs from its carcass while insects, bacteria, and fungi work more slowly on what's left. Within a few months, most cadavers are so thoroughly devoured, trampled, sun-beaten, or rain-soaked that nothing is left to become a fossil.

A tiny fraction of the organisms that die each year are protected from this oblivion. In some cases, they happen to fall into a still lake, where they are rapidly covered in sediment before scavengers can tear them apart. Mineral-rich water percolating through the sediment can fill the tiny spaces within the latticework of bone or shell, or even the insides of cells. Minerals precipitating out of the water accumulate in these spaces, gradually filling them with stone. The bone or shell around them eventually dissolves away, but new minerals leach into the spaces they leave behind. Over thousands of years, the organism and its surrounding sediment turn to rock. A fossil is formed (**Figure 3.5**).

Figure 3.5 Fossils form after organisms die. In some cases, their bodies are covered by sediment and then gradually turn to minerals. (Data from Prothero 2007)

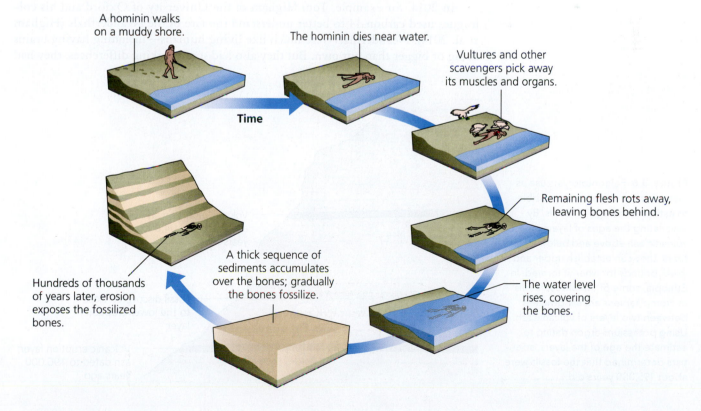

A hominin walks on a muddy shore.

The hominin dies near water.

Vultures and other scavengers pick away its muscles and organs.

Time

Remaining flesh rots away, leaving bones behind.

The water level rises, covering the bones.

A thick sequence of sediments accumulates over the bones; gradually the bones fossilize.

Hundreds of thousands of years later, erosion exposes the fossilized bones.

Millions of years later, the rock may be lifted and eroded, exposing the fossil. Wind and rain can then erode the fossil, destroying it forever. The only chance for paleontologists to study it occurs during the narrow window of time between its exposure and its destruction.

When paleontologists discover a fossil, they attempt to determine its age. Fossils typically don't contain isotopes like strontium and rubidium that make radiometric dating possible. As a result, it's usually not possible to directly date a fossil this way. Often, scientists look for layers of rock chronologically close to a fossil that can serve as time brackets.

In 1967, for example, scientists discovered fossils of humans at a site in the Omo River Valley in southern Ethiopia. The scientists knew the fossils were old, but it was hard to determine just how old they were. Almost three decades later, a team of scientists went back to the site to take a closer look. They discovered two layers of volcanic ash, one above the rocks where the fossils had been found and another right below them (**Figure 3.6**).

The argon in the upper layer yielded an age of approximately 104,000 years. The lower layer was 196,000 years old with a 2000-year margin of error. Thus the fossils had formed sometime between 196,000 and 104,000 years ago. A careful study of the sediments between those two layers indicated that the fossils were located closer to the older boundary than to the younger one. The scientists concluded that the fossils were as old as 195,000 years (McDougall et al. 2005). They were far older than any other fossils of our species known at the time. Since then, older fossils have pushed back the fossil record of *Homo sapiens* to 300,000 years ago (Hublin et al. 2017).

In certain circumstances, scientists can determine the age of organic material itself. As we saw earlier, carbon-14 is continually generated in the atmosphere. As plants take up carbon dioxide from the atmosphere, a small proportion of the carbon they accumulate in their tissues is carbon-14. Animals, likewise, accumulate carbon-14 by eating plants or other animals. Once an organism dies, the carbon-14 in its remains steadily breaks down for thousands of years. Because carbon-14 has a half-life of only 5730 years, it allows scientists to determine only the ages of fossils less than about 50,000 years old. Within that window, however, carbon-14 can serve as a powerful tool.

In 2014, for example, Tom Higham of the University of Oxford and his colleagues used carbon-14 to better understand the fate of the Neanderthals (Higham et al. 2014). Neanderthals were much like living humans—including having brains as big or bigger than our own. But they also had some striking differences: they had

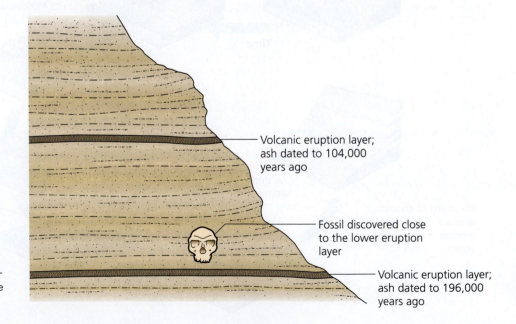

Figure 3.6 Paleontologists use as many lines of evidence as possible to estimate the age of fossils. By calculating the ages of layers of volcanic ash above and below a fossil, they can establish upper and lower bounds for when it formed. In Ethiopia, some of the oldest fossils of *Homo sapiens* are sandwiched between two layers of volcanic ash. Using potassium-argon dating to estimate the age of the layers, scientists determined that the fossils were about 195,000 years old.

Volcanic eruption layer; ash dated to 104,000 years ago

Fossil discovered close to the lower eruption layer

Volcanic eruption layer; ash dated to 196,000 years ago

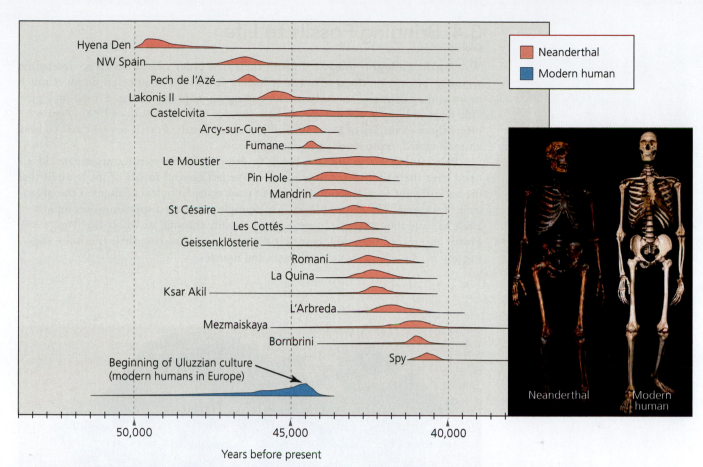

Figure 3.7 Neanderthal carbon dating. Neanderthals (inset, left) and anatomically modern humans (inset, right) descended from a common ancestor that lived about 600,000 years ago. Anatomically modern humans came into contact with Neanderthals in Europe shortly before the Neanderthals disappeared. Carbon-14 dating provided a range of ages for the youngest Neanderthal bones and tools at sites across Europe. This dating indicated that the last Neanderthal sites were no younger than 40,000 years old, implying that Neanderthals disappeared only a few thousand years after the first anatomically modern humans first arrived in Europe. (Data from Higham et al. 2014; inset: Blaine Maley)

a stocky build, had massive brows, and did not leave behind evidence of painting or other kinds of complex artwork. Researchers have long puzzled over how and why the Neanderthals disappeared. Part of the trouble has been the challenge of determining the age of the last Neanderthals in different parts of their range.

Higham and his colleagues extracted organic carbon from Neanderthal bones and made precise measurements of the different isotopes that the carbon comprised. They also looked for organic carbon in shells and charcoal in the same layer in which the Neanderthal bones were found. Based on the carbon isotopes they found, the researchers determined the likely range of the age of the Neanderthals at each site.

The results are displayed in **Figure 3.7.** Neanderthals last lived at different sites at different times. But none lived any later than about 40,000 years ago. Higham and his colleagues also measured carbon from the oldest sites in Europe where anatomically modern humans have been found. Those sites are between about 43,000 and 45,000 years old. As a result, the scientists concluded, modern humans and Neanderthals only coexisted for a few thousand years before Neanderthals disappeared. In Chapter 17, we'll consider what this timing—revealed by carbon dating—tells us about how Neanderthals became extinct and why we survived.

• The fossil record will never be complete because most organisms don't fossilize. • **Key Concept**

3.4 Bringing Fossils to Life

When paleontologists unearth a fossil, their first order of business is to determine where to place it on the tree of life (Chapter 4). But they can often discover much more as they continue to study it. Depending on its preservation, a fossil may provide clues to how an extinct species behaved, for example (Boucot 1990). **Figure 3.8** offers three examples of behavior inferred from fossils. Fossils reveal clues to how animals mated, reproduced, and obtained food.

Paleontologists can also use fossils to determine how extinct organisms developed over their lifetimes. Some species have left behind fossils of individuals that died at different ages. Paleontologists can track morphological changes as the organisms developed. They can observe, for example, whether a species grows rapidly to adult size and then stops, or instead grows steadily through its whole life (**Figure 3.9**). Fossils also record how the organisms themselves aged during their lifetimes, showing any marks left on them by illnesses and injuries.

Figure 3.8 Fossils can preserve clues about the behavior of extinct animals. A: Two turtles dating back about 47 million years died while mating in a lake (Joyce et al. 2012). B: A marine reptile called an ichthyosaur gave birth to a live offspring, rather than an egg. C: A fish was fossilized in the midst of eating a pterosaur. (A: Senckenberg, Messel Research Department, Frankfurt a. M. (Germany); B: The Natural History Museum / Alamy Stock Photo; C: Dirk Wiersma / Science Source)

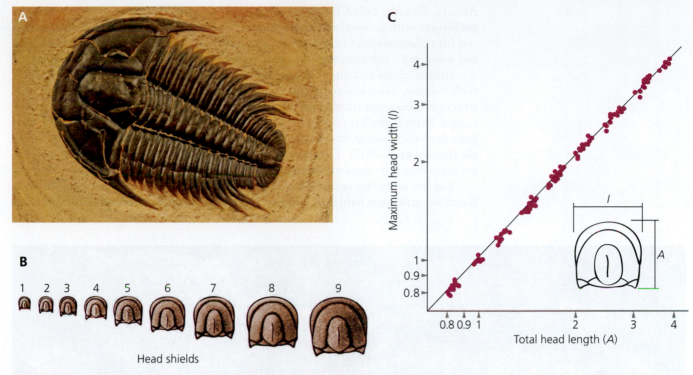

Figure 3.9 A: Trilobites were among the most common animals in the ocean between 541 and 252 million years ago, when they suddenly went extinct. B: Variation in the size of the head shield found in a trilobite species. C: By comparing the width and length of the head shield of trilobite fossils, paleontologists have been able to chart how the animal grew. (A: ScottOrr/Getty Images; B and C: Data from Hunt 1967)

Applying Physics to Paleontology

The biology of living species can inspire hypotheses about extinct ones, but the comparison is rarely straightforward. Consider the mighty carnivorous dinosaur *Tyrannosaurus rex*. It measured about 12 meters in length, weighed up to 8000 kilograms, and stood on its hind legs. It was profoundly different from any living animal today. As a result, paleontologists have long puzzled over how *T. rex* moved. Did it run like an ostrich or was it a lumbering giant?

One way of addressing a question like this is to apply the principles of physics to the fossil record. In 2002, John Hutchinson and Mariano Garcia, then at the University of California, Berkeley, developed a biomechanical model of running animals, estimating how much force leg muscles of a given size could generate (Hutchinson and Garcia 2002). They tested their model by seeing how well it could estimate the running speed of living animals. Hutchinson and Garcia chose two of the closest relatives of dinosaurs alive today—alligators and birds (see Chapter 4). In both cases, the predictions from the models matched the measurements from the real animals.

Next, Hutchinson and Garcia made a model of *T. rex*. To estimate the size of the dinosaur's muscles, they studied the sites on its bones where the muscles attached. They concluded that *T. rex* could not run quickly. Its muscles were simply not powerful enough.

As engineers develop new technologies, paleontologists apply them to fossils to gain new insights. Scanning electron microscopes, for example, allow paleontologists to examine the cellular structure of fossils. Caleb Brown of the Royal Tyrrell Museum of Paleontology and his colleagues have used these microscopes to look at fossilized skin and scales on an exceptionally well-preserved ankylosaur from

Alberta, Canada, called *Borealopelta* (**Figure 3.10**). Ankylosaurs were large terrestrial herbivores with protective armor and knobs on their backs. The fossil that Brown and his colleagues studied came from an animal measuring more than 5 meters long and weighing 1300 kilograms.

Brown examined the fossil with a scanning electron microscope, as well as with another, even newer imaging technique called energy-dispersive X-ray spectroscopy. This inspection revealed a dark pigment called melanin in the fossilized tissues. Brown and his colleagues found more melanin on the animal's dorsal surfaces (its back) than its undersurfaces—the insides of its legs and its belly, for example (Brown et al. 2017). This color pattern, known as countershading, is widespread among living terrestrial mammals, from mice to antelope.

Studies on living animals have revealed that countershading serves a valuable function: making it harder for predators to spot prey. When sunlight falls on, say, a

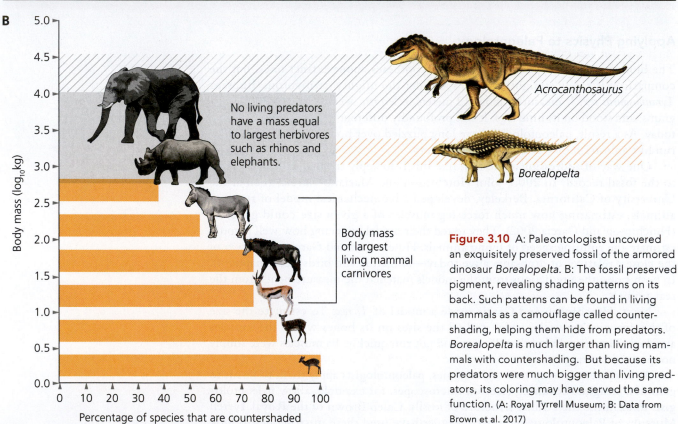

Figure 3.10 A: Paleontologists uncovered an exquisitely preserved fossil of the armored dinosaur *Borealopelta*. B: The fossil preserved pigment, revealing shading patterns on its back. Such patterns can be found in living mammals as a camouflage called countershading, helping them hide from predators. *Borealopelta* is much larger than living mammals with countershading. But because its predators were much bigger than living predators, its coloring may have served the same function. (A: Royal Tyrrell Museum; B: Data from Brown et al. 2017)

jackrabbit, it brightens the animal's back but casts a shadow that darkens its underside. This darkening helps predators recognize the three-dimensional shape of prey. When predators hunt for prey, they often search for the shadows cast by the animals. The countershading on prey species such as jackrabbits cancels out the darkening that shadows create, making it harder for predators to see them.

Big mammals such as moose and elephants lack countershading, probably because they're so big that they aren't threatened by predators like wolves and lions. Thus, natural selection doesn't favor light coloring on their underside. This pattern among living mammals makes the countershading on *Borealopelta*—which is twice the size of a moose—truly remarkable. Its coloration may be the result of the giant dinosaur predators of the Cretaceous.

Even medical technologies can be used to study fossils. Computed tomography (CT) scans were invented to give doctors detailed, three-dimensional views of the insides of patients' bodies. In 2009, David Evans, of the Royal Ontario Museum in Toronto, and Lawrence Witmer and Ryan Ridgely, of Ohio University, used a CT scanner to probe the skulls of a particularly bizarre group of dinosaurs known as hadrosaurs. These plant-eating dinosaurs grew extravagantly long crests on their heads in a diversity of species-specific shapes.

Paleontologists have speculated that the crests served as some kind of signal—possibly to competing rivals or to potential mates. (We'll discuss such sexual displays in more detail in Chapter 11.) The hollow crests are connected to the nasal opening of hadrosaurs, which led to proposals that the dinosaurs moved air through the crests, where it would resonate and produce sounds (**Figure 3.11**; Hopson 1975; Weishampel 1981).

To test this hypothesis, Evans and his colleagues took CT scans of hadrosaur skulls, getting detailed images of the interior spaces. The paleontologists looked at the braincase and observed that the region for interpreting smells was small. This finding suggested that the crests were not an adaptation for enhancing sensitivity to odors.

Evans and his colleagues also looked at the ear regions of the hadrosaur skulls. The shape of an animal's ear bones determines which frequencies it is most sensitive to. Evans and his colleagues found that the hadrosaur's ears were tuned to the frequencies that would have been produced by the crests. These results are compelling evidence that the dinosaurs used the crests to make species-specific sounds (Evans et al. 2009).

A Wealth of Data

Some of the most exciting clues to extinct life-forms come from rare sites where fossils preserve the impressions of muscle and other soft tissues. An exceptionally well-preserved fossil deposit is called a **Lagerstätte**. In almost every Lagerstätte, the fossilized animals were swept into anoxic (or low-oxygen) pools, lagoons, or bays. In these lifeless environments, bacteria and other scavengers did not exist and thus could not destroy the animals' bodies. Instead, the deceased animals were trapped in fine sediments, which preserved even their most delicate tissues in stone. Soft-tissue fossils are exquisitely important to scientists because they preserve an incredible amount of detail.

One of the most important Lagerstätten in the history of paleontology was discovered by Charles Doolittle Walcott in 1909, high in the mountain slopes of British Columbia (**Figure 3.12**). Quarries of the **Burgess Shale** have now yielded more than 65,000 specimens of mostly soft-bodied animals representing at least 93 species.

Around 505 million years ago, a rich community of marine animals thrived in and on shallow underwater mud banks that formed as sediments accumulated on the outer margins of a reef. The reef was located adjacent to a steep escarpment, where periodically the mud banks would collapse, hurling these animals into the abyss below. There, anoxic conditions prevented tissue decomposition, and after the mudslide

Lagerstätte (plural, **Lagerstätten**) is a site with an abundant supply of unusually well-preserved fossils—often including soft tissues—from the same period of time.

Burgess Shale is a Lagerstätte in Canada in which there is a wealth of preserved fossils from the Cambrian period.

Figure 3.11 A: Hadrosaurs, a group of dinosaur species, had striking crests and nasal cavities. B: By taking CT scans of hadrosaur skulls, scientists can reconstruct the structure of the cavities in different species. C: Researchers have made computer models of these cavities, such as the one shown here, to test hypotheses about their function. These studies suggest that hadrosaurs used their nasal cavities and hollow crests to make species-specific sounds. (A: Carl Buell; B: WitmerLab at Ohio University; C: Sandia National Laboratories)

the clouds of sediment in the murky waters settled down and around the bodies of these animals, preserving them intact (Briggs et al. 1995). This process appears to have occurred repeatedly, gradually building a thick sequence of fossil-rich rock.

A Lagerstätte like the Burgess Shale is important not only because it preserves the soft tissues of animals but also because it acts like a snapshot of an entire ecosystem that has long since vanished. As we'll see later in this chapter, the Burgess

Figure 3.12 A: A fossil site in the Canadian Rockies called the Burgess Shale has yielded vast numbers of fossils of animals dating back 505 million years. B: A reconstruction of *Opabinia* based on the Burgess Shale fossil shown in C. D: A reconstruction of *Hallucigenia* based on the fossil shown in E. (A: L. Newman & A. Flowers/Science/ardea.com; B: Quade Paul; C, E: The Smithsonian Institution/Chip Clark; D: Carl Buell)

Shale dates back to a pivotal period in animal evolution when a great diversity of life was emerging. The diversity was so great, in fact, that it included some truly bizarre creatures with names that reflect their strange morphology—for example, *Hallucigenia* was a creature that we might imagine inhabiting a feverish dream (see Figure 3.12D and E). In Chapter 14, we'll see how scientists are integrating their paleontological studies of the Burgess Shale with studies on embryos and ecology to understand the evolution of animal diversity.

- Technology allows scientists to gain new insights into the natural history, behavior, and appearance of extinct species by examining their fossils. ●

Key Concept

3.5 Traces of Vanished Biology

A fossil is not the only trace that an organism can leave behind. A lump of coal, for example, is actually the remains of dead plants. About 300 million years ago, giant swamps spread across many of the continents. When plants died in such conditions, they did not immediately decay. Instead, they fell into the swamps and were rapidly buried in sediment. Bacteria then began to break them down. Eventually, the swamps were drowned by rising oceans and then buried under vast amounts of marine sediment. The plant material was transformed yet again, under tremendous pressure and heat, into coal. Little pockets, called coal balls, sometimes form inside pieces of coal, where leaves and branches can remain preserved.

In certain cases, it's possible to identify the individual molecules of organisms that lived billions of years ago (Gaines 2008). To recognize these so-called **biomarkers**, geochemists must be able to determine that they were formed through biological processes. Amino acids, for example, are produced by organisms, but they can also be produced abiotically. Astronomers have even detected amino acids in interstellar clouds.

But some molecules bear clear hallmarks of their biological origins. Geochemists will sometimes find large molecules that can form only through a long series of enzymatic reactions. Some biomarkers are so distinctive that geochemists can determine which group of species produced them.

Jochen Brocks, of the Australian National University, and his colleagues have found a number of these biomarkers in 1.64-billion-year-old rocks in Australia. One of these biomarkers, called okenane, is derived from pigments made by purple sulfur bacteria (*Chromatiaceae*).

Knowing that the ancient Australian rocks contain high levels of okenane tells us a number of important things about the history of life. Because no abiotic process is known to produce okenane, we can hypothesize that purple sulfur bacteria were present on Earth 1.64 billion years ago. The presence of purple sulfur bacteria also reveals important clues about the chemistry of the oceans at the time. Today, purple sulfur bacteria are rare, found only in extreme environments with low levels of oxygen and high levels of sulfur. Their abundance 1.64 billion years ago supports the hypothesis that the oceans at the time were toxic—at least to organisms like us (Brocks and Banfield 2009).

Even the individual atoms in rocks can offer scientists clues about ancient life. As we saw in Chapter 1, the oxygen atoms in the water consumed by whales end up in the composition of their teeth. The ratio of different oxygen isotopes in whale fossil teeth indicates whether they lived in freshwater, seawater, or water with an intermediate level of salinity.

Isotopes offer clues to the habitats where organisms lived, and they also provide clues to their metabolism. Plants, for example, obtain their carbon from the atmosphere, incorporating a mixture of carbon-12 and carbon-13 isotopes into their biomass. Because carbon-13 is heavier than carbon-12, the plants have more difficulty absorbing it. As a result, the ratio of carbon-13 to carbon-12 is lower in plants than it is in the atmosphere. By analyzing the ratio of carbon isotopes in rocks, scientists can determine whether their carbon was derived from a biological source.

Geologists can also use carbon isotopes for clues about what some extinct animals were eating. That's because different plants have slightly different ratios of carbon isotopes, depending on how they carry out photosynthesis. Most plant species carry out C_3 photosynthesis, so named because it incorporates carbon dioxide into a molecule with three carbon atoms. Grasses and certain other plants have evolved a different way to photosynthesize, incorporating carbon dioxide into four-carbon molecules. This process, called C_4 photosynthesis, helps them to grow rapidly.

Chemists have found that C_4 plants have higher levels of carbon-13 than C_3 plants do. Analysis of plant fossils has revealed this same difference between extinct C_3 and C_4 plants. Because these two types of plants have distinct isotopic signatures,

A **biomarker** is molecular evidence of life in the fossil record. Biomarkers can include fragments of DNA, molecules such as amino acids, or isotopic ratios.

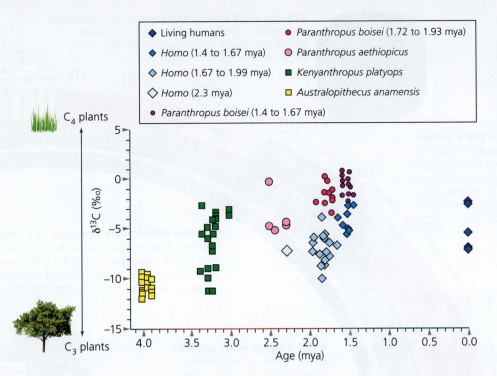

Figure 3.13 Thure Cerling of the University of Utah and his colleagues analyzed carbon isotopes in the teeth of some ancient hominins. The ratio of stable carbon isotopes in early hominins shows that C_3 plants (such as shrubs and trees) played a role in their diet. The ratios in later hominins indicate a shift to C_4 plants such as grasses. Some species probably fed primarily on C_4 plants, whereas others also fed on animals grazing on those plants. (Data from Cerling et al. 2013)

scientists can use the ratio of carbon isotopes in animal tissues to infer the types of plants that they ate. Cows and horses that graze on C_4 grasses, for example, have a higher carbon-13 : carbon-12 ratio than giraffes or elephants, which browse on the leaves of C_3 plants.

Scientists have also found this isotopic signature in the fossils of extinct animals. For example, Thure Cerling of the University of Utah and his colleagues have analyzed the carbon isotopes of our **hominin** forerunners. (We'll discuss human evolution in more detail in Chapter 17.)

Cerling and his colleagues measured carbon isotopes in the tooth enamel from hominin fossils in East Africa dating from 4.2 million years ago to 1.5 million years ago. They found that the earliest hominins had a relatively low ratio of carbon-13 to carbon-12 (**Figure 3.13**). This ratio reflects a diet rich in C_3 plants. Cerling and his colleagues observed that the ratio is similar to the one found today in the teeth of chimpanzees, which feed on fruits and leaves. Plant fossils from the same sites where these teeth were discovered show that these early hominins lived in grassy woodlands where they easily could have found C_4 plants. Thus it appears that they were actively selecting C_3 plants for their diet.

Cerling and his colleagues found that, starting about 3.5 million years ago, the $C_3 : C_4$ ratio shifted. The hominin diet drew more on C_4 plants. Some hominin species, such as *Paranthropus,* appear to have a specialized diet of grasses and other tough plants, judging from their large jaws and teeth. But other hominins—including our own lineage, *Homo*—may have acquired a C_4 signature in their teeth by eating the meat of grazing mammals (Cerling et al. 2013).

The **hominin** group includes humans as well as all species more closely related to humans than to chimpanzees. Within this group, humans are the only surviving members.

• Isotopes and biomarker molecules carry information about the history of life. •   **Key Concept**

3.6 Setting the Stage for Life on Earth

Now that we've surveyed the methods scientists use to extract information about the history of life, we'll trace that history, as we currently understand it, from the geological record (**Figure 3.14**).

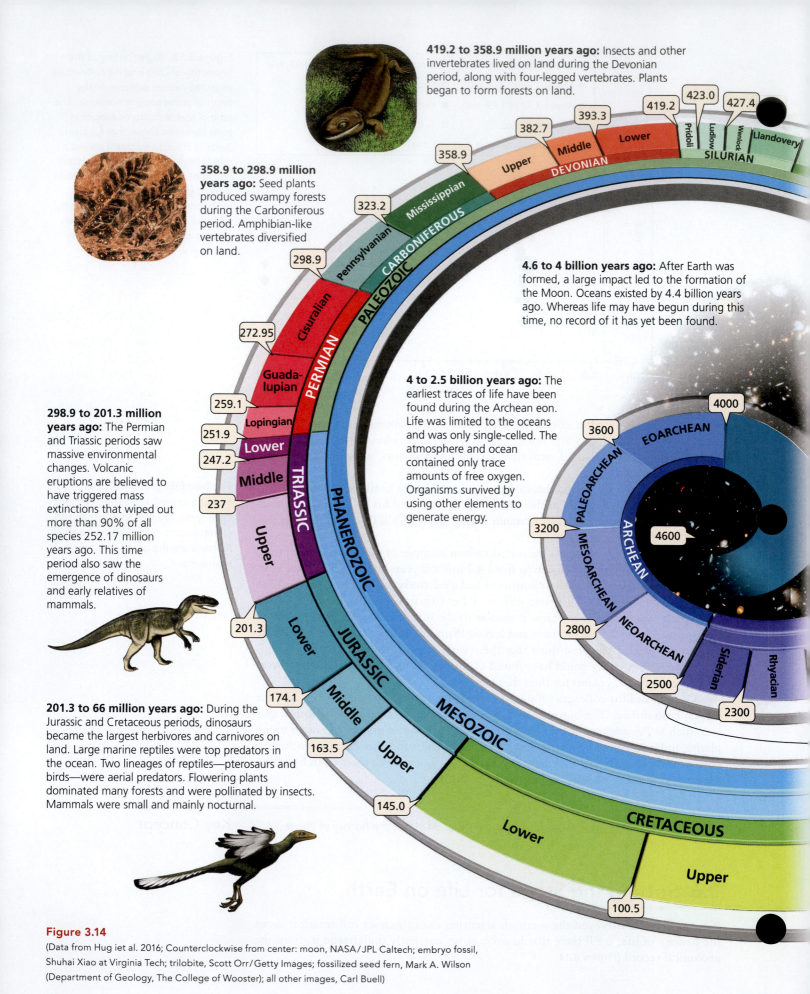

419.2 to 358.9 million years ago: Insects and other invertebrates lived on land during the Devonian period, along with four-legged vertebrates. Plants began to form forests on land.

358.9 to 298.9 million years ago: Seed plants produced swampy forests during the Carboniferous period. Amphibian-like vertebrates diversified on land.

4.6 to 4 billion years ago: After Earth was formed, a large impact led to the formation of the Moon. Oceans existed by 4.4 billion years ago. Whereas life may have begun during this time, no record of it has yet been found.

4 to 2.5 billion years ago: The earliest traces of life have been found during the Archean eon. Life was limited to the oceans and was only single-celled. The atmosphere and ocean contained only trace amounts of free oxygen. Organisms survived by using other elements to generate energy.

298.9 to 201.3 million years ago: The Permian and Triassic periods saw massive environmental changes. Volcanic eruptions are believed to have triggered mass extinctions that wiped out more than 90% of all species 252.17 million years ago. This time period also saw the emergence of dinosaurs and early relatives of mammals.

201.3 to 66 million years ago: During the Jurassic and Cretaceous periods, dinosaurs became the largest herbivores and carnivores on land. Large marine reptiles were top predators in the ocean. Two lineages of reptiles—pterosaurs and birds—were aerial predators. Flowering plants dominated many forests and were pollinated by insects. Mammals were small and mainly nocturnal.

Figure 3.14

(Data from Hug iet al. 2016; Counterclockwise from center: moon, NASA/JPL Caltech; embryo fossil, Shuhai Xiao at Virginia Tech; trilobite, Scott Orr/Getty Images; fossilized seed fern, Mark A. Wilson (Department of Geology, The College of Wooster); all other images, Carl Buell)

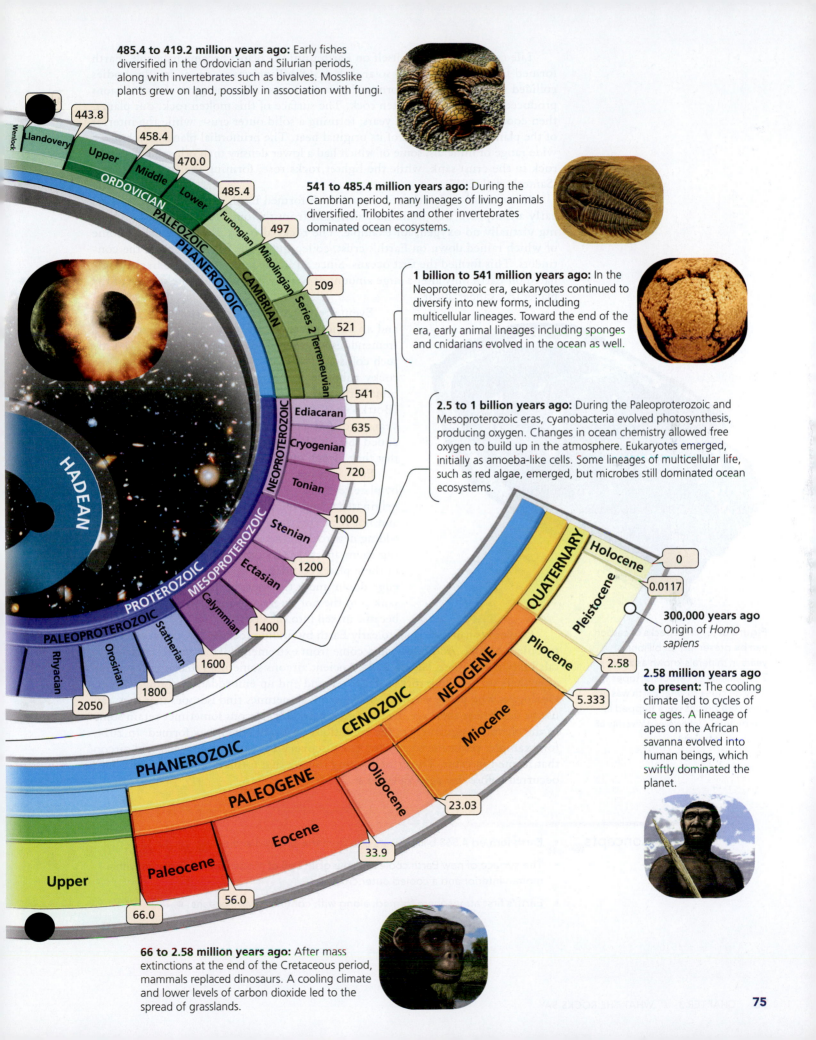

485.4 to 419.2 million years ago: Early fishes diversified in the Ordovician and Silurian periods, along with invertebrates such as bivalves. Mosslike plants grew on land, possibly in association with fungi.

541 to 485.4 million years ago: During the Cambrian period, many lineages of living animals diversified. Trilobites and other invertebrates dominated ocean ecosystems.

1 billion to 541 million years ago: In the Neoproterozoic era, eukaryotes continued to diversify into new forms, including multicellular lineages. Toward the end of the era, early animal lineages including sponges and cnidarians evolved in the ocean as well.

2.5 to 1 billion years ago: During the Paleoproterozoic and Mesoproterozoic eras, cyanobacteria evolved photosynthesis, producing oxygen. Changes in ocean chemistry allowed free oxygen to build up in the atmosphere. Eukaryotes emerged, initially as amoeba-like cells. Some lineages of multicellular life, such as red algae, emerged, but microbes still dominated ocean ecosystems.

300,000 years ago
Origin of *Homo sapiens*

2.58 million years ago to present: The cooling climate led to cycles of ice ages. A lineage of apes on the African savanna evolved into human beings, which swiftly dominated the planet.

66 to 2.58 million years ago: After mass extinctions at the end of the Cretaceous period, mammals replaced dinosaurs. A cooling climate and lower levels of carbon dioxide led to the spread of grasslands.

HADEAN

PHANEROZOIC
PALEOZOIC
ORDOVICIAN
443.8
Llandovery
Wenlock
Upper 458.4
Middle 470.0
Lower 485.4
CAMBRIAN
Furongian 497
Miaolingian 509
Series 2 521
Terreneuvian 541

PROTEROZOIC
NEOPROTEROZOIC
Ediacaran 635
Cryogenian 720
Tonian 1000
MESOPROTEROZOIC
Stenian 1200
Ectasian 1400
Calymmian 1600
PALEOPROTEROZOIC
Statherian 1800
Orosirian 2050
Rhyacian

PHANEROZOIC
CENOZOIC
PALEOGENE
Paleocene 66.0
Eocene 56.0
Oligocene 33.9
NEOGENE
Miocene 23.03
Pliocene 5.333
2.58
QUATERNARY
Pleistocene
Holocene 0
0.0117

Upper

Life could not establish itself on Earth until the planet became habitable. Earth formed from the primordial solar disk 4.568 billion years ago, as smaller bodies collided with each other to form a planet-sized mass. The energy of their collisions produced a giant ball of molten rock. The surface of this molten rock, our planet, then cooled over millions of years, forming a solid outer crust, while the interior of the planet retained much of its original heat. The primordial planet developed a wide range of minerals, some of which had a lower density than others. The denser rock in the crust sank, while the lighter rocks rose, forming the first continents (Santosh et al. 2017).

Meanwhile, rocks released gases, which formed Earth's first atmosphere. This early atmosphere was profoundly different from the one we breathe today, having virtually no oxygen, for example. The rocks also released water vapor, some of which rained down on Earth's crust, collecting in the basins between the continents. This formed the first oceans. Since all life requires water, life likely originated on Earth only after a large amount of water was present on the surface of the planet (Alexander 2016).

Even after Earth developed continents, oceans, and an atmosphere, it still occasionally experienced tremendous impacts (Bottke and Norman 2017). One such collision 4.4 billion years ago was so big that the rocky rubble thrown up from the impact began to orbit Earth and eventually coalesced to form the Moon (Bottke et al. 2015). It wasn't until about 3.8 billion years ago that this period of heavy bombardments ended (Sleep 2010). These impacts were so powerful that they melted some or all of Earth's original crust.

Early rocks on Earth's surface were also destroyed by plate tectonics. Plate tectonics is a mechanism that originated when the planet's crust broke into plates, and hot rock rose into some of the resulting cracks, adding new rock to the margins of the plates and pushing them apart. Meanwhile, the advancing margins of these plates collided with other plates, driving one edge down under the crust of the other. As this rock sank into the hot interior of the planet, it melted and became mixed into the deeper layers.

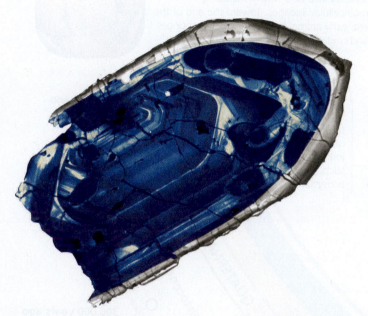

Figure 3.15 Tiny specks of carbon can be preserved for billions of years in minerals known as zircons. The balance of carbon isotopes can provide clues to what Earth was like when the zircons were trapped in the mineral. (John Valley, University of Wisconsin–Madison)

Geologists can gain clues to early Earth by finding rocks that have survived this turbulent history. Additional clues come from extremely durable microscopic crystals known as zircons (**Figure 3.15**). Ancient zircons sometimes manage to survive after their surrounding rock is destroyed and end up embedded in younger rocks. Geologists who probe the interior of zircons sometimes find isotopes that they can use to determine their age. Even more remarkably, zircons sometimes retain some of the chemical signatures of the conditions on Earth when they formed. In 2014, for example, researchers reported the discovery of a 4.4-billion-year-old zircon that formed in crust. The crust-melting impact that formed the Moon must have occurred before that age (Valley et al. 2014).

Key Concepts

- Earth formed 4.568 billion years ago from the primordial disk.

- The surface of new Earth cooled from a giant ball of molten rock to a planet with a molten interior and a cooled outer crust.

- Earth's first atmosphere formed, along with continents and oceans. ●

3.7 Life's Early Signatures

The earliest potential signs of life are 4.1 billion years old. These signs are not fossils, as you might expect. Instead, they are isotopic signatures hidden in ancient zircons.

Before life began, all of the carbon on Earth would have come from lifeless sources, like volcanoes. But once life emerged, it produced abundant amounts of organic carbon, which gradually became incorporated into sedimentary rocks. Organisms have a higher fraction of the lighter carbon-12 than carbon from volcanoes. After the origin of life, sedimentary rocks should have begun to record this shift (Gaines 2008).

In 2015, Mark Harrison, a geologist at the University of California, Los Angeles, and his colleagues reported finding evidence of this shift. They extracted zircons from rocks from Australia and used uranium-lead dating to determine that one zircon found was 4.1 billion years old. Inside the zircon were two tiny, carbon-rich inclusions.

Harrison and his colleagues measured the carbon isotopes in the inclusions and found that they had a ratio of carbon-12 to carbon-13 similar to that of living organisms. Their study suggests that life was already widespread 4.1 billion years ago (Bell et al. 2015). This study, though provocative, is preliminary. It is based on just two inclusions in one zircon. Scientists will need to find many more zircons of that age to see if the same biological signature appears in a larger sample.

When Darwin wrote *The Origin of Species*, the oldest known fossils dated back only to the Cambrian period, which began 541 million years ago. Those fossils belonged to a wide diversity of animals. If Darwin's theory was right, then life must have been evolving long beforehand. "During these vast periods the world swarmed with living creatures," Darwin wrote.

Yet Darwin recognized that no fossils of those creatures had yet been found. "To the question why we do not find rich fossiliferous deposits belonging to these assumed earliest periods prior to the Cambrian system, I can give no satisfactory answer," he wrote.

Today we know the answer: the fossils had yet to be discovered. Since Darwin's day, paleontologists have pushed back the fossil record more than 3 billion years earlier than the Cambrian period.

The oldest fossils, not surprisingly, have inspired the most debate. The earliest forms of life were microbes, and sometimes nonbiological processes can create microbe-like structures in rocks. Making the search even more challenging, the rocks containing these putative fossils have become deformed by heating and pressure over billions of years.

In the 1980s in Australia, J. William Schopf of the University of California, Los Angeles, discovered what he proposed were 3.5-billion-year-old fossils of bacteria. Martin Brasier of the University of Oxford has challenged Schopf's results, arguing that what appeared to be fossils were instead formed by tiny blobs of mineral-rich fluids (Brasier et al. 2006).

To better understand the early history of life, scientists are continuing to scour ancient rocks. Abigail Allwood and her colleagues discovered their strange, egg-carton-like rocks in some of the oldest geological formations on Earth. The researchers then found striking microscopic similarities between the rocks and large mounds built today by colonies of bacteria. These mounds, known as **stromatolites,** grow on the floors of lakes and shallow seas.

A **stromatolite** is a layered structure formed by the mineralization of bacteria.

Stromatolites form when biofilms of microorganisms, especially cyanobacteria, trap and bind sediments to form layered accretionary structures. Sediments and minerals accumulate on the bacteria in thin layers, and more bacteria grow on top of the sediments and minerals. These structures gradually enlarge into cabbage-like structures or even meter-high domes.

On modern Earth, stromatolites are fairly rare. In most environments, microscopic animals and other organisms graze on bacterial mats, preventing them from growing into large, stable structures. Modern stromatolites occur in only a very few extreme environments, such as saline lakes and hot, shallow lagoons where the high salinity keeps grazers away. In Precambrian rocks (older than 541 million years), however, stromatolite fossils are abundant. That's because animals and other grazers were rare or nonexistent before then.

Other researchers who examined the fossils found by Allwood and her colleagues have generally agreed that they are stromatolites produced by microbes. Since then, additional researchers working in Australia and Canada have found what they claim to be even older microbial fossils, dating back perhaps as far as 4 billion years. But these findings are now the subject of fierce debate. Such arguments are to be expected—and welcomed—as scientists push back to the dawn of life on Earth. (See Dodd et al., 2017; Allwood et al., 2018; and Van Zuilen 2018.)

Key Concepts

- Early Earth was a tumultuous place, from which very little original rock remains.

- Extremely ancient rocks provide valuable clues to the environments in which life first arose. ●

3.8 The Rise of Life

Although the earliest putative fossils are still the subject of debate, indisputable fossils emerge in the fossil record about 3.5 billion years ago (Knoll 2015). These fossils are so much like living organisms that it's implausible to claim that they are merely the products of nonliving processes.

But exactly what sort of organisms these fossils were is hard to say. Living microbes can be difficult to distinguish from each other by their appearance alone. Seemingly identical species may use profoundly different forms of biochemistry to harvest energy. That's because they are the product of billions of years of separate evolution.

Figure 3.16 illustrates the large-scale phylogeny of life, based on the analysis of DNA from living species. (See Chapters 4 and 8 for more details about how scientists investigate phylogeny with DNA.) The figure shows how the history of life was dominated by three great branchings. As a result of these early branchings, living things can be divided into three domains: **Bacteria, Archaea,** and **Eukarya**.

The earliest signs of life—such as the stromatolites found by Allwood and colleagues—strongly resemble living bacteria. Yet it's also possible that they're the vestiges of an extinct branch of life. More recent fossils can be more confidently interpreted as bacteria. Researchers have found fossils in 2.6-billion-year-old rocks, for example, that bear a striking resemblance to one particular lineage of bacteria known as cyanobacteria. Living cyanobacteria carry out the type of photosynthesis that produces free oxygen. That's exactly when the first evidence of atmospheric oxygen appears in the fossil record; between 2.45 and 2.32 billion years ago, oxygen increased dramatically. The rise in oxygen was likely due to the emergence of cyanobacteria, as they released oxygen during photosynthesis. Though oxygen levels increased dramatically during this time, they were still very low compared to today. As a result, purple sulfur bacteria were still abundant 1.6 billion years ago, as reflected by the presence of the biomarker okenane.

Some researchers have proposed that Archaea had already diverged from Bacteria by 3.5 billion years ago (Ueno et al. 2006; Schopf et al. 2017). They have analyzed putative microbial fossils from that age, measuring their carbon isotopes. The researchers detected an unusual ratio of these isotopes in methane-producing archaea. Among the places these organisms live today is the digestive tract of cows; thanks to these organisms, cow belches contain methane. Some other researchers

Bacteria is one of the two prokaryotic domains of life. Domain Bacteria includes organisms such as *Escherichia coli* and other familiar microbes.

Archaea is one of the two prokaryotic domains of life. Archaea superficially resemble bacteria, but they are distinguished by a number of unique biochemical features.

Eukarya is the third domain of life, characterized by traits that include membrane-enclosed cell nuclei and mitochondria. Domain Eukarya includes animals, plants, fungi, and protists (a general term for single-celled eukaryotes).

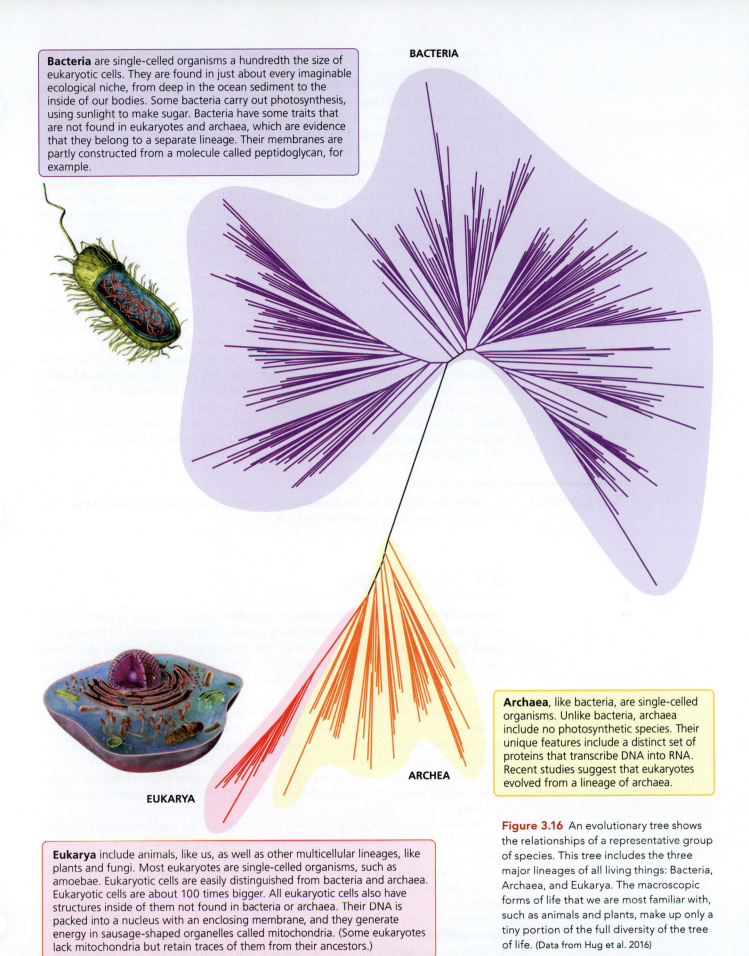

Bacteria are single-celled organisms a hundredth the size of eukaryotic cells. They are found in just about every imaginable ecological niche, from deep in the ocean sediment to the inside of our bodies. Some bacteria carry out photosynthesis, using sunlight to make sugar. Bacteria have some traits that are not found in eukaryotes and archaea, which are evidence that they belong to a separate lineage. Their membranes are partly constructed from a molecule called peptidoglycan, for example.

BACTERIA

ARCHEA

EUKARYA

Archaea, like bacteria, are single-celled organisms. Unlike bacteria, archaea include no photosynthetic species. Their unique features include a distinct set of proteins that transcribe DNA into RNA. Recent studies suggest that eukaryotes evolved from a lineage of archaea.

Eukarya include animals, like us, as well as other multicellular lineages, like plants and fungi. Most eukaryotes are single-celled organisms, such as amoebae. Eukaryotic cells are easily distinguished from bacteria and archaea. Eukaryotic cells are about 100 times bigger. All eukaryotic cells also have structures inside of them not found in bacteria or archaea. Their DNA is packed into a nucleus with an enclosing membrane, and they generate energy in sausage-shaped organelles called mitochondria. (Some eukaryotes lack mitochondria but retain traces of them from their ancestors.)

Figure 3.16 An evolutionary tree shows the relationships of a representative group of species. This tree includes the three major lineages of all living things: Bacteria, Archaea, and Eukarya. The macroscopic forms of life that we are most familiar with, such as animals and plants, make up only a tiny portion of the full diversity of the tree of life. (Data from Hug et al. 2016)

have contested these conclusions, however, questioning whether the fossils are actually fossils at all.

Eukarya emerge in the fossil record only about 1.6 billion years ago (Knoll and Nowak 2017). Their first fossils are single-celled organisms measuring about 100 micrometers across. Although these organisms would have been invisible to the naked eye, they marked a giant leap in size, measuring about 100 times bigger than a typical bacterium. These early eukaryotes had ridges, plates, and other structures that are similar to those of living single-celled eukaryotes. Over the next billion years, the diversity of these single-celled eukaryotes increased—some lineages acquired cyanobacterial symbionts and became able to carry out aerobic photosynthesis, whereas others preyed on bacteria or grazed on their photosynthetic relatives (Knoll et al. 2006).

If you could travel back in time to 1.5 billion years ago, the world would look like a desolate place. On land there were no trees, no flowers, not even moss. In some spots, a thin varnish of single-celled organisms grew. In the ocean, there were no fishes or lobsters or coral reefs. Yet the ocean teemed with microbial life, from the organisms that lived around hydrothermal vents on the seafloor to free-floating bacteria and photosynthetic eukaryotes at the ocean's surface. Along the coasts, microbial mats stretched for miles in the shallow waters.

Today our attention may be distracted by animals and plants, but the world remains dominated by microbes. By weight, microbes make up the bulk of Earth's biomass. Microbes live in a tremendous range of habitats that would kill the typical animal or plant—from Antarctic deserts to the bottom of acid-drenched mine shafts. The genetic variation among single-celled life also far exceeds that of animals or plants. Most genes on the planet belong to microbes or their viruses. It's a microbial world, in other words, and we just happen to live in it.

Key Concept

- The earliest signs of life are microbial, and microbes still constitute most of the world's biomass and genetic diversity. ●

3.9 Life Gets Big

One of the most dramatic transitions in evolution was the origin of multicellular organisms like ourselves. The human body is made of approximately 30 trillion cells glued together with adhesive molecules and differentiated into organs and tissues that work together (Milo and Phillips 2015). Only a minuscule fraction of cells in the human body—the sperm or eggs—have the potential to pass on their genetic material to future generations.

The transition from single-celled microbes to multicellular organisms was a profound one. But the tree of life shows that multicellularity did not evolve just once. Instead, multicellularity evolved on dozens of different occasions. Animals, for example, are closely related to fungi, which can develop mushrooms and other multicellular structures. But animals have lots of even closer relatives that are single-celled, which means that multicellularity likely evolved independently in animals and fungi.

Studies on living organisms offer some clues to what the precursors of multicellular lineages may have looked like. Although bacteria are considered single-celled, for example, they often live together in films that line surfaces ranging from our intestines to rocks on the seafloor. In these biofilms, they send signals to one another that regulate their growth and activity. Among eukaryotes, a model organism used to study the steps leading to multicellularity is *Dictyostelium discoideum*, a slime mold that lives most of its life as a single-celled predator on bacteria (**Figure 3.17**). When its prey runs out, it joins with thousands of other *D. discoideum* to form a sluglike "body" that crawls through the soil. Eventually, it

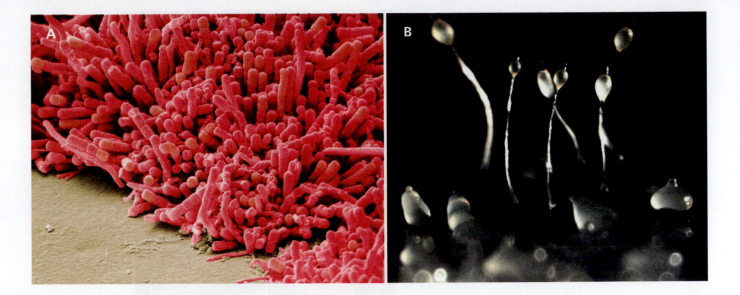

stops, and some of the cells produce cellulose to build a stalk, on top of which a ball of spores forms (we discuss *Dictyostelium* social behavior in Chapter 16).

With so many examples of living multicellular organisms, it's plausible that some extinct lineages independently evolved multicellularity as well. In the West African country of Gabon, for example, researchers have discovered a baffling collection of more than 400 macroscopic fossils dating back about 2.1 billion years (Albani et al. 2010; El Albani et al. 2014). The fossils, measuring up to 17 centimeters across, come in a variety of shapes, from scalloped disks to twisting tubes to bumpy plates. The researchers can't determine which of the main branches of the tree of life the fossils belong to. Adding to the mystery is the age of the fossils. The researchers found them in a series of layers of shale rock, which formed over millions of years in a shallow sea. But as the geologists moved their way up the geological column, the fossils simply disappeared. It's likely that this strange experiment in multicellularity disappeared, leaving behind no living descendants (**Figure 3.18**).

The oldest known fossils of clearly recognizable multicellular eukaryotes are half a billion years younger, dating back 1.6 billion years. In India, for example, researchers have found fossils of red algae (**Figure 3.19**; Bengston et al. 2017). But almost another billion years would elapse before our own lineage of multicellular organisms—the animals—began leaving behind their fossils as well.

Figure 3.17 A: Bacteria can grow and divide individually, but they can also form multicellular aggregates, such as gelatinous sheets called biofilms. B: *Dictyostelium discoideum* (a type of slime mold), a soil eukaryote, is typically unicellular. But *D. discoideum* individuals can come together to form a sluglike mass that can crawl away and form a stalk of spores. These structures are more elaborate than bacterial biofilms and offer clues to how multicellularity first evolved. (A: Science Photo Library/Superstock; B: Owen Gilbert)

Figure 3.18 In 2010, scientists published details of enigmatic fossils measuring up to 12 centimeters across. Dating back 2.1 billion years, they are the oldest fossils of multicellular life. Left: A three-dimensional rendering of one of the fossils generated from a CT scan. Right: A slice through the interior of the fossil. (Data from Albani et al. 2010; © Abderrazak El Albani/Arnaud Mazurier/ CNRS Photo Library)

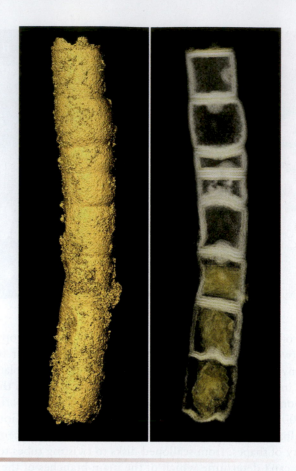

Figure 3.19 A fossil of red algae, known as *Bangiomorpha*, dating back 1.2 billion years. [Bengtson S, Sallstedt T, Belivanova V, Whitehouse M (2017) Three-dimensional preservation of cellular and subcellular structures suggests 1.6 billion-year-old crowngroup red algae. PLoS Biol 15(3)]

Key Concept

- The transition to multicellular life began at least 2.1 billion years ago, but multicellularity evolved independently in a number of lineages. ●

3.10 The Dawn of the Animal Kingdom

Animals are different from other multicellular eukaryotes in a number of ways, but arguably the most important of these differences is how animals capture energy. Green plants, green algae, brown algae, and red algae all use photosynthesis to harvest sunlight. Fungi release enzymes to break down plant or animal matter, which they can then absorb. Animals, on the other hand, evolved bodies that could swallow other organisms.

Today, animals include familiar groups such as mammals and birds, along with less familiar ones as well, like sponges. Sponges may not seem particularly animal-like. They lack a brain, eyes, and even a mouth. They get their food neither by swimming after prey nor by grazing on plants. Instead, they trap organic particles that drift through the water into the pores of their bodies. Yet sponges share thousands of genetic markers with other animals that are not found in non-animal species. And sponges appear to mark the earliest appearance of animals in the fossil record. In 2015, Zongjun Yin of the Nanjing Institute of Geology and Palaeontology and his colleagues discovered an exquisitely preserved, 600-million-year-old fossil sponge the size of a pinhead (**Figure 3.20**; Yin et al. 2015).

Some critics are skeptical that Yin and colleagues have found a sponge. These critics observe that this single putative fossil came to light only after the researchers examined many samples that contained no fossils at all. If the sponge was part of a thriving ecosystem, then you would expect to find more fossils. But there's more evidence of early sponges in the form of molecular signatures.

Sponges produce a distinctive kind of cholesterol molecule that can serve as a biomarker of their presence in rocks. Gordon Love, a geochemist at the University

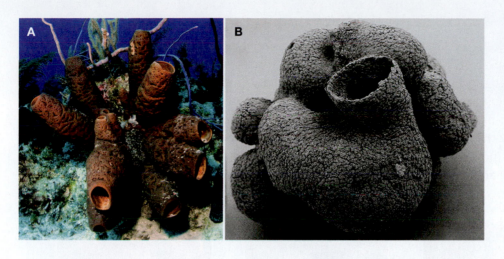

Figure 3.20 In 2015, scientists in Australia described 600-million-year-old fossils, which they interpreted as sponges. A: Living marine sponges. B: A fossil sponge. (A: Stocktrek Images, Inc./Alamy Stock Photo; B: Yin, Z., M. Zhu, E. H. Davidson, D. J. Bottjer, F. Zhao, and P. Tafforeau. 2015. Sponge Grade Body Fossil with Cellular Resolution Dating 60 Myr Before the Cambrian. Proceedings of the National Academy of Sciences 112: E1453–E1460)

of California, Riverside, and his colleagues found this biomarker of sponges in 635-million-year-old rocks in Oman, as well as rocks of similar age in Siberia (Love et al. 2009; Kelly et al. 2011; Gold et al. 2016). In 2011, they reported finding the same molecules in rocks of a similar age in Siberia (Kelly et al. 2011). These biomarkers, as well as early sponge fossils, demonstrate that animals had already evolved tens of millions of years before the start of the Cambrian period.

Sponges, as well as other multicellular animals that emerged before the Cambrian period, were mostly sedentary species, anchored to the ocean floor. But evidence shows that about 585 million years ago, animals that could move across the ocean floor emerged. These moving animals didn't leave behind fossils of their bodies, however. Instead, they appear to have left behind their tracks. In Ecuador, Ernesto Pecoits of the University of Alberta and his colleagues found 585-million-year-old rocks with troughs in them that bear a striking resemblance to the tunnels made by burrowing worms today (**Figure 3.21**). Such animals would have been markedly different from sponges. They must have had muscles and nerves, for example (Pecoits et al. 2012).

Younger rocks reveal more evidence of animals, such as tiny shells measuring 1 to 2 millimeters wide (Budd 2008). And starting at 570 million years ago, a host of much big-

Figure 3.21 Paleontologists recently found this fossil, and many others like it, in Ecuador. They concluded the marks were made by wormlike animals 2 to 3 millimeters in diameter. Dating back at least 585 million years, these fossils are the oldest evidence of animals that can move. (Data from Pecoits et al. 2012; Ernesto Pecoits)

ger organisms appear in the fossil record. Measuring more than a meter in length in some cases, these fossils were bizarre. Some looked like fronds, others like geometrical disks, and still others like blobs covered with tire tracks. They squeezed together on the microbial mats that covered some parts of the seafloor, with thousands of individual organisms sometimes living in a single square meter. Collectively, these enigmatic species are known as the **Ediacaran fauna** (named for a region in Australia where paleontologists first recognized that these kinds of fossils dated back before the Cambrian period).

Paleontologists have compared Ediacaran fossils to living species to determine their place in the tree of life. Some fossils share many traits with living groups of animals. *Kimberella*, for instance, has a rasp-shaped feeding structure found today in mollusks, a group that includes clams and snails. But many Ediacarans have proven far more difficult to decipher. Some fossils may be animals, but they are only distantly related to living lineages. Others may be extinct lineages of different organisms that independently evolved multicellularity (**Figure 3.22**; Droser et al. 2017).

Ediacaran fauna is a group of animal species that existed during the Ediacaran period, just before the Cambrian. The oldest Ediacaran fossils are about 570 million years old. Ediacarans included diverse species that looked like fronds, geometrical disks, and blobs covered with tire tracks. They appear to have become extinct 540 million years ago, although some researchers argue that certain Ediacaran taxa belong to extant animal clades.

EDIACARAN FAUNA

The fossil record allows paleontologists to do more than just identify individual species. It also allows them to reconstruct entire ecosystems, infer how they functioned, and track their change through time. For example, the oldest evidence for abundant multicellular life occurred between about 570 and 540 million years ago, during the Ediacaran period. The reconstruction on the left shows what this ecosystem probably looked like. It was dominated by frondlike organisms (possibly related to animals) that probably fed by filtering seawater. Some small animals lived on the ocean floor, which was covered in microbial mats.

Charniodiscus

Dickinsonia

Aspidella

Kimberella

Spriggina

Tribrachidium

Charni

P. reticulata

During the Cambrian period, starting 541 million years ago, ocean ecosystems became radically reorganized. Animals grazed on microbial mats and burrowed into the ocean floor, oxygenating the sediment. More active animals evolved, including a hierarchy of predators.

Figure 3.22
(Illustration: Quade Paul; *Charniodiscus, Dickinsonia, Spriggina, Tribrachidium,* The Museum Board of South Australia 2014. Photographer: Dr. Jim Gehling; *trilobite,* ScottOrr/Getty Images; *Opabinia, Hallucigenia, Ottoia,* The Smithsonian Institution/Chip Clark)

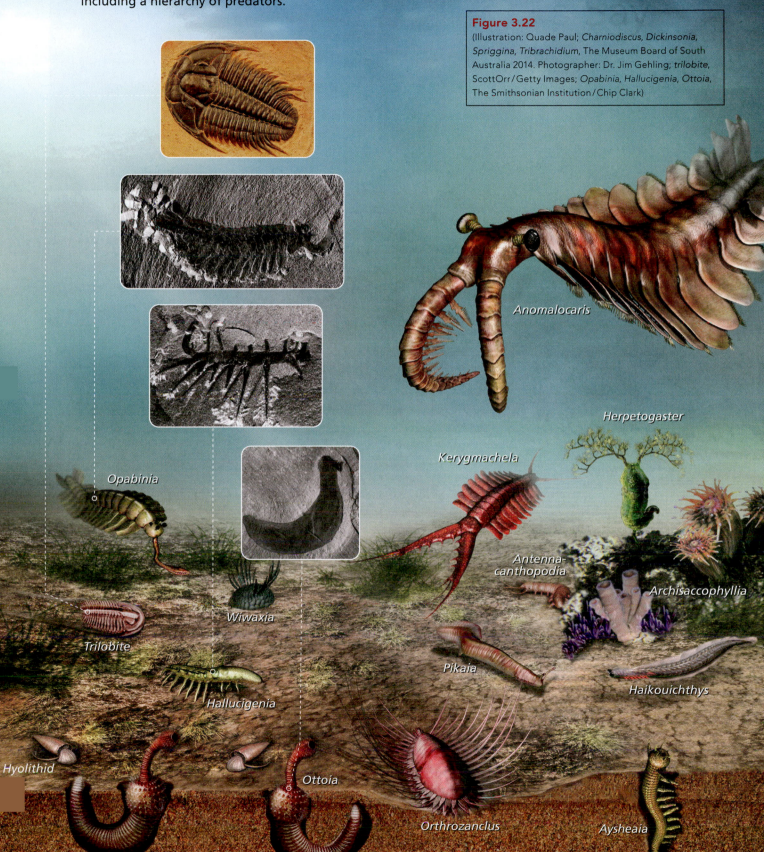

Anomalocaris

Herpetogaster

Kerygmachela

Opabinia

Antenna-
canthopodia

Archisaccophyllia

Wiwaxia

Trilobite

Pikaia

Haikouichthys

Hallucigenia

Hyolithid

Ottoia

Orthrozanclus

Aysheaia

A **chordate** is a member of a diverse phylum of animals that includes the vertebrates, lancelets, and tunicates. As embryos, chordates all have a notochord (a hollow nerve cord), pharyngeal gill slits, and a post-anal tail. Many present-day chordates lose or modify these structures as they develop into adults.

A **notochord** is a flexible, rod-shaped structure found in the embryos of all chordates. Notochords served as the first "backbones" in early chordates, and in extant vertebrates the embryonic notochord becomes part of the vertebral column.

A **trilobite** was a marine arthropod that diversified during the Cambrian period. Trilobites gradually died out during the Devonian period.

Figure 3.23 We humans belong to the chordate lineage that first appeared in the fossil record during the Cambrian period and gave rise to vertebrates. A: *Haikouichthys* was a small, fishlike animal with some traits found only in chordates, such as a brain, a stiffening rod (called a notochord) running next to its spinal cord, and arches that may have supported gills. B: By 370 million years ago, large vertebrate predators had evolved, such as *Dunkleosteus*, which grew up to 6 meters (18 feet) long. (Carl Buell)

Many of these enigmatic Ediacaran forms had disappeared by the beginning of the Cambrian period, 541 million years ago, and the rest became extinct in the following few million years. In the meantime, some of the earliest recognizable members of living animal lineages had emerged (Erwin and Valentine 2013). The Cambrian period, from 541 million to 485 million years ago, is divided into four stages, and each stage saw more first appearances of living groups than the previous one did. We belong to the **chordates,** for example. This group of animals, which as embryos all have a **notochord,** pharyngeal gill slits, and a post–anal tail, first appears in fossil-rich rocks in China called the Chengjiang Formation, which dates back 515 million years (**Figure 3.23**; Shu et al. 1999).

Not all of the major groups of animals that emerged during the Cambrian period can be found on Earth today. After **trilobites** (see Figure 3.9) emerged during the Cambrian period, for example, they flourished until 252 million years ago. The last trilobite species disappeared at around the same time that about 90% of all other species vanished. (For more on the causes and effects of such mass extinctions, see Section 14.9.)

Key Concepts

- Although early Ediacaran fossils were highly diverse and had unique body plans, only a fraction of them share traits with living species. Nearly all Ediacaran species disappear from the fossil record within 40 million years of their appearance.

- Nearly all living animal lineages, including chordates, evolved during the Cambrian period. ●

3.11 Climbing Ashore

Following the rise of multicellular life, another major transition documented in the fossil record is the transition of life from the ocean to land (Labandeira 2005). As life evolved in the sea, dry land remained bare. The earliest hints of terrestrial life come from **prokaryotes**. In South African rocks dating to 2.6 billion years ago, scientists have found remains of microbial mats that grew on land. Fungi, plants, and animals did not arrive on land until much later. In Oman, scientists have found 475-million-year-old fossils of spores that appear to have embedded originally in plant tissues—the oldest plant fossils found so far. The earliest land plants resemble mosses and liverworts (Wellman et al. 2003). Over the next 100 million years, the fossil record documents the rise of larger plants in a greater diversity of forms (Delwiche and Cooper 2015). Eventually, complex forests developed (**Figure 3.24**).

A **prokaryote** is a microorganism lacking a cell nucleus or any other membrane-bound organelles. Prokaryotes comprise two evolutionarily distinct groups, the Bacteria and the Archaea.

Figure 3.24 The earliest land plants resembled mosses (A) and liverworts (B). The oldest treelike plant (C), known as *Wattieza*, lived 385 million years ago and stood 8 meters (26 feet) tall. D: Depiction of ancient trees. (A: Argument/Getty Images; B: Todd Boland/Shutterstock; C: Reprinted by permission from Macmillan Publishers Ltd: Nature, "Giant cladoxylopsid trees resolve the enigma of the Earth's earliest forest stumps at Gilboa" © 2007 (reconstruction, left) Frank Mannolini, NY State Museum, Albany, NY; (photograph, right) South Mountain Trunk, William Stein, State University of NY at Binghamton, NY; D: Peter Giesen)

Today, land plants live in intimate association with fungi. Some fungi feed on dead plants, helping to convert them into soil. Others cause plant diseases, such as chestnut blight, which wiped out almost all American chestnut trees in the early twentieth century. Still others help plants, supplying nutrients to their roots in exchange for organic carbon that the plants create in photosynthesis. The oldest fungi fossils date back 440 million years, suggesting that fungi and plants moved in a coordinated fashion from water onto land (Berbee and Taylor 2007; Smith 2016; see Chapter 15 for more on how different species form intimate partnerships).

Animals left only tentative marks on the land at first. Rocks dating back about 480 million years display tracks that appear to have been made by invertebrate animals—probably ancient relatives of insects and spiders. The tracks were made on a beach dune; whether the animal that made them could actually have lived full-time on land is a mystery. The oldest known fossil of a fully terrestrial animal is more than 60 million years younger than the first trackways: a 414-million-year-old relative of today's millipedes, found in Scotland in 2004 by a bus driver who hunts for fossils in his free time (**Figure 3.25**; Wilson and Anderson 2004, Suarez et al. 2017). The oldest known trackways left by a vertebrate date back 390 million years (**Figure 3.26**;

Figure 3.25 The oldest known fossil of a land animal is the 428-million-year-old millipede *Pneumodesmus newmani*, shown here in an artist's reconstruction. (Carl Buell)

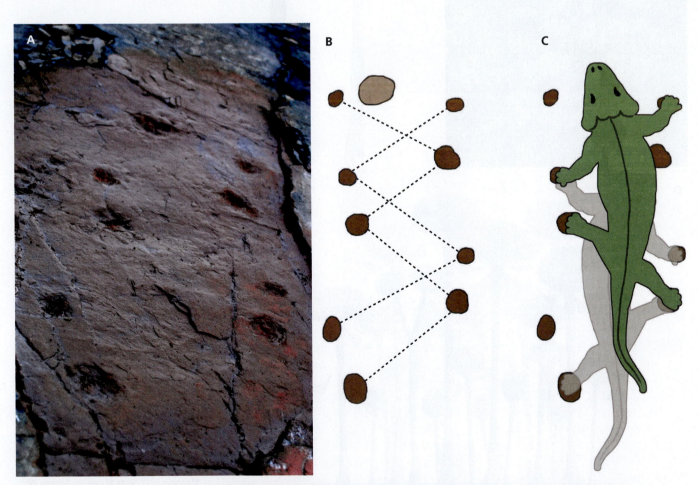

Figure 3.26 A: In 2010, paleontologists reported a 390-million-year-old animal trackway. B: The spacing of the tracks suggests they were made by an animal with an alternating gait. C: It's possible that this animal was an early tetrapod. (A: Grzegorz Niedźwiedzki)

Figure 3.27 *Silvanerpeton* was one of the oldest terrestrial vertebrates (known as tetrapods) that left fossils. (Carl Buell)

Niedźwiedzki et al. 2010), and the oldest known fossils of vertebrates with legs—known as **tetrapods**—are about 370 million years old (**Figure 3.27**). In Chapter 4, we'll look at the origin of tetrapods in more detail.

A **tetrapod** is a vertebrate with four limbs (or, like snakes, descended from vertebrates with four limbs). Living tetrapods include mammals, birds, reptiles, and amphibians.

- The transition from life in the oceans to life on land marked another major change in the fossil record. ●

Key Concept

3.12 Recent Arrivals

An important lesson from the fossil record is that some very familiar kinds of life existing today did not emerge until relatively recently. Most species of fish on Earth today, for example, belong to a group known as the **teleosts**. They include many of the most familiar fishes, such as tuna, salmon, and goldfish. But 350 million years ago there were no teleosts at all. Likewise, 350 million years ago there were no mammals, which today are dominant vertebrates on land. Some 15,000 species of birds fly overhead today, but not a single bird existed 350 million years ago.

Before today's most common groups of species emerged, the planet was dominated by other lineages. Before the rise of teleost fishes, for example, some of the ocean's top predators were giant sea scorpions, which measured up to 6 feet long. On land, 280 million years ago, the dominant vertebrates were relatives of today's mammals: ungainly, sprawling creatures called **synapsids** (**Figure 3.28**). The first synapsids to evolve into something looking even remotely like today's mammals emerged about 200 million years ago. It was not until 160 million years ago that the first members of the living groups of mammals evolved (Luo et al. 2011). In

A **teleost** is a lineage of bony fishes that comprises most living species of aquatic vertebrates. They can be distinguished from other fishes by unique traits, such as the mobility of an upper jawbone called the premaxilla.

A **synapsid** is a lineage of tetrapods that emerged 300 million years ago and gave rise to mammals. Synapsids can be distinguished from other tetrapods by the presence of a pair of openings in the skull behind the eyes, known as the temporal fenestrae.

Chapter 4, we'll examine the evolution of mammals from primitive synapsids in more detail.

Meanwhile, new lineages of reptiles were also evolving. One of the most successful was the dinosaur branch. Dinosaurs emerged about 240 million years ago and steadily grew more diverse. Their ranks included giant long-necked sauropods that were the largest animals ever to walk on Earth as well as fearsome predators. Dinosaurs dominated ecosystems on land until they disappeared in a pulse of mass extinctions 66 million years ago. The only survivors of this lineage today are the birds, which branched off from other dinosaurs about 160 million years ago (Chiappe 2007). In Chapter 4, we'll examine the origin of birds more closely.

Most of the plants we see around us today are also relatively new in the history of life. As we saw earlier, the earliest plants for which we have a fossil record resembled living liverworts and mosses. They likely formed low, ground-hugging carpets of vegetation. The evolution of tough compounds in plants, such as lignin, allowed some lineages to grow stems, stalks, and trunks. Starting in the Carboniferous period 360 million years ago, a number of large-sized plant lineages began to appear. Many lineages later became extinct, but some—such as ferns and gingko trees—have survived until

Figure 3.28 Mammals are descended from sprawling, reptile-like vertebrates called synapsids that first emerged 320 million years ago. (Carl Buell)

today. They are no longer the dominant plant lineages, however. Today, most ecosystems are instead dominated by flowering plants.

The oldest well-accepted fossils of flowering plants date back only 136 million years—some 300 million years younger than the oldest known plant fossils. During the Cretaceous period, flowering plants became more abundant and diverse. In Chapter 15, we'll consider one explanation for their rise: their coevolution with pollinating insects (**Figure 3.29**).

Today, grasses are one of the most widespread forms of flowering plants. Grasses cover the great expanses of savannas and prairies of the world; they thrive in suburban lawns and city parks. Farmers plant domesticated grasses—such as wheat and corn—over much of the planet's arable surface. Yet grass fossils are also late arrivers. The first evidence of grasses in the fossil record is tiny bits of tissue in the 100-million-year-old droppings of dinosaurs (Prasad et al. 2011). Grasses remained relatively rare

Figure 3.29 The oldest insect fossils are 400 million years old. But many of the largest groups of living insect species evolved much later. The first flies, for example, evolved about 250 million years ago. This fly (a gall midge) was trapped in amber (fossilized tree resin) about 30 million years ago. (NHPA/Superstock, Inc.)

BOX 3.2

The Present and the Past in Science

Predictions are essential to science. When scientists devise a hypothesis explaining a natural phenomenon, they can use it to generate predictions about what will happen under a certain set of conditions. To evaluate a hypothesis, scientists will often set up an experiment to see if its predictions are met. Evolutionary biologists, for example, can make predictions about how natural selection will act on a population and then design an experiment to see if they're correct. (See Section 6.6 for an example of such an evolutionary experiment.)

As we've seen in this chapter, though, evolutionary biologists also study things that happened millions or billions of years ago. They generate hypotheses about these events. They can't run experiments on extinct species, but this doesn't mean the past can't be studied scientifically. To study the past, researchers make predictions about what further research about those past events will yield.

Scientists used this method to test Kelvin's hypothesis that Earth was less than 20 million years old. When it became possible to study isotopes in rocks, scientists could make a prediction based on Kelvin's hypothesis, namely, that it should contain short-lived isotopes. This prediction failed.

Evolutionary biologists have many other ways to make predictions about the past. In Chapter 4, we'll look at how scientists construct phylogenetic hypotheses based on the morphology of fossils and living species. Each of these hypotheses makes a prediction that other lines of evidence will support the proposed phylogeny. As we'll see in Chapter 9, scientists can also now examine the evidence encoded in DNA. In a number of cases, the results of these molecular studies match the predictions from fossil evidence. In cases where the results don't match the predictions, further study can often resolve the discrepancies.

Another example of the predictions that evolutionary biologists make comes from one of the greatest catastrophes in the history of life. About 252 million years ago, over 90% of all species became extinct (Section 14.9). Based on the chemistry of rocks formed at that time, some researchers proposed that much of the ocean's oxygen had been depleted, starting in deep water and spreading out to the coasts. That hypothesis led to a prediction: if paleontologists found a good fossil record, they'd find that deepwater species became extinct first, and then shallow-water ones.

In the fossils of bryozoans, coral-like animals that anchor themselves to the seafloor, Catherine Powers and David Bottjer at the University of Southern California found just such a record. Many deepwater bryozoan species went extinct first during mass die-offs, and then shallow ones disappeared later on, just as predicted (Powers and Bottjer 2007).

Three hundred thousand years—the age of our species—is such a vast span of time that it's hard for the human mind to fathom. And yet, as we've seen in this chapter, it is just a tiny fraction of the 3.7 billion or more years that life has existed on Earth. If you were to shrink the time that life existed down to a day, our species would have emerged 5 seconds before midnight.

The geological record of fossils, biomarkers, isotopes, and other traces of past life is clear evidence that life on Earth is immensely old. It also documents a profound transformation. For more than 1.5 billion years, only single-celled organisms inhabited the planet. Over the next 2 billion years, multicellular life-forms also emerged. To understand how these sorts of changes occurred, scientists do not simply catalog lists of bones and stromatolites. They also determine how different species—either alive today or long extinct—are related to one another. By measuring these relationships, scientists can form hypotheses about the processes and patterns of evolution. How they discover life's kinship is the subject of the next chapter.

---

long after their emergence. Only about 20 million years ago did widespread grasslands emerge.

The fossil record is filled with many examples of such long-term shifts in biological diversity and abundance, and, as we'll see in Chapter 14, a major goal of evolutionary biology is to test explanations for these changes. In the case of grasses, many researchers have argued that a gradual shift in the chemistry of the atmosphere has been responsible. Over the past 50 million years, carbon dioxide has been declining gradually in the atmosphere. Grasses, which undergo C_4 photosynthesis, are more efficient at extracting carbon dioxide from the atmosphere than C_3 plants (Piperno and Sues 2005; Soltis et al. 2008).

At about the same time that flowering plants were diversifying, the modern lineages of mammals were also becoming established. Only after the dinosaurs

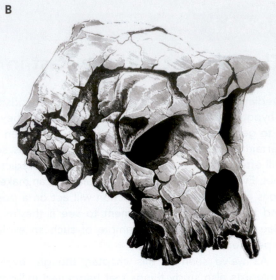

Figure 3.30 A: A reconstruction of the oldest known bipedal hominin, *Sahelanthropus*. B: This fossil, discovered on the edge of the Sahara Desert in 2001, is estimated to have lived 7 million years ago. (A: Carl Buell; B: From "Two new Mio-Pliocene Chadian hominids enlighten Charles Darwin's 1871 prediction" by Michel Brunet. Philosophical Transactions of the Royal Society B: Biological Sciences, 27 October 2010: Vol. 365, No. 1556)

Figure 3.31 This jaw, the oldest known fossil of our own species, discovered in Morocco, is estimated to be at least 300,000 years old. (Jean-Jacques Hublin, MPI EVA Leipzig)

were gone did mammals start evolving into dramatically new forms. Starting around 50 million years ago, for example, the ancestors of whales evolved from land mammals into the ocean's top predators (Chapter 1). At about the same time, bats evolved as the only flying mammals. The first fossils of primates are of the same age (Springer et al. 2012).

The first primates were small, lemur-like creatures, but they shared many traits found in all living primates (including ourselves), such as forward-facing eyes and dexterous hands (Ni et al., 2013). The oldest fossils on the human lineage (known as hominins) are much more recent. They include *Sahelanthropus*, which was discovered in 2001 in Chad and dates back about 7 million years (**Figure 3.30**; Brunet et al. 2002). Early hominins were similar to chimpanzees in some ways, including their body and brain size. The oldest known fossils matching our own stature emerged only about 2 million years ago, and the oldest fossils that clearly belong to our own species, found in Morocco, are estimated to be 300,000 years old (**Figure 3.31**; Hublin et al. 2017). In Chapter 17, we'll investigate some of the key genetic and behavioral changes over the course of hominin evolution.

Key Concept
- Many of the most diverse animal and plant species alive today have undergone relatively recent adaptive radiations. ●

TO SUM UP . . .

- Geologists use the breakdown of radioactive isotopes to estimate the age of rocks.

- Estimates of Earth's age, made using various tools, converge on a precise age of 4.568 billion years.

- Organisms only rarely become fossils.

- The ratio of isotopes in fossils can give hints about the diets and ecology of extinct species.

- Biomarkers, such as molecules from cell walls, can be preserved for hundreds of millions of years.

- Paleontologists can test their predictions about fossils against new evidence as it is discovered.

- The oldest putative chemical traces of life have been found in rocks about 4.1 billion years old.

- Stromatolites and other fossils of microbes date back about 3.7 billion years.

- The three main branches of life—Bacteria, Archaea, and Eukarya—diverged at some point after the first living organisms appeared.

- Multicellular eukaryotic fossils date back as far as 1.6 billion years.

- Biomarkers of animals date back as far as 650 million years.

- The Ediacaran fauna is a puzzling collection of animals that existed between about 570 and 540 million years ago.

- Some of the first members of living groups of animals appeared during the Ediacaran period, and more appeared during the Cambrian period.

- Plant fossils date back 475 million years. Invertebrate animals may have walked on land by then.

- The oldest fossils of vertebrates with four legs (tetrapods) date back about 370 million years.

- The oldest known fossils of animals that looked similar to living mammals are 200 million years old.

- The oldest known fossils of our own species are about 300,000 years old.

MULTIPLE CHOICE QUESTIONS   Answers can be found at the end of the book.

1. What evidence did Darwin use to predict the age of Earth?

   a. Darwin didn't predict the age of Earth.
   b. Darwin used mathematical formulas to calculate the age of Earth.
   c. Darwin used processes he could observe, such as erosion and sedimentation, to predict that Earth must be hundreds of million years old.
   d. It doesn't matter, because Lord Kelvin refuted Darwin's evidence.

2. What is an isochron?

   a. The ratio of rubidium (Rb) to strontium (Sr).

   b. The rate of radioactive decay of an isotope.

   c. The slope of the line describing the ratio of ^{87}Sr to ^{86}Sr.

   d. A line on a graph of isotope ratios that indicates mineral samples formed at a similar time.

3. Which isotope would be useful for dating a fossil found in relatively recent sediments?

   a. An isotope with a moderately high probability of decay.
   b. An isotope with a low probability of decay.
   c. Rubidium.
   d. Strontium.

4. How did the fossils of the Burgess Shale likely form?

   a. The animals fell to the bottom of a deep lake and over thousands of years turned to rock.
   b. The animals dropped into anoxic ocean depths and were covered by fine sediment.
   c. The animals were rapidly covered by ash falling from a volcano.
   d. Shallow seas disappeared, exposing mud sediments to the air.

5. How did scientists determine that *Tyrannosaurus rex* could not run very fast?

   a. They compared skeletal structures of *Tyrannosaurus rex* to modern animals to determine the size of *T. rex*'s muscles.
   b. They used living animals to test a model they had developed on the biomechanics of running.
   c. They used evolutionary theory to determine the most closely related living organisms to *T. rex*.
   d. All of the above.

6. Which group is not considered one of the major lineages of all living organisms?

   a. Bacteria.
   b. Prokarya.
   c. Archaea.
   d. Eukarya.

7. Which of the following helped scientists determine that multicellular life arose more than once?

   a. Animals are more closely related to single-celled eukaryotes than to fungi.
   b. Fungi can produce multicellular structures.
   c. Bacteria live as multicellular groups called biofilms.
   d. Scientists have no idea if multicellular life arose more than once.

8. Define prokaryote.

   a. A descriptive grouping for microorganisms that lack membrane-bound organelles.
   b. A grouping useful in the classification of early microorganisms.
   c. A grouping often used instead of Archaea.
   d. A single-celled eukaryote.

9. Why is a notochord an important adaptation for understanding the evolution of humans?

   a. The development of a notochord occurred in early fishes.
   b. A notochord is a backbone.
   c. A notochord is characteristic of chordates.
   d. A notochord serves to distinguish the Ediacaran fauna from the trilobites.

10. What allows us to know that synapsids were tetrapods?

    a. They lived on land.
    b. They were the dominant land animals.
    c. They evolved into mammals.
    d. They had four legs that they used for walking.

INTERPRET THE DATA   Answer can be found at the end of the book.

11. Which outcome would you predict if you compared the isotopes of fossils of two species of human ancestors and found high ratios of carbon-13 : carbon-12 in one and low ratios in the other based on the illustration at the right?

    a. The species with low ratios likely ate grains from grasses or meat from animals that grazed on grasses.

    b. The species with low ratios likely ate a mixed diet containing berries, leaves, and tubers.

    c. The species with high ratios likely ate a mixed diet.

    d. It would depend on the type of human fossils.

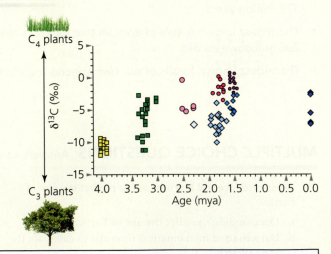

◆ Living humans
◆ *Homo* (1.4 to 1.67 mya)
◇ *Homo* (1.67 to 1.99 mya)
◇ *Homo* (2.3 mya)
● *Paranthropus boisei* (1.4 to 1.67 mya)
● *Paranthropus boisei* (1.72 to 1.93 mya)
○ *Paranthropus aethiopicus*
■ *Kenyanthropus platyops*
□ *Australopithecus anamensis*

SHORT ANSWER QUESTIONS Answers can be found at the end of the book.

12. What evidence refutes Kelvin's claim that Earth is only 20 million years old?

13. Why are fossils rare?

14. How can scientists understand the behavior of extinct animals?

15. How do biomarkers add to our understanding of the history of life?

16. What are stromatolites, and why are they evidence of very early life?

17. Why do scientists believe that plants and fungi may have been integral to each other's colonization of dry land?

ADDITIONAL READING

Conway Morris, S. 1998. *The Crucible of Creation: The Burgess Shale and the Rise of Animal Life.* Oxford: Oxford University Press.

Erwin, D. H., and J. W. Valentine. 2013. *The Cambrian Explosion: The Construction of Animal Biodiversity.* Greenwood Village, CO: Roberts and Company.

Gould, S. J. 1990. *Wonderful Life: The Burgess Shale and the Nature of History.* New York: W. W. Norton.

Knoll, A. H. 2003. *Life on a Young Planet: The First Three Billion Years of Evolution on Earth.* Princeton, NJ: Princeton University Press.

Prothero, D. R. 2007. *Evolution: What the Fossils Say and Why It Matters.* New York: Columbia University Press.

Schopf, W. 1999. *Cradle of Life: The Discovery of Earth's Earliest Fossils.* Princeton, NJ: Princeton University Press.

PRIMARY LITERATURE CITED IN CHAPTER 3

Albani, A. E., S. Bengtson, D. E. Canfield, A. Bekker, R. Macchiarelli, et al. 2010. Large Colonial Organisms with Coordinated Growth in Oxygenated Environments 2.1 Gyr Ago. *Nature* 466 (7302): 100–104.

Alexander, C. M. O'D. 2016. The Origin of Inner Solar System Water. *Philosophical Transactions of the Royal Society A* 375:20150384.

Allwood, A. C., M. T. Rosing, D. T. Flannery, J. A. Hurowitz, and C. M. Heirwegh. 2018. Reassessing Evidence of Life in 3,700-Million-Year-Old Rocks of Greenland. *Nature* (October 17, 2018). doi:10.1038/s41586-018-0610-4.

Bell, E. A., P. Boehnke, T. M. Harriosn, and W. L. Mao. 2015. Potentially Biogenic Carbon Preserved in a 4.1 Billion-Year-Old Zircon. *Proceedings of the National Academy of Sciences USA* 112:14518–21.

Bengston, S., T. Sallstedt, V. Belivanova, and M. Whitehouse. 2017. Three-Dimensional Preservation of Cellular and Subcellular Structures Suggests 1.6 Billion-Year-Old Crown-Group Red Algae. *PLoS Biology* 15 (3): e2000735.

Berbee, M. L., and J. W. Taylor. 2007. Rhynie Chert: A Window into a Lost World of Complex Plant-Fungus Interactions. *New Phytologist* 174 (3): 475–79.

Bottke, W. F., and M. D. Norman. 2017. The Late Heavy Bombardment. *Annual Review of Earth and Planetary Sciences* 45:619–47.

Bottke, W. F., D. Vokrouhlický, S. Marchi, T. Swindle, E. R. D. Scott, et al. 2015. Dating the Moon-Forming Impact Event with Asteroidal Meteorites. *Science* 348 (6232): 321–23.

Boucot, A. J. 1990. *Evolutionary Paleobiology of Behavior and Coevolution.* New York: Elsevier Science.

Brasier, M., N. McLoughlin, O. Green, and D. Wacey. 2006. A Fresh Look at the Fossil Evidence for Early Archaean Cellular Life. *Philosophical Transactions of the Royal Society of London, Series B: Biological Sciences* 361 (1470): 887–902.

Briggs, D. E. G., D. H. Erwin, and F. J. Collier. 1995. *The Fossils of the Burgess Shale.* Washington, DC: Smithsonian.

Brocks, J. J., and J. Banfield. 2009. Unravelling Ancient Microbial History with Community Proteogenomics and Lipid Geochemistry. *Nature Reviews Microbiology* 7:601–9.

Brown, C. M., D. M. Henderson, J. Vinther, I. Fletcher, A. Sistiaga, et al. 2017. An Exceptionally Preserved Three-Dimensional Armored Dinosaur Reveals Insights into Coloration and Cretaceous Predator-Prey Dynamics. *Current Biology* 27:2514–21.

Brunet, M., F. Guy, D. Pilbeam, H. T. Mackaye, A. Likius, et al. 2002. A New Hominid from the Upper Miocene of Chad, Central Africa. *Nature* 418:145–51.

Budd, G. E. 2008. The Earliest Fossil Record of the Animals and Its Significance. *Philosophical Transactions of the Royal Society B: Biological Sciences* 363 (1496): 1425–34.

Cerling, T. E., F. K. Manthi, E. N. Mbua, L. N. Leakey, M. G. Leakey, et al. 2013. Stable Isotope-Based Diet Reconstructions of Turkana Basin Hominins. *Proceedings of the National Academy of Sciences USA* 110:10501–6.

Chiappe, L. M. 2007. *Glorified Dinosaurs: The Origin and Early Evolution of Birds.* Hoboken, NJ: John Wiley & Sons.

Dalrymple, G. B. 1991. *The Age of the Earth.* Stanford, CA: Stanford University Press.

Delwiche, C. F., and E. D. Cooper. 2015. The Evolutionary Origin of a Terrestrial Flora. *Current Biology* 25:R899–R910.

Dodd, M. S., D. Papineau, T. Grenne, J. F. Slack, M. Rittner, et al. 2017. Evidence for Early Life in Earth's Oldest Hydrothermal Vent Precipitates. *Nature* 543 (7654): 60–64.

Droser, M. L., L. G. Tarhan, and J. G. Gehling. 2017. The Rise of Animals in a Changing Environment: Global Ecological Innovation in the Late Ediacaran. *Annual Review of Earth and Planetary Sciences* 45:593–617.

El Albani, A., S. Bengston, D. E. Canfield, A. Riboulleau, C. Rollion, et al. 2014. The 2.1 Ga Old Francevillian Biota: Biogenicity, Taphonomy and Biodiversity. *PLoS ONE* 9 (6): e99438.

England, P. C., P. Molnar, and F. M. Richter. 2007. Kelvin, Perry, and the Age of the Earth. *American Scientist* 95:342–49.

Erwin, D. H., and J. W. Valentine. 2013. *The Cambrian Explosion: The Construction of Animal Biodiversity*. Greenwood Village, CO: Roberts and Company.

Evans, D. C., R. Ridgely, and L. M. Witmer. 2009. Endocranial Anatomy of Lambeosaurine Hadrosaurids (Dinosauria: Ornithischia): A Sensorineural Perspective on Cranial Crest Function. *Anatomical Record* 292 (9): 1315–37.

Gaines, S. M. 2008. *Echoes of Life: What Fossil Molecules Reveal About Earth History*. New York: Oxford University Press.

Gold, D. A., J. Grabenstatter, A. de Mendoza, A. Riesgo, I. Ruiz-Trillo, et al. 2016. Sterol and Genomic Analyses Validate the Sponge Biomarker Hypothesis. *Proceedings of the National Academy of Sciences USA* 113:2684–89.

Higham, T., K. Douka, R. Wood, C. B. Ramsey, F. Brock, et al. 2014. The Timing and Spatiotemporal Patterning of Neanderthal Disappearance. *Nature* 512:306–9.

Hopson, J. A. 1975. The Evolution of Cranial Display Structures in Hadrosaurian Dinosaurs. *Paleobiology* 1 (1): 21–43.

Hublin, J-J., A. Ben-Neer, S. E. Bailey, S. E. Freidline, S. Neubauer, et al. 2017. New Fossils from Jebel Irhoud, Morocco and the Pan-African Origin of *Homo sapiens*. *Nature* 546:289–92.

Hunt, A. S. 1967. Growth, Variation, and Instar Development of an Agnostid Trilobite. *Journal of Paleontology* 41 (1): 203–8.

Hutchinson, J. R., and M. Garcia. 2002. Tyrannosaurus Was Not a Fast Runner. *Nature* 415:1018–21.

Joyce, W. G., N. Micklich, S. F. K. Schall, and T. M. Scheyer. 2012. Caught in the Act: The First Record of Copulating Fossil Vertebrates. *Biology Letters* 8:846–48.

Kelly, A. E., G. D. Love, J. E. Zumberge, and R. E. Summons. 2011. Hydrocarbon Biomarkers of Neoproterozoic to Lower Cambrian Oils from Eastern Siberia. *Organic Geochemistry* 42:640–54.

Knoll, A. H. 2015. Paleobiological Perspectives on Early Microbial Evolution. *Cold Spring Harbor Perspectives in Biology* 7:a018093.

Knoll, A. H., E. J. Javaux, D. Hewitt, and P. Cohen. 2006. Eukaryotic Organisms in Proterozoic Oceans. *Philosophical Transactions of the Royal Society B: Biological Sciences* 361:1023–38.

Knoll, A., and M. Nowak. 2017. The Timetable of Evolution. *Science Advances* 3:e1603076.

Labandeira, C. C. 2005. Invasion of the Continents: Cyanobacterial Crusts to Tree-Inhabiting Arthropods. *Trends in Ecology & Evolution* 20 (5): 253–62.

Lanphere, M. 2001. Radiometric Dating. In R. A. Meyers (ed.), *Encyclopedia of Physical Science and Technology* (pp. 721–30). New York: Academic Press.

Love, G. D., E. Grosjean, C. Stalvies, D. A. Fike, J. P. Grotzinger, et al. 2009. Fossil Steroids Record the Appearance of Demospongiae during the Cryogenian Period. *Nature* 457 (7230): 718–21.

Luo, Z-X., C-X. Yuan, Q-J. Meng, and Q. Ji. 2011. A Jurassic Eutherian Mammal and Divergence of Marsupials and Placentals. *Nature* 476:442–45.

Maloof, A. C., C. V. Rose, R. Beach, B. M. Samuels, C. C. Calmet, et al. 2010. Possible Animal-Body Fossils in Pre-Marinoan Limestones from South Australia. *Nature Geoscience* 3:653–59.

Mattinson, J. M. 2013. Revolution and Evolution: 100 Years of U–Pb Geochronology. *Elements* 9 (1): 53–57.

McDougall, I., F. H. Brown, and J. G. Fleagle. 2005. Stratigraphic Placement and Age of Modern Humans from Kibish, Ethiopia. *Nature* 433:733–36.

Miller, K. R. 1999. *Finding Darwin's God: A Scientist's Search for Common Ground Between God and Evolution*. New York: Cliff Street Books.

Milo, R., and R. Phillips. 2015. *Cell Biology by the Numbers*. New York: Garland Science, Taylor & Francis Group, LLC.

Minster, J. F., J. L. Birck, and C. J. Allegre. 1982. Absolute Age of Formation of Chondrites Studied by the ^{87}Rb–^{87}Sr Method. *Nature* 300:414–19.

Ni, X., D. L. Gebo, M. Dagosto, J. Meng, P. Tafforeau, et al. 2013. The Oldest Known Primate Skeleton and Early Haplorhine Evolution. *Nature* 498:60–64.

Niedźwiedzki, G., P. Szrek, K. Narkiewicz, M. Narkiewicz, and P. E. Ahlberg. 2010. Tetrapod Trackways from the Early Middle Devonian Period of Poland. *Nature* 463 (7277): 43–48.

Nutman, A. P., V. C. Bennett, C. R. L. Friend, M. J. Van Kranendonk, and A. R. Chivas. 2016. Rapid Emergence of Life Shown by Discovery of 3,700-Million-Year-Old Microbial Structures. *Nature* 537:535–38.

Pace, N. R. 2009. Mapping the Tree of Life: Progress and Prospects. *Microbiology and Molecular Biology Review* 73:565–76.

Pecoits, E., K. O. Konhauser, N. R. Aubet, L. M. Heaman, G. Veroslavsky, et al. 2012. Bilaterian Burrows and Grazing Behavior at 585 Million Years Ago. *Science* 336:1693–96.

Piperno, D. R., and H. D. Sues. 2005. Dinosaurs Dined on Grass. *Science* 310 (5751): 1126–28.

Powers, C. M., and D. J. Bottjer. 2007. Bryozoan Paleoecology Indicates Mid-Phanerozoic Extinctions Were the Product of Long-Term Environmental Stress. *Geology* 35:995–98.

Prasad, V. C., A. E. Strömberg, A. D. Leaché, B. Samant, R. Patnaik, et al. 2011. Late Cretaceous Origin of the Rice Tribe Provides Evidence for Early Diversification in Poaceae. *Nature Communications* 2:480.

Prothero, D. R. 2007. *Evolution: What the Fossils Say and Why It Matters*. New York: Columbia University Press.

Santosh, M., T. Aria, S. Maruyama. 2017. Hadean Earth and Primordial Continents: The Cradle of Prebiotic Life. *Geoscience Frontiers* 8:309–27.

Schopf, J. W., A. B. Kudryavtsev, J. T. Osterhout, K. H. Williford, K. Kitajima, et al. 2017. An Anaerobic ~3400 Ma Shallow-Water Microbial Consortium: Presumptive Evidence of Earth's Paleoarchean Anoxic Atmosphere. *Precambrian Research* 299:309–18.

Shu, D-G., H-L. Luo, S. C. Morris, X-L. Zhang, S-X. Hu, et al. 1999. Lower Cambrian Vertebrates from South China. *Nature* 402:42–46.

Sleep, N. H. 2010. The Hadean–Archaean Environment. *Cold Spring Harbor Perspectives in Biology* 2 (6): a00252.

Smith, M. R. 2016. Cord-Forming Palaeozoic Fungi in Terrestrial Assemblages. *Botanical Journal of the Linnean Society* 180 (4): 452–60.

Soltis, D. E., C. D. Bell, S. Kim, and P. S. Soltis. 2008. Origin and Early Evolution of Angiosperms. *Annals of the New York Academy of Sciences* 1133:3–25.

Springer, M. S., R. W. Meredith, J. Gatesy, C. A. Emerling, J. Park, et al. 2012. Macroevolutionary Dynamics and Historical Biogeography of Primate Diversification Inferred from a Species Supermatrix. *PLoS ONE* 7 (11): e49521.

Suarez, S. E., M. E. Brookfield, E. J. Catlos, and D. F. Stöckli. 2017. A U–Pb Zircon Age Constraint on the Oldest-Recorded Air-Breathing Land Animal. *PLoS ONE* 12 (6): e0179262.

Ueno, Y., K. Yamada, N. Yoshida, S. Maruyama, and Y. Isozaki. 2006. Evidence from Fluid Inclusions for Microbial Methanogenesis in the Early Archaean Era. *Nature* 440:516–19.

Valley, J. W., A. J. Cavosie, T. Ushikubo, D. A. Reinhard, D. F. Lawrence, et al. 2014. Hadean Age for a Post-Magma-Ocean Zircon Confirmed by Atom-Probe Tomography. *Nature Geoscience* 7:219–23.

Van Zuilen, M.A. 2018. Proposed Early Signs of Life Not Set in Stone. *Nature* (October 17, 2018). doi:10.1038/d41586-018-06994-x.

Weishampel, D. B. 1981. Acoustic Analyses of Potential Vocalization in Lambeosaurine Dinosaurs (Reptilia: Ornithischia). *Paleobiology* 7 (2): 252–61.

Wellman, C. H., P. L. Osterloff, and U. Mohiuddin. 2003. Fragments of the Earliest Land Plants. *Nature* 425:282–85.

Wilson, H. M., and L. I. Anderson. 2004. Morphology and Taxonomy of Paleozoic Millipedes (Diplopoda: Chilognatha: Archipolypoda) from Scotland. *Journal of Paleontology* 78 (1): 169–84.

Wood, B. 2011. The Formation and Differentiation of Earth. *Physics Today* 64:40–45.

Yin, Z., M. Zhu, E. H. Davidson, D. J. Bottjer, F. Zhao, and P. Tafforeau. 2015. Sponge Grade Body Fossil with Cellular Resolution Dating 60 Myr Before the Cambrian. *Proceedings of the National Academy of Sciences USA* 112:E1453–60.

The Tree of Life

How Biologists Use Phylogeny to Reconstruct the Deep Past

Learning Objectives

- Construct a simple phylogeny and identify the different components.
- Explain how different lines of evidence can lead to different conclusions about species' taxonomical relationships.
- Discuss how scientists can determine the timing of branching events.
- Explain how phylogenies can be used to develop hypotheses about the evolution of tetrapods.
- Explain how the bones of the middle ear can be used to trace the evolution of mammals.
- Discuss exaptations and explain their relevance in the evolution of birds.
- Name three examples of convergent evolution.

In November 2014, Steven Brusatte stepped into Beijing's bustling central railway station. Even though he spoke virtually no Mandarin, he managed to find a train headed for the city of Jinzhou. For the next four hours, Brusatte and a colleague, Junchang Lü, from the Chinese Academy of Geological Sciences, traveled north, almost as far as the border with North Korea. They had a date to see a dinosaur.

Brusatte, who teaches at the University of Edinburgh, has dug up dinosaur fossils in many parts of the world (**Figure 4.1**). He has studied those fossils to address many questions about dinosaurs—how they first evolved some 240 million years ago, how they rose to dominate terrestrial ecosystems, and how they were decimated in a mass extinction 66 million years ago. But there are still thousands of species of dinosaurs with us today (look into the branches of any tree and you may see some), and Brusatte and Lü hoped that the fossil they were traveling to see would shed light on how dinosaurs managed to endure.

Zhenyuanlong was a ground-running dinosaur with feathers that lived about 125 million years ago. The phylogeny of such dinosaurs reveals that they were the ancient relatives of birds and helps us understand how the unique body plan of birds gradually evolved.

Steve Brusatte

Figure 4.1 Stephen Brusatte hunts for dinosaur fossils. He and his colleagues analyzed the *Zhenyuanlong* fossil for clues to the evolution of dinosaurs and birds. (Steve Brusatte)

Figure 4.2 A: Brusatte and Junchang Lü with the fossil of *Zhenyuanlong*. B: The fossil shows exceptional preservation of anatomical details, including long feathers on the arms and tail. (Steve Brusatte)

The fossil, embedded in a massive slab of rock, had been unearthed by a Chinese farmer and ended up in a small museum on the outskirts of Jinzhou. City officials accompanied Brusatte and Lü from the train station to the museum, where they led the two scientists to a side room. There the slab rested on a small table. It was, Brusatte would later recall, one of the most beautiful fossils he had ever seen (**Figure 4.2A**).

The entire dinosaur measured two meters long, nearly the size of a horse. Its skeleton had been pressed flat on the slab, probably by the weight of ash from a volcanic eruption 125 million years ago. As a result, just about every bone in the dinosaur's body was preserved. Its skeleton clearly showed that this was a type of two-legged dinosaur called a theropod. It had the same giant teeth and deadly claws seen on its close relative the *Velociraptor*, made famous by the *Jurassic Park* movies. But the fossil of the Jinzhou dinosaur—which Brusatte and Lü would later name *Zhenyuanlong suni*—showed that real theropods were a lot more bizarre-looking than anything in the movies. It was a dinosaur with wings (Lü and Brusatte 2015).

On a first inspection, Brusatte and Lü could see that its clawed arms had layers of quilled feathers. After months of study on the rest of the fossil, Brusatte and Lü documented long feathers sprouting from its tail, like a pheasant, as well as downy fuzz on other parts of its body (**Figure 4.2B**).

If it sounds like *Zhenyuanlong* was a cross between a dinosaur and a bird, that's because it was, after a fashion. Living birds, capable of flight, evolved from feathered dinosaurs that could only run on the ground. This profound insight about birds came through the study of phylogeny. Brusatte, Lü, and other researchers have drawn the evolutionary tree of birds and found that dinosaurs like *Zhenyuanlong* are among their closest relatives (Brusatte 2018).

In this chapter, we will learn how scientists like Brusatte go about reconstructing the tree of life and use it to learn about the course of evolution.

We'll start by considering Darwin's powerful metaphor of evolution as a tree of life, whose branches are the diverging species. We will then learn how to use treelike representations to examine the relationships of living things and to develop a system of taxonomy based on evolutionary history. Next, we'll consider how we can recover the pattern of these branches by comparing traits in species. And then we'll see how these insights enable us to test hypotheses about important evolutionary questions. In this chapter, we'll use tree-based approaches to consider three questions: How did our ancestors move from ocean to land? What is the origin of the mammalian ear? When did feathers evolve?

The concepts you learn in this chapter are essential for understanding many aspects of evolution that we address in later chapters. Throughout the rest of the book, we will examine a number of additional evolutionary trees that illuminate themes, including the spread of animals and plants across the planet, the origin of diseases, the origin of new adaptations, and the origin of our own species. ●

4.1 Tree Thinking

In the 1830s, Charles Darwin began to picture evolution as a tree growing new branches (see Figure 2.16). In *The Origin of Species,* he put this image into words:

> As buds give rise by growth to fresh buds, and these, if vigorous, branch out and overtop on all sides many a feebler branch, so by generation I believe it has been with the great Tree of Life, which fills with its dead and broken branches the crust of the earth, and covers the surface with its ever branching and beautiful ramifications. (Darwin 1859)

Evolutionary biologists have found that Darwin's metaphor is a powerful and useful way to think about the history of life. Organisms replicate their genetic material when they reproduce. We can think of the succession of generations in a population as the growing tip of a branch. A population may become subdivided into two populations that no longer exchange genes, making it possible for them to diverge into two separate species (Chapter 13). Evolutionary biologists represent this splitting as a single **branch** dividing into two or more new branches. Those new branches can, in time, split into new ones as well (**Figure 4.3**). Depicting the evolution of a lineage into branches is known as a **phylogeny**.

A **branch** is a lineage evolving through time that connects successive speciation or other branching events.

Phylogeny is a visual representation of the evolutionary history of populations, genes, or species.

Constructing Phylogenetic Trees

Evolutionary biologists represent this divergence with phylogenetic trees. Typically, they will reconstruct the relationship of a group of species as a set of branches. A phylogenetic tree can convey a huge amount of information, but only if it is read correctly.

To learn how to make sense of a phylogenetic tree, it helps to compare it to family trees drawn by genealogists (Gregory 2008). Let's say you have a sister and two cousins. You and your sister are more closely related to one another than either of you are to your cousins, so we represent that kinship with a pair of branches joined to a **node** that represents a parent (**Figure 4.4A**). Your cousins are more closely related to each other than either is to you or your sister, so we represent their relationship as another pair of branches joined together at a node that represents your aunt. And your parent and your aunt can, in turn, be joined to an older common ancestor: your grandparent. (For the sake of this analogy, we'll set aside the fact that we have two parents.)

A **node** is a point in a phylogeny where a lineage splits (a speciation event or other branching event, such as the formation of subspecies).

Figure 4.3 A, B: Individual organisms mate to produce offspring, which reproduce in turn. C: Although individuals are born and die, the genetic information they pass from parent to offspring endures over time. D: Populations can be considered aggregations of the genetic lineages of individuals they contain. E: Species, made up of linked populations, represent an enduring mesh of interwoven strands of genetic information, transmitted from generation to generation through time. F: Species sometimes become divided into separate groups of populations that stop interbreeding. This splitting is akin to a branch of an actual tree splitting in two. The two new tips continue to grow as the isolated groups diverge from one another. G: The simple branching structure of phylogenetic trees distills a vast amount of biology into a simple image. (Data from Baum and Smith 2012)

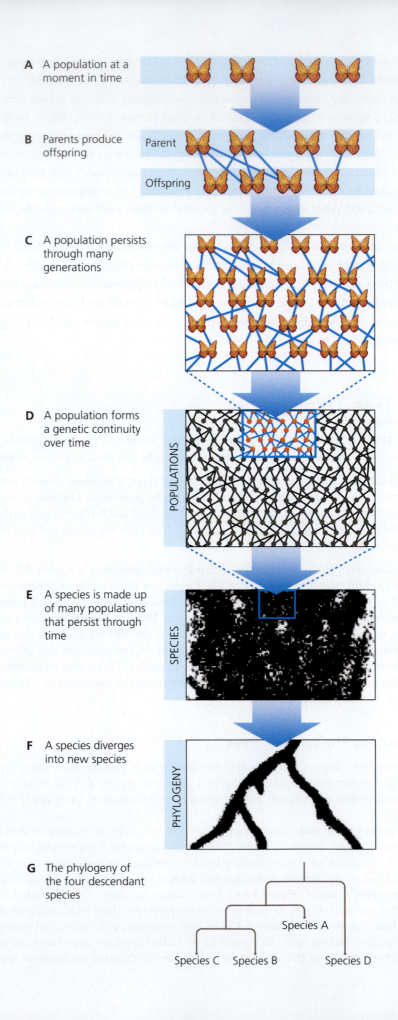

A A population at a moment in time

B Parents produce offspring

Parent

Offspring

C A population persists through many generations

POPULATIONS

D A population forms a genetic continuity over time

SPECIES

E A species is made up of many populations that persist through time

PHYLOGENY

F A species diverges into new species

G The phylogeny of the four descendant species

Species A

Species C   Species B   Species D

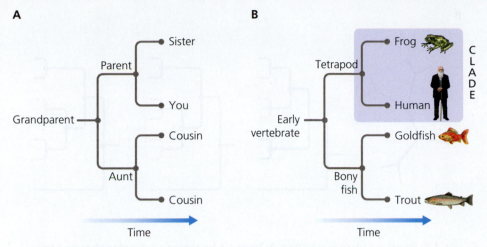

A

Grandparent — Parent — Sister

Parent — You

Aunt — Cousin

Aunt — Cousin

Time

B

Early vertebrate — Tetrapod — Frog

Tetrapod — Human

Bony fish — Goldfish

Bony fish — Trout

CLADE

Time

Figure 4.4 A: A simple genealogical tree shows the relationship between you, a sister, and two cousins. You and your sister share a common ancestor that your cousins do not (your parent)—just as your cousins share a common ancestor that you and your sister do not (your aunt). But you and your cousins also share a more distant ancestor (your grandparent). B: Here we replace the individuals with species. Humans and frogs share a closer common ancestor (a tetrapod) that goldfish and trout do not share. But all four species share an older common ancestor—an early vertebrate. (Data from Gregory 2008)

In **Figure 4.4B**, we have switched the four people at the **tips** of the branches with four species. Humans and frogs share a common ancestry represented by branches and a node that represents tetrapods. Goldfish and trout likewise descend from the common ancestor bony fish. And tetrapods and bony fish, in turn, descend from the common ancestral species early vertebrates. The nodes representing tetrapods, bony fish, and early vertebrates are known as **internal nodes** because they are located within the phylogeny representing ancestral populations or species that have long since disappeared. We can organize the species in this tree according to their relationships. We refer to a node and all its descendants as a **clade.** As we see in the figure, smaller clades are nested within larger ones. Humans and frogs belong to the tetrapod clade, which is nested within a clade that includes early vertebrates and all of their descendants.

A **tip** is the terminal end of an evolutionary tree, representing species, molecules, or populations being compared.

An **internal node** is a node that occurs within a phylogeny and represents ancestral populations or species.

A **clade** is a single "branch" in the tree of life; each clade represents an organism and all of its descendants.

Reading Phylogenetic Trees

When a phylogenetic tree shows only the relationship among species, we refer to it as a cladogram (**Figure 4.5**). The branches do not precisely measure the period of time it took between speciation events. But they do offer some information about the timing of events. In Figure 4.5, for example, divergence from the blue clade occurred before the purple and orange clades diverged. Later in this chapter, we'll show how we can display more information about the passage of time when we draw detailed phylogenetic trees.

There are many ways to visualize the same set of phylogenetic relationships (**Figure 4.6**). This flexibility can be a great strength because evolutionary biologists can select the representation that's most relevant to the question they're investigating. But it can also give rise to misconceptions. Depending on the paper you read, you may encounter phylogenetic trees with branches that form right angles, curves, or diagonal lines. There's no difference in the evolutionary relationship these trees are intended to represent. (In this book, we use right-angled branches, and time is represented as moving from left to right.)

We can also rotate branches around their nodes, much like swinging the arms of a mobile that hangs from a ceiling. **Figure 4.7A** shows a cladogram of six vertebrate species. If we swing around some of the branches, we end up with **Figure 4.7B**. Both cladograms represent exactly the same phylogenetic relationships.

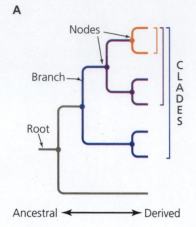

A

Nodes

Branch

Root

CLADES

Ancestral ← → Derived

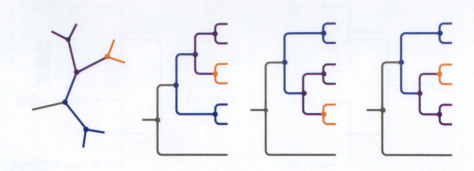

B

Figure 4.5 Reading a phylogenetic tree. A: Trees can be rooted or unrooted. A rooted tree includes the focal group of species being studied as well as a more distantly related outgroup species. Rooted trees contain information about the ancestral states of characters, and they can be used to trace the evolutionary history of groups of organisms through time. Speciation events in ancestral populations are depicted by nodes, and evolution of populations between speciation events is shown by branches. Trees consist of a nested series of clades, each comprising a node and all of its descendant lineages. Rooted trees in this book are depicted so that earlier (ancestral) character states are shown to the left, and later (derived) character states are shown to the right. Note that ancestral and derived characters are relative to specific locations on the phylogeny. In this case, dark blue is a derived character state for the dark blue clade but an ancestral character state for the purple clade. B: Alternative, equally valid portrayals of the same phylogenetic tree shown in A. The first of these trees is shown unrooted. The others are rooted but shown with taxa rotated about their respective nodes.

This equivalence is very important to appreciate when we examine evolutionary trees. It's easy to look at a tree like the one in Figure 4.7A and see it not as a branching process but as a continuum. We might wrongly conclude that humans evolved from a distant goldfish ancestor, through a frog intermediate, and so on until we reached our final form. In fact, the tree shows us that the common ancestor we share with goldfish is older than our common ancestor with cats. After our lineage and the goldfish lineage split from each other, the goldfish lineage underwent its own evolutionary changes that produced the goldfish we know today. As we'll soon see, we can compare species at different positions in a phylogeny to infer what their common ancestors were like.

Evolutionary biologists can also choose how much phylogenetic information they want to represent in a tree. **Figure 4.8A** shows a large tree with seven tips. If we remove species A, C, and D, we can represent the remaining species in the tree shown at the right in Figure 4.8A simply by straightening the remaining branches. These two trees are compatible, because they represent parts of the same underlying phylogeny. This flexibility allows scientists to compare different clades without having to include all the species they contain.

A useful way to condense a phylogenetic tree is to merge an entire clade into a single tip. In **Figure 4.8B**, the species marked D through G all descend from a common ancestor and therefore belong to the same clade, which we've marked H. We can collapse those branches into a single branch. Imagine, for example, that

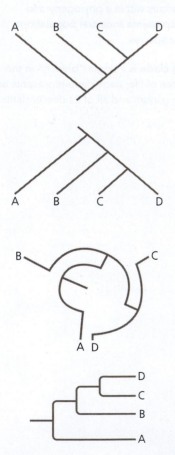

Figure 4.6 Depending on the journal or book you read, you may encounter phylogenies represented in different styles. These four representations are all equivalent. In this book we use rectangular branches, and we show time flowing from left to right (see the tree on the bottom). (Data from Baum and Smith 2012)

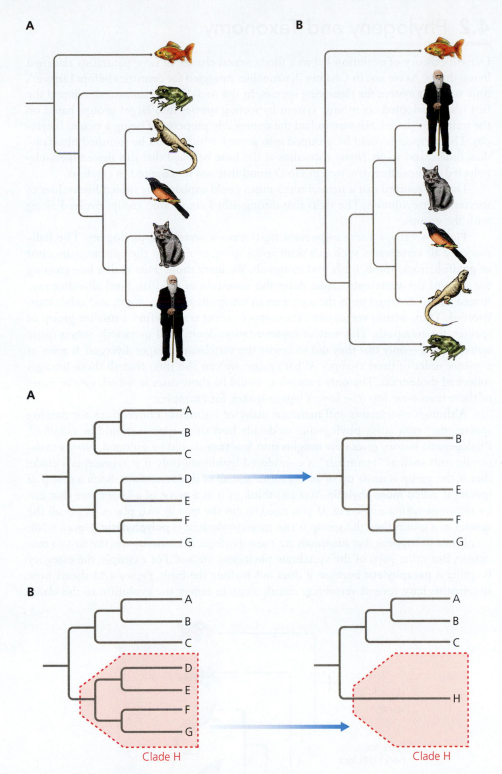

A

B

A

B

Clade H

Clade H

Figure 4.8 Within a clade, we can choose how many taxa we want to include in a phylogenetic tree. A: If we leave out taxa A, C, and D from the full tree on the left, we get the smaller tree on the right. These trees are in agreement about the underlying phylogeny. B: To simplify a tree and make its overall pattern clearer, we can collapse an entire set of taxa into one tip if they all have a recent common ancestor. (Data from Baum and Smith 2012)

branches A through C are reptiles and that branches D through G are different species of mammals. We could simply collapse all of the mammal species into one lineage marked "mammal." If there are unequal amounts of information about different clades, such simplifications can make deeper relationships easier to see.

• Phylogenetic trees represent the branching pattern of evolution over time. •

Key Concept

4.2 Phylogeny and Taxonomy

Darwin's theory of evolution led to a fundamental change in how naturalists classified living things. As we saw in Chapter 2, naturalists struggled for centuries before Darwin's time to find a system for classifying species. In the mid-1700s, Linnaeus developed the first widely adopted taxonomic system by sorting species into larger groups based on the traits they shared. His system had the remarkable property of being a nested hierarchy. That is, species could be grouped into genera, which could be grouped into families, orders, and so on. Many naturalists at the time believed that this nested hierarchy reflected a preexisting structure in God's mind that was represented in creation.

Darwin realized that a natural mechanism could explain why nested hierarchies of species form: evolution. The traits that distinguished taxonomic groups evolved along with the groups.

Figure 4.9 maps a few important traits onto a vertebrate phylogeny. The hallmarks of all vertebrates, such as a skull and a spine, evolved in the common ancestor of goldfish, frogs, lizards, birds, and mammals. We mark those traits with a line crossing the root of the vertebrate clade. After the ancestors of goldfish (and all other ray-finned fishes) diverged from the ancestors of tetrapods, the legs, digits, and other traits evolved. Thus, within vertebrates, these more recent traits define a smaller group of species: the tetrapods. This sort of representation doesn't tell us exactly when those traits evolved—only that they did so before the vertebrate lineages diverged. It gives us a *relative* order of these changes. What's more, we can also infer that all those lineages inherited those traits. The only exception would be those cases in which one or more of these traits were lost (the loss of legs in snakes, for example).

Although systematists still maintain many of Linnaeus's conventions for naming species, they now study phylogenies to decide how those species should be classified. Phylogenetic history gives new insights into *how* **taxa** should be grouped. Now, a taxonomic unit such as "mammals" is considered legitimate only if it represents a clade: that is, the group is made up of an organism and all of its descendants. Such a group of species is called **monophyletic**. You can think of it as a piece of a larger tree that can be removed with a single cut. If you need to cut the tree in two places to get all the species in a group, then the group is not monophyletic; it is **polyphyletic** (**Figure 4.10**).

It just so happens that mammals are monophyletic, as Linnaeus was the first to recognize. But other parts of the vertebrate phylogeny are not. For example, the category Reptilia is **paraphyletic** because it does not include the birds. **Figure 4.11** shows how systematists have revised vertebrate classification to reflect the evolution of the clade.

A **taxon** (plural, **taxa**) is a group of organisms that a taxonomist judges to be a cohesive taxonomic unit, such as a species or order.

Monophyletic describes a group of organisms that form a clade.

Polyphyletic describes a taxonomic group that does not share an immediate common ancestor and therefore does not form a clade.

Paraphyletic describes a group of organisms that share a common ancestor, although the group does not include all the descendants of that common ancestor.

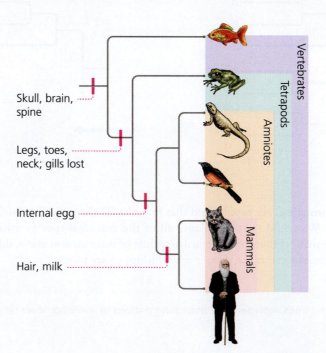

Figure 4.9 Systematists classify species using phylogenies. The nested hierarchy of Linnaean classification is the result of traits evolving along different lineages with new traits appearing in some lineages. All vertebrates have a skull, brain, and spine, for example. But only mammals have hair and milk.

Skull, brain, spine

Legs, toes, neck; gills lost

Internal egg

Hair, milk

Vertebrates
Tetrapods
Amniotes
Mammals

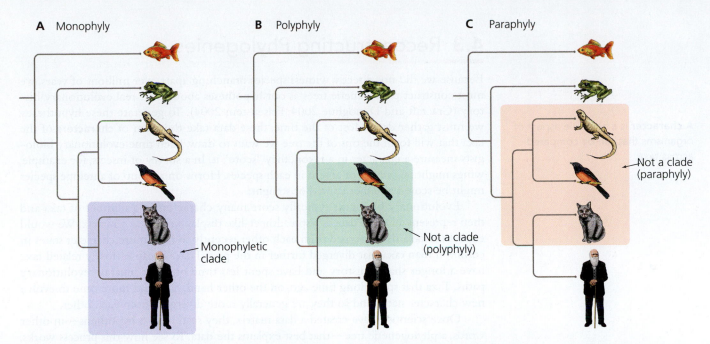

A Monophyly   **B** Polyphyly   **C** Paraphyly

Monophyletic clade

Not a clade (polyphyly)

Not a clade (paraphyly)

Figure 4.10 A clade is considered monophyletic if it can be "snipped" off from the larger tree with a single cut. If two cuts have to be made, then it is not monophyletic. In other words, a clade is monophyletic if it includes an organism and all of its descendants. (Data from Baum and Smith 2012)

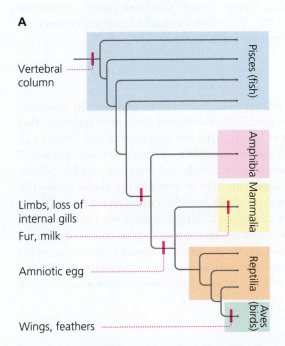

A

Vertebral column

Limbs, loss of internal gills

Fur, milk

Amniotic egg

Wings, feathers

Pisces (fish) · Amphibia · Mammalia · Reptilia · Aves (birds)

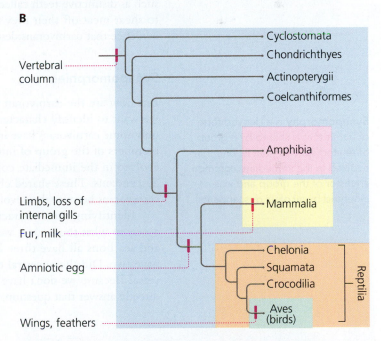

B

Vertebral column

Limbs, loss of internal gills

Fur, milk

Amniotic egg

Wings, feathers

Cyclostomata
Chondrichthyes
Actinopterygii
Coelacanthiformes
Amphibia
Mammalia
Chelonia
Squamata
Crocodilia
Aves (birds)
Reptilia

Figure 4.11 A: Modern taxonomists seek to classify species according to their evolutionary history, creating groups that represent monophyletic clades. In some cases, this means revising old classifications. Reptiles, for example, traditionally do not include birds, even though birds are within the same clade as all the species we refer to as reptiles. B: A modern classification of vertebrates can lead to some controversial renamings. For example, "fish" is no longer a valid name—unless it includes tetrapods, such as reptiles and birds. These invalid groups are called paraphyletic. Thus "fish" is now subdivided into four monophyletic clades. (Data from Baum and Smith 2012)

- • Different conclusions can result from different lines of evidence. Compared with the tools of scientists today, Linnaeus built his taxonomic system on a relatively basic understanding of structural similarities. Evolutionary theory has added new evidence for these relationships and new insight to the meaning of taxonomic groupings. •

Key Concept

4.3 Reconstructing Phylogenies

Because we did not directly witness species branching apart over millions of years, we must construct phylogenetic trees as our hypotheses about their real evolutionary history (Cracraft and Donoghue 2004; Felsenstein 2004). To generate these hypotheses, we must gather data. Most of the time, these data take the form of **characters** of the taxa that will form the tips of the tree we want to draw. Each time evolutionary biologists measure a character in a taxon, they "score" it. In a survey of insects, for example, wings might be present or absent in each species. Horns on a group of antelope species might be scored as either curved or straight.

Evolutionary biologists typically score many characters in a number of taxa and then represent all those data as a spreadsheet-like display known as a matrix. We would expect the taxa that diverged from each other recently to have more character states in common than taxa that diverged further in the past. That's because closely related taxa have a longer shared history and have spent less time on their separate evolutionary paths. Taxa that split a long time ago, on the other hand, have had more time to evolve new character states, and so they are generally more divergent from each other.

Once scientists have created a data matrix, they search for a hypothesis—in other words, a phylogenetic tree—that best explains the data. To see how this process works, let's build a matrix from some real—and familiar—species: dogs, cats, and their close mammal relatives, such as bears, raccoons, walrus, and seals.

Scientists have long recognized that these species belong to a mammalian order called Carnivora. Carnivorans share common traits that aren't found in other mammals, such as distinctive teeth called carnassials. These enlarged side teeth enable carnivorans to shear meat off their prey (**Figure 4.12**). Traits like carnassial teeth are compelling evidence that carnivorans descend from a common ancestor.

Synapomorphies

But how are the carnivoran species related to each other? To explore this question, we want to identify character states—specifically the shared derived characters—that only some carnivorans have in common. Some characters are ancestral and found in all members of the group of interest, but **synapomorphies**, or shared derived characters, evolved in the immediate common ancestor of a clade and were inherited by all the descendants. These shared character states can help us identify the clades *within* the carnivorans because they evolved after these clades branched off from other ones.

Identifying these character states can be a challenge because we have no direct access to what those states were in the common ancestor of carnivorans. Seals, walrus, and sea lions all have three lower incisors, for example, whereas other carnivorans have two. Did the ancestral carnivoran have two, and a third later evolved? Or vice versa? Because we don't have a fossil of the common ancestor of carnivorans, we can't directly answer that question.

A **character** is a heritable aspect of organisms that can be compared across taxa.

Synapomorphy is a derived form of a trait that is shared by a group of related species (that is, one that evolved in the immediate common ancestor of the group and was inherited by all of its descendants).

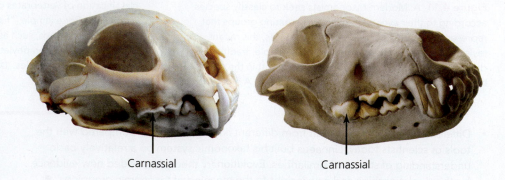

Figure 4.12 Two carnivorans, a bobcat (left) and a Mexican gray wolf. The arrows point to carnassials, teeth for shearing meat that are found only in carnivorans. Synapomorphies such as this show that carnivorans form a monophyletic clade. (Left: Argonaut/ Shutterstock.com; right: Matt Cooper/ Shutterstock.com)

Carnassial

Carnassial

We can, however, infer those states by comparing carnivorans to another species outside the clade. A species chosen for such a comparison is known as an **outgroup**. For this study, we'll use primates—in this case, lemurs—as an outgroup. Carnivorans did not descend *from* primates. Instead, carnivorans and primates diverged from a shared ancestor—an ancestor that lived even earlier than the common ancestor of the carnivorans. After the primates branched off, carnivorans evolved unique derived traits (like carnassial teeth) that are not present in primates. We can therefore use primates to approximate what the ancestral traits looked like before the carnivorans became carnivorans.

A lemur doesn't perfectly represent the early ancestor of carnivorans because primates have been evolving, too. But in cases like these, outgroups give us a way to discern what traits looked like before they became specialized in a particular clade or lineage. They allow us to infer which character states are primitive, or ancestral, and which are derived.

Figure 4.13 shows the characters we will use to create a data matrix. We use lemurs to infer the character states of the ancestral carnivoran. We denote the primitive state for each character with a 0 and the derived state with a 1. Lemurs have a long tail, for example, so 0 denotes the presence of a long tail and 1 the presence of a short tail. **Figure 4.14** shows the character state matrix for ten carnivorans, along with lemurs as an outgroup.

We can then evaluate possible hypotheses for the historical relationships among the carnivorans. Each offers a different version of the past. On each phylogenetic tree,

An **outgroup** is a group of organisms (for example, a species) that is outside of the monophyletic group being considered. In phylogenetic studies, outgroups can be used to infer the ancestral states of characters.

Characters and character states for an analysis of carnivorans

Figure 4.13 In this analysis we choose 12 characters and assign character states for each one.

# Character	State
1 Complexity of the cooling surfaces in the nose	0 = minimally branched 1 = highly branched
2 Bony spur in ear region	0 = straight and projecting 1 = bony spur in ear region
3 Number of lower incisors	0 = 2 lower incisors 1 = 3 lower incisors
4 Upper molar 1	0 = present 1 = absent
5 Bone within the penis	0 = present 1 = absent
6 Tail	0 = elongated 1 = short
7 Hallux (fifth digit, or dew claw, on hind leg)	0 = prominent 1 = reduced or absent
8 Claws	0 = nonretractable 1 = retractable
9 Prostate gland size and shape	0 = small and simple 1 = large and bilobed
10 Kidney structure	0 = simple 1 = conglomerate
11 External ear (pinna)	0 = present 1 = absent
12 Testis position	0 = scrotal 1 = abdominal

Morphological data matrix for Carnivora

Taxa	Character-state scoring											
#	1	2	3	4	5	6	7	8	9	10	11	12
Lemur (outgroup)	0	0	0	0	0	0	0	0	0	0	0	0
Cat	0	1	0	1	0	0	1	1	1	0	0	0
Hyena	0	1	0	1	1	0	1	0	1	0	0	0
Civet	0	1	0	0	0	0	0	0	1	0	0	0
Dog	1	0	0	0	0	0	0	0	0	0	0	0
Raccoon	1	0	0	0	0	0	0	0	0	0	0	0
Bear	1	0	0	0	0	1	0	0	0	1	0	0
Otter	1	0	0	0	0	0	0	0	0	1	0	0
Seal	1	0	1	0	0	1	0	0	0	1	1	1
Walrus	1	0	1	0	0	1	0	0	0	1	1	1
Sea lion	1	0	1	0	0	1	0	0	0	1	0	0

Figure 4.14 We can create a data matrix of species scores for the list of characters in Figure 4.13. The matrix then enables us to generate phylogenetic trees in which particular patterns of character states emerge as species descend from a common ancestor.

we can map the order in which the character states evolved. **Figure 4.15** shows four such trees. The branches in each tree have marks indicating every time one of the characters would have had to change to give rise to the combinations of traits present in living taxa. For example, in Figure 4.15A, the ancestor of cats, hyenas, and civets evolved a cupped bony spur in its ear region.

Homoplasy

All four trees in Figure 4.15 can account for the distribution of character states present in modern taxa. But they require different numbers of evolutionary changes to get there. As you can see from the figures, the evolutionary sequence required to produce the data matrix is different in the four phylogenies. In Figure 4.15A, each of the characters evolved just a single time, and accounting for modern distributions of these character states requires 12 evolutionary "steps." In Figure 4.15B, on the other hand, character 3 evolved twice—the number of lower incisor teeth increased from two to three independently on each of two separate branches. This kind of similarity, known as **homoplasy**, often occurs in the history of life, but we do not expect it to be this common.

Homoplasy can result from the independent evolution of the same trait in two or more lineages—it is not due to shared descent. Birds, bats, and beetles all have wings that can be used for powered flight, for example. This pattern is known as **convergent evolution**. We'll examine the sources of convergent evolution in more detail in Chapter 9.

For the phylogeny shown in Figure 4.15B to account for the distribution of character states that we observe in modern taxa, characters 1, 3, 6, and 10 all would have had to evolve multiple times. In addition, characters 2, 4, 7, 9, and 10 would have had to change from their derived states back to their ancestral state on some of the branches, by a process known as **evolutionary reversal**. All told, the tree topology in Figure 4.15B requires 24 evolutionary steps to explain the data. The trees shown in Figure 4.15C and D require 20 and 21 evolutionary steps, respectively.

Homoplasy describes a character state similarity *not* due to shared descent (for example, produced by convergent evolution or evolutionary reversal).

Convergent evolution is the independent origin of similar traits in separate evolutionary lineages.

Evolutionary reversal describes the reversion of a derived character state to a form resembling its ancestral state.

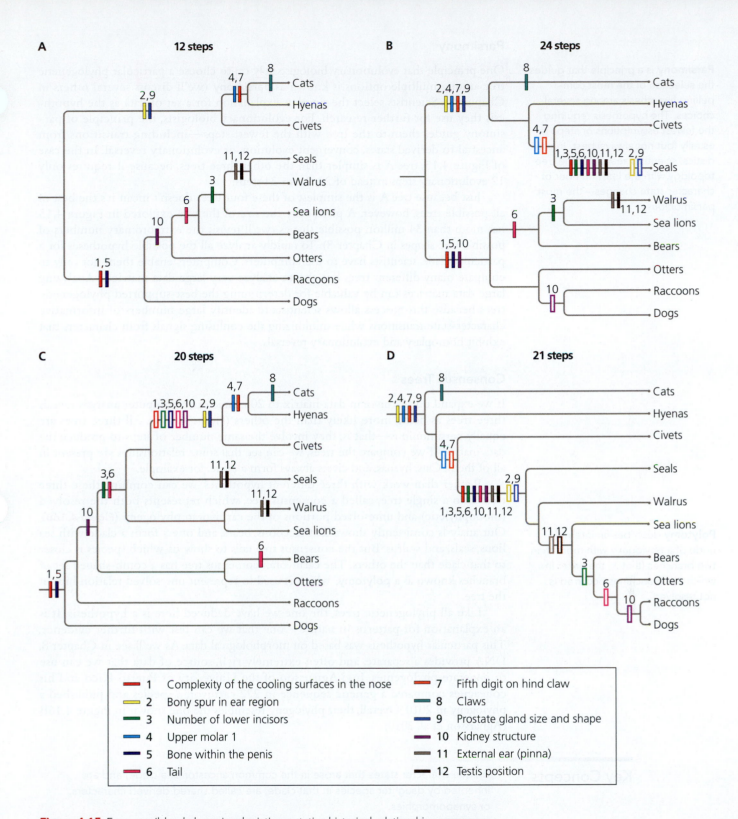

Figure 4.15 Four possible phylogenies depicting putative historical relationships among carnivoran species using the characters in Figure 4.14. Each tree represents a hypothesis for the branching patterns that gave rise to the modern species, and all of them can account for the distribution of character states present in modern taxa. But they require different numbers of evolutionary changes to get there. After "rooting" the tree using character states from lemurs, we can trace the evolution of each of the 12 characters, identifying branches where they changed from the ancestral to their derived state (solid bars) as well as instances where they reverted from their derived state to their ancestral state (open bars). Tree A requires fewer evolutionary steps to explain the data than do trees B–D, so it's considered the most parsimonious of the set.

Parsimony

Parsimony is a principle that guides the selection of the most compelling hypothesis among several choices. The hypothesis requiring the fewest assumptions or steps is usually (but not always) best. In cladistics, scientists search for the tree topology with the least number of character-state changes—the most parsimonious.

One principle that evolutionary biologists rely on to choose a particular phylogenetic tree among multiple options is known as **parsimony** (we'll discuss several others in Chapter 8). Scientists select the simplest explanation for a set of data as the hypothesis they use for further research. For evolutionary biologists, the principle of parsimony guides them to the tree with the fewest steps—including transitions from ancestral to derived states, convergent evolution, or evolutionary reversal. In the case of Figure 4.15, tree A is simpler than the other three trees, because it requires only 12 evolutionary steps instead of 20, 21, or 24 steps.

Just because tree A is the simplest of these four trees doesn't mean it's the best of all possible trees, however. A phylogeny the size of the one explored in Figure 4.15 has more than 34 million possible shapes (we'll revisit the extraordinary numbers of possible tree shapes in Chapter 8). To rapidly analyze all the possible hypotheses for a particular clade, scientists have to use computers. Computers enable them not only to compare many different trees but also to analyze very large data matrices. Analyzing large data matrices can be valuable for determining the best-supported phylogenetic trees because this process allows scientists to identify large numbers of informative character-state transitions while minimizing the confusing signals from characters that exhibit homoplasy and evolutionary reversals.

Consensus Trees

If we expand our carnivoran data matrix to 20 characters, a computer analysis reveals three trees as being more likely than the others (**Figure 4.16A**). All three trees are equally parsimonious—that is, they involve the same number of steps to produce the data matrix. If we compare the trees, we can see that some relationships are present in all of them. Cats, hyenas, and civets always form a clade, for example.

Polytomy describes an internal node of a phylogeny with more than two branches (that is, the order in which the branchings occurred is not resolved).

Rather than work with three different hypotheses, we can combine these three trees into a single tree, called a consensus tree, which represents both the resolved (monophyletic) and unresolved portions of the carnivoran phylogeny (**Figure 4.16B**). Our analysis consistently shows that raccoons, bears, and otters form a clade with sea lions, seals, and walrus. But the consensus tree fails to show us which species is closer to that clade than the others. The carnivoran consensus tree has a comb-shaped set of branches known as a **polytomy**, which is used to represent unresolved relationships in the tree.

Like all phylogenetic trees, the one we have deduced here is a hypothesis. It is an explanation for patterns in nature—one that we can test with further evidence. This particular hypothesis was based on morphological data. As we'll see in Chapter 8, DNA provides a separate, and often extremely rich, source of data that we can use to generate phylogenies. Ingi Agnarsson of the University of Puerto Rico and his colleagues compared a genetic sequence in 222 carnivoran species and published a phylogeny in 2010. Overall, their phylogeny resembles the one shown in Figure 4.16B (Agnarsson et al. 2010).

Key Concepts

- Unique character states that arose in the common ancestor of a clade, and are inherited by daughter species in that clade, are called shared derived characters, or synapomorphies.

- Phylogenetic trees are most accurate when they are constructed using shared derived characters.

- Phylogenetic trees are hypotheses that describe the relationships among taxa based on the best available evidence. Phylogenies help scientists identify questions that can be tested with additional evidence.

- There are a staggering number of possible arrangements of branches (or tree topologies) that could describe any group of taxa. Scientists use approaches like parsimony to evaluate which topologies are best supported by their data. ●

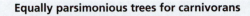

A Equally parsimonious trees for carnivorans

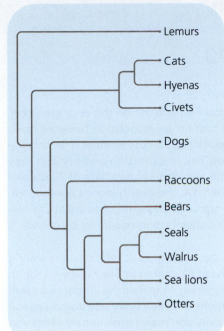
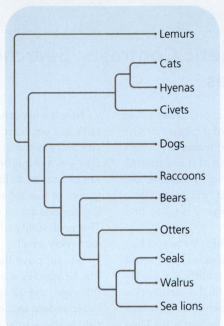
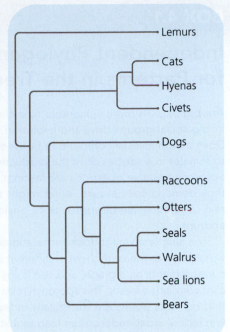

B Consensus tree

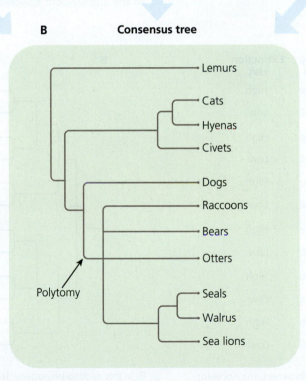

Figure 4.16 A: If we examine 20 characters in carnivorans instead of 12, and we use computers to contrast millions of possible tree shapes instead of four, we find three equally parsimonious trees. All three require 31 steps. The trees differ in some respects, but they all contain some identical relationships. B: We can combine all three trees into one consensus tree. The comb-like branches are instances where species or clades descend from a common ancestor, but we can't determine which members of these branches are most closely related to each other based on the data we analyzed. More detailed studies on much larger data sets agree with this tree's major features in all but a few minor ways. Tree building often reveals smaller details of phylogenies and resolves branch orders as the size of the data set increases.

BOX 4.1

Independent Phylogenetic Contrasts: Searching for Patterns in the Trees

Are big animals more vulnerable to extinction? Does living in big social groups drive the evolution of a bigger brain? Does a high level of competition between males for access to females in a species drive the production of more sperm? Such comparative questions are fascinating to explore, but they demand special care. What might at first seem like a real correlation can disappear once phylogeny is taken into account.

The simplest way to look for relationships between two variables is to measure them in different species, plot those measurements on a graph, and see if any statistically significant correlation exists. This approach rests on the assumption that our data points are statistically independent, however, and lack of independence can lead to a statistical error. Data points in a phylogeny lack that independence. It's not pure coincidence, for example, that humans and chimpanzees have five fingers on their hands. Both species inherited them from a five-fingered ancestor.

Here's a hypothetical example of how this error can occur. Let's say we want to test for an association between body size and extinction risk. We could note whether different species are at high or low risk of extinction and whether they are big or small. In the hypothetical database of 12 species shown in **Box Figure 4.1.1A**, the distribution is nonrandom. Big species are all at high risk, and small ones are at low risk. The probability of this distribution occurring by chance is extremely small.

But suppose that the phylogenetic relationships among the 12 species are those illustrated in **Box Figure 4.1.1B**. In this case, the 12 species we examined no longer represent independent associations between body size and extinction risk. In this scenario, only one major evolutionary event was related to body size. The lineage split into two daughter lineages, one with small body size and low extinction risk and one with large body size and high extinction risk. All of the subsequent species simply inherited their body size

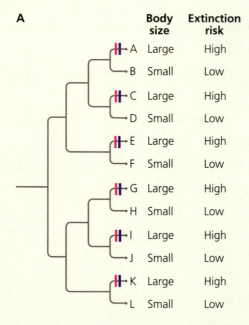

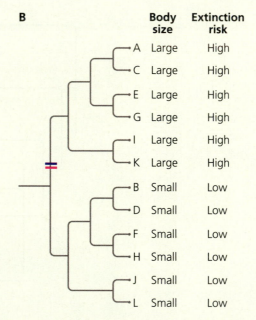

Box Figure 4.1.1 In both of these phylogenies, the ancestor had a small body size and low extinction risk, and the evolution of large body size is associated with a transition to high extinction risk. But this identical result is the product of two different histories. A: In this phylogeny, the origin of large body size presumably coincides with the origin of high extinction risk six separate times.

B: In this second phylogeny, large body size evolved only once. In the first case, evolution of large body size is correlated with evolution of high extinction risk. We can't say anything about such a correlation in the second case, because the pattern arose from just a single event.

A

Species	Trait	
	X	Y
A	20	7
B	24	9
C	40	20
D	30	14

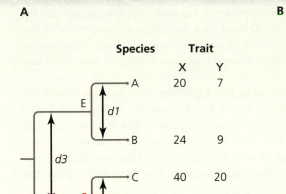

B

Table of independent contrasts

	X	Y
d1	4	2
d2	10	6
d3	13	9

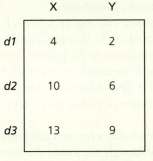

C

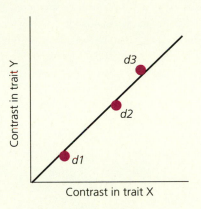

Contrast in trait Y

Contrast in trait X

Box Figure 4.1.2 A: Using phylogenetically independent contrasts, we compare pairs of lineages without using any one lineage more than once. (We estimate the value of traits for internal nodes.) B: Next, we calculate the difference between the pairs, as shown in the table. C: Finally, we can graph these values to see if there is a significant correlation even when we take phylogeny into account. (Data from Avise 2006)

and extinction risk from these common ancestors. Instead of many different evolutionary changes in body size, each associated with a corresponding change in extinction risk, we have only one. We can no longer draw any robust conclusion about the evolutionary relationship between these traits. With only one evolutionary event contributing to our pattern, we cannot reject the possibility that the association between body size and extinction risk is pure coincidence.

One way to minimize these errors is a method called independent contrasts. We make a series of comparisons between nodes and tips in our phylogeny, making sure we don't use any part of the tree more than once. In **Box Figure 4.1.2**, we can compare two values in species A and B and in C and D. We can also compare the internal nodes, E and F, by estimating their value. The simplest way to do so is to take the mean of the extant taxa that descend from them (in this case for trait X, the means are 22 and 35 for E and F, respectively). Next, we calculate the difference for each trait in each pairwise comparison. We can then plot these differences on a graph and see whether there is a significant correlation between them. (This example is adapted from Avise 2006.) Imagine that measurements of body size

are indicated by X and extinction rate is indicated by Y. We would conclude there is a correlation based on these data.

The independent contrast method has become an important tool in a wide range of studies in biology. Conservation biologists use it for their work on endangered species, for example. They need to determine which species are at greatest risk of extinction so they can make the most effective use of their limited resources. If they don't take phylogeny into account, they could make devastating errors.

In 2005, Marcel Cardillo and colleagues at Imperial College London conducted a large-scale analysis of extinction risk in mammals. They used independent contrasts to compare 4000 nonmarine mammal species, ranging from bats that weigh 2 grams to elephants that weigh 4000 kilograms (Cardillo et al. 2005). They found a statistically significant correlation between body size and extinction risk. Their analysis revealed some of the reasons for this link: Big animals need bigger ranges than small ones do. They are also more likely to be hunted, and they have a slow reproductive cycle.

In later chapters, we'll examine a number of other ways that scientists use phylogeny to address questions about biology.

4.4 Fossils, Phylogeny, and the Timing of Evolution

In the previous section, we reconstructed carnivoran phylogeny based on a comparison of extant species (species that are still represented by living individuals). Incorporating fossils of extinct species can allow us to extract more information from phylogenies.

For example, fossils enable us to determine some aspects of the timing of evolution. Consider the phylogeny of five living species in **Figure 4.17A**. We can say that the common ancestor of A, B, and C (species Z) lived after the common ancestor of all five species (species X), but we can't say *when* either common ancestor lived. To constrain the range of time in which these branches diverged, we can determine how fossil taxa are related to extant ones.

Let's say you dig up a fossil, F, and carry out a phylogenetic analysis that shows it is related to A, B, and C, as shown in **Figure 4.17B**. Isotopic dating reveals that it is approximately 50 million years old. By combining these data, we can conclude that Y, the common ancestor of A, B, C, and F, lived before 50 million years ago. After all, a descendant can't live before its own direct ancestor. Likewise, X, the common ancestor of all five extant taxa and the fossil taxa, must have lived even earlier than that.

There's still a lot we can't know about the timing of this phylogeny based on one fossil. Although we know the minimum possible age for Y and Z, one fossil alone cannot tell us what their maximum age is. Fossil Y could be 50.1 million years old

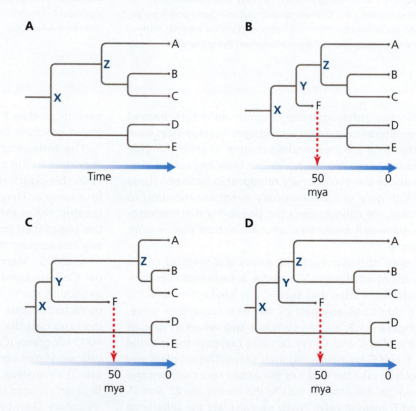

Figure 4.17 Fossils can constrain the timing of branching events in a phylogeny. A: A phylogeny of five living species shows their relationships. According to this hypothesis, we know the order of some branching events. For example, the species at node Z lived after the one at node X. But on its own, this phylogeny cannot give us a precise age for X or Z. B: A fossil of a known age—in the case of F, 50 million years old—helps constrain the age of the phylogeny. Fossil Y, the common ancestor of A, B, C, and F, must be older than 50 million years. Fossil X must be older as well. C, D: Two of the possible temporal patterns that can be accommodated by the discovery of fossil F.

or 501 million years old. We also can't determine from fossil F how old the common ancestor of A, B, and C is. It could be older than F or younger (**Figure 4.17C** and **D**).

Fortunately, paleontologists can frequently study several fossils from the same clade. They can provide constraints to the clade's history. And with the advent of molecular phylogenetics—using DNA instead of morphological traits to reconstruct phylogenies—scientists now have another way to constrain the timing of evolution. As we'll see in Chapter 8, they can use a "molecular clock" along with fossils of known age to estimate the ages of nodes in phylogenies.

• Combining evidence from fossils with morphological evidence from extant species can offer insight to the timing of branching events. ●

Key Concept

4.5 Charting the Major Transitions of Evolution with Phylogenies: From Sea to Land

Phylogenies make it possible to do more than classify species taxonomically. We can examine their pattern of branching to develop hypotheses about the major transitions in evolution, for example.

In Chapter 3, we observed that the oldest known aquatic vertebrates appear in the fossil record about 500 million years ago. Only after about 150 million years do terrestrial vertebrates—tetrapods—appear. All lineages of tetrapods share a complex of traits that supports them out of water. Most obviously, they have four limbs that they can use to move on dry land. Tetrapods can also use lungs to inhale oxygen to fuel those muscles and use skin to protect them from desiccation. Some tetrapods lack some of these traits—but only because their ancestors lost them through evolution. As we saw in Chapter 1, whales today lack legs, but their ancestors had them.

A typical tetrapod is thus profoundly different from an aquatic vertebrate such as a shark or a barracuda. We are left to marvel at how one kind of animal evolved into the other. The phylogeny of tetrapods and their aquatic relatives provides a number of enlightening clues.

A wide range of morphological and molecular studies all point to a handful of aquatic vertebrate species as the closest living relatives of tetrapods. These species include coelacanths, which live in the deep sea off the eastern coast of Africa as well as in the waters around Indonesia (**Figure 4.18**). Another group of closely related species,

Figure 4.18 Coelacanths are rare aquatic vertebrates that live off the coast of East Africa and Indonesia. They are among the closest living relatives of tetrapods. (Hoberman Collection/Superstock, Inc.)

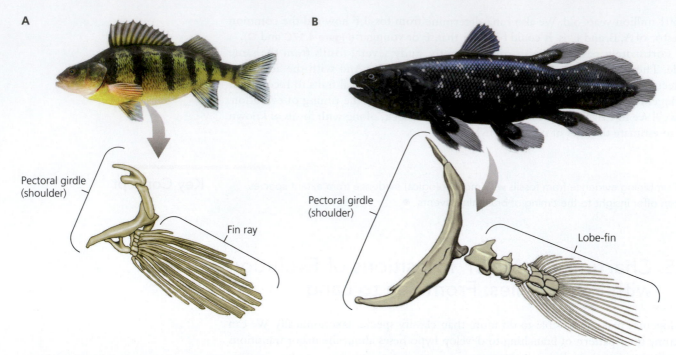

Figure 4.19 A: The ray-finned fishes, known as teleosts, have fins that are made up mostly of slender rays. B: Coelacanths, on the other hand, have a short chain of bones that anchor powerful muscles. This appendage is homologous to the tetrapod limb. Our own limbs evolved from a related "lobe-fin." Coelacanths have been filmed "walking" on the ocean bottom.

Figure 4.20 A: Neil Shubin and Ted Daeschler are part of the team that discovered an early tetrapod fossil they named *Tiktaalik*. B: Their team is shown here digging for fossils in the Arctic. (Courtesy of Neil Shubin)

called lungfishes, live in rivers and ponds in Brazil, Africa, and Australia. Many biologists refer to the clade that includes tetrapods, coelacanths, and lungfishes as lobe-fins (Zimmer 1998).

Tetrapods and lobe-fins are united by a number of homologous traits not shared by other aquatic vertebrates. Some of the most interesting traits are found in the limbs. Instead of the flexible, webbed pectoral fins of goldfish or salmon, coelacanths and lungfishes have fleshy lobes with stout bones inside (**Figure 4.19**). These bones are homologous to the long bones of the arm. But we can gain only limited insights into the origin of tetrapods from studying coelacanths and lungfishes. The ancestor we share with them lived more than 400 million years ago, and since then, the lungfish and coelacanth lineages evolved on separate trajectories. Living lungfishes have since adapted to a freshwater niche, whereas coelacanths have adapted to the deep ocean. To find more clues about the transition from water to land, we need to add the branches belonging to extinct taxa. The only way to do that is to find fossils that are more closely related to tetrapods than to living lobe-fins.

Paleontologists found the first of these fossils in the late 1800s. *Eusthenopteron*, which lived about 385 million years ago, had a stout bone extending from its shoulder girdle and two more bones extending farther out. Over the course of the twentieth century, a few more fossils were found, also dating back to the Middle Devonian epoch. In Greenland, for example, Jennifer Clack of the University of Cambridge found the remains of a 365-million-year-old tetrapod called *Acanthostega* (Clack 2002). It had additional derived tetrapod characters, including digits.

Discoveries like these led Neil Shubin, a University of Chicago paleontologist, and his colleagues to develop a hypothesis about where other transitional fossils might be found (**Figure 4.20**). All of the known fossils marking this transition dated from the Upper Devonian epoch, from about 383 to 359 million years ago. So Shubin reasoned that that time period was probably the one most likely to yield other early tetrapods (**Figure 4.21A**). Fossils of early tetrapods and their closest lobe-fin relatives had been discovered in rocks that had formed in coastal wetlands and river deltas. Shubin hypothesized that it was in these ecosystems that the transition took place.

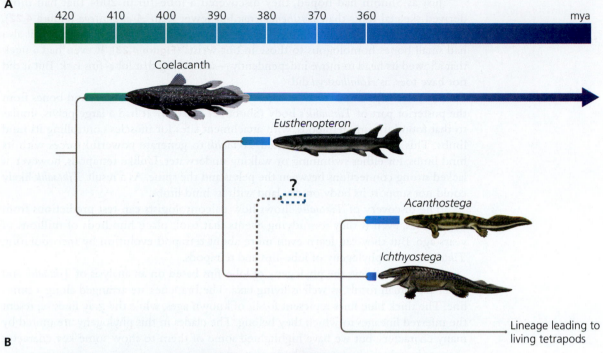

A

420   410   400   390   380   370   360   mya

Coelacanth

Eusthenopteron

?

Acanthostega

Ichthyostega

Lineage leading to
living tetrapods

B

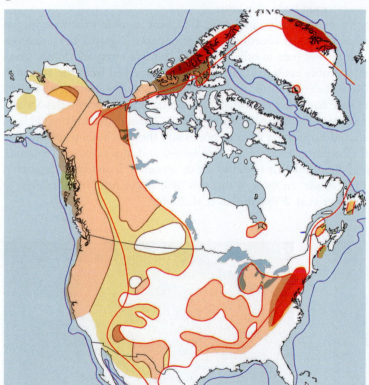

Figure 4.21 Scientists can use phylogenies to generate hypotheses about evolution. A: Here is a simplified phylogeny of tetrapods as of 2000. Based on the distribution of fossils known at the time, Shubin and colleagues reasoned that fossils of transitional tetrapods (closely related to our early tetrapod ancestors) might be found in sedimentary rocks of similar age. B: By examining maps of North American geology, Shubin and his colleagues found three regions where such rocks were found (solid red). One of them, in the Canadian Arctic, had not yet been searched for tetrapod fossils. The scientists went to that formation, where they found *Tiktaalik*. (Data from Dott and Batten 1988)

He therefore narrowed his search further, to Upper Devonian sedimentary rocks that had formed in those environments.

As we saw in Chapter 3, geologists have been mapping geological formations around the world for more than two centuries. Working with geologists, Shubin and his colleagues were able to examine stratigraphic maps to help them look for rocks most likely to hold new early tetrapod fossils. They noticed a large area of Upper Devonian sedimentary rocks in northern Canada (**Figure 4.21B**). When they discovered that no paleontological expedition had ever searched those formations for tetrapod fossils, they decided to head north.

Just as Shubin had hoped, they discovered a lobe-fin in 2004 that had more derived skeletal traits than *Eusthenopteron* but fewer than *Acanthostega* (Figure 4.22). The newfound lobe-fin, which they dubbed *Tiktaalik*, had long limb bones, and it also had small bones homologous to those in our wrists (Figure 4.23). It even had a neck that allowed its head to move independently—something that lobe-fins lack. But it did not have toes, as *Acanthostega* did.

On later expeditions to Canada, Shubin and his colleagues also found bones from the posterior part of *Tiktaalik's* body (Shubin et al. 2014). It had a large pelvis, similar to that found in early tetrapods, with attachment sites for muscles controlling its hind limbs. This anatomy likely permitted *Tiktaalik* to generate powerful forces with its hind limbs, for either swimming or walking underwater. Unlike tetrapods, however, it lacked strong connections between the pelvis and the spine. As a result, *Tiktaalik* likely could not support its body on dry land with its hind limbs.

The discovery of *Tiktaalik* shows how paleontologists can test predictions from hypotheses, even if they're studying events that took place hundreds of millions of years ago. But they can learn even more about tetrapod evolution by incorporating *Tiktaalik* into a phylogeny of lobe-fins and tetrapods.

Figure 4.24 shows the phylogeny of lobe-fins based on an analysis of *Tiktaalik* and other Devonian fossils, as well as living taxa. The branches are arranged along a timeline. The thick blue lines represent fossils of known ages, while the gray lines represent the inferred lineages to which they belong. The clades in this phylogeny are united by many characters, but we have highlighted some of them to show some key character changes by which the tetrapod body plan arose.

When you look at the illustration in Figure 4.24, bear in mind that these species do not form a continuous line of ancestors and descendants. *Tiktaalik* has a number of unique features that are not seen in other lobe-fins and that tetrapods do not share. These peculiar traits probably evolved after its ancestors diverged from the ancestors of other lobe-fins. Nevertheless, this tree gives us insights about the evolution of tetrapods that we'd never have had if not for fossils. The common ancestor of tetrapods and their closest living relatives had stout, paddle-shaped fins. *Tiktaalik* probably could have lifted its head and shoulders, judging from its bones and the attachments for muscles. Digits evolved in the common ancestor of *Acanthostega* and more derived tetrapods. *Acanthostega* had eight digits; other species had six or seven. All living land vertebrates have five or fewer. Only fossils can show us that this "five-finger rule" took millions of years to emerge (Daeschler et al. 2006; Shubin et al. 2006).

Figure 4.22 A: Shubin and his colleagues found several partial skeletons of *Tiktaalik*. B: The team used these fossils to reconstruct the anatomy of *Tiktaalik*, which had some—but not all—of the traits shared by tetrapods. (Data from Shubin et al. 2014) C: *Tiktaalik* lived 375 million years ago. It possessed some traits of living tetrapods, such as a wrist and neck, but it likely lived entirely underwater. (A: Corbin17/Alamy Stock Image; B: John Westlund; C: Carl Buell)

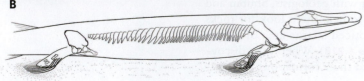

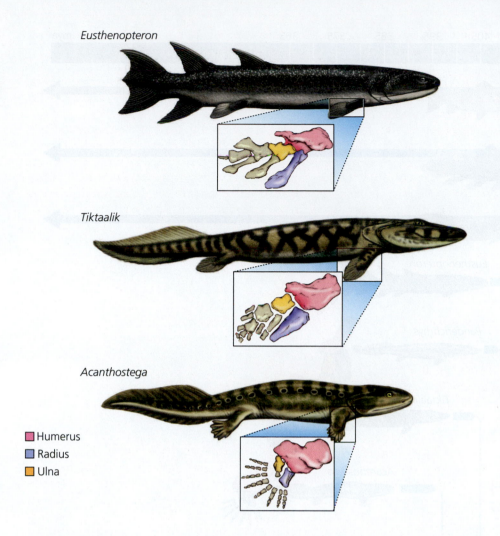

Eusthenopteron

Tiktaalik

Acanthostega

- 🟥 Humerus
- 🟦 Radius
- 🟧 Ulna

Figure 4.23 Fossils of lobe-fins and early tetrapods reveal the homologies in their limb bones. *Eusthenopteron* had bones that were homologous to the long bones (the humerus, radius, and ulna) of our arms. *Tiktaalik* shared more homologies, including wrist bones. *Acanthostega*, an early tetrapod, had distinct digits at the ends of its limbs. All tetrapods today have only five or fewer digits, but *Acanthostega* had eight. (Data from Friedman et al. 2007)

This evolutionary tree also lets us test hypotheses about the selective pressures that led to the origin of the tetrapod body plan. In the early 1900s, Alfred Romer developed an influential theory based on the fact that the oldest tetrapods at the time were found in rocks that appeared to have formed during a time of severe droughts. He envisioned fishlike vertebrates living in rivers and ponds; when these water bodies dried up, the animals had to make their way to remaining bodies of water or die. Mutations that led to more leglike fins would have enabled them to move more quickly over land.

The phylogeny in Figure 4.24 disproves this hypothesis, however. Even after tetrapods had evolved fully formed tetrapod limbs, they were poorly suited for life on land. *Acanthostega* had bones for supporting gills that it may have used to get oxygen. Its shoulder and pelvic bones were so slender that they probably couldn't have supported its weight on land. *Acanthostega*'s tail was still lined with delicate fin rays that would have been damaged by being dragged along the ground. Yet *Acanthostega* was a fairly derived tetrapod that had limbs, digits, and other traits associated with walking. These early tetrapods did not live in the harsh, arid landscape Romer envisioned as their environment; instead, they lived in lush coastal wetlands. Scientists now hypothesize that *Acanthostega* used its digits to move around underwater, perhaps holding onto underwater vegetation or clambering over submerged rocks. Not until millions of years later in tetrapod evolution do fully terrestrial taxa emerge in the fossil record.

- Phylogenies illustrate relationships among *known* species—not all species. Combining different lines of evidence can lead scientists to new fossil discoveries and to insights into why lineages may have evolved. ●

Key Concept

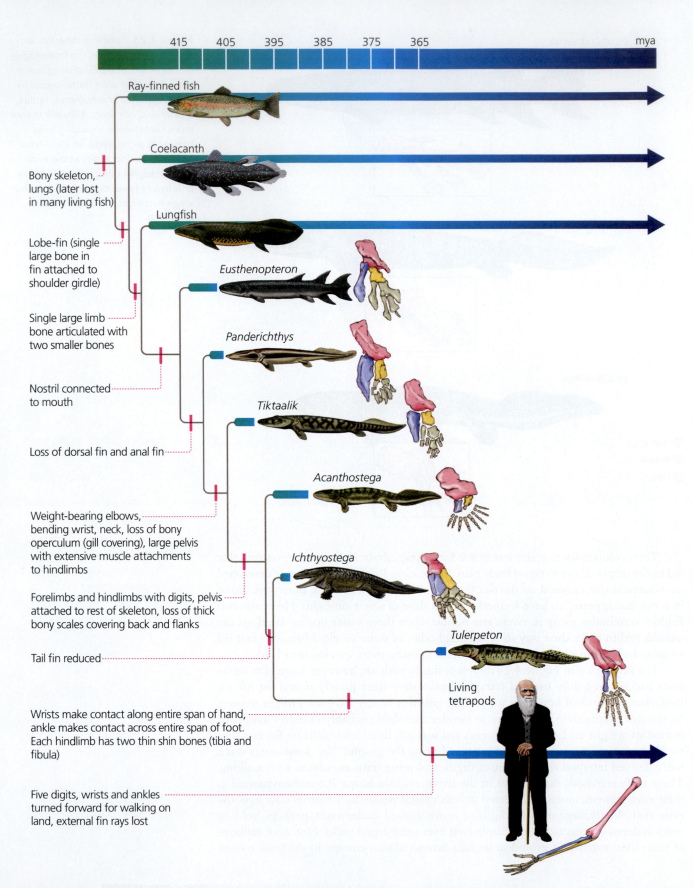

415   405   395   385   375   365   mya

Ray-finned fish

Coelacanth

Bony skeleton, lungs (later lost in many living fish)

Lungfish

Lobe-fin (single large bone in fin attached to shoulder girdle)

Eusthenopteron

Single large limb bone articulated with two smaller bones

Panderichthys

Nostril connected to mouth

Tiktaalik

Loss of dorsal fin and anal fin

Acanthostega

Weight-bearing elbows, bending wrist, neck, loss of bony operculum (gill covering), large pelvis with extensive muscle attachments to hindlimbs

Ichthyostega

Forelimbs and hindlimbs with digits, pelvis attached to rest of skeleton, loss of thick bony scales covering back and flanks

Tulerpeton

Tail fin reduced

Living tetrapods

Wrists make contact along entire span of hand, ankle makes contact across entire span of foot. Each hindlimb has two thin shin bones (tibia and fibula)

Five digits, wrists and ankles turned forward for walking on land, external fin rays lost

Figure 4.24 This tree shows the relationship of lobe-fins to tetrapods and how new tetrapod traits evolved over time. The tetrapod "body plan" evolved gradually, over perhaps 40 million years. The earliest tetrapods probably still lived mainly underwater. This tree includes only a few representative species; paleontologists have discovered many others that provide even more detail about this transition from sea to land.

BOX 4.2

The Tree of Life—or Perhaps the Web of Life?

The metaphor of the tree of life is based on the assumption that genetic information is transferred vertically from parents to offspring, and not horizontally from one branch directly to another. This structural approach is applicable to most multicellular organisms (although hybridization events can occasionally result in the merging of branches).

The treelike representation of history does not necessarily apply to microorganisms, however. Comparisons of the genomes of prokaryotes reveal extensive gene transfer between lineages—that is, across what were traditionally considered distinct branches (**Box Figure 4.2.1**). This transfer is apparent even between distantly related lineages.

Biologists propose considering the relationships among unicellular life-forms as phylogenetic "networks" or "webs" instead of trees due to the prevalence of **horizontal gene transfer** (Syvanen 2012; McInerney and Erwin 2017). We discuss this gene transfer in greater detail in Chapters 8, 9, 11, 13, and 15.

Horizontal gene transfer describes the transfer of genetic material to another organism, sometimes a distantly related one, without reproduction (that is, other than from parent to offspring). Once this material is added to the recipient's genome, it can be inherited by descent.

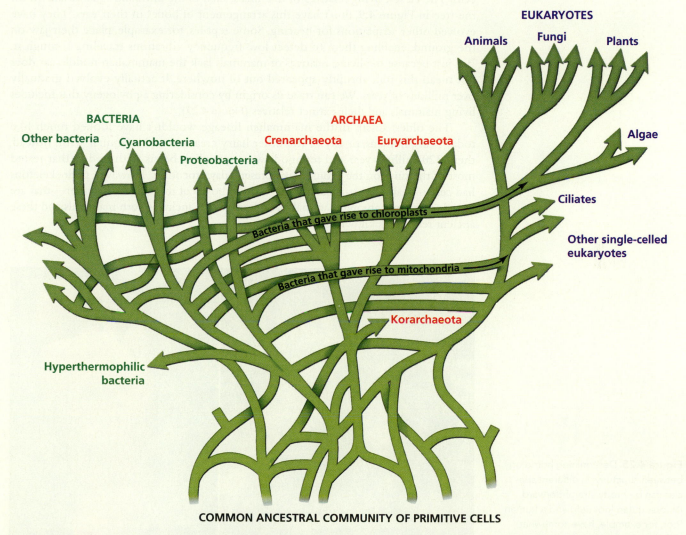

Box Figure 4.2.1 A version of the tree of life that is revised to include horizontal gene transfer—an important mechanism of gene transfer that does not result from genes shared between parents and offspring. (From Doolittle, W. Ford. Uprooting the Tree of Life. Scientific American Vol. 282, No. 2 (February 2000), pp. 90–95. Reproduced with permission. Copyright © 2000 Scientific American, a division of Nature America, Inc. All rights reserved.)

4.6 Homology as a Window Into Evolutionary History

Sometimes the homology between traits in different species is obvious. Anyone can see the similarities between a human foot and an orangutan foot, for example (**Figure 4.25**). Orangutans use their feet to grasp tree branches, and thus selection has resulted in feet with long, curved toes. Humans use their feet for walking and running, and selection resulted in human feet with short toes at the front and long arches. Still, the two types of feet are clearly homologous structures. They have the same bones and muscles, arranged nearly identically. In other cases, homology can be harder to recognize. Phylogenetic analysis can reveal it by tracking the evolution of structures over time.

The Mammalian Ear and Its Homologies

The mammalian ear is a remarkably delicate and sophisticated adaptation. It consists of a chain of tiny bones that vibrate when the tympanic membrane is struck by airborne sound waves (**Figure 4.26**). The bones then transmit those vibrations to nerve cells. The closest living relatives of mammals, such as the birds and iguanas shown on the tree in Figure 4.9, don't have this arrangement of bones in their ears. They have evolved other adaptations for hearing. Some reptiles, for example, place their jaw on the ground, enabling them to detect low-frequency vibrations traveling through it. But just because the living relatives of mammals lack the mammalian middle ear does not mean this trait abruptly appeared out of nowhere. It actually evolved gradually over millions of years. We can trace its origin by considering a phylogeny that includes living mammals and their extinct relatives (**Figure 4.27**).

The oldest fossils in the mammalian lineage wouldn't have looked much like today's cats, humans, or any of the other hairy creatures we're familiar with. Instead, these 320-million-year-old tetrapods were sprawling beasts with bodies that rested most of the time on the ground, like present-day crocodiles. However, their skeletons had certain features—particularly the way the bones of the skull fit together—that are found today only among mammals. The lineage that includes both mammals and these ancient relatives is known as synapsids.

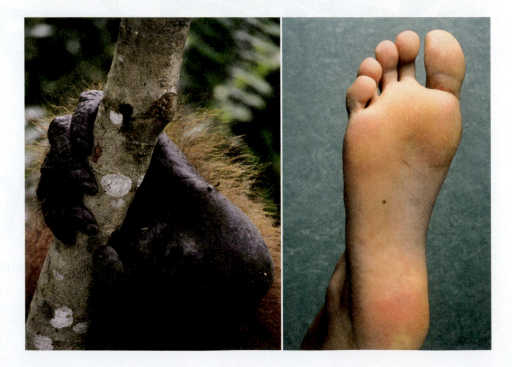

Figure 4.25 Determining homology between structures in different species can be pretty straightforward. An orangutan foot (left) and a human foot, for example, have somewhat different shapes because they're adapted for different functions. (Left: Theo Allofs/Danita Delimont/Alamy Photo Stock; right: Aaron Haupt/Science Source)

For their first 100 million years, synapsids took many different forms. *Dimetrodon*, for example, had a strange, sail-shaped back. Other synapsids looked like turtles with fangs. Still others had peg-shaped teeth for grinding up plants. One lineage of synapsids, known as the cynodonts, evolved a more upright stance and other mammal-like traits not found in other synapsids. By about 200 million years ago, the basic skeletal body plan seen in mammals today had evolved.

About 170 million years ago, the common ancestor of all living mammals diverged. Today, the living descendants of that ancestor belong to three lineages. The deepest branch of living mammals produced the monotremes, which include the duck-billed platypus and the echidna. Like other living mammals, monotremes produce milk, but they secrete it through a network of glands. Their young feed on the milk that coats their fur near the secretion points rather than nursing from a nipple. Like birds and many reptiles, monotremes lay eggs.

All the remaining species of living mammals, known as therians, bear live young. Therians split into two branches of living mammals. One branch, the marsupials, includes opossums, kangaroos, and koalas. Marsupial young crawl into a pouch on the mother's belly after they're born, where they can be carried until they're big enough to survive on their own. The other branch, the eutherian mammals, includes us and all other mammals that develop a placenta to feed embryos in the uterus. (Although therians have traits that monotremes lack, that does not mean monotremes are "lower" or "primitive." See **Box 4.3** for an explanation.)

In Figure 4.27 we highlighted the lower jaw and surrounding bones. Comparing them in a phylogenetic context allows us to see that the bones of the

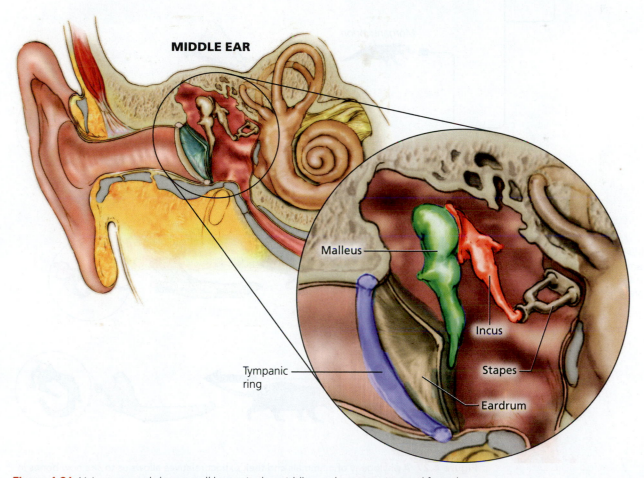

MIDDLE EAR

Malleus

Incus

Tympanic ring

Stapes

Eardrum

Figure 4.26 Living mammals have small bones in the middle ear that transmit sound from the eardrum to the inner ear. Living reptiles and birds have no such structures in their ears, raising the question of how the mammalian ear evolved.

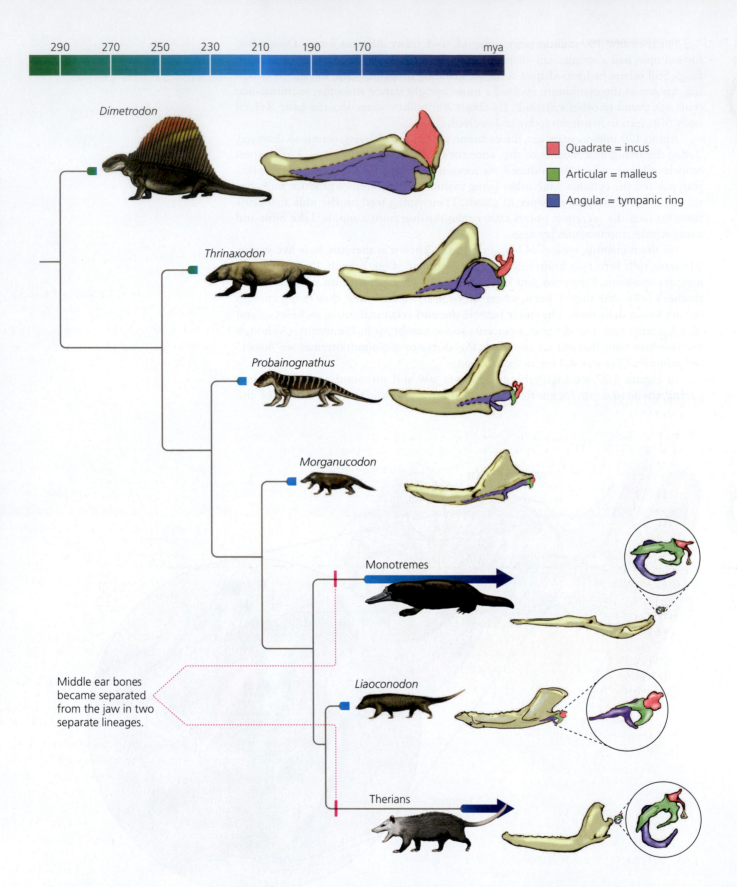

290 270 250 230 210 190 170 mya

Quadrate = incus
Articular = malleus
Angular = tympanic ring

Dimetrodon

Thrinaxodon

Probainognathus

Morganucodon

Monotremes

Middle ear bones became separated from the jaw in two separate lineages.

Liaoconodon

Therians

Figure 4.27 A phylogeny of mammals and their extinct relatives allows us to see how bones in the jaw of synapsids evolved into middle ear bones. Over millions of years, three bones at the rear of the jaw and skull shrank and separated, becoming specialized for transmitting sounds in the ear. The colors indicate homologous bones in different species.

mammalian middle ear are homologous to a set of much larger bones in the synapsid skull (**Figure 4.28**).

In early synapsids, the jaw was made up of a collection of interlocking bones. The front-most bone, the dentary, held many teeth, and the bones in the rear formed a hinge against the back of the skull. In the mammalian lineage, the dentary gradually became larger and larger, ultimately developing its own hinge. As the dentary was expanding, the bones in the back of the jaw were shrinking. Some of these bones eventually disappeared, and the remaining ones became partially separated from the lower jaw. At first these small bones remained tethered to the jaw, but later they became entirely free (Luo 2011; Anthwal et al. 2013; Luo et al. 2016).

By studying living tetrapods, scientists have developed hypotheses for the selective pressures that might have driven these changes. The mammal lineage may have evolved a larger dentary as an adaptation for chewing. A single large jawbone could provide more strength while chewing hard food than a group of smaller bones.

This shift initially may have been driven by the diet of early mammals, but it also opened up a new way of hearing (**Figure 4.29**). Synapsids may have used their lower jaw to detect low-frequency sound traveling through the ground. As the rear jawbones grew smaller and less connected to the dentary, they also began to vibrate in response to airborne sound. At first, this sense of airborne sound was weak. But once the bones were no longer essential for supporting the jaw, they could adapt to a new function. Over millions of years, these former jawbones became a series of levers that amplify faint high-frequency sounds. This transformation may have enabled mammals to occupy new ecological niches.

As paleontologists discover more fossils of mammals, they are revising their hypotheses for how the mammalian ear took its current form. In 2011, for example, a team of Chinese and American scientists published the details of a skeleton belonging

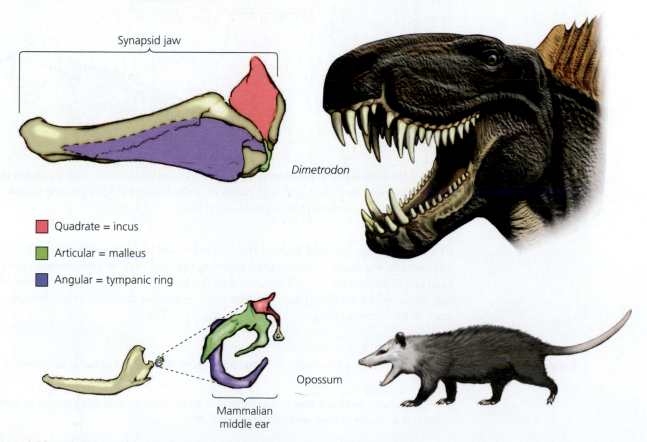

Synapsid jaw

Dimetrodon

- 🟥 Quadrate = incus
- 🟩 Articular = malleus
- 🟦 Angular = tympanic ring

Opossum

Mammalian
middle ear

Figure 4.28 A phylogeny of mammals and their extinct relatives helps us to recognize the homologies between a synapsid jaw and the mammalian middle ear.

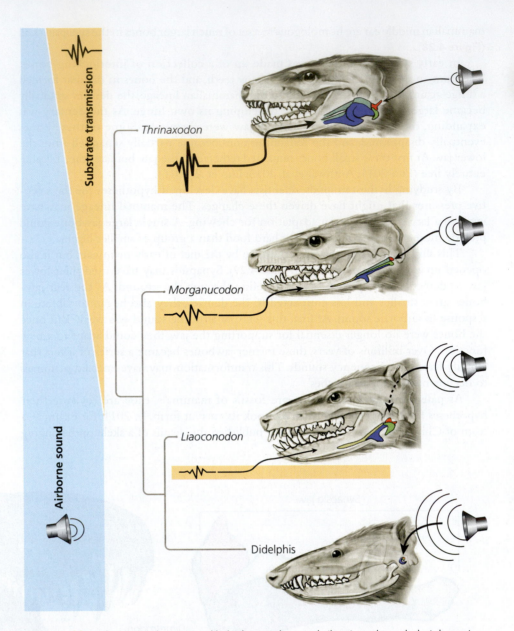

Substrate transmission

Airborne sound

Thrinaxodon

Morganucodon

Liaoconodon

Didelphis

Figure 4.29 Early mammal ancestors likely detected ground vibrations through their lower jaw. The gradual evolution of the mammalian middle ear shifted their hearing to airborne sounds. (From April I. Neander/University of Chicago from Luo et al. 2016)

to a 125-million-year-old therian fossil called *Liaoconodon hui* (Meng et al. 2011). *Liaoconodon* was more closely related to living therians than to monotremes, yet the bones of its middle ear are still connected to the lower jaw. The researchers concluded that the middle ear fully separated at least twice—once in the monotreme lineage and once in the therian lineage.

Key Concepts

- The mammalian ear is made up of modified parts of the lower jaw and surrounding skull.

- Phylogenetic trees can reveal the patterns of small changes and adjustments to traits over the course of their evolutionary history. ●

The Myth of the "Primitive"

For millennia, naturalists organized living things into "lower" and "higher" forms, envisioning life arrayed along a ladder-like scale. Darwin's metaphor of a tree of life is fundamentally opposed to this view of biology. Everything alive today, from bacteria to jellyfish to humans, is the product of a 3.7-billion-year-old lineage. There is nothing intrinsically superior about animals, let alone humans. Indeed, if we want to judge by numbers alone, the most successful organisms are viruses, which have an estimated population size of 10^{31} (Zimmer 2011).

When we look at phylogenies, we must be very careful not to slip into ladderlike thinking. Consider, for example, the mammal tree in Figure 4.27. We've arranged the phylogeny to make it easy for you to follow the series of steps by which the modern mammal body—including the middle ear—evolved. But we could just as easily have swung the nodes around in a different way, so that the tree seemed to lead to platypuses instead of eutherians like ourselves.

Because platypuses have so few close relatives, their position off by themselves on the side of the mammal tree may make them seem like they must be "primitive." It is certainly true that platypuses have some traits that are ancestral to those found in therian mammals. Platypuses lay eggs, for example, whereas marsupial and placental mammal embryos develop in the uterus, attached to a placenta. Platypuses produce milk from mammary glands, but they lack teats (**Box Figure 4.3.1**).

Still, these ancestral traits are only part of platypus biology. Platypuses have a unique bill, which they sweep through water to detect electrical signals from their prey. This bill evolved only after the ancestors of platypuses and the ancestors of the eutherian clade split from each other. Platypuses also have webbing on their feet and produce venom. All of these traits are likely derived in the platypus

Box Figure 4.3.1 Platypuses are mammals that lay eggs. That does not mean they are "primitive," however. (age fotostock/Superstock, Inc.)

lineage (**Box Figure 4.3.2**). These are just the tip of the iceberg when it comes to derived traits in platypuses. In 2008, an international team of scientists sequenced the platypus genome and found that it contained thousands of derived segments of DNA, including protein-coding genes that likely were involved in adaptations to their particular ecological niche (Warren et al. 2008).

We can certainly gain some clues to the nature of the common ancestor of platypuses and other mammals by comparing their biology and mapping the acquisition of traits on their phylogeny. But we cannot treat living platypuses as a stand-in for that ancestor. Indeed, when it comes to electroreceptors and other derived platypus traits, we humans are the "primitive" mammals (Omland et al. 2008).

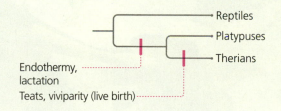

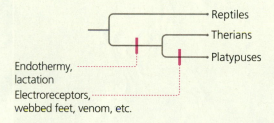

Box Figure 4.3.2 For some traits, platypuses retain the ancestral state relative to therian mammals. But platypuses also have derived traits not found in therian mammals. We could build a similar tree using a "platypus-centric" set of characters and give the impression that our own species has not evolved much.

4.7 How Feathered Dinosaurs Took Flight

An **exaptation** is a trait that initially carries out one function and is later co-opted for a new function. The original function may or may not be retained.

When we examine a trait in a living species, we can't assume it had the same function when it initially evolved that it has today. That's because traits often change their function over the course of evolution. Phylogenies allow us to track this transformation by linking homologous traits in different taxa. Evolutionary biologists refer to "borrowed" traits like these as **exaptations**. You can see one of the most striking exaptations any time you walk down a city street: the feathers on birds.

In 1861, just after *The Origin of Species* was published, German quarry workers discovered the fossil of a bird like nothing alive today (**Figure 4.30**). It had feathers, the impressions of which were preserved thanks to the stagnant, oxygen-free swamp the animal fell into when it died. But the bird, which would turn out to be about 150 million years old, also had teeth in its beak; claws on its wings; and a long, reptilian tail. Scientists dubbed it *Archaeopteryx* ("ancient wing"; Shipman 1998).

Before the discovery of *Archaeopteryx,* birds seemed profoundly different from all other animals. They shared many unique traits, such as feathers and fused arm bones, not found in other living tetrapods. *Archaeopteryx* offered clues to how birds had evolved from extinct reptile ancestors. But *Archaeopteryx* alone left many questions unanswered. Did feathers evolve to aid flight, or did they have other functions before flight? Which reptile ancestor did birds evolve from? For a century, those questions remained open. But over the past 50 years, thanks to new fossil discoveries and careful comparisons of fossils and living birds, a consensus has emerged. Birds, paleontologists now agree, are dinosaurs (Brusatte et al. 2015; Chiappe and Qingjin 2016). **Figure 4.31** shows how birds are related to dinosaurs and other reptiles.

In the 1970s and 1980s, paleontologists observed that the skeletons of birds share many traits with those of one group of dinosaurs in particular, a group known as the theropods. Theropods were bipedal meat-eating dinosaurs whose ranks included *Tyrannosaurus rex* and *Velociraptor.*

Figure 4.30 A: In 1861, German quarry workers discovered a fossil of a bird with reptile traits such as teeth and claws on its hands. Known as *Archaeopteryx*, it lived 150 million years ago. B: A reconstruction of what *Archaeopteryx* may have looked like in life. (A: imageBROKER/Superstock, Inc.; B: Carl Buell)

And in the late 1990s, paleontologists began to discover exquisitely preserved fossils of theropods that allowed them to be linked to birds by a new trait beyond their skeletons: feathers.

These feathered theropods lacked many of the synapomorphies that unite living birds, and many were clearly incapable of powered flight. Their arms were relatively

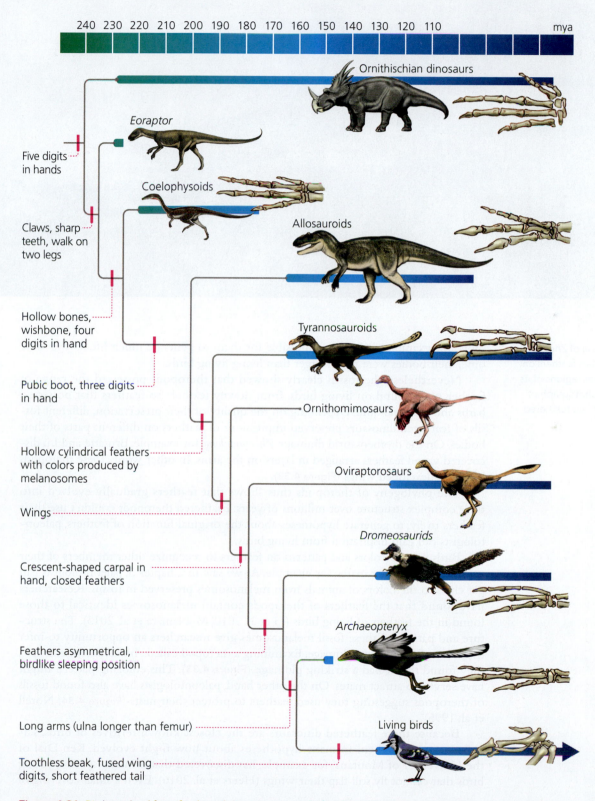

Figure 4.31 Birds evolved from feathered dinosaurs. Long before birds could fly, increasingly complex feathers evolved, as well as other traits that are found today only in birds.

short, for example, making it impossible for them to generate much lift. At the same time, their bodies were much larger than living flying birds.

Nevertheless, the fossils clearly showed that theropods possessed the range of feather types found on living birds, from downy feathers to feathers that possessed barbs and a central quill. Depending on the quality of their preservation, different fossils of feathered dinosaurs preserved impressions of feathers on different parts of their bodies. On the dromeosaurid dinosaur *Zhenyuanlong*, for example, Brusatte and Lü discovered vaned feathers arranged in layers on the arms. In other words, these theropods had well-developed wings (**Figure 4.32**).

The phylogeny of theropods thus shows that feathers gradually evolved into their complex structure over millions of years. Feathered theropods couldn't use their feathers to fly; to generate hypotheses about the original function of feathers, paleontologists can gain inspiration from living birds.

Birds use the colors and patterns on feathers to recognize other members of their species and to attract mates, for example. As we saw in Chapter 3, paleontologists can get clues to the colors of animals from melanosomes preserved in fossils. Researchers have found that the feathers of theropods contain melanosomes identical to those found in the feathers of living birds (Li et al. 2010; McNamara et al. 2013). The structure and pattern of these fossil melanosomes give researchers an opportunity to infer the color of this ancient plumage. Examining a theropod called *Anchiornis*, for example, they found that it had a striking plumage (**Figure 4.33**). This coloring pattern might have served to attract mates. On the other hand, paleontologists have also found fossils of theropods suggesting they used feathers to protect their nests (**Figure 4.34**; Norell et al. 1995).

Because these feathered dinosaurs are the closest known relatives of birds, scientists can study them to make hypotheses about how flight evolved. Ken Dial of the University of Montana, for example, has discovered that in many species, young birds that cannot fly still flap their wings (Heers et al. 2016). Even a small covering of

Figure 4.33 A: Some fossils of a 150-million-year-old dinosaur called *Anchiornis huxleyi* include well-preserved feathers. B: The feathers retain cellular structures called melanosomes that help produce color. C: The melanosomes produced a complex pattern of colors on *Anchiornis*. The size, shape, and organization of the melanosomes allowed paleontologists to reconstruct the dinosaur's original color. (A: Jakob Vinther; B: Republished with permission of the American Association for the Advancement of Science, from L. Quanguo, et al., "Plumage Color Patterns of An Extinct Dinosaur," Vol. 327, Issue 5971, 2010, pp. 1369–1372; permission conveyed through the Copyright Clearance Center; C: Carl Buell)

Figure 4.34 A: Birds sit on their eggs to keep them warm as they develop. B: In Mongolia, paleontologists discovered nesting fossils of dinosaurs that died when they were in a similar posture. This discovery suggests that the nesting behavior seen today in living birds evolved more than 150 million years ago in feathered dinosaurs that could not fly. (A: John Cancalosi/ Oxford Scientific/Getty Images; B: Mick Ellison/American Museum of Natural History)

feathers allows them to generate a downward force, which gives them extra traction while running up inclines. Juvenile theropods might have used the first, primitive flight stroke to help them navigate on the ground. It's possible that some lineages of small, feathered theropods such as *Anchiornis* began to use this stroke to move through the air. Although paleontologists are still debating which theropods were the first fliers, they generally agree that species such as *Archaeopteryx* were capable of true powered flight.

The three examples we've discussed—tetrapods, mammals, and birds—show how scientists use phylogenetic trees to reconstruct evolutionary history. We can see how complex body plans emerge gradually from older ones. Evolutionary trees offer support for Darwin's argument that homology, in all its guises of adaptation, is the result of common ancestry. But like all insights in science, phylogenies also raise new questions. What genetic changes produced the tetrapod body plan, for example, or gave rise to new structures, such as feathers? The answers, as we'll see in the next chapter, lie in the molecules that make heredity possible.

Key Concepts

- Feathers are an exaptation; they evolved originally for functions other than flight.

- Birds are dinosaurs. ●

TO SUM UP . . .

- A phylogenetic tree is a branching diagram that shows possible evolutionary relationships among groups of organisms.

- New evidence has changed how scientists think about taxonomic groupings. Historical groupings, such as reptiles and fish, are no longer considered valid because they are not monophyletic.

- A cladogram is constructed by grouping taxa into nested hierarchies using information from synapomorphies, characters that some members of the clade share because of common ancestry.

- Character states in different species may be similar because they were derived from a common ancestor of the species (synapomorphy) or because the species each converged independently on the same character state (homoplasy).

- Scientists reconstruct evolutionary trees to develop and test hypotheses about how major evolutionary transformations took place.

- Tetrapods belong to a clade called lobe-fins, which includes coelacanths and lungfishes.

- Some tetrapod traits, such as toes and legs, evolved while the ancestors of tetrapods still lived in water.

- The bones of the mammalian ear evolved from bones of the lower jaw of ancestors to mammals.

- Sometimes traits that arise for one function are later co-opted for a different function (exaptations), and phylogenies can help reveal this.

- Feathers evolved before flight. Dinosaurs used these early feathers for other functions, perhaps for insulation, courtship display, and nest brooding.

MULTIPLE CHOICE QUESTIONS Answers can be found at the end of the book.

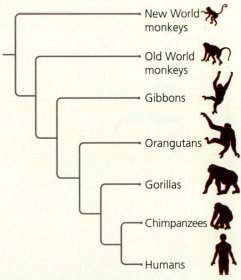

(Data from Gregory 2008)

1. Two species that branch off from the node of a phylogenetic tree are analogous to

   a. siblings on a family tree.

   b. cousins on a family tree.

   c. grandparents on a family tree.

   d. parents on a family tree.

2. Which of the following statements is depicted by the phylogeny at the right?

   a. The ancestors of humans became gradually more "humanlike" over time.

   b. Old World monkeys share a common ancestor with humans.

   c. Humans represent the end of a lineage of animals whose common ancestor was primate-like.

   d. Humans evolved from chimpanzees.

3. If you were looking at a phylogeny of living bird species, where could you find the name of a species of non-theropod dinosaur?

   a. At the tip of a branch, as an outgroup.

   b. At the root of the tree.

   c. Either a or b.

   d. Nested within the bird lineages.

4. Which of the following phylogenies does not indicate the same relationship among whales and other groups?

   a.

   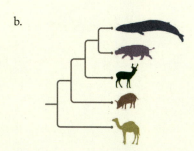

   b.

   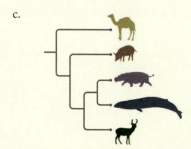

   c.

   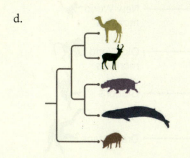

   d.

5. Which of the following is *not* a synapomorphy?

   a. The ability to swim in dolphins and sharks.

   b. The production of milk in humans and cats.

   c. The ability to fly in eagles and pigeons.

   d. The laying of eggs with shells in snakes and lizards.

6. Which of the following is an example of homoplasy?

   a. The reversion of a derived character state to its ancestral state.

   b. The independent origin of similar traits in separate lineages.

   c. The evolution of wings in both birds and bats.

   d. All of the above.

7. According to Figure 4.24, what homologies do *Tiktaalik* and *Acanthostega* share?

   a. Tail fin reduced and weight-bearing elbows.

   b. Weight-bearing elbows and forelimbs and hindlimbs with digits.

   c. Nostril connected to mouth and tail fin reduced.

   d. Bony skeleton and weight-bearing elbows.

8. Why are bird feathers considered an exaptation?

   a. Because they are a shared derived character found in most birds.

   b. Because they are traits that have independently evolved in separate lineages.

   c. Because they first evolved for functions other than flight.

   d. Because they are an evolutionary reversal to an ancestral character state.

9. Scientists have discovered that middle ear bones evolved independently in

   a. monotremes and marsupials.

   b. monotremes and therians.

   c. marsupials and therians.

   d. dinosaurs and birds.

10. Synapomorphies among a group of species is evidence that

   a. the species share an immediate common ancestor.

   b. the species are paraphyletic.

   c. the traits in question evolved independently (convergence).

   d. the traits in question originally evolved for a different function (exaptations).

11. Judging by the phylogenetic tree that follows, which are most closely related?

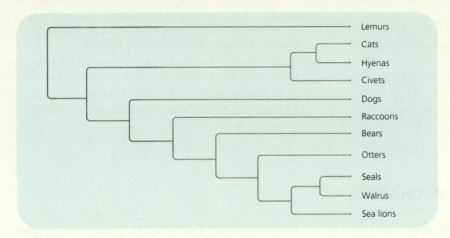

a. Raccoons and sea lions.

b. Dogs and raccoons.

c. Walrus and sea lions.

d. Walrus and seals.

SHORT ANSWER QUESTIONS Answers can be found at the end of the book.

12. Define a clade, and explain how clades are depicted in a phylogenetic tree.

13. What does the order of terminal nodes in a phylogenetic tree tell us? Why does changing the order of terminal nodes affect the way someone might interpret the phylogeny?

14. How can including fossils in phylogenies of extant (living) taxa affect the conclusions scientists can draw?

15. Do you consider *Tiktaalik* to be a missing link in the evolution of tetrapods? Why or why not?

16. How can scientists determine the timing of branching events in a phylogenetic tree?

17. Do you agree with how time is portrayed in the phylogeny illustrated below? Why or why not?

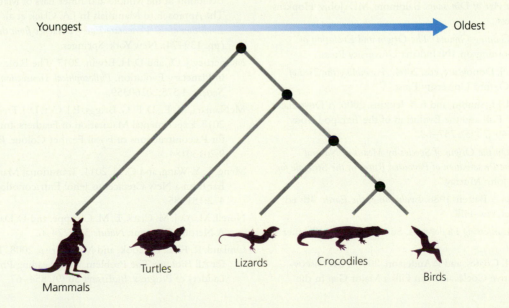

ADDITIONAL READING

Brooks, D. R., and D. A. McLennan. 1991. *Phylogeny, Ecology, and Behavior: A Research Program in Comparative Biology.* Chicago: University of Chicago Press.

Coyne, J. A. 2009. *Why Evolution Is True.* New York: Viking.

Dawkins, R. 2004. *The Ancestor's Tale: A Pilgrimage to the Dawn of Evolution.* Boston: Houghton Mifflin.

Gee, H. 1999. *In Search of Deep Time: Beyond the Fossil Record to a New History of Life.* New York: Free Press.

Gregory, T. 2008. Understanding Evolutionary Trees. *Evolution: Education and Outreach* 1:121–37.

Lecointre, G. 2006. *The Tree of Life: A Phylogenetic Classification.* Cambridge, MA: Belknap Press of Harvard University Press.

Omland, K. E., L. G. Cook, and M. D. Crisp. 2008. Tree Thinking for All Biology: The Problem with Reading Phylogenies as Ladders of Progress. *BioEssays* 30 (9): 854–67.

Shubin, N. 2008. *Your Inner Fish: A Journey into the 3.5-Billion-Year History of the Human Body.* New York: Pantheon Books.

Xu, X., Z. Zhou, R. Dudley, S. Mackem, C. M. Chuong, et al. 2014. An Integrative Approach to Understanding Bird Origins. *Science* 346 (6215): 1253293.

PRIMARY LITERATURE CITED IN CHAPTER 4

Agnarsson, I., M. Kuntner, and L. J. May-Collado. 2010. Dogs, Cats, and Kin: A Molecular Species-Level Phylogeny of Carnivora. *Molecular Phylogenetics and Evolution* 54 (3): 726–45.

Anthwal, N., L. Joshi, and A. S. Tucker. 2013. Evolution of the Mammalian Middle Ear and Jaw: Adaptations and Novel Structures. *Journal of Anatomy* 222:147–60.

Avise, J. C. 2006. *Evolutionary Pathways in Nature: A Phylogenetic Approach.* Cambridge: Cambridge University Press.

Baum, D., and S. Smith. 2012. *Tree Thinking: An Introduction to Phylogenetic Biology.* Greenwood Village, CO: Roberts and Company.

Brusatte, S. L. 2018. *The Rise and Fall of the Dinosaurs: A New History of a Lost World.* New York: William Morrow.

Brusatte, S. L., J. K. O'Connor, and E. D. Jarvis. 2015. The Origin and Diversification of Birds. *Current Biology* 25:R888–98.

Cardillo, M., G. M. Mace, K. E. Jones, J. Bielby, O. R. P. Bininda-Emonds, et al. 2005. Multiple Causes of High Extinction Risk in Large Mammal Species. *Science* 309 (5738): 1239–41.

Chiappe, L. M., and M. Qingjin. 2016. *Birds of Stone: Chinese Avian Fossils from the Age of Dinosaurs.* Baltimore, MD: Johns Hopkins University Press.

Clack, J. A. 2002. *Gaining Ground: The Origin and Evolution of Tetrapods.* Bloomington, IN: Indiana University Press.

Cracraft, J., and M. J. Donoghue, eds. 2004. *Assembling the Tree of Life.* Oxford: Oxford University Press.

Daeschler, E. B., N. H. Shubin, and F. A. Jenkins. 2006. A Devonian Tetrapod-Like Fish and the Evolution of the Tetrapod Body Plan. *Nature* 440 (7085): 757–63.

Darwin, C. 1859. *On the Origin of Species by Means of Natural Selection, or, the Preservation of Favoured Races in the Struggle for Life.* London: John Murray.

Dott, R. H., and R. L. Batten. 1988. *Evolution of the Earth.* 4th ed. New York: McGraw-Hill.

Felsenstein, J. 2004. *Inferring Phylogenies.* Sunderland, MA: Sinauer Associates.

Friedman, M., M. I. Coates, and P. Anderson. 2007. First Discovery of a Primitive Coelacanth Fin Fills a Major Gap in the Evolution of Lobed Fins and Limbs. *Evolution & Development* 9 (4): 329–37.

Gregory, T. R. 2008. Understanding Evolutionary Trees. *Evolution: Education and Outreach* 1 (2): 121–37.

Heers, A. M., D. B. Baier, B. E. Jackson, and K. P. Dial. 2016. Flapping Before Flight: High Resolution, Three-Dimensional Skeletal Kinematics of Wings and Legs During Avian Development. *PLoS ONE* 11 (4): e0153446.

Li, Q., K-Q. Gao, J. Vinther, M. D. Shawkey, J. A. Clarke, et al. 2010. Plumage Color Patterns of an Extinct Dinosaur. *Science* 327 (5971): 1369–72.

Lü, J., and S. L. Brusatte. 2015. A Large, Short-Armed, Winged Dromaeosaurid (Dinosauria: Theropoda) from the Early Cretaceous of China and Its Implications for Feather Evolution. *Scientific Reports* 5:11775.

Luo, Z-X. 2011. Developmental Patterns in Mesozoic Evolution of Mammal Ears. *Annual Review of Ecology, Evolution, and Systematics* 42:355–80.

Luo, Z-X., J. A. Schultz, and E. G. Ekdale. 2016. Chapter 6: Evolution of the Middle and Inner Ears of Mammaliaforms: The Approach to Mammals In J.A. Clack et al. (eds.), *Evolution of the Vertebrate Ear—Evidence from the Fossil Record* (pp. 139–74). New York: Springer.

McInerney, J. O., and D. H. Erwin. 2017. The Role of Public Goods in Planetary Evolution. *Philosophical Transactions of the Royal Society A* 375:20160359.

McNamara, M. E., D. E. G. Briggs, P. J. Orr, D. J. Field, and Z. Wang. 2013. Experimental Maturation of Feathers: Implications for Reconstructions of Fossil Feather Colour. *Biology Letters* 9:20130184.

Meng, J., Y. Wang, and C. Li. 2011. Transitional Mammalian Middle Ear from a New Cretaceous Jehol Eutriconodont. *Nature* 472:181–85.

Norell, M. A., J. M. Clark, L. M. Chiappe, and D. Dashzeveg. 1995. A Nesting Dinosaur. *Nature* 378:774–6.

Omland, K. E., L. G. Cook, and M. D. Crisp. 2008. Tree Thinking for All Biology: The Problem with Reading Phylogenies as Ladders of Progress. *BioEssays* 30 (9): 854–67.

Shipman, P. 1998. *Taking Wing: Archaeopteryx and the Evolution of Bird Flight*. New York: Simon & Schuster.

Shubin, N. H., E. B. Daeschler, and F. A. Jenkins. 2014. Pelvic Girdle and Fin of *Tiktaalik roseae*. *Proceedings of the National Academy of Sciences USA* 111:893–99.

Shubin, N. H., B. D. Edward, and A. J. Farish. 2006. The Pectoral Fin of *Tiktaalik roseae* and the Origin of the Tetrapod Limb. *Nature* 440:764–71.

Syvanen, M. 2012. Evolutionary Implications of Horizontal Gene Transfer. *Annual Review of Genetics* 46:341–58.

Warren, W. C., L. W. Hillier, J. A. M. Graves, E. Birney, C. P. Ponting, et al. 2008. Genome Analysis of the Platypus Reveals Unique Signatures of Evolution. *Nature* 453 (7192): 175–83.

Zimmer, C. 1998. *At the Water's Edge: Macroevolution and the Transformation of Life*. New York: Free Press.

———. 2011. *A Planet of Viruses*. Chicago: University of Chicago Press.

CHAPTER 5

Raw Material

Heritable Variation Among Individuals

Learning Objectives

- Describe the structure that proteins can take.
- Compare and contrast DNA regulation in eukaryotes and in bacteria and archaea.
- Map the events that occur during transcription and translation.
- Discuss mechanisms that influence gene expression.
- Explain the function of coding and noncoding segments of DNA.
- Differentiate between somatic mutations and germline mutations and their roles in variation within a population.
- Explain the roles that independent assortment and genetic recombination play in evolution.
- Discuss vertical and horizontal gene transfer.
- Discuss the complex relationship between genotypes and phenotypes.
- Explain the role of the environment in gene expression.
- Discuss how an organism's genome reflects its evolutionary history.
- Explain why the link between the genotype and phenotypic traits is far more complicated than the genetic polymorphisms that Mendel studied.

Harvard geneticist Joel Hirschhorn studies how tall people are. Height might seem like the simplest thing a biologist could investigate. Hirschhorn doesn't need lasers to probe the inner structure of cells. He doesn't need high-speed video cameras to capture a thousand frames a second of a bat in flight. To measure someone's height, all he needs is a tape measure.

Yet beyond the apparent simplicity of height lies a hidden world of complexity. Humans vary tremendously in the height they reach as adults. In any population of people, some individuals are very short and others are very tall. The average height varies from one population to another. In central Africa,

Bao Xishun (standing 2.36 m tall, right) walks with his wife Xia Shujian. Scientists have identified hundreds of genes that can influence how tall people grow.

pygmies rarely reach more than 1.5 meters. In Denmark, men grow to an average height of 1.8 meters. The average height in many countries has gradually increased over the past century. Explaining all this variation is a big challenge. To address it, Hirschhorn collaborates with hundreds of scientists around the world, studying tens of thousands of individuals. But despite decades of work, they're still a long way from thoroughly explaining height.

It turns out that how tall you are depends on many things. For example, the environment where you grew up influenced your height—including the food you ate and even the chemistry of your mother's uterus while she was pregnant. The genes you inherited from your parents also influenced your height. Short parents tend to have short children, and tall children tend to be born to tall parents.

Hirschhorn searches for the genes that influence height. He and his colleagues examine the DNA of their research participants to find versions of genes that are more likely to be found in tall people or in short ones. In a 2007 study of 4921 people, Hirschhorn (**Figure 5.1**) and his colleagues discovered for the first time a gene with a strong effect on height, called *HMGA2* (Weedon et al. 2007). Carrying two copies of a particular allele of *HMGA2* increases people's height by a centimeter on average. Since then, Hirschhorn and his colleagues have cast a far wider net. As of 2017, they had discovered about 800 loci that influence height (Marouli et al. 2017). Now Hirschhorn and his colleagues are studying the genetic variation they've discovered to see how it gives rise to a range of heights in humans.

This kind of understanding of heredity and variation is a far cry from Charles Darwin's thinking. To make sense of variation, Darwin did the best he could as a naturalist working in the mid-1800s. He skinned rabbits and lined up their bones on his billiard table to measure their lengths. He spent hours in his hothouse observing the variation in orchids and other plants. He hunched down over a microscope to observe the anatomy of barnacles, noting the variations in size and shape within a single species. Despite the technological limitations of his time, Darwin knew enough about heredity and variation to make them cornerstones in his theory of evolution. But since Darwin, scientists have worked out many of the molecular mechanisms involved in heredity and variation, giving evolutionary biology a new foundation of genetics (Zimmer 2018).

In this chapter, we explore this foundation. Some of the topics we discuss will be familiar to you from introductory biology classes, but here we have a different focus: we will examine the aspects of genetics that are most relevant to understanding evolution at the molecular level.

We'll begin by surveying some of the chief components of the cell—proteins, DNA, and RNA. Then we examine how DNA encodes proteins and other molecules and how genes are expressed. Next, we'll take a look at the elements of the genome beyond protein-coding genes, such as regulatory elements, noncoding RNA, and mobile elements.

We'll then examine how genetic information is passed down from one generation to the next and how mutations can alter that information. We consider how this genetic variation gives rise to variation in traits and how genetic variation interacts with the environment. After exploring all of these concepts, we will turn in the next chapter to the ways that variants of genes become more or less common over time—which is the essence of evolution. ●

Figure 5.1 Joel Hirschhorn of Harvard University and his colleagues study DNA from tens of thousands of individuals to figure out why some people are tall and some are short. (Joel Hirschhorn)

5.1 Evolution's Molecules: Proteins, DNA, and RNA

The human body is made up of an estimated 30 trillion cells, each of them made up of millions of molecules (Bianconi et al. 2013). Three kinds of molecules are especially important for evolution: proteins, DNA, and RNA.

Proteins

When you look at a cell through a microscope, you are looking, for the most part, at **proteins**. Most of the dry weight of a cell is composed of proteins. Not only do proteins give the cell much of its structure but they also carry out many of the chemical reactions essential for life. Some proteins act as enzymes, for example, breaking down molecules in the food we eat. Other proteins can carry important molecules, such as hemoglobin, which ferries oxygen in the blood. Proteins can deliver information, relaying signals within a cell or between cells. After you eat a meal, for example, the level of glucose in your blood rises, leading pancreas cells to secrete the protein insulin into the bloodstream.

All told, an estimated 100,000 kinds of proteins are made in the human body. Although proteins come in an astounding number of forms, all of them are constructed from the same building blocks. These building blocks, known as **amino acids**, can be joined together end to end to form long chains (**Figure 5.2**). There are 20 different

A **protein** is an essential macromolecule for all known forms of life. Proteins are three-dimensional biological polymers constructed from a set of 20 different monomers called amino acids.

An **amino acid** is the structural unit that, among other functions, links together to form proteins.

A PRIMARY STRUCTURE   Amino acid sequence

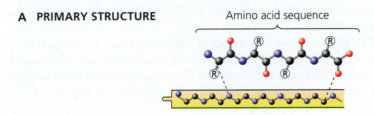

B SECONDARY STRUCTURE (two forms are illustrated)

Beta-pleated sheet   Alpha helix

C TERTIARY STRUCTURE

One subunit of hemoglobin

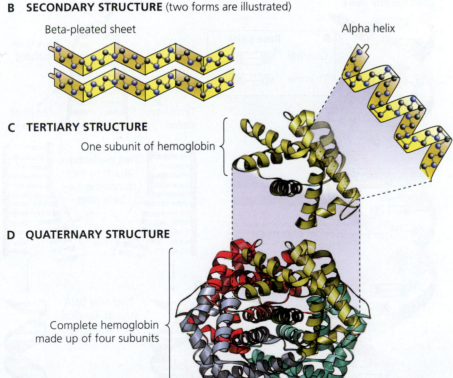

D QUATERNARY STRUCTURE

Complete hemoglobin made up of four subunits

Figure 5.2 Proteins are chains of building blocks known as amino acids. Each protein folds in on itself to produce a new, secondary structure, such as the beta-pleated sheet and alpha helix shown here. These proteins can then bend further into more complex shapes, called the tertiary structure, or join with other proteins into even larger structures (quaternary structure). The structure shown in D is hemoglobin, a protein that carries oxygen in red blood cells.

amino acids that all living things use to build their proteins, and the particular sequence of amino acids in a protein determines the protein's function.

Some amino acids are attracted to surrounding water molecules, whereas others are repelled. These forces cause a chain of amino acids to fold spontaneously into a complex, three-dimensional structure. Depending on a protein's sequence, it may fold into a sheet, a cylinder, or some other shape that allows it to carry out its function. Hemoglobin, for example, folds into a shape that enables it to carry oxygen in red blood cells. The information for assembling the sequence leading to that specific structure is stored in **deoxyribonucleic acid (DNA)**.

DNA

DNA serves as a kind of cookbook for the cell. It stores recipes for each of the cell's proteins as well as for other molecules like **ribonucleic acid (RNA)**. DNA, like a protein, is a linear molecule made up of a limited set of building blocks. However, instead of being made of amino acids, DNA is composed of compounds called **nucleotides**. One end of a nucleotide links to other nucleotides to form a backbone for DNA. The other end of a nucleotide, known as a **base**, helps store the information necessary to build proteins and other molecules.

DNA contains four different bases: adenine, cytosine, guanine, and thymine (A, C, G, and T for short). You can think of these bases as the ingredients that spell out different genetic recipes of life.

It's amazing enough that DNA can encode the information required to build the cells and tissues in an entire organism. But more amazing still is the ability cells have to replicate all that information with almost perfect precision. This fidelity is due to the chemical composition of DNA.

A molecule of DNA consists of two strings of nucleotides that wind together to form a double helix. Each base on a strand binds weakly to a base on the other strand, holding the two together. This pairing follows two simple yet powerful rules: adenine can bind only with thymine, and guanine can bind only with cytosine (**Figure 5.3**).

Deoxyribonucleic acid (DNA) is an essential macromolecule for all known forms of life (along with RNA and proteins). DNA is a double-stranded, helical nucleic acid molecule capable of replicating and determining the inherited structure of a cell's proteins.

Ribonucleic acid (RNA) is an essential macromolecule for all known forms of life (along with DNA and proteins). RNA differs structurally from DNA in having the sugar ribose instead of deoxyribose and in having the base uracil (U) instead of thymine (T).

A **nucleotide** is the structural unit that links together to form DNA (and RNA). Each nucleotide includes a sugar (like deoxyribose or ribose) and a base.

Base refers to one of four nitrogen-containing molecules in DNA: adenine (A), cytosine (C), guanine (G), and thymine (T). (In RNA, uracil (U) replaces T.)

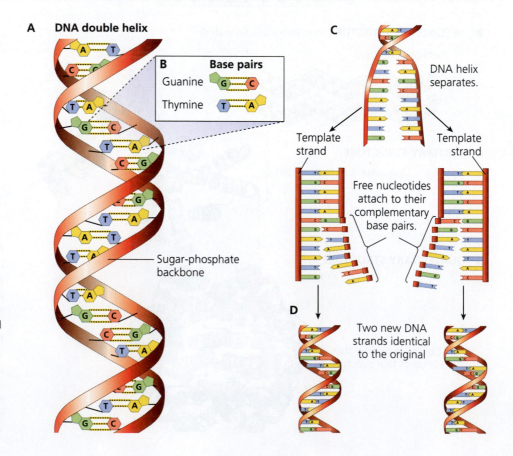

Figure 5.3 DNA structure and replication: A: DNA consists of two phosphate backbones joined by a series of bases. B: Each base can bind to only one other kind (A with T, and G with C). C: When a cell divides, the strands of DNA are separated and new complementary strands are added. D: As a result, the two new DNA strands are identical to the original one.

Thanks to these rules, the sequence of bases along one strand of DNA is perfectly matched to a complementary sequence on the other. If, for example, a part of one of the strands has a base sequence of TGTGCCGATATG, we can be sure that its complement strand has the mirror sequence of ACACGGCTATAC. Although these two strands look different, they contain the same information.

Every time a cell divides, it makes a new copy of its DNA. This process of duplication begins when the two strands of the molecule are pulled apart. Each strand can then serve as a template for the formation of a new complementary strand. When the original strands have been paired with a complementary strand, there are two complete double-stranded copies of the DNA, and the coded information contained in the sequence of bases is preserved.

In some respects, DNA is a lot like the digital storage medium in a computer or a smartphone. In a computer hard drive, information is stored in a string of bits, which can be either one or zero. In DNA, there are four states for each position. As long as an organism can ensure that it copies each position faithfully, it can reproduce the entire genetic cookbook for a new organism.

On rare occasions, however, errors are introduced during copying. These mistakes, known as **mutations**, are then transmitted each time the cell's descendants make new copies of their DNA. Mutations may thus generate heritable changes to the sequence of bases in molecules of DNA. Depending on where a mutation arises and on what form it takes, a mutation can cause a wide range of changes to the organisms that carry it. Mutations can alter the structure, physiology, or behavior of organisms—and they can be deadly or benign. The gradual accumulation of mutations within populations is the ultimate source of heritable genetic variation, the raw material that is essential for evolution. (Later in this chapter, we consider mutations in more detail.)

Just about all living things use DNA as their genetic material. (The only exceptions are RNA viruses, which hijack DNA-bearing cells to replicate themselves.) The fact that DNA is a nearly universal molecule for heredity is a compelling piece of evidence that all living things share a common ancestor. As we'll see in Chapter 8, DNA's universality also allows us to reconstruct some of the relationships between the great lineages of species on Earth. Such studies reveal three main lineages, or domains (see Figure 3.16): eukaryotes, bacteria, and archaea. Eukaryotes have some important differences from bacteria and archaea in the way their DNA is organized, stored, used, and replicated. Let's look at eukaryotic DNA.

Eukaryotic DNA is tightly coiled around a series of histone proteins so that it is arranged much like beads on a string (**Figure 5.4**). This coiling allows a long DNA molecule to be packed inside the nucleus of a cell. (The DNA in a single human cell

A **mutation** is any change to the genomic sequence of an organism.

Figure 5.4 Nuclear DNA is organized into chromosomes in which the molecule is wound around spool-like proteins called histones. By winding and unwinding DNA around histones, cells can expose or hide genes.

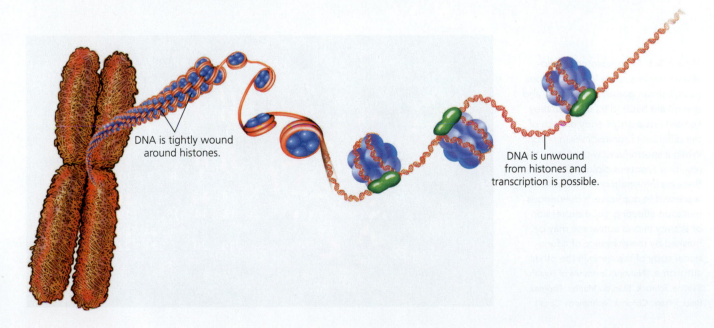

DNA is tightly wound around histones.

DNA is unwound from histones and transcription is possible.

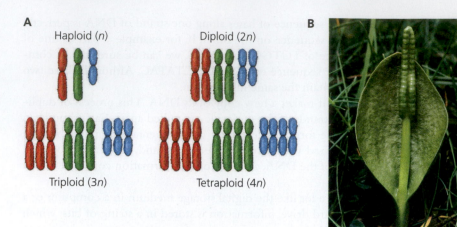

Figure 5.5 A: Chromosomes can occur alone, in duplicate, triplicate, or even higher combinations of homologous sets (ploidy). B: The adder's tongue, *Ophioglossum*, is a relative of ferns that has extreme polyploidy. It can have hundreds of copies of each chromosome. (NHPA/Superstock, Inc.)

A

Haploid (*n*)

Diploid (*2n*)

Triploid (*3n*)

Tetraploid (*4n*)

B

would measure about 2 meters long if it were stretched out completely, and if it were scaled up to the thickness of a human hair, it would extend all the way to the Moon.) The coiling of eukaryotic DNA also affects how cells "read" their genes. Depending on how DNA is wrapped around histone proteins, a particular gene may be hidden away inside a coil, or it may be accessible to the cell's gene-reading proteins.

The information encoded in DNA is stored in multiple copies, providing several levels of redundancy. The double-helix structure of DNA is redundant, for example, because each strand can be used as a template to form a complete copy of the other. But DNA in eukaryotes is also organized into even larger structures, tightly bundled rods known as chromosomes, and chromosomes can come in multiple copies, too (**Figure 5.5**). Chromosomes can occur alone, in duplicate, triplicate, or even higher combinations of homologous sets (**ploidy**). Because information in chromosomes is present in duplicate, a deleterious mutation affecting gene expression or activity in one chromosome may be masked by the presence of a functional copy of the gene in the other.

Almost all human chromosomes belong to nearly identical homologous pairs, providing duplicate copies of almost the entire genome (**Figure 5.6**). In women, all 23 chromosomes form matching pairs, which include two X chromosomes. In men, however, 22 pairs match, but the remaining two chromosomes are a single X chromosome and another one called the Y chromosome. This last pair of chromosomes is called the **sex chromosomes**, whereas the remaining chromosomes are called **autosomes**.

Ploidy refers to the number of copies of unique chromosomes in a cell (*n*). Normal human somatic cells are diploid (*2n*); they have two copies of 23 chromosomes.

A **sex chromosome** is a chromosome that pairs during meiosis but differs in copy number between males and females. For organisms such as humans with XY sex determination, X and Y are the sex chromosomes. Females are the homogametic sex (XX) and males are the heterogametic sex (XY).

An **autosome** is a chromosome that does not differ between sexes.

Figure 5.6 Chromosomes in diploid organisms come in homologous pairs. Human gamete cells (eggs and sperm) are haploid (*n*) because they contain just a single copy of each of the different chromosomes (*n* = 23). When a sperm fuses with an egg, the resulting zygote is diploid (*2n* = 46). Because information in chromosomes is present in duplicate, a deleterious mutation affecting gene expression or activity in one autosome may be masked by the presence of a functional copy of the gene in the other autosome. (National Institutes of Health/ Eveline Schrock, Stan du Manoir, Thomas Reid; Filters: Chroma Technology Corp.)

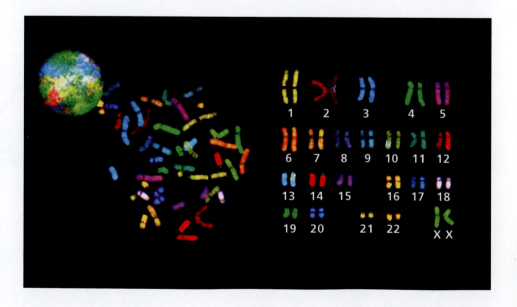

RNA

On the one hand, we have genetic information stored in the bases of DNA. On the other hand, we have thousands of different kinds of proteins, each with a distinctive sequence of amino acids. How do we get from one to the other? This process requires a series of steps that begins with a cell untwisting a segment of DNA that includes a region encoding a protein—in other words, a protein-coding **gene**. The cell can then use the information in the gene to produce a protein in a process known as **gene expression**.

To initiate the process of reading a gene, a number of proteins land on a DNA molecule at a particular spot near the beginning of the gene sequence, which is known as a promoter region. The proteins then move methodically from the "upstream" end of the gene to the "downstream" end. As the proteins travel along the gene, an enzyme called **RNA polymerase** assembles a new string of nucleotides whose sequence matches that of the template DNA in a process called **transcription**. The single-stranded molecule produced by RNA polymerase is known as ribonucleic acid (RNA).

As we'll see later in our discussion, RNA has many functions in the cell. Here we will focus on its role in gene expression. The RNA that is produced by RNA polymerase is known as **messenger RNA** (or **mRNA** for short). It has that name because it carries information from genes to the protein-building factories of the cell like a messenger (**Figure 5.7**).

RNA polymerase continues to build an mRNA molecule until it reaches a specific sequence of bases in the gene that tells it to stop. Once this process is complete, the mRNA molecule can serve as a template for the construction of a protein. This process is called **translation**. Just as a translator converts words from one language to another, cells convert instructions from a language based on bases of DNA to a language based on the amino acids that make up proteins. The set of rules governing the translation of particular base pair sequences into specific amino acids is known as the genetic code (**Figure 5.8**). The four bases in the genetic code can specify 20 different amino acids.

The translation of a gene takes place inside a ribosome, a cluster of proteins and ribosomal RNA molecules. A ribosome binds to the newly formed mRNA molecule and works its way down the sequence of bases. The ribosome grabs three bases at a time. Each trio of bases—which is referred to as a codon—encodes a different

A **gene** is a segment of DNA whose nucleotide sequences code for proteins or RNA or regulate the expression of other genes.

Gene expression is the process by which information from a gene is transformed into a product.

RNA polymerase is the enzyme that builds the single-stranded RNA molecule from the DNA template during transcription.

Transcription is the process that takes place when RNA polymerase reads a coding sequence of DNA and produces a complementary strand of RNA, called messenger RNA (mRNA).

Messenger RNA (mRNA) consists of molecules of RNA that carry genetic information from DNA to the ribosome, where it can be translated into protein.

Translation is the process that takes place when a strand of mRNA is decoded by a ribosome to produce a protein.

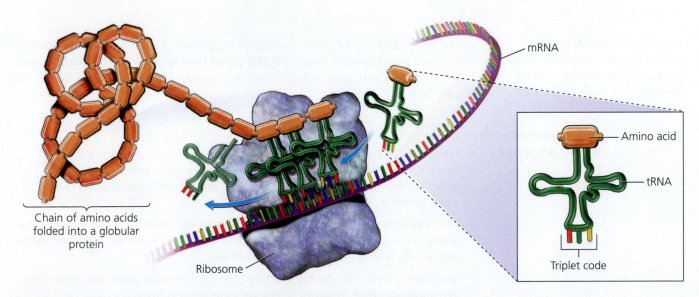

Figure 5.7 Translation. Each messenger RNA molecule acts as a template for building a protein. The ribosome reads three bases at a time (a codon), and a transfer RNA "hooks on" the correct amino acid.

Second base in codon					
	U	**C**	**A**	**G**	
U	UUU = Phenylalanine UUC = Phenylalanine UUA = Leucine UUG = Leucine	UCU = Serine UCC = Serine UCA = Serine UCG = Serine	UAU = Tyrosine UAC = Tyrosine UAA = STOP UAG = STOP	UGU = Cysteine UGC = Cysteine UGA = STOP UGG = Tryptophan	U C A G
C	CUU = Leucine CUC = Leucine CUA = Leucine CUG = Leucine	CCU = Proline CCC = Proline CCA = Proline CCG = Proline	CAU = Histidine CAC = Histidine CAA = Glutamine CAG = Glutamine	CGU = Arginine CGC = Arginine CGA = Arginine CGG = Arginine	U C A G
A	AUU = Isoleucine AUC = Isoleucine AUA = Isoleucine AUG = Methionine START	ACU = Threonine ACC = Threonine ACA = Threonine ACG = Threonine	AAU = Asparagine AAC = Asparagine AAA = Lysine AAG = Lysine	AGU = Serine AGC = Serine AGA = Arginine AGG = Arginine	U C A G
G	GUU = Valine GUC = Valine GUA = Valine GUG = Valine	GCU = Alanine GCC = Alanine GCA = Alanine GCG = Alanine	GAU = Aspartic acid GAC = Aspartic acid GAA = Glutamic acid GAG = Glutamic acid	GGU = Glycine GGC = Glycine GGA = Glycine GGG = Glycine	U C A G

(Left vertical label: First base in codon. Right vertical label: Third base in codon.)

Figure 5.8 The genetic code is used to translate RNA to proteins. "U" stands for uracil, a base in RNA that corresponds to thymine in DNA.

Transfer RNA (tRNA) is a short piece of RNA that physically transfers a particular amino acid to the ribosome.

amino acid, which can be added to the growing protein. Amino acids float around the cell, bound to yet another form of RNA called **transfer RNA** (**tRNA**). Each transfer RNA binds to a particular codon, delivering a particular amino acid to the protein (Figure 5.7).

Key Concepts

- Proteins serve a variety of functions within an organism, and changes to their sequence can affect cell structure, the ability to carry out chemical reactions, the ability to carry information from cell to cell, or even the ability to respond to another signal molecule.

- DNA functions as the basis of the system that encodes and replicates information necessary to build life. Although the replication process is astonishingly faithful, mutations occasionally lead to variation among individuals.

- Mutations to DNA can alter the structure of proteins because genes are transcribed into messenger RNA, which replicates the information in DNA.

- Mutations affecting noncoding forms of RNA, such as ribosomal RNA and transfer RNA, can affect the translation and expression of genes. ●

5.2 Gene Regulation

Humans have about 20,000 protein-coding genes. Scientists have found that about 3800 of those genes are continually expressed in all cells (Eisenberg and Levanon 2013). These so-called housekeeping genes encode proteins, such as RNA polymerase, that all cells need to make at a steady rate to survive. But the remaining 16,200 or so genes are required in some tissues at certain times—though not in others. The cells that produce our hair do not produce hemoglobin; likewise, our blood cells do not express the keratin proteins found in our hair.

Our cells express different combinations of genes thanks to a complex interaction of molecules (Table 5.1). The first type of regulation to be carefully documented targets the transcription of DNA into RNA. As we saw earlier, RNA polymerase and other molecules must attach to a site upstream of the protein-coding region of a gene to commence transcribing it. That upstream section, known as the **gene control region**, may also contain small segments of DNA where molecules can bind. If a protein called a **repressor** attaches to a site in that upstream region, for example, it will block the advance of the transcribing molecules, and the gene will not be expressed. Other proteins, called **transcription factors**, bind to sites called **enhancers** in the gene control region, where they activate gene expression (**Figure 5.9**).

Any given gene control region may contain a number of regulatory regions where different molecules can bind. A cell can thus exercise exquisite control over the precise conditions under which it will express a given gene. Likewise, a cell can also coordinate the expression of many genes at once. Hundreds of genes often carry identical regulatory regions where the same transcription factor can bind. In many cases, those genes encode other transcription factors of their own that switch even more genes on and off. Thus a single protein can trigger the expression of a cascade of genes.

Scientists first discovered gene regulation in the late 1950s. Researchers in France identified a repressor protein that turned off a gene in the bacteria *E. coli*, preventing it from feeding on lactose. And for many years afterward, scientists continued to find other proteins controlling the expression of genes. More recently, however, scientists have come to appreciate that RNA molecules can also regulate gene expression.

Gene control region is an upstream section of DNA that includes the promoter region as well as other regulatory sequences that influence the transcription of DNA.

A **repressor** is a protein that binds to a sequence of DNA or RNA and inhibits the expression of one or more genes.

A **transcription factor** is a protein that regulates the expression of a gene by binding to a specific DNA sequence in association with the gene sequence. A single transcription factor can regulate many genes if they share the same regulatory sequence.

An **enhancer** is a short sequence of DNA within the gene control region where activator proteins bind to initiate gene expression.

Table 5.1 Regulation of Eukaryotic Gene Expression

Stage	Regulatory Mechanism	Consequences
Pre-transcription	**Structural** Coiling or packing of DNA	Can render a gene more or less accessible to RNA polymerase and to cis-regulatory factors necessary for transcription.
Transcription	**Chemical** Methylation of DNA; binding specificities of RNA polymerase, repressors, activators, transcription factors, hormones/signals	Can silence genes by blocking transcription. Can influence when transcription occurs and how much RNA is created.
Post-transcription	Modification of RNA (for example, introns removed, exons spliced)	Can influence how much of the RNA is available for translation.
Translation	Binding of regulatory proteins, antisense RNA (for example, microRNA), or ribosomal subunits	Can influence whether translation is initiated.
Post-translation	Cleavage of amino acid chains, binding of other subunits, phosphorylation	Can alter the structure and function of a protein as well as activate or silence it.

Note: The process of regulating genes in eukaryotes is extremely flexible, and a variety of mechanisms are available at every stage. Mutations affecting any of these regulatory steps can lead to genetic variation for the expression of organismal phenotypes.

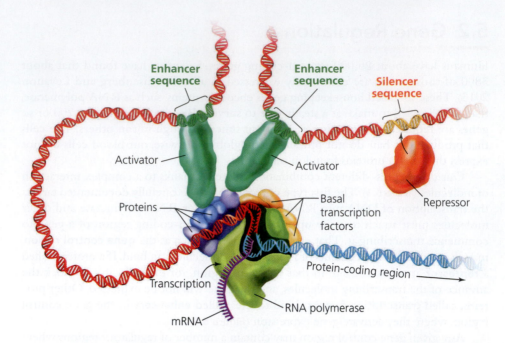

Figure 5.9 Cells can use many mechanisms to regulate when and where particular genes are expressed. Bacteria have small regulatory sequences near protein-coding regions where proteins can bind, promoting or repressing the expression of those genes. Eukaryotic genes have much more complex mechanisms for gene regulation; their activators, transcription factors, and other molecules cooperate to promote a gene's expression.

Enhancer sequence

Enhancer sequence

Silencer sequence

Activator

Activator

Proteins

Basal transcription factors

Repressor

Protein-coding region

Transcription

RNA polymerase

mRNA

MicroRNA describes one group of RNAs that act as post-transcriptional regulators of gene expression. MicroRNAs bind to complementary sequences on specific mRNAs and can enhance or silence the translation of genes. The human genome encodes more than 1000 of these tiny RNAs.

For example, **microRNAs** silence genes by binding to mRNA molecules that would otherwise be translated into proteins (**Figure 5.10**). They act like switches, turning genes on and off in response to changes in the environment. Some RNA molecules help to coordinate the development of embryos. For a human embryo to develop different tissues and organs, for example, certain genes must make proteins in certain cells while other genes must be blocked (Chapter 9).

Transcription factors, microRNAs, and the other regulatory molecules we've considered so far control gene expression within a cell. But other molecules made in one cell can travel to a different cell and control its gene expression. Certain cells release

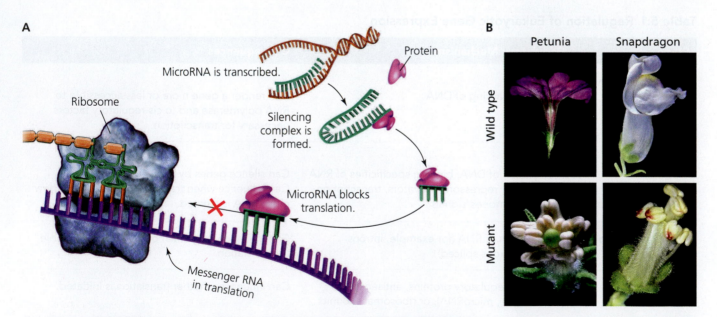

Figure 5.10 A: MicroRNAs are relatively short, but they have powerful effects on gene regulation. They prevent genes from being translated by binding to mRNA, leading to their degradation. MicroRNA molecules can also leave their cell of origin and make their way to distant cells. As a result, they can have widespread effects in animals and plants. B: Shown here are two flowers—petunias and snapdragons—in their normal form (top) and with a mutation to a microRNA gene (bottom). The mutation impairs the flowers' ability to control the identity of its parts. As a result, petals become stamens. (B: MPI for Plant Breeding Research)

hormones, such as adrenalin and estrogen, that travel throughout the body, binding to receptors and triggering a relay of signals that ultimately reach a cell's DNA.

Eukaryotic cells don't just regulate the expression of genes. They can also regulate the kind of protein that is translated from a particular gene. The coding regions of eukaryotic genes are broken up into segments called exons, which are separated from each other by noncoding sections of DNA called introns. After a cell transcribes a gene, it cuts out the introns from the RNA in a process known as **RNA splicing**. The resulting strand of mRNA contains only information from the gene's exons (**Figure 5.11**).

Exons and introns give eukaryotic genes powerful flexibility. Thanks to **alternative splicing**, a eukaryotic cell can combine different subsets of exons from the same gene to produce different combinations and therefore different proteins. In humans, the average gene can be spliced to form one of six different proteins. But some genes can generate thousands of different proteins, each one of them serving a function of its own.

Proteins can be modified even after they are translated. During this post-translational regulation, sugar molecules or lipids may be added to a protein, or it may be cleaved into subunits. It may also undergo chemical reactions that activate it so that it can begin to function.

The regulation of gene expression and protein modification can happen in a matter of seconds. Just think of how quickly a rush of adrenalin affects your entire body. But this regulation can also affect genes and proteins over a much longer timescale. When a single cell develops into an embryo, the cell differentiates into many tissues, inside of which the cells express a distinctive repertoire of genes. The cells these genes give rise to will continue to express their repertoire for decades. Thus our stomach remains a stomach for our lifetime, rather than spontaneously turning into a bone, for example.

Two primary mechanisms are responsible for these long-term changes in gene expression. Methylation involves the addition of methyl groups to DNA. Methyl groups added to an enhancer, for example, can silence a gene. Coiling prevents gene expression because the transcription machinery can't reach the DNA. Once these **epigenetic** modifications are applied to DNA, cells tend to pass them down to daughter cells.

As we'll see in later chapters, mutations altering the regulation of gene expression play a vital role in evolution. For example, changing where or when genes are expressed can lead to dramatic innovations (Chapter 9). Similarly, the evolution of new epigenetic patterns can also lead to different rates of growth in embryos (Chapter 12).

A **hormone** is a molecular signal that flows from cells in one part of the body to cells in other parts of the body. Hormones act directly or indirectly to alter the expression of target genes.

RNA splicing is the process of modifying RNA after transcription but before translation, during which introns are removed and exons are joined together into a contiguous strand.

Alternative splicing is the process of combining different subsets of exons together, yielding different mRNA transcripts from a single gene.

Epigenetic refers to the functional modifications to DNA that don't involve changes to the sequences of nucleotides. Epigenetics is the study of the heritability of these modifications.

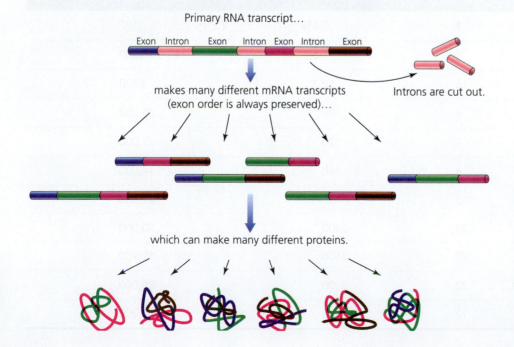

Primary RNA transcript...

Exon Intron Exon Intron Exon Intron Exon

makes many different mRNA transcripts (exon order is always preserved)...

Introns are cut out.

which can make many different proteins.

Figure 5.11 Eukaryotic genes contain introns and exons. A group of proteins called the spliceosome removes introns before translation, and it can remove some exons as well. By regulating which exons are included in the final mRNA transcript, cells can generate many proteins from a single gene. This process is called alternative splicing.

Key Concepts

- Transcription factors, microRNAs, and other regulatory molecules control gene expression within a cell.

- Hormones such as adrenalin and estrogen bind to cell receptors and trigger signals for a cell's DNA.

- Certain genes can generate thousands of different proteins, each serving a different function. ●

5.3 Sizing Up the Genome

A **genome** includes all the hereditary information of an organism. The genome comprises the totality of the DNA, including the coding and noncoding regions.

The entire complement of DNA in a cell is known as its **genome**. Genome size and complexity vary greatly across taxa. Most of the variation comes from differing amounts of mobile genetic elements in the genome (discussed further in Box 5.1). As we can see in **Table 5.2**, the genomes of different species vary tremendously. *Nasuia deltocephalinicola*, a species of bacteria that lives inside of insects, has only 112,000 base pairs in its genome (Bennett and Moran 2013). A spruce tree has 20 billion base pairs of DNA (six times more than in the human genome).

It's easy to assume that a difference in genome size corresponds to a difference in the number of genes it contains. It certainly is true that *N. deltocephalinicola* has very few genes—just about 140. But there's no strong correlation between genome size and gene number across all species. In fact, in many eukaryotes, protein-coding genes make up a tiny portion of the entire genome. Of the 3.5 billion base pairs in the nuclear genome of human DNA, only 1.2% is made up of protein-coding segments (Alexander et al. 2010; van Bakel et al. 2010).

What lies beyond the protein-coding segments? We've already encountered part of the answer to that question. DNA also contains gene control regions, as well as genes for thousands of different RNA molecules including ribosomal RNAs, transfer RNAs, and microRNAs.

Table 5.2 Variation in Genome Size and Complexity

Organism	Number of Chromosomes	Megabases in Genome (Millions of Base Pairs)	Approximate Number of Protein-Coding Genes
N. deltocephalinicola (bacteria)	1	0.112	137
E. coli (bacteria)	1	4.6	4300
S. cerevisiae (yeast)	16	12.1	6700
C. elegans (nematode)	6	100	20,000
A. thaliana (Thale cress)	5	120	27,000
D. melanogaster (fly)	4	180	14,000
N. vectensis (sea anemone)	15	450	27,000
C. familiaris (dog)	39	2400	20,000
M. musculus (mouse)	20	2600	19,900
H. sapiens (humans)	23	3000	20,000
P. abies (Norway spruce)	12	20,000	28,300

Genes and Heredity in Bacteria and Archaea

All living things use DNA or RNA to store genetic information. But eukaryotes are different from bacteria and archaea in the organization of their DNA as well as in how they replicate DNA and pass it down to their offspring (Box Figure 5.1.1).

The name *eukaryote* points to one of the most obvious things that sets eukaryotes apart from other forms of life. In Greek, the term means "true kernel," referring to the nucleus in which eukaryotes keep their DNA tightly coiled. In bacteria and archaea, a single circular chromosome floats within the cell. It is not constrained by a nucleus, nor is it tightly spooled around histones. That does not mean that this DNA is simply a loose tangle. Bacteria and archaea produce proteins that keep sections of DNA organized in twisted loops. Like eukaryotes, bacteria and archaea can regulate gene expression by unwinding and winding their DNA.

Also like eukaryotes, bacteria and archaea have genetic regulatory regions upstream from their genes. Transcription factors can trigger dramatic changes in gene expression through regulatory cascades. Bacteria and archaea can alter their gene expression in response to signals from their environment. When conditions turn stressful, some species respond by making durable spores that can be carried away by water and wind to more favorable places. Others can produce toxins when they sense other microbes competing for resources.

Overall, however, gene regulation is less complex in bacteria and archaea than it is in eukaryotes. Bacteria and archaea lack enhancers, for example, which can be located thousands of base pairs away from genes they control in eukaryotes. Whereas bacteria and archaea have self-splicing introns, eukaryotes have an abundance of introns regulated in a unique process known as alternative splicing. A large group of proteins, called the spliceosome, removes introns from transcripts, potentially producing a vast variety of proteins from the same gene (called isoforms). Bacteria and archaea, by contrast, always produce the same protein from any given gene.

Replication in bacteria and archaea is also simpler. They do not perform mitosis or meiosis. They do not have full-blown sexual reproduction, in which males and females produce gametes that combine in a new offspring. Instead, bacteria and archaea typically grow until they are large enough to divide. They then build a second copy of their circular chromosome, and the two DNA molecules are dragged to either end of the dividing cell. The two daughter cells are identical to the original, except for any mutations that arise during DNA replication.

Bacteria and archaea have many of the same kinds of mutations found in eukaryotes, such as point mutations and insertions. But they cannot acquire genetic variation as a consequence of sexual reproduction the way we see in some eukaryotes (that is, through independent assortment of chromosomes). As we'll see in Chapter 9, this difference

(continued)

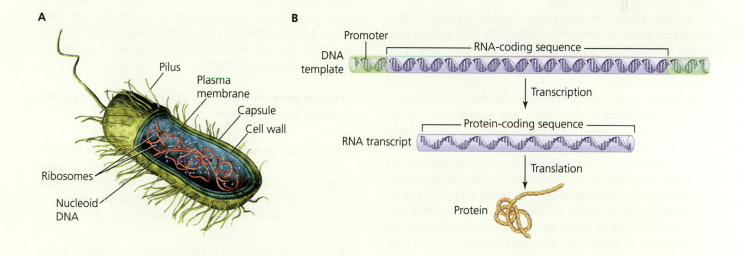

Box Figure 5.1.1 A: Bacteria and archaea differ from eukaryotes, like humans, in that the genetic material in the cell is not contained within a nuclear membrane. B: The process of gene expression and regulation in bacteria and archaea is much simpler than in eukaryotes.

Box 5.1 Genes and Heredity in Bacteria and Archaea (continued)

can have a major effect on how mutations spread through populations.

Beneficial mutations that increase the survival or reproductive rate of bacteria can sweep quickly through a population of microbes, thanks to natural selection. As we'll see in Chapter 6, scientists have used microbes to perform important experiments on evolution, observing natural selection in action. And, as we'll see in Chapter 18, bacteria can rapidly evolve resistance to antibiotics, turning what were once easily treated diseases into serious threats to public health.

One reason antibiotic resistance can spread so quickly is that bacteria are not limited simply to passing down their genes to their descendants (known as **vertical gene transfer**). It's also possible for one individual microbe to "donate" DNA to another, through a process called **horizontal gene transfer** (**Box Figure 5.1.2**).

One mechanism for horizontal gene transfer is via **plasmids**, which are small ringlets of DNA that are separate from the main bacterial chromosome. Under certain conditions, a microbe will translate plasmid genes and assemble a tube called a pilus, which links the cell to a neighboring cell. The donor cell can then pump a copy of the plasmid through the pilus and often pumps a copy of some of its chromosomal DNA as well. In effect, plasmids are genetic parasites, using bacteria and archaea as their hosts and spreading to new hosts through the pili encoded in their own genes. However, they can also carry genes encoding beneficial traits, such as antibiotic resistance, which can provide advantages to their hosts.

Viruses can also carry out horizontal gene transfer. As they replicate, some viruses can accidentally incorporate host genes into their own genome. When they infect a new host, they can insert the genes of the original host into their new host's chromosome.

In many cases, horizontal gene transfer is a dead end. The donated genes are harmful to the recipient cell, which dies or grows too slowly to compete with other individuals. But if a microbe acquires a useful gene, natural selection can favor it. Evidence for successful horizontal gene transfer can be found in studies on the spread of antibiotic resistance: the same gene often turns up in different species. It's also possible to identify cases of horizontal gene transfer that occurred millions of years ago by performing large-scale comparisons of DNA in bacteria and archaea. In a number of cases, scientists have found genes in some species that don't fit with models of vertical gene transfer. The genes are homologous to genes found in distantly related clades but unlike genes found in close relatives. These studies indicate that horizontal gene transfer has been a major element of evolution. In *Escherichia coli*, for example, 80% of all the

genes in its genome show evidence of horizontal gene transfer at some point since the last common ancestor of bacteria (Dagan and Martin 2007). As we saw in Chapter 4, horizontal gene transfer is prompting scientists to revise their concepts of species and of the overall shape of the tree of life.

Compared to bacteria and archaea, eukaryotes appear to have experienced relatively little horizontal gene transfer (Keeling and Palmer 2008). There are a number of possible explanations for this difference. One is opportunity: the complexity of eukaryotic DNA replication may not afford the opportunity to take up foreign genes. In bacteria and archaea, a contiguous set of genes may form a functional unit, called an operon, in which they are all controlled by the same upstream regulatory elements. The entire operon can be inserted into a new host, where it may be able to provide a useful function. Eukaryotes lack operons, however, so that foreign genes are less likely to be useful in a new cell.

More than half of the human genome is composed of neither genes, nor vestiges of human genes, nor regulatory regions. Instead, this segment of the human genome is made up of parasite-like segments of DNA, known as **mobile genetic elements**, with the capacity to make new copies of themselves that can then be reinserted into the genome. Some mobile genetic elements originated as viruses that integrated their genes into the genome of their host. The origins of other mobile genetic elements are more mysterious. Once they become established in their host genome, mobile genetic elements can proliferate into thousands of copies and take up large amounts of space. As seen later in Table 5.2, genomes can vary enormously in size from one species to another, and much of that variation originates from mobile genetic elements.

Vertical gene transfer is the process of receiving genetic material from an ancestor.

Horizontal gene transfer is the transfer of genetic material—other than from parent to offspring—to another organism, sometimes a distantly related one, without reproduction. Once this material is added to the recipient's genome, it can be inherited by descent.

A **plasmid** is a molecule of DNA, found most often in bacteria, that can replicate independently of chromosomal DNA.

A **mobile genetic element** is a type of DNA that can move around in the genome. Common examples include transposons ("jumping genes") and plasmids.

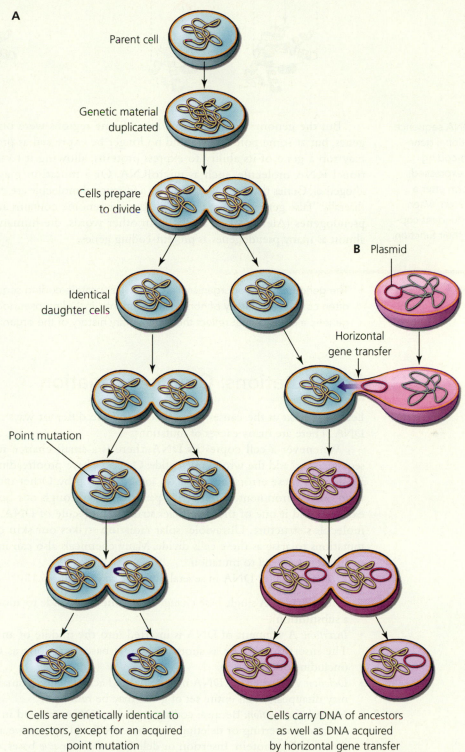

A

Parent cell

Genetic material
duplicated

Cells prepare
to divide

Identical
daughter cells

B Plasmid

Horizontal
gene transfer

Point mutation

Cells are genetically identical to
ancestors, except for an acquired
point mutation

Cells carry DNA of ancestors
as well as DNA acquired
by horizontal gene transfer

Box Figure 5.1.2 Bacteria reproduce by dividing in two. A: As bacteria prepare for division, they separate the two strands of DNA and add a new strand to each one, creating two new DNA molecules, which are typically identical to the original one. B: On rare occasions, however, genes can be passed from one bacterium to another through a process known as horizontal gene transfer.

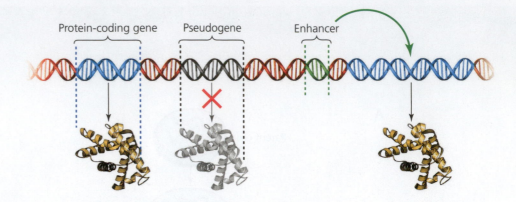

Figure 5.12 Protein-coding genes make up only 1.2% of the human genome. The noncoding portion includes regulatory elements, pseudogenes, and genes encoding functional RNA molecules.

Protein-coding gene   Pseudogene   Enhancer

A **pseudogene** is a DNA sequence that resembles a functional gene but has lost its protein-coding ability or is no longer expressed. Pseudogenes often form after a gene has been duplicated, when one or more of the redundant copies subsequently lose their function.

But the genome contains much more. Some regions were once protein-coding genes, but at some point they could no longer be expressed as proteins. A mutation may rob a gene of its ability to express proteins, allowing it to make only a functional RNA molecule, such as microRNA. Or a mutation may disable the gene altogether. Genes that no longer produce a functional molecule are called **pseudogenes** (literally "false genes"; **Figure 5.12**). The human genome contains an estimated 17,032 pseudogenes (Alexander et al. 2010). In other words, the human genome contains almost as many pseudogenes as protein-coding genes.

Key Concept

- The genomes of most organisms are rarely just protein-coding sequences; they often contain a diversity of noncoding elements, including pseudogenes and mobile genetic elements, that reflect the evolutionary history of the organism. ●

5.4 Mutations: Creating Variation

Let's look next at the causes of mutations and the different ways mutations can alter DNA. There are many causes of mutations.

Whenever a cell copies its DNA, there is a small chance it may misread the sequence and add the wrong nucleotide. Our cells have proofreading proteins that can fix most of these errors, but mistakes sometimes slip by. Other mutations may result from our environment. Radioactive particles pass through our bodies every day, for example, and if one of these particles strikes a molecule of DNA, it can damage the molecule's structure. Ultraviolet solar radiation strikes our skin cells and can cause mutations to arise as these cells divide. Many chemicals also can interfere with DNA replication and lead to mutation.

Mutations alter DNA in several different ways (**Figure 5.13**):

- *Point mutation*: A single base changes from one nucleotide to another (also known as a substitution).
- *Insertion*: A segment of DNA is inserted into the middle of an existing sequence. The insertion may be as short as a single base or as long as thousands of bases (including entire genes).
- *Deletion*: A segment of DNA may be deleted accidentally. A small portion of a gene may disappear, or an entire set of genes may be removed.
- *Frameshift mutation*: Because coding regions of DNA are read in three-base codons, mutations inserting or deleting three bases will add or remove an amino acid from the resulting protein. Insertion or deletion of *one* or *two* bases, on the other hand, will shift the reading frame of the entire sequence, changing the identity of all amino acids coded after—downstream from—the location of the mutation.
- *Duplication*: A segment of DNA is copied a second time. A small duplication can produce an extra copy of a region inside a gene. Entire genes can be duplicated. In some cases, even an entire genome can be duplicated.

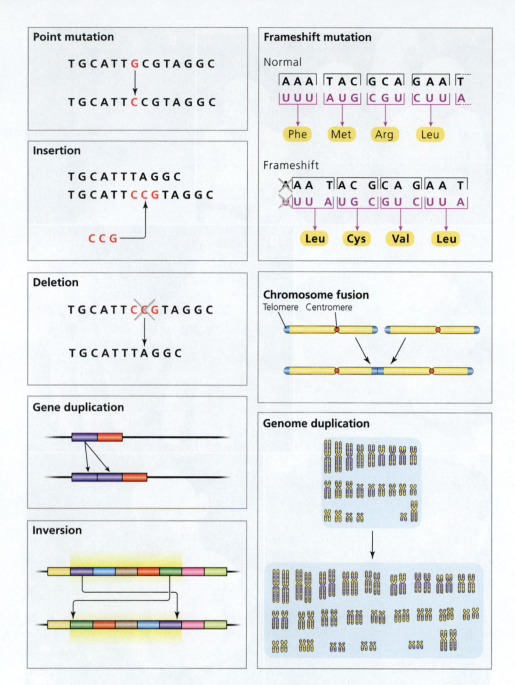

Point mutation

T G C A T T G C G T A G G C

↓

T G C A T T C C G T A G G C

Insertion

T G C A T T T A G G C
T G C A T T C C G T A G G C

C C G

Deletion

T G C A T T C C G T A G G C

↓

T G C A T T T A G G C

Gene duplication

Inversion

Frameshift mutation

Normal

| A A A | T A C | G C A | G A A | T |
| U U U | A U G | C G U | C U U | A |

Phe   Met   Arg   Leu

Frameshift

| A A | T A C | G C A | G A A | T |
| U U | A U G | C G U | C U U | A |

Leu   Cys   Val   Leu

Chromosome fusion

Telomere   Centromere

Genome duplication

Figure 5.13 DNA can undergo several kinds of mutations, such as point mutations, insertions, deletions, and duplications.

- *Inversion*: A segment of DNA is flipped around and inserted backward into its original position.
- *Chromosome fusion*: Two chromosomes are joined together as one.
- *Aneuploidy*: Chromosomes are duplicated or lost, leading to an abnormal number of chromosomes.
- *Genome duplication*: Entire genomes are duplicated, leading to increased ploidy (for example, 4*n*, 8*n*).

Mutations are not necessarily harmful—many are beneficial or even neutral—but all of them result in genetic variation among individuals in a population. Mutations can alter the DNA within the protein-coding region of a gene. An altered coding region may lead to a protein with a different sequence of amino acids, which may fold into a different shape. This change could cause the protein to perform its original activity at a faster or slower rate, or the protein may acquire a different activity (**Figure 5.14**).

Mutations can also have important effects without altering the product of a gene. Instead, they can simply change how much of a protein is made, or they can change

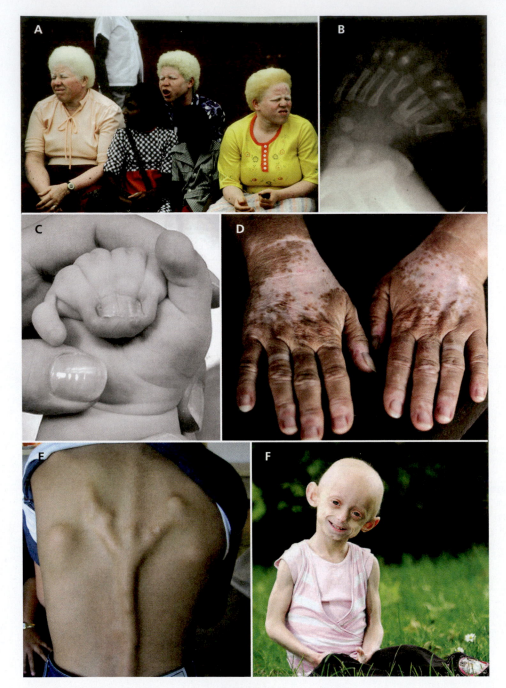

Figure 5.14 Point mutations in human protein-coding genes can have phenotypic effects that range from the benign (for example, eye color) to the severe. Shown here are a variety of striking (but rare) mutations. A: A mutation in the *FGFR3* gene that replaces the amino acid proline with serine at one position is responsible for albinism. (Oetting and King 1993; Carl Purcell / Science Source) B: A single C-to-T transition in the *LMBR1* gene leads to triphalangeal thumb polydactyly. (Wang et al. 2007; Dr. M. A. Ansary / Science Source) C: A missense mutation in one of the exons of the *GLI3* gene, replacing the amino acid proline with serine, leads to Greig's cephalopolysyndactyly. ("The Greig cephalopolysyndactyly syndrome" by Leslie G. Biesecker. Orphanet Journal of Rare Diseases, 2008) D: A missense substitution in the *KIT* gene replacing the amino acid arginine with glycine at one position leads to piebaldism. (Sánchez-Martín et al. 2003; G. Patrick / Science Source) E: A mutation in the *ACVR1* gene substituting the amino acid arginine with histidine at a single position results in fibrodysplasia ossificans progressiva. (Shore et al. 2006; Extensive heterotopic ossification on back of a FOP patient From "Fibrodysplasia ossificans progressiva: clinical and genetic aspects" by Pignolo, Shore, Kaplan. Orphanet Journal of Rare Diseases. 2011. Photo courtesy Dr. Pignolo) F: A point mutation in the *Lamin A* gene causes Hutchinson–Gilford progeria syndrome. (Eriksson et al. 2003; Worldwide Features / Barcroft Media / Getty Images)

Table 5.3 Sources of Heritable Genetic Variation

Location of Mutation	Type of Mutation	Consequence for Gene Action
Coding region	Substitution, insertion, deletion, duplication	Alters the product of the gene and thus its function or activity.
Cis-regulatory regions	Substitution, insertion, deletion, duplication that alters the binding affinity of promoters, activators, repressors, etc.	Alters the timing, location, or level of expression of the gene. Alters the developmental or environmental context in which the gene is expressed.
Trans-regulatory regions	Mutation to coding regions of trans-acting factor Mutation to cis- or trans-regulatory regions of trans-acting factors	Alters the binding affinity and thus the activity of a promoter, activator, repressor, etc. Alters where, when, or to what extent inhibitory, activating, or other trans-acting regulatory factors are expressed.
Physiological pathways (for example, hormones)	Mutations altering where, when, or how much an endocrine signal is produced	Alters the timing, location, or level of expression of the gene. Alters the developmental or environmental context in which the gene is expressed.

the timing or location of its production. Changes in *levels* of gene expression can alter the behavior of cells or tissues and can have profound consequences for evolution as an additional component of heritable variation. Mutations may cause a transcription factor to bind more strongly than it did before. Or they may prevent a specific transcription activator protein from binding so that the gene no longer can be expressed in a particular kind of tissue.

Transcription factors, hormones, and other regulatory molecules are themselves encoded by genes, which means that mutations that alter their genes ultimately can affect the genes they regulate. As a result of these interactions, a gene can be affected by a mutation that is far away from the gene itself. (Nearby elements—on the same chromosome, for example—that affect gene expression are **cis-acting elements**; far-away ones are **trans-acting elements**; see **Table 5.3**.)

The chance of any given DNA base mutating during a single cell division is very low. To measure the average number of mutations arising in a human baby is challenging for several reasons. First, all of our genes are present in duplicate (diploidy), and this means that many mutations that arise in one of the chromosomes will be hidden or masked by a functional copy of the same gene on the sister chromosome. Second, not all mutations that arise in our bodies are transmitted to our offspring. Any individual cell in our body has a chance of mutating as it divides. If it's a skin cell, then the skin cells that descend from it will continue to carry that mutation. But this lineage of cells will come to an end when we die. Such mutations are known as **somatic mutations** because they occur in the "soma," or body.

If, on the other hand, a mutation arises in the line of cells that gives rise to sperm or egg cells, it may be passed on to offspring. And those offspring, in turn, may pass the mutation down to their own descendants. These mutations are known as **germline mutations**. Only germline mutations give rise to heritable variations within populations.

Until recently, scientists could make only very indirect estimates of the germline mutation rate in humans. One common method was to study the rate of diseases that are caused by a mutation of a single gene. But improvements in DNA sequencing have made it possible for scientists to make far more precise estimates.

In 2016, for example, Kári Stefánsson of deCODE Genetics in Iceland and his colleagues sequenced the entire genomes of 283 Icelanders, their mothers, and their fathers. They could then scan each child's genome for mutations that were not present in either parent's DNA (Besenbacher et al. 2016). Stefánsson and his colleagues found 63 point mutations on average in each child. They estimate that the rate of point mutations is 1.29×10^{-8} per position per generation.

A **cis-acting element** is a stretch of DNA located near a gene— immediately upstream (adjacent to the promoter region), downstream, or inside an intron—that influences the expression of that gene. Cis regions often code for binding sites for one or more transposable factors.

A **trans-acting element** is a sequence of DNA located away from the focal gene (for example, on another chromosome). These stretches of DNA generally code for a protein, microRNA, or other diffusible molecule that then influences expression of the focal gene.

A **somatic mutation** is a mutation that affects cells in the body ("soma") of an organism. These mutations affect all the daughter cells produced by the affected cell and can affect the phenotype of the individual. In animals, somatic mutations are not passed down to offspring. In plants, somatic mutations can be passed down during vegetative reproduction.

A **germline mutation** is a mutation that affects the gametes (eggs, sperm) of an individual and can be transmitted from parents to offspring. Germline mutations create the heritable genetic variation that is relevant to evolution.

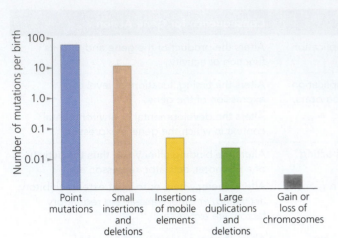

Figure 5.15 Small-scale mutations are much more common than big ones. But even a rare large mutation, like aneuploidy, can change many more base pairs than a lot of point mutations. A: Average number of mutations of each type of variant per birth.

B: Average number of mutated bases contributed by each type of variant per birth. The *y* axis is log10 scaled in both A and B. (Data from Campbell and Eichler 2013)

As **Figure 5.15** shows, point mutations have the highest rate of all categories of mutations in humans. A large duplication (more than 100,000 bases) occurs only once in every 42 live births, for example. But although these mutations may be rarer, their larger size means that they introduce more mutated bases into a population (Campbell and Eichler 2013).

Key Concepts

- Mutations (in germline cells) are not common, especially mutations that affect large parts of the genome, but they can accumulate gradually over time. Within populations, mutations make up the ultimate source of heritable genetic variation, the raw material that is essential for evolution.

- Because information in human chromosomes is present in duplicate, a deleterious mutation affecting gene expression or activity in one chromosome may be masked by the presence of a functional copy of the gene in the other chromosome. (This is not true for haploid organisms like bacteria, which have only a single copy of their genome.)

- Genetic changes in gene expression arise when mutations outside of the protein-coding regions affect where, when, or how much a gene is transcribed.

- Changes in levels of gene expression can have profound consequences for evolution by adding another component to heritable genetic variation. ●

5.5 Heredity

When new mutations arise, organisms can pass this mutated DNA to their offspring along with their unmutated DNA. Bacteria and archaea reproduce by dividing and making a new copy of their genome for each daughter cell (see Box 5.1). In sexually reproducing, multicellular eukaryotes, on the other hand, heredity is more complex. For one thing, only germline cells can pass on genes to offspring, as we saw earlier. For another, the development of sex cells introduces new genetic variation, so sexually reproducing organisms produce genetically unique offspring instead of clones.

One reason that parents do not produce families of identical children is that each parent's own paired chromosomes are not identical. One chromosome may have one version of a gene (an **allele**) and the other chromosome may have an allele with a slightly different DNA sequence. Differences between the chromosomes in each pair generate variation among the gamete cells that are produced, so no two sperm or egg cells are identical.

Gametes of sexually reproducing organisms are produced through a distinctive kind of cell division known as **meiosis**. Meiosis can generate a stunning amount of genetic variation. During meiosis, paired chromosomes come together, and each pair may cross over and exchange segments of DNA in a process known as **genetic recombination** (Figure 5.16). By the time these chromosomes are packed into gamete cells, many of them have already been rearranged so that they differ from either of the parental chromosomes. Recombination occurs randomly every time a germ cell divides to form gametes, so the specific chunks of swapped DNA will differ from

An **allele** refers to one of any number of alternative forms of the DNA sequence of the same locus.

Meiosis, which occurs only in eukaryotes, is a form of cell division that cuts the number of chromosomes in half.

Genetic recombination is the exchange of genetic material between paired chromosomes during meiosis. Recombination can form new combinations of alleles and is a source of heritable variation.

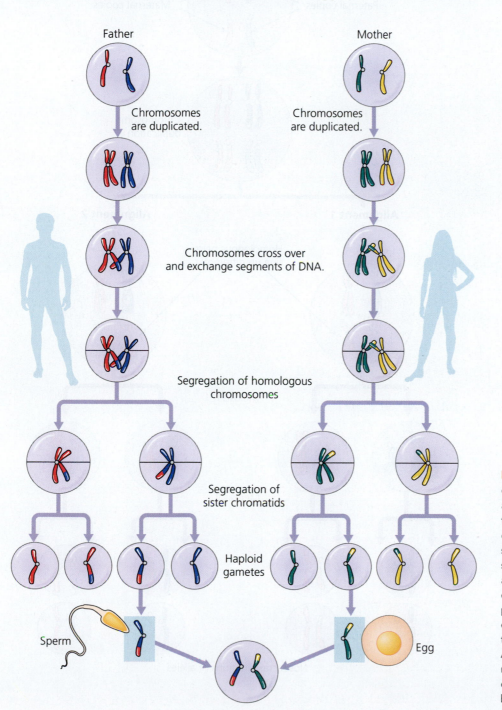

Figure 5.16 Among sexually reproducing organisms, like humans, males and females combine their gametes to reproduce. During the production of gametes, each pair of chromosomes crosses over and exchanges segments of DNA in a process known as genetic recombination. Each gamete receives only one copy from each pair of chromosomes. As a result, each child carries a unique combination of the DNA of his or her parents. Along with independent assortment, recombination increases genetic diversity by producing novel combinations of alleles.

gamete to gamete. A human male, for example, undergoes billions of meiotic cell divisions to generate sperm cells, and all of them will have unique combinations of maternal and paternal DNA along the lengths of their chromosomes.

In addition, during the final stage of meiosis, each pair of chromosomes separates, so that only a single copy ends up in each daughter cell. *Which* copy ends up in each cell is random, and it is random for each of the different chromosomal pairs (**Figure 5.17**). This means that a single sperm cell may inherit the maternal copy of one chromosome and the paternal copy of another. Because *each* chromosome has

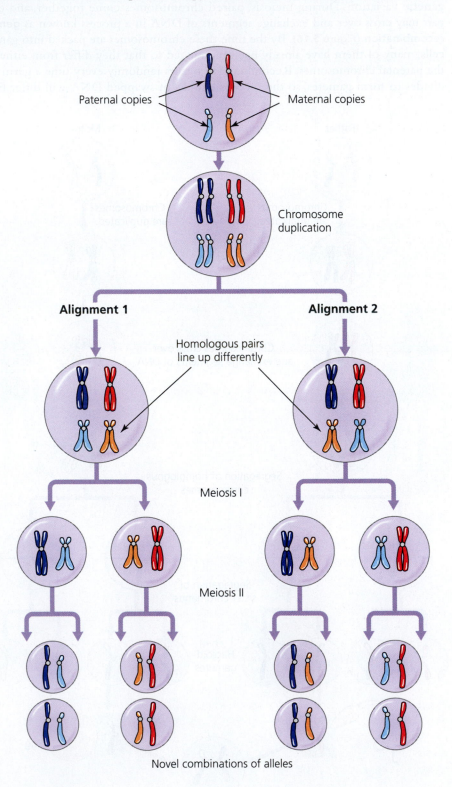

Paternal copies   Maternal copies

Chromosome duplication

Alignment 1   **Alignment 2**

Homologous pairs line up differently

Meiosis I

Meiosis II

Novel combinations of alleles

Figure 5.17 Independent assortment occurs during meiosis in sexually reproducing eukaryotic organisms and produces gametes with a mixture of maternal and paternal chromosomes.

a $50:50$ chance of donating the maternal or paternal copy to an egg or sperm, the number of possible combinations of maternal and paternal chromosomes is 2^N, where N is the number of chromosome pairs. Humans have 23 pairs of chromosomes, and the **independent assortment** of each separate chromosome can result in almost 9 million possible combinations of maternally inherited and paternally inherited genes ($2^{23} = 8,888,608$). Consequently, the processes of recombination and independent assortment can, through the rearrangement of alleles, generate a staggering number of possible gamete genotypes.

This mixing of alleles occurs in the formation of gametes of both the male and the female. Sexual reproduction brings the chromosomal forms of each parent together, creating yet another combination of alleles. Consequently, when a human sperm cell fertilizes an egg, the chromosomes combine to produce a new set of 23 pairs. But this new set is drawn from a rich pool of genetic variants reflecting millions of possible combinations of alleles inherited from both the father and the mother. It is almost impossible for the particular combination that fuses to form the new diploid individual to occur twice. This is why siblings from the same parents can differ in their inherited characteristics. (An obvious exception to that rule is identical twins, which always develop from a single fertilized egg.)

The best estimate of the differences between our paired chromosomes comes from Craig Venter, a genome-sequencing pioneer. In 2007, he and his colleagues published the complete sequence of his own genome (Levy et al. 2007). They compared each pair of chromosomes and tallied the differences. The researchers identified 3.2 million places where a single nucleotide in one chromosome did not match the corresponding nucleotide in its partner. The scientists also found about a million segments of DNA on one chromosome that were missing from its partner or that had been inserted.

Independent assortment is the random mixing of maternal and paternal copies of each chromosome during meiosis, resulting in the production of genetically unique gametes.

Key Concepts

- Because of genetic recombination and the independent assortment of chromosomes, meiosis can generate extraordinary genetic diversity among gametes in sexually reproducing organisms.

- The fusion of egg and sperm results in great genetic diversity among offspring, even in those from the same parents. ●

5.6 The Complex Link Between Most Genotypes and Phenotypes

Scientists draw a distinction between the genetic material in an organism and the traits that the genetic material encodes. The genetic makeup of an organism is known as its **genotype**, and the manifestation of the genotype is known as the **phenotype**. Organisms do not inherit a phenotype; they inherit alleles, which together constitute a genotype, which then gives rise to a phenotype.

Determining how phenotypes emerge from genotypes is no easy task. A trait does not come with a label on it, detailing all the genes that helped to build it and the specific role played by each of the genes. Instead, scientists rely on a variety of methods to explore how genes and gene expression contribute to the formation of organismal phenotypes. These methods range from controlled breeding experiments to detailed genetic mapping studies—even to perturbations of expression of focal developmental genes (Chapter 9 describes many of these methods in greater detail).

The traits that Mendel studied in his peas (see **Box 5.2**) have relatively simple, discrete, alternative phenotypic states. The peas were either wrinkled or smooth, for example, and not anything in between. The traits Mendel studied also turned out

Genotype describes the genetic makeup of an individual. Although a genotype includes all the alleles of all the genes in that individual, the term is often used to refer to the specific alleles carried by an individual for any particular gene.

Phenotype is an observable, measurable characteristic of an organism. A phenotype may be a morphological structure (for example, antlers, muscles), a developmental process (for example, learning), a physiological process or performance trait (for example, running speed), or a behavior (for example, mating display). Phenotypes can even be the molecules produced by genes (for example, hemoglobin).

BOX 5.2

Genetics in the Garden

It was in a population of pea plants that the father of genetics, Gregor Mendel, got some of the first clues about how genes work (**Box Figure 5.2.1**). Mendel (1822–84) lived most of his adult life as a monk in a monastery in what is now the Czech Republic. Before entering the monastery, he attended the University of Vienna, where he became fascinated by heredity. After many years of reflection, he concluded that heredity was not a blending of traits, as many naturalists then believed. Instead, he believed that it came about by the combination of discrete factors from each parent.

To test his idea, Mendel designed an experiment to cross different varieties of pea plants and keep track of the color, size, and shape of the new generations of pea plants that they produced. For two years, he collected varieties and tested them to see if they would breed true. Mendel settled on 22 different varieties and chose seven different traits to track. His peas were either round or wrinkled and either yellow or green. Their pods were yellow or green as well, and they were also either smooth or ridged. The plants themselves might be tall or short, and their flowers, which could be violet or white, might blossom at their tips or along their stems.

Delicately placing the pollen of one plant on another, Mendel created thousands of hybrids, which he then interbred to create a new generation. After crossing smooth and wrinkled peas, for example, he shucked the pods a few months later and found that the hybrid peas were all smooth. The wrinkled trait had utterly disappeared from that generation of plants. Mendel then bred these smooth hybrids together and grew a second generation. Although most of the peas were smooth, some were wrinkled—just as deeply wrinkled as their wrinkled grandparents. The wrinkled trait had not been destroyed during the smooth generation, in other words: it had gone into hiding in the hybrids and then reappeared when the hybrids were interbred.

The number of peas that ended up wrinkled would vary on each plant, but as Mendel counted more and more of them, he ended up with a ratio of one wrinkled seed for every three smooth ones. He crossed varieties to follow the fate of other traits, and the same pattern emerged: one green seed for every three yellow ones, and one white flower for every three violet ones. Again and again, the peas produced a three-to-one ratio (3 : 1) of the traits.

Mendel realized that he had found an underlying regularity to the confusion of heredity. He concluded that pea plants must have factors of some sort that they could pass down to their offspring. Each parent of a new plant passed down a copy of the same factor—what we now call alleles. The two alleles might be identical or they might differ in how they influenced the development of the plant. (Diploid

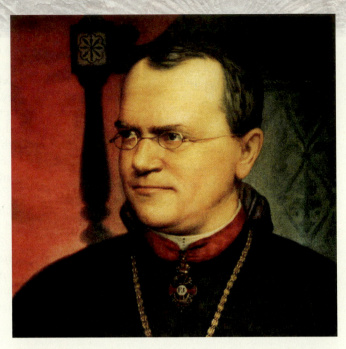

Box Figure 5.2.1 Gregor Mendel (1822–84) first recognized that inherited traits were made possible by factors (now called genes) passed down from parents to offspring. (Pictorial Press Ltd/Alamy Stock Photo)

organisms with a set of identical alleles are homozygotes, and those with different alleles are heterozygotes.)

Homozygous pea plants would produce the traits encoded by their identical alleles. A pea plant with two wrinkled alleles, in other words, would produce wrinkled peas. But a heterozygous pea plant with a wrinkled allele and a smooth allele always produced a smooth pea. Mendel recognized that some alleles could, in effect, exert more influence than their partner allele and produce their trait. These came to be known as **dominant alleles**. The other alleles that exerted less influence in a heterozygous individual, such as the allele for wrinkles, came to be known as **recessive alleles**.

Mendel also developed hypotheses for how these alleles were passed down from parents to offspring. First, he argued, the pairs of heritable factors in plants were separated (or segregated) during the formation of gametes. In a homozygote, all the gametes carried the same allele. In heterozygotes, however, half carried the smooth allele and half carried the wrinkled allele. Because Mendel randomly mixed together the gametes of different heterozygotes, each heterozygote had a 50% chance of passing down either allele to its offspring. **Box Figure 5.2.2** shows the probability that a pea plant will inherit two copies of the smooth allele, two copies of the wrinkled allele, or a combination.

Only 25% of the peas will be wrinkled homozygotes. All the others will either be smooth homozygotes or smooth heterozygotes. Thus they produce Mendel's 3 : 1 ratio of phenotypes.

Most of Mendel's contemporaries—including Darwin—never became aware of his work. But he was actually a pioneer in genetics, a field that didn't even come into formal existence until 16 years after his death. After a hundred years of research, it is now clear why Mendel's peas grew the way they did (Box Figure 5.2.2). The difference between a smooth pea and a wrinkled pea, for example, is determined by a single gene that encodes a protein, called starch-branching enzyme (SBEI), which helps to break down sugar.

At some point in the distant past, a mobile element measuring 800 base pairs long was inserted into the gene for SBEI in a pea plant. Over time, this new recessive allele

(continued)

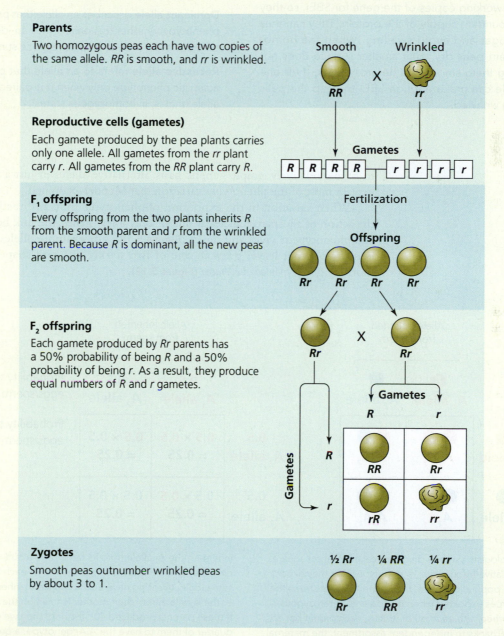

Parents

Two homozygous peas each have two copies of the same allele. *RR* is smooth, and *rr* is wrinkled.

Smooth   Wrinkled

RR   X   *rr*

Reproductive cells (gametes)

Each gamete produced by the pea plants carries only one allele. All gametes from the *rr* plant carry *r*. All gametes from the *RR* plant carry *R*.

Gametes

R R R R — r r r r

F₁ offspring

Every offspring from the two plants inherits *R* from the smooth parent and *r* from the wrinkled parent. Because *R* is dominant, all the new peas are smooth.

Fertilization

Offspring

Rr   *Rr*   *Rr*   *Rr*

F₂ offspring

Each gamete produced by *Rr* parents has a 50% probability of being *R* and a 50% probability of being *r*. As a result, they produce equal numbers of *R* and *r* gametes.

Rr   X   *Rr*

Gametes

R   r

	R	r
R	*RR*	*Rr*
r	*rR*	*rr*

Zygotes

Smooth peas outnumber wrinkled peas by about 3 to 1.

½ *Rr*   ¼ *RR*   ¼ *rr*

Rr   *RR*   *rr*

Box Figure 5.2.2 Mendel learned that heterozygote peas produced offspring with a 3 : 1 ratio of certain traits. This figure shows how this ratio emerges as alleles for wrinkled or smooth peas are passed down through three generations.

Box 5.2 Genetics in the Garden *(continued)*

spread through the population (Bhattacharyya et al. 1990). Because homozygous recessive individuals cannot produce any SBEI, they can't break down sugar effectively. As a result, they become rich in sugar. A sugary seed absorbs extra water as it develops, so it swells to a larger size. Later, when the homozygous recessive pea begins to dry out, it shrinks and its surface folds in on itself, forming wrinkles.

Things go differently in homozygous dominant peas. They have two working copies of the gene for SBEI, so they can make an abundant supply of the protein. They can thus break down sugar and avoid swelling. When the homozygous dominant peas dry, their smaller surface does not wrinkle, leaving them smooth. In heterozygotes, their one functional allele can make enough SBEI to keep the peas from becoming wrinkled.

It would have been impossible for Mendel to observe the patterns he discovered if he had studied only a single pea plant. Only by studying a cross composed of many plants could he see the 3:1 ratio of traits and then infer dominant and recessive alleles. As we'll see in the next chapter, this statement remains true today: only by studying populations can biologists understand how alleles change in frequency and produce evolutionary change.

Dominant allele describes an allele that produces the same phenotype whether it is paired with an identical allele or a different allele (that is, a heterozygotic state).

Recessive allele refers to an allele that produces its characteristic phenotype only when it is paired with an identical allele (that is, in homozygous states).

to have a relatively simple genetic basis: alternative alleles of just a few genes could explain most of the phenotypic variation that Mendel examined.

When variation in the expression of phenotypic traits is directly attributable to the action of alternative alleles at one or only a few genes, the link between genotype and phenotype is vivid and direct (**Figure 5.18**). For example, alleles of a single gene control whether leaves grow in one of two starkly different shapes in the ivy-leaf morning glory, *Ipomoea hederacea* (**Figure 5.19**).

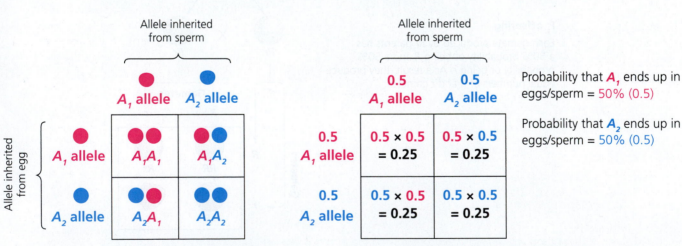

Figure 5.18 Biologists use box diagrams called Punnett squares to predict the relative frequencies of genotypes that they expect for offspring in a population. Here, the Punnett square shows the offspring genotypes expected when two parents heterozygous for the *A* allele breed with each other (that is, each parent has an A_1A_2 genotype). Thanks to independent assortment, the maternal and paternal copies of the chromosome containing the *A* allele have a 50:50 chance of ending up in each egg or sperm that these parents produce. As a result, every offspring has a 50% chance of

inheriting the A_1 allele from its father and a 50% chance of inheriting the A_2 allele (shown across the top of each box). And, by the same logic, offspring have a 50% chance of inheriting either the A_1 or the A_2 allele from their mother (shown on the side of each box). If these parents produced 40 offspring, then we would expect a quarter of them to have the A_1A_1 genotype, a quarter of them to have the A_2A_2 genotype, and half of them to be heterozygotes, with one A_1 and one A_2 allele.

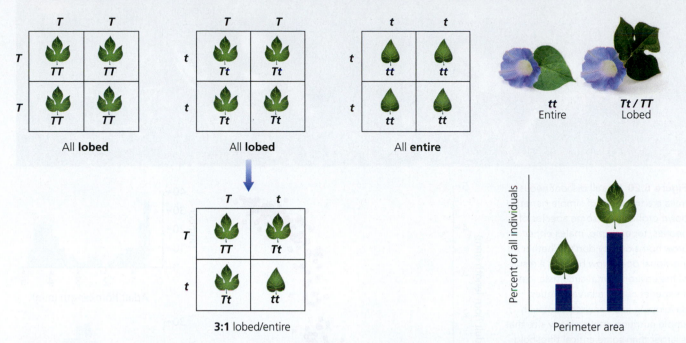

Figure 5.19 Ivy-leaf morning glories possess a genetic polymorphism for the shape of their leaves. Plants with one or two copies of the dominant allele *T* grow lobed leaves. Plants with two copies of the recessive allele *t* produce leaves with a different shape, known as "entire." Natural populations contain individuals with lobed leaves, entire leaves, or mixtures of the two forms, but not a continuous distribution of leaf shapes. Variation in leaf shape (shown here for a sample population) is discrete. (Data from Bright and Rausher 2008)

Simple **genetic polymorphisms** are also responsible for a number of human diseases. Huntington's disease, for example, causes certain kinds of neurons to waste away. Its victims slowly decline, losing the ability to speak and becoming unable to control their body movements. Geneticists have long known that Huntington's is a disorder that runs in families. In 1993, a team of American and British scientists finally pinpointed the cause of the disease: a mutation in a single gene, which produces a protein now known as huntingtin (Gusella et al. 1993; Walker 2007). Scientists still don't know the function of huntingtin, but its link to the disease is clear.

For most phenotypic traits, however, the link to the genotype is far more complicated. Many alternative phenotypes, for example, are not actually associated directly with alternative alleles of a single gene. In fact, organisms often only develop particular phenotypes in response to certain environmental signals. Aphids, for example, can either develop wings or become wingless. There are no "half-winged" aphids. Yet these alternate phenotypes do not correspond to alternative genotypes. The same genotype can produce both phenotypes, depending on the environment in which the aphid develops.

These traits are called **polyphenisms**. Often, the development of polyphenisms incorporates a threshold of sensitivity to the environment, such as the amount of daylight or the temperature to which an organism is exposed. Although these traits may appear to have a simple genetic basis due to their distinct alternative phenotypes, they actually have more complex underlying mechanisms that can involve many genes interacting with each other and with the environment (**Figure 5.20**).

A trait such as height, on the other hand, does not exist in discrete alternative states. Instead, there is a continuous range of values for the trait. In 1885, Francis Galton (Charles Darwin's first cousin) gathered measurements of the height of 1329 men. When he plotted the distribution of heights on a graph, he found that height was distributed symmetrically around a mean (**Figure 5.21**). Most of the men were of average or near-average height. Very tall and very short men were equally rare. Traits with continuous distributions of phenotypic variation are known as **quantitative traits**.

Genetic polymorphism is the simultaneous occurrence of two or more discrete phenotypes within a population. In the simplest case, each phenotype results from a different allele or combination of alleles of a single gene. In more complex cases, the phenotypes result from complex interactions between many different genes and the environment.

Polyphenism is a trait for which multiple, discrete phenotypes can arise from a single genotype, depending on environmental circumstances.

A **quantitative trait** is a measurable phenotype that varies among individuals over a given range to produce a continuous distribution of a phenotype. Quantitative traits are sometimes called complex traits; they're also sometimes called polygenic traits because their variation can be attributed to polygenic effects (that is, the cumulative action of many genes).

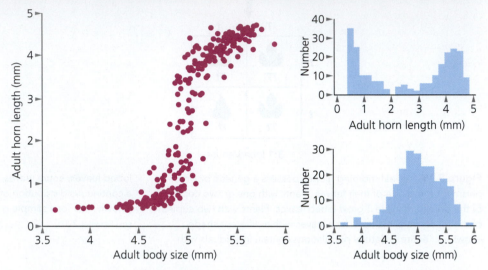

Figure 5.20 Not all discontinuous traits are the result of simple genetic polymorphisms. In some species of beetles, for example, males either grow horns or they don't. Whether an individual does grow horns is a result of the environmental circumstances it experienced as a larva. Beetles developing in an environment with ample nutrition attain a body size that is larger than some critical threshold, and these individuals grow horns. Beetles developing in an environment with insufficient nutrition fail to attain that threshold body size, and in these individuals horn growth is blocked. (Top: Douglas Emlen)

The total variation of a phenotypic trait is the sum of the variation caused by genetic factors and the variation caused by environmental factors. Height, for example, is strongly influenced both by genes and by the environment. If people live in the same environment, tall people will tend to have tall children, and short people will tend to have short children. On the other hand, people who grow up in more affluent families tend to be taller than those who grow up in poorer ones. Mothers who suffer from malnutrition while pregnant tend to give birth to babies who grow up to be shorter than people with mothers who have been properly nourished. Thus, even in the absence of any genetic variation, the environment can create phenotypic variation.

One of the clearest illustrations of the power of the environment comes from the research of Barry Bogin, an anthropologist who teaches at Loughborough University in England. In the 1970s, Bogin began to study the short stature of the Maya people of Guatemala. Some scholars called them the pygmies of Central America because the men averaged only 1.6 meters tall and the women 1.4 meters. The other major

Figure 5.21 A: Francis Galton (1822–1911) studied the continuous variation in human height by collecting measurements from 1329 British men. B: Galton found that most men were around average height and that the proportions of taller and shorter men were the same. Variation in height in this sample has a "normal" distribution. This is a common pattern found in continuous, or "quantitative," phenotypic traits. (Data from Hartl 2007; A: Pantheon/Superstock, Inc.)

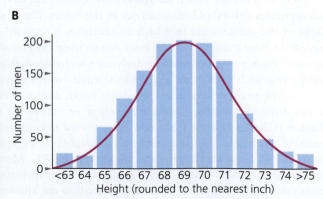

ethnic group in Guatemala is the Ladinos, who are descended from a mix of Maya and Spanish ancestors. Ladinos are of average height.

The biggest factor in the difference between the Ladinos and the Maya was not genetic. It was poverty. The Maya had less food and less access to modern medicine, which caused them to be shorter. During the Guatemalan civil war, which lasted from 1960 to 1996, a million refugees came to the United States. By 2000, Bogin found, American Maya were 10 centimeters taller than Guatemalan Maya, making them the same height as Guatemalan Ladinos. To grow much taller, the so-called pygmies needed improved quality of life, including a better diet and health care, for only a few generations (Bogin et al. 2002). They did not require new alleles.

Yet scientists have long known that genes also help make some people tall and other people short. In 1903, for example, the British statistician Karl Pearson published data on 1100 families and showed that tall fathers tended to have tall children. More recently, David Duffy of the Queensland Institute of Medical Research and his colleagues surveyed twins (**Figure 5.22**; Duffy, personal communication). They compared identical twins (who develop from the same egg and thus share the same set of alleles) to fraternal twins (who develop from separate eggs and thus share only some of the same alleles). Identical twins grow to be much closer in height than fraternal twins. Both sorts of twins grow up in the same environment, so the main source of such a difference must be genetic.

Although these studies demonstrate that variation in height has a genetic component, they do not reveal which genes are involved. One method to find candidate genes for a quantitative trait is known as quantitative trait locus (QTL) analysis. We describe this method in detail in Chapter 7. Briefly, QTL analysis involves searching for genetic markers scattered through the genome that tend to be found in people with high or low values of a particular trait. For example, a genetic marker might tend to be found in very tall people.

This linkage may mean that a gene associated with the trait in question is located near the marker. Recall that in each generation, chromosomes swap homologous segments via recombination. So alleles that once shared the same chromosome

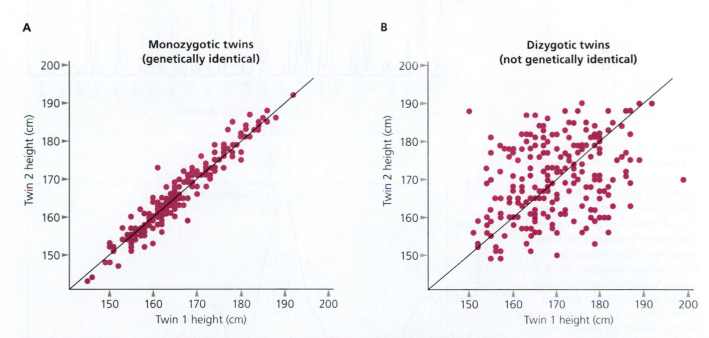

Figure 5.22 A: This graph shows the relationship between the heights of identical twins. One twin's height is marked along the x axis, and his or her twin sibling's height is marked on the y axis. The tight correlation shows that identical twins tend to be of similar height. B: Fraternal twins do not show as strong a tendency to be of similar height. The difference between these two results is due to the strong influence of genes on height. Because identical twins inherit identical sets of genes, they are more likely to be of similar height than fraternal twins, who develop from separate eggs. (Data from David Duffy)

become separated. The closer two alleles are, the more likely they are to remain on the same segment of DNA. Two people both may be tall because they have inherited the same allele from a common ancestor—along with a genetic marker nearby.

Figure 5.23 shows the result of one such survey. In 2004, Dutch researchers searched for QTLs associated with height in a group of 1036 people. They found that height was associated with a few genetic markers up to 100 times more often than you'd expect from chance (Willemsen et al. 2004).

Joel Hirschhorn, whom we met at the beginning of this chapter, and his colleagues have been developing increasingly powerful methods for identifying loci linked to height. One of the big challenges they face is distinguishing between alleles that are more common in tall people not because they promote height but thanks merely to chance. This challenge becomes especially vexing when researchers search for alleles that have an extremely weak influence on height.

Improving DNA-sequencing technology is one way to overcome this challenge. Hirschhorn and his colleagues are analyzing denser arrays of genetic markers across the genome in their new studies. It may even become affordable to sequence entire genomes of large groups of people to study traits like height. Another solution is simply to study a lot of people. Large-scale studies provide the statistical power to confidently identify very weak effects of alleles. But even with 800 or so loci identified

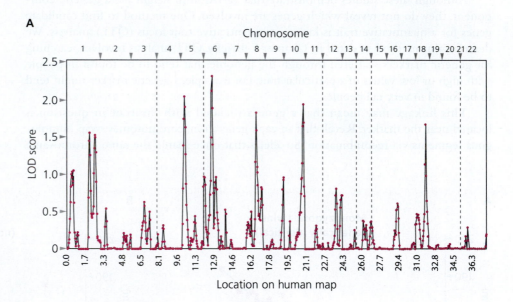

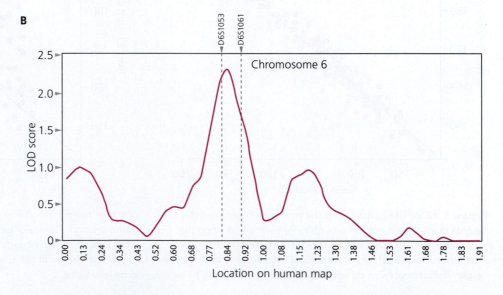

Figure 5.23 These graphs show the results of a genome-wide search for quantitative genetic loci (QTLs) associated with height. The LOD score along the vertical axis refers to "log of odds." An LOD of 2 means that a genetic marker is 100 times more likely to be associated with a height-related gene than not. A: This graph shows the location of strong candidates for QTLs on different chromosomes. B: This graph provides a closer view of the markers on chromosome 6. Once scientists identify such QTLs, they can test nearby genes (shown here as dashed lines) for effects on height. (Data from Willemsen et al. 2004)

for height, Hirschhorn and his colleagues have a long way to go in identifying all the genetic influences on height. So far, all the loci they've identified account for only 27% of the heritability of height (Marouli et al. 2017). Scientists likely have yet to discover hundreds—maybe even thousands—of additional loci contributing to variation in human height.

Key Concepts

- Phenotypes are a result of both an organism's genotype and the environment.

- Polyphenisms often result from developmental threshold mechanisms, whereby organisms respond to a critical level of some environmental cue (for example, photoperiod or temperature) by switching from production of a default phenotype to that of an alternate phenotype.

- Continuously varying traits are called quantitative traits, and the study of their inheritance and evolution is called quantitative genetics.

- Evolutionary biologists study variation in the expression of phenotypic traits. ●

5.7 How Do Genes Respond to the Environment?

The fact that individuals can grow to such different heights depending on their diet, as well as other measures of quality of life, is a striking demonstration of how much the environment can influence the expression of quantitative traits. These differences arise in part because different environments trigger different patterns of gene expression. When an organism experiences stress, for example, it may respond with a change in its level of developmental hormones. The hormones travel throughout the body and bind to regulatory regions of genes. Stress can also trigger changes in the coiling of DNA around histones, which can have further effects on gene expression. These gene expression changes caused by stress ultimately can cause changes in an organism's metabolism, growth, and behavior.

We tend to think of the "environment" as the external conditions around us. But for a gene inside a cell, the environment can be anything that interacts with its promoter region in a way that influences whether the gene is expressed. Products of other genes, like activators or transcription factors, are part of the environment. When they are present in an adequate concentration, the gene is transcribed; when they are in an inadequate concentration, it is not. Signals that come from other cells are also part of a gene's environment.

During development, cells respond to a multitude of signals from their environments. Often this means the immediate surroundings of the cell, such as contact with neighboring cells or exposure to molecular signals from those cells; for example, **morphogens** are molecular signals that move through a field of cells. Genes housed in cells at one end of an insect embryo encounter a different mix of molecular signals than genes housed in cells at the other end. Because of these different molecular environments, cells at the anterior and posterior of insect larvae do different things. Genes turned on in one location may be repressed in the other. The different environments in which cells develop in an animal's body ultimately produce an organism with a complex anatomy and organs, instead of a bag of uniform cells.

Some phenotypic traits are relatively unaffected by differences in these signals. For example, humans are almost always born with two eyes; we do not grow 5 or 10 or 20 eyes depending on how much food we eat. But other traits, such as height, are more sensitive to environmental stimuli. This potential for variation is known as **phenotypic plasticity**.

A **morphogen** is a signaling molecule that flows between nearby cells and acts directly to alter the expression of target genes.

Phenotypic plasticity refers to changes in the phenotype produced by a single genotype in different environments.

It's rare for phenotypic plasticity to be just a random response to the environment. Usually, phenotypic plasticity is a flexible, adaptive strategy that organisms can use to tailor their phenotypes to their surroundings. And as we'll see in Chapter 7, individuals often vary genetically in how they respond to the environment. Thanks to this variation, phenotypic plasticity itself can evolve.

Key Concept

- The expression of genes is often influenced by signals from the environment. This can cause gene activity to be matched with particular developmental or ecological contexts. ●

TO SUM UP . . .

- Genetic information is stored in DNA. DNA is copied into RNA, which has many functions in the cell.

- Proteins are essential tools in the structure and function of an organism; they can function as structural elements, enzymes, signals, or receptors. Some RNA molecules serve as templates for proteins.

- Mutations are changes in the genetic information of an organism—they are the ultimate source of heritable genetic variation, the raw material essential for evolution. Offspring can inherit germline mutations from their parents.

- Mutations can affect the product of a gene, such as a protein, or the levels of expression of that gene product.

- Mutations are rare, but they arise at a roughly steady rate.

- Mutations can have harmful or beneficial effects; most mutations are neutral or mildly deleterious.

- Sexual reproduction adds genetic variation within a population. During meiosis, genetic recombination and independent assortment separate, mix, and combine alleles, resulting in siblings from the same parents inheriting different characteristics.

- The phenotype is the manifestation of the genotype, but the relationship is not always straightforward. Some phenotypic polymorphisms exist because of a mutation to a single gene. Some genotypes can produce multiple phenotypes, depending on the environment.

- Quantitative traits can reflect the simultaneous effects of multiple genes and the environment, so it is misleading to speak of a gene "for" a trait. For instance, there is no "height gene." Any particular gene that affects height probably accounts for only a small amount of variation among the heights of individuals in a population. Other complex traits are also affected by numerous genes, and in some cases environmental circumstances as well.

- The environment can produce large variations in the expression of a trait.

- Phenotypic plasticity is the result of interactions between genotypes and environments.

MULTIPLE CHOICE QUESTIONS Answers can be found at the end of the book.

1. Which of the following is *not* a protein?

   a. Histone.
   b. Insulin.
   c. Nucleotide.
   d. Hemoglobin.

2. What role do histones play in eukaryotes?

   a. Condensing the DNA and controlling the transcription of genes.
   b. Translating RNA and condensing the DNA.
   c. Translating RNA and transcription of genes.
   d. None of the above.

3. According to Figure 5.8, which of the following codons does *not* designate the amino acid serine?

   a. UCU.
   b. UGU.
   c. UCA.
   d. AGU.

4. Which of the following is an important job of a hormone?

   a. To splice alternative exons.
   b. To alter expression of target genes.
   c. To halt transcription of mRNA.
   d. To induce translation of DNA.

5. What is the name of DNA sequences that have lost their protein-coding ability?

   a. Transposons.
   b. Pseudogenes.
   c. Introns.
   d. Exons.

6. Which of the following mutations does *not* affect gene expression?

   a. Mutations to promoter regions.
   b. Mutations to cis-regulatory regions.

   c. Mutations that code for recessive alleles.
   d. Mutations that code for hormones.

7. Which of the following statements about genetic recombination is *false*?

   a. Genetic recombination acts independently of independent assortment.
   b. Genetic recombination occurs during the production of sperm.
   c. Genetic recombination is relatively unimportant to the process of natural selection.
   d. During meiosis, chromosomes can cross over and exchange segments of DNA so that chromosomes of gametes are different from the chromosomes of the parents.

8. Which of these statements about the link between phenotypes and genotypes is *not* true?

   a. The environment often affects how a phenotype will develop.
   b. A single genotype may produce multiple phenotypes.
   c. Human height is controlled by more than one gene.
   d. Risk of Huntington's disease is a quantitative (polygenic) trait.

9. Which of the following is/are part of the environmental control of gene expression?

   a. Signals from outside the body.
   b. Signals from other genes within the cell.
   c. Signals from other cells.
   d. All of the above.

10. Which of the following is *not* true of DNA in eukaryotes?

    a. It is regularly altered through horizontal gene transfer.
    b. It is coiled tightly around a series of histone proteins.
    c. It is reorganized into larger structures known as chromosomes.
    d. It is coiled to allow it to fit within a cell's nucleus.

INTERPRET THE DATA Answer can be found at the end of the book.

11. Which form of DNA mutation is illustrated in the figure below?

    TGCATTTAGGC
    TGCATTCCGTAGGC
    CCG

    a. Point mutation.
    b. Insertion.
    c. Frameshift mutation.
    d. Deletion.

SHORT ANSWER QUESTIONS Answers can be found at the end of the book.

12. What are three roles that ribonucleic acid plays in the eukaryotic cell?

13. How can many different proteins be made from a single gene?

14. What are the possible outcomes to a protein product if the gene that codes for it has a point mutation?

15. What is the difference between trans- and cis-acting elements? How can stress affect a cis-acting element?

16. What are two mechanisms of meiosis that can cause offspring to be genetically different from their parents?

17. Why don't all phenotypic traits occur as discrete, alternative states like Mendel's peas?

ADDITIONAL READING

Campbell, C. D., and E. E. Eichler. 2013. Properties and Rates of Germline Mutations in Humans. *Trends in Genetics* 29:575–84.

Carroll, S. B. 2000. Endless Forms: The Evolution of Gene Regulation and Morphological Diversity. *Cell* 101 (6): 577–80.

———. 2005. *Endless Forms Most Beautiful: The New Science of Evo-Devo.* New York: W. W. Norton.

Henig, R. 2000. *The Monk in the Garden: The Lost and Found Genius of Gregor Mendel, the Father of Genetics.* Boston: Houghton Mifflin.

Leroi, A. 2003. *Mutants: On Genetic Variety in the Human Body.* New York: Viking.

McCutcheon, J. P., and N. A. Moran. 2012. Extreme Genome Reduction in Symbiotic Bacteria. *Nature Reviews Microbiology* 10:13–26.

Stern, D. L. 2000. Perspective: Evolutionary Developmental Biology and the Problem of Variation. *Evolution* 54 (4): 1079–91.

West-Eberhard, M. J. 2003. *Developmental Plasticity and Evolution.* New York: Oxford University Press.

Zimmer, C. 2018. *She Has Her Mother's Laugh: The Power, Perversions, and Potential of Heredity.* New York: Dutton.

PRIMARY LITERATURE CITED IN CHAPTER 5

Alexander, R. P., G. Fang, J. Rozowsky, M. Snyder, and M. B. Gerstein. 2010. Annotating Non-Coding Regions of the Genome. *Nature Reviews Genetics* 11 (8): 559–71.

Bennett, G. M., and N. A. Moran. 2013. Small, Smaller, Smallest: The Origins and Evolution of Ancient Dual Symbioses in a Phloem-Feeding Insect. *Genome Biology and Evolution* 5 (9): 1675–88.

Besenbacher, S., P. Sulem, A. Helgason, H. Helgason, H. Kristjansson, et al. 2016. Multi-nucleotide *de novo* Mutations in Humans. *PLoS Genetics* 12 (11): e1006315.

Bhattacharyya, M. K., A. M. Smith, T. H. Ellis, C. Hedley, and C. Martin. 1990. The Wrinkled-Seed Character of Pea Described by Mendel Is Caused by a Transposon-Like Insertion in a Gene Encoding Starch-Branching Enzyme. *Cell* 60 (1): 115–22.

Bianconi, E., A. Piovesan, F. Facchin, A. Beraudi, R. Casadei, et al. 2013. An Estimation of the Number of Cells in the Human Body. *Annals of Human Biology* 40:463–71.

Bogin, B., P. Smith, A. B. Orden, M. I. Varela Silva, and J. Loucky. 2002. Rapid Change in Height and Body Proportions of Maya American Children. *American Journal of Human Biology* 14 (6): 753–61.

Bright, K. L., and M. D. Rausher. 2008. Natural Selection on a Leaf-Shape Polymorphism in the Ivy-Leaf Morning Glory (*Ipomoea hederacea*). *Evolution* 62:1978–90.

Campbell, C. D., and E. E. Eichler. 2013. Properties and Rates of Germline Mutations in Humans. *Trends in Genetics* 29:575–84.

Cartolano, M., R. Castillo, N. Efremova, M. Kuckenberg, J. Zethof, et al. 2007. A Conserved MicroRNA Module Exerts Homeotic Control over *Petunia hybrida* and *Antirrhinum majus* Floral Organ Identity. *Nature Genetics* 39 (7): 901–5.

Dagan, T., and W. Martin. 2007. Ancestral Genome Sizes Specify the Minimum Rate of Lateral Gene Transfer During Prokaryote Evolution. *Proceedings of the National Academy of Sciences USA* 104 (3): 870–5.

Eisenberg, E., and E. Y. Levanon. 2013. Human Housekeeping Genes, Revisited. *Trends in Genetics* 29: 569–74.

Eriksson, M., W. T. Brown, L. B. Gordon, M. W. Glynn, J. Singer, et al. 2003. Recurrent de Novo Point Mutations in *Lamin A* Cause Hutchinson-Gilford Progeria Syndrome. *Nature* 423 (6937): 293–8.

Gusella, J. F., M. E. MacDonald, C. M. Ambrose, and M. P. Duyao. 1993. Molecular Genetics of Huntington's Disease. *Archives of Neurology* 50 (11): 1157–63.

Hartl, D. L. 2007. *Principles of Population Genetics.* Sunderland, MA: Sinauer Associates.

Keeling, P. J., and J. D. Palmer. 2008. Horizontal Gene Transfer in Eukaryotic Evolution. *Nature Reviews Genetics* 9: 605–18.

Levy, S., G. Sutton, P. C. Ng, L. Feuk, A. L. Halpern, et al. 2007. The Diploid Genome Sequence of an Individual Human. *PLoS Biology* 5 (10): e254.

Marouli, E., M. Graff, C. Medina-Gomez, K. S. Lo, A. R. Wood, et al. 2017. Rare and Low-Frequency Coding Variants Alter Human Adult Height. *Nature* 542 (7640): 186–90.

Oetting, W. S., and R. A. King. 1993. Molecular Basis of Type I (Tyrosinase-Related) Oculocutaneous Albinism: Mutations and Polymorphisms of the Human *Tyrosinase* Gene. *Human Mutation* 2 (1): 1–6.

Sánchez-Martín, M., J. Pérez-Losada, A. Rodríguez-García, B. González-Sánchez, B. R. Korf, et al. 2003. Deletion of the Slug (*SNAI2*) Gene Results in Human Piebaldism. *American Journal of Medical Genetics Part A* 122A (2): 125–32.

Shore, E. M., M. Xu, G. J. Feldman, D. A. Fenstermacher, FOP International Research Consortium, et al. 2006. A Recurrent Mutation in the BMP Type I Receptor ACVR1 Causes Inherited and Sporadic Fibrodysplasia Ossificans Progressiva. *Nature Genetics* 38 (5): 525–7.

van Bakel, H., C. Nislow, B. J. Blencowe, and T. R. Hughes. 2010. Most "Dark Matter" Transcripts Are Associated with Known Genes. *PLoS Biology* 8 (5): e1000371.

Walker, F. O. 2007. Huntington's Disease. *Lancet* 369 (9557): 218–28.

Wang, Z-Q., S-H. Tian, Y-Z. Shi, P-T. Zhou, Z-Y. Wang, et al. 2007. A Single C to T Transition in Intron 5 of *LMBR1* Gene Is Associated with Triphalangeal Thumb-Polysyndactyly Syndrome in a Chinese Family. *Biochemical and Biophysical Research Communications* 355 (2): 312–7.

Weedon, M. N., G. Lettre, R. M. Freathy, C. M. Lindgren, B. F. Voight, et al. 2007. A Common Variant of *HMGA2* Is Associated with Adult and Childhood Height in the General Population. *Nature Genetics* 39 (10): 1245–50.

Willemsen, G., D. I. Boomsma, A. L. Beem, J. M. Vink, P. E. Slagboom, et al. 2004. QTLs for Height: Results of a Full Genome Scan in Dutch Sibling Pairs. *European Journal of Human Genetics* 12 (10): 820–8.

Zimmer, C. 2018. *She Has Her Mother's Laugh: The Power, Perversions, and Potential of Heredity.* New York: Dutton.

CHAPTER 6

The Ways of Change

Drift and Selection

Learning Objectives

- Define population genetics.
- Calculate allele frequencies within a population.
- Use allele frequencies to determine whether a population is in Hardy–Weinberg equilibrium.
- Compare and contrast the effects of genetic drift on large versus small populations.
- Predict the effect of a bottleneck or founder event on allelic diversity.
- Compare and contrast measures of fitness of a phenotype and fitness of an allele within a population.
- Explain how slight differences in fitness can change the frequencies of alleles within a population over time.
- Discuss how pleiotropy affects the response to selection acting on alleles.
- Explain why natural selection cannot drive dominant alleles to fixation within a population.
- Describe how selection can act to either remove or maintain allelic diversity.
- Discuss the effects of inbreeding on an individual's fitness.
- Analyze the influence of drift and inbreeding on the genetics of populations within a landscape.

I n the 1960s, the government of France had a plan. To attract tourists to the Mediterranean coast, they would build new cities from scratch. There was just one problem: the balmy climate that attracted tourists to the French coast also created a splendid environment for mosquitoes (*Culex pipiens*) to breed. Before the tourists came, the mosquitoes would have to go.

The government launched a program to regularly spray mosquito-breeding sites along the coast with pesticides. Starting in 1969, they used organophosphate insecticides, which kill the insects by inhibiting an enzyme called acetylcholinesterase (AChE1) in the mosquito nervous system. At first,

A tourist-filled beach and promenade in Nice, along the French Riviera. Insecticides used to keep the beaches free of mosquitoes drove rapid evolution of insecticide resistance in coastal mosquito populations.

Ken Walsh/Alamy Stock Photo

the treatment was successful. The mosquito population fell, and people got bitten less often. But in 1972, the mosquitoes started coming back.

To find out why they were rebounding, Nicole Pasteur of the University of Montpellier and her colleagues collected mosquito larvae from streams and cisterns (**Figure 6.1**). She took the insects to her lab to expose them to different doses of insecticides. The mosquitoes from just north of the spray zone died when exposed to even a weak dose of insecticide. However, when Pasteur ran the same experiment on mosquitoes from the coast, where spraying was heavy, the insects could resist stronger doses (Pasteur and Sinègre 1975).

Further research revealed the source of this resistance. *C. pipiens* carries a gene called *Ester*, which encodes an enzyme known as esterase. Esterase breaks down a wide range of toxins, including organophosphate insecticides. If a mosquito is exposed to a high level of insecticide, however, it can't produce enough esterase to defend itself, and it dies. Pasteur and her colleagues discovered that the resistant mosquitoes carried a mutation that altered the expression of *Ester*. Mosquitoes that carried this mutant allele, known as *Ester*[1], produced much more esterase, which could eliminate the insecticide before it could kill them. They tended to survive and to reproduce, passing down the *Ester*[1] allele to their offspring (Raymond et al. 1998).

Figure 6.1 Michel Raymond, Mylene Weill, and Claire Berticat of the University of Montpellier collecting mosquito larvae in Tunisia. Their work around the Mediterranean has provided one of the most fine-grained studies of how alleles can spread through wild populations. (Michel Raymond)

The scientists began conducting annual surveys around southern France to find more mosquitoes carrying *Ester*[1]. Before 1972, they had not detected any *Ester*[1] alleles. But by 1973, the allele was present at a frequency of more than 60% in coastal populations. Farther inland, the allele became rarer; it was present at less than 20% in populations just 11 kilometers from the ocean (**Figure 6.2**). As the years passed, the frequency of the *Ester*[1] allele in the coastal populations continued to climb until it reached 100% by 1975. Yet the allele remained rare farther inland. In other words, in about a decade, the genes of entire coastal populations of mosquitoes had changed, giving them resistance to a highly toxic insecticide. The populations had evolved.

The change among France's mosquitoes is a vivid illustration that evolution can take place rapidly and have a major impact on our lives. It's especially instructive because the change itself has been relatively clear. Mutations in the population of mosquitoes raised the reproductive success in the animals that carried them, thanks to a change in the environment. Natural selection then made those mutations more common.

In Chapter 5 we learned about how genes work within individual organisms. We examined how they interact with each other and with their environment to produce phenotypes, how an individual passes down genetic information to its offspring, and how mutations can give rise to new genetic and phenotypic variation—the raw ingredients for evolution.

In this chapter, we will move from an individual perspective to consider genetic variation in populations. After new alleles arise within a population,

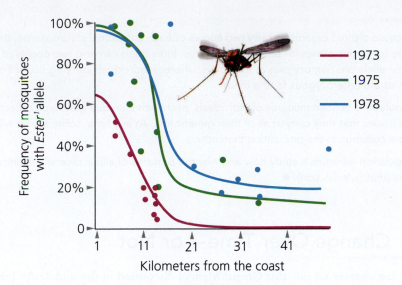

Figure 6.2 This graph shows the spread of the *Ester*[1] allele through mosquito populations around Marseille. By 1973, the *Ester*[1] allele was present in more than half of the mosquitoes along the coast but was absent 21 kilometers inland. By 1975, *Ester*[1] was the only variant for that gene in the coastal populations but continued to be rare in inland populations. In 1978, the allele had become somewhat more common in inland populations but still nowhere as prevalent as on the coast. This difference in selection was due to heavy pesticide spraying on the coast, raising the fitness of *Ester*[1] alleles there. (Data from Raymond et al. 1998)

they either increase or decrease in frequency. This change in the allelic composition of populations is at the very heart of the evolutionary process. As we'll see in the following pages, many interacting factors determine whether a particular allele spreads through an entire population or disappears.

We will also examine the ways in which populations change: the mechanisms of evolution, in other words. These mechanisms include genetic drift, natural selection, migration, and mutation. We will focus in this chapter on two of these mechanisms—drift and selection—and we'll examine the factors that can strengthen and weaken each of them. As we'll see, scientists study these mechanisms through the complementary approaches of observation, experimentation, and mathematical modeling. ●

6.1 The Genetics of Populations

When we refer to a **population**, we are referring to a group of interacting and potentially interbreeding individuals of a species. Some populations span large geographic ranges, whereas others occupy only small ranges and are isolated from other populations of the same species. A population is made up of individuals, and those individuals, in turn, carry alleles. These alleles typically vary from individual to individual; some populations contain large amounts of allelic diversity, and others are made of individuals that are nearly identical, genetically speaking. The study of allele distributions and frequencies is known as **population genetics**. Population geneticists study the patterns of allelic diversity in populations and how and why these patterns change over time.

An individual's genotype includes alleles for all of the millions of genetic loci that it carries in its chromosomes. But when population geneticists study a population, they typically focus their attention on only one or a few loci across a sample of individuals. For this reason, they often use the term *genotype* to refer to the combination of alleles carried by an individual at a particular **genetic locus**, or just a few loci. When scientists study resistance to insecticides in mosquitoes, for example, they typically study the genotype at the esterase locus. Because mosquitoes are diploids, they can potentially be homozygous or heterozygous at this locus. Thus, their genotype can be *Ester*[1]*Ester*[1], *Ester*[1]*Ester*[2], or *Ester*[2]*Ester*[2]. Scientists who study insecticide resistance thus set aside the many other loci in the mosquito genome that are not involved in this particular trait.

A **population** is a group of interacting and potentially interbreeding individuals of a species.

Population genetics is the study of the distribution of alleles within populations and the mechanisms that can cause allele frequencies to change over time.

Genetic locus (plural, **loci**) refers to the specific location of a gene or piece of DNA sequence on a chromosome. When mutations modify the sequence at a locus, they generate new alleles—variants of a particular gene or DNA region. Alleles are mutually exclusive alternative states for a genetic locus.

- Because diploid organisms carry two copies of each autosomal chromosome, they can have up to two alleles for each gene or locus. Individuals carrying two copies of the same allele are homozygous at that locus, whereas individuals carrying two different alleles are heterozygous for the locus.

- Populations contain mixtures of individuals, each with a unique genotype reflecting the alleles that they carry at all of their genetic loci. At any time, some alleles will be more common in the population than others.

- Population geneticists study how and why the patterns of allelic diversity change over time (that is, evolution). ●

6.2 Change Over Time—or Not

In the last chapter we saw that Gregor Mendel discovered in the mid-1800s how two heterozygous pea plants could produce offspring in a 3:1 ratio of traits. Heterozygous smooth peas would produce roughly three smooth peas for every wrinkled one, for example. Mendel also demonstrated how two homozygous plants would produce offspring with the same allelic combinations—all smooth or all wrinkled, for example.

In the early 1900s, geneticists discovered how to extend Mendel's model from parents and their offspring to entire populations. Some of the most important insights from geneticists emerged from mathematical models. With these models, the researchers considered how many parents of each possible genotype were present in a population and how often each of the possible crosses of different pairs occurred. By examining mathematical models, the researchers saw several interesting patterns emerge.

As long as the parents mated with each other randomly (**Box 6.1**), their gametes would mix together in predictable ways. In fact, the frequency of the alternative alleles in the gametes could be used to predict exactly how many offspring should be produced in the population with each possible genotype. Once the alleles were assembled into genotypes, the population would retain these allele and genotype frequencies indefinitely—the frequencies would persist in a state of equilibrium—as long as nothing acted on the population to change the frequencies. (**Box 6.2** explains the difference between allele frequencies and genotype frequencies.)

This persistence is known as Hardy–Weinberg equilibrium, after the British mathematician G. H. Hardy and the German physiologist Wilhelm Weinberg, who independently created similar versions of the same **theorem** in 1908. Hardy and Weinberg demonstrated that in the absence of outside forces (which we describe later), the allele frequencies of a population will not change from one generation to the next. As we'll see, this theorem is a powerful tool for looking for evidence of evolution in populations. But it's important to bear in mind that the theorem rests upon some assumptions. When these assumptions are violated, the result is a change in allele frequencies, or evolution within the population.

One assumption of the model is that a population is infinitely large. Real populations are finite, which means that their allele frequencies can change randomly from generation to generation simply due to chance. (We will explore this mechanism, known as genetic drift, in detail later in the chapter.) Although no real-life population is infinite, of course, very large ones are big enough that they behave quite similarly to the model. In large populations, variation due to chance is inconsequential. As a result, the allele frequencies will not change very much from generation to generation.

Another assumption of the Hardy–Weinberg theorem is that all of the genotypes at a locus are equally likely to survive and reproduce—it assumes there is no selection. If individuals with certain genotypes produce twice as many offspring as individuals with other genotypes, for example, then the alleles that these individuals carry will make up a greater proportion of the total in the offspring generation than would be

A **theorem** is a mathematical statement that has been proven based on previously established theorems and axioms. Theorems use deductive reasoning and show that a statement necessarily follows from a series of statements or hypotheses—the proof. Theorems are not the same as theories. Theories are explanations supported by substantial empirical evidence—the explanations are necessarily tentative but weighted by the quantity of evidence that supports them.

BOX 6.1

What Is Meant by "Random Mating"?

A central assumption of many population genetics models, including the Hardy–Weinberg theorem, is that the populations are made up of sexually reproducing organisms that mate at random. Strictly speaking, this is never actually the case. Species typically evolve strong preferences about their choice of mates, meaning that organisms do not mate randomly. (We'll explore these preferences in depth in Chapter 11.) And yet, oddly enough, nonrandom mating patterns usually don't interfere with population genetics studies.

The solution to this paradox arises from the fact that when population geneticists refer to random mating, they mean random mating *with respect to alleles at their genetic locus*

of interest. If they are studying the locus responsible for smooth versus wrinkled peas, for example, random mating means that pollen with the allele for wrinkled peas is just as likely to fertilize ovules that carry the wrinkled allele as those that carry the smooth allele at that locus. Similarly, for the malaria example we discuss later in this chapter, random mating means that sperm with the A allele are equally likely to fertilize eggs with the A allele as they are to fertilize eggs with the S allele. In other words, mating is assumed to be random *for that genetic locus*. Only rarely do mate choice decisions result in strongly biased (that is, nonrandom) mating at a particular genetic locus under study.

predicted given the Hardy–Weinberg theorem. In other words, selection for or against particular genotypes may cause the relative frequencies of alleles to change and result in evolution.

Another assumption of the Hardy–Weinberg theorem is that no alleles enter or leave a population through migration. This assumption is violated if some individuals disperse out of a population or if new individuals arrive because these events alter the frequencies of alleles. The model also assumes that there is no mutation in the population. New mutations would produce new alleles, changing the overall frequency of alleles in the population.

In each of these four violations of the assumptions of the Hardy-Weinberg theorem, the offspring genotype frequencies will differ from the theorem's predictions about the equilibrium of alleles. Because they alter allele frequencies from one generation to the next, drift, selection, migration, and mutation are all possible mechanisms of evolution.

Key Concepts

- The Hardy–Weinberg theorem states that in the absence of drift, selection, migration, and mutation, allele frequencies at a genetic locus will not change from one generation to the next.

- Mechanisms of evolution, such as genetic drift and natural selection, are processes that can change allele frequencies in a population from one generation to the next. ●

6.3 Evolution's "Null Hypothesis"

The Hardy–Weinberg theorem is useful because it provides mathematical proof that evolution will not occur in the absence of selection, drift, migration, or mutation. By explicitly delineating the conditions under which allele frequencies *do not* change, the theorem serves as a useful **null hypothesis** for studying the ways that allele frequencies *do* change. By studying how populations deviate from Hardy–Weinberg equilibrium, we can learn about the mechanisms of evolution (**Figure 6.3**).

Scientists can analyze populations to see if they are in Hardy–Weinberg equilibrium. In the 1970s, for example, Luigi Luca Cavalli-Sforza of Stanford University and

A **null hypothesis** is a default hypothesis that there is no relationship between two measured phenomena. By rejecting this hypothesis, scientists can provide evidence that such a relationship may exist.

Mechanisms of Evolution: Altering Allele Frequencies

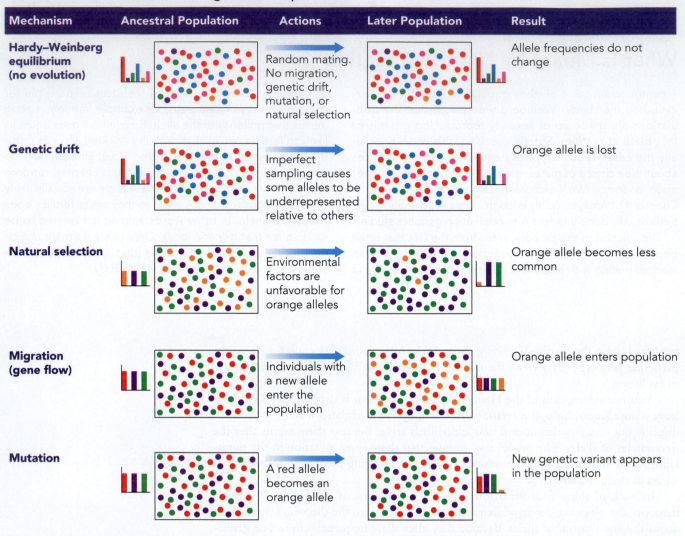

Mechanism	Ancestral Population	Actions	Later Population	Result
Hardy–Weinberg equilibrium (no evolution)		Random mating. No migration, genetic drift, mutation, or natural selection		Allele frequencies do not change
Genetic drift		Imperfect sampling causes some alleles to be underrepresented relative to others		Orange allele is lost
Natural selection		Environmental factors are unfavorable for orange alleles		Orange allele becomes less common
Migration (gene flow)		Individuals with a new allele enter the population		Orange allele enters population
Mutation		A red allele becomes an orange allele		New genetic variant appears in the population

Figure 6.3 Populations can evolve through a variety of mechanisms. This diagram illustrates how some of the most important mechanisms change allele frequencies. Colors represent alternative alleles at a genetic locus, and bars indicate relative allele frequencies.

his colleagues studied variations in the genes that encode hemoglobin, the molecule that transports oxygen through the bloodstream (Stone et al. 2007). They surveyed 12,387 adults in Nigeria, noting whether they had either of two alleles at a locus encoding part of the hemoglobin molecule called β-globin. These alleles are known as *A* and *S*.

Cavalli-Sforza and his colleagues calculated that for their population, the frequency of the *A* allele was 0.877, whereas the frequency of the *S* allele was 0.123. They then used the Hardy–Weinberg theorem to calculate how many people with each genotype they'd expect to find if the population was in Hardy–Weinberg equilibrium. (**Box 6.3** shows these calculations.) Then they compared these expected results with the actual numbers of people they observed with each genotype.

The Nigerian population deviated significantly from the values predicted by the Hardy–Weinberg theorem (Stone et al. 2007). There were fewer *SS* genotypes than expected, and significantly more *AS* and *AA* genotypes than expected. The population was therefore not at Hardy–Weinberg equilibrium.

When scientists discover a population out of Hardy–Weinberg equilibrium, as Cavalli-Sforza discovered in the case of hemoglobin, they can then test the various assumptions to figure out which was violated. In some cases, this search can reveal a mechanism of evolution at work. As we'll see later, the Nigerian population was

BOX 6.2

The Relationship Between Allele and Genotype Frequencies

For a genetic locus where two alternative alleles occur in the population (A_1 and A_2), population geneticists often use p to represent the frequency of one allele and q to represent the frequency of the other. This means that $p + q = 1$. If the frequencies of the alleles are known, then we can use the Hardy–Weinberg theorem to predict what the genotype frequencies should be in the next generation, assuming that individuals mate randomly (see Box 6.1) with respect to their genotype at that locus, and assuming that the population is not evolving.

If we think of the population as a pool of eggs and sperm that join together randomly, then we can predict how many of the offspring formed from these gametes should have each possible genotype at our locus. The total frequency (100%, or 1) of all genotypes in the offspring population will be the sum of the frequencies (f) of individuals with each possible genotype:

$$1 = f(A_1A_1) + f(A_1A_2) + f(A_2A_1) + f(A_2A_2)$$

For example, A_1A_1 individuals will be produced every time an egg with the A_1 allele fuses with a sperm that also has the A_1 allele. The frequency, $f(A_1A_1)$, at which this occurs is simply the product of the probability that the egg will have the A_1 allele (p) multiplied by the probability that the sperm will have the A_1 allele (also p), which is p^2 (**Box Figure 6.2.1**). By the same logic, the offspring will inherit the A_2 allele from both parents with a frequency of q^2. A_1A_2 individuals will be produced when an egg with the A_1 allele fuses with a sperm that has the A_2 allele and also when an egg with the A_2 allele fuses with a sperm that has the A_1 allele. Each of those types

of offspring will occur with a probability of pq, so the total frequency of A_1A_2 in the offspring population will be $2pq$. If we combine these terms, we get an equation for the sum of all genotypes in the offspring population:

$$1 = p^2 + 2pq + q^2$$

Thus, for a randomly mating population at equilibrium, the Hardy–Weinberg theorem gives us the genotype frequencies expected for any possible set of allele frequencies.

Alternatively, we can use *genotype* frequencies to calculate *allele* frequencies whenever the genotype frequencies are known. To do this, we multiply the frequency with which the genotype occurs in the population by the number of copies of the allele carried by each genotype (A_1A_1 individuals have two copies of the A_1 allele, for example, whereas A_2A_2 individuals have zero copies of the A_1 allele) and then divide this by 2 (to determine the frequency of the single allele):

$$f(A_1) = [2 \times f(A_1A_1)] + 1 \times f(A_1A_2) + 1 \times f(A_2A_1) + 0 \times f(A_2A_2)]/2$$

We know that relative frequencies of the A_1A_1, A_1A_2, A_2A_1, and A_2A_2 genotype are p^2, pq, qp, and q^2, respectively, so $f(A_1) = p$.

$$f(A_1) = [(2 \times p^2) + 2pq + (0 \times q^2)]/2$$
$$= [2p^2 + 2pq]/2$$
$$= p^2 + pq$$
$$= p(p + q)$$

Recall that $p + q = 1$. So $f(A_1) = p$.

The frequency of the A_2 allele is calculated in a similar manner:

$$f(A_2) = [0 \times f(A_1A_1)] + 1 \times f(A_1A_2) + 1 \times f(A_2A_1) + 2 \times f(A_2A_2)]/2$$
$$= [(0 \times p^2) + 2pq + (2 \times q^2)]/2$$
$$= [2p^2 + 2pq]/2$$
$$= p^2 + pq$$
$$= q(p + q)$$

And because $p + q = 1$, $f(A_2) = q$.

Sperm allele (and frequency)

		A_1 (p)	A_2 (q)
Egg allele (and frequency)	A_1 (p)	A_1A_1 (p^2)	A_1A_2 (pq)
	A_2 (q)	A_2A_1 (pq)	A_2A_2 (q^2)

Box Figure 6.2.1 The Hardy–Weinberg theorem can be used to determine the frequencies of genotypes based on the frequencies of alleles within a population.

BOX 6.3

Testing Hardy–Weinberg Predictions for the Human β-Globin Locus

The predictions of the Hardy–Weinberg theorem can be tested for each genetic locus. In fact, it may be common for some loci to be in Hardy–Weinberg equilibrium even though other loci in the same population are not. The Hardy–Weinberg theorem can be used to determine whether particular loci of interest are evolving in contemporary populations. Subsequent studies can then explore which mechanisms (mutation, migration, drift, selection) are responsible.

Cavalli-Sforza and colleagues measured allele frequencies for the A and S alleles to be 0.877 and 0.123, respectively. With these data, we can test the prediction of the Hardy–Weinberg theorem for the β-globin locus and calculate the frequencies of AA, AS, and SS genotypes that we would expect if the population were in equilibrium.

We assume that gametes in this population combine at random (with respect to alleles at this locus; see Box 6.1), so the expected frequency of individuals with the AA genotype

is the product of the probability that the egg will have the A allele, p, multiplied by the probability that the sperm will have the A allele, p, which is p^2. Because the measured frequency of the A allele is 0.877, the expected frequency of the AA genotype is 0.877×0.877, or 0.769. By the same logic, the offspring will inherit the S allele from both parents with a frequency of $q^2 (0.123 \times 0.123 = 0.015)$, and heterozygotes will occur with a frequency of $2pq (2 \times 0.877 \times 0.123 = 0.216)$.

But this is not what Cavalli-Sforza and his colleagues found (**Box Figure 6.3.1**). When they measured the genotypes of 12,387 individuals, they observed genotype frequencies that were significantly different from the Hardy–Weinberg expectations. By rejecting the null model of no evolution, they concluded that at least one of the assumptions of the Hardy–Weinberg theorem had been violated. In fact, as we'll see later in this chapter and in upcoming chapters, the β-globin locus is evolving in response to selection because of its role in conferring resistance to malaria.

Genotype	Predicted by Hardy–Weinberg	Observed
AA	9527.2 (76.9%)	9365 (75.6%)
AS	2672.4 (21.6%)	2993 (24.2%)
SS	187.4 (1.5%)	29 (0.2%)
Total	12,387	12,387

Box Figure 6.3.1 Geneticists measured the frequencies of two alleles of a gene that encodes part of hemoglobin, known as A and S. They then determined the frequencies of homozygotes and heterozygotes that would be expected if the alleles were in Hardy–Weinberg equilibrium. As this chart shows, the observed frequencies were significantly different from the expected values. As we'll see later, natural selection is responsible for the difference between these expectations and the actual frequencies. (Data from Cavalli-Sforza 1977)

experiencing natural selection, violating the Hardy–Weinberg assumptions. The genotypes at the β-globin locus alter the reproductive success of people, by either providing resistance to malaria or causing sickle-cell anemia.

Key Concepts

- Because it describes the conditions in which evolution will not occur, the Hardy–Weinberg theorem serves as the fundamental null hypothesis of population genetics.

- For any measured set of allele frequencies, the Hardy–Weinberg theorem predicts the genotype frequencies expected for a population that is not evolving. ●

6.4 A Random Sample

In the 1950s, Peter Buri, then a graduate student at the University of Chicago, carried out an experiment that dramatically illustrates how populations can violate the Hardy–Weinberg theorem's assumptions (Buri 1956). He bred thousands of fruit flies of the species *Drosophila melanogaster* that had two different alleles, called *bw* and *bw*^{75}, for a gene that influences their eye color. A fly with two copies of *bw* had white eyes; a *bw/bw*^{75} heterozygote had light-orange eyes, and the eyes in a fly with two copies of *bw*^{75} were bright red-orange.

Buri established 107 separate fly populations with light-orange-eyed flies (so each fly was heterozygous, with one *bw* and one *bw*^{75} allele). He started each population with eight males and eight females, and he let them reproduce. From the next generation, he randomly selected eight new males and eight new females to breed. Buri bred the flies this way for 19 generations, and in each generation he tallied the number of *bw* and *bw*^{75} alleles in the populations.

If Buri's populations of flies had been infinitely large, the frequency of the alleles should not have changed from one generation to the next. That's because all the rest of the assumptions for a population in Hardy–Weinberg equilibrium had been met: there was no selection (reproducing flies were chosen randomly in each generation), there was no migration (because each population was kept in a separate vial), and the probability of mutations occurring at that locus within the short duration of the experiment was small. Throughout the experiment, half of the alleles should have remained *bw* and half should have remained *bw*^{75}.

But that's not what happened. Instead, as **Figure 6.4** shows, the allele frequencies changed, and they changed differently in different populations. Over the generations, *bw* became rarer in some populations until it disappeared completely, leaving only *bw*^{75}/*bw*^{75} flies with bright-red-orange eyes. In other populations, *bw*^{75} disappeared instead, leaving only white-eyed *bw/bw* flies. The rest of the populations spanned the range between these two extremes. All these different outcomes emerged from populations that were initially identical and housed under identical conditions.

Buri's populations of flies evolved through a process known as **genetic drift**. The name of this process comes from the way allele frequencies "drift" randomly away from their starting value. Genetic drift is a powerful mechanism of evolution. In many species with relatively small populations, genetic drift has eliminated many variations in DNA. For all its importance, however, genetic drift is far less familiar than natural selection as a mechanism of evolution. Its obscurity may be due to the way it produces dramatic changes in allele frequencies thanks purely to chance.

Genetic drift occurs when a random, nonrepresentative sample from a population produces the next generation. A simple analogy illustrates how this happens (**Figure 6.5**). Imagine a bowl of jelly beans. They're all the same size and shape, but half are red and half are white. If you grab some jelly beans from the bowl without looking, there's a 50% chance that each of the jelly beans you pick out will be red. If you were to scoop a big handful of the jelly beans, the proportion of the jelly beans should be close to 50% red and 50% white. You'd be surprised to end up with only red jelly beans. But if you pick out only two jelly beans, it wouldn't be surprising for both of them to be red. The smaller the sample, in other words, the more likely it is that you will observe large deviations from the original frequency of red and white colors. Alleles are just like different-colored jelly beans, and this effect on frequency can have lasting effects on the variation within populations.

In Buri's experiment, he selected only 16 flies from each generation to reproduce. By picking such a small sample, he increased the probability that the allele frequencies in the small set of parents would deviate strongly from the frequencies in the entire population just by chance. In some populations, random deviations caused *bw* to become rarer until only a single heterozygous fly was left carrying a single copy of the allele. And then, in the next generation, that single fly passed down its copy of *bw*^{75}

Genetic drift is evolution arising from random changes in the genetic composition of a population from one generation to the next.

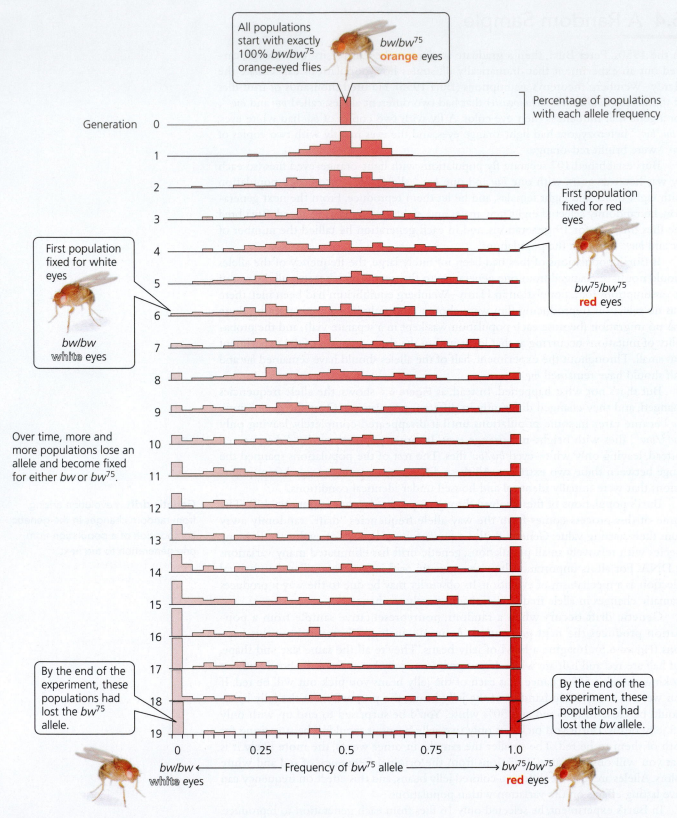

Figure 6.4 This sequence of graphs charts the results of an experiment performed by Peter Buri to measure genetic drift. He began with 107 replicate populations of *Drosophila*, each containing eight flies heterozygous at his marker locus, or *bw*/*bw*^75. The flies in each population interbred to produce the first generation, from which he randomly selected eight males and eight females to found all subsequent generations. Because their population sizes were small, his replicate populations began to diverge from each other due to the random effects of drift. In some, the frequency of the *bw*^75 allele increased, whereas in others it decreased. Over time, populations continued to diverge in their respective allele frequencies, and by the end of the experiment, the *bw*^75 allele had either become fixed or disappeared from most of the populations. (Data from Buri 1956; data from Ayala and Kiger 1984)

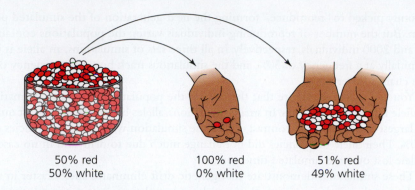

50% red
50% white

100% red
0% white

51% red
49% white

Figure 6.5 Jelly bean analogy of alleles in a population. If you grab only a few jelly beans, there is a good chance that the proportion of colors in your hand will differ from that originally in the bowl. In small populations, alleles may become more or less common from one generation to the next through a similar random process, known as genetic drift.

instead of *bw*, and *bw* was gone from the population for good. The same probabilities left other populations without any copies of *bw*[75]. In each case, biologists would say that one of the alleles had become **fixed** in the population—in other words, all its members now carried only that particular allele.

Scientists can study genetic drift through experiments like those of Buri's. They can also use mathematical models and computer simulations. **Figure 6.6** shows the output of one computer simulation study that demonstrates the effect of population size on the strength of genetic drift. In each generation, a set number of individuals are

Fixed describes an allele that remains in a population when all of the alternative alleles have disappeared. No genetic variation exists at a fixed locus within a population because all individuals are genetically identical at that locus.

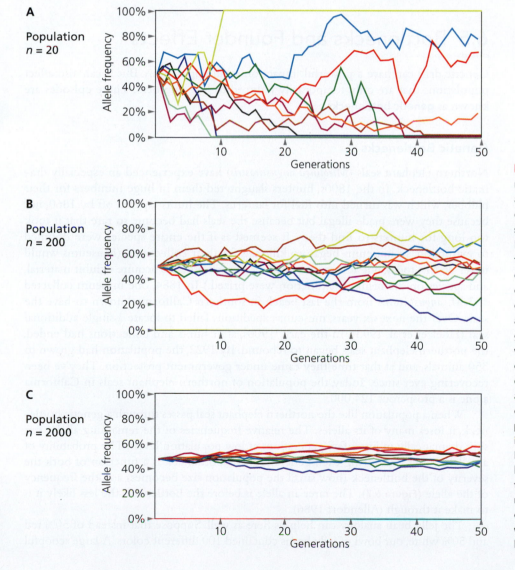

A

Population
n = 20

B

Population
n = 200

C

Population
n = 2000

Figure 6.6 These graphs show the results of computer simulations of allele frequencies within populations of different sizes (each colored line shows the results of one simulation). Each simulation began with an allele frequency of 0.5, and subsequent changes to this frequency arose entirely due to chance—the result of random sampling errors that accumulated with each successive generation. A: When populations were small (20 individuals sampled each generation), they tended to experience strong genetic drift, and the allele either became fixed or disappeared from several of the populations. B: In simulations with 200 individuals, genetic drift was weaker. Each new generation was much more likely to have the same frequency of the allele as the previous one, and in no case did the allele become fixed or disappear. C: The effect of drift was weaker still when the simulated populations contained 2000 individuals.

randomly picked to "reproduce," forming the next generation of the simulated population. But the number of reproducing individuals varies: the populations contain 20, 200, and 2000 individuals, respectively. In all three sets of simulations, an allele is present initially at a frequency of 50%, and the simulations track how the frequency of the allele changed over time.

You can see in the figure that the smaller the population, the more dramatically the allele frequency fluctuates. In small populations, alleles become fixed or lost quickly. The largest populations, by contrast, ended the simulation with allele frequencies close to 50%. Their allele frequencies did not change much due to drift, and in no case was an allele lost over the simulated time period.

These simulations demonstrate that genetic drift eliminates alleles faster in small populations than in bigger ones. But even large populations, given enough time, will eventually lose alleles due to genetic drift. As a result, genetic drift tends to rob populations of their genetic variation.

Key Concepts

- Genetic drift is the random, nonrepresentative sampling of alleles from a population during breeding. Drift is a mechanism of evolution because it causes the allelic composition of a population to change from generation to generation.

- Alleles are lost due to genetic drift much more rapidly in small populations than in large populations. ●

6.5 Bottlenecks and Founder Effects

Genetic drift can have a powerful impact on small populations. But it can also affect populations that are only temporarily reduced to low numbers. These episodes are known as **genetic bottlenecks**.

Genetic Bottlenecks

A **genetic bottleneck** is an event in which the number of individuals in a population is reduced drastically. Even if this dip in numbers is temporary, it can have lasting effects on the genetic variation of a population.

Northern elephant seals (*Mirounga angustirostris*) have experienced an especially dramatic bottleneck. In the 1800s, hunters slaughtered them in huge numbers for their blubber, which was turned into fuel for lanterns. The hunts tapered off by 1860, not because they were made illegal but because the seals had become so rare that it took too long for hunters to find them. It seemed as if the entire species were doomed to extinction, so museums began to kill seals for their collections. (Museums would never do this today, but at that time they raced each other to acquire exhibit material, and species on the brink of extinction were prized.) In 1884, one museum collected 153 *M. angustirostris* from the last beach in southern California known to have the seals. Over the next six years, museum expeditions failed to locate a single additional seal (Hoelzel et al. 1993). In the early 1900s, after hunts and collections had ended, the northern elephant seals began to rebound. By 1922, the population had grown to 350 animals, and at that time they came under government protection. They've been recovering ever since. Today, the population of northern elephant seals in California alone is a prosperous 124,000.

When a population like the northern elephant seal passes through a genetic bottleneck, it loses many of its alleles. The relative frequencies of the remaining alleles can shift dramatically in just a few generations at low population levels. The probability of an allele being lost during each generation of a bottleneck is a function of both the severity of the bottleneck (how small the population size becomes) and the frequency of the allele (**Figure 6.7**). The rarer an allele is before the bottleneck, the less likely it is to make it through (Allendorf 1986).

The jelly bean analogy can help us here as well. Suppose that instead of 50% red and 50% white, our bowl of jelly beans contained 100 different colors. A large scoopful

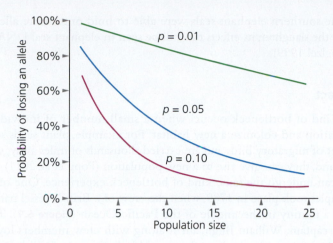

Figure 6.7 The probability of an allele making it through a bottleneck depends on the severity of the bottleneck (population size during the bottleneck) and the allele's frequency within the population before the bottleneck. Rare alleles (for example, $p = 0.01$) are the most likely to be lost. (Data from Allendorf and Luikart 2006)

of jelly beans may retain close to 100 colors. But in a small sample, many colors will be missing (Figure 6.8).

Once a population returns to its former size, its genetic variation will remain low, often for many generations. In the early 1990s, Rus Hoelzel, then at the University of Cambridge, documented this lingering effect on northern elephant seals (Hoelzel et al. 1993). He gathered tissue samples from two northern elephant seal colonies in California, and then he and his colleagues measured the diversity of alleles in the populations.

For their analysis, they chose a 300-base-pair stretch of mitochondrial DNA that is highly variable. It's typical for populations to contain 30 or more variable sites within this segment—an indication that it experiences a high mutation rate. In the case of the elephant seals, however, Hoelzel and his colleagues found only two polymorphic sites in the populations—so the level of genetic variation was extremely low.

Hoelzel found further evidence for a bottleneck when he compared the genetic variation in northern elephant seals with that in southern elephant seals. Southern elephant seals have been luckier than their northern relatives because they rear their pups on isolated islands in the ocean around Antarctica, where they were harder for hunters to find. Whereas Hoelzel estimates that the population of northern elephant seals dwindled to about only 30 animals in the nineteenth century, the number of southern elephant seals probably never dipped below 1000.

That difference is reflected in each species' genetic variation. Hoelzel found 23 different variable sites in southern elephant seals in the same 300-base-pair stretch of mitochondrial DNA that had harbored only 2 variable sites in northern elephant seals. Because their populations never dropped to as small a size (their bottleneck was

Figure 6.8 Northern elephant seals experienced a population bottleneck in the 1800s when their numbers shrank to about 30. They lost much of their genetic diversity, which has not increased much as their population has expanded. (Douglas Emlen)

less severe), the southern elephant seals were able to hold onto more alleles. Some 150 years after the slaughter, its effects can still be seen in elephant seal DNA (Hoelzel et al. 1993; Hoelzel 1999).

Founder Effect

A particular kind of bottleneck occurs when a small number of individuals leave a larger population and colonize a new habitat. For example, plant seeds sometimes stick to the feet of migratory birds and are carried thousands of miles away; when they fall to the ground, they can give rise to a new population (Popp et al. 2011).

Humans can undergo the same kind of bottleneck experience. One of the most dramatic examples took place in 1789, when the crew of a British vessel named HMS *Bounty* staged a mutiny in the middle of the Pacific Ocean (**Figure 6.9**). The mutineers set their captain, William Bligh, adrift along with crew members loyal to him and then searched for an island where they could hide from the British Royal Navy. They stopped at Tahiti, where they picked up supplies as well as 6 Tahitian men and 12 women. They sailed until they discovered Pitcairn Island, a place so remote and tiny that it had been misplaced on the Royal Navy's charts. The mutineers made the island their new home, and about a year later, on January 23, 1790, they burned the *Bounty*, leaving them stuck on the island.

Twenty-seven adults and one baby originally arrived at Pitcairn on the *Bounty*, and together they and their descendants lived in almost complete isolation for decades. By 1856 they were no longer able to support themselves on Pitcairn, so 193 residents moved to Norfolk Island. Today, Norfolk supports a population of 2000, and most residents are descendants of Pitcairn's original founders.

Figure 6.9 This painting by Robert Dodd shows mutineers turning Captain Bligh and his officers and crew adrift from the ship HMS *Bounty* on April 29, 1789. (The Print Collector/Alamy Stock Photo)

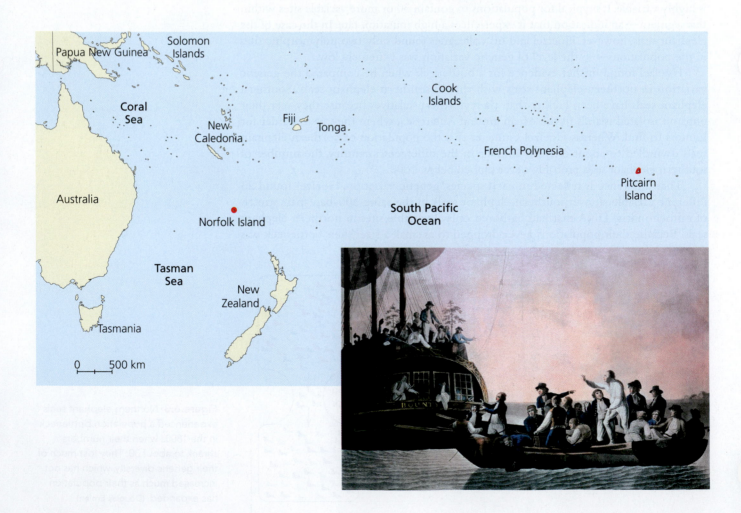

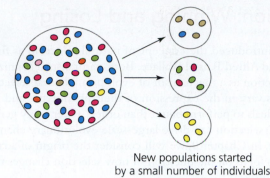

New populations started
by a small number of individuals

Figure 6.10 Genetic variation is lost when a small number of individuals start a new population. Allele frequencies may be very different in these new populations, and allelic diversity may continue to disappear due to drift while the populations remain small.

In such instances, the small number of founding individuals results in a genetic bottleneck: only a small subset of the genetic diversity of the source population is likely to be included in the new population, and the relative frequencies of these alleles may be very different from what they had been before. This is called a **founder effect** (**Figure 6.10**).

Even today, the founder effect still has a powerful influence on the genetic diversity of the descendants of the Pitcairn settlers. A remarkable 25.5% of the Norfolk islanders suffer from migraine headaches, for example, a frequency far higher than in mainland populations. To better understand this affliction, Bridget Maher, a geneticist at Griffith University in Australia, and her colleagues sequenced DNA from 600 residents of the island (Maher et al. 2012). They linked migraines among the islanders to the presence of an allele on the X chromosome (**Figure 6.11**). This allele had gone overlooked until Maher's research, likely because it is rare in mainland populations. The founder effect made the allele more prominent among residents of Norfolk Island. The same rule applies to founder effects around the world, thus making them valuable sources of insight into the human genome.

The **founder effect** is a form of genetic drift. It describes the loss of allelic variation that accompanies the founding of a new population from a very small number of individuals (a small sample of a much larger source population). This effect can cause the new population to differ considerably from the source population.

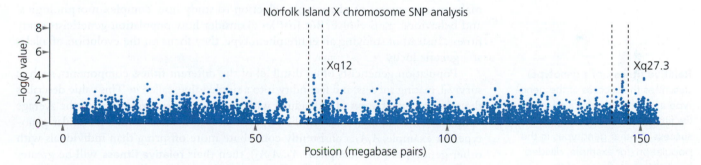

Figure 6.11 Australian researchers analyzed the X chromosome of people from Norfolk Island. They found two loci (marked Xq12 and Xq27.3) where certain alleles were strongly associated with either migraines or a lack of them. Migraine-associated alleles are unusually common among the people of Norfolk Island, thanks to the founder effect. (Data from Maher et al. 2012)

Key Concepts

- Even brief bottleneck events can lead to drastic reductions in the amount of genetic variation within a population, and this loss of allelic diversity can persist for many generations after the event.

- The founder effect is a loss of allelic variation that accompanies the founding of a new population from a very small number of individuals. Founder effects can cause the new population to differ considerably from the source population. ●

6.6 Selection: Winning and Losing

In Chapter 2, we introduced the concept of natural selection as first developed by Charles Darwin and Alfred Russel Wallace. Both naturalists recognized the profound importance of selection as a mechanism of evolution. Natural selection arises whenever (1) individuals vary in the expression of their phenotypes and (2) this variation causes some individuals to perform better than others. Over many generations, Darwin and Wallace argued, selection can drive large-scale evolutionary change, allowing new adaptations to arise. In Chapter 9, we will consider the origin of adaptations in more detail. For now, let's focus on the question of how selection changes the frequencies of alleles in a population.

Fitness

Fitness refers to the success of an organism at surviving and reproducing and thus contributing offspring to future generations.

The reproductive success of a particular phenotype is known as **fitness**, and selection occurs when individuals in a population vary in their fitness. Although this may seem straightforward enough, studying the actual fitness of real organisms is a surprisingly complicated matter. The ideal way to measure fitness would begin with tallying the lifetime reproductive contribution of an individual and then noting how many of its offspring manage to survive to reproduce themselves. In practice, however, it's hardly ever possible to make such a detailed measurement.

Scientists settle instead for reliable proxies of fitness. They sometimes measure the probability that an individual survives to the age of reproduction, for example, or they measure the number of offspring that organisms produce in a specific season. Whatever the actual metric, measuring selection entails comparing these fitness measures for many different individuals within a population and relating variation in fitness with variation in the expression of a phenotype.

Another difficulty when it comes to measuring fitness is the complicated relationship between genotype and phenotype (see Chapter 5). The fitness of an organism is the product of its *entire* phenotype. We'll see in Chapters 7 and 10 how scientists can make measurements of phenotypic selection to study how complex morphological and behavioral traits evolve. But first let's consider how population geneticists study fitness. Instead of studying an entire phenotype, they focus on the evolution of alleles at a genetic locus.

Relative fitness (of a genotype) describes the success of the genotype at producing new individuals (its fitness) standardized by the success of other genotypes in the population (for example, divided by the average fitness of the population).

Population geneticists often distill all of the different fitness components, such as survival, mating success, and fecundity, into a single value, called w. This value describes the relative contribution of individuals with one genotype, compared with the average contribution of all individuals in the population. If individuals with a particular genotype, for example, A_1A_1, consistently contribute more offspring than individuals with other genotypes (for example, A_1A_2, A_2A_2), then their **relative fitness** will be greater than 1. Conversely, if the net contributions of individuals with a genotype are lower than those of other individuals, the relative fitness will be less than 1. Sometimes population geneticists calculate relative fitness by comparing the fitness of all individuals to the fitness of the most successful genotype in the population, rather than to the mean fitness of the population. In such cases, the genotype with the highest fitness has a relative fitness of $w = 1$, and all other genotypes have relative fitnesses that are between 0 and 1. Regardless of which way fitness is measured, selection will always occur if two or more genotypes differ consistently in their relative fitness. The strength of selection will reflect how different the genotypes are in their respective fitnesses.

To understand how selection leads to changes in the frequencies of alleles, we can consider the contributions of a specific *allele*, rather than a genotype, to fitness. But calculating the relative fitness of an allele is more complicated than calculating that of a genotype, for two reasons. First, alleles in diploid organisms don't act alone. They are always paired with another allele to form the genotype (and subsequently the phenotype). If there is, say, a dominance interaction between alleles at a locus (see Box 5.2),

Genotypes, Phenotypes, and Selection

Selection operates directly on phenotypes because phenotypic variation among organisms influences the relative probability of survival and reproduction. Those phenotypes, in turn, are influenced by alleles. Although the relationship between alleles and phenotypes is rarely known and often complex, it is still possible for alleles at genetic loci to experience selection. Population geneticists can sample individuals for their genotype at a locus and compare the fitness of individuals with one genotype (that is, the average fitness of the genotype) with the fitnesses of individuals with other genotypes. When genotypes differ consistently in their fitness, the genetic locus can be said to be under selection. The selection coefficient(s) is used to describe how much the genotypes differ in their fitness.

In many cases, the allelic variation at a particular locus does not influence the phenotype. In such cases, the alleles are "hidden" from the action of selection because they are selectively neutral. Even if an allele does affect the phenotype, it still could be selectively neutral if the change in phenotype has no effect on reproductive success.

---

this interaction will influence the phenotype. Second, selection does not act directly on alleles; it acts on individuals and their phenotypes (**Box 6.4**).

Nevertheless, it is still possible to calculate the net contributions of an allele to fitness. To do so, we must consider the fitness contributions of individuals heterozygous and homozygous for the allele. Then we must weigh how many individuals with each genotype are actually present in the population and contributing offspring to the next generation. **Box 6.5** shows how the net fitness contribution of an allele, called the **average excess of fitness**, is calculated.

The average excess of fitness for an allele can be used to predict how the frequency of the allele will change from one generation to the next:

$$\Delta p = p \times (a_{A_1}/\overline{w})$$

where Δp is the change in allele frequency due to selection, p is the frequency of the A_1 allele, $\overline{w}$ is the average fitness of the population, and a_{A_1} is the average excess of fitness for the A_1 allele. This equation can tell us a lot about the nature of natural selection.

The sign of the average excess of fitness (a_{A_1}), for example, determines whether selection increases an allele's frequency or decreases it. Whenever an allele is present in a population, its frequency is greater than zero; as long as the population exists, its average fitness, $\overline{w}$, is also greater than zero (because $\overline{w}$ is the sum of all individuals with each genotype multiplied by their respective contributions of offspring to the next generation). Because both p and $\overline{w}$ are by definition positive, the sign of Δp must be determined by the average excess of fitness of the allele. Whenever the fitness effects of an allele are positive, selection should increase the frequency of the allele over time; the converse is true when the fitness effects are negative.

This equation also tells us that the speed of change in the frequency of an allele will depend on the strength of selection that it experiences—the magnitude of a_{A_1}. When the average excess of fitness is very large (positive or negative), the resulting change in allele frequency will be greater than when the average excess of fitness is smaller.

Finally, this equation shows us that the effectiveness of selection at changing an allele's frequency depends on how common it is in the population. When an allele is very rare $(p \approx 0)$, the power of selection to act will be low even if the fitness effects of the allele are pronounced.

Average excess of fitness (of an allele) is the difference between the average fitness of individuals bearing the allele and the average fitness of the population as a whole.

BOX 6.5

Selection Changes Allele Frequencies

Let's consider how natural selection changes allele frequencies by starting with a population in Hardy–Weinberg equilibrium at a genetic locus. We will then calculate how selection pulls the population out of equilibrium and, in so doing, shifts the frequencies of the alleles.

We'll use the same locus and alleles that we did in Box 6.2, A_1 and A_2, and starting frequencies of p and q, respectively. We've already seen that for a population in Hardy–Weinberg equilibrium, the frequencies of each possible genotype are

$$f(A_1A_1) = p^2$$

$$f(A_1A_2) = 2pq$$

$$f(A_2A_2) = q^2$$

Selection acts on a genetic locus whenever the genotypes of that locus differ in their relative fitness. In this case, we can assign fitnesses to each genotype as $w_{A_1A_1}$, $w_{A_1A_2}$, and $w_{A_2A_2}$, or, more simply, w_{11}, w_{12}, and w_{22}, respectively. Fitness can act through many components, such as survivorship to the age of reproduction, mating success, and fecundity, but ultimately these all translate into the success of each genotype at contributing offspring to the next generation. Here, we'll let our fitness measures encompass all of these, so that w_{11}, w_{12}, and w_{22} denote the proportional contributions of offspring by individuals with A_1A_1, A_1A_2, and A_2A_2 genotypes, respectively.

To calculate the genotype frequencies after selection (time $t + 1$), we need to multiply the frequency of each genotype by its relative fitness. In essence, this simulates a parental population that reproduces to give rise to an offspring generation with zygote genotype frequencies of p^2, $2pq$, and q^2. (These are the genotype frequencies in the population immediately before selection.) These offspring then experience selection as they develop into reproductively mature adults themselves, who then mate to produce yet another generation of progeny. The relative success of individuals with each genotype at surviving to adulthood, competing successfully for mates, and producing viable offspring is reflected in their respective relative fitness values—selection is acting on this genetic locus. Because of selection, some genotypes will increase in frequency at the expense of others in the next generation.

So, at time $(t + 1)$, the number of individuals with each genotype will be represented by the following:

Genotype:	A_1A_1	A_1A_2	A_2A_2
Numbers:	$p^2 \times w_{11}$	$2pq \times w_{12}$	$q^2 \times w_{22}$

But the total number of individuals in this new generation will not be the same as in the previous generation. Individuals may have produced multiple offspring, or individuals with particular genotypes may have died before breeding, for example. To convert these numbers into new frequencies for each genotype, we need to standardize them by the total number of individuals in the new generation. This new total is just the sum of the numbers of individuals having each possible genotype:

$$\overline{w} = p^2 \times w_{11} + 2pq \times w_{12} + q^2 \times w_{22}$$

The term $\overline{w}$ is also called the *average fitness of the population* because it's the sum of the fitnesses of each genotype multiplied by (that is, weighted by) the frequencies at which they occur. Using the average fitness of the population, we can now turn the relative numbers of individuals with each genotype after selection into the new genotype frequencies at time $t + 1$.

Genotype:	A_1A_1	A_1A_2	A_2A_2
f_{t+1}:	$(p^2 \times w_{11})/\overline{w}$	$(2pq \times w_{12})/\overline{w}$	$(q^2 \times w_{22})/\overline{w}$

And from these results, we can calculate each *allele* frequency in this new generation as the frequency of homozygote individuals plus half the frequency of heterozygotes:

$$p_{t+1} = [(p^2 \times w_{11})/\overline{w}) + (pq \times w_{12})]/\overline{w}$$
$$= (p^2 \times w_{11} + pq \times w_{12})/\overline{w}$$

and

$$q_{t+1} = [(q^2 \times w_{22})/\overline{w}) + (pq \times w_{12})]/\overline{w}$$
$$= (q^2 \times w_{22} + pq \times w_{12})/\overline{w}$$

Natural selection is a mechanism of evolution because it can cause allele frequencies to change from generation to generation. Now that we have applied selection (as differential fitnesses) to our genotypes, let them reproduce, and calculated the new allele frequencies in the offspring generation, the question is, how have the allele frequencies changed?

To calculate the change in frequency of the A_1 allele, Δp, we subtract the starting frequency, p, from the new frequency, p_{t+1}:

$$\Delta p = p_{t+1} - p$$

The starting frequency $p = p^2 + pq$. To express this over the denominator, $\overline{w}$, we multiply it by 1 in the form of $\overline{w}/\overline{w}$ so that $p = (p^2 \times \overline{w} + pq \times \overline{w})/\overline{w}$, Therefore,

$$\Delta p = p_{t+1} - p$$

$$= [(p^2 \times w_{11} + pq \times w_{12})/\overline{w}] - [(p^2 \times \overline{w} + pq \times \overline{w})/\overline{w}]$$

$$= (p^2 \times w_{11} + pq \times w_{12} - p^2 \times \overline{w} - pq \times \overline{w})/\overline{w}$$

$$= p \times (p \times w_{11} + q \times w_{12} - p \times \overline{w} - q \times \overline{w})/\overline{w}$$

$$= (p/\overline{w}) \times (p \times w_{11} - p \times \overline{w} + q \times w_{12} - q \times \overline{w})$$

$$= (p/\overline{w}) \times [p \times (\overline{w}_{11} - \overline{w})] + [q \times (\overline{w}_{12} - \overline{w})]$$

The term $[p \times (w_{11} - \overline{w})] + [q \times (w_{12} - \overline{w})]$ is known as the average excess of fitness for the A_1 allele, written as a_{A_1}. This term represents the difference between the average fitness of individuals having the A_1 allele and the average fitness of the population as a whole (**Box Figure 6.5.1**). In essence, the average excess of fitness is a way to assign fitness values to *alleles*, even though alleles have no phenotypes of their own—they have phenotypes only when they are combined in pairs to form genotypes. Thus, although it is individuals with *genotypes* who experience selection and who differ in their relative contributions to subsequent generations, we can still assign relative fitnesses to *alleles* in the form of their average excess of fitness. This approach allows us to see clearly the relationship between alleles and fitness.

We can now express the change in allele frequencies resulting from selection as

$$\Delta p = (p/\overline{w}) \times a_{A_1}$$

$$\Delta p = p \times (a_{A_1}/\overline{w})$$

where a_{A_1} is the average excess of fitness of the A_1 allele. We can also calculate the average excess of fitness for the A_2 allele as

$$a_{A_2} = [p \times (w_{12} - \overline{w})] + [q \times (w_{22} - \overline{w})]$$

and the predicted change in frequency of the A_2 allele as a result of selection as

$$\Delta q = (q/\overline{w}) \times a_{A_2}$$

$$\Delta q = q \times (a_{A_2}/\overline{w})$$

Two important conclusions can be drawn from these calculations. First, because p and $\overline{w}$ are always greater than or

Frequency of A_1 alleles that are present in A_1A_1 homozygous individuals	Frequency of A_1 alleles that are present in A_1A_2 heterozygous individuals

$$a_{A_1} = [p \times (w_{11} - \overline{w})] + [q \times (w_{12} - \overline{w})]$$

Difference in fitness between A_1A_1 individuals and the mean fitness of the population	Difference in fitness between A_1A_2 individuals and the mean fitness of the population

Box Figure 6.5.1 The average excess of fitness of the A_1 allele (a_{A_1}) is calculated from the frequency of A_1 alleles in homozygotes and heterozygotes, each adjusted by their differences in fitness from the mean of the population.

equal to 0, whether Δp is positive or not depends entirely on the sign of a_{A_1}. This means that whether the allele increases or decreases in frequency from generation to generation depends on whether its average excess of fitness is positive or negative. When the net effect of an allele is an increase in fitness (average excess in fitness is greater than 0), meaning that the allele experiences positive selection, the allele is predicted to increase in frequency. Conversely, when the net effect of an allele is a decrease in fitness (average excess in fitness is less than 0), meaning that the allele experiences negative selection, the allele is predicted to decrease in frequency.

Second, the average excess of fitness depends not only on the fitnesses *but also on the frequencies* of each allele. This means that the effect of selection acting on an allele will depend on the population context in which it is found. For example, two populations with identical fitnesses for each genotype could have very different average excesses of fitness if they have different allele frequencies before selection. Selection may cause rapid changes in allele frequency in one population, but only minor changes in the other. When an allele is very rare (as it would be if it had recently arisen through mutation), selection may be much less effective at changing its frequency than it would be if the allele were more common.

Small Differences, Big Results

Alleles can differ enormously in their effects on fitness. A single mutation can disable an essential protein, for example, leading to a lethal genetic disorder. Such an allele experiences strong negative selection because children who die of the disorder cannot pass on the mutation to offspring. As a result, a typical severe genetic disorder affects only a tiny fraction of the population.

But even when alleles are separated by only a small difference in their average excess of fitness, selection can have big long-term effects. That's because populations grow like investments earning interest.

Let's say you invest $100 in a fund that earns 5% interest each year. In the first year, the fund will increase by $5. In the second, it will increase by $5.25. In every subsequent year, the fund will increase by a larger and larger amount. In 50 years, you'll have more than $1,146. Because of this accelerating growth, even a small change in the interest rate can have a big effect over time. If the interest rate on your fund is 7% instead of 5%, you'll make only an extra $2 in the first year. But in 50 years, the fund will be worth more than $2,945—close to triple what an interest rate of 5% would yield. Slight differences in fitness get magnified in a similar way. Over time, an allele with a slightly higher average excess for fitness can come to dominate a population.

Unlike genetic drift, this compounding power of natural selection is more effective in larger populations than in smaller ones. That's because genetic drift can erode allelic variation in small populations, even eliminating beneficial mutations. In large populations, by contrast, genetic drift has a weaker effect, leaving selection free to alter allele frequencies.

Figure 6.12 shows computer simulations that illustrate the relative impacts of drift and natural selection, in which an allele with a selective advantage of 5% is added to populations of different sizes. In the largest population of 10,000 individuals, the allele becomes more common in all the simulations. In a population of 10 individuals, however, it disappears from half of the simulations. High relative fitness, in other words, is not a guarantee that an allele will spread—or even persist—in a population, because the effects of drift can be stronger than those of selection when populations are very small.

The relative importance of selection and drift spurred fierce debates over much of the twentieth century. The British geneticist Ronald Fisher (1890–1962) (Figure 6.13A), for example, developed some of the first statistical methods for studying natural selection. Fisher held that populations were large and well mixed, making natural selection more effective. On the other hand, the American geneticist Sewall Wright (1889–1988) (Figure 6.13B) considered populations to be networks of smaller groups that sometimes exchanged alleles. In Wright's model, genetic drift played a much more important role. The Japanese geneticist Motoo Kimura (1924–1994) (Figure 6.13C)

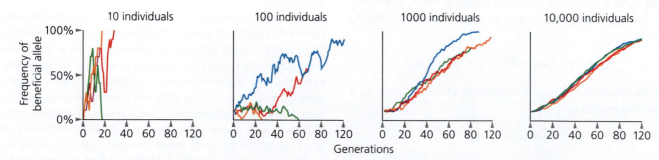

Figure 6.12 Natural selection is ineffective in small populations and effective in large ones. These graphs show the results of computer simulations of a population in which an allele that raises fitness by 5% is added to populations of different sizes (each colored line represents a different simulation). In all cases, the allele starts at a frequency of 0.1 (10%), and subsequent changes in its frequency result from the combined action of selection and drift. In the smallest populations, the allele disappears from half the simulations, even though it has beneficial effects on fitness. But in large populations, the allele becomes more common in all of them. (Data from Bell 2008)

Figure 6.13 A: Ronald Fisher (1890–1962). B: Sewall Wright (1889–1988). C: Motoo Kimura (1924–1994). (A: A. Barrington Brown/ Science Source; B: U.S. Department of Agriculture; C: Annual Review of Genetics, December 1996: Vol. 30:1–5)

noted that much of the molecular variation in species did not influence their phenotypes. As a result, he maintained, genetic drift could account for much of the allelic variation in genomes. These debates were based largely on mathematical models of evolution. In recent years, advances in DNA sequencing have allowed researchers to test models of drift and selection against empirical data.

Patterns of Selection in Time and Space

Selection can produce patterns of surprising complexity. In the next few sections, we'll consider how those patterns are generated, starting with one important fact about mutations: they often have more than one effect on an organism. These multiple effects, known as **pleiotropy**, are an example of the interconnectedness of biology. A single regulatory gene, for example, can influence the expression of many other genes.

> **Pleiotropy** is the condition when a mutation in a single gene affects the expression of more than one different phenotypic trait.

The evolution of resistance in mosquitoes on the coast of France demonstrates how pleiotropy can affect the nature of selection. When the $Ester^1$ allele emerged in the early 1970s, it provided mosquitoes with resistance to insecticides. But it had other effects on the mosquitoes as well. Researchers at the University of Montpellier have found that $Ester^1$ mosquitoes have a higher probability of being caught by spiders and other predators than insecticide-susceptible mosquitoes do, for example (Berticat et al. 2004). A mutation that has improved fitness in one context—by providing resistance to insecticides—has also altered the physiology of these mosquitoes in a way that may well lower their fitness in other contexts. This form of pleiotropy, in which the effects of a mutation have opposite effects on fitness, is known as **antagonistic pleiotropy**.

> **Antagonistic pleiotropy** occurs when a mutation with beneficial effects for one trait also causes detrimental effects on other traits.

The net effect of an allele on fitness is the sum of its pleiotropic effects on the organism in question. Even if an allele has some beneficial effects, it may, on balance, lower reproductive success overall. How the balance tips depends on the environment in which an organism lives. For mosquitoes on the French coast, any protection against insecticides can dramatically raise fitness because susceptible mosquitoes are dying in droves. Even if the extra esterases make the mosquitoes more vulnerable to predators, they still, on balance, make the insects more fit.

Such is not the case for mosquitoes farther inland. There, the $Ester^1$ allele provides no benefit from resistance because there's no insecticide to resist. Instead, the allele lowers fitness by making the insects easier prey. The curves shown in Figure 6.2 are the result of this shift. Selection raised the frequency of $Ester^1$ along the coast while keeping it low inland. This difference was maintained even as mosquitoes were migrating from one site to another and their alleles were flowing across southern France. As soon as copies of $Ester^1$ left the insecticide zone, they were often eliminated by selection.

As **Figure 6.14** shows, the $Ester^1$ allele became common along the coast in the 1970s, but it later became rare. That's because a new allele, known as $Ester^4$, emerged around 1985. $Ester^4$ also led to the overproduction of esterases. Intriguingly, $Ester^4$ became more common as $Ester^1$ was disappearing—even though $Ester^4$ provides slightly *less* protection against insecticides than the older $Ester^1$ allele. A clue to the success of $Ester^4$ comes from the fact that the prevalence of $Ester^4$ does not drop

Figure 6.14 This series of graphs extends the history of resistance alleles in mosquitoes that we encountered in Figure 6.2. After spreading widely in the 1970s, the *Ester*[1] allele gradually became rarer in the 1980s and 1990s, while another allele, known as *Ester*[4], became widespread. The shift may reflect a physiological cost imposed on the insects by *Ester*[1]. *Ester*[4] alleles may confer resistance on mosquitoes without this cost, making its relative fitness higher and driving it to higher frequencies. (Data from Raymond et al. 1998)

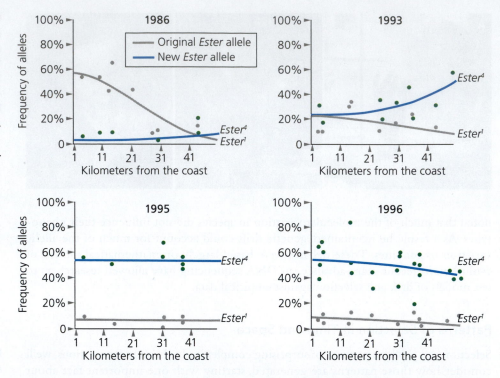

off steeply as you go inland. It's likely that *Ester*[4] does not impose the high cost of increased predation of *Ester*[1]. Selection favors *Ester*[4] on the coast, but mosquitoes don't pay a price for carrying it if they migrate inland (Raymond et al. 1998).

Sixty Thousand Generations of Selection: Experimental Evolution

Some of the most important insights into how selection affects alleles have come from experiments that scientists set up in their laboratories. In such situations, researchers can carefully control the conditions in which organisms grow and reproduce, and they can analyze the entire population under study.

One of the longest running of these experiments is taking place at Michigan State University (Barrick et al. 2009). Richard Lenski started it in 1988 with a single *Escherichia coli* bacterium. He allowed the microbe to produce a small group of genetically identical descendants. From these clones, he started 12 genetically identical populations of bacteria (**Figure 6.15**). Each population lives in a flask containing 10 milliliters (ml) of a solution. The bacteria grow on glucose, but Lenski supplied only a limited concentration. Each day—including weekends and holidays—someone in Lenski's group withdraws 0.1 ml from a culture and transfers that into 9.9 ml of fresh solution. They do this for each of the 12 populations, keeping each population separate from the others. The bacteria grow until the glucose is depleted and then sit there until the same process is repeated the next day. In a single day, the bacteria divide about seven times.

All of the bacteria descended from a single ancestral genotype. As they reproduced, they occasionally acquired new mutations. Alleles that lowered their reproductive success experienced **negative selection**. Any alleles that sped up their growth or boosted their survival rate experienced **positive selection**. The random sample Lenski took each day from each flask reflected these shifting frequencies of alleles.

Every 500 generations, Lenski and his students stored some of the bacteria from each of the 12 lines in a freezer. Freezing did not kill them, so the samples became a frozen "fossil" record that could be resurrected at a later time. When he thawed them out later, Lenski could directly observe how quickly the ancestral and descendant bacteria grew under the same conditions, as a measure of relative fitness. He could thus directly measure their change in average fitness.

Negative selection refers to selection that decreases the frequency of alleles within a population. Negative selection occurs whenever the average excess for fitness of an allele is less than zero.

Positive selection is the type of selection that increases allele frequency in a population. Positive selection occurs whenever the average excess for fitness of an allele is greater than zero.

The experiment has now progressed for 60,000 generations. (It would have taken about 1.5 million years if Lenski were using humans as experimental organisms instead of bacteria.) **Figure 6.16** shows the evolution of Lenski's *E. coli* over the first 50,000 generations (Good et al. 2017). In all 12 populations, the bacteria became more fit in their low-glucose environment than their ancestors had been. The average competitive fitness of the populations increased by approximately 75% relative to the ancestor. In other words, all 12 of the bacterial populations evolved in response to natural selection: they had accumulated mutations that made them more efficient at growing under the conditions that Lenski set up. The rate of increase in fitness has declined in recent years, but the fitness of the bacteria continues to rise and is expected to continue for years to come (Wiser et al. 2013; Good et al. 2017).

Preserving a frozen fossil record doesn't just allow Lenski to compare ancestors against descendants. It also allows him and his colleagues to compare the DNA of those individuals. Because the experiment began with a single microbe, and because the microbe's descendants reproduced asexually without horizontal gene transfer, the researchers can be confident that alleles present in descendants but not in the original ancestor must have arisen through mutation during the experiment itself.

Lenski and his colleagues have been investigating these new mutations, observing how they affect the fitness of the bacteria. In one experiment, they selected a single microbe from generation 10,000 to analyze (Stanek et al. 2009). They transferred 1296 different segments of its DNA into ancestral bacteria from the same line. Then they mixed each kind of engineered bacteria with unmanipulated ancestral ones and allowed them to grow side by side. These trials revealed one evolved segment in particular that increased the fitness of the bacteria. Further analysis allowed Lenski and his colleagues to pinpoint the mutation within the segment that was responsible. A single nucleotide was mutated in a protein-binding site, called *BoxG1*, which regulates a pair of nearby genes. These genes encode proteins called GlmS and GlmU, which help synthesize the bacterial cell membrane. To confirm that the mutation was indeed responsible for increasing bacterial fitness, they inserted the single nucleotide into *BoxG1* in the ancestral bacteria. That tiny insertion raised the relative fitness of the bacteria by 5%.

Having identified this mutation and measured its fitness, Lenski and his colleagues then set out to trace its origin. At some point during the evolution of that particular line of *E. coli*, they hypothesized, the mutation must have emerged and then increased in frequency. They turned to the line's frozen fossil record, selected bacteria from each 500-generation sample, and examined them for the presence of the *BoxG1* mutation. None of the bacteria they examined from generation 500 had the *BoxG1* mutation. So the mutation must have arisen after that point. The bacteria in generation 1000 told a different story: 45% of them carried the mutation. And in generation 1500, the researchers found that 97% of the bacteria had it. This rapid spread is the kind of pattern you'd expect from a mutation that increases fitness.

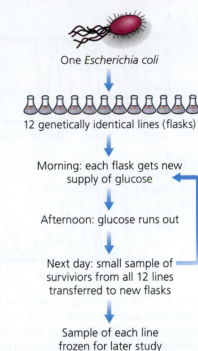

One *Escherichia coli*

12 genetically identical lines (flasks)

Morning: each flask gets new supply of glucose

Afternoon: glucose runs out

Next day: small sample of surviviors from all 12 lines transferred to new flasks

Sample of each line frozen for later study every 500 generations

Figure 6.15 Richard Lenski and his colleagues have bred bacteria for more than 25 years using this method.

Figure 6.16 The bacteria in Lenski's experiment have experienced natural selection. New mutations have caused the descendants to reproduce faster under the conditions of the experiment than their ancestors did. Even after 25 years, the bacteria are continuing to experience natural selection. (Data from Good et al. 2017; luismmolina/Getty Images)

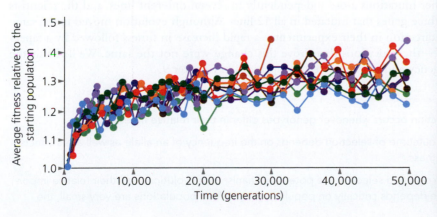

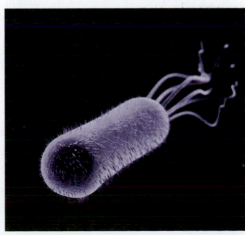

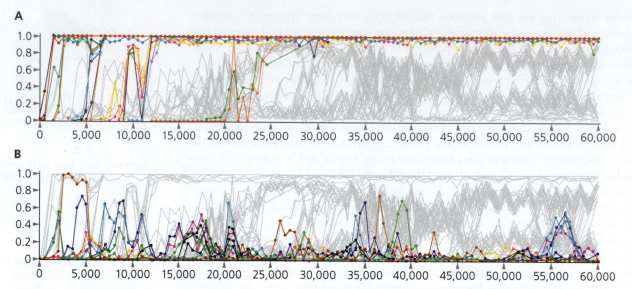

Figure 6.17 By sequencing the genomes of bacteria sampled every 500 generations, Lenski and his colleagues were able to reconstruct a complete picture of the evolution of each of their experimental populations, documenting the appearance of every new mutation and the fates of each of these new alleles. Here we show changes in relative frequency over time for all of the alleles that arose in one of their experimental populations (gray lines). A: Some of these alleles proved to be beneficial, eventually becoming fixed in the population by selection (colored lines). B: Other alleles arose and persisted for a while, but were eventually lost either because their effects were deleterious to the bacteria or simply due to the action of genetic drift. Studies like this provide an unusually comprehensive view of the mechanisms of evolution in action. (Data from Good et al. 2017)

It's not immediately obvious how the mutation benefits the bacteria, but Lenski and his colleagues have some clues. In bacteria with the *BoxG1* mutation, less GlmS and GlmU is expressed. It's possible that the bacteria divert resources from building thick membranes to other functions, speeding up their reproduction.

The *BoxG1* mutation is just one of a growing collection of beneficial mutations that Lenski and his colleagues have identified in their long-term evolution experiment. These mutations arose sequentially in the bacterial lines, building on the increased fitness of previous mutations.

Lenski's team has also documented other alleles that arose, persisted for a period of time, and eventually disappeared. **Figure 6.17** shows the fates of new beneficial and deleterious alleles for one of the 12 experimental populations, providing an unusually clear picture of 60,000 generations of evolution.

Large-scale comparisons of these mutations are revealing lessons about how beneficial mutations interact. Some mutations, for example, are beneficial only when they follow certain other mutations. That's because their effects on the bacteria interact with alleles at other loci in a process known as **epistasis**.

Finally, Lenski's experiment also makes it possible to observe how predictable evolution is at the molecular level. Only one line of *E. coli* evolved the *BoxG1* mutation. But other mutations arose independently in several different lines, and the scientists found three genes that mutated in all 12 lines. Although evolution moved in the same overall direction in their experiment—a rapid increase in fitness followed by a tapering off—the mutations that drove this change were not the same. We'll revisit the contingency and convergence of adaptations in Chapter 9.

Epistasis occurs when the effects of an allele at one genetic locus are modified by alleles at one or more other loci.

Key Concepts

- Selection occurs whenever genotypes differ in their relative fitnesses.

- The outcome of selection depends on the frequency of an allele as well as its effects on fitness.

- Both drift and selection are potent mechanisms of evolution, and their relative importance depends critically on population size. When populations are very small, the effects of drift can enhance the action of selection (for example, by removing harmful alleles that have been driven to low frequency by selection) or oppose it (for example, by removing beneficial alleles). When populations are large, the effects of drift are minimal and selection is the more important force.

- Alleles often affect the fitness of an organism in more than one way. When these fitness effects oppose each other, the balance between them will determine the net direction of selection acting on the allele. This balance may tip one way in one environment and a different way in others.

- Laboratory studies of experimental evolution help reveal how new alleles can arise and spread through a population in response to selection. ●

6.7 Dominance: Allele Versus Allele

Bacteria are useful for running evolution experiments because their haploid genetics are relatively simple. Selection can be more complex in diploid organisms, however, due to interactions between the two alleles at each genetic locus. As we saw in Chapter 5, an allele can act independently of its partner or it can be either dominant or recessive. Each of these states can have different effects on the course of selection.

Let's first consider alleles that act independently. In Chapter 5, we introduced the work of Joel Hirschhorn and his colleagues on the genetics of height. One of the genes they discovered, *HMGA2*, has a strong influence on stature. People who carry one copy of a variant of the gene will grow about half a centimeter taller, on average, than people who lack it. People who are homozygous for the allele get double the effect and grow about a centimeter taller. Such interactions between alleles are called **additive** because the effects of the alleles can be predicted simply by summing the number of copies that are present.

Additive alleles are especially vulnerable to the action of selection. Whenever an additive allele is present, it will affect the phenotype, and selection can act on it. Favorable alleles can be carried all the way to fixation because heterozygous individuals will have higher fitness than individuals lacking the allele, and homozygous individuals will fare even better. Eventually, the population will contain only individuals homozygous for the allele (**Figure 6.18**). Conversely, deleterious additive alleles can be entirely removed from a population. Every time the allele is present, it is exposed to selection, and its bearers suffer lower fitness than other individuals lacking the allele. Here, too, the result will be absolute: selection will remove the allele completely from the population.

Dominant and recessive alleles, on the other hand, are not additive. A dominant allele will overshadow the other allele at the same locus. It will have the same effect on an individual's phenotype whether one copy is present in a heterozygote or two copies are present in a homozygote. A recessive allele, on the other hand, can affect the phenotype only when it is paired with another copy of the same recessive allele—that is, when it occurs in a homozygous recessive individual.

This interaction blunts the power of selection to spread alleles to fixation or to eliminate them from a population. When a mutation gives rise to a new recessive allele, the individual carrying it is, by necessity, a heterozygote. As a result, the new recessive allele will have no effect on the phenotype. The heterozygous individual may or may not pass down the new recessive allele to its offspring; if it does, its offspring will be heterozygotes as well because no other individuals in the population carry the allele (they are all homozygous for the ancestral allele). Even if, by chance, some other member of the population also acquires the same recessive mutation, the odds will be tiny that the two alleles will end up combined in a homozygote. As a result, rare recessive alleles are almost always carried in heterozygous individuals.

Because recessive alleles don't affect the phenotype of heterozygotes, they remain largely hidden from the action of selection. Drift alone determines whether recessive

Additive describes an allele that yields twice the phenotypic effect when two copies are present at a given locus than occurs when only one copy is present. Additive alleles are not influenced by the presence of other alleles (for example, there is no dominance).

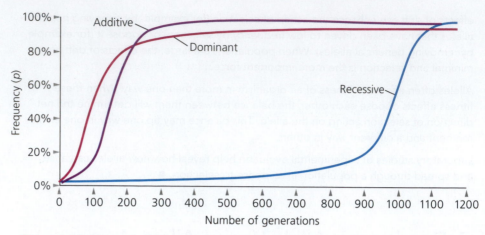

Figure 6.18 Effects of positive selection on additive, recessive, and dominant alleles. Each line shows predicted changes in allele frequency given a selection coefficient of 0.05. Alleles with additive effects on phenotypes (purple line) are always exposed to selection, so they will increase steadily from the moment they arise due to mutation until they are fixed in the population. Recessive alleles (blue line) are not exposed to selection initially because they are likely to occur only in heterozygous genotypes. They may linger for thousands of generations until drift either removes them or increases their frequency. Eventually, if drift increases their frequency sufficiently, homozygous recessive individuals will begin to appear in the population. As soon as this happens, selection will begin to increase the frequency of the allele and swiftly carry it to fixation. Dominant alleles (red line) are exposed to selection immediately, and their frequency will increase rapidly. However, as dominant alleles become increasingly common, the alternative alleles (by definition, recessive) become increasingly rare. As we've just seen, rare recessive alleles are invisible to the action of selection because they are carried in a heterozygous state. Thus selection alone is unlikely to fix a completely dominant allele. (Data from Conner and Hartl 2004)

alleles persist in the population. Eventually, drift may increase a recessive allele's frequency such that heterozygotes become fairly common. At that stage, the odds become more likely that two heterozygotes will encounter each other and mate. Only then do homozygous recessive offspring begin to appear in the population. And only then can selection begin to act on the recessive allele.

If the effects of the allele are positive, selection can quickly increase the frequency of the allele. As the allele spreads, more and more individuals are born with homozygous recessive genotypes. Because the dominant alternative allele performs less well (its average excess of fitness is negative), it declines in frequency. Because these negative effects are present in both heterozygous and homozygous genotypes, this deleterious dominant allele has nowhere to hide. Selection can purge it completely from the population. Thus, after a long period during which the recessive allele experiences only drift, it can rapidly spread to fixation (see Figure 6.18).

Selection has a different effect on deleterious recessive alleles. If drift creates a high frequency of heterozygotes, they will start to produce recessive homozygotes that in turn will suffer lowered fitness. As a result, the allele will become less common. But selection cannot remove the recessive allele completely, despite its low fitness. As soon as the recessive allele's frequency drops again, it occurs mostly in a heterozygous state where it is hidden once more from selection.

Selection has a very different impact on a dominant allele that appears in a population. Right from the start, the new dominant allele is exposed to selection. If its effects are favorable (its average excess of fitness is positive), it spreads rapidly through the population. At first, while the allele is still rare, it is present almost entirely in heterozygous individuals. That's because the rest of the population carries the ancestral allele at that locus. As the dominant allele becomes more common, however, heterozygous individuals begin to pair with other heterozygous individuals and produce homozygous individuals that carry two copies of the new dominant allele. Such individuals experience the same fitness advantage as heterozygotes, and the frequency of the allele continues to climb.

As the frequency of the new dominant allele approaches fixation, the population is increasingly composed of dominant homozygotes. Fewer and fewer individuals are heterozygous. Even fewer offspring that are homozygous for the ancestral allele are produced. Eventually, the ancestral (now recessive) allele becomes so rare that heterozygotes almost never meet and mate. At this point, the recessive allele is present only in heterozygotes. Because the recessive allele has no effect on the phenotype of heterozygotes, there is no longer any difference in fitness among individuals caused by this genetic locus. There is no more selection acting on the allele. Its fate is now governed by drift. Thus, although selection can drive a dominant allele to high frequency very rapidly, it is unlikely to drive the allele all the way to fixation because it cannot eliminate the ancestral recessive allele (Figure 6.18).

These dynamics help explain why populations harbor so much genetic variation, and why so much of this variation is made up of rare recessive alleles with deleterious effects. Whenever mutations generate alleles with dominance interactions, the potential arises for deleterious recessive alleles to hide from selection in a heterozygous state. And as long as they are rare, the deleterious alleles can persist in populations for thousands of generations, until they are eventually lost to drift. We'll see later in this chapter how this variation can rear its ugly head when recessive alleles are flushed out of hiding.

Mutation-Selection Balance

Ultimately, new mutations are the source of genetic diversity. At first, this might seem like a weak force, because the rate of new mutations at any particular genetic locus is typically very low (see Chapter 5). Several recent studies have estimated the human mutation rate to be 1.2×10^{-8} per position per haploid genome (Campbell and Eichler 2013). In other words, a gene would need to be copied for more than 100 million generations before a particular single nucleotide mutated.

But we actually don't have to wait nearly so long for mutations to arise. For one thing, each human genome is huge, containing 3.5 billion base pairs. With such a big target, even a low mutation rate will be guaranteed to produce some mutations. About 61 new point mutations arise in each baby. And because about 130 million babies are born each year, we can estimate that about 7.9 billion new mutations are arising in humans each year. Although the odds of a mutation striking a particular locus as it is being copied in any given individual are extremely low, the rate at which mutations arise in the entire human population is not.

Many of these mutations turn out to be neutral, but a significant number have important phenotypic effects. Cystic fibrosis, for example, is a genetic disorder in which the lungs build up with fluid, leading to pneumonia. The median life expectancy for Americans with cystic fibrosis is 37 years. The disease is caused by mutations to the *CFTR* gene, which encodes a chloride channel in epithelial cells. More than 2000 different disease-causing alleles of the *CFTR* gene have already been identified (Veit et al. 2016).

Mutations are thus an important mechanism of evolution, injecting new alleles into gene pools and changing allele frequencies as a result. Once a new mutation arrives, drift and selection may begin to act on it. If the allele is deleterious, selection will act to reduce its frequency. But other new mutations at the same locus will keep emerging, lifting up the allele's frequency. The production of new alleles and negative selection will act like opposing teams in a tug-of-war. Together, this mutation-selection balance will result in an equilibrium frequency of any new allele (we show how to calculate this equilibrium in **Box 6.6**). Mutation-selection balance helps explain why rare deleterious alleles with recessive effects persist in populations, adding to genetic variation (Crow 1986; Templeton 2006).

BOX 6.6

Mutation-Selection Balance for a Recessive Allele

Selection is not very effective at culling deleterious alleles from populations when they are recessive (see Figure 6.18), especially once they have been driven to a low frequency. At this point, alleles can linger for so long that the mutation rate, μ, becomes much more important as a force influencing their frequency. For example, consider the situation where an allele, A_2, is both rare and recessive, and mutation increases its frequency by changing the ancestral allele, A_1, from its original state to the new state, A_2, at a rate of μ per generation. We can specify the frequency of A_1 and A_2 as p and q, respectively, so that

$$p + q = 1, \quad so \quad p = 1 - q$$

In each generation, the frequency of A_2, q, will increase due to mutation. The change in allele frequency from one generation to the next, Δq, due to mutation will be

$$\Delta q = p \times \mu$$

which is simply the starting frequency of A_1 multiplied by the rate at which mutation converts A_1 to A_2. The mutation rate is multiplied by p because each copy of A_1 in the population has a chance to mutate each generation. However, given that μ is generally very small, the evolutionary change due to mutation alone is likely to be relatively small. (Note that we can also calculate the rate at which the new A_2 alleles mutate back to A_1 alleles, as $q \times \mu$. However, as both q and μ are tiny, and their product even tinier, back-mutation from A_2 to A_1 is generally considered negligible, and for these calculations we'll ignore it.)

If we assume that the new allele A_2 is deleterious and that its effects are recessive to A_1, then we can examine how the frequency of this allele will evolve given the combined effects of mutation and selection. Mutation will be acting to increase the frequency of A_2, while selection will be acting

to decrease it. The combined effects of selection and mutation can be expressed as

$$\Delta q = \Delta q(\text{due to mutation}) + \Delta q(\text{due to selection})$$

We know from Box 6.5 that

$$\Delta q(\text{selection}) = (q/\overline{w}) \times a_{A_2}$$

where a_{A_2} is the average excess of fitness of the A_2 allele, so

$$\Delta q(\text{selection}) = (q/\overline{w}) \times [q \times (w_{22} - \overline{w})] + [p \times (w_{12} - \overline{w})]$$

The net change in q from both mutation and selection combined is

$$\Delta q = \Delta q(\text{due to mutation}) + \Delta q(\text{due to selection})$$
$$= p \times \mu + (q/\overline{w}) \times [q \times (w_{22} - \overline{w})] + [p \times (w_{12} - \overline{w})]$$

Because we've specified that the A_2 allele is deleterious (and that it is recessive), the relative fitnesses of each genotype will be as follows.

Genotype:	A_1A_1	A_1A_2	A_2A_2
Fitness:	w_{11}	w_{12}	w_{22}
	1	1	$1 - s$

(The fitness of A_2A_2 individuals is lower than that of the others because this is a deleterious allele, and the fitnesses of A_1A_1 and A_1A_2 are identical because we assumed that the A_2 allele was recessive.) Because we assume that A_2 is rare and that increases in the frequency of A_2 due to mutation are small, the frequency of A_2 will be negligible in the population ($q \approx 0$), and the frequency of A_1 will be $p \approx 1$.

In this case, the mean fitness of the population is

$$\overline{w} = p^2 \times w_{11} + 2pq \times w_{12} + q^2 \times w_{22}$$
$$\approx 1 \times (1) + 0 \times (1) + 0 \times (1 - s)$$
$$\approx 1$$

Selecting Diversity

Negative frequency-dependent selection takes place when rare genotypes have higher fitness than common genotypes. This process can maintain genetic variation within populations.

We've seen how selection can reduce genetic diversity by driving some alleles to fixation and eliminating others from populations. But under certain conditions, selection actually fosters variation rather than reducing it. In some situations, for example, the relative fitness of a genotype is high when it is rare, but low when it is common. Selection in such cases is known as **negative frequency-dependent selection**.

Before we explain how it works, we want to stress that negative frequency-dependent selection is not like the selection we discussed in Box 6.5, in which the effect of selection depends on the frequency of an allele because Δp is sensitive to p. In that case, the relative fitness of each genotype was always the same, and the constant selection simply drove larger changes when the allele was common than when the

And the change in frequency of A_2 due to the combined action of selection and mutation becomes

$$\Delta q = p \times \mu + (q/\overline{w}) \times [q \times (w_{22} - \overline{w})] + [p \times (w_{12} - \overline{w})]$$

$$= 1 \times \mu + (q/1) \times [q \times (1 - s - 1)] + [(1 \times (1 - 1)]$$

$$= \mu + q \times (-sq)$$

$$= \mu - sq^2$$

We can solve for the equilibrium frequency of A_2, $\hat{q}$ (pronounced Q-hat) under the combined action of mutation and selection by setting $\Delta q = 0$, because q will be at equilibrium only when it is not changing from generation to generation, in other words, when $\Delta q = 0$:

$$0 = \mu - sq^2$$

$$sq^2 = \mu$$

$$q^2 = \mu/s$$

$$\hat{q} = \sqrt{\mu/s}$$

What this tells us is that when the effects of a deleterious allele are recessive, selection is predicted to drive the frequency of the allele down, but not all the way to 0. Populations are predicted to reach an equilibrium state where the rate of influx of alleles through recurrent mutation, μ, *balances* the ability of selection, s, to purge the allele from the population.

Most of the time, the effects of recessive alleles are only partially masked by their dominant counterpart. That is, alleles are partially recessive. For this more general situation, population geneticists use a slightly different equation that includes the extent to which the allele is expressed in the phenotype of a heterozygote (called h, or the coefficient of dominance). With this new variable included, the balance between mutation and selection works out to the equation

$$\hat{q} = \sqrt{\mu/hs}$$

where hs is the reduction in relative fitness of the heterozygote relative to the higher-fitness homozygote. Although this equation is not directly comparable to the example we just worked out for a completely recessive allele (different starting assumptions were used to solve for it), it turns out to be very useful for a broad range of circumstances. Consider, for example, a situation where an allele is *mostly* recessive and deleterious. In this situation, selection acting on the allele is likely to be very weak because the effects on the phenotype are small. Individuals with a heterozygous genotype might perform only slightly less well than their homozygous counterparts — say, $hs = 1/10,000$. In this case, the equilibrium frequency of the allele would be

$$\hat{q} = \mu/hs$$

$$\hat{q} = \mu/1/10,000$$

$$\hat{q} = \mu \times 10,000$$

As we saw earlier for completely recessive alleles, selection on mostly (but not completely) recessive alleles can offset the increase in allele frequency caused by mutation. Acting alone, selection would eventually eliminate the allele ($\mu = 0$, so $\hat{q} = 0$). Similarly, in the absence of negative selection, mutation alone would eventually fix the allele. There would be no equilibrium frequency. Combined, however, these mechanisms of evolution result in a balance—an equilibrium frequency for the allele that is low, but nonzero—and we expect these alleles to persist within populations for many generations.

allele was rare. In negative frequency–dependent selection, fitness itself changes as the genotype frequency changes.

One particularly pretty example of negative frequency–dependent selection can be found in the elderflower orchid (*Dactylorhiza sambucina*), a flower that grows in France. In a single population, some individuals have deep purple flowers, whereas others produce yellow ones. These colors are the result of a genetic polymorphism (Gigord et al. 2001).

The orchids produce packages of pollen, which are attached to the bumblebees that visit the orchids in search of nectar. But the orchids in effect play a trick on the bees: they produce no nectar, thus offering the insects no reward for their pollen-delivery services. This deception saves the orchids the energy required to produce nectar, but it comes with a risk. Bumblebees that visit rewardless orchids can learn to avoid them on later visits.

Luc Gigord of the University of Exeter and his colleagues ran an experiment to measure the effect of flower color on the fitness of orchids. They planted different proportions of yellow and purple flowers in 10 patches in a French meadow. As **Figure 6.19** shows, when yellow flowers were rare, they were more fit: they delivered more of their pollen to bees, and they produced more fruit. Bees were more likely to encounter purple flowers first and may have learned to avoid them. But when yellow flowers made up most of a population, they were less fit than purple ones—presumably because the bees learned to avoid them instead.

As a result of negative frequency-dependent selection, two different colors can coexist in the flower populations. Whenever one color starts to disappear from the population, its fitness relative to the other increases, pulling it back from the brink until it becomes common again. When it becomes too common, the fitness of the other color increases, and it then spreads in the population. Frequency-dependent

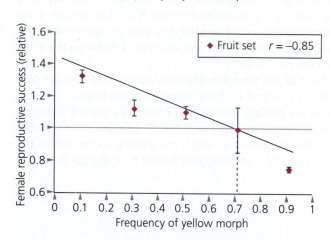

Figure 6.19 A: Elderflower orchids in Europe have a polymorphism for yellow and purple flowers. Luc Gigord and his colleagues set up 10 plots of 50 flowers apiece with proportions ranging from mostly yellow to mostly purple. Each flower color performed best when it was rare (fitness of the yellow genotype is shown in B and C). B: Male fitness was measured as the number of pollen packets attached to visiting bees. C: Female fitness was measured as the number of fertilized seeds. When yellow flowers were rare (for example, frequencies < 0.2), they had high fitness, resulting in positive selection for increased yellow-morph frequency. In contrast, when yellow flowers were abundant (for example,

frequencies > 0.8), they had much lower fitness, resulting in negative selection. The flowers experience negative frequency-dependent selection. Unlike most other flowers, the orchids do not produce nectar for pollinating bumblebees to eat. The bumblebees learn to avoid the orchids with the most common color and are more likely to visit the rare color instead. (A: From "Negative frequency-dependent selection maintains a dramatic flower color polymorphism in the rewardless orchid *Dactylorhiza sambucina*" by Gigord, Macnair and Smithson. PNAS, May 22, 2001: Vol. 98 (11), © 2001 National Academy of Sciences; data from Gigord et al. 2001)

selection in this case keeps both of the flower colors in the population through a cyclical "leapfrogging" of color frequencies.

Natural selection can also maintain allelic variation when heterozygotes have a higher fitness than homozygotes (**heterozygote advantage**). Rather than driving one allele at a locus to fixation, selection can maintain both alleles in the population. Such is the case with the *S* and *A* alleles for hemoglobin we discussed in Section 6.3. In Nigeria, there are very few *SS* homozygote genotypes because the *S* allele gives rise to a deformed hemoglobin molecule. The red blood cells that carry these deformed molecules take on a long, curved shape like the blade of a sickle. This deformity leads to a dangerous condition, known as sickle-cell anemia, in which many red blood cells die and others clump together, damaging blood vessels, organs, and joints.

In affluent countries, intensive medical care can allow people with sickle-cell anemia to survive for decades (the average life span of individuals with the disease in the United States is a little more than 40 years). But in Nigeria and most other countries where sickle-cell anemia is common, most children with the *SS* genotype die before age 5. Thus, the *S* allele experiences strong negative selection when it is paired with the same allele. Yet Nigeria also has fewer people with *AA* genotypes and more people with the heterozygous genotype *AS* than would be expected if the Nigerian population were in Hardy–Weinberg equilibrium (see Box 6.3).

That's because the *S* allele does more than just cause red blood cells to sickle. It also protects people from malaria, a disease that kills 627,000 people a year worldwide and infects an estimated 207 million. Malaria is caused by a single-celled protozoan in the genus *Plasmodium*, which is spread by mosquitoes. When a mosquito bites a victim, the parasite slips into the bloodstream. It eventually invades red blood cells and replicates inside them. The infected red blood cells become sticky and tend to clog small blood vessels, sometimes leading to fatal bleeding. The *S* allele makes infected cells less sticky, thus reducing the risk of death from malaria (Cyrklaff et al. 2011).

A 2018 study determined that the *S* allele arose approximately 7300 years ago in Africa (Shriner and Rotimi 2018). Population expansions and migrations then spread the allele throughout much of the continent and into Europe, the Near East, and India. Despite the negative impact of fitness that sickle-cell anemia had, the *S* allele was favored by selection because people with one copy of the *S* allele enjoy protection against malaria (**Figure 6.20**). As a result, the *AS* genotype is more common in Nigeria (a country with high rates of malaria) than expected by Hardy–Weinberg frequencies, and the *AA* genotype is less common.

At this locus, selection maintains genetic variation in populations, keeping both the *S* and the *A* alleles from disappearing. Heterozygous individuals in regions with malaria have higher fitness than individuals with either of the homozygote genotypes. As a result, neither allele can become fixed at the other allele's expense. **Box 6.7** shows

Heterozygote advantage occurs when selection favors heterozygote individuals over either the dominant homozygote or the recessive homozygote.

Relative fitness of *SS*, *AS*, and *AA* genotypes

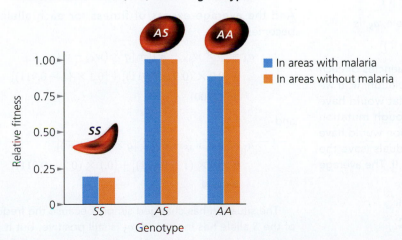

Figure 6.20 People with one copy of the *S* allele are more likely to survive malaria than people who are *AA* homozygotes. People who are *SS* homozygotes suffer sickle-cell anemia and have much lower fitness. The higher fitness of *AS* heterozygotes has the unfortunate effect of raising the frequency of *SS* homozygotes.

BOX 6.7

Calculating the Average Excess of Fitness for *A* and *S* Alleles of the β-Globin Locus

For thousands of years, people living in Central and West Africa have been continuously exposed to malaria, a major parasite-induced disease that today remains a leading cause of death worldwide. About 7300 years ago, a mutation to the β-globin locus generated a new allele, *S*, that arose in an individual in this region. It is called *S* because in homozygous form it causes a deformity to red blood cells responsible for a severe form of anemia ("sickle-cell" anemia). Despite these detrimental effects on fitness, the *S* allele has increased in frequency in many parts of the world. The reason is that although homozygous (*SS*) individuals experience a severe reduction in fitness, heterozygous individuals (*AS*) do not. In fact, heterozygotes have higher fitness than (*AA*) homozygotes in parts of the world where malaria is prevalent because they are protected from the effects of this pathogen. Somehow, then, the beneficial effects of this allele when it is present in *AS* heterozygotes must outweigh the deleterious effects that it has when it is present in *SS* homozygotes. We can better understand how this happens by calculating the average excesses of fitness for the *A* and *S* alleles and by considering what happens as the frequency of the *S* allele changes (Templeton 2006).

In regions where malaria is present, we can characterize the phenotypes and relative fitnesses of each genotype as follows.

Genotype:	*AA*	*AS*	*SS*
Phenotype (anemia):	Normal	Normal	Anemic
Phenotype (malaria):	Susceptible	Resistant	Resistant
Fitness:	w_{AA}	w_{AS}	w_{SS}
	0.9	1.0	0.2

As we saw in Box 6.5, the average excess of fitness for the *A* allele, a_A, is

$$a_A = [p \times (w_{AA} - \overline{w})] + [q \times (w_{AS} - \overline{w})]$$

and the average excess of fitness for the *S* allele, a_S, is

$$a_S = [p \times (w_{AS} - \overline{w})] + [q \times (w_{SS} - \overline{w})]$$

Both of these values depend on the frequencies of each allele and on the average fitness of the population, $\overline{w}$. If we start at the beginning, we can consider what would have happened when the *S* allele first arose through mutation in the population. At that time, the population would have been made up almost entirely of *AA* individuals (save the one *AS* mutant individual), so $p \approx 1$ and $q \approx 0$. The average fitness of the population would have been

$$\overline{w} = p^2 \times w_{AA} + 2pq \times w_{AS} + q^2 \times w_{SS}$$
$$= 1 \times (0.9) + 0 \times (1) + 0 \times (0.2)$$
$$= 0.9$$

So the average excess of fitness for each allele would be

$$a_A = [p \times w_{AA} - \overline{w}] + [q \times (w_{AS} - \overline{w})]$$
$$= [1 \times (0.9 - 0.9)] + [0 \times (1 - 0.9)]$$
$$= 0$$

and

$$a_S = [p \times w_{AS} - \overline{w}] + [q \times (w_{SS} - \overline{w})]$$
$$= [1 \times (1 - 0.9)] + [0 \times (0.2 - 0.9)]$$
$$= 0.1$$

We can see from this calculation that when the *S* allele is rare, its average excess of fitness is positive. Selection acting on this allele should cause it to increase in frequency over time, despite its strong deleterious effect on homozygotes. In a population with random mating (with respect to the *AS* genotype), homozygotic *SS* individuals are likely to be very, very rare. Most individuals in the population still have the *AA* genotype, so whenever carriers of the *S* allele mate, they are likely to produce only heterozygote offspring. (When an *AS* genotype mates with an *AA* genotype, the progeny can be only *AA* or *AS*.) Because no homozygous *SS* individuals are present in the population, the deleterious fitness effects of this genotype are not realized. The fitness advantage resulting from resistance to malaria is all that matters, and this positive selection causes the frequency of the *S* allele to increase.

Now let's consider what happens after the frequency of the *S* allele has increased from $q \approx 0$ to $q = 0.1$. The average fitness of the population is

$$\overline{w} = p^2 \times w_{AA} + 2pq \times w_{AS} + q^2 \times w_{SS}$$
$$= 0.81 \times (0.9) + 0.18 \times (1) + 0.01 \times (0.2)$$
$$= 0.911$$

And the average excess of fitness for each allele now becomes

$$a_A = [p \times w_{AA} - \overline{w}] + [q \times (w_{AS} - \overline{w})]$$
$$= [0.9 \times (0.9 - 0.911)] + [0.1 \times (1 - 0.911)]$$
$$= -0.001$$

and

$$a_S = [p \times w_{AS} - \overline{w}] + [q \times (w_{SS} - \overline{w})]$$
$$= [0.9 \times (1 - 0.911)] + [0.1 \times (0.2 - 0.911)]$$
$$= 0.009$$

The situation has changed simply because the frequency of the *S* allele has increased. a_S is still positive, but it is not as large as it was before. Now there are enough *S* alleles segregating in the population that a few *SS* homozygotes are produced, and these individuals suffer a severe reduction in fitness, bringing down the average excess of fitness for the *S* allele. At the same time, the frequency of heterozygotes has increased, so the impact of the fitness advantage of the *AS* genotype is more pronounced. By this point, the average excess of fitness for the *A* allele is slightly negative, the average excess of fitness for the *S* allele is still positive, and we expect the population to continue to evolve toward more *S* alleles and fewer *A* alleles.

But what happens when the *S* allele becomes even more common? Presumably, the population should reach a point where the *S* allele becomes so common that *SS* homozygotes begin to crop up with regularity. Once this occurs, a_S should plummet. This is exactly what happens. For example, by the time the frequency of *S* reaches 0.2, the average excess of fitness of *S* is negative:

At this point, $p = 0.8$, $q = 0.2$, and

$$\bar{w} = p^2 \times w_{AA} + 2pq \times w_{AS} + q^2 \times w_{SS}$$

$$= 0.64 \times (0.9) + 0.32 \times (1) + 0.04 \times (0.2)$$

$$= 0.904$$

And the average excess for fitness of each allele becomes

$$a_A = [p \times (w_{AA} - \bar{w})] + [q \times (w_{AS} - \bar{w})]$$

$$= [0.8 \times (0.9 - 0.904)] + [0.2 \times (1 - 0.904)]$$

$$= 0.016$$

and

$$a_S = [p \times (w_{AS} - \bar{w})] + [q \times (w_{SS} - \bar{w})]$$

$$= [0.8 \times (1 - 0.904)] + [0.2 \times (0.2 - 0.904)]$$

$$= -0.064$$

As soon as the *S* allele becomes too common, its average excess of fitness drops below zero, and we expect selection to cause a decrease in its frequency. As discussed in the text, this particular mixture of relative fitnesses, called heterozygote advantage, is a form of selection that can maintain genetic variation within populations indefinitely. Whenever *S* becomes sufficiently rare, selection will increase its frequency. But selection cannot drive the allele to fixation because once it gets common, more *SS* individuals are born, and selection begins to decrease its frequency.

For this example, we kept the relative fitnesses of each genotype constant, and changes in the effects of selection were driven entirely by changes in the frequency of the *S* allele. This situation fits well with regions where malaria is prevalent because it is in these environments that the fitness

advantage for *AS* heterozygotes is realized. But what happens to the *S* allele in populations living in areas where malaria is absent?

For these populations, the phenotypes and fitnesses of each genotype are as follows:

Genotype:	AA	AS	SS
Phenotype (anemia):	Normal	Normal	Anemic
Fitness:	w_{AA}	w_{AS}	w_{SS}
	1	1	0.2

Now there is no fitness advantage to the *AS* genotype, and the only effects of the *S* allele will be the deleterious effects that arise in *SS* homozygotes. Given that there is never a benefit of having the *S* allele in this population, the average excess of fitness is always going to be negative. But to work through the calculations, let's assume the frequency of *S* is 0.1.

This means that $p = 0.9$, $q = 0.1$, and

$$\bar{w} = 0.81 \times (1) + 0.18 \times (1) + 0.01 \times (0.2)$$

$$= 0.992$$

And the average excess for fitness of each allele is

$$a_A = [p \times (w_{AA} - \bar{w})] + [q \times (w_{AS} - \bar{w})]$$

$$= [0.9 \times (1 - 0.992)] + [0.1 \times (1 - 0.992)]$$

$$= 0.008$$

and

$$a_S = [p \times (w_{AS} - \bar{w})] + [q \times (w_{SS} - \bar{w})]$$

$$= [0.9 \times (1 - 0.992)] + [0.1 \times (0.2 - 0.992)]$$

$$= -0.072$$

When malaria is absent, selection acts strongly to reduce the frequency of the *S* allele.

Our investigation of β-globin has illuminated several crucial points about selection. First, its effectiveness at increasing or decreasing allele frequencies depends on how common the alleles are in the population. Second, the relative fitnesses of genotypes—and thus the average excess of fitness for the alleles—depends on the environment. In this case, selection favors increases in the *S* allele (up to a point) in environments with prevalent malaria, but it favors the elimination of the *S* allele in environments lacking malaria.

As we'll see in Chapter 18, this pattern of selection has dramatically affected the distribution of the *S* allele worldwide. Maps of the frequency of this allele closely track the geographic distribution of the disease it protects against (see Figure 18.21).

how the average excess of fitness can be calculated for the A and S alleles, and how the effects of selection change with the frequency of the S allele.

Biologists call these special forms of selection **balancing selection**. In the case of S and A alleles, balancing selection makes the population more resistant overall to malaria. Unfortunately, it also leaves millions of people suffering from sickle-cell anemia. If the S allele provided no protection against malaria, it would rapidly become very rare because people with two copies of it would have fewer children than people with one or no copies of the allele. But heterozygotes have enough reproductive success that they raise the number of S alleles circulating in the population. In the process, they also raise the odds of some people being born with two copies of the allele.

Sickle-cell anemia drives home an important truth about the nature of fitness: fitness is not an inherent quality of a genotype. It emerges from the relationship of organisms to their environment. A genotype that increases fitness in some environmental conditions might decrease fitness in other conditions. If malaria were eliminated tomorrow, the AS genotype would immediately lose its high fitness advantage, and the S allele would begin to disappear.

Balancing selection describes the type of selection that favors more than one allele. This process acts to maintain genetic diversity in a population by keeping alleles at frequencies higher than would be expected by chance or mutation alone.

Key Concepts

- Rare alleles are almost always carried in a heterozygous state.

- When recessive alleles exist in heterozygous individuals, they are invisible to the action of selection.

- Selection cannot drive dominant beneficial alleles all the way to fixation because once the alternative (recessive) alleles become rare, they can hide indefinitely in a heterozygous state.

- The mutation rate for any specific locus may be extremely low, but it is much higher when considering an entire genome or population. The gradual accumulation of mutations within populations is the ultimate source of heritable genetic variation.

- Balancing selection actively maintains multiple alleles within a population. Two mechanisms are negative frequency-dependent selection and heterozygote advantage. Negative frequency dependence occurs if the fitness of an allele is higher when that allele is rare than when it is common. Heterozygote advantage occurs when the heterozygotes for the alleles in question have higher fitness than either of the homozygotes. ●

6.8 Inbreeding: The Collapse of a Dynasty

Sickle-cell anemia illustrates how natural selection can, paradoxically, maintain a disease in a population. We'll now look at another way diseases can emerge in a population: by bringing together rare deleterious mutations through inbreeding. There are many examples to choose from, among humans and nonhuman species alike. Probably the most spectacular example concerns the fall of the Spanish Empire.

Charles II: "The Hexed"

Charles II belonged to the great Hapsburg dynasty, which took over Spain in 1516 and dramatically expanded its sphere of power. When Charles II became king of Spain in 1665, the Spanish Empire was the greatest power on the planet. In the New World, its power reached from California down to the tip of Tierra del Fuego. In Europe, Spain possessed half of Italy. It held sway over much of the Caribbean as well as the Philippines. But all that would soon end.

Charles II was crowned at age 4, and from the start it was clear that the boy was an unfortunate monarch. He had a host of deformities, including a jaw that was so large it left him unable to chew and a tongue so big that people could hardly understand his speech. He did not walk until he was 8, and he was such a poor learner that he was never formally educated. He vomited regularly and suffered from diarrhea all his life. By age 30 he looked like an old man (**Figure 6.21**). All this suffering earned Charles II the name *El Hechizado*—"the Hexed"—because he was widely believed to be the victim of sorcery.

Spain suffered under Charles's reign. Its economy shrank as it fought a host of small but draining wars. And worst of all, it became increasingly clear that Charles II would not produce an heir. The Hapsburgs had long feared such an outcome; to hold onto Spanish rule, they had taken to marrying within their family. Charles II not only failed to produce an heir with his two wives but also lacked any brothers or other Hapsburgs who could succeed him. And so, when Charles II died in 1700 at the young age of 39, he left the throne to Philip, a French duke who was the grandson of his half-sister and King Louis XIV of France.

Philip was not just the king of Spain. He was also in the line of succession to the French throne. There was a real chance that he might eventually become king of a united France and Spain—a prospect that terrified the rest of Europe. Soon England, the Netherlands, the Holy Roman Empire, and other European powers declared war to stop Philip from creating a super-empire. They battled not just in Europe but in their colonies as well. The English in the Carolinas waged war with the Spanish of Florida; in Canada, they battled the French and their Indian allies.

The War of the Spanish Succession had claimed hundreds of thousands of lives by the time it ended in 1714. Spain and France were defeated and forced to sign away substantial parts of their empires. Philip forsook the French throne. Spain fell into decline, while England started its rise to become the most powerful empire on the planet.

Any major historical event like the War of Spanish Succession has many causes. But one of the most important causes involved population genetics—specifically, how the genes of Spanish kings made their way down through the generations of the Hapsburg dynasty.

Like many royal dynasties, the Hapsburgs tended to marry within their extended family. It was quite common for first cousins to marry, for example, and uncles even married nieces. Marrying relatives kept power within the dynasty, but it had an unfortunate side effect known as inbreeding.

As we saw earlier in this chapter, rare recessive alleles can be preserved in large populations, even if they're deleterious, because the more common dominant alleles overshadow them. In an inbreeding population, however, rare deleterious alleles can become unmasked in homozygotes. That's because parents in these populations tend to be closely related and are thus much more likely to share rare alleles than are two people picked at random from a large population. The more closely the parents are related to each other, the greater the odds that their children will be homozygous for recessive alleles, including alleles that are deleterious. And often, as was the case for Charles II, inbred individuals will suffer fitness consequences from their resulting phenotypes.

On its own, inbreeding does not change the frequency of alleles in a population. It simply rearranges alleles such that homozygotes for rare recessive alleles become more common. This means that inbreeding on its own is not a mechanism of evolution. But it can create the conditions for evolution to take place. When deleterious rare alleles are combined in homozygotes, they can cause genetic disorders that lower fitness. Selection can then reduce the frequency of these rare alleles, reducing the genetic variation in the population.

Figure 6.21 King Charles II was the victim of centuries of inbreeding, which left him physically and mentally impaired. (The Art Archive / Superstock, Inc.)

Measuring the Impact of Inbreeding

In 2009, Gonzalo Alvarez, a geneticist at the University of Santiago in Spain, and his colleagues measured the impact of inbreeding on the Hapsburgs—and Charles II in particular—by building a detailed genealogy of the dynasty (Alvarez et al. 2009). They charted the kinship of three thousand of Charles II's ancestors and other relatives. To calculate the probability that Charles II was homozygous at any of his loci, Alvarez and colleagues worked their way through his ancestry. His father, King Philip IV, was the uncle of his mother, Mariana of Austria. And Philip and Mariana themselves were also the product of a long history of inbreeding, going back to the early 1500s (**Figure 6.22**). As a result, Charles II was far more inbred than you'd expect if his parents were an uncle and his niece. In fact, Charles II was more inbred than the children of a brother and sister (**Figure 6.23**).

This level of inbreeding must have dramatically increased the number of loci at which the Hapsburgs were homozygous, Alvarez and his colleagues concluded. They could not test this conclusion by examining the DNA of the royal family directly, but they could test it indirectly by measuring the family's rate of infant mortality. The higher the homozygosity, the higher you'd expect rates of lethal genetic disorders to become. And indeed, Alvarez and his colleagues measured an astonishingly high rate of infant mortality in the Hapsburg dynasty (**Figure 6.24**). Despite all the comforts enjoyed by the royal family of Spain, only half the Hapsburg children survived to their first birthday. In Spanish villages at the time, by contrast, four out of five babies survived.

Charles II inherited a genotype that had been ravaged by two centuries of familial inbreeding. Alvarez notes that many of the symptoms historians have chronicled in Charles II could have been produced by two rare genetic disorders, known as

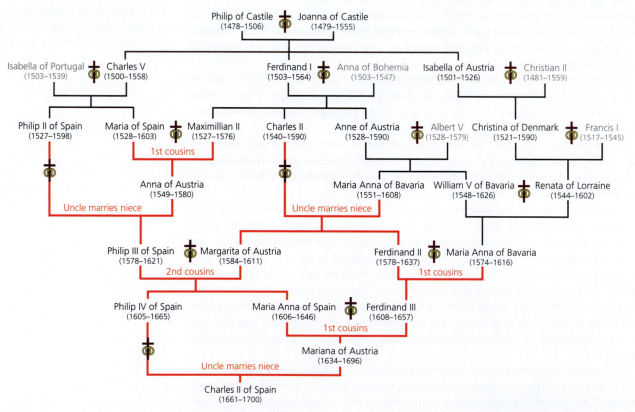

Figure 6.22 The Hapsburg dynasty intermarried to retain power. They also gave their descendants many genetic disorders as a result of inbreeding. Gonzalo Alvarez and his colleagues traced the genealogy of 3000 relatives of Charles II to determine his level of inbreeding. Red lines in this family pedigree show marriages between relatives.

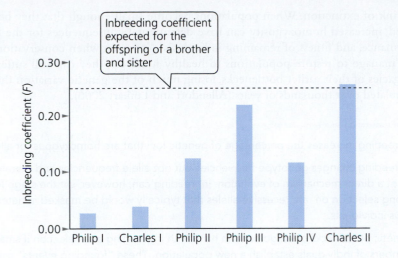

Inbreeding coefficient expected for the offspring of a brother and sister

Figure 6.23 This graph shows the level of inbreeding in the kings of Spain, which tended to increase over subsequent generations. (Data from Alvarez et al. 2009; also see Box 6.8)

combined pituitary hormone deficiency and distal renal tubular acidosis. Inbreeding can also give rise to homozygotes with recessive alleles that cause infertility—probably explaining why Charles II failed to produce an heir. Charles II's level of inbreeding was even higher than that of children born to brothers and sisters, for whom the **inbreeding coefficient (F)** is 0.25. In other words, the fall of the Hapsburg dynasty and the reorganization of the world's great powers of the eighteenth century were due in part to inbreeding.

Inbreeding Depression

Inbreeding is not unique to royal dynasties. If the descendants of a small founding population remain isolated from other populations, their only source of genetic variation is what existed in the founders. Any recessive alleles the founders bring to the new population in a heterozygous state may be combined later in homozygotic offspring. These recessive alleles will be expressed in the phenotypes of the offspring, and if they turn out to be deleterious, they may reduce the fitness of their bearers. These individuals may suffer from genetic disorders or have low rates of fertility. As they produce fewer offspring, their alleles will be removed from the population due to selection. This combination of inbreeding and selection is known as **inbreeding depression**.

Inbreeding depression exacerbates the loss of allelic diversity caused by genetic drift. As a result, inbreeding depression can make harmful mutations more common in populations that pass through bottlenecks, leading to high rates of otherwise rare genetic disorders. Inbreeding depression is of concern to conservation biologists because it pushes endangered populations of animals and plants even closer to

Inbreeding coefficient (F) refers to the probability that the two alleles at any locus in an individual will be identical because of common descent. F can be estimated for an individual, $F_{pedigree}$, by measuring the reduction in heterozygosity across loci within the genome of that individual attributable to inbreeding, or it can be estimated for a population, by measuring the reduction in heterozygosity at one or a few loci sampled for many different individuals within the population.

Inbreeding depression is a reduction in the average fitness of inbred individuals relative to that of outbred individuals. It arises because rare recessive alleles become expressed in a homozygous state where they can affect detrimentally the performance of individuals.

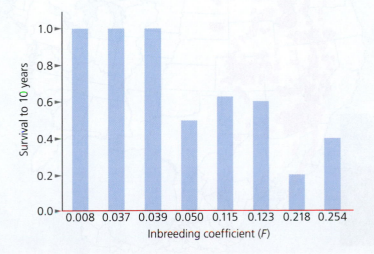

Figure 6.24 High inbreeding tended to be associated with low infant survival rates in the Hapsburg dynasty. As a result, Spanish royalty had few children, and Charles II died without an heir. (Data from Alvarez et al. 2009)

the brink of extinction. When populations become small enough that they begin to inbreed, increased homozygosity can have detrimental consequences for the health, performance, and fitness of remaining individuals. And even when conservation biologists manage to restore populations to healthy numbers, they can still suffer from the legacies of their earlier bottlenecks, losing much of the genetic variation they had accumulated over thousands of years (Allendorf and Luikart 2006).

Key Concepts

- Inbreeding increases the percentage of genetic loci that are homozygous for alleles.

- Inbreeding changes genotype frequencies but not allele frequencies and therefore is not a direct mechanism of evolution. Inbreeding can, however, set the stage for strong selection on rare recessive alleles that typically would be masked in heterozygous individuals.

- Genetic bottlenecks often go hand in hand with inbreeding and selection if small numbers of individuals establish a new population. These "founding events" can be important episodes of rapid evolution because genetic drift has noticeable effects, and the increased homozygosity arising from inbreeding exposes recessive alleles to positive and negative selection. If the new population survives this bottleneck, it may be very different from its parent population. ●

6.9 Landscape Genetics

Landscape genetics is a field of research that combines population genetics, landscape ecology, and spatial statistics.

Real-world populations don't exist as freely mixing jumbles of individuals. They are spread out across landscapes, and those landscapes influence which individuals mate with each other. **Landscape genetics** accounts for real-world populations by looking at population genetics, landscape ecology, and spatial statistics. Bighorn sheep (*Ovis canadensis*), for example, inhabit a range extending from southern Canada to the Baja Peninsula in Mexico (**Figure 6.25**). Their range is not continuous, however; they prefer habitats of steep rocky cliffs, which can be isolated from each other by long stretches of open prairies and harsh deserts. It's also virtually impossible for a bighorn sheep in Canada to wander to Mexico and mate with a sheep there. Instead, the Canadian

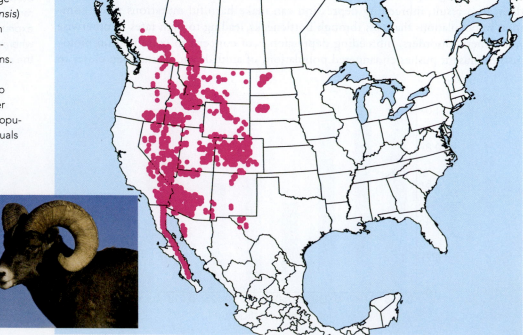

Figure 6.25 Although the range of bighorn sheep (*Ovis canadensis*) extends across much of western North America, animals are subdivided into local subpopulations. Individuals within each local subpopulation are more likely to interact—and mate—with other individuals from the same subpopulation than they are with individuals from more distant populations. (J. T. Chapman/Shutterstock)

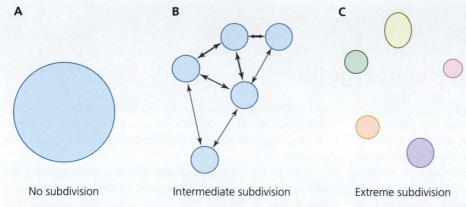

A
No subdivision

B
Intermediate subdivision

C
Extreme subdivision

Figure 6.26 The extent of population subdivision depends on features of the landscape as well as how readily individuals move between locations. A: When landscapes are homogeneous and/or individuals move readily, populations have little or no subdivision. B: Most populations show at least some subdivision, such that movement between locations is less frequent than movement within but still occurs regularly enough to shuttle alleles from place to place. C: At the other extreme, subpopulations may be completely isolated from each other, with almost no movement between them.

animal is far more likely to mate with other individuals that live near it in Canada. When the constraints of landscape and distance restrict the movement of individuals from place to place, the result is called **population structure**, or **population subdivision** (**Figure 6.26**).

Population Structure

Population structure can affect evolution dramatically because it increases the opportunity for genetic drift to change allele frequencies. Consider a hypothetical population of bighorn sheep divided into completely isolated subpopulations (see Figure 6.26C). Suppose we examine a locus with several alleles that are selectively neutral. Even if the sheep were to mate at random with respect to these alleles within each subpopulation, the frequencies of the alleles in different subpopulations would diverge from each other. That's because the subpopulations are smaller than the entire population, and as we've already seen, small populations are especially vulnerable to the vagaries of chance. They lose alleles much faster than large populations do.

Because drift occurs independently in each of the subpopulations, the outcomes will be different from place to place. An allele that increases due to drift in one subpopulation may decrease or be lost in another. Thus, even in the absence of selection, subdivided populations will begin to diverge from each other, just like in Buri's populations of bottled flies (see Figure 6.4). The longer the subpopulations are separated, the more they should diverge in their respective allele frequencies—in other words, the more **genetically distant** they should become (**Figure 6.27**; see Wright 1943, 1951; Nei 1973).

Population structure (or **population subdivision**) refers to the occurrence of populations that are subdivided by geography, behavior, or other influences that prevent individuals from mixing completely. Population subdivision leads to deviations from Hardy–Weinberg predictions.

Genetic distance is a measure of how different populations are from each other genetically. Genetic distance can inform population geneticists about levels of inbreeding within a population or about the historic relationships between populations or species.

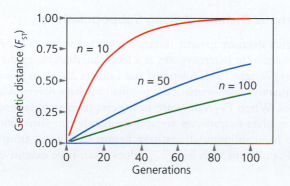

Figure 6.27 Thanks to genetic drift, isolated populations grow more distant from each other over time, as alleles are randomly fixed or lost from place to place. Genetic distance (F_{ST}) increases faster when subpopulations are small. (Data from Allendorf and Luikart 2006)

BOX 6.8

Measuring the Missing Heterozygotes

The level of heterozygosity in a population is an important factor for population geneticists. Inbreeding is a drastic form of nonrandom mating (see Box 6.1) that results in a paucity of heterozygous loci. Related individuals share recent ancestors, and because of this they pass to their offspring alleles that are more similar than would be expected by chance. Another way of saying this is that alleles combined to form an inbred individual are sampled nonrandomly from the base population because they come from parents who are more closely related to each other than parents drawn at random from the larger population.

If we scanned the length of the genome of any diploid individual, we'd find that some of the loci are heterozygous and the remaining proportion are homozygous. Were we to do this for many different individuals, we could calculate the average percentage of heterozygosity for individuals in that population. A highly inbred individual would stand apart from this population average. Highly inbred individuals are unusually likely to inherit identical copies of alleles, so they have more homozygous loci—and, by the same logic, fewer heterozygous loci—than is typical for outbred individuals in their population. Biologists can quantify precisely how inbred any particular individual is by measuring the extent of this deficit in heterozygous loci, a value called *pedigree F*. Specifically,

$$F_{pedigree} = (H_e - H_o)/H_e$$

where H_e is the average proportion of heterozygous loci in outbred individuals of the base population—the expected level of heterozygosity—and H_o is the observed level of heterozygosity in the focal, putatively inbred individual. When there is no difference in the proportion of heterozygous loci between the focal individual and the rest of the population, $F = 0$. On the other hand, when individuals are strongly inbred, the proportion of heterozygous loci may be dramatically reduced and $F_{pedigree} > 0$. In Charles II's case, his $F_{pedigree} = 0.254$ (see Figure 6.23). Because of inbreeding, the loci in Charles II's genome were 25% less likely to be heterozygous (0.254×100). It's no wonder that rare, disease-causing recessive alleles ended up being homozygous in his genome.

Most of the time, population geneticists calculate F for one locus at a time within a population instead of scanning across all the loci of a single individual's genome. For example, they may sample 500 individual black bears and measure whether they are homozygous or heterozygous at a single genetic locus. In these cases, F reflects the paucity in heterozygous individuals relative to homozygous individuals at a particular genetic locus *within a population*. If inbreeding is occurring often, then samples from that population should have fewer heterozygous individuals than would be expected if mating in the base population were random with respect to the marker locus. This discrepancy can be quantified via the same basic equation we used earlier:

$$F = (H_e - H_o)/H_e$$

In this case, though, our measure concerns a single genetic locus sampled across many different individuals (rather than many different loci sampled from within a single individual). Thus H_e is the proportion of heterozygous individuals expected if the population were not inbred, and H_o is the proportion of heterozygous individuals actually observed. If there is no inbreeding—if mating is random with respect to our marker locus—and if there is no selection, then the frequency of heterozygous individuals should equal $2pq$, as predicted by Hardy–Weinberg equilibrium (see Box 6.2). However, if parents within that population mate nonrandomly with close relatives, then fewer of the offspring will be heterozygous at that locus than would be expected by chance.

Population subdivision can reduce the frequency of observed heterozygotes, too. Whenever features of the landscape interfere with movement, they effectively divide the overall population into spatially separated subpopulations.

Measuring Genetic Distance Among Subpopulations

F_{ST} is a measure of genetic distance between subpopulations.

One way biologists measure genetic distance is with a value called F_{ST} (**Box 6.8**). F_{ST} measures the reduction in heterozygotes at a locus attributable to the effects of population subdivision (Wright 1951; Allendorf and Luikart 2006). In practice, F_{ST} provides a useful way to quantify how dramatically populations have diverged in their respective allele frequencies. When F_{ST} values are near zero, then populations have very little genetic structure; allele frequencies are approximately the same from place to place. However, when local subpopulations have begun to diverge from each other, F_{ST} values increase. Populations with high F_{ST} values often have extensive spatial variation

Now when individuals mate, they are more likely to mate with other individuals from within the same local subpopulation. If the subpopulation is small, this could also increase mating among close relatives (inbreeding). However, even without inbreeding per se, individuals within a local population tend to share more recent common ancestry with each other than with individuals from farther away. Within a subpopulation, individuals are more likely to share the same alleles—alleles, for example, that might be common in their local population but rare or absent elsewhere—and this should affect the likelihood that progeny are heterozygous.

Consider our bighorn sheep population from Section 6.9, for example, and our hypothetical genetic marker locus with multiple alternative alleles. Due to the enhanced effects of genetic drift, simply dividing this population into subunits begins a process that erodes genetic variation from within each subpopulation. As the number of alleles at our marker locus dwindles within a local area, more and more of the matings bring together parents with identical alleles at that locus. Fewer and fewer of the offspring will be heterozygous. Eventually, only a single allele will remain within the local population, and from that point forward none of the offspring will be heterozygous.

Similar erosion of allelic diversity will occur in the other subpopulations, too. However, because drift is random, the specific alleles fixed should differ. Were we to survey allelic diversity at our genetic marker across the range of our sheep species, we might find all of the original alleles persisting. But instead of having all alleles circulating in all places, we'd find that each local population had only a few, and more and more of them were becoming fixed for just one allele. Eventually, all of the subpopulations will be fixed for one of the original alleles. (Figure 6.4 illustrates this process as it occurred in Buri's experimental populations of flies. At the end of his experiment, both of the original eye-color alleles remained, just not within the same local populations; each subpopulation had been fixed for one or the other allele.)

What would heterozygosity look like at that point in time? If we ignored the fact that the population was subdivided, and we simply pooled our genetic samples from across the species' range, we could calculate the overall frequencies for each of the alleles. Based on these frequencies, we could then calculate the proportion of sampled individuals we would expect to be heterozygous at this locus using Hardy–Weinberg expectations. But this isn't what we'd actually observe because in reality our "population" of sheep comprises a number of isolated subpopulations.

Mates aren't being chosen randomly from across the species' range; they're picked from within their local subpopulation. To account for this, we'd need to estimate our expected frequencies of heterozygotes separately for each of the different subpopulations. But this would give us very different expected frequencies. In our extreme example, all of the local populations have become fixed for a single marker allele. We actually wouldn't expect there to be any heterozygous individuals at all. Most populations fall somewhere in between, but this example illustrates a critical point: subdividing a population lowers the expected frequencies of heterozygous individuals relative to what we'd predict if the population behaved as a freely mixing whole.

F_{ST} measures the amount of population subdivision—the genetic distance between populations—by quantifying the discrepancy in the expected frequency of heterozygotes, where S stands for "subdivided" population and T stands for the "total" population: $F_{ST} = (H_T - H_S)/H_T$.

In this case, H_T refers to the proportion of heterozygotes we would expect if the population were freely mixing, given the allele frequencies in the collective whole, and H_S is the proportion of heterozygotes we'd expect if the population were subdivided—if we calculated our expected frequencies separately for each local population and then averaged these values together. F_{ST} is one of a series of metrics biologists use to measure the genetic distance between subpopulations. For subdivided populations, $H_S < H_T$, and $F_{ST} > 0$, but when they are mixing, $F_{ST} \approx 0$.

in allele frequencies from place to place, reflecting genetic differences among populations across the landscape (**Figure 6.28**).

Scientists typically estimate F_{ST} values with neutral genetic markers (neutral with respect to selection). These markers therefore reflect genetic structure resulting from the action of random genetic drift. However, selection can also act differently on subpopulations, and divergent selection can drive allele frequencies apart even faster than drift. In fact, biologists have recently begun to use F_{ST} as a handy way to identify candidate genes likely to be involved in local adaptation in response to selection. By estimating F_{ST} for hundreds or even thousands of different genetic loci, it's possible to provide rigorous estimates for the overall effects of population subdivision and genetic

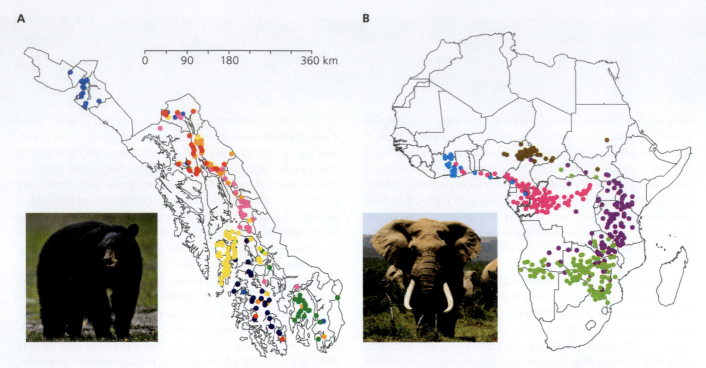

A

0   90   180   360 km

B

Figure 6.28 Populations with extensive subdivision often show heterogeneity in allele frequencies from place to place, called genetic structure, reflecting divergence due to drift. (Selection also can drive subpopulations apart. However, because most genetic markers are selectively neutral, spatial patterns in their relative frequencies reflect drift rather than selection.) For example, populations of black bears in southeastern Alaska can be clustered into distinct subpopulations based on their respective allele frequencies (A), as can populations of African elephants (B). Elephant populations have diverged so much that authorities can use DNA sampled from confiscated ivory to trace the source locations of illegally poached animals. (A: Data from Peacock et al. 2007; Sorin Colac/Shutterstock; B: Data from Wasser et al. 2004; Johan Swanepoel/Shutterstock)

drift, because most of the loci will show similar genetic distances. A few of them, however, may stand out. These outliers have F_{ST} values that are much higher than the rest, meaning that allele frequencies at these loci diverged among subpopulations faster than expected simply due to drift (**Figure 6.29**). The genes containing these loci warrant closer scrutiny, because it's likely they evolved especially rapidly in response to selection.

Whereas isolation of subpopulations causes them to diverge from each other, movement between them has the opposite effect. Whenever individuals from one local population mate with individuals from other populations, the result will be a mixing of alleles in their offspring. This movement of alleles between populations is known as **gene flow** (**Figure 6.30**). Gene flow can counteract the loss of alleles due to drift, restoring genetic variation to local populations. Over time, gene flow can homogenize allele frequencies among subpopulations across the landscape.

Ultimately, the degree of genetic structure of natural populations reflects a balance between divergence due to subdivision and mixing due to gene flow. Often, the biology of particular species contributes to how subdivided they are. For example, species with restricted habitats or limited movement may be much more impeded by landscape features, such as mountain ranges, than are species with more liberal habitat requirements or species that routinely move long distances. For example, **Figure 6.31** illustrates the relationship between geographic distance and genetic distance for four North American mammal species. Over the same geographic range, populations of species with restricted movement, like bighorn sheep, have diverged far more extensively than species that move more widely, like coyotes and lynx.

Gene flow describes the movement, or migration, of alleles from one population to another.

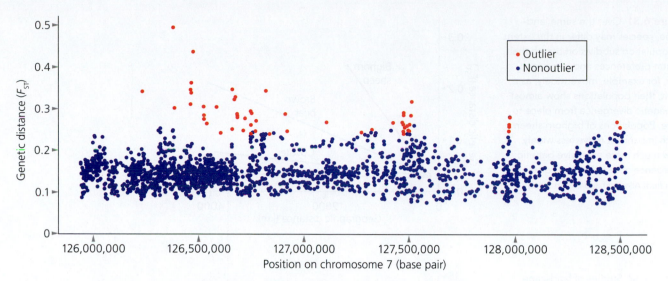

Figure 6.29 Locus-specific estimates of F_{ST} along human chromosome 7. When F_{ST} values are calculated for many thousands of genetic loci, it's possible to detect outliers (red)—specific genetic loci that have diverged faster than expected by drift alone (blue). These markers may indicate genes under strong selection. (Data from Guo et al. 2009; Holsinger and Weir 2009)

Understanding barriers to gene flow has important implications for speciation (Chapter 13) and for conservation (**Figure 6.32**). As we'll discuss in more detail in Chapter 14, many species are currently facing the threat of extinction due to the destruction of their habitat. Populations of species that once existed in large numbers across wide geographic ranges have shrunk down to small fragments. In those fragments, genetic drift can erode the genetic diversity of remaining populations, reducing

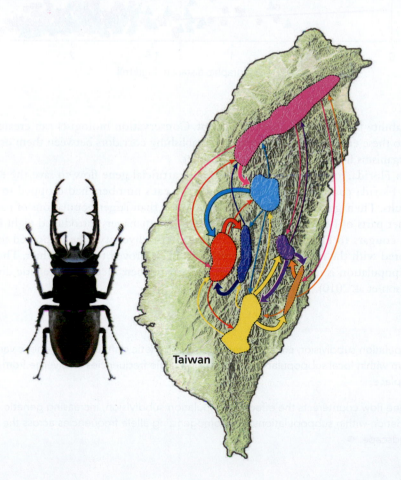

Figure 6.30 Rates of gene flow measured among Taiwanese populations of stag beetle (*Lucanus formosanus*). Colored regions represent subpopulations, and the thickness of arrows reflects intensity of gene flow between them. (Data from Huang and Lin 2010)

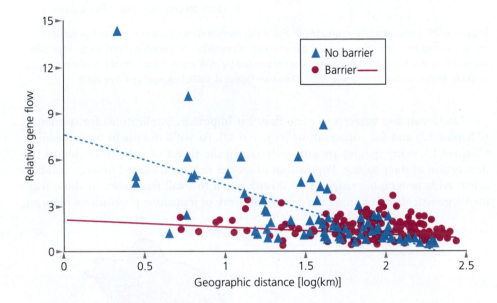

Figure 6.31 Over the same landscape, species may differ in the extent of population subdivision due to their habitat preferences and behavior. Lynx, for example, move widely; as a result, their populations show almost no genetic divergence from place to place. Populations of bighorn sheep, which in contrast move less widely, show much more extensive genetic divergence across the same distances. (Data from Allendorf and Luikart 2006)

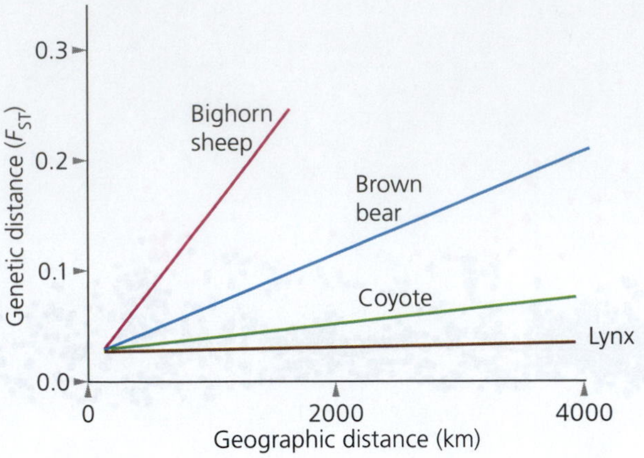

Figure 6.32 Studies of landscape genetics can reveal the impacts of anthropogenic changes to habitat on levels of gene flow and population structure. In desert populations of bighorn sheep, for example, populations separated by barriers such as highways experience less gene flow than populations not separated by major roadways. In as little as 40 years, fragmented and isolated populations of sheep have already lost 15% of their allelic diversity. (Data from Epps et al. 2005)

their ability to adapt to their environment. Conservation biologists can create gene flow to these endangered populations by establishing corridors between them or moving organisms from one population to another.

In Florida, for example, scientists created artificial gene flow to save the endangered Florida panther. By the early 1990s, the cat's numbers had dropped to 20 to 25 adults. Their genetic diversity was much lower than larger populations of panthers in other parts of North America. In 1995, wildlife managers introduced eight female Texas cougars to Florida to increase their genetic diversity. The imported females interbred with the native panthers, and the hybrids proved to be healthier. The panther's population is now growing, thanks in part to their increase in genetic diversity (Johnson et al. 2010).

Key Concepts

- Population subdivision enhances the effects of genetic drift, eroding genetic variation from within local subpopulations and causing allele frequencies to diverge from place to place.

- Gene flow counteracts the effects of population subdivision, increasing genetic variation within subpopulations and homogenizing allele frequencies across the landscape. ●

TO SUM UP . . .

- Population genetics focuses on the distribution of alleles and how allele frequencies change over time.

- Alleles can become more or less common in a population over generations.

- The Hardy–Weinberg theorem offers mathematical proof for how allelic frequencies are maintained within a population. The theorem is based on several assumptions: there is random mating; there is no mutation; and there is no migration into or out of a population, no genetic drift, and no selection.

- The Hardy–Weinberg theorem serves as the fundamental null model of population genetics.

- Genetic drift is the random rise or fall of alleles in a population. Its effects are stronger in smaller populations than in larger ones.

- A slightly higher degree of relative fitness can allow one genotype to outcompete another and replace it over time.

- A mutation in one gene often affects the expression of many phenotypic traits, but the net effect determines the direction of selection acting on the allele.

- Scientists can study the same evolutionary processes that occur in nature in the laboratory. These studies can follow the rise and spread of alleles through populations in response to selection.

- Rare alleles are almost always carried in a heterozygous state. Selection alone is unlikely to drive a dominant allele to fixation because rare recessive alleles can continue to exist in heterozygotes.

- Balancing selection can help maintain genetic variation in a population.

- Although inbreeding alone does not affect the frequency of alleles, it causes an increase in homozygotes for rare recessive alleles within a population. Inbreeding can also significantly affect individual fitness.

- Landscapes influence evolution by altering how individuals interact over time. When populations are subdivided and gene flow is low, genetic drift can influence the genetic variation differently within different subpopulations, ultimately leading to divergence.

MULTIPLE CHOICE QUESTIONS   Answers can be found at the end of the book.

1. If two individuals mate, one of them heterozygous at a locus and the other homozygous for a recessive allele at the same locus, what will be the outcome?

   a. The offspring will be either heterozygous or homozygous for the recessive allele.

   b. The offspring will be homozygous for the dominant allele, heterozygous, or homozygous for the recessive allele.

   c. The offspring will not evolve because they will carry the same alleles as the parents.

   d. The recessive allele eventually will become the dominant allele in the population.

2. The Hardy–Weinberg theorem demonstrates that

   a. dominant alleles are more common than recessive alleles.

   b. in the absence of outside forces, allele frequencies of a population will not change from one generation to the next.

   c. a locus can have only one of two alleles.

   d. evolution is occurring.

3. Which population would be most likely to have allele frequencies in Hardy–Weinberg equilibrium?

   a. A population in a rapidly changing environment.

   b. A population where immigration is common.

   c. A large population that currently is not evolving.

   d. A population that cycles between a very large and a very small number of individuals.

4. Genetic drift

   a. can cause the loss of an allele in a species.

   b. happens faster in large populations than in small ones.

   c. does not occur in large populations.

   d. is a function of Hardy–Weinberg equilibrium.

5. A genetic bottleneck in a population often results in

   a. loss of alleles.

   b. an increase in inbreeding.

   c. an increase in genetic drift.

   d. All of the above.

6. What do population geneticists mean when they refer to the fitness of an allele?

   a. The ability of the allele to survive in a population.

   b. The contribution of an allele to the strength and overall health of a genotype.

   c. The contribution of an allele to a genotype's relative success at producing new individuals.

   d. Whether or not an allele is dominant.

7. If a mutation produces a new deleterious recessive allele in a population, what is least likely to happen to the frequency of that allele?

   a. It will remain at a low frequency within the population for a very long time.

   b. Drift will determine whether it persists in the population.

   c. The allele will be rare enough that it almost never occurs in a homozygous state.

   d. The allele will quickly be purged from the population by selection.

8. If Cavalli-Sforza and colleagues had measured allele frequencies as 0.869 for the A allele and 0.131 for the S allele, how many homozygous genotypes should they have expected to find? Would they have considered the population to be at equilibrium?

   a. 9354 AA and 29 SS. No, they would not have considered the population to be at equilibrium.

   b. 9354 AA and 211 SS. Yes, they would have considered the population to be close to equilibrium.

   c. 9354 AA and 211 SS. No, they would not have considered the population to be at equilibrium.

   d. 2811 AA and 2993 SS. No, they would not have considered the population to be at equilibrium.

9. Which of these statements about inbreeding is *false*?

   a. Inbreeding is not a mechanism of evolution.

   b. Inbreeding can affect the fitness of individuals, but it does not necessarily alter allele frequencies within a population.

   c. Inbreeding increases the probability that two alleles at any locus will be identical because of a shared common ancestor.

   d. Inbreeding alters allele frequencies within a population but does not affect the fitness of individuals.

10. What can measuring genetic distance, or F_{ST}, tell scientists about a group of organisms?

   a. Whether groups have begun to diverge from each other.

   b. Whether genes are under strong selection.

   c. How barriers may be influencing gene flow.

   d. All of the above.

INTERPRET THE DATA Answer can be found at the end of the book.

11. The figure below shows how human-made barriers, such as highways, have affected standing genetic diversity in natural populations of bighorn sheep. Which of the following is *not* a valid conclusion from this graph?

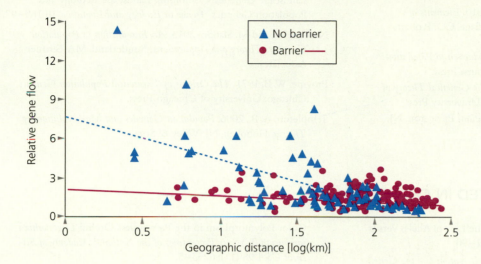

a. Because of highways, some populations that are physically very close to each other look, genetically, just as isolated as if they were far apart.

b. Sheep in populations isolated by barriers like highways do not manage to migrate into other populations in their range.

c. Inbreeding is likely to be more prevalent when the geographic distance illustrated in the graph is greater.

d. In the absence of barriers, the probability that sheep migrate and interbreed decreases as the physical distance between populations increases.

SHORT ANSWER QUESTIONS Answers can be found at the end of the book.

12. Does your urine smell after you've eaten asparagus? A survey through 23andMe.com found that of 4737 individuals of European ancestry, 3002 said they could smell asparagus in their urine and 1735 said they could not. If the *A* allele for odor detection is dominant over the *G* allele for lack of odor detection, and 1027 individuals are heterozygotes, what are the allele frequencies for this locus, assuming random mating? What are the genotype frequencies? Is the population in Hardy–Weinberg equilibrium?

13. Why did the genetic variation of northern elephant seal populations remain low for generations after the bottleneck event? How could genetic drift have played a role in slowing the recovery of genetic diversity?

14. What would happen to the frequency of heterozygous carriers of sickle-cell anemia (with an *AS* genotype) if mosquitoes were completely wiped out in a large region? Explain.

15. Using the average excess of fitness of an allele, explain why when an allele is very rare ($p \approx 0$), the change in allele frequency from one generation to the next due to selection (Δp) will be small even when the fitness effects of the allele are considerable.

16. How can a drastic reduction in population size lead to inbreeding depression?

17. How does pleiotropy affect selection on alleles?

ADDITIONAL READING

Allendorf, F. W., and G. Luikart. 2006. *Conservation and the Genetics of Populations*. Hoboken, NJ: Wiley-Blackwell.

Bohonak, A. J. 1999. Dispersal, Gene Flow, and Population Structure. *Quarterly Review of Biology* 74:21–45.

Charlesworth, B., and D. Charlesworth. 2010. *Elements of Evolutionary Genetics*. Greenwood Village, CO: Roberts and Company.

Crow, J. F., and M. Kimura. 2009. *An Introduction to Population Genetics Theory*. Caldwell, NJ: Blackburn Press.

Fisher, R. A. 1930 (reprinted in 2000). *The Genetical Theory of Natural Selection*. New York: Oxford University Press.

Haldane, J. B. S. 1990. *The Causes of Evolution*. Princeton, NJ: Princeton University Press.

Hartl, D. L. 2007. *Principles of Population Genetics*. Sunderland, MA: Sinauer Associates.

Manel, S., M. K. Schwartz, G. Luikart, and P. Taberlet. 2003. Landscape Genetics: Combining Landscape Ecology and Population Genetics. *Trends in Ecology and Evolution* 18:189–97.

Nielsen, R., and M. Slatkin. 2013. *An Introduction to Population Genetics: Theory and Applications*. Sunderland, MA: Sinauer Associates.

Provine, W. B. 1971. *The Origins of Theoretical Population Genetics*. Chicago: University of Chicago Press.

Templeton, A. R. 2006. *Population Genetics and Microevolutionary Theory*. Hoboken, NJ: Wiley & Sons.

PRIMARY LITERATURE CITED IN CHAPTER 6

Allendorf, F. W. 1986. Genetic Drift and the Loss of Alleles Versus Heterozygosity. *Zoo Biology* 5 (2): 181–90.

Allendorf, F. W., and G. Luikart. 2006. *Conservation and the Genetics of Populations*. Hoboken, NJ: Wiley-Blackwell.

Alvarez, G., F. C. Ceballos, and C. Quinteiro. 2009. The Role of Inbreeding in the Extinction of a European Royal Dynasty. *PLoS ONE* 4 (4): e5174.

Ayala, F., and J. Kiger, Jr. 1984. *Modern Genetics*. 2nd ed. Menlo Park, CA: Benjamin Cummings Publishing Company.

Barrick, J. E., D. S. Yu, S. H. Yoon, H. Jeong, T. K. Oh, et al. 2009. Genome Evolution and Adaptation in a Long-Term Experiment with *Escherichia coli*. *Nature* 461:1243–7.

Bell, G. 2008. *Selection: The Mechanism of Evolution*. Oxford: Oxford University Press.

Berticat, C., O. Duron, D. Heyse, and M. Raymond. 2004. Insecticide Resistance Genes Confer a Predation Cost on Mosquitoes, *Culex pipiens*. *Genetical Research* 83:189–96.

Buri, P. 1956. Gene Frequency in Small Populations of Mutant *Drosophila*. *Evolution* 10 (4): 367–402.

Campbell, C. D., and E. E. Eichler. 2013. Properties and Rates of Germline Mutations in Humans. *Trends in Genetics* 29:575–84.

Cavalli-Sforza, L. L. 1977. *Elements of Human Genetics*. 2nd ed. Menlo Park, CA: Benjamin Cummings Publishing Company.

Conner, J. K., and D. L. Hartl. 2004. *A Primer of Ecological Genetics*. Sunderland, MA: Sinauer Associates.

Crow, J. 1986. *Basic Concepts in Population, Quantitative and Evolutionary Genetics*. New York: W. H. Freeman.

Cyrklaff, M., C. P. Sanchez, N. Kilian, C. Bisseye, J. Simpore, et al. 2011. Hemoglobins S and C Interfere with Actin Remodeling in *Plasmodium falciparum*–Infected Erythrocytes. *Science* 334:1283–6.

Epps, C. W., P. J. Palsboll, J. D. Wehausen, G. K. Roderick, R. R. Ramey II, et al. 2005. Highways Block Gene Flow and Cause a Rapid Decline in Genetic Diversity of Desert Bighorn Sheep. *Ecology Letters* 8:1029–38.

Gigord, L. D. B., M. R. Macnair, and A. Smithson. 2001. Negative Frequency-Dependent Selection Maintains a Dramatic Flower Color Polymorphism in the Rewardless Orchid *Dactylorhiza sambucina* (L.) Soò. *Proceedings of the National Academy of Sciences USA* 98 (11): 6253–5.

Good, B. H., M. J. McDonald, J. E. Barrick, R. E. Lenski, and M. M. Desai. 2017. The Dynamics of Molecular Evolution over 60,000 Generations. *Nature* 551:45–50.

Guo, F., D. K. Dey, and K. E. Holsinger. 2009. A Bayesian Hierarchical Model for Analysis of Single-Nucleotide Polymorphisms Diversity in Multilocus, Multipopulation Samples. *Journal of the American Statistical Association* 104:142–54.

Hoelzel, A. R. 1999. Impact of Population Bottlenecks on Genetic Variation and the Importance of Life-History: A Case Study of the Northern Elephant Seal. *Biological Journal of the Linnean Society* 68 (1–2): 23–39.

Hoelzel, A. R., J. Halley, S. J. O'Brien, C. Campagna, T. Arnbom, et al. 1993. Elephant Seal Genetic Variation and the Use of Simulation Models to Investigate Historical Population Bottlenecks. *Journal of Heredity* 84 (6): 443–9.

Holsinger, K. E., and B. S. Weir. 2009. Genetics in Geographically Structured Populations: Defining, Estimating and Interpreting F_{ST}. *Nature Reviews Genetics* 10:639–50.

Huang, J-P., and C-P. Lin. 2010. Diversification in Subtropical Mountains: Phylogeography, Pleistocene Demographic Expansion, and Evolution of Polyphenic Mandibles in Taiwanese Stag Beetles, *Lucanus formosanus*. *Molecular Phylogenetics and Evolution* 57:1149–61.

Johnson, W. E., D. P. Onorato, M. E. Roelke, E. D. Land, M. Cunningham, et al. 2010. Genetic Restoration of the Florida Panther. *Science* 329:1641–5.

Maher, B. H., R. A. Lea, M. Benton, H. C. Cox, C. Bellis, et al. 2012. An X Chromosome Association Scan of the Norfolk Island Genetic Isolate Provides Evidence for a Novel Migraine Susceptibility Locus at Xq12. *PLoS ONE* 7 (5): e37903.

Nei, M. 1973. Analysis of Gene Diversity in Subdivided Populations. *Proceedings of the National Academy of Sciences USA* 70:3321–3.

Pasteur, N., and G. Sinègre. 1975. Esterase Polymorphism and Sensitivity to Dursban Organophosphorus Insecticide in *Culex pipiens pipiens* Populations. *Biochemical Genetics* 13 (11): 789–803.

Peacock, E., M. M. Peacock, and K. Titus. 2007. Black Bears in Southeast Alaska: The Fate of Two Ancient Lineages in the Face of Contemporary Movement. *Journal of Zoology* 271:445–54.

Popp, M., V. Mirré, and C. Brochmann. 2011. A Single Mid-Pleistocene Long-Distance Dispersal by a Bird Can Explain the Extreme Bipolar Disjunction in Crowberries (*Empetrum*). *Proceedings of the National Academy of Sciences USA* 108 (16): 6520–5.

Raymond, M., C. Chevillon, T. Guillemaud, T. Lenormand, and N. Pasteur. 1998. An Overview of the Evolution of Overproduced Esterases in the Mosquito *Culex pipiens*. *Philosophical Transactions of the Royal Society of London, Series B: Biological Sciences* 353 (1376):1707–11.

Shriner, D., and C. N. Rotimi. 2018. Whole-Genome-Sequence-Based Haplotypes Reveal Single Origin of the Sickle Allele During the Holocene Wet Phase. *The American Journal of Human Genetics* 102:1–11.

Stanek, M., T. Cooper, and R. Lenski. 2009. Identification and Dynamics of a Beneficial Mutation in a Long-Term Evolution Experiment with *Escherichia coli*. *BMC Evolutionary Biology* 9 (1): 302.

Stone, L., P. F. Lurquin, and L. L. Cavalli-Sforza. 2007. *Genes, Culture, and Human Evolution: A Synthesis*. Hoboken, NJ: Wiley-Blackwell.

Templeton, A. R. 2006. *Population Genetics and Microevolutionary Theory*. New York: Wiley-Liss.

Veit, G., R. G. Avramescu, A. N. Chiang, S. A. Houck, Z. Cai, et al. 2016. From CFTR Biology Toward Combinatorial Pharmacotherapy: Expanded Classification of Cystic Fibrosis Mutations. *Molecular Biology of the Cell* 27 (3): 424–33.

Wasser, S. K., A. M. Shedlock, K. Comstock, E. A. Ostrander, B. Mutayoba, et al. 2004. Assigning African Elephant DNA to Geographic Region of Origin: Applications to the Ivory Trade. *Proceedings of the National Academy of Sciences USA* 101:14847–52.

Wiser, M. J., N. Ribeck, and R. E. Lenski. 2013. Long-Term Dynamics of Adaptation in Asexual Populations. *Science* 342 (6164): 1364–7.

Wright, S. 1943. Isolation by Distance. *Genetics* 28:114–38.

———. 1951. The Genetical Structure of Populations. *Annals of Eugenics* 15:323–54.

CHAPTER 7

Beyond Alleles

Quantitative Genetics and the Evolution of Phenotypes

Learning Objectives

- Explain how continuous variation in phenotypes arises.
- Analyze the differences between broad sense and narrow sense heritability.
- Discuss how dominance and epistatic effects of alleles are transmitted.
- Measure narrow sense heritability using an offspring–parent regression.
- Explain the concept of physical linkage of loci on a chromosome and how genetic recombination can erode the resulting linkage disequilibrium over time.
- Discuss three uses of linkage disequilibrium.
- Discuss the value of quantitative trait loci for examining the genetics of phenotypic traits.
- Explain how QTL analysis helps researchers make connections between variations in phenotype and allelic variations in genotypes.
- Compare and contrast plastic and evolutionary changes within a phenotype.
- Explain why a population-level change in plasticity is evolution but the plastic change in phenotype is not.

Each February, Hopi Hoekstra does what many residents of New England do: she gets on a plane and flies south to spend a couple of weeks on a Florida beach. But for Hoekstra, a biologist at Harvard University, the journey is not a vacation. On the beaches of Florida, she and her colleagues can discover the molecular foundations of biological adaptations. Thanks to advances in technology and our understanding of biology, Hoekstra can make discoveries that Darwin could only dream of.

Walking along the snow-white dunes, Hoekstra and her students lay small metal boxes in the sand, each baited with food. Later, they return to find visitors trapped in some of them: oldfield mice (*Peromyscus polionotus*). The mice live in tunnels under

An Alabama beach mouse, Peromyscus polionotus ammobates, *hiding in the dunes.*

the sand dunes, coming out at night to gather seeds. They typically have white bellies and flanks, along with a narrow band of light tan down their backs (Hoekstra 2010).

Oldfield mice also live inland, where they dig their tunnels in the dark loamy soil of abandoned farm fields and open woodlands. The pelage (fur coat) of these mainland mice is markedly different from that of beach mice. Mainland mice are dark brown on top, and they have a smaller patch of white limited to their underside.

These two color patterns are strikingly well matched to the environments in which the mice live: the light coat of the beach mice blends into the white background of the dunes, whereas the dark coat of the inland mice is similar to the dark soil of fields and wooded areas. What makes this match even more intriguing is that the Florida beaches and barrier islands inhabited by beach mice are just 4000 to 6000 years old. This means that the dark mainland mice that colonized these new habitats evolved a new coat color in a matter of millennia.

In the early 2000s, Hoekstra recognized in these mice a promising opportunity to investigate how organisms adapt to their environment. Their populations diverged only recently, and yet the mice have evolved clearly distinguishable phenotypes. And the different phenotypes are heritable. When scientists breed oldfield mice in captivity, parents with dark coats generally produce offspring with dark coats, and the same is true for light coats. Dorsal (back-side) pelage is transmitted faithfully from parents to offspring. Individual differences in pelage color, therefore, must be attributable at least partially to genetic variation that can be transmitted via eggs or sperm from parents to offspring.

Hoekstra was also drawn to the animals because geneticists have been studying another species of mice for a century. Shortly after the rediscovery of Mendel's work on the genetics of peas in the early 1900s, some researchers began looking for a mammal they could breed in a similar fashion. They settled on the house mouse (*Mus musculus*), and one of the first traits in the mice they began to study was coat color. Over the past hundred years, researchers have studied the functions of thousands of mouse genes, and in 2002 researchers published the entire *M. musculus* genome. The lineages of the house mouse and the oldfield mouse diverged from a common ancestor roughly 25 million years ago. They're similar enough that Hoekstra can use the vast store of research on the house mouse as the foundation for her own research (**Figure 7.1**).

Despite all these advantages, studying the color of mice is not a simple task. In the previous chapter, we saw how a single allele conferred insecticide resistance to mosquitoes in France, and how scientists could track selection

Figure 7.1 Hopi Hoekstra, an evolutionary biologist at Harvard University, has uncovered the genetic basis of coat-color differences in oldfield mice. (Lynn Johnson/National Geographic Creative)

on that allele over time. Variations in mouse color, on the other hand, are not due to alleles at a single locus. To uncover the genetic basis of the colors of oldfield mice, Hoekstra would need another set of tools.

In Chapter 6, we investigated how to study evolution from the bottom up, measuring the fitness of individual alleles and charting the rise and fall of their frequency due to selection and drift. In this chapter, we will move from the top down, from complex phenotypic traits to the genes that influence them. Together, Chapters 6 and 7 present a range of methods scientists can use to study the mechanisms of evolution in wild populations. In Chapter 10, we'll see how scientists use these methods to document evolution in our own time, among organisms including weeds, birds, snakes, and Hoekstra's own oldfield mice. ●

7.1 Genetics of Quantitative Traits

Many traits that are important for survival and reproductive success have a complicated genetic basis. They are polygenic—that is, their expression is influenced by many genetic loci. They may depend on nonadditive interactions between alleles at those loci (epistasis). They may also be shaped by the interaction of these alleles with the environment (phenotypic plasticity). Because polygenic traits are influenced by alleles at many loci, the traits often vary continuously among individuals, just as we observed with human height in Chapter 5. The phenotypic difference among individuals is therefore typically quantitative rather than qualitative: it's a difference of degree rather than a difference of kind (**Figure 7.2**). The study of the evolution of these complex phenotypic traits is known as evolutionary **quantitative genetics**. This discipline involves building models that include genetic variation, environmental variation, and the interaction between genetics and the environment (Falconer and Mackay 1996; Lynch and Walsh 1998).

The quantitative genetics approach to evolution is different from the population genetics approach discussed in Chapter 6 in that the two methods start at opposite sides of the genotype–phenotype relationship. Population genetics starts with alleles at genetic loci and builds from the genotype up. Quantitative genetics gives scientists a top-down approach that starts with distributions of phenotypic values in a population. Scientists can use information gathered with this approach to then figure out how mechanisms like selection cause these distributions of phenotypes to change over time.

Quantitative genetics is the study of continuous phenotypic traits and their underlying evolutionary mechanisms.

Variance

One of the most important pieces of information quantitative geneticists need is an estimate of the amount of variation for a trait in a population, known as its **variance**. Traits that vary widely have a larger variance than traits that are practically identical in every individual (**Figure 7.3**).

To calculate the variance of a trait, quantitative geneticists first measure the value of that trait in a large sample of individuals from a population. Let's say that scientists are interested in human height, a trait we considered in Chapter 5. The researchers would first calculate how far each of these measured values deviates from the mean of the population. They would square each of those deviations, and then they would divide the sum of the squares by the number of individuals in the sample.

Variance is a statistical measure of the dispersion of trait values about their mean.

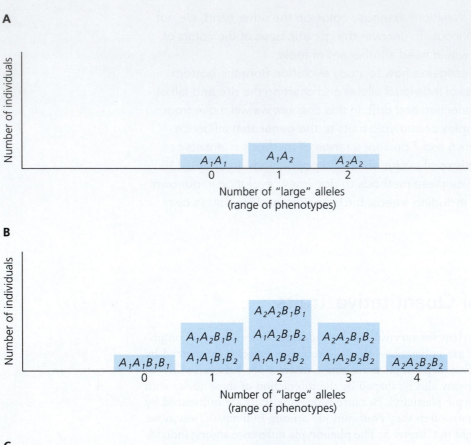

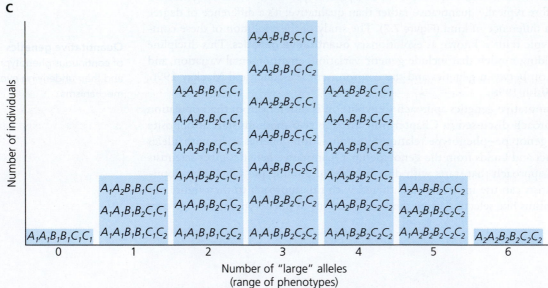

Figure 7.2 The Hardy-Weinberg theorem can be extended to situations where traits are influenced by multiple genetic loci. As loci are added, the number of possible genotypes increases, and the number of phenotypes increases as well. For example, if we consider loci with two alleles and additive effects on body size, and give each copy of the "small" allele (A_1, B_1, etc.) a value of 0 and each copy of the "large" allele (A_2, B_2, etc.) a value of 1, then two alleles at a single locus results in 3 genotypes and 3 different phenotypes (A). Two alleles at *two* genetic loci results in 9 genotypes and a total of 5 different phenotypes (B). Two alleles at each of three genetic loci results in 27 genotypes and 7 different phenotypes (C). As the number of genetic loci increases, the number of possible trait values also increases, resulting in a continuous distribution of phenotypes. (Data from Plomin et al. 2009)

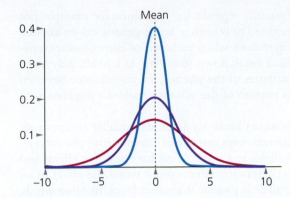

Figure 7.3 Complex phenotypic traits typically vary continuously within a population, yielding a "normal" distribution of trait values around the population mean. Here, three hypothetical populations are shown, each with a mean phenotypic value of 0, and with variances ranging from 1 (tallest curve) to 3 (lowest curve). The variance of each distribution is a measure of how widely dispersed trait values are from the mean (mathematically, the variance, σ^2, is equal to the square of the standard deviation). As the variance of a sample increases, more and more individuals have trait values located far away from the mean. (Data from Conner and Hartl 2004)

The total variance in a phenotypic trait in a population (known as V_P) is actually the sum of several different kinds of variances. One important source of this variance is the genetic difference among individuals, V_G. Another is the difference arising from the environmental conditions in which the individuals developed, V_E (Figure 7.4). We can express this relationship in a simple equation:

$$V_P = V_G + V_E$$

For some traits, the environment plays a much bigger role in generating phenotypic variation than do alleles. In these cases, $V_E > V_G$. In other cases, the environment has little influence on the variation of a trait, and thus the alleles at the relevant genetic loci are mostly responsible. If the environment has no effect ($V_E \approx 0$), then $V_p \approx V_G$. Scientists use the term *heritability* to refer to the proportion of the total phenotypic variance that is attributable to genetic variation among individuals.

Heritability

There are several different ways to express the concept of heritability mathematically, and each works best for solving specific problems. For example, quantitative geneticists sometimes calculate **broad sense heritability (H^2)** as follows:

$$H^2 = \frac{V_G}{V_P} = \frac{V_G}{V_G + V_E}$$

Broad sense heritability is a useful way to measure the relative importance of genetic and environmental effects on trait expression. However, the problem with broad sense heritability is that it represents all genetic variance as a single value. Real genetic

Broad sense heritability (H^2) is the proportion of the total phenotypic variance of a trait that is attributable to genetic variance, where genetic variance is represented in its entirety as a single value (that is, genetic variance is not broken down into different components).

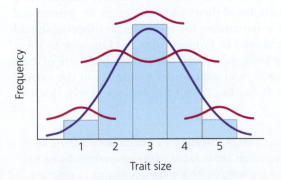

Figure 7.4 Combining V_G with V_E demonstrates how a continuous phenotypic distribution can result from only two genetic loci with two alleles each. In this illustration, bars show the five phenotypic trait values produced from additive combinations of alleles at the two loci (as in Figure 7.2B). This variation among genotypes is V_G. The small normal curve over each bar represents the distribution of phenotypes produced by each genotype caused by the environment (V_E). Because the small distributions overlap, the result approximates a smooth normal curve (blue). That is, the combination of genetic and environmental effects yields a continuous distribution of phenotypes. (Data from Conner and Hartl 2004)

variance is far more complicated. In sexually reproducing organisms, for example, not all of an individual's genotype is transmitted to offspring. Some genetic effects are lost. Combinations of alleles change during meiosis when each pair of chromosomes separates and the associations between alleles break down (**Box 7.1**). As a result, only some of the genetic variation actually contributes to the phenotypic resemblance between offspring and their parents. Only this portion of the variance enables a population to evolve in response to selection.

For this reason, scientists will sometimes break V_G down into smaller components. These components result from the different ways that alleles can influence phenotypes. As we saw in Chapter 6, sometimes alleles at a single locus have additive effects and sometimes they have dominant effects. Epistatic effects result when the effect of an allele at one locus depends on which allele is present at another locus. In other words, epistasis occurs when alleles at loci interact *nonadditively* to affect phenotype expression. For example, an allele may affect the phenotype in one way if that individual has a *B* allele at a second locus and in a different way if it has an *A* allele at that second locus.

We can include these distinctions in our equation by breaking down V_G into three different sources of genetic variation: V_A, Vwhich represents the additive genetic variance; V_D, which represents the variance due to dominance effects of alleles; and V_I, which represents the variance attributable to epistatic interactions among alleles at the various genetic loci. (The *I* in V_I stands for "interactions." This value is used for epistatic effects because V_E is already used for the environmental effects on phenotypic variance.) The relationship of these sources of genetic variation can be represented with the following equation:

$$V_G = V_A + V_D + V_I$$

We can then update our equation for the components of phenotypic variation:

$$V_P = V_G + V_E$$

$$V_P = V_A + V_D + V_I + V_E$$

> **Narrow sense heritability (h^2)** is the proportion of the total phenotypic variance of a trait attributable to the *additive effects of alleles* (the additive genetic variance). This is the component of variance that causes offspring to resemble their parents, and it causes populations to evolve predictably in response to selection.

V_A, the variance from additive effects of genes, is especially important in the study of evolution. It's the reason why relatives resemble each other—because they have more alleles in common than nonrelatives do. And as we'll see later in this chapter, V_A is what causes a population to evolve predictably in its phenotypic distribution following selection. Biologists find it useful, therefore, to calculate the proportion of the total variance of a trait attributable to only the additive effects of alleles.

This more nuanced representation of heritability is known as the **narrow sense heritability (h^2)** of a trait, and it can be expressed in this way:

$$h^2 = \frac{V_A}{V_P} = \frac{V_A}{V_G + V_E} = \frac{V_A}{V_A + V_D + V_I + V_E}$$

Scientists have developed a number of ways to estimate narrow sense heritability. If they want to estimate h^2 for the body mass of fishes, for example, they can randomly pair fishes together in laboratory aquaria, breed them, and then weigh the parents and offspring. The greater the narrow sense heritability, the more similar offspring should be to their parents. Large fishes should tend to produce large fry. One way to measure this similarity is to plot offspring body size against the average body size of their parents. The slope of the regression will yield a quantitative estimate of the narrow sense heritability of that trait (**Figure 7.5**). See **Box 7.2** for a closer look at why this is so.

Key Concepts
- When the components of variation act independently, their effects are additive, so variation attributable to genes and variation attributable to the environment sum to yield the total phenotypic variance of the sample. This allows biologists to estimate the relative contributions of different sources of variation to the phenotypic distribution observed.

- The heritability of a trait is the proportion of phenotypic variance that is due to genetic differences among individuals.

- Broad sense heritability (H^2) reflects all of the genetic contributions to a trait's phenotypic variance, including additive, dominant, and epistatic gene effects. It also includes influences of the parent phenotype (maternal and paternal effects) on the environment of offspring that can cause siblings to resemble each other.

- Narrow sense heritability (h^2) more accurately reflects the contributions of specific, additive components of genetic variation to offspring. ●

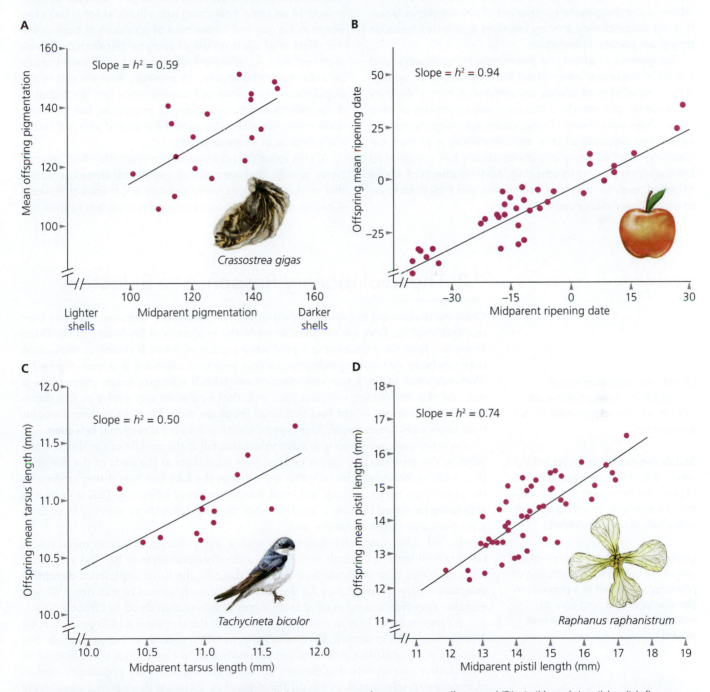

Figure 7.5 One way to estimate narrow sense heritability is to measure trait values in parents and their offspring. The slope of the regression is equivalent to the narrow sense heritability. Shown are regressions for (A) shell pigmentation (darkness) in Pacific oysters, (B) relative date of flowering in apple trees, (C) length of the tarsus bone in tree swallows, and (D) pistil length In wild radish flowers. All four traits display some narrow sense heritability. (Data in A from Evans et al. 2009; B from Tancred and Zeppa 1995; C and D from Conner and Hartl 2004)

BOX 7.1

Why Doesn't h^2 Include Dominance or Epistasis?

For a phenotypic trait to evolve in response to selection, it must be heritable. The heritability of a trait can be thought of as differences among individuals (variation) that can be transferred from one generation to the next. At its simplest, heritability is just the genetic component of phenotypic variance. It is the resemblance among relatives that arises because they share genetic information.

But genes can affect the phenotype in many ways, and not all of these are transmitted from parents to offspring. When the effects of alleles are additive (Chapter 6), they influence the phenotype in the same way regardless of what other alleles are present. These alleles act independently of each other. Because of this, any differences in phenotype resulting from additive effects of alleles get transmitted faithfully from parents to offspring. Additive effects of alleles cause relatives to resemble each other, and they contribute to evolutionary responses to selection.

Dominance and epistatic effects of alleles are not transmitted in the same way. Dominance and epistasis result from interactions among alleles—either an allele interacting with its sister allele (on the other chromosome in the case of dominance) or an allele interacting with alleles at other loci elsewhere in the genome (in the case of epistasis). In both cases, the effect of an allele on the phenotype *depends on what it is paired with*. An allele influencing body size might cause an individual to be smaller, on average, than the rest of the population if it is paired with another copy just like it (that is, if the individual is homozygous for the locus), but the same allele might not affect size at all if it is paired with a different allele (that is, in a heterozygote).

Alleles exhibiting dominance or epistasis affect phenotypes in ways that are context (genotype) dependent. But this context dissolves every generation, thanks to meiosis (Chapter 5). When chromosome pairs separate to give rise

7.2 The Evolutionary Response to Selection

Once we understand the sources of variance in a quantitative trait, we can study how that trait evolves. Let's say we want to study the evolution of body size in the fishes living in a lake. We examine the reproductive success of fishes. If there's an association between body size and reproductive success, selection exists for that trait. **Figure 7.6** illustrates some of the forms this selection can take. If selection favors phenotypes at one end of a distribution of values for a trait, the population may evolve in that direction. In our lake, we might find that small fishes are more likely to survive droughts than larger ones, for example. This type of selection is called **directional selection**.

In other cases, selection may favor values that fall at the middle of the distribution, whereas the reproductive fitness of organisms with traits at the ends of the distribution is lower. We might discover that the fishes in the lake fare best if they're close to the population mean, and big and small fishes have fewer offspring. This is known as **stabilizing selection** because it tends to keep the population from moving away from a narrow range of values for the trait.

In still other cases, the individuals with a trait value close to the mean might fare less well than individuals at the ends of the distribution; very big and very small fishes do better than medium-sized fishes. In this case, the fishes experience **disruptive selection**. **Figure 7.7** and **Figure 7.8** illustrate population responses to selection. (We will examine empirical examples of all three forms of selection in detail in Chapter 10.)

It's important to bear in mind that selection of the sort shown in Figure 7.6 is *not* synonymous with evolution. Evolution happens when there is a change in allele frequencies in a population. Selection can *potentially* lead to evolution if the difference in reproductive success is tied to genetic variation. How quickly the population evolves in response to selection depends on the amount of variation there is in a phenotypic trait in the population and how much of the variation in that trait is heritable (h^2).

To calculate the evolutionary response to selection, quantitative geneticists must measure the selection on a phenotypic trait. We saw in Chapter 6 how population geneticists measure the strength of selection as the selection coefficient: the amount, s,

Directional selection favors individuals on one end of the distribution of phenotypes present in a population.

Stabilizing selection favors individuals in the middle of the distribution of phenotypes present in a population (for example, by acting against individuals at either extreme).

Disruptive selection favors individuals at the tails of the distribution of phenotypes present in a population (for example, by acting against individuals with intermediate trait values).

to haploid gametes, all effects arising from interactions with a sister allele vanish. Likewise, when chromosomes independently assort and recombine, associations between alleles at different loci break down. This means that although dominance and epistasis contribute importantly to genetic variation, their effects disappear each generation. Because neither can be transmitted to offspring, they will not cause relatives to resemble each other, and they cannot contribute to evolutionary responses to selection.

For this reason, the narrow sense heritability (h^2) includes only the additive effects of alleles—the effects that contribute to the phenotypic resemblance between parents and their offspring—and this is the metric most relevant for predicting evolutionary responses to selection in diploid, sexually reproducing populations.

However, there are circumstances where interactions among alleles *are* transmitted from parents to offspring, and in these situations dominance and epistatic effects of alleles can be included in heritability. Bacteria do not undergo meiosis, for example. Neither do asexually reproducing plants and animals. Epistatic interactions between alleles in these species can thus have an important effect on progeny phenotypes.

Sometimes even when alleles are shuffled during meiosis, dominance and epistatic effects can cause relatives to resemble each other. In highly inbred populations, for example, so many of the alleles are homozygous because of the inbreeding (Chapter 6) that interactions among alleles in offspring are likely to be the same as they were in the parents. In both of these situations, the broad sense heritability (H^2) will predict the response of the population to selection.

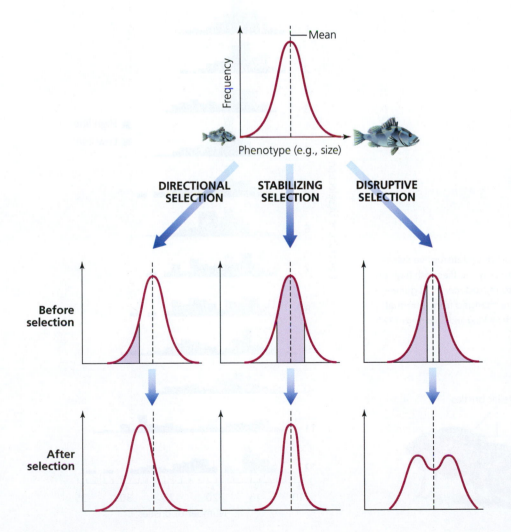

Figure 7.6 Selection can act in different ways on a population. Directional selection (left) favors individuals at one end of a trait distribution, such as animals with small body size. As illustrated, small individuals have higher fitness than larger individuals (positive selection for small sizes is indicated by the shaded area). After selection, and provided the trait is heritable, the distribution of phenotypes should shift to the left, toward a smaller mean body size. Stabilizing selection (middle) favors individuals with a trait near the population mean. In this case, fishes with intermediate sizes have the highest fitness (shaded area), and in the generation after selection, the variance of the population (but not the mean) should be smaller than it was in the preceding generation. Disruptive selection (right) favors individuals at either end of the distribution (shaded area). Here, if selection is strong enough, populations may begin to diverge in phenotype (that is, they may become dimorphic).

Selection differential (S) is a measure of the strength of phenotypic selection. The selection differential describes the difference between the mean of the reproducing members of the population that contribute offspring to the next generation and the mean of all members of a population.

by which the fitness of a genotype is reduced relative to the most-fit genotype in the population. Quantitative geneticists use a different method. They measure selection for a trait as the difference in the mean of a trait of reproducing individuals and the mean of the trait for the general population (the **selection differential, S; Figure 7.9**). Directional selection occurs whenever the mean phenotype of breeding individuals ($\overline{X}_B$) differs from the mean phenotype of all the individuals in the parents' generation ($\overline{X}_P$). If the difference is large, selection is strong.

Let's say that in the lake we're studying, fishes with big body sizes are much more successful than smaller ones at reproducing when conditions are harsh. But under mild

Figure 7.7 In 1896, researchers in Illinois began the world's longest-running scientific experiment in directional selection—an experiment that continues today. Out of a stand of several thousand corn plants, they sampled 163 ears and measured their oil content. They selected the 24 ears with the highest oil content to create one line of corn and the 24 ears with the lowest oil content to create another line. Each year, they selected the highest oil producers from the high strain and the lowest from the low. As you can see in this graph, the average oil content in each line of corn has changed steadily. After 100 years of selection, the oil content of each line was far different from that in the original plants. (Data from Moose et al. 2004)

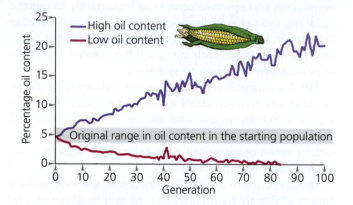

Figure 7.8 Thoday and Gibson (1962) produced disruptive selection in a population of flies. They allowed only the flies with high or low numbers of bristles on their thorax to reproduce. In 12 generations, the distribution of bristle numbers changed from a normal distribution to two isolated peaks. (Data from Klug and Cummings 1997; Roblan/Shutterstock)

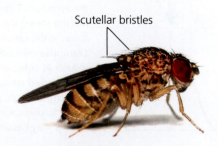

Scutellar bristles

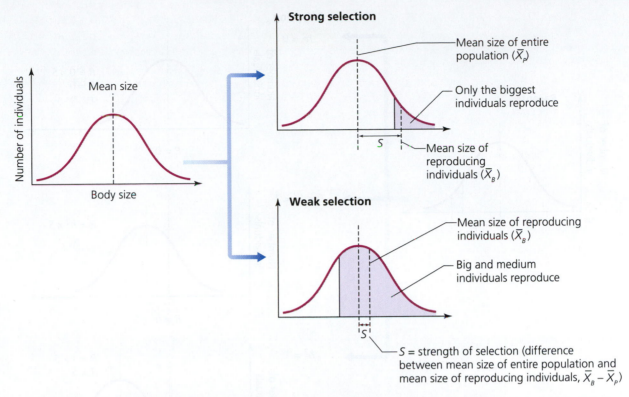

Figure 7.9 The graph on the left shows the range of body sizes in a hypothetical population. Top right: If only the very biggest individuals reproduce, the population experiences strong selection for large body size. The mean size of the reproducing individuals is much bigger than the mean of the entire population. Bottom right: If big and medium individuals reproduce, selection is less strong, and the mean size of the reproducing individuals is much closer to the mean size of the entire population.

conditions, smaller fishes also survive and reproduce. The graph in Figure 7.9 shows the two possibilities of fish body size. Under harsh conditions, selection for large bodies is strong ($\overline{X}_B \gg \overline{X}_P$). In milder conditions, selection is weaker ($\overline{X}_B > \overline{X}_P$).

Selection is present in both of these examples because a nonrandom subset of fishes is producing more offspring than average. But will that selection lead to evolution? That depends on how much of the phenotypic variation in body size is attributable to additive genetic differences among the individuals—that is, on the narrow sense heritability of the trait (h^2).

If differences in the body size of fishes depend solely on the environment—the temperature of the water, for example, or how much food a fish larva finds—then h^2 will be zero. The offspring body sizes in this population will not closely resemble the sizes of their parents. The next generation of fishes will grow into adults that have the same distribution of body sizes that the entire population had before. The population will *not* evolve, despite the presence of selection (**Figure 7.10**, top).

At the other extreme, when $h^2 = 1$, all of the phenotypic variation is due to allelic differences among the individuals. In this case, offspring sizes exactly track the sizes of their parents, regardless of the water temperatures or food supplies they encounter. Selection on body size translates into an increase in the average size of fishes in the next generation. In this case, the mean body size of the next generation of fishes will be the same as the mean size of the selected parents. The evolutionary response to selection, in other words, will equal the strength of selection imposed (Figure 7.10, bottom).

We can now see that the evolutionary response of a population to selection depends on both the strength of selection on a trait and the heritability of that trait. In fact, we can calculate the evolutionary response (R) with a remarkably simple equation, known as the breeder's equation, using these two variables:

$$R = h^2 \times S$$

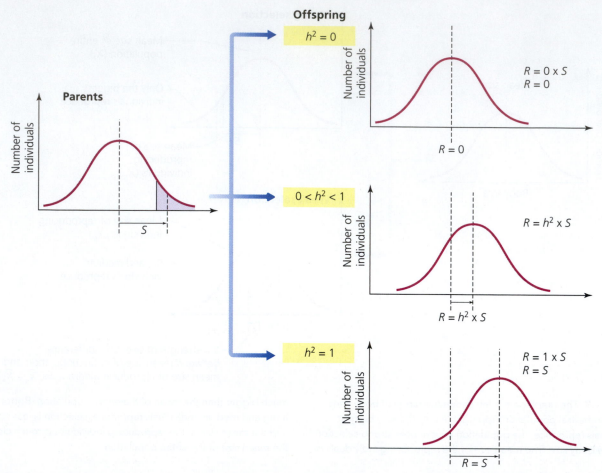

Figure 7.10 A population's response to selection (R) depends in part on the heritability of the trait being selected. Left: Large individuals in a hypothetical population are selected. Top right: In this population, body size is not heritable. In other words, the size of parents is not correlated with the size of their offspring. Despite experiencing strong selection for body size, the population's mean size does not change in the next generation. The response is zero.

Middle right: At intermediate levels of heritability, the offspring are intermediate in size between the mean size of the parental population as a whole and the mean size of the selected individuals. The response to selection is equal to the strength of selection times the heritability of the trait ($R = h^2 \times S$). Lower right: If body size is completely heritable, the response is equal to selection.

The two terms on the right side of the equation reflect the conditions Darwin first recognized as necessary for evolution in response to selection (Chapter 2): phenotypic variation that influences fitness (S) and the ability to transmit those phenotypic characteristics to offspring (h^2). If selection is strong, a population can respond even if a trait is only weakly heritable. And even weak selection can lead to significant evolutionary change if a trait's heritability is high. But the most rapid evolutionary responses occur when both selection and heritability are large.

Key Concepts

- Selection can shape populations by favoring individuals with trait values at one end of a distribution (directional selection), by favoring individuals with trait values near the middle of a distribution (stabilizing selection), or by favoring individuals with trait values at both ends of a distribution (disruptive selection).

- Selection and evolution are *not* the same thing. Populations can experience selection even if they cannot evolve in response to it.

- The evolutionary response (R) is a product of the strength of selection (S) and the extent to which offspring resemble their parents for that trait (the heritability of the trait, h^2). ●

7.3 Linkage: Mixing and Matching Alleles

When selection acts on a trait, it does not necessarily select only those individuals carrying one particular allele. Many traits are complex, meaning that they are influenced by many genes. As we saw in Chapter 5, for example, human height has been linked to hundreds of loci, with many more loci likely left to be discovered. Different combinations of alleles can produce the same quantitative phenotype favored by natural selection. To make sense of the evolution of complex traits, we need to equip ourselves with new analytical tools.

To start, it's important that we take into account the way alleles get mixed and matched over the generations. As we saw in Chapter 5, eukaryotes undergo meiosis during their cycle of sexual reproduction. In a diploid species such as our own, each pair of chromosomes shuffles some of its DNA through genetic recombination before the individual chromosomes are packaged in haploid gametes (eggs or sperm). And because of independent assortment (Figure 5.17), there is a 50:50 chance that the copy of any given chromosome that ends up in a gamete is maternal (or paternal).

If genetic recombination and independent assortment did not occur, the distribution of alleles in human populations would be profoundly different. Without recombination, for example, every chromosome would remain intact from one generation to the next. All its alleles would remain physically linked together, like beads on a string.

In reality, our alleles are far less predictable. If we look across a population of humans, we will find that some alleles tend to be inherited together, but others are not. The ways in which gametes are generated are the reasons for these differences.

Let's first consider independent assortment. Each time a pair of chromosomes is pulled apart and packaged into gametes, the particular version of the chromosome (maternal or paternal) included is random. There is a 50% chance that the maternal copy of chromosome 1 will end up in a particular sperm, and a 50% chance for chromosome 2, and for chromosome 3, and so on. As a result, the maternal versions of any two loci located on different chromosomes have only a 50% chance of being packaged together in the same gamete. We refer to this relationship between loci on different chromosomes as a state of **linkage equilibrium**. Chromosomes are mixed and matched so readily that the chances of maternal and paternal alleles being inherited together are random.

But what about loci located on the same chromosome? If chromosomes were always passed down unchanged from one generation to the next like beads on a string, then their state would be the opposite of linkage equilibrium: **linkage disequilibrium**. The presence or absence of any allele at any locus would be dependent on that of all the other alleles on the same chromosome.

But even under such circumstances, because of meiosis and genetic recombination, alleles sometimes get moved from one copy of a chromosome to another—like moving a bead from one string to another (Figure 5.16). This transfer takes place in the development of the thousands of eggs carried by each woman and the millions of sperm generated by each man. The process is repeated in every generation. The result of all this movement is that pairs of alleles on a given chromosome can range from complete linkage disequilibrium to linkage equilibrium.

One crucial factor in this linkage is how close a given pair of loci are physically located to each other on a chromosome. Recombination can break and rejoin a chromosome pair at any spot along the length of the chromosome. For most chromosomes in most eukaryotic species, this is a big target, and it means that loci located far apart on the chromosome have a high probability of being swapped by recombination. The probability is so high that sampled across a population, maternal and paternal alleles at distant loci behave just as if they were located on different chromosomes.

When loci are closer together, on the other hand, there is less chance that recombination will divide them, simply because there is less DNA separating them

Linkage equilibrium exists for any two loci if the occurrence of an allele at one of the loci is independent of the presence or absence of an allele at the other locus.

Linkage disequilibrium exists for any two loci if the occurrence of an allele at one of the loci is nonrandomly associated with the presence or absence of an allele at the other locus.

BOX 7.2

Using an Offspring–Parent Regression to Measure the Narrow Sense Heritability, h^2

Narrow sense heritability (h^2) is the proportion of phenotypic variance that is transmitted from parents to offspring (see Box 7.1). For this reason, it's the variance that causes a population to evolve in response to selection. One way to measure h^2 is with a so-called offspring–parent regression (Falconer and Mackay 1996). Scientists measure the phenotypic resemblance of a trait between parents and their offspring in a number of different families. They then regress the mean trait value of offspring against the mean trait value of the parents. (The parent mean is often called the midparent value because there are only two parents.)

When these values are plotted for many different families, the relationship is an indication of how closely the offspring trait values resemble those of their parents. If there is a significant positive relationship (that is, the slope is greater than zero), then parents with unusually large trait values tend to produce offspring who also have unusually large trait values (and vice versa for parents with smaller trait values). Graphs A and B in **Box Figure 7.2.1** show two such relationships.

Let's say that we're examining the body size of two species of fish. In each of the graphs in panels A and B, the parent values are plotted along the x axis, and offspring trait values are plotted on the y axis. (Both plots are drawn so that the x and y axes intersect at the mean phenotypic value of each population—the origin of the plot is the mean of both offspring and parental trait values.) As you can see, both graphs have a positive slope, but the top graph displays a steeper slope than the bottom graph ($a = 0.8$ and 0.2, respectively). In other words, the offspring in the top graph have a stronger resemblance to their parents than the offspring in the bottom graph (at least when it comes to body size).

Let's now consider how the phenotypic trait changes from one generation to the next. In the absence of selection or another mechanism of evolution, a population should not evolve. Offspring phenotypes should be similar to parental phenotypes, even if the trait in question is heritable. (This is the phenotypic manifestation of the Hardy–Weinberg theorem we saw in Chapter 6.)

But consider what happens if we apply selection to this population. We've already seen how selection can be a powerful mechanism of evolution because it can change allele frequencies from one generation to the next (see Box 6.5). Because additive effects of alleles cause relatives to resemble each other in their phenotypes (that is, because of h^2), selection also can be a powerful mechanism of phenotypic evolution. Selection can cause the distribution of phenotypes to change from one generation to the next.

Let's say that selection favors big body size in both populations of fishes. We'll illustrate this selection by shaded circles to represent individuals that were able to reproduce (individuals that failed to breed are shown in open circles). These parents alone contribute offspring to the next generation of the population. This situation results in positive directional selection on the phenotypic trait because the mean trait value in selected parents (vertical dashed blue line) is greater than the starting mean of the parental population (vertical dashed black line). The difference between these means is the selection differential (S).

To predict how much the offspring generation will evolve in response to this selection, we use the parent-offspring regression. To do this, follow the mean value (X) of selected parents on the x axis up until it intersects with the regression line. From this point, read across to the corresponding value on the offspring (y) axis (horizontal dashed blue lines). This is the new mean trait value expected for the offspring. If this new value differs from the mean of the population before selection—in this case, if the new mean lies above the origin—then the offspring distribution will have shifted toward larger trait sizes. The population will have evolved in response to selection, as shown in the curves to the right of each graph. (The difference between the new mean of offspring and the starting mean is a measure of the population response to selection, R.)

When the slope of the offspring–parent regression is steep, the offspring phenotype distribution will experience a big shift. When the slope of the regression is shallow, the offspring phenotypes will shift less. The slope of the regression, then, determines how much a given population will evolve in response to selection. As we know already, the component of variance that causes a population to evolve in response to selection is, by definition, the narrow sense heritability (h^2). So the slope of the offspring–parent regression must equal h^2.

We can reach this same conclusion by looking at the regression equation itself. A linear regression will take the form $y = a \times x + b$, where the slope of the relationship, a, is equal to the change in trait value along the y axis, ΔY, divided by the change in trait value along the x axis, ΔX:

$$a = \Delta Y / \Delta X$$

In an offspring–parent regression, $\Delta Y = R$ and $\Delta X = S$, so

$$a = \Delta Y / \Delta X = R / S$$

and we know from the breeder's equation (see Section 7.2) that $R = h^2 \times S$ and therefore that $h^2 = R/S$. So the slope of the regression, a, equals h^2.

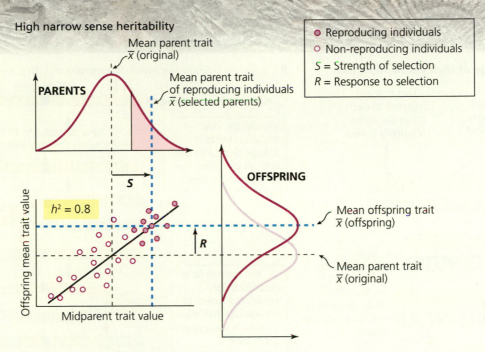

A   High narrow sense heritability

Mean parent trait
$\bar{x}$ (original)

PARENTS

Mean parent trait
of reproducing individuals
$\bar{x}$ (selected parents)

● Reproducing individuals
○ Non-reproducing individuals
S = Strength of selection
R = Response to selection

S

$h^2 = 0.8$

OFFSPRING

Mean offspring trait
$\bar{x}$ (offspring)

R

Mean parent trait
$\bar{x}$ (original)

Offspring mean trait value

Midparent trait value

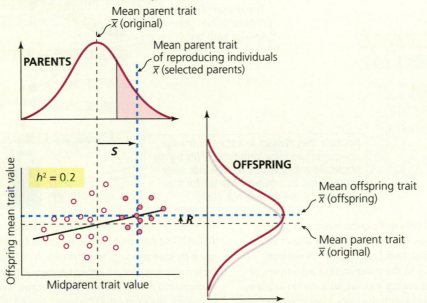

B   Low narrow sense heritability

Mean parent trait
$\bar{x}$ (original)

PARENTS

Mean parent trait
of reproducing individuals
$\bar{x}$ (selected parents)

S

$h^2 = 0.2$

OFFSPRING

Mean offspring trait
$\bar{x}$ (offspring)

R

Mean parent trait
$\bar{x}$ (original)

Offspring mean trait value

Midparent trait value

Box Figure 7.2.1 Panels A and B show parent–offspring regressions for two populations. Phenotypic distributions for the parents and offspring are shown above and to the right of each graph, respectively. Shaded circles indicate the parents in the first generation that reproduced to create the second (offspring) generation, and the vertical dashed blue line indicates the mean trait value of these selected parents. The difference between the average value of the reproducing individuals [$\bar{x}$ (selected parents)] and the average value of the trait in the entire population before selection [$\bar{x}$ (original)] is the strength of selection (S). The response to selection (R) depends on the narrow sense heritability of the trait, represented here by the slope of the parent–offspring regression. R is calculated as the difference between the offspring mean trait value (horizontal blue dashed line) and the average value of the trait in the population prior to selection (horizontal dashed black line). In the top example, the evolutionary response is large because narrow sense heritability is high; in the bottom example, the response is small because narrow sense heritability is low, perhaps because of high environmental variation or high epistasis.

(**Figure 7.11**). Recombination is still possible and, given enough meiotic events, it will eventually happen. But its low probability means that it takes much longer. For a while, at least, a maternal allele at one locus is likely to be inherited in tandem with other maternal alleles at other loci nearby. The closer any two loci are to each other on the string, the less likely recombination is to split them apart, and the more likely they are to be inherited together—in a state of linkage disequilibrium.

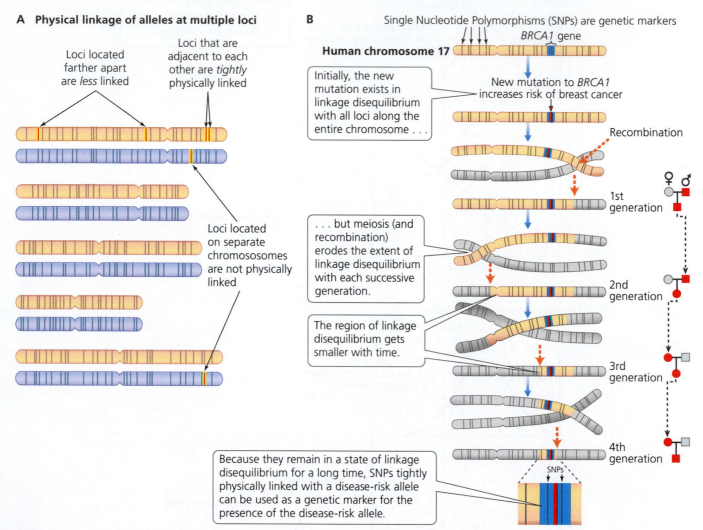

A   Physical linkage of alleles at multiple loci

Loci located farther apart are *less* linked

Loci that are adjacent to each other are *tightly* physically linked

Loci located on separate chromososomes are not physically linked

B   Single Nucleotide Polymorphisms (SNPs) are genetic markers

BRCA1 gene

Human chromosome 17

Initially, the new mutation exists in linkage disequilibrium with all loci along the entire chromosome . . .

New mutation to *BRCA1* increases risk of breast cancer

Recombination

1st generation

. . . but meiosis (and recombination) erodes the extent of linkage disequilibrium with each successive generation.

2nd generation

The region of linkage disequilibrium gets smaller with time.

3rd generation

4th generation

SNPs

Because they remain in a state of linkage disequilibrium for a long time, SNPs tightly physically linked with a disease-risk allele can be used as a genetic marker for the presence of the disease-risk allele.

Figure 7.11 A: The fact that eukaryotic genomes are organized into discrete chromosomes has implications for how alleles at different loci are mixed and matched. Loci housed on separate chromosomes are shuffled due to the independent assortment of maternal and paternal strands during meiosis, so a maternal allele on one chromosome is equally likely to be packaged into a gamete alongside the maternal or the paternal allele at a locus on another chromosome. When two loci exist on the same chromosome, their alleles may still be inherited independently thanks to genetic recombination. However, it takes much longer, on average, for recombination to strike between loci that are located very close to each other than it does between loci that are farther apart. As a result, loci that are physically near to each other on a chromosome are said to be linked, and they may be inherited together more often than expected by chance (linkage disequilibrium). B: When a new mutation arises, it begins in a state of perfect linkage disequilibrium along the entire length of its chromosome, shown here

for a new mutation in the breast cancer tumor suppressor gene *BRCA1*. The new mutation is indicated by the red line, the *BRCA1* gene by blue shading, and the region of linkage disequilibrium by tan shading. Because they happened to be part of the genetic background on the particular chromosome where the mutation occurred, alleles at other "marker" loci on the same chromosome (red lines) co-segregate with the new mutation—they are nonrandomly associated with the disease-causing allele. But successive generations of meiosis and genetic recombination mix and match these alleles with corresponding alleles from other versions of the chromosome (gray shading). This tends to occur in distant regions first, eroding the region of linkage disequilibrium in the process. Marker loci located very close to the original mutation, however, may persist in a state of linkage disequilibrium for hundreds or even thousands of generations. For this reason, doctors can screen for the presence of the marker allele and use this as an inexpensive proxy for the disease-causing allele.

Physical linkage—the adjacency of two loci on a chromosome strand—isn't required for two loci to be in a state of linkage disequilibrium. Technically, linkage disequilibrium is a statistical state we detect whenever some combinations of alleles at two loci occur together more often than expected by chance (**Figure 7.12**), and this can sometimes occur even when loci are located very far apart or on different chromosomes. But linkage disequilibrium very often does result from loci being physically adjacent simply because it takes many generations for meiosis to uncouple loci that are located very close to each other on a chromosome.

Biologists often think of linkage disequilibrium as an accident of history— a pattern in the variation that decays predictably over time. When a new mutation arises

> **Physical linkage** is the adjacency of two or more loci on the same chromosome.

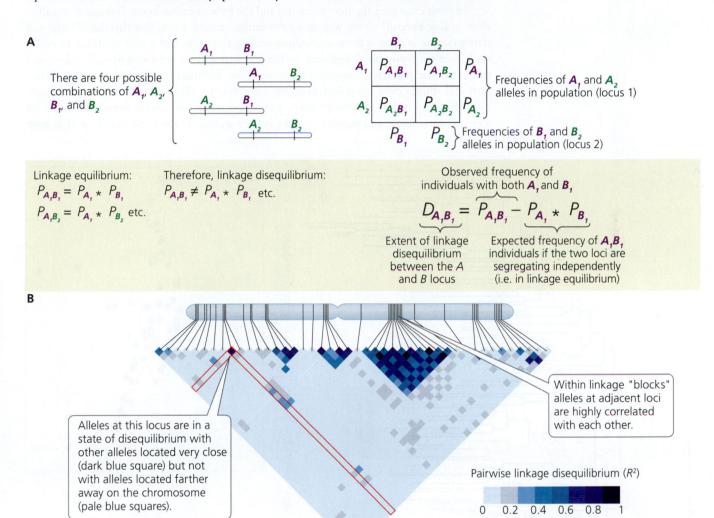

Linkage equilibrium:
$$P_{A_1B_1} = P_{A_1} * P_{B_1}$$
$$P_{A_1B_2} = P_{A_1} * P_{B_2} \text{ etc.}$$

Therefore, linkage disequilibrium:
$$P_{A_1B_1} \neq P_{A_1} * P_{B_1} \text{ etc.}$$

Observed frequency of individuals with both A_1 and B_1

$$D_{A_1B_1} = \overbrace{P_{A_1B_1}}^{} - \underbrace{P_{A_1} * P_{B_1}}_{}$$

Extent of linkage disequilibrium between the A and B locus

Expected frequency of A_1B_1 individuals if the two loci are segregating independently (i.e. in linkage equilibrium)

Alleles at this locus are in a state of disequilibrium with other alleles located very close (dark blue square) but not with alleles located farther away on the chromosome (pale blue squares).

Within linkage "blocks" alleles at adjacent loci are highly correlated with each other.

Pairwise linkage disequilibrium (R^2)

0   0.2   0.4   0.6   0.8   1

Figure 7.12 Detecting linkage disequilibrium. A: For any two loci it's possible to ask whether an allele at one of the loci, say A_1, co-occurs with an allele at the other locus, B_1, more (or less) often than expected by chance. If A_1 and B_1 are in a state of linkage equilibrium, then the presence of A_1 on a particular chromosome tells us nothing about whether we are likely to find B_1 or B_2 at the other locus, since the two loci are segregating independently. If the frequency of A_1 is p_{A_1}, and the frequency of B_1 is p_{B_1}, then simply by chance we'd expect to see A_1 and B_1 occur together on the same chromosome with a frequency of $p_{A_1} \times p_{B_1}$. If, on the other hand, alleles at these two loci are *not* segregating independently— say individuals with both A_1 and B_1 are unusually common—then we'd detect this because the frequency of A_1B_1 individuals ($p_{A_1B_1}$) would be greater than $p_{A_1} \times p_{B_1}$. Biologists quantify the extent of linkage disequilibrium between any two loci, $D_{A_1B_1}$, as the difference between the observed frequency of individuals with both A_1 and B_1 and the frequency with which these alleles are expected to co-occur by chance. B: With increasing amounts of genomic information, it's become possible to estimate D for hundreds or thousands of pairs of loci simultaneously, assembling "heat maps" for the extent of linkage disequilibrium along the length of a chromosome. Studies like this are revealing exciting new patterns, such as "blocks" of linked loci, indicated here in blue for a linkage map of chromosome 3 of the common bean (*Phaseolus vulgaris*). (Data from Perseguini et al. 2016)

on a chromosome, it first appears on a specific genetic background (Figure 7.11B). The new allele sits amid a string of beads defined by the alleles that happened to exist at that moment on that particular chromosome. At that instant, the new allele is in a perfect state of linkage disequilibrium. There is only one copy, and that one copy only exists alongside that particular string of beads. Similarly, if two individuals from different species mate to produce a hybrid offspring, the alleles in that hybrid exist in perfect linkage disequilibrium. Every fixed difference that has accumulated between the two species now sorts perfectly along the two chromosomes: all of the alleles unique to the maternal species occur on the maternal copy of the chromosome, and all of the alleles unique to the paternal species exist on the paternal copy.

In both examples, the new mutation and the hybridization event, linkage disequilibrium at first extends all the way along the entire chromosome. But this initial state will start to decay as soon as these individuals undergo meiosis, and it will continue to erode with every subsequent generation. In time, only loci that are close to each other will still show the pattern of nonrandom association. Eventually, meiosis and recombination will shuffle even these loci, bringing neighboring alleles into linkage equilibrium.

Thinking of linkage disequilibrium as a statistically detectable signal that arises by chance, starts large, and steadily shrinks is very useful. There is no question that new

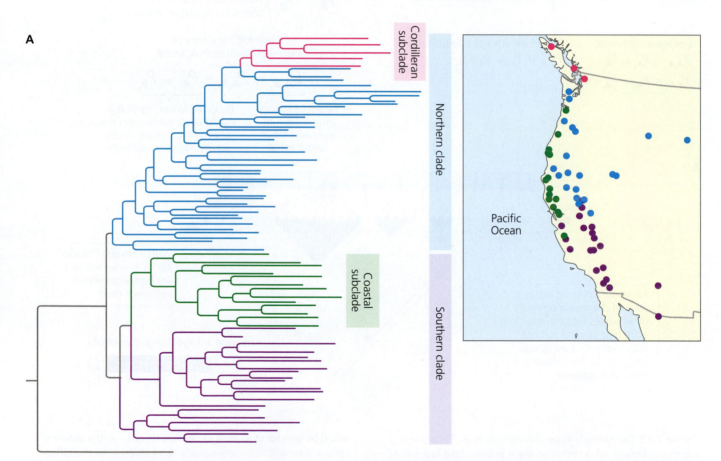

Figure 7.13 A: The yellow monkeyflower (*Mimulus guttatus*) inhabits much of the inland Pacific Northwest, where it encounters hot, dry summer conditions. But at least twice this plant has expanded into coastal areas, where it now lives in milder, moister environments. B: Inland and coastal populations have diverged in a suite of traits including whether they are annual or perennial and whether they flower early or late in the season, and in traits related to the relative amount of growth the plants experience before they begin to flower, such as stem thickness and internode stem length. These traits are controlled by alleles in several different genes, and selection favors particular—inland or coastal—combinations of

alleles at these loci. Recombination would break apart these adaptive combinations of alleles, causing plants to inherit unfortunate mixtures of inland- and coastal-adapted traits. However, the loci responsible for these traits are clustered very close to each other in a "supergene," and this region of the chromosome has been inverted in the coastal populations. When coastal plants mate with other coastal plants, recombination works just fine, even if it occurs within the inverted region. Because these plants already have both the inversion and all of the coastal alleles (A_2, B_2, C_2, etc.), recombination does not break apart the favorable "coastal" combination of these traits. Similarly, when two inland plants mate, recombination

mutations arise all the time, randomly creating new blocks of linkage disequilibrium. But sometimes patterns of linkage disequilibrium are more than just chance associations among adjacent loci. Sometimes selection favors specific combinations of alleles. When this happens, selection and recombination act in opposition because recombination breaks even favorable combinations of alleles apart. If selection for particular combinations of alleles is strong enough, it can favor structural changes to chromosomes that help lock favorable combinations of alleles in place. Chromosomal rearrangements that bring functionally related loci close together, for example, increase their physical linkage and reduce the rate of recombination between them (suites of functionally related genes located close enough to each other that they segregate as a single unit are called **supergenes**; Schwander et al. 2014; Thompson and Jiggins 2014). Mutations that invert a segment of a chromosome containing functionally linked genes are an even better way to preserve favorable combinations of alleles at different loci because inversions suppress genetic recombination in the inverted region (**Figure 7.13**; Kirkpatrick 2010).

A **supergene** is a group of functionally related genes located close enough together that they segregate as a single unit.

Let's consider one example: The yellow monkeyflower *Mimulus guttatus* inhabits a wide range of environments extending across the Pacific Northwest of the United States and Canada (Lowry and Willis 2010). Inland populations are locally adapted to

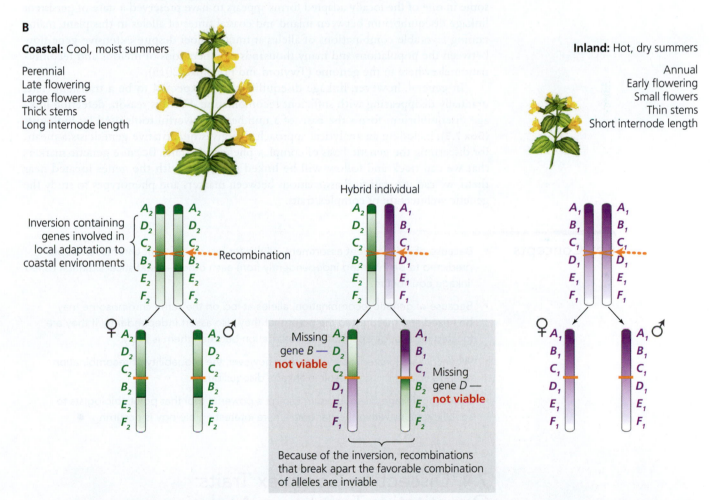

B

Coastal: Cool, moist summers

Perennial
Late flowering
Large flowers
Thick stems
Long internode length

Inversion containing genes involved in local adaptation to coastal environments

Recombination

Hybrid individual

Missing gene B — **not viable**

Missing gene D — **not viable**

Because of the inversion, recombinations that break apart the favorable combination of alleles are inviable

Inland: Hot, dry summers

Annual
Early flowering
Small flowers
Thin stems
Short internode length

works just fine. Neither plant has the chromosome inversion, and they each already possess all of the inland alleles (A_1, B_1, C_1, etc.), so mixing and matching will not undo the favorable "inland" combination of traits. However, when an inland plant mates with a coastal plant—the sort of mating that could break apart the favorable combinations of locally adapted traits—recombination fails. Only one of the plants has the chromosome inversion, and

recombination in this region results in offspring that are inviable since some genes get duplicated and others are lost entirely. Because these hybrid recombinants die, the chromosome inversion suppresses recombination within the region of the inversion, locking the favorable combinations of traits in place. (Data from Lowry and Willis 2010; Twyford and Friedman 2015)

dry soils and hot summer conditions. To escape the heat, these plants flower early in the year—before summer temperatures have baked the soils. And they die after flowering (that is, they are annuals), leaving their tough seeds buried in the soil until rains return the following year. Up and down the coast, however, plants encounter cooler, moister conditions. Coastal populations can wait longer before flowering because there is less risk of baking in hot sun. And they can survive through the whole year. By flowering later in the summer, coastal plants can allocate more resources for growth early in the year, producing larger plants that live for several years (perennials).

Local adaptation to coastal and inland conditions involves alleles at many loci, including genes affecting the timing of flowering, whether plants allocate resources early in the year for growth or for reproduction, and whether the plants are annual or perennial. And mixing and matching inland and coastal alleles for these different traits is likely to be disastrous. When David Lowry and John Willis, biologists at Duke University, compared the genomes of inland and coastal populations of *M. guttatus*, they found that the genes for these traits were located very close to each other on the same chromosome (that is, they formed a supergene), and they discovered that this region had inverted in the coastal populations (Figure 7.13; Lowry and Willis 2010). Clustering of related loci into a supergene and inverting this region of the chromosome in one of the locally adapted forms appears to have preserved a state of persistent linkage disequilibrium between inland and coastal suites of alleles in this plant, maintaining favorable combinations of alleles at multiple loci despite extensive gene flow between the populations and many thousands of generations of meiosis and recombination elsewhere in the genome (Twyford and Friedman 2015).

In general, however, linkage disequilibrium is expected to be a transient state, gradually disappearing with sufficient recombination. For this reason, detecting linkage disequilibrium forms the basis of a number of powerful tools used by biologists (**Box 7.3**), including an analytical approach that gives quantitative geneticists a means for discerning the genetic basis of complex phenotypic traits. Because genetic markers that we can track and follow will be linked physically with the genes located near them, we can use statistical associations between markers and phenotypes to study the genetic architecture of complex traits.

Key Concepts

- Because of independent assortment, alleles for loci on different chromosomes are predicted to be inherited independently from each other—a statistical situation called linkage equilibrium.

- Because of genetic recombination, alleles at loci on the same chromosome may be mixed and matched to the point that they also assort independently if they are located far enough apart that recombination between them is likely.

- When loci are located closer together, however, the probability of recombination decreases, and this can result in linkage disequilibrium.

- Detecting linkage disequilibrium can be a powerful tool that points biologists to specific regions within the genome where interesting biology is occurring. ●

7.4 Dissecting Complex Traits: Quantitative Trait Locus Analysis

Quantitative trait locus (QTL) is a stretch of DNA that is correlated with variation in a phenotypic trait. These regions contain genes, or are linked to genes, that contribute to population differences in a phenotype.

Linking the evolution of quantitative traits to their genetic basis requires a scientific form of detective work. Researchers have to hunt for the connections between variations in phenotype and allelic variations in genotype. One of the most powerful ways to conduct this sleuthing is a method known as **quantitative trait locus (QTL)** analysis.

Let's say a team of scientists wants to use QTL analysis to find the genetic basis of body size in the fishes in a lake. They can begin their search by breeding two lines of

BOX 7.3

The Many Uses of Linkage Disequilibrium

When new mutations arise, they begin in a state of perfect, chromosome-wide linkage disequilibrium (LD) (Figure 7.11B). They sit on a string of beads that comprises the full length of the chromosome on which they arose. From that point forward, however, meiosis and genetic recombination will mix and match this string of beads with other strings, and the chunk of flanking DNA associated with the new mutation will get smaller and smaller. Distant regions tend to get shuffled first, whereas regions immediately adjacent to the new mutation are likely to remain in a state of LD for much longer. This gradual decay of LD is a very useful signal that is now routinely exploited by biologists in a number of ways.

The extent of LD provides information about the age of an allele, for example. New mutations are likely to be associated with larger blocks, or regions, of LD than are older alleles, for the simple reason that they have experienced fewer recombination events. When alleles of interest are discovered in a genome, one of the first things scientists can do is look at large samples of individuals, examine their genotypes at lots of markers (for example, single nucleotide polymorphisms, or SNPs) along the chromosome containing the allele, and estimate the relative size of the flanking region that is still in LD. By modeling the rate of decay of LD, they can then work backward to provide an estimate of how long ago the allele first appeared. Andrew Clark, a geneticist at Cornell University, and his colleagues used this approach to calculate that the Med allele of the glucose-6-phosphate dehydrogenase gene (*G6PD-Med*) likely arose in Europeans between 1600 and 6640 years ago. Like the β-globin allele we examined in Chapter 6, the *G6PD-Med* allele confers some resistance to infection by the malaria-causing *Plasmodium* parasite. And, like that β-globin allele, the benefits of *G6PD-Med* are balanced by steep costs—in this case, enzyme deficiencies due to this mutation impair glucose metabolism in roughly 400 million people worldwide (Tishkoff et al. 2001).

Thanks to physical linkage, loci that are located very close to each other on a chromosome can persist in a state of LD for hundreds of generations. This simple observation means it is possible for doctors, and even companies like 23andMe, to offer inexpensive tests for inherited risk of numerous diseases. Several mutations to the human tumor suppressor gene *BRCA1*, for example, increase the risk of acquiring breast and ovarian cancers. A few of these mutations, when they arose, happened to be sitting next to a rare allele at another locus—a SNP. The fact that these adjacent alleles were rare is important because not many people carried the alleles to begin with, and now these alleles are nonrandomly associated with the cancer-causing mutation. The cancer-causing allele and the rare SNP allele co-occur together more often than expected by chance—they exist in a state of LD—and this means that by testing for the presence of the marker allele, doctors can get a good idea of whether a person also carries the disease risk allele. LD makes it possible for inexpensive markers to serve as proxies for alleles related to all sorts of traits, ranging from risk of various cancers and other heritable diseases to whether a person's skin is likely to flush after drinking alcohol, whether they are likely to have dimples, freckles, blue eyes, or red hair.

Genetic markers in LD with biologically relevant loci also make it possible for quantitative geneticists to find the genes responsible for phenotypic variation in complex traits. Quantitative trait locus (QTL) mapping and genome-wide association (GWA) studies (discussed later in the chapter) each rely on genotyping lots of individuals at many marker loci sprinkled throughout the genome and testing for associations between specific marker alleles and the expression of particular phenotypic traits. By first finding markers whose alleles are correlated with the trait, biologists can then search nearby within the genome to try to find the genes. Thanks to physical linkage and the characteristic patterns of decay of LD, any markers closely associated with expression of a phenotypic trait are likely to sit very close to the actual alleles influencing trait expression.

Finally, these marker-based approaches can be used to search within genomes for "footprints" of selection—patterns lurking within the standing genetic variation of a population that point to the recent action of natural selection. As we've seen, when new mutations arise, they persist in a state of LD with alleles at neighboring loci for hundreds of generations. If one of these mutations happens to enhance the fitness of its carriers, then selection can increase its frequency within the population quickly (Chapter 6). As the new allele becomes more common, it carries with it nearby beads on its string. Alleles at neighboring loci "hitchhike" along for the ride. They all sweep to high frequency together, and this creates a signal—a local region with unusually little genetic variation—that biologists can detect. We'll discuss selective sweeps, as well as several related approaches that also leverage LD, in Chapter 10.

fish, selecting large size in one and small size in the other (**Figure 7.14A**). With enough inbreeding, fishes in each line will become homozygous at many loci associated with body size. The scientists then cross a homozygous male from one line with a homozygous female from the other to produce offspring that are heterozygous across loci (this first generation of offspring is known as F_1). Then researchers breed these heterozygotes with each other to produce a second generation of offspring (F_2).

The scientists end up with a population of F_2 offspring with a wide range of body sizes. That's because genetic recombination will have swapped pieces of the chromosomes from each of the original parental types. Recombination mixes and matches segments from the original parental types into new arrangements of alleles. Now offspring will differ in whether they have alleles from the large or small parental lines at each of their many loci (**Figure 7.14B**). Some of the fishes will end up carrying a high number of alleles for large body size, whereas others will inherit more of the alleles for small body size.

Next, the scientists examine the DNA of as many of these F_2 fishes as they can, looking for loci that correlate with variation in the trait of interest—body size, in our case. To do this, scientists return to each parental line and identify a large number of short segments of DNA distinct to each lineage. These markers are often

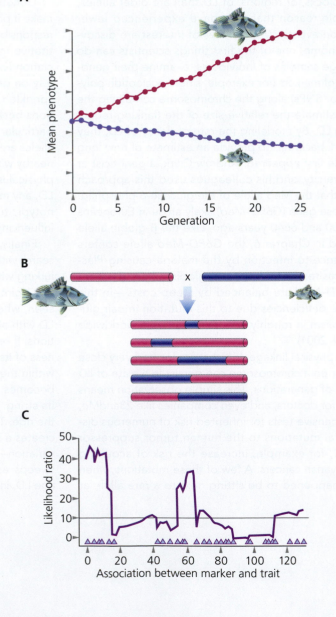

Figure 7.14 A: Quantitative trait loci (QTL) mapping often begins with two purebred strains that have been selected over many generations for different values of a phenotypic trait. B: Scientists then cross these two lines to create various combinations of their genotypes. After two generations of interbreeding, the researchers then search in the F_2 individuals for unique genetic markers found in each parental strain. They test each marker for a statistically significant association with a quantitative trait. C: The triangles represent markers along a chromosome. Two regions, centered on positions 10 and 60, are strongly associated with the trait. Once scientists find these quantitative trait loci, they can conduct further studies to identify the genes in these regions that are influencing the trait in question. (Data from Mackay 2001)

single-nucleotide polymorphisms (called SNPs), but they may also include simple sequence repeats or transposable elements. Any readily identifiable fragment of DNA that differs among individuals will work. Scientists choose these markers so that the alleles at each one are different, depending on whether they're inherited from the original maternal or paternal chromosome. In our example of measuring fish size, this means that each marker can be scored for whether it came from the large-body-size copy of its chromosome or the small-body-size copy.

The scientists then examine the F2 fishes for these genetic markers and test whether the genotype at particular markers is associated with the variation in the phenotypic trait of interest (**Figure 7.14C**). Certain genetic markers may turn out to be associated with the body-size phenotype. Fishes with paternal alleles at a marker may consistently have larger body sizes than fishes that inherited maternal alleles at the marker. Finding this association does not mean that the genetic marker itself has a direct effect on the phenotype of the animals. Instead, the loci that explain part of the phenotypic variance in body size may be nearby. By identifying these markers, scientists narrow down the regions they need to search for genes and other DNA elements (Miles and Wayne 2008). Markers that have a statistically significant association with the expression of a phenotypic trait are called quantitative trait loci (QTLs).

QTL mapping studies require a tremendous amount of work. Scientists first raise hundreds of F2 offspring and measure their phenotypes. Then they carry out large-scale comparisons of huge numbers of genetic markers to find significant associations. But such effort has been amply rewarded: QTL mapping has helped scientists to uncover regions of the genome responsible for the evolution of complex phenotypes (**Figure 7.15**; Mackay 2001). Plant geneticists, for example, can use QTL techniques to identify genes that are crucial for traits in crops and then use those insights to develop

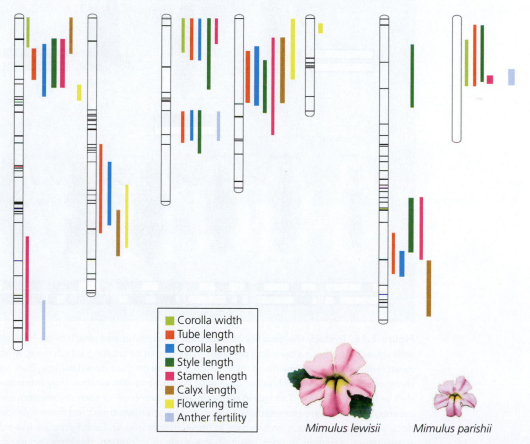

Legend:
- Corolla width
- Tube length
- Corolla length
- Style length
- Stamen length
- Calyx length
- Flowering time
- Anther fertility

Mimulus lewisii   *Mimulus parishii*

Figure 7.15 QTL analysis can help scientists to identify the genes contributing to variation in several different traits. This figure shows the result of one study on two closely related species of the flowering plant *Mimulus*. The analysis revealed loci associated with variations in several components of flower shape, as well as flowering time. (Data from Fishman et al. 2015)

new strains that are resistant to drought, salt, and pests. Medical geneticists can use the same techniques to identify genes that play important roles in conditions such as diabetes and high blood pressure.

QTL Mapping in Mouse Coat Color

Hoekstra and her colleagues have used QTL mapping to pinpoint some of the genes underlying the variation in oldfield mouse coat color (Nachman et al. 2003; Steiner et al. 2007). They began by interbreeding the beach and mainland populations of mice. Starting with three mice from each population, they carried out two rounds of crosses between inbred lineages from each population to produce 465 F_2 progeny. The F_2 mice had higher variation in coat color than either parental population—ranging from nearly white to mostly brown (**Figure 7.16**).

Hoekstra's team created a map of 124 markers in the *Peromyscus* genome and then looked for associations between these markers and variation in the color patterns on the face, back, rump, and tail. The comparison revealed two markers where alleles from each population tended to show up in mice that were either very light or very dark (**Figure 7.17**).

This discovery allowed Hoekstra and her colleagues to narrow their search to just two regions of the mouse genome. These regions were still very large—large enough to include many different genes. So the scientists hunted through each region

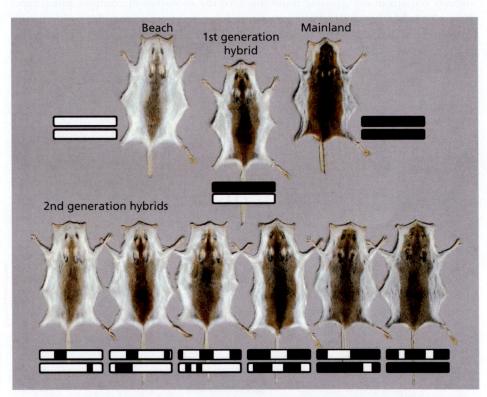

Figure 7.16 To study the genetics of coloration in mainland and beach mice, Hoekstra and her colleagues carried out a breeding program. They began by crossing beach mice, which have mostly white coats, with mainland mice, which are dark on their dorsal surface. Bars represent chromosomes color-coded black and white to correspond with whether they originated from the dark or light population. The hybrid offspring inherit one copy of each chromosome from their parents, and they have intermediate coat color. The scientists then bred these hybrids to produce a second generation (shown here as museum specimens). Each F_2 mouse has a different combination of genetic regions, resulting in a range of coloration. Hoekstra and her colleagues were then able to search the genomes of the F_2 mice for genetic markers correlated with light or dark coloration. (From "Adaptive Variation in Beach Mice Produced by Two Interacting Pigmentation Genes" by Steiner, Weber and Hoekstra. PLoS Biology 08/14/2007)

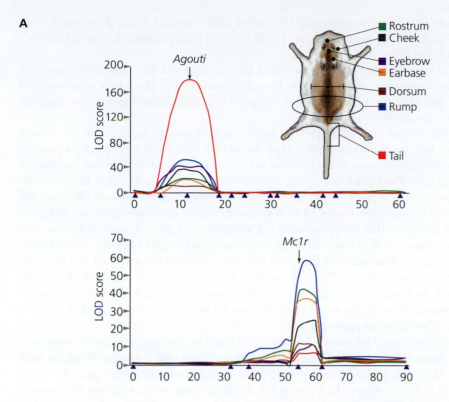

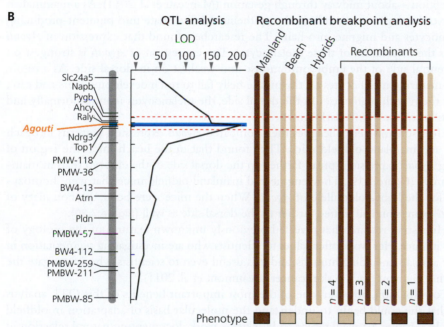

Figure 7.17 Hoekstra and her colleagues conducted a QTL analysis on F₂ crosses between beach and mainland populations of oldfield mice. They examined the color of each mouse's coat at seven locations, as shown in the skin diagram in A, and searched for correlations with alleles. A: They found two loci with significant associations. Each region contained a gene known to be involved in coat color: *Agouti* and *Mc1r*. (LOD refers to the logarithm of the odds score, a statistical estimate of whether two loci [for example, a marker and a gene influencing a focal trait] are likely to lie next to each other on a chromosome.) B: In a subsequent study, the researchers looked more closely at the genetic region containing *Agouti* (cM = centimorgans, units of distance along a chromosome). They crossed beach and mainland mice (represented here by light and dark bars, respectively). They were able to narrow down the size of the segment that correlates with color. In this analysis, *Agouti* in particular revealed the strongest correlation with coat color compared to nearby genes. (Data from Steiner et al. 2007; Manceau et al. 2011)

for genes that might be responsible for color differences. To focus their search, they started with the 100 genes that were already known from previous studies to influence coloration in mice.

Hoekstra and her colleagues found that each of the QTL regions contained a known coat-color gene. Together, the allelic variation in these two genes—called *Agouti* and *Mc1r* (for melanocortin-1 receptor)—could explain most of the variation in coat color in the F₂ mice.

Hoekstra and her colleagues then looked at each gene to break down the genetic variance into smaller components. They found that *Agouti* and *Mc1r* produce proteins that are critical elements of the pathway for synthesizing the dark pigment, melanin, in growing hair. The lighter coat color of Gulf Coast beach mice resulted from a mutation that changes a single amino acid in the *Mc1r* protein. The mutation decreases the activity of the receptor (**Figure 7.18**). A second mutation increases expression of a gene, known as *Agouti*, that interferes with the expression of *Mc1r*. Combined, these two genetic changes reduce levels of melanin synthesis and result in lighter-colored mice.

This study demonstrated that even the interactions of just two loci can be complex (Steiner et al. 2007). These researchers found that *Agouti* and *Mc1r* do not act additively to produce pale color. Instead, the *Agouti* allele found in beach mice interacts epistatically with the other gene: it must be present for the *Mc1r* allele to have a measurable effect on pigmentation.

Hoekstra's team followed up on this 2007 study by figuring out exactly where and when *Agouti* is expressed in beach mice and mainland oldfield mice. Working with Marie Manceau, a postdoctoral researcher in her lab, and other colleagues, Hoekstra found that *Agouti* began to be expressed in the developing mouse embryos 12 days after conception—about midway through gestation (Manceau et al. 2011). As a mammalian embryo develops, certain cells in the epithelium differentiate into pigment-producing melanocytes and migrate into hairs. The researchers found that expression of *Agouti* delays the maturation of these melanocytes. The expression of *Agouti* is strongest on the ventral side of the embryonic mouse and weakest on its dorsal side. As a result, slow-maturing melanocytes on the mouse belly fail to reach developing hairs and can't color them with pigments. On the dorsal side, the melanocytes mature normally and can still turn the hairs a dark brown.

Hoekstra and her colleagues then compared *Agouti* expression patterns in beach mice and mainland oldfield mice. They found that in the beach mice, the region of strong *Agouti* expression spread farther up the dorsal side of the embryo than in mainland mice (**Figure 7.19**). They engineered mainland oldfield mice that were homozygous for the light-color allele of *Agouti*. When the mice were born, the boundary of *Agouti* expression had shifted farther to the dorsal side as well (**Figure 7.20**).

Hoekstra's research has revealed previously unknown features of the biology of color in mice. Her work offers clues to scientists who are investigating pigmentation in other animals, including humans, and it is useful even to scientists who investigate the role that pigment plays in skin cancer (Beaumont et al. 2011).

For Hoekstra, however, one of the most important benefits of this QTL analysis was that she could begin to understand the molecular basis of adaptation in oldfield mice. In Chapter 10, we'll return to Hoekstra's work documenting natural selection in the wild.

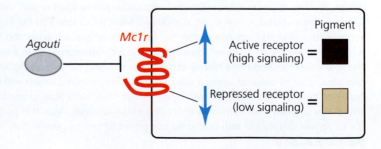

Figure 7.18 *Mc1r* encodes a receptor that triggers the production of pigment. One allele causes high signaling from the receptor (and the production of dark pigment), and another causes low signaling (and the production of light pigment). *Agouti* also causes the production of light pigment, but it does so by encoding a repressor that shuts down the *Mc1r* receptor, causing low signaling to the pigment-producing cells.

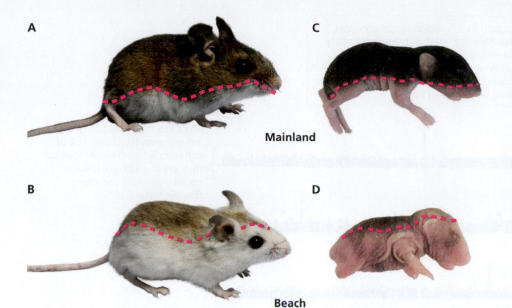

Figure 7.19 In adult mainland old-field mice, the boundary between the dark dorsal pelage and light ventral pelage is relatively low (dashed line in A). In adult beach mice, this line is much closer to the dorsal side of the animal (dashed line in B). These shifting boundaries help the mice blend into their surroundings. The pelage boundary is already marked in mouse embryos before they develop hairs (dashed lines in C and D). (Data from Manceau et al. 2011; photos by Maria Manceau / Hoekstra Lab)

Genome-Wide Association Mapping

In the early 2000s, researchers developed a new twist on QTL mapping, called **genome-wide association (GWA)** mapping. GWA studies require detailed maps of the entire genome of a species, as well as genotype information for very large numbers of individuals. For this reason, they have proven most effective for humans and a few unusually well-studied genetic model species (for example, the plant *Arabidopsis thaliana*; Aranzana et al. 2005).

Instead of crossing individuals from genetically divergent populations, GWA studies begin with large numbers of individuals sampled from within a single population. These individuals are grouped according to their phenotype—presence or absence of a particular disease, for example. Then they're compared, marker for marker, across the genome in an attempt to identify alleles whose presence is correlated with the trait of interest.

In one of the first big successes for this approach, researchers at the Wellcome Trust, a biomedical research foundation, used a GWA approach to screen for alleles associated with seven major diseases. For each disease, the team compared the genomes of 2000 individuals who had the disease with another 3000 individuals who

Genome-wide association (GWA) mapping involves scanning through the genomes of many different individuals, some with, and others without, a focal trait of interest, to search for markers associated with expression of the trait.

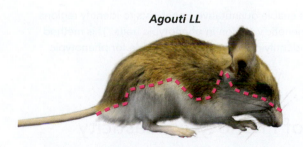

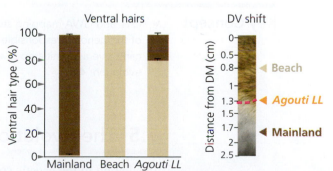

Figure 7.20 To test their hypothesis about *Agouti* patterning's influence on the two populations of oldfield mice, Hoekstra and her colleagues replaced the *Agouti* alleles in mainland mice with those of beach mice. The engineered mice (*Agouti LL*) developed a higher boundary between light belly and dark dorsal coat color. DV shift stands for the shift of the coloration boundary along the dorsoventral axis. DM stands for dorsal midline. (Photo by Maria Manceau / Hoekstra Lab; data from Manceau et al. 2011)

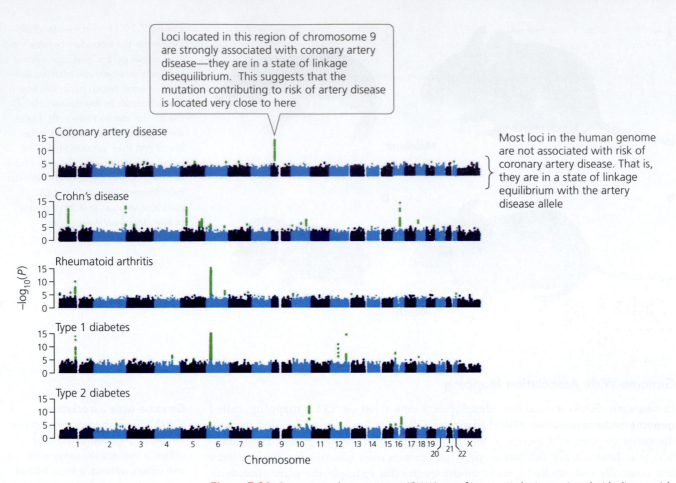

Figure 7.21 Genome-wide association (GWA) scan for genetic loci associated with disease risk. The term $-\log_{10}(P)$ is a metric for the extent of association between SNPs and each disease. Green dots indicate loci likely to be associated with inherited risk of disease. (Data from Burton et al. 2007)

lacked the disease. The Wellcome team were able to pinpoint genetic loci likely to be involved with inherited risk for coronary artery disease, Crohn's disease, rheumatoid arthritis, and both type 1 and type 2 diabetes (**Figure 7.21**; Burton et al. 2007). Since then, many GWA studies have been published, and they are increasingly leading researchers to insights about the functional differences between variants (Gallagher and Chen-Plotkin 2018).

Key Concept

- QTL and GWA mapping studies enable quantitative geneticists to identify regions of the genome responsible for genetic variation in phenotypic traits. This method can serve as a first step toward identifying the genes responsible for phenotypic evolution. ●

7.5 The Evolution of Phenotypic Plasticity

In addition to the genetic component (V_G) of phenotypic variance, the environmental component (V_E) is important as well. The environmental component of variation in a trait includes all of the external factors that can influence its development, from the temperature in which an organism develops to the food it eats to the infections it suffers. But it also includes how organisms *respond* to the environment. Traits that are shaped by these responses are known as plastic (from the original sense of the word, meaning "able to be molded"). Phenotypic plasticity is the capacity for a genotype to express more than one phenotype, depending on the environment.

In many cases, phenotypic plasticity represents a sophisticated, flexible adaptation that enables organisms to survive in different environments. Plants, for example, may grow under different amounts of sunlight. Some seeds may wind up in an open meadow, whereas others sprout below a thick canopy of taller plants casting shadows on the ground. Genetically identical plants may respond to different light levels by growing on dramatically different trajectories (**Figure 7.22**; Sultan 2000). A plant that grows in abundant sunlight will grow small leaves but produce large amounts of biomass in other structures. A genetically identical plant growing in low light will grow large leaves that capture more light, but it will produce only a tiny fraction of the biomass found in plants growing in abundant light.

These different responses each maximize the fitness of the plant. To capture sunlight, they need to build leaves. But the more tissue they dedicate to leaves, the less they have available to build other essential structures, such as roots for extracting water and nutrients from the soil, and seeds for reproduction. In dim light, plants grow large leaves to capture more energy, but they must make a trade-off, investing fewer resources in other tissues. Plants respond to bright sunlight by diverting those resources from their leaves.

Phenotypic plasticity allows organisms not only to respond to a range of unpredictable changes but also to respond to regularly occurring ones as well. Snowshoe hares, for example, have brown fur during the summer. But as they sense the shortening of the day in autumn, they molt and grow a coat of white fur, which matches the

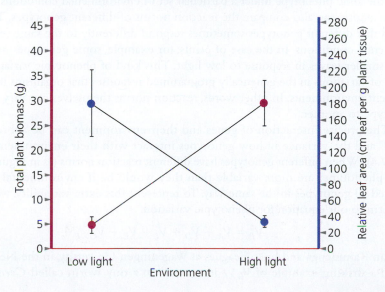

Figure 7.22 The environment strongly influences plant development. This photo shows two genetically identical plants that were reared in dim versus bright light, with their total mass (red circles) and relative leaf area (blue circles). The same genotypes produce dramatically different phenotypes in response to different conditions. (Photo by Sonia E. Sultan/The Sultan Lab; data from Sultan 2000)

Figure 7.23 Snowshoe hares change coat color to match their background. This annual change is the result of phenotypic plasticity, not evolution. (L. Scott Mills Research Photo)

Reaction norm refers to the pattern of phenotypic expression of a single genotype across a range of environments. In a sense, reaction norms depict how development maps the genotype into the phenotype as a function of the environment.

snow that arrives in winter. In spring, they molt again, turning brown (**Figure 7.23**). These cyclical changes in coat color are life-saving because hares that blend into their background may avoid being killed by predators.

We can visualize the responses of organisms to the environment by using a diagram called a **reaction norm** (**Figure 7.24**). Any one organism may live in only a single environment and therefore produce only one phenotype out of a wide range of phenotypes that the reaction norm can generate. Reaction norms allow individuals across an entire population to produce phenotypes that match a wide range of environmental conditions.

The simplest way to study reaction norms is to rear genetically similar organisms (for example, clones, inbred lines, or siblings) under different environmental conditions. **Figure 7.25** shows a hypothetical experiment in which scientists grow plants under different light levels and then measure the total leaf area per gram of all tissue. The plants exhibit phenotypic variation (V_P) in all cases (A–D), but the specific components contributing to this variation (V_G, V_E, $V_{G \times E}$) differ. In A, for example, the phenotypic variation has no genetic component because all the plants are genetically identical ($V_G = 0$; $V_P = V_E$). The effects of the environment on the phenotype are not random, however, because scientists find that the same genotype will repeatedly produce the same phenotype under a particular set of environmental conditions.

Scientists can also compare the reaction norms of different genotypes. They have found that different genotypes sometimes respond differently to the same set of environmental conditions. In the case of plants, for example, some genotypes might produce smaller leaves in response to low light. This kind of phenotypic variation is the result of variations in the genetically programmed responses that organisms produce in different environments. In other words, reaction norms themselves can vary from one genotype to another.

The precise interaction of genes and their environment can be abbreviated as $G \times E$ and the variance in how genotypes interact with their environment as $V_{G \times E}$ (**Box 7.4**). When different genotypes have different reaction norms (as in Figure 7.23D), their phenotypes are more variable than they would be if environmental variation affected all genotypes in the same way. To represent this extra variability, we can add the term to our equation for phenotypic variation:

$$V_P = V_A + V_D + V_I + V_E + V_{G \times E}$$

Jan Kammenga and his colleagues at Wageningen University, in the Netherlands, found a striking example of $V_{G \times E}$ in a study on a tiny worm called *Caenorhabditis*

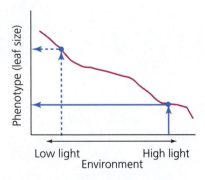

Figure 7.24 This diagram illustrates a hypothetical relationship between variation in the environment (low to high ambient light) and the phenotype (leaf size) expressed by a genotype grown in that environment. The blue arrows show how the reaction norm can be used to predict the phenotype value (leaf size) that will be produced by that genotype for any specific environment.

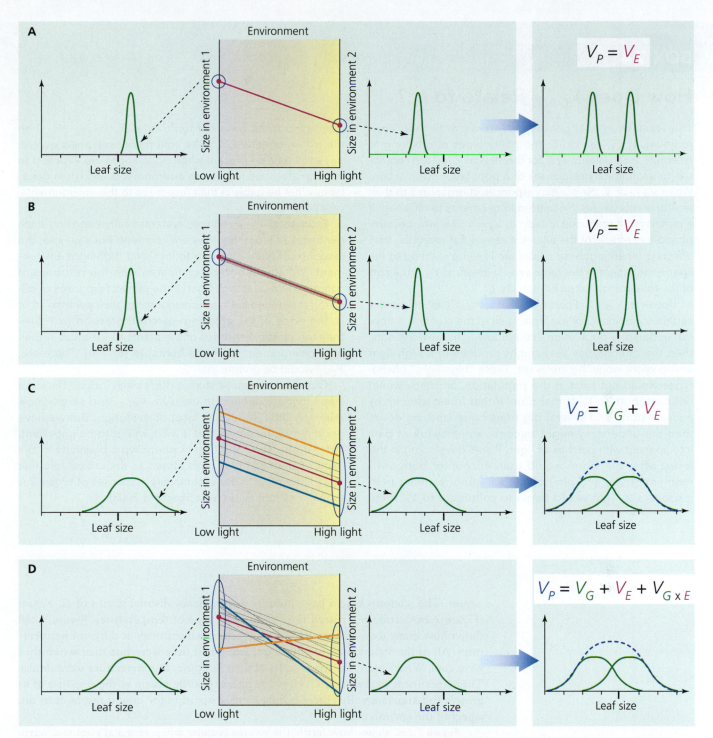

Figure 7.25 Components of phenotypic variation in a phenotypically plastic trait (leaf size). A: Simplified reaction norm for a single plant genotype grown in environments with low and high light. This genotype is plastic because it produces smaller leaves in conditions of high light than in low light. Phenotypic variation is entirely due to the light environments. B: Reaction norms now shown for multiple plant genotypes, all of which respond to ambient light in the same fashion (the slopes are parallel). There are no differences among genotypes for leaf size, so, as before, phenotypic variation is all due to the environment. C: Genotypes differ in the leaf sizes that they produce. Within each light environment, there is genetic variation for leaf size. Across environments, all genotypes produce smaller leaves in high light, so there is also variation caused by the environment. A leaf with a genotype for being big could still turn out smaller than a leaf with a genotype for being smaller if it finds itself in an environment that is less suitable for growth. D: Genotypes differ in how they respond to the environment. Overall, there is a net trend toward smaller leaves in high-light environments, so there is environmental variation, but the genotypes producing the smallest leaves in the low-light environment (orange) are not the same as the genotypes producing small leaves in the high-light environment (turquoise). Statistically, there is an interaction between genotypes and the environment ($V_{G \times E}$). Such an interaction represents heritable variation for phenotypic plasticity in trait expression.

BOX 7.4

How Does $V_{G \times E}$ Relate to h^2?

The variance in how genotypes interact with their environment, $V_{G \times E}$, can cause the phenotypes of offspring to resemble their parents, and for this reason $V_{G \times E}$ can contribute to evolutionary responses of a population to selection. When $V_{G \times E} > 0$, the plastic responses of organisms to their environments will be transmitted from parents to offspring. If a parent plant responds to light in a particular way because of additive effects of the alleles it carries, for example, then offspring inheriting these alleles are likely to respond to the light environment in the same way. In essence, $V_{G \times E}$ is a part of the total narrow sense heritability, h^2.

However, $V_{G \times E}$ is a part of h^2 that reveals itself only under certain conditions. For example, if all of the plants in a tropical understory population lived in low-light environments, then the phenotypes they might produce in a high-light environment would be irrelevant. These "high-light" phenotypes would not exist in the population, and they would not contribute to V_P at that time. If that forest were hit by a major storm and several big trees came crashing down, then the understory might become a patchwork of dark interspersed with patches of light. Plants developing in this forest after the storm would then encounter both environment types, and phenotypic variation resulting from responses to light would begin to contribute to V_P. If the plant responses were heritable (that is, $V_{G \times E} > 0$), then the extent or nature of these light responses could evolve. $V_{G \times E}$ will thus be present when populations are exposed to appropriately heterogeneous environments and when genotypes differ heritably in their responses to this environmental variation.

Estimating $V_{G \times E}$ requires systematically exposing each genotype to more than one environment. For example, the response of plant offspring to low- and high-light environments would need to be compared with the response of parents to low and high light. If the plastic responses of offspring resembled the plastic responses of their parents—that is, the slope of the offspring–parent regression ($a > 0$)—then the plastic responses of the plants to variation in their light environment would be heritable ($h^2 > 0$). Therefore, $V_{G \times E}$ would be greater than 0.

Quantitative genetic studies don't often include the extra steps that are needed to measure $V_{G \times E}$, and so we know relatively little about this facet of evolution. But we have good reason to expect that it will turn out to be important. Indeed, the near ubiquity of phenotypic plasticity in the expression of complex traits attests to the critical role that $V_{G \times E}$ likely plays in the evolutionary responses of organisms to their environments (West-Eberhard 2003).

elegans. The scientists bred a large number of genetically distinct strains of *C. elegans* (**Figure 7.26A**) and then reared the strains at a range of temperatures. **Figure 7.26B** shows how many days each strain took to reach sexual maturity at different temperatures. All of the strains matured more quickly when they were hot than when they were cool. What's more, the magnitudes of their responses to temperature were similar. This experiment showed that *C. elegans* is phenotypically plastic in the response of its growth to temperature. But there is little genetic variation for how it responds to this aspect of the environment.

Figure 7.26C shows how fertile the worms became when raised at cool and warm temperatures. The data illustrated present a very different picture than the reaction norm for sexual maturity. In some strains, a warm environment made the worms more fertile than a cool one. But in other strains, there was no response to environmental variation. And in still other strains, the response was the opposite: warmer temperatures made them less fertile. Thus *C. elegans* has a high $V_{G \times E}$ for fertility (Gutteling et al. 2007).

When scientists find a genetic basis for plasticity, it's sometimes possible for them to use QTL analysis to hunt for the genes responsible for the variance in plasticity. Kammenga and his colleagues crossed purebred lines of *C. elegans* and then searched for correlations between their responses to temperature and genetic markers. They found five QTLs associated with plasticity for various traits. **Figure 7.26D** shows one

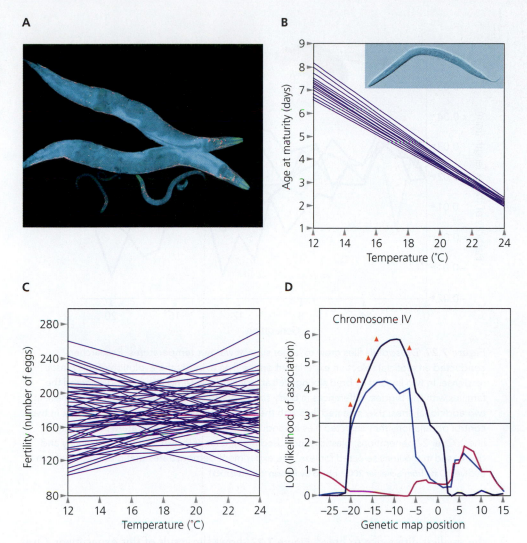

Figure 7.26 Scientists tested the plasticity of different strains of the worm *Caenorhabditis elegans* (A). Each line in these graphs represents the response of one strain to a range of temperatures. B: Worms took less time to reach maturity at high temperatures than low ones. There was little difference in the response of the strains, suggesting no $V_{G \times E}$ for this trait. C: The fertility of different strains responded in different ways to temperature, increasing with temperature in some strains and declining in others. For this trait, $V_{G \times E}$ is high. D: QTL analysis reveals a locus on chromosome IV strongly associated with levels of plasticity in fertility. (A: Heiti Paves/Shutterstock; data from Gutteling et al. 2007)

QTL they found for plasticity in fertility on chromosome IV. Currently, they're testing the genes around these QTLs to learn more about the genetic encoding of plasticity.

When $V_{G \times E} > 0$, plastic responses of organisms to their environments can evolve. Scientists can study the evolution of plasticity with artificial selection experiments. Samuel Scheiner and Richard Lyman, then at Northern Illinois University, conducted one such experiment on *Drosophila* flies. Like *C. elegans*, *Drosophila* develops differently at different temperatures. At low temperatures, the flies end up with a small body size, and at high temperatures, the body size becomes larger.

Scheiner and Lyman reared families of flies and then reared siblings at either 19°C or 25°C. The difference in the size of the warm- and cold-reared siblings varied from family to family. In two separate populations, the scientists allowed only flies with the greatest size difference between temperatures to breed—that is, the most plastic genotypes were allowed to breed. In two other populations, they allowed only the flies with

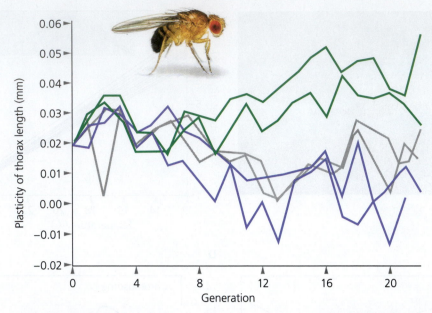

Figure 7.27 *Drosophila* flies grow to larger sizes at warmer temperatures. Researchers conducted an artificial selection experiment on the flies to study the evolution of this plastic response. In two lines, they bred sibling flies at high and low temperatures and selected the families with the biggest differences in body size to breed the next generation (green lines). In two additional lines, they selected flies with the smallest difference (purple lines). Finally, in two control populations they selected flies at random with respect to temperature plasticity (gray lines). Over 22 generations, plasticity increased in the lines selected for greater plasticity and decreased in the lines selected for less. The experiment demonstrates that plasticity itself can evolve. (Data from Scheiner 2002; photo by Roblan/Shutterstock)

the smallest difference to breed. **Figure 7.27** shows the result of this experiment. Over 22 generations, plasticity increased in the first two populations and decreased in the remaining two (Scheiner and Lyman 1991). In other words, the amount of phenotypic plasticity had evolved in response to the selection that the scientists had imposed.

Experiments such as these allow scientists to observe the evolution of phenotypic plasticity in fine detail. It's also possible to study this form of evolution in the wild. As we'll see in Chapter 10, snowshoe hares have been experiencing natural selection on their response to changing photoperiod and temperature over the past several thousand years. And it's even possible that their long-term survival will depend on the continued evolution of this plastic trait.

Key Concept

- Organisms may differ in how they react to environmental situations, and these differences may be heritable. When this occurs, the responses themselves can evolve, leading to the evolution of adaptive phenotypic plasticity. (Think of this as genetic changes to the underlying physiological and developmental response mechanisms.) ●

TO SUM UP . . .

- Scientists can study the evolution of traits even when the traits' mechanisms of inheritance are complex or unknown. Statistical approaches can be used to estimate the contribution of genetic variation to the expression of phenotypic traits based on the degree of resemblance among relatives (for example, parents and their offspring).

- Variation in phenotypes can be broken down into variation due to genetic components and variation due to the environment. Broadly defined, heritability is the genetic component of variation. In its narrow sense, heritability is the phenotypic resemblance among relatives that arises as a result of the additive effects of alleles.

- The rate of adaptive evolution depends both on the strength of selection and on the heritability of traits, both of which can be measured from quantitative studies of the phenotypes of organisms in natural populations.

- Quantitative trait loci (QTLs) are stretches of DNA that can be isolated and correlated with variation in a phenotypic trait. QTL analysis is a powerful tool for identifying genes that have contributed to the evolution of traits important in adaptation.

- With phenotypically plastic traits, it is the reaction norms, rather than the traits themselves, that evolve in response to selection.

- Understanding the evolution of phenotypic plasticity can be difficult because the plastic change in a particular phenotype is not evolution—a hare turns from brown to white in response to the environment. A population-level change in the amount or nature of plasticity, however, is evolution—populations of hares become more or less sensitive to day length as a cue for the color change.

MULTIPLE CHOICE QUESTIONS   Answers can be found at the end of the book.

1. Phenotypic traits often have a continuous distribution because they are

   a. a result of dominance interactions.
   b. not related to genotypes.
   c. influenced only by the environment.
   d. often polygenic.

2. Which of these statements about narrow sense heritability (h^2) is *true*?

   a. The numerator of narrow sense heritability includes additive, dominant, and epistatic gene effects.
   b. The numerator of narrow sense heritability includes only the additive effects of alleles.
   c. Narrow sense heritability includes only the epistatic effects of alleles.
   d. Narrow sense heritability can be estimated by comparing quantitative trait loci among offspring using regression.

3. The breeder's equation incorporates two of the conditions Darwin identified that must be met for evolution by natural selection to take place. Which two?

   a. Greater survival (S) and reproduction (R) of phenotypes with specific alleles.

   b. Variation in phenotypic traits (R) and heritability of additive alleles (h^2).

   c. Differences in phenotypes that influence the probability of survival or reproduction (S) and differences in phenotypic traits that must be at least partially heritable (h^2).

   d. Heritability of additive alleles (h^2) and the evolutionary response of the population (R).

4. How can scientists determine what constitutes a quantitative trait locus?

   a. They painstakingly examine the genotypes of hundreds of individuals and look for genes that are consistently similar.

   b. They examine nucleotide sequences and count the repeated segments that they feel are important.

   c. They hybridize species and compare how genetic markers recombine in the offspring.

   d. They select for different traits in lineages of an organism, cross-breed the lineages for two generations, and search for genetic markers that are correlated with expression of the trait.

5. If the age of sexual maturation is a phenotypically plastic trait, what relationship(s) would you expect to find?

   a. Genotypes differ in the age at which they reproduce.

   b. Environmental conditions (such as nutrition) affect the age at which individuals begin reproducing.

   c. Body size affects the age at which different genotypes reproduce.

   d. All of the above.

6. Which of the following is an example of the process of evolution?

   a. A population of snowshoe hares has a different frequency of alleles than the previous generation.

   b. Trees drop their leaves in the fall.

   c. A man becomes immune to a strain of virus that caused him to have a cold when he was younger.

   d. A female bird lays more eggs one season than she did the three previous seasons combined.

7. The rate of adaptive evolution depends on

   a. the strength of selection.

   b. heritability of traits.

   c. both a and b.

   d. phenotypes of organisms in natural populations.

8. Which of the following is responsible for the heritability of a trait?

   a. Influences on the genotype by additive environmental components.

   b. The proportion of phenotypic variance that is due to genetic differences among individuals.

   c. All of the genetic contributions to a trait's phenotype.

   d. Influences of the parental phenotype on the environment of the offspring.

9. Genetic variance among individuals can be broken down into all of the following categories *except*

   a. total variance in phenotypic trait in a population.

   b. additive genetic variance.

   c. variance due to dominance effects of alleles.

   d. variance attributable to epistatic interactions among alleles at various genetic loci.

10. Evolution happens when

    a. there is a change in allele frequencies in a population.

    b. there is a change in allele frequencies in a family.

    c. there is selection of any kind.

    d. All of the above.

INTERPRET THE DATA Answer can be found at the end of the book.

11. What type of selection is depicted in the graph at the right?

    a. Stabilizing.

    b. Directional.

    c. Disruptive.

    d. None of the above.

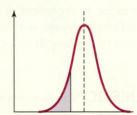

SHORT ANSWER QUESTIONS Answers can be found at the end of the book.

12. What is the difference between the selection differential and the evolutionary response to selection?

13. Why don't dominance effects and epistatic interactions among alleles generally contribute to phenotypic resemblance of relatives?

14. How can reaction norms be used to understand phenotypic plasticity?

15. How might selection drive the evolutionary response of snowshoe hare populations experiencing warming in the far north due to climate change?

16. Why are QTL mapping studies so difficult?

17. Under what circumstances will directional selection occur?

ADDITIONAL READING

Conner, J. K., and D. L. Hartl. 2004. *A Primer of Ecological Genetics.* Sunderland, MA: Sinauer Associates.

Endler, J. A. 1986. *Natural Selection in the Wild.* Princeton, NJ: Princeton University Press.

Falconer, D. S., and T. F. C. Mackay. 1996. *Introduction to Quantitative Genetics.* 4th ed. New York: Longman.

Fox, C. W., and J. B. Wolf, eds. 2006. *Evolutionary Genetics: Concepts and Causes.* Oxford: Oxford University Press.

Hallgrímsson, B., and B. K. Hall, eds. 2005. *Variation: A Central Concept in Biology.* New York: Elsevier.

Lynch, M., and B. Walsh. 1998. *Genetics and Analysis of Quantitative Traits.* Sunderland, MA: Sinauer Associates.

Roff, D. A. 1997. *Evolutionary Quantitative Genetics*. New York: Chapman & Hall.

Schlicting, C. D., and M. Pigliucci. 1998. *Phenotypic Evolution: A Reaction Norm Perspective*. Sunderland, MA: Sinauer Associates.

Stearns, S. C. 1992. *The Evolution of Life Histories*. Oxford: Oxford University Press.

West-Eberhard, M. J. 2003. *Developmental Plasticity and Evolution*. Oxford: Oxford University Press.

PRIMARY LITERATURE CITED IN CHAPTER 7

Aranzana, M. J., J. A. S. Kim, K. Zhao, E. Bakker, M. Horton, et al. 2005. Genome-Wide Association Mapping in *Arabidopsis* Identifies Previously Known Flowering Time and Pathogen Resistance Genes. *Public Library of Science Genetics* 1:531–9.

Beaumont, K. A., S. S. Wong, S. A. Ainger, Y. Y. Liu, M. P. Patel, et al. 2011. Melanocortin MC_1 Receptor in Human Genetics and Model Systems. *European Journal of Pharmacology* 660 (1): 103–10.

Burton P. R., D. G. Clayton, L. R. Cardon, N. Craddock, P. Deloukas, et al. 2007. Genome-Wide Association Study of 14,000 Cases of Seven Common Diseases and 3,000 Shared Controls. *Nature* 447:661–78.

Conner, J. K., and D. L. Hartl. 2004. *A Primer of Ecological Genetics*. Sunderland, MA: Sinauer Associates.

Evans, S., M. D. Camara, and C. J. Langdon. 2009. Heritability of Shell Pigmentation in the Pacific Oyster, *Crassostrea gigas*. *Aquaculture* 286:211–16.

Falconer, D. S., and T. F. C. Mackay. 1996. *Introduction to Quantitative Genetics*. 4th ed. New York: Longman.

Fishman, L., P. M. Beardsley, A. Stathos, C. F. Williams, and J. P. Hill. 2015. The Genetic Architecture of Traits Associated with the Evolution of Self-Pollination in *Mimulus*. *New Phytologist* 205 (2): 907–17.

Gallagher, M. D., and A. S. Chen-Plotkin. 2018. The Post-GWAS Era: From Association to Function. *The American Journal of Human Genetics* 102 (5): 717–30.

Gutteling, E. W., J. A. G. Riksen, J. Bakker, and J. E. Kammenga. 2007. Mapping Phenotypic Plasticity and Genotype–Environment Interactions Affecting Life-History Traits in *Caenorhabditis elegans*. *Heredity* 98:28–37.

Hoekstra, H. E. 2010. From Mice to Molecules: The Genetic Basis of Color Adaptation. In J. B. Losos (ed.), *In the Light of Evolution: Essays from the Laboratory and Field*. Greenwood Village, CO: Roberts and Company.

Jones, M. R., L. S. Mills, P. C. Alves, C. M. Callahan, J. M. Alves, et al. 2018. Adaptive Introgression Underlies Polymorphic Seasonal Camouflage in Snowshoe Hares. *Science* 360 (6395): 1355–8.

Kirkpatrick, M. 2010. How and Why Chromosome Inversions Evolve. *PLoS Biology* 8 (9): e1000501.

Klug, W. S., and M. R. Cummings. 1997. *Concepts of Genetics*. 5th ed. Englewood Cliffs, NJ: Prentice-Hall.

Lowry, D. B., and J. H. Willis. 2010. A Widespread Chromosomal Inversion Polymorphism Contributes to a Major Life-History Transition, Local Adaptation, and Reproductive Isolation. *PLoS Biology* 8 (9): e1000500.

Lynch, M., and B. Walsh. 1998. *Genetics and Analysis of Quantitative Traits*. Sunderland, MA: Sinauer Associates.

Mackay, T. F. C. 2001. Quantitative Trait Loci in *Drosophila*. *Nature Reviews Genetics* 2 (1): 11–20.

Manceau, M., V. S. Domingues, R. Mallarino, and H. E. Hoekstra. 2011. The Developmental Role of *Agouti* in Color Pattern Evolution. *Science* 331:1062–5.

Miles, C., and M. Wayne. 2008. Quantitative Trait Locus (QTL) Analysis. *Nature Education* 1 (1): 208.

Moose, S. P., J. W. Dudley, and T. R. Rocheford. 2004. Maize Selection Passes the Century Mark: A Unique Resource for 21st Century Genomics. *Trends in Plant Science* 9 (7): 358–64.

Nachman, M. W., H. E. Hoekstra, and S. L. D'Agostino. 2003. The Genetic Basis of Adaptive Melanism in Pocket Mice. *Proceedings of the National Academy of Sciences USA* 100:5268–73.

Perseguini, J. M. K. C., P. R. Oblessuc, J. R. B. F. Rosa, K. A. Gomes, A. F. Chiorato, et al. 2016. Genome-Wide Association Studies of Anthracnose and Angular Leaf Spot Resistance in Common Bean (*Phaseolus vulgaris L.*). *PLoS One* 11 (3): e0150506.

Plomin, R., C. M. A. Haworth, and O. S. P. Davis. 2009. Common Disorders Are Quantitative Traits. *Nature Reviews Genetics* 10 (12): 872–8.

Scheiner, S. M. 2002. Selection Experiments and the Study of Phenotypic Plasticity. *Journal of Evolutionary Biology* 15 (6): 889–98.

Scheiner, S. M., and R. F. Lyman. 1991. The Genetics of Phenotypic Plasticity. II. Response to Selection. *Journal of Evolutionary Biology* 4:23–50.

Schwander, T., R. Libbrecht, and L. Keller. 2014. Supergenes and Complex Phenotypes. *Current Biology* 24 (7): R288–94.

Steiner, C. C., J. N. Weber, and H. E. Hoekstra. 2007. Adaptive Variation in Beach Mice Produced by Two Interacting Pigmentation Genes. *PLoS Biology* 5 (9): e219.

Sultan, S. E. 2000. Phenotypic Plasticity for Plant Development, Function and Life History. *Trends in Plant Science* 5 (12): 537–42.

Tancred, S. J., and A. G. Zeppa. 1995. Heritability and Patterns of Inheritance of the Ripening Date of Apples. *HortScience* 30 (2): 325–8.

Thoday, J. M., and J. B. Gibson. 1962. Isolation by Disruptive Selection. *Nature* 193 (4821): 1164–6.

Thompson, M. J., and C. D. Jiggins. 2014. Supergenes and Their Role in Evolution. *Heredity* 113 (1): 1.

Tishkoff, S. A., R. Varkonyi, N. Cahinhinan, S. Abbes, G. Argyropoulos, et al. 2001. Haplotype Diversity and Linkage Disequilibrium at Human *G6PD*: Recent Origin of Alleles That Confer Malarial Resistance. *Science* 293 (5529): 455–62.

Twyford, A. D., and J. Friedman. 2015. Adaptive Divergence in the Monkey Flower Mimulus Guttatus Is Maintained by a Chromosomal Inversion. *Evolution* 69 (6): 1476–86.

West-Eberhard, M. J. 2003. *Developmental Plasticity and Evolution*. Oxford: Oxford University Press.

CHAPTER 8

The History in Our Genes

Learning Objectives

- Explain how gene trees reconstruct the historical relationships among alleles within and between populations.
- Compare and contrast gene trees and species trees.
- Discuss how coalescence helps explain phylogenetic relationships between closely related species.
- Explain how the distribution of coalescent events can be used to detect historical changes in the size of a population.
- Describe the methods scientists use to construct phylogenetic trees.
- Discuss the kinds of evidence used to determine the origin of tetrapods, humans, and HIV.
- Discuss how the neutral theory of evolution is used to deduce the timing of evolutionary events and the history of natural selection.
- Explain how phylogenetic approaches can assist researchers in identifying disease-causing genes.
- Compare and contrast the evolution of genome size in bacteria and eukaryotes.

Sarah Tishkoff has been traveling from one end of Africa to the other for well over a decade. She took her first trip to Africa as a graduate student in genetics at Yale University, and she still returns now that she's a professor at the University of Pennsylvania (**Figure 8.1**). She has bounced along cratered roads in Tanzania, and she has traveled aboard hand-cranked ferries in the jungles of Cameroon. On her journeys, Tishkoff carries syringes, vials, and centrifuges. Her expansive goal for all this travel is to create a genetic portrait of the 1 billion people who live in Africa. She and her colleagues still have a long way to go toward reaching that goal, but they have collected DNA from more than 7000 people from more than 100 ethnic groups.

Tishkoff hopes to learn many things from this genetic portrait she is putting together. She and her colleagues are beginning to identify alleles that make some Africans more vulnerable to certain diseases and resistant to others. But she also travels to Africa to understand history—not just the history of Africans, but the history of all humans. Tishkoff and her colleagues are creating a detailed genealogy of the human race. Their research has been crucial to our

Scientists are using DNA to determine how the peoples of the world are related to one another.

Figure 8.1 Sarah Tishkoff of the University of Pennsylvania gathers genetic samples to study human diversity in Africa. (Sarah Tishkoff)

current understanding of how our species evolved. Thanks to the work of Tishkoff and other researchers, we now know that our species evolved in Africa for hundreds of thousands of years. Only after thousands of generations did a small group of Africans migrate out of the continent, interbreed with Neanderthals and other human populations that are now extinct, and ultimately spread across Asia, Europe, and the New World (Campbell and Tishkoff 2008; Nielsen et al. 2017; Chapter 17).

In Chapter 4, we saw how scientists use morphological traits to construct evolutionary trees. Until the 1990s, these data were the only kind of evidence readily available to evolutionary biologists seeking to understand phylogenies. But since then, researchers like Tishkoff have started using powerful computers and DNA-sequencing technology to unlock an extraordinary historical archive stored in the genomes of all living things. They've combined this insight with an understanding of how allele frequencies of a population can change over time through mechanisms of evolution such as drift and selection (Chapter 6 and Chapter 7).

As we shall see in this chapter, we can reconstruct the history of genes as well as the species that carried them. We can even detect episodes of genetic drift and natural selection that occurred millions of years ago. ●

8.1 Gene Trees

Conceptually, molecular phylogenies are similar to morphological ones of the sort we discussed in Chapter 4. In each case, evolutionary biologists compare homologous characters in a group of organisms to reconstruct their relationships. In a morphological phylogeny of birds and other dinosaurs, those characters might include the presence or absence of feathers (see Figure 4.31). In the case of molecular phylogenies, the characters might be the presence of adenine at a certain position in a certain gene. Biologists can use both morphological and molecular characters to identify clades. The common ancestry of humans and bats thus explains not only the homology of their bones but also the homology of many regions of their genomes.

But there are also some important differences in how scientists construct phylogenies from molecules versus morphology. To appreciate them, let's take a closer look at how genetic variation arises in populations and how, over millions of years, speciation affects the distribution of that variation.

For this discussion, let's consider the gene *BRCA1*, which we first discussed in Chapter 7. Located on chromosome 17, *BRCA1* normally functions as a tumor-suppressor gene. Mutations that disrupt it can dramatically increase a woman's risk of developing breast or ovarian cancer. Although the *BRCA1* gene is more than 126,000 base pairs long, it takes only a single mutation to a single nucleotide to create this risk. Scientists have identified hundreds of mutations in *BRCA1* that can lead to cancer, but here we'll just consider one: a rare mutation that converts a single G nucleotide to T.

People who carry this particular G-to-T mutation can theoretically end up with it in one of two ways. It may have arisen in the egg or sperm that combined to produce their zygote. Or they may have inherited it from one of their parents. Their parent may have inherited the mutation in turn from one of their parents, and so on back through history. But at some point, the gene must have undergone the spontaneous mutation from G to T in one of their ancestors.

For example, the T variant could have arisen in an egg that was later successfully fertilized with a sperm carrying the G variant of *BRCA1*. The zygote now carried one benign G variant and one pathogenic T variant. If this zygote then developed into a woman, she would be at greater risk of breast and ovarian cancer. And if she had children, approximately half would inherit her T version of *BRCA1*. (Recall that thanks to segregation, only half of the eggs or sperm will carry this copy of the gene.) Her children who inherited the pathogenic T variant might then pass it down to some of their own children.

Figure 8.2A represents this process in a simplified form. Each circle in the column on the far left represents a *BRCA1* allele in a population. (For simplicity's sake, we'll only consider ten alleles.) The columns to the right each represent *BRCA1* alleles in the subsequent generations. From one generation to the next, some alleles are replicated, whereas others fail to be transmitted. The black lines highlight the descendants of a single allele. The orange line represents the transition from G to T in a single *BRCA1* allele. In the final generation, on the right-hand side, there are now two copies of the pathogenic allele in the population.

The two orange circles and the black circle in the final generation are related to each other because they share common ancestors. The two orange circles are more closely related to each other than they are to the black circle, as evidenced by the G-to-T mutation that they share. If we want to represent their relationship, we can draw a genetic version of a family tree, akin to the pedigree we drew for Charles II of Spain in Chapter 6 (Figure 6.22). In that tree, each node and branch tip represented a human being. Now, to trace the spread of the pathogenic version of *BRCA1*, we can make these nodes and tips represent alleles (**Figure 8.2B**; Rosenberg and Nordborg 2002).

When a pathogenic T variant of *BRCA1* arises in one person, it is initially very rare in a population, existing side by side with more common variants. As the new variant is passed down to descendants, it can become more common. In Chapter 6, we discussed how mechanisms such as selection and drift can drive variants to fixation or eliminate them from a population. In this chapter, we will look at this process from a genealogical perspective. This perspective applies not just to one pathogenic variant of *BRCA1* but to every allele of every gene we carry. Each allele has a *history*, a path of descent that we can trace backward through time. **Figure 8.3** places the origin of the T variant in a broader population.

We can see from this figure that when G becomes T in one lineage, it becomes phylogenetically informative. In other words, it is a **synapomorphy**, because individuals with a T are similar to each other given they each inherited this character state from a recent common ancestor. In Chapter 4, we saw how evolutionary biologists use many morphological synapomorphies to determine the relationship between species and clades. The same holds true for gene trees. We have focused on a single site in the *BRCA1* gene in this section, but there are many sites across the entire gene that are variable from person to person.

Each of these variable sites can be treated as a separate character in a phylogenetic analysis, and from these clues biologists can reconstruct the long-term evolutionary history of the gene. The branching path of descent with modification that describes the genealogy of a gene is called a **gene tree**.

Figure 8.4A shows a hypothetical gene tree for *BRCA1*, reconstructed from 20 copies of the gene sampled from the same population. Constructing gene trees for alleles sampled *within* a single population can reveal clues about the past—whether the

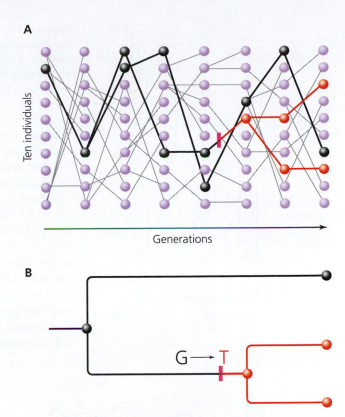

Figure 8.2 Alleles have their own genealogies. A: Paths of descent of alternative copies of the *BRCA1* gene for eight successive generations. The black circles and lines represent one sample lineage of alleles. The red dash indicates the change from G at one position in the gene to T, giving rise to a cancer-causing mutation. The red lines and circles track the replication of this new variant in later generations. B: This figure represents the relationship between the two red circles and black circle in the final generation in a simplified form. (Data from Rosenberg and Nordborg 2002)

Synapomorphy is a derived form of a trait that is shared by a group of related species (that is, one that evolved in the immediate common ancestor of the group and was inherited by all its descendants).

Gene tree refers to the branched genealogical lineage of homologous alleles that traces their evolution back to an ancestral allele.

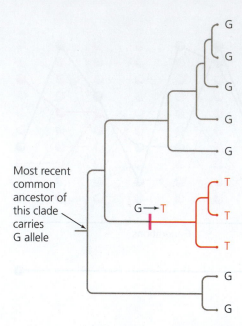

Figure 8.3 In this figure, we expand the *BRCA1* gene tree in Figure 8.2B to show the relationship of ten sampled alleles at one base pair. The hypothetical cancer risk mutation is indicated as a change at this site from a G to T, occurring in one branch and creating a polymorphism that now coexists with the ancestral allele in the population. (Data from Rosenberg and Nordborg 2002)

Most recent common ancestor of this clade carries G allele

population went through a bottleneck event, for example, or whether the gene recently experienced directional or stabilizing selection (**Box 8.1**).

But gene trees constructed from individuals from *different* populations or species can illuminate events that occurred much further back in time. For example, **Figure 8.4B** shows a gene tree constructed from 2622 base pairs of the *BRCA1* gene sampled across 57 species of mammals (Fleming et al. 2003). The common ancestor of these mammals lived about 100 million years ago. When Melissa Fleming of Johns Hopkins University and her colleague created this gene tree, they could trace the rise of mutations over this vast period of time. They determined that some regions of the *BRCA1* gene have accumulated very few mutations in that period, suggesting that they may be especially vulnerable to negative fitness effects. As we will see later, such surveys can help cancer biologists better understand how mutations to *BRCA1* affect human health.

A

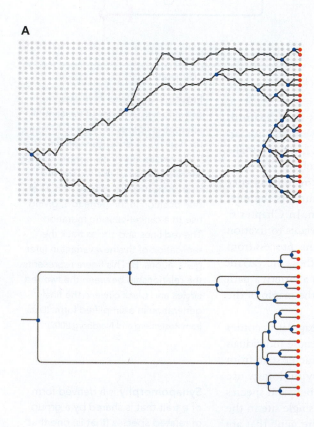

B

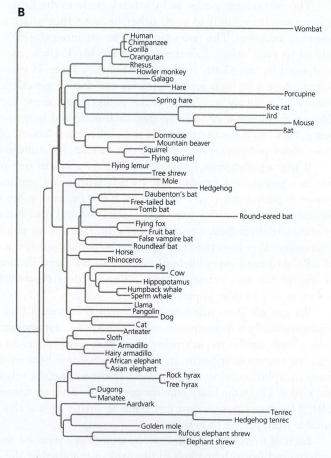

Figure 8.4 A: A hypothetical gene tree for *BRCA1*, showing the historical relationships among 20 alleles sampled from a single population. Because diploid individuals carry two copies of each gene, and they each have their own unique history, studies like this focus on sampled *alleles* rather than individuals. Often, thousands of base pairs of sequence are collected for each sampled allele. When these sequences are aligned, polymorphisms such as the one depicted in Figure 8.3 provide informative characters for reconstructing their historical relationships. B: Alleles can also be sampled from different populations or species, revealing their phylogenies over much greater spans of time. This shows a gene tree for the *BRCA1* gene in mammals descending from a common ancestor approximately 160 million years ago. (Data from Fleming et al. 2003)

BOX 8.1

Clues from Coalescence

Coalescent events can tell us a great deal about the history of a population. They can tell us if the population has experienced any major changes in size or if it has undergone natural selection (Nielsen and Slatkin 2013).

To see how we can do this, let's move forward in time through a gene tree (Box Figure 8.1.1A), counting the splitting events as they occur. One branch splits into two, and these each split to produce four, and so on. The number of splitting events increases exponentially.

To examine coalescent events, on the other hand, we just read the tree in the reverse direction, beginning with the branch tips and working our way back to base of the tree. As tips coalesce into their most recent common ancestors, the number of branches decreases. We expect a constant rate of coalescence to lead to an exponential *decrease* in the number of coalescent events as we move farther back through time. Most alternative alleles should coalesce soon after we move back from the tips of the tree, and only a few should coalesce deeper back in time.

The rate at which branches coalesce depends on the size of the population. We saw in Chapter 6 that neutral alleles

(continued)

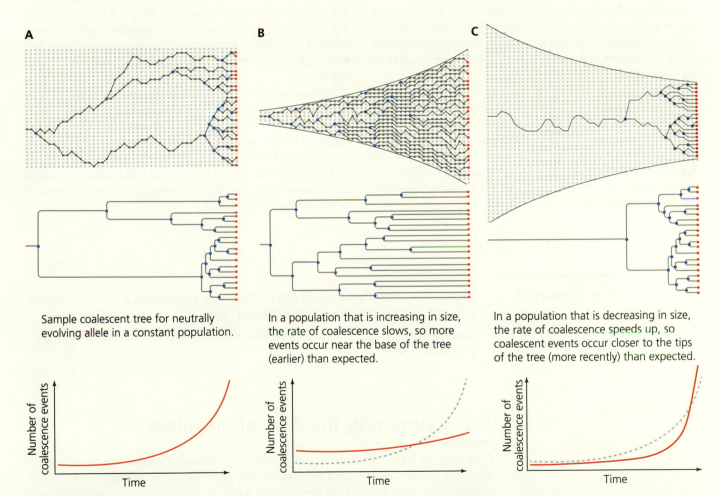

Sample coalescent tree for neutrally evolving allele in a constant population.

In a population that is increasing in size, the rate of coalescence slows, so more events occur near the base of the tree (earlier) than expected.

In a population that is decreasing in size, the rate of coalescence speeds up, so coalescent events occur closer to the tips of the tree (more recently) than expected.

Box Figure 8.1.1 A: Over time, a gene tree splits, generating an increasing number of alleles. B: If the rate of coalescence deviates from null expectations, this deviation may have one of several possible causes. If the size of a population dramatically increases, the rate of coalescence slows. As a result, more coalescent events occur near the base of the tree than expected, and fewer occur near the tips. C: By the same logic, populations that have decreased in size over time should show the reverse pattern. The rate of coalescence speeds up as the population gets smaller, so observed gene trees will have fewer coalescent events than expected deep in the tree, and more coalescence than expected right at the tips.

Box 8.1   Clues from Coalescence *(continued)*

are either lost or fixed faster in small populations than in larger ones. If we look at a gene tree of a small population, we expect its genealogies to converge onto a common ancestor in fewer generations than in large populations. For any two copies of a gene sampled from a haploid population, the average time to coalescence should be $2N$ generations, where N is the number of individuals in the population. In a diploid population, the time would be $4N$ generations.

We can predict the distribution of coalescent events for a population if we can estimate its population size and generate thousands of randomly generated tree topologies for the number of alleles in question (Box Figure 8.1.1A shows just one out of many possible topologies). This prediction serves as our null hypothesis. If actual gene trees violate these predictions, this can point to interesting properties of the population.

One way for a population to violate these predictions is to expand dramatically over time. If it started out very small, then it will have more accumulated coalescent events early in its history than we'd expect (**Box Figure 8.1.1B**). Shrinking populations will produce the opposite violation of the null hypothesis. Coalescent events will occur slower than predicted deep in the tree, and faster than expected near the tips (**Box Figure 8.1.1C**).

We can also detect even finer scale patterns. For example, a population that was large, experienced an acute bottleneck event, and then expanded once again will leave a mark of this history on the topology of its tree. It will have an unexpectedly high number of nodes clustered in the *middle* region of the tree.

Selection can cause observed gene trees to deviate from the null expectations, too. Strong positive selection, for example, can quickly pull an allele to high frequency. Reconstructing the gene tree for this gene would show coalescence times that are shallower—closer to the tips of the tree—than expected by chance. Balancing selection, on the other hand, like we observed with the A and S alleles of the β-globin locus in Chapter 6, can maintain multiple alleles within a population indefinitely. Coalescent events on this sort of gene tree should occur much deeper on the tree than expected.

Since changes in population size and selection can each generate similar patterns in the distribution of coalescent events, how can biologists distinguish between them? One way is to look at several different genes. Neutrally evolving genes should provide a clear picture of changes to overall population size—*and they should all provide the same picture*—whereas particular genes under specific types of selection should stand out.

Key Concepts

- Alternative copies of a gene exist side by side within populations, and they each have a lineage that traces their history back through time.

- Gene trees reconstruct the historical relationships among alleles within and between populations. ●

8.2 Estimating the Age of an Allele

Now that we have followed these alleles forward through the generations, let's travel back in time. When we march forward, we talk about nodes as splitting events. When we march backward, nodes are the points at which two lineages converge, or coalesce, into a single ancestral lineage. These events of **coalescence** occur at the most recent common ancestor of any two alleles. For closely related alleles like the branches depicted in Figure 8.2, we don't have to trace back many generations before the alleles coalesce. For more distantly related alleles we might have to trace back the generations much further. As we saw in Chapter 6, some alternative alleles may persist side by side in populations for many thousands of generations. Gene trees like the one in Figure 8.4A can be used to estimate the times of coalescence for all possible pairwise combinations of the different alleles in a population.

Coalescence is the process in which the genealogy of any pair of homologous alleles merges in a common ancestor.

The fact that all extant copies of a gene eventually coalesce does not mean that the population originally consisted of only a single individual with that ancestral allele. It just means that a particular individual's allele was the one that, out of all the alleles present at that time, later became fixed in the population. If we were to examine the extant copies of another gene, they would probably coalesce in a different ancestral individual (and at a different time).

The timing of coalescence can vary a great deal. It depends on factors such as whether alleles are under selection or not. Positive selection can accelerate the rise in frequency in an allele, for example, shortening the time to fixation—and leading to a short coalescence. Two alleles that experience little selection may coexist for a longer period of time. The more generations in which multiple alleles persist side by side, the further back in time we need to travel before they converge on a common ancestral allele. Powerful statistical approaches enable scientists to use information, like the frequencies of the alleles and the size of the population, to make accurate estimates of the likely times to coalescence for alleles in particular genes (Box 8.1).

Consider, for example, mitochondrial DNA. All humans carry DNA in their mitochondria, which they inherit solely from their mothers, without meiotic recombination. Certain mitochondrial mutations are more common in certain ethnic groups. That pattern would suggest that they arose after the initial divergence of the lineages of alleles found in living humans. How far back in human ancestry must we go to find the common ancestral allele?

We can't find the answer directly. We don't have a complete record of all the mitochondrial DNA in every human alive today. Nor do we have all the ancestors of humans going back millions of years. But we can narrow the range of likely answers by sampling the current distribution of alleles and applying statistical models to them.

Alan Templeton, a population geneticist at Washington University in St. Louis, used one of these models to estimate the time to coalescence of human mitochondrial DNA. He found a range of 152,000 to 473,000 years. **Figure 8.5** shows the results of his analysis of mitochondrial DNA as well as a number of autosomal genes. The time to coalescence varies tremendously, from tens of thousands of years to millions of years. (For the mathematics of coalescent theory, see Templeton 2006; Wakeley 2008.)

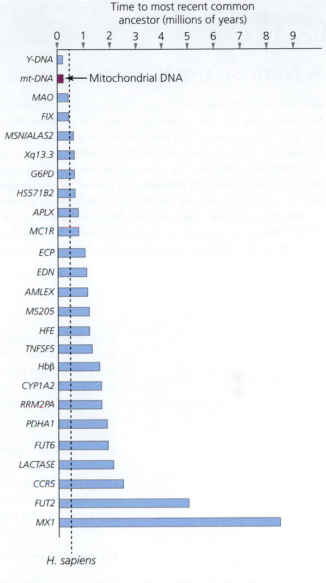

Figure 8.5 Alan Templeton estimated the coalescence of alleles in a number of human genes. The times differ greatly and in some cases are longer than the age of *Homo sapiens*. (Data from Templeton 2006)

- It is possible to trace the genealogies of genes back through time, reconstructing when mutations generated new alleles and how these alleles subsequently spread. •

Key Concept

8.3 Gene Trees and Species Trees

Sometimes a population will split into two or more reproductively isolated populations. As we'll explore in more detail in Chapter 13, this isolation can eventually lead to the origin of new species. It can take thousands or even millions of years for isolated populations to become clearly delineated species. During that process of speciation, their alleles will be passed from generation to generation, accumulating mutations and forming branched genealogical lineages—gene trees—along the way. After

A Note on the Names of Genes and Proteins

BRCA1 was originally discovered because mutations to the gene raise the risk of breast cancer. Thus, researchers named it *Breast Cancer 1*; it was then abbreviated to *BRCA1*. Genes are typically italicized (*BRCA1*), whereas the proteins they encode have the same name but in roman type (BRCA1). Discussions of **orthologous genes**, or **orthologs**, can be confusing because the scientists who identify them in different species may not give them the same name. *BRCA1* refers to the human ortholog, the mouse ortholog is *Brca1*, the African clawed frog ortholog is *brca1*, and the pig ortholog is *PIGBRCA1*. The confusion can be even greater when the scientists who discover and name a gene do not recognize their orthology to genes in other species. As a result, many orthologous genes have entirely different names. In Chapter 9, for example, we will learn how *Dpp* and *BMP4* are orthologs in invertebrates and chordates, respectively.

An **ortholog** is one of two or more homologous genes separated by a speciation event.

---

populations become reproductively isolated, they continue to accumulate changes, and their evolutionary trajectories will diverge. (See **Box 8.2** for a discussion of how genes and proteins are named, and see **Box 8.3** for a discussion of how genes in different species are related to one another.)

Gene trees often resemble the phylogeny of the species in which they are embedded. Over long periods of time, a species can acquire fixed alleles at many different loci, and collectively these differences can distinguish it from other species. If that species then gives rise to new species, the subsequent species will also inherit these loci. As a result, DNA can provide a clear signal of the phylogeny of these species, much like the morphological traits we discussed in Chapter 4.

But the history of genes is not always the same as the history of species. Here we will consider two important mechanisms that can create this mismatch.

Introgression

Introgression describes the movement of alleles from one species or population to another.

Sometimes individuals from one species interbreed with individuals from another. If the hybrid offspring survive and they end up mating with individuals from either of the original species, then gene copies from one of the species can be introduced into the genomes of the other. This introduction of genes through hybridization is known as **introgression**.

If these gene copies happen to carry beneficial variation, then they may be favored by selection and retained within the genomes of the recipient species. A gene tree reconstructed from these particular regions of the genome would appear more closely related to gene copies from the donor species than to other gene copies sampled from the recipient species.

As we'll see in later chapters, introgression is proving to be a significant factor in molecular evolution. In Chapter 10, we'll see how introgression has introduced an allele that's allowing snowshoe hares to adapt to warmer climates. And in Chapter 17, we'll see how Neanderthal DNA has been introgressed into the human genome.

Incomplete Lineage Sorting

Gene trees can sometimes fail to match species trees even without introgression. To understand why, we need to take a closer look at the fate of a gene as a population diverges. Initially, the population will have many alleles of the gene. And when the population splits, several alleles may be carried together into both of the resulting species. If one of these lineages splits again, some of the same alleles will be carried through once more.

But eventually some alleles will be lost to genetic drift. If we then sample alleles of the gene from each of the daughter species, we might pluck a different one of our original alleles from each of them. Depending on which of the alleles happened to persist in each of the species, the gene tree we reconstruct might not reflect the actual branching history of the species.

Figure 8.6 illustrates how this mismatch can happen. Both panels of the illustration show three species with the same phylogenetic relationship. Embedded in each species tree is a gene tree that relates all alleles for a particular gene to each other.

In Figure 8.6A, alleles of the gene present in species 1 and species 2 diverged more recently from one another than either did from the allele present in species 3. The same relationship holds true for the species themselves. In Figure 8.6B, on the other hand, the coalescent time of the alleles reaches back before the divergence of the species. All three alleles were already present by the time of the first speciation event, and which copy ultimately persisted in each species was the result of drift. In this case, the alleles and the species have different histories, a situation called **incomplete lineage sorting**. The allele sampled from species 2 happens to be more closely related to the one we sampled from species 3.

If we were to rely on this one gene alone to determine the phylogeny of the three species, we'd end up concluding that 2 and 3 are closely related sister species when, in fact, they are not. And if we used the coalescence time of the gene to estimate the ages of the species, we would make another error because, in this case, the coalescent time is much older than the species themselves.

> **Incomplete lineage sorting** occurs when a genetic polymorphism persists through several speciation events. When fixation of alternative alleles eventually occurs in the descendent species, the pattern of retention of alleles may yield a gene tree that differs from the true phylogeny of the species.

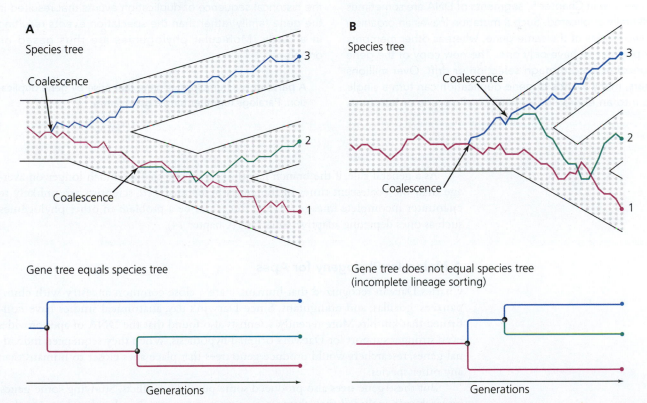

Figure 8.6 Two phylogenies of three species are depicted by the thick branches. Within these three lineages, the dots and lines represent the alleles of a gene carried by individuals. If we sample the same gene in one individual from each species, we can trace back their genealogies until they coalesce in an ancestral allele. A: Here, the gene tree we derive matches the species tree. That is, the copies of the gene in species 1 and 2 share a more recent ancestry than either does with the gene in species 3. B: In this case, the relationship between the species is the same as in A, but the coalescent time is long compared to the time it took for the species to branch. As a result, alleles in the ancestral species were sorted in the descendant species to produce a pattern that is discordant with the species tree. If we were to rely on this particular gene to reconstruct the phylogeny of these species, we might conclude erroneously that species 2 and 3 have a closer common ancestor. (Data from Rosenberg and Nordborg 2002)

BOX 8.3

Homology in Genes

As we saw in Chapter 4, wings and arms are homologous structures because their similarities are the result of a common ancestry. Genes can also be homologous, but the nature of their homology includes some important differences from that of anatomical traits. Those differences are reflected in the names scientists use to refer to homologous genes.

When a species splits into two or more new species (a process we discuss more fully in Chapter 13), mutations begin to accumulate independently in each of the descendant species. A gene passed down to these new lineages will begin to diverge as mutations gradually alter its sequence through time. The more time that passes since the speciation event, the more different the sequences of the gene are likely to become. Although versions of the gene carried by the two species are no longer identical, they can still retain an overwhelming similarity due to common ancestry. As we noted earlier, homologous genes that are separated by speciation events are known as orthologs.

As we saw in Chapter 5, segments of DNA are sometimes accidentally duplicated. Such a mutation leaves an organism with two copies of the same gene, whereas other members of its population have only one. The new copy of the gene can become fixed through selection or drift. Over millions of years, many rounds of gene duplication can turn a single gene into an entire "gene family" of dozens or hundreds of homologous genes. Homologous genes resulting from duplication events are called **paralogs**.

Once a single gene evolves into two or more paralogs, each paralog begins to evolve along its own trajectory. A mutation can produce a new allele of one paralog, while a different mutation to another paralog creates its own allele. The genes diverge *even though they lie side by side within the same genome.*

As we'll see in Chapter 9, paralogs play an important role in the evolution of new traits because new copies of genes often evolve new functions. When genes diverge before the speciation event, comparing an allele of one paralog in a species to an allele of another paralog in another species will not reflect the history of those two species. The coalescence of the paralogs extends back before the common ancestor of the two species, all the way back to the original gene duplication that produced the two paralogs. Reconstructing molecular evolution among paralogs reflects the historical sequence of duplication events that resulted in the gene family rather than the speciation events resulting in a clade. Molecular phylogenies are thus based on orthologs.

A **paralog** is a homologous gene that arise by gene duplication. Paralogs together form a gene family.

As a general rule, if the branch lengths of a species tree are much longer on average than the coalescent times of the genes being analyzed, scientists are unlikely to encounter incomplete lineage sorting. But it can be a problem in other phylogenies, such as ones depicting adaptive radiations (Chapter 14).

A Molecular Phylogeny for Apes

Charles Darwin recognized that humans share a close common ancestry with chimpanzees, gorillas, and orangutans. Since Darwin's day, anatomical studies have confirmed that kinship. More recently, scientists also found that the DNA of apes provides even stronger support for Darwin's original hypothesis. When they sequenced individual genes, researchers would produce gene trees that place apes closer to humans than any other species.

But these gene trees also produced some puzzling conflicts. Studying some genes, researchers concluded that chimpanzees were more closely related to humans than were gorillas or orangutans. But studies on other genes pointed to gorillas as our closest relatives. These studies, it's now clear, were bedeviled by incomplete lineage sorting.

Thanks to advances in DNA-sequencing technology, scientists can now compare not just individual genes but entire genomes. In 2018 a team of scientists led by Evan Eichler at the University of Washington assembled reconstructions of genomes of humans, chimpanzees, gorillas, and orangutans (Kronenberg et al. 2018). Whereas earlier technologies could allow scientists to reconstruct genomes from only short fragments a few hundred bases long, Eichler's team read fragments that were thousands of base pairs long or longer. As a result, they could compare the genomes with far greater precision than before.

The researchers surveyed the single-nucleotide polymorphisms (SNPs) in the genomes of humans and other apes. To do so, they compared every corresponding 1000-base-pair segment in each species and determined how much each segment differed from species to species. Across the entire genome, they found that chimpanzees and humans differ by 1.27%. In terms of SNPs, we are 98.8% identical to chimpanzees. Gorillas are only slightly more divergent, differing on average by 1.61%. Orangutans, on the other hand, differ on average by 3.12%.

These percentages reflect the evolutionary history of the great apes. Using their new genomes, the researchers determined that the ancestors of orangutans branched off early in the history of the clade from our own ancestors. Chimpanzees and gorillas are more closely related, and as a result they differ by a smaller amount.

But when the researchers generated phylogenies based only on each 1000-base-pair segment, they sometimes ended up with different trees. As shown in **Figure 8.7**, 64.4% of their trees produced the clade with chimpanzees as our closest relatives. But

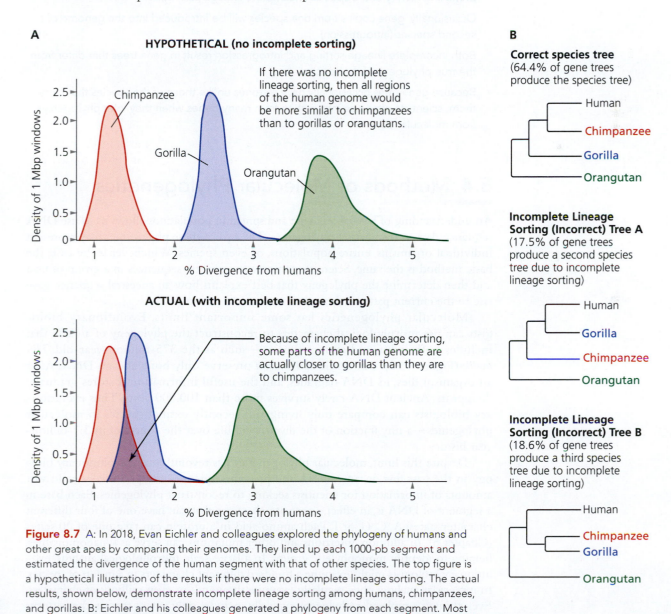

Figure 8.7 A: In 2018, Evan Eichler and colleagues explored the phylogeny of humans and other great apes by comparing their genomes. They lined up each 1000-pb segment and estimated the divergence of the human segment with that of other species. The top figure is a hypothetical illustration of the results if there were no incomplete lineage sorting. The actual results, shown below, demonstrate incomplete lineage sorting among humans, chimpanzees, and gorillas. B: Eichler and his colleagues generated a phylogeny from each segment. Most produced the correct result (top), but small fractions produced incorrect ones due to incomplete lineage sorting. (Data from Kronenberg et al. 2018)

in 17.5% of the trees, gorillas were closest. And in 18.6% of the trees, gorillas and chimpanzees were closer to each other than either was to humans.

The interpretation that best explains these results is that orangutans split off from other great apes a long time ago, and their relationship to us is not masked by incomplete lineage sorting. By contrast, the ancestors of gorillas, chimpanzees, and humans split from each other over a relatively brief period of time. As a result, a substantial number of variants did not sort completely. Their evolutionary history is different from the evolutionary history of great ape species.

This evolutionary history explains why early studies on ape phylogeny, based on only single genes, produced confusing results. Later, when Eichler and his colleagues took into account *all* the available evidence, they could see that chimpanzees are our closest living relatives.

Key Concepts

- Because alternative alleles sometimes persist side by side for a very long time, they may be passed down to daughter species in a fashion that does not reflect the actual branching history of the species (incomplete lineage sorting).

- Occasionally, gene copies from one species will be introduced into the genome of a second species (introgression).

- Both incomplete lineage sorting and introgression result in gene trees that differ from the true phylogeny of the species.

- Because gene trees occasionally have histories unlike those of the species that carry them, scientists often use information from many genes when they infer phylogenies from molecular data. ●

8.4 Methods of Molecular Phylogenetics

An understanding of how alleles arise and spread in populations allows us to use DNA sequence data to reconstruct phylogenies. The tips of these phylogenies may represent individual organisms, entire populations, or even species. Whichever is the case, the basic method is the same. Scientists compare the genetic sequences in a group of taxa and then determine the phylogeny that best explains how an ancestral sequence gave rise to the current patterns through mutation and fixation.

Molecular phylogenetics has some important limits. Evolutionary biologists can use morphological characters to reconstruct the phylogeny of a clade that includes both living and extinct species—such as the 375-million-year-old *Tiktaalik* (Figure 4.22). But fossils of *Tiktaalik* preserve only bone and no DNA. After an organism dies, its DNA degrades, and the useful information it stores gradually disappears. Ancient DNA rarely survives more than 100,000 years. Thus evolutionary biologists can compare only living and recently extinct species in molecular phylogenies—a tiny fraction of the diversity of life over the course of its 3.7-billion-year history.

Despite this limit, molecular phylogenetics has revolutionized evolutionary biology in the past few decades. DNA and protein sequences can potentially yield vast amounts of information for scientists seeking to reconstruct phylogenies. Each base in a segment of DNA is, in effect, a separate character that can have one of four different character states: A, C, G, or T. Each amino acid in a protein can take one of 20 states (Chapter 5). In addition, insertions and other mutations can serve as synapomorphies, helping to identify monophyletic groups. In a large-scale morphological analysis, evolutionary biologists might examine a hundred different characters in a group of species. In molecular analyses, they regularly examine thousands or even millions of characters. Even though a molecular phylogeny typically includes only extant species, the patterns found within and among taxa still yield important clues about the nature of the common ancestors of those species that lived hundreds of millions of years ago.

To take advantage of this rich trove of information, evolutionary biologists must contend with the special challenges posed by molecular evolution. We've already seen how species trees and gene trees don't always match. Another source of error is homoplasy. As we saw in Chapter 4, separate lineages can independently arrive at the same character state. Morphological homoplasy—such as the hydrodynamic shape of dolphins and tuna (Chapter 1)—is often the product of mutations to many interacting genes. Molecular homoplasy, on the other hand, can evolve much more easily. Because each base in a segment of DNA can exist in only one of four states, the probability that separate lineages will independently arrive at the same character state can be high. It's also possible for a site to mutate to a new nucleotide and then mutate again back to the original state. Instead of providing a stronger phylogenetic signal, such reversals erode it.

Thus any attempt to reconstruct a branching pattern for the past is likely to incorporate data with a jumble of signals. Some data provide true signals that accurately reflect the branching history of the group, and some provide false signals arising from homoplasy or reversals. Scientists use a number of analytical approaches to select the phylogeny that best approximates the actual history of a group.

The **maximum parsimony** method rests on the logic that we explored in Chapter 4's discussion of morphological phylogenies, namely, that the simplest solution is also the most reasonable one. When scientists use parsimony methods, they examine the distribution of characters among taxa on a number of trees with different topologies. They calculate how often those characters would have changed if a candidate phylogeny is correct. The tree with the fewest number of character state changes is considered the most parsimonious (Kitching et al. 1998; Swofford 2002).

Homoplasy can present a misleading picture of the most parsimonious tree, but scientists can use statistical methods to reduce its effect. They give extra weight to informative portions of genomes and less weight to ones that are more prone to homoplasy. Exon regions of protein-coding genes, for instance, are useful for reconstructing phylogenies of distantly related species. That's because they typically evolve very slowly when they are under strong **purifying selection**. The changes that do occur in exon regions are typically functional and are thus conserved for long periods of time. These types of slow substitutions can provide clear signals for constructing phylogenies because they reflect the patterns of ancestry of the lineages.

Introns and intergene regions of DNA are often effectively neutral with respect to selection. Such noncoding regions of DNA have more variable sequences—more information to use in building a tree—but also more homoplasy due to random convergence of base pairs. By giving less weight to rapidly changing characters such as noncoding bases, scientists can minimize the conflicting signal arising from homoplasy in the DNA (Williams and Fitch 1990; Maddison and Maddison 1992). Scientists may reverse these choices, however, if they are studying recently diverged lineages. Slow-evolving regions of DNA may experience too little change to be useful, whereas fast-evolving regions will offer more information.

When scientists use maximum parsimony methods, their analysis may present them with a single tree that's the most parsimonious. Or they may end up with a group of trees that are equally parsimonious, which can be combined into a single consensus tree that includes only branching patterns that appear in all of the most-parsimonious trees. In either case, scientists can then analyze the final tree with statistical methods to test how strongly the available data support it.

One of these tests is called **bootstrapping**. In this process, researchers select a random sample of characters from their full data set, much like drawing characters from a hat—with the caveat that each time a character is pulled it is immediately replaced in the hat, so that some characters may be sampled multiple times and others not at all. Scientists keep sampling the characters until they reach the same number as in the original set. Then they create a new data matrix and use it to generate a potential phylogeny. They repeat the process, randomly selecting characters and creating another phylogeny. After generating thousands of these potential trees, they can then compare them to each other. If the trees are very different from each other, it means

Maximum parsimony is a statistical method for reconstructing phylogenies that identifies the tree topology that minimizes the total amount of change, or the number of steps, required to fit the data to the tree.

Purifying selection (also called negative selection) removes deleterious alleles from a population. It is a common form of stabilizing selection (Chapter 7).

Bootstrapping is a statistical method that allows for assigning measures of accuracy to sample estimates; scientists use it for estimating the strength of evidence that a particular node in a phylogeny exists.

the data offer poor support for the original tree. On the other hand, if the bootstrap test reveals that the trees are all very similar, it indicates stronger support.

Scientists can use bootstrapping to evaluate how reliable the support is for each branching event (node) in an evolutionary tree. They will often print the bootstrap values above each of the nodes in a phylogeny so that the relative strength of support for each part of the tree can be evaluated.

Other methods for generating phylogenetic trees don't rely on the assumption of parsimony. For example, they may group taxa based on how different they are from each other. These **distance-matrix methods** convert DNA or protein sequences from different taxa into a pairwise matrix of the evolutionary distances (dissimilarities) between them. These methods predict that closely related species will have more similarities than more distantly related species, and this information can be used to group species into clades. These methods also can be used to estimate the lengths of the branches in the tree by equating the genetic distance between nodes with the length of the branch.

For example, **neighbor joining** is a distance-matrix method in which scientists pair together the two least-distant species by joining their branches at a node (**Figure 8.8**). They then join this node to the next-closest sequence, and so on. By joining neighboring species in round after round of this procedure, scientists can find the tree with the smallest possible distances—and the shortest possible branch lengths—between species (Saitou and Nei 1987).

One big advantage to using molecular data for phylogenetic studies is that we actually know a lot about the way DNA and protein sequences evolve (Chapter 6). For example, we've already seen that noncoding regions of DNA are likely to accumulate mutations faster than coding regions do. Within coding regions, substitutions that don't affect the amino acid coded for by a site (Chapter 5) are likely to accumulate more rapidly than substitutions that do. This means that not all base pairs in a DNA sequence are expected to change at the same rate or in the same way. It also means that there's a lot of additional information lurking in the sequences of species.

Maximum likelihood methods can use this extra information, and as a result they have become some of the most powerful and prevalent approaches for inferring phylogenies (Felsenstein 1981, 2004; Huelsenbeck and Crandall 1997). Maximum likelihood methods begin with an explicit model of evolution at the molecular level.

These models force some substitution rates to be equal. For example, the rate from A to C and from A to T might be set to have the same value. But the models allow other rates to vary. For example, the rate at one gene might be allowed to differ from the rate at another gene. Allowing less restricted models improves the realism of the model but comes at the cost of more difficulty in estimating the parameters.

These models of molecular evolution can be very complex, and they can be tailored to particular genes or regions of the sequences included. For example, one model can be used for coding regions and another model for noncoding regions. Scientists can also optimize these models to best explain the data they're studying (for example, by incorporating mutation rates that are estimated from their data for the particular species being studied).

Once scientists have specified their model for molecular evolution, they can use it to calculate the probability of observing their data set, given the tree and the model. This result is known as the "likelihood." In this approach to testing molecular phylogenies, the better trees are those for which the data are most probable. By comparing these likelihood scores for many different possible trees, scientists can identify the tree that best fits their data and their model for molecular evolution.

Another group of tests is known as **Bayesian methods** (Yang and Rannala 1997; Huelsenbeck and Ronquist 2001). They are similar to maximum likelihood in that they also use statistical models of the way DNA or protein sequences evolve. Maximum likelihood methods determine the probability of the data, given an evolutionary model and a hypothetical tree. Bayesian methods, on the other hand, determine the probability of a tree, given an evolutionary model and a particular data set.

A **distance-matrix method** is a procedure for constructing phylogenetic trees by clustering taxa based on the proximity (or distance) between their DNA or protein sequences. These methods place closely related sequences under the same internal branch, and they estimate branch lengths from the observed distances between sequences.

Neighbor joining is a distance method for reconstructing phylogenies. Neighbor joining identifies the tree topology with the shortest possible branch lengths given the data.

Maximum likelihood is an approach used to estimate parameter values for a statistical model. Maximum likelihood and the similar Bayesian method are used in phylogeny reconstruction to find the tree topologies that are most likely, given a precise model for molecular evolution and a particular data set.

Bayesian methods refer to tests that are similar to maximum likelihood; they use statistical models to determine the probability of data given an evolutionary model and a phylogenetic tree.

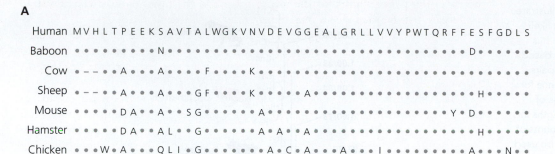

A

Human	M V H L T P E E K S A V T A L W G K V N V D E V G G E A L G R L L V V Y P W T Q R F F E S F G D L S
Baboon	• • • • • • • • • N • D • • • • • •
Cow	• – – • • A • • • A • • • • F • • • • K •
Sheep	• – – • • A • • • A • • • G F • • • • K • • • • • • A • • • • • • • • • • • • • • • • • • H • • • • •
Mouse	• • • • • D A • • A • • S G • • • • • A • • • • • • • • • • • • • • • • • • Y • D • • • • • • •
Hamster	• • • • • D A • • A L • • G • • • • • A • A • • A • • • • • • • • • • • • • • • • H • • • • • •
Chicken	• • • W • A • • • Q L I • G • • • • • • • • A • C • A • • • A • • • I • • • • • • • • • • A • • • N • •

B

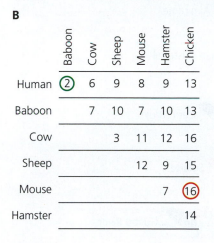

	Baboon	Cow	Sheep	Mouse	Hamster	Chicken
Human	②	6	9	8	9	13
Baboon		7	10	7	10	13
Cow			3	11	12	16
Sheep				12	9	15
Mouse					7	⑯
Hamster						14

C

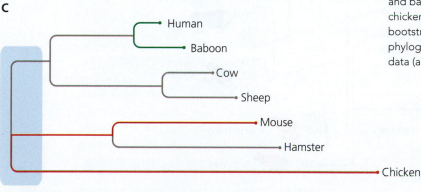

Figure 8.8 This figure shows how the neighbor-joining method can be used to generate an evolutionary tree. A: The analysis starts with sequence data—in this case, the amino acids in the β-globin protein. (Dots represent amino acids identical to those of humans at the same position. Dashes indicate deleted positions.) B: The sequence data are converted into a distance matrix. Each number in the matrix represents the number of differences in the sequence of each species pair. C: Finally, through a process of neighbor joining, scientists find the species that have the shortest distance between them. Over the course of the analysis, they end up with a tree that minimizes the total length of the tree. (The branch lengths in the tree are proportional to the distance between the nodes.) The red and green branches illustrate two examples. Humans and baboons differ by only two positions, while mice and chickens differ by 16. Each node can then be tested with bootstrapping. The blue shading shows the parts of the phylogeny that scientists can't resolve with the available data (a polytomy; see Chapter 4). (Data from Hartl 2011)

Scientists start with a possible tree and then make small changes to its topology. They evaluate the new topology, given the data and the model, and then change the tree again. They repeat this procedure thousands of times and in the process generate a probability distribution for the different possible trees. Eventually, they converge on a set of the most likely trees. Rather than returning a single best tree and set of parameters, Bayesian methods seek to estimate the probabilities of a wide range of trees and parameters and thus give an estimate of the most probable history as well as the uncertainty of the results.

Over the past two decades, scientists have argued vigorously about the merits of each method we've presented here. What scientists agree on is that no single method for inferring molecular phylogenies is superior to all the rest. Although maximum likelihood methods have proven to be very powerful, for example, they also demand far more computer processing than, say, neighbor joining. In practice, scientists compare the results of several of these methods to explore the relationships of species and populations (**Figure 8.9**).

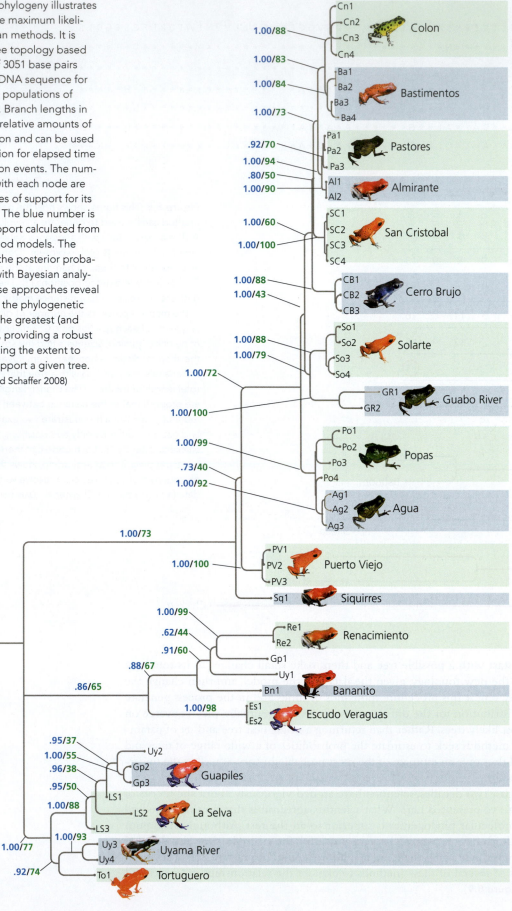

Figure 8.9 This phylogeny illustrates how biologists use maximum likelihood and Bayesian methods. It is the most likely tree topology based on comparison of 3051 base pairs of mitochondrial DNA sequence for Central American populations of poison dart frogs. Branch lengths in this figure reflect relative amounts of molecular evolution and can be used as an approximation for elapsed time between speciation events. The numbers associated with each node are statistical measures of support for its grouping of taxa. The blue number is the bootstrap support calculated from maximum likelihood models. The green number is the posterior probability calculated with Bayesian analysis. Together, these approaches reveal which portions of the phylogenetic hypothesis have the greatest (and weakest) support, providing a robust metric for evaluating the extent to which the data support a given tree. (Data from Wang and Schaffer 2008)

- Phylogenetic trees are hypotheses about the relationships among species or groups of individuals.

- Statistical models help scientists sift through volumes of molecular evidence to determine the best hypothesis or hypotheses that explain the data. ●

8.5 Three Case Studies in Molecular Phylogeny

As we saw in Chapter 4, scientists can use morphological characters to reconstruct phylogenies to test hypotheses about major transitions in evolution. Molecular phylogenetics also allows scientists to address these evolutionary questions. In this section we will look at three case studies, stretching across a huge range of taxa, that illustrate the power of this approach.

The Origin of Tetrapods

Molecular phylogeny enables scientists to reevaluate phylogenies that were developed based on morphology alone. Paleontologists, for example, have long argued that the closest living relatives of tetrapods were lobe-finned fishes, a group that today includes only lungfishes and coelacanths (Section 4.5). That was a fairly precise prediction, because there are around 30,000 species of fishes alive today. Of all those fishes, paleontologists predicted that only a half dozen should share a close common ancestry with tetrapods.

In 2017, Naoko Takezaki of Kagawa University and Hidenori Nishihara of the Tokyo Institute of Technology carried out a study of vertebrate phylogeny to test this hypothesis (Takezaki and Nishihara 2017). They compared DNA of a variety of tetrapods to that of lungfish and coelacanths. As an outgroup, they examined gars, which are a primitive lineage of ray-finned fishes. The phylogeny they reconstructed is shown in **Figure 8.10**. Their analysis shows lungfishes as the closest relative to tetrapods and coelacanths as the next-closest sister clade to tetrapods and lungfishes. Thus scientists studying one line of evidence—DNA—have confirmed a hypothesis originally developed from another line of evidence, the anatomy of fossil and living species.

How Did *Homo Sapiens* Evolve?

Earlier in this chapter, we saw how molecular phylogenetics illuminates our common ancestry with apes. As we'll discuss in more detail in Chapter 17, fossil and molecular evidence suggests that the last common ancestor of humans and chimpanzees lived about 6 million years ago. Over the past 6 million years, our lineage has produced perhaps 20 different species, known collectively as hominins. One of the most important questions about hominin evolution that scientists have tried to answer has to do with our own species: how did *Homo sapiens* evolve?

Until the 1980s, the evidence that scientists could use to address that question was almost entirely limited to morphology. Based on the fossil record, a number of paleoanthropologists concluded that *Homo sapiens* had evolved gradually across the entire Old World from an older hominin species over the past 1 million years. This hypothesis came to be known as the "multiregional model" of human evolution.

In the 1980s, Chris Stringer of the National History Museum in London and other paleoanthropologists put forward a competing model. They argued that *Homo sapiens* evolved in Africa alone, and that other hominin fossils from the past million years were extinct branches. For evidence, they pointed to the fact that the earliest fossils with clearly modern anatomical traits are found in Africa, dating back as far as 300,000 years (Section 3.12). The oldest clearly modern fossils outside of Africa, found in Israel, date back only 100,000 years, and the fossil record of modern humans outside

Figure 8.10 Scientists studying the anatomy of fossil and living species proposed that lobe-fins are the closest living relatives of tetrapods. A maximum-likelihood phylogeny based on their DNA, shown here, supports this hypothesis. The support for the nodes was 100% in all cases except for the elephant-armadillo clade, which had 58% support. (Data from Takezaki and Nishihara 2017)

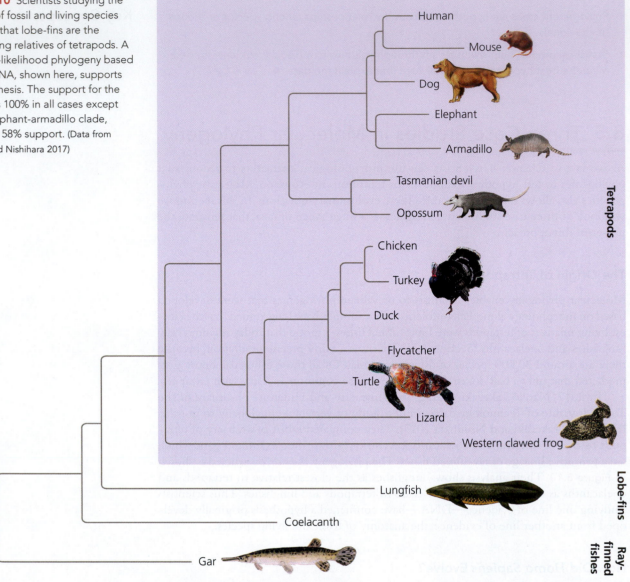

A **microsatellite** is a noncoding stretch of DNA containing a string of short (one to six base pairs), repeated segments. The number of repetitive segments can be highly polymorphic, and for this reason microsatellites are valuable genetic characters for comparing populations and for assigning relatedness among individuals (DNA fingerprinting).

of Africa becomes strong only 50,000 years ago. Stringer and his colleagues proposed that *Homo sapiens* evolved in Africa and that some populations expanded to other continents much later (Stringer 2012).

This hypothesis generates a clear prediction: all major ethnic groups of humans—Africans, Europeans, and Asians—are derived from recent African ancestry. Once geneticists began gathering DNA from a wide range of human populations, they began to put that hypothesis to the test.

Sarah Tishkoff has been in a particularly good position to do this research because she has gathered so much genetic information about people in Africa, where Stringer and others proposed humans originated. Tishkoff and her colleagues analyzed DNA from Africans and compared their genetic sequences with those of people from other parts of the world.

The results of one study are shown in **Figure 8.11** (Tishkoff et al. 2009). Tishkoff and her colleagues studied 121 African populations, 4 African American populations, and 60 non–African populations. They identified patterns of variation at 1327 genetic loci. Some of these loci, known as nuclear **microsatellites**, are stretches of repeating DNA that have a very high mutation rate. The researchers also looked at loci where DNA had been either inserted or deleted. Using the neighbor-joining method,

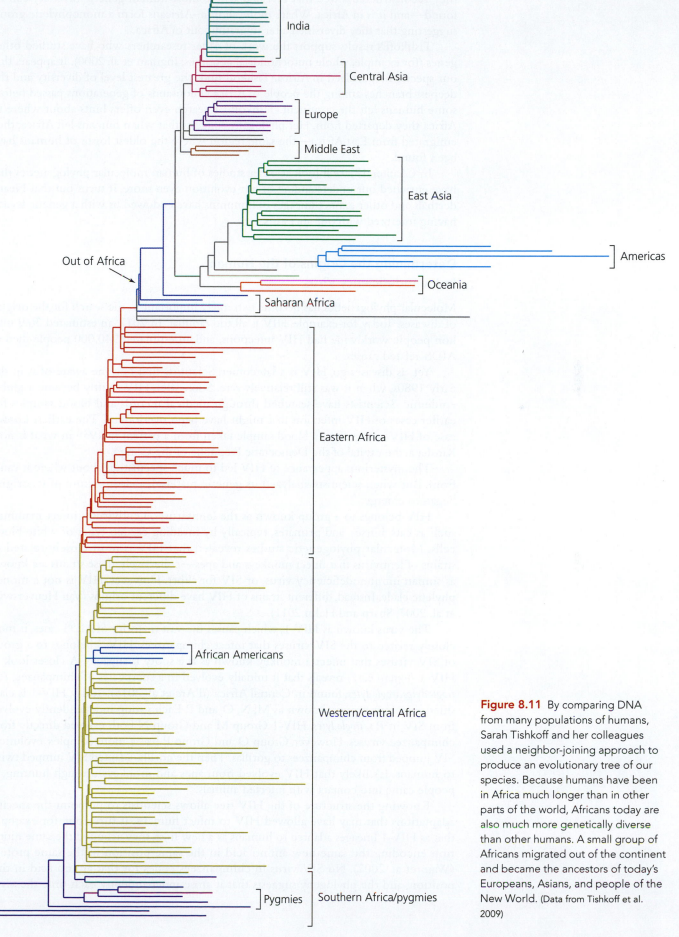

India

Central Asia

Europe

Middle East

East Asia

Out of Africa

Americas

Oceania

Saharan Africa

Eastern Africa

African Americans

Western/central Africa

Pygmies

Southern Africa/pygmies

Figure 8.11 By comparing DNA from many populations of humans, Sarah Tishkoff and her colleagues used a neighbor-joining approach to produce an evolutionary tree of our species. Because humans have been in Africa much longer than in other parts of the world, Africans today are also much more genetically diverse than other humans. A small group of Africans migrated out of the continent and became the ancestors of today's Europeans, Asians, and people of the New World. (Data from Tishkoff et al. 2009)

they reconstructed a tree that revealed where most human genetic diversity can be found—and it is in Africa. What's more, all non–Africans form a monophyletic group, suggesting that they diversified after migrating out of Africa.

Tishkoff's results support the work of other researchers who have studied other genes (for example, whole mitochondrial genomes; Ingman et al. 2000). It appears that our species first evolved in Africa. Tishkoff finds the greatest level of diversity and the deepest branches among the people of Africa. Thousands of generations passed before some humans left the continent. Tishkoff's research even offers hints about where in Africa they departed from. Her phylogeny suggests that when humans left Africa, they emigrated from East Africa—the same region where the oldest fossils of humans have been found.

In Chapter 17, we'll look at other studies of human molecular phylogenetics that have enriched our understanding of our evolution even more. It turns out that Neanderthals and other extinct lineages of hominins have endowed us with a genetic legacy, having interbred with our ancestors.

Determining the Origins of the Human Immunodeficiency Virus

Molecular phylogenetics has become extremely important in the search for the origins of diseases. Today, for example, HIV is all too familiar. In 2017, an estimated 36.9 million people worldwide had HIV infections, and an estimated 940,000 people died of AIDS-related causes.

Yet, as diseases go, HIV is a latecomer. Scientists first became aware of it in the early 1980s, when it was still relatively rare. Soon after, HIV swiftly became a global epidemic. Scientists have searched through medical records and blood samples for earlier cases of HIV infection that might have been overlooked. The earliest known case of HIV comes from a blood sample taken from a patient in 1959 in what is now Kinshasa, the capital of the Democratic Republic of the Congo.

The mysterious appearance of HIV led to much speculation about where it came from. But when scientists analyzed its genetic material, a clear picture of its origins began to emerge.

HIV belongs to a group known as the lentiviruses. Lentiviruses infect mammals such as cats, horses, and primates, typically by invading certain types of white blood cells. Molecular phylogenetic studies revealed that HIV is most closely related to strains of lentivirus that infect monkeys and apes—collectively, these strains are known as simian immunodeficiency virus, or SIV for short. However, HIV is not a monophyletic clade. Instead, different strains of HIV have different origins (Van Heuverswyn et al. 2007; Sharp and Hahn 2011).

The virus known as HIV-1, which causes the vast majority of AIDS cases, is most closely related to the SIV viruses that infect chimpanzees. HIV-2 belongs to a group of SIV viruses that infect a monkey known as the sooty mangabey. A closer look at HIV-1 (Figure 8.12) reveals that it initially evolved in a subspecies of chimpanzee, *Pan troglodytes troglodytes*, found in Central Africa (d'Arc et al. 2015). Today, HIV-1 is classified into four groups, known as M, N, O, and P. Each group independently evolved from SIV in *P. t. troglodytes*. HIV-1 Group M and Group N both evolved directly from chimpanzee viruses. However, Group O and Group P had a more complex evolution. SIV jumped from chimpanzees to gorillas. Then the gorilla-adapted SIV jumped twice to humans. It's likely that HIV evolved from apes and monkeys through hunting, as people came into contact with infected animals.

Knowing the structure of the HIV tree allows scientists to pinpoint the specific adaptations that may have allowed HIV to infect humans. It turns out, for example, that as HIV-1 lineages adapted to humans as a new host, they acquired the same mutation encoding the same new amino acid in the same position in the same protein (Wain et al. 2007). No SIV virus in chimpanzees codes for that amino acid in that position, and the phylogeny suggests that it arose independently each time the virus

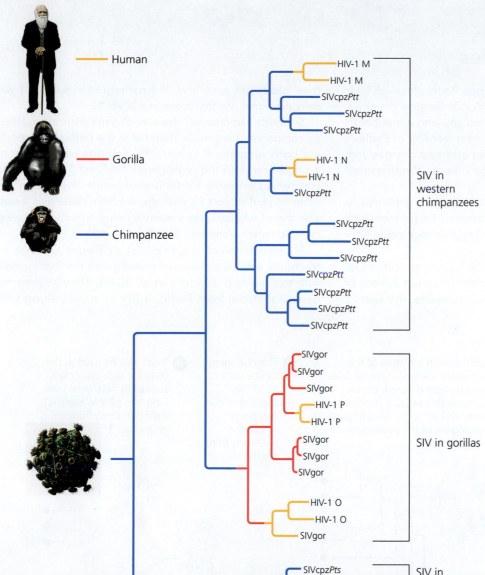

HOST OF VIRUS

Human

Gorilla

Chimpanzee

HIV-1 M
HIV-1 M
SIVcpz*Ptt*
SIVcpz*Ptt*
SIVcpz*Ptt*

HIV-1 N
HIV-1 N
SIVcpz*Ptt*

SIVcpz*Ptt*
SIVcpz*Ptt*
SIVcpz*Ptt*
SIVcpz*Ptt*
SIVcpz*Ptt*
SIVcpz*Ptt*
SIVcpz*Ptt*

} SIV in western chimpanzees

SIVgor
SIVgor
SIVgor
HIV-1 P
HIV-1 P
SIVgor
SIVgor
SIVgor
HIV-1 O
HIV-1 O
SIVgor

} SIV in gorillas

SIVcpz*Pts*
SIVcpz*Pts*
SIVcpz*Pts*

} SIV in eastern chimpanzees

Figure 8.12 The evolutionary tree of HIV-1 reveals how the virus hopped from chimpanzee hosts to humans four times. This tree topology was generated using a maximum likelihood approach. (Data from Sharp and Hahn 2011; virus: Russell Kightley Media)

adapted to its new human hosts. This mutation altered a gene encoding the shell of the virus, and experiments suggest that the mutation was crucial to the success of the new HIV viruses in humans. It's possible that the mutation allowed the virus to better manipulate its hosts into building new copies of itself.

Studies like these allow scientists to better understand the evolution of human disease and may help to better predict the emergence of new pathogens. Molecular phylogenies have also helped solve legal cases involving reckless and criminally negligent transmission of HIV (**Box 8.4**), and they've played a pivotal role in detecting the origins of a recent outbreak of Ebola virus (Gire et al. 2014). We'll explore the relevance of evolution to medicine in greater detail in Chapter 18.

• Constructing phylogenies is often a process of evaluating evidence. Scientists can test the predictions of phylogenetic hypotheses developed with one line of evidence by using other, independent lines of evidence to draw conclusions. •

Key Concept

BOX 8.4

Forensic Phylogenies

In 2005, a Texas man named Philippe Padieu learned from his doctor that he was infected with HIV. Despite the news, Padieu went on to have unprotected sex with a number of women, who later contracted HIV. After learning of Padieu's reckless behavior, the local district attorney charged him in 2007 with six indictments for the offense of aggravated assault with a deadly weapon.

The prosecutors marshaled many lines of evidence to show that Padieu had knowingly infected his partners with HIV. And some of the most compelling evidence came from evolutionary biology.

When a pathogen such as HIV infects an individual (subject 1), it rapidly reproduces within him or her. In each of the lineages, distinctive mutations accumulate. If subject 1 then infects someone else, the pathogens in subject 2 will carry the mutations that arose in subject 1.

Scientists can discover this trail of genealogical evidence by comparing the genetic material in the pathogens infecting both subjects. If subject 2's pathogens are all from a single branch in the phylogeny of subject 1, that can be compelling evidence for transmission from 1 to 2. On the other hand, if subject 2's pathogens form a clade that is outside that of subject 1, the evidence points to subject 2 being infected by someone else (**Box Figure 8.4.1**).

Michael Metzker, a geneticist at Baylor College of Medicine, and his colleagues investigated the Padieu case with this method (Scaduto et al. 2010). They sequenced genetic material from Padieu's HIV infection, along with

❶ As pathogens spread from person to person, their genomes mutate.

❷ To reconstruct the history of this transmission, scientists take samples of pathogens from different people. They sequence highly variable regions of the genome and use these data to produce the phylogeny of the pathogens.

❸ This phylogeny can be used to reject the hypothesis that one person acquired a pathogen from another.

❹ It can also be used as the basis for molecular clock studies to determine how long ago people acquired their pathogen from an individual.

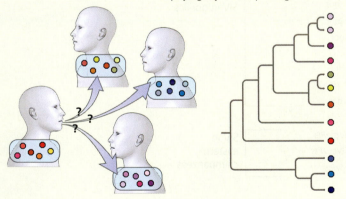
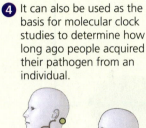
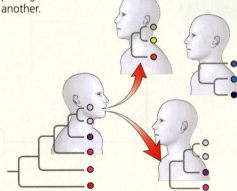

Box Figure 8.4.1 Molecular phylogenies provide insights that can allow scientists to reconstruct how pathogens spread from one person to another. This evidence is now being used in criminal trials. (Data from Bhattacharya 2014)

8.6 Natural Selection Versus Neutral Evolution

In the last section, we saw how molecular phylogenies help scientists solve specific mysteries about particular taxa. They can also help scientists answer broad questions about the process of evolution. One of the biggest questions is how much of life's diversity can be explained by evolutionary mechanisms other than selection.

Evolutionary biologists agree that natural selection is critically important to the evolution of complex morphology and behavior, because these phenotypes directly

viruses from the six infected women. The result was the phylogeny shown in **Box Figure 8.4.2**. The HIV from all six women formed branches nested within Padieu's clade. Each woman was infected by a different virus, in other words, but all the viruses evolved within Padieu's body.

Padieu was convicted and sentenced to 45 years in prison. The case is one of a growing number of examples of "forensic phylogenetics" making its way into the courtroom (Bhattacharya 2014).

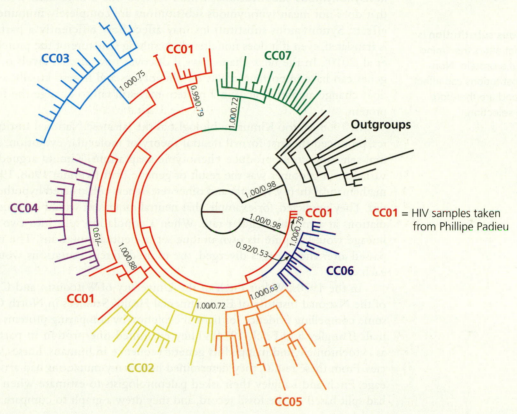

Box Figure 8.4.2 HIV viruses sampled from Philippe Padieu are marked CC01. All of the samples from Padieu's HIV-infected partners (CC02–CC07) were nested within the CC01 clade. Such a pattern is consistent with the six women getting infected by Padieu. Because this was used as evidence in an assault trial, the strength of support for the tree topology was critical. As we saw with Figure 8.9, scientists often indicate how strongly their data support particular nodes in a phylogenetic tree by placing the node support results right next to the nodes. Here, the numbers depict Bayesian posterior probabilities (1.0 equals 100% probability), and maximum likelihood bootstrap values (% of sampled trees retaining the node). (Data from Scaduto et al. 2010)

affect the fitness of individuals. But as we saw in Chapter 6, neutral mutations can also spread to fixation due solely to processes such as genetic drift. As scientists have mapped the genomes of numerous organisms, they've found that much of the variation in their DNA is hardly affected by selection at all (Nei 2005).

Substitutions

One way that genetic mutations can escape the action of selection is by not affecting an organism's phenotype (see Box 6.4). As we saw in Chapter 5, much of the noncoding DNA—including most pseudogenes—has no known function. As far as we know, these regions of DNA are simply transmitted as baggage from one generation to the next. Mutations to nonfunctional swaths of sequence are not likely to affect the phenotypes of the individuals that carry them, and as such, they are not likely to be exposed to selection.

Even mutations to protein-coding genes can sometimes escape the action of selection. Mutations to a protein-coding gene may fail to change a protein thanks to the redundancy built into the genetic code. Several different codons may encode the same amino acid (see Figure 5.8). A mutation may switch one codon to another without changing the corresponding amino acid. Scientists call this type of mutation a **synonymous** (or "silent") **substitution**.

As we'll see, synonymous substitutions are much less subject to selection than **nonsynonymous substitutions**, which replace one amino acid with another. But that does not mean synonymous substitutions are completely immune to selection's effects. Synonymous substitutions may affect how efficiently a particular protein is translated, even if it does not alter the resulting structure of the protein itself (Tuller et al. 2010). In the next chapter, we'll discover how altered levels of expression of genes can have important effects on phenotypes and fitness. Finally, a mutation that does change an amino acid in a protein may still fail to change the function of the protein.

In 1968 Motoo Kimura, a biologist at the Japanese National Institute for Genetics, produced the first formal neutral theory of molecular evolution. Although natural selection could produce phenotypic adaptations, Kimura argued, much of the variation in genomes was the result of genetic drift (Kimura 1968, 1983). From this mathematical theory, Kimura and other researchers constructed hypotheses they could test. They predicted, for example, that neutral mutations would become fixed in populations at a roughly regular rate. When a population split into two lineages, each lineage would acquire its own unique set of neutral mutations. The more time that passed after the lineages diverged, the more different mutations would be fixed in each one.

In the 1970s, Walter Fitch of the University of Wisconsin and Charles Langley of the National Institute of Environmental Health Sciences in North Carolina found some compelling evidence for neutral evolution by comparing proteins from 17 mammals (Langley and Fitch 1974). They examined one protein in particular, known as cytochrome *c*, and mapped its genetic sequence in humans, horses, and other species. From these results, they determined how many mutations had arisen in each lineage. Fitch and Langley then asked paleontologists to estimate when those lineages had split based on the fossil record, and they drew a graph to compare the two sets of results.

As **Figure 8.13** shows, they discovered that the more distantly related two species were, the more mutations had accumulated in each lineage since they split from a common ancestor. The graph was especially striking because the relationship was so linear. Mutations became fixed in the lineages with almost clocklike regularity.

In recent years, scientists have been able to carry out large-scale surveys of the rate of nucleotide substitutions in different classes of DNA sequences. As shown in **Figure 8.14**, different types of sites experience different rates of molecular evolution. Pseudogenes evolve far faster than nonsynonymous sites in protein-coding genes, for example. These differences also support the neutral theory of evolution because most pseudogenes no longer encode proteins or RNA molecules that are important to the fitness of an organism.

The precise relationship between natural selection and neutral evolution is a complex one that scientists are still exploring. But it is clear that neutral evolution has played a major role in how genomes got to be the way they are today. Neutral theory also provides evolutionary biologists with powerful tools for investigating other aspects of evolution. In the following sections, we will consider two of these applications: a "molecular clock" for determining the age of branching events and a null hypothesis for recognizing cases of natural selection.

A **synonymous substitution** is a substitution that does not alter the amino acid sequence of a protein. Because synonymous substitutions do not affect the protein an organism produces, they are less prone to selection and often free from selection completely.

A **nonsynonymous substitution** is a substitution that alters the amino acid sequence of a protein. Nonsynonymous substitutions can affect the phenotype and are therefore more subject to selection.

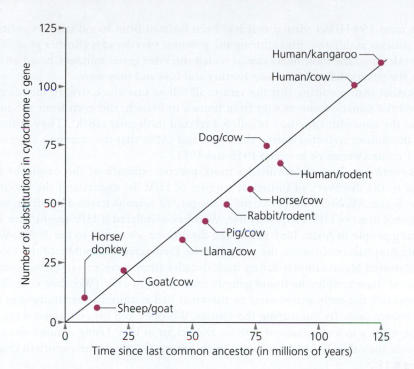

Figure 8.13 DNA fixes substitutions at a roughly clocklike rate. This graph shows how distantly related pairs of species have a large number of different substitutions in the cytochrome *c* gene. (Data from Moore and Moore 2006)

The Molecular Clock

Because base pair substitutions accumulate at a roughly clocklike rate, scientists can use mutations to tell time. By counting the number of base pair substitutions in a species' cytochrome *c* gene, for example, it's possible to estimate how long ago its ancestors branched off from our own. Scientists refer to this method of tracking time as the **molecular clock**.

To use molecular clocks, scientists must first calibrate them. If they can date an event relevant to the history of a group, such as the origin of an island or the age of a fossil, then they can calculate the rate at which nucleotides are being substituted over time for the genes and lineages they are studying.

Scientists must also select the most appropriate genetic material for a given molecular clock study (Moorjani et al. 2016). As Figure 8.14 illustrates, different segments of DNA evolve at different rates. To measure the divergence of species separated by hundreds of millions of years, a slow-evolving segment of DNA will provide greater accuracy than a fast-evolving one because it's likely to have accumulated less noise due to homoplasy (Section 8.4). Faster evolving regions, on the other hand, permit scientists to date events that unfolded over much shorter time scales—in some cases, mere decades.

As we saw earlier, scientists have used molecular phylogenetics to trace the origin of HIV to viruses that infect apes in central Africa. Once the researchers had a robust phylogeny, they could begin using it to estimate when the viruses shifted to human hosts. In 2000, researchers based at Los Alamos National Laboratory compared the

A **molecular clock** is a method used to determine time based on base pair substitutions. Molecular clocks use the rates of molecular change to deduce the divergence time between lineages in a phylogeny, for example. They work best when they can be "calibrated" with other markers of time, such as fossils with known ages and placements.

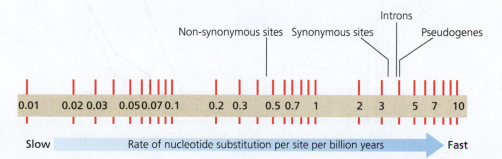

Figure 8.14 Different types of DNA segments evolve at different rates. Pseudogenes, for example, have a far faster rate of nucleotide substitution than do nonsynonymous (replacement) sites in protein-coding genes. This variation is a prediction of the neutral theory of evolution. (Data from Hartl 2011)

RNA from 159 HIV-1 viruses that had been isolated from blood samples collected from patients at different times during the previous two decades (Korber et al. 2000). They calculated the most likely rate at which the virus genes mutated, based on how much the viruses differed from one another and how old they were.

Rather than requiring that the viruses all follow one clock strictly, the scientists allowed the mutation rate to vary from branch to branch, and even from site to site within the genes (this method is called a relaxed molecular clock). They estimated from the isolates collected during the 1980s and 1990s that the common ancestor of HIV-1 existed sometime between 1915 and 1941.

Researchers later determined a more precise estimate of the origin of HIV, thanks to the discovery of historical samples of HIV. To understand the evolution of the disease, Michael Worobey of the University of Arizona traveled to Africa, where the animal hosts of HIV's ancestors live. Worobey wondered if HIV might have been infecting people in Africa long before the disease was identified in the 1980s. While visiting hospitals in Kinshasa, the capital of the Democratic Republic of the Congo, he discovered blood samples dating back decades (they were preserved in paraffin). In one of those samples, he found genetic material from HIV (Worobey et al. 2008). As expected, the early generations of the virus had acquired fewer mutations than more recent ones. By comparing the viruses, Worobey and his colleagues were able to determine a more precise estimate for the origin of HIV. Using a molecular clock approach, they showed that the disease likely emerged early in the twentieth century (**Figure 8.15**).

These molecular clock studies, combined with the phylogeny described earlier in this chapter, help us understand how HIV evolved. Hunters in West Africa have a long tradition of killing primates to eat or to sell in village markets. The hunters occasionally would have been exposed to various SIV strains. Some of the viruses that infected them might have replicated slowly, but they soon died out.

In the early 1900s, however, things changed. French and Belgian colonists established railways and extracted timber and other resources from deep within African forests. People had more contact with each other in the region, and the growing population drove an increased demand for bushmeat. In this new environment, SIV crossed over into humans and established itself as a new, devastating human pathogen.

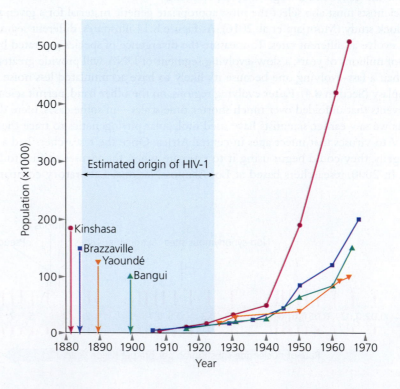

Figure 8.15 Growth of major settlements in central Africa coincide with the emergence of HIV-1. Michael Worobey and colleagues used early samples of HIV-1 discovered in hospitals in Kinshasa, the Democratic Republic of the Congo, to calibrate estimates of the rate of molecular evolution of HIV-1. Using a molecular clock approach, they showed that the disease likely emerged early in the twentieth century, a period coinciding with the beginnings of population growth in the region. (Data from Worobey et al. 2008)

- Neutral mutations accumulate in a clocklike fashion in genomes.
- Scientists can use molecular clocks to estimate the origin of diseases and major clades.

8.7 Footprints of Selection

Natural selection leaves behind traces in genomes, like footprints in wet sand. Scientists can use a number of different methods to throw a light on those footprints, long after these episodes of selection have passed. **Table 8.1** summarizes these methods. We'll take a closer look in this section at a few of these methods and some examples of the insights they can provide.

Table 8.1 General Approaches and Timing of Detecting Selection in Genome-Wide Selection Studies

Approaches	Signatures	Scope of the Comparison	Selection Detected	Time Frame (years)
Comparative				
Divergence rate	Reduction in the interspecific sequence divergence around a selected region relative to divergence of homologous regions genome-wide (Mayor et al. 2000; Ovcharenko et al. 2004) or when compared with a third species (Tajima 1993)	Between species	Positive, purifying	Greater than 1 million
Increased function-altering substitution rates	Elevated ratio of nonsynonymous (dN) to synonymous (dS) changes (dN/dS) in coding regions of selected genes (Nielsen and Yang 1998; Yang and Nielsen 1998)	Within a species	Positive	Greater than 1 million
Interspecies divergence versus intraspecies polymorphism	Reduction in the ratio of intraspecific diversity to interspecific divergence (Hudson et al. 1987; McDonald and Kreitman 1991)	Between species	Positive	Greater than 1 million
Population based				
Local reduction in genetic variation	A significant decrease in genetic variation (often measured as heterozygosity) around the selected site relative to its chromosomal neighborhood or genome-wide (Oleksyk et al. 2008)	Within a population	Positive	Less than 200,000
Differentiating between populations (F_{ST})	An increase or decrease in population differentiation in genomic regions under selection relative to the rest of the genome (Akey et al. 2002; Beaumont and Balding 2004)	Between populations	Positive, balancing	Less than 80,000
Extended linkage disequilibrium segments	Extended linkage disequilibrium producing remarkably long haplotypes around the beneficial single-nucleotide phase polymorphism (Tishkoff et al. 2001; Sabeti et al. 2002; Voight et al. 2006)	Within a population	Positive	Less than 30,000

(Data from Oleksyk et al. 2010)

Linkage Disequilibrium

Selective sweep describes the situation in which strong selection can "sweep" a favorable allele to fixation within a population so fast that there is little opportunity for recombination. In the absence of recombination, alleles in large stretches of DNA flanking the favorable allele will also reach high frequency.

Genetic hitchhiking occurs when an allele increases in frequency because it is physically linked to a positively selected allele at a nearby locus.

When a neutral mutation arises, especially in a large population, it may take a very long time for it to reach a high frequency through drift. By the time these alleles wander to fixation, recombination has had plenty of opportunity to mix and match the surrounding stretches of the chromosome, eroding any physical linkage among the new mutation and alleles at adjacent loci. Only the actual mutation becomes fixed, having reached linkage equilibrium.

When an allele experiences strong natural selection, on the other hand, it can spread quickly through a population. When this sweep occurs faster than recombination can separate the allele from the nearby regions of the genome, it leaves a footprint of natural selection known as a **selective sweep**. Alleles that happened to be sitting on the same chromosome when the mutation occurred get pulled along for the ride—as the new mutation becomes more common, so too do these other alleles—in a phenomenon known as **genetic hitchhiking**. Strongly selected alleles frequently will be found in a population surrounded by the same set of alleles at neighboring locations (**Figure 8.16**).

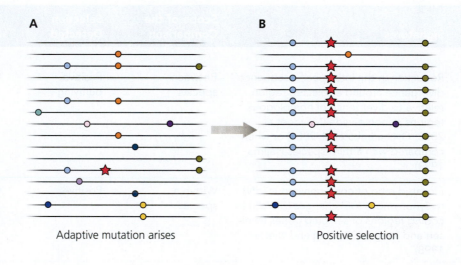

Adaptive mutation arises · Positive selection

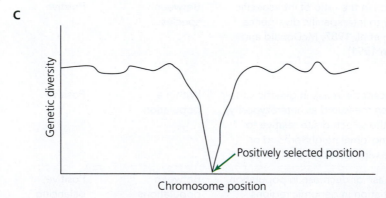

Figure 8.16 Scientists can detect the signature of natural selection in an allele by comparing its neighboring alleles in different individuals. Strong natural selection favoring one allele will spread its entire neighborhood to high frequencies in a population before recombination can separate the adjacent loci. A: Each line represents a segment of DNA of one individual in the population. Circles represent nucleotide bases unique to that individual. A new mutation (red star) arises in one individual and raises its fitness. B: The same population, a number of generations later. Individuals who inherited a segment of DNA with the new mutation had higher fitness. The mutation increased in frequency, carrying along its neighboring DNA. As a result, this particular recombinant will be unusually abundant in the population. C: Scans across the genome examining patterns of genetic variation will detect such a selective sweep as a "valley" in the amount of standing genetic variation. Loci adjacent to the selected allele will be less variable than expected when compared with samples of loci from other parts of the genome.

Figure 8.17 Humans domesticated cattle in both East Africa (left) and Northern Europe (right) several thousand years ago. In both places, mutations that enabled humans to digest milk as adults spread rapidly because they raised fitness in those contexts. (Left: Ton Koene/age fotostock/Superstock; right: Bernard/imagebroker/Superstock)

Linkage disequilibrium has revealed a striking case of natural selection in humans: the ability of some people to digest milk as adults (Figure 8.17).

Humans are mammals, and one of the hallmarks of living mammals is the production of milk. Milk is rich in a sugar called lactose, and young mammals produce an enzyme called lactase to break it down into simpler sugars they can digest. Around the time young mammals are weaned, they typically stop producing lactase in their guts because they stop drinking milk. Natural selection should favor this shift because it means that mammals don't waste energy making an enzyme with no advantage.

About 70% of humans also stop producing lactase in their intestinal cells during childhood. As a result, they can digest milk when they're young, but they have a difficult time with it when they're adults. Lactose builds up in their guts, spurring the rapid growth of bacteria that feed on the sugar. The waste released by the bacteria causes indigestion and gas. In about 30% of people, however, cells in the gut continue to produce lactase into adulthood. These people can consume milk and other dairy products without any discomfort because they can break down the lactose, leaving less of the sugar for the gas-generating bacteria to feed on. The difference between lactose-tolerant and lactose-intolerant people is largely due to alleles of the lactase gene, *LCT* (Swallow 2003).

To understand how 30% of people ended up with alleles for lactose tolerance, we must take a look at the history of cattle. Starting about 10,000 years ago, humans began to domesticate cattle in northwest Europe, East Africa, and certain other regions, leading to a dramatic change in their diet. Now energy-rich milk and milk-based foods were available well into adulthood.

The geography of lactose tolerance matches the geography of domestication fairly well (Figure 8.18). An *LCT* allele for lactose tolerance (called *LCT*P*) is most common today in northwest Europe—where cattle were domesticated 10,000 years ago—and rarest in southeast Europe, the farthest point in Europe from that origin. Scientists have also compared the frequency of the allele in traditional milk-drinking societies and non-milk-drinking ones in the same countries. *LCT*P* is generally much more common in the milk drinkers (Swallow 2003). If *LCT*P* had spread due merely to genetic drift, we would not expect such a strong association between the presence of both cattle herding and the allele. Instead, this pattern points strongly to natural selection.

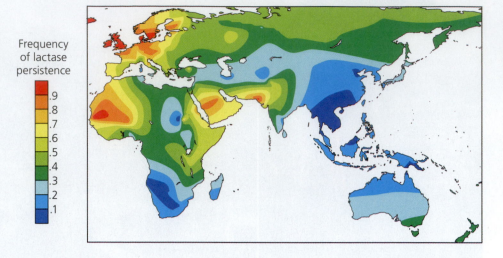

Figure 8.18 The ability to digest milk as an adult is known as the lactase persistence phenotype. This map shows the estimated frequency of the phenotype in the world. (Data from Itan et al. 2010)

Sara Tishkoff and her colleagues have found another line of evidence in favor of natural selection (versus random genetic drift) by comparing the DNA of individuals in milk-drinking societies and searching for a selective sweep. Specifically, she looked for this signal of natural selection around the *LCT* gene in two milk-drinking populations: East Africans and Europeans. The results are shown in **Figure 8.19**. Strong natural selection has preserved large segments of homologous DNA around *LCT*. However, a different allele was favored by natural selection in each population of humans. In other words, a mutation arose independently in each population that conferred lactose tolerance and then spread rapidly in both continents (Tishkoff et al. 2007).

We can combine this evidence to come up with a hypothesis for the origin of lactose tolerance. Originally, we humans had an *LCT* allele that stopped producing lactase when we outgrew nursing. Sometimes mutations gave rise to *LCT* alleles conferring lactose tolerance in adults, but they did not raise fitness because feeding on milk as adults was rare. In cattle-herding cultures, however, milk was plentiful, and the ability to digest milk brought huge benefits. People who could get protein and other nutrients from milk were more likely to survive and to pass on their mutant copy of *LCT* to their offspring. We will return to this *LCT* story in Chapter 18, to see how ancient DNA in fossils is enriching our understanding of its evolution.

F_{ST} Outlier Methods

We saw in Chapter 6 that most natural populations are subdivided into populations dispersed across the landscape. The frequencies of alleles in these populations are the product of opposing forces. Gene flow among the populations works to homogenize their allele frequencies. Meanwhile, drift and selection act within particular populations, causing allele frequencies to diverge from one population to the next.

In Box 6.8, we discussed how population geneticists characterize the extent of subdivision among populations using F_{ST}. F_{ST} ranges from 0 (fully homogenized) to 1 (fully segregated). Sequencing a few dozen loci allows scientists to use F_{ST} to calculate a reliable estimate of gene flow.

Initially, scientists used estimates of F_{ST} to measure gene flow between populations. For this research, they sampled a few dozen loci. Now, thanks to the genomics revolution, scientists can estimate F_{ST} for thousands of loci. This wealth of information allows scientists to use F_{ST} to address an entirely new question: how natural selection acts on populations.

When scientists calculate F_{ST} for so many loci, they can determine the overall differentiation between two populations. In some cases, one locus may be an outlier

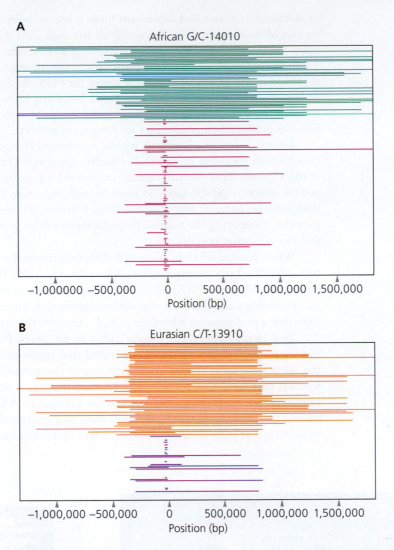

A

African G/C-14010

B

Eurasian C/T-13910

Position (bp)

Figure 8.19 A: Sarah Tishkoff of the University of Pennsylvania and her colleagues examined genetic linkage in Africans to detect natural selection around LCT. They compared 123 people from Kenya and Tanzania with an allele for lactose tolerance and one for intolerance. The people with the lactose tolerance allele (green lines) share much larger segments of homologous DNA around the gene, represented by the length of lines, than those with the alternative lactose-intolerant allele (red lines). B: Working with 101 people of European and Asian descent, Joel Hirschhorn of Harvard and his colleagues also found that an allele for lactose persistence was surrounded by large swaths of homologous DNA (orange lines). Note that the mutations for these two alleles are located in different parts of the same gene. They are lined up in this graph simply to show the different sizes of the hitchhiking regions. (Data from Tishkoff et al. 2007)

compared to the others. In other words, the frequency of an allele at that locus differs between populations to a greater degree than expected by chance. Natural selection is the only mechanism that can produce such extreme outliers. Searching across the genome for such F_{ST} **outliers** can point researchers to regions of the genome under strong contemporary selection.

Scientists used F_{ST} to discover one such case of intense selection in Tibet. At least 30,000 years ago, humans migrated onto the Tibetan Plateau, a vast uplifted swath of Central Asia often called the "Roof of the World." Spanning more than 2,500,000 square kilometers and averaging 4500 meters in elevation, the Tibetan Plateau is home to large herds of yak, gazelle, and antelope, as well as locally adapted populations of wolf and pika.

All of these species faced the same challenge when they colonized this high, barren landscape: there wasn't enough oxygen. The partial pressure of oxygen in air drops

The **F_{ST} outlier** method detects loci with allele frequencies that are more different than expected between populations. These outlier loci are likely to be near to regions of the genome experiencing strong selection.

as elevation increases, and this means there is less in every breath available for animals to use. Mountain climbers know this all too well, and this is the reason they acclimate at high-elevation base camps for several weeks before a significant ascent. Over a period of a week or two their bodies acclimate, upregulating synthesis of cofactors that bind to the hemoglobin in their blood, for example, making it more readily accessible to their oxygen-starved tissues. Faster breathing rates can bring in more oxygen, too. But these are just short-term strategies for coping with low oxygen. Natural selection has produced other solutions for long-term survival.

Several teams of researchers set out to identify the genes involved with adaptation to high elevation using the F_{ST} outlier approach. One team, led by Shuhua Xu at the Chinese Academy of Sciences, conducted a genome-wide screen of 1,000,000 genetic markers (SNPs) sampled from 46 high-elevation Tibetans and compared this with samples from 92 low-elevation Han Chinese. They then marched through the genome, comparing the relative frequencies of alleles at each locus across the high- and low-elevation populations.

When Xu and his team compared allele frequencies for Han Chinese and Tibetans, they identified two strong outliers, located next to the *EPAS1* and *EGLN1* genes, respectively (**Figure 8.20**; Xu et al. 2010). Both of these genes are known to affect oxygen physiology, and the particular alleles identified in Tibetans have since been shown to confer a performance advantage at high elevation (Moore 2017).

Two twists have recently been added to this story. First, a detailed study of the history of the Tibetan *EPAS1* allele found that regions of the genome surrounding this allele were more similar to the genomes of Denisovans, a hominin sister species to humans that we'll revisit in Chapter 17, than they were to *EPAS1* alleles from other human populations. The gene tree for the Tibetan *EPAS1* allele did not match our species tree. It now appears that Denisovans first possessed the *EPAS1* allele favored by selection at high elevation. The ancestors of Tibetans must have encountered and

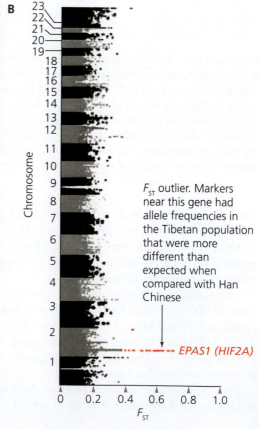

Figure 8.20 A: Both Tibetans and their domesticated Mastiff dogs have had to adapt to the extremely low concentrations of oxygen on the Tibetan Plateau. B: Shuhua Xu and his colleagues used an F_{ST} outlier approach to search for footprints of selection within the genomes of Tibetan and Han Chinese people. Each point in this graph depicts the F_{ST} value estimated for one of the sampled marker loci. Markers adjacent to the *EPAS1* and the *EGLN1* genes stood out. A related study used this same approach to identify F_{ST} outlier loci in the genomes of Tibetan and lowland Chinese dogs (Wang et al. 2014). (A: B&M Noskowski / E+ / Getty Images)

interbred with the Denisovans, and their *EPAS1* allele was introgressed into the human genome as a result (Huerta-Sánchez et al. 2014).

The second twist involves studies of Tibetan mastiff dogs. Guo-Dong Wang, also at the Chinese Academy of Sciences, and colleagues used the F_{ST} outlier approach to compare allele frequencies in the genomes of domesticated dogs sampled from the Tibetan Plateau and from the Chinese lowlands. Not only did they also discover several candidate genes likely to be associated with local adaptation to high elevation, but one of them, *EPAS1*, was the same gene as in humans (Wang et al. 2014). It turns out that Tibetan domesticated dogs also likely acquired their beneficial *EPAS1* allele through introgression—in this case from Tibetan gray wolves (Miao et al. 2016).

dN/dS

Both selective sweeps and F_{ST} outlier methods are able to detect recent selection because the footprints of selection depend on there still being linkage disequilibrium between the mutation under selection and markers located at nearby loci. But, like tracks on a beach, these signals decay with time. Even strongly selected mutations become uncoupled from alleles at adjacent loci through recombination, causing the footprint of selection to disappear.

Here we will consider other methods, based on molecular phylogeny, that scientists can use to detect natural selection that took place millions of years ago.

These methods also use neutral evolution as their null hypothesis. Scientists start out by assuming that any variations they find in homologous segments of DNA are the result of neutral evolution. If they test that hypothesis and reject it, the result supports the interpretation that selection is responsible. (This approach is similar to the way scientists use the Hardy–Weinberg equilibrium as a null hypothesis to detect evidence of selection in genotype frequencies; see Section 6.3.)

One way to test DNA is to compare the substitutions that occur in nonsynonymous sites to those that occur in synonymous sites. We would expect the difference between these two kinds of substitutions to reflect the balance between selection and neutral evolution because only those occurring in nonsynonymous sites would be consistently subject to natural selection. Let's consider a neutrally evolving pseudogene. It encodes no useful protein or RNA molecule. As a result, selection cannot act on any mutation that it acquires. We would expect that synonymous and nonsynonymous mutations are equally likely to become fixed through genetic drift. To estimate these probabilities, we can calculate the number of nonsynonymous substitutions per nonsynonymous site in the pseudogene (known as *dN*) and the number of synonymous substitutions per synonymous site in the pseudogene (known as *dS*). Under neutral evolution, we would expect that *dN* = *dS* (**Figure 8.21**; Bell 2008).

We would reject this hypothesis if we found that synonymous and nonsynonymous mutations did *not* occur equally as predicted by neutral evolution. Consider, for instance, a gene that undergoes strong positive selection (Chapter 6). It acquires a replacement (nonsynonymous) mutation that alters the structure of the encoded protein in a way that improves its performance. Because this allele is beneficial, it increases in frequency faster than synonymous alleles, the frequencies of which change only due to chance. As mutations continue to arise, beneficial ones will be pulled to fixation by selection, whereas neutral mutations will not. Positive selection will thus produce a gene in which there are more nonsynonymous mutations than would be expected through neutral evolution (*dN* > *dS*).

Another deviation from the null hypothesis occurs if nonsynonymous mutations are fewer than expected. This purifying selection results when a segment of DNA plays an essential role that is easily disrupted by mutations. A gene may encode a protein, for example, that cannot function if even a single amino acid is altered. Natural selection will eliminate alleles of such a gene with harmful mutations that lower fitness. Synonymous mutations, on the other hand, will remain hidden from selection because they don't alter the protein. Alleles with synonymous mutations will become more frequent, thanks to genetic drift. In other words, *dS* > *dN*.

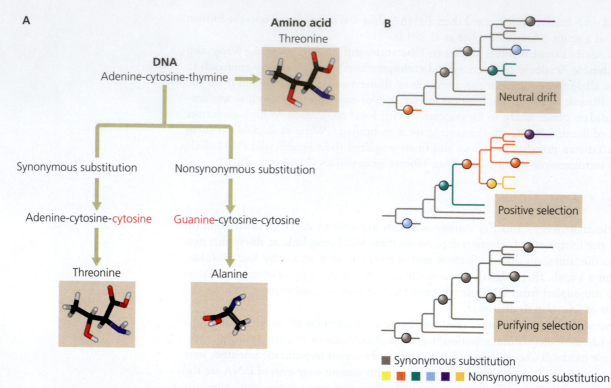

Figure 8.21 One sign of positive selection is the accumulation of an unusually high level of substitutions that change the structure of proteins. A: Synonymous substitutions alter a codon without changing the amino acid it encodes. Nonsynonymous substitutions change the amino acid; thus, they can potentially change the way a protein functions. B: Scientists can look for evidence of selection by comparing substitutions of different genes. (*Top*) Nucleotides in this gene are equally likely to acquire synonymous or nonsynonymous substitutions. This is a sign of neutral drift. (*Middle*) This gene acquires nonsynonymous substitutions that raise fitness. They are favored by natural selection and become fixed at a greater rate than synonymous substitutions. (*Bottom*) Under purifying selection, nonsynonymous substitutions in a particular gene lower the fitness of organisms, so they are lost from populations. Synonymous substitutions are thus far more common.

Evolutionary biologists have found that selection has to be very strong to create a clear signal in the difference between dS and dN (Charlesworth and Charlesworth 2010). Under certain conditions, they can use other tests to detect weaker footprints of selection. One of these is called the McDonald-Kreitman test, or MK test for short (McDonald and Kreitman 1991). To carry out the MK test on a gene in a particular species, evolutionary biologists compare its alleles within that species, and they also compare it to the homologous gene in other species. If the gene has experienced neutral evolution, then the ratio of nonsynonymous to synonymous substitutions across species should be the same as the ratio of nonsynonymous to synonymous polymorphic loci within the species.

If, on the other hand, the gene has experienced positive selection in the recent history of the focal species, then beneficial mutations will have rapidly increased and become fixed. These beneficial mutations are more likely to be nonsynonymous substitutions that improve the structure or function of the corresponding protein. As a result, scientists will find a higher ratio of nonsynonymous substitutions to synonymous ones between the focal species and a closely related species.

Selection will have a different effect on alleles within the focal species. Most of the standing genetic variation within the species will be the result of synonymous substitutions. As we saw earlier, selection will lead to the fixation of beneficial nonsynonymous substitutions. As a result, the ratio of nonsynonymous to synonymous fixed substitutions compared across species will be higher, whereas the same ratio calculated for polymorphic loci within the species will be lower. Such a pattern can be interpreted as evidence for positive selection fixing beneficial mutations between the species.

The *BRCA1* gene offers a fascinating case study in natural selection revealed by nonsynonymous mutations. As we discussed earlier, *BRCA1* is best known as a gene associated with breast cancer. But when it isn't causing cancer, it serves a number of vital functions. It oversees repairs to damaged DNA, and it helps control the complex molecular events that allow one cell to divide into two.

To explore the evolutionary history of this medically important gene, University of Texas biologist Sara Sawyer and her colleagues studied the synonymous and nonsynonymous mutations in *BRCA1* (Lou et al. 2014). They compared orthologs in 23 species of primates.

On many branches, the researchers estimated that the *dN/dS* ratio was well below 1. This could be the result of negative selection eliminating nonsynonymous mutations that disrupted *BRCA1*'s function. But on a few branches, the *dN/dS* value was well above 1, indicating positive selection. Intriguingly, one branch had the highest value of all. **Figure 18.22** shows the results of Sawyer's analysis for humans and our closest ape relatives.

Sawyer and her colleagues found that we have acquired 22 nonsynonymous substitutions since our ancestors split from the ancestors of chimpanzees and bonobos more than 7 million years ago. By contrast, humans have only three fixed synonymous mutations in the *BRCA1* gene. What makes this discovery all the more intriguing is that many of these changes to *BRCA1* result in a significantly elevated risk for breast cancer.

Sawyer and her colleagues speculate that viruses are the solution to this paradox. When cells divide and make new copies of their DNA, some viruses can slip their own genetic material into our cellular machinery to create new copies of themselves. Mutations to *BRCA1* may allow the gene to shut viruses out of this process. But viruses may then evolve new adaptations to evade the BRCA1 protein. As we'll see in Chapter 15, this open-ended back-and-forth form of evolution is quite common in parasites and their hosts. The benefits of this positive selection may be so great that they outweigh the increased risk of cancer, much like the trade-off between malaria resistance and sickle-cell anemia that we explored in Chapter 6.

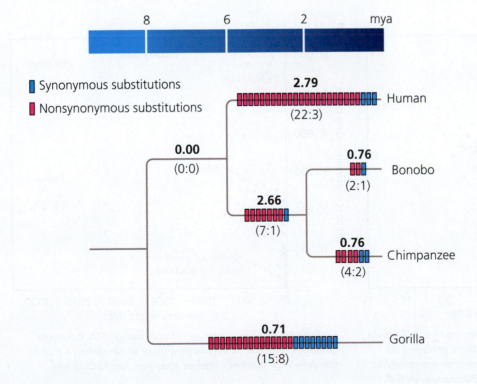

Figure 8.22 Certain mutations to the *BRCA1* gene create a significant risk for hereditary breast cancer. Surprisingly, the gene has undergone strong positive selection in our own lineage, as well as the lineage giving rise to chimpanzees and bonobos. The number on the top of each branch is the *dN/dS* ratio. On the bottom of each branch are the number of nonsynonymous and synonymous substitutions estimated to have been fixed on each lineage. (Data from Lou et al. 2014)

- The neutral theory of molecular evolution describes the pattern of nucleotide sequence evolution under the forces of mutation and random genetic drift in the absence of selection.

- The neutral theory predicts that neutral mutations will yield nucleotide substitutions in a population at a rate equivalent to the rate of mutation, regardless of the size of the population.

- As long as mutation rates remain fairly constant through time, neutral variation should accumulate at a steady rate, generating a molecular signature that can be used to date events in the distant past.

- Positive selection and purifying selection both leave distinctive signatures in nucleotide or amino acid sequences that can be detected using statistical tests. ●

8.8 Genome Evolution

Some evolutionary biologists examine the evolution of individual genes; other researchers look at the genome as a whole. As we saw in Chapter 5, genomes can vary tremendously in size. Many evolutionary factors are at play in determining the genome size in each species. As **Figure 8.23** shows, the size of bacterial genomes is proportional to the number of genes in each species. That's because bacterial genomes are mostly composed of genes.

Bacteria increase their genomes by gaining new genes. They have several mechanisms for doing so. An accidental duplication of a segment of DNA can create an extra copy of a gene. As we'll discuss in more detail in the next chapter, duplicated genes can diverge, acquiring new functions. Bacteria also acquire new genes from other bacteria through horizontal gene transfer.

Bacterial genomes can shrink as well. Accidental deletions of DNA may eliminate genes; if the loss of those genes doesn't reduce the fitness of bacteria, they may disappear entirely from a population. Scientists have found that certain kinds of niches favor bacterial genomes of different sizes. Free-living bacteria typically have relatively large

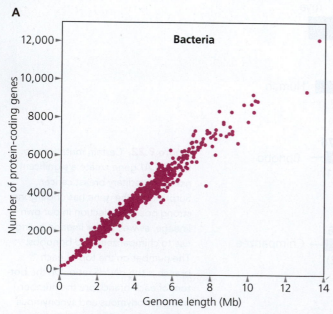

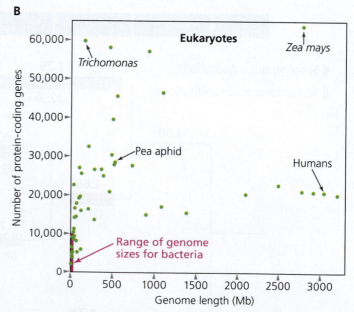

Figure 8.23 Species vary enormously in genome size. A: Bacteria have relatively small genomes. Because bacterial genomes are made up mostly of protein-coding genes, genome size correlates strongly with the total number of genes. B: Eukaryotes, on the other hand, have large amounts of noncoding DNA. Eukaryote genomes can become very large without accumulating a proportionately large number of genes. (Data from John McCutcheon)

genomes, for example, with a large repertoire of genes to cope with a variety of conditions. Some bacteria have become restricted to living in hosts, either as pathogens or as symbionts. With a reliable supply of amino acids and other nutrients from their hosts, many genes become less essential. Mutations that create pseudogenes are less likely to lower fitness. Subsequently, these noncoding regions may be deleted altogether, shrinking the genome (**Figure 8.24**). As we'll see in Chapter 15, the genomes of bacteria that exist inside hosts for many millions of years can shrink drastically to merge with the cells they inhabit.

Eukaryotes, on the other hand, don't have genomes that correlate tightly in size with their number of genes. Genomes with roughly the same number of genes can have vastly different sizes. Deletions can shrink the size of eukaryote genomes, but many processes can increase them. Like prokaryotes, eukaryotes can experience gene duplication. Even their entire genome can be duplicated.

Eukaryotic genomes also contain large amounts of non-protein-coding DNA. Mobile elements (Chapter 5), for example, proliferate in a parasitic fashion. Some researchers think that the dangers of these mobile elements are so great that eukaryotes have evolved defense systems to prevent mobile elements from replicating, thus protecting vital parts of the genome from the insertion of new copies of mobile elements (Slotkin and Martienssen 2007).

Why the genome structure of eukaryotes is so different from that of bacteria is not yet clear. Michael Lynch of Indiana University has proposed that the key factors are that the cells of eukaryotes were much larger than those of bacteria, and that the populations of eukaryotes were smaller (Lynch 2007). Genetic drift in those small populations could have allowed the first mobile elements to invade their genomes, despite their deleterious effects. Ever since that happened, mobile elements and their hosts have been locked in an evolutionary battle.

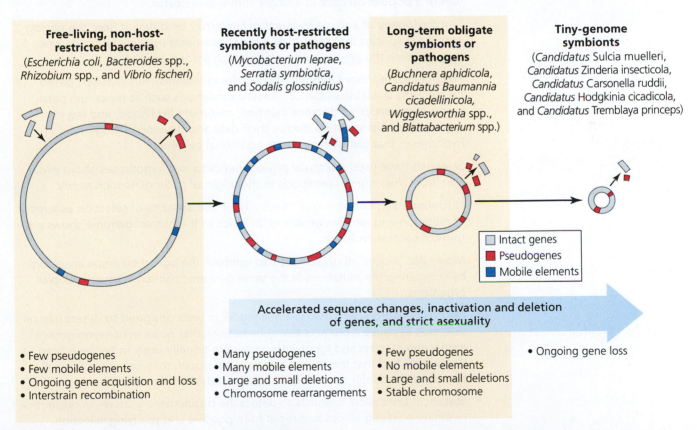

Free-living, non-host-restricted bacteria
(*Escherichia coli*, *Bacteroides* spp., *Rhizobium* spp., and *Vibrio fischeri*)

- Few pseudogenes
- Few mobile elements
- Ongoing gene acquisition and loss
- Interstrain recombination

Recently host-restricted symbionts or pathogens
(*Mycobacterium leprae*, *Serratia symbiotica*, and *Sodalis glossinidius*)

- Many pseudogenes
- Many mobile elements
- Large and small deletions
- Chromosome rearrangements

Long-term obligate symbionts or pathogens
(*Buchnera aphidicola*, *Candidatus* Baumannia cicadellinicola, *Wigglesworthia* spp., and *Blattabacterium* spp.)

- Few pseudogenes
- No mobile elements
- Large and small deletions
- Stable chromosome

Tiny-genome symbionts
(*Candidatus* Sulcia muelleri, *Candidatus* Zinderia insecticola, *Candidatus* Carsonella ruddii, *Candidatus* Hodgkinia cicadicola, and *Candidatus* Tremblaya princeps)

- Ongoing gene loss

☐ Intact genes
🟥 Pseudogenes
🟦 Mobile elements

Accelerated sequence changes, inactivation and deletion of genes, and strict asexuality

Figure 8.24 Bacteria can experience drastic reduction in genome size as they evolve into symbionts that live inside hosts. Deletions remove the many genes that are no longer essential. This reduction can continue for tens of millions of years. (Data from McCutcheon and Moran 2012)

Genomics is the study of the structure and function of genomes, including mapping genes and DNA sequencing. The discipline unites molecular and cell biology, classical genetics, and computational science.

The variation in genome sizes of eukaryotes is striking and puzzling. Why, for example, do genome sizes in animals vary 6650-fold (Gregory 2014)? Salamanders have some of the largest animal genomes, which can be up to 40 times the size of the human genome (Sun et al. 2012). Researchers have offered a number of hypotheses to explain these differences in eukaryotes. Some have suggested that animals with larger cells have larger genomes, for example. But no single hypothesis has gained strong support. As research in **genomics** advances, scientists may be able to untangle answers to these important questions.

Key Concepts

- Bacteria typically have relatively small genomes made up mostly of genes, whereas eukaryotes have genomes that vary greatly in size.

- As more and more genomes are sequenced, our understanding of genome evolution is changing rapidly. Scientists are discovering ways to answer important questions about the role that genome size and architecture play in the origins of our species as well as in other eukaryotes. ●

TO SUM UP . . .

- The relationships among species and other lineages can be inferred from different lines of evidence, such as their DNA, as well as their morphology. These different lines of evidence may yield different phylogenetic hypotheses.

- Scientists use coalescence to trace the alleles of a gene shared by all individuals of a population back to a single common ancestor.

- The phylogeny of a single segment of DNA may be different from the phylogeny of the species that carry it. As a result, scientists use several genes when they examine the phylogenetic relationships among species.

- Phylogenies represent hypotheses describing historical relationships based on currently available evidence. Statistical methods such as maximum parsimony, bootstrapping, distance matrixes, maximum likelihood, and Bayesian approaches help scientists resolve their data and develop phylogenetic hypotheses that can be tested with additional evidence.

- Scientists have used molecular phylogenetics to test hypotheses about evolution, from the origin of tetrapods to the origin of pathogens such as HIV.

- Nucleotide sequences evolve "neutrally" in the absence of selection as a result of mutation and random genetic drift. Much of the human genome shows patterns of neutral evolution.

- Molecular "clocks" allow scientists to estimate the age of common ancestors by comparing the mutations in the same genomic context across a group of organisms.

- Neutral evolution can serve as a null model in tests designed to detect natural selection because it predicts that substitutions that occur in nonsynonymous (replacement) sites and synonymous sites are equally likely to become fixed. If scientists find that the substitutions are not equal, they can reject the null hypothesis and infer that selection is responsible for the differences.

- Statistical tests allow scientists to detect the distinctive signatures in nucleotide or amino acid sequences that result from positive and purifying selection.

- Although scientists have hypothesized that genome size may be related to cell size, understanding the diversity of genome sizes and its role in evolution are fertile areas of study in evolutionary biology.

MULTIPLE CHOICE QUESTIONS Answers can be found at the end of the book.

1. Why is understanding coalescence important when developing molecular phylogenies?
   a. Because scientists can't know the true genealogy of a lineage without coalescing phylogenies to determine which is the most parsimonious.
   b. Because scientists can sample only a limited portion of the history of any allele.
   c. Because alleles that change over time are not valuable to developing phylogenies.
   d. Because scientists can't possibly determine the genealogy of a lineage from the limited samples available to them.

2. Why don't all gene trees reflect the phylogeny of species?
   a. Because the branch lengths of a species tree are usually much longer on average than the coalescence times of the genes being analyzed.
   b. Because coalescence of specific genes can occur before speciation events.
   c. Because speciation events can sometimes be very rapid.
   d. Both b and c.

3. Why might scientists use a statistical tool, such as Bayesian or maximum likelihood analyses, when reconstructing phylogenies?
   a. Because otherwise scientists can easily misinterpret the outcome.
   b. Because scientists can specify the parameters of a statistical model and test the capacity of the tool to produce comparable trees.
   c. Because molecular data can provide both true and false signals of the branching history, and statistical tools can reveal important patterns in the changes that occurred.
   d. Both b and c.

4. Molecular phylogenies indicate which of the following about HIV?
   a. The same mutation evolved in three separate lineages of HIV; in each instance, the mutation improved the ability of the virus to infect humans.
   b. HIV came from a monkey virus that was introduced into people by contaminated vaccinations.
   c. HIV is a monophyletic strain of lentivirus that infects both humans and chimpanzees.
   d. The common ancestor of simian immunodeficiency virus and human immunodeficiency virus came from horses.

5. The theory of neutral evolution describes
   a. the rate of mutation at a site that results from purifying selection, regardless of the size of the population.
   b. the rate of fixation of alleles at a site in the absence of selection.
   c. the competition between genetic drift and natural selection within the genome.
   d. Both a and b.

6. Which of these is a *true* statement about molecular clocks?
   a. Molecular clocks use neutral theory to date events within a phylogeny.
   b. Molecular clocks can be calibrated using fossils of known age.
   c. Molecular clocks can be affected by the segments of DNA being examined and relative sizes of the populations.
   d. All of the above.

7. When $dN > dS$,
   a. scientists would reject the null hypothesis of neutral evolution because the number of replacement substitutions is greater than expected.
   b. scientists would accept the hypothesis that the population is undergoing purifying selection because more replacement mutations were found than expected.
   c. scientists would reject the hypothesis that natural selection took place millions of years ago and is no longer relevant.
   d. scientists would accept the hypothesis that neutral evolution took place millions of years ago.

8. Which of the following is *not* true of coalescence?
   a. The timing of coalescence can depend on whether or not alleles are under selection.
   b. Positive selection can accelerate the rise in frequency in an allele, leading to a short coalescence.
   c. Two alleles that experience little selection may coexist longer, and thus the farther back coalescence occurs.
   d. In some instances, it is impossible to trace the genealogies of two homologous alleles in a population to a common ancestral allele, even if it is possible to trace their histories indefinitely.

9. Why do scientists use several genes when they examine the phylogenetic relationships among species?
   a. Synonymous substitutions are more likely to be present in multiple genes.
   b. The phylogeny of a single segment of DNA may be different from the phylogeny of the species that carry it.
   c. Purifying selection can remove deleterious alleles from a population, and a single segment of DNA may be missing from one gene.
   d. Scientists are aiming to increase their chances of finding microsatellites, which can be valuable genetic characters for comparing populations.

10. Which of the following is *not* true of the neutral theory of molecular evolution?
    a. When neutral variation accumulates at a steady rate, the molecular signature generated is an unreliable measure to date events in the distant past.
    b. The neutral theory of molecular evolution describes the pattern of nucleotide sequence evolution under the forces of mutation and random genetic drift in the absence of selection.
    c. The neutral theory predicts that neutral mutations will yield nucleotide substitutions in a population at a rate equivalent to the rate of mutation, regardless of the size of population.
    d. As long as mutation rates remain constant, neutral variation is expected to accumulate at a steady rate.

11. The figure below shows the estimated times of coalescence for alleles of 24 human autosomal genes and the mitochondrial genome. Most of the genes appear to coalesce at roughly the same time, between 0.5 and 1.5 million years ago. But the coalescent time of *MX1* is much deeper. Why might alleles of this gene be so much older than alleles of the other genes?

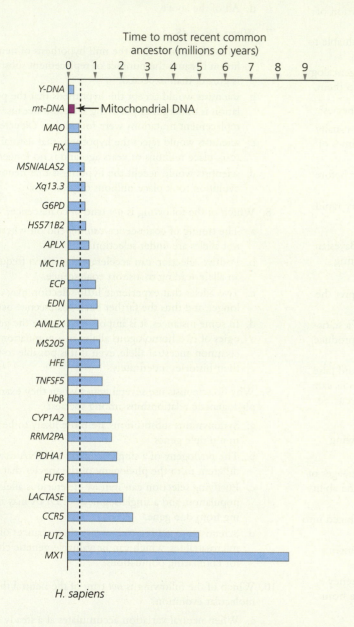

a. Because human populations likely went through a bottleneck event when they dispersed out of Africa.

b. Because one or more of the alleles of this gene were introgressed into human populations from a divergent lineage, such as the Neanderthals or Denisovans.

c. Because alleles of this gene are experiencing positive selection.

d. Because human populations have undergone a dramatic recent explosion in numbers.

SHORT ANSWER QUESTIONS Answers can be found at the end of the book.

12. Why are different statistical methodologies important for developing phylogenies?

13. How is the theory of neutral evolution different from the theory of evolution by natural selection? How is it the same?

14. Why might humans and dogs have used the same genes when they locally adapted to high elevation populations?

15. How did Sarah Tishkoff and her colleagues test the hypothesis that all modern humans are derived from recent African ancestry?

16. Why are conserved genes important when developing phylogenies?

17. Discuss the differences between a gene tree and a species tree.

ADDITIONAL READING

Baum, D., and S. Smith. 2010. *Tree Thinking: An Introduction to Phylogenetic Biology.* Greenwood Village, CO: Roberts and Company.

Carrol, S. 2006. *The Making of the Fittest.* New York: Norton.

Felsenstein, J. 1988. Phylogenies from Molecular Sequences: Inference and Reliability. *Annual Review of Genetics* 22:521–65.

———. 2003. *Inferring Phylogenies.* 2nd ed. Sunderland, MA: Sinauer Associates.

Grauer, D., and W-H. Li. 2000. *Fundamentals of Molecular Evolution.* Sunderland, MA: Sinauer Associates.

Hall, B. G. 2007. *Phylogenetic Trees Made Easy: A How-To Manual.* 3rd ed. Sunderland, MA: Sinauer Associates.

Hartl, D. 2011. *Essential Genetics: A Genomics Perspective.* Sudbury, MA: Jones and Bartlett.

Maddison, W. P. 1997. Gene Trees in Species Trees. *Systematic Biology* 46: 523–36.

Nei, M., and S. Kumar. 2000. *Molecular Evolution and Phylogenetics.* Oxford: Oxford University Press.

Oleksyk, T. K., M. W. Smith, and S. J. O'Brien. 2010. Genome-Wide Scans for Footprints of Natural Selection. *Philosophical Transactions of the Royal Society of London. Series B: Biological Sciences* 365 (1537): 185–205.

Templeton, A. R. 2006. *Population Genetics and Microevolutionary Theory.* New York: Wiley-Liss.

Wakeley, J. 2008. *Coalescent Theory: An Introduction.* Greenwood Village, CO: Roberts and Company.

Wallace, D. C. 2010. *Inside the Human Genome.* Oxford: Oxford University Press.

PRIMARY LITERATURE CITED IN CHAPTER 8

Akey, J. M., W. J. Swanson, J. Madeoy, M. Eberle, and M. D. Shriver. 2006. TRPV6 Exhibits Unusual Patterns of Polymorphism and Divergence in Worldwide Populations. *Human Molecular Genetics* 15:2106–13. (doi:10. 1093/hmg/ddl134)

Beaumont, M. A., and D. J. Balding. 2004. Identifying Adaptive Genetic Divergence among Populations from Genome Scans. *Molecular Ecology* 13:969–80 (doi:10.1111/j.1365- 294X. 2004.02125.x)

Bell, G. 2008. *Selection: The Mechanism of Evolution.* Oxford: Oxford University Press.

Bhattacharya, S. 2014. Disease Detectives. *Nature* 506:424–6.

Campbell, M. C., and S. A. Tishkoff. 2008. African Genetic Diversity: Implications for Human Demographic History, Modern Human Origins, and Complex Disease Mapping. *Annual Review of Genomics and Human Genetics* 9:403–33.

Charlesworth, B., and D. Charlesworth. 2010. *Elements of Evolutionary Genetics.* Greenwood Village, CO: Roberts and Company.

D'arc, M., A. Ayouba, A. Esteban, G. H. Learn, V. Boué, et al. 2015. Origin of the HIV-1 Group O Epidemic in Western Lowland Gorillas. *Proceedings of the National Academy of Sciences USA* 112:E1343–52.

Felsenstein, J. 1981. Evolutionary Trees from DNA Sequences: A Maximum Likelihood Approach. *Journal of Molecular Evolution* 17 (6): 368–76.

———. 2004. *Inferring Phylogenies.* Sunderland, MA: Sinauer Associates.

Fleming, M. A., J. D. Potter, C. J. Ramirez, G. K. Ostrander, and E. A. Ostrander. 2003. Understanding Missense Mutations in the *BRCA1* Gene: An Evolutionary Approach. *Proceedings of the National Academy of Sciences USA* 100 (3): 1151–6.

Gire, S. K., A. Goba, K. G. Andersen, R. S. G. Sealfon, D. J. Park, et al. 2014. Genomic Surveillance Elucidates Ebola Virus Origin and Transmission During the 2014 Outbreak. *Science* 345 (6202): 1369–72.

Gregory, T. R. 2014. Animal Genome Size Database. http://www.genomesize.com.

Hartl, D. L. 2011. *Essential Genetics: A Genomics Perspective.* 5th ed. Sudbury, MA: Jones and Bartlett.

Hudson, R. R., M. Kreitman, and M. Aguade. 1987. A Test of Neutral Molecular Evolution Based on Nucleotide Data. *Genetics* 116:153–9.

Huelsenbeck, J. P., and K. A. Crandall. 1997. Phylogeny Estimation and Hypothesis Testing Using Maximum Likelihood. *Annual Review of Ecology and Systematics* 28 (1): 437–66.

Huelsenbeck, J. P., and F. Ronquist. 2001. MRBAYES: Bayesian Inference of Phylogenetic Trees. *Bioinformatics* 17 (8): 754–5.

Huerta-Sánchez, E., X. Jin, Z. Bianba, B. M. Peter, N. Vinckenbosch, et al. 2014. Altitude Adaptation in Tibetans Caused by Introgression of Denisovan-like DNA. *Nature* 512 (7513): 194.

Ingman, M., H. Kaessmann, S. Pääbo, and U. Gyllensten. 2000. Mitochondrial Genome Variation and the Origin of Modern Humans. *Nature* 408:708–13.

Itan, Y., B. L. Jones, C. J. Ingram, D. M. Swallow, and M. G. Thomas. 2010. A Worldwide Correlation of Lactase Persistence Phenotype and Genotypes. *BMC Evolutionary Biology* 10:36.

Kimura, M. 1968. Evolutionary Rate at the Molecular Level. *Nature* 217 (5129): 624–6.

———. 1983. *The Neutral Theory of Molecular Evolution*. Cambridge: Cambridge University Press.

Kitching, I., P. Forey, C. Humphries, and D. Williams. 1998. *Cladistics: The Theory and Practice of Parsimony Analysis*. Oxford: Oxford University Press.

Korber, B., M. Muldoon, J. Theiler, F. Gao, R. Gupta, et al. 2000. Timing the Ancestor of the HIV-1 Pandemic Strains. *Science* 288:1789–96.

Kronenberg, Z. N., I. T. Fiddes, D. Gordon, S. Murali, S. Cantsilieris, et al. 2018. High-Resolution Comparative Analysis of Great Ape Genomes. *Science* 360:eaar6343.

Langley, C. H., and W. M. Fitch. 1974. An Examination of the Constancy of the Rate of Molecular Evolution. *Journal of Molecular Evolution* 3 (3): 161–77.

Lou, D. I., R. M. McBee, U. Q. Le, A. C. Stone, G. K. Wilkerson, et al. 2014. Rapid Evolution of *BRCA1* and *BRCA2* in Humans and Other Primates. *BMC Evolutionary Biology* 14:155.

Lynch, M. 2007. *The Origins of Genome Architecture*. Sunderland, MA: Sinauer Associates.

Maddison, W. P., and D. R. Maddison. 1992. *MacClade: Analysis of Phylogeny and Character Evolution*. Sunderland, MA: Sinauer Associates.

Mayor, C., M. Brudno, J. R. Schwartz, A. Poliakov, E. M. Rubin, et al. 2000. VISTA: Visualizing Global DNA Sequence Alignments of Arbitrary Length. *Bioinformatics* 16:1046–7.

McCutcheon, J. P., and N. A. Moran. 2012. Extreme Genome Reduction in Symbiotic Bacteria. *Nature Reviews Microbiology* 10:13–26.

McDonald, J. H., and M. Kreitman. 1991. Adaptive Protein Evolution at the *Adh* Locus in *Drosophila*. *Nature* 351 (6328): 652–4.

Miao, B., Z. Wang, and Y. Li. 2016. Genomic Analysis Reveals Hypoxia Adaptation in the Tibetan Mastiff by Introgression of the Gray Wolf from the Tibetan Plateau. *Molecular Biology and Evolution* 34 (3): 734–43.

Moore, L.G. 2017. Human Genetic Adaptation to High Altitudes: Current Status and Future Prospects. *Quaternary International* 461:4–13.

Moore, J., and R. Moore. 2006. *Evolution 101*. Westport, CT: Greenwood Press.

Moorjani, P., C. E. Amorim, P. F. Arndt, and M. Przeworski. 2016. Variation in the Molecular Clock of Primates. *Proceedings of the National Academy of Sciences USA* 113 (38): 10607–12.

Nei, M. 2005. Selectionism and Neutralism in Molecular Evolution. *Molecular Biology and Evolution* 22:2318–42.

Nielsen, R., J. M. Akey, M. Jakobsson, J. K. Pritchard, S. Tishkoff, et al. 2017. Tracing the Peopling of the World Through Genomics. *Nature* 541 (7637): 302.

Nielsen, R., and M. Slatkin. 2013. *An Introduction to Population Genetics: Theory and Applications*. Sunderland: Sinauer Associates.

Nielsen, R., and Z. Yang. 1998. Likelihood Models for Detecting Positively Selected Amino Acid Sites and Applications to the HIV-1 Envelope Gene. *Genetics* 148:929–36.

Oleksyk, T. K., M. W. Smith, and S. J. O'Brien. 2010. Genome-Wide Scans for Footprints of Natural Selection. *Philosophical Transactions of the Royal Society B: Biological Sciences* 365 (1537): 185–205.

Oleksyk, T. K., K. Zhao, F. M. De La Vega, D. A. Gilbert, S. J. O'Brien, et al. 2008. Identifying Selected Regions from Heterozygosity and Divergence Using a Light-Coverage Genomic Dataset from Two Human Populations. *PLoS One* 3:e1712.

Ovcharenko, I., M. A. Nobrega, G. G. Loots, and L. Stubbs. 2004. ECR Browser: A Tool for Visualizing and Accessing Data from Comparisons of Multiple Vertebrate Genomes. *Nucleic Acids Research* 32:W280–6.

Pfeffer, C. M., B. N. Ho, and A. T. Singh. 2017. The Evolution, Functions and Applications of the Breast Cancer Genes *BRCA1* and *BRCA2*. *Cancer Genomics-Proteomics* 14:293–8.

Rosenberg, N. A., and M. Nordborg. 2002. Genealogical Trees, Coalescent Theory and the Analysis of Genetic Polymorphisms. *Nature Reviews Genetics* 3 (5): 380–90.

Sabeti, P. C., P. Varilly, B. Fry, J. Lohmueller, E. Hostetter, et al. 2002. Detecting Recent Positive Selection in the Human Genome from Haplotype Structure. *Nature* 419:832–7.

Saitou, N., and M. Nei. 1987. The Neighbor-Joining Method: A New Method for Reconstructing Phylogenetic Trees. *Molecular Biology and Evolution* 4 (4): 406–25.

Scaduto, D. I., J. M. Brown, W. C. Haaland, D. J. Zwickl, D. M. Hillis, et al. 2010. Source Identification in Two Criminal Cases Using Phylogenetic Analysis of HIV-1 DNA Sequences. *Proceedings of the National Academy of Sciences USA* 107:21242–47.

Sharp, P. M., and B. H. Hahn. 2011. Origins of HIV and the AIDS Pandemic. *Cold Spring Harbor Perspectives in Medicine* 1:a006841.

Slotkin, R. K., and R. Martienssen. 2007. Transposable Elements and the Epigenetic Regulation of the Genome. *Nature Reviews Genetics* 8:272.

Stringer, C. 2012. *Lone Survivors: How We Came to Be the Only Humans on Earth*. New York: Henry Holt.

Sun, C., D. B. Shepard, R. A. Chong, J. L. Arriaza, K. Hall, et al. 2012. LTR Retrotransposons Contribute to Genomic Gigantism in Plethodontid Salamanders. *Genome Biology and Evolution* 4:168–83.

Swallow, D. M. 2003. Genetics of Lactase Persistence and Lactose Intolerance. *Annual Review of Genetics* 37:197–219.

Swofford, D. L. 2002. *PAUP*. Phylogenetic Analysis Using Parsimony (*and Other Methods).* Version 4. Sunderland, MA: Sinauer Associates.

Tajima, F. 1993. Simple Methods for Testing Molecular Clock Hypothesis. *Genetics* 135:599–607.

Takezaki, N., and H. Nishihara. 2017. Support for Lungfish as the Closest Relative of Tetrapods by Using Slowly Evolving Ray-Finned Fish as the Outgroup. *Genome Biology and Evolution* 9:93–101.

Templeton, A. R. 2006. *Population Genetics and Microevolutionary Theory.* New York: Wiley-Liss.

Tishkoff, S. A., F. A. Reed, A. Ranciaro, B. F. Voight, C. C. Babbitt, et al. 2007. Convergent Adaptation of Human Lactase Persistence in Africa and Europe. *Nature Genetics* 39:31.

Tishkoff, S. A., F. A. Reed, F. R. Friedlaender, C. Ehret, A. Ranciaro, et al. 2009. The Genetic Structure and History of Africans and African Americans. *Science* 324 (5930): 1035–44.

Tishkoff, S. A., R. Varkonyi, N. Cahinhinan, S. Abbes, G. Argyropoulos, et al. 2001. Haplotype Diversity and Linkage Disequilibrium at Human G6PD: Recent Origin of Alleles that Confer Malarial Resistance. *Science* 293:455–62.

Tuller, T., Y. Y. Waldman, M. Kupiec, and E. Ruppin. 2010. Translation Efficiency Is Determined by Both Codon Bias and Folding Energy. *Proceedings of the National Academy of Sciences USA* 107 (8): 3645–50.

Van Heuverswyn, F., Y. Li, E. Bailes, C. Neel, B. Lafay, et al. 2007. Genetic Diversity and Phylogeographic Clustering of SIVcpz*Ptt* in Wild Chimpanzees in Cameroon. *Virology* 368 (1): 155–71.

Voight, B. F., S. Kudaravalli, X. Wen, and J. K. Pritchard. 2006. A Map of Recent Positive Selection in the Human Genome. *PLoS Biology* 4:e72.

Wain, L. V., E. Bailes, F. Bibollet-Ruche, J. M. Decker, B. F. Keele, et al. 2007. Adaptation of HIV-1 to Its Human Host. *Molecular Biology and Evolution* 24:1853–60.

Wakeley, J. 2008. *Coalescent Theory: An Introduction.* Greenwood Village, CO: Roberts and Company.

Wang, G. D., R. X. Fan, W. Zhai, F. Liu, L. Wang, et al. 2014. Genetic Convergence in the Adaptation of Dogs and Humans to the High-Altitude Environment of the Tibetan Plateau. *Genome Biology and Evolution* 6 (8): 2122–8.

Wang, I. J., and H. B. Schaffer. 2008. Rapid Color Evolution in an Aposematic Species: A Phylogenetic Analysis of Color Variation in the Strikingly Polymorphic Strawberry Poison-Dart Frog. *Evolution* 62:2742–59.

Williams, P. L., and W. M. Fitch. 1990. Phylogeny Determination Using Dynamically Weighted Parsimony Method. *Methods in Enzymology* 183:615–26.

Worobey, M., M. Gemmel, D. E. Teuwen, T. Haselkorn, K. Kunstman, et al. 2008. Direct Evidence of Extensive Diversity of HIV-1 in Kinshasa by 1960. *Nature* 455 (7213): 661–4.

Xu, S., S. Li, Y. Yang, J. Tan, H. Lou, et al. 2010. A Genome-Wide Search for Signals of High-Altitude Adaptation in Tibetans. *Molecular Biology and Evolution* 28 (2): 1003–11.

Yang, Z., and R. Nielsen. 1998. Synonymous and Non-synonymous Rate Variation in Nuclear Genes of Mammals. *Journal of Molecular Evolution* 46:409–18.

Yang, Z., and B. Rannala. 1997. Bayesian Phylogenetic Inference Using DNA Sequences: A Markov Chain Monte Carlo Method. *Molecular Biology and Evolution* 14 (7): 717–24.

CHAPTER 9

From Genes to Traits

The Evolution of Genetic Networks and Development

Learning Objectives

- Explain how mutations to regulatory networks affect development of an organism.
- Explain how novel traits can arise when existing genes are expressed in new contexts.
- Name two examples in which proteins with a given function were co-opted for other functions.
- Discuss two ways scientists have discovered in which gene duplication plays a part in evolution.
- Explain the importance of *Hox* genes as part of the genetic toolkit.
- Explain how location and timing of the expression of developmental genes influence development.
- Describe three important steps in the evolution of the vertebrate eye.
- Define "pleiotropic effects" of a mutation.
- Explain why pleiotropic effects might differ depending on whether the regulatory or coding region of a gene is affected.
- Explain how imperfections of a complex adaptation can still confer advantages to fitness.
- Compare and contrast homology, convergent evolution, and parallel evolution.

To study evolution, Bryan Fry puts his life on the line. Fry, a biologist at the University of Melbourne, studies the evolution of snake venom, and that means he has to make intimate contact with some of the most lethal animals in the world, from deadly sea snakes to king cobras (Figure 9.1). Fry is an admitted adrenaline junkie, but he's not reckless. He prepares himself for every encounter so that he comes home safe and sound. It helps that he understands snake behavior. Fry knows, for example, that a king cobra

The eyelash viper, Bothriechis schlegelli—so named because of the scales above the eyes—is a venomous snake species found in Central and South America.

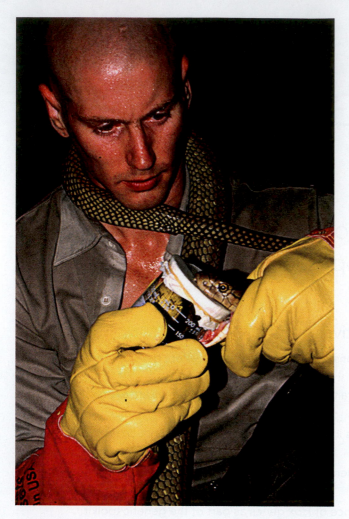

Figure 9.1 Bryan Fry investigates the evolution of this complicated adaptation by analyzing the genes for venom. (Bryan Fry)

signals dominance over another king cobra by rising up and touching its head to the top of its rival's. So to trap a king cobra, Fry first shows it who's boss by tapping it on the head. The cobra briefly bows down in submission, and Fry takes the opportunity to slip it into a bag.

Back in his lab at the University of Melbourne, Fry gets a close look at the biology that makes venomous snakes so deadly. At the back of their mouths, these animals have glands that produce venom. The cells in the glands express a set of venom genes, build the corresponding proteins, and pump the venom into the surrounding fluid. A king cobra opens its jaws and stabs its fangs into its victim. Muscles squeeze down on the glands, and the venom shoots down a pair of tubes leading into the fangs and then squirts out through holes at their tips, delivering the fatal concoction.

In their venom, king cobras and other species of snakes produce a veritable cocktail of molecules, each helping to subdue the prey. Some snakes make venom that relaxes the walls of the victim's aorta, which results in its blood pressure dropping until it blacks out. Other venoms lock onto receptors in neurons, causing paralysis. Yet others interfere with the biochemistry inside muscle cells, causing them to break down rapidly.

Fry doesn't just want to catalog these venoms. He wants to understand how these molecules—along with the rest of the venom-delivery apparatus in snakes—first evolved. Only some species of snakes use venom to kill their prey. Others, like pythons, suffocate their victims instead. Lizards, the closest living relatives of snakes, mostly lack venom. As we'll see later in this chapter, Fry and his colleagues are uncovering remarkable insights into venom evolution—such as the surprising fact that snake venom evolved before snakes themselves.

Nature brims with similarly complex adaptations, and reconstructing the details of their evolution is one of the richest lines of inquiry in biology. To explain the origin of a structure like the eye, for example, it's necessary to go beyond examining eyes that already exist. We instead must discover how animals with eyes evolved from animals with no eyes whatsoever.

In earlier chapters, we began to explore complex adaptations by examining their phenotypes. In Chapter 4, for example, we learned about the origin of powered flight in birds by looking at the fossil record. In this chapter, we will dive into the genetic foundations of these complex adaptations. We'll pay particularly close attention to how networks of genes guide the development of organisms and discover how evolutionary tinkering with this development can give rise to new complex adaptations and dramatically modify existing ones. Finally, we'll see how the pattern of evolution is remarkably similar in complex adaptations across the tree of life, from bacteria to plants to animals. ●

9.1 Regulatory Networks Across Nature

In Chapter 5, we learned how DNA is composed of a number of different elements. Some elements are genes, encoding proteins, or RNA molecules. Some are **gene control regions**, which can bind **repressors** or **transcription factors** that regulate the expression of nearby genes. Every **complex adaptation** is encoded and controlled by a large number of such genetic elements. These elements produce the proteins required by a complex adaptation and determine where and when they're made.

We tend to think of bacteria as being simple organisms, but the fact is that they have many complex adaptations of their own. And because bacteria are easier to study in the laboratory than animals or plants, they provide some of the best understood examples of complex adaptations. For instance, many species of bacteria survive harsh conditions by making spores. Spores are special types of cells encased in tough coats that can remain dormant for years. They can be carried off by wind or water to distant places. If the environment is right and conditions are improved enough, the spores can turn into viable cells once more. Understanding the biology of spore formation is important not just to microbiology but to human health. A species of bacteria called *Bacillus anthracis* can produce abundant spores in the soil. If these spores get into a person's body, they can germinate and cause the deadly disease anthrax.

Figure 9.2 shows the "blueprint" for making spores. Molecular sensors both inside and outside a microbe can detect signals of danger. These signals may reveal that the microbe has suffered DNA damage or that external conditions have become threatening. If a microbe detects a certain combination of these signals, the microbe can trigger the expression of a few of its regulator genes. The proteins from these regulatory genes then activate or repress more than 300 other genes.

Gene control region refers to an upstream section of DNA that includes the promoter region as well as other regulatory sequences that influence the transcription of DNA.

A **repressor** is a protein that binds to a sequence of DNA or RNA and inhibits the expression of one or more genes.

A **transcription factor** is a protein that binds to specific DNA sequences and acts like a light switch by turning all the sequences on or off simultaneously.

A **complex adaptation** is a suite of coexpressed traits that together experience selection for a common function. Phenotypes are considered complex when they are influenced by many environmental and genetic factors and when multiple components must be expressed together for the trait to function.

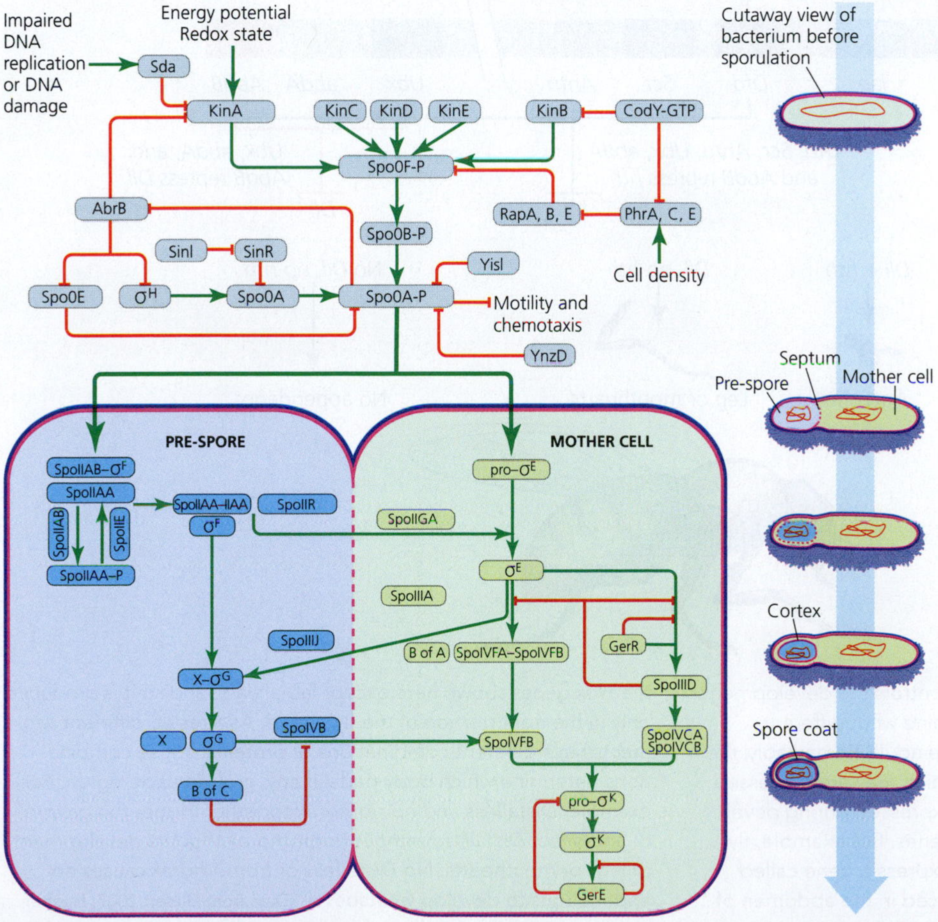

Figure 9.2 Many genes are organized in hierarchies, allowing the expression of just a few genes to trigger the expression of hundreds of other genes. This diagram shows the hierarchy of genes that controls the formation of spores in many species of bacteria. Some proteins or signals promote the expression of genes (green arrows), whereas others reduce expression (red lines). (Data from Paredes et al. 2005)

Together, these genes lead to the production of a special cell called a "pre-spore." Instead of splitting off on its own, the pre-spore remains inside its mother cell, where regulatory genes continue to direct the development of its tough coat. At the same time, the mother cell's own activity is controlled by regulatory genes, leading it to help the pre-spore turn into a true spore ready to survive on its own (de Hoon et al. 2010).

The genes that govern the formation of a spore form what's known as a **regulatory network**. In all cellular organisms, regulatory networks have a similar organization. Many of them are hierarchical in structure, with just a few "master" regulators at the top receiving external signals. They in turn control a large number of genes at lower levels. Regulatory networks also include genes for repressors. They keep the expression of other genes in the network from getting too high and can shut them down once they've finished constructing a complex adaptation like a spore.

In our own cells, strikingly similar regulatory networks govern complex adaptations as well. But unlike bacteria, we develop from a single cell into more than 30 trillion cells. From a single genotype, our bodies generate hundreds of different cell types and tissues. Animals and plants use regulatory networks of gene hierarchies to guide their development.

Figure 9.3 illustrates one set of important developmental genes in animals, known as *Hox* genes. They encode transcription factors that determine the identity of body

A **regulatory network** is a system of interacting genes, transcription factors, promoters, RNA, and other molecules. It functions like a biological circuit, responding to signals with output that controls the activation of genes during development, the cell cycle, and the activation of metabolic pathways.

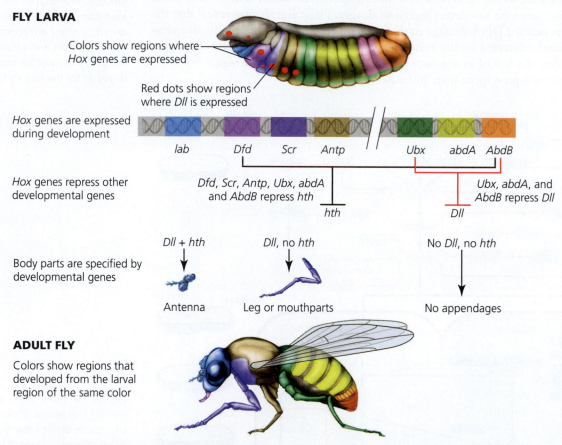

Figure 9.3 Hierarchical gene regulation controls the development of animal embryos. *Hox* genes help determine what different segments in a fly embryo will become in the adult. Remarkably, the genes are arranged in the same order in which they are expressed in the embryo. Some *Hox* genes that are expressed during development shut down other developmental genes. For example, fly larva cells in the head and thorax regions express a gene called *Distalless*, or *Dll*. But the *Hox* genes produced in the abdomen of the larva (*Ubx*, *abdA*, and *AbdB*) all repress the *Distalless* gene. Another gene, called *homothorax* (or *hth*), is repressed by all of the *Hox* genes shown here except labial (*lab*), and so it is produced only in the front portion of the fly's head. As a result, different segments have different combinations of proteins. Those combinations determine which body parts, if any, grow in each region. For example, *Distalless* and *homothorax* together trigger the growth of antennae. *Distalless* without *homothorax* triggers development of legs or mouthparts. No *Distalless* or *homothorax* causes no appendages to develop whatsoever. (Data from Gilbert 2007; Hueber and Lohmann 2008)

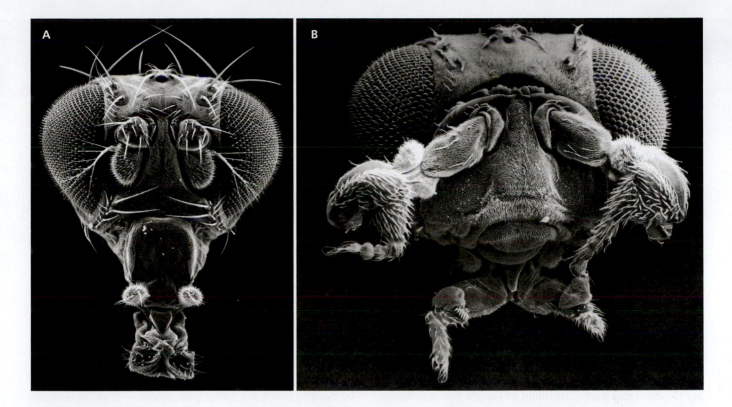

Figure 9.4 Mutations to genes at the top of the hierarchy in developmental networks can have drastic phenotypic effects. A: The head of a normal fly. (Thom Kaufman/Flybase) B: A mutation to a *Hox* gene called *Antennapedia* leads to the growth of legs where antennae normally grow on fly heads. (Thom Kaufman/Flybase)

parts along the head-to-tail body axis. The regulatory genes in the *Hox* gene network regulate such fundamental networks that mutations to them can lead to startling deformations. A leg may grow from a fly's head where its antenna would normally be found (Figure 9.4). These freakish deformities occur when *Hox* genes assign an incorrect identity to the cells in a region of the fly embryo.

As scientists make more detailed maps of these hierarchies, they're discovering how gene networks offer opportunities for evolutionary change, including large evolutionary shifts in body plan or novel biological functions achieved via the repurposing of ancient gene networks. For more details about how gene control regions can generate complex gene expression patterns in embryos, see Box 9.1.

Key Concepts

- Genes rarely function in isolation—the expression of a gene is influenced by a network of regulatory elements and interactions with RNA and other gene products. These hierarchies are important building blocks in the evolution of complex adaptations.

- Novel traits can arise when existing genes or processes are expressed in new developmental contexts (for example, activating a genetic pathway at a new time or in a new tissue during development). ●

9.2 Generating Innovations

For new adaptations to evolve, existing proteins and RNA molecules must take on new functions. At first this might sound impossible. If a mutation successfully switches a protein to a new function, you might assume that the protein must also stop performing its original function. And if that function is essential, this mutation should lead to a loss of fitness, rather than a gain. But many lines of research have resolved this paradox, revealing several ways in which genetic innovations can fuel the evolution of adaptations.

BOX 9.1

The Importance of Cis-Regulation

In a developing embryo, gene expression can be remarkably complex. Consider, for example, a gene known as *even-skipped* (*eve* for short). In a fly embryo, the cells that express *eve* form a series of seven distinct stripes spread across the embryo, from the anterior to the posterior ends (Box Figure 9.1.1).

How can a gene be turned on in these bands of cells and yet remain shut off in the cells in the intervening stripes? The answer lies in how cells regulate their gene expression. Gene control regions located on the same chromosome as the gene, called **cis-regulatory regions**, control where and when a gene is expressed. As proteins and RNA molecules bind to specific cis-acting elements during development, they can provide information to a cell about its location within the embryo. The blend of transcription factors and other proteins at the anterior end of an embryo is different from that found in posterior regions. Cells inside a band of *eve* expression are exposed to a particular blend of signals that is sufficient to activate a cis-acting element and initiate expression of the *eve* gene. Cells in the adjacent regions experience blends of signals that fail to initiate expression.

What's truly remarkable about the striped pattern of *eve* expression is that the precise combination of signals in any given stripe is different from the other stripes. How can a single gene switch on under all these different combinations? By having separate cis-regulatory regions precisely tuned to each of the seven regions in the embryo that become stripes. This organization is common in eukaryotic organisms. Genes that have multiple functions have multiple cis-acting elements, each corresponding to a context in which the gene must be expressed. The sophisticated control of gene expression sometimes gives rise to cis-regulatory regions that are actually longer than the genes they regulate. **Box Figure 9.1.2** shows the geography of cis-regulation of the *eve* gene.

Many scientists maintain that this organization of gene regulation has profoundly affected the evolution of multicellular organisms (Stern 2000; Carroll 2008). Genes involved in development are often pleiotropic—they have many functions in different cell types. A mutation that occurs in the coding region of such a gene may be beneficial in one context but hugely detrimental in another. These kinds of

Box Figure 9.1.1 Expression of the embryonic patterning gene *even-skipped* (*eve*) in *Drosophila* embryos is confined to seven distinct bands along the anterior–posterior axis. Each band is regulated by a different cis-acting element within the promoter region of this gene. (Republished with permission of Oxford University Press from Alexander Spirov, Mikhail Borovsky, Olesya Spirova, "HOX Pro DB: the functional genomics of hox ensembles", Nucleic Acids Research, 2002, v. 30 (1): 351–353; permission conveyed through Copyright Clearance Center, Inc.)

Horizontal Gene Transfer: A Primordial Source of Innovation

Bacteria and archaea have an astounding variety of adaptations. Even closely related strains may vary drastically in their biology; one strain may live on decaying matter in the soil whereas the other makes a living as a deadly pathogen. In recent decades, it's become clear that horizontal gene transfer (Box 5.1) has been essential to the evolution of these new adaptations for billions of years (Hall et al. 2017).

Bacteria and archaea can acquire new genes by many routes. Viruses can move microbial genes from one host to another. Plasmids can be transferred. Some species

mutations will experience strong purifying selection. Mutations in a cis-acting element, by contrast, affect gene expression only in a single developmental context. These kinds of mutations will likely have less catastrophic effects on the resulting phenotype.

As a result, a gene's expression can evolve in different directions in different contexts. A mutation can block expression of a gene in one context while allowing it to be expressed in others. Insertions can add new cis-acting elements to a gene, recruiting it for new functions without altering its performance in other conditions. This relative independence allows, for example, our arms and legs to develop different shapes. Limb development genes not only can receive information that they are in a limb bud, but also can "sense" that they are in a limb bud that will become either an arm or a leg.

A **cis-regulatory region** is a portion of the gene control region located on the same chromosome as the gene. It often lies near the promoter regions of a gene, but it can also be located inside an intron or downstream of the gene.

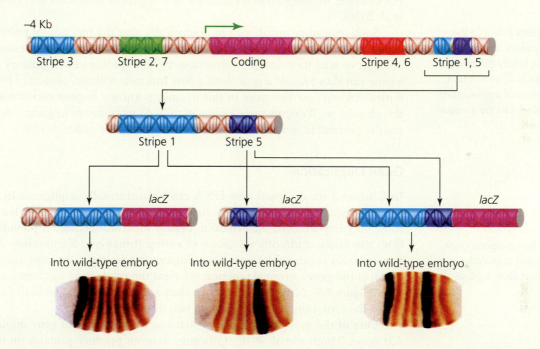

Box Figure 9.1.2 Regulation of gene expression is often compartmentalized. Specific cis-acting regulatory elements correspond with particular contexts or tissues in which the gene is expressed. For example, the cis-regulatory regions responsible for expression of the *Drosophila eve* gene in stripe 1 (yellow) are different from those responsible for stripe 5 (blue). When individual cis regions are coupled with a reporter gene (*lacZ*) and injected into *Drosophila* embryos, only cells in the corresponding stripe are affected (black staining). (Data from Gilbert 2007; Photos republished with permission of The Company of Biologists Ltd. from M. Fujioka, Y. Emi-Sarker, G.L. Yusibova, T. Goto, and J.B. Jaynes, "Analysis of an even-skipped rescue transgene reveals both composite and discrete neuronal and early blastoderm enhancers, and multi-stripe positioning by gap gene repressor gradients," Development 1999: 126, 2527–2538.)

can even take up loose DNA that spills out of dying microbes. Sometimes the new genes produce proteins that interfere with a microbe's existing biology and lower its fitness. But in other cases, the new genes provide a benefit.

In the most extreme of these cases, microbes can acquire entire sets of genes that together carry out an important function, such as breaking down complex molecules into metabolites. In other cases, individual genes can become incorporated into existing pathways, improving their performance or giving them capabilities. In a recent study on the evolution of *Escherichia coli* strains, Tin Yau Pang and Martin Lercher of the University of Dusseldorf estimated that 10% of the new adaptations and 19% of the strong improvements of *E. coli* were the result of horizontal gene transfer (Pang and Lercher 2015).

As we'll see in Chapter 18, the evolution of new adaptations through horizontal gene transfer poses a direct threat to our health. It's responsible for the rapid rise of new strains of bacteria that are resistant to almost every antibiotic known to science. And although horizontal gene transfer appears to be most common in bacteria and archaea, it has also had a significant influence on our eukaryote lineage. As we'll see in Chapter 15, a dramatic form of horizontal gene transfer took place some 2 billion years ago, when an entire bacterial cell took up residence inside another cell, giving rise to mitochondria.

Gene Recruitment

Many genes are tightly regulated by transcription factors and RNA molecules, ensuring that they are expressed only at certain times and in certain tissues (Box 9.1). The genetic control region for genes can mutate just as readily as the genes themselves. For example, mobile elements can insert new copies of enhancers near genes (Chuong et al. 2016).

Such regulatory mutations can dramatically alter the function of proteins without changing their structure. They can prevent a gene from being expressed in certain contexts or wire them into new genetic networks. Adding new regulatory elements to a gene can thus provide a gene with a new function without detracting from its other duties. Adding new functions in this manner is known as **gene recruitment**. Later in this chapter, we'll discuss some examples of gene recruitment in greater detail, including the proteins in our lenses and the venoms that make snakes deadly.

Gene recruitment refers to the co-option of a particular gene or network for a totally different function as a result of a mutation. The reorganization of a preexisting regulatory network can be a major evolutionary event.

Gene Duplication

In Chapter 5 we discussed how DNA can be accidentally duplicated. In some cases, these duplications will create new copies of genes. Gene duplication has turned out to be very important to the evolution of new functions because it provides an escape from the fitness trade-offs we discussed earlier (Innan and Kondrashov 2010). Once there are two copies of the same gene (known as **paralogs**), one copy can continue to perform the gene's original function, whereas the other copy can acquire new mutations (**Figure 9.5**). Natural selection can then act independently on both paralogs, leading to the evolution of two genes with two distinct functions.

A **paralog** is a homologous gene that arises by gene duplication. Paralogs together form a gene family.

One of the most striking demonstrations of the power of gene duplication is in our noses (Hughes et al. 2018). Olfactory neurons produce proteins on their surface to bind odor-bearing molecules in the air. The precise shape of an olfactory receptor determines which group of odorants it can detect. Scientists have identified 855 genes for olfactory receptors in the human genome. These genes are variations on a theme. In fact, scientists can organize them into a phylogeny of "gene families." This is exactly the pattern you'd expect from repeated rounds of gene duplication producing new olfactory receptors.

Figure 9.5 When genes duplicate, they can take on new functions. A mutation can affect the regulatory region, altering the gene's function or even halting it completely.

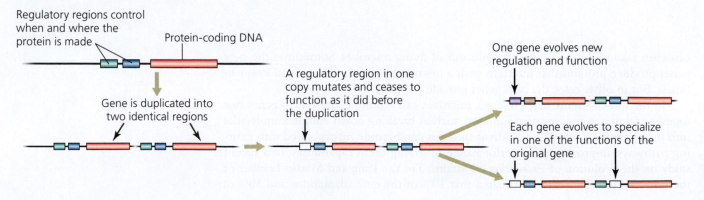

Regulatory regions control when and where the protein is made

Protein-coding DNA

Gene is duplicated into two identical regions

A regulatory region in one copy mutates and ceases to function as it did before the duplication

One gene evolves new regulation and function

Each gene evolves to specialize in one of the functions of the original gene

Even more remarkably, only 390 of the human olfactory receptor genes remain functional. The other 465 are pseudogenes, carrying mutations that prevent their successful expression. To get a better understanding of this evolutionary legacy, scientists have compared human olfactory receptor genes to those in other species. We share some functional genes in common with closely related species, along with some of our pseudogenes.

These patterns give us a fascinating glimpse into the long, dynamic history of gene duplication. When an olfactory receptor gene first duplicates, the two new copies still encode identical receptors. Over time, they may diverge, potentially allowing a species to tell more odors apart. This heightened sensitivity can increase the fitness of animals if they can, say, better detect the fragrance of fruit they eat or pick up the scent of a predator. But if these ecological conditions change, some olfactory receptors will stop providing these benefits. Now the animals pay no penalty for mutations that turn their genes into pseudogenes. Eventually, mutations that delete DNA can eliminate these pseudogenes altogether.

Compared to other mammals, we humans have a mediocre repertoire of olfactory receptors. Dolphins have the smallest repertoire of olfactory receptors, with only 156 genes in total, and just 58 that are functional. But other species have far more copies than we do. Elephants currently hold the record, with 4230 olfactory receptor genes in total, including 2514 functional genes and 1716 pseudogenes. This finding fits well with what scientists know about the elephant's keen sense of smell. Reportedly, elephants can lift their trunks, sniff the air, and smell ponds or rivers that are miles away. The role of gene duplication in evolution helps us understand the origin of this remarkable adaptation.

Promiscuous Proteins

In the case of olfactory receptor evolution, the genes duplicate first and new functions arise later. In other cases, the order is reversed. That's because many proteins can have two functions at once. Say, for example, that a protein cleaves one molecule. It may also be able to weakly cleave a second one as well. Scientists refer to such moonlighting molecules as **promiscuous proteins** (Khersonsky and Tawfik 2010; Andersson et al. 2015).

The promiscuity of proteins opens up several avenues for evolutionary change. If there's more benefit to a protein cleaving the second molecule than the first, natural selection may alter its specificity. Gene duplication can create two copies of a promiscuous protein, allowing each one to specialize on a different target. Alternatively, the two copies can undergo evolution in their regulation, recruiting them into new networks.

A **promiscuous protein** is a protein capable of carrying out more than one function, such as catalyzing reactions of different substrates.

Key Concepts

- Genes can acquire new functions through evolution by overcoming the trade-offs between old functions and new ones.

- Regulatory evolution can add genes into new genetic networks—a process known as gene recruitment.

- Gene duplication facilitates divergence in gene function because redundant copies can take on new and different tasks. Because they are "extra" copies, duplicated genes are released from purifying selection acting on the performance of the original gene. They are able to accumulate mutations faster and are often co-opted to serve new functions.

- Proteins are said to be promiscuous when they are able to perform functions other than their primary task. This ability can serve as a starting point for the evolution of novel traits if proteins that evolved primarily for one function become recruited to another function. ●

9.3 The Origin of New Adaptations in Microbes

Some of the most important insights into gene recruitment, gene duplication, and promiscuous proteins have come from experiments on bacteria. Here we will discuss two experiments that are particularly revealing when it comes to the evolution of new functions.

Promiscuous Proteins in Action

Joakim Näsvall and his colleagues at Uppsala University in Sweden ran an experiment to observe the evolution of promiscuous proteins in *Salmonella* (Näsvall et al. 2012). These bacteria have a number of genes for enzymes to build amino acids. The scientists altered a gene called *hisA*, which encodes an enzyme that produces histidine. The mutation that they introduced allowed the HisA enzyme to also help synthesize a second amino acid, tryptophan.

Although this promiscuous HisA enzyme could now do two jobs, it did neither of them well. If the scientists withheld histidine from these bacteria, they grew slowly because they struggled to make their own supply. To see how well the mutant HisA enzymes synthesized tryptophan, they disabled the gene that normally carries out this job. Now the bacteria could only make tryptophan very slowly.

The researchers then ran an evolutionary experiment on these mutant bacteria, similar to the one led by Richard Lenski that we discussed in Chapter 6. They predicted that if they grew the bacteria on a medium lacking both tryptophan and histidine, then natural selection would strongly favor mutations to improve their production of these amino acids. Just as they predicted, the bacteria began multiplying much faster. By sequencing the bacteria's DNA, the researchers were able to uncover the genetic basis of this adaptation.

In fewer than 50 generations, the bacteria gained extra copies of the mutant *hisA* gene. These extra copies allowed the bacteria to make more HisA enzymes, speeding up their creation of both amino acids. Then these new copies began to mutate. In some populations of the bacteria, one copy of the gene evolved an improved activity making tryptophan while becoming unable to synthesize histidine. Meanwhile, a second copy of the gene evolved in the opposite direction, making histidine but not tryptophan. Thus, in less than a year, natural selection generated a new gene adapted for producing tryptophan.

Rewiring Metabolic Networks

When Richard Lenski began his experiment on *E. coli* in 1988, he set up 12 genetically identical *E. coli* lines and fed them a meager diet of glucose. As the lines reproduced, they mutated and adapted to their new conditions.

But in one of those lines, something remarkable happened: a new trait evolved. The bacteria acquired the ability to eat in a new way.

The first sign of this remarkable event was in 2003, when Lenski and his students noticed that, overnight, one of the 12 flasks had become much cloudier than the others. In a microbiology lab, cloudiness is a surefire sign that the bacteria in a flask have experienced a population explosion.

At first, the team suspected that some other species of bacteria had slipped into the flask and was overgrowing the native *E. coli*. But they found that the bacteria making the medium cloudy was their own *E. coli* and not some accidental contamination—the flask population were descendants of the original ancestor that Lenski had used to start the experiment. Somehow, the bacteria in this one flask had evolved a way to grow more than the bacteria in the other 11 flasks.

The scientists determined that the strain of *E. coli* that evolved in the cloudy flask had found an extra food supply. It was feeding on citrate, a compound that was present in the medium. The scientists had added citrate to the medium to adjust its chemistry so that *E. coli* could take up iron more easily and grow faster. But initially the bacteria could not feed on citrate itself.

Normally, *E. coli* can feed on citrate only in the absence of oxygen. It's a strategy that they rely on when they can't use aerobic metabolism to grow on higher-quality sources of energy, such as glucose.

When *E. coli* senses that it is in an oxygen-free environment with citrate, it responds by expressing a set of genes known collectively as *citT*. The genes encode a protein that ends up in the membrane of the microbe. There, the protein pumps out a compound called succinate while it is pumping in citrate. *E. coli* can then harvest the energy in the citrate to power its metabolism. If the oxygen level in its environment rises again, *E. coli* silences the *citT* genes and stops feeding on citrate.

E. coli's inability to grow on citrate in the presence of oxygen is a hallmark of the species. For example, when microbiologists are trying to determine the species of bacteria causing an infection, they culture the microbes on a petri dish that contains only citrate. If bacteria start to grow on it, they know that they're dealing with a species other than *E. coli*. It was thus a surprise to Lenski and his students to find that the bacteria feeding on citrate were *E. coli*.

On closer inspection, the scientists realized that the *E. coli* had undergone a new adaptation. The ability to metabolize citrate in the presence of oxygen increased the fitness of bacteria compared to bacteria that lacked it. Christina Borland, a postdoctoral researcher in Lenski's lab, began investigating this adaptation, and the project was later taken over by Zachary Blount—then a graduate student in Lenski's lab and now a postdoctoral researcher there (Blount et al. 2012; Blount 2017). To chart the origin of the adaptation, the researchers first thawed out ancestors of the citrate-feeding bacteria.

For the first 31,000 generations after Lenski started the experiment, the bacteria could not grow on citrate. Then the citrate-feeding bacteria emerged. At first, they grew only slowly. Over the next few thousand generations, their performance improved gradually, and around generation 33,000, they suddenly dominated their population (**Figure 9.6**).

To identify the mutations responsible for citrate feeding, Lenski and his colleagues sequenced the genomes of representative generations along this lineage. Although they identified many candidate mutations, one in particular stood out. At around 31,500 generations, a crucial gene duplication event occurred.

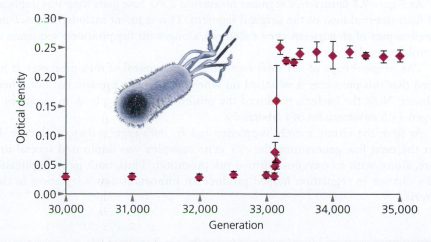

Figure 9.6 As Richard Lenski's bacteria adapted to the laboratory environment, one lineage evolved a new trait just after generation 33,000—the ability to feed on citrate. As the bacteria evolved to eat more citrate, they grew to greater numbers, which researchers identified because their flask grew cloudy. Optical density on this graph is a measure of this cloudiness. (Data from Blount et al. 2008)

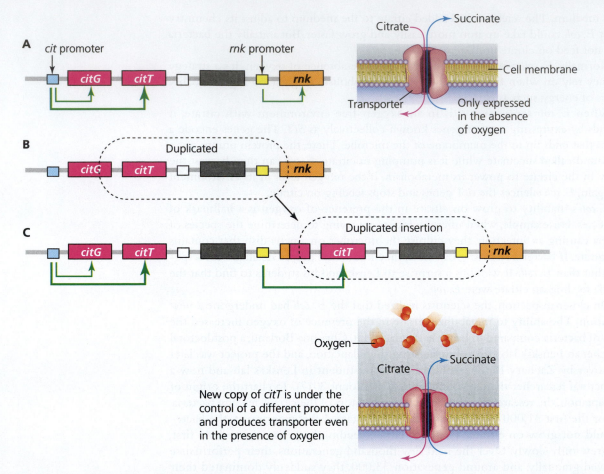

Figure 9.7 A key step in the evolution of citrate feeding in *E. coli* was an accidental duplication of a segment of its DNA. A: The segment contained a gene called *citT*. It encodes a transporter protein called *citT* that draws citrate into the cell by exchanging succinate and other small molecules. The gene is expressed only in the absence of oxygen. B: The region in the dashed line was duplicated. It included *citT* as well as the promoter region of an adjacent gene, *rnk*. C: The new copy of the DNA segment was inserted next to the old one. The duplicated copy of *citT* now came under the control of the duplicated *rnk* promoter, which expressed *citT* in the presence of oxygen. As a result, the bacteria could now feed on the citrate in the laboratory flask. (Data from Blount et al. 2012)

As **Figure 9.7** illustrates, a segment measuring 2933 base pairs long was duplicated and then inserted next to the original segment. This segment included the *citT* gene as well as part of an adjacent gene called *rnk*, along with the promoter sequence that controls expression of *rnk*.

The original copy of *citT* now came under the control of *rnk*'s promoter. It happened that this promoter is switched on when oxygen was present, but not when it is absent. Now the bacteria produced the proteins required to feed on citrate in the oxygen-rich environment of a laboratory.

At first, the citrate merely supplemented *E. coli*'s regular diet of glucose. But over the next few generations, the *citT* gene complex was duplicated several times more, along with its oxygen-sensitive *rnk* promoter. Thus, both gene duplications and a change in regulation helped produce an important new adaptation in these bacteria.

Key Concept

- Microbes are ideal organisms for use in examining the evolution of complex adaptations because they are so diverse and they reproduce rapidly. Studying these organisms gives scientists valuable insight into how the co-option of promiscuous proteins, gene duplication, and changes in gene regulation can lead to novel traits. ●

9.4 The History of Venom: Evolving Gene Networks in Complex Organisms

In animals and plants, gene networks produce different tissues and cell types that can interact in complex ways. The same factors that drive the evolution of new adaptations in microbes—such as promiscuous proteins, gene duplication, changes in gene regulation, and gene recruitment—also enable new adaptations to emerge in complex organisms as well.

Scientists generally can't observe the emergence of a new adaptation in an animal or a plant the way they can in bacteria. Instead, they infer the history of these adaptations by comparing the anatomy and genes of related species. That's just what Bryan Fry and his colleagues are doing to trace the evolution of snake venoms over the past 200 million years (Casewell et al. 2013; Fry 2015).

To start their investigation, the scientists isolated genes for venom in a wide range of snakes. They collected cells from the venom glands of snakes and determined which of their genes were active. About half of the active genes were ordinary housekeeping genes that carry out basic processes in all cells. The remaining active genes encoded venom molecules. Fry and his colleagues sequenced these venom genes and compared them to figure out how they evolved from ancestral genes.

Each species produced its own distinctive cocktail of venoms, but each venom gene typically showed a close kinship with venom genes in closely related snakes. The pattern suggested that venomous snakes inherited genes for venom from a common ancestor—but, after the lineages diverged, the venom genes were shaped differently by natural selection.

Fry and his colleagues reconstructed the history of these venom genes with gene trees. **Figure 9.8** shows one of these trees for a muscle-destroying venom called crotamine. Fry compared crotamine genes in snakes with those in different species, as well as with other genes in snakes and other vertebrates. The tree reveals that the closest relatives of crotamine genes are defensin genes, which are expressed in the pancreas of

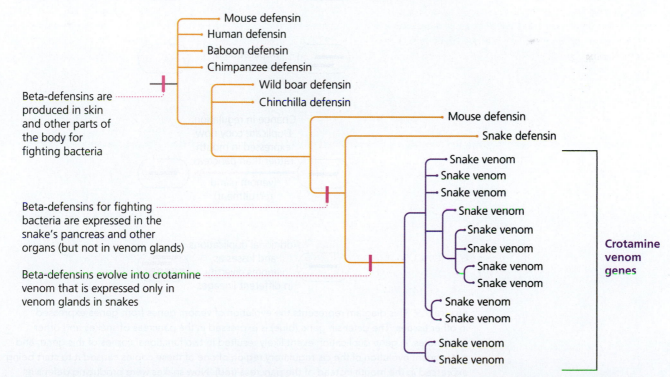

Figure 9.8 Bryan Fry and his colleagues sequenced genes for a venom called crotamine in snakes and compared them to homologous genes in other animals. This evolutionary tree shows how crotamine genes are closely related to beta-defensins, the bacteria-fighting molecules found in many vertebrates. (Data from Fry 2005)

snakes. Snakes use defensins to fight infections. Pigs, mice, and humans make defensins for the same purpose.

These results support the hypothesis that the defensins originally evolved in a common ancestor of snakes and mammals. As new lineages of animals split off from that ancestor, they inherited that defensin gene. Gene duplication produced extra copies of the gene, which became specialized for attacking different pathogens in the pancreas.

The ancestors of today's snakes inherited those duplicated genes, and one of them experienced a regulatory mutation that affected where the protein was made. Instead of producing it in the pancreas, snakes began to produce defensin in their mouths. When the snakes bit their prey, they now released this defensin into the wound. Fry proposes (Figure 9.9) that further mutations to the duplicated defensin gene changed its protein's shape so that it began to take on a new function. Instead of fighting pathogens, it began to damage muscles. Further mutations made defensin increasingly deadly, and subsequent gene duplications eventually gave rise to an entire family of these venom genes.

Fry and his colleagues have drawn similar trees for more than two dozen venom genes. Some evolved from genes expressed in the heart, whereas others were expressed initially in the brain, in white blood cells, and in many other places in the snake body (Figure 9.10). An Australian snake known as the inland taipan produces a venom that causes its victims to "black out," for example. The gene for this venom evolved from proteins that slightly relax the muscles around the heart. Once these proteins evolved into venom, this response changed from a slight relaxation to a rapid drop in blood pressure. In each case, venoms evolved through gene duplication, changes in gene

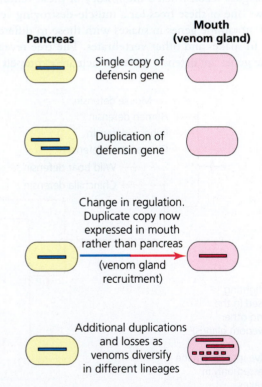

Figure 9.9 This diagram represents the evolution of venom genes from genes expressed in other tissues. The defensin gene (blue) is expressed in the pancreas of snakes and other vertebrates. A gene duplication event likely resulted in two functional copies of this gene, and subsequent evolution of the cis-regulatory region of one of these copies caused it to start being expressed in the mouth instead of the pancreas (red). Now snakes were producing defensins in both their mouth and their pancreas, and defensin proteins began to function as venoms. A novel trait, venom, arose from the expression of an existing gene in a new part of the body. (Data from Casewell et al. 2013)

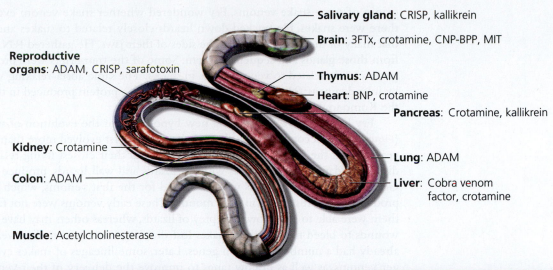

Salivary gland: CRISP, kallikrein

Brain: 3FTx, crotamine, CNP-BPP, MIT

Reproductive organs: ADAM, CRISP, sarafotoxin

Thymus: ADAM

Heart: BNP, crotamine

Pancreas: Crotamine, kallikrein

Kidney: Crotamine

Colon: ADAM

Lung: ADAM

Liver: Cobra venom factor, crotamine

Muscle: Acetylcholinesterase

Snake toxin	Effects
3FTx	Neurotoxin
Acetylcholinesterase	Disruption of nerve impulses, causing heart and respiratory failure
ADAM	Tissue decay
BNP	Acute low blood pressure
CNP-BPP	Acute low blood pressure
Cobra venom factor	Anaphylactic shock
CRISP	Paralysis of peripheral smooth muscle, hypothermia
Crotamine	Muscle decay and neurotoxicity
Kallikrein	Acute low blood pressure, shock, destruction of blood-clotting factors
MIT	Constriction of intestinal muscles, resulting in cramping, increased perception of pain
Sarafotoxin	Acute high blood pressure

regulation leading to gene recruitment, and mutations that, through natural selection, fine-tuned the genes themselves.

The venom-delivery system in a taipan is, of course, more than just the venom genes. It also includes the venom glands and the fangs for delivering the venom deep into the flesh of its prey. It may be hard to imagine how this complex system evolved because all the parts seem to depend on each other. Yet when Fry looked closer at the venom system, he made a major discovery about how it evolved.

Fry wondered if gene recruitment for venom production took place independently in each lineage of venomous snakes, or if some venom genes evolved before their lineages split and before the evolution of complex venom-delivery systems. He discovered that some venom genes are quite ancient in origin, having evolved in the common ancestor of *all* snakes—even snakes not previously recognized as venomous. Garter snakes, for example, lack hollow fangs and high-pressure delivery systems. But Fry discovered that these relatively harmless snakes make some of the same venoms that are found in rattlesnakes, and those venoms are just as potent, molecule for molecule. Garter snakes may not have fangs, but they can use their tiny teeth to puncture a frog's delicate skin, through which the venom can enter their prey's body. The venom of garter snakes had gone undiscovered until Fry's investigation simply because humans didn't get hurt by their bite.

Fossil evidence and studies on reptile DNA indicate that snakes evolved about 60 million years ago (Yi 2017). Their closest living relatives include iguanas and monitor lizards, such as the Komodo dragon, the biggest lizard alive today. Given the ancient

Figure 9.10 Venom genes have been recruited from genes expressed in many organs of snakes.

origin of some snake venoms, Fry wondered whether snake venom evolved before there were snakes. He tracked down lizards closely related to snakes and discovered that many of them had glands on the sides of their jaws. He gathered RNA transcripts from those glands and sequenced them. Some of the transcripts were produced from genes closely related to venom genes in snakes. The same aorta-relaxing venom made by inland taipans, for example, is closely related to a protein produced in the mouth of the Komodo dragon.

Fry has proposed a surprising new hypothesis for the evolution of snake venom (**Figure 9.11**). According to this hypothesis, venom first evolved more than 200 million years ago in the common ancestor of snakes and their closest living relatives. (More distantly related lizards, such as whiptail lizards and wall lizards, produce no venom.) Genes with other functions were recruited for the first venoms, which early lizards produced in mucus glands in their mouths. These early venoms were not fatal. Some of them were able to slow down the prey of lizards, whereas others may have caused their wounds to bleed more. When snakes lost their legs, about 60 million years ago, they already had a number of venom genes. Later, some lineages of snakes evolved stronger venoms, as well as hollow fangs to improve the delivery of their venom. Rather than simply slowing down their prey, now the snakes' venom could kill them outright.

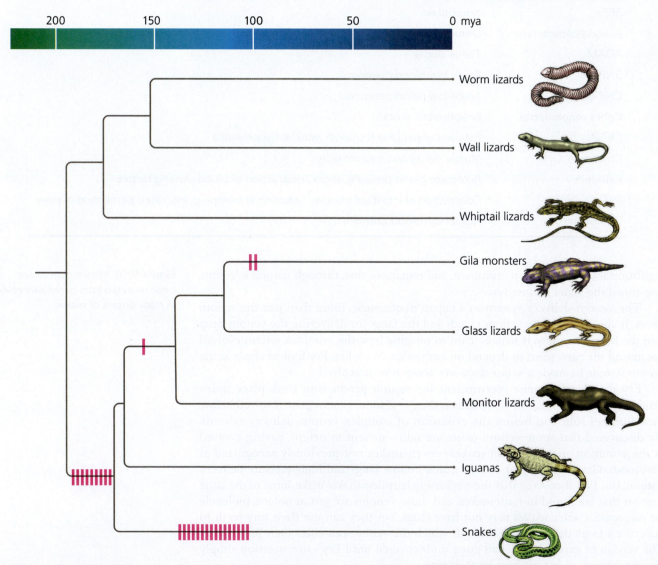

Figure 9.11 Fry and his colleagues discovered that some lizards have genes closely related to snake venom genes. Each pink bar shows when one of these genes evolved. This discovery suggests that the complex venom system in snakes began evolving millions of years before snakes evolved. (Data from Fry et al. 2006)

What looks to us today like a complex adaptation made up of parts that can't work on their own was actually assembled, bit by bit, over many millions of years.

If Fry is right, there's an unexpected answer to the question, "Which came first, the venom or the fang?" The answer is the venom. In fact, the venom may have come even before the snake.

Key Concept

- Many novel traits (such as snake venoms) seem so complex that it can be hard to imagine how the individual components could have evolved. But research has shown they can evolve through a series of duplication events and co-option of proteins originally involved with other body functions. The complex adaptations we see today are the result of the combination of these processes with natural selection and time. ●

9.5 The Genetic Toolkit for Development

As we saw in Chapter 2, Charles Darwin recognized that the emergence of many new adaptations involved changes in the development of embryos. A bat, a human, and a seal all share a common mammalian ancestor. The forelimbs of that ancestral mammal diverged in each lineage so that they developed into wings, hands, or flippers.

Darwin based this inference on the homology of mammal forelimbs. But when later scientists compared mammals to more distantly related animals such as insects or worms, it became harder to identify homologous structures. Insect legs, for example, are drastically different from our own. They lack internal skeletons, and their muscles and blood are located within an exoskeleton. The main nerve of a fly's body runs along the bottom of its abdomen, and the main structures of its digestive system run along the top of its back—the reverse of the arrangement in mammals. A fly feeds with three pairs of segmented mouthparts, and its eyes are made up of many hexagonal columns, each capturing an image of a tiny fragment of its surroundings. Neither organ has any counterpart among mammals.

In the late 1900s, however, a crucial new source of information about animal development emerged. Scientists finally began to pinpoint some of the genes that guide the development of an embryo from a fertilized egg. They started out by studying the best-understood model organisms, such as mice, chickens, and flies. The scientists were shocked by what they found. The same underlying network of genes governed the development of all of these model organisms, no matter how different they looked. This discovery led to a new understanding of how adaptations in animals evolved (Panganiban et al. 1997; Shubin et al. 1997, 2009). Here we will focus on how developmental genes play a role in animal evolution. In **Box 9.2**, we discuss some striking similarities with the evolution of plants.

Same Genes, Different Bodies

As we saw in Figure 9.3, *Hox* genes assign different developing sections of the body of a fly along the head-to-tail (anterior–posterior) axis to different body parts. (See **Figure 9.12** for a guide to axes of animal bodies.) Surprisingly, mice have a related set of *Hox* genes—four

Figure 9.12 This diagram shows some of the names that biologists use to refer to different axes of an organism's body.

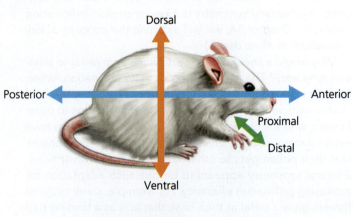

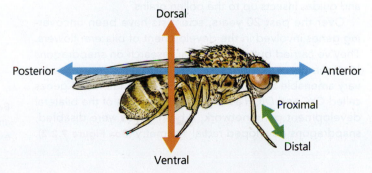

BOX 9.2

Evolution of Plant Morphology

Elsewhere in this chapter, we consider the evolution of complex adaptations such as eyes and limbs in animals. Plants have their own suite of complex adaptations, too, such as leaves and flowers.

The parallels between animals and plants are deep: animals and plants both develop from individual cells, growing into multicellular bodies made up of many tissues and cell types. As in animals, networks of genes guide the development of structures in plants. And scientists are also finding that evolution of adaptations in plants has some striking similarities to animal evolution. Evolution produces new plant forms through the tweaking or redeploying of ancient, conserved regulatory networks (Shepard and Purugganan 2002; Friedman et al. 2004; Della Pina et al. 2014).

Thanks to the gene hierarchy that controls plant development, changes to a limited number of genes can trigger large phenotypic changes. Flowers, for example, typically produce one of two arrangements of petals. Some flowers produce petals like the spokes on a wheel—a form known as radial symmetry. Other flowers look as if their two sides are reflections in a mirror—this is known as bilateral symmetry (**Box Figure 9.2.1**). Phylogenetic studies have revealed that the first flowering plants had radial symmetry. Bilateral symmetry evolved from radial symmetry at least 70 separate times (Hileman 2014). In most cases, this transition led to a greater number of species in the bilateral clade than in its radial sister clade. This overall increase in diversity suggests that bilateral symmetry is a key innovation in flowering plants. (In Chapter 14, we will examine the concept of key innovations in more detail.)

Why would a lineage of flowers switch from radial to bilateral symmetry? The answer appears to be pollination. When birds or insects visit flowers to feed on nectar, they become dusted with pollen. They can then deliver that pollen to other flowers, where it can fertilize the plant's ovules. Flowers have evolved many adaptations to lure pollinators and to ensure that their pollen gets to other flowers of their own species. Bilateral symmetry appears to be one such adaptation for increasing pollination efficiency. For example, some bilateral flowers grow a petal at their base that acts as a landing pad and guides insects up to the pollen grains.

Over the past 20 years, scientists have been uncovering genes involved in the development of bilateral flowers. They've carried out most of their research on snapdragons (*Antirrhinum majus*) because that species has proven to be very amenable to lab work. Scientists have identified genes called *CYC* and *DICH* as important regulators of the bilateral development gene network. If both genes were disabled, snapdragons developed radial symmetry (**Box Figure 9.2.2**).

Box Figure 9.2.1 Some flowers have radial symmetry—the petals radiate from the center like spokes in a wheel. Others have bilateral symmetry, meaning the petals are arranged in a mirror image.

Subsequent studies revealed that *CYC* and *DICH* are members of a gene network that sets up the left-right symmetry of snapdragon flower petals and also differentiates the upper and lower petals on each side of the divide.

Once scientists had deciphered snapdragon genetics, they wondered how other bilateral flowers controlled their development. Not surprisingly, they've found that the *CYC/DICH* pathway controls symmetry in other lineages as well. This research suggests that during the evolution of flowering plants, evolution has recruited the same pathway again and again to produce the same phenotype (**Box Figure 9.2.3**).

Scientists have also documented a number of cases in which bilaterally symmetrical flowers have evolved back to radial symmetry. In those lineages, the *CYC/DICH* pathway is no longer being expressed (Hileman 2014). Reversals like these elegantly confirm the importance of this pathway in the evolution of flowers.

Box Figure 9.2.2 A: In snapdragons, two genes known as *CYC* and *DICH* control bilateral development. B: Mutants that don't express either gene develop radial symmetry. (Enrico Coen)

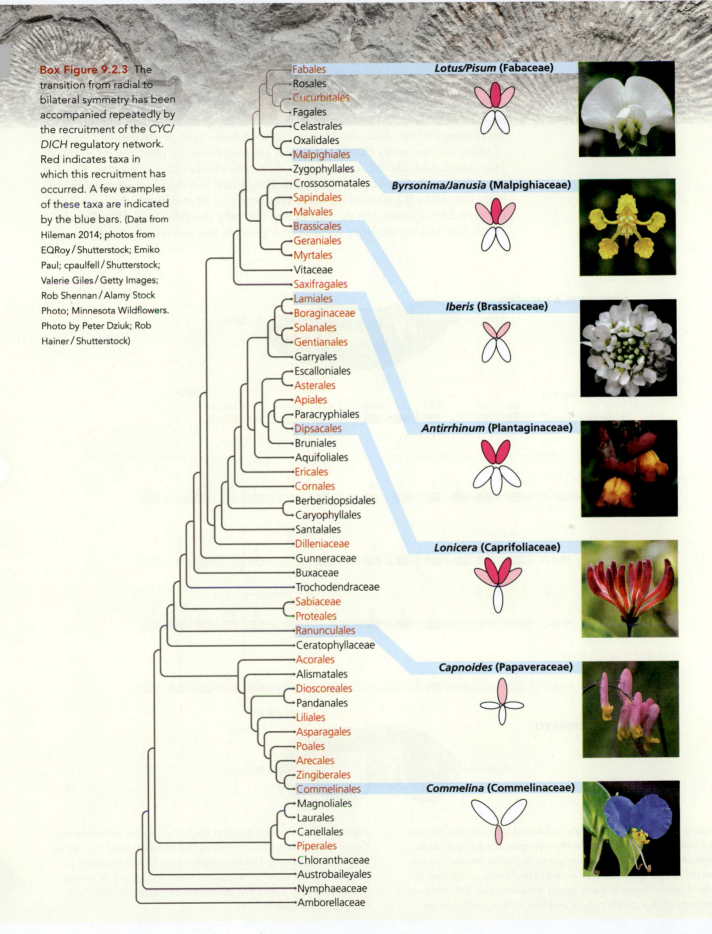

Box Figure 9.2.3 The transition from radial to bilateral symmetry has been accompanied repeatedly by the recruitment of the *CYC/DICH* regulatory network. Red indicates taxa in which this recruitment has occurred. A few examples of these taxa are indicated by the blue bars. (Data from Hileman 2014; photos from EQRoy/Shutterstock; Emiko Paul; cpaulfell/Shutterstock; Valerie Giles/Getty Images; Rob Shennan/Alamy Stock Photo; Minnesota Wildflowers. Photo by Peter Dziuk; Rob Hainer/Shutterstock)

Fabales
Rosales
Cucurbitales
Fagales
Celastrales
Oxalidales
Malpighiales
Zygophyllales
Crossosomatales
Sapindales
Malvales
Brassicales
Geraniales
Myrtales
Vitaceae
Saxifragales
Lamiales
Boraginaceae
Solanales
Gentianales
Garryales
Escalloniales
Asterales
Apiales
Paracryphiales
Dipsacales
Bruniales
Aquifoliales
Ericales
Cornales
Berberidopsidales
Caryophyllales
Santalales
Dilleniaceae
Gunneraceae
Buxaceae
Trochodendraceae
Sabiaceae
Proteales
Ranunculales
Ceratophyllaceae
Acorales
Alismatales
Dioscoreales
Pandanales
Liliales
Asparagales
Poales
Arecales
Zingiberales
Commelinales
Magnoliales
Laurales
Canellales
Piperales
Chloranthaceae
Austrobaileyales
Nymphaeaceae
Amborellaceae

Lotus/Pisum (Fabaceae)

Byrsonima/Janusia (Malpighiaceae)

Iberis (Brassicaceae)

Antirrhinum (Plantaginaceae)

Lonicera (Caprifoliaceae)

Capnoides (Papaveraceae)

Commelina (Commelinaceae)

sets of them, to be exact (**Figure 9.13**). Each set of *Hox* genes is arrayed along a mouse chromosome in the same order in which it is expressed from head to tail in the body. So are the genes in the fly genome (Mallo 2018).

Our understanding of the similarities between the *Hox* genes in flies and in mice has increased since scientists first began studying them. In 2004, for example, David Bartel of the Massachusetts Institute of Technology (MIT) and his colleagues discovered microRNA molecules that map to the *Hox* gene cluster in mice. These RNA molecules keep some of the other *Hox* genes shut down in certain regions of the developing embryo. Flies also have microRNA molecules nearly identical to those of mice, and the genes that encode them are located in the same region in the fly *Hox* cluster. And, like the mouse RNAs, fly RNAs silence other *Hox* genes. The *Hox* genes function so similarly in mice and in flies, in fact, that their sequences are literally interchangeable. If a scientist silences a *Hox* gene in a fly and inserts the corresponding gene from a mouse, the fly will develop normally (Bachiller et al. 1994; Lutz et al. 1996). The striking similarity between *Hox* genes in flies and mice indicates that they

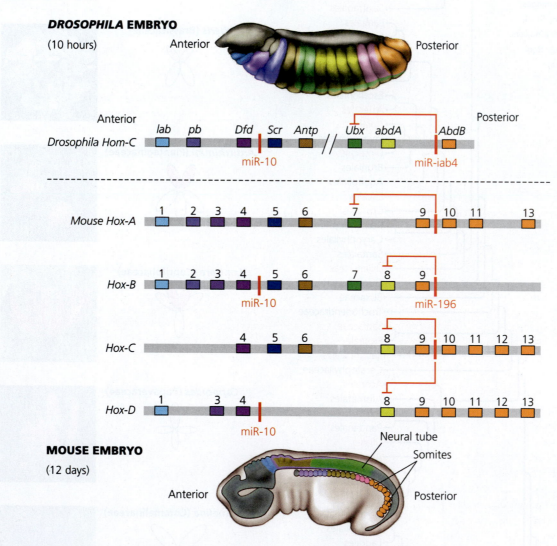

Figure 9.13 Flies and mice are separated by more than 570 million years of evolution, yet the development of each species is controlled by homologous *Hox* genes. In the vertebrate lineage, the entire *Hox* gene cluster was duplicated twice, producing four sets of genes. Some of these genes were later lost. Still, the overall similarity of *Hox* genes in mice and flies is clear. Both animals even have homologous genes in the same location within the *Hox* cluster that encode RNA molecules that regulate other *Hox* genes (marked here as miR). The best explanation for this structural similarity is that the common ancestor of flies and mammals already had a set of *Hox* genes that controlled development. (Data from Gilbert 2007; De Robertis 2008)

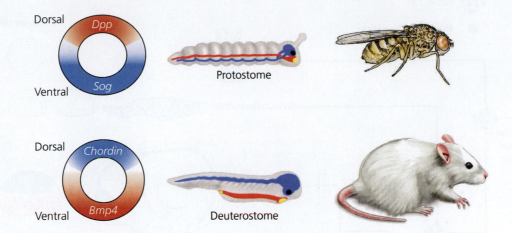

Dorsal
Dpp
Ventral
Sog

Protostome

Dorsal
Chordin
Ventral
Bmp4

Deuterostome

inherited this set of genes from a common ancestor, only later adapting it for building different structures from head to tail.

An equally intriguing pattern emerges when we look at the dorsal–ventral axis of mice and flies. In a developing mouse, cells along the belly (the ventral side) express the gene *Bmp4*. Flies express a homologous gene called *Dpp*, but they express it along the back (the dorsal side; **Figure 9.14**). The side of the body that produces proteins from *Bmp4/Dpp* becomes the site of the digestive system. The side without it is where the main axis of the nervous system develops (De Robertis 2008). In other words, flies and mice use the same developmental genes to mark where their nervous system will develop, but they do so in a mirror image of each other.

Deep History of the Genetic Toolkit

As scientists have found more similarities between the developmental genes in animals as different as mice and flies, they've wondered what their common ancestors looked like. Did the common ancestor of mice and flies have a nerve cord running from its head to tail, for example, patterned by *Hox* genes? If so, did it run along the dorsal or ventral side of its body?

These questions cannot be answered solely by studies on mice and flies, however. These animals belong to just two twigs on the animal tree of life. Recently José M. Martín-Durán, Kevin Pang, and their colleagues at the University of Bergen in Norway took a far broader look at the animal kingdom (Martín-Durán et al. 2018). Their findings are summarized in the evolutionary tree shown in **Figure 9.15**.

Mice and flies belong to a large group of animal species called bilaterians. They're named for having left and right sides to their bodies; in addition, they also have clearly defined anterior and posterior ends of their bodies. The ancestors of bilaterians split off from the ancestors of cnidarians, the group that includes jellyfish and corals. Cnidarians have a radial symmetry to their bodies, rather than a left–right symmetry. Instead of a nerve cord as found in mice or flies, their nervous system is a diffuse net.

The deepest branch of bilaterians is a group of little-known ocean-dwelling worms called xenacoelomorphs. They express *Hox* genes along their anterior–posterior axis. They also express *BMP* along their dorsal side, like flies do. Whereas some xenacoelomorphs have a nerve cord, many species produce nerve nets like cnidarians.

Among bilaterians, a nerve cord is present in mice and other vertebrates. Insects, as well as their relatives, also have one. So do some species of a group called annelids, which includes earthworms. But Martín-Durán, Pang, and their colleagues failed to find any dorsal–ventral patterning in the closest relatives of annelids, such as mollusks and flatworms. Even some annelids lacked the patterning.

Taken together, these findings hint that the ancestor of bilaterians still had a diffuse net of nerves. In three separate lineages, these nerve nets evolved into nerve cords

Figure 9.14 Left: A simplified cross section of fly and mouse embryos. Flies and mice use homologous genes for dorsal-ventral patterning. (*Bmp4* is homologous to *Dpp*, and *Chordin* is homologous to *Sog*.) But the genes are expressed in an opposite, mirrorlike pattern. Middle: These expression patterns lead to the development of a nerve cord on the dorsal side of the mouse embryo and the gut on the ventral side. Flies develop in an opposite pattern.

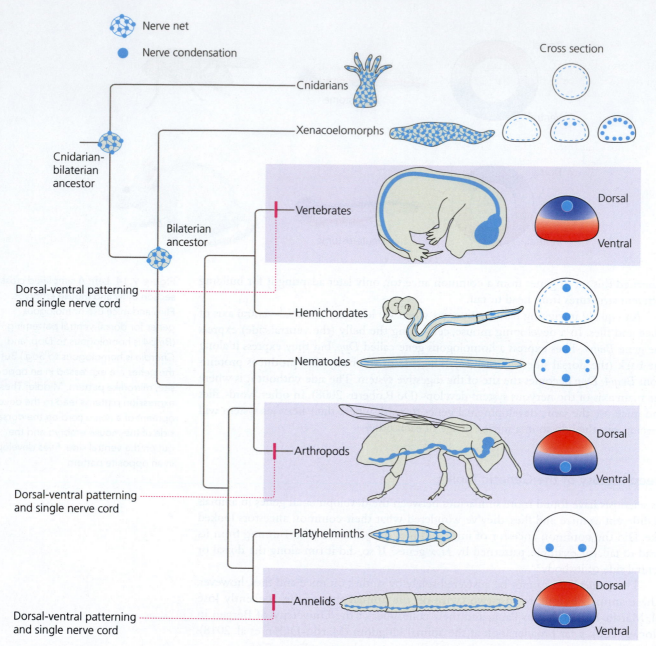

Figure 9.15 The phylogeny of animals sheds light on how the genetic toolkit was deployed in animals. Early animals likely had nervous systems arranged in a diffuse net, much like living jellyfish and other cnidarians. A large clade of animals, known as bilaterians, is characterized by left-right symmetry. The common ancestor of bilaterians may have had a nerve net, judging from the fact that xenacoelomorphs still do. It's possible that animals evolved a single nerve cord in at least three independent lineages. In each case, evolution recruited the same genetic toolkit for its development. (Republished with permission of Nature Publishing Group, from "Convergent evolution of bilaterian nerve cords", José M. Martín-Durán et al. Nature. 2018 Jan 4;553(7686):45–50, permission conveyed via Copyright Clearance Center, Inc.)

running the length of the body. In each case, evolution recruited the same *BMP/Dpp* patterning system to lay down the coordinates where the nerve cord would develop. Evolution, in other words, picked out the same tool from an ancestral toolkit over and over again.

Key Concept

- *Hox* genes and other patterning genes participate in regulatory networks that demarcate the geography of developing animals, determining the relative locations and sizes of body parts. These networks make up an ancient genetic toolkit inherited by all animals with bilateral body symmetry. •

9.6 The Deep History of Limbs

Limbs provide one of the best case studies for the way the shared genetic toolkit has been deployed again and again, generating diversity over hundreds of millions of years.

The Origin(s) of Limbs

Today, bilaterian appendages come in a vast range of forms—from lobster claws to butterfly wings to horse hooves. They develop in radically different ways. A mouse leg, for example, begins as a bulge known as a limb bud, which grows steadily outward from the body, creating the bones and other parts of the limb along the way. The mouse leg first builds the proximal structures like the femur, and only later forms the most distal elements like the toes and claws. A fly leg, in contrast, starts out as a disk, which unfurls only when the fly molts from a larva to a pupa. Unlike the bones of a mouse leg, all of the segments in the fly's leg develop at the same time. The fly's femur forms as the outermost of a series of concentric rings in the disk; the tibia and tarsi are stacked closer to the center. The "bull's-eye" of the disk of cells corresponds to the distal-most tip of the leg—the part that will extend the farthest out from the body once the structure has been unfolded.

Yet despite these differences in growth and form, mouse legs and fly legs develop in response to the same network of regulatory genes (Panganiban et al. 1997; Shubin et al. 1997; Tabin et al. 1999). Not only are the gene sequences similar to each other, but the roles of these genes, and the ways they interact with each other, are the same (**Figure 9.16**).

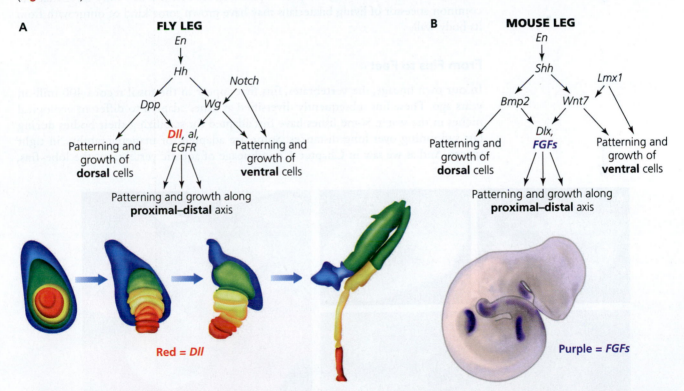

Figure 9.16 A, top: This diagram shows the hierarchical patterning network of genes involved in fly limb development. It is responsible for delineating anterior-posterior, dorsal-ventral, and proximal-distal axes within the developing limb and for initiating and controlling growth of the structure. A, bottom: The disk of cells that will form a fly leg resembles a bull's-eye with concentric rings mapping to segments in the final appendage. Cells expressing the patterning gene *Dll* are shown in red. B, top: Orthologous genes in mice form a patterning network. Many of these genes have different names (such as *Hh*, *Shh*, *Dpp*, and *Bmp2*) because they were discovered independently by researchers studying mice, flies, nematodes, or other taxa. Only relatively recently did it become apparent that many developmental genes shared striking similarities in structure, regulation, and function across taxa that could be attributed to their derivation from genes in a common ancestor. B, bottom: In this image of a mouse embryo, cells that express *FGFs* appear in purple. These cells will give rise to limb buds.

An **ortholog** is one of two or more homologous genes separated by a speciation event (as opposed to paralogs—homologous genes, produced by gene duplication, that are both possessed by the same species).

For example, the *Engrailed* (*En*) gene is crucial for defining the posterior portion of the limb bud in mice (the portion of the limb that is naturally oriented toward the back of the animal). Its **ortholog** in flies has an identical role in the leg disc. In flies, engrailed proteins turn on the expression of the signaling gene *Hedgehog* (*Hh*) in these posterior cells. The Hh protein then diffuses into a strip of cells, in the adjacent portion of the disc, that lack the En protein. This combination of signals—the high concentrations of Hh and an absence of En—causes these cells to begin expressing two additional signaling genes, *Decapentaplegic* (*Dpp*) and *Wingless* (*Wg*).

The *Dpp* gene is expressed in the dorsal part of this strip of cells, and *Wg* is expressed in the ventral portion. Together, these signals interact to initiate expression of yet another tier of transcription factors (including *Distalless*, *Dll*, and the *Epidermal Growth Factor Receptor*, *EGFR*). This cascade of genetic signaling delineates a clear pattern within the developing limb. These signals map out the anterior-posterior, dorsal-ventral, and proximal-distal axes of the forming structure.

The same genetic cascade unfolds in the developing limbs of mice. It involves the same homologous set of genes and the same gene–gene interactions. Thus, despite superficial differences, insect and mammal limbs have a deep similarity, and this similarity reflects common ancestry and homology of the developmental regulatory gene networks. Mutations that block this gene network lead to birth defects in which babies are born without limbs (Figure 9.17).

The finding that two distantly related bilaterians use the same genetic network to develop limbs suggests that the common ancestor of all bilaterians already had this network. But it would be wrong to assume that it used that network to grow limbs in particular. A number of different outgrowths, including butterfly wings and gill arches, also use the same development pathway. At best, we can only infer that the common ancestor of living bilaterians may have grown some kind of outgrowth from its body wall.

From Fins to Feet

In our own lineage, the vertebrates, fins first appear in the fossil record 400 million years ago. These fins subsequently diversified as fishes adapted to different ecological niches in the water. Some fishes have fins adapted for stabilizing their bodies during fast swimming over long distances. Some are adapted for maneuverability in tight spaces. And as we saw in Chapter 4, one lineage of aquatic vertebrates, the lobe-fins,

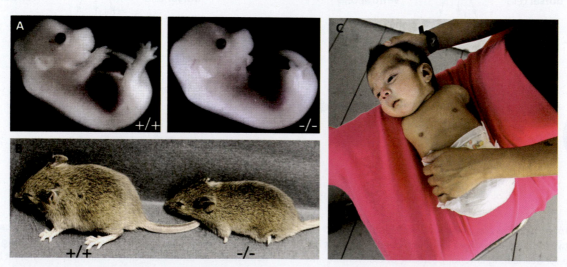

Figure 9.17 Blocking the expression of *Sonic hedgehog* (*Shh*), one of the three vertebrate orthologs of the insect gene *Hh*, can truncate activity of the limb-patterning pathway and block growth of the limbs. Mutant mice develop with only rudimentary legs (A, B) (Sagai et al. 2005). Similar interruption to the limb pathway may contribute to limbless mutations in humans (C). (A, B: Permission from "Elimination of a long-range cisregulatory module causes complete loss of limb-specific Shh expression and truncation of the mouse limb." T. Sagai, M. Hosoya, Y. Mizushina, M. Tamura, T. Shiroishi, Development 2005 132: 797–803; C: Taylor Jones / Palm Beach Post / ZUMA Press)

evolved stout appendages that eventually evolved into the limbs that tetrapods use to walk on land.

Fossils can provide crucial clues to this tremendous transition from sea to land. But the development of fins and limbs in living vertebrates can also add additional pieces to our understanding. Early in their development, all tetrapod limbs and fish fins are remarkably similar. They start out as buds bulging out from the body wall.

Later, their development diverges. In fishes, mesodermal tissue produces a small cluster of skeleton bones that develops near the base of the fin. A set of fin rays, produced by ectodermal tissues, grows into a large, stiff flap.

In tetrapods, by contrast, the skeletal bones derived from the mesoderm become much larger, forming the long bones of the limb. Later, the smaller bones of wrists, ankles, and digits develop. Tetrapods develop no fin rays at all.

Remarkably, tetrapods and fishes use *Hox* genes to produce these different phenotypes (Nakamura et al. 2016). In both cases, many of the same genes that help determine the anterior-posterior axis of the entire body are expressed again in developing appendages. Although the genes are the same in both fishes and tetrapods, the timing and location of their expression differ.

These different developmental paths generate a hypothesis about the evolution of tetrapod limbs from fins. It's possible that an important element in this transition was a shift in the regions dedicated to different tissues in the developing bud. As tetrapod limbs evolved, the region where bones developed expanded. Meanwhile, the region where fin rays developed shrank.

Over the past 20 years, scientists have begun identifying some of the genes that set up the patterns of development in fish fins and tetrapod limbs. Many of the genes in the genetic toolkit, such as *Hox* genes, are expressed in the developing appendages of both fishes and tetrapods (**Figure 9.18A**). *Hox* genes establish zones where bones, fin rays, and other elements of the appendages will develop.

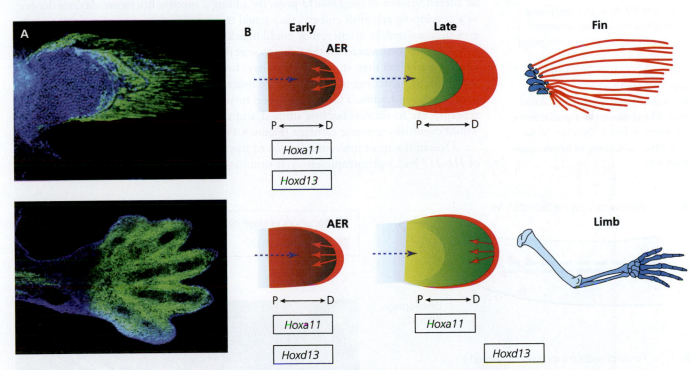

Figure 9.18 A: The expression patterns of *Hoxd13* in embryonic limbs of a zebrafish and a mouse, shown here, share some striking similarities (Nakamura et al. 2016). B: In fish fins, domains of expression of *Hoxa11*, and *Hoxd13* overlap. The region known as the apical ectodermal ridge (AER) stimulates distal growth of the limb bones in the mesodermal tissue (blue shading). When expression of the *Hox* genes ceases, fin rays begin to grow from ectodermal tissue (red shading). In tetrapod limbs *Hox* genes become active a second time later in development, and their respective domains of expression no longer overlap (boxes). This modified pattern of expression leads to a longer period of bone growth, rather than the development of fin rays. This suggests that the evolution of tetrapod limbs involved changes to the timing and location of expression of *Hox* genes. (A: Neil Shubin; data from Freitas et al. 2012, 2014)

One *Hox* gene in particular, known as *Hoxd13*, proved especially important in the evolution of limbs. Scientists found striking differences in its expression in fish fins and tetrapod limbs. In the developing fish fin, *Hoxd13* produces proteins along the outer rim early in development. The proteins bind to other genes and switch them on, triggering a cascade of activity—some genes are expressed, others repressed—that helps guide the development of the fin. The *Hoxd13* gene then becomes silent in fish embryos.

The *Hoxd13* gene also becomes active in tetrapods during the early development of limbs. As in fishes, its expression also subsides. But after a few days of development, cells on the outer edge of the limb bud start to express *Hoxd13* yet again. This second wave of *Hoxd13* marks a new set of coordinates that guides the development of digits and wrist bones (**Figure 9.18B**).

Scientists proposed that this development process might be an important clue to how limbs evolved. Did mutations in our tetrapod ancestors cause *Hoxd13* to turn on again late in development? And could this new expression pattern have added new structures to the ends of fin buds—structures that would eventually evolve into the hands of primates?

If these mutations did in fact occur in the ancestors of tetrapods, it happened more than 370 million years ago, during the age of lobe-finned fossils with tetrapod-like limbs. And yet this hypothesis is testable today. If it were true, it would mean that our finned ancestors already had some of the genetic tools needed to develop a primitive hand. The only evolutionary change required would be to alter the timing and location of the expression of some of the genes. If a scientist could artificially alter the genes in a living fish, it might be possible to produce a fin that looked more like a tetrapod limb.

At Universidad Pablo de Olavide in Spain, Renata Freitas and her colleagues set out to try to unlock that potential (Freitas et al. 2012). They engineered zebrafish with an altered version of the *Hoxd13* gene. By adding a specific hormone, dexamethasone, to a developing zebrafish embryo, they could trigger the expression of the gene whenever they wanted. In essence, they could hijack the genetic cascade.

The scientists waited for the fishes to start developing their fin. The fishes expressed *Hoxd13* early and then stopped. A few days later, the scientists added the hormone to the embryos, switching on *Hoxd13* once more. In this way, they mimicked what happens in tetrapod limbs. This manipulation turned out to have a dramatic effect on the zebrafish fin. Its fin rays became stunted, and the end of its fin swelled with cells that would eventually become cartilage (**Figure 9.19**).

One of the most interesting results of this experiment is that the extra expression of *Hoxd13* had a pleiotropic effect. It simultaneously shrank the outer area of the fin

Figure 9.19 A: In an experiment testing this hypothesis, scientists expressed *Hoxd13* late in the development of a fish fin. B: The result was an appendage with a striking resemblance to a tetrapod limb. It's possible that a second phase of *Hoxd13* expression was crucial to the evolution of limbs. (Data from Freitas et al. 2012, 2014; photo by Juniors/Super-stock, Inc.)

A   Normal developing zebrafish fin

B   Fin with second expression of *Hoxd13*

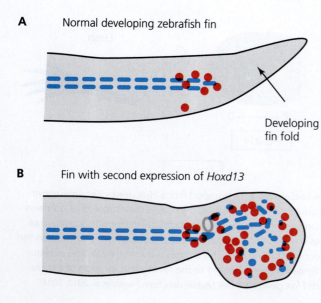

Developing fin fold

■ Precursors of bone cells
■ Fast-growing cells

where fin rays develop and expanded the region where bone grew. In the evolution of the hand, Freitas's experiment suggests, these two changes might have occurred at the same time.

From Hands to Wings

Once the tetrapod limb evolved, it diversified into many new forms. The fossil record provides crucial evidence about how these transitions occurred. As we saw in Chapter 1, the hindlimbs in ancient whales became gradually reduced, and the forelimbs became flippers adapted for swimming. In Chapter 4 we saw how in the dinosaur ancestors of birds, the forelimbs became reduced and their digits were lost, giving rise to feathered wings. Just as the genetic toolkit was involved in the origin of limbs, recent studies have shown that it was also involved in their diversification.

One of the most carefully studied of these transitions is the origin of bat wings. Bats evolved from small, rodent-like animals more than 60 million years ago. The fossil record of their origin is not yet as rich as that of birds, but even a comparison of living bats to other living mammals gives us some clues. They still retain all five digits of the mammalian hand, but the digits have become drastically elongated (**Figure 9.20**).

The bat wing starts out its development much like the arm of a mouse. The humerus forms first, followed by the ulna and radius. At this point, however, arm development in mice and bats diverges. The cells in a mouse limb bud stop proliferating once they've developed relatively small digits. Bat cells, on the other hand, begin to grow at a faster rate. As a result, the digits of the bat take on the shape of chopsticks.

In 2006, Karen Sears of the University of Illinois and her colleagues identified for the first time a key difference in the gene expression in mouse hands and bat wings (Sears et al. 2006). The wing's acceleration is controlled by the expression of a gene known as *Bmp2* (see Figure 9.16). Cells that will form the distal tip of the limb (in a part of the limb bud called the apical ectodermal ridge, which is analogous to the bull's-eye in the center of developing insect legs) produce high levels of *Bmp2*, driving more cell division in the developing digits.

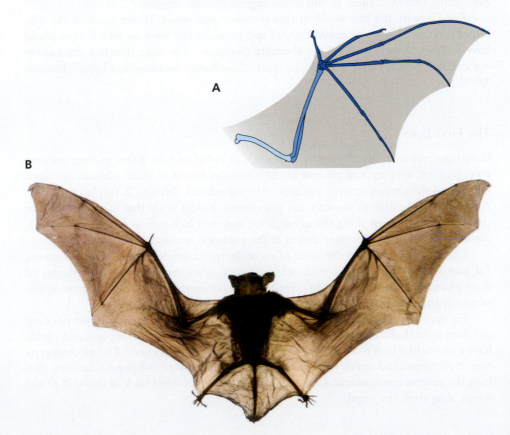

A

B

Figure 9.20 Subtle shifts in the levels of expression of patterning genes can alter the shape or size of limbs. Increased expression of *Bmp2*, for example, stimulates extra growth in the finger bones of bat forelimbs, contributing to their unusual length. Although changes in levels of expression of "upstream" patterning genes like *Bmp2* can have dramatic effects, these aren't the only genes involved in adaptation. The wings of bats evolved from more familiar mammalian digits through the altered expression of thousands of limb-related developmental genes. (spxChrome/Getty Images)

That doesn't mean that *Bmp2* alone is responsible for the unique trajectory of bat wing development, however. More recently, Katherine Pollard of the University of California, San Francisco, and her colleagues took a much broader survey of the activity of genes in bat wings (Booker et al. 2016; Eckalbar et al. 2016). Instead of comparing the bat wing to the mouse hand, they compared bat wings to their small feet. Pollard and her colleagues identified more than 7000 genes that were expressed at significantly different levels in bat forelimbs and hindlimbs. The researchers also identified 2796 segments of bat DNA that appear to be genetic control regions that experienced natural selection. Intriguingly, many of these regions are close to limb-associated genes. Rather than evolving entirely new genes, bats may have evolved by changing expression of thousands of genes in their forelimbs to produce wings.

Key Concept

- Subtle changes in the patterns of expression of developmental genes can dramatically alter phenotypes, such as turning fins into limbs, or ordinary mammal limbs into bat wings. ●

9.7 Evolving Eyes

Of all the complex adaptations in nature, the eye has held a special place since Darwin's day. "The eye to this day gives me a cold shudder," he once wrote to a friend. If his theory of evolution was everything he thought it was, a complex organ such as the human eye could not lie beyond its reach. And no one appreciated the beautiful construction of the eye more than Darwin—from the way the lens was perfectly positioned to focus light onto the retina to the way the iris adjusted the amount of light that could enter the eye.

In *The Origin of Species*, he wrote that the idea of natural selection producing the eye "seems, I freely confess, absurd in the highest possible degree."

For Darwin, the key word in that sentence was *seems.* If you look at the different sorts of eyes in the natural world and consider the ways in which they could have evolved, Darwin realized, the absurdity disappears. The objection that the human eye couldn't possibly have evolved, he wrote, "can hardly be considered real" (Darwin 1859).

The First Eyes

If we map eyes onto the phylogeny of animals, as in **Figure 9.21**, a few striking patterns emerge. First, eyes are not present in the common ancestor of all animals. Sponges, considered the earliest derived lineages of living animals (Section 3.10), have no eyes. Another basal lineage of animals, the placozoans, have a body that is a thin sheet of cells. Placozoans creep along the ocean floor, and they lack eyes as well. Thus, we can confidently infer that eyes were absent in the common ancestor of animals.

Eyes are found only in the clade of more derived animals, known as eumetazoans. All animal eyes capture light with cells known as photoreceptors. But the shapes of these photoreceptors take very different forms in different clades, and they grow inside of eyes with an even greater diversity of shapes (Oakley and Pankey 2008).

Fly eyes are built out of columns. Scallops have a delicate chain of eyes peeking out from their shells. Flatworms have simple light-sensitive spots. Octopuses and squids have camera-like eyes like we do, but with some major differences. The photoreceptors of octopuses and squids point out from the retina, toward the pupil. Our own eyes have the reverse arrangement. Our photoreceptors are pointed back at the wall of the retina, away from the pupil.

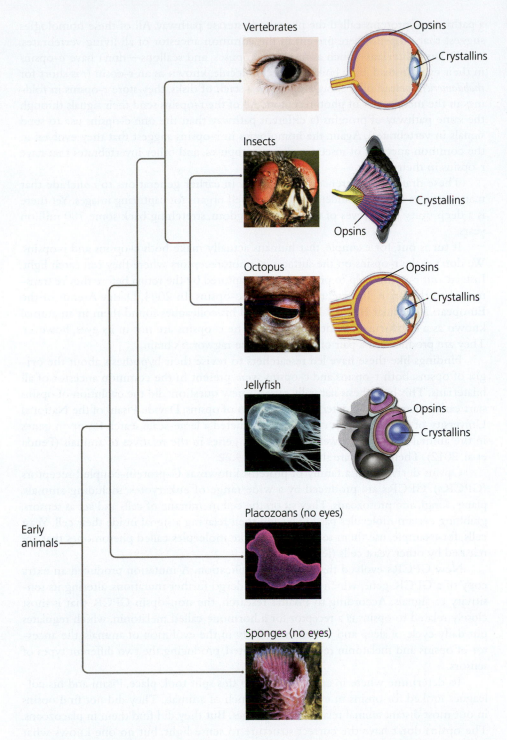

Vertebrates
— Opsins
— Crystallins

Insects
— Crystallins
— Opsins

Octopus
— Opsins
— Crystallins

Jellyfish
— Opsins
— Crystallins

Placozoans (no eyes)

Early animals —

Sponges (no eyes)

Figure 9.21 Complex eyes have evolved in several different lineages of animals. Each kind of eye contains crystallins for directing incoming light and opsins for capturing it. But the particular molecules each kind of eye uses as opsins and crystallins are different from the others. (monika3steps/Shutterstock; Stana/Shutterstock; Kerry L. Werry/Shutterstock; ANT Photo Library/Science Source; Schierwater Lab/Karolin von der Chevallerie; Andrew J. Martinez/Science Source)

The Role of Opsins

When light enters your eye, it strikes a molecule known as an opsin that sits on the surface of photoreceptor cells. Photons trigger a series of chemical reactions that cause the photoreceptor to send an electrical message toward the brain. Biologists have long known that all vertebrates carry the same basic kind of opsin in their eyes, known as a c-opsin (*c* is short for *ciliary*). All c-opsins have the same basic molecular shape, whether they're in the eye of a shark or the eye of a hummingbird. All c-opsins are stored in a stack of disks, each of which grows out of a hairlike extension of the retina called a cilium. In all vertebrates, c-opsins relay their signal from the stack of disks through a pathway of proteins called the phosphodiesterase pathway. All of these homologies suggest that c-opsins were present in the common ancestor of all living vertebrates.

Other bilaterians—such as insects, octopuses, and scallops—don't have c-opsins in their eyes. Instead, they build another molecule, known as an r-opsin (*r* is short for *rhabdomeric*). Instead of keeping r-opsins in a stack of disks, they store r-opsins in foldings in the membranes of photoreceptors. All of the r-opsins send their signals through the same pathway of proteins (a different pathway than the one c-opsins use to send signals in vertebrates). Again, the homologies in r-opsins suggest that they evolved in the common ancestor of insects, scallops, octopuses, and other invertebrates that have r-opsins in their eyes.

These dramatic differences led scientists in earlier generations to conclude that many lineages of animals independently evolved organs for capturing images. Yet there is a deep unity to the eyes of the animal kingdom, stretching back some 700 million years.

It turns out, for example, that humans actually make both c-opsins and r-opsins. We don't make r-opsins on the surfaces of photoreceptors where they can catch light. Instead, our r-opsins help to process images captured by the retina before they're transmitted to the brain. Invertebrates also have c-opsins. In 2004, Detlev Arendt of the European Molecular Biology Laboratory and his colleagues found them in an animal known as a ragworm (Arendt et al. 2004). The c-opsins are not in its eyes, however. They are produced in a pair of organs atop the ragworm's brain.

Findings like these have led researchers to revise their hypothesis about the origin of opsins: both r-opsins and c-opsins were present in the common ancestor of all bilaterians. This hypothesis naturally raised a new question: did the evolution of opsins start earlier than that? To determine the origin of opsins, Davide Pisani of the National University of Ireland and his colleagues conducted a large-scale search for opsin genes in the animal kingdom, as well as for related genes in the relatives of animals (Feuda et al. 2012). Their results are shown in **Figure 9.22**.

Opsins derive from a family of proteins known as G-protein-coupled receptors (GPCRs). GPCRs are produced by a wide range of eukaryotes, including animals, plants, fungi, and protozoans. They sit on the cell membrane of cells and act as sensors, grabbing certain molecules passing by and then relaying a signal inside their cell. Yeast cells, for example, use them to detect odor-like molecules called pheromones that are released by other yeast cells (Jékely 2003).

New GPCRs evolved through gene duplication. A mutation produced an extra copy of a GPCR gene, which could then undergo further mutations, altering its sensitivity to signals. According to Pisani's research, the non-opsin GPCR that is most closely related to opsins is a receptor for a hormone called melatonin, which regulates our daily cycle of sleep and wakefulness. Early in the evolution of animals, the ancestor of opsins and melatonin receptors duplicated, producing the two different types of sensors.

To determine where in animal evolution this split took place, Pisani and his colleagues looked for opsins in every major branch of animals. They did not find opsins in our most distant animal relatives, the sponges. But they did find them in placozoans. The opsins don't have the correct structure to sense light, but no one knows what other signal the opsins of placozoans are sensing.

After placozoans split off from our own ancestors, opsins duplicated two more times, giving rise to the r-opsins and c-opsins, which can be found not just in bilaterians but also in cnidarians—a group that includes jellyfishes, corals, and anemones. In both lineages, these opsins took on the duty of catching light. But in some lineages, one type or the other evolved to take on other functions, such as the r-opsins in our own eyes.

This research is helping us appreciate just how far back the homology in the molecular biology of eyes goes. The earliest animals probably produced these opsins in simple light-sensitive eyespots. Such eyespots are found today on many animals, which use them only to sense changes from light to dark. But opsins are only part of the story

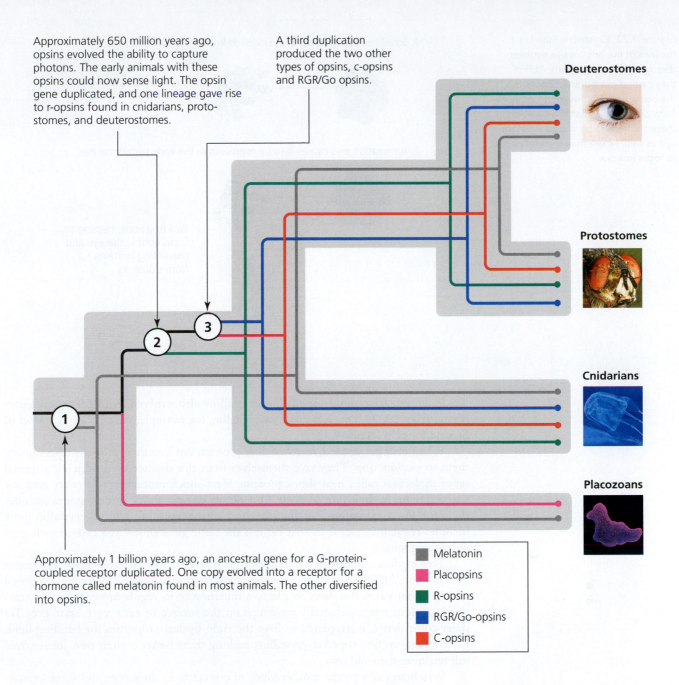

Approximately 650 million years ago, opsins evolved the ability to capture photons. The early animals with these opsins could now sense light. The opsin gene duplicated, and one lineage gave rise to r-opsins found in cnidarians, protostomes, and deuterostomes.

A third duplication produced the two other types of opsins, c-opsins and RGR/Go opsins.

Deuterostomes

Protostomes

Cnidarians

Placozoans

Approximately 1 billion years ago, an ancestral gene for a G-protein-coupled receptor duplicated. One copy evolved into a receptor for a hormone called melatonin found in most animals. The other diversified into opsins.

■ Melatonin
■ Placopsins
■ R-opsins
■ RGR/Go-opsins
■ C-opsins

Figure 9.22 Recent studies on opsins show that they evolved from a single common ancestor perhaps a billion years ago. Gene duplication gave rise to different types of opsins that have taken on different functions in the eyes and nervous systems of animals. (Photos: monika3steps/Shutterstock, Stana/Shutterstock, Daleen Loest/Shutterstock, Schierwater Lab/Karolin von der Chevallerie data from Feuda et al. 2012)

of eye evolution. To produce useful sensory input, opsins need to be part of larger light-gathering organs.

The Role of Crystallins

The evolution of spherical eyes allowed some animals to form images. Crucial to these image-forming eyes was the evolution of lenses that could focus light. Lenses are made of remarkable molecules called crystallins, which are among the most specialized proteins in the body. Different clades of animals produce different types of crystallins (vertebrate eyes, for example, produce one form known as α–crystallins). But all types of crystallins have many things in common. They are transparent, to begin with. As light passes through crystallins, the proteins alter the path of incoming light, making it possible to focus an image on the retina. All crystallins are incredibly rugged. They are among the most stable proteins in the body and keep their structure for decades. (Cataracts are caused by crystallins clumping late in life.)

Figure 9.23 Crystallins (blue) in the lens of the human eye evolved through gene recruitment. At first they carried out other functions in the body, such as preventing proteins from clumping. Mutations caused these proteins to be produced in the eye as well, where they then helped to focus images.

1. Small heat-shock protein is expressed in muscles and other tissues.

Original function: preventing proteins from clumping

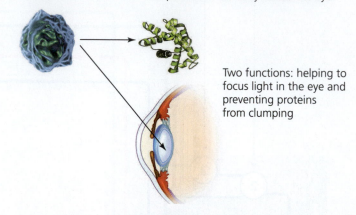

2. A mutation also causes it to be expressed in the early vertebrate eye.

Two functions: helping to focus light in the eye and preventing proteins from clumping

Scientists have determined that crystallins also evolved from recruited genes (Piatigorsky 2007). The vertebrate α-crystallin, for example, started out as a kind of first aid for cells (Figure 9.23).

When cells get stressed, many of their proteins can lose their proper shape, causing them to malfunction. They save themselves from this disaster by producing a special set of molecules called heat-shock proteins. Heat-shock proteins cradle other proteins, keeping them in their proper shape. One family of heat-shock protein genes includes the gene for α-crystallin. But the genetic control region for the α-crystallin gene includes regulatory elements that express the same gene in stressed cells as well as in developing lenses.

This evidence points to a hypothesis for how α-crystallin evolved in vertebrates (Cvekl et al. 2017). An ancestral version of the α-crystallin gene was only expressed in cells under stress. But then, thanks to a mutation to its regulatory region, this heat-shock protein was accidentally produced on the surface of early vertebrate eyes. By sheer coincidence, it happened to have the right optical properties for bending light. Later mutations fine-tuned α-crystallins, making them better at their new job—while still retaining their old job.

Vertebrates also produce other kinds of crystallins in their eyes, and some crystallins are limited to only certain groups, such as birds or lizards. And invertebrates with eyes, such as insects and squid, make entirely different crystallins. Scientists are gradually discovering the origins of all these crystallins. Evidently, many different kinds of proteins have been recruited, and they all proved to be good for bending light.

Evolution of the Eye Lens

Figure 9.24 summarizes many lines of research into a hypothesis about the evolution of the vertebrate eye (Lamb et al. 2008). The forerunners of vertebrates produced light-sensitive eyespots on their brains that were packed with photoreceptors carrying c-opsins. These light-sensitive regions ballooned out to either side of the head and later evolved an inward folding to form a cup. Early vertebrates could then do more than merely detect light: they could get clues about where the light was coming from. The ancestors of hagfish branched off at this stage of vertebrate eye evolution, and today their eyes offer some clues to what the eyes of our own early ancestors would have looked like (Lamb et al. 2008).

1. Early chordates had light-sensitive eyespots expressing photoreceptor genes.

Eyespot

Figure 9.24 One recent hypothesis for how the vertebrate eye evolved. It started out as a simple light sensor and then gradually evolved into a precise, image-forming organ. Living vertebrates and relatives of vertebrates offer clues to how this transformation took place. (Data from Lamb et al. 2008)

2. Light-sensitive regions bulge outward to the sides of the head.

Light-sensing cells with opsins

3. Patch folds inward into a cup, beneath unpigmented skin (lens placode).

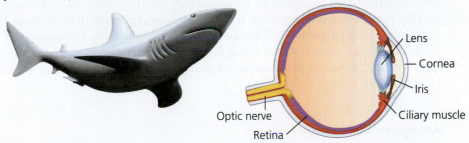

Optic cup — Lens placode
— Retina

4. Surface becomes transparent, and lens evolves ability to focus an image.

Lens

Optic nerve

5. Eyes become spherical and evolve greater acuity.

Lens
— Cornea
— Iris

Optic nerve — Ciliary muscle
Retina

After hagfish diverged from the other vertebrates, a thin patch of tissue evolved on the surface of the eye. Light could pass through the patch, and crystallins were recruited into it, leading to the evolution of a lens. At first the lens probably focused light quite crudely. But even a crude image was better than none. A predator could follow the fuzzy outline of its prey, and its prey could flee at the fuzzy sight of its attackers. Mutations that improved the focusing power of the lens were favored by natural selection, leading to the evolution of a spherical eye that could produce a crisp image.

The evolution of the vertebrate eye did not stop there. Double lenses evolved in some fishes, which allowed them to see above and below the water's surface at the same time. The ability to see in ultraviolet light evolved in birds. But all subsequent vertebrate eyes would be variations on the basic theme established half a billion years ago.

Key Concept

- Very complex adaptations, such as the vertebrate eye, can evolve through a series of small steps. Opsins link vertebrates to organisms that lived 650 million years ago, and so, too, do the mechanisms for co-opting regulatory networks. ●

9.8 Constraining Evolution

In this chapter, we've been considering some of the many remarkable adaptations that have evolved over the history of life. But it's also intriguing to consider the adaptations that *haven't* evolved. Why aren't there any hawk-sized dragonflies? Why are there no nine-toed tetrapods? One explanation for the absence of such forms is that life evolves within constraints. There may be some directions in which evolution rarely—if ever—travels (Maynard Smith et al. 1985).

Obeying the Laws of Physics

The laws of physics can impose constraints that evolution cannot overcome. Animals such as elephants and sharks can reach enormous sizes, for example, and yet animals that make up the most diverse clade, the insects, are relatively small. The reason there are no gigantic insects alive today, a number of researchers have argued, is that the concentration of oxygen in today's atmosphere makes it impossible.

Insects get oxygen through tiny tubes penetrating their exoskeletons. As the insects become bigger, they need more tubes to supply enough oxygen to support their larger bodies. But as an insect gets bigger, the tubes become a less efficient way to deliver oxygen to its tissues. As a result, the tubes themselves must get bigger in order to deliver sufficient oxygen to power the insect's muscles, and there's less room for muscles and other structures that support the animal.

It's possible to test this hypothesis by looking back through the history of life to times when the oxygen concentrations were higher. If insects can absorb oxygen more efficiently, they don't require larger tubes, and they can thus evolve to larger body sizes. During the Carboniferous period, for example, oxygen concentrations were much higher than they are today (35% compared to the current 21%). Paleontologists have found remarkably large fossils of flying insects from that period—one dragonfly-like species called *Meganeura*, for example, reached the size of a seagull (Kaiser et al. 2007; Harrison et al. 2010).

Antagonistic Pleiotropy

Antagonistic pleiotropy is the condition that occurs when a mutation with beneficial effects for one trait also causes detrimental effects on other traits.

In Chapter 6, we learned about pleiotropy—the ability that genes have to exert more than one phenotypic effect. A mutation can also be pleiotropic, altering an organism in more than one way. In some cases, the changes may all lower fitness, or all raise it. But in other cases, mutations can have opposite effects—a condition known as **antagonistic pleiotropy**. Antagonistic pleiotropy can constrain the evolution of adaptations. A mutation that might seem to be favored by natural selection may actually lower fitness when you take into account all its effects on the phenotype.

Antagonistic pleiotropy can be especially powerful in the evolution of multicellular organisms with complex developmental programs. A single gene, such as a *Hox* gene, may help guide the development of many different structures in an embryo. The harmful effects of mutations on developmental genes may explain some of the limits

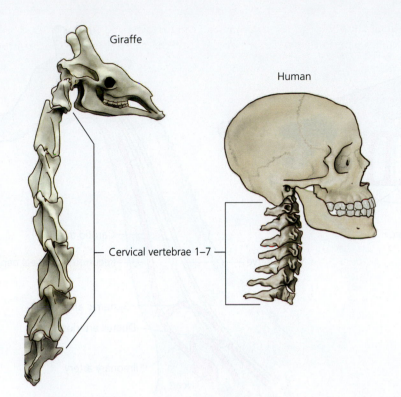

Giraffe

Human

Cervical vertebrae 1–7

Figure 9.25 Almost all mammals have only seven cervical vertebrae in their neck. This rule applies even to species with long necks, such as giraffes. This pattern may be the result of evolutionary constraint. Mutations that would lead to more or fewer cervical vertebrae may have deleterious effects as well.

found in complex organisms. Almost all mammals, for example, have exactly seven cervical vertebrae in their necks: the only exceptions are sloths and manatees. It's hard to see how natural selection would favor seven cervical vertebrae over six or eight. It seems particularly unlikely when you consider that even giraffes, with their extremely long necks, also have seven cervical vertebrae (albeit seven very long ones; **Figure 9.25**).

Frietson Galis of Leiden University and her colleagues wondered if some type of strong antagonistic pleiotropy was responsible. They searched through medical records in Dutch hospitals, identifying hundreds of cases in which human fetuses developed an abnormal number of cervical vertebrae. They found that such fetuses were more likely to be stillborn and that children born with an abnormal number of cervical vertebrae are 120 times more likely to develop pediatric cancers (Galis et al. 2006; Varela-Lasheras et al. 2011). Galis and her colleagues have also found fitness costs for abnormal numbers of cervical vertebrae in mice and other animals.

Two mammals that break the rule—sloths, with eight cervical vertebrae, and manatees, with six—have very low metabolisms for a mammal. Low metabolic rates have been linked to low rates of cancer. Thus, the fitness cost of evolving more or less cervical vertebrae may disappear in slow-moving mammals like sloths and manatees.

- Antagonistic pleiotropy can constrain the directions of evolution by causing some phenotypes to be unsuccessful even when mutations may be beneficial. ●

Key Concept

9.9 Building on History: Imperfections In Complex Adaptations

As we have seen, new adaptations do not evolve from scratch. They are modifications of previously existing structures, pathways, and other traits. As a result, even traits that are impressive in their complexity often have deep flaws. The vertebrate eye, for example, has photoreceptors pointing away from the light. The photoreceptors all send projections to the optic nerve on the surface of the retina, and the nerve actually

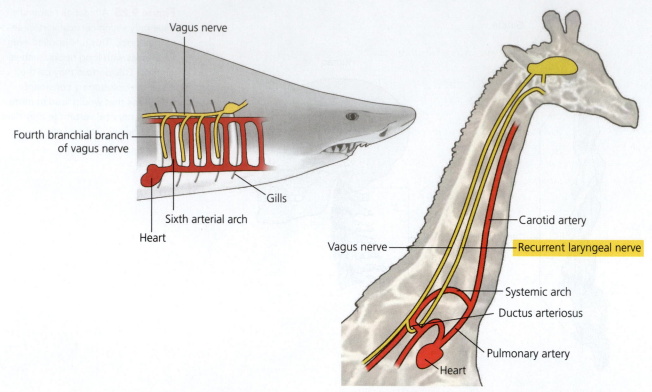

Figure 9.26 All tetrapods evolved from aquatic vertebrate ancestors. These ancestors had nerves that extended back from the brain to the gill arches, crossing over blood vessels. As tetrapods adapted to life on land, they lost functional gills but retained the gill pouch organization in their neck. As the structures homologous to gill pouches evolved to be in the neck, the nerves continued to loop around blood vessels. In giraffes, which have the longest neck of any living tetrapod, this evolutionary constraint led to a bizarre anatomical organization. The recurrent laryngeal nerve travels all the way down the giraffe's neck, where it loops around a blood vessel and then returns all the way back up, to a spot just a few inches from its origin. (Data from Dawkins 2009)

plunges back through the retina in order to travel back to the brain. The place where the optic nerve passes through the retina forms a blind spot in our vision. To get a complete picture of what we see, our eyes constantly dart around so that our brains can fill in the blind spot with extra visual information.

The shortcomings arise because complex traits are not built from scratch. Instead, evolution reworks what already exists. Fishes, for example, grow a series of nerve branches from their spinal cord that extend into their gill pouches. Tetrapods evolved from lobe-fins, inheriting this basic arrangement of nerves. But as the tetrapod neck evolved, the ancestral gill arches shifted their positions and their sizes. The nerves migrated with the arches, stretching into peculiar paths (**Figure 9.26**).

The most spectacular of these detoured nerves, the recurrent laryngeal nerve, extends down the neck to the chest, loops around a lung ligament, and then runs back up the neck to the larynx. In long-necked giraffes, the nerve grows to a length of 20 feet in order to make this U-turn, although 1 foot of nerve would have done quite nicely. But a 1-foot nerve never arose in giraffes, because this type of variation never existed. Ancestral giraffe populations contained individuals with nerves that developed according to their ancient, inherited developmental mechanism, and this mechanism was incapable of producing a simple, 1-foot-long nervous connection to the larynx. History in this case dictated what permutations to the giraffe phenotype were possible, and selection could operate only on the variation that was available at the time.

Key Concept

- Natural selection often retools the form and function of characters present within a population, leading to complex adaptations that are far from perfect but that still provide fitness advantages. ●

9.10 Convergent Evolution

The cougars and wolves that roam parts of North America are placental mammals. In other words, female cougars and wolves carry their developing fetuses in their uterus, where an organ called the placenta helps the fetuses grow. Cougar-like and wolf-like animals also once roamed Australia, but they were of a different sort (Feigen et al. 2018). They were marsupial mammals, which do not grow placentas. Instead, their young crawl out of the uterus and into a pouch to finish developing. Marsupial and placental mammals diverged from a common ancestor about 130 million years ago, and both lineages eventually produced species that looked remarkably similar and occupied the same ecological niches (Figure 9.27).

The independent evolution of two lineages into a similar form is known as convergent evolution (see Chapter 4). Although the evolution of similarity among two distantly related lineages has occurred quite often, the process does not always follow the same routes. Butterflies and birds both have wings that they use to fly, but butterflies evolved from terrestrial insects without wings, whereas birds evolved from flightless dinosaurs. Both lineages independently evolved wings. Bird wings evolved from dinosaur arms, which did not exist on the flightless ancestors of butterflies. Although wings evolved independently on butterflies and birds, they converged on similar structures. Both animals have flat surfaces, which they can flap to generate lift.

In some cases multiple populations evolve the same trait in parallel, shifting independently from the same ancestral state to the same derived state. Surface populations

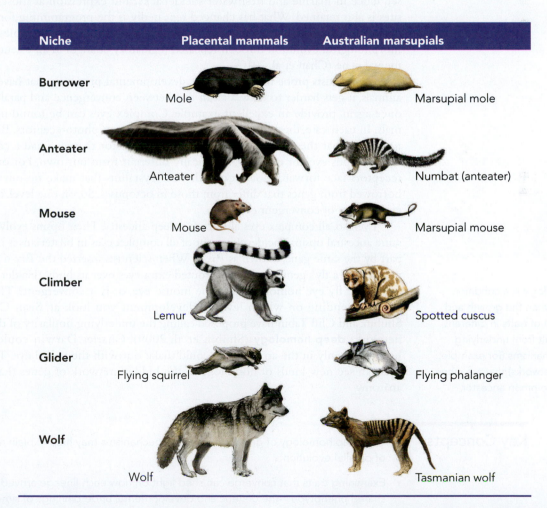

Niche	Placental mammals	Australian marsupials
Burrower	Mole	Marsupial mole
Anteater	Anteater	Numbat (anteater)
Mouse	Mouse	Marsupial mouse
Climber	Lemur	Spotted cuscus
Glider	Flying squirrel	Flying phalanger
Wolf	Wolf	Tasmanian wolf

Figure 9.27 Marsupials and placental mammals diverged from a common ancestor 130 million years ago. But they have repeatedly evolved body plans that show remarkable convergence. This convergence is the result of adaptation to the same ecological niches, such as burrowing underground or attacking large prey.

of the ray-finned teleost fish *Astyanax*, for example, have repeatedly colonized underground Mexican caves. Selection in dark, underground pools is very different from selection in surface waters, and cave populations evolved the same suite of derived traits in cave after cave, including pale body coloration and reduced eyes (Jeffery 2009). Populations undergoing such **parallel evolution** are like replicated experiments, providing scientists with opportunities to observe how selection shapes the same suite of adaptations over and over. In some cases, lineages have evolved identical phenotypes by independently acquiring mutations on the same genes.

In Chapter 10, we discuss a striking example of parallel evolution in three-spined stickleback fishes (Section 10.5). In many lakes, sticklebacks have lost some of their armor as they have adapted to a predator-free life. David Kingsley, a biologist at Stanford University, and his colleagues have found that selection has repeatedly favored an ancient mutation in a particular armor-controlling gene in freshwater populations; but in other populations, an independent mutation to the same gene (known as *Eda*) caused the loss of armor (Colosimo et al. 2005).

Further research has revealed that the parallels among the sticklebacks actually run even deeper. Kingsley and his colleagues discovered another gene, called *Pitx1*, that is also essential for the development of pelvic spines. In marine sticklebacks, *Pitx1* is expressed in the spines as well as in the developing nose, thymus gland, and sensory neurons. The gene's array of regulatory elements allows it to be expressed in all of these tissues. But when sticklebacks invaded freshwater lakes, the specific regulatory region controlling expression of *Pitx1* only in spines was disabled, through independent deletions in many different lakes. The coding sequence of *Pitx1* has an identical sequence in marine and freshwater sticklebacks, and expression at most other body sites is also retained. What has changed repeatedly is the programming for where that gene should be expressed; thus, sticklebacks provide a striking example of repeated morphological evolution through repeated regulatory changes in an essential developmental gene (Chan et al. 2010).

As scientists probe deeper into the developmental programs that have evolved in animals, it gets harder to draw a clean line between convergence and parallelism. Eyes, once again, provide an excellent example. Complex eyes can be found in many animals. In each case, they contain lenses to focus light on photoreceptors. But that does not mean that these animals have a common ancestor that also had a complex eye. The octopus eye, for example, is radically different from our own. For one thing, its receptors face forward, not backward. The crystallins that make up our lenses were borrowed from genes that differ from those in octopuses. So, on one level, the complex eye is a case of convergent evolution.

Even so, all complex eyes also share a deep ancestry. Their opsins evolved from the same ancestral opsin. The development of all complex eyes in bilaterians is controlled in part by the same gene, known as *Pax-6*. When scientists inserted the *Pax-6* gene from a mouse into a fly's genome, the fly sprouted extra eyes over its body (Halder et al. 1995).

Is the fly eye homologous to the mouse eye, or is it convergent? The answer is both, depending on which level of development you look at. Sean Carroll, Neil Shubin, and Cliff Tabin have proposed calling the underlying similarity of these genetic networks **deep homology** (Shubin et al. 2009). Charles Darwin could recognize homology only in the anatomy he could make out with the naked eye. Today, scientists can see new kinds of homology in the hidden network of genes that build that anatomy.

Parallel evolution refers to independent evolution of similar traits, starting from a similar ancestral condition.

Deep homology is a condition that occurs when the growth and development of traits in different lineages result from underlying genetic mechanisms (for example, regulatory networks) that are inherited from a common ancestor.

Key Concepts

- The deep homology of genetic regulatory mechanisms may help explain many cases of parallel evolution.

- Examining traits that converge can shed light on how each lineage arrived at the shared phenotype—the genetic and developmental underpinnings of similar adaptations. It can reveal multiple solutions to a common challenge as well as the limitations imposed by history within each lineage. ●

TO SUM UP . . .

- Genes at the top of regulatory hierarchies influence many other genes involved in development. Mutations in a few of these regulatory genes can produce far-reaching changes in development and ultimately in an organism's morphology.

- Promiscuous proteins may be recruited for new functions or into new regulatory networks fairly easily.

- Gene duplication is an important kind of mutation for the evolution of new functions.

- Genes are sometimes recruited or co-opted to become active in new organs or in new genetic networks.

- Animals as different as mammals and insects use modified versions of the same genetic toolkit of regulatory genes during development.

- During the evolution of animal eyes, genes for photoreceptors and crystallins were recruited and duplicated. Simple light-sensing eyes gradually evolved into complex camera-like eyes.

- Pleiotropy and physical constraints can restrict the possible forms into which a species can evolve.

- Adaptations are often imperfect due to co-option of the regulatory networks available. Natural selection can operate only on the available options.

- In convergent evolution, two separate lineages independently evolve similar adaptations (for example, bird wings and bat wings, or the body forms of whales and fishes). In parallel evolution, the same trait evolves independently in two separate lineages from the same ancestral state.

MULTIPLE CHOICE QUESTIONS Answers can be found at the end of the book.

1. How do novel traits arise?
   a. Through an existing trait that reverts to a previous version.
   b. Through mutation leading to new phenotypes.
   c. Through genes that are inherited from the previous generation.
   d. Through a new combination of traits already existing in the individual.

2. What is a paralog?
   a. A protein that has taken on a new function.
   b. A protein that can have two different functions at once.
   c. A gene that serves a similar function in a related species.
   d. A gene that has taken on a new function through gene duplication.

3. What is an ortholog?
   a. Two genes in two different species that serve the same function and were inherited from a common ancestor.
   b. Two genes in two different species that serve the same function but were inherited from different ancestors.
   c. Two genes in the same species that serve different functions but were duplicated from the same original gene.

   d. Two genes in the same species that serve the same function but were inherited separately from different ancestors.

4. Which of the following processes did *not* lead to the evolution of snake venom?
   a. Gene recruitment.
   b. Natural selection.
   c. Gene duplication.
   d. Broad-sense heritability

5. What do you predict would happen to the hind legs of a fly if the gene *Dpp* were artificially overexpressed?
   a. The legs would grow longer.
   b. The legs would not grow as long.
   c. The growth of the legs would be completely suppressed.
   d. The early leg cells would be reabsorbed into the body.

6. What role does *Bmp2* play in the expression of bat wing phenotypes?
   a. It is the only protein responsible for the shape of the wing.
   b. It is one of two proteins that determine the shape of the wing.

c. It is a regulatory gene that turns on many other genes.

d. It is turned on by the regulatory protein known as calmodulin.

7. Which organism does *not* use crystallins in its eyes to focus light?

a. Humans.

b. Octopuses.

c. Fishes.

d. Ragworms.

8. What is *not* an example of antagonistic pleiotropy?

a. Genes that create insect breathing tubes cause decreased production of growth hormones.

b. Mutations that confer resistance to *Bacillus thuringiensis* (Bt) also make insects more vulnerable to natural plant defensive chemicals.

c. Developmental genes that code for greater or fewer than seven cervical vertebrae have negative fitness costs.

d. Carriers of a recessive gene for cystic fibrosis have greater resistance to tuberculosis.

9. Why do giraffes have a recurrent laryngeal nerve that is 19 feet longer than the shortest route possible?

a. Their ancestors were fishes and did not have a larynx.

b. The route of that nerve was inherited from their ancestors.

c. The nerve makes more connections to their lungs than in their ancestors.

d. The route that the nerve takes is the result of random processes.

10. What do you predict would happen if you inserted a *Pax-6* gene from a mouse into the genome of an octopus?

a. No eyes would develop in the octopus.

b. The normal two eyes would develop, but they would not work.

c. Extra eyes would develop, but with mouse crystallins.

d. Extra eyes would develop, but with octopus crystallins.

INTERPRET THE DATA   Answer can be found at the end of the book.

11. The figure below shows evolutionary changes (amino acid substitutions) to venom genes in snakes and their relatives. If the scientists were to use the *dN/dS* method introduced in Chapter 8 to test for the action of natural selection on this gene, what pattern would you expect them to find? (N and S stand for nonsynonymous and synonymous, respectively.)

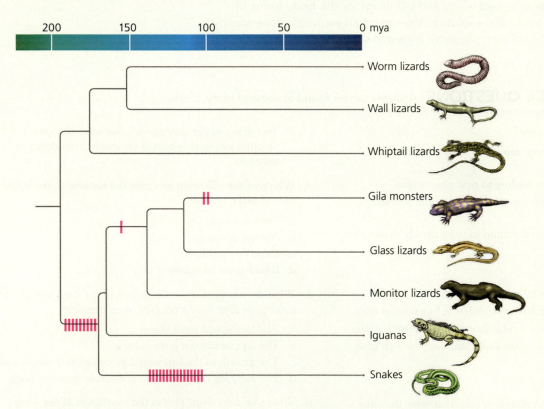

a. *dN > dS* when snakes are compared with iguanas.

b. *dS > dN* when snakes are compared with monitor lizards.

c. *dS > dN* when snakes are compared with whiptail lizards.

d. *dN = dS* when snakes are compared with worm lizards.

SHORT ANSWER QUESTIONS Answers can be found at the end of the book.

12. How can horizontal gene transfer give rise to a new adaptation?

13. What are the differences and similarities between gene duplication and gene recruitment?

14. Describe how a *defensin* gene was co-opted for predation in snakes.

15. Why do mice and flies have the same genetic toolkit of patterning genes? Explain in general terms how these patterning genes guide the development of a fly's limb.

16. Describe the role of opsins and crystallins in the evolution of the vertebrate eye.

17. Consider similar traits in both sharks and dolphins. Describe the evidence that would allow you to identify two traits as being homologous and two different traits as being convergent.

ADDITIONAL READING

Carroll, S. B. 2006. *Endless Forms Most Beautiful: The New Science of Evo Devo.* New York: Norton.

Friedman, W. E., R. C. Moore, and M. D. Purugganan. 2004. The Evolution of Plant Development. *American Journal of Botany* 91:1726–41.

Gerhart, J., and M. Kirschner. 1997. *Cells, Embryos, and Evolution: Towards a Cellular and Developmental Understanding of Phenotypic Variation and Evolutionary Variability.* Malden, MA: Blackwell Science.

Gilbert, S., and D. Epel. 2009. *Ecological Developmental Biology: Integrating Epigenetics, Medicine, and Evolution.* Sunderland, MA: Sinauer Associates.

Hoekstra, H. E., and J. A. Coyne. 2007. The Locus of Evolution: Evo Devo and the Genetics of Adaptation. *Evolution* 61:995–1016.

Losos, J. B. 2017. *Improbable Destinies: Fate, Chance, and the Future of Evolution.* New York: Riverhead Books.

Shubin, N., C. Tabin, and S. Carroll. 2009. Deep Homology and the Origins of Evolutionary Novelty. *Nature* 457:818–23.

Stern, D. L. 2011. *Evolution, Development, and the Predictable Genome.* Greenwood Village, CO: Roberts and Company.

PRIMARY LITERATURE CITED IN CHAPTER 9

Andersson, D. I., J. Jerlström-Hultqvist, and J. Näsvall. 2015. Evolution of New Functions De Novo and from Preexisting Genes. *Cold Spring Harbor Perspectives in Biology* 7:a017996.

Arendt, D., K. Tessmar-Raible, H. Snyman, A. W. Dorresteijn, and J. Wittbrodt. 2004. Ciliary Photoreceptors with a Vertebrate-Type Opsin in an Invertebrate Brain. *Science* 306 (5697): 869–71.

Bachiller, D., A. Macías, D. Duboule, and G. Morata. 1994. Conservation of a Functional Hierarchy Between Mammalian and Insect *Hox/Hom* Genes. *EMBO Journal* 13 (8): 1930–41.

Blount, Z. D. 2017. Replaying Evolution. *American Scientist* 105:156.

Blount, Z. D., J. E. Barrick, C. J. Davidson, and R. E. Lenski. 2012. Genomic Analysis of a Key Innovation in an Experimental *Escherichia coli* Population. *Nature* 489: 513–8.

Blount, Z. D., C. Z. Borland, and R. E. Lenski. 2008. Historical Contingency and the Evolution of a Key Innovation in an Experimental Population of *Escherichia coli. Proceedings of the National Academy of Sciences USA* 105 (23): 7899–906.

Booker, B. M., T. Friedrich, M. K. Mason, J. E. VanderMeer, J. Zhao, et al. 2016. Bat Accelerated Regions Identify a Bat Forelimb Specific Enhancer in the HoxD Locus. *PLoS Genetics* 12 (3): e1005738.

Carroll, S. B. 2008. Evo-Devo and an Expanding Evolutionary Synthesis: A Genetic Theory of Morphological Evolution. *Cell* 134 (1): 25–36.

Casewell, N. R., W. Wüster, F. J. Vonk, R. A. Harrison, and B. G. Fry. 2013. Complex Cocktails: The Evolutionary Novelty of Venoms. *Trends in Ecology and Evolution* 28 (4): 219–29.

Chan, Y. F., M. E. Marks, F. C. Jones, G. Villarreal, M. D. Shapiro, et al. 2010. Adaptive Evolution of Pelvic Reduction in Sticklebacks by Recurrent Deletion of a *Pitx1* Enhancer. *Science* 327 (5963): 302–5.

Chuong, E. B., N. C. Elde, and C. Feschotte. 2016. Regulatory Evolution of Innate Immunity Through Co-option of Endogenous Retroviruses. *Science* 351:1083–7.

Colosimo, P. F, K. E. Hosemann, S. Balabhadra, G. Villarreal Jr., M. Dickson, et al. 2005. Widespread Parallel Evolution in Sticklebacks by Repeated Fixation of Ectodysplasin Alleles. *Science* 307 (5717): 1928–33.

Cvekl, A., Y. Zhao, R. McGreal, Q. Xie, X. Gu, et al. 2017. Evolutionary Origins of Pax6 Control of Crystallin Genes. *Genome Biology and Evolution* 9:2075–92.

Darwin, C. 1859. *On the Origin of Species by Means of Natural Selection, or, the Preservation of Favored Races in the Struggle for Life.* London: John Murray.

Dawkins, R. 2009. *The Greatest Show on Earth.* New York: Free Press.

de Hoon, M. J., P. Eichenberger, and D. Vitkup. 2010. Hierarchical Evolution of the Bacterial Sporulation Network. *Current Biology* 20 (17): R735–45.

Della Pina, S., E. Souer, and R. Koes. 2014. Arguments in the Evo-Devo Debate: Say It with Flowers! *Journal of Experimental Botany* 65 (9): 2231–42.

De Robertis, E. M. 2008. Evo-Devo: Variations on Ancestral Themes. *Cell* 132 (2): 185–95.

Eckalbar, W. L., S. A. Schlebusch, M. K. Mason, Z. Gill, A. V. Parker, et al. 2016. Transcriptomic and Epigenomic Characterization of the Developing Bat Wing. *Nature Genetics* 48 (5): 528.

Feigin, C. Y., A. H. Newton, L. Doronina, J. Schmitz, C. A. Hipsley, et al. 2018. Genome of the Tasmanian Tiger Provides Insights into the Evolution and Demography of an Extinct Marsupial Carnivore. *Nature Ecology and Evolution* 2 (1): 182.

Feuda, R., S. C. Hamilton, J. O. McInerney, and D. Pisani. 2012. Metazoan Opsin Evolution Reveals a Simple Route to Animal Vision. *Proceedings of the National Academy of Sciences USA* 109: 18868–72.

Freitas, R., C. Gómez-Marín, J. M. Wilson, F. Casares, and J. L. Gómez-Skarmeta. 2012. *Hoxd13* Contribution to the Evolution of Vertebrate Appendages. *Developmental Cell* 23: 1219–29.

Freitas, R., J. L. Gómez-Skarmeta, and P. N. Rodrigues. 2014. New Frontiers in the Evolution of Fin Development. *Journal of Experimental Zoology Part B: Molecular and Developmental Evolution* 322 (7): 540–52.

Friedman, W. E., R. C. Moore, and M. D. Purugganan. 2004. The Evolution of Plant Development. *American Journal of Botany* 91 (10): 1726–41.

Fry, B. G. 2005. From Genome to "Venome": Molecular Origin and Evolution of the Snake Venom Proteome Inferred from Phylogenetic Analysis of Toxic Sequences and Related Body Proteins. *Genome Research* 15:403–20.

Fry, B., ed. 2015. *Venomous Reptiles and Their Toxins: Evolution, Pathophysiology and Biodiscovery.* Oxford: Oxford University Press.

Fry, B. G., N. Vidal, J. A. Norman, F. J. Vonk, H. Scheib, et al. 2006. Early Evolution of the Venom System in Lizards and Snakes. *Nature* 439:584.

Galis, F., T. J. van Dooren, J. D. Feuth, J. A. Metz, A. Witkam, et al. 2006. Extreme Selection in Humans Against Homeotic Transformations of Cervical Vertebrae. *Evolution* 60 (12): 2643–54.

Gilbert, S. F. 2007. *Developmental Biology.* 8th ed. Sunderland, MA: Sinauer Associates.

Halder, G., P. Callaerts, and W. J. Gehring. 1995. Induction of Ectopic Eyes by Targeted Expression of the *Eyeless* Gene in *Drosophila. Science* 267 (5205): 1788–92.

Hall, J., M. A. Brockhurst, and E. Harrison. 2017. Sampling the Mobile Gene Pool: Innovation via Horizontal Gene Transfer in Bacteria. *Proceedings of the Royal Society B: Biological Sciences* 372:20160424.

Harrison, J. F., A. Kaiser, and J. M. VandenBrooks. 2010. Atmospheric Oxygen Level and the Evolution of Insect Body Size. *Proceedings of the Royal Society B: Biological Sciences* 277: 1937–46.

Hileman, L. C. 2014. Bilateral Flower Symmetry—How, When and Why? *Current Opinion in Plant Biology* 17:146–52.

Hueber, S. D., and I. Lohmann. 2008. Shaping Segments: *Hox* Gene Function in the Genomic Age. *BioEssays* 30 (10): 965–79.

Hughes, G. M., E. S. M. Boston, J. A. Finarelli, W. J. Murphy, D. G. Higgins, et al. 2018. The Birth and Death of Olfactory Receptor Gene Families in Mammalian Niche Adaptation. *Molecular Biology and Evolution* 35:1390–1406.

Innan, H., and F. Kondrashov. 2010. The Evolution of Gene Duplications: Classifying and Distinguishing Between Models. *Nature Reviews Genetics* 11 (2): 97–108.

Jeffery, W. R. 2009. Regressive Evolution in *Astyanax* Cavefish. *Annual Review of Genetics* 43:25–47.

Jékely, G. 2003. Small Gtpases and the Evolution of the Eukaryotic Cell. *BioEssays* 25 (11): 1129–38.

Kaiser, A., C. J. Klok, J. J. Socha, W-K. Lee, M. C. Quinlan, et al. 2007. Increase in Tracheal Investment with Beetle Size Supports Hypothesis of Oxygen Limitation on Insect Gigantism. *Proceedings of the National Academy of Sciences USA* 104 (32): 13198–203.

Khersonsky, O., and D. S. Tawfik. 2010. Enzyme Promiscuity: A Mechanistic and Evolutionary Perspective. *Annual Review of Biochemistry* 79:471–505.

Lamb, T., E. Pugh, and S. Collin. 2008. The Origin of the Vertebrate Eye. *Evolution: Education and Outreach* 1 (4): 415–26.

Lutz, B., H. C. Lu, G. Eichele, D. Miller, and T. C. Kaufman. 1996. Rescue of *Drosophila* Labial Null Mutant by the Chicken Ortholog *Hoxb-1* Demonstrates That the Function of *Hox* Genes Is Phylogenetically Conserved. *Genes and Development* 10 (2): 176–84.

Mallo, M. 2018 Reassessing the Role of Hox Genes During Vertebrate Development and Evolution. *Trends in Genetics* 34:209.

Martín-Durán, J. M., K. Pang, A. Børve, H. Semmler Lê, A. Furu, et al. 2018. Convergent Evolution of Bilaterian Nerve Cords. *Nature* 553:45.

Maynard Smith, J., R. Burian, S. Kauffman, P. Alberch, J. Campbell, et al. 1985. Developmental Constraints and Evolution: A Perspective from the Mountain Lake Conference on Development and Evolution. *Quarterly Review of Biology* 60 (3): 265–87.

Nakamura, T., A. R. Gehrke, J. Lemberg, J. Szymaszek, and N. H. Shubin. 2016. Digits and Fin Rays Share Common Developmental Histories. *Nature* 537:225–8.

Näsvall, J., L. Sun, J. R. Roth, and D. I. Andersson. 2012. Real-Time Evolution of New Genes by Innovation, Amplification, and Divergence. *Science* 338:384–7.

Oakley, T., and M. Pankey. 2008. Opening the "Black Box": The Genetic and Biochemical Basis of Eye Evolution. *Evolution: Education and Outreach* 1 (4): 390–402.

Pang, T. Y., and M. Lercher. 2015. The Adaptive Acquisition of Single DNA Segments Drives Metabolic Evolution Across *E. coli* Lineages. *arXiv:*1509.06667.

Panganiban, G., S. M. Irvine, C. Lowe, H. Roehl, L. S. Corley, et al. 1997. The Origin and Evolution of Animal Appendages. *Proceedings of the National Academy of Sciences USA* 94 (10): 5162–6.

Paredes, C. J., K. V. Alsaker, and E. T. Papoutsakis. 2005. A Comparative Genomic View of Clostridial Sporulation and Physiology. *Nature Reviews Microbiology* 3:969–78.

Piatigorsky, J. 2007. *Gene Sharing and Evolution: The Diversity of Protein Functions.* Cambridge, MA: Harvard University Press.

Sagai, T., M. Hosoya, Y. Mizushina, M. Tamura, and T. Shiroishi. 2005. Elimination of a Long-Range Cis-Regulatory Module Causes Complete Loss of Limb-Specific *Shh* Expression and Truncation of the Mouse Limb. *Development* 132 (4): 797–803.

Sears, K. E., R. R. Behringer, J. J. Rasweiler, and L. A. Niswander. 2006. Development of Bat Flight: Morphologic and Molecular Evolution of Bat Wing Digits. *Proceedings of the National Academy of Sciences USA* 103 (17): 6581–6.

Shepard, K. A., and M. D. Purugganan. 2002. The Genetics of Plant Morphological Evolution. *Current Opinion in Plant Biology* 5 (1): 49–55.

Shubin, N., C. Tabin, and S. Carroll. 1997. Fossils, Genes and the Evolution of Animal Limbs. *Nature* 388 (6643): 639–48.

———. 2009. Deep Homology and the Origins of Evolutionary Novelty. *Nature* 457 (7231): 818–23.

Stern, D. L. 2000. Perspective: Evolutionary Developmental Biology and the Problem of Variation. *Evolution* 54 (4): 1079–91.

Tabin, C. J., S. B. Carroll, and G. Panganiban. 1999. Out on a Limb: Parallels in Vertebrate and Invertebrate Limb Patterning and the Origin of Appendages. *American Zoologist* 39 (3):650–63.

Varela-Lasheras, I., A. Bakker, S. van der Mije, J. Metz, J. van Alphen, et al. 2011. Breaking Evolutionary and Pleiotropic Constraints in Mammals: On Sloths, Manatees and Homeotic Mutations. *EvoDevo* 2 (1): 11.

Yi, H. 2017. How Snakes Came to Slither. *Scientific American* 318:70.

CHAPTER 10

Natural Selection

Empirical Studies in the Wild

Learning Objectives

- Compare and contrast the factors leading to directional and stabilizing selection.
- Discuss the outcomes of directional and stabilizing selection.
- Explain how predators can act as agents of selection.
- Explain how selection can vary across a species' range.
- Explain how natural selection can act on an extended phenotype.
- Analyze the role of natural experiments in our understanding of evolutionary change in response to selection.
- Explain how selective sweeps can be detected within genomes.
- Evaluate the evidence for the role of humans as selective agents in the evolution of plants and animals.

Charles Darwin managed to visit only a few of the Galápagos Islands in 1835 while on his journey aboard the *Beagle*. Among the many islands he didn't get to explore was a tiny volcanic cone known as Daphne Major.

Even today, it is not an easy place to visit. To set foot on Daphne Major, you have to approach a steep cliff in a small boat and then take an acrobatic leap onto a tiny ledge. There are no houses on Daphne Major and no supply of water. In fact, just about the only things to see on Daphne Major are low scrubby plants and the little birds that eat their seeds.

In 1973, a British-born couple named Peter and Rosemary Grant came to Daphne Major and lived on the island for months. They returned every year for more than four decades, bringing with them a team of students and all the supplies they needed for a lengthy stay: tents, coolers, jugs of water, cooking fuel, clothes, radios, binoculars, and notebooks. This dedication has allowed the Grants—who

The beak sizes of Darwin's finches on the Galápagos Islands are influenced by natural selection.

Peter and Rosemary Grant

Figure 10.1 Peter and Rosemary Grant collect body measurements and place colored leg bands on wild-caught birds (A). The tiny island of Daphne Major (B), which is accessible only by scrambling up the surrounding cliffs (C), provides an isolated and unusually pristine environment for this study. (A: Peter and Rosemary Grant; B: Bjoern Backe/Alamy Stock Photo; C: Dr. Martin Wikelski/Max-Planck-Institut fur Ornithologie)

for much of their career have been biologists at Princeton University—to make one of the most extensive studies of natural selection in the wild (**Figure 10.1**).

As we saw in Chapter 6, some scientists study natural selection by conducting laboratory experiments. Richard Lenski, for example, has tracked more than 60,000 generations of evolution in *Escherichia coli*. Thanks to his carefully designed experiments, he and his colleagues can measure natural selection in the bacteria with great precision. He knows that all 12 lines of *E. coli* that he rears descend from a common ancestor and that they all have experienced precisely the same controlled conditions since the experiment began. He can even thaw out frozen ancestors to compare them to their evolved descendants.

But Darwin's theory encompassed far more than experiments that scientists carry out in their laboratories. He claimed that natural selection shaped all life on Earth. An important area of research in evolutionary biology is the study of selection in natural populations. In this chapter, we explore how scientists like the Grants conduct this challenging research.

To measure selection in wild populations, scientists need to observe the components that make selection possible. As we saw in Chapter 5 and Chapter 7, those ingredients are genetic variation, heritability of traits, and nonrandom survival of offspring. Although it's not possible to measure those values perfectly in wild populations, scientists can make robust estimates.

Measuring natural selection in the wild is important for reasons other than simply documenting its existence. As we'll see in this chapter, natural selection produces remarkably complicated patterns across time and space. Selection changes over time as the environment changes. From one location to another, the physical conditions and competition from other species may vary, creating a patchwork of selection pressures. Many different selection pressures can act on a population at the same time.

As scientists uncover this complexity of natural selection in the wild, they can delve even deeper into evolution. They can identify the genes under selection and document changes in allele frequency. And, as we'll see at the end of this chapter, scientists are discovering that we humans have become a major selection pressure for wild populations. ●

10.1 Evolution in a Bird's Beak

Any examination of the study of natural selection in the wild has to begin with Peter and Rosemary Grant, because they were the first scientists to make rigorous long-term measurements of the phenomenon (Weiner 1994; Grant and Grant 2014).

The Grants study Darwin's finches, birds that Darwin himself collected on his visit to the Galápagos Islands. Studies on the DNA of Darwin's finches indicate their ancestors arrived on the islands roughly a million years ago and then diversified into

Figure 10.2 Diversity in Darwin's finches. Over the past 3 million years, these birds have specialized for feeding on cactus flowers (A); for using twigs as tools to pry insects from bark (B); and for eating eggs (C), leaves (D), blood (E), and ticks (F). (A: Valeranda Media/ Shutterstock.com; B: Mary Plage/Oxford Scientific/Getty Images; C: Minden Pictures/Superstock; D: robertharding/Alamy Stock Photo; E: Ardea/D. Parer & E. Parer-Cook/Diomedia; F: John Abbott/Nature Picture Library)

as many as 18 species. The birds rapidly evolved into many different forms, adapting to the many different opportunities the islands offered for finding food.

The cactus finch, for example, nests in cactus, sleeps in cactus, mates in cactus, drinks cactus nectar, and eats the flowers, pollen, and seeds of cactus. Two different species of finches use tools: they pick up a twig or a cactus spine, trim it to shape with their beaks, and then poke it into bark on dead branches to pry out larvae. Some finches eat green leaves, which is practically unheard of for birds to do. Still other finches perch on the backs of Nazca boobies and peck at their wings and tails until they draw blood, which they then drink. And some finches even ride on the backs of iguanas and eat their ticks (**Figure 10.2**).

The Medium Ground Finch

The bird that captured most of the Grants' attention is the medium ground finch (*Geospiza fortis*), a species on Daphne Major that primarily eats seeds (**Figure 10.3**). Despite the inaccessibility of the island—indeed, precisely because of it—Daphne Major is an outstanding place to measure selection in the wild. The island remains relatively pristine. No one has ever tried to farm on the island. No one has introduced goats or other invasive species. As far as the Grants can tell, no species on Daphne Major have become extinct since the arrival of humans.

The island also has the added advantage of being ecologically simple. There aren't many plant species on Daphne Major, so the Grants were able to identify and measure every type of seed that the island's finches eat. The island is small, and so is its population of birds. On Daphne Major, only a few hundred ground finches may be hatched in a given year, and most spend their entire lives there, thus permitting the Grants to mark and follow every individual in the population. Emigrant finches rarely leave the island, and immigrants rarely arrive. As a result, the Grants can be confident that migrations have a negligible effect, at best, on the changes in the allele frequencies of the island population.

The Grants survey every bird on Daphne Major, measuring vital statistics such as body mass and beak width. They trace families, determining how many offspring each

Figure 10.3 Ground finches on Daphne Major differ in the thickness of their bills, and this variation causes some individuals to be more efficient at processing hard seeds. (Peter and Rosemary Grant)

bird had, and how many offspring their offspring had. From year to year, the Grants also compare individual finches to their offspring to determine how strongly inherited each kind of variation was.

The Grants' team has found that beak size is heritable. Roughly 65% of the phenotypic variance in beak length, and as much as 90% of the variance in beak depth, is attributable to additive genetic effects of alleles ($h^2 = 0.65$ and 0.90, respectively; Boag 1983; Grant and Grant 1993). In other words, big-beaked birds tend to produce chicks with big beaks, and small-beaked birds tend to produce chicks with small beaks. With such high heritability, we can use the breeder's equation ($R = h^2 \times S$) to find that the average beak size on Daphne Major has the potential to evolve rapidly in response to natural selection (Chapter 7).

But how much natural selection do the birds actually experience? The Grants reasoned that the size of a bird's beak could affect how it ate seeds, so they investigated the kinds of food available to the birds on Daphne Major. They measured the sizes and hardness of each of the seeds produced by two dozen species of plants on the island. They took samples of the seeds to see when and where they were available to the birds. They dug up soil samples and counted all of the seeds that they contained.

The Grants were able to measure precisely how much food was available to the birds, including the relative amounts of each kind of seed. The Grants and their colleagues also closely observed the birds as they ate, noting which kinds of seeds they chose and the time it took birds to process seeds of each type. During that first season alone, they observed more than four thousand meals.

When the Grants started their historic study, they were surprised to find that different species did not specialize on different kinds of seeds. In addition to the medium ground finch, Daphne Major is also home to the small ground finch (*Geospiza fuliginosa*), which has a narrower, pointier bill. Despite the different shapes of their beaks, both species of birds fed on the same soft, small seeds that were abundant on the island at that time. Even species that weren't seed specialists, such as cactus finches, were also eating the seeds.

Selection Pressures on Daphne Major

When the Grants returned 6 months later, however, the island was transformed. The dry season had begun, and the island had not gotten a drop of rain for 4 months. Many of the plants on Daphne Major had died, leaving behind a barren landscape. The small, soft seeds were all gone. Now the birds on the island were no longer all eating the same kind of food. They had become specialists. The Grants discovered that even within each species, individuals selected different kinds of seeds. Their choice, it turned out, depended on subtle differences in the shapes of their beaks.

The medium ground finches could choose from two kinds of seeds: small seeds from a plant known as spurge (*Chamaesyce amplexicaulis*) and hard, woody seeds from the plant *Tribulus cistoides*, commonly called caltrop. Finches with big beaks (11 millimeters deep) could crack open the caltrop seeds in 10 seconds. Finches with beaks 10.5 mm deep needed 15 seconds. If a bird's beak was 8 mm deep or less, it took so long to crack a caltrop seed that the bird gave up on it altogether. Instead, it ate only small spurge seeds.

The Grants found that the size of a finch's beak could make the difference between life and death. In 1977, Daphne Major was hit by a major drought. Most of the spurge plants died, leaving the medium ground finches without any small seeds to eat. Many of the medium ground finches died, most likely because the birds with smaller beaks were unable to crack open the big seeds from caltrops and starved.

A few years later, the Grants discovered that the population of finches had recovered. But now the average size of their beaks was deeper. Before the drought, the population ranged in beak size from 8 to 11 mm, with an average depth of 9.2 mm. After the drought, the average beak size had shifted half a millimeter to 9.7 mm, or about 15% of the range of variation (Grant 1986). The shift occurred because finches with bigger beaks had a better chance of surviving the drought. They could therefore produce a bigger fraction of the next generation. In other words, natural selection caused the average size of the beaks of medium ground finches to increase within the population (**Figure 10.4**).

Five years later, the Grants were able to observe natural selection at work again. At the end of 1985, heavy rains came to the islands. Spurge bloomed, producing lots of small seeds. Now small-beaked birds had the advantage. They could eat small seeds more efficiently than the big-beaked birds could, allowing them to grow faster and have more energy for producing offspring. In just a few generations after the heavy rains, the average size of beaks decreased by 2.5% (about a tenth of a millimeter).

The Grants had made a historic observation. It was the first time scientists had measured the effects of natural selection in a wild population as they were unfolding.

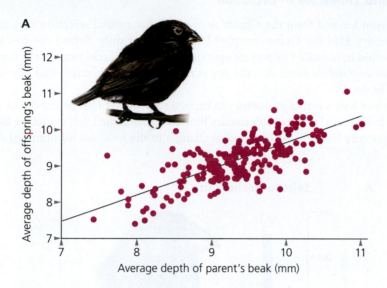

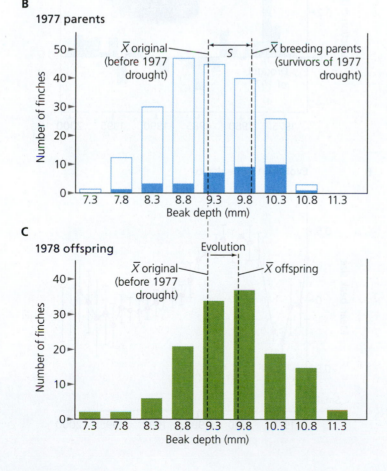

Figure 10.4 A: The size of beaks is heritable in medium ground finches. B: During a drought in 1977, birds with large, deep beaks had more chicks than did birds with small, narrow beaks. (The white bars show the total number of medium ground finches on Daphne Major with beaks in each size class before the drought. The blue bars show the number of birds with beaks in each size class that survived the drought and subsequently reproduced.) C: The average beak size increased in the offspring produced by birds surviving the drought. The dashed vertical lines show the average bill size from one year to the next. (Data from Grant and Grant 2002)

The Grants had measured the heritability of beak size (h^2), they had measured the strength and direction of several episodes of natural selection acting on beak size, and they had measured what happened to the population for several generations after the episode of selection (the evolutionary response, R). They documented how, precisely, selection caused evolution in the finches.

The Grants have continued to return to Daphne Major, and their persistence has paid off (**Figure 10.5**). Their research now offers many deep insights into how natural selection works (Grant and Grant 2002, 2011; Grant et al. 2017). (See **Box 10.1** for a discussion on character displacement.)

Intensity and Timescale of Evolution

The first lesson learned from the Grants' research is that natural selection may itself vary in intensity. Had the Grants sampled Daphne Major only during the wet seasons, they would have missed important episodes of directional selection. It was during the dry seasons—and in particular, the dry seasons of drought years—that selection favored big beaks.

The second lesson is that evolution can happen surprisingly fast. Before the Grants conducted their research, many evolutionary biologists maintained that evolution likely occurred over very long timescales. Gradual changes in the fossil record unfolded over

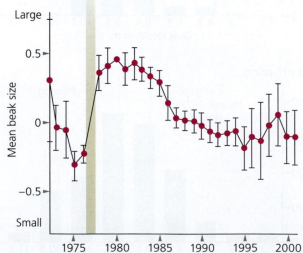

Figure 10.5 The Grants have measured natural selection on finch beaks for almost 40 years. During this period, the strength and direction of selection fluctuated. The Grants plotted the difference between the mean beak size of the population and the mean beak size of the individuals producing offspring in the next generation (the selection differential) for each year from 1972 to 2001 (A). In some years, birds with large beaks were favored (bars above the horizontal line); in other years, birds with small beaks were (bars below horizontal line). In still other years, selection on beak size was minimal. The finch population evolved in response to these episodes of selection, with the result that beak size fluctuated in tandem with the directions of selection (B). (Data from Grant and Grant 2002)

millions of years, suggesting that the strength of natural selection was probably very weak—weaker, for example, than the **artificial selection** imposed on populations of domesticated plants and animals. But the Grants were able to observe evolutionary changes in a natural population that were every bit as fast as those resulting from artificial selection. The selection they measured was strong, and their populations evolved in a matter of generations.

The third, and perhaps most important, lesson is that the pattern of selection can change over time. The Grants measured selection favoring big, deep beaks in some years and small, narrow beaks in others. Both the strength of selection, and its direction, fluctuated several times over the course of their study.

- During a severe drought on Daphne Major, subtle differences in beak depth among medium ground finches affected who lived and who died. Because beak depth is highly heritable, natural selection could lead to rapid evolution of beak size.

- Long-term studies of natural selection often show fluctuations in the direction and magnitude of selection. ●

Artificial selection is similar to natural selection, except that it results from human activity. When breeders nonrandomly choose individuals with economically favorable traits to use as breeding stock, they impose strong artificial selection on those traits.

Key Concepts

10.2 Mice in Black and White

The studies of Peter and Rosemary Grant are exceptional because of their 40-year span. But they are now just one example of hundreds of studies documenting selection in wild populations. These studies reveal different aspects of selection's complexity. And in some cases, researchers are even able to zero in on the specific genes that selection is altering.

In Chapter 7, we met Hopi Hoekstra, who studies oldfield mice (*Peromyscus poliopnotus*) in the southeastern United States. As we learned, Hoekstra and her colleagues are interested in the variation in coat color in different populations. They used quantitative trait locus (QTL) mapping to identify a few key genes involved in determining the color on the coats of the mice. This discovery raises the possibility that the difference in coloration between the beach mice and the mainland mice Hoekstra studies is the result of selection. It shows us that the trait possesses two of the three requirements for natural selection to lead to evolution—phenotypic variation and a genetic basis that can be passed down from parents to offspring.

The third requirement is that variation in the trait leads to differential reproductive success among individuals in a population. To explore this possibility, Hoekstra and her colleagues studied how the color of mice affects their chances of getting killed by predators. To catch a mouse, a bird or another predator has to see it. The oldfield mice make themselves difficult to see by foraging mainly on dark, cloudy nights. It was possible that the color of the mice also helped them become harder to see (Kaufman 1974). Oldfield mice that live on the mainland tend to be dark, matching the dark, loamy soils they walk on. Beach mice, which live on white sand, are much lighter.

To test the hypothesis that natural selection produced this variation, Hoekstra and her colleague, Sacha Vignieri, then at Harvard, conducted a simple field experiment. They made hundreds of life-sized clay models of mice and put them in each type of habitat. Half of the imitation mice were dark and half were light. Hoekstra and Vignieri then waited for predators to attack the models.

The predatory birds and mammals attacked some of the imitation mice, but then quickly discarded them once they realized their prey wasn't real. Hoekstra and Vignieri then gathered all the models and tallied the ones that had been damaged by predators. They discovered that predators are much more likely to attack mismatched phenotypes. In the light-colored sands of the beach habitats, predators attack primarily dark

BOX 10.1

Ecological Character Displacement

When two species live in the same area and compete for the same resource, selection can push their phenotypes in opposite directions. This divergence, known as **ecological character displacement**, is considered an important process in the diversification of life (Grant 2013; Stuart and Losos 2013).

In 1956, William Brown and Edward O. Wilson of Harvard proposed a model for how ecological character displacement occurs (Brown and Wilson 1956). When two species overlap, individuals that are similar to the other species will experience intense competition for limited food and nesting sites. But the individuals that are the least similar to the other species will experience less competition—and thus have more reproductive success. Over time, Brown and Wilson argued, natural selection favors dissimilar individuals so much that the two populations will diverge.

One way to study this process is to find opportunities to observe it occurring in real time. As has been discussed, on the island of Daphne Major, the Grants have been able to observe ecological character displacement in

medium ground finches, for example (Grant and Grant 2006; Lamichhaney et al. 2016).

The Grants found that the size of a medium ground finch's beak influenced the kinds of food it consumed. The individuals with large beaks would sometimes crack open hard woody fruits, but the ones with smaller beaks preferred feeding on smaller seeds.

In 1982, Daphne Major was invaded by a few large ground finches (*Geospiza magnirostris*), which flew in from another island. Large ground finches have very big, strong beaks. In 2004, a drought brought the large ground finches into intense competition with the big members of the medium ground finch population. Meanwhile, the small members of the medium ground finch population didn't experience this competition. The average beak size in the medium ground finch population decreased in the next generation.

More recently, Yoel Stuart, then at Harvard University, and Todd Campbell, then at the University of Tennessee, set up an experiment to see if they could trigger ecological

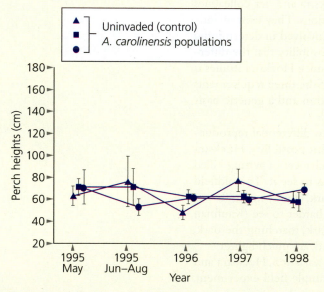

A Only *A. carolinensis* occurs on each island

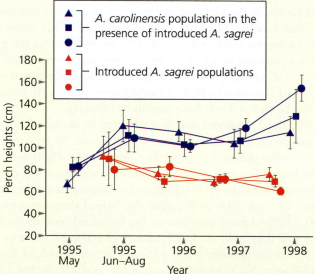

B Both species occur together

Box Figure 10.1.1 *Anolis carolinensis* is native to the southeastern United States, whereas *Anolis sagrei* was introduced from Cuba in 1880. In 1995, scientists introduced *A. sagrei* to islands in Florida and observed a rapid shift in the behavior of the native lizards (A, B). Revisiting the islands 15 years later, the scientists found that the same behavior change had persisted (C). What's more, the lizards had adapted to their higher perch with changes to their toes, which they use to grip the trees (D). (Data from Stuart et al. 2014; photos: top, Ambika Kamath; bottom, jillyafah/iStock/Getty Images)

character displacement in a species of anole lizards called *Anolis carolinensis* (Stuart et al. 2014).

The lizards are native to the southeastern United States, where they forage on a range of substrates—on the ground, on the trunks of trees, and in the lower canopy. Until the early 1900s, the lizards had no competition in this ecological niche. But then an invasive species, the Cuban brown anole lizard (*Anolis sagrei*), was introduced to the southeastern United States and began to compete directly with *A. carolinensis* for the same substrates.

Today, *A. sagrei* has yet to reach some tiny islands just off the east coast of Florida. Stuart and Campbell realized that these islands could act as laboratories for testing ecological character displacement. They introduced *A. sagrei* to three islands, where the lizards began to compete with the native *A. carolinensis*. As they observed the competition on these islands, Stuart and Campbell also observed the native *A. carolinensis* on three other islands. These invader-free islands served as their controls.

The results of their experiments are shown in **Box Figure 10.1.1**. On islands where *A. carolinensis* occurred by itself, the lizards preferred to perch on trees almost a meter from the ground (A). But on the islands where invaders were introduced (B), *A. carolinensis* began to concentrate their activity higher in the canopy, where they experienced less competition.

Fifteen years later, the scientists came back to the islands to revisit the lizards. *A. sagrei* had managed to reach more islands, but there were still islands that were home only to *A. carolinensis*. On islands with competition, *A. carolinensis* remained higher in trees (C). This behavioral shift also led to an evolutionary change in morphology. The lizards that shifted to higher in the trees evolved larger toe pads with more ridges (D). In only 20 generations, natural selection appears to have driven this change, in a striking case of ecological character replacement.

Ecological character displacement refers to evolution driven by competition between species for a shared resource (for example, food). Traits evolve in opposing directions, minimizing overlap between the species.

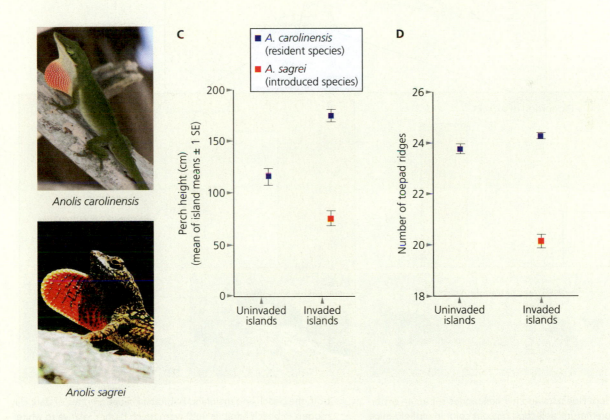

Anolis carolinensis

Anolis sagrei

■ *A. carolinensis* (resident species)

■ *A. sagrei* (introduced species)

individuals, whereas in the more complex and darker backgrounds of inland habitats, they attack primarily light individuals (**Figure 10.6**; Vignieri et al. 2010).

Experiments like these allowed Hoekstra and her colleagues to develop a detailed hypothesis for the evolution of coat color in oldfield mice—a hypothesis that can address both the ecological factors driving natural selection and the genetic basis that makes it possible. In mainland populations, there is genetic variation for coat color, based on different alleles for genes involved in pigmentation. Predators are quick to kill off mice with alleles that produce light coats, keeping the frequency of those alleles very low in the population. As a result, a disproportionate number of brown mice survive long enough to breed. Because coat color is heritable to some degree, later generations will also tend to be brown.

Several thousand years ago, some oldfield mice colonized Gulf Coast beaches and barrier islands. The dark mice stood out and were more likely to be killed. Natural selection favored genetic variants that produce pale coats, leading to a drastic shift in the average phenotype of the beach population. On the Gulf Coast, it turns out, these lighter coats were the results of mutations to several genes involved in the pathway for pigmentation. One mutation changed a single amino acid in the melanocortin-1 receptor (*Mc1r*), decreasing the sensitivity of the receptor to signals that would otherwise lead to the production of dark pigmentation. A second mutation increased expression of a gene known as *Agouti,* which interferes with the signaling of *Mc1r.*

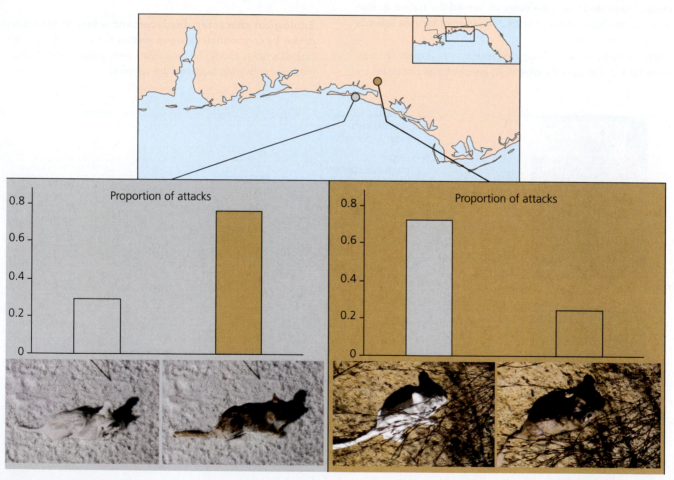

Figure 10.6 Hopi Hoekstra and her colleagues set up an experiment to measure natural selection on coat color in oldfield mice. Clay models of mice were painted to resemble beach or mainland forms and placed in either mainland or beach habitat in Florida. Blending into the background effectively reduced predation rate in both the beach and mainland habitats. Predation rates of dark clay models in beach habitats (left) were much higher relative to white models, and predation rates of light models in mainland habitats (right) were much higher relative to dark models. (Data from Vignieri et al. 2010; photos: Sacha Vignieri)

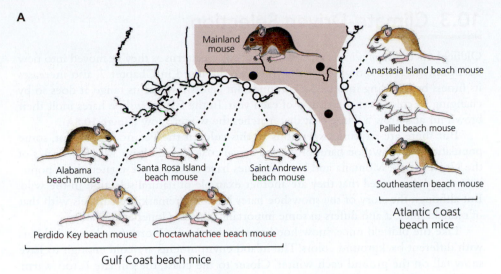

A

Mainland mouse

Anastasia Island beach mouse

Pallid beach mouse

Southeastern beach mouse

Atlantic Coast beach mice

Alabama beach mouse

Santa Rosa Island beach mouse

Saint Andrews beach mouse

Perdido Key beach mouse

Choctawhatchee beach mouse

Gulf Coast beach mice

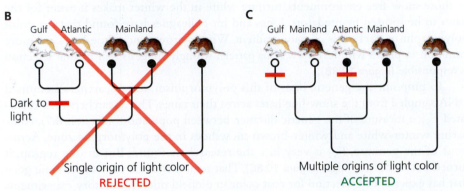

B

Gulf Atlantic Mainland

Dark to light

Single origin of light color
REJECTED

C

Gulf Mainland Atlantic Mainland

Multiple origins of light color
ACCEPTED

Figure 10.7 A: Separate pale beach mouse subspecies occur along both the Gulf Coast and the Atlantic Coast of Florida. Two hypotheses for their evolution are illustrated in B and C. In B, the pale beach mice from both coasts share a pale common ancestor. In C, two lineages of mainland mice independently evolved pale coloration. Studies on the DNA of the mice reject hypothesis B and support hypothesis C (see text).

Combined, these two genetic changes resulted in reduced levels of melanin synthesis and lighter overall coat color (Figure 7.18).

The Gulf Coast is not the only place where you can find white oldfield mice. On Florida's Atlantic Coast, they exist as well. It's unlikely that one population descended from the other because that would have required white mice traveling across 300 kilometers of dark soils. Instead, it's much more likely that the two populations evolved from mainland mice independently. Hoekstra and her colleagues compared the genetic basis of coat color in the two coastal populations and got an intriguing result: the light-colored mice on the Atlantic Coast lacked the Gulf Coast alleles of *Mc1r*. The altered phenotype of the Atlantic Coast mice appears to be produced by mutations to other genes (Steiner et al. 2009). This turns out to be a common pattern in natural selection: closely related populations under the same selective pressures often evolve the same phenotype in parallel. But the populations can realize that phenotype through different mutations (**Figure 10.7**).

Key Concepts

- When oldfield mice moved into new coastal habitats, the nature of selection they experienced changed. Camouflaged individuals still survived best, but the colors conferring the best camouflage shifted from brown to white.

- Atlantic and Gulf Coast mouse populations each evolved white fur, but the specific mutations responsible are different. ●

10.3 Climate-Driven Selection

Oldfield mice (*Peromyscus*) have evolved new color patterns as they've moved into new environments. The snowshoe hare, which we discussed in Chapter 7, also increases its fitness by blending into its background. But across most of its range, it does so by changing its color over the course of each year. In the fall, snowshoe hares molt their brown fur and grow a white coat that matches the winter snow (**Figure 10.8A**).

But there are striking exceptions to this rule. In the Pacific Northwest, some populations of snowshoe hares produce a *brown* coat after molting. Matthew Jones of the University of Montana and his colleagues have recently investigated these populations and discovered that they are another example of natural selection in the wild. But although the story of the snowshoe hares has some remarkable parallels with that of oldfield mice, it also differs in some important respects (Jones et al. 2018).

Like the oldfield mice, snowshoe hares in the Pacific Northwest live in habitats with different background colors. The inland environments are cold enough to have snow fall on the ground each winter. Closer to the coast, the climate is too warm. In those snow-free environments, turning white in the winter makes it easier for the hares to be spotted by predators. Jones and his colleagues have found that the color polymorphism tracks this climate gradient. Winter-brown snowshoe hares are more common in places with less snow. This pattern strongly suggests that natural selection is responsible (**Figure 10.8B**).

To pinpoint the genetic basis of this polymorphism, the researchers examined DNA sampled from the snowshoe hares across their range. The researchers then calculated F_{ST}, a measure of the genetic distance between populations (Section 8.7), comparing winter-white and winter-brown snowshoes in the polymorphic zone. Across most of the genome, F_{ST} is very low, the researchers found. But in one region, it turned out to be very high (**Figure 10.8C**). That region contains *Agouti*, the same gene that has experienced selection for coat color in oldfield mice. Laboratory experiments on snowshoe hare cells revealed that in winter-white animals, the expression of *Agouti* is increased in the fall.

To reconstruct the history of this allele, the researchers studied the phylogeny of snowshoe hares and their close relatives. They compared SNPs across the entire genome to build a species tree, which showed that winter-brown snowshoe hares from Washington are closely related to winter-white snowshoe hares from Utah and Montana. But when the scientists only used the region of DNA that included the *Agouti* gene, they ended up with a different tree. On this tree, the winter-brown snowshoe hare belonged to a clade with black-tailed jackrabbits, which never turn white in the winter (**Figure 10.8D**).

The best explanation for this pattern is that snowshoe hares in the Pacific Northwest were originally winter-white. During the last ice age, glaciers covered part of the region. Thanks to the cold climate of the time, the parts that were ice-free were covered each winter by snow. Natural selection strongly favored hares that could match that white background.

From time to time, snowshoe hares interbred with black-tailed jackrabbits. During the ice age, any offspring inheriting the winter-brown allele experienced lower fitness because they were easier to spot against a snowy background. But once the glaciers retreated at the end of the ice age, the situation flipped. Now the winter-brown allele from the black-tailed jackrabbits was beneficial, and it experienced a selective sweep.

Understanding examples of natural selection in response to past climate change can offer important clues to the future. As we will discuss in further detail in Chapter 14, we humans are rapidly altering the climate by emitting greenhouse gases. Computer models indicate that large regions of the current range of snowshoe hares will receive less snowfall, lowering the fitness of the winter-white phenotype once more. It's possible that winter-brown alleles will be favored by natural selection in these regions as

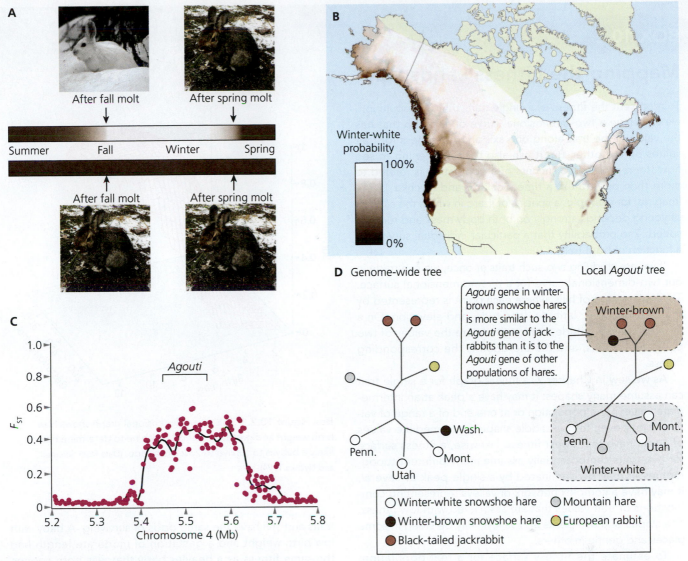

Figure 10.8 A: Across most of their range, snowshoe hares produce a white coat after their fall molt. But in some places, they remain brown. B: The probability that snowshoe hares will turn white is close to 100% across most of their range. But at the warm fringes of their range, they are more likely to remain brown. C: Winter-brown and winter-white snowshoe hares in Washington state have little genetic divergence, except in a tiny region of their genome that includes the *Agouti* gene. D: Genome-wide analysis of rabbit DNA (left) shows that winter-brown snowshoe hares in Washington state are most closely related to winter-white snowshoe hares from other parts of North America. But the winter-brown *Agouti* allele has a different evolutionary history (right). This pattern suggests that the winter-brown allele was introgressed into the snowshoe hare genome when hares hybridized with black-tailed jackrabbits, and the allele spread rapidly in the Pacific Northwest due to natural selection. (Data from Jones et al. 2018; photos: L. Scott Mills)

well. On the other hand, it's also possible that hare populations will not evolve fast enough to adapt to rapid climate change, which may drive down their numbers.

Key Concepts

- Natural selection can lead to variation in space—across habitats or environments—just as dramatically as it can lead to variation in a single habitat over time.
- The effect of natural selection can change in response to the changing climate.
- Natural selection can rapidly spread an allele introduced into a species by hybridization. ●

BOX 10.2

Mapping the Fitness Landscape

To represent the fitness of a single trait within a population, we can draw a two-dimensional curve on a graph that has the value of the trait along one axis and the fitness for trait values on the other. But it can also be enlightening to see how the reproductive success of a population is related to more than one trait at a time (Schluter and Nychka 1994). Imagine, for example, a species of lizard in which the survival of young depends strongly on both body mass and running speed. The probability that a particular individual survives is a function of both traits.

If we are studying two such traits at once, we can trade in our two-dimensional curve for a three-dimensional surface. Think of a range of hills, where each point is represented by three coordinates: latitude, longitude, and elevation. On a fitness surface, latitude and longitude are the values of two phenotypic traits, and the elevation is the corresponding fitness.

As we saw in Chapter 7, a fitness graph for a single trait can assume many shapes: it may have a peak at an intermediate value in the population or at one end of a range of values; it may even form a saddle shape if intermediate values of a trait have the lowest fitness. Likewise, a fitness surface for two traits can potentially assume many different topographies. It may be dominated by a single peak or several; it may have a complex network of valleys representing combinations of trait values that are associated with low fitness. The slopes of these peaks and valleys may be steep in some places and gentle in others.

To estimate the fitness surface for a real population, we must take measurements of the two traits in a large sample of individuals and then find the topography that best fits the data. There are a number of methods for doing so; one of the most influential was developed in 1994 by Dolph Schluter and Douglas Nychka, based on a curve-fitting technique known as cubic splines. In one example, they examined medical records from 7307 babies. For each child, Schluter and Nychka compared two traits—birth weight and maternal gestation period—and also noted whether the child survived the first 2 weeks after birth.

Box Figure 10.2.1 shows their result: a dome-shaped topography. The steepness of the dome shows the strength of selection. Selection acts most strongly against small babies with short gestation periods. But babies could have different combinations of intermediate values of the two

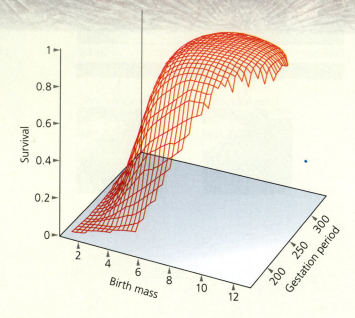

Box Figure 10.2.1 This three-dimensional graph shows how birth weight and gestation period combine to determine how likely a baby is to survive to 2 weeks of age. (Data from Schluter and Nychka 1994)

traits and still have the same odds of surviving. A baby with low birth weight and a gestation of moderate length had the same fitness as a heavier baby that was born sooner. The topography gains a dome shape because babies with the largest body size and longest gestation periods were less likely to survive. Thus, this topography reveals that the selection human babies experience is mainly directional selection but also weakly stabilizing (Schluter and Nychka 1994).

Of course, it's also possible for more than two traits to be strongly selected in a population. Unfortunately, our brains cannot visualize graphs in four or more dimensions, but it is possible to use the same methods we've considered here to analyze these more complex interactions. The notion of an evolutionary landscape can also be a useful metaphor for thinking about how populations change over time. For extensive discussions of these issues, see Schluter (2000) and Svensson and Calsbeek (2012).

10.4 Predators Versus Parasitoids

More than one agent of selection can act on a trait (see **Box 10.2**). And sometimes, scientists have found, those agents drive a population in two different directions at once. At first it can be hard to see these multiple agents at work, just as a rope may become motionless during a game of tug-of-war as two teams are pulling in opposite directions. It takes careful experiments to tease apart the effects of the different agents. One of the most striking cases of this evolutionary tug-of-war is well documented in the gallfly (*Eurosta solidaginis*; **Figure 10.9**).

Gallflies and Goldenrod

Female gallflies lay eggs into the growing tips of goldenrod (*Solidago* spp.), a plant that thrives in old farm fields. After the eggs hatch, each larva bores into the bud tissues to feed. The larva secretes fluids containing proteins and other molecules that change the gene expression of cells in the plant. The plant cells grow into a bulbous, tumorlike structure, known as a gall, that is hard on the outside and soft on the inside. Cradled at the center of the gall, the gallfly larva can feed on the plant's fluids.

The gall is made of plant cells, and yet its growth is controlled by the flies. It can thus be considered an **extended phenotype** of the flies. Biologists Arthur Weis and Warren Abrahamson found that the final size of galls varies, and at least some of this variation is due to inherited differences among the fly larvae, not the plants. The scientists were able to demonstrate this link with a two-part experiment: they allowed several different female gallflies to lay eggs on goldenrod, and they let each female lay eggs on several different goldenrod plants. (To remove the effects of genetic differences among the plants, they had all of the females lay eggs on clones—plants with the same genotype.) Abrahamson and Weis let all of the offspring of each female make galls, and

An **extended phenotype** is a structure constructed by an organism that can influence its performance or success. Although it is not part of the organism itself, its properties nevertheless reflect the genotype of each individual. Animal examples include the nests constructed by birds and the galls of flies.

Figure 10.9 The gallfly (*Eurosta solidaginis*) lays her eggs into the stems of goldenrod (A). When the fly larvae hatch, they secrete chemicals that induce the plant to form a gall (B), which serves as both a food source and protection for the developing larva (C). (Warren G. Abrahamson)

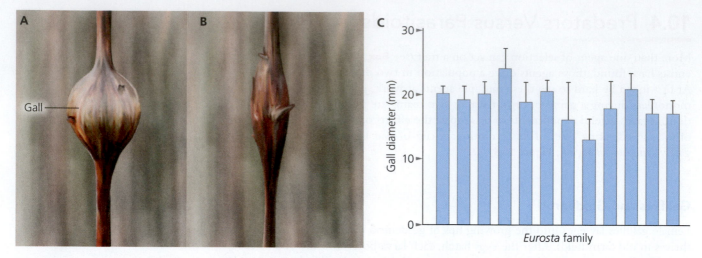

Gall

Eurosta family

Gall diameter (mm)

Figure 10.10 Fly larvae differ in the diameter of the galls they produce (A, B), and some of this variation is heritable. Larvae from the same mother produce similar-sized galls, and some families produce much larger galls than others (C). (Data from Abrahamson and Weis 1997; photos: Warren G. Abrahamson)

they compared the average sizes of the galls that were produced. They found that fly families differed significantly in the gall sizes that they produced (**Figure 10.10**).

This difference pointed to an inherited component in the variation in how the flies induced galls to form in their host plants. Galls met two of the conditions for natural selection—variation in populations and an inherited component of that variation (Weis and Abrahamson 1986).

A Tug-of-War on Fitness

For the second part of their experiment, Weis and Abrahamson considered the third condition for evolution by natural selection: whether an inherited phenotypic trait influenced fitness. When they investigated whether the size of galls affected the survivorship of larval gallflies in natural populations, gall size turned out to matter a lot. Galls give the larvae physical protection from two major sources of mortality: predatory birds and parasitoid wasps. Predatory birds tear into the galls and pull out the larvae, but parasitoid wasps pose a different problem. Female parasitoid wasps drill their ovipositor—the organs used to lay eggs—into the galls to lay eggs beside the fly larvae. Once the parasitoid eggs hatch, the wasp larvae develop very fast—faster than the fly larva—and they eat both the gall tissues and the fly larva as they grow.

The likelihood of each of these sources of mortality is influenced by the size of the gall, but in different ways. Bird predation, Abrahamson and Weis found, selects very strongly for small gall sizes. During the winter, when vegetation has died back, bigger galls are easier for the birds to find. As a result, the fly larvae in large galls get eaten more often than the fly larvae in smaller galls. Abrahamson and Weis observed this same pattern of selection at several different sites and during multiple years. In all these cases, predation by birds favored the evolution of small, inconspicuous gall sizes (**Figure 10.11**).

The parasitoids also cause strong selection on the galls, but their effect is opposite that of bird predators. Parasitoid female wasps must reach into the center of the gall to place their eggs on the surface of the gallfly larvae. Although the wasps have unusually long ovipositors, some galls in a population are so big that the parasitoids cannot reach the fly larva inside. As a result, the larvae in the largest galls often escape being parasitized. Parasitoid wasps thus favor the evolution of large galls, and this pattern, too, was observed across multiple populations and many years.

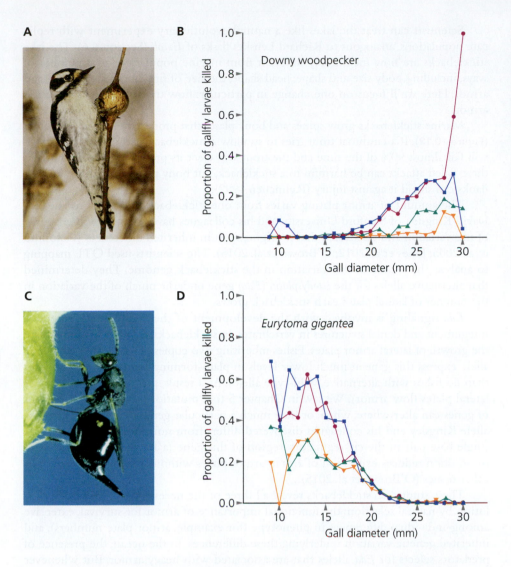

Figure 10.11 A: Downy woodpeckers feed on goldenrod galls during the winter, when oldfield vegetation has died back and the galls are most visible. B: They find primarily large galls and thus exert negative directional selection on gall size. C: Tiny parasitoid wasps inject their eggs into galls and kill larvae in the smallest galls. D: They produce positive directional selection on gall size. The result is a balance: natural selection favors flies that produce intermediate gall sizes. Colored lines represent different sampled populations. (Data from Weis et al. 1992; A: Read, Marie / Animals Animals; C: Warren G. Abrahamson)

Taken together, the studies of Abrahamson and Weis reveal a balance. When galls are too large, larvae are likely to be eaten by birds. When galls are too small, larvae are likely to die from parasitoids. The result is a trade-off with stabilizing selection for intermediate-sized galls (Weis et al. 1992).

- When agents of selection act in opposition, the net effect can be a balance: stabilizing selection for an intermediate trait value. ●

Key Concept

10.5 Replicated Natural Experiments

Three-spined sticklebacks (*Gasterosteus aculeatus*) are small fishes that live across much of the Northern Hemisphere. Some populations live as adults off the coasts of North America, Europe, and Asia. Many of these fishes enter streams and lakes to spawn, and their offspring swim back to the sea. Other stickleback populations live their entire lives in lakes. Their ancestors colonized newly formed freshwater lakes as the glaciers withdrew at the end of the last ice age about 12,000 years ago. Some individuals remained in the lakes, founding resident populations. In each of these lakes, the isolated sticklebacks experienced a new set of selection pressures.

Scientists can treat the lakes like a natural evolutionary experiment with replicate populations, analogous to Richard Lenski's flasks of *E. coli* (Section 6.6). The lake sticklebacks are now measurably different from marine populations in a number of ways, including body size and shape, head shape, the size of fins, numbers of teeth, and armor. Here we'll focus on one change in particular: how the sticklebacks lost their armor.

Marine sticklebacks grow spines and bony plates that protect them from predators (**Figure 10.12**). If a cutthroat trout tries to swallow a stickleback with spines, the attack will fail almost 90% of the time and the trout will eject its prey from its mouth. Even these failed attacks can be harmful to a stickleback, but bony plates that grow along its flanks can shield it against injury (Reimchen 1992).

The amount of armor plating varies from one stickleback individual to another. David Kingsley, of Stanford University, and his colleagues have investigated the source of this variation and discovered that it originates from inherited differences (Colosimo et al. 2005; Jones et al. 2012; O'Brown et al. 2015). The scientists used QTL mapping to analyze the source of this variation in the stickleback genome. They determined that alternative alleles for the *ectodysplasin* (*Eda*) gene underlie much of the variation in the number of lateral plates each stickleback grows.

Eda signaling is involved with the development of the adult outer protective integument and dental structures in vertebrates. In sticklebacks, *Eda* appears to regulate the growth of lateral armor plates. Fishes inheriting two copies of a recessive "low" *Eda* allele express this gene at much lower levels in plate-forming regions of their bodies than do fishes with alternative, wild-type alleles. As a result, they develop with fewer lateral plates (low armor). We saw in Chapter 5 that mutations to regulatory regions of genes can alter where, when, or how much a particular gene is expressed. The low allele Kingsley and his colleagues discovered differs from normal *Eda* alleles at just a single base pair in the cis-regulatory region of this gene (a T changed to a G). This substitution reduces expression of *Eda* in armor plates without affecting expression in other tissues (O'Brown et al. 2015).

These studies on sticklebacks reveal all three of the necessary conditions for evolution by natural selection: the functional importance of armor for survival, extensive among-individual differences in phenotype (for example, armor plate numbers), and inherited genetic variation underlying these differences. In the ocean, the presence of predators selects for *Eda* alleles that are associated with heavy armor. But whenever stickleback fishes became isolated in lakes with few or no predators, the selection pressures they experience changed. Elaborate defenses no longer raised their fitness.

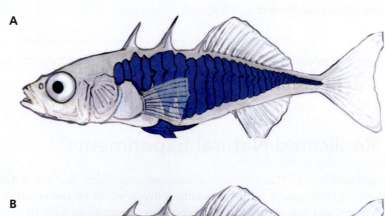

Figure 10.12 A: Marine three-spined sticklebacks protect themselves from predators with long dorsal and pelvic spines and with a row of tough lateral plates. B: In freshwater lakes, the expression of these defensive structures can be greatly reduced.

Sticklebacks have made the transition from the ocean to freshwater lakes repeatedly for millions of years, as sea level has risen and fallen. Michael Bell, of Stony Brook University, and his colleagues have found a remarkable set of freshwater stickleback fossils from a geological formation in Nevada dating back 10 million years (Bell et al. 2006). The fossil record is so dense that they have reconstructed 110,000 years of evolutionary history in 250-year slices. Bell and his colleagues measured the armor on the stickleback fossils in order to estimate the long-term history of selection on the animals.

For the first 93,000 years, Bell found sticklebacks with just a few small protective spines. But then this ancestral phenotype was joined by more heavily armored sticklebacks with spines that were longer and more numerous. Bell suspects that this influx of fishes came from nearby streams, possibly driven by the disappearance of phytoplankton they fed on in the streams. At first the two stickleback forms coexisted in the fossil record. But then the early fishes with few spines disappeared.

Over the next 17,000 years, the defensive structures in the new, more heavily armored fish regressed (**Figure 10.13**). Step by step, its spines got shorter and disappeared. By the end of this period, the new stickleback had come to resemble the earlier form that it had replaced.

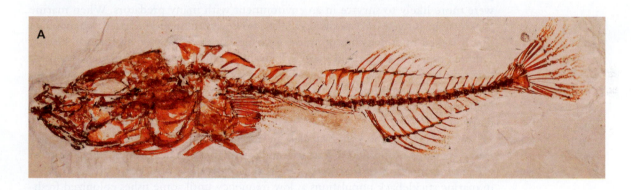

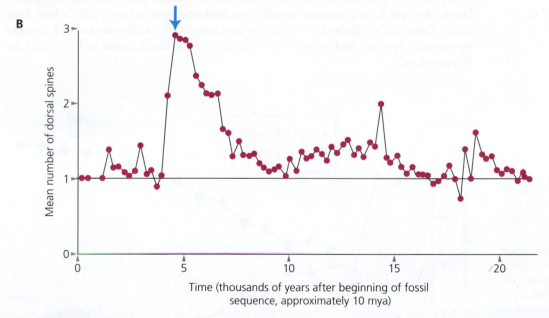

Time (thousands of years after beginning of fossil sequence, approximately 10 mya)

Figure 10.13 A: Sticklebacks have an unusually comprehensive fossil record, and this has let biologists reconstruct the gradual evolution of their form. B: A site in Nevada preserves more than 100,000 years of stickleback evolution in a lake from 10 million years ago. This graph shows the mean number of dorsal spines on fossils from the last 20,000 years of the sequence. The arrow indicates the sudden arrival of highly armored fishes, which most likely came from nearby streams where predators favored larger numbers of protective spines. These fishes replaced the earlier population of sticklebacks in the lake and then gradually lost their spines as well. (Data from Bell et al. 2006; A: Michael A. Bell)

The same pattern has occurred among sticklebacks that became isolated in lakes after the last ice age. Heavily armored fishes spread into the freshwater lakes, but they varied, to some extent, in the extent of their armor. Individuals wielding less armor had more offspring than the heavily armored ones. The mean level of armor plating thus dropped over time. Today, the sticklebacks in freshwater lakes have fewer spines and drastically reduced lateral plating compared to their closest marine relatives. This evolutionary reduction of defensive weaponry occurred repeatedly in lake after lake.

Dolph Schluter, of the University of British Columbia, and his colleagues are studying living populations of sticklebacks in Canadian lakes to better understand how natural selection can erode defenses. Their data show that without predatory fishes in the lakes, the energy investment to grow armor no longer benefits the fishes. In fact, it's very energy-expensive to produce armor in lakes because freshwater has low concentrations of the ions necessary for bone growth. As a result, fishes with low–*Eda* alleles have an advantage in freshwater. They grow to be larger as juveniles, which helps more of them survive the cold winters, and begin breeding sooner than fishes that have the "complete armor" version of the *Eda* allele (Barrett et al. 2008).

Taken together, these studies allowed scientists to reconstruct the recent history of natural selection on sticklebacks in freshwater lakes. The low allele for the *Eda* gene was rare in populations of marine sticklebacks because heavily armored fishes were more likely to survive in an environment with many predators. When marine sticklebacks moved into freshwater lakes, however, their environment changed so that it lacked high densities of predatory fishes. In these new environments, heavily armored fishes no longer had a survival advantage, and the cost of growing spines and plates meant that they could not grow in mass as fast as other fishes. The low allele for the *Eda* gene, which once lowered fitness in marine sticklebacks, now raised it in the predator-free lakes. The allele spread, and the average number of lateral armor plates dropped.

By comparing the low–*Eda* allele in different stickleback populations, Kingsley found that it is quite old. At least two million years ago, the low–*Eda* allele arose in the marine ancestors of freshwater sticklebacks. This allele is recessive, and it lingered in marine stickleback populations at low frequency until some fishes colonized freshwater habitats. (We saw in Chapter 6 that when recessive alleles are rare in a population, they are largely invisible to selection, enabling them to persist for a very long time.) Once the sticklebacks were in the new habitat, the allele was favored strongly by natural selection, leading to parallel evolution of reduced armor in lake after lake (**Figure 10.14**).

Figure 10.14 Rapid evolution of stickleback armor in Loberg Lake. In 1982, wildlife managers poisoned all the fish in Loberg Lake near Cook Inlet, Alaska, to improve lake conditions for introduced recreational fish. Before the extermination, stickleback fishes in the lake had all been of the "low-armor" form (that is, fishes with low-*Eda* alleles). In 1988 an influx of marine sticklebacks from Cook Inlet colonized the lake, and these fishes had a mixture of high- and low-armor forms. Over the next two decades, selection drove rapid increases in the frequency of the low-armor alleles. (Data from Bell and Aguirre 2013)

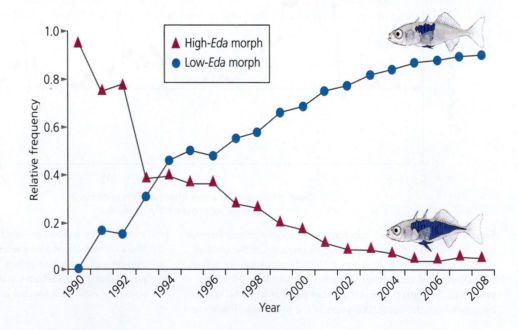

One final piece completes this story: an exception that, in effect, proves the rule. The sticklebacks in Lake Washington, near Seattle, have full armor. Daniel Bolnick, a University of Texas biologist, and his colleagues wondered why these sticklebacks should differ so much from those in other lakes (Kitano et al. 2008). They discovered that the increase in armor in Lake Washington sticklebacks happened very recently. Fishes collected in the late 1950s had the reduced armor typical of other lake sticklebacks. Within the following 40 years, however, the sticklebacks in Lake Washington changed. They rapidly evolved back toward a fully armored form.

Bolnick argues that the solution to this paradox lies in the government regulations that have reduced pollution in Lake Washington over the past few decades. Trout that have been introduced into the lake can now see the sticklebacks more easily in the clearer water and can attack them. This increase in predation can be dated to the 1970s. It also coincides with the beginning of the evolutionary reversal in stickleback morphology. Thus, humans may have reversed the direction of selection.

- Sometimes multiple populations independently experience the same change in their selection environment. These populations are ideal for evolutionary studies because they are replicated natural experiments. The nature of the evolutionary response can be observed for each population and compared across the different populations. •

Key Concept

10.6 Humans as Agents of Selection

The evolution of sticklebacks in Lake Washington is just one of many examples of human-driven selection. Human-driven selection had its first huge impact on the world about 10,000 years ago, when we first started to domesticate plants and animals (Doebley 2006).

The early stages of domestication may have begun inadvertently. Wild wheat plants, for example, grow seeds that break away through a process called shattering. In the wild, a mutation that makes wheat fail to shatter is deleterious because the seeds remain trapped on their parent plants and germinate less often. When people began to gather wheat plants, they preferred the ones that failed to shatter because the seeds were still attached. They may have planted some of the seeds near their settlements. As a result, those people inadvertently began to select for reduced shattering. Thousands of years later, people started raising the plants on large-scale farms and consciously selecting certain plants to breed (Diamond 2002; Doebley 2006).

Regardless of whether these early episodes of artificial selection were incidental or deliberate, the impacts were dramatic. Domesticated wheat plants have many traits not found in their wild relatives. Their seeds ripen simultaneously, grow in tight bunches at the end of branches, and don't shatter—all traits that make them easier to harvest (Zeder et al. 2006).

Domestic animals underwent a similar transformation. Humans selected behavioral traits in their livestock, such as increased tolerance to penning, increased sexual precocity, and reduced wariness and aggression (Clutton-Brock 1999). Sometimes the same wild species was subsequently selected in many different directions (**Figure 10.15**). Wild cabbage (*Brassica oleracea*) was selected for its leaves (cabbage, kale), stems (kohlrabi), flower shoots (broccoli, cauliflower), and buds (brussels sprouts).

Artificial Selection in Maize

Archaeologists and geneticists are reconstructing the evolutionary steps that some crops took from their wild ancestors (Bruford et al. 2003; Zeder et al. 2006; Kantar

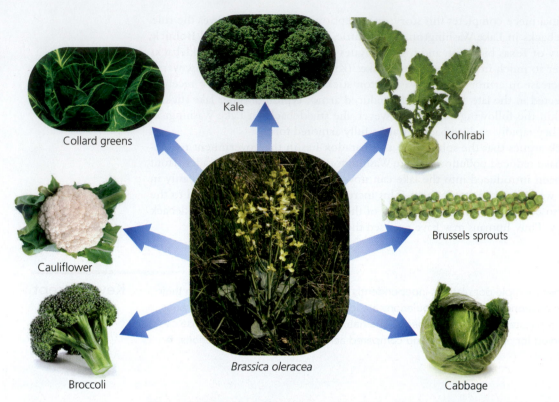

Figure 10.15 Artificial selection on wild cabbage (*Brassica oleracea*) resulted in the evolution of diverse plant forms, including broccoli, cauliflower, brussels sprouts, cabbage, collard greens, kale, and kohlrabi. (Clockwise from left: Valentyn Volkov / Shutterstock; Pukach / Shutterstock.com; chantel / Shutterstock; Maria Meester / Shutterstock; Soyka / Shutterstock; marilyn barbone / Shutterstock; JIANG HONGYAN / Shutterstock; FloralImages / Alamy Stock Photo)

et al. 2017), as shown in **Figure 10.16**. One of the best characterized of these events is the evolution of maize (or what's known as corn in English-speaking countries). Approximately 9000 years ago, farmers in the Balsas River Valley of southern Mexico began selectively planting and harvesting individuals of a streamside plant called teosinte. Teosinte was taller and broader leaved than most grasses, and people collected its seed heads for food. All modern maize appears to have descended from this original domestication event (Doebley 2004).

After domesticating teosinte, farmers continued to select for advantageous traits for thousands of years. Archaeologists have documented the evolutionary response of maize to artificial selection by unearthing ancient cobs (**Figure 10.17**). By 5500 years ago, cobs had already increased in the number of rows of kernels and in kernel size.

Some ancient cobs still retain fragments of DNA. The sequences of these fragments contain important clues about how this crop evolved. Researchers have managed to extract DNA from a 5310-year-old cob from Mexico (Ramos-Madrigal et al. 2016). This maize plant already had many of the derived alleles found in modern maize breeds, but not found in teosinte. Yet the cob also had some primitive alleles, indicating that the domestication process had many years left to go.

Analysis of allelic diversity in these early cobs suggests that by 4400 years ago, early maize had lost almost 30% of the allelic diversity originally present in wild teosinte populations. Such loss of variation is indicative of strong selection and a genetic bottleneck (Chapter 6), as would be expected if selective harvesting entailed breeding only a small subset of the wild population (Jaenicke-Després et al. 2003).

John Doebley of the University of Wisconsin and his colleagues identified mutations of major effect in three genes that contribute to the evolved morphology of maize. One of them (*teosinte branched 1*) carries a maize variant that represses the growth of lateral meristems. It helps give rise to fewer branches on maize compared to

A Corn

WILD   DOMESTICATED

B Tomatoes

WILD   DOMESTICATED

C Sunflowers

WILD   DOMESTICATED

Figure 10.16 Domesticated crops and their closest wild relatives are separated by dramatic differences in morphology. A: Teosinte (left), from which maize was developed, grows multiple stalks and long branches. Maize, by contrast, grows only a single stalk. The ears on teosinte and maize plants are also different. Kernels grow naked on the surface of maize, whereas teosinte grains are enclosed in a triangular casing. B: A massive cultivated tomato and the small fruit of its wild progenitor. C: A wild sunflower plant (left) has many small heads borne on multiple slender stalks, whereas a cultivated sunflower plant (right) has a single large head borne on a thick stalk. (Data from Doebley et al. 2006; A: John Doebley; HUIZENG/Shutterstock.com; B: Photo by Kent Loeffler/Cornell University; C: Trong Nguyen/Shutterstock.com; toa55/Depositphotos)

teosinte. A second gene, *prolamin box binding factor*, is involved in the production of seed storage proteins in the kernels. A third gene, *sugary 1*, encodes an enzyme that alters the properties of starch in the kernels in ways that affect the textural properties of food produced from the kernels, such as tortillas (Jaenicke-Després et al. 2003).

5500 years ago

4400 years ago

Figure 10.17 Archaeologists have uncovered discarded maize cobs from sequential periods of occupation of Tehuacán, Mexico. These specimens document the gradual evolutionary increase of kernel number and cob size. (Mesoamerican Maise: 5000 years of maize evolution from tiny wild corn to a modern example ca. 1500 C.E. from Prehistory of the Tehaucan Valley, Vol. 1, 1967. Fig. 122 © Robert S. Peabody Museum of Archaeology, Phillips Academy, Andover, Massachusetts)

Artificial Selection in Dogs

A similarly impressive history of human-caused evolution occurred with the domestication of dogs. Studies on dog fossils and genes indicate they were domesticated from gray wolves in Eurasia about 25,000 years ago (Freedman and Wayne 2017). Early domestication appears to have involved primarily selection for behavior. Dogs, unlike wolves, can understand humans surprisingly well. They can, for example, recognize that a pointed finger indicates something they should pay attention to.

A more recent phase of domestication occurred in the past few centuries as people began selecting for a variety of morphological and physiological traits associated with hunting or recreational tasks. Today, there are more than four hundred recognized breeds, and dogs exhibit more phenotypic variation than any other species (Vilà et al. 1997; Cruz et al. 2008). As in the case of maize, researchers are beginning to identify the genetic changes responsible for these remarkable evolutionary transformations in form (Figure 10.18).

As dog breeders selected for certain alleles, many deleterious mutations got swept along for the ride. Under natural circumstances, these mutations would likely have reduced individual performance and fitness. As a result, purebred dogs today are faced with an inordinate frequency of genetic maladies (Cruz et al. 2008; Freedman et al. 2016).

Figure 10.18 Centuries of artificial selection have influenced the size, shape, and behavior of domesticated dogs. Recent genetic studies are identifying some of the genes that appear to have contributed to diversity in dog morphology. A: Sutter et al. (2007) showed that an allele of the *IGF1* gene contributes to small body size. B: Akey et al. (2010) found that *HAS2* is associated with skin wrinkling. C: Shearin and Ostrander (2010) identified alleles of three genes that affect coat properties: *RSPO2* is associated with wiry hair and moustaches, *FGF5* alleles cause long or short fur, and *KRT71* alleles lead to curly or straight hair. [A: Eric Isselee/Shutterstock; B: Yuri Kevhiev/Alamy; C: Photos by Mary Bloom, American Kennel Club, From "Canine Morphology: Hunting for Genes and Tracking Mutations" by Shearin and Ostrander. Plos Biology (2010)]

Chemical Warfare

When humans domesticated crops, they created a new food supply not only for themselves but also for huge hordes of insects. The very traits that farmers favored in plants—a failure to shatter, large seeds, and a tightly synchronized life history—made their crops an ideal source of nutrition for many species of pests. These insects already had an impressive capacity for rapid growth and reproduction, and once we provided them with a banquet of crops, their numbers exploded. Swarms of pests besieged the fields, laying waste to entire farms. The battle between humans and insects was on.

Farmers searched for ways to fight off pests. Some of their attempts seem laughable today. Roman farmers believed that rubbing trees with green lizard gall could repel caterpillars and that nailing a toad to a barn door could scare weevils away from stored grain. But early farmers also stumbled across chemicals that were effective at warding off insects. For example, 4500 years ago in the ancient empire of Sumer, farmers put sulfur on their crops. Early Europeans learned to extract chemicals from plants, and by the nineteenth century, farmers had a fairly extensive arsenal of pesticides for killing insects.

Around 1870, a tiny Chinese insect turned up in farm fields around the city of San Jose, California. The creature would inject a syringe-like mouthpart into a plant and suck up the juices. The San Jose scale, as the insect came to be known, spread quickly through the United States and Canada, leaving ravaged orchards in its path. Eventually, farmers found that a mixture of lime and sulfur was most effective against the scale. After a few weeks of spraying, the San Jose scale would disappear. By 1900, however, the lime-sulfur cure was failing. Here and there, the San Jose scale returned to its former abundance.

An entomologist at the time named A. L. Melander found some San Jose scales living happily under a thick crust of dried lime-sulfur spray. Melander embarked on a widespread experiment, testing out lime-sulfur on orchards across Washington state (Melander 1914). He found that in some orchards, the pesticide wiped out the insects completely. In other orchards, as many as 13% of the scales survived.

Melander wondered why some populations of scales became able to resist pesticides. Could the lime-sulfur spray trigger a change in their biology, the way manual labor triggers the growth of calluses on our hands? Melander doubted it. After all, 10 generations of scales lived and died between sprayings. The resistance must be hereditary, he reasoned. He sometimes would find families of scales still alive amid a crowd of dead insects.

Attributing the resistance observed to heredity was a radical idea at the time. Biologists had only recently rediscovered Mendel's laws of heredity (Box 5.2). They talked about genes being passed down from one generation to the next, although they didn't know what genes were made of yet. Still, they did recognize that genes could spontaneously change—mutate—and in so doing, alter traits permanently.

In the short term, Melander suggested that farmers switch to fuel oil to fight scales, but he warned that the scales would eventually become resistant to fuel oil as well. In fact, the best way to keep the scales from becoming entirely resistant to pesticides was, paradoxically, to do a bad job of applying those pesticides. By allowing some susceptible scales to survive, farmers would keep the susceptible genes in the scale population.

Unfortunately, Melander's prophetic words appear to have fallen on deaf ears. Today, 12% of all the ice-free land on Earth is farmed, and farmers apply pesticides and herbicides across this vast expanse of cropland (Figure 10.19). When farmers apply a new chemical pesticide to a field, they kill a large proportion of its vulnerable population of pests. This die-off produces strong selection on the insect.

Individual insects with mutations for biochemical mechanisms enabling them to survive, to somehow detoxify the chemical poison, do very well. They live, whereas most of their competitors do not. These survivors now have more food to eat, boosting their survival and fecundity. As they propagate themselves, they populate subsequent generations of the pest population with offspring who are also resistant to the

Figure 10.19 When farmers spray their fields with either pesticides or herbicides, these chemicals act as agents of selection on the affected populations. (Federico Rostagno/Shutterstock)

pesticide, and alleles conferring resistance spread. The large size of insect populations can produce substantial genetic variation. When the intense selection of pesticides is applied to the insects, resistance can evolve rapidly.

It takes only a few years, in fact, for resistance to a new pesticide to emerge (Gould et al. 2018). More than five hundred species of pest insect are now known to be resistant to at least one pesticide (**Figure 10.20**). Farmers often have to apply more of a pesticide to control resistant pests over subsequent years; today, farmers in the United States spend $12 billion on pesticides (Palumbi 2001a). Many species are now resistant to so many pesticides that they are impossible to control, and up to a third of farm production is lost to pest damage (Palumbi 2001b). The evolution of resistance also poses a risk to public health because the high concentrations of pesticides can contaminate groundwater and streams.

Insects are not the only organisms that can make life difficult for farmers. Weeds can invade farm fields and outcompete crop plants for space. On large farms, pulling weeds out of the ground is simply not practical. So farmers fight weeds by spraying their fields with chemicals known as herbicides, which can kill plants. Once the weeds are dead, the farmers can plant their crops. Yet time and again, weeds have

Figure 10.20 Evolution of resistance to insecticides by houseflies on Danish farms. The width of each bar reflects the extent of use, the inverted triangle symbol the date of the first confirmed case of resistance of a given insecticide, and the letter *R* the date when most populations were resistant. (Data from Wood and Bishop 1981)

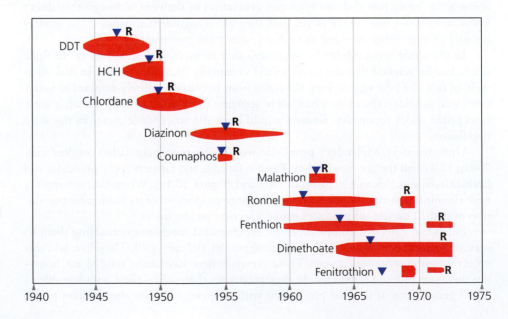

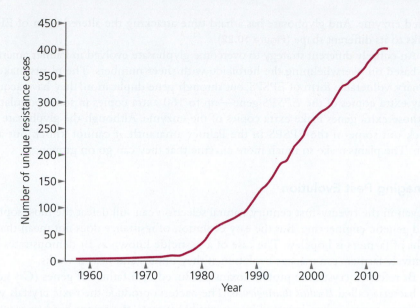

Figure 10.21 Weeds around the world have become more resistant to a variety of herbicides. This graph shows the global increase in the number of herbicide-resistant weeds since 1955. (Data from Heap 2014)

evolved resistance to herbicides, just as insects have evolved resistance to pesticides (**Figure 10.21**; Powles and Yu 2010; Heap 2014).

Evolution in the Age of Genetically Modified Crops

One of the latest failures of herbicides involves the chemical glyphosate, which the company Monsanto sells under the brand name Roundup. Glyphosate kills weeds by blocking the construction of amino acids that are essential for the survival of plants. It attacks an enzyme called EPSPS that only plants use, so it's harmless to people, insects, and other animals. And unlike other herbicides that wind up in groundwater, glyphosate stays where it's sprayed, degrading within weeks (Powles and Yu 2010).

In 1986, Monsanto scientists improved the performance of glyphosate on weeds by engineering crop plants to be resistant to glyphosate. They did so by inserting genes from bacteria that could produce amino acids even after a plant was sprayed with herbicides. In the 1990s Monsanto began to sell glyphosate-resistant corn, cotton, sugar beets, and many other crops. The crops proved hugely popular. Instead of applying a lot of different herbicides, farmers found they could hit their fields with a modest amount of glyphosate alone, which wiped out weeds without harming their crops. Studies indicate that farmers who used these transgenic crops used fewer herbicides than those who grew regular plants—77% less in Mexico, for example—while getting a significantly higher yield from their fields.

For a while, it seemed as if glyphosate would avoid Melander's iron rule that pests evolve resistance. Monsanto scientists ran tests that showed no evidence that weeds could withstand glyphosate. But as farmers converged on glyphosate as their herbicide of choice, more and more weeds were exposed to it. Glyphosate became an ever more potent agent of selection on weed populations, and after a few years farmers began to notice horseweed and morning glory encroaching once more into their fields. Some farmers had to cut down fields of cotton rather than harvest them due to infestations of a weed called Palmer amaranth. Other farmers had to abandon glyphosate and turn back to older, more toxic herbicides. As of 2014, 24 weed species had evolved resistance to glyphosate in 21 countries (Heap 2014).

A century ago, Melander could study the evolution of resistance only by observing which insects lived and died. Today, scientists can pop the lid off the genetic toolbox that insects and weeds use to resist chemicals. What's striking is how many different ways weeds have found to overcome glyphosate. Scientists had thought that glyphosate was invincible in part because the enzyme it attacks, EPSPS, is similar in all plants. That uniformity suggests that plants can't tolerate mutations to it. But it turns out that one mutation, which has independently turned up in many populations of ryegrass and goosegrass, changes a single amino acid in EPSPS. The plant can still survive with this altered enzyme. And glyphosate has a hard time attacking the altered form of EPSPS thanks to its different shape (Figure 10.22).

An entirely different strategy to overcome glyphosate evolved in Palmer amaranth: one based on overwhelming the herbicide with sheer numbers. The plants make the ordinary, vulnerable form of EPSPS, but through gene duplication, they have acquired many extra copies of the *EPSPS* gene—up to 160 extra copies in some populations. All those extra genes make extra copies of the enzyme. Although the glyphosate may knock out some of the EPSPS in the Palmer amaranth, it cannot knock out all of them. The plants make so much more enzyme that they can go on growing.

Managing Pest Evolution

So even in the twenty-first century, natural selection can still defeat the most sophisticated genetic engineering. But the easy evolution of resistance does not mean that the plight of farmers is hopeless. The case of a pesticide known as Bt demonstrates how effectively farmers can manage evolution, if they understand how it works.

Bt refers to crystalline protein toxins produced by a family of genes (*Cry* genes) in a bacteria called *Bacillus thuringiensis*. The bacteria produce the toxic crystals when they form spores; when ingested by susceptible insects, the toxins bind to receptors in the insects' gut and make them sick. For decades farmers have sprayed Bt on crops. Among Bt's attractions is its short life. It rapidly breaks down in sunlight, so it does not create dangerous groundwater pollution. More recently, scientists developed genetically modified crops that carried the *Bt* gene. When farmers plant these crops, the plants make their own pesticide.

When Bt was applied to cotton and to other crops, Bruce Tabashnik, of the University of Arizona, and other researchers warned that insects might evolve a resistance to the toxin (Tabashnik et al. 2013). In a field planted with Bt-treated crops, insects that could resist Bt were able to flourish. But the scientists pointed out that farmers could slow the rise of resistance by creating Bt-free "refuges" on their farms.

Tabashnik and his colleagues based this prediction on the fact that resistance mutations come at a cost. As we saw with sticklebacks, insects have a finite supply of resources that they can invest in physiological processes. If an insect is genetically programmed to put extra resources into resisting a pesticide, it has fewer resources to invest in other activities, such as growth and reproduction. In a field without Bt, a Bt-resistant insect is therefore at a disadvantage compared to susceptible insects. In a field with Bt, the cost of Bt resistance is outweighed by its benefits, and the resistant insects take over.

Figure 10.22 The plants (A) ryegrass, (B) goosegrass, and (C) Palmer amaranth evolved mechanisms to overcome the glyphosate in Roundup. In ryegrass and goosegrass, an altered form of the EPSPS enzyme evolved. Palmer amaranth, on the other hand, produces more of the original form of EPSPS. (A, B: Nigel Cattlin/Alamy Stock Photo; C: Bill Barksdale/AGE Fotostock)

If farmers planted nothing but Bt-producing crops, they could drive the rapid evolution of Bt-resistant insects and make their genetically modified crops useless. Tabashnik and his colleagues suggested that farmers plant a few of their fields with ordinary crops instead. In these refuges, Bt-resistant insects would be outcompeted by other insects that didn't invest so much in detoxifying Bt. Insects from the Bt-producing fields and the refuges would interbreed, and their offspring would inherit some genes for Bt susceptibility.

Several years after Bt crops were introduced, Bt-resistant insects began to appear in significant numbers. In 2008, Tabashnik and his colleagues surveyed the rise of resistance. In states with large areas of refuge, resistance evolved much more slowly than in states with small areas of refuge. The farmers had carried out a giant experiment in evolution, and it had turned out as the evolutionary biologists predicted. Today, farmers using Bt corn are required to set aside 20% of their crop area as a Bt-free refuge, and farmers planting Bt cotton must set aside 50% of their crop area as refuge (Cullen et al. 2008).

Altered Environments and Invasive Species

Along with domestication and chemical resistance, humans have also influenced selection on many other kinds of species. By building cities, for example, we have favored animals and plants that can survive in urban environments instead of the rural ones that existed beforehand. In southern France, scientists have documented this urban selection acting on a small flowering plant called *Crepis sancta* (Cheptou et al. 2008). Populations of the flower grew in the countryside, whereas others grew in busy Marseille, colonizing the patches of ground around trees planted along the streets. The scientists examined the plants that grew in a part of the city that had been paved in the early 1990s.

C. sancta can make two different kinds of seeds—one that can drift off in the wind and another that simply drops to the ground. The scientists hypothesized that in Marseille, wind-carried seeds would be a burden to plants because they would be likely to land on the pavement instead of the ground. Dropped seeds would have a better chance of surviving because they'd fall onto the patch of ground where the parent plants grew.

To test their hypothesis, the scientists raised *C. sancta* from Marseille in a greenhouse alongside *C. sancta* from the countryside. Under the same conditions, the scientists found that the city plants were making 4.5% more nondispersing seeds than the ones in the countryside.

The scientists estimated that about 25% of the variation in the ratio of the two types of seeds is controlled by genetic differences. With these levels of heritability and selection, it should have taken about 12 generations to produce the observed change of 4.5% in the seed ratio. As predicted, about 12 generations of plants have lived in Marseille since the sidewalks were built. Without intending it, humans created a new environment for these plants, which are now adapting to it. As more time passes, the city plants may continue to make more dropping seeds and fewer windborne ones.

Besides changing the habitats of many species, we humans can also move species to new habitats. In some cases, we move them intentionally. Potatoes, for example, were domesticated 5000 years ago in Peru and then introduced into Europe in the sixteenth century. In other cases, the introductions to new habitats occur by accident. Ships take up ballast water when they begin their voyages and then dump it when they arrive at their destination. In that ballast water may be a vast number of exotic animals, algae, and bacteria.

Most relocated species die off. Others, however, are able to persist and, in some cases, to spread. Invasive species often experience strong directional selection within their new habitat. Both the invader species and the native species in their path may rapidly evolve. For example, cane toads (*Bufo marinus*) were introduced to Australia in the 1930s to control insect pests in sugarcane fields. The introduction was a disaster

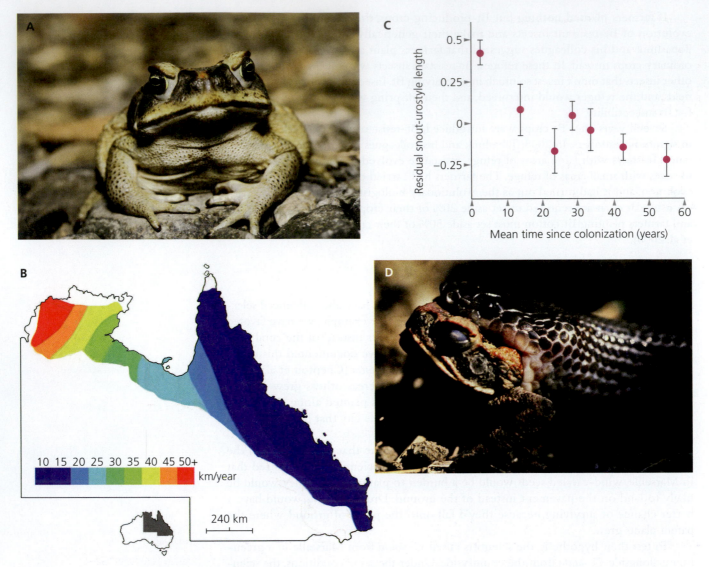

Figure 10.23 A: Cane toads (*Bufo marinus*) were introduced to eastern Australia in the 1930s. B: The toads have expanded steadily and increasingly rapidly since that time. C: Toads are rapidly evolving to smaller body sizes in these new habitats. D: Native snakes are killed by toxins in skin glands of cane toads, and these predators also are evolving rapidly in response to this introduced species. (Data from Phillips and Shine 2005; Rollins et al. 2015; A: Dirk Ercken/Shutterstock; D: Greg Watson/Photographers Direct)

(Phillips and Shine 2005). Instead of controlling farm pests, the toads fed on harmless animals—even small mammals. Native predators that attacked the new prey got an awful surprise. Cane toads exude a milky poison from large glands behind their eyes, and it is toxic to many animals, including humans and dogs.

Ben Phillips and Richard Shine, two biologists at the University of Sydney in Australia, have shown that toad lineages responded to altered patterns of selection in this new environment by rapidly evolving smaller body sizes and smaller gland sizes (**Figure 10.23**; Phillips and Shine 2005). Relatively long legs also evolved. With these longer legs, the cane toads moved faster, and they expanded their range in Australia at a faster rate (Phillips et al. 2006).

The cane toads are also strong agents of selection on their predators—native Australian snakes, overall, have become larger since cane toads were introduced. Phillips and Shine propose that larger body size raises the fitness of the snakes because it lowers the concentration of the toxins they ingest when they attack a cane toad. Bigger snakes are thus more likely to survive a given dose of toxins. But they also found evidence that narrower mouths (measured as gape width) evolved in the snakes. Snakes with smaller gape widths cannot swallow the biggest toads—which are also the most toxic and thus most likely to kill the snakes (Shine 2018). Because small snakes have the smallest mouths, cane toads result in disruptive selection on the size of their snake predators.

Hunting and Fishing as Agents of Selection

To feed ourselves, we humans farm much of the world's arable land, hunt wild animals on land, and catch fishes at sea. As the world's human population has grown, and as technology has grown more sophisticated, our harvest has increased dramatically. As we'll see in Chapter 14, hunting and fishing are endangering a number of species. But they're also exerting selection on many populations. That's because this harvesting of wild animals is not random. Individuals with certain traits are more likely to be killed than others.

Hunting and fishing have an evolutionary effect that's the opposite of domestication. Farmers select individual plants to breed because they have desirable traits. But when animals are heavily hunted, it's the undesirable individuals that can survive and pass on their traits to the next generation (**Table 10.1**; Allendorf and Hard 2009).

Trophy hunters of big game almost universally prefer to kill the largest, most ornamented males of deer, elk, moose, and bighorn sheep. David Coltman of the University of Sheffield has analyzed records of big game animals and discovered evidence of selection. The preference of hunters has led to the rapid evolution of smaller horn and body sizes (**Figure 10.24**; Coltman et al. 2003).

This recent evolution may be altering how these big game animals choose mates. As we'll see in Chapter 11, male bighorn sheep and other game species use their horns to compete with other males and to attract females. The size of their horns is also linked to high quality in males most preferred by females. So hunters are killing off the very individuals that would normally have the highest breeding success. When trophy hunting was finally banned in Coltman's study population in 1996, for example, males with big horns started doing better once again and male horn sizes increased (Pigeon et al. 2016).

Table 10.1 Traits That Experience Selection Due to Human Hunting and Fishing

Trait	Selective Action	Response(s)	Remedy
Age and size at sexual maturation	Increased mortality	Sexual maturation at earlier age and size, reduced fertility	Reduce harvest mortality or modify selectivity of harvest.
Body size or morphology, sexual dimorphism	Selective harvest of larger or more distinctive individuals	Reduced growth rate, attenuated phenotypes	Reduce selective harvest of large or distinctive individuals.
Sexually selected weapons (horns, tusks, antlers, etc.)	Trophy hunting	Reduced weapon size or body size	Implement hunting regulations that restrict harvest based on size or morphology of weapons under sexual selection.
Timing of reproduction	Selective harvest of seasonally early or late reproducers	Altered distribution of reproduction (truncated or altered seasonality)	Harvest throughout reproductive season.
Behavior	Harvest of more active, aggressive, or bolder (more vulnerable to predation) individuals	Reduced boldness in foraging or courtship behavior, potentially reduced productivity	Implement harvest methods less likely to impose selection on activity or aggressive behavior.
Dispersal/migration	Harvest of individuals with more predictable migration patterns	Altered migration routes	Interrupt harvest with key time and area closures tied to primary migration routes.

(Data from Allendorf and Hard 2009)

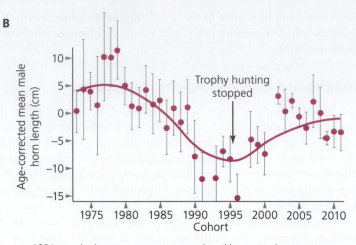

Figure 10.24 A: Bighorn sheep (*Ovis canadensis*) have experienced selection from hunters who prefer large males with long horns. B: Over the period from 1975 to 1996, this "unnatural" selection resulted in the evolution of shorter male horns. After 1996, trophy hunting was prevented and horn size began to evolve toward larger sizes again (Data from Pigeon et al. 2016; A: photo by J. T. Chapman/Shutterstock)

Fishes are experiencing strong selection from hunting as well. In some salmon populations, 90% are caught by fishermen (Hard et al. 2008). But these catches are not random samples of fish populations. Fishermen tend to catch more large fishes than small ones—gillnets, for example, let smaller fishes swim through. The result of this practice has been strong artificial selection for salmon with smaller body sizes (Allendorf and Hard 2009). In salmon, as in many other heavily fished species, the fecundity of individuals increases exponentially with body size (Marteinsdottir and Begg 2002). The small fishes favored by this new pattern of selection have dramatically reduced reproductive potential, potentially fueling the utter collapse of these harvested populations (**Figure 10.25**; Belgrano and Fowler 2013).

Figure 10.25 A: Atlantic cod have experienced decades of selection for smaller body size as a result of selective harvest by fisheries. B: This has led to the evolution of a life history trait, the age of maturity, so that fishes today reach sexual maturity at significantly smaller sizes than fishes 50 years ago. (Data from Kuparinen and Festa-Bianchet 2017; A: Juniors/Superstock, Inc.)

- The speed of evolution is a product of the amount of available genetic variation and the strength of selection. Weed and pest populations can be highly variable, and herbicides and pesticides can impose extremely strong selection. The result: rapid evolution of resistance.

- An understanding of evolutionary biology can lead to novel management practices that slow the evolution of resistance in pest populations or minimize undesirable evolutionary consequences of harvesting. ●

TO SUM UP . . .

- Evolution in response to natural selection is the inevitable outcome whenever three conditions are met: (1) individuals differ in their expression of a trait; (2) this variation is at least partially heritable; and (3) because of these differences, individuals differ in their reproductive success.

- Specific features of the environment can generate natural selection on a trait. These agents of selection can be events, such as storms or droughts, or environmental factors like predators or diet.

- Episodes of natural selection may be associated with particular seasons or events (such as droughts or floods). This means that selection need not be visible or measurable all the time.

- Natural selection can at times be strong and lead to rapid evolution that is observable in wild populations. Even infrequent episodes of strong selection can have important effects on the evolution of populations.

- The strength and the direction of natural selection can change over time.

- Selection may be similarly heterogeneous over space, so that individuals in different parts of a species' range encounter very different patterns of selection.

- During their lifetime, organisms experience many different sources and types of selection. Often, these agents of selection act in opposing directions, and this can generate a net balance. When this balance occurs, the combined effects result in stabilizing selection for intermediate trait values.

- Artificial selection can be such a powerful force that alleles can be swept to fixation in a population, carrying genetically linked alleles along with them. For this reason, many domesticated lineages such as purebred dogs are prone to genetic disorders.

- Humans have dramatically altered their environments, and this has resulted in novel types of selection on many organisms.

- Domestication, application of pesticides or herbicides, translocation of species, and hunting and fishing all have led to rapid and recent evolution of affected populations.

- In some cases, an understanding of the principles of evolution can lead to new strategies for mitigating unintended consequences of human activity.

1. Which is *not* one of the three conditions that must be met for evolution by natural selection to take place?

   a. Variation in phenotypic traits must exist in the population.

   b. Differences in phenotype influence the probability of survival or reproduction.

   c. One extreme of the phenotype leads to greater survival.

   d. Differences in phenotypic traits must be at least partially heritable.

2. Which experiment would allow you to test whether coat color affects oldfield mouse fitness?

   a. Using dark and light models of oldfield mice to determine predation rates in forest and beach habitats.

   b. Trapping oldfield mice in both forest and beach habitats and counting whether there were more dark mice in forest habitats or in beach habitats.

   c. Following oldfield mice with dark coats in beach habitats to determine whether they reproduced.

   d. Conducting late-night surveys in both forest and beach habitats to determine whether predators could see mice with dark or light coats better in either habitat.

3. In the Pacific Northwest, snowshoe hares molt from summer brown to winter brown instead of the typical summer brown to winter white. Which of the following statements is incorrect?

   a. The evolutionary loss of coat color plasticity (brown to white) resulted from an allele that snowshoe hares borrowed from another species that is not plastic (that is, through introgression).

   b. Jones et al. (2018) used an F_{ST} outlier approach to identify a mutation in the *Agouti* gene as the likely genetic basis for the shift to "winter-brown" pelage.

   c. The gene tree for the *Agouti* gene matched the genome-wide (that is, the "species") tree for the hares.

   d. All of these statements are correct.

4. The text describes the galls of flies as examples of extended phenotypes. What are extended phenotypes?

   a. Phenotypes that are shared by multiple generations.

   b. Behaviors that influence the survival of offspring.

   c. Morphological features that affect reproductive output.

   d. Structures constructed by organisms that can influence their performance or success.

5. How have selective sweeps affected domesticated animals such as purebred dogs?

   a. Variation in flanking loci was reduced.

   b. Alleles at flanking loci "hitchhiked" to high frequency.

   c. Deleterious mutations leading to disease became more common.

   d. All of the above.

6. Which of the following is *not* a potential agent of selection?

   a. Human fishing.

   b. Genetic drift.

   c. A flood.

   d. A predator.

7. What is the best course of action for a farmer who wants to slow the evolution of resistance of a pest population feeding on her crops?

   a. Allowing some nonresistant pests to survive.

   b. Decreasing the amount of pesticide, but increasing the concentration.

   c. Increasing the amount of pesticide, but decreasing the concentration.

   c. Planting genetically modified crops that make their own pesticide.

8. A selective sweep describes which of the following situations?

   a. A favorable allele is pushed to fixation within a population so quickly that recombination is not likely.

   b. An unfavorable allele is pushed to fixation within a population so quickly that recombination is not likely.

   c. A favorable or unfavorable allele is pushed to fixation within a population so quickly that recombination is frequent.

   d. Alleles flanking a favorable allele will appear in DNA infrequently.

9. Which of these statements about selection is *false*?

   a. During their lifetime, plants may experience many different sources of selection.

   b. Insects often experience different types of selection as larvae than they do as adults.

   c. Birds can experience different directions of selection in different years.

   d. Selection in mammals always operates more strongly on survival than on reproduction.

10. What is the most likely reason that Atlantic cod populations have evolved to become sexually mature at younger ages?

    a. Warming ocean temperatures favor faster growth.

    b. Removal of their predators from harvesting relaxed selection on juvenile cod.

    c. Smaller fishes are more likely to escape gill nets and therefore have a survival advantage.

    d. All of the above.

INTERPRET THE DATA Answer can be found at the end of the book.

11. Which of the following statements about the graph that follows is *not true*?

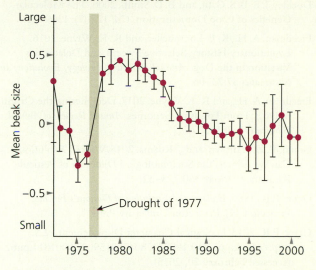

Evolution of beak size

a. The drought of 1977 caused a dramatic increase in beak size.

b. Mean beak size remained constant for medium ground finches from 1975 to 1985.

c. By the early 2000s, mean beak sizes were close to pre-drought sizes.

d. Mean beak size in ground finches fluctuated from 1990 to 2000.

SHORT ANSWER QUESTIONS Answers can be found at the end of the book.

12. What information did Peter and Rosemary Grant's team need to measure or record in order to demonstrate the effect of natural selection on the beak size of finches in the Galápagos?

13. What are the differences and similarities between directional and stabilizing selection?

14. For evolution by natural selection to occur, why is it important for the coat color of oldfield mice to be variable and at least partly heritable? What would happen if the variation or heritability were reduced?

15. Is domesticated corn (maize) better adapted to its environment than teosinte, the wild plant it evolved from? Why or why not?

16. Why is the evolution of resistance so rapid? How do farmers and scientists attempt to slow the evolution of resistance in pest populations?

17. What were the steps involved for snakes to evolve a smaller gape width in Australia after the introduction of cane toads?

ADDITIONAL READING

Abrahamson, W. G., and A. E. Weis. 1997. *Evolutionary Ecology Across Three Trophic Levels: Goldenrods, Gallmakers, and Natural Enemies.* Princeton, NJ: Princeton University Press.

Allendorf, F. W., and J. J. Hard. 2009. Human-Induced Evolution Caused by Unnatural Selection through Harvest of Wild Animals. *Proceedings of the National Academy of Sciences USA* 106 (Supplement 1): 9987–94.

Barrett, R. D. H., S. M. Rogers, and D. Schluter. 2008. Natural Selection on a Major Armor Gene in Threespine Stickleback. *Science* 322 (5899): 255–7.

Diamond, J. 2002. Evolution, Consequences and Future of Plant and Animal Domestication. *Nature* 418 (6898): 700–707.

Endler, J. A. 1986. *Natural Selection in the Wild.* Princeton, NJ: Princeton University Press.

Grant, P. R., and B. R. Grant. 2006. Evolution of Character Displacement in Darwin's Finches. *Science* 313 (5784): 224–6.

Linnen, C. R., E. P. Kingsley, J. D. Jensen, and H. E. Hoekstra. 2009. On the Origin and Spread of an Adaptive Allele in Deer Mice. *Science* 325 (5944): 1095–8.

Palumbi, S. R. 2001. Humans as the World's Greatest Evolutionary Force. *Science* 293 (5536): 1786–90.

PRIMARY LITERATURE CITED IN CHAPTER 10

Abrahamson, W. G., and A. E. Weis. 1997. *Evolutionary Ecology Across Three Trophic Levels: Goldenrods, Gallmakers and Natural Enemies. Princeton Monographs in Population Biology.* Princeton, NJ: Princeton University Press.

Akey, J. M., A. L. Ruhe, D. T. Akey, A. K. Wong, C. F. Connelly, et al. 2010. Tracking Footprints of Artificial Selection in the Dog Genome. *Proceedings of the National Academy of Sciences USA* 107 (3): 1160–5.

Allendorf, F. W., and J. J. Hard. 2009. Human-Induced Evolution Caused by Unnatural Selection Through Harvest of Wild Animals. *Proceedings of the National Academy of Sciences USA* 106 (1): 9987–94.

Barrett, R. D. H., S. M. Rogers, and D. Schluter. 2008. Natural Selection on a Major Armor Gene in Threespine Stickleback. *Science* 322:255–7.

Belgrano, A., and C. W. Fowler. 2013. How Fisheries Affect Evolution. *Science* 342:1176–7.

Bell, M. A., and W. E. Aguirre. 2013. Contemporary Evolution, Allelic Recycling, and Adaptive Radiation of the Threespine Stickleback. *Evolutionary Ecology Research* 15:377–411.

Bell, M. A., M. P. Travis, and D. M. Blouw. 2006. Inferring Natural Selection in a Fossil Threespine Stickleback. *Paleobiology* 32 (4): 562–77.

Boag, P. T. 1983. The Heritability of External Morphology in Darwin's Ground Finches (*Geospiza*) on Isla Daphne Major, Galápagos. *Evolution* 37 (5): 877–94.

Brown, W. L., and E. O. Wilson. 1956. Character Displacement. *Systematic Zoology* 5:49–64.

Bruford, M. W., D. G. Bradley, and G. Luikart. 2003. DNA Markers Reveal the Complexity of Livestock Domestication. *Nature Reviews Genetics* 4 (11): 900–910.

Cheptou, P. O., O. Carrue, S. Rouifed, and A. Cantarel. 2008. Rapid Evolution of Seed Dispersal in an Urban Environment in the Weed *Crepis sancta. Proceedings of the National Academy of Sciences USA* 105 (10): 3796–9.

Clutton-Brock, J. 1999. *A Natural History of Domesticated Mammals.* Cambridge, MA: Cambridge University Press.

Colosimo, P. F., K. E. Hosemann, S. Balabhadra, G. Villarreal, M. Dickson, et al. 2005. Widespread Parallel Evolution in Sticklebacks by Repeated Fixation of Ectodysplasin Alleles. *Science* 307 (5717): 1928–33.

Coltman, D. W., P. O'Donoghue, J. T. Jorgenson, J. T. Hogg, C. Strobeck, et al. 2003. Undesirable Evolutionary Consequences of Trophy Hunting. *Nature* 426 (6967): 655–8.

Cruz, F., C. Vilà, and M. T. Webster. 2008. The Legacy of Domestication: Accumulation of Deleterious Mutations in the Dog Genome. *Molecular Biology and Evolution* 25 (11): 2331–6.

Cullen, E., R. Proost, and D. Volenberg. 2008. *Insect Resistance Management and Refuge Requirements for Bt Corn.* Vol. A3857: Nutrient and Pest Management Program. Madison: University of Wisconsin–Extension.

Diamond, J. 2002. Evolution, Consequences and Future of Plant and Animal Domestication. *Nature* 418 (6898): 700–708.

Doebley, J. 2004. The Genetics of Maize Evolution. *Annual Review of Genetics* 38 (1): 37–59.

———. 2006. Unfallen Grains: How Ancient Farmers Turned Weeds into Crops. *Science* 312 (5778): 1318–9.

Doebley, J. F., B. S. Gaut, and B. D. Smith. 2006. The Molecular Genetics of Crop Domestication. *Cell* 127 (7): 1309–21.

Freedman, A. H., K. E. Lohmueller, and R. K. Wayne. 2016. Evolutionary History, Selective Sweeps, and Deleterious Variation in the Dog. *Annual Review of Ecology, Evolution, and Systematics* 47:73–96.

Freedman, A. H., and R. K. Wayne. 2017. Deciphering the Origin of Dogs: From Fossils to Genomes. *Annual Review of Animal Biosciences* 5:281–307.

Gould, F., Z. S. Brown, and J. Kuzma. 2018. Wicked Evolution: Can We Address the Sociobiological Dilemma of Pesticide Resistance? *Science* 360:728–32.

Grant, P. R. 1986. *Ecology and Evolution of Darwin's Finches.* Princeton, NJ: Princeton University Press.

Grant, P. R. 2013. Ecological Character Displacement. *eLS.* http://dx.doi.org/10.1002/9780470015902.a0001811.pub2 (accessed February 20, 2015).

Grant, B. R., and P. R. Grant. 1993. Evolution of Darwin's Finches Caused by a Rare Climatic Event. *Proceedings of the Royal Society of London, Series B: Biological Sciences* 251 (1331): 111–7.

Grant, P. R., and B. R. Grant. 2002. Unpredictable Evolution in a 30-Year Study of Darwin's Finches. *Science* 296 (5568): 707–11.

Grant, P. R., and B. R. Grant. 2006. Evolution of Character Displacement in Darwin's Finches. *Science* 313:224–6.

Grant, P. R., and B. R. Grant. 2011. *How and Why Species Multiply: The Radiation of Darwin's Finches.* Princeton, NJ: Princeton University Press.

Grant, P. R., and B. R. Grant. 2014. *40 Years of Evolution: Darwin's Finches on Daphne Major Island.* Princeton, NJ: Princeton University Press.

Grant, P. R., B. R. Grant, R. B. Huey, M. T. Johnson, A. H. Knoll, et al. 2017. Evolution Caused by Extreme Events. *Philosophical Transactions of the Royal Society of London, Series B: Biological Sciences* 372 (1723): 20160146.

Hard, J. J., M. R. Gross, M. Heino, R. Hilborn, R. G. Kope, et al. 2008. Evolutionary Consequences of Fishing and Their Implications for Salmon. *Evolutionary Applications* 1 (2): 388–408.

Heap, I. 2014. Global Perspective of Herbicide-Resistant Weeds. *Pest Management Science* 70 (9): 1306–15.

Jaenicke-Després, V., E. S. Buckler, B. D. Smith, M. T. P. Gilbert, A. Cooper, et al. 2003. Early Allelic Selection in Maize as Revealed by Ancient DNA. *Science* 302:1206–8.

Jones, F. C., M. G. Grabherr, Y. F. Chan, P. Russell, E. Mauceli, et al. 2012. The Genomic Basis of Adaptive Evolution in Threespine Sticklebacks. *Nature* 484:55–61.

Jones, M. R., L. S. Mills, P. C. Alves, C. M. Callahan, J. M. Alves, et al. 2018. Adaptive Introgression Underlies Polymorphic Seasonal Camouflage In Snowshoe Hares. *Science* 360:1355–8.

Kantar, M. B., A. R. Nashoba, J. E. Anderson, B. K. Blackman, and L. H. Rieseberg. 2017. The Genetics and Genomics of Plant Domestication. *BioScience* 67 (11): 971–82.

Kaufman, D. W. 1974. Adaptive Coloration in *Peromyscus polionotus*: Experimental Selection by Owls. *Journal of Mammalogy* 55 (2): 271–83.

Kitano, J., D. I. Bolnick, D. A. Beauchamp, M. M. Mazur, S. Mori, et al. 2008. Reverse Evolution of Armor Plates in the Three-spine Stickleback. *Current Biology* 18 (10): 769–74.

Kuparinen, A. and Festa-Bianchet, M. 2017. Harvest-Induced Evolution: Insights from Aquatic and Terrestrial Systems. *Philosophical Transactions of the Royal Society of London, Series B: Biological Sciences* 372 (1712): 20160036.

Lamichhaney, S., F. Han, J. Berglund, C. Wang, M. S. Almén, et al. 2016. A Beak Size Locus in Darwin's Finches Facilitated Character Displacement During a Drought. *Science* 352 (6284): 470–4.

Marteinsdottir, G., and G. A. Begg. 2002. Essential Relationships Incorporating the Influence of Age, Size and Condition on Variables Required for Estimation of Reproductive Potential in Atlantic Cod *Gadus morhua*. *Marine Ecology Progress Series* 235:235–56.

Melander, A. L. 1914. Can Insects Become Resistant to Sprays? *Journal of Economic Entomology* 7:167–73.

O'Brown, N. M., B. R. Summers, F. C. Jones, S. D. Brady, and D. M. Kingsley. 2015. A Recurrent Regulatory Change Underlying Altered Expression and Wnt Response of the Stickleback Armor Plates Gene EDA. *eLife* 4:e05290.

Palumbi, S. R. 2001a. *The Evolution Explosion: How Humans Cause Rapid Evolutionary Change.* New York: Norton.

———. 2001b. Humans as the World's Greatest Evolutionary Force. *Science* 293 (5536): 1786–90.

Phillips, B. L., G. P. Brown, J. K. Webb, and R. Shine. 2006. Invasion and the Evolution of Speed in Toads. *Nature* 439 (7078): 803.

Phillips, B. L., and R. Shine. 2005. The Morphology, and Hence Impact, of an Invasive Species (the Cane Toad, *Bufo marinus*): Changes with Time Since Colonisation. *Animal Conservation* 8 (4): 407–13.

Pigeon, G., M. Festa-Bianchet, D. W. Coltman, and F. Pelletier. 2016. Intense Selective Hunting Leads to Artificial Evolution in Horn Size. *Evolutionary Applications* 9 (4): 521–30.

Powles, S. B., and Q. Yu. 2010. Evolution in Action: Plants Resistant to Herbicides. *Annual Review of Plant Biology* 61 (1): 317–47.

Ramos-Madrigal, J., B. D. Smith, J. V. Moreno-Mayar, S. Gopala-krishnan, J. Ross-Ibarra, et al. 2016. Genome Sequence of a 5,310-Year-Old Maize Cob Provides Insights into the Early Stages of Maize Domestication. *Current Biology* 26:3195–201.

Reimchen, T. E. 1992. Injuries on Stickleback from Attacks by a Toothed Predator (*Oncorhynchus*) and Implications for the Evolution of Lateral Plates. *Evolution* 46 (4): 1224–30.

Rollins, L. A., M. F. Richardson, and R. Shine. 2015. A Genetic Perspective on Rapid Evolution in Cane Toads (*Rhinella marina*). *Molecular Ecology* 24:2264–76.

Schluter, D. 2000. *The Ecology of Adaptive Radiation.* Oxford: Oxford University Press.

Schluter, D., and D. Nychka. 1994. Exploring Fitness Surfaces. *American Naturalist* 143:597–616.

Shearin, A. L., and E. A. Ostrander. 2010. Canine Morphology: Hunting for Genes and Tracking Mutations. *PLoS Biology* 8 (3): e1000310.

Shine, R. 2018. *Cane Toad Wars.* Oakland: University of California Press.

Steiner, C. C., H. Römpler, L. M. Boettger, T. Schöneberg, and H. E. Hoekstra. 2009. The Genetic Basis of Phenotypic Convergence in Beach Mice: Similar Pigment Patterns but Different Genes. *Molecular Biology and Evolution* 26 (1): 35–45.

Stuart, Y. E., T. S. Campbell, P. A. Hohenlohe, R. G. Reynolds, L. J. Revell, et al. 2014. Rapid Evolution of a Native Species Following Invasion by a Congener. *Science* 346 (6208): 463–6.

Stuart, Y. E., and J. B. Losos. 2013. Ecological Character Displacement: Glass Half Full or Half Empty? *Trends in Ecology and Evolution* 28:402–8.

Sutter, N. B., C. D. Bustamante, K. Chase, M. M. Gray, K. Zhao, et al. 2007. A Single *IGF1* Allele Is a Major Determinant of Small Size in Dogs. *Science* 316 (5821): 112–5.

Svensson, E., and R. Calsbeek. 2012. *The Adaptive Landscape in Evolutionary Biology.* Oxford: Oxford University Press.

Tabashnik, B. E., T. Brévault, and Y. Carrière. 2013. Insect Resistance to Bt Crops: Lessons from the First Billion Acres. *Nature Biotechnology* 31:510–21.

Vignieri, S. N., J. G. Larson, and H. E. Hoekstra. 2010. The Selective Advantage of Crypsis in Mice. *Evolution* 64 (7): 2153–8.

Vilà, C., P. Savolainen, J. E. Maldonado, I. R. Amorim, J. E. Rice, et al. 1997. Multiple and Ancient Origins of the Domestic Dog. *Science* 276 (5319): 1687–9.

Weiner, J. 1994. *The Beak of the Finch: A Story of Evolution in Our Time.* New York: Knopf.

Weis, A. E., and W. G. Abrahamson. 1986. Evolution of Host-Plant Manipulation by Gall Makers: Ecological and Genetic Factors in the Solidago-Eurosta System. *American Naturalist* 127 (5): 681–95.

Weis, A. E., W. G. Abrahamson, and M. Andersen. 1992. Variable Selection on *Eurosta*'s Gall Size, I: The Extent and Nature of Variation in Selection. *Evolution* 46 (6): 1674–97.

Wood, R., and J. Bishop. 1981. Insecticide Resistance: Populations and Evolution. In J. A. Bishop and L. M. Cook (eds.), *The Genetic Basis of Man-Made Change* (pp. 97–127). New York: Academic Press.

Zeder, M. A., E. Emshwiller, B. D. Smith, and D. G. Bradley. 2006. Documenting Domestication: The Intersection of Genetics and Archaeology. *Trends in Genetics* 22 (3): 139–55.

CHAPTER 11

Sex

Causes and Consequences

Learning Objectives

- Compare and contrast the genetic consequences of sexual and asexual reproduction.
- Explain how differential investment by males and females in sexual reproduction can lead to sexual selection.
- Compare and contrast the leading hypotheses developed to explain mate choice.
- Discuss the costs and benefits to males and females of different mating systems.
- Explain how sexual selection can act on traits that function after mating.
- Discuss how sexual conflict could lead to antagonistic coevolution.

Scientists do some pretty strange things. For Patricia Brennan, an evolutionary biologist at Mount Holyoke College, those strange things include inspecting a duck's genitals (**Figure 11.1**).

Normally, a male duck keeps his phallus (the avian equivalent of a human penis) retracted inside his body. Only during mating does he extend it into view. Measuring the length of a duck phallus is therefore a two-person job. Brennan has a colleague grab a bird and hold it upside down, its legs sticking out in the air—a maneuver that, if done with care, doesn't disturb the duck. Brennan gently presses around a small dome of muscle below the bird's tail, and after a little coaxing the phallus emerges. Brennan grabs a ruler.

The measurement she records will vary depending on the time of year that she does her research. When the breeding season ends, a duck's phallus shrinks to a fraction of its former size. When the next breeding season approaches, the phallus grows large again. Shaped like a spiraling corkscrew, the duck phallus can grow to astonishing lengths. In some species of ducks, it grows as long as the bird's entire body.

A male wood duck (right) swims alongside a female.

Figure 11.1 Patricia Brennan has discovered that an evolutionary arms race between male and female birds can produce gigantic sexual organs. (Bernard J. Brennan)

The length of a duck's phallus is all the more remarkable when you consider that only 3% of all bird species have phalluses at all. (In the other bird species, the male has only a simple opening, called a cloaca, that he positions against a similar opening in the female.) Brennan and her team want to understand why duck phalluses are so elaborate, especially when other bird species have none. To discover why, she has embarked on a study of the forces driving their evolution.

A male duck's phallus twists counterclockwise. But Brennan found that in ducks and other waterfowl, the female's reproductive tract—called an oviduct—twists clockwise. To see the effect of these mismatched twists, Brennan decided to watch duck phalluses in action. So she designed twisting glass tubes of the same size and shape as duck oviducts. She took her bizarre sculptures to a California duck farm, where the workers are skilled at collecting the sperm from prize male ducks to use for breeding. There, she coaxed the male ducks to mate with the glass tubes.

First she had them insert their phalluses into glass tubes that had the same counterclockwise twist as the duck phalluses. Brennan found that the duck phalluses expanded to the end of the correspondingly shaped tube in just a third of a second. Next, she tried out glass tubes that twisted clockwise, the way real duck oviducts do. Now the male ducks could only push their phalluses into the base of the tube. This information revealed that the female's reproductive tract was actually hampering mating, not helping it (Brennan et al. 2010).

In ducks, the female's reproductive tract thwarts the efforts of males to fertilize their eggs. Brennan's research echoes the work of other biologists who study sex in other species. Adaptations often evolve in males and females that put one sex in conflict with the other.

The reason for the conflict in ducks appears to lie in their mating system. More than one male duck may mate with a female, and so she will end up with the sperm of several males. Due to the mismatched twists of the female's oviduct, males can insert their sperm, but only at its opening. In other animals, scientists found that females can store sperm from several males in different pouches and then can later control which male's sperm will fertilize their eggs. Brennan discovered that the duck's oviducts have rows of pouches along their length. Could ducks be doing the same thing?

The results from another study that Brennan carried out suggest that this is indeed the case. She traveled to Alaska, where she caught 16 species of ducks and other waterfowl that migrate there for the summer. After measuring the phalluses on the male birds, Brennan then turned her attention to the females. She found a striking correspondence. In species where males have longer phalluses, the females have oviducts with more pouches and more coils. Her research suggests that the sexual conflict between males and females is more intense in some species than others, and that intensity drives the evolution of extreme genitalia in both sexes (Brennan and Prum 2012).

Brennan's ducks are a stunning demonstration that life is much more than just finding food and surviving cold, heat, and other environmental challenges. To pass down their genes to future generations, sexually reproducing organisms must find mates, overcome obstacles to mating, and then produce healthy offspring, which in turn must survive until they can have offspring of their own.

In this chapter, we will examine the evolution of sex and the influence of sex on evolution. We will begin by considering why sex exists in the first place and discuss how it may have originated. For organisms that do reproduce sexually, a new set of evolutionary mechanisms comes into play. We will look at how sexual reproduction drives the evolution of different roles for males and females. We will also consider how males evolve weapons for competition for females and showy ornaments to attract potential mates. Finally, we will examine how sexual conflict over reproduction leads to some of the most bizarre features of the animal kingdom, such as the sexual organs studied by Brennan and her colleagues. As we'll see in later chapters, understanding the influence of sex on evolution is important for understanding many other aspects of the history of life, from the origin of new species to human behavior. ●

11.1 Evolution of Sex

A discussion of sex must start with the existence of sex itself. We humans can reproduce only sexually, by combining gametes from males and females. But this is hardly a universal rule among other species. In fact, the means by which organisms reproduce are surprisingly diverse (Tree of Sex Consortium 2014).

Why Sex?

Some species of whiptail lizards in the southwestern United States turn out to be remarkable: they are made up entirely of females. Biologists have carefully studied their reproductive biology to determine how they have ended up so different from most other animals. They do not require additional DNA from sperm to reproduce. Instead, the chromosomes inside developing lizard eggs duplicate and then undergo recombination through meiosis. (Crews et al. 1986; Dias and Crews 2008; Lutes et al. 2010). The lizards thus use one component of sex—meiosis—to produce gametes, but they no longer use a second: male gametes. (This mode of sexual reproduction is called **parthenogenesis**.)

Bacteria don't even *have* males and females; they produce offspring simply by cell division. They reproduce asexually—without sex—because they have neither meiosis nor fusion of haploid gametes through fertilization. (The whiptails, on the other hand, have both.)

For still other species, sex is optional. A strawberry plant, for example, has ovules (the plant version of eggs) that can be fertilized with pollen (grains containing the plant version of sperm) from another plant. Or it can send out runners, which produce new plants that are genetically identical to it. Many species of fungi, such as bread mold, are capable of sex, but they can also produce spores asexually (**Figure 11.2**).

Some animals and plant species are **hermaphrodites**. An individual may develop both male and female organs (**Figure 11.3**). Many species of hermaphroditic flatworms, for example, can receive sperm from other flatworms and also inseminate their partners—a form of sexual reproduction. But they don't have to mate to reproduce.

Parthenogenesis is a mode of reproduction in which female sex cells undergo meiosis but are not fertilized by sperm; in parthenogenetic species, females produce only daughters.

A **hermaphrodite** is an individual that produces both female and male gametes.

Figure 11.2 For many species, reproduction is always sexual. But for other species, it's optional. A: Strawberry plants reproduce sexually when ovules are fertilized with pollen. Or they can grow runners that produce genetically identical daughter plants. B: Many kinds of fungi, such as bread mold, typically reproduce by producing spores that contain genetically identical individuals. But they have different sexes and sometimes will reproduce sexually.

A

Runner (stolon)

Crown

Roots

Daughter plants

B

Spores

Mold

They can simply fertilize their own eggs with their own sperm (self-fertilization). Many flowering plants are hermaphrodites because they incorporate both female and male gametes in a single flower. For example, ovules embedded in the styles and pollen produced on the surfaces of stamens. In other cases, a single plant may produce both male flowers and female flowers separately.

This survey makes clear that sexual reproduction involving males and females is far from an absolute law in the natural world. Evolution has produced a wide range of reproductive strategies. Each of these strategies is made possible by genes that are inherited by each generation, and those genes are subject to natural selection.

And therein lies a puzzle. As an evolutionary strategy, sexual reproduction imposes huge costs on organisms—costs that could well make sex a recipe for extinction.

To appreciate the toll that sexual organisms pay, imagine a population of lizards in which some don't need to mate and some do. Each asexual female can produce many daughters, and each of them can produce daughters of their own. Meanwhile, the sexually reproducing females must mate with males before they can reproduce. About half of their offspring will be males, which cannot produce young themselves. Sex, in other words, effectively cuts the reproducing population of these lizards in half.

Figure 11.3 Some animal species are hermaphrodites: each individual has both male and female sex organs. Banana slugs, shown here, can receive sperm from other slugs, deliver sperm to them, or fertilize their own eggs. After some matings, banana slugs may devour the phallus of their partner. (Doug Steakley/ Getty Images)

Figure 11.4 English biologist John Maynard Smith introduced the concept of the "twofold cost of sex." Asexual lineages multiply faster than sexual lineages because all progeny are capable of producing offspring. In sexual lineages, on the other hand, half of the offspring are males who cannot themselves produce offspring. This effectively halves the rate of replication of sexual species and should favor the invasion and spread of asexual genotypes within sexual populations. Maynard Smith's model is based on two assumptions: (1) asexual and sexual females produce the same number of offspring; (2) the offspring of asexual and sexual females have the same fitness. The fact that sex exists must be explained through a violation of either or both of these assumptions.

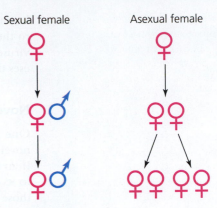

For the same amount of energy invested, asexual females produce twice as many copies of their genotype as sexual females do. This means that asexual genotypes should increase in frequency, replacing the sexual ones in the population. That is, natural selection *should* favor the evolution of asexual reproduction. Because two asexual females can produce twice as many offspring as one female and one male, the late British evolutionary biologist John Maynard Smith called this disadvantage the **twofold cost of sex** (**Figure 11.4**; Maynard Smith 1978).

Yet sex is common. In a recent study of insects and other arthropods, researchers estimated that 99.96% reproduce sexually, whereas 0.04% are parthenogenetic (van der Kooi et al. 2017). The researchers found that parthenogenesis evolved independently hundreds of times, yet this reproductive strategy remains remarkably rare. To solve this paradox, evolutionary biologists have developed and explored hypotheses, some of which we show in **Table 11.1**. Many of these hypotheses have focused on the fact that sexual reproduction makes recombination possible, which generates genetic variation. (Recombination is also a component of reproduction in parthenogenic and

Twofold cost of sex refers to the disadvantages of being a sexual rather than an asexual organism. Asexual lineages multiply faster than sexual lineages because all progeny are capable of producing offspring. In sexual lineages, half of the offspring are males who cannot themselves produce offspring. This limitation effectively halves the rate of replication of sexual species.

Table 11.1 Consequences of Sexual Reproduction

Disadvantages of Sex	Advantages of Sex
Twofold Cost of Sex Asexual lineages have an intrinsic capacity to grow more rapidly in each generation because all progeny can produce offspring. In sexual populations, males cannot themselves produce offspring.	**Combining Beneficial Mutations** By combining alleles of genes from two different individuals, sexual reproduction can bring separate beneficial mutations together in a single individual faster than would be expected if they had to arise spontaneously in the same genome.
Search Cost Males and females must locate each other in order to mate. This can involve time, energy, and risk of predation.	**Generation of Novel Genotypes** Through recombination, meiosis provides an opportunity for paired chromosomes to cross over, creating gametes with unique combinations of alleles.
Reduced Relatedness Sexually reproducing organisms pass only half of their alleles to their offspring because meiosis generates gametes that are haploid. This halves the relatedness between parents and their progeny.	**Faster Evolution** Offspring of sexual parents will be more genetically variable than offspring of asexually reproducing parents. This can speed the evolutionary response to selection of sexual populations and is critical for maintaining resistance to parasites (the Red Queen effect).
Risk of Sexually Transmitted Infections Mating between males and females provides an effective means of transmission for many pathogens. Asexual populations do not mate and so avoid this risk.	**Clearance of Deleterious Mutations** Sexual populations can purge themselves of harmful mutations because recombination can generate individuals with allelic combinations that exclude deleterious mutations. Asexual populations cannot do so, and they steadily and irreversibly accumulate mutations until a lineage is driven extinct (Muller's ratchet).

hermaphroditic species, but the offspring have decreased genetic variation compared to the offspring of sexually reproducing animals. Asexually reproducing bacteria and viruses also have some capacity for mixing alleles.) Here we will consider two hypotheses that have emerged as particularly promising.

Novelty

One hypothesis explains the benefit of sex as a way for parents to rapidly generate novel genotypes in their offspring, allowing for quick adaptation to their environment. If an asexual organism picks up a beneficial mutation, it can pass the mutation only to its offspring. If two asexual individuals each acquire a different beneficial mutation, those beneficial mutations will persist in isolation within their respective lineages.

Sexual reproduction, on the other hand, splits up genotypes and shuffles them into new combinations (see Chapter 5). If two sexual organisms each acquire a different beneficial mutation, sexual reproduction can combine these alleles into a single descendant individual much faster than if both mutations were to arise spontaneously in the same genome. Beneficial mutations can be combined with other beneficial mutations, and they can be separated from harmful ones.

When asexual organisms acquire deleterious mutations that offset earlier, beneficial mutations, the two types of mutations can't be separated. The bad mutations continue to accumulate (through a process called **Muller's ratchet**), and the burden that these mutations place on the fitness of individuals (their **genetic load**) gradually increases.

Sex releases populations from this constraint. Recombination shuffles the alleles of the two parents and creates new combinations in their offspring. Sometimes harmful mutations from both parents will end up in the same individuals. These individuals perform poorly and are removed from the population by selection. Other offspring may inherit only the beneficial alleles of their parents, and when these favorable new genotypes spread, they effectively purge the population of the deleterious alternative alleles. Over many generations, sexual lineages are expected to yield fitter individuals faster than asexual lineages (faster evolution).

Michael McDonald, a biologist at Harvard University, and his colleagues demonstrated these genetic effects of sex using a clever experiment in brewer's yeast, *Saccharomyces cerevisiae*. They took advantage of the fact that yeast can switch between sexual and asexual reproduction. For the most part, yeast reproduce by budding off daughter cells. These cells, the product of mitosis, are genetically identical to their mother cells.

Occasionally, however, yeast can also reproduce using sex. This process begins with a diploid parent cell undergoing meiosis to produce two types of haploid spores, which scientists refer to as a and α. The yeast release their spores into the environment. Later, if opposite spore types encounter each other, they can fuse to form a fertilized diploid zygote.

As we saw in Chapter 6, single-celled microbes such as *Escherichia coli* can allow scientists to observe selection in action in their labs. Yeast offer similar experimental advantages. Under optimal conditions, yeast can reproduce every 100 minutes, making it possible to measure fitness of different genotypes and observe the origin and spread of new mutations. And in order to measure the effect of sex on evolution, McDonald and his colleagues figured out how to control how they reproduced.

McDonald set up 12 experimental populations that could only reproduce asexually and another six that he forced to reproduce sexually once every 90 generations. He let his populations evolve for 1000 generations, adapting to the medium in which they grew. Then he looked to their genomes to see how they had changed (McDonald et al. 2016).

New mutations arose spontaneously in all of his populations, and he could follow their fates through time. Some of these new alleles proved to be beneficial, and

Muller's ratchet describes the process by which the genomes of an asexual population accumulate deleterious mutations in an irreversible manner.

Genetic load is the burden imposed by the accumulation of deleterious mutations.

selection pulled them to fixation. Other alleles were deleterious. But the fate of these alleles was different in the sexual and asexual populations.

In the asexual populations, deleterious mutations could not be uncoupled from the beneficial ones because the yeast could not undergo genetic recombination. As a result, many of the deleterious alleles were pulled to fixation alongside their beneficial neighbors through "hitchhiking" (**Figure 11.5**). The yeast accumulated a substantial genetic load that slowed the speed of their adaptation.

Sexual populations, on the other hand, did not suffer from these problems. They underwent recombination every 90 generations. By breaking apart the linkage between alleles, sex uncoupled the fates of the beneficial and deleterious alleles. Beneficial alleles were fixed without the harmful baggage of linked deleterious neighbors, and the deleterious alleles were purged from the populations.

The Red Queen Effect: Running in Place

Parasites and pathogens may also help explain the benefits of sex. All cellular organisms risk becoming the victim of parasites and pathogens. Being infected can lower the fitness of an individual in many ways—a parasite can kill its host outright, it can make its host too sick to find a mate, or it can even damage its host's reproductive organs. Hosts that can resist parasites better than other members of their population should be favored by natural selection. As we'll discuss in more detail in Chapter 15, the selection

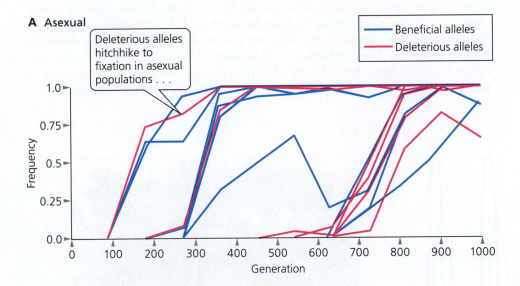

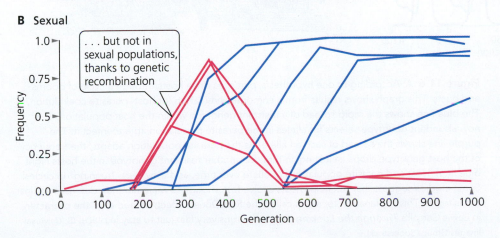

Figure 11.5 Yeast can reproduce either asexually or sexually. Michael McDonald and colleagues set up experimental asexual and sexual populations and let them evolve in the laboratory, tracing the fates of new mutations (alleles) that arose in each lineage. A: In asexual populations, deleterious mutations (red) often hitchhiked to fixation with beneficial mutations (blue). B: In sexual populations, on the other hand, beneficial alleles could be fixed by selection, whereas deleterious alleles were rapidly purged from the population. (Data from McDonald et al. 2016)

pressure from parasites has major impacts both on the course of evolution and on human health.

As strong defenses against parasites evolve in hosts, countermeasures may evolve in parasites that allow them to overcome those defenses. In the 1970s, several evolutionary biologists suggested that this evolution of parasites and their hosts occurs in cycles. Parasites strike the most common genotypes in the population of their hosts. Those hosts reproduce less or even die off, whereas other more resistant host genotypes become more common. The most common parasite is now no longer well adapted to the most common host genotype. Over time new parasite genotypes evolve that can exploit this new common host, eventually killing them, and the cycle begins again (**Figure 11.6A**). This model is known as the **Red Queen effect** (Van Valen 1973), named after the character of the Red Queen in Lewis Carroll's book *Through the Looking-Glass*, who takes Alice on a run that never seems to get them anywhere. "Now here, you see, it takes all the running you can do to keep in the same place," the Red Queen explained (**Figure 11.6B**).

The Red Queen effect predicts an evolutionary arms race between hosts and their parasites. As a result, hosts constantly have to evolve defenses at a rapid rate just to "stay in the same place"—that is, to survive. Through the Red Queen effect, parasites may drive the maintenance of sex in their hosts. Genetic variation within parasites, coupled with rapid reproduction and short generation times, allows them to adapt quickly to the genotypes of their hosts. It may be even easier for them to do so with asexual hosts because the only new variation in asexually reproducing hosts comes from rare mutations. Sexual reproduction, on the other hand, creates lots of genetic variation in every generation of the hosts by shuffling alleles into new combinations. Among those new variations may be some that are very resistant to the dominant parasite genotype. These individuals will be favored by natural selection, and their descendants will thrive. Sex is beneficial because it makes the host a moving target for parasites. (See **Box 11.1** for a discussion of long-lived clades of asexual animals.)

The **Red Queen effect** describes a phenomenon seen in coevolving populations—to maintain relative fitness, each population must constantly adapt to the other.

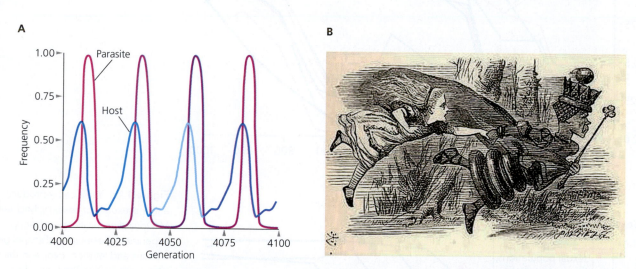

A

B

Figure 11.6 A: According to one hypothesis, parasites drive hosts through a cycle of booms and busts. This graph shows results from a computer simulation for host–parasite coevolution. The blue line shows the rapid spread of a resistant genotype within the host population. This now abundant genotype selects for alleles in the parasite population able to infect it. The purple line shows the spread of such an allele. As the parasite population adapts, the frequency of the host genotype drops, selecting, in turn, for another resistant genotype in the hosts (light blue line). Host and parasite populations oscillate over time, as if they were "running" in circles. Some studies suggest that sex helps hosts evolve fast enough to maintain their defenses against parasites. B: This explanation for sex is called the Red Queen effect, named after the character in Lewis Carroll's *Through the Looking-Glass* who runs very fast just to stay in place. (B: Universal Images Group/Superstock)

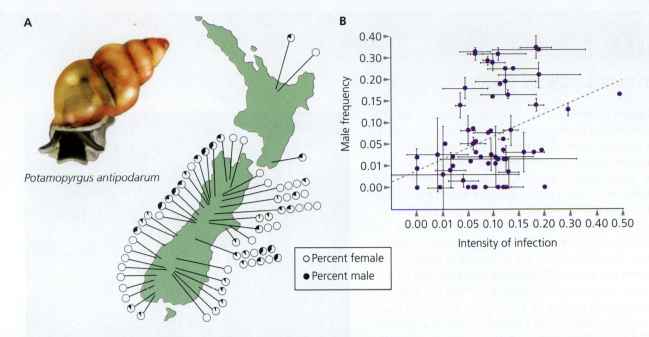

Figure 11.7 A: The New Zealand freshwater snail *Potamopyrgus antipodarum* can reproduce either sexually or asexually through parthenogenesis, making it an ideal organism for testing when the benefits of sex outweigh the costs (see Table 11.1). Curt Lively of Indiana University sampled 66 snail populations, tallying the asexual and sexual forms. His results are shown here as the proportion of snails that are males. B: Variation in the proportion of sexual individuals correlated across Lively's study sites with the intensity of infection of snails by trematode parasites. This correlation is consistent with selection from parasites driving the evolution of sex. (Data from Lively 1992)

Over many generations, this adaptive edge may cause asexual individuals in a population to become increasingly rare, whereas sexual individuals become increasingly common. That's what mathematical models of the Red Queen effect suggest, and a number of studies on real organisms offer support as well (Lively 2010). **Figure 11.7** illustrates one of the best-studied examples of the Red Queen effect, on sexual and asexual snail populations in New Zealand.

Key Concepts

- Sex is the combining and mixing of chromosomes during the formation of offspring. It involves two main processes: (1) meiosis halves the number of chromosomes during gamete formation, and (2) fertilization restores the original chromosome number as two gametes fuse to form a zygote.

- Self-fertilization occurs in hermaphroditic organisms when the two gametes fused in fertilization come from the same individual. Although this is still sexual reproduction (because it incorporates meiosis and fertilization), it yields offspring that are lacking in genetic variation.

- Sexual reproduction may have evolved because sex creates new genetic variation within populations by mixing parental alleles into novel offspring genotypes while at the same time enabling the purging of deleterious alleles.

- The Red Queen effect is used to explain the evolution of sexual reproduction because sexually reproducing organisms are likely to fare better in the continual evolutionary arms races that can arise between species (for example, between parasites or pathogens and their hosts).

BOX 11.1

A Curious Lack of Sex

The Red Queen effect offers a promising explanation for why sex has become so widespread despite its apparent cost. But some species seem to challenge such explanations—their ancestors abandoned sex millions of years ago, and they have continued to adapt and diversify. Among the most striking of these asexual groups are the bdelloid rotifers: 400 species of invertebrates that live in freshwater (Box Figure 11.1.1). Their closest relatives are all sexual, but almost 100 million years ago, the lineage switched to asexual reproduction. Bdelloid rotifers are all female, in other words, and daughters are genetically identical to their mothers.

If sex offers so many advantages, how have the bdelloid rotifers managed to survive as a lineage? Fascinating clues emerged from studies on their genomes. Some segments of their DNA were inserted through horizontal gene transfer from bacteria, fungi, and plants. These genes can enter the bdelloid rotifer genome thanks to their peculiar life cycle.

Bdelloid rotifers survive harsh conditions by entering a state of metabolic dormancy and complete dehydration; they can later rehydrate when conditions improve. During desiccation and rehydration, their membranes rupture. As they repair their membranes, the rotifers sometimes ingest fragments of foreign DNA and incorporate them into their genomes. This extra supply of genes may allow bdelloid rotifers to escape some of the deleterious genetic consequences of a loss of sex (Mark Welch et al. 2004; Gladyshev et al. 2008).

Desiccation may also help bdelloid rotifers live without sex in another way. Parasites and pathogens cannot survive in their dehydrated bodies. Thus, bdelloid rotifers may be better able to withstand diseases, and they may partially escape the relentless coevolution of parasites predicted by the Red Queen effect (Wilson and Sherman 2010).

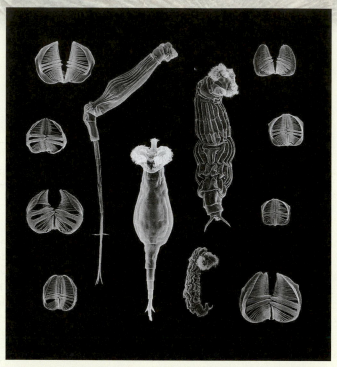

Box Figure 11.1.1 Bdelloid rotifers abandoned sexual reproduction almost 100 million years ago, yet they have adapted to diverse freshwater environments and diverged into 400 successful species. They reproduce parthenogenetically, yielding daughter genotypes identical to their mothers, and they completely lack both males and meiosis. Recent sequencing of their genomes revealed several surprises, including DNA originating from bacteria, fungi, and plants. It now appears that these rotifers may escape some of the deleterious genetic consequences of a loss of sex by incorporating fragments of foreign DNA into their genomes. (Data from Mark Welch et al. 2004; Gladyshev et al. 2008; photo by Public Library of Science)

11.2 Sexual Selection

Once sex evolves, it changes the evolutionary landscape in dramatic ways—ways that we will explore in the rest of this chapter.

Cheap Sperm and Costly Eggs

Anisogamy refers to sexual reproduction involving the fusion of two dissimilar gametes; individuals producing the larger gamete (eggs) are defined as female, and individuals producing the smaller gamete (sperm) are defined as male.

One particularly dramatic change in the evolutionary landscape is in the size of gametes. In many species, eggs are large, whereas sperm are tiny. Kiwis, flightless birds that live in New Zealand, take this imbalance—known as **anisogamy**—to an extreme. Whereas a male kiwi's sperm are microscopic in size, a single egg produced by a female kiwi may weigh as much as a quarter of her own weight (Figure 11.8). There is a vast difference in the resources required by each sex to make these gametes. A male kiwi can produce trillions of sperm with the same resources a female needs to make one egg.

Figure 11.8 Eggs can demand huge resources from females. The X-ray image of this kiwi bird shows an egg that weighs approximately a quarter of her body weight. (Otorohanga Kiwi House, NZ)

Males can produce many, many more gametes than females because sperm are much smaller than eggs. Anisogamy has far-reaching effects on the evolution of males and females because it means that the best strategy for reproductive success for males is not the same as the best one for females. Males and females both face limits to the number of offspring they can produce, but those limits have different sources.

Because sperm are so abundant, the reproductive success of females is almost never limited by the availability of male gametes. A man, for example, makes tens of millions of new sperm every day. Any man could, in his lifetime, theoretically fertilize all the eggs of every woman on Earth. Females, on the other hand, are limited by how many eggs they can produce and provision—a trait known as **fecundity**. Females differ in their fecundity, and that difference can determine which individual females contribute the most copies of their genome to future generations. In many populations, the most successful females are the ones who provision more eggs than other females. This is not true for males, because the most successful males are not necessarily those that produce the most sperm.

Males face a different limit to their reproductive success: access to females. In an anisogamous population, males that can gain access to more eggs can produce more offspring than other males. As a result, there is often a direct relationship between the number of females that a male mates with and the number of offspring that he sires. More matings with more females translates into more fertilized eggs. Consequently, when it comes to males, the ones who mate with many females (specifically, who mate with more females than other males) generally have the highest fitness (**Figure 11.9**).

Fecundity describes the reproductive capacity of an individual, such as the number and quality of eggs or sperm. As a measure of relative fitness, fecundity refers to the number of offspring produced by an organism.

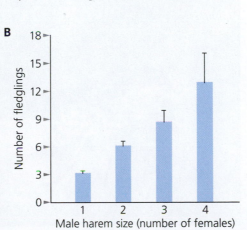

Figure 11.9 A: Male reed warblers guard harems of females. B: Males that have bigger harems have more offspring. (Data from Hasselquist 1998; A: BogdanBoev/Shutterstock)

Unequal Investment Results in Sexual Selection

Another consequence of anisogamy is that females often benefit when they perform additional parental care. Females already invest more than males by virtue of their large, yolk-rich gametes, and their reproductive success is most limited by the number and quality of offspring they can produce. A female that continues to invest in the nourishment or protection of her offspring may be able to raise her reproductive success (Trivers 1972; Clutton-Brock 1991).

One reason maternal care is so often beneficial is that when a mother invests in her offspring, the beneficiaries of her care are very likely to actually be *her* offspring. The same cannot be said for males. We'll see later in this chapter that when a male mates with a female, it does not necessarily mean he is the father of the offspring she produces. His **certainty of paternity** may be much lower than a female's certainty of maternity. As a result, when offspring require extra care, the females usually provide it (Kokko and Jennions 2008).

In many species, females pack extra nutrients into their eggs. This strategy can help raise the odds of their offspring surviving. Females may spend time looking for safe places to lay their eggs or to protect their broods once they hatch. A particularly safe place to rear the young is inside the mother's own body. In a wide range of species, from flies to mammals, the female parent carries the pair's offspring in a cavity in its body where the offspring are nurtured as well as protected (**Figure 11.10**).

Males typically don't experience such strong selection for investing additional care in offspring. For them, the best strategy will often be to leave females after mating and use all their available time and resources in the pursuit of other chances to mate.

Because of these two different strategies, reproductive rates can be very high in males and are generally low in females. It may take just a few minutes after mating for

Certainty of paternity is the probability that a male is the genetic sire of the offspring his mate produces.

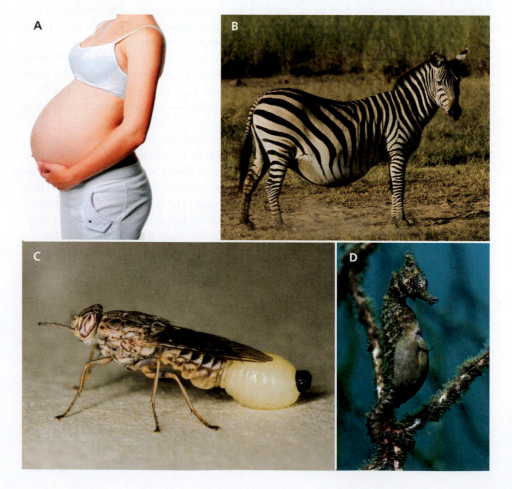

Figure 11.10 Many species keep their young inside their bodies as they develop, providing both nourishment and protection. Examples include (A) humans, (B) zebras, (C) tsetse flies, and (D) seahorses. We will see in Chapter 12 that seahorses are unusual because pregnancy occurs in males rather than females. (A: In Green/Shutterstock; B: ClassicStock.com/Superstock; C: www.raywilsonbirdphotography.co.uk; D: NaturePL/Superstock, Inc.)

a male to be ready to breed again with a different female. Females, by contrast, often need time to nurture the offspring in some way before they can reproduce again. A female elephant, for example, spends about 20 months pregnant with a single calf and then nurses it for another 18 months. All told, 3 or 4 years may pass before a female elephant can get pregnant again.

These different reproductive rates mean that at any moment, there will be more members of one sex in a population ready to mate than there are potential partners from the other sex. Typically, a population has more available males than receptive females. The ratio of males to females available for mating at any one time is called the **operational sex ratio** (OSR; Emlen and Oring 1977). The OSR may not be obvious to a casual observer. If you simply tally the number of males and females, the ratio will typically be close to one to one. But in many species, the OSR is very biased because there are so few receptive females per available male.

When the OSR is strongly male biased, males may face stiff competition for the opportunity to mate with a reproductive female. This competition creates the opportunity in males for selection for traits that are primarily concerned with reproduction. (See **Box 11.2** for further discussion of the strength of selection in males and females. As we'll see in the next chapter, the process is reversed when the OSR is skewed toward females; in such cases, these traits arise in females.)

This type of selection is called **sexual selection**. As with many concepts in evolutionary biology, sexual selection was first recognized by Darwin (1859, 1871). He defined it as "the advantage which certain individuals have over others of the same sex and species, in exclusive relation to reproduction."

In many species, a male-biased OSR leads to sexual selection favoring males that can outcompete other males for access to females. Different traits may evolve in males of different species. Some males use weapons to fight with each other for access to females (**Figure 11.11**). Ornaments make females more likely to choose some males over other males (Darwin 1871; Andersson 1994). Occasionally, the same trait

Operational sex ratio (OSR) is the ratio of male to female individuals who are available for reproducing at any given time.

Sexual selection refers to differential reproductive success resulting from the competition for fertilization, which can occur through competition among individuals of the same sex (intrasexual selection) or through attraction to the opposite sex (intersexual selection).

Figure 11.11 In many species, males have weapons that they use to compete with other males for access to females and for territory. (Mark Bridger/Shutterstock)

BOX 11.2

Which Sex Will Experience the Stronger Sexual Selection?

Because every egg is fertilized by a sperm, the number of offspring sired by all of the males in a population must equal the number of offspring produced by females. Similarly, the mean reproductive success of males must equal the mean reproductive success of females. What matters for sexual selection is how this reproductive success is distributed among individuals of each sex. For example, two populations could each have a mean reproductive success of five offspring per male. In one, every male sires exactly five offspring, and the variance among males is 0. In the other, two males each sire 30 and the rest sire 0. Here, the variance among males is much greater. The variance in reproductive success is critical because it shows how much better winners fare than losers. In the first population there are no winners or losers because all males sire the same number of offspring. As a result, there is no selection resulting from competition over access to females. In the second population, on the other hand, winners do much better than losers, and the difference between them is stark. Any traits that enabled a male to sire 30 offspring rather than 0 would be favored very strongly by sexual selection. The opportunity for selection will be greater in the second population because the variance in reproductive success in that population is larger.

This same logic can be used to compare the opportunity for selection in males and females of a single population. They, too, share the same mean reproductive success, so any differences between them in the relative strength of sexual selection will lie in their respective variances. Stated simply, *the sex with the greater variance in reproductive success will be the one experiencing the stronger sexual selection* (Box Figure 11.2.1).

Often, the variance in reproductive success is much greater for males than it is for females. In these populations, sexual selection is likely to be stronger in males than it is in females, and we expect extravagant ornaments or weapons to evolve in males rather than females.

However, this need not always be the case. In Chapter 12 we will look at populations where the variance in reproductive success is greater for females. In these "role-reversed" species, sexual selection is stronger in females than males, and bright colors, aggression, and weapons evolve in females rather than males.

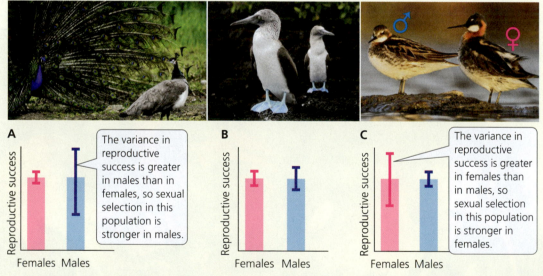

A: The variance in reproductive success is greater in males than in females, so sexual selection in this population is stronger in males.

C: The variance in reproductive success is greater in females than in males, so sexual selection in this population is stronger in females.

Box Figure 11.2.1 Which sex will experience the strongest sexual selection? Males and females in a population have the same mean reproductive success (colored bars), but they may differ in the reproductive variance among members of each sex (error bars). A: When males experience a greater opportunity for sexual selection than females (higher variance in reproductive success), sexual selection is likely to be stronger in males than in females. Peacocks, for example, do not help peahens with parental care, and some of the males have much more reproductive success than others. In such a case, elaborate male traits may evolve. B: When males contribute to offspring care, the variance in reproductive success will be more similar for males and females, and extreme sexual dimorphism is not expected. In many seabird species, such as this pair of blue-footed boobies, males and females are monogamous and raise their chicks together. Not surprisingly, the males and females of many of these species are practically indistinguishable. C: When males provide all the parental care, the variance in reproductive success will be higher in females. As a result, sexual selection can be stronger in females than in males, and females may evolve ornaments or weapons. Red-necked phalarope females compete for access to males who provide all the parental care. Females are larger, brighter, and more aggressive than males, as shown here. (Data from Brennan 2010; A: Dave Blackey/All Canada Photos/Superstock, Inc.; B: Photography by Jessie Reeder/Getty Images; C: Glenn Bartley/All Canada Photos/Superstock, Inc.)

functions as both an ornament and a weapon (Berglund et al. 1996). Sexual selection can thus be divided into two forms. **Intrasexual selection** occurs when members of the less limiting sex compete with each other for access to the limiting sex. **Intersexual selection**, on the other hand, occurs when the members of the limiting sex discriminate among potential mates from the other sex.

Male–Male Competition: Horns, Teeth, and Tusks

Struggles among males can take many forms, from mountain sheep slamming their horns against each other to fiddler crabs flipping each other with their oversized claws. In species ranging from horses to flies, males guard groups of females, known as harems, from rival males. Fights for control over these harems can be intense and frequent. If one male manages to guard 20 females, then 19 other males will likely get access to none.

The difference between winners and losers is especially stark among the southern elephant seals that breed on Sea Lion Island, one of the Falkland Islands in the South Atlantic Ocean. For a number of years, A. Rus Hoelzel, a biologist at Durham University in England, and his colleagues surveyed all of the seals on Sea Lion Island. They snipped small pieces of skin from the adult elephant seals as well as from 192 baby seals that were born from 1996 to 1997. They extracted DNA from each piece of skin and found that they could identify the father of almost every seal pup. That's because the females were mating only with males that were lying around on Sea Lion Island.

Among the males, 72% failed to have any offspring at all. The remaining 28% did not share reproductive success equally. Many had only one or two pups, whereas a few managed to have many offspring. One particularly successful male seal fathered 32 pups (**Figure 11.12**; Fabiani et al. 2004).

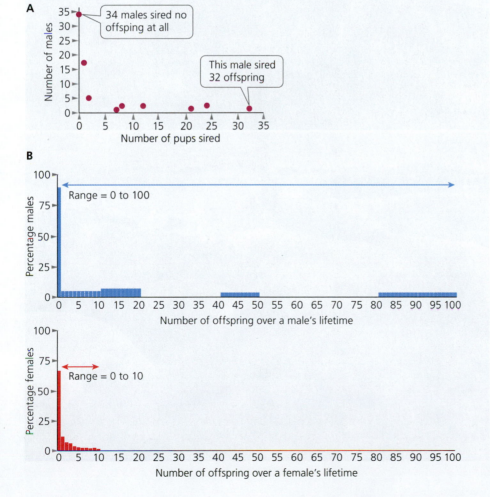

Intrasexual selection occurs when members of the less limiting sex (generally males) compete with each other over reproductive access to the limiting sex (generally females). Often called male–male competition.

Intersexual selection occurs when members of the limiting sex (generally females) actively discriminate among suitors of the less limited sex (generally males). Often called female choice.

Figure 11.12 Extreme variance in reproductive success among male elephant seals. A: Fertilization success was measured for seven different harems on Sea Lion Island during one breeding season. A small number of dominant males (the harem holders) were able to achieve disproportionate mating success (for example, 21, 24, and 32 pups sired), whereas most males sired no pups at all. (Data from Fabiani et al. 2004) B: Over their lifetimes, male (blue) and female (red) seals varied in their reproductive success, but this variance was much higher for males than it was for females. Such high variance in male reproductive success generates intense sexual selection for traits that enable males to win in these contests. (Data from Le Boeuf and Reiter 1988)

Sexual dimorphism is a difference in form between males and females of a species, including color, body size, and the presence or absence of structures used in courtship displays (elaborate tail plumes, ornaments, pigmented skin patches) or in contests (antlers, tusks, spurs, horns).

Success among male southern elephant seals is not random. Males fight each other in order to mate with females, which gather together in harems. Males rear up on their flippers and throw their tremendous bodies against each other, regularly gashing each other with their teeth (**Figure 11.13**). The losers slink away, to lurk at the edges of the colony. Hoelzel found that 90% of the seals that fathered pups were the heads of harems. The other 10% of the seals that fathered pups were lurkers who managed to sneak in and mate with a female while the dominant male was distracted by a fight with another male.

Sexual selection may explain why male elephant seals are several times larger than female elephant seals (**sexual dimorphism**). Bigger males tend to win fights with smaller ones, so bigger males tend to hold onto harems. The big males thus have more offspring. Female elephant seals, on the other hand, don't fight with each other, so extremely large females don't have a reproductive advantage over smaller ones. In other words, sexual selection for body size is much stronger in male elephant seals than in females.

Figure 11.13 A: Male elephant seals fight each other to mate with large numbers of females. B: The female elephant seals gather together in harems. (A: David Osborn/Shutterstock; B: Comstock/Exactostock-1557/Superstock, Inc.)

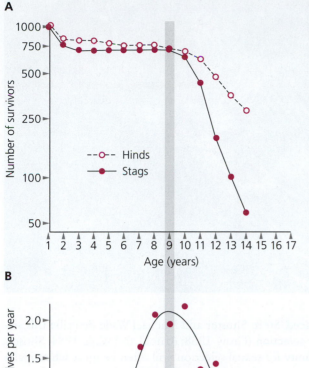

A

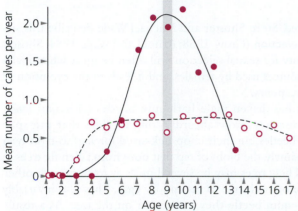

B

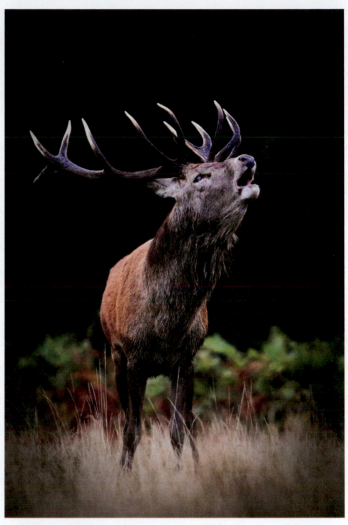

Figure 11.14 The costs of fighting. As with elephant seals, male red deer (stags) fight for possession of harems of females (hinds). A few successful males realize high reproductive success, but getting to the top is dangerous and costly. A: In a long-term study, Timothy Clutton-Brock of the University of Cambridge and his colleagues found that as soon as males reached reproductive age, they began to die at a much faster rate than females. B: This dangerous battle for harem ownership was reflected in the distribution of reproductive success across deer life spans. Most females reproduced; once they began breeding, they produced calves at a fairly steady rate for many years. In contrast, most males failed to breed at all; those that did were not able to begin breeding until they were much older and larger than females and able to fight to the top of the pack. If the males were able to get to the top of the pack, they could hold that position for only a brief period before being killed or replaced by new males. (Data from Clutton-Brock 1988; C: DamianKuzdak/Getty Images)

In species such as bighorn sheep and red deer, battles over harems have driven the evolution of elaborate horns and antlers (**Figure 11.14**). These weapons help males in their fights. Males that are in the best condition grow the largest weapons, and these males mate with the largest numbers of females. Here, too, sexual selection has led to the evolution of male traits that aid them in contests over access to reproduction (Kruuk et al. 2002; Preston et al. 2003; Malo et al. 2005; Vanpé et al. 2007; Emlen 2008).

In some species, males don't fight directly with each other for females. Instead, they fight over the resources the females need, such as shelter or food. If the resources are patchy or rare enough, dominant males can drive off other males and thus have the opportunity to mate with a disproportionate number of females. Males that are very successful at guarding resources can mate with many different females. Other males in the same population are less successful, and many males have no reproductive success at all. This large variance in reproductive success means that the strength of sexual

Figure 11.15 A: Male harlequin beetles fight with long forelegs to control territories where females feed on tree sap. B: Tiny male pseudoscorpions ride on the backs of these males. Instead of long forelegs, they have evolved large appendages called pedipalps, which they use to fight other males for access to females. (A: Danita Delimont/Getty Images; B: Stefan Sollfors/Alamy Stock Photo)

Opportunity for selection refers to the variance in fitness within a population. When there is no variance in fitness, there can be no selection; when there is large variance in fitness, there is a great opportunity for selection. In this sense, the opportunity for selection constrains the intensity of selection that is possible.

selection can be powerful indeed. Steve Shuster and Michael Wade describe this situation as the **opportunity for selection** (Crow 1958; Arnold and Wade 1984; Shuster and Wade 2003). The opportunity for sexual selection will often be great when males can guard females or critical resources used by females, and it leads to the evolution of enlarged or exaggerated male weaponry.

A struggle over resources has driven male harlequin beetles (*Acrocinus longimanus*) to extravagant extremes: they grow absurdly long forelegs that can span 16 inches (**Figure 11.15A**). The beetles use their chopstick-sized forelegs to fight over the places where females feed: namely, the blobs of sap that ooze from fallen fig trees in the tropical forests of Panama. These trees may be several miles apart from each other, and many females will converge on each tree to eat and lay eggs. They also invariably mate with whatever male harlequin beetle they encounter on the tree. As a result, males guard their sap by battling intruders with their forelegs. David and Jeanne Zeh, of the University of Nevada, have found that males with longer forelegs and larger body sizes tend to win these fights and thus tend to have more offspring than other males (**Figure 11.16**; Zeh et al. 1992).

The Zehs have also discovered an even stranger twist to this story. Along with harlequin beetles, a species of pseudoscorpion (*Cordylochernes scorpioides*; Figure 11.15B) also feeds on the sap of fallen fig trees. Unlike harlequin beetles, however, the pseudoscorpions lack wings. To get from one tree to another, they ride on the backs of the beetles.

Figure 11.16 Male harlequin beetles with (A) large body sizes and (B) extra-long forelegs have the highest mating success, indicating strong sexual selection due to male–male competition. The dashed lines indicate standard error confidence limits. (Data from Zeh et al. 1992)

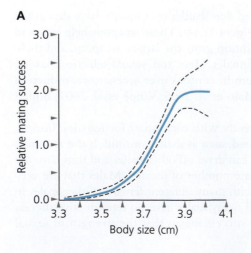

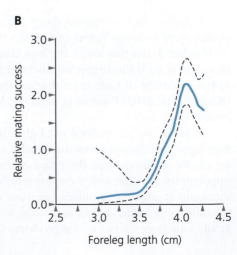

A ride on the back of a beetle is a limited resource for the pseudoscorpions, and as a result, the males fight to guard this airborne territory. Dominant males force subordinate ones off the backs of the harlequin beetles. These dominant male pseudoscorpions then mate with females while the beetle flies to the next fig tree. On the way from one fig tree to another, successful males can mate with 20 or more females. Just as in the case of the harlequin beetles, these pseudoscorpions have experienced sexual selection for imposing male weapons. Male pseudoscorpions have enlarged, claw–like appendages called pedipalps that they use in fights with rival males (Zeh and Zeh 1992).

Males Bearing Gifts

Females do not merely sit and wait passively for males to fight over them. Some make choices about which males they will and will not mate with. By preferentially mating with particular males, females may derive any number of reproductive benefits. In general, these benefits come in two forms, direct and indirect. **Direct benefits** of female choice include food and protection (**Table 11.2**). Females who make prudent

A **direct benefit** is a benefit that affects a particular female directly, such as food, nest sites, or protection.

Table 11.2 Direct Benefits of Female Choice

Direct Benefit	Examples
Protection When a female risks being trampled or injured by competing males, she may benefit by preferentially mating with large (dominant) males that guard her from harassment. Females in these species often have a "last male" fertilization advantage that selects for mate-guarding behavior on the part of the males.	**Protection from lethal injuries by other males** • Elephant seals (Le Boeuf and Mesnick 1990) • Dung flies (Sigurjónsdóttir and Parker 1981) **Male gives female toxic chemicals that she uses to protect her eggs** • Rattlebox moths (Dussourd et al. 1991; Iyengar and Eisner 1999)
Territories and/or nests In some species females preferentially mate with males who have already constructed an elaborate nest, saving them the time and effort. By choosing males with high-quality nests or optimal nest locations, females may improve the likelihood of survival of their offspring relative to offspring of other, less choosy females.	**Access to good egg-laying sites** • Dragonflies (Campanella and Wolf 1974) • Frogs (Howard 1978) **Suitable vegetation for nesting** • Red-winged blackbirds (Yasukawa 1981) **Nests the male has manufactured** • Weaverbirds (Collias and Victoria 1978) **Territories with good forage** • Antelope (Balmford et al. 1992)
Food Females of most animal species are fecundity limited: eggs and young are expensive to produce and to provide for. By preferentially mating with males that offer food rewards, females can increase the resources available to their young, improving their fecundity relative to other, less choosy females. Food rewards for mating are often called nuptial gifts; extreme examples involve sexual cannibalism.	**Prey food items** • Scorpionflies (Thornhill 1976, 1983) **Female eats fleshy hindwings of male** • Crickets (Eggert and Sakaluk 1994) **Spermatophores** • Katydids and bushcrickets (Gwynne 1988; Wedell and Arak 1989) **Female eats the male** • Redback spiders (Andrade 1996)
Help raising young Females preferentially mate with males who will invest more parental effort in their offspring (for example, by choosing males with courtship displays signaling high parental ability).	**Females choose males who are better egg guardians** • Sculpins (Downhower and Brown 1980) **Females choose males better at fanning eggs** • Sticklebacks (Östlund and Ahnesjö 1998) **Females choose males better at feeding nestlings** • Pied flycatchers (Saetre et al. 1995)
Reduced risk By choosing males with bright ornaments or vigorous displays, females pick healthy males who are less likely to transmit mites, fleas, lice, or sexually transmitted infections to the female.	**Females choose males with large, bright red combs** • Junglefowl; these males have healthier immune systems (Zuk et al. 1990; Zuk and Johnsen 1998)

mate choices increase their fitness directly because males give them those benefits in exchange for mating.

In some species, males entice females to mate with them by bearing gifts. The most common of these so-called nuptial gifts is food. Because a female's fecundity is limited by the resources available to her, a gift of food can mean a great deal to her reproductive success. Many male insects, for example, don't just deliver sperm into the reproductive tracts of their mates. They also deliver blobs known as spermatophores, which contain lots of proteins and lipids.

When females receive spermatophores, they may eat them or just absorb them directly into their reproductive tissues. In either case, the females can use the nutrients to "yolk up" their eggs, so that their offspring have more nutrients to feed on while they develop. Darryl Gwynne, at the University of Toronto, has studied the evolution of nuptial gifts in katydids. Gwynne found that the more nuptial gifts a female received, the larger and more numerous were her eggs (**Figure 11.17**; Gwynne 1988).

Praying mantises and some species of spiders have pushed the practice of nuptial gift giving to the extreme: in the middle of copulation, the female eats the male himself (**Figure 11.18**; Lelito and Brown 2006; Barry et al. 2008). Such sexual cannibalism tends to evolve in species that are already predators. Females may simply look upon males as yet more prey. And because females are almost never limited by the number of males they have access to, females can afford to devour their mates.

For males, sexual cannibalism can obviously have a devastating impact on their reproductive success. As might be expected, they try to avoid this outcome whenever possible. In some species, the males are very small compared to the females, and some researchers have argued that this diminutive size evolved as a defense against being devoured. The males may be so small that they're not worth the effort to eat. On the other hand, some males appear to gain reproductive success through cannibalism. Even as the female slowly dines on the male, he continues to mate with her and his body's nutrients enhance her fecundity (**Figure 11.19**; Andrade 1996).

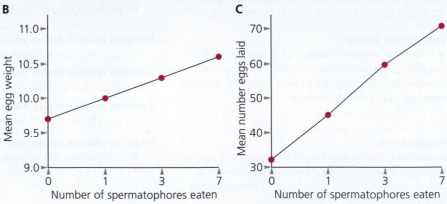

Figure 11.17 When katydids mate, males give females nutrient-rich spermatophores to eat. A: A female feeds on a spermatophore. B: The more spermatophores a female eats, the greater the mean weight of her eggs will be. C: Likewise, an increase in the number of spermatophores leads to a greater number of eggs laid. (Data from Darryl Gwynne, personal communication; A: Darryl Gwynne)

Figure 11.18 Some species of invertebrates engage in sexual cannibalism. Here, a female praying mantis devours a male during mating. (Tobias Bernhard/Getty Images)

By preferentially mating with males who bear gifts, a female can derive direct material benefits that improve her survivorship and enhance her fecundity relative to other females who are less discriminating. This process leads to the evolution of ever stronger female mate choice behavior and increasingly significant (expensive) male gifts.

Indirect benefits, on the other hand, enhance the quality of a female's offspring (they are called *indirect* benefits because they bypass the female and instead increase the fitness of her offspring). As we'll see in the next section, such benefits arise when females are able to choose high-quality males as sires of their offspring, because the

An **indirect benefit** is a benefit that affects the genetic quality of a particular female's offspring, such as male offspring that are more desirable to females.

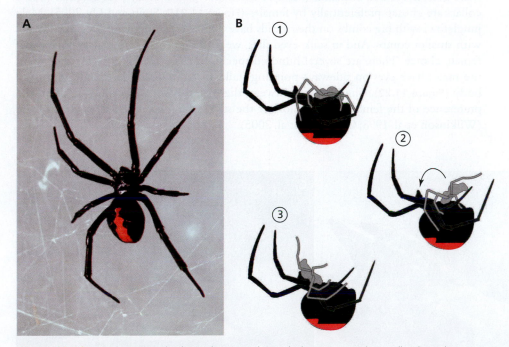

Figure 11.19 Maydianne Andrade studies sexual cannibalism in Australian redback spiders and has found that males in this species have a very low probability of finding and mating with a second female. A: A female Australian redback spider. B: In this extreme case, as Andrade has demonstrated, the males actually benefit from being eaten because they can prolong copulation as the female eats them. The male's pedipalps also often break off in the female's reproductive tract, acting as a copulatory plug to prevent other males from copulating with the female. The numbers in this figure show the sequence of this self-sacrifice. While mating, a male arches himself upward and backward, placing his abdomen in front of the female's jaws. (A: Peter Waters/Shutterstock)

sperm from these males contains more favorable alleles than does the sperm from males of poorer quality.

Dancing Males and Showy Ornaments

In many species, scientists have found strong preferences among the females for certain ornamental traits in males (**Figure 11.20**). These preferences can be remarkably precise. Male blue-crowned manakins (*Lepidothrix coronata*; **Figure 11.21A**), for example, perform a dancing display in the understory of Ecuadorian tropical forests to attract females. The males may dance in isolation or in groups—called **leks**—of up to seven simultaneously displaying males. Female blue-crowned manakins will visit several displaying males before accepting a male as a mate. The females are very consistent in their choice behavior: males displaying in leks are more likely to attract females than males displaying alone, and each lek generally includes a single male who sires most of the offspring. In blue-crowned manakins, females typically select the male that displays the fastest (Durães et al. 2009).

Male long-tailed manakins (*Chiroxiphia linearis*; **Figure 11.21B**) dance in pairs. Two males team up and bounce side by side in a series of leapfrog hops, punctuated by wing flaps known as butterfly flight. Two males may remain dance partners for many years. The females, meanwhile, assess these performances. They can judge several dances at once, because males will gather in the same small area of forest. After a female has observed the dances, she will approach one pair and mate with the alpha male. Long-tailed manakin females are also remarkably consistent in their dance preferences, and it is not unusual for one or a few males to receive most of the matings in a population (Trainer and McDonald 1995).

Females judge males not only on what they do but also on how they look. Male golden-collared manakins (*Manacus vitellinus*; **Figure 11.21C**) with brighter yellow collars are chosen preferentially by females (**Figure 11.21D**; Stein and Uy 2005). Male junglefowl with big combs on their heads have higher reproductive success than males with smaller combs. And in stalk-eyed flies, we find one of the most surreal effects of female choice. There are several hundred species of these flies, and in most of them the males have eyes on sideways-pointing stalks that can be longer than their entire body (**Figure 11.22**). The distance between the males' eyes has a strong effect on the preference of the female flies: the longer the stalks, the more likely they are to mate (Wilkinson et al. 1998; Chapman et al. 2005).

A **lek** is an assemblage of rival males who cluster together to perform courtship displays in close proximity.

Figure 11.20 Male displays can take many forms, from bright plumage to neck flaps to loud, croaking calls. (A: ZUMA Press, Inc./Alamy Stock Photo; B: David Byron Keener/Shutterstock; C: Rike_/Getty Images)

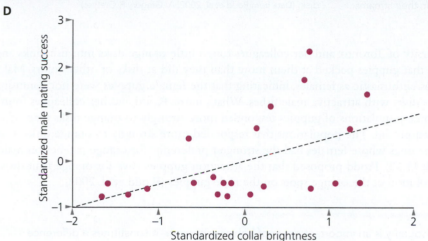

Figure 11.21 Male manakins perform elaborate displays alongside rival males to attract females. A: In the blue-crowned manakin, females prefer males with the highest display rates. B: In the long-tailed manakin, males team up for a joint display, even though only the dominant male will mate. C: In the golden-collared manakin, males clear display arenas of debris, enhancing the contrast of their plumage against this background. D: Golden-collared manakin females choose males with the brightest and most conspicuous plumage. (Data from Stein and Uy 2005; A: Roy Toft/Getty Images; B: Glenn Bartley/Getty Images; C: Gualberto Becerra/Shutterstock)

But why do females exhibit the preferences that they do? Some studies suggest that females have ancient preferences for certain shapes or colors, and these sensory biases trigger sexual selection in males (Ryan 1990). On the Caribbean island of Trinidad, for example, male guppies sport bright-orange patches, and females prefer to mate with the males with the biggest, brightest patches. But the guppies also eat bright-orange fruits that plop into their streams. Helen Rodd, a biologist at the

Figure 11.22 The stalk-eyed fly gets its name from the long stalks on which its eyes develop. The distance between a fly's eyes can be longer than its entire body. Female flies prefer to mate with male flies with longer stalks rather than shorter ones. (Bazzano Photography/Alamy Stock Photo)

Figure 11.23 A: Female guppies prefer to mate with males with bright-orange spots. Their preference may be linked to an attraction to orange-colored fruits that sometimes land in their streams.

B: In one experiment, scientists found that females with stronger preferences for orange in males would peck more at an orange disk. (Data from Rodd et al. 2002; A: Gregory F. Grether)

University of Toronto, and her colleagues tossed little orange disks into fish tanks and found that guppies pecked at them more than they did at disks of other colors. Males were as enthusiastic as females, indicating that the female guppies were not confusing orange disks with attractive male fishes. What's more, Rodd and her colleagues found that some populations of guppies responded more strongly to orange disks than other populations did. The populations that responded more strongly to orange disks were also the ones whose females had the strongest preference for orange patches on males (**Figure 11.23**). Rodd proposed that the attraction guppies have for orange fruit drove the evolution of the orange spots on the male guppies (Rodd et al. 2002).

Key Concepts

- Anisogamy is an important driver of behavior because it constitutes a difference between the sexes in their relative investment in offspring. Because of anisogamy, males and females maximize their reproductive success in distinct ways.

- A female's reproductive success is often limited by the number of eggs that she can produce and provision. Females with access to the most resources generally achieve the highest egg (and offspring) numbers and have the highest fitness relative to other females.

- Male reproductive success is often limited by the number of matings the male can obtain. Males who mate with the most females generally sire the greatest number of offspring and have the highest fitness relative to other males.

- Biased operational sex ratios can generate strong sexual selection because the abundant sex (typically males) must compete over access to the limiting sex (typically females).

- Males often fight with each other over access to females, and this behavior can generate strong sexual selection for large body size, weapons, and aggression.

- Sexual selection affects the sex with the greater variance in reproductive success (usually males).

- Depending on the species, a female may select males based on characteristics that benefit her directly (such as nutrients, nest sites, protection, parental care) or indirectly (for example, high genetic quality transmitted to her offspring).

- Females often choose males based on appearance or details of their courtship behavior.

- The variance in reproductive success of males and females can influence the opportunity for sexual selection. ●

11.3 The Rules of Attraction

By preferentially mating with high-quality males, a female may improve the genetic quality of her offspring. In other words, she may get indirect benefits from female choice. To gain these benefits, however, she has to be able to choose males of the highest possible quality. Males present a spectacular diversity of badges, plumes, crests, dances, songs, and smells to females in their efforts to mate. What makes these traits so informative?

Several hypotheses explore why some male traits are more informative than others (Table 11.3). Tests of these hypotheses can explain the types of ornamental male traits that females prefer.

Arbitrary Choice

In 1915, the British biologist Ronald Fisher developed the first detailed model of the evolution of exaggerated male ornaments through sexual selection. He proposed that once females developed a preference for certain males, a runaway process of sexual selection was triggered. In Fisher's model, a population contained choosy females that preferred males with some ornament. It also included unchoosy females that were equally likely to mate with any male. In such a population, all the choosy females and even some of the unchoosy females mate with the showy males, whereas only some of the unchoosy females mate with the unshowy males.

As a result, having an attractive ornament raises the odds that a male will reproduce. At the same time, females that choose males with ornaments have more reproductive success as well, because they give birth to showy sons that also attract females

Table 11.3 Indirect Benefits of Female Choice

Arbitrary Choice ("Fisher's Positive Feedback Cycle")

Certain male traits are advantageous not because they indicate good quality, but simply because they are attractive to females. Offspring of females choosing males with attractive traits inherit alleles influencing the expression of both the preference (from their mother) and the trait (from their father). The resulting association between preference and trait can lead to a positive feedback cycle of ever stronger preferences and larger display traits. Ornaments can evolve to such extremes that their severe costs balance the reproductive advantages of having the trait (Fisher 1930; Lande 1981; Kirkpatrick 1982; Iwasa and Pomiankowski 1994; Pomiankowski and Iwasa 1998).

Good Genes

Elaborate or bright male ornaments signal underlying genetic quality (good genes) such as efficient metabolism, body condition, or resistance to parasites or disease. This hypothesis assumes that among-male variation in the expression of ornaments reliably signals individual differences in overall quality of the males. It predicts that choosy females will produce offspring with higher survivorship or in better condition than less choosy females. Ornaments may indicate that males are able to successfully wield costly "handicaps" (Zahavi 1975; Iwasa et al. 1991), are resistant to parasites (Hamilton and Zuk 1982; Folstad and Karter 1992), or are in top physiological condition (Grafen 1990; Price et al. 1993; Iwasa and Pomiankowski 1999; Bonduriansky and Day 2003).

Fisher–Good Genes Process (Combination)

Alleles for female preferences can coevolve with alleles influencing expression of costly indicator traits in males. Although genetic correlations between female preference for a male ornament and its degree of expression have been found, these ornaments often are costly indicators of male quality. Thus, a Fisher process of linkage disequilibrium between ornament and preference applies to a "good genes" indicator trait (Kokko et al. 2002; Mead and Arnold 2004; Kokko and Jennions 2008).

(for this reason, Fisher's model is often called the sexy son hypothesis). Their offspring inherit the genes for choosiness, which are expressed in females, as well as the genes for showiness, which are expressed in males. A positive feedback cycle thus emerges in Fisher's model as female preferences and male ornaments continue to coevolve (Fisher 1930): as female preference becomes more discerning, male ornaments become more exaggerated.

Fisher proposed that this cycle of escalation would be stopped only when a sexually selected trait started to seriously threaten a male's ability to survive. Male guppies have a trait that illustrates this concept—their orange patches. These patches don't merely attract female guppies; they also attract predatory fishes. In streams with predators, the orange patches on male guppies are drab compared with those of the vibrant guppies that live in predator-free waters—just as Fisher would have predicted (Endler 1987; Endler and Houde 1995).

In Fisher's original model, the male ornament was arbitrary. Any blaze of color or tuft of feathers could work, as long as females preferred it. Only if the ornament was heritable, and female preferences were heritable, too, could Fisher's runaway process kick in. Since Fisher's original research, a number of theoretical studies all have shown that Fisher's process can drive the evolution of ornaments that are otherwise arbitrary (Iwasa and Pomiankowski 1994; Pomiankowski and Iwasa 1998). The simple fact that females prefer the trait is enough to make it beneficial, and choosy females get indirect benefits when they pass the trait to their sons (Ryan and Wilczynski 1988; Prum 1997; Funk et al. 2009).

Good Genes

There is a risk, however, in relying on such arbitrary signals in choosing a mate. Because these signals contain no information about the overall genetic quality of a male, attractive males need not be of any higher quality than unattractive males. Females could do much better if they chose males based on traits that were also reliable signals of male quality. In that case, they could derive indirect benefits from choice for two reasons instead of just one: their sons would still be attractive (because they inherited the ornament), but now they would also be healthier and stronger than other males because their fathers were of unusually high quality. If females can accurately pick sires with good genes, then they can pass this genetic quality to their offspring.

Mathematical models suggest that one way that a sexual display can become a reliable or "honest" signal is for an animal to have to pay a cost to display it (Zahavi 1975; Kokko et al. 2002; Mead and Arnold 2004). That cost might be the energy it takes to build antlers or to croak all night long. Strong animals can afford that extra energy, whereas weaker ones cannot. Such costly signals ought to be particularly difficult for weaker animals to make when they're under other kinds of stress, such as sickness or starvation (Lande 1981; Hamilton and Zuk 1982; Kirkpatrick 1982; Johnstone 1995).

These models have inspired a number of biologists to search for evidence of honest signals in animals. Sarah Pryke, of the University of New South Wales in Australia, has tested the hypothesis by studying red-collared widowbirds in Africa. Male red-collared widowbirds have tail feathers that measure around 22 centimeters (cm) long—longer than their entire body. During the mating season, they establish territories and fly around in them, fanning out their tail feathers so that they flap in the breeze. Female red-collared widowbirds spend this time choosing where to make their nests and then mating with the male whose territory they've picked.

Males with longer tails, Pryke found, attracted more females to their territories. To see how strong this preference was, Pryke snipped the tails of 120 male red-collared widowbirds. On 60 birds, she snipped their tails down to 20 cm; on the other 60, she cut the tails down to 12 cm. Then she compared how well the males fared in the mating game. (Although 12 cm is short for a red-collared widowbird tail, it's still within the natural range of the birds.)

Pryke found some dramatic differences between the two groups of birds (**Figure 11.24**). A long tail costs male red-collared widowbirds extra energy because

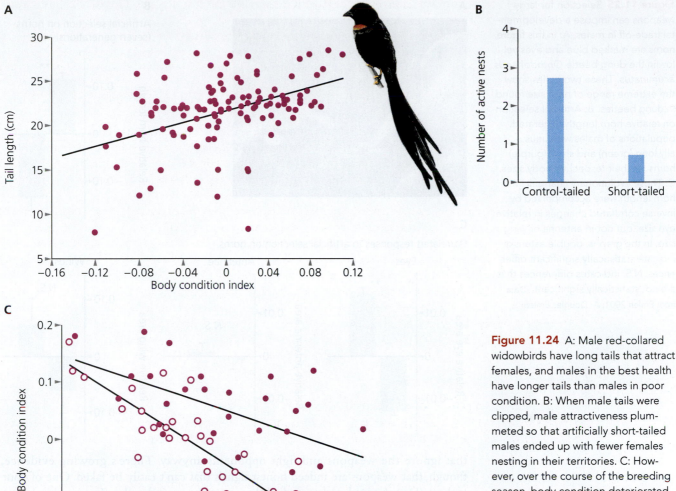

Figure 11.24 A: Male red-collared widowbirds have long tails that attract females, and males in the best health have longer tails than males in poor condition. B: When male tails were clipped, male attractiveness plummeted so that artificially short-tailed males ended up with fewer females nesting in their territories. C: However, over the course of the breeding season, body condition deteriorated much faster in control males with long tails than it did in the males with artificially shortened tails, indicating that the long tails were costly for males to bear. (Data from Pryke et al. 2001; Pryke and Andersson 2005)

of the drag it creates in the air. Birds with clipped tails were able to spend more time flying around, attracting females. Yet that extra time did not translate into more reproductive success. The birds with the longest tails ended up with three times more nesting females in their territories than the short-tailed birds (Pryke and Andersson 2005).

A long tail, in other words, comes at a cost to male red-collared widowbirds. That cost suggests that the trait is an honest signal to females. What clues does it give the females? Pryke has found that males with naturally long tails are in better condition than males with naturally short tails. By selecting long-tailed mates, female red-collared widowbirds may be picking males that are strong enough to hold their territories and to give their offspring genes for good health.

The evolution of honest signals may account not only for female preferences but also for the horns, tusks, antlers, and other extravagant weapons that male animals can grow. These weapons may be not just for fights, but for avoiding fights. Battling another male is very risky and potentially leaves both animals wounded or dead. In many species, males reduce this risk by sizing up their opponents from a distance. A big weapon may serve as a clear signal of the strength of the male that carries it, and weaker males may back away at just the sight of it.

Of course, such a signal cannot survive very long if it's easy to fake. If weak males can make big weapons, cheating the system, then natural selection may favor males

Figure 11.25 Selection for large weapons can impose a developmental trade-off in males. A: In this figure, horns are marked blue and eyes yellow in the dung beetle *Onthophagus acuminatus*. These two males show the extreme range of horn size found in dung beetles. B: Artificial selection on relative horn length generated populations of males with unusually long (green) and short (purple) horns for their respective body sizes. C: These evolutionary changes in horn length were accompanied by inverse correlated changes in relative eye size, but not in antenna or wing size. In the graphs, double asterisks indicate statistically significant differences; N.S. indicates differences that are not statistically significant. (Data from Emlen 2001; A: Douglas Emlen)

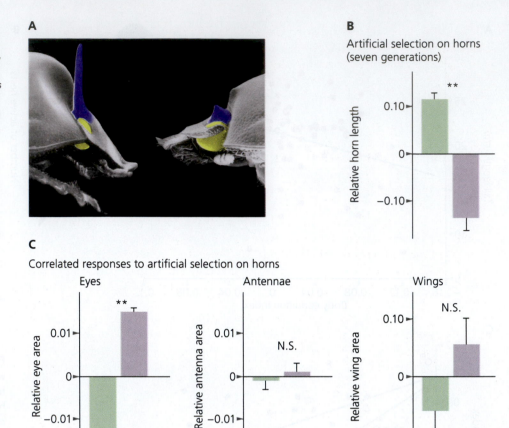

that ignore the weapons and fight opponents anyway. There's growing evidence, though, that weapons are indeed honest signals that can't easily be faked. One of your authors (Douglas Emlen) has studied the cost of horns in beetles. To grow big horns, male beetle larvae have to allocate resources to the cells on their heads that grow into those organs. That means they have fewer resources remaining to develop into neighboring organs, such as eyes or antennae, which they need to survive (**Figure 11.25**; Emlen 2001).

Key Concepts

- Females who choose males with attractive traits pass to their offspring alleles influencing the expression of both the preference and the trait. The resulting genetic correlation between preference and trait can lead to an escalating positive feedback cycle of coevolution of stronger preferences and larger display traits.

- Although the traits that strike the females' fancy may originally be arbitrary, they often allow females to choose males based on characteristics that reliably distinguish the highest-quality, healthiest individuals from the rest of the lot.

- Costly signals or display traits will be the most reliable because they are difficult or impossible for those of poor genetic quality to bluff. ●

11.4 The Evolution of Mating Systems

In the animal kingdom, many different kinds of mating systems can be defined based on sexual behavior. The males and females of the mimic poison dart frogs (*Ranitomeya imitators*) will pick mates and then remain absolutely loyal. Together, a male and a female will work to raise their young in tiny pools of water that form in bromeliads lining the branches of tropical trees. This mating system, where a male and female form a stable pair and mate only with each other, is known as **monogamy**. Among elephant seals, by contrast, a single male mates with many females—a system known as **polygyny**. Central American birds known as wattled jacanas are at the other extreme, **polyandry**, in which each female may mate with several males. Researchers have observed that in some species (for example, chimpanzees), a combination of systems may be at play.

Sexual monogamy is very rare and occurs when each male mates only with a single female, and vice versa. Social monogamy occurs when a male and female form a stable pair bond and cooperate to rear the young, even if either or both partners engage in extra-pair copulations. Social monogamy occurs in a few fish, insect, and mammal species and in almost 90% of bird species.

Until the late 1900s, naturalists could study mating systems only by observing animals. Yet looks are often deceiving. Many pair-bonding birds are not as loyal to one another as they may appear. When scientists started using DNA to analyze the parentage of the chicks in nests of pair-bonding birds, they often found that a large fraction of the eggs did not carry the DNA of their mother's apparent partner. The mothers were mating with other males (extra-pair copulation), and their partners were doing the hard work of raising the chicks (**Figure 11.26**).

To make sense of these intricate mating systems, evolutionary biologists consider how different strategies boost the reproductive success of males and females. The low cost of sperm, for example, can make polygyny a good strategy for males because they can fertilize many females, a situation that can result in many offspring. But polygyny has its downsides. Each male has to compete with lots of other males, and even if he succeeds in mating with females, the females can sometimes control which sperm they use to fertilize their eggs. This can be bad for a male if females end up using sperm from other males.

Under some conditions, a male actually may be better off mating with a single female, trying to fight off other males, and helping to raise the offspring. Whereas most male mammals are not monogamous, for example, male California mice are (Ribble 1991). If a male California mouse helps rear his pups, more than twice as many of them will survive than if the mother rears them alone (Gubernick and Teferi 2000).

Polyandry can benefit females in a number of ways. By mating with a number of males, females may be able to get the highest-quality genes possible for their offspring. In essence, they "hedge their bets" in case sperm from any one of the males performs poorly. Each female superb fairy-wren forms a bond with one male, for example, but then mates with other males. The males they pick for these extra-pair copulations show signs of being in particularly good condition, suggesting that these females are "upgrading" the quality of the sperm they use to conceive their chicks (Double and Cockburn 2000).

But there may be other explanations for polyandry. One possibility is that females may be benefiting not from good genes but from genes that are simply different from their own. If animals mate with partners that are too similar genetically, their inbred

Monogamy is a mating system in which one male pairs with one female.

Polygyny is a mating system where males mate (or attempt to mate) with multiple females.

Polyandry is a mating system where females mate (or attempt to mate) with multiple males.

Figure 11.26 Female superb fairy-wrens (A) form long-term bonds with males (B) but nevertheless mate with other male wrens. (A: Houshmand Rabbani/Shutterstock; B: Katarina Christenson/Shutterstock)

offspring may have lower fitness (Chapter 6). A second explanation is that mating with multiple males may allow females to boost the health of some of their offspring by giving them a defense against a wider range of pathogens. Alternatively, females may mate with multiple males simply because the costs of resisting courtship overtures are higher than the costs of mating multiple times. In addition, in gift-giving species, females may benefit from acquiring more resources from the nuptial gifts of multiple mates.

Key Concept

- Mating systems evolve because of the benefits and costs they confer on males and females. Who mates with whom can depend on the opportunities that arise through both inter- and intrasexual selection. In some species, a combination of systems may be at play. ●

11.5 The Hidden Dimension of Sexual Selection

So far in this chapter, we've discussed the behaviors that males and females display *before* mating that can increase their fitness—in the case of males, competition and display, and in the case of females, choosing mates or engaging in polyandry. But competition and choice can also take place *after* mating is over (Eberhard 2009).

Sperm Competition

Because females often mate with more than one male, they can end up with sperm from many males inside them at once. In effect, these sperm compete to fertilize eggs. If one male has a mutation that makes it more likely that his sperm will fertilize a female's eggs than the sperm of other males, he will have more reproductive success. As a result, **sperm competition** has led to the evolution of a number of strategies that males (or their sperm) use to raise their success in fertilization.

Sperm competition is a form of sexual selection that arises after mating, when males compete for fertilization of a female's eggs.

One way males can increase their reproductive success is by getting rid of the sperm females may be carrying from previous matings. Seed beetles (*Callosobruchus maculatus*) have sharp spines on their penises. When a male pairs with a female who has already mated with another male, the spines on his penis enable him to remove sperm from the earlier male. The longer the spines, the more success the male seed beetles have at fertilizing eggs (**Figure 11.27**; Hotzy and Arnqvist 2009).

In some species, males have sperm that can outcompete sperm from other males also present in a female's reproductive tract. One such strategy is simply to swamp the opposition with sperm. Among primates, for example, many species with strong sperm competition have bigger testicles than the species in which males and females tend to

A

B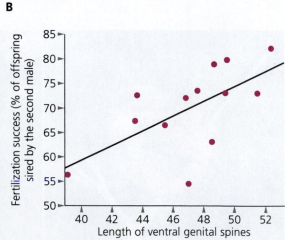

Figure 11.27 Female seed beetles may mate with multiple males. A: Male seed beetles have spines on their copulatory organ, which may help to remove sperm left by other males during mating. B: The longer the spines on a male, the more success he has fertilizing eggs when he is second in line—that is, when the female has already mated with another male. (Data from Hotzy and Arnqvist 2009; A: Johanna L. Ronn)

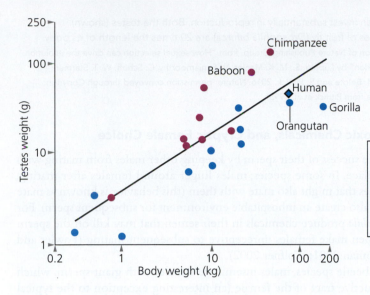

Figure 11.28 In primates, male testes are proportionately larger in species where females regularly mate with multiple males and sperm competition is high. (Data from Harcourt et al. 1981)

mate monogamously. Bigger testicles supply more sperm, and more sperm can raise a male's odds of fertilizing a female's eggs when sperm from other males are present (**Figure 11.28**).

In deer mice (genus *Peromyscus*), a male's sperm will join together to form aggregates that can swim faster together than they can individually. Heidi Fisher and Hopi Hoekstra of Harvard University, who discovered this cooperation, also found that the sperm can recognize other sperm from the same male, so that the sperm preferentially aggregate with their kin (**Figure 11.29**). They also discovered that males of a promiscuous species (*P. maniculatus*) with a history of strong sperm competition were more likely to form aggregates with related sperm—sperm from the same male—than were sperm from a related monogamous species (*P. polionotus*), which formed aggregates indiscriminately. This suggests that sperm competition was responsible for the evolution of kin recognition in cooperating sperm (Fisher and Hoekstra 2010; Fisher et al. 2014).

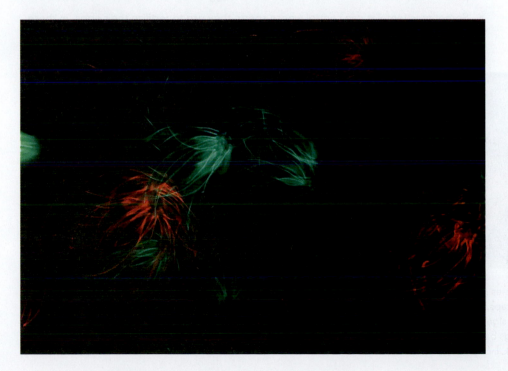

Figure 11.29 Sperm in some species of mice join together as they swim through a female's reproductive tract. Swimming cooperatively, they can travel faster than sperm swimming individually. In *Peromyscus maniculatus*, a species experiencing sperm competition, sperm preferentially aggregate with genetically related sperm from the same sire as they race to the eggs. In this image, sperm from one male are stained red, and sperm from the other are in green. (Data from Fisher and Hoekstra 2010; photo by Heidi Fisher)

Testes

Figure 11.30 Males can invest substantially in reproduction. Both the testes (shown) and sperm of this species of fruit fly (*Drosophila bifurca*) are 20 times the length of its body. (Republished with permission of Nature Publishing Group. from "How sexual selection can drive the evolution of costly sperm ornamentation" by Lüpold, S., M. K. Manier, N. Puniamoorthy, C. Schoff, W. T. Starmer, S. H. Buckley Luepold, J. M. Belote and S. Pitnick. 2016. Nature; permission conveyed through Copyright Clearance Center, Inc.; data from Pitnick et al. 1995)

Male Guarding, Toxic Chemicals, and Cryptic Female Choice

Males can increase the success of their sperm by keeping other males from mating with females in the first place. In some species, males linger around females after mating, chasing off other males that might also mate with them (this behavior is known as mate guarding). Males can also create an inhospitable environment for subsequent sperm. For example, male *Drosophila* produce chemicals in their semen that may kill off the sperm of other males and then make females unreceptive to subsequent mating (Fowler and Partridge 1989; Chapman 2001; Wolfner 2002).

In some fly and beetle species, males inseminate females with giant sperm, which fill the entire reproductive tract of the female (an interesting exception to the typical rule that males have tiny gametes). This may prevent sperm introduced from subsequent copulations from even entering the female (**Figure 11.30**; Pitnick et al. 1995).

Females do not simply wait passively as the sperm from males compete to fertilize their eggs, however. A number of studies show that females of many species can exercise some influence over which males succeed by selecting sperm from among the males they have mated with. This **cryptic female choice** explains a number of peculiar features of mating (Peretti and Aisenberg 2015). In some species, for example, males sometimes continue to perform courtship rituals *after* they've finished mating.

In a species of fly called *Dryomyza anilis*, the male taps the female's body for two minutes after withdrawing his sexual organ (Otronen 1990). Scientists have found that as the number of taps increases, the percentage of eggs the female fertilizes with that male's sperm increases as well. The female may be evaluating the male on the basis of this post-mating courtship (**Figure 11.31**).

Cryptic female choice refers to a form of sexual selection that arises after mating, when females store and separate sperm from different males and thus bias which sperm they use to fertilize their eggs.

Key Concepts

- When females mate with multiple males, sperm from the different males must compete for opportunities to fertilize the females' eggs.

- Sometimes females can control whose sperm they use to fertilize their eggs. In these species, males may continue to court females even after they have finished mating. ●

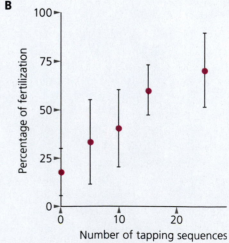

Percentage of fertilization

Number of tapping sequences

Figure 11.31 A: Male flies of the species *Dryomyza anilis* tap females after they have finished mating. B: The longer a male taps, the more eggs he fertilizes. It's possible that the taps are post-copulatory courtship signals. The female flies may be choosing how many sperm to use to fertilize their eggs based on the courtship. (Data from Otronen 1990; A: Chua Wee Boo/age fotostock/Superstock, Inc)

11.6 Sexual Conflict and Antagonistic Coevolution

As we've seen in this chapter, both males and females have evolved a number of adaptations to increase their fitness in preparation for mating, as well as afterward. Some strategies that are optimal for a male have direct negative fitness effects on females, and vice versa. As noted earlier, male *Drosophila* flies inject chemicals into their mates to ward off other males. Some of these chemicals are quite toxic (in fact, they're structurally related to snake venoms). In addition to warding off sexual competitors, these toxic chemicals shorten the life span of female flies. Although they can lower the fitness of females, these chemicals impose no fitness costs on the males, because males don't form long-term bonds with individual females.

Sexual Conflict

Scientists refer to such an evolutionary conflict of interests as **sexual conflict** (Chapman et al. 2003; Arnqvist and Rowe 2005). A number of experiments have shown that sexual conflict is a potent evolutionary engine. If a male's reproductive strategy lowers the fitness of females, then there will be strong selection for females that can counter their strategy. And males then come under selective pressure to overcome female defenses. Males and females thus get caught in a coevolutionary arms race.

Patricia Brennan's findings on the reproductive organs of ducks and other waterfowl, for example, are best explained by sexual conflict (Brennan and Prum 2015). In many of these species, the males and females bond for an entire breeding season. Males that fail to find a mate of their own at the start of a season often harass females, trying to force them to mate. And when a bonded male's partner is busy incubating his eggs, the male partner may also search for other females and force them to mate with him. All told, about a third of all matings are forced in some species of waterfowl. The harassment from male ducks can get so intense that some females drown as a result.

Although forced matings are common in ducks, the unwanted males father only about 3% of the ducklings each year. Brennan suspects that the female birds are controlling the sperm in their bodies, favoring the sperm from their partners over that of the intruders.

Just like the evolution of measures and countermeasures that enhance survival in the face of parasites, sexual selection favors the evolution of measures and countermeasures that enhance reproductive fitness. Because males and females benefit from different strategies, when defenses evolve in females, selection favors countermeasures from the males. In the case of ducks, longer, more flexible phalluses may evolve in males, and in response, even more twisted oviducts with more side pouches may evolve in females. Sexual conflict can explain why male and female sexual organs are so twisted in the birds Brennan has studied (**Figure 11.32**).

The mating system of *Drosophila* flies also exhibits sexual conflict. Females defend themselves against male mating strategies with their own chemicals. Female *Drosophila* produce proteins that can destroy some of the toxic compounds in male semen. But as these defenses evolve, selection in turn favors males who are able to incorporate new proteins into their semen that can overcome the defenses. Scientists have detected strong selection in the genes for both kinds of proteins, suggesting that male and female *Drosophila* are locked in an arms race.

William Rice, at the University of California at Santa Barbara, and his colleagues have been able to document this sexual conflict by manipulating evolution experimentally. He reasoned that the mating system of *Drosophila* was driving their evolution and that if he were to alter the mating system, he would alter the evolution of the flies. He and his colleagues reared pairs of males and females so that they had no choice but to be monogamous for 30 generations (Holland and Rice 1999). These flies experienced no sperm competition, and as a result, producing costly toxins was selected against because there was no rival male sperm.

> **Sexual conflict** is the evolution of phenotypic characteristics that confer a fitness benefit to one sex but a fitness cost to the other.

Figure 11.32 A: Patricia Brennan and her colleagues found that in bird species where males had long phalluses, the females had more pouches in their oviducts. B: The researchers also found a similar correlation between phallus length and the number of spirals in the oviduct. Both patterns are consistent with sexual conflict between male and female birds. (Data from Brennan et al. 2007; photo by Bernard J. Brennan)

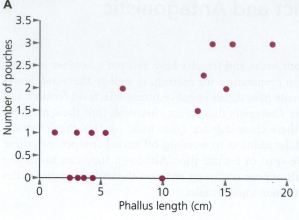

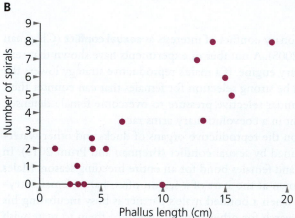

Monogamy removed the "antagonism" between males and females because without new toxins there was no longer selection for female countermeasures. Furthermore, a male's fitness was tied with that of a single female. If a male harmed his mate under the experimental conditions, he too would suffer, because any future offspring produced by the female would also have been his. After 30 generations of monogamy, males no longer harmed the females with their seminal fluids. Females that mated lived just as long as females who did not mate, and what once was an antagonistic interaction became a mutualistic one instead.

Rice then tested the resistance of these monogamous females to the seminal fluids of wild-type males. He bred them with male flies from a polygynous population, which experienced sperm competition and strong sexual conflict. The monogamous females fared very badly. They laid fewer than half as many eggs as females from the polygynous population, and they also died earlier. These results showed that monogamy had relaxed selection in female flies for resistance to the toxins in male sperm. Females from the monogamous population had "let down their guard," and after this experimental evolution they suffered drastically when they were exposed to the seminal proteins of males from the polygynous population (Figure 11.33).

The Genes Behind Sexual Conflict

Other insect species also enable scientists to uncover the genetic changes that drive the evolution of these conflict-related traits. Locke Rowe, of the University of Toronto, and his colleagues study the genetic basis of sexual conflict in water striders. Male water striders do not bother with singing or other forms of courtship. Instead, they just leap onto females and try to mate with them. In some species, the males may prolong the mating for hours (Figure 11.34).

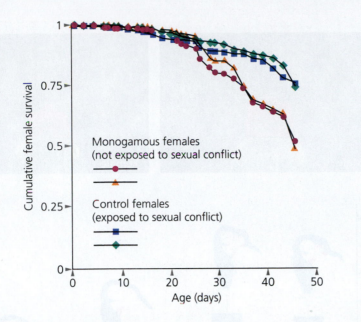

Figure 11.33 *Drosophila* were experimentally forced to mate monogamously. After more than 30 generations of evolution without any sperm competition, male semen became less toxic to females, and the females lost their defenses against seminal proteins. When these females were later forced to mate with males reared in natural, competitive environments (red, orange), they died much sooner than females that had evolved with the sexual conflict (blue, green). (Data from Holland and Rice 1999)

These attacks reduce a female's opportunity to exercise mate choice. Instead of mating with multiple males, she is held down by one male. The attacks can also lower a female's fitness by making her more vulnerable to predators. Research by Rowe and others suggests that female water striders have evolved a defense to these ambushes. First, they try to skate away from approaching males. If the males succeed in grabbing them, the females perform somersaults to throw them off (Rowe 1994).

Males that can overcome these female counterattacks will have greater reproductive success. In some species of water striders, this selection pressure has given rise to bizarre modifications of male water strider anatomy. Their legs are twisted and grow a row of projections that help them grip the female's body. At their front end, their antennae have evolved peculiar extensions that allow them to lock onto the head of a female, like a wrench locking onto a nut.

To understand the evolution of this morphology, Rowe and his colleagues screened the genes of water striders (Khila et al. 2012). They found that the gene *Distalless* (*dll*)

Figure 11.34 Water striders exhibit intense sexual conflict. Males will jump onto females and grip them tightly to mate. Females respond by trying to flip the males off their bodies. (blickwinkel/Alamy Stock Photo)

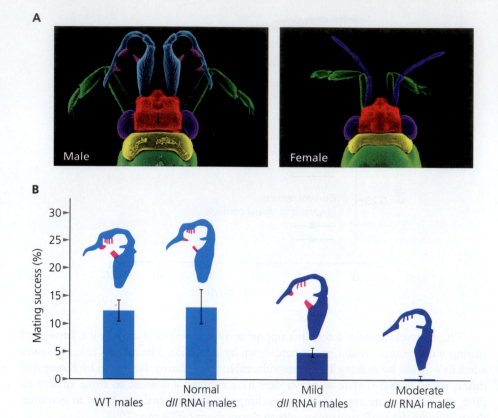

Figure 11.35 A: Males of some water strider species have elaborate antennae. Scientists at the University of Toronto hypothesized that these were adaptations for mating that had evolved through sexual conflict. When a male jumped on a female, he could use his antennae to grip her head, making it harder for her to shake him off. B: To test this hypothesis, the scientists used a method called RNAi to lower the expression of a developmental gene in the structure. Males with smaller antennae structures fertilized fewer eggs. This result supports the hypothesis that the antennae had evolved through sexual conflict. (Data from Khila et al. 2012; A: Khila Abderrahman)

controls the development of the wrench-like extensions. We saw in Chapter 9 that *dll* is crucial in the development of many appendages, such as legs, wings, and antennae (Figure 9.3). Indeed, the scientists found that *dll* is expressed in both male and female water strider antennae. Later in development, males express *dll* again, but only in specific regions of their antennae that then develop into its mating "wrenches."

Rowe and his colleagues hypothesized that *dll* was co-opted to produce this novel trait in response to sexual conflict. They tested their hypothesis by reducing the expression of *dll* in male water strider larvae during the stage when they developed their female-gripping appendages. When the males matured, they had smaller wrenches.

Rowe and his colleagues allowed these modified males to attempt to mate with females. The males were more likely to lose their grip on their mates when the females flipped. As a result, they had less mating success. The lower the expression of *dll* was, the lower the mating success (**Figure 11.35**).

This experiment is a vivid demonstration of how evolutionary biologists can weave together different lines of research to better understand life's diversity. The importance of *dll* to animal evolution was discovered by developmental biologists who were surprised by how powerful these genes were in animal embryos—and even more surprised to find orthologs in distantly related species. Now scientists who study a seemingly different aspect of evolution—sexual conflict—are discovering how it co-opts this same genetic toolkit in the race between males and females.

In this chapter, we've seen how mating opens up the opportunity for the evolution of an astonishing diversity of morphology and behavior. Once the mating is over, however, and the eggs have been fertilized, the opportunity for selection remains. In the next chapter, we'll investigate the evolution of parental care.

Key Concept

- Sexual conflict can lead to antagonistic coevolution and arms races between males and females. Mutations advancing the reproductive interests of males select for countermeasures in the females, and vice versa. ●

TO SUM UP . . .

- Sex is widespread in nature, but its success is puzzling due to the twofold reproductive cost it induces compared to asexual reproduction. Benefits that outweigh this cost include allowing populations to adapt faster to fluctuating environments and keeping deleterious mutations from accumulating.

- The Red Queen effect predicts a cyclical coevolution between hosts and their parasites because hosts have to evolve defenses quickly just to survive in the presence of parasites. Sexual recombination creates additional genetic variation and may allow populations to evolve rapidly.

- Females invest heavily in a limited number of eggs, whereas males can produce vast numbers of sperm. This imbalance may be the source of sexual selection and sexual conflict in many species.

- The sex with the greater variance in mating success will more likely experience sexual selection. Intrasexual selection is the competition for mates between rivals (usually males), and intersexual selection is the choice of mates one sex (usually females) makes based on appearances, displays, songs, and other forms of courtship.

- Females stand to gain directly by choosing males that provide food or protection as well as indirectly by choosing males with high-quality genes.

- In many species, females prefer males with certain traits over others. Those traits, in some cases, signal good genes in the males.

- Females in many species mate with more than one partner. As a result, sperm competition has evolved among males in these species.

- Conflicts of interest over the control of fertilization can result in antagonistic coevolutionary "arms races" between males and females.

MULTIPLE CHOICE QUESTIONS   Answers can be found at the end of the book.

1. Which of the following is *not* a major reason why sexual reproduction can speed the spread of adaptations in a population?

   a. Sexual reproduction encourages species to travel to find mates, which allows for individuals to spread their genes over the largest range.

   b. Through recombination, meiosis provides an opportunity for paired chromosomes to cross over, creating gametes with unique combinations of alleles.

   c. Independent assortment during meiosis can mix and match the maternal and paternal copies of chromosomes, leading to novel combinations of alleles in offspring.

   d. Beneficial mutations can be combined and harmful mutations can be purged.

2. The Red Queen effect refers to the fact that

   a. parasites often kill their hosts and therefore act as potent agents of selection on host populations.

   b. host immune systems evolve continuously and quickly in an arms race with parasite populations that are also evolving to evade their defenses.

   c. parasites evolve faster than their hosts.

   d. social insect colonies often have reproductive queens who actively suppress the reproductive capacity of worker females in the colony.

3. Which is an example of anisogamy?

   a. Male red deer have large antlers, but females do not have antlers at all.

   b. Female jacanas are larger in size than the males.

   c. Female fiddler crabs have two small claws, but males have one small claw and one huge claw.

   d. The eggs of a female kiwi are large, but the sperm of the males are small.

4. Are unequal gamete sizes relevant for explaining adult behavior? Why or why not?

   a. No. Divergent gamete sizes are a consequence of sexual reproduction, but they are largely irrelevant for understanding adult behavior.

   b. Yes. There typically are insufficient eggs to go around; males end up having to compete for access to them.

   c. Yes. Females sometimes build elaborate nests or burrows in which to place their eggs.

   d. Yes. Both a male and a female gamete are needed to produce viable offspring.

5. Conspecific females who differ in their *fecundity* differ in what?

   a. The number of offspring they produce that survive to successfully reproduce themselves.
   b. The number of eggs they produce at one time.
   c. The number of times a female breeds during her lifetime.
   d. The number of mates a female has during her lifetime.

6. Which of these situations offers the lowest opportunity for selection for males?

   a. Locations that females need access to in order to reproduce are rare.
   b. Every female chooses one male to mate with for life.
   c. Males fight each other for access to groups of females.
   d. Females choose mates with a fancy ornament.

7. Why are traits like the bright colors of a male manakin considered to be honest indicators of male genetic quality?

   a. The colors stimulate sensory preferences or biases that are intrinsic.
   b. The colors are very costly for males to produce.
   c. Bright-colored ornaments evolve extremely rapidly and often diverge in form among populations or closely related species.
   d. The colors accurately reveal unpalatability or distastefulness to predators, so that predators can quickly learn to avoid them.

8. Which of the following is *not* an example of the *costs* relevant to honest expression of male sexually selected ornaments or weapons?

   a. Metabolic or energetic resources that a male must allocate to the growth of ornaments or weapons.

   b. The inability of males to invest simultaneously in ornament production and other functions, such as defenses against parasites.
   c. The risk a male faces because ornaments slow him down or because bright colors make him more conspicuous to predators.
   d. The production of brighter colors in males inheriting particular combinations of alleles than in males with other combinations of alleles.

9. Which of the following is *not* a strategy used by male insects to increase the likelihood that their sperm will fertilize a female's eggs?

   a. Mate with more females than rival males.
   b. Guard females after mating.
   c. Physically remove sperm by inflating the penis or using a penile flagellum.
   d. Inject toxic chemicals that induce females to avoid re-mating.

10. In the context of sexual selection, *antagonistic coevolution* refers to

   a. the fact that male genitalia sometimes pierce or tear the insides of females in ways detrimental to the females.
   b. the arms races that can result as males and females compete for control over fertilization.
   c. the fact that suboptimal males will sometimes trick females into mating with them by sneaking up to them and mating while the dominant males are busy fighting or mating with another female.
   d. the coercion behaviors that males sometimes adopt during mating, such as tapping or stroking the back of the females, that appear to increase the chances of a female using that male's sperm to fertilize her eggs.

INTERPRET THE DATA   Answer can be found at the end of the book.

11. In the figure below showing average reproductive success as a function of age in red deer, why do males begin breeding so much later than females?

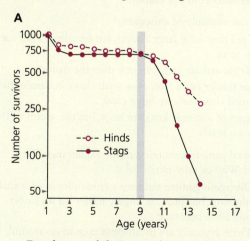

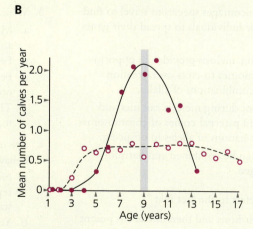

   a. Females spread their reproductive effort over more of their lifetime.
   b. It takes males longer to reach sexual maturity than it does females.
   c. Females don't invest as much as males do in reproduction.
   d. Males have to battle for access to harems, and it takes several years of experience and growth before males become dominant enough to win.

SHORT ANSWER QUESTIONS   Answers can be found at the end of the book.

12. Why is sexual reproduction so widespread when the costs of reproducing sexually are so high?

13. What is the difference between sexual selection and natural selection? Why is sexual selection an outcome of the evolution of sex?

14. Would sexual cannibalism be more likely in a species with a mating system of polygyny or polyandry? Why? Upon what would the likelihood of sexual cannibalism depend?

15. The three house finch males pictured below differ in their expression of a sexually selected ornament, a patch of red feathers on the face and breast. Why might selection favor females who choose the male on the right, over the others, as a mate?

16. What is polyandry? Describe some of the advantages of polyandry for a female. How could this system of mating increase the chances of survival of her offspring?

17. Why might male *Drosophila* benefit from seminal fluid containing chemicals that are harmful to a female?

(Left to right: Arto Hakola/Shutterstock; Tim Zurowski/All Canada Photos/Superstock, Inc.; Steve Byland/Shutterstock)

ADDITIONAL READING

Birkhead, T. R. 2000. *Promiscuity: An Evolutionary History of Sperm Competition*. Cambridge, MA: Harvard University Press.

Burt, A. 2006. *Genes in Conflict: The Biology of Selfish Genetic Elements*. Cambridge, MA: Belknap Press of Harvard University Press.

Clutton-Brock, T. 2007. Sexual Selection in Males and Females. *Science* 318:1882–5.

Darwin, C. 1871. *The Descent of Man, and Selection in Relation to Sex*. New York: Appleton.

Emlen, D. J. 2008. The Evolution of Animal Weapons. *Annual Review of Ecology, Evolution, and Systematics* 39:387–413.

Hadany, L., and J. M. Comeron. 2008. Why Are Sex and Recombination So Common? *Annals of the New York Academy of Sciences* 1133:26–43.

Judson, O. 2002. *Dr. Tatiana's Sex Advice to All Creation*. New York: Metropolitan Books.

Maynard Smith, J. 1976. Sexual Selection and the Handicap Principle. *Journal of Theoretical Biology* 57 (1): 239–42.

Ryan, M. J. 2018. *A Taste for the Beautiful: The Evolution of Attraction*. Princeton, NJ: Princeton University Press.

Simmons, L. W. 2001. *Sperm Competition and Its Evolutionary Consequences in the Insects*. Princeton, NJ: Princeton University Press.

Wilkinson, G. S., F. Breden, J. E. Mank, M. G. Ritchie, A. D. Higginson, et al. 2015. The Locus of Sexual Selection: Moving Sexual Selection Studies into the Post-genomics Era. *Journal of Evolutionary Biology* 28 (4): 739–55.

Williams, G. C. 1966. *Adaptation and Natural Selection: A Critique of Some Current Evolutionary Thought*. Princeton, NJ: Princeton University Press.

Andersson, M. B. 1994. *Sexual Selection: Monographs in Behavior and Ecology.* Princeton, NJ: Princeton University Press.

Andrade, M. C. B. 1996. Sexual Selection for Male Sacrifice in the Australian Redback Spider. *Science* 271 (5245): 70–2.

Arnold, S. J., and M. J. Wade. 1984. On the Measurement of Natural and Sexual Selection: Theory. *Evolution* 38 (4): 709–19.

Arnqvist, G., and L. Rowe. 2005. *Sexual Conflict.* Princeton, NJ: Princeton University Press.

Balmford, A., A. M. Rosser, and S. D. Albon. 1992. Correlates of Female Choice in Resource-Defending Antelope. *Behavioral Ecology and Sociobiology* 31 (2): 107–14.

Barry, K. L., G. I. Holwell, and M. E. Herberstein. 2008. Female Praying Mantids Use Sexual Cannibalism as a Foraging Strategy to Increase Fecundity. *Behavioral Ecology* 19 (4): 710–5.

Berglund, A., A. Bisazza, and A. Pilastro. 1996. Armaments and Ornaments: An Evolutionary Explanation of Traits of Dual Utility. *Biological Journal of the Linnean Society* 58:385–99.

Bonduriansky, R., and T. Day. 2003. The Evolution of Static Allometry in Sexually Selected Traits. *Evolution* 57 (11): 2450–8.

Brennan, P. 2010. Sexual Selection. *Nature Education Knowledge* 1 (8): 24.

Brennan, P. L. R., C. J. Clark, and R. O. Prum. 2010. Explosive Eversion and Functional Morphology of the Duck Penis Supports Sexual Conflict in Waterfowl Genitalia. *Proceedings of the Royal Society B: Biological Sciences* 277:1309–14.

Brennan, P. L., and R. O. Prum. 2012. The Limits of Sexual Conflict in the Narrow Sense: New Insights from Waterfowl Biology. *Philosophical Transactions of the Royal Society B: Biological Sciences* 367 (1600): 2324–38.

Brennan, P. L., and R. O. Prum. 2015. Mechanisms and Evidence of Genital Coevolution: The Roles of Natural Selection, Mate Choice, and Sexual Conflict. *Cold Spring Harbor Perspectives in Biology* 7 (7): a017749.

Brennan, P. L. R., R. O. Prum, K. G. McCracken, M. D. Sorenson, R. E. Wilson, et al. 2007. Coevolution of Male and Female Genital Morphology in Waterfowl. *PLoS ONE* 2 (5): e418.

Campanella, P. J., and L. L. Wolf. 1974. Temporal Leks as a Mating System in a Temperate Zone Dragonfly (Odonata: Anisoptera) I: *Plathemis lydia* (Drury). *Behavioural Brain Research* 51 (1–2): 49–87.

Chapman, T. 2001. Seminal Fluid-Mediated Fitness Traits in *Drosophila. Heredity* 87:511–21.

Chapman, T., G. Arnqvist, J. Bangham, and L. Rowe. 2003. Sexual Conflict. *Trends in Ecology and Evolution* 18 (1): 41–7.

Chapman, T., A. Pomiankowski, and K. Fowler. 2005. Stalk-Eyed Flies. *Current Biology* 15 (14): R533–5.

Clutton-Brock, T. 1988. *Reproductive Success. Studies of Individual Variation in Contrasting Breeding Systems.* Chicago: University of Chicago Press.

———. 1991. *The Evolution of Parental Care.* Princeton, NJ: Princeton University Press.

Collias, N. E., and J. K. Victoria. 1978. Nest and Mate Selection in the Village Weaverbird (*Ploceus cucullatus*). *Animal Behaviour* 26 (2): 470–9.

Crews, D., M. Grassman, and J. Lindzey. 1986. Behavioral Facilitation of Reproduction in Sexual and Unisexual Whiptail Lizards. *Proceedings of the National Academy of Sciences USA* 83 (24): 9547–50.

Crow, J. F. 1958. Some Possibilities for Measuring Selection Intensities in Man. *Human Biology* 30 (1): 1–13.

Darwin, C. 1859. *On the Origin of Species by Means of Natural Selection, or, the Preservation of Favored Races in the Struggle for Life.* London: John Murray.

———. 1871. *The Descent of Man, and Selection in Relation to Sex.* New York: Appleton.

Dias, B. G., and D. Crews. 2008. Regulation of Pseudosexual Behavior in the Parthenogenetic Whiptail Lizard, *Cnemidophorus uniparens. Endocrinology* 149:4622–31.

Double, M., and A. Cockburn. 2000. Pre-Dawn Infidelity: Females Control Extra-Pair Mating in Superb Fairy-Wrens. *Proceedings of the Royal Society B: Biological Sciences* 267 (1442): 465–70.

Downhower, J. F., and L. Brown. 1980. Mate Preferences of Female Mottled Sculpins, *Cottus bairdi. Animal Behaviour* 28 (3): 728–34.

Durães, R., B. A. Loiselle, P. G. Parker, and J. G. Blake. 2009. Female Mate Choice Across Spatial Scales: Influence of Lek and Male Attributes on Mating Success of Blue-Crowned Manakins. *Proceedings of the Royal Society B: Biological Sciences* 276:1875–81.

Dussourd, D. E., C. A. Harvis, J. Meinwald, and T. Eisner. 1991. Pheromonal Advertisement of a Nuptial Gift by a Male Moth (*Utetheisa ornatrix*). *Proceedings of the National Academy of Sciences USA* 88 (20): 9224–7.

Eberhard, W. G. 2009. Postcopulatory Sexual Selection: Darwin's Omission and Its Consequences. *Proceedings of the National Academy of Sciences USA* 106 (Suppl. 1): 10025–32.

Eggert, A-K., and S. K. Sakaluk. 1994. Sexual Cannibalism and Its Relation to Male Mating Success in Sagebrush Crickets, *Cyphoderris strepitans* (Haglidae: Orthoptera). *Animal Behaviour* 47 (5): 1171–7.

Emlen, D. J. 2001. Costs and the Diversification of Exaggerated Animal Structures. *Science* 291 (5508): 1534–6.

———. 2008. The Evolution of Animal Weapons. *Annual Review of Ecology, Evolution, and Systematics* 39:387–413.

Emlen, S. T., and L. W. Oring. 1977. Ecology, Sexual Selection, and the Evolution of Mating Systems. *Science* 197 (4300): 215–23.

Endler, J. A. 1987. Predation, Light Intensity and Courtship Behavior in *Poecilia reticulata* (Pisces: Poeciliidae). *Animal Behaviour* 35 (5): 1376–85.

Endler, J. A., and A. E. Houde. 1995. Geographic Variation in Female Preferences for Male Traits in *Poecilia reticulata. Evolution* 49 (3): 456–68.

Fabiani, A., G. Filippo, S. Simona, and A. R. Hoelzel. 2004. Extreme Polygyny Among Southern Elephant Seals on Sea Lion Island, Falkland Islands. *Behavioral Ecology* 15 (6): 961–9.

Fisher, H. S., L. Giomi, H. E. Hoekstra, and L. Mahadevan. 2014. The Dynamics of Sperm Competition in a Competitive Environment. *Proceedings of the Royal Society B: Biological Sciences* 281:20140296.

Fisher, H. S., and H. E. Hoekstra. 2010. Competition Drives Cooperation Among Closely Related Sperm of Deer Mice. *Nature* 463:801–3.

Fisher, R. A. 1930. *The Genetical Theory of Natural Selection*. Oxford: Clarendon Press.

Folstad, I., and A. J. Karter. 1992. Parasites, Bright Males, and the Immunocompetence Handicap. *American Naturalist* 139 (3): 603–22.

Fowler, K., and L. Partridge. 1989. A Cost of Mating in Female Fruitflies. *Nature* 338:760–1.

Funk, W. C., D. C. Cannatella, and M. J. Ryan. 2009. Genetic Divergence Is More Tightly Related to Call Variation Than Landscape Features in the Amazonian Frogs *Physalaemus petersi* and *P. freibergi*. *Journal of Evolutionary Biology* 22:1839–53.

Gladyshev, E. A., M. Meselson, and I. R. Arkhipova. 2008. Massive Horizontal Gene Transfer in Bdelloid Rotifers. *Science* 320:1210.

Grafen, A. 1990. Sexual Selection Unhandicapped by the Fisher Process. *Journal of Theoretical Biology* 144 (4): 473–516.

Gubernick, D. J., and T. Teferi. 2000. Adaptive Significance of Male Parental Care in a Monogamous Mammal. *Proceedings of the Royal Society B: Biological Sciences* 267:147–50.

Gwynne, D. T. 1988. Courtship Feeding and the Fitness of Female Katydids (Orthoptera: Tettigoniidae). *Evolution* 42 (3): 545–55.

Hamilton, W. D., and M. Zuk. 1982. Heritable True Fitness and Bright Birds: A Role for Parasites? *Science* 218 (4570): 384–7.

Harcourt, A. H., P. H. Harvey, S. G. Larson, and R. V. Short. 1981. Testis Weight, Body Weight and Breeding System in Primates. *Nature* 293 (5827): 55–7.

Hasselquist, D. 1998. Polygyny in the Great Reed Warbler: A Long-Term Study of Factors Contributing to Male Fitness. *Ecology Letters* 79:2376–90.

Holland, B., and W. R. Rice. 1999. Experimental Removal of Sexual Selection Reverses Intersexual Antagonistic Coevolution and Removes a Reproductive Load. *Proceedings of the National Academy of Sciences USA* 96 (9): 5083.

Hotzy, C., and G. Arnqvist. 2009. Sperm Competition Favors Harmful Males in Seed Beetles. *Current Biology* 19 (5): 404–7.

Howard, R. D. 1978. The Influence of Male-Defended Oviposition Sites on Early Embryo Mortality in Bullfrogs. *Ecology Letters* 59 (4): 789–98.

Iwasa, Y., and A. Pomiankowski. 1994. The Evolution of Mate Preferences for Multiple Sexual Ornaments. *Evolution* 48:853–67.

———. 1999. Good Parent and Good Genes Models of Handicap Evolution. *Journal of Theoretical Biology* 200 (1): 97–109.

Iwasa, Y., A. Pomiankowski, and S. Nee. 1991. The Evolution of Costly Mate Preferences: II. The "Handicap" Principle. *Evolution* 45 (6): 1431–42.

Iyengar, V. K., and T. Eisner. 1999. Female Choice Increases Offspring Fitness in an Arctiid Moth (*Utetheisa ornatrix*).

Proceedings of the National Academy of Sciences USA 96 (26): 15013–6.

Johnstone, R. A. 1995. Sexual Selection, Honest Advertisement and the Handicap Principle: Reviewing the Evidence. *Biological Reviews* 70 (1): 1–65.

Khila, A., E. Abouheif, and L. Rowe. 2012. Function, Developmental Genetics, and Fitness Consequences of a Sexually Antagonistic Trait. *Science* 336 (6081): 585–9.

Kirkpatrick, M. 1982. Sexual Selection and the Evolution of Female Choice. *Evolution* 36 (1): 1–12.

Kokko, H., R. Brooks, J. M. McNamara, and A. I. Houston. 2002. The Sexual Selection Continuum. *Proceedings of the Royal Society B: Biological Sciences* 269:1331–40.

Kokko, H., and M. D. Jennions. 2008. Parental Investment, Sexual Selection and Sex Ratios. *Journal of Evolutionary Biology* 21 (4): 919–48.

Kruuk, L. E. B., J. Slate, J. M. Pemberton, S. Brotherstone, F. E. Guinness, et al. 2002. Antler Size in Red Deer: Heritability and Selection but No Evolution. *Evolution* 56:1683–95.

Lande, R. 1981. Models of Speciation by Sexual Selection on Phylogenetic Traits. *Proceedings of the National Academy of Sciences USA* 78 (6): 3721–5.

Le Boeuf, B. J., and S. L. Mesnick. 1990. Sexual Behavior of Male Northern Elephant Seals: I. Lethal Injuries to Adult Females. *Behavior* 116 (1–2): 143–62.

Le Boeuf, B. J., and J. Reiter. 1988. Lifetime Reproductive Success in Northern Elephant Seals. In T. Clutton-Brock (ed.), *Reproductive Success: Studies of Individual Variation in Contrasting Breeding Systems* (pp. 344–62). Chicago: University of Chicago Press.

Lelito, J. P., and W. D. Brown. 2006. Complicity or Conflict over Sexual Cannibalism? Male Risk Taking in the Praying Mantis *Tenodera aridifolia sinensis*. *American Naturalist* 168 (2): 263–9.

Lively, C. M. 1992. Parthenogenesis in a Freshwater Snail: Reproductive Assurance Versus Parasitic Release. *Evolution* 46 (4): 907–13.

———. 2010. A Review of Red Queen Models for the Persistence of Obligate Sexual Reproduction. *Journal of Heredity* 101 (Suppl. 1): S13–20.

Lutes, A. A., W. B. Neaves, D. P. Baumann, W. Wiegraebe, and P. Baumann. 2010. Sister Chromosome Pairing Maintains Heterozygosity in Parthenogenetic Lizards. *Nature* 464 (7286): 283–6.

Malo, A. F., E. R. S. Roldan, J. Garde, A. J. Soler, and M. Gomendio. 2005. Antlers Honestly Advertise Sperm Production and Quality. *Proceedings of the Royal Society B: Biological Sciences* 272 (1559): 149–57.

Mark Welch, J. L., D. B. Mark Welch, and M. Meselson. 2004. Cytogenetic Evidence for Asexual Evolution of Bdelloid Rotifers. *Proceedings of the National Academy of Sciences USA* 101 (6): 1618–21.

Maynard Smith, J. 1978. *Evolution of Sex*. Cambridge: Cambridge University Press.

———. 1985. Sexual Selection, Handicaps, and True Fitness. *Journal of Theoretical Biology* 115 (1): 1–8.

McDonald, M. J., D. P. Rice, and M. M. Desai. 2016. Sex Speeds Adaptation by Altering the Dynamics of Molecular Evolution. *Nature* 531 (7593): 233.

Mead, L. S., and S. J. Arnold. 2004. Quantitative Genetic Models of Sexual Selection. *Trends in Ecology and Evolution* 19 (5): 264–71.

Östlund, S., and I. Ahnesjö. 1998. Female Fifteen-Spined Sticklebacks Prefer Better Fathers. *Animal Behaviour* 56 (5): 1177–83.

Otronen, M. 1990. Mating Behavior and Sperm Competition in the Fly, *Dryomyza anilis*. *Behavioral Ecology and Sociobiology* 26:349–56.

Peretti, A., and A. Aisenberg, eds. 2015. *Cryptic Female Choice in Arthropods*. Cham, CH: Springer.

Pitnick, S., T. A. Markow, and G. S. Spicer. 1995. Delayed Male Maturity Is a Cost of Producing Large Sperm in *Drosophila*. *Proceedings of the National Academy of Sciences USA* 92 (23): 10614–8.

Pomiankowski, A., and Y. Iwasa. 1998. Runaway Ornament Diversity Caused by Fisherian Sexual Selection. *Proceedings of the National Academy of Sciences USA* 95:5106–11.

Preston, B. T., I. R. Stevenson, and K. Wilson. 2003. Overt and Covert Competition in a Promiscuous Mammal: The Importance of Weaponry and Testes Size to Male Reproductive Success. *Proceedings of the Royal Society B: Biological Sciences* 270 (1515): 633–40.

Price, T., D. Schluter, and N. E. Heckman. 1993. Sexual Selection When the Female Directly Benefits. *Biological Journal of the Linnean Society* 48 (3): 187–211.

Prum, R. O. 1997. Phylogenetic Tests of Alternative Intersexual Selection Mechanisms: Trait Macroevolution in a Polygynous Clade (Aves: Pipridae). *American Naturalist* 149 (4): 668–92.

Pryke, S. R., and S. Andersson. 2005. Experimental Evidence for Female Choice and Energetic Costs of Male Tail Elongation in Red-Collared Widowbirds. *Biological Journal of the Linnean Society* 86 (1): 35–43.

Pryke, S. R., S. Andersson, and M. J. Lawes. 2001. Sexual Selection of Multiple Handicaps in the Red-Collared Widowbird: Female Choice of Tail Length but Not Carotenoid Display. *Evolution* 55 (7): 1452–63.

Ribble, D. O. 1991. The Monogamous Mating System of *Peromyscus californicus* as Revealed by DNA Fingerprinting. *Behavioral Ecology and Sociobiology* 29 (3): 161–6.

Rodd, F. H., K. A. Hughes, G. F. Grether, and C. T. Baril. 2002. A Possible Non-sexual Origin of Mate Preference: Are Male Guppies Mimicking Fruit? *Proceedings of the Royal Society B: Biological Sciences* 269 (1490): 475–81.

Rowe, L. 1994. The Costs of Mating and Mate Choice in Water Striders. *Animal Behaviour* 48 (5): 1049–56.

Ryan, M. J. 1990. Sexual Selection, Sensory Systems and Sensory Exploitation. *Oxford Surveys in Evolutionary Biology* 7:157–95.

Ryan, M. J., and W. Wilczynski. 1998. Coevolution of Sender and Receiver: Effect on Local Mate Preference in Cricket Frogs. *Science* 240:1786–8.

Saetre, G-P., T. Fossnes, and T. Slagsvold. 1995. Food Provisioning in the Pied Flycatcher: Do Females Gain Direct Benefits from Choosing Bright-Coloured Males? *Journal of Animal Ecology* 64 (1): 21–30.

Shuster, S. M., and M. J. Wade. 2003. *Mating Systems and Strategies*. Princeton, NJ: Princeton University Press.

Sigurjónsdóttir, H., and G. A. Parker. 1981. Dung Fly Struggles: Evidence for Assessment Strategy. *Behavioral Ecology and Sociobiology* 8 (3): 219–30.

Stein, A. C., and J. A. C. Uy. 2005. Plumage Brightness Predicts Male Mating Success in the Lekking Golden-Collared Manakin, *Manacus vitellinus*. *Behavioral Ecology* 17 (1): 41–7.

Thornhill, R. 1976. Sexual Selection and Nuptial Feeding Behavior in *Bittacus apicalis* (Insecta: Mecoptera). *American Naturalist* 110 (974): 529–48.

———. 1983. Cryptic Female Choice and Its Implications in the Scorpionfly *Harpobittacus nigriceps*. *American Naturalist* 122 (6): 765–88.

Trainer, J. M., and D. B. McDonald. 1995. Singing Performance, Frequency Matching and Courtship Success of Long-Tailed Manakins (*Chiroxiphia linearis*). *Behavioral Ecology and Sociobiology* 37 (4): 249–54.

Tree of Sex Consortium. 2014. Tree of Sex: A Database of Sexual Systems. *Scientific Data* 1:140015.

Trivers, R. L. 1972. Parental Investment and Sexual Selection. In B. Campbell (ed.), *Sexual Selection and the Descent of Man 1871–1971* (pp. 136–79). Chicago: Aldine.

van der Kooi, C. J., C. Matthey-Doret, and T. Schwander. 2017. Evolution and Comparative Ecology of Parthenogenesis in Haplodiploid Arthropods. *Evolution Letters* 1 (6): 304–16.

Vanpé, C., J-M. Gaillard, N. Morellet, P. Kjellander, A. Mysterud, et al. 2007. Antler Size Provides an Honest Signal of Male Phenotypic Quality in Roe Deer. *American Naturalist* 169 (4): 481–93.

Van Valen, L. 1973. A New Evolutionary Law. *Evolutionary Theory* 1:1–30.

Wedell, N., and A. Arak. 1989. The Wartbiter Spermatophore and Its Effect on Female Reproductive Output (Orthoptera: Tettigoniidae, *Decticus verrucivorus*). *Behavioral Ecology and Sociobiology* 24 (2): 117–25.

Wilkinson, G. S., H. Kahler, and R. H. Baker. 1998. Evolution of Female Mating Preferences in Stalk-Eyed Flies. *Behavioral Ecology* 9 (5): 525–33.

Wilson, C. G., and P. W. Sherman. 2010. Anciently Asexual
Bdelloid Rotifers Escape Lethal Fungal Parasites by Drying
Up and Blowing Away. *Science* 327 (5965): 574–6.

Wolfner, M. F. 2002. The Gifts That Keep on Giving: Physiological
Functions and Evolutionary Dynamics of Male Seminal
Proteins in *Drosophila*. *Heredity* 88 (2): 85–93.

Yasukawa, K. 1981. Male Quality and Female Choice of Mate in
the Red-Winged Blackbird (*Agelaius phoeniceus*). *Ecology Letters*
62 (4): 922–9.

Zahavi, A. 1975. Mate Selection—A Selection for a Handicap.
Journal of Theoretical Biology 53 (1): 205–14.

Zeh, D. W., and J. A. Zeh. 1992. Dispersal-Generated Sexual Selection in a Beetle-Riding Pseudoscorpion. *Behavioral Ecology
and Sociobiology* 30 (2): 135–42.

Zeh, D. W., J. A. Zeh, and G. Tavakilian. 1992. Sexual Selection
and Sexual Dimorphism in the Harlequin Beetle *Acrocinus
longimanus*. *Biotropica* 24 (1): 86–96.

Zuk, M., and T. S. Johnsen. 1998. Seasonal Changes in the Relationship Between Ornamentation and Immune Response in Red
Jungle Fowl. *Proceedings of the Royal Society B: Biological Sciences*
265 (1406): 1631–5.

Zuk, M., R. Thornhill, K. Johnson, and J. D. Ligon. 1990. Parasites
and Mate Choice in Red Jungle Fowl. *American Zoologist*
30:235–244.

CHAPTER 12

After Conception

The Evolution of Life History and Parental Care

Learning Objectives

- Use your understanding of natural selection and trade-offs to explain why organisms do not always produce as many offspring as they can.

- Compare and contrast the investments males and females make in reproduction.

- Compare and contrast the life history trade-offs associated with different reproductive strategies.

- Explain how different parental care strategies can lead to conflicts between parents and between parents and their offspring.

- Compare and contrast mechanisms of sex ratio adjustment.

- Explain how parental conflict can influence gene expression.

- Describe how senescence may arise as a result of life history trade-offs.

- Discuss the evidence scientists have used to explain the evolution of menopause in humans.

Steven Austad decided to become a scientist while lying in a hospital bed, wondering how much longer he had to live. Before his stay in the hospital, he had worked in Hollywood, taking care of lions on movie sets. One day, a lion unexpectedly turned on him. As Austad recovered from the injury, he thought about his future. If he made any more mistakes in his lion-taming job, he realized, his life would be too short to bother with long-term plans.

Once Austad was released from the hospital, he turned his love of animals into a career in zoology, conducting fieldwork in remote jungles in countries such as Venezuela and New Guinea. Austad spent much of that time studying opossums (**Figure 12.1**). Austad found that once opossums reached about 18 months of age, their health declined rapidly. They quickly

Opossums (Didelphis virginiana) give birth to seven or eight babies in a single litter, nurturing them with milk until they can live on their own. Natural selection maximizes the lifetime fitness of organisms by adjusting the size of litters and the interval between them.

STEVE MASLOWSKI/Science Source

Figure 12.1 Steven Austad discovered that the longevity and age of sexual maturity can evolve rapidly in opossums. (Steven Austad)

Life history refers to the pattern of investment an organism makes in growth and reproduction. Life history traits include an organism's age at first reproduction, the duration and schedule of reproduction, the number and size of offspring produced, and life span.

developed cataracts, arthritis, and other symptoms of old age. Most were dead by 2 years of age.

Austad grew interested in the timing and duration of the key events in the lives of the opossums. Theoretical studies had suggested that natural selection could change these events in response to changes in the environment. To test these ideas, Austad searched for a natural experiment. He was looking for populations of neighboring opossums living in different conditions.

Austad, who now teaches at the University of Alabama at Birmingham, discovered just such a natural experiment in the southeastern United States. On the mainland of Georgia, predators killed 80% of the opossums. But on Sapelo Island, just off the coast, there are opossums but no predators. This odd environment developed after the end of the ice age. Before then, the island had been joined to the mainland, but when the glaciers melted, sea levels rose. Opossums became isolated when the island was cut off from the mainland. But perhaps simply by chance, no predators were on the island when it was separated. For the past five thousand years, opossums living there have been free of predation.

Austad compared the **life history** of mainland opossums to the Sapelo opossums and found that the island opossums live 25% longer than their mainland cousins (Austad 1993). This discovery was as momentous as if Austad had discovered an isolated tribe of people who regularly lived well past 100 years. The animals not only lived longer, but also enjoyed good health for a longer time. Their tendons, for example, remained springier far later in life than the tendons of opossums on the mainland. These striking differences evolved rapidly in the five thousand years since Sapelo Island was cut off from the mainland.

In Chapter 11, we saw how species can evolve a wide range of mating strategies, depending on factors such as environmental conditions, sexual selection, and sexual conflict. However, selection does not stop with the fertilization of eggs. An organism's fitness is determined by the number of offspring it produces—and the number that survive to maturity—over its *entire* lifetime. Many organisms reproduce more than once during their lives, and they must balance how they allocate resources to these successive bouts of reproduction. Producing too many offspring at once, or providing them with too many resources, might reduce an organism's ability to reproduce as effectively later on. As a result, selection can shape how organisms invest in reproduction over their lifetimes. In fact, it can even shape the life span of a species (**Figure 12.2**). Austad's studies are part of a growing body of research, called life history theory, that illuminates how quickly species reach sexual maturity, how much care they invest in their offspring, and how quickly they age.

Key Concept

- Life history theory explores how the schedule and duration of key events in an organism's lifetime are shaped by natural selection. It helps explain variation in the age at which organisms begin reproducing, the size and number of offspring produced, the amount and type of parental care invested, and the onset of senescence. ●

Figure 12.2 Life history traits vary spectacularly among species. A: For example, some insects such as mayflies live for only a matter of hours as adults, and individual females lay thousands of eggs. B: At the other extreme, female African elephants (*Loxodonta africana*) typically don't begin reproducing until they are 10–12 years old. They give birth to only a single calf every 5 years, but they can be reproductive for more than 50 years. C: Plants vary over even greater extremes. Many short-lived plants such as the mustard (*Brassica rapa*) reproduce only a single time in their 2-month lifetimes. D: Other species such as bristlecone pines reproduce repeatedly over a period of more than 4000 years. (A: Stana/Shutterstock.com; B: Mogens Trolle/Shutterstock.com; C: Snowbelle/Shutterstock.com; D: Joy Stein/Shutterstock.com)

12.1 Selection Across a Lifetime

In his book *Adaptation and Natural Selection* (Williams 1966), the American evolutionary biologist George Williams argued that scientists should expand their concept of fitness. He argued that they should not focus simply on particular traits that raise reproductive success: they instead should think of fitness across an organism's entire lifetime.

Antagonistic Pleiotropy over a Lifetime

To see why this matters, consider a population of opossums. Individuals are steadily killed off by predators, accidents, and diseases. How long any one opossum lives may be partly a matter of chance—whether it happens to be walking in the path of a falling tree, for example. But across the entire population, older opossums are rarer than younger ones. You can use these data to calculate the probability that an opossum lives to any given age and observe that, for opossums, that probability declines with time. (The rate at which external events, such as predation, lead to death in a population is known as the extrinsic mortality rate.)

If a mutation is deleterious only in old age, it may have no effect on an individual opossum's fitness because the animal is likely to die from external causes before the allele can cause any harm. On the other hand, mutations that are deleterious early in life can have a much stronger effect on fitness because the odds of an opossum being alive are much higher. If an opossum dies before leaving its mother's pouch due to an early-acting mutation, it has lost a lifetime's opportunity to reproduce.

The same contrast also applies to beneficial mutations at different times in life. A mutation that maintains health in a very old opossum may not raise fitness if the likelihood is high that most individuals have already been killed by predators by that age. And a mutation that leads to better health in a juvenile opossum will be strongly favored by selection because it can lead to many more offspring.

As we saw in Chapter 9, some mutations that are beneficial in one context can be harmful in another. Antagonistic pleiotropy plays an important role in selection on mutations that affect an organism's life history. Consider a mutation that is beneficial early in life, but then is harmful later in life. Because fitness effects are more important earlier in life, natural selection may favor this mutation despite the harm it causes in old age.

These mutations can affect the life history of an opossum. Mutations may change the time it takes for an animal to reach sexual maturity and how many resources it invests in its young. Williams noted that organisms do not have an infinite amount of energy to bring to these functions. Consider a female opossum. Over the course of her lifetime, she may have several litters. She could potentially increase her fitness by giving birth to larger litters. But larger litters require more resources from her during both pregnancy and nursing. If she cannot find enough food to nurture her extra young, they all may become malnourished and run a greater risk of dying before maturity. What's more, the experience of raising a larger litter can take a toll on her body. She may become more vulnerable to disease and predation or have difficulty bearing another litter later in life.

Now consider a female opossum that gives birth to only a few small litters. Because fewer offspring require fewer resources, she would be able to rear much healthier offspring and put herself at less risk than if she produced big litters. But producing too few offspring may mean a decline in her relative fitness because other opossums will outbreed her.

Williams argued that the lifetime fitness of an organism is determined in part by the trade-offs it experiences between competing demands. Selection favors the optimal strategy, the one that maximizes the number of offspring that survive to maturity over the course of an organism's entire life. If individuals are likely to die at an early age, then there is much greater benefit from increased reproduction early in life—even if it comes at the price of fewer offspring late in life. Populations can accumulate mutations that raise fitness early in life, even if those mutations have antagonistic pleiotropic effects later in life. This trade-off, Williams argued, could explain why species like the opossum deteriorate and die at an early age even when they are housed in captivity, away from extrinsic threats like predators: they have inherited mutations that have strongly deleterious effects in old age.

The Limits of Self-Repair

In the late 1970s, Thomas Kirkwood, an evolutionary biologist at Newcastle University in England, extended life history theory by proposing mechanisms underlying the trade-offs proposed by Williams and others (Kirkwood and Holliday 1979; Kirkwood and Austad 2000). Kirkwood noted that organisms of any age continually need to repair their cells. To survive, they have to fix DNA replication errors, replace deformed proteins with new copies, and produce molecules that can shield cells from damaging oxidation.

This continual repair demands energy, and Kirkwood argued that natural selection should favor levels of self-repair that are good enough to keep an organism in sound condition only for as long as it has a reasonable chance of reproducing. For example, more than 90% of mice in the wild die in their first year, so, theoretically, a mouse that invests in mechanisms for survival beyond its first year has only a 10% chance of experiencing any benefit from its self-repair. So mice that invest extra energy into producing lots of pups in their first year should outcompete mice that invest that energy into repair to keep their cells in good working order for a decade.

Studies on mice are consistent with this hypothesis. Mice take only 6 to 8 weeks to reach sexual maturity after birth. A female mouse's gestational period is just 6 weeks, and within hours of delivering her litter, she becomes fertile again.

Testing Life History Theories

The work of theoreticians like Williams and Kirkwood prompted Austad to test the predictions with the opossum. He made empirical observations on the animals and checked whether the patterns he observed matched the predictions from life history theory. Given the high rate of predation on the mainland, the theory predicted that selection would favor earlier sexual maturity and a larger number of offspring per litter. This life history would maximize the reproductive success of the mainland opossums in the short time they had to live. Antagonistic pleiotropy might lead them to suffer harmful effects from these traits later in life, but the effects would have less impact on fitness due to the high extrinsic mortality rate.

On Sapelo Island, the lack of predators changed the extrinsic mortality rate. Because the opossums were more likely to survive to older ages, individuals that invested in cell repair would be able to have more offspring throughout their life span. Mutations that were beneficial in young animals would not be favored as much by selection because their negative effects in old age would be greater. Over the past five thousand years, this shift in selection pressures might lead to the evolution of measurable differences in the opossums' life histories.

During any given time interval, Austad found, the Sapelo Island opossums were less likely to die than mainland opossums (**Figure 12.3**). As a result, the Sapelo opossums lived markedly longer on average. This was not simply due to predation: the

A

Comparison between Sapelo and mainland opossums			
	Mainland	Island	P value
Longevity (mo)	20.0	24.5	0.002
Litter size (1st yr)	7.6	5.9	0.002

B

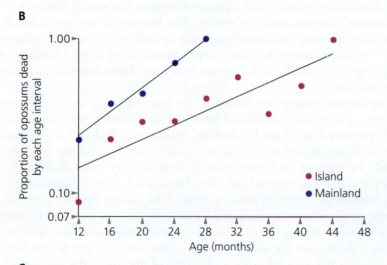

C

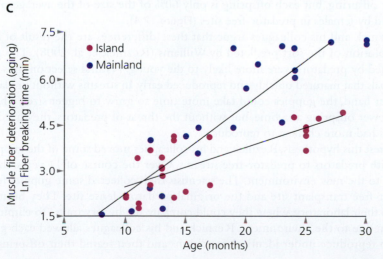

Figure 12.3 Steven Austad compared opossums living on Sapelo Island, off the coast of Georgia, to mainland opossums. A: He found a number of differences in life history traits in the two populations. The island opossums lived longer and had smaller litters. B: As this graph shows, the death rate increased slower on Sapelo Island than on the mainland. C: Austad also found that the island opossum aged more slowly than the mainland opossum. (Data from Austad 1993)

mainland opossums deteriorated faster than the ones on the island. Austad documented this deterioration by examining the collagen in the tendons of the opossums. Tendons are made of collagen fibers, which slide past each other to allow the tendons to stretch. As animals get older, cross-links form between the fibers, making them less flexible.

Austad measured the cross-linking in the island and mainland opossums. To do so, he hung a weight from each tendon and placed it in a solution that gradually dissolved the cross-links between the collagen fibers. As Figure 12.3C shows, the island opossum tendons broke under the strain of the weight faster than the mainland opossum tendons did. That fast breaking indicates that the island opossums' tendons had fewer cross-links. In other words, the island opossums were aging more slowly than their mainland counterparts.

Along with exhibiting slower aging, the Sapelo opossums produced fewer offspring in each litter, as would be expected if investment in survival mechanisms came at the expense of reproduction—that is, if there were a trade-off between survival and reproduction. This life history strategy was beneficial on the island. Even though females on Sapelo produce fewer offspring in their first year than opossums on the mainland, more of these females lived to breed again in their second year. Because more opossums on Sapelo Island survived longer, a slower—yet successful—reproductive strategy evolved.

An Experiment in Life History

More support for life history trade-offs has come from research on guppies (*Poecilia reticulata*) that live in the streams of the Caribbean island of Trinidad. These guppies do not lay fertilized eggs; instead, the eggs develop into live young within the female fishes, absorbing nutrients from the mother. The amount of resources a female provisions for each offspring will influence its size at birth, which in turn influences its odds of survival to adulthood. A female guppy's overall fitness is also a function of how many offspring it produces in each litter and how many litters it produces over its lifetime (Reznick 2011).

Like opossums, guppies on Trinidad also face differing predation pressures. In some sections of streams, killifish and other predators attack them. Other sections of streams are predator-free. David Reznick, an evolutionary biologist at the University of California, Riverside, and his colleagues hypothesized that different mortality rates would select for different life history traits. To find out, they compared the life histories of the populations in each habitat. They found that in guppy populations that do not face predators, the offspring grow up slowly. The females produce large offspring but few offspring per litter. In the guppy populations that are menaced by predators, however, a different strategy has evolved. The males in high-predation sites are ready to mate much earlier than the males at predator-free sites. The females produce twice as many offspring, but each offspring is only 60% of the size of the average offspring produced by females in predator-free sites (Figure 12.4).

Reznick and his colleagues argue that these differences are the result of life history evolution of the sort predicted by Williams (Reznick et al. 2008). The guppies threatened by predators were more likely to die young. Natural selection thus favored individuals that matured quickly and reproduced early. In streams without predators, on the other hand, the guppies could take more time to grow to bigger sizes. They produced fewer offspring at a time, but without the threat of predators, they lived longer and thus had more chances to reproduce.

To test this hypothesis, Reznick and his colleagues moved some of the guppies that lived with predators to predator-free streams. Over the course of 11 years, the fishes adapted to the new environment. The scientists then collected some guppies from the predator-free transplant site and the original predator-dense site. They brought the fishes to their laboratory, where they could compare life history traits. To eliminate any variation due to the environment, Reznick and his colleagues allowed each group of fishes to reproduce under identical conditions and then reared their offspring under

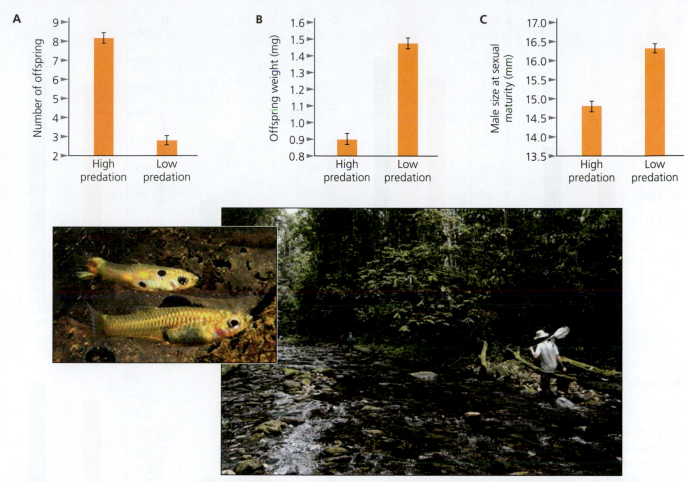

Figure 12.4 David Reznick and his colleagues compared guppies (inset photo) in streams with many predators to those in streams with few. As these graphs show, the guppies that faced few predators produced fewer offspring (A) and bigger offspring (B) than the guppies menaced by many predators. The guppies also grew to larger sizes by the time they were sexually mature (C). Natural selection shaped the life history of the guppies in different ways, depending on whether they were more or less likely to be killed by predators. (Data from Reznick et al. 2006; photo: Andrew Hendry; photo inset: Paul Bentzen)

identical conditions (this is called a common garden experiment). They then allowed the lab-reared guppies to reproduce again (Reznick et al. 2008).

As illustrated in **Figure 12.5**, the scientists found significant differences in the life history traits of the two populations of guppies. In the absence of predators, female guppies produced smaller broods, and their offspring took longer to reach maturity. However, when they did reach maturity, they were bigger than guppies from populations adapted to living with predators.

More recently, Terry Dial, of Brown University, and his colleagues discovered another advantage that comes when Trinidadian guppies evolved to larger size. The bigger guppies could open their mouths wider (Dial et al. 2017). The researchers found that as the guppies evolve to bigger sizes, their jaw muscles undergo a proportionately larger increase than their overall body. As a result, the muscles can provide a larger gape. The big guppies may thus gain a competitive advantage at feeding compared with smaller guppies.

These experiment on guppies illustrate just how quickly populations can respond to selection on life history trade-offs. It only takes a matter of years—not millennia.

The Case of Brown Anolis Lizards

Sometimes scientists can document these life history trade-offs within the lifetime of an individual animal. For example, Ryan Calsbeek and Robert Cox ran an experiment

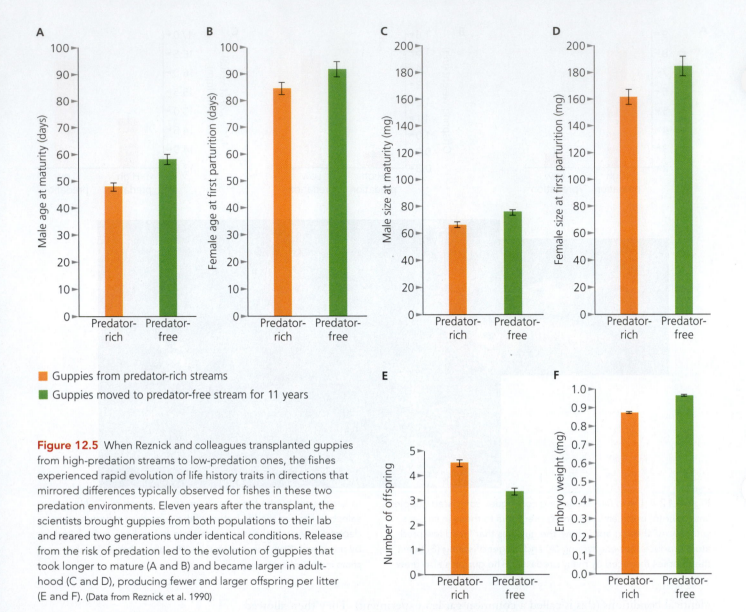

Figure 12.5 When Reznick and colleagues transplanted guppies from high-predation streams to low-predation ones, the fishes experienced rapid evolution of life history traits in directions that mirrored differences typically observed for fishes in these two predation environments. Eleven years after the transplant, the scientists brought guppies from both populations to their lab and reared two generations under identical conditions. Release from the risk of predation led to the evolution of guppies that took longer to mature (A and B) and became larger in adulthood (C and D), producing fewer and larger offspring per litter (E and F). (Data from Reznick et al. 1990)

Legend:
- Guppies from predator-rich streams
- Guppies moved to predator-free stream for 11 years

in which they removed the ovaries from brown anolis lizards (**Figure 12.6**; Cox and Calsbeek 2010). They then performed the same surgery on another group of females without removing the ovaries, in order to create a set of controls. The control females invested resources to grow eggs. The females without ovaries, on the other hand, grew faster because they had free resources they could invest in growth.

Key Concepts

- Trade-offs arise when allocation of resources to one life history trait reduces investment in another trait.

- Investment in reproduction often comes at the expense of growth or body maintenance.

- Selection may favor mutations that are beneficial early in life, even if those same mutations are harmful later on.

- Investment in reproduction early in life often reduces an individual's ability to breed later in life.

- A change in the selective environment can bring about rapid evolution of life history traits. ●

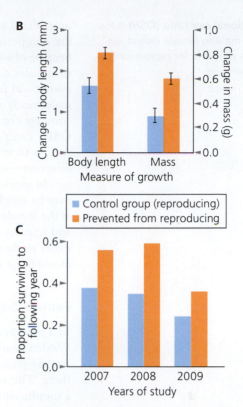

Figure 12.6 Sometimes life history trade-offs are visible within an individual's lifetime. For example, when Ryan Calsbeek and Robert Cox prevented some brown anolis lizards (A) from reproducing (by surgically removing their ovaries), treated nonreproductive females grew faster (B) and were more likely to survive (C) than were sham-operated control females who produced offspring. Results like these are consistent with a resource allocation trade-off between growth and reproduction. (A: "Experimental evidence for physiological costs underlying the trade-off between reproduction and survival," *Functional Ecology* 2010, 24, 1262–1269. Figure 1 Photo by Ryan Calsbeek; data from Cox and Calsbeek 2010)

12.2 Parental Investment

One of the most important life history traits is the effort that organisms put into rearing offspring. Investment by a parent into current offspring may come at the expense of other reproductive options (such as producing additional young later or seeking other mating opportunities). This trade-off is similar in many ways to the one that governs the evolution of gametes, which we discussed in Chapter 11. Females produce a limited number of large, nutrient-rich eggs. As a result, they are likely to benefit the most from providing additional investments of parental care: anything a female can do that increases the probability of survival, or the quality, of her offspring compared with those of other females is likely to be favored by selection. A male producing many sperm invests less to begin with, and he usually will lose less if he elects to abandon the young.

Females also have the advantage of certainty that their offspring are their own. Males, on the other hand, have much less certainty about whether a partner's offspring are his because females often mate with multiple males. Resources he invests in rearing offspring that turn out not to be his reduce his fitness.

Thus, in most animal species in which parents care for their young, it is the mothers who provide that care. In placental mammal species, for example, fertilized eggs develop in the mother's uterus, receiving abundant nutrients from the mother's bloodstream. After giving birth, the mother continues to provide her offspring with nutrients in the form of milk.

However, there are exceptions to this rule: in some species, it's the males who do most of the work of rearing young. These role-reversed species offer exciting insights into sexual selection. Recall that the nature of sexual selection—members of one sex competing over opportunities to mate with members of the other sex—arises because of a biased **operational sex ratio (OSR)**. Typically, females invest so much more than males do in their offspring that they take longer to recycle between reproductive events, and fewer females are available for mating at any point in time. The excess of males ready to mate sets the stage for sexual selection on traits that help males outcompete rivals and gain access to females (Chapter 11). Highly skewed (male-biased) OSRs can result in strong sexual selection on males. But the OSR can vary from species to species, and when this ratio changes, so does the nature—and even the direction—of sexual selection.

In species where both males and females cooperate in the care of young, the OSR may be much less male-biased. If males spend the same amount of time with young as the females do, then they are not going to recycle any faster than the females. Males and females are equally "tied down" with parental care. As a result, the number of males ready to breed will be comparable to that of females. Sexual selection is predicted to be much weaker in these populations.

Female-Biased Operational Sex Ratios

In a few species, the OSR is reversed—more females are ready to mate than males. Wattled jacanas (*Jacana jacana*) are an incredible example of such a species (**Figure 12.7**). Female jacanas lay their clutches of eggs into the nests of males and then abandon them. The male protects the eggs, and later the chicks, and will spend more than a month with his brood. Because females abandon their eggs and males take on the parental care, females are able to recycle faster than males. The result is that these populations have an OSR skewed toward females.

Such role-reversed taxa provide elegant tests of the predictions of sexual selection theory, because they are the exceptions that prove the rule. Nothing about sexual selection requires that males must compete for access to females. This just happens to be the most prevalent manifestation of sexual selection, thanks to anisogamy and the associated asymmetry of investment by females and males. In principle, either sex could have to compete. In role-reversed taxa like the jacanas, it is the females who compete over a limited number of reproductively ready males.

Female jacanas fight with rival females for the possession of territories that include male nests. Dominant females are often able to hold territories that include several different males, and in this way they are able to mate with all of them. As we might expect for a species with "reversed" operational sex ratios, female jacanas are larger than males. They are also more aggressive than males, and they grow spurs on their wings that they use as formidable weapons to fight with rival females (Emlen and Wrege 2004).

Figure 12.7 A: Males incubate eggs and protect chicks after they hatch. B: Females guard harems of multiple males. Females are larger than males and have larger weapons. (A: Joe Petersburger/National Geographic/Getty Images; B: Marie Read/Science Source)

Perhaps the most extreme case of parental role reversal can be found in seahorses, pipefishes, and other species in the family Syngnathidae. In syngnathid species, the males effectively get pregnant (**Figure 12.8**). When syngnathid fishes mate, the female transfers unfertilized eggs to the male. The male stores the eggs, sometimes inside a fleshy pouch, where he fertilizes them with his sperm. Whereas sperm in other animals may be champion swimmers that can make the long journey up a female's reproductive tract, syngnathid sperm barely move at all.

Because the eggs develop in males, the embryos get some of their energy from each parent. Some of it comes from the yolk their mother provided them, and some of it comes from their fathers. The pouch of some species changes shape, taking on

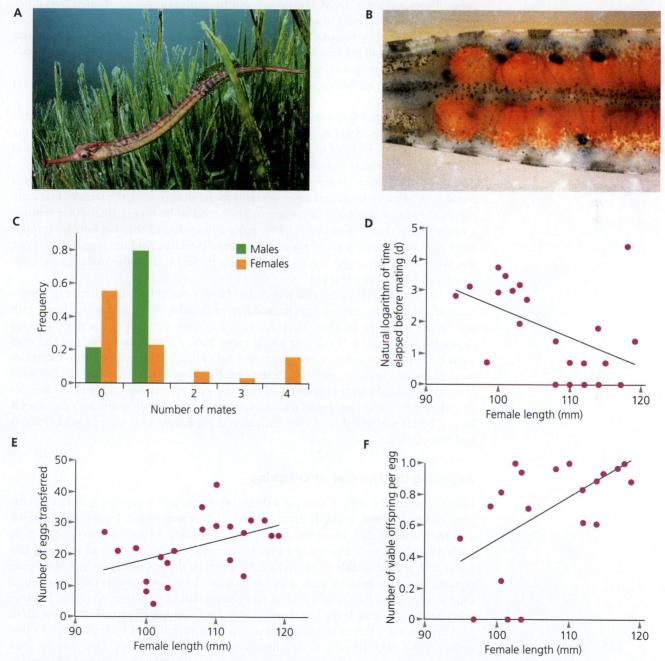

Figure 12.8 A: Male gulf pipefishes carry fertilized eggs in pouches (B). Because "pregnant" males make a bigger investment in rearing offspring, mate choice is more important in males. C: In an experiment, only a few females mated with most males. D: Males prefer to mate with large females, accepting their advances and mating more quickly than with smaller females. E, F: Eggs from larger females are more likely to be successfully transferred to males and develop into viable offspring. (B: Kimberly Ann Paczolt, Wilkinson Lab University of Maryland, Dept. of Biology; data from Jones et al. 2001; Paczolt and Jones 2010)

a complex anatomy that brings each fish embryo into intimate contact with the father's blood supply so that he can give them nutrients. Eventually the baby fishes wiggle out and are ready for life on their own.

In this mating system, the females still produce the eggs, but they don't have to put the time and effort into rearing them. Sexual selection theory suggests that the females would be better off looking for as many males as possible to take their eggs. And with all those females swimming around in search of males, they're going to face some fierce competition.

Female–Female Competition

Adam Jones, of Texas A&M University, and his colleagues have studied gulf pipe-fishes (*Syngnathus scovelli*) to see whether this reversed competition in fact occurs. In one experiment, they found that most females failed to find any mates at all, whereas a small proportion of the females managed to mate with four males (compare this finding to the poor mating success of most male elephant seals as discussed in Chapter 11). Most of the male pipefishes, on the other hand, mated once during Jones's experiment (Jones et al. 2001).

Male syngnathid fishes are not limited by the number of females they can fertilize. As a result, they don't benefit from making lots of sperm. And that explains the remarkably scant supply of sperm these fishes produce. Instead of making millions of sperm every day, a male seahorse's testes may carry just 150 sperm.

The competition of females opens up the opportunity for males, rather than females, to be picky. Jones and other scientists found that, indeed, the females that mate the most have certain traits in common. They tend to be bigger than other females, and they have fancier fins and brighter colors. Jones found that big females transfer more eggs into the pouches of males than do small females. And the bigger the female, the more likely each egg was to survive. The preferences of the males cause large females to have more offspring.

Jones and his colleagues also wondered if males controlled the amount of investment they put into rearing eggs from different females. They had males mate with one female and then another. They discovered that the survival of the second brood depended on the first. If the first brood came from a big female, fewer of the eggs from the second brood survived. Jones and his colleagues concluded that the males were likely giving more resources to the eggs from big females, leaving less for small females they might later encounter. If they did end up mating with smaller females first, they gave these eggs fewer resources, storing more for themselves so that they'd be in a better position should they encounter a big female next time around (Paczolt and Jones 2010).

Adjusting the Number of Offspring

Investments can be risky. If you put a thousand dollars into a promising start-up company, that money may multiply many times over, or it may vanish. Parents face a similar uncertainty when they invest resources in their offspring. Mutations can arise, causing birth defects. A female may carry a fetus for months, only to produce an offspring that ultimately will be unable to reproduce. Changes to the environment can also bring reproductive risks. In some years, a female bird may be able to provide large amounts of food to her chicks thanks to abundant rain and growth of seed-bearing plants. In other years, a drought may leave her struggling to find enough food to keep them alive.

Evolutionary theory suggests that natural selection should favor adaptations that reduce these risks, thereby raising fitness over an entire lifetime. One way to cope with the uncertainty of reproduction is to respond flexibly to different challenges. Empirical studies reveal that parents in many species do just this. In some cases, they can adjust the number of their offspring to raise their long-term fitness; in other cases, they can manipulate the ratio of sons to daughters in their offspring.

Miscarriages are sometimes the result of these flexible strategies. When fertilized eggs begin to develop, they may carry harmful mutations. In some cases, for example, they have extra copies of chromosomes, leading to Down syndrome and other disorders in humans. Scientists have found that 90% of human embryos with abnormal chromosome numbers result in miscarriages in the first trimester of pregnancy. On the other hand, 93% of embryos with normal chromosome numbers continue to term (Quenby et al. 2002). A number of scientists have argued that these miscarriages are not simply the failure of an embryo to develop. Instead, the mother's body may be using chemical cues to assess offspring quality and spontaneously abort embryos that show signs of genetic abnormalities (Forbes 1997). Miscarriages are a source of great psychological suffering, but they are favored by natural selection because they can reduce 9 months of investment to a month or less for embryos that are unlikely to survive or reproduce on their own.

Spontaneous abortion may be an effective strategy for maximizing fitness by regulating parental investment, but it only works in species that control the development of their offspring inside their bodies after fertilization. It is not an option for many other species, but related strategies may have evolved in species such as the sand goby (*Pomatoschistus minutus*), a European species of fish. The male sand goby builds a nest in an empty shell that he excavates. After successful courtship, a female goby releases her eggs, which are fertilized by the male. The male sand goby then tends the nest, covering it with sand, guarding it from predators, and cleaning off any algae. Sand goby fathers even fan the eggs to provide them with a fresh supply of water as they develop.

Male sand gobies will sometimes do something that seems unthinkable: they dig up their nests and devour their own eggs. Hope Klug, a biologist at the University of Helsinki, found an evolutionary logic behind this cannibalism (Klug et al. 2006). She experimentally raised and lowered levels of oxygen in the nests. The number of eggs the father ate depended on how low the oxygen levels dropped (**Figure 12.9**). Klug concluded that by selectively eating some of the eggs, the fathers were increasing the survivorship of the remaining eggs. Supporting this conclusion was the fact that sand goby fathers also adjusted the extent of their cannibalism to the density of eggs in their nests. The denser the eggs in a nest, the more eggs the father was likely to eat. In another study, males even detected whether eggs had been infected with a water mold. By preferentially eating only eggs exposed to the mold, male sand gobies protected the rest of their clutches from infection (Vallon et al. 2016). Rather than being a random act of destruction, this cannibalism appears to be a response the males make to certain changes in their environment that results in increased egg survivorship.

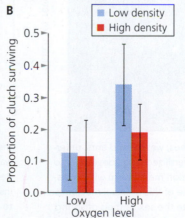
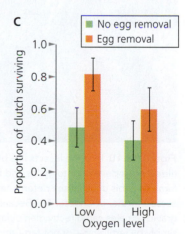

Figure 12.9 Male sand gobies (A) adjust the size of their brood by eating eggs. When egg densities are high, for example, or when oxygen levels are low, males reduce the number of eggs that they care for (B). These adjustments increase the proportion of offspring surviving (C). (A: Cigdem Sean Cooper/Shutterstock.com; data from Klug et al. 2006)

Adjusting the Sex Ratio of Offspring

In addition to adjusting the number of developing offspring, animals can maximize their fitness by adjusting the ratio of sons and daughters away from the typical one-to-one proportions. Ronald Fisher offered an elegant argument for why this balance is normally in place (Fisher 1930). Imagine that mutations arise in a population that lead to more female births than male births. The imbalance gives males an advantage; a male is more likely to find a mate than a female. If some individuals produce more sons, they will be favored by natural selection. But as the numbers of males come to equal the numbers of females in each new generation, the advantage of being male dwindles. The same process would work under the opposite conditions, with more males than females. The sex ratio of the population balances itself at one to one (1:1), an example of frequency-dependent natural selection.

In 1973, Robert Trivers and Dan Willard, both then at Harvard University, argued that natural selection could drive sex ratios away from one to one under certain conditions (Trivers and Willard 1973). Consider, for example, a polygynous species in which a few males in good condition mate with most of the females. If a female is in good condition herself, she may be able to boost her reproductive success by having more sons than daughters. Her sons, in good condition themselves, will mate with many partners and give her more grandchildren than daughters would.

On the other hand, if a female is in poor condition, Trivers and Willard argued, she may be better off having more daughters than sons. Sons in poor condition may fail to attract any mates at all, and may therefore leave their mother without any grandchildren. Daughters, on the other hand, will probably have at least some offspring, even if they are in poor condition.

In their hypothesis, Trivers and Willard proposed that a female could alter the sex ratio of her offspring to suit her condition (**Figure 12.10** and **Figure 12.11**). Many mammals do just this: females in prime physiological condition at the time of conception are more likely to produce sons than are females in poorer condition (Clutton-Brock

A

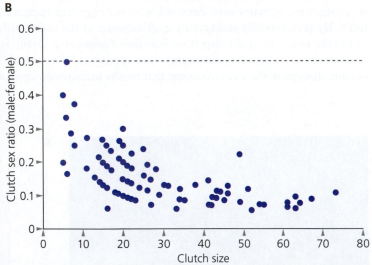

B

Figure 12.10 In some insects like bees, wasps, and ants, unfertilized eggs develop into males and fertilized eggs develop into females. (This unusual sex determination mechanism will be discussed further in Chapter 16.) Because of this unusual mechanism, gravid females can directly manipulate the sex of their offspring simply by fertilizing—or not fertilizing—their eggs before laying. Sex ratios in hymenopteran insects vary widely, and they provide superb tests of theories for sex allocation (Hamilton 1967; Charnov 1982; Herre 1987). A: Fig wasp (*Liporrhopalum tentacularis*) females lay from 10 to 80 eggs into ovules of fig flowers lining the inner surfaces of inflorescences. When these eggs hatch, males wander through the fig and mate with their sisters as they emerge. Because only a few males are needed to fertilize all of the females inside a fig, females benefit by adjusting their offspring sex ratios to produce just a few sons and as many daughters as possible. B: A 2008 study by Shazia Raja and her colleagues showed that females adjust the sex ratios of their broods still further, and the largest clutch sizes have more extreme female bias than smaller clutches. (A: Robert F. Sisson/Getty Images; data from Raja et al. 2008)

Figure 12.11 The different types of sexual selection experienced by males and females can influence the evolution of sex ratios in another way. In many species of coral reef fishes, individuals switch sexes during their lifetimes (Warner et al. 1975; Warner 1984). In bluestreak cleaner wrasse (*Labroides dimidiatus*), large males fight to guard harems of females, and they monopolize matings within their harem. As might be predicted from Trivers–Willard logic, these wrasse begin life as females, and they breed as females when they are young and relatively small. Only when an individual becomes the largest in its group will it switch from female to male—that is, when the benefits to being a male outweigh those of being a female. When this male dies, the next largest (female) individual will switch to become the new male (Robertson 1972). (Howard Chew/iStock/Getty Images)

et al. 1986; Cameron 2004). But in some species, females manipulate sex ratios in the opposite direction. Such is the case for the Seychelles warbler (*Acrocephalus sechellensis*), a bird that lives on the Seychelles Islands in the Indian Ocean (Uller et al. 2007). When a mother warbler's eggs hatch, the male and female chicks can look forward to different lives. The male birds tend to fly away from their natal home in search of female warblers and other territories. Young female birds tend to stay behind, helping their mother incubate her eggs. A mother benefits from this help because she is able to rear more chicks over her lifetime with the aid of her daughters (Brouwer et al. 2012).

Jan Komdeur, a biologist at the University of Groningen in the Netherlands, and his colleagues have been carefully chronicling the lives of all two thousand or so Seychelles warblers that live on the islands (**Figure 12.12**; Komdeur 2003). They discovered that female Seychelles warblers can adjust the balance of male and female chicks in their broods in response to their environment. An unassisted female living on a patch of land with abundant food may produce a brood that's as much as 88% daughters. But Komdeur has found that female warblers that live in places where food is scarce may produce broods in which as few as 23% of the chicks are daughters.

Komdeur hypothesized that the birds were adjusting the balance of daughters and sons to maximize their reproductive success. A female bird that lives in a high-quality territory can use the help of her daughters to produce more chicks. A female that is stuck living in a low-quality territory will be better off producing sons that can search for greener pastures. Komdeur tested his hypothesis by moving birds from low-quality territories to high-quality ones. And as he predicted, the birds switched from mostly sons to mostly daughters.

Figure 12.12 Female Seychelles warblers can adjust their ratio of sons to daughters to boost their reproductive success. (Rene van Bakel/ASAblanca/Getty Images)

There is, however, such a thing as too much help. When Seychelles warbler mothers living in high-quality territory have more than three female helpers, trouble arises. The helpers eat too much food, and they may crack the mother's eggs as they clamber around the nest. In response, the Seychelles warblers adjust the sex ratio yet again. They produce more sons that will soon fly away and not be such a burden. To test his hypothesis in a new way, Komdeur took away the helpers in some of the warbler nests. The sex ratio changed again, and in exactly the way he predicted: the unassisted birds started producing more daughters again. (See **Box 12.1** for a discussion of mechanisms of sex ratio adjustments.)

Key Concepts

- Females generally benefit more than males from parental care of offspring. Reversals in the operational sex ratio, when they occur, can also cause a reversal in the more typical roles of the sexes, providing exciting tests of sexual selection theory.

- Frequency-dependent selection can maintain variation within populations if the fitness of an allele or phenotype decreases when it is common and increases as it becomes rare. This may contribute to the persistence of two sexes (males, females) within populations.

- The Trivers–Willard hypothesis predicts greater investment in male offspring by parents in good condition and greater investment in female offspring by parents in poor condition. ●

12.3 Family Conflicts

Parental conflict occurs when parents have an evolutionary conflict of interest over the optimal strategy for parental care.

So far in this chapter, we've examined how an individual's parental strategies can raise its fitness. But the best strategy for one parent sometimes lowers the fitness of other members of its family. In Chapter 11, we saw how different mating strategies can lead to sexual conflict (Section 11.6). Now we will consider conflicts within families.

Parental Conflict

One of the most striking examples of **parental conflict** is displayed by a bird called the Eurasian penduline tit (*Remiz pendulinus*; **Figure 12.13**). Unmated males build elaborate nests and sing to attract a female. After a male and female mate, they both get to work enlarging the nest. Hanging from a bough, the nest has a narrow-mouthed opening for the mother to fly into and lay the eggs. Either the mother or the father will take on the job of incubating the eggs and then feeding the chicks once they hatch. Despite this care, a third of breeding pairs desert their eggs, leaving them to die.

Tamás Székely, a biologist at the University of Bath in England, and his colleagues have designed experiments to discover the source of this perplexing behavior (Pogány et al. 2008). They've concluded that parental conflict is responsible. Each parent can boost its total number of offspring by abandoning the nest to its partner and finding

Figure 12.13 Among penduline tits, mothers and fathers cooperate to build nests and bring food to their offspring. But their interests conflict over who should provide care the longest. Each sex has the highest reproductive success if it can abandon the nest first, leaving their mate to care for the young. But leaving too soon can be risky, because the fledglings may die without enough food. The result is a tenuous alliance replete with deceit and trickery. (nbgbgbg/Getty Images)

BOX 12.1

How Do They Do It? Mechanisms of Sex Ratio Adjustment

Although it is clear that avian and mammalian females often *do* adjust the sex of their offspring, for now it is far from clear *how* they do it. Several mechanisms appear likely, however (Krackow 1995; Pike and Petrie 2003; Cameron 2004; Uller and Badyaev 2009). There are many steps from copulation to fertilization, and each step affords females opportunities to affect how they use sperm from different males (see Chapter 11 for more details on cryptic female choice). The same opportunities arise with ovulation. Several events along the path from follicle to fertilization must unfold; **Box Figure 12.1.1** shows some of these events for a chicken. Many of these events are under the control of maternal hormones, and this control could influence the likelihood that the eggs develop into male or female.

For example, sex is determined just before a follicle is released and engulfed by the oviduct funnel. Unlike mammals, female birds determine the sex of their offspring because females are the heterogametic sex (females are ZW and males are ZZ). In birds, the chromosome segregation that results in sex determination takes place during the first stage of meiosis, which generates haploid daughter cells from a single diploid mother cell and determines whether the oocyte will receive a Z or W chromosome. During meiosis, numerous mechanisms intervene to balance segregation of the Z and W chromosomes because the size of the W chromosome is so much smaller than the Z chromosome. It now appears, however, that several of these mechanisms may be influenced by maternal hormones, providing a nongenetic means whereby a mother could bias which of her chromosomes is incorporated into the developing ova (the other ends up in a polar body), in a form of segregation distortion (Uller and Badyaev 2009).

Once the sex of an ovum is determined, the processes of yolk acquisition and growth also appear to be sensitive to maternal hormones in ways that could bias the growth of male versus female ova (Pike and Petrie 2003; Badyaev et al. 2005). Sometimes females reabsorb the yolk from eggs, effectively truncating their development, and selective reabsorption could also alter the likelihood of a male or a female ova being released (Emlen 1997). Changing hormones are responsible for triggering the release itself, and speeding up or delaying this step could determine how well developed particular ova are at the time of their release, thus possibly influencing how likely they are to be successfully fertilized (Pike and Petrie 2003). Facilitating, impeding, or even withholding stored sperm as it passes into the oviduct funnel is yet another possible means of selectively fertilizing ova.

All of these putative mechanisms could potentially influence the sex of offspring before successful fertilization. After fertilization, additional maternal investments, such as production of the shell, could be enhanced or withheld in a sex-specific manner. Birds could also engage in behaviors after fertilization that could alter the sex ratio. They might lay eggs outside of their nest, for example (called "dump laying"). Because each of these steps is sensitive in one way or another to levels of maternally released hormones, they are all viable candidates for mechanisms regulating the sex of offspring in vertebrates.

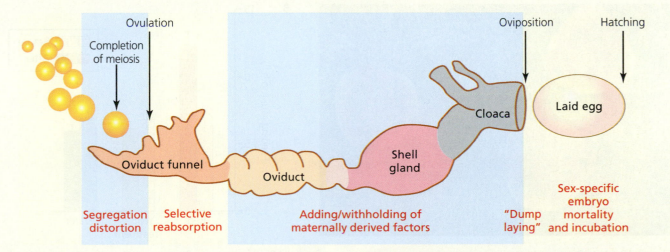

Box Figure 12.1.1 A chicken egg travels through several organs before it is ready to be laid. At each stage, the mother can potentially intervene in its development, influencing its sex or adjusting its investment.

a new bird to mate with. But deserting a nest also has its risks. If a father abandons a mother too soon, she may not be able to lay her eggs successfully. Another risk of an early departure is that another male bird will visit the mother and fertilize her remaining eggs. However, staying around too long also has its risks. If the father waits too long, he may fail to find additional free female birds left to mate with. The mother may even abandon the nest before he does, leaving him to raise the chicks by himself.

To leave at the best possible moment (for him), a male needs to keep careful track of the female's preparations for laying her eggs. In turn, strategies that make it difficult for males to keep such careful track of egg-laying preparation evolved in females. If a male sticks his head into the nest while she's laying eggs, she'll fight him off—in some cases, even kill him. As a female lays her eggs, she hides some of them in the bottom of her nest. If the male does manage to slip into the nest, he'll get the impression that the female needs more time to finish laying all her eggs.

Experimental results are consistent with the hypothesis that females hide their eggs due to parental conflict. When scientists uncover the eggs, for example, a female becomes far more aggressive in keeping the male away from the nest. Nevertheless, the male usually deserts her that very day. Now that he can see how far things have progressed, he moves off to find other females.

Parent–Offspring Conflict

In addition to the conflicts mothers may have with fathers, they can have evolutionary conflicts with their own offspring. To maximize her lifetime fitness, for example, a female bird may not provide the maximum resources to earlier broods. A chick from one of those early broods might be more likely to survive to maturity if its mother did provide it with more food, but that would lower the survival rate of siblings from later broods. Natural selection can lead to this conflict because siblings are not genetically identical—a subject we'll revisit in greater detail in Chapter 16. The bottom line: optimal allocation of parental effort differs depending on perspective. What's most beneficial for mom is often not the same as what's most beneficial to offspring (Trivers 1974; Hinde and Kilner 2007).

This **parent–offspring conflict** can help us to better understand many of the features of animal family life. Consider the way chicks beg for food. Chicks will put considerable energy into calling out to their parents, straining their bodies, and opening their mouths wide—even putting their entire families at risk of being attacked by predators (**Figure 12.14A**).

Parent–offspring conflict occurs when parents benefit from withholding parental care or resources from some offspring (for example, a current brood) and invest in other offspring (for example, a later brood). Conflict arises because the deprived offspring would benefit more if they received the withheld care or resources.

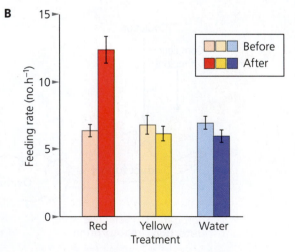

Figure 12.14 A: Chicks within a nest compete for the attention of their parents, and this can lead to loud begging calls and bright, gaping mouths and throats, as in these skylark chicks. B: Parents in many bird species prefer to feed chicks with the brightest red mouths. When Nicola Saino, a biologist at the Università degli Studi di Milano, and her colleagues applied droplets of red food coloring to the mouths of barn swallow chicks, parental feeding rates increased—a pattern not observed with yellow food coloring or when chicks were treated simply with drops of water. (A: Gabor Pozsgai/Fotolibra/ZUMA Press/Newscom; data from Saino et al. 2000)

Chicks manipulate the intensity of their begging to maximize the food they get from their parents. The intensity of begging increases as the chicks get close to fledging, for example, which is exactly when parents are likely to benefit most if they begin to pull back on their efforts.

Offspring–Offspring Conflict

Begging also allows chicks to compete against members of their own brood in an example of **offspring–offspring conflict**. If parents preferentially feed chicks with the loudest calls or those with the biggest, brightest, gaping mouths, then selection should favor chicks able to outshine their nest mates (Macnair and Parker 1979). In barn swallows, for example, chicks display the insides of their mouths and throats to their parents when they beg. These tissues are red, and their brightness turns out to be an honest indicator of health. Chicks fighting infections or those with compromised immune functions have duller mouths than their healthier nest mates (Saino et al. 2000). When given a choice, parents prefer to give their food to nestlings with the brightest throats (**Figure 12.14B**; Saino et al. 2000).

Competition for parental efforts has become especially intense among birds called coots, because parents routinely produce many more offspring than they can effectively feed. The faces of juvenile coots are surrounded by bright red feathers, enhancing the stimulus provided by their gaping mouths (**Figure 12.15**). Like swallows, coot parents prefer to feed their brightest chicks. Bruce Lyon, a biologist at the University of California, Santa Barbara, trimmed the bright tips from feathers of some of the chicks, causing them to be black instead of red. In response, the parents avoided feeding the black chicks and instead gave their food to the red nest mates beside them (Lyon et al. 1994).

Offspring–offspring conflict (sibling rivalry) occurs when siblings compete for parental care or limited resources.

- Parental care creates the opportunity for many kinds of conflict—between parents, among siblings, and between parents and offspring. ●

Key Concept

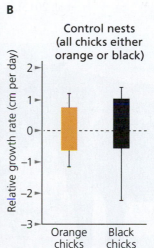
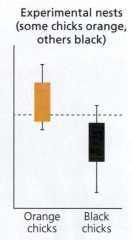

Figure 12.15 A: Sibling competition for parental attention reaches an extreme in American coots (*Fulica americana*), where chicks complement their colorful mouths with bright orange-red plumage surrounding their faces. Here, too, parents prefer to feed the brightest chicks in the nest. B: By trimming the bright tips of the feathers Bruce Lyon could cause chicks to appear black instead of orange. If he trimmed all of the chicks in a nest, parental feeding rates did not change. But when he trimmed only some of the chicks, their parents most often fed the brighter siblings beside them; the dull-colored chicks ate less often, resulting in slower growth rates. (A: Bruce Lyon, Ecology & Evolutionary Biology, University of California at Santa Cruz; data from Lyon et al. 1994)

12.4 Conflicts Within the Genome

Parent-of-origin effect describes an effect on the phenotype of an offspring caused by an allele inherited from a particular parent.

Genomic imprinting occurs when genes inherited from one or the other parent are silenced due to methylation. Imprinting can result in offspring who express either the maternal or paternal copy of the gene, but not both.

Methylation is the process by which methyl groups are added to certain nucleotides. Methylation is associated with altered gene expression, thereby reducing or eliminating the production of proteins or RNA molecules.

Parental conflict may help explain a long-standing mystery first discovered by mule breeders three thousand years ago. Mules are hybrids, produced from a cross between a horse and a donkey. But mule breeders must take care to use a male donkey and a female horse. If they use a female donkey and a male horse instead, they will get a different hybrid, known as a hinny, which has a thick mane, stout back legs, and short ears (**Figure 12.16**). Since that long-ago discovery was made, animal breeders and scientists have found many other cases of this **parent-of-origin effect** (Mott et al. 2014).

At first, this mismatch seems to defy the rules of genetics. Genes should have the same phenotypic effects on an organism, no matter which parent it inherits them from. But some genes are an exception to the rule. In a phenomenon known as **genomic imprinting**, the two copies of an imprinted gene behave differently from one another. In a particular organ, the father's copy may be silenced while the mother's is expressed. In other cases, the reverse is true (Barlow and Bartolomei 2014).

Imprinting

Imprinting occurs through **methylation**, a process we first discussed in Chapter 5. In sperm and eggs, methyl groups are attached to certain cytosine nucleotides in the DNA. After fertilization, many genes maintain the differential pattern of methylation across the maternal and paternal copies of genes. For the most part, both copies of a gene share the same pattern of methylation. But scientists estimate that several hundred genes have differing methylation patterns between the sexes (DeVeale et al. 2012).

Scientists are only beginning to link specific imprinted genes to specific parent-of-origin effects. Even though there are probably only a few hundred imprinted genes, they appear to have a widespread influence on many traits (Mott et al. 2014). That's probably because imprinted genes interact with many other nonimprinted ones. Furthermore, many imprinted genes play a role in development, where they can have long-term effects on an organism. The differences between mules and hinnies, for example, have been linked to the activity of imprinted genes in the placenta (Wang et al. 2013).

Figure 12.16 Mules (A) and hinnies (B) are both the hybrid offspring of horses and donkeys. But their appearance differs because the species that serves as the father and mother are reversed. (Mules have donkeys for fathers; hinnies have horses.) In each species, different sets of genes are imprinted in males and females, leading to different patterns of gene expression in their hybrid offspring. (A: Royer/Getty Images; B: Zoonar GmbH/Alamy Stock Photo)

Another sign of the importance of imprinted genes is how devastating their disruption can be. If a mother's copy of a gene is normally silenced, the deletion of the father's copy can leave an offspring with no active copy of the gene whatsoever. As we'll discuss in more detail in Chapter 18, a range of genetic disorders have been traced to disruptions in imprinted genes (Ishida and Moore 2013).

Evolutionary biologists are developing several theories to explain how some genes became imprinted (Patten et al. 2014). David Haig, of Harvard University, argues one of these theories, arguing that it is a molecular version of parental conflict in placental mammals (Wilkins and Haig 2003).

Like mothers in other species, female mammals can maximize their lifetime fitness by moderating the amount of investment they put into individual offspring. In the case of mammals, this investment includes nutrients mothers provide to embryos through the placenta. It also includes milk they produce to nurse their offspring after birth. If a female invests too much of these resources on one offspring, she may put her own health and future reproductive success at risk. Haig proposed that natural selection favors adaptations that let mothers rein in the nutrition they supply to their offspring.

A father, on the other hand, benefits if his mate puts lots of energy into her current pregnancy. Chances are good that her current offspring are his, but there's little guarantee that her future offspring will be his as well. So even if the mother's health is harmed, the father benefits from more energy going to his offspring. They'll be healthier as a result, and more likely to survive until adulthood.

Haig proposed that differential methylation can alter the growth of embryos in a way that raises the fitness of either fathers or mothers. This is because embryo growth is regulated by a network of genes in which some genes stimulate growth and others slow it (see Chapter 9).

A mother can theoretically rein in the growth of her embryos by passing down a silenced copy of a growth-promoting gene. Instead of two working copies of the gene, the embryo will have only one. As a result, it produces only half the growth-promoting proteins it would have if both copies of the gene were active. Theoretically, the opposite tactic would be better for fathers. If they passed down a silenced copy of a growth-inhibiting gene, they would cut the supply of the growth-inhibiting proteins in embryos and accelerate the growth of their offspring.

Igf2 and Igf2r

The best way to test Haig's hypothesis is to look at how imprinted genes affect the fitness of mothers and fathers. Biologists have uncovered these details for only a few genes. Two of these genes provide compelling support for Haig's hypothesis.

One of these gene products, called insulin-like growth factor 2 (*Igf2*), is produced by fetus-derived cells that invade the lining of the uterus, where they extract nutrients from the mother. Normally, only the father's copy is active; the female's is silenced. To understand the gene's function, scientists disabled the father's copy in the placenta of fetal mice. Without the *Igf2* gene to help draw nutrients from their mothers, the mice were born weighing 40% below average. It's possible that the mother's copy of *Igf2* is silent because turning it off helps to slow the growth of a fetus (**Figure 12.17**).

On the other hand, mice carry another gene called *Igf2r*, which interferes with the growth-spurring activity of *Igf2*. This gene may have evolved to provide defense to the mother, reducing the damage to her body from excessive growth of offspring. In the case of *Igf2r*, it is the father's gene that is silent, perhaps as a way for fathers to speed up the growth of their offspring. If the mother's copy of *Igf2r* is disabled, mouse pups are born 125% heavier than average.

These experiments confirm the predictions of Haig's hypothesis. But other researchers have developed alternative explanations. Troy Day, of Queens University in Canada, and Russell Bonduriansky, of the University of New South Wales in Australia, have developed a model that they've dubbed the sexual antagonism hypothesis (Day and Bonduriansky 2004; Patten et al. 2014).

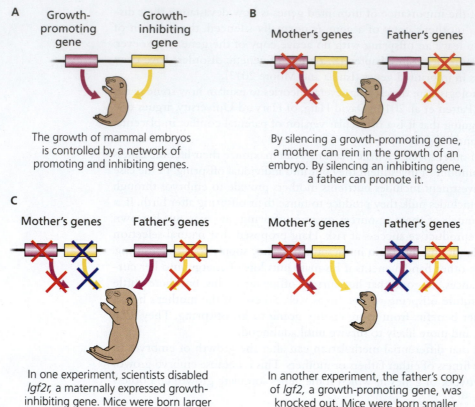

A

Growth-promoting gene | Growth-inhibiting gene

The growth of mammal embryos is controlled by a network of promoting and inhibiting genes.

B

Mother's genes | Father's genes

By silencing a growth-promoting gene, a mother can rein in the growth of an embryo. By silencing an inhibiting gene, a father can promote it.

C

Mother's genes | Father's genes

In one experiment, scientists disabled *Igf2r*, a maternally expressed growth-inhibiting gene. Mice were born larger than normal.

Mother's genes | Father's genes

In another experiment, the father's copy of *Igf2*, a growth-promoting gene, was knocked out. Mice were born smaller than normal.

Figure 12.17 Several hundred genes in our genome are imprinted. In other words, the copy from one parent is silenced, whereas the other is active. David Haig and other researchers have theorized that imprinted genes evolve through an evolutionary conflict between parents, which is resolved in their offspring. A: Haig's hypothesis focuses on genes involved in promoting or inhibiting the growth of embryos and infants. B: By contributing a silenced copy of a growth-inhibiting gene, a father can theoretically accelerate the growth rate of his offspring. Likewise, a female can benefit if she contributes a silenced copy of a growth-promoting gene because that will rein in the growth of the embryos. C: Experiments on genes called *Igf2* and *Igf2r* are consistent with their hypothesis. Red Xs mark genes silenced by imprinting. Blue Xs mark genes knocked out in experiments. (Data from Wilkins and Haig 2003)

Intralocus sexual conflict is a conflict between the fitness effects of alleles of a given locus on males and females.

They base their hypothesis on the fact that males and females of a given species depend on different traits to raise their fitness. At some loci, certain alleles can raise the fitness of females, whereas other alleles may raise the fitness of males. Likewise, a female-fitness allele can *lower* the fitness of males, and vice versa. This is yet another evolutionary conflict of interest, known as **intralocus sexual conflict** (Bonduriansky and Chenoweth 2009).

If a female passes down a high female-fitness allele to her sons, she may lower his fitness. The reverse is true for daughters and their fathers. As a result, natural selection should favor loci that can silence maternally inherited alleles in males and other loci that silence paternally inherited alleles in females.

We know more about imprinted genes today than people did back when mules and hinnies were first bred. Still, it will take a lot more research before scientists can determine why those genes evolved.

Key Concepts

- Parents sometimes have conflicting interests when it comes to how much a mother should invest in her offspring. Males benefit when they can maximize a female's parental contribution to current offspring, but females benefit when they save some of their resources for later offspring.

- Parental conflict can lead to battles over the control of gene expression in offspring through genomic imprinting. ●

12.5 Searching for Immortality Genes

Life history theory can guide scientists in the search for answers to a wide range of questions. One of the most intriguing of those questions is what determines how long we live. Animals have a wide range of life spans: fruit flies live for only a few weeks, whereas 500-year-old clams have been discovered off the coast of Iceland. Yet the aging process is strikingly similar throughout the animal kingdom. As animals get older, for example, cells accumulate malformed proteins. The immune system becomes less effective at fighting infections. Rates of noninfectious diseases like cancer go up. These parallel declines hint that the same evolutionary factors are shaping the **senescence**, or deterioration associated with aging, of all species.

To explore the evolution of senescence, scientists have also searched for ways to extend the life spans of animals (Bartke 2011). They have discovered that if they dramatically reduce the calories in an animal's diet, it often lives much longer. Some experiments suggest that restricting an animal's diet triggers a special response in its cells. The cells begin to produce proteins that can repair the damage caused by the stress of not getting enough to eat.

This stress-fighting response appears to be an ancient strategy, given that the same genes can be found in animals ranging from nematode worms to mice. These genes may have evolved as a way to cope with famines and droughts. Scientists who study nematode worms have found that they can double the life span of the animals with mutations that keep these stress-response genes switched on. The genes may be able to repair damage to cells continually, fighting the effects of aging. Restricting calories may have the same effect, by keeping the genes switched on permanently.

Other genes may provide lifesaving functions early in life but contribute to higher rates of cancers and other noninfectious diseases later. The p53 tumor-suppressor protein is effective at defending against cancer in humans and other vertebrates, for example. It responds to stress inside cells, particularly to damaged DNA, which may signal the first steps toward cancer, and it can cause a cell to die or to stop dividing. In either case, p53 prevents the cell from possibly growing into a tumor, but it takes a toll. As the years pass, p53 can kill or stunt so many cells that tissues can no longer renew themselves. By forcing cells into early retirement, p53 may prevent them from becoming tumors, but the cells may damage surrounding tissue and even release abnormal proteins that stimulate the growth of other cancer cells (Rodier et al. 2007).

In other words, p53 is an effective stopgap defense against cancer. It helps keep young animals relatively cancer free. But it also damages the body in the process. The damage accumulates slowly, over the course of many years. By the time p53 has this detrimental impact, animals are so old that in general they have already left offspring, so natural selection cannot act to eliminate this protein. Senescence, then, is a trade-off.

We humans value long life. But Nicole Jenkins, a biologist at the Buck Institute for Age Research in California, and her colleagues have found that longevity-extending mutations lower the early reproductive fitness of nematodes (Jenkins et al. 2004). The scientists put 50 of the long-lived worms in a dish with 50 normal worms and let them breed. Jenkins and her colleagues then randomly picked out 100 of the eggs and used them to rear the next generation. The scientists found that the long-lived worms experienced a reduction in fertility shortly after they reached sexual maturity. Later in life, there was no difference in fertility. Yet this difference in early reproduction had a huge effect on overall fitness. Within just a few generations, the long-lived worms had vanished from the dish. They had been outcompeted by the shorter-lived worms. Natural selection, once again, did not favor long life simply for long life's sake.

Senescence refers to the deterioration in the biological functions of an organism as it ages.

- Senescence is thought to be a by-product of selection for alleles that enhance growth and reproduction early in life. It arises because of a trade-off between investing in reproduction and investing in body maintenance. ●

Key Concept

12.6 Menopause: Why Do Women Stop Having Children?

Humans are remarkably long lived for primates. Chimpanzees, our closest living relatives, begin to show signs of old age in their mid-30s, and the oldest confirmed chimpanzee lived for 59.4 years. The average life span of women in the United States is 80 years, and the record for any human is more than 122 years (de Magalhães and Costa 2009). Yet after about age 50, women can no longer reproduce—a phenomenon known as menopause. Scientists have long puzzled over the lengthy post-reproductive life of women. In chimpanzees, the fertility of female chimpanzees declines as well, but only at the end of life when their entire bodies are deteriorating due to old age (**Figure 12.18**).

A number of evolutionary biologists have investigated menopause, seeking to use life history theory to explain this phenomenon (Austad 1994). Menopause is not the result of better hygiene, nutrition, or medicine made possible by modern civilization. Women in hunter-gatherer societies experience menopause as reliably as women living in affluent cities. The universality of menopause strongly suggests that it is a biological feature of our species; it evolved at some point after our ancestors branched off from other apes some seven million years ago.

Evolutionary biologists have proposed several different hypotheses to explain our species' unusual life history (Kirkwood and Shanley 2010). In 1957, George Williams first proposed what came to be one of the most influential of these hypotheses. Human babies, he observed, command especially high parental investment from mothers because they are born helpless but require a steady, energy-rich diet for their developing brains. (We'll discuss human brain evolution in more detail in Chapter 17.) Williams pointed out that a woman in her late 40s would have a harder time caring for a newborn than she would in her 20s. What's more, she would have to divide her limited resources between rearing babies and continuing to care for her older children. Williams suggested that natural selection favored mutations that reduced a woman's fertility so that she could focus on raising her older children instead (Williams 1957). Other researchers have elaborated on the so-called mother hypothesis (Pavard et al. 2008). They argued that older women who became pregnant faced a much more direct threat to their reproductive success: they were more likely to die in childbirth, leaving their other children at grave risk of dying as well.

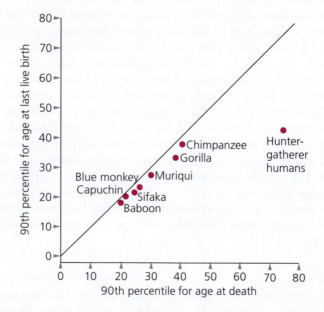

Figure 12.18 Female monkeys and apes typically don't live long after their last live birth. But as this graph shows, humans are an exception to this rule. Women undergo menopause and may live for decades more. (Data from Alberts et al. 2013)

But other researchers have tested the mother hypothesis and found it wanting. Virpi Luumaa, of the University of Sheffield, and her colleagues analyzed historical records from Canada and Finland to measure the costs older women pay for child-bearing. The records from Finland covered the years from 1741 to 1908, and those from Canada stretched from the nineteenth century into the twentieth. The mothers covered in these records generally lived on farms without access to modern medicine. Although Luumaa and her colleagues found that the risk of death from childbirth was real, it was small. At age 50, women faced only a 1%–2% risk of dying during labor. And even when women did die, the effect on their surviving children was also small. Only children who were not yet weaned faced an increased risk of death if their mothers died in giving birth to another child. Because humans have traditionally lived in extended families, the scientists concluded, children who lose their mother can get sufficient care from other family members (Lahdenperä et al. 2011).

Luumaa and her colleagues suggested a variant on the mother hypothesis—known as the grandmother hypothesis—that may explain prolonged life span but not neces-sarily menopause. For most of human history, many of the women who were still alive in their late 40s had become grandmothers. By helping their children raise their grandchildren, older women could increase their total genetic contribution to future generations (a concept called inclusive fitness, which we will discuss in Chapter 16). In their studies of historical records, Luumaa and her colleagues found that children were more likely to survive until adulthood if their grandmothers were still alive (Lahdenperä et al. 2004).

Some researchers are trying to understand the evolution of menopause by looking beyond humans to the other species that experience it. So far, they've identified only two other species: fin whales and killer whales. Darren Croft and Lauren Brent, of the University of Exeter, have followed groups of killer whales for years and reviewed decades of observations by other scientists to find evolutionary advantages that females experience by going through menopause (Brent et al. 2015).

Intriguingly, they've found that in each group of killer whales, the menopausal females are more likely to take the lead. The other whales follow in their path. Killer whale families stick together for decades, so many of the whales that follow the meno-pausal females are their own offspring. When food is hard to find, the odds that a group of killer whales is led by menopausal females go up. It's possible, Brent and Croft speculate, that female killer whales that outlive their reproductive years can still help their descendants by making use of the knowledge of the ocean that they've acquired over decades. Perhaps human grandmothers offered the same valuable resource: information.

Menopause can be a major factor in the health of older women, often leading to bone loss and other adverse symptoms. As medical researchers search for ways to improve the well-being of postmenopausal women, evolutionary biologists hope to aid their efforts by understanding the evolutionary history that gave rise to this feature of our life histories. As we'll see in Chapter 18, life history theory can help us better understand many medical conditions.

- Among primates, menopause is a unique biological feature of humans that likely evolved at some point after our ancestors branched off from other apes some seven million years ago. ●

Key Concept

TO SUM UP . . .

- The timing and duration of key events in an organism's lifetime are traits that can evolve through natural selection.

- Trade-offs can arise when the timing of one event, such as age at first breeding, reduces the duration of another event, such as life span.

- The number of offspring that females produce and the timing of their reproduction are also subject to selection.

- When males invest more resources into offspring than females do, the sex roles can reverse, supporting sexual selection theory and highlighting evolutionary trade-offs.

- Females in some species adjust the ratio of sons to daughters they produce in ways that may maximize the fitness they derive from their parental investment.

- Males and females each have optimal strategies for reproductive success. Sometimes those strategies come into conflict, as when birds abandon a nest to find another mate.

- In sexually reproducing organisms, parents are equally related to all their offspring (when there are no extra-pair matings), but siblings are at most half related. These differences may lead to parent–offspring conflicts, where siblings compete among each other for parental care.

- Genomic imprinting may have evolved in response to conflict between males and females during reproduction.

- As a result of life history trade-offs, the biological functions of an organism change as the organism ages, leading to senescence. Because these trade-offs appear relatively late in an organism's life, after reproduction, the capacity for natural selection to influence their effects is limited.

- Unlike other primates, human females experience menopause. Whether menopause is adaptive is still up for debate.

MULTIPLE CHOICE QUESTIONS   Answers can be found at the end of the book.

1. Which statement is the best way to explain why an understanding of life history trade-offs is important when examining the fitness of an organism?

   a. Individuals need to live longer, so they will not produce the maximum number of offspring each season.

   b. When individuals need offspring, they can choose whether to reproduce or not.

   c. Individuals may not be producing the maximum number of offspring they possibly can in any particular breeding season.

   d. In any particular breeding season, individuals may need to produce fewer offspring to enhance their own survival.

2. What was Austad's prediction regarding opossums on Sapelo Island?

   a. Opossums on Sapelo Island would mature later and have fewer offspring per season than opossums on the mainland.

   b. Opossums on Sapelo Island would have more offspring per season than opossums on the mainland because there were no predators.

   c. Natural selection would favor opossums with more stretchable muscle fibers on Sapelo Island.

   d. Opossums on Sapelo Island would have higher fitness than opossums on the mainland.

3. Use Figure 12.5 to predict the evolution of life history traits that a population of lake trout (a popular food fish) might experience after a new lakeside fishing resort was built.

   a. Because the fishermen act like predators, the lake trout should produce more offspring that are bigger in size.

   b. Because the fishermen act like predators, the lake trout will spend less time breeding and should produce fewer offspring.

   c. Because the fishermen act like predators, the lake trout should produce more offspring that are smaller in size and the trout will mature at smaller sizes.

   d. Nothing will happen to the lake trout.

4. Why do male gobies dig up and devour eggs in their nests?

   a. The male needs the nutrition, and he can always mate again later.

   b. If there is not enough oxygen in the water, the young won't survive anyway.

   c. Male gobies that devour some of their eggs in response to poor environmental conditions tend to have more surviving offspring than male gobies that do not.

   d. The male eats the young for the betterment of the species.

5. Which of these statements about gene imprinting is *true*?

   a. Gene imprinting is a male reproductive strategy that can reduce the lifetime reproductive success of females.

   b. Gene imprinting is a female reproductive strategy that can reduce the lifetime reproductive success of males.

   c. Gene imprinting is found only in placental mammals.

   d. Both a and b.

6. How does an evolutionary perspective help us understand the aging process?

   a. The theory of evolution allows scientists to compare closely related species that differ in how they invest in reproduction and in body maintenance to test for effects on the aging process.

   b. The theory of evolution generates predictions about how individuals with certain suites of life history characteristics might fare in the aging process when compared to other individuals without those characteristics.

   c. The theory of evolution provides insight into the historical development of adaptations, such as the number of offspring an individual has, that may contribute to life history trade-offs, such as aging.

   d. All of the above.

7. Do all scientists agree that menopause is adaptive?

   a. Yes. Menopause has to be adaptive because life history theory predicts that women should not outlive their reproductive capacity, so menopause must have some function.

   b. Yes. Menopause has to be adaptive because humans, not chimpanzees, experience menopause, and humans are evolutionarily more advanced than chimpanzees.

   c. No. Several hypotheses for menopause as an adaptation have been proposed, but they have not been tested.

   d. No. Several hypotheses for menopause as an adaptation have been proposed, but more evidence is necessary to support any one of them.

8. Investment in reproduction can come at the expense of

   a. growth.

   b. body maintenance.

   c. future fertility.

   d. All of the above.

9. The Trivers–Willard hypothesis predicts which of the following?

   a. Greater investment in male offspring by parents in good condition.

   b. Greater investment in female offspring by parents in good condition.

   c. Equal investment in male and female offspring by parents in good condition.

   d. Equal investment in male and female offspring by parents in poor condition.

10. Senescence describes all of the following conditions *except*

   a. aging cells accumulating malformed proteins.

   b. a less effective immune system.

   c. the onset of Alzheimer's disease.

   d. higher rates of cancer.

11. The figures that follow show data on the distribution of reproductive success (number of offspring sired, or number of mates) across individuals within a population. Which sex has the greater variance in reproductive success in each case? Why is this relevant to the strength and direction of sexual selection likely to be acting in each species?

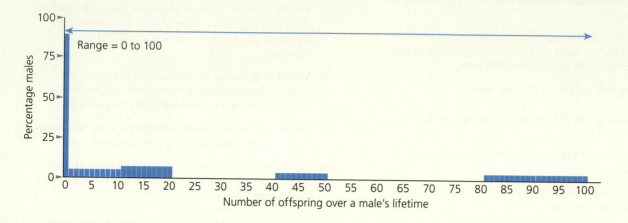

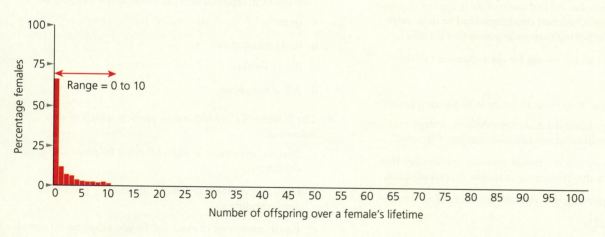

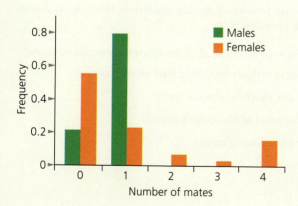

SHORT ANSWER QUESTIONS   Answers can be found at the end of the book.

12. How does natural selection act to optimize reproductive fitness of individuals in light of life history trade-offs?

13. Discuss how sexual conflict may shape the reproductive strategies of male and female gulf pipefishes.

14. From your understanding of the mechanisms of sex ratio adjustment, how might you test whether natural selection may be operating to drive sex ratios in Seychelles warblers? What is your hypothesis?

15. Based on your understanding of genomic imprinting, how might you explain the parent-of-origin effect observed when horses and donkeys are hybridized?

16. What is senescence? Give an example of how evolutionary factors are affecting senescence in a species.

17. Does the following statement accurately reflect the current understanding of the evolution of life histories? Why or why not? "An individual female chooses how many pups to raise because she needs to have high fitness over her lifetime."

ADDITIONAL READING

Charnov, E. 1982. *The Theory of Sex Allocation*. Princeton, NJ: Princeton University Press.

Flatt, T., and A. Heyland. 2011. *Mechanisms of Life History Evolution: The Genetics and Physiology of Life History Traits and Trade-Offs*. London: Oxford University Press.

Hamilton, W. D. 1967. Extraordinary Sex Ratios. *Science* 156:477–88.

Roff, D. A. 1992. *The Evolution of Life Histories, Theory and Analysis*. New York: Chapman and Hall.

———. 2002. *Life History Evolution*. Sunderland, MA: Sinauer Associates.

Royle, N. J., P. T. Smiseth, and M. Kölliker, eds. 2012. *The Evolution of Parental Care*. Oxford: Oxford University Press.

Stearns, S. C. 1992. *The Evolution of Life Histories*. London: Oxford University Press.

Williams, G. C. 1966. *Adaptation and Natural Selection: A Critique of Some Current Evolutionary Thought*. Princeton, NJ: Princeton University Press.

PRIMARY LITERATURE CITED IN CHAPTER 12

Alberts, S. C., J. Altmann, D. K. Brockman, M. Cords, L. M. Fedigan, et al. 2013. Reproductive Aging Patterns in Primates Reveal That Humans Are Distinct. *Proceedings of the National Academy of Sciences USA* 110:13440–5.

Austad, S. N. 1993. Retarded Senescence in an Insular Population of Virginia Opossums (*Didelphis virginiana*). *Journal of Zoology* 229 (4): 695–708.

———. 1994. Menopause: An Evolutionary Perspective. *Experimental Gerontology* 29 (3-4): 255–63.

Badyaev, A. V., H. Schwabl, R. L. Young, R. A. Duckworth, K. J. Navara, et al. 2005. Adaptive Sex Differences in Growth of Pre-ovulation Oocytes in a Passerine Bird. *Proceedings of the Royal Society B: Biological Sciences* 272:2165–72.

Barlow, D. P., and M. S. Bartolomei. 2014. Genomic Imprinting in Mammals. *Cold Spring Harbor Perspectives in Biology* 6 (2): a018382.

Bartke, A. 2011. Single-Gene Mutations and Healthy Ageing in Mammals. *Philosophical Transactions of the Royal Society B: Biological Sciences* 366 (1561): 28–34.

Bonduriansky, R., and S. F. Chenoweth. 2009. Intralocus Sexual Conflict. *Trends in Ecology and Evolution* 24 (5): 280–8.

Brent, L. J., D. W. Franks, E. A. Foster, K. C. Balcomb, M. A. Cant, et al. 2015. Ecological Knowledge, Leadership, and the Evolution of Menopause in Killer Whales. *Current Biology* 25 (6): 746–50.

Brouwer, L., D. S. Richardson, and J. Komdeur. 2012. Helpers at the Nest Improve Late-Life Offspring Performance: Evidence from a Long-Term Study and a Cross-Foster Experiment. *PLoS One* 7 (3): e33167.

Cameron, E. Z. 2004. Facultative Adjustment of Mammalian Sex Ratios in Support of the Trivers–Willard Hypothesis: Evidence for a Mechanism. *Proceedings of the Royal Society B: Biological Sciences* 271 (1549): 1723–8.

Charnov, E. 1982. *The Theory of Sex Allocation*. Princeton, NJ: Princeton University Press.

Clutton-Brock, T. H., S. D. Albon, and F. E. Guinness. 1986. Great Expectations: Dominance, Breeding Success and Offspring Sex Ratios in Red Deer. *Animal Behaviour* 34 (2): 460–71.

Cox, R. M., and R. Calsbeek. 2010. Severe Costs of Reproduction Persist in *Anolis* Lizards Despite the Evolution of a Single-Egg Clutch. *Evolution* 64:1321–30.

Day, T., and R. Bonduriansky. 2004. Intralocus Sexual Conflict Can Drive the Evolution of Genomic Imprinting. *Genetics* 167 (4): 1537–46.

de Magalhães, J. P., and J. Costa. 2009. A Database of Vertebrate Longevity Records and Their Relation to Other Life-History Traits. *Journal of Evolutionary Biology* 22 (8): 1770–4.

DeVeale, B., D. Van Der Kooy, and T. Babak. 2012. Critical Evaluation of Imprinted Gene Expression by RNA-Seq: A New Perspective. *PLoS Genetics* 8 (2012): e1002600.

Dial, T. R., L. P. Hernandez, and E. L. Brainerd. 2017. Morphological and Functional Maturity of the Oral Jaws Covary with Offspring Size in Trinidadian Guppies. *Scientific Reports* 7 (1): 5771.

Emlen, S. T. 1997. When Mothers Prefer Daughters over Sons. *Trends in Ecology and Evolution* 12 (8): 291–2.

Emlen, S. T., and P. H. Wrege. 2004. Size Dimorphism, Intrasexual Competition, and Sexual Selection in Wattled Jacana (*Jacana jacana*), a Sex-Role-Reversed Shorebird in Panama. *The Auk* 121 (2): 391–403.

Fisher, R. A. 1930. *The Genetical Theory of Natural Selection*. Oxford: Clarendon Press.

Forbes, L. S. 1997. The Evolutionary Biology of Spontaneous Abortion in Humans. *Trends in Ecology and Evolution* 12 (11): 446–50.

Hamilton, W. D. 1967. Extraordinary Sex Ratios. *Science* 156 (3774): 477–88.

Herre, E. A. 1987. Optimality, Plasticity and Selective Regime in Fig Wasp Sex Ratios. *Nature* 329 (6140): 627–9.

Hinde, C. A., and R. M. Kilner. 2007. Negotiations Within the Family over the Supply of Parental Care. *Proceedings of the Royal Society B: Biological Sciences* 274:53–60.

Ishida, M., and G. E. Moore. 2013. The Role of Imprinted Genes in Humans. *Molecular Aspects of Medicine* 34 (4): 826–40.

Jenkins, N. L., G. McColl, and G. J. Lithgow. 2004. Fitness Cost of Extended Lifespan in *Caenorhabditis Elegans*. *Proceedings of the Royal Society B: Biological Sciences* 271 (1556): 2523–6.

Jones, A. G., D. Walker, and J. C. Avise. 2001. Genetic Evidence for Extreme Polyandry and Extraordinary Sex-Role Reversal in a Pipefish. *Proceedings of the Royal Society B: Biological Sciences* 268 (1485): 2531–5.

Kirkwood, T. B., and S. N. Austad. 2000. Why Do We Age? *Nature* 408 (6809): 233.

Kirkwood, T. B. L., and R. Holliday. 1979. The Evolution of Ageing and Longevity. *Proceedings of the Royal Society B: Biological Sciences* 205 (1161): 531–46.

Kirkwood, T. B. L., and D. P. Shanley. 2010. The Connections Between General and Reproductive Senescence and the Evolutionary Basis of Menopause. *Annals of the New York Academy of Sciences* 1204 (1): 21–9.

Klug, H., K. Lindström, and C. M. S. Mary. 2006. Parents Benefit from Eating Offspring: Density-Dependent Egg Survivorship Compensates for Filial Cannibalism. *Evolution* 60 (10): 2087–95.

Komdeur, J. 2003. Daughters on Request: About Helpers and Egg Sexes in the Seychelles Warbler. *Proceedings of the Royal Society B: Biological Sciences* 270:3–11.

Krackow, S. 1995. Potential Mechanisms for Sex Ratio Adjustment in Mammals and Birds. *Biological Reviews* 70 (2): 225–41.

Lahdenperä, M., V. Lummaa, S. Helle, M. Tremblay, and A. F. Russell. 2004. Fitness Benefits of Prolonged Post-Reproductive Lifespan in Women. *Nature* 428 (6979): 178–81.

Lahdenperä, M., A. F. Russell, M. Tremblay, and V. Lummaa. 2011. Selection on Menopause in Two Premodern Human Populations: No Evidence for the Mother Hypothesis. *Evolution* 65 (2): 476–89.

Lyon, B. E., J. M. Eadie, and L. D. Hamilton. 1994. Parental Choice Selects for Ornamental Plumage in American Coot Chicks. *Nature* 371:240–3.

Macnair, M. R., and G. A. Parker. 1979. Models of Parent-Offspring Conflict. III. Intra-brood Conflict. *Animal Behaviour* 27:1202–9.

Mott, R., W. Yuan, P. Kaisaki, X. Gan, J. Cleak, et al. 2014. The Architecture of Parent-of-Origin Effects in Mice. *Cell* 156 (1): 332–42.

Paczolt, K. A., and A. G. Jones. 2010. Post-copulatory Sexual Selection and Sexual Conflict in the Evolution of Male Pregnancy. *Nature* 464 (7287): 401–4.

Patten, M. M., L. Ross, J. P. Curley, D. C. Queller, R. Bonduriansky, et al. 2014. The Evolution of Genomic Imprinting: Theories, Predictions and Empirical Tests. *Heredity* 113:119–28.

Pavard, S., C. J. E. Metcalf, and E. Heyer. 2008. Senescence of Reproduction May Explain Adaptive Menopause in Humans: A Test of the "Mother" Hypothesis. *American Journal of Physical Anthropology* 136 (2): 194–203.

Pike, T. W., and M. Petrie. 2003. Potential Mechanisms of Avian Sex Manipulation. *Biological Reviews* 78 (4): 553–74.

Pogány, A., S. István, K. Jan, and T. Székely. 2008. Sexual Conflict and Consistency of Offspring Desertion in Eurasian Penduline Tit *Remiz pendulinus*. *BMC Evolutionary Biology* 8 (1): 242.

Quenby, S., G. Vince, R. Farquharson, and J. Aplin. 2002. Recurrent Miscarriage: A Defect in Nature's Quality Control? *Human Reproduction* 17 (8): 1959–63.

Raja, S., N. Suleman, S. G. Compton, and J. C. Moore. 2008. The Mechanism of Sex Ratio Adjustment in a Pollinating Fig Wasp. *Proceedings of the Royal Society B: Biological Sciences* 275 (1643): 1603–10.

Reznick, D. 2011. Guppies and the Empirical Study of Adaptation. In J. Losos (ed.), *In the Light of Evolution: Essays from the Laboratory and the Field* (pp. 205–32). Greenwood Village, CO: Roberts and Company.

Reznick, D. N., M. Bryant, and D. Holmes. 2006. The Evolution of Senescence and Post-reproductive Lifespan in Guppies (*Poecilia reticulata*). *PLoS Biology* 4 (1): e7.

Reznick, D. N., H. Bryga, and J. A. Endler. 1990. Experimentally Induced Life-History Evolution in a Natural Population. *Nature* 346 (6282): 357–9.

Reznick, D. N., C. K. Ghalambor, and K. Crooks. 2008. Experimental Studies of Evolution in Guppies: A Model for Understanding the Evolutionary Consequences of Predator Removal in Natural Communities. *Molecular Ecology* 17 (1): 97–107.

Robertson, D. R. 1972. Social Control of Sex Reversal in a Coral-Reef Fish. *Science* 177 (4053): 1007–9.

Rodier, F., J. Campisi, and D. Bhaumik. 2007. Two Faces of p53: Aging and Tumor Suppression. *Nucleic Acids Research* 35 (22): 7475–84.

Saino, N., P. Ninni, S. Calza, R. Martinelli, F. De Bernardi, et al. 2000. Better Red Than Dead: Carotenoid-Based Mouth Coloration Reveals Infection in Barn Swallow Nestlings. *Proceedings of the Royal Society B: Biological Sciences* 267:57–61.

Trivers, R. L. 1974. Parent-Offspring Conflict. *American Zoologist* 14:249–64.

Trivers, R. L., and D. E. Willard. 1973. Natural Selection of Parental Ability to Vary the Sex Ratio of Offspring. *Science* 179:90–2.

Uller, T., and A. V. Badyaev. 2009. Evolution of "Determinants" in Sex Determination: A Novel Hypothesis for the Origin of Environmental Contingencies in Avian Sex-Bias. *Seminars in Cell and Developmental Biology* 20 (3): 304–12.

Uller, T., P. Ido, W. Erik, W. B. Leo, and K. Jan. 2007. The Evolution of Sex Ratios and Sex-Determining Systems. *Trends in Ecology and Evolution* 22 (6): 292–7.

Vallon, M., N. Anthes, and K. U. Heubel. 2016. Water Mold Infection but Not Paternity Induces Selective Filial Cannibalism in a Goby. *Ecology and Evolution* 6 (20): 7221–9.

Wang, X., D. C. Miller, R. Harman, D. F. Antczak, and A. G. Clark. 2013. Paternally Expressed Genes Predominate in the Placenta. *Proceedings of the National Academy of Sciences USA* 110 (26): 10705–10.

Warner, R. R. 1984. Mating Behavior and Hermaphroditism in Coral Reef Fishes. *American Scientist* 72 (2): 128–36.

Warner, R. R., D. R. Robertson, and E. G. Leigh. 1975. Sex Change and Sexual Selection. *Science* 190 (4215): 633–8.

Wilkins, J. F., and D. Haig. 2003. What Good Is Genomic Imprinting: The Function of Parent-Specific Gene Expression. *Nature Reviews Genetics* 4 (5): 359–68.

Williams, G. C. 1957. Pleiotropy, Natural Selection, and the Evolution of Senescence. *Evolution* 11 (4): 398–411.

———. 1966. *Adaptation and Natural Selection; A Critique of Some Current Evolutionary Thought*. Princeton, NJ: Princeton University Press.

CHAPTER 13

The Origin of Species

Learning Objectives

- Compare and contrast the phylogenetic, biological, and general lineage species concepts.
- Discuss the influence on gene flow of geographic and reproductive isolating barriers.
- Compare and contrast allopatric, parapatric, and sympatric speciation.
- Explain how isolating barriers contribute to varying models of speciation.
- List the types of evidence scientists use to evaluate different models of speciation.
- Discuss the challenges of studying speciation.
- Explain why the rate of speciation may vary among organisms.
- Describe the significance of the discovery of an example of a cryptic species.
- Discuss the challenges of applying species concepts to bacteria and archaea.

Eline Lorenzen (**Figure 13.1**) is fascinated by the origin of species. But there's one species whose origins she is particularly interested in: *Ursus maritimus*, otherwise known as the polar bear.

It only takes a visit to the zoo for anyone to recognize that polar bears are very different from other species of bears. Other bears are black or brown, for example, but polar bears are white—or, to be more precise, they have transparent hairs that scatter sunlight, much as transparent snowflakes do. Polar bears are also different from other bears in their amazing swimming ability, spending much of their time chasing after ringed seals or traveling from one Arctic Ocean ice floe to another. One polar bear was recorded swimming a mind-boggling journey of 426 miles.

Lorenzen, a professor at the Museum of Natural History of Denmark, wants to understand how such a remarkable species arose. How did the polar bear split away from other bears and become so distinct from them? To investigate that question, she needs a means of looking through the history of *U. maritimus*, all the way back to the time when the ancestors of polar bears had not yet diverged from other bear populations.

To get that glimpse of history, Lorenzen regularly travels with other polar bear researchers to Greenland. Although the government of Greenland protects polar bears, they make an exception for Inuits, who harvest a small number of the animals to continue their tradition of subsistence hunting.

Polar bears are a distinct species (Ursus maritimus). Studies on their DNA have revealed how their species originated within the last few hundred thousand years from a population of brown bears (Ursus arctos) that became isolated in the Arctic.

Vladone/iStock/Getty Images

Figure 13.1 Eline Lorenzen travels to Greenland to collect polar bear tissue. By analyzing the genes of polar bears, she and her colleagues can reconstruct the history of the species. (Rune Dietz)

Lorenzen travels to Greenland in February during hunting season, when the sea ice starts melting and the polar bears move closer to land. Stationed at a remote Inuit village, she waits for word of a polar bear kill and then hops on a snowmobile to race to the hunting site. There, Lorenzen and her colleagues collect blood, muscle, and other tissues, preserving them in chemicals to be sure they survive the journey to universities around the world.

For Lorenzen, the most precious thing the researchers bring back is polar bear DNA. Back at the university, she and her colleagues can sequence all the genes in the polar bear genome. She can then use a computer to compare the DNA in individual polar bears to measure the genetic variation within the species. And she can also compare the DNA of polar bears to that of closely related species, such as brown bears (also known as grizzly bears in North America).

Those trips to Greenland have been rewarded with a surprising picture of how the species *U. maritimus* evolved (Liu et al. 2014). As we'll explore in greater detail later in this chapter, Lorenzen's research indicates that polar bears split off from brown bears less than 500,000 years ago and rapidly evolved their distinctive features. And yet they are still so closely related to brown bears that the two species sometimes interbreed. That capacity for interbreeding makes the distinctiveness of polar bears as a species all the more striking.

In this chapter, we will examine how species—from polar bears to monkeyflowers to *Escherichia coli* bacteria—originate. This process is known as speciation. To explore speciation, we have to start by considering what it means to be a species. As we'll see, species concepts have evolved over the centuries, and scientists continue to refine them today. One crucial feature of a species is that it is distinct from other species. We'll consider the ways in which species become distinct from each other and then remain so. In many cases, barriers to gene flow play an important role. These barriers can affect all stages of reproduction, from mating to fertilization to the fitness of hybrid offspring.

As we'll see, barriers can evolve between populations of the same species. Over time, these barriers can allow the populations to diverge along separate evolutionary trajectories. Eventually, they may become distinct enough to warrant being considered separate species. But the process of speciation is remarkably dynamic; gene flow may even continue between well-established species.

Understanding speciation is important for many reasons. It was the fundamental question that inspired the title of Charles Darwin's groundbreaking book, *The Origin of Species*. It's also important for understanding the threat of diseases, because different microbial species can affect our health in different ways. And it's important for conserving the world's biodiversity. By seeing how new species evolve and are maintained, we can better appreciate the threats to their continued existence. ●

13.1 What Is a Species?

Long before the dawn of science, humans were naming species. To be able to hunt animals and gather plants, people had to know what they were dealing with. In the 1600s, taxonomy emerged as a science and then came into its own in the next century, thanks largely to the work of Swedish naturalist Carolus Linnaeus.

Carolus Linnaeus

Linnaeus invented a system to sort living things into groups, inside which were smaller groups (see Figure 2.4). Every member of a particular group shared certain key traits. Humans belonged to the mammal class, and within that class the primate order, and within that order the genus *Homo*, and, within that genus, the species *Homo sapiens*. To Linnaeus, species were fixed for the most part ever since creation. "There are as many species as the Infinite Being produced diverse forms in the beginning," he wrote (Wilkins 2018).

Linnaeus's system made the work of taxonomists much easier, but trying to draw the lines between species often proved frustrating. Two apparently distinct species might produce hybrids from time to time, for instance. Within a species there was confusion as well. The willow ptarmigans in Ireland, for example, have a slightly different plumage than the willow ptarmigans in Finland, which differ in turn from the ones in Norway. Naturalists during Linnaeus's time could not agree about whether they belonged to different ptarmigan species or were just varieties—subsets, in other words—of a single species.

Darwin, for one, was amused by these struggles. "It is really laughable to see what different ideas are prominent in various naturalists' minds, when they speak of 'species,'" he wrote in 1856. "It all comes, I believe, from trying to define the indefinable" (Wilkins 2018).

Species, Darwin argued, were not fixed since creation. They had evolved. Each group of organisms that we call a species starts out as a variety of an older species. Over time, forces such as natural selection and genetic drift transformed them independently in their separate environments. When these varieties have diverged sufficiently from each other, Darwin argued, we see them as distinct species in their own right. In this way, new species arise over time, just as others become extinct.

"I look at the term 'species' as one arbitrarily given, for the sake of convenience, to a set of individuals closely resembling each other," Darwin declared.

Phylogenetic Species Concept

Ever since Darwin, biologists have recognized that species are indeed the product of evolution. But they have fiercely debated just what it means to be a species. According to one recent tally, biologists have put forward 25 different competing definitions (Wilkins 2018). Their disagreement is partly a matter of philosophy and partly a matter of research methods. Different concepts can be more useful in different branches of evolutionary biology, as we will explore in this chapter. Paleontologists use fossils to identify extinct species, for example, so they need a concept that can let them distinguish between species based only on the anatomical features of their fossils. Molecular phylogeneticists (biologists who study DNA to reconstruct the relationship of organisms) value a concept that can help them use DNA to recognize species.

The **phylogenetic species concept** is especially useful for systematists because it focuses on the phylogenetic history of organisms. It defines a species as the smallest aggregation of populations that can be diagnosed by a unique combination of character states. In essence, species are the tufts of leaves at the tips of a phylogenetic tree (Chapter 4).

The **phylogenetic species concept** is the idea that species are the smallest possible groups whose members are descended from a common ancestor and who all possess defining or derived characteristics that distinguish them from other such groups.

Biological Species Concept

The **biological species concept** holds that species are groups of actually (or potentially) interbreeding natural populations that are reproductively isolated from other such groups.

Another widely used species concept is the **biological species concept**, championed by the ornithologist Ernst Mayr. Mayr defined a species as a group of "actually or potentially interbreeding populations which are reproductively isolated from other such groups" (Mayr 1942). This definition proved useful because it focused on the process by which species form. Other species concepts tended to emphasize how you recognize a species once it comes into existence. Mayr's definition focuses on what might keep populations apart. But the biological species concept works only for organisms that reproduce sexually, where there is potential gene flow, and it is practical only in situations where it's possible to study the extent to which candidate populations can or would interbreed.

General Lineage Species Concept

A **metapopulation** is a group of spatially separated populations of the same species that interact at some level (for example, exchange alleles).

The **general lineage species concept** is the idea that species are metapopulations of organisms that exchange alleles frequently enough that they comprise the same gene pool and therefore the same evolutionary lineage.

Although different concepts may differ in the criteria required to designate a set of organisms as a species, most modern views of species share the same basic concept of what true species are (de Queiroz 2005). Species are **metapopulations** of organisms that evolve independently from other metapopulations. This view recognizes that the various populations within a species (that is, within the larger metapopulation) exchange alleles frequently enough that they can be treated as members of the same evolutionary lineage. Individuals within the metapopulation all share the same gene pool, which thus remains fairly consistent over time. The systematist Kevin de Queiroz has called this the **general lineage species concept**.

Today, biologists generally agree that no single definition will ever fit all taxa (Coyne and Orr 2004). But as tricky as species may be to define, most biologists agree that there is something unique about them. Within a species, alleles flow among populations, so that evolutionary forces affecting one population also affect other populations, to some extent. These populations remain parts of the same whole, evolving as one lineage. Between species, on the other hand, often there is no such cohesion. Different species behave like independent evolutionary units, following separate trajectories (Dobzhansky 1937; de Queiroz 2005).

Key Concept

- As scientists gain more insight into the mechanism of gene flow and the relationships between genotype and phenotype across the diversity of life on our planet, they are developing more precise ways of defining species. ●

13.2 Barriers to Gene Flow: Keeping Species Apart

For the practical task of identifying separate species, the general lineage species concept is arguably the most useful. This is especially true given the recent surge in availability of genomic data and advances in phylogenetic methods for handling large genetic data sets, which we'll discuss in more detail later in this chapter. For these studies, it makes sense to define species based on measures of the genetic differences among populations as well as on their independent phylogenetic histories.

For studying how new species form, on the other hand, the biological species concept has been instrumental. That's because this species concept focuses on the things that keep species apart. By examining what causes individuals from one population to be unlikely to breed with individuals from another, biologists can study how, precisely, new species evolve.

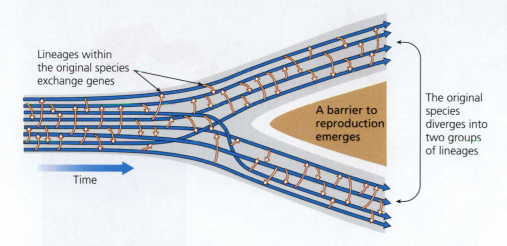

Figure 13.2 A key step in the evolution of new species is the splitting in two of an original population. A geographic barrier can divide it, but scientists have identified other factors that can also act as barriers. Genetic exchange within a gene pool is symbolized by white arrows. (Data from Doolittle 2008)

Lineages within the original species exchange genes

A barrier to reproduction emerges

The original species diverges into two groups of lineages

Time

For sexually reproducing species to remain distinct, they require **isolating barriers** to block gene flow with other species (**Figure 13.2**). In this section, we'll explore some of the many ways that gene flow between populations can be interrupted. We'll then turn our attention to how these barriers emerge within species and drive the process of **speciation**: the evolution of new species.

Isolating barriers can be extrinsic properties, such as geographic barriers, that reduce the opportunity for gene flow. Or they can be intrinsic features of the organisms themselves, such as reproductive barriers, that minimize interbreeding between populations. When new isolating barriers arise, they cause populations to begin behaving as distinct evolutionary entities.

Given sufficient time, populations separated by isolating barriers should begin to diverge. Genetic differences can accumulate due simply to genetic drift or because selection favors different alleles in different populations. At some point, the populations may become so different from each other that they no longer would or could interbreed, even if they were given the opportunity. In such instances, they have become reproductively isolated from each other, and most biologists would agree that from that point forward they are now separate species. According to most species concepts, a pair of genetically distinct, reproductively isolated sister taxa are separate species (Mallet 2008).

An **isolating barrier** refers to an aspect of the environment, genetics, behavior, physiology, or ecology of a species that reduces or impedes gene flow from individuals of other species. Isolating barriers can be geographic or reproductive.

Speciation is the evolutionary process by which new species arise. Speciation causes one evolutionary lineage to split into two or more lineages (cladogenesis).

Geographic Barriers

The most common factor that prevents species from interbreeding is geographic separation, or **allopatry**. Elk of North America, for example, cannot cross the Atlantic to interbreed with their close relatives, the red deer of Europe and Russia (**Figure 13.3**).

Yet, according to the biological species concept, allopatry alone is not enough to warrant dividing two populations into separate species. Populations must also be unable or unlikely to interbreed, even if they were given the opportunity. Whereas elk and red deer do not interbreed on their own, they can when they are brought together in zoos. That's because they're still very closely related. Even though they are geographically isolated, they have not evolved any intrinsic barriers to reproduction. About seven million years ago, the ancestors of both populations spread from Central Asia into Europe, East Asia, and over the Bering Land Bridge into the New World. The North American animals continued to interbreed with their counterparts in East Asia for millions of years, until a rise in sea level nine thousand years ago cut off the two continents. In this instance, nine thousand years of geographic isolation appears not to have been enough time for **reproductive isolation** to evolve. Some scientists have argued that the two populations should be lumped together in a single species,

Allopatry occurs when populations are in separate, nonoverlapping geographic areas (that is, they are separated by geographic barriers to gene flow).

Reproductive isolation occurs when reproductive barriers prevent or strongly limit reproduction between populations. The result is that few or no genes are exchanged between the populations.

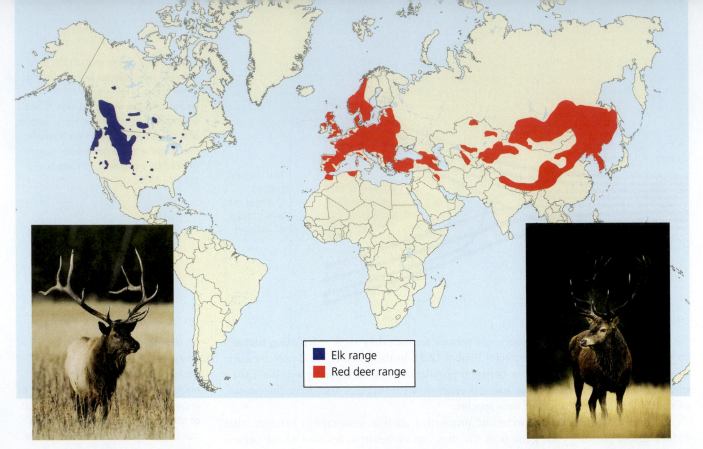

Figure 13.3 Elk (lower left) and red deer (lower right) are closely related and can reproduce when they are brought together in sympatry. Before nine thousand years ago, the ancestral populations of red deer and elk had no geographic boundaries preventing them from interbreeding. Since then, changes in sea level have geographically isolated them. Scientists are debating whether to consider them separate species. (Data from iucnredlist.org; left: Paul Tessier / Getty Images; right: DamianKuzdak / Getty Images)

Cervus elaphus. Others have argued that because they never interbreed in the wild, each population should be considered a species in its own right.

According to most species concepts, species must be able to remain distinct from each other even when they occur in **sympatry**—in the same geographic area. Coexistence in sympatry depends on the existence of reproductive isolating barriers. Unlike geographic barriers, reproductive barriers are intrinsic properties of the species: they are differences in phenotypes that reduce the chances of interbreeding between populations. These reproductive barriers fall into three main categories, reflecting the three main stages of reproduction: finding a mate, forming a zygote (fertilization of egg and sperm), and successfully developing reproductively viable offspring. We'll explore each of the three categories in turn.

> **Sympatry** occurs when populations are in the same geographic area.

Reproductive Barriers: Finding a Mate

Sexually reproducing organisms have to find each other (or at least their gametes do). The ecology of different species can reduce the opportunities two species have to mate with each other. For example, many species of fruit flies seek out a particular kind of fruit to find a mate. The females then lay their eggs on the fruit, so that when the eggs hatch they have a ready supply of food. As a result, for example, the fruit flies that specialize on hawthorns will rarely mate with fruit flies that specialize on blueberries (Bush 1969; Diehl and Prokopy 1986; Feder et al. 1989).

Timing

Time can also create reproductive barriers. Nancy Knowlton, a marine biologist at the Smithsonian Institution, and her colleagues have documented a striking case of isolation by time in corals. Coral reefs are actually giant skeletons secreted by colonies of tiny animals related to jellyfishes and sea anemones. To mate, corals release sperm and eggs that drift through the water. On rare occasions, two gametes meet and produce a fertilized egg. Each population of corals that Knowlton studies releases its gametes in a pulse—lasting just 15 to 30 minutes—that occurs at precisely the same time each evening. The odds that a sperm and egg will meet are highest during this spawning event because the density of gametes is high. With time, the gametes get dispersed by ocean currents, lowering their concentration, and the probability of sperm–egg encounters goes down. Knowlton and her colleagues found that the spawning times of closely related, sympatric corals are separated by 1.5 to 3 hours (**Figure 13.4**). By the time the second species releases its gametes, currents will have carried away most of the gametes from the first species. The density of the gametes from the first spawning species is thus so low when the second one spawns that the two species are not likely to cross-fertilize (Levitan et al. 2004).

Pollinators

Like corals, plants stay in one place and send their gametes abroad. Many species depend on pollinators, such as bees and birds, to carry pollen (which contains sperm) to other plants, where the sperm can fertilize the eggs in a flower's ovules. But these pollinators can create pre-mating reproductive barriers between plants.

One striking case of this kind of isolation is found in monkeyflowers (*Mimulus*) that grow in the western United States. In 1998, Douglas Schemske and H. D. Bradshaw, then at the University of Washington, studied two species of monkeyflowers (*M. cardinalis* and *M. lewisii*), whose ranges overlap in the Sierra Nevada in California. Schemske and Bradshaw spent weeks watching the monkeyflowers and observing which animals visited them to feed on their nectar. *M. cardinalis* was visited 97% of the time by hummingbirds. *M. lewisii* was visited 100% of the time by bees (Schemske and Bradshaw 1999).

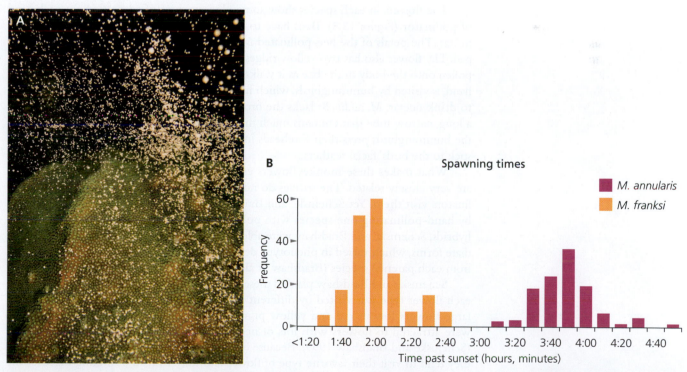

Figure 13.4 A: Corals reproduce by releasing gametes into the water. B: This graph shows how two species of *Montastraea* corals remain reproductively isolated: by spawning at different times after sunset. (A: Universal Images Group/Superstock, Inc.; data from Levitan et al. 2004)

Mimulus lewisii *Mimulus cardinalis* *Mimulus lewisii* F₁ HYBRIDS *Mimulus cardinalis*

Pollinator   Pollinator

F₂ HYBRIDS

Figure 13.5 Two species of partially sympatric monkeyflowers are pollinated by different animals, *Mimulus lewisii* by bees (left) and *M. cardinalis* by hummingbirds (middle). Right: First generation hybrids of *M. lewisii* and *M. cardinalis* have intermediate forms, and when crossed to each other, produce second generation (F₂) hybrids with a wide range of flower forms. (Data from Schemske and Bradshaw 1999; photos by Douglas Schemske and Toby Bradshaw)

The flowers in each species show some obvious signs of adaptation for each kind of pollinator (**Figure 13.5**). Bees have to land on flowers before they can reach the nectar. The petals of the bee–pollinated *M. lewisii* thrust forward, serving as a landing pad. The flower also has two yellow ridges of brushy hairs that probably serve to guide pollen onto the body of the bee as it walks toward the nectar. *M. cardinalis*, on the other hand, is visited by hummingbirds, which hover in front of flowers and insert their beak to drink nectar. *M. cardinalis* lacks the broad landing pads of *M. lewisii*; instead, it has a long, narrow tube that contains much more nectar. It also positions its pollen so that the hummingbirds press their foreheads into the grains as they drink, causing them to stick to the birds' facial feathers.

What makes these monkeyflowers especially fascinating is that the two species are very closely related. The species do not hybridize in nature, because different pollinators visit them. Yet Schemske and Bradshaw were able to produce hybrid flowers by hand-pollinating one species with pollen from the other. By then crossing these hybrids, Schemske and Bradshaw were able to generate a remarkable range of intermediate forms, which varied in phenotype depending on how many alleles they inherited from each parental species (Bradshaw et al. 1995; see Figure 13.5).

Schemske and Bradshaw planted these hybrids in the field and observed how often each flower type was visited by different pollinators. They found that bees preferred large flowers low in red and yellow pigments. Hummingbirds, on the other hand, were attracted to flowers with lots of nectar and rich in red anthocyanin pigments (Schemske and Bradshaw 1999). Because bees and birds have distinct color preferences, they tend to visit their favorite type of flower faithfully. Instead of mixing together the genes of the two plant populations, the pollinators dramatically reduced the probability that the pollen from one type of flower could end up on the other. As a result, the specialization of monkeyflowers for different pollinators has become a reproductive barrier keeping the species apart.

Courtship Rituals

Many animal species have to go through certain courtship rituals in order to mate. Each species may have a distinctive ritual, making it unlikely for a male from one species to get a chance to mate with a female from another. Fireflies use flashes of light for courtship, and in a single field in Massachusetts, as many as six species of fireflies may carry out their courtships at the same time. The males of each species produce a distinctive series of flashes, and the females respond only to the ones they find attractive. What's more, the males approach only the females that wait a suitable and species-specific period of time before responding with a flash of their own. Distinct male courtship signals, combined with corresponding female preferences, act in these fireflies as a pre-mating reproductive barrier between species (Lewis and Cratsley 2008).

Reproductive Barriers: Forming a Zygote

Some reproductive barriers are brought about by the very act of mating itself. Consider the beetles *Carabus maiyasanus* and *C. iwawakianus*, which live in sympatry in Japan. When biologists bring the two species together in the lab, they will mate readily. But after these cross-species matings, the females often die, and the survivors produce only a few eggs.

It turns out that it's harmful to the females' health to mate with males of the other species. Male beetles deliver their sperm with an ornate structure called an aedeagus. It fits into a specialized pouch inside the female. Within each species, the fit is good, but between species, it's not. The males get stuck, and the aedeagus can break off inside the female, tearing her reproductive organs in the process and preventing fertilization (Sota and Kubota 1998).

Many closely related species can mate without suffering the ugly fate of carabid beetles. But very often those matings fail to give rise to fertilized eggs. As it turns out, there are many reproductive barriers on the road from mating to fertilization.

In many species of insects, for example, the male cannot simply deliver his sperm and leave. He must engage in copulatory courtship (Eberhard 1991). Some females require males to flap their wings to produce a distinctive buzz as they mate. For other species, copulatory courtship involves stroking the female with special appendages at the right frequency. If a male from a different species carries out the wrong copulatory courtship while mating with a female, she's less likely to fertilize her eggs with the sperm he has deposited inside her.

After mating, sperm must make the long journey to the female's eggs. In birds, for example, sperm must first pass from the vagina to storage tubes, where females store sperm until ovulation. Scientists have found that relatively few sperm from other species can survive this first leg of the journey. When female birds draw sperm from the tubes to fertilize their eggs, sperm from other species have less success in reaching the eggs. And when the sperm come into contact with the egg itself, the ones from other species often fail to penetrate and deliver their DNA.

A lot of this failure lies in the biochemistry of sperm and the female reproductive tract. Some studies suggest that the immune cells in a female's reproductive tract may attack foreign sperm, as they would an invading microbe. Even if sperm survive and are able to reach an egg, they have to attach to its surface with proteins that function like a key in a lock. Closely related species often have locks of different shapes.

Such **gametic incompatibilities** are especially important in marine organisms, like corals and urchins, because these species broadcast their gametes over such a large area and because they, in turn, are likely to encounter gametes from many different species (Palumbi 1994; Howard 1999). In sea urchins, binding proteins on the surface of sperm help with the attachment of the sperm to eggs. The structure of the proteins differs

Gametic incompatibility occurs when sperm or pollen from one species fails to penetrate and fertilize the egg or ovule of another species.

slightly from species to species; the differences are sufficient to ensure that sperm from a different species don't end up fertilizing the eggs (Palumbi and Metz 1991; Metz et al. 1994). In abalone, the lysin protein accomplishes this task. It lies at the tip of the sperm, and its job is to dissolve a hole in the protective surface of the egg so that the sperm can contact the lipid membrane of the egg directly. Even closely related abalone species differ in the amino acid sequences of their lysin proteins; the result is that conspecific (within-species) sperm are better at fertilizing eggs than heterospecific (different-species) sperm (Lee and Vaquier 1992).

Reproductive Barriers: Viable Offspring

All of the reproductive barriers we've looked at so far are known as **prezygotic reproductive barriers** because they occur before the formation of the zygote (the fertilized egg). Yet even after a zygote develops, reproductive barriers (called **postzygotic reproductive barriers**) can still arise and isolate species from each other.

As a first example of a postzygotic reproductive barrier, embryos with parents from two different species may fail to develop. There are many reasons that hybrid embryos may fail. Even if two species have all of the same genes, they may have alleles of those genes that cannot function together.

Hybrids do sometimes develop successfully, but in some cases they are born with deformities that leave them in poor health. Healthy hybrids can sometimes live long lives, but in many cases they are sterile. Hybrid male *Drosophila* flies, for example, produce defective sperm. Mules, which are the hybrid offspring of female horses and male donkeys, are almost never able to breed with each other, nor can they breed with horses or donkeys (Chapter 12). Thus, the genes from the horses cannot flow into the donkey population, and vice versa. Even though mules are viable animals, their sterility means that horses and donkeys remain distinct species.

In 1909, the British geneticist William Bateson proposed that genetic incompatibility was the cause of hybrid sterility—and thus species isolation (Bateson 1909). The idea was then developed by Theodosius Dobzhansky (1936) and by H. J. Muller (1942). Collectively, these genetic incompatibilities are known as **Bateson-Dobzhansky-Muller incompatibilities** (Orr 1996). These incompatibilities occur when alleles at different loci can no longer cooperate. Consider an ancestral population with the genotype *aabb*. In one population, allele *A* arises and sweeps to fixation. In the second, allele *B* fixes. Now the two populations have genotypes *AAbb* and *aaBB*. If these two populations interbreed, they will create hybrids with mixed genotypes, such as *AaBb*. Allele *A* may perform perfectly well with *b*, and *B* may work just fine with *a*. But if *A* and *B* are not compatible with each other, then these hybrids will suffer lower fitness. Epistatic incompatibilities can thus act as postzygotic barriers to the exchange of genes across populations (Orr and Presgraves 2000).

One example of Bateson-Dobzhansky-Muller incompatibility occurs in monkeyflowers. Approximately 300,000 years ago, a central Californian population of the monkeyflower *Mimulus guttatus* got separated from the rest of the flowers in its range (Brandvain et al. 2014). This isolated population began to evolve independently from *M. guttatus*, and today it is known as *M. nasutus*. Modern populations of these closely related species overlap in places, and where they do, they occasionally interbreed. But male flowers produced by these interspecies crosses either produce pollen with reduced viability or are completely sterile.

Andrea Sweigart and colleagues at Duke University used quantitative trait locus (QTL) mapping to compare the level of pollen viability in hybrids with the presence of certain alleles. They identified a single incompatibility between two loci. *M. guttatus* and *M. nasutus* populations differ in the alleles they carry at each of these loci, called *hybrid male sterility 1* (*Hms1*) and *hybrid male sterility 2* (*Hms2*). Within each species, the alleles at *Hms1* and *Hms2* interact without causing the flower any harm. But whenever an *M. guttatus* allele at *Hms1* combines with an *M. nasutus* allele at *Hms2*, the result is harmful: the pollen produced by these hybrids is inviable, so the males are sterile (**Figure 13.6**; Fishman and Willis 2001; Sweigart et al. 2006).

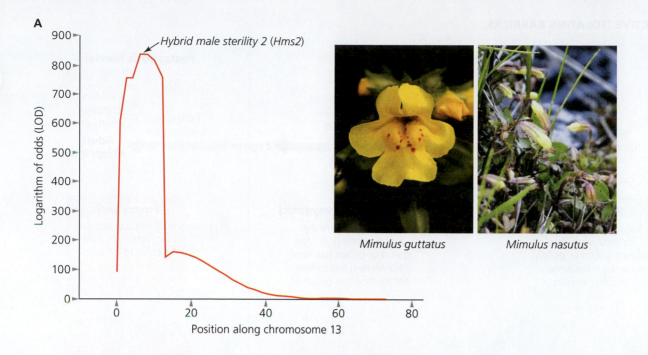

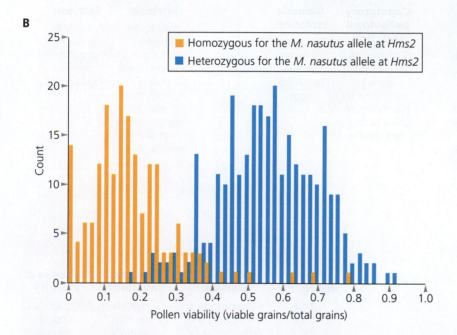

Figure 13.6 A: Naturally occurring hybrids between the monkeyflowers *Mimulus guttatus* and *M. nasutus* produce pollen with reduced viability. Some crosses can result in complete sterility. This reproductive barrier reduces gene flow between the two species. Lila Fishman and her colleagues found genes involved in producing this sterility by rearing hybrids and using QTL mapping to compare levels of sterility with the presence of certain alleles. The peaks pointed the biologist Andrea Sweigart to two genes: *hybrid male sterility 1* (*Hms1*), located on chromosome 6, and *hybrid male sterility 2* (*Hms2*), on chromosome 13 (shown). B: The scientists then sorted the flowers based on which alleles they inherited at these loci and recorded the viability of pollen produced by their hybrid offspring. Plants inheriting *M. guttatus* alleles for *Hms1* produce viable pollen if they also inherit at least one *M. guttatus* allele at *Hms2* (that is, if they are heterozygous for the *M. nasutus* allele at *Hms2*; blue), but not if they inherit only *M. nasutus* alleles at *Hms2* (orange). This epistatic interaction is an example of Bateson-Dobzhansky-Muller incompatibility. (Data from Sweigart et al. 2006; left photo: Juergen + Christine Sohns/Picture Press/Getty Images; right photo: Amanda Kenney)

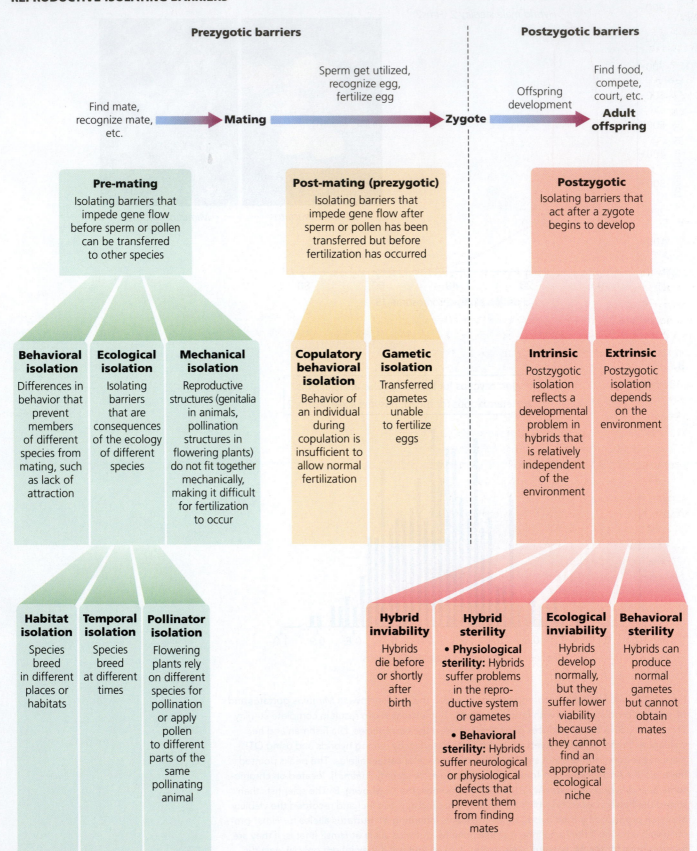

Figure 13.7 Reproductive barriers to gene flow can take many forms. (Data from Coyne and Orr 2004)

Barriers to gene flow can thus arise at different stages of reproduction. See **Figure 13.7** for a summary of the diverse ways species remain distinct.

Key Concepts

- Geographic barriers to gene flow are features of the environment that physically separate populations from each other. They are important for all types of species, regardless of the species concept that applies.

- Reproductive barriers to gene flow are intrinsic properties of organisms that reduce the likelihood of interbreeding between individuals of different populations.

- Separation of populations in time can reduce the likelihood they will exchange gametes. In these cases, divergent behavior (for example, habitat preference, spawning time) acts as a reproductive barrier to gene flow.

- Floral traits can act as reproductive barriers to gene flow when their divergence causes them to attract different pollinators.

- Genetic incompatibilities between eggs and sperm can be important barriers to gene flow between species, causing conspecific sperm to be more likely to fertilize eggs than heterospecific sperm.

- Genetic incompatibilities can be important barriers to gene flow between populations if they cause hybrid offspring to be sterile or to perform poorly relative to other individuals in the population. ●

13.3 Models of Speciation

The evolution of new barriers to reproduction between populations typically occurs gradually, in a series of stages (Mallet 2008; Hendry 2009).

The first stage begins with a population of organisms whose genetic variation is continuous, and there are no barriers to reproduction. In the second stage, geographic or minor reproductive barriers begin to emerge, so that offspring more often tend to be produced by matings within each group rather than between them. At this stage, the genetic variation becomes partially discontinuous between the two new groups, meaning that the organisms within each of the new groups are more genetically similar to each other than they are, on average, to the individuals in the other group.

In the third stage of speciation, the reproductive barriers become even stronger, and the two groups become even more genetically discontinuous. However, at this stage the reproductive barriers still can be reversed if environmental conditions change. In the fourth stage, reproductive isolation between the two groups has become complete and irreversible.

One way to study this process is to build a theoretical model and analyze it. Scientists build these models by drawing on insights from fields such as population genetics and ecology. As the scientists probe these models, they can develop hypotheses for the conditions under which speciation can occur and which mechanisms are most important for the process. Armed with these insights, scientists can then design studies to carry out on real populations at different stages of speciation. Here we will explore speciation models, and in the next section we'll consider some empirical examples of speciation.

Allopatric Speciation

One crucial factor in these models is the geographic context in which speciation takes place. For example, imagine a change in the landscape that splits a population in two

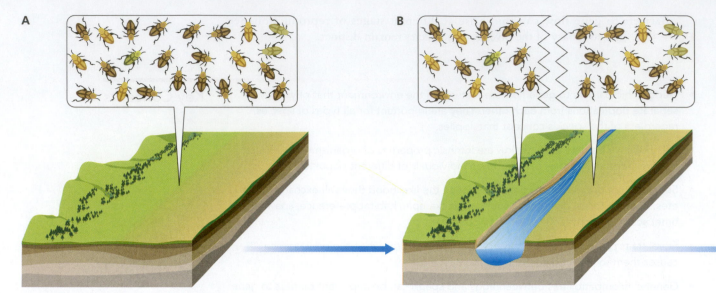

A

B

Figure 13.8 Allopatric speciation is the result of geographic iso-lation. A: Here, a population begins with a continuous geographic range. It contains genetic variation, but gene flow ensures that new mutations can spread across the range once they arise. B: A river divides the population into two subpopulations. C: The change of allele frequencies in the two subpopulations is no longer linked. They become increasingly divergent. D: The river later dries up, allowing the two subpopulations to make contact. During their separation, reproductive barriers may evolve, reducing the gene flow between the two subpopulations.

Allopatric speciation is the evolution of new species after populations have been separated geographically, preventing gene flow between them.

(Figure 13.8). A river might slice across the range of a species of terrestrial beetle, for example. A flock of birds might be swept out to sea and settle on an island, cut off from the rest of their species on the mainland. Speciation that arises from geographic separa-tion of populations is called **allopatric speciation**.

While the two populations are physically isolated, they will continue to evolve due to drift and selection. During this time apart, the two populations may accumulate genetic differences. Existing alleles may become more common in one population and not in the other, and mutations arising in one population are not likely to be present in the other. Sometimes these alleles create reproductive barriers such as incompatibil-ities between loci, or differences in where or when individuals breed. Alleles that make males attractive to females in one population might make them unattractive to females from the other.

Such incompatibilities are irrelevant as long as the populations stay separated because individuals from the divergent populations don't encounter each other. But they can cause problems later on, if the geographic context changes. The populations may come back into contact. The river dividing two populations of beetles may dry up, allowing them to rejoin each other. The descendants of birds that colonized an island may return to the mainland. These populations will now have the opportu-nity to interbreed, and any barriers that have accumulated will affect the likelihood of this occurring. If the habitats or courtship displays they prefer have become different enough, individuals from the two populations will be unlikely to breed. Any genetic incompatibilities they've accumulated will start to take effect in the hybrid offspring they produce if they do interbreed.

In models of allopatric speciation, what happens next depends on how strong the reproductive barriers have become. If the barriers are weak, the rejoined populations will interbreed easily. Mutations will be free to spread throughout the population, and any accumulated genetic differences will quickly disappear. The populations will lapse back toward the first stage of speciation.

If the reproductive barriers are very strong, on the other hand, the populations may not interbreed at all, showing that the process of speciation is complete. In other

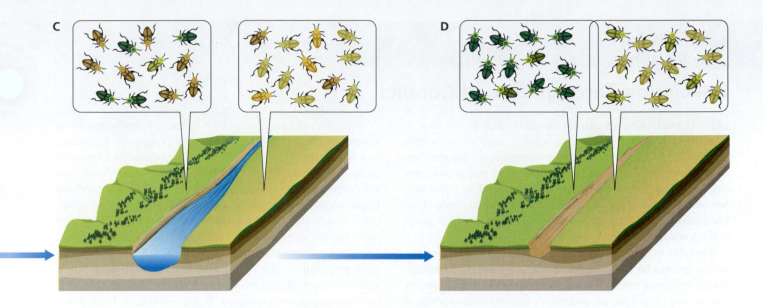

cases, the reproductive barriers may not be complete, but greater reproductive isolation continues to evolve in the two populations. Selection may favor prezygotic isolation mechanisms that reduce the odds that individuals will produce hybrids with low fitness. This process is called **reinforcement** (Servedio and Noor 2003).

Sympatric Speciation

At the other end of the spectrum, scientists have investigated the possibility that speciation can occur in populations that are *not* geographically disrupted—a process known as **sympatric speciation** (Bolnick and Fitzpatrick 2007). These models start with a population of interbreeding individuals occupying the same range—say, a species of fish living in a lake in an extinct volcano crater. In these models, individuals begin to mate nonrandomly: they mate more often with individuals that are genetically or phenotypically similar to themselves. The original population diverges into two or more populations, divided by increasingly effective reproductive barriers.

Biologists have also modeled speciation in other geographic contexts that fall between these two extremes. In some cases, for example, a geographic barrier emerges that only partially isolates two populations. The gene flow is reduced, but not eliminated. When populations diverge in such an arrangement, it's known as **parapatric speciation**.

Even with no geographic barriers at all, a species can undergo divergence if its range is big enough. If individuals disperse in only a limited distance, gene flow will become reduced across the entire range, and genetic drift can produce variation between populations that grows larger over greater distances (Chapter 6). This variation across the range of a species is known as **isolation by distance**.

Selection can also drive sympatric speciation if populations in different habitats within a species' range experience significantly different conditions. The populations of a species at the northern end of its range may face bitter winters, for example, while populations at the southern end endure blazing summers. These populations can still breed with their neighbors, enabling some alleles to flow between the north and the south. But alleles that promote survival in the cold may not be favored by natural selection in the south the way they are in the north. As long as the northern and southern populations are evolving away from each other *faster* than gene flow across the extent of the range can homogenize them, the populations will continue to diverge.

Reinforcement refers to the increase of reproductive isolation between populations through selection against hybrid offspring.

Sympatric speciation is the evolution of new species within a contiguous population that still has extensive gene flow.

Parapatric speciation is the evolution of new species within a spatially extended population that still has some gene flow.

Isolation by distance is a pattern in which populations that live in close proximity are genetically more similar to each other than populations that live farther apart.

BOX 13.1

Speciation Through Sexual Conflict

Male crickets sing to attract females. But as we saw in Chapter 11, sexual selection does not stop once mating begins. Females often choose among sperm from different males that have mated with them. And males compete with one another via the sperm and semen they inject into females. Just as pre-mating sexual selection can help drive speciation, scientists are discovering cases in which post-mating sexual selection is also at work.

Females in many species mate with more than one male, so sex creates an evolutionary conflict of interest. If a female mates in a short time with two different males, the first male's fitness may drop if the second male's sperm fertilizes some of her eggs. As a result, male strategies have evolved to make it more likely that his sperm, rather than the sperm of his competitors, will fertilize the eggs. In some species, a male's semen includes compounds that make females less receptive to other males. Or it may cause females to lay all their eggs immediately, rather than waiting until they've received sperm from several males.

These strategies can raise a male's fitness, but they can also lower a female's fitness. Females that become unreceptive to mating may not be able to mate with a higher-quality male if they encounter one later. Dumping her eggs all at once may rob a female of the chance to improve her offspring's survival or may compromise her ability to reproduce later in her life. In some *Drosophila* species, the compounds that males use to manipulate females are actually toxic. Mating shortens the life span of the females, and the more often a female mates, the sooner she dies (Chapter 11).

These challenges to females have favored the evolution of counterstrategies. And in every new generation, the strategies of males and females continue to evolve. As we saw in Chapter 11, females can counteract the effects of the toxic proteins in *Drosophila* semen. But it takes only a few generations before females become less resistant to seminal toxins when they are forced to mate in a monogamous relationship (Rice and Salt 1988). Loss of resistance can cause serious harm to females if they come back into contact with a polygynous population.

Antagonistic coevolution of males and females can drive speciation by creating a post-mating, prezygotic reproductive barrier to gene flow. When two populations become geographically isolated, the males and females in each one will continue to experience sexual conflict. They will continue to accumulate new mutations, some that allow males to overcome female defenses and others that allow females to exert more control over their own reproductive success. The two populations may move rapidly in different directions. Within each population, the males and females will remain in a stalemate, with almost every attack protein counterbalanced by a female countermeasure. But if individuals from those two populations mate, the females may not have sufficiently specific defenses, and as a result, successful interbreeding will be rare.

Various studies have confirmed that sexual conflict can create major barriers to gene flow. Sexual conflict can drive populations apart so quickly that a number of scientists have investigated whether it is an engine for speciation—akin to the coevolutionary arms race between insects and host plants (Chapter 15). Göran Arnqvist, of Uppsala University in Sweden, tested this hypothesis by comparing a wide range of insects (Arnqvist et al. 2000). He identified clades of insects in which the females were monandrous—they mated only once in their lifetime, providing little opportunity for sexual conflict to cause populations to diverge. Each monandrous clade had a sister group that was polyandrous—in other words, the females mated with more than one male over their lifetime, opening up the opportunity for sexual conflict.

Arnqvist and his colleagues compared 25 pairs of these monandrous and polyandrous taxa to see if there were an equal number of species in the two clades in each pair. After all, the species all descend from a recent common ancestor, and new species have had the same time to evolve within the two lineages in each pair. But that wasn't what Arnqvist found. Speciation was four times faster in polyandrous insects than in monandrous insects, suggesting that sexual conflict may be a potent driver of speciation.

Ecological Speciation

Ecological speciation is the evolution of reproductive barriers between populations by adaptation to different environments or ecological niches.

Scientists building speciation models often take into consideration the evolution of ecological adaptations. Two populations of predators can specialize in attacking different kinds of prey, for example. Two populations of plants may become specialized for growing on different types of soil. In **ecological speciation**, selection for different ecological traits creates reproductive barriers. It may cause populations to come into

Other studies suggest that sexual conflict may also drive the evolution of reproductive isolation in flowering plants (Bernasconi et al. 2004). Pollen grains grow "pollen tubes" as they burrow through the fleshy tissue of the flower style en route to ovules (Box Figure 13.1.1). As they plow through the style, the pollen tubes face two challenges: they must race rival pollen tubes, and they must cope with the molecular environment of the female style. As a result, flowering plants have the opportunity to experience sexual conflict similar to that experienced by animals. In species where the pollen from several plants lands on a style at the same time, selection can favor adaptations that raise the odds that a pollen grain can beat its competitors. Those adaptations may be harmful to the plant being fertilized. In contrast, sexual conflict is not expected to evolve in selfing species—species that fertilize themselves.

Mark Macnair, of the University of Exeter, explored this hypothesis by comparing two closely related species of monkeyflower, *Mimulus nasutus* and *M. guttatus*. *M. nasutus* is a selfing plant, and in *M. guttatus* plants can receive pollen grains from many other plants (see Box Figure 13.1.1). Macnair placed pollen grains from each species on the style of the other and observed how successful they were in fertilizing ovules. The scientists applied a 50-50 mix of pollen grains from both species to the selfing *M. nasutus*. Half of the resulting seeds were hybrids, suggesting that the pollen from each species was equally successful. But when they applied the same mix to *M. guttatus* styles, none of the seeds were fertilized by *M. nasutus* (Diaz and Macnair 1999).

These findings suggested that selfing in *M. nasutus* resulted in a lack of sexual conflict, and its pollen lacked adaptations for competing with other pollen. Macnair found support for this hypothesis by measuring how fast the pollen tubes grew. The selfing *M. nasutus* pollen tubes grew slowly, whereas *M. guttatus* pollen tubes grew quickly. Recent studies by Lila Fishman, at the University of Montana, and her colleagues have identified at least eight different genetic loci that contribute to this

Box Figure 13.1.1 Sexual conflict may drive speciation in plants. A: When outcrossed pollen is deposited on the stigma of a flower, it competes with pollen from rival individuals as it burrows through the style to reach the ovules (pollen tubes shown for *Arabidopsis thaliana*). Traits enabling pollen to perform well in this race may be harmful to the reproductive interests of females, selecting for alleles expressed in the style that interfere with pollen tube growth. B: In the monkeyflower, *Mimulus guttatus*, a history of outcrossing appears to have led to rapid evolution of pollen and style resulting from sexual conflict. The related species *M. nasutus* is self-fertilized, alleviating any conflicts of interest between males and females. These differences in the direction and intensity of sexual conflict help explain why pollen from *M. nasutus* is unsuccessful at competing with *M. guttatus* pollen when fertilizing flowers of *M. guttatus*. (A: From "Evolutionary Ecology of the Prezygotic Stage" by Bernasconi et al. Science, 2004. 303 (5660): 971–5. Reprinted by permission of Dr Juan M. Escobar Restrepo and Dr Amal J Johnston; B: Juergen + Christine Sohns / Picture Press / Getty Images)

prezygotic barrier to reproduction, and all but one appear to exert their influence in the styles of *M. guttatus* flowers—exactly what you'd expect if antagonistic coevolution were driving pollen and style evolution in this species (Fishman et al. 2008).

less contact, for example, and reduce their opportunity to mate. It may also lower the fitness of their hybrid offspring, which may not be able to compete for the resources that the two diverging populations are adapted for. It can even potentially influence mating behavior. Natural selection may favor an avoidance of partners from the other population because such a preference would ultimately lead to lower fitness.

Key Concepts

- New species can form when two or more populations become geographically isolated from each other (allopatry). Allopatric populations will begin to diverge genetically because they independently experience mutation, selection, and drift. Eventually, they may become sufficiently divergent that they no longer would or could interbreed, even if the physical barrier disappeared.

- Reproductive isolation can arise even when populations live in sympatry—they are not geographically separated from each other and can thus exchange migrants. New alleles accumulate independently in these populations because individuals rarely interbreed.

- Isolation by distance occurs because individuals tend to mate with individuals from the same or nearby populations, resulting in imperfect mixing of alleles across the geographic range of a species.

- Speciation can arise as a by-product of ecological adaptation of populations to different habitats or resources. ●

13.4 Testing Speciation Models

Models of speciation allow scientists to generate hypotheses about real populations, which they can then test. Here we will look at a series of studies to illustrate this kind of research.

Splitting an Ocean

Scientists have found a great deal of evidence for allopatric speciation. In these studies, scientists typically examine the history of the environment where species evolved and explore how geographic barriers influenced their speciation. Before 15 million years ago, for example, the Isthmus of Panama did not exist. North America and South America were separated, and the Pacific and Atlantic were joined.

In those times, marine animals could move back and forth between the oceans. But then the passageway began to close. About 3 million years ago, the two oceans were entirely cut off from each other. Species that previously had ranges that straddled the two oceans were now split by a major terrestrial barrier.

Nancy Knowlton and her colleagues study the animals that live today in the waters on either side of Panama. In the 1990s, the scientists set out to study how the formation of the isthmus influenced the speciation of shrimp. To do so, they worked out the phylogeny of shrimp species, some on the Atlantic side of the isthmus and some on the Pacific. The phylogeny revealed that for many species, their sister taxon was on the other side of Panama rather than in their own ocean (**Figure 13.9**).

To see how these clades behaved in sympatry, Knowlton brought some of these Pacific and Atlantic species pairs to her laboratory. When she placed them in tanks together, they did not interbreed (Knowlton et al. 1993). These results are best explained by the hypothesis that the formation of the Isthmus of Panama led to the allopatric speciation of the shrimp.

"Magic Traits" and Ecological Speciation

The shrimp on either side of Panama are far along in the process of speciation because they have been separated for millions of years. Their genomes were free to diverge without the homogenizing influence of gene flow and genetic recombination. In cases like this we expect that divergence will have arisen gradually through the independent accumulation of genetic differences in each species. A simple test for genetic divergence (such as F_{ST}, which we discussed in Chapter 8) among the sister species

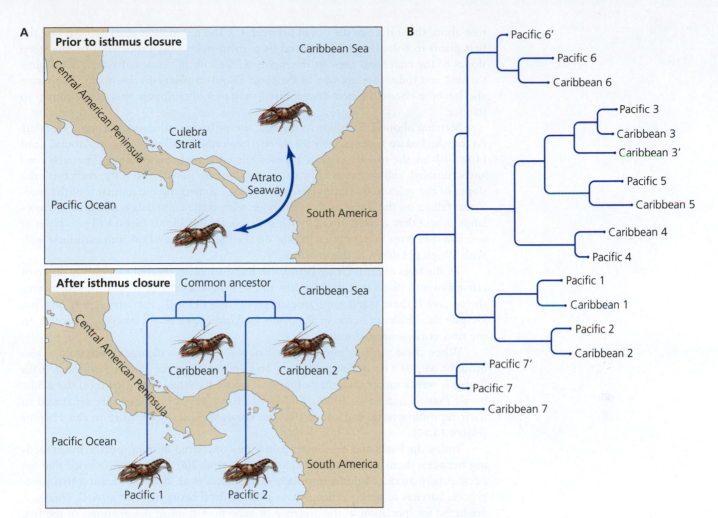

Figure 13.9 A: Until about three million years ago, the Atlantic and Pacific Oceans were joined by the Atrato Seaway. Some species of shrimp had ranges that extended into both oceans. When the Isthmus of Panama formed, it isolated Atlantic and Pacific populations of some of these species of shrimp. B: Genetic studies show that the closest relatives of some shrimp species in the Atlantic are not other Atlantic shrimp, but Pacific species. These studies indicate that these species underwent allopatric speciation, or the formation of new species by geographic barriers. (Data from Knowlton et al. 1993)

pairs in Figure 13.9 should show fairly uniform divergence across their respective genomes.

In contrast, speciation that occurs in the presence of gene flow—in parapatric or sympatric populations, for example—is likely to leave a very different signature of divergence within the genome. In these cases, divergence may begin to evolve when some individuals start to exploit a new resource or habitat. In the process of adapting to this new challenge, certain alleles may permit individuals to perform well. Those same alleles may also cause them to become less likely to interbreed with the original population. If selection spreads these alleles, they can initiate the process of speciation.

Traits that simultaneously confer divergent local adaptation and act as reproductive barriers are called **magic traits** (Servedio and Kopp 2012). In the early stages of speciation, incipient species should differ only in regions of the genome containing the genes for the magic traits. The rest of their genome remains otherwise largely homogenous (Feder et al. 2012; Wolf and Ellegren 2017). Only further in the process of speciation will genetic differences accumulate throughout the genome.

> A **magic trait** is a trait that simultaneously confers ecological divergence and reproductive isolation.

Adapting to a New Type of Soil

Lord Howe Island is a speck in the middle of the Tasman Sea, sitting by itself some 580 kilometers off the eastern coast of Australia. It formed when a volcanic shield rose above the surface of the ocean between 6.4 and 6.9 million years ago. Among the first plants to colonize the new island were palm trees, which took root on the steep slopes of the crater and grew in the highly acidic soils of volcanic basalt. These palms thrived, and today they are one of the most abundant plants on the island. They have also become one of the best-documented examples of sympatric speciation known to science.

Starting about 2.5 million years ago, the polar ice caps and glaciers expanded. As they locked up water in their ice sheets, they caused sea levels to drop. Around Lord Howe Island, the receding ocean exposed the coral reefs encircling it. The reefs died and crumbled, and the wind blew calcium carbonate powder from the reefs onto the slopes of the volcano. Sea levels later rebounded, submerging the remnants of the reefs. Rains falling on the slopes of the volcanoes washed the coral debris down to the lowlands, where they created calcium-rich soils. Now Lord Howe Island had two types of soil: low-pH crust from volcanic basalts on the high slopes and calcium carbonate soils with a high pH down in the lowlands.

To the trees of Lord Howe Island, this transformation created two starkly different environments. Plants that grow in low pH soils have a suite of adaptations to survive the acidity. If these plants are transplanted to high pH soils, they struggle to survive. Despite this challenge, some of the palms on Lord Howe Island evolved to survive on the new coral-generated soils.

When these palms adjusted to their new soil pH, this change in physiology also brought with it a barrier to gene flow. Trees growing on the calcium carbonate soils flower 6 weeks earlier than they otherwise would have had they grown on the acidic basalt (Savolainen et al. 2006). The Lord Howe palms have become separated in their reproductive timing, much like the corals we discussed earlier in this chapter (**Figure 13.10**).

Today, the basalt and calcareous palms are considered distinct species. Interbreeding between them almost never occurs (Babik et al. 2009). The few hybrids that are occasionally produced perform poorly (Hipperson et al. 2016), indicating that postzygotic barriers to gene exchange have now evolved between them as well. Finally, as predicted for speciation in the presence of gene flow, scans of the genomes of the two species show just a few loci with strong signatures of divergence against an otherwise similar genetic background (Savolainen et al. 2006). The tiny size of the island makes it unlikely that the palm populations were ever isolated in allopatry, making the case for sympatric speciation compelling.

There are few documented cases of sympatric speciation as compelling as that of the Lord Howe palms. The low number of reports doesn't necessarily mean that sympatric speciation is rare; it may simply be hard to document. Making matters even more complex, it appears that some species have evolved through a combination of allopatric and sympatric speciation. For example, they may have diverged initially in allopatry but did not evolve in complete reproductive isolation through the process of reinforcement until after they later became sympatric. In these cases, researchers might end up studying contemporary populations that live in sympatry without realizing that the initial divergence between them actually had occurred while they were geographically separated. Today, biologists generally agree that speciation in sympatry is possible, based on theoretical models of population genetics, although the process is more complex than speciation in allopatry.

A Song of Speciation

Kerry Shaw, of Cornell University, studies a particularly spectacular group of new species—the 37 species of swordtail crickets found only in Hawaii.

To understand the origin of these species, Shaw and her colleagues gathered DNA samples and reconstructed their phylogeny (**Figure 13.11**). This phylogeny enabled the researchers to map the route that the insects took from island to island. The pattern

A

B

Adjusting to calcareous soils caused trees to flower earlier

Frequency of flowering

- - - *Howea belmoreana*
——— *Howea forsteriana*

Weeks

Howea belmoreana   *Howea forsteriana*

Figure 13.10 A: Lord Howe Island and its two species of *Howea* palm. Molecular clock estimates suggest that these species diverged from each other well after the island formed, and the island is too small (14.5 km² total area) for the palms to have ever experienced allopatry. Instead, they diverged in sympatry as they began to inhabit two chemically distinct soil types, an acidic olcanic basalt on the sides of the volcano and a calcareous sediment (basic pH) that collected in the lowlands. B: The stark shifts in physiology required to grow on the calcareous soil also caused trees to flower 6 weeks earlier than they otherwise would have; as a result, trees on the calcareous soils finished flowering before trees on the volcanic basalt soils had started. Males and females are indicated by solid and dashed lines, respectively. (A: Top: Gary Bell/Oceanwidelmages.com; bottom: William J. Baker, Royal Botanic Gardens, Kew; data from Savolainen et al. 2006)

was clear: the oldest lineages are found on the westernmost islands. Later, the swordtail crickets hopped eastward from island to island (Mendelson and Shaw 2005).

This pattern is strikingly similar to the origin of the islands themselves. Like Lord Howe Island, the Hawaiian Islands owe their existence to molten rock upwelling from a mantle plume beneath the floor of the Pacific Ocean. Initially, this magma built a volcano that rose up to the ocean's surface, forming the first island of the Hawaiian archipelago. Earth's crust was moving slowly to the west, however, and so over time the island drifted away from the upwelling.

Without the magma, the island no longer grew; instead, the underlying rock cooled and the island sank. Eventually the island disappeared underwater. Meanwhile, the plume had begun to build a new island to the east. This cycle of growth and disappearance has repeated itself many times over, producing Hawaii's chain of islands. The youngest island, not surprisingly, is the largest and most volcanically active: the big island of Hawaii, which is only about 430,000 years old (**Figure 13.12**).

Shaw's research suggests that swordtail crickets first colonized Hawaii's oldest islands and then expanded to the newer islands as these islands emerged from the ocean. These movements apparently have been rare because Shaw and her colleagues haven't found genetic evidence of crickets moving back east and interbreeding. Once the crickets came to a new island, it seems, they were isolated from populations on

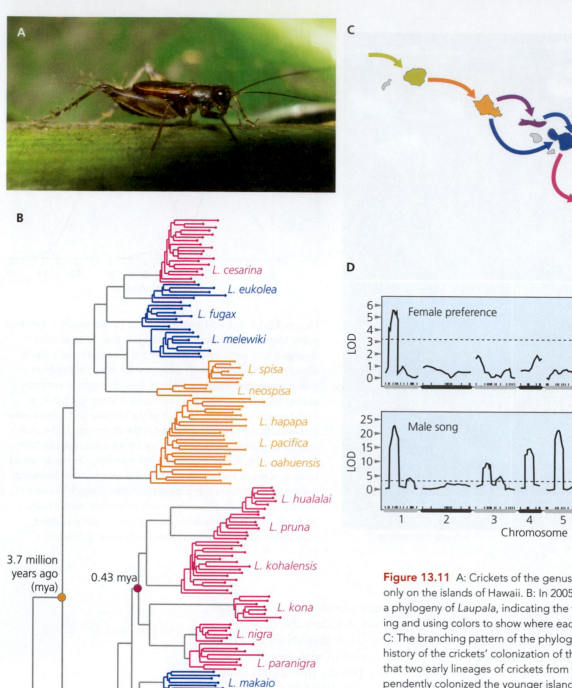

Figure 13.11 A: Crickets of the genus *Laupala* are found only on the islands of Hawaii. B: In 2005, scientists published a phylogeny of *Laupala*, indicating the timing of the branching and using colors to show where each species is found. C: The branching pattern of the phylogeny reflects the history of the crickets' colonization of the islands, indicating that two early lineages of crickets from Oahu (orange) independently colonized the younger islands of the chain (purple, blue, pink). (Data from Mendelson and Shaw 2005) D: Populations on a single island often differ in features of both the male song (pulse rate) and the female preference for those songs. Kerry Shaw and Sky Lesnick used QTL mapping to identify loci responsible for song differences between the Big Island species *L. kohalensis*, which sings songs with a fast pulse rate, and *L. paranigra*, which sings a much slower song. They found that loci contributing to changes in pulse rate mapped to the same location (chromosome 1) as loci contributing to the evolution of female song preferences. The co-localization of these loci may enable the rapid speciation of the crickets. LOD refers to the logarithm of the odds score. An LOD greater than the value indicated by the dashed line means that parental alleles at that genetic marker are significantly associated with the trait in question. (A: Kerry L. Shaw; data from Shaw and Lesnick 2009)

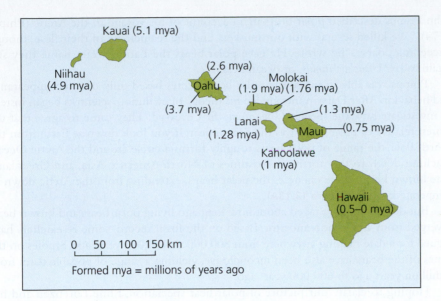

Figure 13.12 The island chain of Hawaii was formed by the motion of Earth's crust over a large underlying hotspot. Many of the older islands in the chain have sunk under the ocean's surface. The largest islands are also the youngest.

other islands and then diverged to the point that they no longer recognize crickets on other islands as potential mates.

It's important to note that speciation did not simply produce a single new species per island. Shaw's research shows that when the insects arrived on a new island, they rapidly radiated into several new species. The rate of speciation on the big island, where six species have evolved in the past 430,000 years, is especially striking—more than ten times higher than the average rate of speciation in arthropods.

What makes this evolutionary explosion so remarkable is the lack of evidence that the crickets have diverged ecologically (Mullen and Shaw 2014). Unlike the Lord Howe Island palms, they have not shifted to occupy a new habitat, nor have they begun feeding on new foods. Still, there is one striking phenotypic difference between the crickets: their courtship songs.

Male swordtail crickets rub their legs on their bodies to produce a rapid series of chirps. Each species produces these chirps in a distinct pattern, and the females are keenly sensitive to tiny variations in their songs. Shaw and her colleagues have also found that even within species, the cricket songs have diverged along with the female preference for different songs.

These findings led Shaw to do what an increasing number of biologists are doing: search for the genes involved in speciation. Given the importance of courtship songs for species barriers in swordtail crickets, she investigated their genetic basis. In one study, she and her colleagues hybridized two closely related species from the big island of Hawaii. The male hybrids displayed a range of courtship song patterns, and the females displayed a range of preferences. Using QTL analysis (Section 7.4), Shaw and her colleagues discovered the loci that were associated with each trait. Remarkably, they found that the same loci were associated with both male song production and female song preference (Shaw and Lesnick 2009; Wiley et al. 2011).

As the cricket species expand their range, Shaw's research suggests, genetic variations cause populations to diverge in both their male songs and female preferences. These shifts rapidly introduce reproductive barriers between the populations, opening the path to speciation in the presence of gene flow. Given that the new species of swordtail crickets evolve at an astonishing rate, Shaw's research hints that sexual selection may play a powerful role in many cases of speciation (see Chapter 11).

Polar Bears: A Tale of Ecological Speciation and Interspecies Gene Flow

When European naturalists first set eyes on the polar bear, they immediately recognized it was a separate species. In 1773, the British naval officer Constantine

John Phipps described polar bears in an account of his voyage to the Arctic (Phipps 1774). "We killed several with our muskets, and the seamen ate of their flesh, though exceeding coarse," he wrote. He gave polar bears the Latin species name they still retain today: *Ursus maritimus*, or ocean bear.

Phipps was able to judge the polar bear species based only on their appearance and behavior. After Darwin introduced his theory of evolution, scientists began instead to investigate how the species *U. maritimus* first evolved. They came to agree that the closest relative of polar bears was the brown bear. If you look down at Earth from the North Pole, the range of polar bears roughly forms a circle around the Arctic Ocean, reaching down to the northern coastlines of North America, Asia, and Greenland. The brown bear's range encircles the polar bear's, extending from the Arctic down to temperate latitudes (**Figure 13.13A**).

But debate has long raged about how long ago living polar bears and brown bears diverged from a common ancestor. Based on the fossil record, some researchers have argued for a date ranging anywhere from 800,000 to 150,000 years ago. Studies on the genes of the bears have also been inconclusive, yielding a range of possible dates from 5 million years ago to 600,000 years ago.

Hoping to clarify our picture of polar bear speciation, Eline Lorenzen and her colleagues traveled to Greenland to get polar bear DNA. They sequenced the entire genome of a number of individual polar bears, and they also sequenced the genomes of brown bears from Alaska, Montana, and Scandinavia (Liu et al. 2014). This unprecedented amount of data enabled the scientists to reconstruct many details of polar bear evolution, such as their divergence from brown bears and their population sizes at different times in their history (see Box 8.1).

The scientists estimate that polar bears diverged from brown bears relatively recently—somewhere between 479,000 and 343,000 years ago. At that time, the planet was in a warm period between ice ages. Much of Greenland was covered in a dense spruce forest. Lorenzen and her colleagues suspect that this warm period made it possible for brown bears to expand their range northward into regions that previously were uninhabitable.

At first, the northern populations of brown bears would have survived as they had at lower latitudes, feeding on land on a variety of food, from fishes to berries. But then the climate changed, replacing high Arctic forests with tundra and ice sheets. The northern brown bears became isolated from the brown bears and began to evolve in allopatry.

Unlike the swordtail crickets, polar bears experienced strong selection for rapid divergence in ecological traits. Fossils of polar bears dating back 110,000 years ago clearly demonstrated that by then, *U. maritimus* had evolved from a terrestrial bear to a bear specializing in hunting marine mammal prey at sea. Based on the estimate that Lorenzen and her colleagues came up with for the divergence of polar bears and brown bears, this transformation may have taken place in as little as 230,000 years or less—about 20,500 generations.

To understand this adaptation better, Lorenzen and her colleagues looked for molecular traces of positive selection (Section 8.7) in polar bears after their split with brown bears. The researchers first looked for genes that were different in brown and polar bears; these divergent genes may explain some of the differences between the species. Then they looked for genes that had low variation *within* polar bear populations. Recall that strong selection can sweep alleles to fixation, carrying with them flanking stretches of DNA. Thus, genes with low variation may be needed to maintain adaptive polar bear phenotypes. Using this information (along with other measures of selection; see Table 8.1), the scientists found that, consistent with their rapid adaptation to a harsh new environment, polar bears experienced stronger selection than brown bears.

Many of the genes that have experienced the strongest selection in polar bears show clear links to their new ecological niche. One gene, for example, called *APOB*, encodes a protein that binds cholesterol. Because polar bears feed on seals, their diet

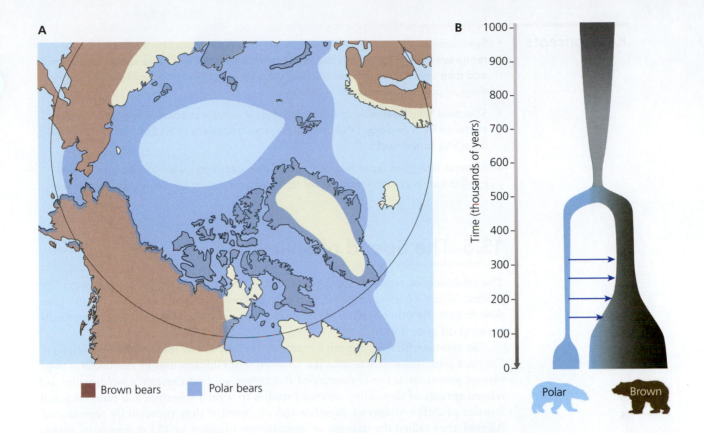

A

B

Brown bears   Polar bears

Polar   Brown

Figure 13.13 Polar bears evolved into a distinct species starting about 500,000 years ago. They diverged from brown bears, likely becoming isolated in the Arctic. Yet even after adapting to their new habitat, they continued to interbreed with brown bears. The result was sporadic gene flow. (Data from Liu et al. 2014)

is rich in fatty acids, leading to a high level of cholesterol in their blood. A drastic change to their APOB proteins may enable them to reduce the harm that comes with such a diet, such as heart disease. Lorenzen and her colleagues also found that genes involved in heart function have evolved rapidly in polar bears. They may be under selection to cope with the polar bear diet as well, or perhaps with their demanding swims through frigid water.

Other rapidly evolving genes in polar bears include genes involved in pigmentation. As polar bears shifted from living in forests to living on ice floes, the production of pigment in their hairs decreased, leading to a white coat that helped them blend into their new background.

Once a pocket of northern brown bears diverged and became polar bears, the new study suggests, they rapidly spread around the Arctic Ocean. The cycle of ice ages continued. During cold periods, the polar bears' range expanded, and during warm periods, it shrank. The climate changes also affected the brown bears, and sometimes the two species came back into contact with each other.

Lorenzen and her colleagues (as well as other researchers; see Cahill et al. 2015) discovered evidence that polar bears and brown bears have interbred since they diverged. Intriguingly, the genomes of the bears indicate that this gene flow has traveled only one way, from polar bears to brown bears (**Figure 13.13B**). On rare occasion, scientists have found hybrids between the two species—nicknamed "pizzlies" or "grolar bears." Such interbreeding has been going on for hundreds of thousands of years, Lorenzen's research suggests, and the hybrids have been able to contribute their DNA to future generations.

This finding raises some intriguing questions. Why, for example, don't polar bears and brown bears collapse back into a single species? Although they can and do interbreed, it's possible that most alleles from one species are selected against in the other, so that the species remain distinct in the face of gene flow. As we'll see in Chapter 17, we humans experienced a similar flow of genes in the past 50,000 years as our ancestors interbred with Neanderthals and other extinct groups of humans.

- Speciation is often a complex process, and models of speciation are based on currently available evidence. As scientists understand more and more about populations and their interactions, they are gaining valuable insight into the processes leading to these important divergences in lineages.

- Scientists have been able to test hypotheses about the process of speciation using evidence from geology, DNA, and phylogenetic analyses as well as with mating and breeding experiments.

- Because it triggers rapid and local coevolution between males and females, sexual conflict can lead to the evolution of reproductive barriers between populations. ●

13.5 The Speed of Speciation

The evolution of reproductive barriers of the sort we've described so far is a gradual process. When a barrier first evolves, it may be very weak, only slightly reducing gene flow. Eventually other barriers evolve, and over time they become stronger, gradually choking off gene flow.

To estimate how long it takes reproductive barriers to lead to new species, Jerry Coyne, a population geneticist at the University of Chicago, and H. Allen Orr, a population geneticist at the University of Rochester, studied *Drosophila melanogaster* and related species of flies. They reviewed studies in which scientists had put males and females of different species together and observed if they successfully reproduced. All told, they tallied the strength of reproductive isolation of 171 species pairs, giving them a score of 0 to 1. Species that never had hybrid matings scored a 1. If species mated freely with each other, they scored 0. Most pairs fell somewhere in between (Figure 13.14). Coyne and Orr then used the molecular clock (see Section 8.6) to calculate how long ago each pair of species had descended from a common ancestor. They found it can take as long as several hundred thousand years for two populations of flies to become isolated enough to be considered true species by the biological species concept (Coyne and Orr 1997). Interestingly, when Coyne and Orr looked at the

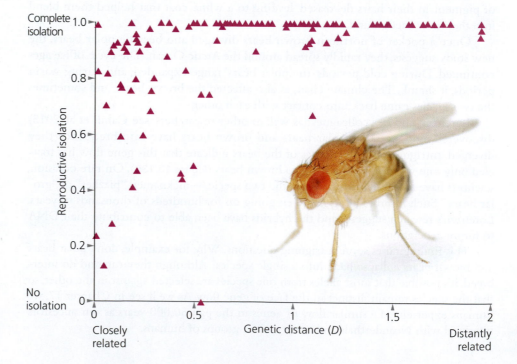

Figure 13.14 This graph shows how reproductive isolation evolved among species of *Drosophila*. The genetic distance (*D*) between two species increases with time. It takes roughly a million years for *D* to reach a value of 1. By then, a typical pair of *Drosophila* species no longer interbreeds. (Data from Coyne and Orr 2004; photo by Antagain / E+ / Getty Images)

different forms of reproductive isolation in the flies, they found that prezygotic isolation often evolved much faster than postzygotic isolation.

Although speciation sometimes takes hundreds of thousands of years, in other cases it can take just a few generations. And many species of plants have a biology that enables them to speciate within a generation. If a pollen grain lands on a flower of a different species, it may succeed in fertilizing an ovule, producing a hybrid offspring. In many cases, the hybrid is sterile. In other cases, the hybrid can breed with one of its parental species. But in still other cases, the hybrid can evolve into a new species. **Figure 13.15** shows how this can happen. A species with two pairs of chromosomes mates with a species with three. The hybrid cannot produce viable gametes because it

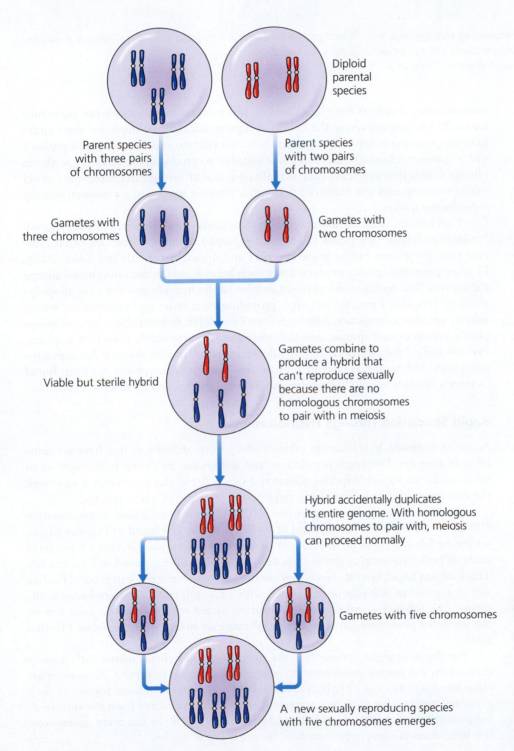

Diploid parental species

Parent species with three pairs of chromosomes

Parent species with two pairs of chromosomes

Gametes with three chromosomes

Gametes with two chromosomes

Viable but sterile hybrid

Gametes combine to produce a hybrid that can't reproduce sexually because there are no homologous chromosomes to pair with in meiosis

Hybrid accidentally duplicates its entire genome. With homologous chromosomes to pair with, meiosis can proceed normally

Gametes with five chromosomes

A new sexually reproducing species with five chromosomes emerges

Figure 13.15 Allopolyploidy is a common form of speciation that occurs when two species hybridize. If the resulting hybrid offspring have an odd number of chromosomes, they cannot reproduce sexually. A duplication event in the hybrid's asexually produced offspring can double the number of chromosomes, thus allowing the new species to reproduce sexually.

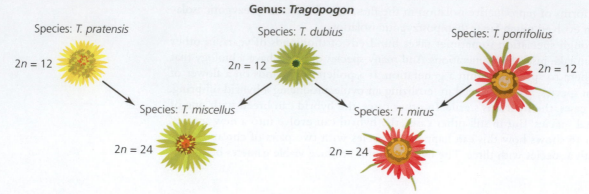

Genus: *Tragopogon*

Species: *T. pratensis*

$2n = 12$

Species: *T. dubius*

$2n = 12$

Species: *T. porrifolius*

$2n = 12$

Species: *T. miscellus*

$2n = 24$

Species: *T. mirus*

$2n = 24$

Figure 13.16 Examples of recent speciation by allopolyploidy in salsify flowers (*Tragopogon*). *T. dubius* hybridized with two other species of salsify, once with *T. pratensis* to produce *T. miscellus* and another time with *T. porrifolius* to produce *T. mirus*. (Data from Soltis and Soltis 2009)

cannot evenly divide its five chromosomes. But in some cases, hybrids can reproduce asexually. On rare occasions, the hybrid's offspring accidentally duplicate their entire genome, resulting in ten chromosomes. Now the chromosomes can pair and produce viable gametes. Those gametes can come together to produce new offspring with ten chromosomes that can reproduce sexually. But the offspring can't interbreed easily with the two species that produced it. In other words, it has become a distinct, sexually reproducing species.

This process is called **allopolyploidy**. Scientists sometimes call allopolyploidy "instant speciation" (see **Figure 13.16** for an example), although studies on plants indicate that the process can be more complex and drawn out (Soltis and Soltis 2009). In some cases, two species produce many such hybrids, which then interbreed among themselves. The multiple independent origins of the hybrids give the new allopolyploid species added genetic variation, providing even more raw material on which natural selection can operate. Although there's much left to learn about how allopolyploidy produces new species, one thing is clear: it's surprisingly common in plants. Perhaps half of the estimated 300,000 flowering plant species evolved through allopolyploidy. Although it is rarer in animals than plants, allopolyploidy has been found in insects, crustaceans, mollusks, fishes, amphibians, reptiles, and even some mammals.

Allopolyploidy refers to polyploidy (more than two paired chromosomes) resulting from interspecific hybridization. (If polyploidy arises within a species, it's called autopolyploidy.)

Rapid Speciation Through Hybridization

As we've discussed, hybridization often erodes genetic differences that have accumulated in allopatry. Divergent populations and species can be drawn back together to homogenize into a single species (Grant and Grant 2014). But sometimes it can work the other way around: hybridization can lead to the origin of a new species.

In 1981, a young male cactus finch (*Geospiza conirostris*) flew approximately 100 kilometers from his home island of Española to the tiny island of Daphne Major, where he landed right in the middle of Peter and Rosemary Grant's long-term study of beak evolution in ground finches (*Geospiza fortis*) (discussed in Chapter 10). This bird was larger, but otherwise very similar in shape to the resident ground finches, and by the end of that year he had mated with a *G. fortis* female and sired several offspring. The following year one of his offspring mated with another *G. fortis* female, but from that point forward none of the hybrids ever mated with the ground finches again.

For the next eight generations, the newly formed hybrids mated only among themselves, and despite extensive inbreeding, they managed to survive. Now there are eight breeding pairs and 23 hybrid individuals living on the island, and surveys of their genomes confirm that they are completely reproductively isolated from the surrounding population of ground finches (Lamichhaney et al. 2018). In just three generations, this hybridization event appears to have spawned a new species.

- The speed of speciation can vary enormously among taxa. In plants, interspecific hybridization and allopolyploidy can generate new species rapidly. In most birds and mammals, however, reproductive isolation can take millions of years.

- Although hybridization often erodes any genetic differences that have accumulated among species, it can occasionally produce a new species that is reproductively isolated from either of the parent species. ●

13.6 Uncovering Hidden Species

The studies we've discussed so far all focus on how new species form. But it's also important to recognize the boundaries that divide current species. Policies intended to preserve species require a way to tell which species a given organism belongs to. It's less important to determine the reasons that two groups of organisms became genetically distinct than to have a reliable way to discern their current differences. For conservation biologists, the general lineage species concept is therefore more practically useful than the biological species concept.

As scientists accumulate more types of genetic information, they can use this knowledge to do a better job of classifying the world's biological diversity. In Costa Rica, for example, Daniel Janzen, of the University of Pennsylvania, and his colleagues have been trying to document the true extent of insect diversity. In one study, he has examined the neotropical skipper butterfly, *Astraptes fulgerator*. This handsome blue and black insect was first described as a species in 1755, and it has been recorded living as far north as the United States and as far south as Argentina.

Over 25 years, Janzen has reared thousands of skipper caterpillars, and he noticed that they feed on a wide range of plants. Caterpillars are usually very fussy about which plants they feed on, because they produce chemicals that are finely tuned for overcoming defensive chemicals in plants. It seemed odd that the neotropical skipper butterfly would be able to feed on so many different plants. A closer look revealed that caterpillars that prefer the same plants tend to have the same color patterns. Janzen and his colleagues also found subtle differences among the adult forms into which the caterpillars developed. Janzen hypothesized that *A. fulgerator* might be six or more species rather than just one.

To test whether these so-called **cryptic species** exist, Janzen joined forces with Paul Hebert, a biologist at the University of Guelph in Canada. To distinguish between species, Hebert and other scientists identified short segments of fast-evolving DNA contained within the mitochondria of cells that scientists can examine with relatively little effort. The divergence of these hypervariable segments of DNA is not likely to have caused speciation, of course, but it can act as a marker for a broader divergence over the entire genome. This method, known as DNA barcoding, can help discern whether populations are behaving as genetically distinct, independently evolving lineages.

Hebert, Janzen, and their colleagues used DNA barcoding to measure the genetic divergence of the butterflies. In 2004 they reported that the insects formed ten distinct genetic clusters, each of which produced a distinct color pattern as caterpillars (**Figure 13.17**; Hebert et al. 2004).

Many of the caterpillars that Janzen has captured over the years have turned out to be infected with parasitoid wasps. Adult female wasps inject the eggs into caterpillar hosts, where they hatch and develop as larvae, feeding on the still-living caterpillars. Janzen wondered if the wasps might form cryptic species as well. Janzen, Hebert, and their colleague James Whitfield, from the University of Illinois, analyzed 2597 different wasps (Smith et al. 2008). Upon inspecting the wasps visually, the scientists recognized six different genera of wasps parasitizing the caterpillars. On closer examination, they were able to sort them into 171 provisional species. But DNA barcoding revealed

Cryptic species are lineages that historically have been treated as one species because they are morphologically similar but that are later revealed to be genetically distinct.

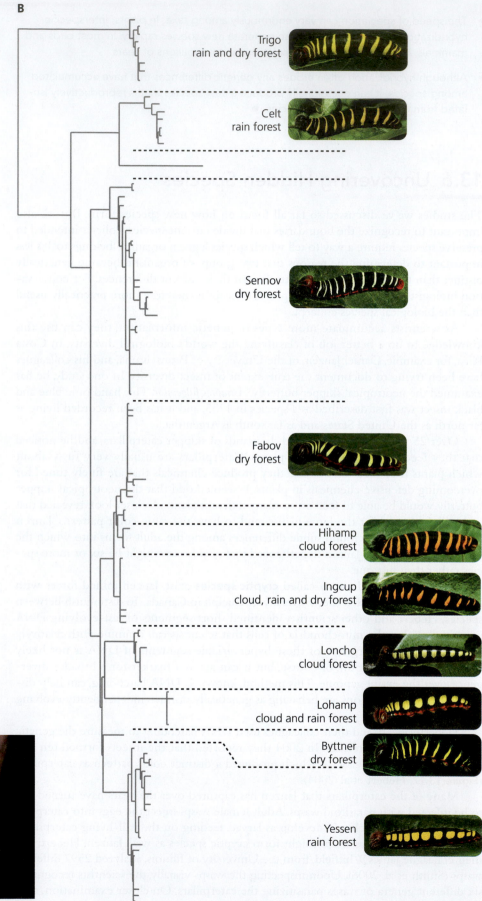

Figure 13.17 A: Skipper butterflies in Costa Rica have a wide range of color patterns as caterpillars. B: An analysis of their DNA reveals that populations have diverged from one another, indicating a diversity of species that had previously gone unnoticed. (Photos by Dan Janzen. "Ten species in one: DNA barcoding reveals cryptic species in the neotropical skipper butterfly *Astraptes fulgerator*" by Herbert et al. PNAS, 2004: Vol. 101, No. 41, pp.14812–14817; data from Hebert et al. 2004)

142 more species, bringing the total to 313. One species, a tiny black wasp known as *Apanteles leucostigmus*, turned out to be 32 different cryptic species!

Although results like this are exciting, Janzen and other scientists also know that identifying species can't happen too soon. Biodiversity, especially in tropical rain forests, is facing extraordinary threats from humans, including logging operations and growing cities (see Chapter 14 for more on extinctions). Janzen and his colleagues want to know how many species there are in places like Costa Rica so that they can figure out how to save as many of them as possible.

Key Concepts

- Cryptic species may diverge from ancestral populations without evolving easily distinguishable morphologies.

- Identifying cryptic species is important to measures of biodiversity. ●

13.7 The Puzzle of Microbial "Species"

In Linnaeus's grand scheme of classification, he never named a species of bacteria. Only in the nineteenth century did microbiologists start cataloging the diversity of microbial life. And when they did so, they followed Linnaeus's example, giving microbes proper Latin names like *Bacillus anthracis* and *Escherichia coli*. To distinguish one species from others, microbiologists examined its observable traits—its appearance under a microscope, whether it took up a particular stain, what nutrients it needs to survive, and so on.

In the twentieth century, zoologists and botanists reconsidered the concept of a species in the light of evolution. They developed the biological species concept and other concepts. But these concepts proved difficult to apply to microbes. Bacteria and archaea, for example, do not exist as males and females that reproduce sexually. Instead, they mainly reproduce asexually, producing identical or nearly identical clones. The biological species concept as envisioned by Mayr has no relevance for them.

The Phylogenetic Species Concept Applied to Microbes

With the advent of molecular biology, microbiologists could use the phylogenetic species concept. One widespread method involves examining a gene for a ribosomal RNA molecule called *16S rRNA*. This gene, which is found throughout the tree of life, evolves slowly enough that it is phylogenetically informative even when scientists are comparing lineages that descend from a common ancestor that lived billions of years ago.

Microbiologists compared the variations in *16S rRNA* genes from individual microbes that, according to the traditional methods of diagnosing a microbial species, belonged to the same species. They found that the gene varied by up to 3% within a traditional species. To emulate earlier methods to diagnose species, these researchers decided that if two microbes were at least 97% identical in the *16S rRNA* gene, they belonged to the same taxonomic group. They referred to such a group of microbes as a species, or sometimes as an operational taxonomic unit, or OTU (Stackebrandt and Goebel 1994).

By this standard, the diversity of microbes is vast. A single spoonful of soil may have 10,000 different species of bacteria, most of which cannot be grown and characterized in pure culture. To put that in perspective, there are only 6399 species of mammals on the entire planet (Burgin et al. 2018). Each survey microbiologists carry out—whether it is in a lake, in the Arctic tundra, or in the human body—continues to yield many DNA sequences that do not closely match those of anything found before. What's more, scientists have come to recognize that many microbial species are so rare that conventional surveys are likely to miss them (Sogin et al. 2006). In 2016, Kenneth Locey and Jay Lennon, two biologists at Indiana University, extrapolated from current studies to estimate how many species of microbes there are on Earth. They settled on an astonishing number: one trillion species (Locey and Lennon 2016).

However, scientists also recognize that the 97% threshold is somewhat arbitrary. To appreciate this arbitrariness, consider a pair of *Streptococcus* species that live on our bodies, *Streptococcus pneumoniae* and *S. mitis*. These microbes were identified as separate species before the advent of molecular biology. Microbiologists simply observed their strikingly different ecological niches in our bodies: *S. pneumoniae* is a pathogen that can cause pneumonia by growing in our airway. *S. mitis*, by contrast, lives harmlessly on teeth. Yet the *16S rRNA* sequences of these two "species" turn out to be more than 97% identical. In fact, they're 99% identical. If scientists relied on only their genetic similarity, then these two profoundly different types of organisms would be considered a single species (Eren et al. 2013).

As if that wasn't confusing enough, microbes make species concepts even more challenging because they rampantly share genes across lineages.

Horizontal Gene Transfer Among Microbes

Microbes can acquire genes from other species by several mechanisms. In homologous recombination, a microbe that is repairing a damaged segment of DNA can replace it with a closely matching version from another microbe. Viruses can incorporate DNA from a host microbe belonging to one species and then insert it into the genome of another host belonging to an entirely different species. Plasmids can engineer their passage from one microbe to another.

Although the chances of any one microbe acquiring genes by horizontal transfer are small, the cumulative impact of the process on microbes has been immense. One of the most impressive demonstrations has been made in the study of *E. coli*. Scientists have long recognized a number of strains of *E. coli*; some are pathogenic and others are harmless or even beneficial in the human gut (Zimmer 2008).

When researchers began comparing the genomes of different *E. coli* strains, they initially expected to find a few differences between them—some point mutations on genes, for example, or a few homologous recombination events. They were sorely mistaken.

In 2002, a team of researchers compared the three genomes of *E. coli* strains. Only a small "core" of genes was shared by all three of them. The total number of genes found in all three strains—the so-called pan-genome—far outnumbered the *E. coli* core genome (Welch et al. 2002). In the years that followed, many teams of researchers have followed suit. By 2015, they had sequenced 2085 strains. The core genome shrank further, stabilizing at about 3188 genes. But the pan-genome grew, reaching roughly 90,000 genes (Land et al. 2015). In other words, only about 3% of *E. coli* genes are essential for all strains of *E. coli* (**Figure 13.18**).

Findings such as these raise serious questions about the usefulness of the phylogenetic species concept for microbes (Doolittle and Zhaxybayeva 2009). The

Figure 13.18 A: After the genomes of three strains of *E. coli* were sequenced in 2002, scientists discovered that they share only a limited "core" of genes. (Data from Welch et al. 2002) B: By 2015, researchers had sequenced 2085 genomes of *E. coli* strains. The core genome is now made of only 3188 genes, whereas the pan-genome has expanded to as many as 90,000 genes. (Data from Land et al. 2015)

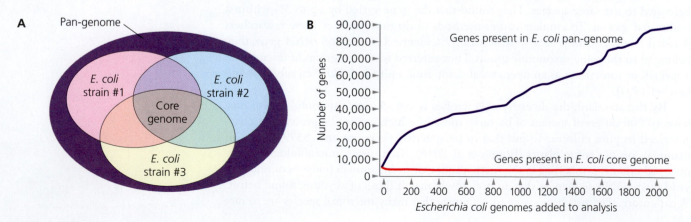

phphylogenetic species concept is useful only if organisms reliably pass down their genes to their offspring. If genes are moving frequently from one lineage to another, life seems to blur into a jumble of mosaic-like genomes (**Figure 13.19**).

Moreover, larger-scale studies demonstrate that this pattern applies not just to *E. coli* but also across all bacteria and archaea. Tal Dagan, a biologist at the University of Düsseldorf, and her colleagues carried out a large-scale search for horizontal gene transfer by analyzing the genomes of 181 distantly related species of bacteria. All told, their analysis encompassed 539,723 genes (Dagan et al. 2008). By constructing phylogenies for the genes, they could reconstruct their history, identifying how they were passed down by vertical descent from ancestors to descendants as well as by horizontal transfer to other species. The scientists concluded that, on average, 80% of the genes in each of the species they studied had experienced horizontal gene transfer in the past.

If the tree of life is supposed to represent the vertical descent of genes from one generation to the next, then it's necessary to join its branches together with weblike strands to show the rampant horizontal gene transfer that has moved genes from species to species for billions of years (**Figure 13.20**).

Does this mean that species do not exist among microbes? Are they simply fleeting aggregations of genes that will be shuffled into new configurations in future generations? A number of microbiologists don't think so, and they're developing new species concepts that can explain the ecological and evolutionary patterns of microbes.

Stable Ecotype Model

Frederick Cohan, an evolutionary geneticist at Wesleyan University, has offered an influential species concept for microbes known as the **stable ecotype model** (Cohan 2006). In essence, Cohan argues that microbes undergo ecological speciation like animals and plants do.

In studying the dense ecosystem of microbes that live around hot springs in Yellowstone National Park, Cohan and his colleagues noticed that the bacteria and archaea could be sorted into genetic and ecological clusters. Each genetically related group of microbes lives in a certain niche in the hot springs—enjoying a certain

Stable ecotype model is a species concept constructed for microbes in particular. In this model, a species is a long-lived population of genetically related individuals that share a stable set of adaptations for the same ecological niche.

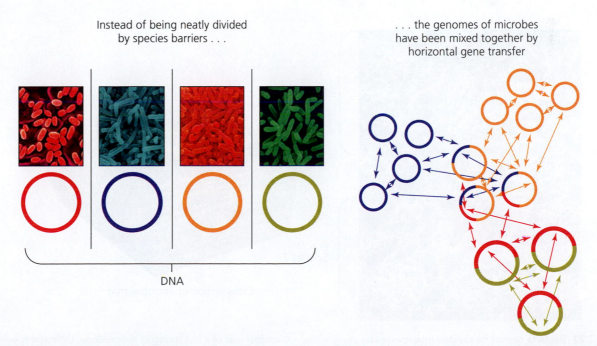

Instead of being neatly divided by species barriers . . .

. . . the genomes of microbes have been mixed together by horizontal gene transfer

DNA

Figure 13.19 The distribution of homologous genes in bacteria and archaea suggests that rampant horizontal gene transfer has taken place over billions of years, transforming microbial genomes into mosaics. This mixing of genes makes it difficult, if not impossible, to delineate species boundaries. (Photos by Dennis Kunkel Microscopy/Science Source; data from Doolittle and Papke 2006)

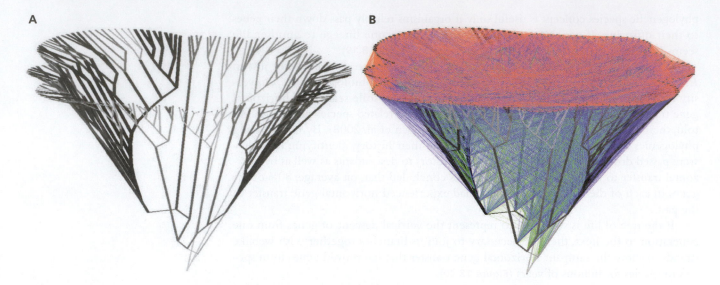

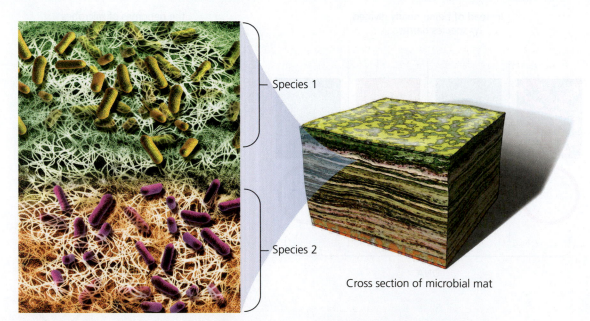

Figure 13.20 A: An evolutionary tree of 181 species of bacteria, calculated from some of their genes. B: Over billions of years, many genes have moved from one branch to another. Horizontal gene transfer events are indicated here by colored lines. (Republished with permission of The Royal Society from Dagan T, Martin W. Getting a better picture of microbial evolution en route to a network of genomes. Philos Trans R Soc Lond B Biol Sci. 2009;364 (1527):2187–96; permission conveyed through Copyright Clearance Center, Inc.)

temperature, for example, or requiring a certain amount of sunlight. The genetic and ecological clusters thus correspond to each other (**Figure 13.21**).

Cohan argues that when lineages of microbes adapt to a particular ecological niche—one that's distinct from the niche of other lineages—it's appropriate to call them a species. These species endure over time because selection strongly favors mutations that enable them to exploit their niche more effectively. When a beneficial new mutation arises in a microbe, its descendants swiftly outcompete other members of the species. Selection thus prunes a species' side branches, maintaining a strong genetic similarity among its members.

Variation on the Stable Ecotype Model

B. Jesse Shapiro, of the Université de Montréal, and his colleagues have more recently developed a more complex model of microbial speciation that takes into account horizontal gene transfer. Their model emerged out of research they did on *Vibrio*

Figure 13.21 Bacteria cannot be divided into species by using the rules that apply to animals. A different way to define bacterial species is based on the way they are adapted to narrow ecological niches. This figure shows how bacteria are distributed in the

microbial mat of a hot spring in Yellowstone. Different species are adapted to particular temperatures and concentrations of different minerals and other nutrients.

Species 1

Species 2

Cross section of microbial mat

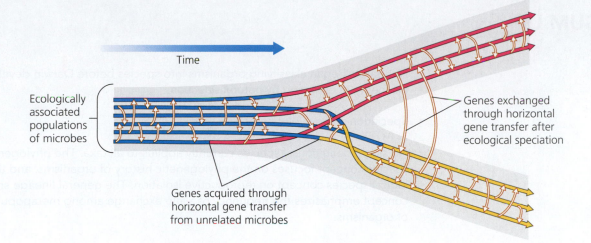

Time

Ecologically associated populations of microbes

Genes exchanged through horizontal gene transfer after ecological speciation

Genes acquired through horizontal gene transfer from unrelated microbes

cyclitrophicus, a bacteria that they collected in seawater off the coast of Massachusetts (Shapiro et al. 2012). They found that some of the bacteria grew in biofilms on small animals and algae, and others floated in the open water. Here were two groups of closely related bacteria, each of which had undergone natural selection for a different ecological niche. According to the stable ecotype model, you would expect that selective sweeps would have caused the genomes of the two new species to diverge. The mutations that distinguished the two species would be distributed across the entire genome.

That's not what the scientists found, however. They identified 725 point mutations that partitioned the bacteria into the two ecological groups. But those mutations were concentrated into 11 "islands" of DNA. Three of those islands contained 80% of the mutations. Outside of those islands, the scientists found many stretches of DNA that had been swapped between the two groups due to homologous recombination.

Shapiro and his colleagues proposed a variation on the stable ecotype model to explain these results, one that takes horizontal gene transfer into account (Shapiro 2014). The *V. cyclitrophicus* population started out as an ecologically uniform group of bacteria in which homologous recombination occurred frequently. Then some members of the group acquired adaptive alleles from other bacteria through homologous recombination. These alleles became more common, either through selection or through homologous recombination. As a result, the once uniform population of bacteria began to diverge into two groups, each with a separate ecological adaptation (Figure 13.22).

As the groups diverge, selection favors mutations that improve their performance in their own niche. Homologous recombination continues to occur within each group, but it becomes less common between the groups. That's partly because the two groups occupy different places in the ecosystem, so there's less opportunity for them to trade genes. But selection begins to eliminate the genes from the other group because they lower a microbe's fitness.

These new models are helping scientists come to terms with the world's staggering microbial diversity. And they are also helping the medical community better understand the pathogens that make us sick. Doctors still rely on outmoded traditional species names to classify pathogens. In years to come, a new classification may emerge that recognizes the actual evolutionary history of these disease-causing microbes.

Figure 13.22 Some scientists have developed speciation models for bacteria that take horizontal gene transfer into account. Horizontal gene transfer is frequent within ecologically associated populations of microbes. Genes acquired from other lineages can be important for the emergence of lineages with new adaptations.

Key Concepts

- Microbial organisms present particular challenges to species concepts developed for eukaryotes because of the variation in rates and kinds of exchange of genetic material.

- New species concepts in microbiology may provide valuable insight into the evolutionary history of disease-causing organisms. ●

TO SUM UP . . .

- Naturalists began classifying organisms into species before Darwin developed his theory of evolution by natural selection.

- Speciation is the evolution of new species.

- No single definition of species currently applies to all taxa. The phylogenetic species concept focuses on the phylogenetic history of organisms, and the biological species concept on reproductive isolation. The general lineage species concept emphasizes the frequency of allele exchange among metapopulations of organisms.

- Gene flow can be interrupted by geographic and reproductive barriers. Populations can be physically separated from each other by features of the environment, and intrinsic properties of the organisms themselves can reduce the likelihood of interbreeding.

- Reproductive barriers can evolve between populations isolated by space or time. Prezygotic reproductive barriers reduce the likelihood that a zygote will form, either before mating or before fertilization. Postzygotic reproductive barriers prevent fertilized zygotes from successfully developing and reproducing themselves.

- Allopatric speciation occurs when geographically isolated populations evolve independently as a result of drift, mutation, and selection.

- Speciation can also take place without geographic isolation. Parapatric speciation, sympatric speciation, and allopolyploidy are examples.

- Complete reproductive isolation can take millions of years to evolve between two species of birds, for example; but in plants, allopolyploidy can lead to new species quite rapidly.

- Genetic studies sometimes reveal that what appears to be a single species may be several species.

- Species of bacteria and archaea are difficult to identify using concepts developed for sexually reproducing species.

1. In which case would the biological species concept *not* be useful?

   a. Different kinds of birds occurring sympatrically with very different appearances.

   b. A group of lizards reproducing asexually.

   c. Polar bears and grizzly bears living in the Arctic.

   d. Both b and c.

2. Why is defining the concept of species such a difficult task?

   a. Species are constantly evolving.

   b. Species are often defined in relation to research methods.

   c. Species are fixed taxonomic units—the difficulty arises from asexually reproducing organisms.

   d. Both a and b.

3. Which of the following is an example of gametic incompatibility?

   a. A male cat's reproductive organ does not fit a female's reproductive tract.

   b. A male shark deposits his sperm in a female shark, but those sperm fail to attach to her eggs.

   c. A male abalone produces faster-swimming sperm than another male abalone.

   d. A male coral releases sperm at a different time of day than another male coral.

4. Which is *not* an example of a pre-mating isolating barrier?

   a. Flowers that make pollen available to pollinators at different times of the day.

   b. Frogs that sing in ponds at different elevations.

   c. Females of a fish species that prefer one color of male fish over another color.

   d. The hybrid offspring of two mice that cannot produce viable gametes.

5. Which is an example of a postzygotic isolating barrier?

   a. The genitalia of a male duck that do not fit properly with the genitalia of females of another population.

   b. Females of one kind of fly that are not attracted to the buzz of another kind of male.

   c. Two species of bats breeding in different habitats.

   d. The hybrid offspring of two species of crocodiles that can produce normal gametes but cannot obtain a mate.

6. Which is the most likely order of events that could lead to allopatric speciation?

   a. Geographic separation, then genetic divergence, then reproductive isolation.

   b. Genetic divergence, then geographic separation, then reproductive isolation.

   c. Genetic divergence, then reproductive isolation, then geographic separation.

   d. Geographic separation, then reproductive isolation, then genetic divergence.

7. Why did Mendelson and Shaw examine the geological history of the Hawaiian Islands to test their hypothesis about cricket speciation?

   a. They predicted that the evolution of the cricket lineages matched the evolution of the islands.

   b. Their phylogeny was inconclusive and they needed more evidence.

   c. They wanted to find evidence to support the theory of continental drift.

   d. They predicted that allopatric speciation was more important in crickets than sympatric speciation.

8. Which of these statements about allopolyploidy is *false*?

   a. Allopolyploidy occurs only in plants.

   b. Allopolyploidy can involve the doubling of chromosomes as a result of hybridization.

   c. Allopolyploidy can quickly lead to speciation.

   d. In a phylogeny of *Tragopogon*, an allopolyploidy event would be represented as a merging of two branches.

9. Which of these statements about the speed of speciation is *false*?

   a. Speciation can take millions of years.

   b. Speciation can happen in a single generation.

   c. Speciation in plants is rarely due to hybridization.

   d. Speciation often happens faster in flowering plants than in animals.

10. Why are genetic tests a good way to discover cryptic species?

    a. They allow you to compare extinct individuals with living individuals.

    b. Grouping similar genotypes might reveal populations that do not or cannot interbreed.

    c. The alleles of a species will be identical in each individual.

    d. Cryptic species will have more mutations.

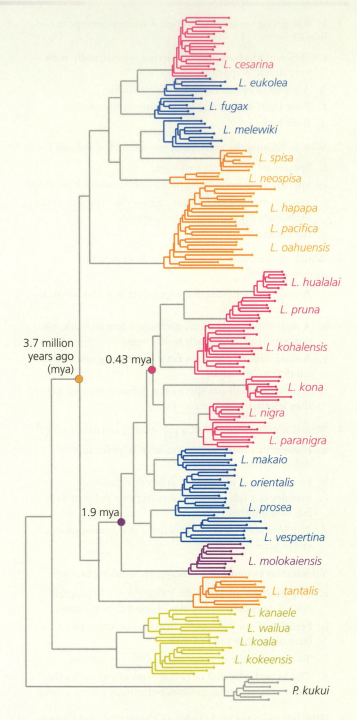

INTERPRET THE DATA   Answer can be found at the end of the book.

11. In the phylogeny at the right, Mendelson and Shaw reconstructed the historical patterns of relatedness among 25 species of Hawaiian cricket. They noted that often the closest relative of a species was another species that lived on the same island. They interpreted this as evidence that

   a. new species of cricket had formed after the animals colonized each island.

   b. new species formed primarily when crickets migrated from one island to another.

   c. crickets from the same island had adapted to different ecological habitats on each island.

   d. many of the cricket "species" were not actually true species.

SHORT ANSWER QUESTIONS   Answers can be found at the end of the book.

12. Do geographic isolating barriers function equally under the phylogenetic, biological, and general lineage species concepts?

13. Why would different kinds of bacteria and archaea be difficult to identify as different species using the biological species concept?

14. What are the similarities and differences between allopatric and sympatric speciation?

15. Under what ecological and evolutionary conditions is sympatric speciation most likely?

16. Which species concept would you apply to plants that hybridize relatively frequently, such as salsify flowers (*Tragopogon*), often resulting in allopolyploidy?

17. What does the plethora of cryptic species being discovered using genetic tools say about our understanding of what a species is?

ADDITIONAL READING

Coyne, J. A., and H. A. Orr. 2004. *Speciation*. Sunderland, MA: Sinauer Associates.

Eberhard, W. G. 1991. Copulatory Courtship and Cryptic Female Choice in Insects. *Biological Reviews* 66:1–31.

Howard, D. J., and S. H. Berlocher. 1998. *Endless Forms: Species and Speciation*. London: Oxford University Press.

Lowry, D. B., J. L. Modliszewski, K. M. Wright, C. A. Wu, and J. H. Willis. 2008. Review. The Strength and Genetic Basis of Reproductive Isolating Barriers in Flowering Plants. *Philosophical Transactions of the Royal Society B: Biological Sciences* 363 (1506): 3009–21.

Panhuis, T. M., R. Butlin, M. Zuk, and T. Tregenza. 2001. Sexual Selection and Speciation. *Trends in Ecology and Evolution* 16:364–71.

Price, T. 2008. *Speciation in Birds*. Greenwood Village, CO: Roberts and Company.

Rieseberg, L. H., and J. H. Willis. 2007. Plant Speciation. *Science* 317:910–4.

Ritchie, M. G. 2007. Sexual Selection and Speciation. *Annual Review of Ecology, Evolution, and Systematics* 38:79–102.

Schluter, D. 2009. Evidence for Ecological Speciation and Its Alternative. *Science* 323:737–41.

Soltis, P. S., and D. E. Soltis. 2009. The Role of Hybridization in Plant Speciation. *Annual Review of Plant Biology* 60:561–88.

Uy, J. A. C., D. E. Irwin, and M. S. Webster. 2018. Behavioral Isolation and Incipient Speciation in Birds. *Annual Review of Ecology, Evolution, and Systematics* 49:1–24.

Wu, C-I., and C-T. Ting. 2004. Genes and Speciation. *Nature Reviews Genetics* 5:114–22.

PRIMARY LITERATURE CITED IN CHAPTER 13

Arnqvist, G., M. Edvardsson, U. Friberg, and T. Nilsson. 2000. Sexual Conflict Promotes Speciation in Insects. *Proceedings of the National Academy of Sciences USA* 97 (19): 10460–4.

Babik, W., R. K. Butlin, W. J. Baker, A. S. Papadopulos, M. Boulesteix, et al. 2009. How Sympatric Is Speciation in the *Howea* Palms of Lord Howe Island? *Molecular Ecology* 18 (17): 3629–38.

Bateson, W. 1909. Heredity and Variation in Modern Lights. In A. C. Seward (ed.), *Darwin and Modern Science* (pp. 85–101). Cambridge: Cambridge University Press.

Bernasconi, G., T. L. Ashman, T. R. Birkhead, J. D. D. Bishop, U. Grossniklaus, et al. 2004. Evolutionary Ecology of the Prezygotic Stage. *Science* 303 (5660): 971–5.

Bolnick, D. I. and B. M. Fitzpatrick. 2007. Sympatric Speciation: Models and Empirical Evidence. *Annual Review of Ecology, Evolution, and Systematics* 38:459–87.

Bradshaw, H. D., S. M. Wilbert, K. G. Otto, and D. W. Schemske. 1995. Genetic Mapping of Floral Traits Associated with Reproductive Isolation in Monkeyflowers (*Mimulus*). *Nature* 376 (6543): 762–5.

Brandvain, Y., A. M. Kenney, L. Flagel, G. Coop, and A. L. Sweigart. 2014. Speciation and Introgression Between *Mimulus nasutus* and *Mimulus guttatus*. *PLoS Genetics* 10 (6): e1004410.

Bush, G. L. 1969. Sympatric Host Race Formation and Speciation in Frugivorous Flies of the Genus *Rhagoletis* (Diptera, Tephritidae). *Evolution* 23 (2): 237–51.

Burgin, C. J., J. P. Colella, P. L. Kahn, and N. S. Upham. 2018. How Many Species of Mammals Are There? *Journal of Mammalogy* 99 (1): 1–11.

Cahill, J. A., I. Stirling, L. Kistler, R. Salamzade, E. Ersmark, et al. 2015. Genomic Evidence of Geographically Widespread Effect of Gene Flow from Polar Bears into Brown Bears. *Molecular Ecology* 24 (6): 1205–17.

Cohan, F. M. 2006. Towards a Conceptual and Operational Union of Bacterial Systematics, Ecology, and Evolution. *Philosophical Transactions of the Royal Society B: Biological Sciences* 361 (1475): 1985–96.

Coyne, J. A., and H. A. Orr. 1997. "Patterns of Speciation in *Drosophila*" Revisited. *Evolution* 51 (1): 295–303.

———. 2004. *Speciation*. Sunderland, MA: Sinauer Associates.

Dagan, T., Y. Artzy-Randrup, and W. Martin. 2008. Modular Networks and Cumulative Impact of Lateral Transfer in Prokaryote Genome Evolution. *Proceedings of the National Academy of Sciences USA* 105 (29): 10039–44.

De Queiroz, K. 2005. A Unified Concept of Species and Its Consequences for the Future of Taxonomy. *Proceedings of the California Academy of Sciences* 56 (18):196–215.

Diaz, A., and M. R. Macnair. 1999. Pollen Tube Competition as a Mechanism of Prezygotic Reproductive Isolation Between *Mimulus nasutus* and Its Presumed Progenitor *M. guttatus*. *New Phytologist* 144 (3): 471–8.

Diehl, S. R., and R. J. Prokopy. 1986. Host-Selection Behavior Differences Between the Fruit Fly Sibling Species *Rhagoletis pomonella* and *R. mendax* (Diptera: Tephritidae). *Annals of the Entomological Society of America* 79 (1): 266–71.

Dobzhansky, T. 1936. Studies on Hybrid Sterility. II. Localization of Sterility Factors in *Drosophila pseudoobscura* Hybrids. *Genetics* 21 (2): 113–35.

———. 1937. *Genetics and the Origin of Species*. New York: Columbia University Press.

Doolittle, W. F. 2008. Microbial Evolution: Stalking the Wild Bacterial Species. *Current Biology* 18 (13): R565–7.

Doolittle, W. F., and R. T. Papke. 2006. Genomics and the Bacterial Species Problem. *Genome Biology* 7 (9): 116.

Doolittle, W. F., and O. Zhaxybayeva. 2009. On the Origin of Prokaryotic Species. *Genome Research* 19 (5): 744–56.

Eberhard, W. G. 1991. Copulatory Courtship and Cryptic Female Choice in Insects. *Biological Reviews* 66 (1): 1–31.

Eren, A. M., L. Maignien, W. J. Sul, L. G. Murphy, S. L. Grim, et al. 2013. Oligotyping: Differentiating Between Closely Related Microbial Taxa Using 16S rRNA Gene Data. *Methods in Ecology and Evolution* 4:1111–9.

Feder, J. L., C. A. Chilcote, and G. L. Bush. 1989. Are the Apple Maggot, *Rhagoletis pomonella* and Blueberry Maggot Fly, *R. mendax,* Distinct Species? Implications for Sympatric Speciation. *Entomologia Experimentalis et Applicata,* 51 (2): 113–32.

Feder, J. L., S. P. Egan, and P. Nosil. 2012. The Genomics of Speciation-with-Gene-Flow. *Trends in Genetics* 28 (7): 342–50.

Fishman, L., J. Aagaard, J. C. Tuthill, and M. Rausher. 2008. Toward the Evolutionary Genomics of Gametophytic Divergence: Patterns of Transmission Ratio Distortion in Monkeyflower (*Mimulus*) Hybrids Reveal a Complex Genetic Basis for Conspecific Pollen Precedence. *Evolution* 62 (12): 2958–70.

Fishman, L., and J. H. Willis. 2001. Evidence for Dobzhansky-Muller Incompatibilities Contributing to the Sterility of Hybrids Between *Mimulus guttatus* and *M. nasutus. Evolution* 55 (10): 1932–42.

Grant, P. R., and B. R. Grant. 2014. Speciation Undone. *Nature* 507:178.

Hebert, P. D. N., E. H. Penton, J. M. Burns, D. H. Janzen, and W. Hallwachs. 2004. Ten Species in One: DNA Barcoding Reveals Cryptic Species in the Neotropical Skipper Butterfly *Astraptes fulgerator. Proceedings of the National Academy of Sciences USA* 101 (41): 14812–7.

Hendry, A. P. 2009. Ecological Speciation! Or the Lack Thereof? *Canadian Journal of Fisheries and Aquatic Sciences* 66 (8): 1383–98.

Hipperson, H., L. T. Dunning, W. J. Baker, R. K. Butlin, I. Hutton, et al. 2016. Ecological Speciation in Sympatric Palms: 2. Pre- and Post-Zygotic Isolation. *Journal of Evolutionary Biology* 29 (11): 2143–56.

Howard, D. J. 1999. Conspecific Sperm and Pollen Precedence and Speciation. *Annual Review of Ecology and Systematics* 30 (1): 109–32.

Knowlton, N., L. A. Weigt, L. A. Solorzano, D. K. Mills, and E. Bermingham. 1993. Divergence in Proteins, Mitochondrial DNA, and Reproductive Compatibility Across the Isthmus of Panama. *Science* 260 (5114): 1629–32.

Lamichhaney, S., F. Han, M. T. Webster, L. Andersson, B. R. Grant, et al. 2018. Rapid Hybrid Speciation in Darwin's Finches. *Science* 359 (6372): 224–8.

Land, M., L. Hauser, S. R. Jun, I. Nookaew, M. R. Leuze, et al. 2015. Insights from 20 Years of Bacterial Genome Sequencing. *Functional and Integrative Genomics* 15 (2): 141–61.

Lee, Y. H., and V. D. Vacquier. 1992. The Divergence of Species-Specific Abalone Sperm Lysins Is Promoted by Positive Darwinian Selection. *Biological Bulletin* 182 (1): 97–104.

Levitan, D. R., H. Fukami, J. Jara, D. Kline, T. M. McGovern, et al. 2004. Mechanisms of Reproductive Isolation Among Sympatric Broadcast-Spawning Corals of the *Montastraea annularis* Species Complex. *Evolution* 58 (2): 308–23.

Lewis, S. M., and C. K. Cratsley. 2008. Flash Signal Evolution, Mate Choice, and Predation in Fireflies. *Annual Review of Entomology* 53:293–321.

Liu, S., E. D. Lorenzen, M. Fumagalli, B. Li, K. Harris, et al. 2014. Population Genomics Reveal Recent Speciation and Rapid Evolutionary Adaptation in Polar Bears. *Cell* 157 (4): 785–94.

Locey, K. J., and J. T. Lennon. 2016. Scaling Laws Predict Global Microbial Diversity. *Proceedings of the National Academy of Sciences USA* 113 (21): 5970–5.

Mallet, J. 2008. Hybridization, Ecological Races and the Nature of Species: Empirical Evidence for the Ease of Speciation. *Philosophical Transactions of the Royal Society B: Biological Sciences* 363 (1506): 2971–86.

Mayr, E. 1942. *Systematics and the Origin of Species.* New York: Columbia University Press.

Mendelson, T. C., and K. L. Shaw. 2005. Sexual Behaviour: Rapid Speciation in an Arthropod. *Nature* 433 (7024): 375–6.

Metz, E. C., R. E. Kane, H. Yanagimachi, and S. R. Palumbi. 1994. Fertilization Between Closely Related Sea Urchins Is Blocked by Incompatibilities During Sperm-Egg Attachment and Early Stages of Fusion. *Biological Bulletin* 187 (1): 23–34.

Mullen, S. P., and K. L. Shaw. 2014. Insect Speciation Rules: Unifying Concepts in Speciation Research. *Annual Review of Entomology* 59:339–61.

Muller, H. J. 1942. Isolating Mechanisms, Evolution, and Temperature. *Biological Symposium* 6:71–125.

Orr, H. A. 1996. Dobzhansky, Bateson, and the Genetics of Speciation. *Genetics* 144 (4): 1331–5.

Orr, H. A., and D. C. Presgraves. 2000. Speciation by Postzygotic Isolation: Forces, Genes and Molecules. *BioEssays* 22 (12): 1085–94.

Palumbi, S. R. 1994. Genetic Divergence, Reproductive Isolation, and Marine Speciation. *Annual Review of Ecology and Systematics* 25 (1): 547–72.

Palumbi, S. R., and E. C. Metz. 1991. Strong Reproductive Isolation Between Closely Related Tropical Sea Urchins (Genus *Echinometra*). *Molecular Biology and Evolution* 8 (2): 227–39.

Phipps, C. J. 1774. *A Voyage Towards the North Pole Undertaken by His Majesty's Command, 1773.* London: W. Bowyer and J. Nichols for J. Nourse.

Rice, W. R., and G. W. Salt. 1988. Speciation via Disruptive Selection on Habitat Preference: Experimental Evidence. *American Naturalist* 131 (6): 911–7.

Savolainen, V., M. C. Anstett, C. Lexer, I. Hutton, J. J. Clarkson, et al. 2006. Sympatric Speciation in Palms on an Oceanic Island. *Nature* 441 (7090): 210.

Schemske, D. W., and H. D. Bradshaw. 1999. Pollinator Preference and the Evolution of Floral Traits in Monkeyflowers (*Mimulus*). *Proceedings of the National Academy of Sciences USA* 96 (21): 11910–5.

Servedio, M. R., and M. Kopp. 2012. Sexual Selection and Magic Traits in Speciation with Gene Flow. *Current Zoology* 58 (3): 510–6.

Servedio, M. R., and M. A. F. Noor. 2003. The Role of Reinforcement in Speciation: Theory and Data. *Annual Review of Ecology, Evolution, and Systematics* 34:339–64.

Shapiro, B. J. 2014. Signatures of Natural Selection and Ecological Differentiation in Microbial Genomes. *Advances in Experimental Medicine and Biology* 781:339–59.

Shapiro, B. J., J. Friedman, O. X. Cordero, S. P. Preheim, S. C. Timberlake, et al. 2012. Population Genomics of Early Events in the Ecological Differentiation of Bacteria. *Science* 336:48–51.

Shaw, K. L., and S. C. Lesnick. 2009. Genomic Linkage of Male Song and Female Acoustic Preference QTL Underlying a Rapid Species Radiation. *Proceedings of the National Academy of Sciences USA* 106 (24): 9737–42.

Smith, M. A., J. J. Rodriguez, J. B. Whitfield, A. R. Deans, D. H. Janzen, et al. 2008. Extreme Diversity of Tropical Parasitoid Wasps Exposed by Iterative Integration of Natural History, DNA Barcoding, Morphology and Collections. *Proceedings of the National Academy of Sciences USA* 05: 12359–64.

Sogin, M. L., H. G. Morrison, J. A. Huber, D. M. Welch, S. M. Huse, et al. 2006. Microbial Diversity in the Deep Sea and the Underexplored "Rare Biosphere." *Proceedings of the National Academy of Sciences USA* 103 (32): 12115–20.

Soltis, P. S., and D. E. Soltis. 2009. The Role of Hybridization in Plant Speciation. *Annual Review of Plant Biology* 60 (1): 561–88.

Sota, T., and K. Kubota. 1998. Genital Lock-and-Key as a Selective Agent Against Hybridization. *Evolution* 52 (5): 1507–13.

Stackebrandt, E., and B. M. Goebel. 1994. Taxonomic Note: A Place for DNA-DNA Reassociation and 16S rRNA Sequence Analysis in the Present Species Definition in Bacteriology. *International Journal of Systematic Bacteriology* 44 (4): 846–9.

Sweigart, A. L., L. Fishman, and J. H. Willis. 2006. A Simple Genetic Incompatibility Causes Hybrid Male Sterility in *Mimulus*. *Genetics* 172:2465–79.

Welch, R. A., V. Burland, G. Plunkett, P. Redford, P. Roesch, et al. 2002. Extensive Mosaic Structure Revealed by the Complete Genome Sequence of Uropathogenic *Escherichia coli*. *Proceedings of the National Academy of Sciences USA* 99 (26): 17020–4.

Wiley, C., C. K. Ellison, and K. L. Shaw. 2012. Widespread Genetic Linkage of Mating Signals and Preferences in the Hawaiian Cricket *Laupala*. *Proceedings of the Royal Society of London B: Biological Sciences* 279:1203–9.

Wilkins, J. 2018. *Species: The Evolution of the Idea*. Boca Raton, FL: CRC Press.

Wolf, J. B. W., and H. Ellegren. 2017. Making Sense of Genomic Islands of Differentiation in Light of Speciation. *Nature Reviews Genetics* 18 (2): 87–100.

Zimmer, C. 2008. What Is a Species? *Scientific American* 298: 72–9.

CHAPTER 14

Macroevolution

The Long Run

Learning Objectives

- Compare and contrast the processes involved in macroevolution and microevolution.
- Compare and contrast the patterns that result from macroevolution and microevolution.
- Evaluate the effects on total species diversity when origination and extinction rates vary.
- List the kinds of evidence needed to distinguish between dispersal events and vicariance events in the fossil record.
- Describe how paleontologists analyze the fossil record to reconstruct macroevolutionary patterns.
- Explain what kinds of opportunities can give rise to adaptive radiations.
- Distinguish between background extinctions and mass extinctions.
- Describe two abiotic factors potentially responsible for mass extinctions.
- Evaluate the evidence for human influence on biotic and abiotic factors affecting biodiversity.
- Discuss whether human influence on biotic and abiotic factors may lead to another mass extinction.

Anthony Barnosky can often be found in a cave, searching for fossils (**Figure 14.1**). He's spent decades unearthing bones of North American mammals that lived during the Pleistocene, from 2.58 million years ago to 11,700 years ago. Barnosky, a paleontologist at Stanford University, and his colleagues observe when new species first appear in the fossil record, and also take note of their last occurrence. Even as recently as 15,000 years ago, Barnosky and his team found, the biodiversity of western North America was markedly different from today. Most strikingly, it was home to many large mammals—from mastodons to giant ground sloths to saber-toothed cats—that are now extinct.

As recently as 12,000 years ago, North America was home to a wide range of large mammals such as this saber-toothed tiger (Smilodon fatalis) and giant bison (Bison latifrons). These and many other species of large North American mammals have become extinct, due at least in part to the arrival of human hunters on the continent.

Carl Buell

Figure 14.1 Anthony Barnosky, of Stanford University, explores caves to reconstruct the changing biodiversity of North America over the past 30 million years. (Jessica Blois)

Macroevolution is evolution occurring above the species level, including the origination, diversification, and extinction of species over long periods of evolutionary time.

Microevolution is evolution occurring within populations, including adaptive and neutral changes in allele frequencies from one generation to the next.

The diversity of life on Earth is staggeringly vast—and staggeringly hard to make sense of. Traditionally, scientists only recognize a new species once a detailed description has been published for it, along with a Latin name. So far, scientists have described and named 1.8 million living species. But they also name thousands of new species each year. Researchers have attempted to extrapolate from this pace of discovery to estimate how many species in total, described and undescribed, there are on Earth.

In a 2011 study (Mora et al. 2011), a team of researchers put that number at 8.7 million. Since then, others have argued that this figure is too low. One team has estimated that animals alone total 163 million species (Larsen et al. 2017). And these estimates may seriously neglect the vast diversity of microbes, which may total a *trillion* species (Locey and Lennon 2016).

The vast number of species in existence today raises a host of questions. Why are there many more species per square mile in the tropics than near the poles? Why are there 1.5 million species of beetles (Stork 2018) and just one species of gingko tree (Crane 2013)? But these questions only scrape the surface, because they only address life as we see it today. Life also has a history, one that stretches back at least 3.7 billion years. During that time, the fossil record documents, many species evolved, endured, and then disappeared. In fact, extinct species vastly outnumber species that exist today. As we'll see in this chapter, the diversity of life has gone up and down over millions of years.

Such large-scale processes and patterns of evolution are known as **macroevolution**. The term stands in contrast to **microevolution**, the change of allele frequencies within populations caused by mechanisms such as genetic drift and natural selection.

As we saw in Chapter 6, microevolution occurs on such a small scale that scientists can observe it in wild populations and conduct experiments to test evolutionary hypotheses. Scientists cannot observe macroevolution as closely, for the simple reason that they don't live for millions of years. But that does not make the study of macroevolution any less scientific than the study of microevolution. Scientists simply have to use a different set of methods to study it. They explore the evidence of macroevolutionary processes left behind in the fossil record, in the genetic patterns revealed through molecular phylogenetics, or in the current distribution of living species. (See Box 3.2 for a further discussion of how evolutionary biologists study life's deep history.)

To explore macroevolution in this chapter, we will begin with its geographic patterns—why certain clades are found only in some parts of the world, for example, or why some regions have more species than others. Macroevolution produces patterns through time as well as space; we'll look at how scientists measure the rates of origination and extinction in the fossil record and examine the causes of changes in global biodiversity. We will consider the factors that drive some clades to diversify into many new species whereas other clades remain species poor (**Figure 14.2**).

Macroevolution also includes abrupt drops in diversification rates caused by the rapid extinction of many species. Mass extinctions offer a sobering warning for our own future. As we'll see in this chapter, the research of scientists like Barnosky has revealed that we humans are driving species extinct at a worrying rate, a rate that could potentially increase to truly catastrophic levels in the years to come (Barnosky et al. 2017). ●

Figure 14.2 There are an estimated 1.5 million species of beetles. Studies on their evolution suggest that this high number is the result of new species originating faster than old species becoming extinct. (imv/Getty Images)

14.1 Biogeography: Darwin and Wallace's Pioneering Work

When we look at a map of the world, we can see some striking patterns in biological diversity. Some regions, such as the tropics, are more species rich than others (**Figure 14.3**). Certain groups of species are common in some places and rare in others. Almost every species of marsupial, for example, is native to Australia, whereas placental mammals dominate the other continents.

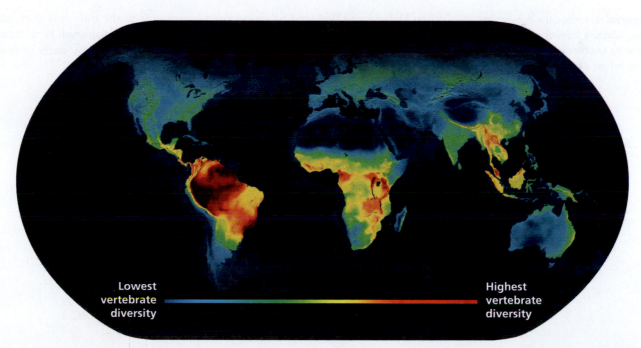

Lowest vertebrate diversity

Highest vertebrate diversity

Figure 14.3 The tropics contain a wealth of unique species. This map shows levels of vertebrate diversity. Some scientists have proposed that the tropics are rich in species because of the high speciation rate there. Others have argued that extinction rates are low. It's possible that a combination of factors is responsible. (Jenkins C.N. et al. Global patterns of terrestrial vertebrate diversity and conservation. Proc. Natl. Acad. Sci. U.S.A. 2013; 110: E2602–E2610)

Biogeography is the study of the distribution of species across space (geography) and time.

The study of these patterns of diversity is known as **biogeography**. Before the nineteenth century, many naturalists assumed that biogeographic patterns existed because the Creator had chosen to put species in places to which they were well adapted. But in his travels aboard the HMS *Beagle*, Charles Darwin observed many serious flaws in this explanation (Chapter 2). Environmental conditions such as climate alone could not completely account for the distribution of species. Llamas are found in the Andes, for example, and yet they are missing from the Rockies, despite the environmental similarities of these two regions.

Darwin was also struck by the similarity between some groups of species despite their living in different environments. The rodents he encountered in the Andes showed many signs of being closely related to the rodents that live on the South American plains—as opposed to the rodents of the environmentally similar Rockies. Darwin argued that the rodents of South America had adapted to different environments over time.

Islands also presented intriguing patterns to Darwin on his travels. Island ecosystems, he recognized, tended to be dominated by species that were good dispersers. He encountered many species of bats and birds on islands, for example, but few amphibians or elephants. This distribution of island species made sense as the result of species colonizing the islands from the mainland and then adapting to those habitats.

Alfred Russel Wallace, co-discoverer of evolution by natural selection, also made important observations about biogeography. Rather than examining islands and other small-scale features of biogeography, he looked at the planet as it was understood by naturalists in the nineteenth century. He recognized six major biogeographic provinces, each one having its own distinctive balance of species (**Figure 14.4**). The boundary between two of these provinces—southeast Asia on one side and Australia and New Zealand on the other—is particularly striking. It actually runs through the middle of the Indonesian Archipelago. Islands on both sides of the boundary have very similar environments but very different species assemblages. The science of biogeography seeks to explain such intriguing patterns of species distribution.

Dispersal and Vicariance

Dispersal describes the movement of populations from one geographic region to another with very limited, or no, return exchange.

When Darwin and Wallace first investigated biogeography, the only way they could envision species getting to where they are now was through **dispersal**. Birds flew to islands; elephants walked over mountain ranges; seeds were carried by water to distant shores.

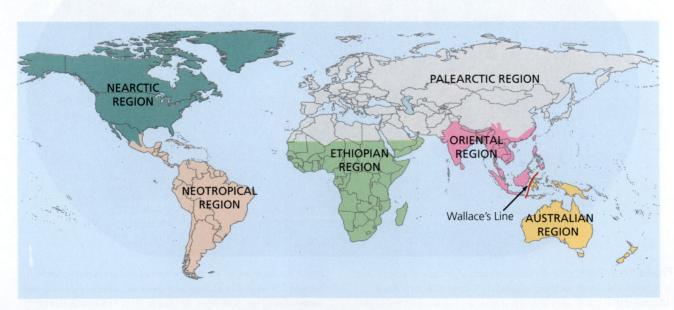

Figure 14.4 Alfred Russel Wallace drew the first global map of biogeographic regions.

But dispersal fell short in explaining some patterns of biogeography—why, for example, marine invertebrates on the east and west coasts of Panama are so remarkably similar (a shrimp today would have to swim all the way around the continent of South America to get from one of these populations to the other). As geologists learned more about how the surface of Earth changes, it became clear that the formation of barriers to migration and gene flow can also shape biogeographic patterns (Chapter 13). This process is known as **vicariance**. Many geological processes can drive vicariance. Rising sea levels can turn peninsulas into islands. Mountain ranges can rise up, splitting a population. Continents can break up and drift apart.

As shown in **Figure 14.5**, the distribution and phylogeny of species can provide clues about vicariance. In this hypothetical example, a continent breaks in two, and those two landmasses then split. If a clade does not disperse during this time, its phylogeny should reflect the history of the continent's breakup. The most closely related clades should be found on the most recently separated landmasses.

The history of marsupials is a spectacular example of how both vicariance and dispersal produce complex patterns of biodiversity. Although today most living species of marsupials are found in Australia, the oldest marsupial fossils come from China and North America. Yet there is only one species of marsupial in all of North America today (the Virginia opossum), and none at all in China. How did we get to this puzzling situation?

We can understand this pattern by integrating several separate lines of evidence: from studies on the molecular phylogeny of living marsupials, from the fossil record, and from understanding the history of plate tectonics. In 2015, for example, Maria Nilsson, of the Senckenberg Nature Research Society in Germany, and her colleagues published a molecular phylogeny of marsupials (Gallus et al. 2015). They concluded that all Australian marsupials form a single clade. This clade, in turn, is nested within a clade of South American species. That tree topology indicates that a single dispersal event may have brought marsupials from South America to Australia (**Figure 14.6**).

A molecular phylogeny can show only the relationships of species from which scientists can obtain DNA. A morphological phylogeny, on the other hand, can include fossil species as well (although it's typically based on fewer characters; see Chapter 4). The addition of fossil evidence complements the molecular evidence for the evolution of marsupials. A phylogeny of the major marsupial lineages known from fossils, based on morphological characters (Figure 14.6B), shows the same monophyly of Australian marsupials. It also reveals that marsupials lived in Antarctica when it was warmer and

Vicariance is the formation of geographic barriers to dispersal and gene flow, resulting in the separation of once continuously distributed populations.

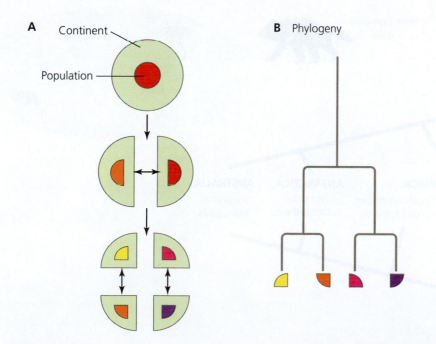

Figure 14.5 Clades can become isolated when geographic barriers emerge, in a process called vicariance. A: Here, a single continent drifts apart into four fragments. B: The phylogeny of the clade reflects the geological history.

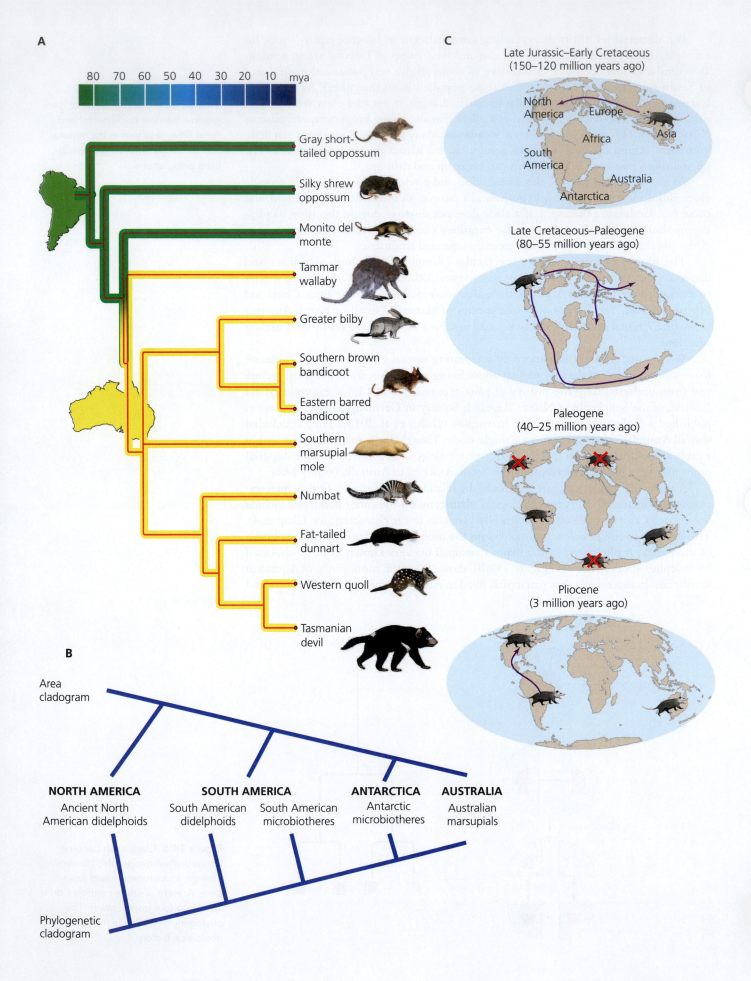

A

80 70 60 50 40 30 20 10 mya

Gray short-
tailed opossum

Silky shrew
opossum

Monito del
monte

Tammar
wallaby

Greater bilby

Southern brown
bandicoot

Eastern barred
bandicoot

Southern
marsupial
mole

Numbat

Fat-tailed
dunnart

Western quoll

Tasmanian
devil

B

Area
cladogram

NORTH AMERICA
Ancient North
American didelphoids

SOUTH AMERICA
South American
didelphoids

South American
microbiotheres

ANTARCTICA
Antarctic
microbiotheres

AUSTRALIA
Australian
marsupials

Phylogenetic
cladogram

C

Late Jurassic–Early Cretaceous
(150–120 million years ago)

North
America   Europe
Asia
Africa
South
America
Australia
Antarctica

Late Cretaceous–Paleogene
(80–55 million years ago)

Paleogene
(40–25 million years ago)

Pliocene
(3 million years ago)

that these extinct Antarctic marsupials were more closely related to living Australian marsupials than to South American ones. Finally, it shows that extinct North American marsupials belong to the deepest of the branches. This branching pattern parallels the order in which these continents separated from each other. North America split off first; next, South America separated from Australia and Antarctica; and lastly, Australia and Antarctica became separate landmasses.

These separate lines of evidence combine to offer a detailed scenario for the evolution of marsupials (Springer et al. 2011). The fossil record provides the earliest details about marsupials because many deep branches of the marsupial tree are now extinct.

Marsupial-like mammals were living in China by 150 million years ago, the age of the oldest fossils yet found. By 120 million years ago, marsupials had dispersed into North America, which at the time was linked to Asia. Many new lineages of marsupials evolved there over the next 55 million years. From North America, some marsupials dispersed to Europe, even reaching as far as North Africa and Central Asia. All of these Northern Hemisphere marsupials eventually died out in a series of extinctions between 30 and 25 million years ago (Beck et al. 2008).

Living marsupials descend from a lineage of North American marsupials that dispersed to South America around 80 million years ago. Living South American marsupials, belonging to deep branches in Nilsson's molecular phylogeny, descend from those migrants that headed south. At the time, South America was joined to Antarctica, allowing some marsupials to disperse there—and to continue expanding to Australia, which was attached to the other side of Antarctica. The oldest Australian fossils of marsupials are 55 million years old. But molecular clock studies (Mitchell et al. 2014) indicate that marsupials may have arrived in Australia 10 million years earlier, around the end of the Cretaceous period.

After marsupials arrived in Australia, the three linked continents split apart, each carrying with it a population of marsupials (vicariance). The fossil record shows that marsupials were still in Antarctica 40 million years ago. But as the continent moved nearer to the South Pole, it grew colder. Eventually it became unsuitable for tetrapods of any kind. Antarctica's marsupials became extinct.

Meanwhile, marsupials in South America diversified into a wide range of species, including catlike marsupial saber-tooths. But these large, carnivorous species became extinct, along with many other unique South American marsupials, when the continent reconnected to North America a few million years ago.

Competition with placental mammals dispersing from North America may have been an important factor driving this extinction. However, there are still many different species of small and medium-sized marsupials living in South America today. One South American marsupial, the Virginia opossum, even expanded back into North America, where marsupials had disappeared millions of years before (Dias and Perini 2018).

Australia, meanwhile, drifted in isolation for more than 40 million years. The fossil record of Australia is currently too patchy for paleontologists to say whether there were any placental mammals in Australia during that time. Abundant Australian fossils date back to about 25 million years ago, when all of the therian mammals in Australia were marsupials. They evolved into a spectacular range of forms, including kangaroos and koalas. It was not until 15 million years ago that Australia moved close enough to

Figure 14.6 (left) Marsupials evolved through a mix of vicariance and dispersal. A: Molecular phylogeny shows that Australian marsupials form a clade nested with South American marsupials. The only living North American marsupial, the Virginia opossum (*Didelphis*) belongs to the clade Didelphimorphia, which also originated in South America. (Data from Gallus et al. 2015) B: Separate lines of evidence also add insight into marsupial biogeography. A cladogram showing the ranges of marsupial lineages (known as an area cladogram) shows that marsupial phylogeny tracks the major separations of continents (top) and tracks a phylogenetic cladogram of known fossil lineages based on morphological characters (bottom). C: By combining this evidence, we can construct a scenario for the evolution of marsupials.

Asia to allow placental mammals—rats and bats—to begin colonizing the continent. These invaders diversified into many ecological niches, but there's no evidence that they displaced a single marsupial species that was already there (Christopher Norris, personal communication).

Key Concepts

- Biogeography is a highly interdisciplinary field that explores the roles of geography and history in explaining the distributions of species in space and time.

- Dispersal and vicariance explain distribution patterns of taxa. Dispersal occurs when a taxon crosses a preexisting barrier, like an ocean. Vicariance occurs when a barrier interrupts the preexisting range of the taxon, preventing gene flow between the now separated populations. ●

14.2 The Drivers of Macroevolution: Speciation and Extinction

The fossil record and DNA provide clues not only about biogeography but also about the changing levels of diversity over time. To estimate diversity, evolutionary biologists have adapted a method from population ecology (Marshall 2017).

Population ecologists chart the growth of populations through time by a formula. They start with a population's current size, add births and immigrations, and subtract deaths and emigrations of individuals. In the formula shown here, N_1 stands for the current size of a population, and N_2 stands for the size of the population in the next time step.

$$N_1 + \text{births} + \text{immigrations} - \text{deaths} - \text{emigrations} = N_2$$

When scientists study macroevolution, they can adapt the formula, making species or higher clades their units of analysis instead of individual organisms. Likewise, they can substitute the originations and extinctions of species for the birth and death of individuals. The number of species in a region can also be altered through immigration and emigration. Here, D stands for the diversity—that is, the total number of species in a particular clade.

$$D_1 + \text{originations} + \text{immigrations} - \text{extinctions} - \text{emigrations} = D_2$$

In many cases, scientists study macroevolution on a global scale, rather than in one particular region. In these cases, we can eliminate the immigration and emigration terms from the equation:

$$D_1 \text{ (diversity)} + \text{originations} - \text{extinctions} = D_2 \text{ (new diversity)}$$

A **fauna** is an assemblage of many different species that live together. A fauna may comprise the species in a single ecosystem, in a region, or across the entire planet.

This simple equation brings into focus a fundamental idea about macroevolution: changes in diversity through time can be studied by looking at the interplay between origination and extinction. We can look at these processes in entire **faunas**. We can also look at them within a single clade, tracking its fluctuations in diversity through time.

We can illustrate this method with a hypothetical example, shown in **Figure 14.7**. A horizontal bar represents the temporal range of fossils belonging to each species. We will assume that the oldest fossil of a species marks its origin, and the youngest marks its extinction. The fossils were deposited in four consecutive geological stages marked here as A, B, C, and D. If we add up the fossils present at any point during stage A, we find a total of 24 species. During stage B, 10 new ones evolved, bringing the total to 34. But 6 species did not survive beyond B; the total in C is thus 36—only 2 species more than in B, despite the emergence of 8 new species. During D, another combination of originations and extinctions dropped the total to 30 species.

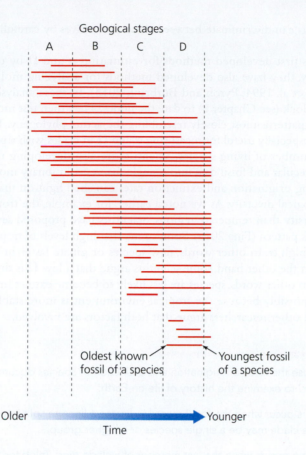

Geological stages

A   B   C   D

Oldest known
fossil of a species

Youngest fossil
of a species

Older ——————————► Younger

Time

Figure 14.7 This diagram of a hypothetical scenario shows how species are preserved in the fossil record. Paleontologists can calculate rates of processes such as origination and extinction from these kinds of data.

The total number of originations and extinctions in a given interval of time is known as **turnover**. In the example shown in Figure 14.7, a total of 20 turnovers occur in stage C (8 originations plus 12 extinctions).

Turnover rates can tell us many important things about extinct clades. Paleontologists have found that some clades have especially high turnover rates, whereas others have low ones. Trilobites, for example, had very high turnover rates. Each trilobite species had only a brief duration in the fossil record, but the trilobite clade persisted because new species continued to arise at least as fast as earlier ones disappeared. Other groups, such as clams and snails, appear to have had relatively low turnover rates. Their species last much longer, and new species appear less frequently.

These shifts in diversity result from changes in both the rate at which new species evolve and the rate at which old species become extinct. We can represent the rate at which new species evolve, or originate, in a clade by the Greek letter α (alpha) and the rate at which species go extinct by Ω (omega). In Figure 14.7, the mean origination rate for the three stages B, C, and D is 8 new species per stage, whereas the mean extinction rate is 9 species per stage. Although our sample is not very large, we can see that 12 extinctions in stage C is more than usual and that on average the extinction rate is slightly greater than the origination rate.

In the example shown in Figure 14.7, we measure the rate of origination and extinction in terms of species per stage. Radiometric dating of stages makes it possible to measure these rates by using years instead. If the stages shown in Figure 14.7 were each 6 million years long, for example, then α would be 1.3 species per million years, and Ω would be 1.5 species per million years.

The origination and extinction rates in a clade determine the total diversity at any given time—a value known as **standing diversity**. Over long periods of time, differences in α and Ω can lead to large changes in the diversity of a clade. If Ω becomes higher than α, the diversity of a clade will decline. But diversity can also decline if α drops. Likewise, either a drop in Ω or a rise in α will lead to an increase in diversity. Changes in both of these rates can lead to superficially similar patterns, but

Turnover refers to the disappearance (extinction) of some species and their replacement by others (origination) in studies of macroevolution. The turnover rate is the number of species eliminated and replaced per unit of time.

Standing diversity is the number of species (or other taxonomic unit) present in a particular area at a given time.

paleontologists can discriminate between these alternatives by carefully inspecting the fossil record.

Scientists first developed methods for estimating α and Ω by using fossils, but more recently, they have also developed methods for analyzing molecular phylogenies (Harvey et al. 1994; Pyron and Burbrink 2013). In these analyses, scientists use a molecular clock (see Chapter 8) to date the nodes. They then test models of α and Ω that produce patterns most closely resembling the actual phylogeny. These molecular methods are especially useful for measuring α and Ω in clades with a poor fossil record but a large number of living species. Scientists are also developing new methods to combine molecular and fossil data into a single macroevolutionary model.

Measuring origination and extinction rates can shed light on many striking patterns of biological diversity. As we noted earlier, for example, the tropics have higher levels of diversity than temperate regions. Scientists have proposed several hypotheses to explain this pattern (Fine 2015). Some argue that high levels of tropical diversity are the result of high α. In other words, new species originate faster in the tropics than elsewhere. On the other hand, some scientists argue that a low Ω is the cause of tropical diversity. In other words, species are less likely to become extinct in the tropics than elsewhere—possibly because the tropical environment is more stable over the long term. But still other researchers argue that both factors are involved.

Key Concepts

- Scientists use the rates of origination and extinction of species documented in the fossil record to examine the history of life on Earth.

- Originations occur when the fossil record indicates a lineage split into two distinct clades. The clade may be a single species or a higher group.

- An extinction occurs when the last member of a clade dies. Trilobites, for example, were a clade containing many species; the entire trilobite clade became extinct 252 million years ago. ●

14.3 Charting Life's Rises and Falls

Charting the diversity of life poses two challenges. One challenge is the task of distinguishing species based on their fossils alone. Another challenge is that the fossil record is far from a complete picture of past biodiversity. For example, biologists have described 61,000 living vertebrate species, but paleontologists have described only about 12,000 fossil vertebrate species from the past 540 million years. There must be many, many other fossil species we have yet to find (International Institute for Species Exploration 2010).

A pioneering paleobiologist, David Raup, of the University of Chicago, took on these challenges by amassing a tremendous database of marine invertebrate fossils (Raup 1972). For the purposes of his study, these animals had several advantages over other taxa: they make mineralized tissues such as shells and exoskeletons that fossilize well, and many species produced huge populations that left an abundant record from the Cambrian onward. In the 1960s, Raup started to build a database that by 1970 included 144,251 species with information about each species, such as the geological age in which it lived. He then calculated changes in diversity from the Cambrian to the present, finding intriguing rises and falls in the total number of genera in marine clades. Ever since, other researchers have been building on that database and starting others, as well as applying new statistical methods to get a more accurate understanding of macroevolutionary patterns (Alroy 2008; Jablonski 2010).

In the late 1970s, for example, Jack Sepkoski of the University of Chicago built on Raup's work by comparing the changes in diversity of different taxonomic groups in the marine fossil record (Sepkoski 1981). Sepkoski found that the diversity of groups through time was not random. In fact, certain groups rose and fell together with other groups with remarkable consistency. He could explain more than 90% of the data by grouping taxa into three great "evolutionary faunas" that succeeded each other through time (**Figure 14.8**).

These three faunas are called the Cambrian, Paleozoic, and Modern faunas. (The names refer to when the faunas reached their peak diversity.) The Cambrian fauna arose at the beginning of the Paleozoic. It was dominated by trilobites, inarticulate brachiopods, and coil-shelled mollusks known as monoplacophorans. Following a quick start, this fauna went into decline. Today, only one group each of inarticulate brachiopods and monoplacophorans survives.

By the Ordovician, Sepkoski found, the Cambrian fauna was overshadowed by the Paleozoic fauna, which was rich in various brachiopods, echinoderms, corals, crustaceans, ammonite mollusks, and many other groups. These groups thrived until the end of the Permian, 251.9 million years ago, when the biggest mass extinction in the fossil record wiped out up to 96% of marine species (we'll discuss this extinction in more detail later in this chapter). Most of these groups never recovered, and some major groups became extinct.

The Modern fauna has its roots in the Cambrian era, although its members were of relatively low diversity at the time. After the Cambrian, diversity increased gradually. The Modern fauna, which has been dominant since the end–Permian extinction, consists mainly of gastropods (snails) and bivalves (clams).

This research, although important, only illuminates macroevolution among fossil-producing animals over the past half-billion years. As we saw in Chapter 3, by the beginning of the Cambrian, life had already existed in microbial form for more than three billion years. Even today, as we discussed earlier in this chapter, microbes likely make up the vast majority of species.

Microbes only rarely fossilize, leaving much of their evolution largely a mystery. But their DNA serves as a chronicle of their diversification and extinction. Stilianos Louca, of the University of British Columbia, and his colleagues have analyzed the genes of living microbes to infer the patterns of macroevolution in their history.

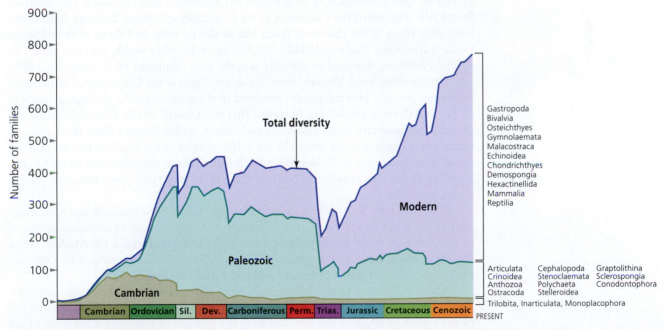

Figure 14.8 Jack Sepkoski identified three faunas in the fossil record since the Cambrian. The Cambrian fauna arose at the beginning of the Paleozoic and quickly declined. The Paleozoic fauna arose in the Ordovician. The so-called Modern fauna has its roots in the Cambrian but came to dominate the planet after the end of the Permian, 251.9 million years ago. (Data from Sepkoski 1981)

The researchers compared hundreds of thousands of microbial DNA sequences, organizing them into phylogenies. They tested out different rates of speciation and extinction to find the model that best explained the evolutionary history of the microbes over the past billion years. They found that tens of thousands of lineages become extinct every million years. As a result, most lineages of microbes found in the early fossil record are likely missing from the planet today. That implies that we must not assume that we can infer what early microbes were like based on living species. On the other hand, the researchers also found that the diversification rate of microbes has remained slightly higher overall than their extinction rate. As a result, the world is now home to many more microbe species than a billion years ago (Louca et al. 2018).

Key Concepts

- Large-scale studies of fossils help chart the rise and fall of assemblages of species.

- Sometimes groups of species change together in ways that distinguish them as discrete evolutionary faunas.

- Genomic studies are permitting researchers to study macroevolutionary patterns in species that don't fossilize well, such as microbes. ●

14.4 The Drivers of Macroevolution: Changing Environments

When we see large-scale patterns of macroevolution such as Sepkoski's three faunas, we can test hypotheses to explain them. Scientists have found evidence for two types of factors involved in macroevolutionary change. Intrinsic factors, such as the physiology of clades, can play a role. Extrinsic factors in the environment can as well.

Sepkoski, for example, proposed that the transition between faunas may have been driven by their differences in origination (α) and extinction (Ω) rates. The invertebrates that dominated the Cambrian fauna (especially trilobites) had very high turnover rates, those of the Paleozoic fauna had moderate ones, and those of the Modern fauna (particularly clams and snails) had low ones. In other words, when it comes to animal evolution, slow and steady may win the race (Sepkoski 1981; Valentine 1989).

On the other hand, Shanan Peters, a paleontologist at the University of Wisconsin, has found possible physical factors involved in these changes: the geological context of Sepkoski's three faunas. He observed that most fossils of the Paleozoic fauna are found in sedimentary rocks known as carbonates, which formed from the bodies of microscopic organisms that settled to the seafloor. Most of the Modern fauna fossils are found in rocks known as silicoclastics, which formed from the sediments carried to the ocean by rivers. The ecosystems built on these rocks may favor preservation of certain clades over others.

Over the past 540 million years, carbonate rocks have become rarer, whereas silicoclastic rocks have become more common, possibly because of sediment delivered to the oceans by rivers. Peters proposed that as the seafloor changed, the Modern fauna could expand across a greater area, whereas the Paleozoic fauna retreated to a shrinking habitat where it suffered almost complete extinction (Peters 2008).

Another physical factor drives long-term changes in biodiversity: the climate. Earth is warmed by incoming radiation from the Sun (**Figure 14.9**). Changes in the planet's orbit or angle to the Sun can alter the amount of radiation it receives. Once this energy reaches Earth, it can be stored in the atmosphere or ocean, or it may be bounced back into space. The chemical composition of the atmosphere can change the amount of heat it traps. The concentration of heat-trapping gases, such as carbon

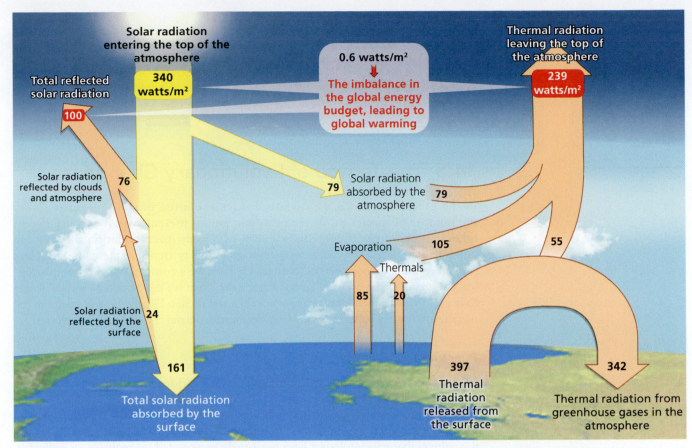

Figure 14.9 Earth receives incoming radiation from the Sun. It warms the atmosphere and surface of the planet before being released back into space. Changes in the concentration of greenhouse gases can lead to warming or cooling of the planet. Currently, human emissions are responsible for a net imbalance in the planet's overall energy budget. (Data from USGCRP 2017)

dioxide and methane, is influenced by many factors, from the absorption of carbon dioxide by photosynthesizing plants and algae to the eruptions of carbon dioxide-rich plumes from volcanoes.

Geologists can reconstruct the history of Earth's climate by looking at the chemistry of its rocks. Warm water has a higher concentration of the isotope oxygen-18 than cool water does, for example, and so rocks that form in warm water will lock in those isotopes as well. These records have demonstrated that Earth's climate has indeed fluctuated over the planet's history. One major source of this variation is the amount of carbon dioxide in the atmosphere. For example, when certain types of volcanoes erupt more, they deliver more of these heat-trapping greenhouse gases to the atmosphere.

Peter Mayhew, of the University of York, examined how these changes in climate might affect the diversity of life. As we saw earlier, differences in climate have been proposed to explain the different levels of diversity found today in the tropics and in temperate zones. Mayhew and his colleagues looked to see if changes to the entire planet's climate could lead to changes in diversity. By carefully comparing the fossil record with the climate record, Mayhew and his colleagues did indeed find a correlation. After correcting for sampling biases in the fossil record, they found that periods with warmer ocean temperatures also had increased standing diversity of marine invertebrates (**Figure 14.10**; Mayhew et al. 2012).

Figure 14.10 Taxonomic diversity (number of genera) is positively correlated with global mean ocean temperature. Over the past 500 million years, warmer periods have had higher global levels of standing diversity than cooler periods. (Data from Mayhew et al. 2012)

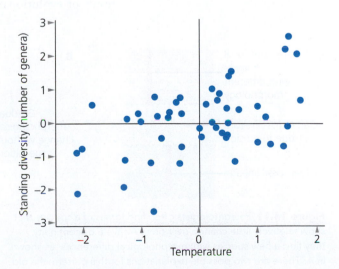

14.5 The Tempo of Evolutionary Change

In the previous sections, we examined macroevolution broadly, measuring it by the number of species or other taxonomic groups present worldwide at different times in evolutionary history. But scientists who study macroevolution also zoom in on individual clades, to observe long-term evolutionary trends in these lineages. Here we will consider three aspects of this research. First, we'll consider whether morphological changes in clades occur in bursts or gradually. Then we'll look at how some lineages have rapidly diversified, occupying new ecological niches. Finally, we will examine the macroevolutionary patterns at the dawn of animals, during the Cambrian explosion.

Punctuated Equilibria

In Chapter 10 we saw how paleontologists measured long-term phenotypic change in stickleback fishes by examining their fossils over millions of years (Figure 10.13). Although the fossil record of these fishes is impressive, it's far from complete. At any moment during this period of time, thousands of these fishes existed. But the fossils of the animals that paleontologists have recovered are separated by centuries or millennia. Most of the rest of the fossil record is even less complete.

Here's a typical experience for paleontologists, which we've illustrated schematically in **Figure 14.11**. They find an outcrop bearing clam fossils and begin chipping away at the rock. The older rock contains a series of morphologically similar fossils, which we'll represent here as a vertical column. The outcrop is not a continuous series of layers because erosion destroyed some of them. In the younger layers that still survive, the paleontologists find clams that are morphologically similar to the older ones—but not identical. Instead, they're distinct enough to be considered a separate species.

Traditionally, paleontologists confronted with these findings would infer that the imperfect fossil record reflected a period of smooth, gradual evolutionary change (Newell 1956). The fossil record illustrated in Figure 14.11A could plausibly be the result of evolutionary change shown in Figure 14.11B. The species undergoes a

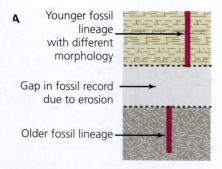

A

Younger fossil lineage with different morphology

Gap in fossil record due to erosion

Older fossil lineage

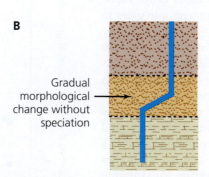

B

Gradual morphological change without speciation

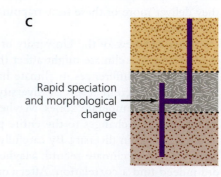

C

Rapid speciation and morphological change

Figure 14.11 Paleontologists often find fossils of a species that show relatively little change over time. Above a discontinuity, they find a new species with morphological differences, as shown in A. There are two possible explanations for this pattern: the old species underwent anagenesis, gradually evolving a new morphology (B), or a new species may have branched off the old one, rapidly evolving morphological differences before entering its own stasis (C).

gradual change, reaching a recognizably different morphology, before stabilizing in its new phenotype. This change is known as **anagenesis**.

In 1972, paleontologists Niles Eldredge and Stephen Jay Gould argued that scientists should not assume that anagenesis underlies morphological change documented by an imperfect fossil record (Eldredge and Gould 1972). Cladogenesis is also a plausible explanation (Figure 14.11C). A population of clams belonging to the species in the older layer might become isolated and undergo evolutionary change, branching off into a new species. In the time that is unrecorded in the rocks, both the old species and the new one coexisted. The older species then became extinct, leaving only the new species in the fossil record.

Eldredge and Gould proposed a new picture of macroevolution, which they dubbed **punctuated equilibria** (**Figure 14.12**). In this pattern, morphological transformations happen relatively rapidly, during the origin of new clades, and are followed by long periods of stasis (the *equilibria* of punctuated equilibria).

Anagenesis refers to wholesale transformation of a lineage from one form to another. In macroevolutionary studies, anagenesis is considered to be an alternative to lineage splitting or speciation.

Punctuated equilibria is a model of evolution that proposes that most species undergo relatively little change for most of their geological history. These periods of stasis are punctuated by brief periods of rapid morphological change, often associated with speciation.

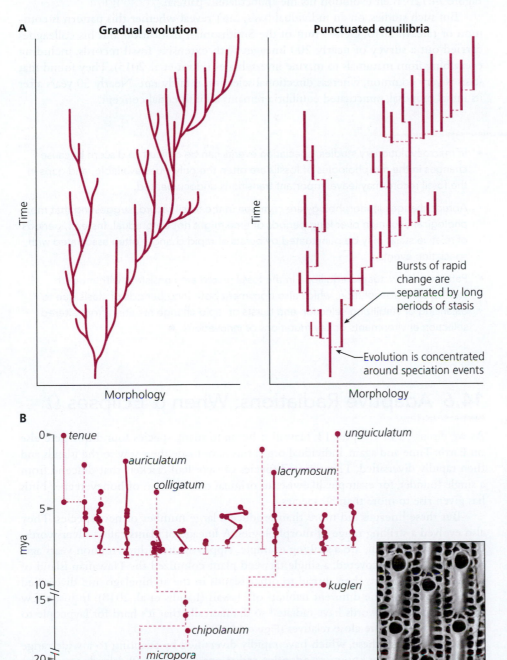

Figure 14.12 A: Gradual evolution through anagenesis is expected to lead to speciation arising against a backdrop of relatively continuous morphological change in lineages. The concept of punctuated equilibria, on the other hand, envisions evolution as comprising long periods of stasis interrupted by concentrated bursts of morphological change that occur during speciation. B: Paleontologists have found some patterns in the fossil record that best fit the punctuated equilibria model. This diagram shows how a lineage of bryozoans (*Metrarabdotos*) evolved rapidly into new species, but changed little once those species were established. (Data from Benton 2003)

The concept of punctuated equilibria is consistent with the mechanisms of evolution we discussed earlier in the book. Stabilizing selection can keep a lineage from deviating far from a given phenotype, for example, making it possible to produce long periods of stasis.

As for punctuations, Eldredge and Gould did not invoke some kind of creationist appearance of new species out of nowhere. They observed that thousands—or even hundreds of thousands—of years could vanish from the fossil record due to erosion. As we saw in Chapter 13, speciation can take place within such an interval, made possible by well-understood mechanisms of evolution.

The hypothesis of punctuated equilibria has spurred a number of scientists to put it to the test. Jeremy Jackson, a marine biologist, and Alan Cheetham, a paleontologist, did so by examining a collection of fossils of bryozoans, a colony-forming marine animal. They charted the emergence of new species in the fossil record, while also charting their morphological divergence (Jackson and Cheetham 1994). As shown in Figure 14.12B, their evolution fits the "punctuated" pattern.

But such studies, on an individual basis, can't reveal whether this pattern is common or rare. In 2015, Gene Hunt of the Smithsonian Institution and his colleagues carried out a survey of nearly 200 lineages with extensive fossil records, including everything from mammals to marine invertebrates (Hunt et al. 2015). They found that stasis is very common, whereas directional selection is fairly rare. Nearly 50 years after its initial proposal, punctuated equilibria remains an influential concept.

Key Concepts

- In macroevolutionary studies, speciation events can be difficult to discern because changes in the morphologies of fossils are often the only clues available, and gaps in the fossil record may leave important transitions undocumented.

- Abrupt changes in morphology are common in the fossil record, suggesting that morphological evolution over long periods of time might not be gradual. Instead, periods of relative stasis may be punctuated by bursts of rapid change, often associated with speciation events.

- Patterns of punctuated equilibria in the fossil record are consistent with modern studies of microevolution, which also document both long periods of stasis due to balancing or stabilizing selection and bursts of rapid change resulting from altered selection environments or key mutations or innovations. ●

14.6 Adaptive Radiations: When α Eclipses Ω

As we discussed in Chapter 13, Hawaii is home to many species found nowhere else on Earth. Time and again, individual organisms have made their way to the islands and then rapidly diversified. There are 37 species of swordtail crickets that descend from a single founder, for example; likewise, an original colonization of honeycreeper birds has given rise to more than 50 species.

But these lineages did more than produce a large number of new species. They also evolved a striking range of morphologies. A lineage of plants called silverswords, found only in Hawaii, are a striking example. Approximately three million years ago, researchers have discovered, a single tarweed plant colonized the Hawaiian island of Kauai. Its descendants then spread to other islands in the archipelago and diversified as they adapted to the different habitats of Hawaii (Landis et al. 2018). In just a few million years, silverswords have radiated so dramatically that it's hard for laypeople to recognize that they are close relatives (**Figure 14.13**).

Clades such as these, which have rapidly diversified by adapting to a wide range of resource zones, are known as **adaptive radiations** (Losos 2010). Islands are not the only places where adaptive radiations take place. The Great Lakes of East Africa are

An **adaptive radiation** is an evolutionary lineage that has undergone exceptionally rapid diversification into a variety of lifestyles or ecological niches.

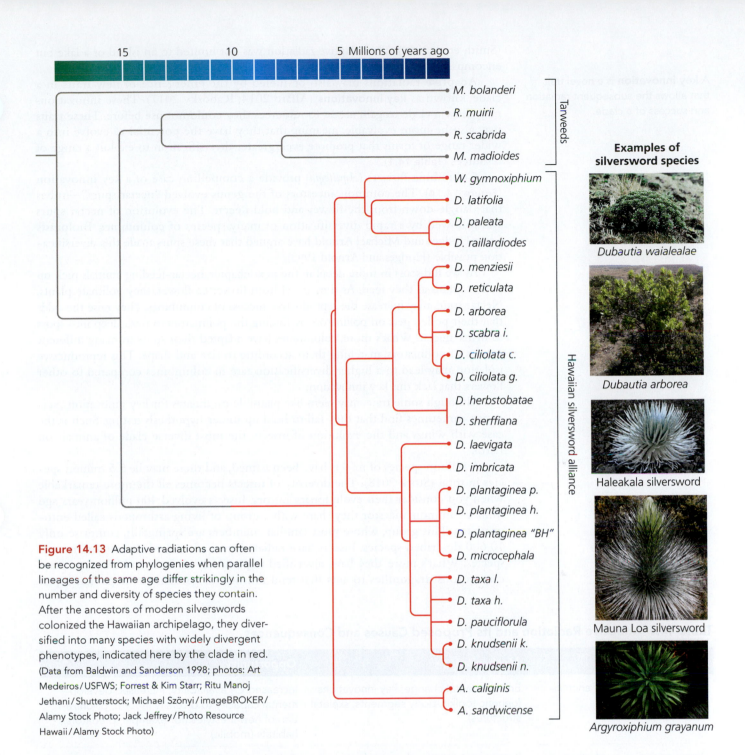

15 **10** **5 Millions of years ago**

M. bolanderi ⎤
R. muirii ⎥
R. scabrida ⎥ Tarweeds
M. madioides ⎦

W. gymnoxiphium
D. latifolia
D. paleata
D. raillardiodes
D. menziesii
D. reticulata
D. arborea
D. scabra i.
D. ciliolata c.
D. ciliolata g.
D. herbstobatae
D. sherffiana
D. laevigata
D. imbricata
D. plantaginea p.
D. plantaginea h.
D. plantaginea "BH"
D. microcephala
D. taxa l.
D. taxa h.
D. pauciflorula
D. knudsenii k.
D. knudsenii n.
A. caliginis
A. sandwicense

Hawaiian silversword alliance

Figure 14.13 Adaptive radiations can often be recognized from phylogenies when parallel lineages of the same age differ strikingly in the number and diversity of species they contain. After the ancestors of modern silverswords colonized the Hawaiian archipelago, they diversified into many species with widely divergent phenotypes, indicated here by the clade in red. (Data from Baldwin and Sanderson 1998; photos: Art Medeiros/USFWS; Forrest & Kim Starr; Ritu Manoj Jethani/Shutterstock; Michael Szönyi/imageBROKER/Alamy Stock Photo; Jack Jeffrey/Photo Resource Hawaii/Alamy Stock Photo)

Examples of silversword species

Dubautia waialealae

Dubautia arborea

Haleakala silversword

Mauna Loa silversword

Argyroxiphium grayanum

geologically very young, in many cases having formed in just the past few hundred thousand years (Sturmbauer et al. 2001). Once the lakes formed, cichlid fishes moved into them from nearby rivers. The fishes then diversified explosively into hundreds of new species. Along the way, the cichlids adapted to making a living in an astounding range of ways—from crushing mollusks to scraping algae to eating the scales off other cichlids (Kocher 2004; Salzburger et al. 2005).

Adaptive radiations may occur when clades evolve to occupy ecological niches in the absence of competition. These opportunities can arise with the emergence of a new island or lake. But they can arise in other ways as well. When extinctions remove certain species from an ecological resource zone, other lineages can evolve that take their place. Such appears to be the case for mammals. When large dinosaurs became extinct at the end of the Cretaceous, large mammals rapidly evolved and diversified

(Smith et al. 2010). This adaptive radiation was not limited to an island or a lake but encompassed the world.

Adaptive radiations may also be fueled by the emergence of new traits in a clade, known as **key innovations** (Alfaro 2014; Rabosky 2017). These innovations may allow a clade to gain access to resources they could not use before. These traits may also be more evolvable, meaning that they have the potential to evolve into a wider range of forms that promote even greater diversification to exploit a range of resources (Table 14.1).

Columbine flowers (*Aquilegia*) provide a compelling case of a key innovation (Figure 14.14). The common ancestors of the genus evolved "nectar spurs"—tubes that dangle down from the flower and hold nectar. The evolution of nectar spurs was followed by a rapid diversification of many species of columbines. Biologists Scott Hodges and Michael Arnold have argued that these spurs made this diversification possible (Hodges and Arnold 1995).

As we'll discuss in more detail in the next chapter, nectar-feeding animals pick up pollen grains as they feed. As they travel from flower to flower, they pollinate plants. Nectar spurs may increase the reproductive success of columbines. They raise the odds that their pollen gets on pollinators by forcing the pollinators to push deep into spurs to drink nectar. What's more, columbines have adapted their spurs to many different species of pollinators, matching them according to size and shape. This reproductive isolation may lead to a higher diversification rate in columbines compared to other flowers that lack this key innovation.

Although some traits may seem like plausible candidates for key innovations, scientists sometimes find that they fail to hold up under hypothesis testing. Such is the case with wings and the evolution of insects, the most diverse clade of animals on Earth.

A million species of insects have been named, and there may be 5.5 million species in total (Stork 2018). The diversity of insects becomes all the more remarkable when you consider their evolutionary history. Insects evolved 400 million years ago from a common ancestor they share with a group of living arthropods called entognathans. This group, whose most familiar members are springtails, comprise only 10,600 described species. Insects have radiated into a hundred times more named species. What's more, they have diversified impressively in form and behavior: from carnivorous dragonflies to ants that tend mushroom gardens to wasps that inject

A **key innovation** is a novel trait that allows the subsequent radiation and success of a clade.

Table 14.1 Adaptive Radiation and Its Proposed Causes and Consequences

Taxa/Clade	Mechanism	Opportunity
Cambrian radiation of animals	Environmental change; key innovations (genetic toolkit, body segments, skeletal structures)	Increased O_2 availability; increased developmental capacity to diversify in form; colonization of new lifestyles (for example, predators), habitats (mobile)
Devonian radiation of plants	Key innovations (seeds, vascular tissue)	Colonization of terrestrial environments
Cretaceous radiation of angiosperms	Key innovation (flowers)	Initiation of mutualistic coevolution with insects
Cenozoic radiation of mammals	Extinction of dinosaurs, large reptiles	Undercontested resources/niches
Radiation of Darwin's finches	Colonization of Galápagos archipelago	Undercontested resources/niches
Radiation of silverswords, fruit flies, honeycreepers	Colonization of Hawaiian archipelago	Undercontested resources/niches
Radiation of columbines	Key innovation (nectar spurs)	Increased reproductive success and reproductive isolation

Figure 14.14 A: Columbine flowers evolved deep cones for holding nectar, known as nectar spurs. Here an insect drinks from a nectar spur—and also picks up pollen from the flower. B: Nectar spurs have been proposed as a key innovation, enabling the clade to rapidly diversify. Here is a range of nectar spurs from columbines, adapted to different pollinators. C: Some of the morphological diversity in columbine flowers. (A: Scott Hodges; B, C: Republished with permission of The Royal Society from "Aquilegia as a model system for the evolution and ecology of petals," Kramer and Hodges, Philos Trans R Soc Lond B Biol Sci. 2010 Feb 12; 365(1539): 477–490. © 2010 The Royal Society; B, C: Data from Kramer and Hodges 2010)

their eggs into living hosts. Meanwhile, entognathans remain very similar to one another after hundreds of millions of years of evolution.

What accounts for the staggering success of insects? One of the traits that distinguishes them from entognathans is wings. This new adaptation allowed insects to occupy ecological roles unavailable to flightless invertebrates. Wings enabled insects to occupy new adaptive zones and also enabled them to colonize new habitats (Grimaldi and Engel 2005; Mayhew 2007; Nicholson et al. 2014).

In 2016, Fabien Condamine and colleagues tested whether wings qualified as a key innovation by examining the rate of diversification in insects, using both their fossil record and molecular phylogenies (Condamine et al. 2016). Surprisingly, the researchers found no measurable increase in diversity after insects evolved wings.

Instead, the study indicates that the great diversity of insects may be due to the independent evolution of several key innovations *within* several insect clades. Beetles, for example, have unusual wings. They only use their hindwings to fly because their front pair of wings have hardened into structures called elytra. Studies on beetles have demonstrated that elytra protect the insect against a range of stresses, from cold temperatures to predation to desiccation (Linz et al. 2016). Elytra also protect their delicate hindwings from tearing or abrasion. Although the first appearance of insect wings may not have been a key innovation, uncoupling the development of beetle fore- and hindwings led to an explosion of diversity.

Key Concepts

- Most adaptive radiations have a common theme: the absence of established competitors for the resources within an environment. Undercontested resources permitted ancestral populations to flourish and adapt to increasingly specialized and localized subsets of those available resources and/or habitats, leading to diversification and speciation.

- Sometimes intrinsic properties of a lineage create ecological opportunity. Key innovations can transform how organisms interact with their environments in ways that take them into new and undercontested habitats or permit them to exploit novel ways of life. These opportunities can trigger explosive subsequent diversification and adaptive radiation. ●

14.7 The Cambrian Explosion: Macroevolution at the Dawn of the Animal Kingdom

As spectacular as the insect radiation has been, the sudden appearance of animal body forms in the fossils record at the dawn of the Cambrian period 541 million years ago was far more profound. Insects belong to the arthropod phylum, along with crustaceans, centipedes, and many other groups of animals. The earliest fossils of arthropods have been found in early Cambrian rocks—along with the earliest fossils of most other phyla of living animals, including the vertebrates, starfish, corals, comb jellyfish, sponges, roundworms, segmented worms, mollusks, and rotifers, as well as several other phyla that no longer exist today.

The near simultaneous appearance of such a diversity of body forms has been called the "Cambrian explosion," and it stands as one of the most important macroevolutionary events in the history of life (Erwin and Valentine 2013; Sperling and Stockey 2018). The Cambrian explosion is also the subject of intense study, as scientists seek to understand all the factors—both environmental and developmental—that triggered it.

The Cambrian explosion was already apparent to paleontologists in the mid-1800s. The oldest remains of animals that scientists found reached back only to the Cambrian period. When paleontologists looked at rocks that formed before the Cambrian, they found nothing. This suggested that all of the animal phyla arose at once, in a pulse of diversification captured in the sudden appearance of the fossils.

Darwin cautioned that the fossil record is spotty. This meant that the suddenness of the radiation of animal body forms might also have been an artifact. Darwin predicted that older fossils of simpler organisms predating the animals would someday emerge. He was right (Schopf 2000). As we saw in Chapter 3, paleontologists have found fossils of microbes dating back some 3.4 billion years, and multicellular life emerges more than 2.1 billion years ago. Researchers have even found some evidence of animals from before the Cambrian period. Biomarkers of sponges date back to 635 million years ago, for example, and macroscopic animal fossils (the enigmatic Ediacarans) date back to 570 million years ago.

Molecular phylogenies offer a second line of evidence showing that animals first evolved long before the Cambrian period (Cunningham et al. 2017). **Figure 14.15** shows the range of molecular clock dates from recent studies. According to these studies, the common ancestor of metazoans—living animals—lived approximately 750 million years ago, nearly 200 million years before the Cambrian explosion. The first major split in animal evolution was the divergence of sponges and all other animals. The ancestors of cnidarians and bilaterians diverged about 640 million years ago. The major lineages of living bilaterians diverged from each other between about 630 and 600 million years ago. These studies indicate that some animal phyla likely

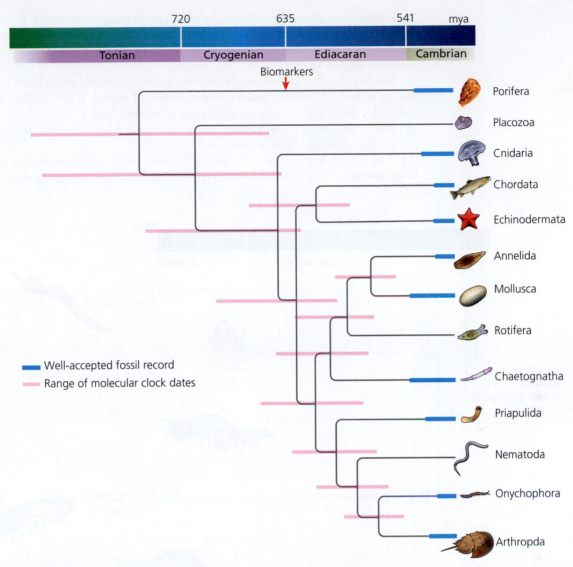

Figure 14.15 Molecular phylogenetic and molecular clock studies indicate that the animal kingdom arose about 750 million years ago, and bilaterians about 630 million years ago. The Cambrian explosion thus occurred long after animals had begun to diversify. (Data from Cunningham et al. 2017)

had already arisen before the "explosion" in the Cambrian fossils. By the same logic, some of their body plan evolution probably occurred gradually over this earlier period.

Nevertheless, a great profusion of diverse body plans does appear to have evolved over a very brief period (approximately 25 million years) right at the beginning of the Cambrian. Scientists are piecing together clues to test whether this pulse of rapid evolution resulted from novel traits—key innovations—from novel combinations of those traits or from changes in the climate or environment occurring at that time.

Ongoing discoveries of new animal fossils are allowing scientists to better understand the step-by-step anatomical transformations that produced the body plans of Cambrian animals. Living arthropods, for example, are united by a unique combination of traits—including a hardened exoskeleton made of chitin—that isn't seen in any other group of animals. Some of the earliest Cambrian arthropod fossils, such as trilobites, share these key synapomorphies with living species.

But that does not mean that arthropods leaped into existence in the Cambrian fully formed. Researchers have identified early relatives of arthropods that display some of the traits found in all living arthropods, but lack others (Caron and Aria 2017).

Their phylogeny shows the sequence by which these traits accumulated. Because these Cambrian relatives of arthropods became extinct long ago, living arthropods appear to be more distinct from other groups than they might have seemed early in animal evolution (**Figure 14.16**).

Despite these insights, many fascinating questions remain about the early evolution of animals. The oldest biochemical evidence of animals is 635 million years old, and molecular clock studies push the origin of animals back to 700 million years ago or more. And yet the oldest known animal fossils—Ediacarans—only appear around 570 million years ago. What explains that absence?

The Ediacarans (Figure 3.22) pose their own puzzles: how were they related to living animals? They produced a wide range of phyla before becoming extinct around

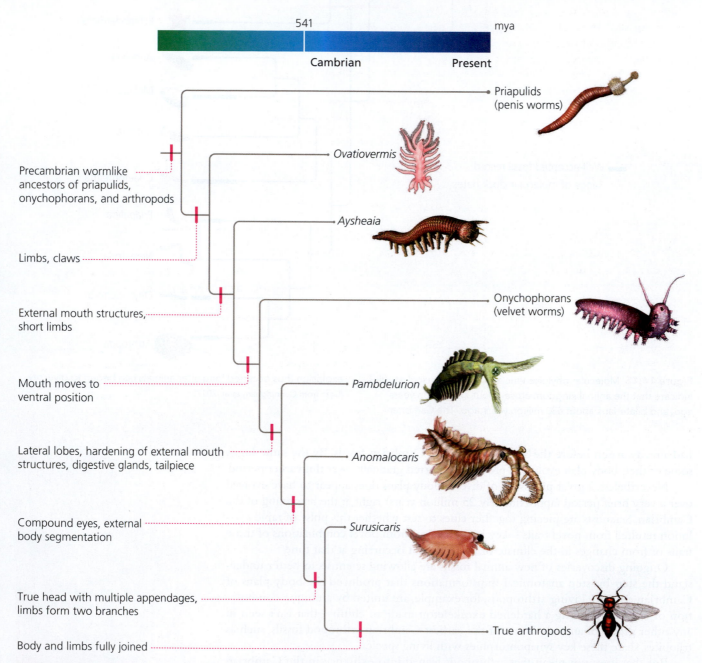

Figure 14.16 The fossil record documents how animal phyla emerged during the Cambrian. Arthropods—a phylum that includes insects, spiders, and crustaceans—share a number of synapomorphies, such as jointed exoskeletons. Some Cambrian fossils are related to today's arthropods, having some synapomorphies but lacking others. This phylogeny shows the relationships of a selection of these species. (Data from Cedric Aria)

540 million years ago. Paleontologists have found a very small range of metazoan fossils from that time. But in the beginning of the Cambrian period, as Ediacarans disappear, the metazoan phyla turn up in astonishing numbers. Why did it take perhaps 150 million years after the origin of animals for this fossil "explosion" to occur?

The full answer to these questions will demand insights from many scientific fields, including geochemistry, oceanography, ecology, and developmental biology. In 2013, Paul Smith, of the University of Oxford, and David Harper, of Durham University, put forward a detailed scenario, taking into account these different lines of evidence. They see the Cambrian explosion as a having had a very long evolutionary fuse (Smith and Harper 2013).

The first step toward the Cambrian explosion may have been the evolution of the developmental toolkit we discussed in Chapter 9. It may have enabled lineages of animals to evolve dramatically new body plans with relatively modest mutations to developmental genes. But that toolkit alone was not enough to trigger an explosion. Early bilaterians, which presumably had this toolkit, seem to have been mainly small worm-shaped animals for tens of millions of years.

The evolution of these animals may have been constrained by the chemistry of the Precambrian ocean. Oxygen levels were very low, for example, making it difficult for animals to swim or burrow or do other energetically demanding activities. Sponges and Ediacarans appear to have had much lower metabolic demands because they were anchored to the seafloor.

Around the beginning of the Cambrian, geological research indicates, the oceans underwent some major changes that could have finally lit the fuse. Tectonic activity caused the sea level to rise, submerging large swaths of coastal regions. Marine animals swiftly colonized these new habitats. Phosphates, eroded from the submerged land, served as fertilizer that spurred the animals' growth.

But the sea level rise also raised levels of calcium in the ocean to potentially toxic levels. At that time, Smith and Harper note, an abundance of small shells appear in the fossil record. They don't think this is a coincidence. Shells might have evolved originally as a defense against calcium poisoning, allowing animals to safely remove the mineral from their tissues. But eventually, the mineralization of animals took on new functions such as hardened weapons like mandibles and claws for predators and thick defenses for prey (**Figure 14.17**).

At the same time, oxygen levels in the oceans were rising for reasons that are not yet entirely clear (Sahoo et al. 2012). The extra oxygen was a great boon to bilaterian animals. For one thing, animals need to burn fuel to make collagen, a protein that binds cells together in their bodies. And as animals began to move around in the ocean, powering their muscles demanded even more energy.

The genetic toolkit, Smith and Harper propose, enabled bilaterians to rapidly evolve into new forms to take advantage of all the new ecological niches that were opening up. And their biological evolution altered the chemical evolution of the oceans. Some bilaterians evolved into burrowers, and for the first time in the history of the oceans, the seafloor became shot through with tunnels. The sediments on the seafloor became oxygenated, enabling many more animals to move into this vast habitat.

Figure 14.18 illustrates the intricate web of causes that Smith and Harper propose to explain the Cambrian explosion. Together, these processes drove the expansion of habitats for animals and spurred the increased complexity of the food web. Once animals began to evolve rapidly, they may have become caught in a feedback loop. Bigger predators evolved to eat smaller ones, for example. Both predators and prey may have evolved new sensory organs, like eyes, to detect their prey and their enemies. Once the Cambrian explosion was finally triggered, it didn't take very long (geologically speaking, of course) for most of the modern groups of animals to emerge.

Figure 14.17 During the Cambrian, the ecology of the ocean changed dramatically. Scientists have found 550-million-year-old fossils of an animal called *Cloudina* that bear holes bored by a predator—one of the earliest signs of animal predation in the fossil record. (Republished with permission of American Association for the Advancement of Science from "Predatorial Borings in Late PreCambrian Mineralized Exoskeletons" by Bengtson and Yue. Science, 1992: Vol. 257, pp. 367–369; permission conveyed through Copyright Clearance Center, Inc.; data from Erwin and Valentine 2013)

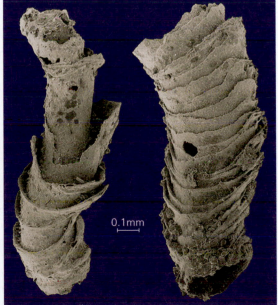

0.1mm

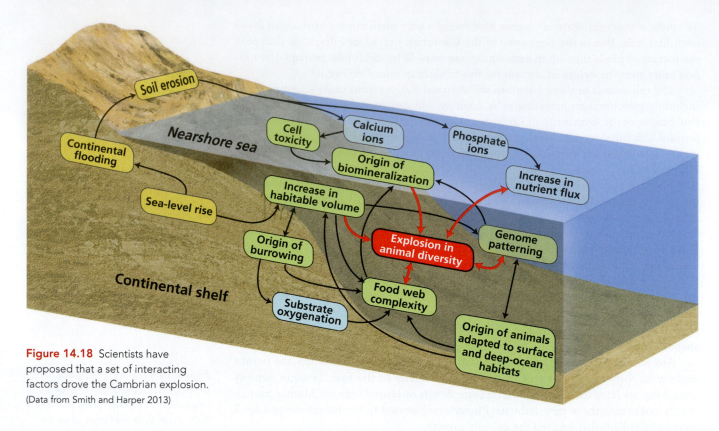

Figure 14.18 Scientists have proposed that a set of interacting factors drove the Cambrian explosion. (Data from Smith and Harper 2013)

Like any hypothesis, Smith and Harper's explanation of the Cambrian explosion needs to be tested against future evidence that could explain the timing of geochemical changes and the onset of new ecological features of the Cambrian oceans.

Key Concepts

- Adaptive radiations are high rates of originations that occur in an area in a relatively short period of geological time. The Cambrian explosion likely resulted because a developmental innovation at the microevolutionary level allowed lineages to radiate and occupy a tremendous diversity of new ecological opportunities.

- Although several animal phyla were already present long before the beginning of the Cambrian, a combination of environmental changes (for example, increased availability of calcium and oxygen), a developmental capacity for rapid body plan evolution (the genetic toolkit), and the creation of new ecological niches (tunneling, predation) contributed to a sudden and spectacular adaptive radiation of animal body plans. ●

14.8 Extinctions: From Background Noise to Mass Die-Offs

The small-scale mechanisms of microevolution underlie the large-scale patterns of macroevolution. To understand how new species emerge into the fossil record, we can integrate the lessons we gain from studying speciation in living organisms, which we explored in Chapter 13. Here we will consider the flip side of macroevolution—extinction—and consider what studies on living organisms can tell us about it.

A species is a lineage made up of linked populations. It can endure for a million years or more, although the total number of individuals in the species may fluctuate wildly over time—booming when a new source of food becomes available or shrinking under attack from a parasite. A species can endure even if some populations completely disappear, because other populations survive and can expand their range to replace the missing ones. But if the total number of individuals in a species shrinks too much, it faces the risk of disappearing altogether.

Once a species falls below this threshold, any number of different factors may drive it extinct. If a lizard species drops to just 50 individuals that all live on a single tiny island, a hurricane can kill them all in one fell swoop. Small populations also face threats from their own genes. As we saw in Chapter 6, they can become vulnerable to genetic drift, which can fix harmful mutations, lowering the population's average reproductive fitness. Small populations also have less genetic variation, which can leave them less prepared to adapt quickly to a changing environment.

Scientists have documented all these processes in living species, and they've even managed to document some actual extinctions. When Dutch explorers arrived on the island of Mauritius in the 1600s, for example, they discovered a big, flightless bird called the dodo (**Figure 14.19**). The explorers killed dodos for food and also inadvertently introduced rats to Mauritius. The rats began eating the dodos' eggs and so drove down their numbers even further. As adult and young dodos alike were killed, the population shrank until only a single dodo was left. When it died, the species was gone forever (Rijsdijk et al. 2015).

Simply killing off individuals is not the only way to drive a species toward extinction. Habitat loss—the destruction of a particular kind of environment where a species can thrive—can also put a species at risk. The Carolina parakeet once lived in huge numbers in the southeastern United States. Loggers probably hastened its demise in the early 1900s by cutting down the old-growth forests where the parakeets made their nests in hollow logs. A smaller habitat supported a smaller population, until the entire species collapsed (Saikku 1990).

These two extinctions were both caused by human activity, as were many others in recent centuries. But extinction is not unique to this age. Long before our own species evolved, other species were becoming extinct. Studies on fossils and living populations indicate that some species likely became extinct through competition with other species. Others were unable to withstand stressful changes in their local physical environment.

If we look at the extinction of a single species, we can think of it as analogous to the death of an individual. Each person's longevity depends on the particulars of his or her own history. But if we statistically study the longevity of everyone, some clear patterns emerge. Worldwide, the average human life span has gradually increased over the past century, but it remains higher in some countries than in others due to factors like nutrition and clean water. Likewise, the longevity of any individual species depends on its own particular history of competition and environmental change. And yet, if we look across species, we can see global patterns.

To estimate the longevity of a given species, paleontologists measure the time between its oldest and youngest known fossils. Over the past 65 million years, North American mammal species have had a mean longevity of 1.6 million years. By contrast, ferns and other vascular plants without seeds (pteridophytes) have a mean longevity of 12 million years (Marshall 2017).

There's another parallel between death and extinction that's valuable for studying macroevolution. Even though the day that any one of us dies may be a matter of chance, the overall death rate remains relatively steady from year to year. (In 2015, 56,657,000 people died worldwide—a rate of 7.7 per 1000 people.) Extinctions follow a similar pattern. Over geological stretches of time, the probabilistic nature of extinctions for individual species produces a fairly steady rate of extinctions. The typical rate of extinctions is called **background extinction**.

A

B

Figure 14.19 A: The dodo became extinct in the late 1600s, probably due to hunting and predation by introduced species. B: The Carolina parakeet became extinct in the early 1900s, due in part to logging, which removed the logs where it built its nests. (A: Encyclopaedia Britannica/UIG/Getty Images)

Background extinction refers to the normal rate of extinction for a taxon or biota.

A clade can endure only if its origination rate (α) of new species is greater than the background rate of extinction. If species in a clade become extinct faster than new ones emerge, the entire clade will become extinct. This can happen if the rate of species extinctions (Ω) rises above α. But a clade can also become extinct if α drops below Ω.

Trilobites were one of the dominant animal clades during the Cambrian period. But after reaching a peak about 500 million years ago, they dropped dramatically at the end of the Ordovician period 443.8 million years ago. They then slowly tapered off for almost 200 million years, finally vanishing for good at the end of the Permian period, 251.9 million years ago.

Mass extinction describes a statistically significant departure from background extinction rates that results in a substantial loss of taxonomic diversity.

What makes the decline of trilobites all the more striking is the simultaneous decline of many other clades. The end of the Ordovician and Permian periods are unusual events in the fossil record known as **mass extinctions**. They are analogous to spikes in the human death rate, when wars, famines, or diseases abruptly kill off many people at once.

Figure 14.20 shows the percentage of marine animal genera that became extinct during geological intervals since the end of the Cambrian (Hull 2015). There are a number of small pulses of increased extinction rates. But above all of these, five peaks stand out as mass extinction events—often nicknamed the "Big Five."

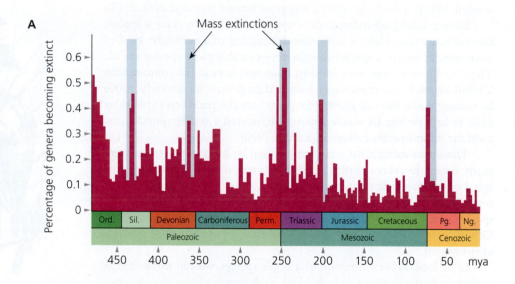

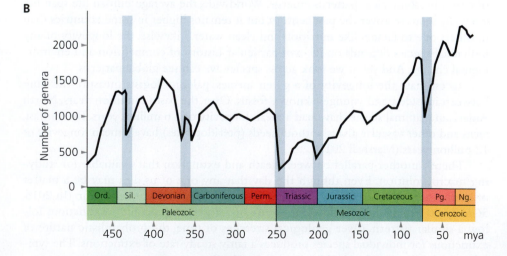

Figure 14.20 A: Studies on marine animal fossils reveal five mass extinction events in the past 500 million years, marked by gray bars. Here, extinctions are measured by the percentage of genera that became extinct in a given time interval. B: Here, the mass extinction events are plotted over a graph of the total diversity of genera. (Data from Hull 2015)

Although the Big Five all pass the same statistical test for being mass extinctions, they had different impacts. Some were more extreme than others, for one thing. The mass extinction at the end of the Permian was the biggest of all, claiming 96% of all species on Earth (Benton 2018).

Mass extinctions have been harder on some clades than others. The end-Ordovician mass extinction took the heaviest toll on trilobites, for example. The end-Permian extinction caused the greatest losses to brachiopods, crinoids, and anthozoan corals. At the end of the Cretaceous, all dinosaurs became extinct, with the exception of birds. Small tetrapods, such as mammals and frogs, suffered extinctions as well but managed to survive.

- Extinction is a common event in the history of life, well documented in the fossil record. Scientists can determine the regularity of this process and use that background extinction rate to examine how departure from that rate affects the diversity of life on Earth. ●

Key Concept

14.9 The Causes of Mass Extinctions

In the search for the cause of mass extinctions, many researchers have focused their attention in recent years on volcanoes (Bond and Grasby 2017; Ernst and Youbi 2017). Given that a volcano erupts somewhere on Earth about once a week, volcanism might seem an unlikely candidate for rare, catastrophic losses of biodiversity. But every 20 or 30 million years, an extraordinary form of volcanism occurs—one that we humans have never witnessed.

In these events, Earth cracks open, pushing vast amounts of molten rock to the surface. These eruptions last for thousands of years—sometimes even a few million. The vast stretches of fresh rock in these so-called Large Igneous Provinces can total millions of cubic kilometers in volume.

The chronology of Large Igneous Provinces bears a striking similarity to the chronology of extinctions. Four out of the five mass extinctions occur very close to the occurrence of Large Igneous Provinces. The end-Permian extinction, for example, took place in less than 60,000 years—right in the middle of a 500,000-year period of volcanic activity in Siberia, in a Large Igneous Province known as the Siberian Traps. A large number of minor extinction pulses also coincided with Large Igneous Provinces.

How could Large Igneous Provinces cause mass extinctions? Some clues come from the chemistry of rocks that formed at these times. Marine sedimentary rocks from these intervals often show signs of having formed in oceans with abnormally low levels of oxygen, for example. Carbon isotopes from these times also suggest that carbon dioxide gas was at extremely high concentrations in the atmosphere.

Figure 14.21 shows how the Siberian Traps could trigger such tremendous mass extinctions. The first step would be a catastrophic change to the atmosphere. The eruptions would inject huge amounts of carbon dioxide, sulfur dioxide, nitrous oxides, and chlorine into the air, along with toxic metals such as mercury. As molten rock spread across the land, it would heat up the underlying soil and rock, unlocking methane and other greenhouse gases.

The toxic metals might be enough to kill organisms on land and in the ocean. Acid rain would form from sulfur dioxide and other gases. These changes to the environment would stress terrestrial ecosystems, and also affect the ocean. Acid rain speeds up the weathering of rocks, delivering minerals and other nutrients to the ocean that can feed oxygen-hungry organisms—lowering the oxygen levels throughout the ocean.

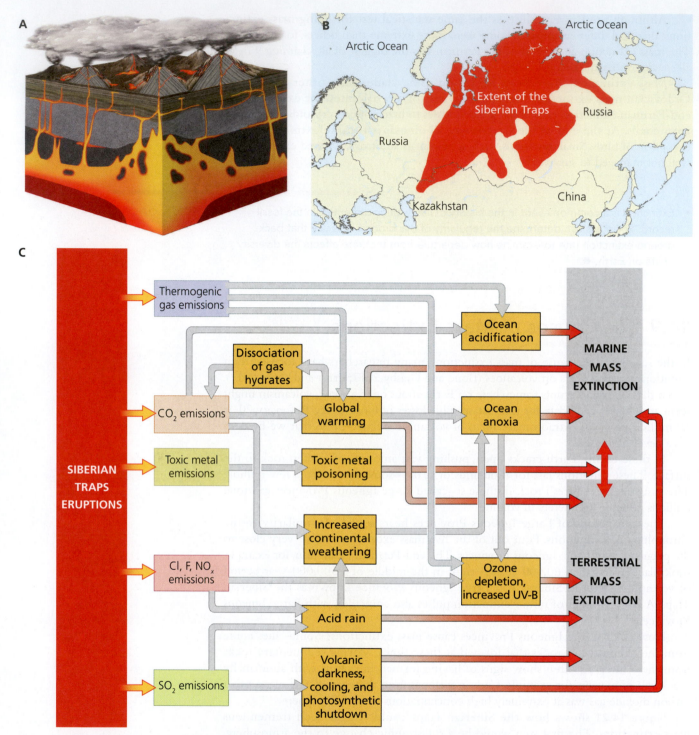

Figure 14.21 A: Large Igneous Provinces are produced by massive amounts of volcanic activity that can last for thousands or millions of years. B: Over the course of about a million years, a Large Igneous Province known as the Siberian Traps spread across much of Asia. During this time, 252 million years ago, the Permian-Triassic mass extinction took place. C: Studies on late-Permian rocks suggest that the Siberian Traps eruptions caused a planet-wide environmental crisis, driving species extinct on land and in the ocean. (Data from Bond and Grasby 2017)

In the atmosphere, these gases can wreak havoc, too. They can destroy the ozone layer, which shields the planet from ultraviolet rays. Without an ozone layer, ultraviolet radiation can reach the surface of the planet, triggering mutations leading to cancer and birth defects. Sulfur dioxide can also form a dense haze that cools the atmosphere and blocks sunlight, casting the planet into shadow and preventing plants and algae from photosynthesizing.

Carbon dioxide causes its own host of troubles. It dissolves into the ocean, lowering the pH of seawater. Ocean acidification can prevent marine animals from making shells and cause other disruptions in their biology. In the atmosphere, carbon dioxide absorbs radiation, leading to global warming. Geochemical evidence suggests that at the time of the Siberian Traps, the climate of the planet swung from cold to warm. A sudden increase in temperature may create too much stress for some species to survive. The warmer oceans become, the less oxygen they can hold. Thus, the oxygen level would drop even further. And so a single event could cause a host of planetary assaults leading to the extinction of almost every species on Earth.

It is possible that other major disturbances to the biosphere can also cause mass extinctions. The last mass extinction event, at the end of the Cretaceous period 66 million years ago, also occurred just after the emergence of another Large Igneous Province, one in India known as the Deccan Traps. But the extinctions also coincided with a visit from outer space. A 10-mile-wide asteroid crashed into the planet.

The first clues of this impact emerged in the 1970s. Walter and Luis Alvarez and their colleagues discovered that rocks at the boundary between the Cretaceous and Paleogene periods had unusually high levels of an element called iridium (**Figure 14.22**). They proposed that an iridium-rich object struck Earth, and the impact distributed the iridium around the world. The Alvarezes and their colleagues proposed that the impact was the cause of the end-Cretaceous extinction (Alvarez et al. 1980).

This provocative hypothesis led other researchers to look for additional evidence. They found many such signs, including quartz crystals that had experienced massive shocks—the kind of shocks you'd expect from the pressure of an impact. Subsequently, a team of geologists found vestiges of the crater itself, off the coast of Mexico (**Figure 14.23**).

Excited by this discovery, other scientists explored how such an impact would have affected the entire planet. By striking a coastal environment, the asteroid would have lofted not just seawater from the ocean, but dust from the adjacent land. The dust clouds could have spread around the planet, blocking out sunlight for weeks, months, or years. A variety of compounds injected into the atmosphere could have had an effect similar to that of Large Igneous Provinces, causing acid rain and succeeding periods

Figure 14.22 A: The rock layer that marks the end of the Cretaceous period 66 million years ago is easy to spot thanks to its bright white color. B: In Italy, researchers discovered that the boundary layer had a high concentration of iridium. They attributed it to an extraterrestrial impact. (A: U.S. Geological Society)

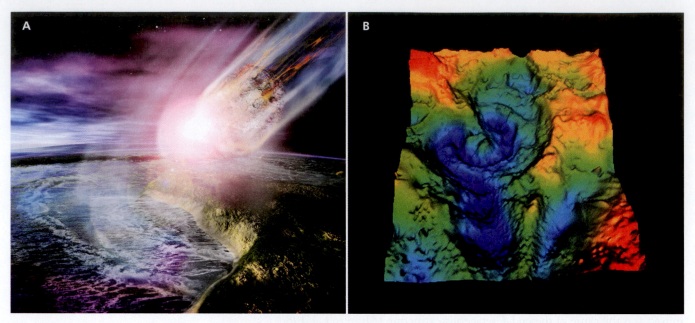

Figure 14.23 A: An artist's conception of the impact of an asteroid 66 million years ago. B: The asteroid's crater was discovered in the seafloor off the coast of Mexico. The colors here show different elevations in the structure. Spanning 100 kilometers, it contains a sheet of melted rock several kilometers thick. (NASA (Chicxulab asteroid impact) Courtesy, V. L. Sharpton/LPI/NASA)

of extreme heat and cold. Soot deposits found near the boundary layer suggested that the energy of the impact triggered wildfires around the world as well (Alvarez 2008).

Both the impact and the volcanism had a profound effect on the planet at the end of the Cretaceous. But determining which of them was responsible for the mass extinctions is challenging. Deccan volcanism occurred over a stretch of 300,000 years at the end of the Cretaceous, whereas the asteroid impact was instantaneous. Paleontologists have searched for outcrops from this precise period that are rich with fossils to see if the extinction rate increased with the Deccan volcanism or only coincided with the impact.

The only place on Earth where paleontologists have conducted a systematic survey of dinosaurs at the end of the Cretaceous, for example, is the Hell Creek Formation in the northwestern United States. The top 3 meters of rock, recording the very end of the Cretaceous, is almost entirely missing dinosaur fossils. Some researchers have argued that this absence means that dinosaurs died out well before the impact—whereas others maintain there are more dinosaur fossils yet to be found (Brusatte et al. 2015).

Paleontologists are searching the planet for more formations where they can get a closer look at the fossil record at the end of the Cretaceous for the many clades that suffered mass extinctions. In 2015, researchers with the British Antarctic Survey reported the discovery of fossil mollusks on Seymour Island, off the coast of Antarctica (Witts et al. 2015). As shown in **Figure 14.24**, the extinction rate spiked abruptly at the very end of the Cretaceous period. The authors of the study conclude from this pattern that the mollusks suffered mass extinctions entirely due to the impact, and not due to the volcanism.

Scientists also have to consider the possibility that the two events, happening so close in geological time, worked together to drive down biological diversity. Researchers have found that in the one million years leading up to the impact temperatures rose and fell several times, and sea levels also changed drastically. Perturbations caused by the Deccan Traps might have put stress on the global ecosystem, and the impact may have then delivered a fatal blow (Schoene et al. 2019).

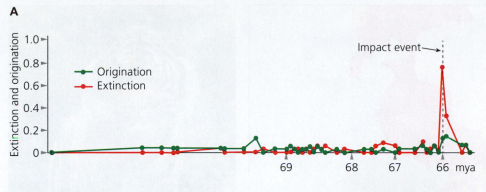

A

B

Antarctic
Peninsula

Seymour
Island

Antarctica

C

Figure 14.24 Researchers have discovered a rock formation on Seymour Island, off the coast of Antarctica, that is rich with mollusk fossils from the end of the Cretaceous. They find a sharp spike of extinctions, consistent with the asteroid impact, as the cause of the Cretaceous-Paleogene mass extinctions 66 million years ago. (Data from Witts et al. 2016; C: © Auckland Museum CC BY)

- The Big Five mass extinctions had different causes and affected different kinds of organisms. There is no single mechanism that explains all mass extinctions.

- Extreme periods of large-scale volcanism can drastically affect climate, leading to toxic gases and rain, depletion of Earth's protective ozone layer, and greenhouse effects. All of these may contribute to mass extinction on a global scale.

- Impacts from asteroids also can cause catastrophic changes to climate and may act either alone or in tandem with volcanism to trigger mass extinction. ●

Key Concepts

14.10 Macroevolution and Our "Sixth Mass Extinction"

The dodo and the Carolina parakeet are far from the only species to have become extinct since the emergence of human civilization. An organization of biologists known as the International Union for Conservation of Nature (IUCN) is systematically evaluating the status of the world's species. By 2018 they had assessed 93,577 species. They found that 872 were extinct, and many of the remaining species were at risk of extinction in the future. The scientists determined that 26,197 species are

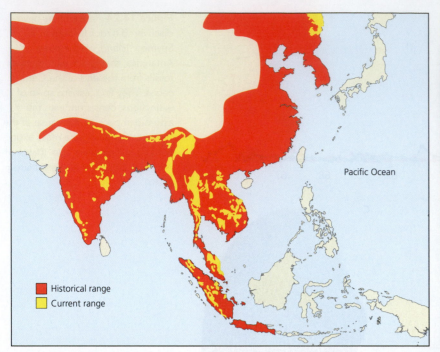

Historical range
Current range

Pacific Ocean

Figure 14.25 Tigers have become an endangered species largely because they have lost most of their historical habitat. Across Asia, the forests where they lived have been logged and converted to farmland and urban settlements. Only about 3800 tigers survive in about 7% of their original habitat. (Data from Barnosky et al. 2017; photo: Colette3/Shutterstock)

threatened by human activities such as deforestation, overfishing, and the introduction of alien species (**Figure 14.25**).

Are these recent extinctions any different from the background extinctions that have occurred for billions of years? Or have the Big Five become the Big Six? Barnosky and his colleagues have used their research on the fossil record to gauge the scale of the current crisis (Ceballos et al. 2015). For mammals, they estimated that the background extinction rate is two extinctions per 100 years per 10,000 species. Using IUCN data, Barnosky and his colleagues determined that over the past century, the average extinction rate of vertebrates is up to 114 times higher than the background rate (**Figure 14.26**). They conclude that a sixth extinction is indeed already underway.

And yet as sobering as these results may be, they may actually underestimate the threat of extinction. Barnosky and his colleagues took into account only the extinctions that have already occurred due to well-established factors such as hunting and

Figure 14.26 A study of vertebrate extinctions over the past few centuries revealed that the average vertebrate extinction rate is far above the background extinction rate. Conservation biologists fear we have entered the sixth mass extinction. (Data from Ceballos et al. 2015)

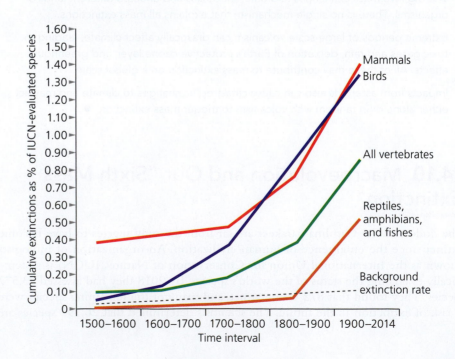

habitat loss. But humans are also altering the atmosphere, and the effects of that change are just now starting to be felt.

Every year, humans release more than 7 billion metric tons of carbon dioxide into the atmosphere. Over the past two centuries, humans have raised the concentration of carbon dioxide in the air, from 280 parts per million (ppm) in 1800 to 405 ppm in 2018. Depending on how much coal, gas, and oil we burn in the future, levels of carbon dioxide could reach 1000 ppm in a few decades (**Figure 14.27**).

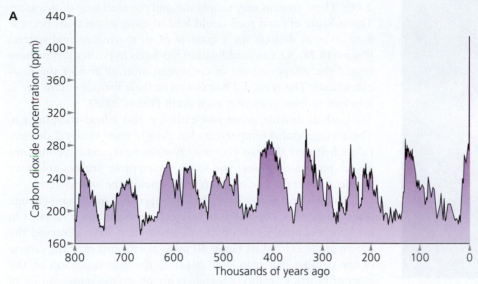

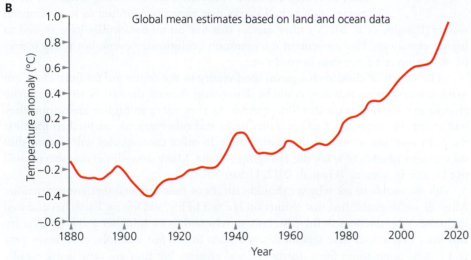

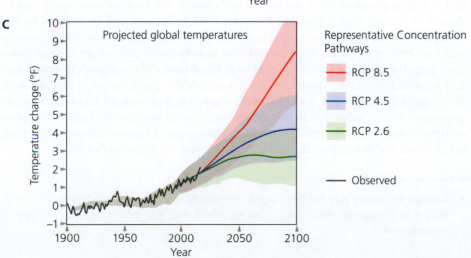

Figure 14.27 A: Human activity has already dramatically raised the concentration of carbon dioxide in the atmosphere. B: As a result, the average temperature of the planet has shown a warming trend for the past century. C: Computer projections consistently show that the planet will warm much more in the next century if the concentration of atmospheric carbon dioxide continues to increase. This graph shows three scenarios. In RCP 2.6, greenhouse gas emissions peak by 2020 and then substantially drop after that. In RCP 4.5, emissions peak around 2040 before declining. And in RCP 8.5, emissions rise throughout the twenty-first century. Rapid climate change may raise the extinction rate by reducing suitable habitat for species.

Figure 14.28 In addition to raising the global temperature, atmospheric carbon dioxide is lowering the pH level of the oceans. This change of chemistry could be devastating to coral reefs, which are home to a quarter of the ocean's biodiversity. (David Burdick / NOAA)

As this carbon dioxide enters the oceans, it is making the water more acidic, with potentially huge impacts on marine life. Bärbel Hönisch, a paleoceanographer at Columbia University's Lamont-Doherty Earth Observatory, and her colleagues have found that the ocean is acidifying faster now than at any point in the past 300 million years (Hönisch et al. 2012). As the pH of seawater drops, the additional hydrogen ions interfere with the growth of coral reefs and shell-bearing mollusks, such as snails and clams (Zeebe et al. 2008). These animals may simply die, and the reefs may disintegrate. The collapse of coral reefs could lead to more extinctions because they serve as shelters for a quarter of all marine animal species (**Figure 14.28**). Ocean acidification has been hypothesized to have caused the disappearance of coral reefs after all five of the mass extinctions. The reefs did not return to their former extent for at least four million years after each event (Veron 2008).

Carbon dioxide, as we saw earlier, is also a heat-trapping gas. The average global temperature has already risen nearly 2 degrees Fahrenheit over the past century. Over the next century, computer models project, the planet will warm several more degrees unless we can slow down the rise of greenhouse gases in the atmosphere.

Animals and plants have already responded to climate change (Parmesan 2006). Thousands of species have shifted their ranges. On land, for example, British butterfly researchers have observed the brown argus expand its range 40 miles northward over the course of two decades (Pateman et al. 2012). In the Atlantic Ocean off the coast of North Carolina, researchers are observing more and more species such as bull sharks that were once limited to more tropical waters (Bangley et al. 2017). Other species that live on mountainsides have shifted to higher elevations. This movement is a common evolutionary event, but today it may be occurring at a faster than normal rate.

The effects of climate change on biodiversity in the future are far from clear, but many scientists warn that they could be devastating. Among the first victims of climate change may be mountain-dwelling species. As they move to higher elevations, they will eventually run out of refuge. Polar bears and other animals adapted to life near the poles may also see their habitats melt away. In other cases, species will have to shift their ranges quickly to track the changing climate. Many slow-dispersing species will not be able to keep up (Hannah 2012; Urban 2015).

It's reasonable to ask why we should care about these impending mass extinctions. After all, we've established that extinction is a fact of life, and life on Earth has endured big pulses of extinctions in the past, only to rebound to even higher levels of diversity. But mass extinctions have serious repercussions for us. For example, mangroves protect coastal populations from storms and soil erosion, but they are now being rapidly destroyed. People who depend on fishes for food or income will be harmed by the collapse of coral reefs, which provide shelter for fish larvae. Bees and other insects pollinate billions of dollars of crops, and now, as introduced diseases are driving down their populations, farmers will suffer as well. Biodiversity also sustains the ecosystems that support human life, whether they are wetlands that purify water or soil in which plants grow. In some cases, a single species can disappear without much harm to an ecosystem. But, as we've seen in this chapter, mass extinctions can lead to the complete collapse of ecosystems for millions of years. The lessons of macroevolution do not just tell us about the distant past: they offer a warning for the future of civilization.

Key Concept

- A single extinction may not have significant effects on an ecosystem, but we should be concerned about the cascading effects that may result from a sixth mass extinction. ●

TO SUM UP . . .

- Microevolution describes the evolution of alleles and processes such as selection and drift. Macroevolution describes evolution at a much larger scale; it is evolution applied above the species level, including the origination, diversification, and extinction of species and clades over long periods of time.

- Biogeography is the interdisciplinary study of the distribution of species around the world.

- Dispersal and vicariance are important processes in species distributions. Dispersal describes the movement of organisms from their place of origin—for example, a seed being transported to an island by wind. Vicariance is the process of barrier formation, such as the barriers that develop through plate tectonics (for example, mountains).

- Changes in biodiversity can be examined using models that account for the number of originations, immigrations, extinctions, and emigrations over a specific time period.

- Lineages can produce new species at fast or slow rates. Lineages can also experience stasis and bursts of change.

- Extinction can be caused by predation, loss of habitat, or other factors that reduce a species' population.

- Rates of origination (α) and extinction (Ω) can be calculated using changes in their numbers over some unit of time. The difference between α and Ω determines the fate of a particular clade.

- Adaptive radiations occur when lineages diversify into many different species with disparate lifestyles, behaviors, and morphologies.

- Opportunities for adaptive radiation can arise if a key innovation transforms how an organism interacts with its environment. These key innovations can allow the organism to exploit new and undercontested habitats or novel ways of life.

- The early evolution of animals was a major radiation, possibly triggered by worldwide environmental changes. The struggle between predators and prey accelerated the diversification of animal lineages.

- The global extinction rate has varied over time. At least five mass extinctions have occurred, but no clear ecological or taxonomic signal or cause unites all five.

- Periods of high extinction rates often coincide with major environmental changes, including sea level change, asteroid impacts, volcanic eruptions, and global warming.

- Scientists have gathered a compelling amount of evidence indicating that humans are a major contributing factor to a new pulse of extinctions.

1. How are extinctions related to biodiversity?

    a. Earth's biodiversity is a result of the relationship between α and Ω.

    b. Extinctions are less important to biodiversity within a specific geographic area than immigration and emigration.

    c. Extinctions always lead to a decrease in biodiversity because extinctions are negative.

    d. Extinctions can decrease standing diversity but not biodiversity.

2. Wallace's boundary is significant in biogeography because it

    a. indicates continental barriers to dispersal.

    b. divides two very distinctive faunas that are geographically very close together.

    c. separates Australia from South America.

    d. divides two geographic areas whose faunas are almost identical.

3. Which of these statements about vicariance is *true*?

    a. Plate tectonics is a primary mechanism of vicariance.

    b. Vicariance led to Australian and South American lineages of marsupials.

    c. Vicariance prevents dispersal.

    d. Both a and b.

4. Which of the following is *not* a hypothesis about the conditions that can lead to adaptive radiations?

    a. Adaptive radiations occur as a result of the absence of competition for ecological resources.

    b. Adaptive radiations occur as a result of island formation.

    c. Adaptive radiations occur as a result of key innovations.

    d. All are hypotheses about the conditions that can lead to adaptive radiations.

5. The typical tempo of extinctions within a particular taxon is called

    a. background extinction.

    b. mass extinction.

    c. total extinction.

    d. episodic extinction.

6. Can the Big Five extinctions all be attributed to a single cause? If so, what caused them?

    a. Yes. The Big Five extinctions were caused by asteroids that had major impacts on habitats when they hit Earth.

    b. Yes. The Big Five extinctions resulted from plate tectonics that changed the quantity and quality of available habitats.

    c. No. The Big Five extinctions were caused by various abiotic and biotic factors that affected different taxa differently.

    d. No. The Big Five extinctions resulted from low origination rates that resulted from a variety of biotic factors.

7. Which of the following is *not* a dispersal event?

    a. Plant seeds falling in rivers and being carried downstream.

    b. Insect populations separated by a cataclysmic volcanic event.

    c. Elk crossing mountain ranges.

    d. Seagulls flying from island to island.

8. Which of the following best explains the model of punctuated equilibria?

    a. The average life span of a species is about one million years.

    b. A large population gradually splits into two distinct lineages.

    c. Most species change very little for most of their history.

    d. Periods of stasis are punctuated by brief periods of rapid morphological change.

9. Which of the following statements about climate change is true?

    a. Photosynthesis helps mitigate the effects of carbon dioxide on the atmosphere.

    b. The chemical composition of the atmosphere can alter the amount of heat it traps.

    c. Eruptions of carbon dioxide from volcanoes contributes to the concentration of gases in the atmosphere.

    d. All of the above.

10. Studies show that the Cambrian explosion

    a. occurred after about 200 million years of animal evolution.

    b. was a time when evolution proceeded at a relatively slow pace.

    c. was a result of one of the Big Five extinctions.

    d. happened because the oceans underwent a massive drop in sea level.

11. What is the turnover rate in stage B shown in the figure at the right?

    a. 20 (34 total species minus 2 that originated and went extinct within the stage, minus 4 more extinctions, minus 8 more originations).

    b. 16 (6 extinctions plus 10 originations).

    c. 28 (34 total species minus 6 extinctions).

    d. 4 (10 originations minus 6 extinctions).

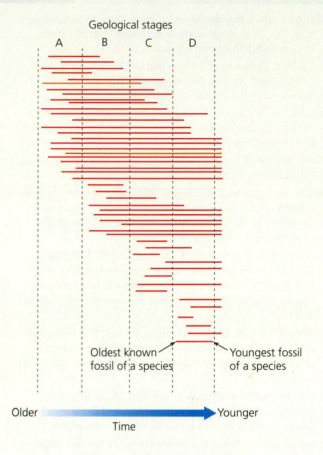

Geological stages

A   B   C   D

Oldest known fossil of a species — Youngest fossil of a species

Older ➡ Younger

Time

SHORT ANSWER QUESTIONS Answers can be found at the end of the book.

12. What are the differences between microevolution and macroevolution? What are the similarities? Which has more evidentiary support: microevolution or macroevolution?

13. What lines of evidence have macroevolutionary biologists used to determine the origin of marsupials? How are these lines distinct?

14. What is the difference between an abiotic and a biotic factor, and how might these factors have contributed to extinctions? Provide examples.

15. Should humans be concerned about the pace of extinctions of organisms that are not directly related to our survival?

16. How do paleontologists use fossils to reconstruct macroevolution?

17. What are the major proposed causes of the Big Five mass extinctions?

ADDITIONAL READING

Barnosky, A. D. 2014. *Dodging Extinction: Power, Food, Money, and the Future of Life on Earth*. Berkeley: University of California Press.

Benton, M. J., and D. A. T. Harper. 2008. *Introduction to Paleobiology and the Fossil Record*. Hoboken, NJ: Wiley.

Erwin, D. H., and J. W. Valentine. 2013. *The Cambrian Explosion*. Greenwood Village, CO: Roberts and Company.

Hunt, G. 2010. Evolution in Fossil Lineages: Paleontology and the Origin of Species. *American Naturalist* 176 (S1): 61–76.

Kocher, T. D. 2004. Adaptive Evolution and Explosive Speciation: The Cichlid Fish Model. *Nature Reviews Genetics* 5:288–98.

Schluter, D. 2000. *The Ecology of Adaptive Radiation*. Oxford: Oxford University Press.

Sepkoski, D. 2012. *Rereading the Fossil Record: The Growth of Paleobiology as an Evolutionary Discipline*. Chicago,: University of Chicago Press.

Vrba, E., and N. Eldredge, eds. 2005. *Macroevolution: Diversity, Disparity, Contingency: Essays in Honor of Stephen Jay Gould*. Lawrence, KS: Paleontological Society.

Alfaro, M. E. 2014. Key Evolutionary Innovations. In D. A. Baum et al. (eds.), *The Princeton Guide to Evolution* (pp. 594–600). Princeton, NJ: Princeton University Press.

Alroy, J. 2008. Dynamics of Origination and Extinction in the Marine Fossil Record. *Proceedings of the National Academy of Sciences USA* 105 (Suppl. 1): 11536–42.

Alvarez, L. W., W. Alvarez, F. Asaro, and H. V. Michel. 1980. Extra-terrestrial Cause for the Cretaceous Tertiary Extinction. *Science* 208 (4448): 1095–108.

Alvarez, W. 2008. T. rex *and the Crater of Doom*. 2nd ed. Princeton, NJ: Princeton University Press.

Baldwin, B. G., and M. J. Sanderson. 1998. Age and Rate of Diversification of the Hawaiian Silversword Alliance (Compositae). *Proceedings of the National Academy of Sciences USA* 95 (16): 9402–6.

Bangley, C. W., L. Paramore, D. S. Shiffman, and R. A. Rulifson. 2018. Increased Abundance and Nursery Habitat Use of the Bull Shark (*Carcharhinus leucas*) in Response to a Changing Environment in a Warm-Temperate Estuary. *Scientific Reports* 8:6018.

Barnosky, A. D., E. A. Hadly, P. Gonzalez, J. Head, P. D. Polly, et al. 2017. Merging Paleobiology with Conservation Biology to Guide the Future of Terrestrial Ecosystems. *Science* 355 (6325): eaah4787.

Beck, R. M. D., H. Godthelp, V. Weisbecker, M. Archer, and S. J. Hand. 2008. Australia's Oldest Marsupial Fossils and Their Biogeographical Implications. *PLoS ONE* 3 (3): e1858.

Benton, M. J. 2003. *When Life Nearly Died: The Greatest Mass Extinction of All Time*. New York: Thames & Hudson.

Benton, M. J. 2018. Hyperthermal-Driven Mass Extinctions: Killing Models During the Permian–Triassic Mass Extinction. *Philosophical Transactions of the Royal Society A* 376 (2130): 20170076.

Bond, D. P., and S. E. Grasby. 2017. On the Causes of Mass Extinctions. *Palaeogeography, Palaeoclimatology, Palaeoecology* 478:3–29.

Brusatte, S. L., R. J. Butler, P. M. Barrett, M. T. Carrano, D. C. Evans, et al. 2015. The Extinction of the Dinosaurs. *Biological Reviews* 90:628–42.

Caron, J. B., and C. Aria. 2017. Cambrian Suspension-Feeding Lobopodians and the Early Radiation of Panarthropods. *BMC Evolutionary Biology* 17:29.

Ceballos, G., P. R. Ehrlich, A. D. Barnosky, A. García, R. M. Pringle, et al. 2015. Accelerated Modern Human–Induced Species Losses: Entering the Sixth Mass Extinction. *Science Advances* 1 (5): e1400253.

Condamine, F. L., M. E. Clapham, and G. J. Kergoat. 2016. Global Patterns of Insect Diversification: Towards a Reconciliation of Fossil and Molecular Evidence? *Scientific Reports* 6:19208.

Crane, P. R. 2013. *Ginkgo: The Tree That Time Forgot*. New Haven, CT: Yale University Press.

Cunningham, J. A., A. G. Liu, S. Bengtson, and P. C. Donoghue. 2017. The Origin of Animals: Can Molecular Clocks and the Fossil Record Be Reconciled? *BioEssays* 39:1–12.

Dias, C., and F. A. Perini. 2018. Biogeography and Early Emergence of the Genus *Didelphis* (Didelphimorphia, Mammalia). *Zoologica Scripta* 47:645–54.

Eldredge, N., and S. J. Gould. 1972. Punctuated Equilibria: An Alternative to Phyletic Gradualism. In T. J. M. Schopf (ed.), *Models in Paleobiology* (pp. 82–115). San Francisco: Freeman, Cooper and Co.

Ernst, R. E., and N. Youbi. 2017. How Large Igneous Provinces Affect Global Climate, Sometimes Cause Mass Extinctions, and Represent Natural Markers in the Geological Record. *Palaeogeography, Palaeoclimatology, Palaeoecology* 478:30–52.

Erwin, D. H., and J. W. Valentine. 2013. *The Cambrian Explosion*. Greenwood Village, CO: Roberts and Company.

Fine, P. V. A. 2015. Ecological and Evolutionary Drivers of Geographic Variation in Species Diversity. *Annual Review of Ecology, Evolution, and Systematics* 46:369–92.

Gallus, S., A. Janke, V. Kumar, and M. A. Nilsson. 2015. Disentangling the Relationship of the Australian Marsupial Orders Using Retrotransposon and Evolutionary Network Analyses. *Genome Biology and Evolution* 7:985–92.

Grimaldi, D., and M. Engel. 2005. *Evolution of the Insects*. Cambridge: Cambridge University Press.

Hannah, L., ed. 2012. *Saving a Million Species*. Washington, DC: Island Press.

Harvey, P. H., E. C. Holmes, A. O. Mooers, and S. Nee. 1994. Inferring Evolutionary Process from Molecular Phylogenies. In R. W. Scotland, D. J. Siebert, and D. M. Williams (eds.), *Models in Phylogeny Reconstruction* (pp. 312–33). Oxford: Oxford University Press.

Hodges, S. A., and M. L. Arnold. 1995. Spurring Plant Diversification: Are Floral Nectar Spurs a Key Innovation? *Proceedings of the Royal Society of London B* 262 (1365): 343–8.

Hönisch, B., A. Ridgwell, D. N. Schmidt, E. Thomas, S. J. Gibbs, et al. 2012. The Geological Record of Ocean Acidification. *Science* 335 (6072):1058.

Hull, P. 2015. Life in the Aftermath of Mass Extinctions. *Current Biology* 25 (19): R941–52.

Hunt, G., M. J. Hopkins, and S. Lidgard. 2015. Simple Versus Complex Models of Trait Evolution and Stasis as a Response to Environmental Change. *Proceedings of the National Academy of Sciences USA* 112 (16): 4885–90.

International Institute for Species Exploration. 2010. *State of Observed Species: A Report Card on Our Knowledge of the Earth's Species*. Tempe, AZ: Arizona State University.

Jablonski, D. 2010. Macroevolutionary Trends in Space and Time. In P. R. Grant and B. R. Grant (eds.), *In Search of the Causes of Evolution: From Field Observations to Mechanisms* (pp. 25–43). Princeton, NJ: Princeton University Press.

Jackson, J. B. C., and A. H. Cheetham. 1994. Phylogeny Reconstruction and the Tempo of Speciation in Cheilostome Bryozoa. *Paleobiology* 20:407–23.

Kocher, T. D. 2004. Adaptive Evolution and Explosive Speciation: The Cichlid Fish Model. *Nature Reviews Genetics* 5:288–98.

Kramer, E. M., and S. A. Hodges. 2010. Aquilegia as a Model System for the Evolution and Ecology of Petals. *Philosophical Transactions of the Royal Society of London B: Biological Sciences* 365 (1539): 477–90.

Landis, M. J., W. A. Freyman, and B. G. Baldwin. 2018. Retracing the Hawaiian Silversword Radiation Despite Phylogenetic,

Biogeographic, and Paleogeographic Uncertainty. *Evolution* 72:2343–59.

Larsen, B. B., E. C. Miller, M. K. Rhodes, and J. J. Wiens. 2017. Inordinate Fondness Multiplied and Redistributed: The Number of Species on Earth and the New Pie of Life. *Quarterly Review of Biology* 92:229–65.

Linz, D. M., A. W. Hu, M. I. Sitvarin, and Y. Tomoyasu. 2016. Functional Value of Elytra Under Various Stresses in the Red Flour Beetle, *Tribolium castaneum*. *Scientific Reports* 6:34813.

Locey, K. J., and J. T. Lennon. 2016. Scaling Laws Predict Global Microbial Diversity. *Proceedings of the National Academy of Sciences USA* 113 (21): 5970–5.

Losos, J. B. 2010. Adaptive Radiation, Ecological Opportunity, and Evolutionary Determinism. *American Naturalist* 175 (6): 623–39.

Louca, S., P. M. Shih, M. W. Pennell, W. W. Fischer, L. W. Parfrey, and M. Doebeli. 2018. Bacterial Diversification Through Geological Time. *Nature Ecology and Evolution* 2:1458.

Marshall, C. R. 2017. Five Palaeobiological Laws Needed to Understand the Evolution of the Living Biota. *Nature Ecology and Evolution* 1:0165.

Mayhew, P. J. 2007. Why Are There So Many Insect Species? Perspectives from Fossils and Phylogenies. *Biological Reviews* 82 (3): 425–54.

Mayhew, P. J., M. A. Bell, T. G. Benton, and A. J. McGowan. 2012. Biodiversity Tracks Temperature over Time. *Proceedings of the National Academy of Sciences USA* 109 (38): 15141–5.

Mitchell, K. J., R. C. Pratt, L. N. Watson, G. C. Gibb, B. Llamas, et al. 2014. Molecular Phylogeny, Biogeography, and Habitat Preference Evolution of Marsupials. *Molecular Biology and Evolution* 31:2322–30.

Mora, C., D. P. Tittensor, S. Adl, A. G. B. Simpson, and B. Worm. 2011. How Many Species Are There on Earth and in the Ocean? *PLoS Biology* 9 (8): e1001127.

Newell, N. D. 1956. Fossil Populations. In P. C. Sylvester-Bradley (ed.), *The Species Concept in Palaeontology No. 2* (pp. 123–37). London: Systematics Association Publication.

Nicholson, D. B., A. J. Ross, and P. J. Mayhew. 2014. Fossil Evidence for Key Innovations in the Evolution of Insect Diversity. *Proceedings of the Royal Society B: Biological Sciences* 281 (1793): 20141823.

Parmesan, C. 2006. Ecological and Evolutionary Responses to Recent Climate Change. *Annual Review of Ecology, Evolution, and Systematics* 37 (1): 637–69.

Pateman, R. M., J. K. Hill, D. B. Roy, R. Fox, and C. D. Thomas. 2012. Temperature-Dependent Alterations in Host Use Drive Rapid Range Expansion in a Butterfly. *Science* 336:1028–30.

Peters, S. E. 2008. Environmental Determinants of Extinction Selectivity in the Fossil Record. *Nature* 454:626–9.

Pyron, R. A., and F. T. Burbrink. 2013. Phylogenetic Estimates of Speciation and Extinction Rates for Testing Ecological and Evolutionary Hypotheses. *Trends in Ecology and Evolution* 28 (12): 729–36.

Rabosky, D. L. 2017. Phylogenetic Tests for Evolutionary Innovation: The Problematic Link Between Key Innovations and Exceptional Diversification. *Philosophical Transactions of the Royal Society B: Biological Sciences* 372 (1735): 20160417.

Raup, D. M. 1972. Taxonomic Diversity During the Phanerozoic. *Science* 177 (4054): 1065–71.

Rijsdijk, K. F., J. P. Hume, P. G. B. de Louw, H. J. Meijer, et al. 2015. A Review of the Dodo and Its Ecosystem: Insights from a Vertebrate Concentration Lagerstätte in Mauritius. *Journal of Vertebrate Paleontology* 35 (Suppl. 1): 3–20.

Sahoo, S. K., N. J. Planavsky, B. Kendall, X. Wang, X. Shi, et al. 2012. Ocean Oxygenation in the Wake of the Marinoan Glaciaton. *Nature* 489 (7417): 546–9.

Saikku, M. 1990. The Extinction of the Carolina Parakeet. *Environmental History Review* 14 (3): 1–18.

Salzburger, W., T. Mack, E. Verheyen, and A. Meyer. 2005. Out of Tanganyika: Genesis, Explosive Speciation, Key-Innovations and Phylogeography of the Haplochromine Cichlid Fishes. *BMC Evolutionary Biology* 5 (1): 17.

Schopf, J. W. 2000. Solution to Darwin's Dilemma: Discovery of the Missing Precambrian Record of Life. *Proceedings of the National Academy of Sciences USA* 97 (13): 6947–53.

Schoene, B., M. P. Eddy, K. M. Samperton, C. B. Keller, G. Keller, et al. 2019. U-Pb Constraints on Pulsed Eruption of the Deccan Traps Across the End-Cretaceous Mass Extinction. *Science* 363:862–6.

Sepkoski Jr., J. J. 1981. A Factor Analytic Description of the Phanerozoic Marine Fossil Record. *Paleobiology* 7 (1): 36–53.

Smith, F. A., A. G. Boyer, J. H. Brown, D. P. Costa, T. Dayan, et al. 2010. The Evolution of Maximum Body Size of Terrestrial Mammals. *Science* 330 (6008): 1216–9.

Smith, M. P., and D. A. Harper. 2013. Causes of the Cambrian Explosion. *Science* 341 (6152): 1355–6.

Sperling, E. A., and R. G. Stockey. 2018. The Temporal and Environmental Context of Early Animal Evolution: Considering All the Ingredients of an 'Explosion'. *Integrative and Comparative Biology* 58:605–22.

Springer, M. S., R. W. Meredith, J. E. Janecka, and W. J. Murphy. 2011. The Historical Biogeography of Mammalia. *Philosophical Transactions of the Royal Society B: Biological Sciences* 366 (1577): 2478–502.

Stork, N. E. 2018. How Many Species of Insects and Other Terrestrial Arthropods Are There on Earth? *Annual Review of Entomology* 63:31–45.

Sturmbauer, C., S. Baric, W. Salzburger, L. Rüber, and E. Verheyen. 2001. Lake Level Fluctuations Synchronize Genetic Divergences of Cichlid Fishes in African Lakes. *Molecular Biology and Evolution* 18 (2): 144–54.

Urban, M. C. 2015. Accelerating Extinction Risk from Climate Change. *Science* 348:571–3.

U. S. Global Change Research Program (USGCRP). 2017. Climate Science Special Report: Fourth National Climate Assessment. Washington, DC: U.S. Global Change Research Program. https://science2017.globalchange.gov/ (accessed November 15, 2018).

Valentine, J. W. 1989. How Good Was the Fossil Record? Clues from the California Pleistocene. *Paleobiology* 15 (2): 83–94.

Veron, J. E. 2008. Mass Extinctions and Ocean Acidification: Biological Constraints on Geological Dilemmas. *Coral Reefs* 27:459–72.

Witts, J. D., R. J. Whittle, P. B. Wignall, J. A. Crame, J. E. Francis, et al. 2016. Macrofossil Evidence for a Rapid and Severe Cretaceous–Paleogene Mass Extinction in Antarctica. *Nature Communications* 7:11738.

Zeebe, R. E., J. C. Zachos, K. Caldeira, and T. Tyrrell. 2008. Carbon Emissions and Acidification. *Science* 321 (5885): 51–2.

CHAPTER 15

Intimate Partnerships

How Species Adapt to Each Other

Learning Objectives

- Explain how coevolution is driven by natural selection.
- Explain the process of reciprocal selection in terms of the geographic mosaic theory of coevolution.
- Compare and contrast the outcomes of coevolution when antagonistic relationships are between two species versus among many species.
- Differentiate between negative and positive frequency-dependent selection and how they function in coevolutionary relationships.
- Compare and contrast the coevolution dynamics of Müllerian and Batesian mimicry.
- Explain how coevolution can promote diversification.
- Describe an example of endosymbiosis.
- Explain how parasites affect the fitness of their hosts and vice versa.

Edmund Brodie Jr. first heard the story of the poisonous coffeepot in the early 1960s. His biology professor at Western Oregon University, Kenneth Walker, told him about three hunters found dead at their campsite in the Oregon Coast Range. There was no sign of struggle, no wounds. The only strange thing about the campsite was the coffeepot. It contained a boiled rough-skinned newt (*Taricha granulosa*). Brodie was hungry for a research project, and so he decided to find out if the newts were poisonous (Brodie 2011).

Brodie set up a makeshift lab and went to nearby ponds to collect the rough-skinned newts. He ground the skin of the newts with a mortar and pestle and mixed it with water. Brodie then injected the resulting solution into mice. Even at minute concentrations, he discovered, the skin could kill a mouse in minutes. Brodie then tried the skin solution on bigger animals, and found it was potent enough to kill them, too.

*A rough-skinned newt (*Taricha granulosa*) produces deadly toxins in its skin. It shows off its bright underside to warn off predators, but some predators have evolved resistance to the toxin.*

Paul Freed / Animals Animals—Earth Scenes

While Brodie was studying the newts, a team of Stanford University researchers published a paper in which they identified the toxic compound in the newt skin. It was a compound called tetrodotoxin, or TTX for short. Several lineages of animals, including pufferfish and poisonous snails, have independently evolved the ability to produce TTX. TTX locks onto a type of receptor on neurons, leading to fatal paralysis.

Brodie was crestfallen at being scooped. But there was still much he wanted to know about the newts. Why should a single rough-skinned newt produce such an overwhelming amount of poison? One day while Brodie was collecting newts, he made a discovery. He caught newts in "pit traps"— a combination of metal drift fences and 5-gallon buckets set up at local ponds. When newts encountered one of Brodie's fences, they crawled alongside it to find a way around and then tumbled into a bucket.

Sometimes the buckets Brodie retrieved also contained snakes, much to his surprise. The snakes weren't trapped in the buckets; when Brodie disturbed them, they slithered quickly away. The snakes were intentionally entering the pit traps. And in one bucket, Brodie discovered why: a common garter snake was actually eating a rough-skinned newt. Brodie then caught some more common garter snakes and injected them with a preparation from newt skin. He discovered that the snakes could withstand levels of TTX that would kill a far bigger animal.

Brodie went on to become a professor at Utah State University and studied snakes and amphibians in many parts of the world. His son, Edmund Brodie III, followed in his father's footsteps, becoming a biologist at the University of Virginia (Figure 15.1). While working in the Oregon Coast Range in the 1980s, Edmund III was studying the diet of snakes. He would squeeze the snakes to force up their recent meals. The main species he was studying, the northwestern garter snake (*Thamnophis ordinoides*), had a fairly dull diet of worms and slugs. But the common garter snake (*T. sirtalis*), he discovered, was loaded with interesting meals, including frogs, birds, and, most surprisingly, a rough-skinned newt.

Edmund III knew that his father had researched the toxins of rough-skinned newts, but he had forgotten that his father had found common garter snakes eating them as well. When he shared his discovery with his father, they began to talk about the mysteries of this strange relationship between predator and prey. Out of that conversation came a research project that has spanned the past 25 years and has revealed many astonishing details about the evolutionary history of the rough-skinned newt and the common garter snake.

The newt and the snake are partners in an evolutionary dance known as **coevolution**. Coevolution is defined as a process of reciprocal evolutionary change between ecologically intimate species driven by natural selection. In this chapter, we will look at the dynamics of coevolution and examine some of the striking adaptations it has produced, such as the toxic overkill of rough-skinned newts. We'll explore the tremendous importance that coevolution has in our everyday life: viruses and other pathogens have coevolved with

Figure 15.1 Edmund Brodie Jr. (left) and his son Edmund Brodie III (right) study the coevolution of rough-skinned newts and common garter snakes. In this photo, they are catching newts in British Columbia with the help of Edmund III's son, Fisher. (Susan Brodie)

Coevolution is reciprocal evolutionary change between interacting species, driven by natural selection.

hosts like ourselves, whereas many of the foods we eat come from plants that have coevolved with insects and other animals that pollinate their flowers and spread their seeds. In fact, our own bodies are the product of coevolution: our cells are amalgams of different species that have come together to form a new collective. ●

15.1 The Web of Life

Coevolution is a major feature of evolution because every species exists in a dense web of interactions with other species. Let's consider a zebra (**Figure 15.2**). The zebra does not generate its own organic carbon; it grazes on grasses to fulfill this need. And once the zebra swallows a mouthful of plant matter, it does not digest this food by itself. Trillions of bacteria and archaea in its digestive tract produce enzymes necessary to break down the food. The grasses, in turn, cannot use photosynthesis to grow without help from another species. They depend on a partnership with underground fungi, called mycorrhizae. The fungi supply nutrients to grass roots in exchange for carbon compounds generated through photosynthesis. Likewise, the zebra is prey for lions and other predators. The zebra may also become infected with deadly viruses and other pathogens, and ticks and other ectoparasites suck on its blood. Still other species, such as oxpecker birds, may visit the zebra to pick off the ticks (Nunn et al. 2011).

Each of these interactions can potentially affect the fitness of either the zebra or one of its ecological partners, or both. The zebra may decrease the fitness of grass

Figure 15.2 Each species exists in a web of ecological interactions. Each interaction can potentially raise or lower an organism's fitness and therefore can be subject to coevolution.

by eating seeds that would otherwise have been able to sprout and reproduce. The oxpeckers may increase the zebra's fitness by removing potentially harmful ectoparasites. And if a lion consumes the zebra, the effect on the zebra's future fitness is obvious.

We can organize such coevolutionary relationships into categories based on the effect these relationships have on each partner's fitness (Table 15.1). **Mutualism** is a positive/positive relationship between species that raises each other's fitness. **Commensalism** is a positive/neutral relationship in which one species benefits but the other suffers no loss of fitness. Other relationships (negative/positive) can lead to significant fitness loss for one species but benefit the other, as in the case of viruses and their hosts.

Key Concept

• Each species in a coevolutionary relationship exerts selective pressures on the others, thereby affecting each other's evolution. ●

15.2 Variation and Selection: The Building Blocks of Coevolution

Coevolved adaptations arise from the same mechanisms that give rise to the other adaptations we've discussed in previous chapters. The first requirement for the evolution of an adaptation, of course, is that heritable variation must be present in the traits relevant to ecological interactions of species. Scientists have documented this kind of variation in a number of different species.

Heritable Variation in Parasitoid Wasps

While at Cornell University, Heather J. Henter documented heritable variation in parasitoid wasps that lay their eggs inside aphids (Henter 1995). The eggs hatch and the wasp larvae feed on the internal organs of the aphids, even as their hosts continue to live. Aphids, however, have defenses against the wasp larvae. Their immune cells coat the larvae, suffocating them. Although this defense can kill some wasps, others manage to survive. Henter set out to measure the heritable variation in the survival of the wasps under attack from the immune systems of their aphid hosts.

She began by producing a genetically uniform stock of aphids, rearing them from a single clone. These aphids would all produce an identical immune response to the wasps. Henter then reared a stock of wasps to infect them with—not an identical set of clones, but ones that varied genetically. She collected female wasps from the wild and produced offspring from them. Then she used these offspring to infect the aphids.

When Henter measured the survival of the wasps, she found that related wasps had similar survival rates (Figure 15.3). Based on her observations, Henter estimated the narrow sense heritability, h^2 (Chapter 7), for successful parasitization to be 0.26.

Henter also ran a mirror-image experiment, in which she used genetically identical wasps to infect aphids with genetic variation. In that experiment, she found that genetic variation within the aphid population affected how susceptible they were to the wasps (Henter and Via 1995). Thus, both the host and parasite have genetic variation for coevolutionary traits. In other words, they have the potential for reciprocal selection.

Heritable Variation in Soapberry Bugs

This variation can fuel rapid evolution. One striking case was documented by Scott Carroll, of the University of California, Davis, and his colleagues when they studied how Australian insects shifted to feed on a newly introduced plant (Carroll et al. 2005).

Table 15.1 Species Can Evolve a Range of Relationships with Other Species

Effect on Fitness	Examples	Coevolved Adaptations
Positive/Positive Mutualism A relationship between species that raises each other's fitness  *Plants and rhizobia*	**Pollination:** Insects and other animals visit flowers to gather nectar. Plants benefit because the animals spread their pollen, allowing them to reproduce. **Seed dispersal:** Birds and mammals eat fleshy fruits. The seeds pass through their digestive tracts and are released in their droppings. The plant benefits from being dispersed across a wide range. **Nutrient exchange between mycorrhizae and plants:** Fungi in soil deliver minerals and other nutrients to plant roots. Plants deliver organic carbon to fungi. **Farming:** Some species of ants rear "mushroom gardens" in their nests. **Animals and microbiota:** Humans and other animals depend on microbes to help digest food and synthesize vitamins. **Cleaners:** Some species of fish eat ectoparasites on the skin of other fishes.	Bright colors on flowers attract insects and birds. Hummingbirds insert slender bills into flower tubes.  *Pollinating birds and flowers* Farmed fungi can grow only inside ant nests. Ants harbor antibiotic-producing bacteria on skin that kill pathogenic fungi that sometimes invade gardens.
Positive/Neutral Commensalism A relationship in which one species benefits but the other suffers no loss of fitness	**Scavengers:** Remora fishes attach to larger fishes and detach to feed on the prey killed by the larger fishes.	Remora fishes have structures to keep them attached to another fish.  *Remora on shark*
Negative/Positive  *Virus*	**Predators and prey:** A relationship in which an animal ingests another animal. **Herbivores and plants:** A relationship in which an animal feeds on plants. **Deceptive pollination:** A flower tricks an insect into visiting it, without providing nectar in exchange. **Host and parasite:** A virus, intestinal worm, or other organism lives in or on another organism, often causing disease or death.	Prey produce toxins to deter predators; predators evolve defenses against toxins.  *Garter snake and rough-skinned newt* Plants produce sticky latex to stop insects; insects avoid triggering latex production. Flowers produce pheromones and grow structures that appear like female insects. Male insects learn to avoid flowers. Many species of parasites castrate hosts or alter their behavior. Host immune systems attack the pathogens.

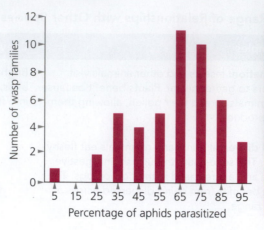

The insect in question is the soapberry bug. It has a long mouthpart that it uses to drill into fruits to reach the seeds at their center. In the 1960s, a species of balloon vine was introduced from the United States and began to spread invasively across eastern Australia. The balloon vine has a larger fruit than the native species that the soapberry previously fed on. Researchers observed that as the balloon vine became abundant, the soapberry bugs began attempting to feed on its fruits. But because their mouthparts were adapted to the smaller, native fruits, they only sometimes succeeded in feeding on the seeds of the larger balloon vine fruits.

Carroll and his colleagues wondered if natural selection led soapberry bugs to grow larger mouthparts to better deal with the larger balloon vine fruits. To detect this shift, they compared the soapberry bug populations that feed on invasive balloon vines to populations of the insect that still feed on native fruits, as well as to museum specimens from the early 1900s before the introduction of the balloon vine (**Figure 15.4**). They discovered that the insects that have shifted to the balloon vine have 10%–15% longer beaks (Carroll et al. 2005). This change led to a marked rise in feeding success. Carroll and his colleagues offered balloon vine fruits to soapberry bugs in their laboratory. They found that longer-beaked bugs were able to reach the seeds within the balloons at almost twice the rate of the shorter-beaked soapberry bugs that still feed on native fruits.

Reciprocal Selection

Reciprocal selection describes selection that occurs in two species due to their interactions with one another. Reciprocal selection is the critical prerequisite of coevolution.

As one species adapts to another, its partner may evolve as well. This two-way evolution results from **reciprocal selection**. Bacteria, for example, are attacked by an enormous variety of viruses known as bacteriophages. The viruses can kill the bacteria, and as a result they can act as an agent of selection on the bacteria, favoring resistance mechanisms. They may evolve receptors that are difficult for viruses to bind to and enter, for example. Other defenses include enzymes that can recognize incoming virus DNA and cleave it into harmless fragments. But as the bacteria become resistant, they favor adaptations in the viruses that let them overcome their hosts' defenses. As each population continues to adapt to the other, the reciprocal nature of the selection increases.

The strength and the direction of reciprocal selection can differ from population to population, and even when they experience similar selection, populations may differ in how they evolve in response to it. Some populations may be very small, for example, so that selection is overwhelmed by drift. Some populations may experience selection on other traits that have pleiotropic effects on coevolved traits. In other words, based on population genetics, we would expect that coevolved traits will display geographic variation in their evolution.

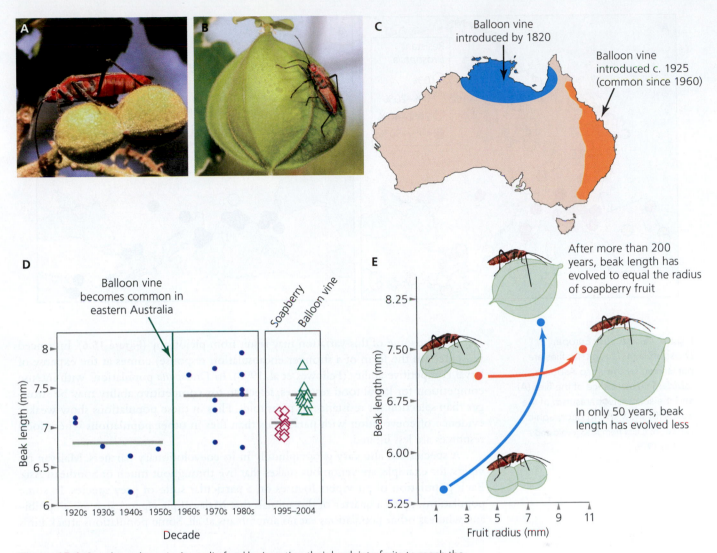

Figure 15.4 Soapberry bugs in Australia feed by inserting their beak into fruits to reach the seeds inside. A: A soapberry bug feeds on a native woolly rambutan fruit. B: A soapberry bug feeds on the nonnative balloon vine, which produces a much larger fruit. C: Balloon vines have colonized two regions of Australia. In northern Australia they became established two centuries ago, but in eastern Australia they were not common until the 1960s. D: In eastern Australia, the beaks of Australian soapberry bugs that feed on balloon vines have evolved to longer lengths in three decades. Blue circles in this graph represent historic museum specimens. Gray bars mark the mean beak length in the specimens before and after the 1960s, when balloon vines became common in eastern Australia. Diamonds show beak length of insects collected in 2004 from native host plants, and triangles show insects collected from balloon vine plants. E: Much longer beaks have evolved in soapberry bugs in northern Australia, where the balloon vines have been established longer. (A, B: Scott P. Carroll; data from Carroll et al. 2005)

And that, indeed, is what scientists do find. **Figure 15.5** shows the geographic variation across Europe in the interacting traits of both the fly *Drosophila melanogaster* and a parasitoid wasp, *Asobara tabida* (Kraaijeveld and Godfray 1999). The map on the left shows the percentage of flies in each area that can defend against the parasite by encapsulating wasp eggs. The map on the right shows the percentage of wasps that can resist this encapsulation and grow anyway. Both species show variations in the expression of coevolved traits across their range—and, interestingly, the geographic patterns are somewhat different.

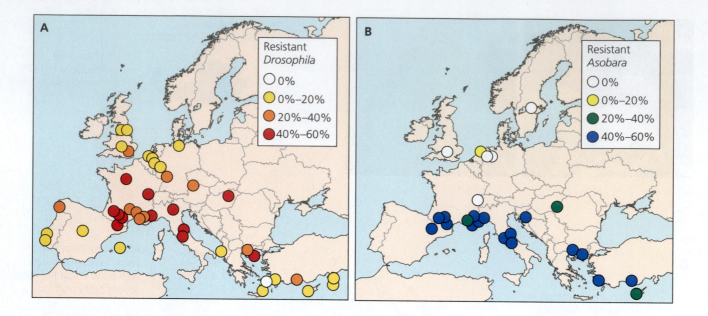

Figure 15.5 Across Europe, *Drosophila melanogaster* flies are parasitized by the wasp *Asobara tabida*. The resistance of the flies (A) and the successful parasitism of the wasps (B) show strong geographic variation. (Data from Kraaijeveld and Godfray 1999)

At least some of this variation may result from pleiotropy (**Figure 15.6**). Enhanced resistance (in the form of a stronger encapsulation response) comes at the expense of larval competitive ability (Fellowes et al. 1998). In *Drosophila* populations with intense competition for larval food resources, selection for competitive ability may be stronger than selection for resistance to parasitoids. Flies in these populations show weaker evidence of coevolution with parasitoids than flies in other populations where food resources are less limited.

A species can also vary geographically in its coevolutionary partners. Malayan pit vipers, for example, are venomous snakes that live throughout much of Southeast Asia. Each population of pit vipers focuses on a particular suite of prey species. In some populations, up to a quarter of the animals that Malayan pit vipers catch are amphibians, whereas other populations eat no amphibians at all. Some populations attack birds

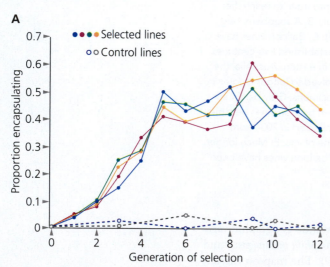

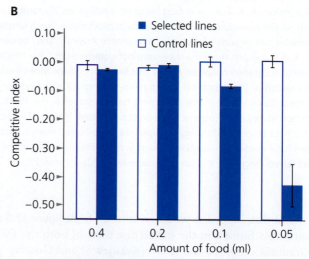

Figure 15.6 One source of variation in *Drosophila* resistance to parasitoid wasps is the cost that strong resistance incurs. A: H. C. J. Godfray and colleagues ran a selection experiment to produce four lines of highly resistant flies. B: They then put the resistant flies in competition with wild-type flies at four different levels of food. They measured the relative success of the flies with a competitive index. (The index is log[e/(r + 1)], where *e* is the number of experimental larvae and *r* is the number of reference larvae that successfully eclose in each replicate.) The performance of control larvae is shown in the open bars, and the selected larvae in blue bars. The control larvae outperform the selected ones at low levels of food. (Data from Fellowes et al. 1998)

and mammals preferentially, whereas others attack only reptiles. These specializations, in turn, select for variations in the venom the vipers produce. Venom evolves differently from population to population because the partners in the coevolutionary interactions differ. Today, each population of pit vipers produces a combination of venoms adapted to their particular prey (Daltry et al. 1996; see Chapter 9 for more on the evolution of snake venom).

Geographic Mosaic Theory of Coevolution

John Thompson, an evolutionary biologist at the University of California, Santa Cruz, and his colleagues have sought to formalize the components of variation in evolving interactions and build an ecological explanation of the coevolutionary process in what he has dubbed the **geographic mosaic theory of coevolution** (Thompson 2005, 2009, 2010). Thompson hypothesizes that the evolution of interactions among species is shaped by three factors. The first is the geographic variation in the type of selection on the interacting species. In some places the interaction may be under mutualistic selection, for example, whereas it may be under antagonistic selection in other places. It's also possible for the interaction to be antagonistic everywhere, but selection may be directional in some environments and frequency-dependent in other environments. The second factor in the theory is the geographic variation in the strength of reciprocal selection resulting from these interactions. And the third is the geographic variation in the distribution of traits that evolve in response to this selection.

In this model, coevolution plays out differently in different populations, leading to a variety of coevolutionary outcomes. You can think of the geographic distribution of these outcomes as a mosaic made up of differently colored tiles. At some locations, populations will become coevolutionary "hotspots" where reciprocal selection is intense. Others will become "coldspots" where there is no selection. Gene flow, genetic drift, and other processes will remix these coevolving traits across landscapes. The result is spatial variation in coevolutionary traits.

The geographic mosaic theory predicts that coevolution can give rise to several different local dynamics (**Figure 15.7**). Generally, these dynamics can be grouped as antagonistic or mutualistic. Some examples will help clarify the variety of interactions that can evolve.

The **geographic mosaic theory of coevolution** proposes that the geographic structure of populations is central to the dynamics of coevolution. The direction and intensity of coevolution vary from population to population, and coevolved genes from these populations mix together as a result of gene flow.

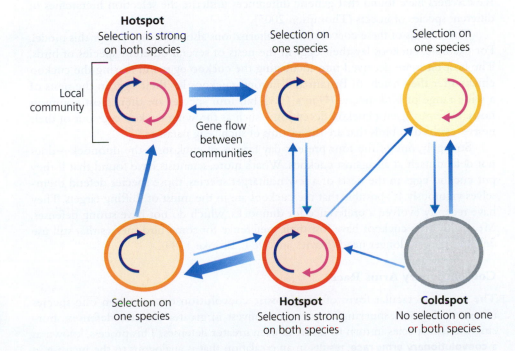

Figure 15.7 The geographic mosaic theory explains coevolution as a complex pattern of reciprocal selection unfolding in a linked group of populations. Some populations are coevolutionary "hotspots," where selection is strong on both species. Others are "coldspots," where one or both species experience no selection. Gene flow delivers coevolved genes from one population to another. (Data from Thompson 2010)

Antagonists

When two species are in an antagonistic relationship—such as parasites and their hosts—they can generate negative frequency-dependent selection on each other. Parasites that are adapted for the most common host genotype in a population will have the highest fitness. As these parasite genotypes multiply, the fitness of the common hosts goes down. Rare host genotypes then have higher fitness than the common ones because they suffer from fewer parasite infections. Over time, these host genotypes will become more prevalent. At that point, the most common parasite genotype will be selected against, whereas rare parasite genotypes better adapted to the new dominant host genotype are favored. As we saw in Chapter 11, such reciprocal coevolution may favor sexual reproduction (the Red Queen effect), and it can maintain genetic variation for host resistance and pathogen virulence within local populations (Dodds and Thrall 2009; Lively 2010).

The strongest antagonistic coevolution results from pairwise interactions between two species. But many antagonistic relationships have a more complex ecological structure. For example, a single species may have an antagonistic relationship with several partner species. Lions eat zebras, for example, but they can also select other prey species to attack. A zebra, meanwhile, may graze on several different plant species.

In such interactions, we wouldn't expect strict, one-on-one coevolutionary dynamics to evolve. One possible alternative is that a predator focuses on the least-defended of several possible target species. For a time, this paired interaction is strong, though not necessarily reciprocal. Predator pressure would select for better defenses in the focal prey species. Eventually, the focal prey population may become rarer relative to other locally available prey species. Or a more successful strategy may evolve in the prey, enabling it to better escape the predator. In either case, the predator may switch over to a new species before counter-adaptations to the original prey species evolve. This shifting strategy, known as **coevolutionary alternation**, results in sequential or alternating bouts of pairwise coevolution between many different species (**Figure 15.8**).

Plant-feeding insects fit this model of coevolutionary alternation. In many species, female insects lay eggs on one of several different species of plants. They often prefer one species over all of the others; if they can't find their top preference, they choose a less preferable species, and so on down a hierarchy. This preference hierarchy results in variation in the selection pressure exerted by the insects on different plants. Researchers have found that genetic differences underlie the selection hierarchies of different species of insects (Thompson 2005).

The victims of these coevolutionary alternations also offer support for this model. For example, cuckoos lay their eggs in the nests of several different species of birds. The host birds are deceived into incubating the cuckoo eggs and rearing the cuckoo chicks after they hatch. In Britain and Europe, cuckoos lay their eggs in the nests of a wide range of bird species (**Figure 15.9**), but four species are their most common hosts. Defenses against cuckoo deception—such as throwing cuckoo eggs out of their nests—evolved in birds that are the victims of such "egg parasites."

Strangely, one of the four present-day targets of cuckoos—the dunnock—does not defend itself at all against cuckoos. What's more, scientists have found that if they put cuckoo eggs in the nests of a few nontarget species, those species defend themselves vigorously. It's possible that the cuckoos are in the midst of shifting targets. They have recently evolved a preference for dunnocks, which do not have strong defenses. Meanwhile, the cuckoos have lost their preference for some bird species that still use defenses they no longer need (Davies and De L. Brooke 1989).

Coevolutionary Arms Races

The most spectacular form of antagonistic coevolution occurs when one species invests in defenses, spurring its partner to invest in greater counter-defenses, spurring the first species in turn to invest in even greater defenses. This process, known as a **coevolutionary arms race**, results in an escalation that is analogous to the increase in

Coevolutionary alternation occurs when one species coevolves with several other species by shifting among the species with which it interacts over many generations.

A **coevolutionary arms race** occurs when species interact antagonistically in a way that results in each species exerting reciprocal directional selection on the other. As one species evolves to overcome the weapons of the other, it, in turn, selects for new weaponry in its opponent.

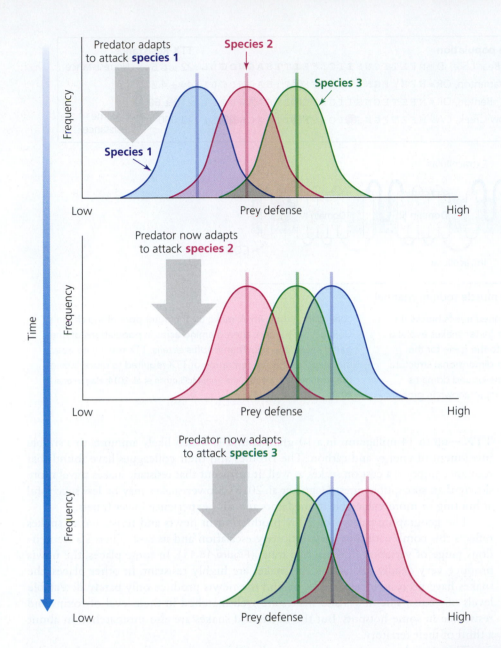

Predator adapts to attack **species 1**

Species 2

Species 3

Species 1

Frequency

Low   Prey defense   High

Predator now adapts to attack **species 2**

Frequency

Low   Prey defense   High

Predator now adapts to attack **species 3**

Frequency

Low   Prey defense   High

Time

Figure 15.8 Coevolutionary alternation occurs when one species is in an antagonistic relationship with several other species. In this diagram, a predator attacks three prey species. In each prey species, individuals vary in the extent of their defenses against predation. By specializing on the species with the lowest defenses (here marked species 1), the predator can have the greatest success. But the predator thereby imposes selection on species 1, which can drive it to evolve better predatory defenses. As the defensive capabilities of this species increase, a new prey species emerges as the least defended and, therefore, the more profitable prey. The predator can then switch to species 2, which now has the weakest defenses, and so on.

nuclear weapons in both the United States and the former Soviet Union during the Cold War (see **Box 15.1**).

Thanks to the work of the Brodies and their colleagues, the rough-skinned newt and common garter snake have become one of the best-understood cases of coevolutionary escalation. After the ability to produce high levels of TTX evolved in rough-skinned newts, resistance to the toxin evolved in common garter snakes. The Brodies and their colleagues have identified mutations in the snakes that alter the shape of the receptor to which TTX normally binds (**Figure 15.10**). In populations where common garter snakes can resist TTX and eat the newts, selection then favors newts with even greater production of TTX.

The adaptations that arise during coevolutionary escalation can impose a cost on each species. The most toxic newts carry a lot of

Figure 15.9 Cuckoos lay eggs in the nests of other species. The defenses exhibited by their hosts show signs of being the result of coevolutionary alternation. (John Mason / Pantheon / Superstock)

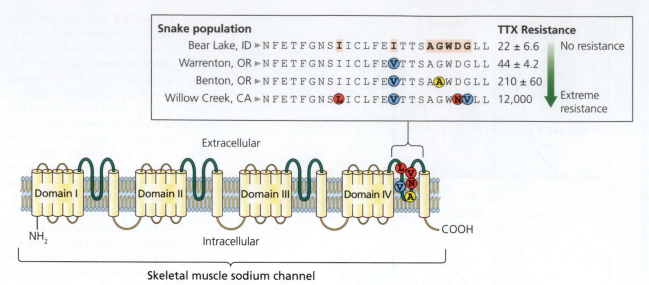

Snake population		TTX Resistance	
Bear Lake, ID ▶ N F E T F G N S **I** I C L F E **I** T T S **A G W D G** L L	22 ± 6.6		No resistance
Warrenton, OR ▶ N F E T F G N S I I C L F E **V** T T S A G W D G L L	44 ± 4.2		
Benton, OR ▶ N F E T F G N S I I C L F E **V** T T S A **A** W D G L L	210 ± 60		
Willow Creek, CA ▶ N F E T F G N S **I** I C L F E **V** T T S A G W **N** **V** L L	12,000		Extreme resistance

Figure 15.10 The TTX toxin of rough-skinned newts blocks the skeletal muscle sodium channel. Common garter snakes evolved resistance to the toxin through mutations to the gene for the sodium channel. This figure shows the two-dimensional structure of part of the protein, illustrating membrane-bound domains (DI–DIV) and the polypeptide chains ("p-loops" shown in green) connecting them, which make up the outer pore of the channel. Colored circles mark altered amino acids in populations of one garter snake species, *Thamnophis sirtalis*. TTX resistance was measured as the concentration of TTX required to block 50% of the sodium channels. (Data from McGlothlin et al. 2014; Hague et al. 2017)

TTX—up to 14 milligrams in a 10-gram animal—which likely amounts to a sizable investment of energy and carbon. The Brodies and their colleagues have found that resistance imposes a cost on snakes as well. It turns out that resistant snakes travel more slowly than susceptible ones (Hague et al. 2018). Slower snakes may be less successful at hunting or more vulnerable to predators and thus experience lower fitness.

The geographic pattern of TTX production in newts and resistance in snakes reflects this combination of coevolutionary escalation and its cost. There's a tremendous range of variation in these two traits (**Figure 15.11**). In some places, the newts produce very deadly toxins, and the snakes are highly resistant. In other places, the snakes have no resistance to speak of, and the newts produce only barely detectable levels of toxins. The newts and snakes are well matched in their level of toxins and resistance in some hotspots. But the newts and snakes are also mismatched in about a third of their territory.

This complex pattern is the sort predicted by the geographic mosaic theory. It's possible that this mismatch evolves because snakes can quickly evolve strong resistance to the newts, whereas it's harder for the newts to evolve a more potent toxin (McGlothlin et al. 2014; Hague et al. 2017). It takes only the change of a single amino acid for the neuron channels of the snakes to resist the newt toxin. On the other hand, newts appear to require a series of several mutations to arise to become more deadly. Edmund Brodie III and his colleagues argue that when a population of snakes becomes resistant enough, the arms race is essentially over (Brodie 2011).

Coevolutionary escalation is by no means universal. Parasites, for example, do not always evolve to cause their hosts greater and greater harm. (The degree of harm caused by a parasite in this process is called its virulence.) Instead, different levels of virulence evolve in different parasites (Bull and Lauring 2014). About half of people who are infected by the Ebola virus experience a swift, horrific death, for example, whereas a cold virus can produce billions of descendants inside of us while making us feel only a bit under the weather.

Starting in the 1980s, scientists developed an explanation of how parasites can evolve different levels of virulence, and they called it the trade-off hypothesis (Alizon et al. 2009). They argued that virulence represents a trade-off between selection within hosts for rapid replication and selection for transmission between hosts.

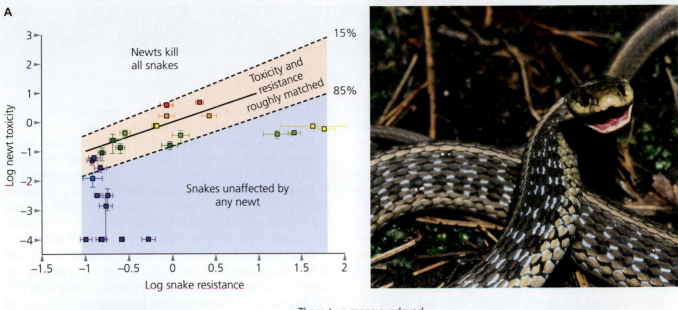

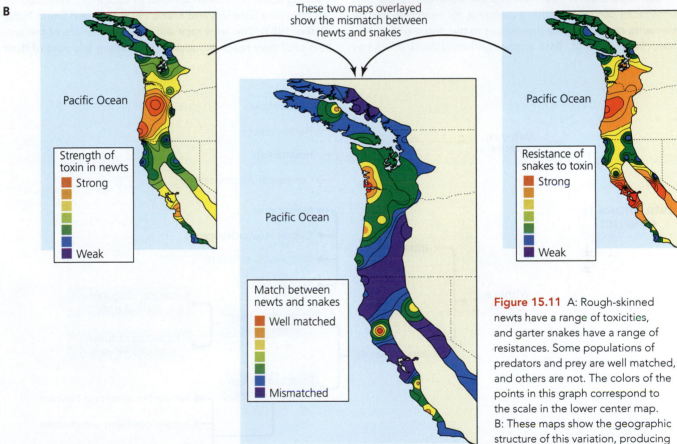

Figure 15.11 A: Rough-skinned newts have a range of toxicities, and garter snakes have a range of resistances. Some populations of predators and prey are well matched, and others are not. The colors of the points in this graph correspond to the scale in the lower center map. B: These maps show the geographic structure of this variation, producing a geographic mosaic of coevolution. (Data from Geffeney et al. 2005; Hanifin et al. 2008; photo: Nature's Images/Science Source)

Within an infected host, there may be many different strains of a pathogen. They may arise from a single ancestor through mutations, or they may independently infect the same host. In either case, these strains compete. Virulent strains will tend to dominate because they use the host's resources to reproduce more rapidly. Competition within a host favors increased virulence.

If a host dies before transmitting any pathogen, however, these fast-reproducing pathogens will die, too. Selection for transmission to new hosts thus favors *reduced* virulence. These two opposing agents of selection together shape the optimal virulence of a pathogen.

BOX 15.1

Biological Arms Races

Arms races crop up in all sorts of interesting places, and whenever they do, the effect is profound. They can dramatically accelerate the speed of evolution and alter subsequent directions of evolution of the affected species.

The crucial prerequisites of an evolutionary arms race are conflict and reciprocal coevolution: the agents of selection must themselves evolve. In the case of antagonistic interactions among different species, this means that when one species changes, it acts as an agent of selection on the other. This selection then favors a change in the second species, which subsequently acts as an agent of selection on the first, and so on.

Box Figure 15.1.1 categorizes the different forms that arms races can take. Within a genome, for example, mobile elements can propagate themselves to the detriment of their host (Section 15.4). That impact on host fitness produces reciprocal selection and coevolution between the mobile elements and the host (Burt and Trivers 2006).

Mary Jane West-Eberhard (1983) noted that many social situations can also lead to reciprocal coevolution. The fitness of an individual often depends on its size or status relative to other individuals in the population. As the population evolves, an individual's fitness may change. If male elk with the largest antlers consistently win contests for females, then mutations causing an increase in antler size will spread through the population to fixation. The mean antler size of the population thus increases, effectively resetting the baseline and favoring yet another increase in antler size. This "sliding scale" of fitness can lead to escalatory evolution, or an arms race (Dawkins and Krebs 1979; Maynard Smith 1982; Parker 1983). The arms race will increase the size of the antlers until they reach an upper limit, where the cost of their

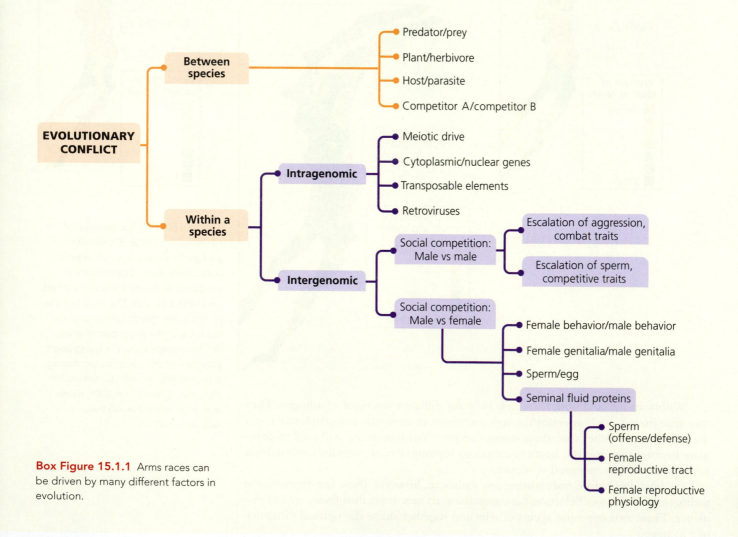

Box Figure 15.1.1 Arms races can be driven by many different factors in evolution.

size is greater than the advantage they provide in competition with other males.

In elk, the arms race arises among individuals of a single sex. Another form of arms race can arise between the sexes of a species, for each sex benefits when it can usurp control of fertilization from the other (Rice and Holland 1997). A new tactic that arises in males selects for a countertactic in the females, which then selects for yet another tactic in the males. Antagonistic coevolutionary arms races are thought to have led to the evolution of extreme genitalia in male and female ducks (Chapter 11), as well as rapid evolution of male genital claspers and female abdominal spines in water striders (Perry and Rowe 2012).

Many of the best-understood arms races arise between reciprocally interacting (coevolving) species. But here also,

the crucial prerequisite is that the agents of selection relevant to a population evolve themselves. Predators and prey are locked in classic evolutionary arms races, as are parasites and their hosts (for example, the Red Queen effect; Chapter 11). Even mimic species can enter an arms race with the species that they copy. When a mimic converges too closely with the model species and its numbers grow, the model suffers a cost because predators switch from aversion to attraction and begin tasting the mimic and model alike. This shift selects for any new trait—a blaze of color on a butterfly wing or a new spot on the side of a frog—that will permit the predator to discern model from mimic. Now selection acts on the mimic to converge again on the new model phenotype. Such rapid coevolutionary arms races can generate spectacular diversity in forms (**Box Figure 15.1.2**).

Three toxic species	Papilio dardanus

Box Figure 15.1.2 Toxic butterflies evolve bright coloration to warn off birds and other predators. Batesian mimics, which are themselves harmless, gain protection by evolving to look similar to the toxic species. The toxic butterflies can, in turn, benefit by evolving away from the mimic. Thus, the butterflies can become swept up in a coevolutionary arms race. This figure shows, on the left, three different toxic species in the family Danaidae, and on the right, three different mimetic forms of the harmless species *Papilio dardanus*.

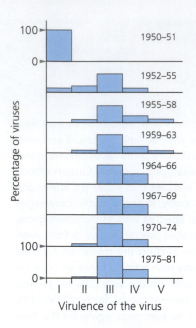

Figure 15.12 Myxoma virus was introduced into Australia to control the exploding rabbit population. This graph shows the virulence of the virus over time, as measured by "virulence grades" (I is the deadliest, V is the mildest). Initially the virus was highly lethal, and later it became less virulent. Less virulent strains were able to spread more effectively than more virulent ones. A similar shift occurred when myxoma virus was introduced into France to control rabbits there. (Data from Kerr et al. 2012; photo: Popperfoto/Getty Images)

When this balance leads to the evolution of lower virulence, the strength of reciprocal selection can diminish for both host and pathogen. What begins as an intensely antagonistic interaction gradually attenuates until each species exerts only weak selection on the other, and the speed of coevolution slows. On the other hand, if transmission becomes easier, then selection may tip in favor of within-host replication and increased virulence.

Attenuated Coevolution: Rabbits in Australia

One of the best-documented examples of attenuated coevolution has unfolded in Australia over the past 60 years (Kerr et al. 2012). In 1859 rabbits were introduced to Australia by Thomas Austin, an immigrant farmer, so that he could have game to shoot. Without predators to control them, the rabbits exploded across the continent, eating so much vegetation that they began to cause serious soil erosion. In the 1950s, scientists deployed a biological counteroffensive, known as rabbit myxoma virus (**Figure 15.12**).

The rabbit myxoma virus, which was discovered in South America, causes deadly infections in hares on that continent. The virus was introduced to Australia in the hopes that it would be just as deadly against the imported rabbits there. At first, it lived up to those hopes: scientists found that 99.8% of infected Australian rabbits died. But within a few years, the virus's mortality rate dropped to between 70% and 95%. That's still fairly deadly, but not deadly enough to keep the fast-breeding rabbits in check. Because the virulence of the myxoma virus evolved to an intermediate, less deadly state, rabbits continue to be an ecological blight on Australia today.

In 1952, the same experiment took place in Europe, where another myxoma virus strain was imported to France to control rabbits. As in Australia, the virus turned from highly deadly to moderately deadly. The two strains of myxoma acquired different sets of mutations over the past 60 years. But despite their differences, they converged on the same intermediate level of virulence as a result of the same trade-off between rapid reproduction within hosts and transmission between hosts. (In Chapter 18, we will return to the evolution of virulence and explore its importance for medicine.)

Mutualists

Coevolution can also produce a variety of mutualisms, in which the alleles that are favored by selection benefit, rather than harm, another species. Here, too, the strongest selection and most rapid coevolution tend to occur when species participate in intimate pairwise interactions with one other species. But reality is often more complex.

Pairwise interactions may crop up for a while and then disappear, or they may be confined to local hotspots on an otherwise patchy mosaic of species interactions.

When pairs of species interact mutualistically, each species may experience an increase in the frequency of alleles that allow it to better complement the alleles of its partner species. If a mutation results in plants supplying mycorrhizal fungi with more carbon, for example, the plant can also benefit because the fungi will be able to grow faster and provide more nutrients from the soil. In these situations, new alleles that enhance the mutualistic interaction are favored by selection, and they begin to spread.

But unlike antagonistic coevolution, where the performance of alleles drops off as they become common, here the fitness of these alleles only increases as they spread through the population. In this case, the frequency-dependent selection is positive. As individuals with the new alleles become common, the more likely they are to encounter the other species, and the more likely they are to reap the benefits of the mutualistic interaction. (This assumes the most common genotype of one species is most likely to interact with the most common genotype in the other species.)

Positive frequency-dependent selection sweeps these alleles to fixation. When another mutually beneficial mutation arises, it will be favored by selection, too, and its fitness also will increase the more common it becomes. With each step, the intimacy of the partnership grows, and the fitness benefits of mutualism increase. This escalating interaction can generate rapid coevolution of both species.

Long-Tongued Flies

In South Africa, biologists Steven Johnson and Bruce Anderson have documented a geographic mosaic pattern of coevolution in long-tongued flies (*Prosoeca ganglbaueri*) and the flowers they pollinate. The tongues of the flies are actually tubes that they dangle behind them as they fly. When they find a flower that they want to drink nectar from, they fold their tongue forward until it extends before them a distance several times longer than their entire body (**Figure 15.13**).

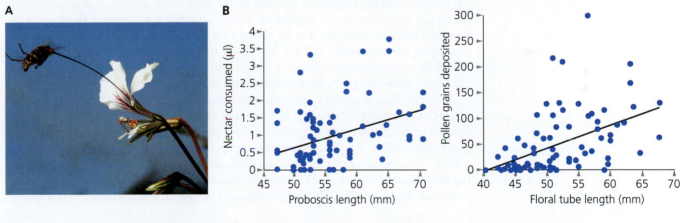

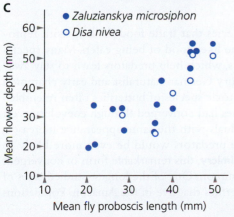

Figure 15.13 A: Long-tongued flies in South Africa feed on flowers with correspondingly long tubes. These extremes are the result of the coevolution of the flies and flowers. B: The longer a fly's tongue, the more nectar it can drink (left). But the longer a flower's tube, the more effort a fly must make to reach the nectar. As a result, the fly deposits more pollen grains on the flower (right). C: This graph shows the length of the proboscis of the fly *Prosoeca ganglbaueri* compared to the flowers it feeds on. *Z. microsiphon* is the fly's main source of nectar, whereas *D. nivea* is a deceptive orchid that mimics *Z. microsiphon* but provides no food. In different populations of the flies and the flowers, different matching lengths have coevolved. (A: Steven Johnson; data from Johnson and Anderson 2010)

These flies use their extraordinary tongues to feed on the nectar of long-necked flowers, especially the flowers of a species called *Zaluzianskya microsiphon*. As they try to push their tongues to the bottom of the deep nectar receptacles, the flies rub their heads on the flowers, picking up pollen. When the flies visit another flower, the pollen from the first plant can fertilize the second one's ovules (Pauw et al. 2009; Johnson and Anderson 2010).

Anderson and Johnson traveled to 16 sites in South Africa, where they caught hundreds of long-tongued flies. They measured the length of each fly's tongue, along with many other traits of the flies. At the same 16 sites, they also gathered *Zaluzianskya* flowers, measuring them as well. The scientists found a striking pattern. At some sites, the flies have tongues as long as 50 millimeters (2 inches), closely matching the depth of the *Zaluzianskya* flower tubes at those sites. At other sites, the flies have tongues only half that length. Their shorter tongues are matched by shorter flower tubes at the same sites (Johnson and Anderson 2010).

Anderson and Johnson argue that natural selection must be creating the local tongue lengths. They suspect that *Zaluzianskya* flowers at some sites dominate the sources of nectar for the flies, resulting in strong pairwise reciprocal selection between the species. Natural selection favors deeper tubes for nectar in these places because it forces the flies to pick up pollen as they struggle to reach their tongues into the tubes. As the flower tubes get deeper, natural selection favors longer tongues, which in turn selects for deeper flower tubes. But in other populations, the flowers may have to compete with smaller flowers and shorter tubes, thus diluting the strength of reciprocal selection. In such cases, if a fly can't reach to the bottom of a *Zaluzianskya* flower, it can always drink from a flower of another species, and having too long a tongue may make this more difficult. At these sites, natural selection may favor flies with shorter tongues as well as *Zaluzianskya* flowers with shorter tubes.

Mutualisms Encompassing Many Species

Whereas some ecological interactions involve only a pair of species, mutualisms between free-living organisms typically encompass many species. A large number of species of flowering plants depend on fruit-eating birds to disperse their seeds, for example. Each plant species may be visited by a diversity of bird species. Likewise, any given bird species usually will not limit itself to a single species of fruit. Indeed, that would be a poor strategy because a bird would have food for only a short time each year. Both birds and plants form a mutualistic network.

Despite the complexity of mutualistic networks, they can also coevolve. A group of species that play the same role in a network will converge on similar traits. In Costa Rica, for example, hawk moths pollinate many species of trees, vines, and shrubs (Haber and Frankie 1989). All these plants have independently evolved traits that increase the odds of their being visited by hawk moths—white flowers, sucrose nectar, and opening their flowers at dusk. Each species of plant does not depend on a single species of pollinator, however. They all can be pollinated by several different hawk moth species.

Mimicry

Mutualistic networks are not limited to species that trade food for help with reproduction. Some of these networks reduce the likelihood of being eaten. Many species with toxins also have bright warning colors, which help predators learn to stay away from them. Fritz Müller, a nineteenth-century German naturalist and early champion of Darwin, noticed in Brazil that unrelated toxic species of butterflies often resembled one another. He proposed that the butterflies had converged through coevolution on the same pattern. If the number of individuals with the same appearance increased, unrelated species could all benefit because predators would be even more likely to learn to avoid them. Known as **Müllerian mimicry**, this remarkable form of convergent coevolution turns out to be surprisingly common. One of the longest-studied cases of Müllerian mimicry involves *Heliconius* butterflies that live in the Amazon. Populations

Müllerian mimicry occurs when several harmful or distasteful species resemble each other in appearance, facilitating the learned avoidance of predators.

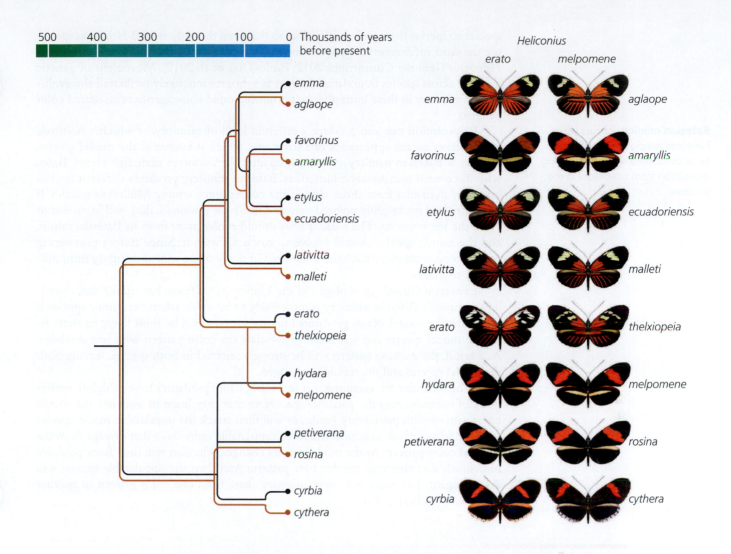

Figure 15.14 *Heliconius* butterflies produce compounds that are toxic to birds. Their bright colors enable birds to quickly learn to avoid them. Despite extraordinary variation in wing coloration from place to place, local populations of *H. erato* and *H. melpomene* always converge on similar wing patterns. Coevolution caused these populations to diversify in tandem, resulting in parallel phylogenies. (Data from Cuthill and Charleston 2012; photo: Republished with permission of American Association for the Advancement of Science from "Optix Drives the Repeated Convergent Evolution of Butterfly Wing Pattern Mimicry" by Reed et al. Science, 2011: Vol. 333, No. 6046 pp. 1137–141; permission conveyed through Copyright Clearance Center, Inc.)

of several species—for example, *Heliconius melpomene* and *H. erato*—have co-diversified, resulting in dozens of local color forms (**Figure 15.14**). From place to place, the particular warning patterns differ, but within each location, the colors of the two species match. Reconstructions of their recent history show that changes in each species tracked changes in the other as independent lineages of butterflies coevolved (Cuthill and Charleston 2012).

Although scientists have known about *Heliconius* mimicry for more than 130 years, they've only recently begun to work out the molecular underpinnings of this remarkable example of convergent evolution (Reed et al. 2011; Martin et al. 2014). A diversity of red-striped patterns have repeatedly evolved on *Heliconius* wings, for example, and coevolving species appear to have independently converged on each of the local stripe patterns. Robert Reed and Riccardo Pappa, of the University of California, Irvine, and their colleagues found that a single developmental gene was responsible for the red coloration in a number of species. What's more, the gene, called *optix*, produces an identical protein in all of the butterflies. The different color patterns are the result of different *optix* expression patterns in the developing butterfly wings.

This convergence is similar to the evolution of stickleback armor we discussed in Chapter 10, in which the same gene was repeatedly modified in different fish populations to produce the same phenotype. It also helps solve the puzzle of how readily Müllerian mimicry can evolve: a shared developmental gene can be deployed easily to produce the same pattern in different species. In fact, it now appears that the regulatory region responsible for spatial patterning of *optix* expression was shuttled from species to species through hybridization, so that even distantly related *Heliconius* species use the same mechanism to generate red-colored patterns on their wings (**Figure 15.15**; *Heliconius* Genome Consortium 2012; Pardo-Diaz et al. 2012). Movement of genetic material across species boundaries, known as introgression, likely facilitated the evolution of mimicry in these butterflies by permitting rapid convergence onto shared color patterns.

Coevolution can also produce a different kind of mimicry, in which a nontoxic species takes on the appearance of a toxic one, which is known as the model species. Known as **Batesian mimicry**, it was named after the Victorian naturalist Henry Bates, who discovered it in Amazon butterflies. Batesian mimicry produces different coevolutionary dynamics from those seen in the convergence among Müllerian mimics. If predators feed on brightly colored prey and don't get poisoned, they will learn not to avoid the toxic species. The toxic species should evolve away from its Batesian mimic, and the mimic species should evolve to catch up with it. Since Bates's pioneering work, Batesian mimicry has been discovered in many other animals, ranging from millipedes to snakes.

Lawrence Gilbert, an ecologist at the University of Texas, has argued that coevolution through Batesian mimicry is most likely to be stable when the mimic species is rarer than the model. Naive predators hunting for prey will be most likely to encounter the model species and will learn to associate the color pattern with unpalatability. As a result, the existing pattern will be strongly selected in both species, leaving both the model species and the mimic unchanged.

When mimics are common, on the other hand, predators have a higher probability of encountering the palatable species, so that they learn to associate the shared color pattern with palatability. Predators will then attack the unpalatable model species as well, and as a result, selection will favor unpalatable individuals that diverge from the original color pattern. As the model species changes, selection will then favor palatable individuals that converge on that new pattern. And then the unpalatable species will diverge again. The result is a coevolutionary chase from one color pattern to another to another (Gilbert 1975).

Batesian mimicry occurs when harmless species resemble harmful or distasteful species, deriving protection from predators in the process.

Key Concepts

- The intensity and specificity of interactions among coevolving species may vary across landscapes, and this spatial mosaic of interactions can shift over time.

- Host and parasite (or pathogen) populations may generate negative frequency-dependent selection on each other, a reciprocal interaction that can maintain genetic variation within both populations.

- Predator populations sometimes switch between several different prey populations, resulting in sequential or alternating bouts of pairwise coevolution between multiple species.

- Escalated and rapid coevolution of weapons and defenses (an "arms race") results when species interact antagonistically so that each generates directional selection on the other.

- Interactions among species that begin as strongly antagonistic sometimes become less antagonistic over time, as each species adapts in ways that reduce the intensity of the interaction for the other.

- When mutualistic species exert positive frequency-dependent selection on each other, the rapid coevolution that results can be very similar to an arms race, only without the antagonism.

- Mutualistic interactions are vulnerable to the invasion of cheaters. Cheaters can spread rapidly within a population, resulting in the collapse of the mutualism or the evolution of defenses, such as sanctions. ●

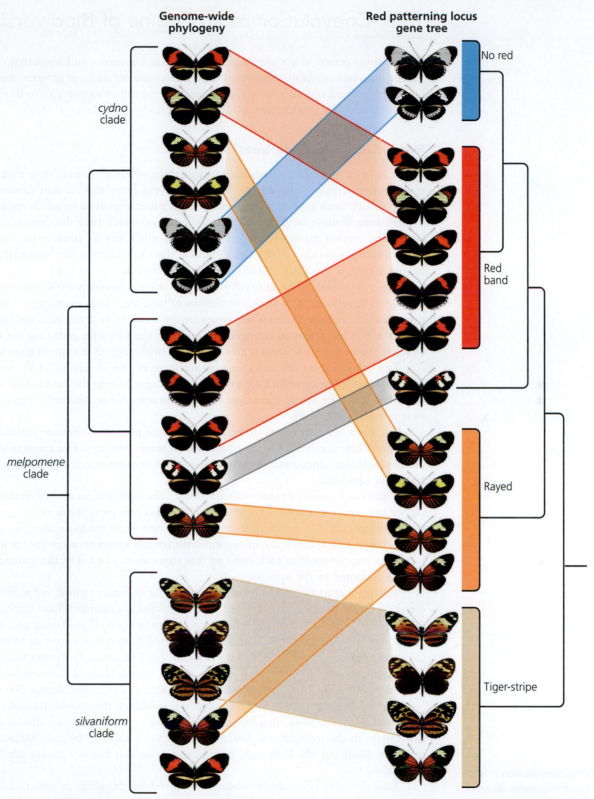

Figure 15.15 Genomic studies of *Heliconius* evolution suggest that the regulatory region controlling *optix* expression introgressed repeatedly during the recent history of these butterflies. The phylogeny on the left shows the actual history of the species (the "species tree"), as inferred from comparison of genome-wide DNA sequence data. In contrast, the phylogeny of the regulatory region of the *optix* gene suggests a different evolutionary history, shown on the right. Species with similar wing patterns cluster together on this tree, even though many of them are distantly related. This pattern would be expected if the segment of DNA containing the *optix* regulatory region had been horizontally transferred between clades through hybridization and introgression. Across-species transfer of this regulatory module may have facilitated the rapid evolutionary convergence of wing patterns in these coevolving butterfly species. (Data from Kronforst 2012; *Heliconius* Genome Consortium 2012; Pardo-Diaz et al. 2012)

15.3 Coevolution as an Engine of Biodiversity

Coevolution results in a wide range of ecological outcomes and traits that can vary among populations. As species coevolve over thousands or millions of years, the coevolutionary process can also drive speciation. Coevolution can even leave its mark on macroevolution by altering rates of diversification.

Red Crossbills and Pine Trees

In the Rocky Mountains, birds known as red crossbills are coevolving with several species of pine trees (**Figure 15.16**). The birds have a distinctive bill that crosses over at the tip, which enables them to pry seeds out of pinecones. Red crossbills spend much of their time feeding on the distal tips of the cones, which have the densest supply of seeds. The smaller the scales and cones, the easier it is for the birds to pry out seeds. Crossbills avoid the cones with thick scales at their tips because the benefit they'd get from the seeds is not worth the cost of prying them out.

Craig Benkman, an evolutionary ecologist at the University of Wyoming, and his colleagues have found that some populations of pine trees have adaptations that defend them from the foraging birds. Their cones are short and wide at the base, and they grow thick scales. This defensive arrangement impedes the crossbills, reducing the number of seeds lost to the birds. Trees with this cone morphology thus enjoy higher fitness.

The evolution of the trees, in turn, influences the evolution of the crossbills. Forests where cones are thick select for longer, larger bills in the birds. Benkman and his colleagues have found that in areas where pine trees have thicker cones, the birds have unusually large bills.

But Benkman has found other populations of pine trees that are maladapted to crossbills. Their cones produce thinner scales at their distal tips. The cause of this mismatch, Benkman discovered, was the presence of another animal with a taste for the seeds: red squirrels.

Scales can't protect a pinecone's seeds from the squirrels. Instead of feeding at the tips of the cones, red squirrels start at the base and pull away the scales to get to the seeds underneath. As a result, pine trees whose cones have thicker scales gain no extra fitness when red squirrels are in the environment. A better strategy for such trees is to produce more seeds in each cone, so that some seeds can fall to the ground before getting devoured by the squirrels.

The trees thus face a trade-off between their defenses against red squirrels and crossbills. Selection for increased seed number results in cones that have thinner scales. Thus, pine trees that are more efficient at escaping squirrel predation end up with cones that are more accessible to the crossbills. These cones give rise to selection for shorter bills in the crossbills in these locations. The result, Benkman found, was a geographic mosaic of coevolution in which pine trees and crossbills coevolved differently depending on whether squirrels were present or absent (Benkman 2003, 2010).

Squirrels are not the only species that can influence the coevolution of crossbills and pines. Other seed-feeding birds, such as Clark's nutcrackers, can also create local differences in the strength and direction of selection imposed by each species on the other. In Montana, the lodgepole pine borer moth also exerts a strong selection on pinecones.

This complex geographic mosaic has caused red crossbills to experience **diversifying coevolution**, a process in which coevolution leads to significant differences between populations. Such rapid diversification can help push populations apart (Figure 15.16D).

Joshua Trees and Yucca Moths

Coevolution can also help drive some populations to evolve into distinct species. Researchers have found evidence for this splitting in one of the few tree species to

Diversifying coevolution refers to an increase in genetic diversity caused by the heterogeneity of coevolutionary processes across the range of ecological partners.

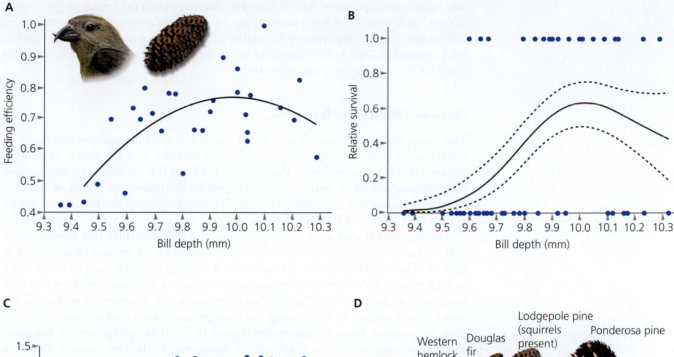

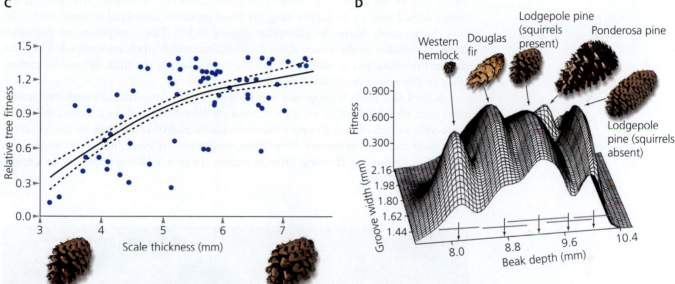

Figure 15.16 A: Crossbills feed on pinecones. Because the seeds are nestled deep in the cones, the birds' bill depth influences the efficiency of their feeding. B: Birds with the most efficient beak depth have the highest probability of survival. The circles along the top of the figure represent uniquely marked individuals recaptured or resighted up to 3 years after initial capture. The circles along the bottom represent individuals not seen a year or more after initial capture. These data come from South Hills, Idaho. C: The fitness of pine trees, in turn, is determined by their scale thickness, which protects their seeds from birds. These data are taken from a site in the western Pyrenees. D: A fitness surface for five crossbill taxa based on foraging data from the laboratory that were converted into fitness using the relationship between feeding efficiency and survival. The peaks in the graph correspond with the diversity of beak shapes that have evolved in the Rocky Mountains, allowing the birds to feed efficiently on differently shaped pinecones. (Data from Benkman 2003, 2010)

survive in the Mojave Desert, the Joshua tree (*Yucca brevifolia*). In the western part of its range, the tree is pollinated by a species of yucca moth called *Tegeticula synthetica*; in the east, it is pollinated by another species called *T. antithetica*.

Olle Pellmyr, a biologist at the University of Idaho, and his colleagues reconstructed the evolution of the tree and the moths on which it depends (Godsoe et al. 2008). They found that the common ancestor of both moth species pollinated Joshua trees as well. When the two species of moth diverged, each specialized on only one population of the trees. In response, the scientists found, canals in the flowers of the two Joshua tree populations have diverged so that they match the length of the tongues of their "own" species of yucca moth. If the trees and moths continue to diverge, there will be less and less opportunity for pollen to move from one population of Joshua trees to another. This pattern may lead to the building of a reproductive barrier that could set the trees on the path to speciation.

Milkweed Plants and Butterflies

The examples of red crossbills and yucca moths involve timescales of thousands of years. But what effects does coevolution have on diversity over the course of millions of years? In 1964, two biologists, Paul Ehrlich and Peter Raven, addressed this question by publishing one of the most influential papers on the long-term effects of coevolution. They argued that antagonistic coevolution could drive up the number of species in a clade (Ehrlich and Raven 1964).

Ehrlich and Raven used milkweed plants and the butterflies that feed on them as their case study. An array of defenses against their insect enemies has evolved in milkweed plants. They grow hairs that make it difficult for the insects to reach down to their tissues. If a caterpillar does manage to bite into a milkweed, sticky white fluid bursts out of the plant (the "milk" that gives milkweed its name). And even if the insect should manage to keep eating, the plant produces a cocktail of toxic molecules that can seriously harm the caterpillar (**Figure 15.17**). The caterpillars, on the other hand, have many of their own defenses, as Raven and Ehrlich recognized. They can disarm the toxins in the milkweed, and they can sabotage the milk defense by cutting holes in the vessels that it flows through.

Raven and Ehrlich proposed that milkweed plants periodically acquired mutations that allowed them to escape from insects. Liberated by the mutation, the plants were able to spread and diversify into previously empty ecological niches, producing a new radiation of species. Over time, adaptations to overcome the new defenses evolved in the insects, allowing them to exploit the new milkweed species. As a result, the insects radiated.

Figure 15.17 A caterpillar chews on a milkweed plant. In response, the plant oozes out sticky latex, along with toxic chemicals. Defenses against these weapons, in turn, have evolved in caterpillars. For example, by chewing a ring this caterpillar has blocked the toxic latex from flowing into the tissue where it feeds. (Anurag Agrawal)

Flowering Plants and Ants

Ehrlich and Raven have inspired generations of biologists to investigate the conditions under which coevolution spurs long-term increases in biodiversity (Yoder and Nuismer 2010). Just like the antagonistic interactions of milkweeds and insects, mutualisms may be able to boost diversity as well. Szabolcs Lengyel, a biologist at the Hungarian Academy of Sciences, and his colleagues analyzed the diversity that arises from the coevolution of flowering plants and the ants that spread their seeds (Lengyel et al. 2009). About 11,000 known plant species around the world grow fleshy handles, called elaiosomes, on their seeds. After ants bring the seeds to their nests, they eat the elaiosomes and discard the seeds in a special space in their colony. There the seeds sprout, protected from being eaten by other animals.

Elaiosomes have evolved independently at least 101 times. The researchers also found that ant-dispersing lineages contain more than twice as many species as the most closely related lineages of plants. Ants may foster this plant diversity by protecting the seeds, which increases the plant's fitness. In addition, the plants end up growing in a small range around specific ant colonies, causing them to become geographically isolated from other plant populations.

Coevolution's Role in Extinction

As we saw in Chapter 14, the world's biodiversity is under serious threat, and we may be sliding into the sixth mass extinction to have occurred over the past 500 million years. Coevolution fosters biodiversity, but it may play a role in extinction as well (Toby Kiers et al. 2010). If one species depends on another one for its survival, then it will not be able to endure after the other species becomes extinct. If it can shift to a new partner, however, it may be able to survive.

Mass extinctions in the past offer some clues about how coevolution makes species vulnerable. Some species of corals live mutualistically with algae; others do not. In the last major mass extinction, 66 million years ago (Chapter 14), mutualistic corals suffered about four times more extinctions than nonmutualistic corals. These extinctions coincided with a huge asteroid impact that blocked out the light of the Sun for months. It's possible that this blackout killed off photosynthetic algae as well as the corals that depended on them for survival (Kiessling and Baron-Szabo 2004).

A number of scientists are investigating the role that coevolution plays in the current extinction crisis. Sandra Anderson, of the University of Auckland, and her colleagues, for example, are studying the effect of the introduction of cats and other mammal predators to New Zealand (Anderson et al. 2011). These introduced predators have helped to drive 49% of New Zealand's native bird species extinct. Anderson and her colleagues wondered what happened to the plants that depended on the extinct birds for pollination.

To find out, they studied the plants on the upper portion of the North Island, comparing them to the plants on the small adjacent islands. On the mainland, two of the three native pollinating birds have become extinct. But on the nearby islands, all three of those pollinators—the bellbird, stitchbird, and tui—still survive. The scientists hypothesized that the two populations of plants would have different levels of reproductive success.

They tested this hypothesis on the native flower *Rhabdothamnus solandri*, which grows on both the mainland and the islands. They placed bags over some flowers so that they could not be pollinated at all. They pollinated other plants by hand. And they left a third group of the plants to be pollinated naturally by birds. On both the mainland and the islands, Anderson found, bagged flowers produced almost no fruit at all. Plants pollinated by hand produced equally abundant fruits in both sites. But the scientists discovered a big difference between the islands and mainland when they examined the naturally pollinated flowers. The island flowers produced about as much fruit on their own as they did when pollinated by hand. The mainland flowers, by contrast, produced far fewer.

Figure 15.18 A: New Zealand and its surrounding islands are home to many endemic species, such as the flower *Rhabdothamnus solandri* (B). C: Many of these plants depend on endemic species of birds for pollination, such as the tui (*Prosthemadera novaeseelandiae*). Many endemic New Zealand birds are now endangered or extinct, raising the question of what effect their disappearance will have on plants. D: New Zealand researchers gauged the effect of bird extinctions by measuring how much fruit *R. solandri* produced. They studied a mainland population that has lost two of its three main pollinators and compared it to flowers on islands where all three birds are still present. In one trial, they put bags over the flowers, preventing pollination. In another, they pollinated the flowers by hand to ensure that all of them were pollinated. In the third trial, they measured the fruit set produced by natural pollination. The scientists found that island and mainland plants set similar numbers of fruit in each of the artificial treatments, but naturally pollinated mainland flowers produced much less fruit than did their island counterparts due to the local extinction of pollinators. (A, B: Dave Kelly; C: Tessa Palmer/Shutterstock; data from Anderson et al. 2011)

These results suggest that native flowers on the mainland of New Zealand are suffering pollination failure because they have lost their pollinators. Because *R. solandri* is a slow-growing plant, it may take a long time for it to disappear from the mainland. But without its coevolutionary partners, it may be doomed (**Figure 15.18**). Scientists suspect that many other plant species will also suffer from the disruption of their mutualisms (Aslan et al. 2013).

Key Concepts

- Coevolutionary interactions can increase biodiversity as genetic diversity diverges among populations of each species within a geographic mosaic.

- Highly specialized coevolutionary interactions make species highly dependent on each other. If one species becomes extinct, the other may be more likely to disappear as well. ●

15.4 Endosymbiosis: How Two Species Become One

The aster leafhopper (*Macrosteles quadrilineatus*, **Figure 15.19**) is a nightmare for many farmers. The insect feeds on crops such as carrots, tomatoes, and onions by piercing their stems and drinking their sap. As it feeds, an aster leafhopper can also inject bacteria that can sicken the plants. The bacteria cause a disease called aster yellows, named for the way an infected plant's leaves turn yellow. The insects can nearly wipe out a farmer's entire crop, and the only way to control the disease is to kill infected plants as soon as they're discovered.

As devastating as aster leafhoppers may be, they are also marvels of coevolution. The sap they extract from plants lacks many of the amino acids they need for growth, and so it should be impossible for the insects to survive on it. Their secret to survival is two species of bacteria, known as *Nasuia* and *Sulcia*. The bacteria can live only inside the insects, which build special organs to house them. Inside these shelters, the bacteria can convert compounds in the plant sap into the amino acids that the insects need to build proteins (Bennett and Moran 2013).

Mutualists that must live inside coevolutionary partners are known as **endosymbionts**. The evolutionary dynamics of endosymbionts differ in some important ways from the dynamics of free-living mutualists. For example, they have the potential to enter into long-lasting relationships that may stretch over millions or even billions of years. One way to reconstruct this long history is to compare the phylogenies of endosymbionts and their hosts (see **Box 15.2**).

An **endosymbiont** is a mutualistic organism that lives within the body or cells of another organism.

Evolution of Endosymbionts

Gordon Bennett and Nancy Moran, of the University of Texas, have traced the evolution of the aster leafhopper's endosymbionts. The relatives of aster leafhoppers, such as cicadas and glassy-winged sharpshooters, also carry amino-acid-synthesizing endosymbionts in their bacteriomes. And Bennett and Moran found that many of these bacteria are closely related to the aster leafhopper's *Nasuia* and *Sulcia*. The scientists concluded that the common ancestor of these insects acquired two species of bacteria, and the descendants of those bacteria still live inside aster leafhoppers and related hosts. The scientists estimate that the bacteria first became endosymbionts 260 to 280 million years ago.

The evolution of these endosymbionts has been nothing short of astonishing. Each species has lost the vast majority of its genome. The *Nasuia* species that lives in aster leafhoppers has a genome that's only 112,000 base pairs long, making it the smallest nonviral genome yet discovered—it contains just 137 protein-coding genes. Both *Nasuia* and *Sulcia* have lost most of the genes required for free-living bacteria to survive. Instead, they depend on their host for most of their nutrients. But the endosymbionts still retain genes for synthesizing the amino acids the aster leafhoppers need to survive. To carry out this essential task, the bacteria have divided the work. *Nasuia* synthesizes only two of the amino acids the insects require. *Sulcia* has retained genes for synthesizing all the other amino acids (Bennett and Moran 2013).

As **Figure 15.20** shows, the evolution of endosymbionts in the relatives of aster leafhoppers is more complex than a simple case of cospeciation. In some lineages, new species of bacteria have replaced the original ones. Once this replacement occurred, the new endosymbionts cospeciated with their hosts and lost most of their genes. And in some species, it appears that the leafhoppers have lost their endosymbionts altogether. Understanding this dynamic history can potentially lead to new ways of fighting the crop-threatening diseases spread by these insects.

Figure 15.19 Aster leafhoppers feed on crops, spreading a disease known as aster yellows. They can survive on nutrient-poor plant sap by harboring two species of bacteria in special organs called bacteriomes. The bacteria synthesize amino acids that they provide to the insects. (Gordon M. Bennett)

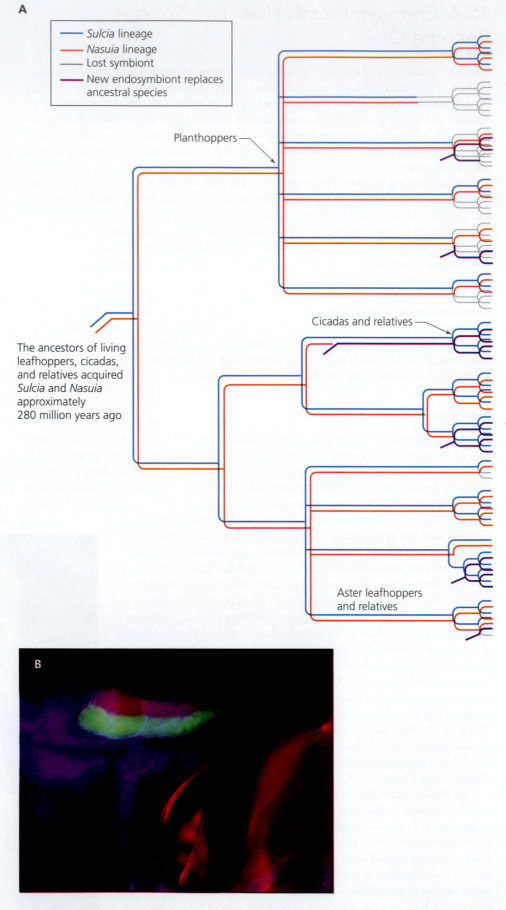

Figure 15.20 A: The endosymbionts found in aster leafhoppers today descend from ancient bacteria that were taken up by insects 280 million years ago. Since then, the bacteria have become adapted to their new environment, losing most of the genes required to survive outside of their host. Some relatives of aster leafhoppers have bacteria descending from the same ancestors, whereas others have lost their endosymbionts or acquired new ones. B: This photo shows an organ called the bacteriome, where leafhoppers store endosymbionts. *Sulcia* is stained in green, *Nasuia* in magenta, and host cell nuclei in blue. (Data from Bennett and Moran 2013; B: Gordon M. Bennett)

A

— *Sulcia* lineage
— *Nasuia* lineage
— Lost symbiont
— New endosymbiont replaces ancestral species

Planthoppers

Cicadas and relatives

The ancestors of living leafhoppers, cicadas, and relatives acquired *Sulcia* and *Nasuia* approximately 280 million years ago

Aster leafhoppers and relatives

BOX 15.2

Comparing Phylogenies to Reconstruct Coevolution

To test coevolutionary hypotheses, biologists sometimes use phylogenies to analyze the deep history of the partners. Species of gophers, for example, carry particular species of lice. When biologists compare the evolutionary trees of the gophers and their lice, they see a mirrorlike symmetry (**Box Figure 15.2.1**). This symmetry tells us something about the coevolution of the two clades. When a population of gophers becomes isolated from the rest of its species and evolves into a new species, its lice can become a new species as well (Hafner and Page 1995). This process is known as **cospeciation**.

Although biologists sometimes find cases where clades have cospeciated, they also find branches that do not match. For example, in Chapter 8, we looked at the phylogeny of

HIV. If HIV and related viruses had cospeciated with primates, we would expect that all HIV strains would form a monophyletic clade. That's not what scientists find, however. Some strains of HIV are more closely related to strains of simian immunodeficiency virus (SIV) that infect chimpanzees, gorillas, and monkeys than they are to other human strains. When branches don't match, it can be evidence that species have switched coevolutionary partners. In the case of HIV, primate strains have become human strains several times over the past century.

Cospeciation occurs when a population speciates in response to, and in concert with, another species.

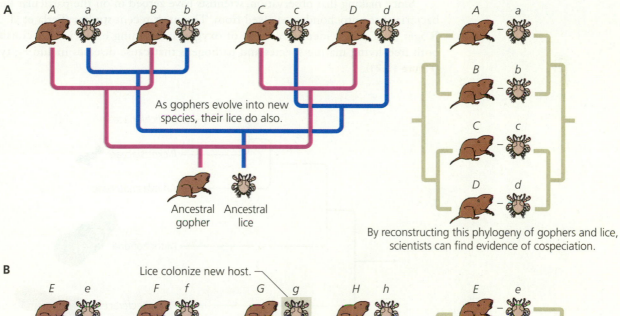

As gophers evolve into new species, their lice do also.

Ancestral gopher Ancestral lice

By reconstructing this phylogeny of gophers and lice, scientists can find evidence of cospeciation.

Lice colonize new host.

Lice become extinct.

Very often, coevolutionary partners do not cospeciate perfectly. Partner species may become extinct or evolve partnerships with new species.

Extinction and host colonization can make phylogenies mismatched.

Box Figure 15.2.1 A: As a species of gopher splits into new species, the lice that live on it also diverge into new species. This cospeciation is recorded in the evolutionary trees of the parasite and its host. B: The lice on gopher species *G* become extinct, and some lice from species *F* colonize them. This extinction and species-hopping breaks the mirror symmetry of the evolutionary trees.

The Origin of Mitochondria

Humans also harbor endosymbionts. Today, they're so well integrated into our cells that until relatively recently cell biologists did not recognize them as the descendants of free-living bacteria. They are mitochondria, the membrane-bound structures that use oxygen, sugar, and other molecules to produce fuel for eukaryotic cells. Mitochondria also carry out other important jobs such as building clusters of iron and sulfur atoms that are then attached to certain proteins. All eukaryotes contain mitochondria in their cells. (Some protozoans have highly reduced forms known as mitosomes.)

After discovering mitochondria in the late 1800s, cell biologists were deeply puzzled by them. They seemed like little cells within our cells. Mitochondria are surrounded by two membranes and carry their own DNA, which replicates as they divide. At the dawn of the twentieth century, Russian biologists proposed a solution to the mystery: mitochondria were once free-living, oxygen-consuming bacteria that entered a single-celled host (Sapp 1994).

The proposal was generally forgotten until the 1960s, when Lynn Margulis, a biologist at the University of Massachusetts, revived it. In the 1970s, scientists were able to test her hypothesis by examining bits of mitochondrial DNA. This DNA closely resembled neither human genes nor the genes of any animals. It did not even resemble the genes of eukaryotes. The closest matches came from bacteria.

Since making that observation, scientists have zeroed in on the particular kind of bacteria that mitochondria evolved from. The most recent studies (Ferla et al. 2013; Roger et al. 2017) identify a clade of oxygen-consuming bacteria that consists of both free-living marine species and pathogens that cause diseases including typhus (**Figure 15.21**).

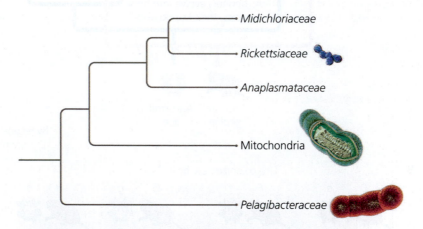

Figure 15.21 Mitochondria generate energy for our cells. These sausage-shaped structures were originally free-living bacteria that were later engulfed in our single-celled ancestors. They are now present in the cells of most eukaryotes. This evolutionary tree shows the relationship of mitochondria to their closest bacterial relatives, based on a study of their DNA. (Data from Ferla et al. 2013; bottom: Dr. David Furness, Keele University / Science Source)

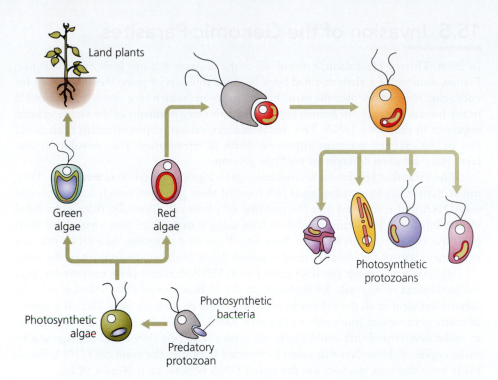

Figure 15.22 Plants can use sunlight to grow with the help of bacterial partners. A single-celled ancestor of plants ("predatory protozoan") engulfed photosynthetic bacteria. The descendants of these photosynthetic cells evolved into several living lineages, including red algae and green algae. One lineage of green algae evolved into land plants. One lineage of red algae was engulfed by another host, which evolved into a variety of living photosynthetic protozoans. (Data from Hohmann-Marriott and Blankenship 2011)

The most compelling hypothesis that accounts for these results is that mitochondria arose when oxygen-consuming bacteria took up residence inside some free-living cells some two billion years ago. In one scenario, they started out as parasites that consumed their host's ATP (Wang and Wu 2014). Gradually, the relationship shifted from parasitic to mutualistic. Early hosts exploited the energy provided by their new resident, and they gradually abandoned their own energy-generating proteins. All of the lineages of living eukaryotes evolved from those early hosts.

Mitochondria today are not typical bacteria by any measure. They would not survive for a second outside cells. As the early mitochondria replicated inside their host cells, their DNA mutated. Some of those mutations deleted genes that were no longer necessary to their survival, now that they were protected inside another cell. Other mutations caused genes to be transferred from the mitochondria to the DNA in the nucleus of the cell.

Another coevolutionary merger took place hundreds of millions of years later, in the ancestors of plants (**Figure 15.22**). Plants descended from single-celled eukaryotes that originally were most likely microscopic predators, feasting on bacteria. At some point, however, they became hosts to bacteria that could carry out photosynthesis. At first they may have gotten energy both from eating other organisms and from capturing sunlight. Some photosynthetic eukaryotes live this way today in the ocean. Others shifted to relying entirely on photosynthesis. Some of these marine species became photosynthetic by engulfing photosynthetic eukaryotes, through a process known as secondary endosymbiosis. Green algae colonized land, using their endosymbiotic bacteria to grow in the sunlight. Living plants still carry remnants of these photosynthetic bacteria, called plastids. Every patch of living green you see, from a blade of grass to a forest of redwoods, got its start with a merger of two species (Hohmann-Marriott and Blankenship 2011).

• Mitochondria and plastids are endosymbiotic bacteria that coevolved with their hosts until they became organelles, rather than free-living organisms. ●

Key Concept

15.5 Invasion of the Genomic Parasites

In 2006, Thierry Heidmann, a researcher at the Gustave Roussy Institute in Villejuif, France, resurrected a virus that had been dead for millions of years. Heidmann and his colleagues did not discover the virus buried in ice or hidden in a cave—they found it in the human genome. All human beings on Earth carry remnants of the virus's genetic sequence in their own DNA. This spectacular revival has helped scientists understand one of the strangest yet most important forms of coevolution: the coevolution that takes place between different parts of our genome.

The virus that Heidmann revived belongs to a group known as **retroviruses**. They infect their hosts by creating an RNA copy of their genome, which is then turned into DNA that is inserted into the genome of a host cell. Typically, these embedded viruses hijack the biochemistry of their host, using it to produce new viruses that then burst out of the cell. But scientists have also discovered retrovirus-like DNA that is a permanent part of the human genome, passed down from one generation to the next.

It's likely that these viruslike stretches of DNA descend from retroviruses that infected sperm or egg cells. An organism produced from one of those infected sex cells carried the virus in all the cells of its body, including its own sex cells. Over the course of many generations, mutations to this viral DNA robbed the viruses of their ability to make new viruses that could escape their host. The best they could manage was to make copies of themselves that could be inserted back into the same cell's DNA. Eventually even that trait was lost, and the virus's DNA became inert (**Figure 15.23**).

To test this idea, Heidmann tried to revive one of these retroviruses into its once-active form. He and his colleagues selected a viruslike segment of DNA found only in humans. They found slightly different versions of the segment in different people. These differences presumably arose as the original retrovirus mutated in different lineages of humans. Heidmann and his colleagues compared the variants to determine what the original sequence had been. They built a piece of DNA that matched the original sequence and inserted it into a colony of human cells reared in a petri dish. Some of the cells produced new viruses that could infect other cells. Heidmann named the virus Phoenix, for the mythical bird that rose from its own ashes (Dewannieux et al. 2006).

Phoenix has been found only in humans, which indicates that the virus infected our ancestors after they had branched off from the apes some 7 million years ago. But we do share other endogenous retroviruses with other apes, as well as with monkeys in Africa and Asia. These viruses must be much older because the common ancestor of all these primates lived about 30 million years ago. After the ancestors of those viruses infected early primates, they continued to replicate and to insert new copies back into the genomes of their hosts. In your own genome, making up about 8% of your DNA, there are almost 100,000 fragments of endogenous retroviral DNA. They take up about four times more of the human genome than the 20,000 genes that encode proteins (Johnson 2010).

"Jumping Genes"

Endogenous retroviruses were not the first "jumping genes" scientists discovered in the genome. In the 1950s, biologist Barbara McClintock was studying the genes that control the color of corn kernels. She discovered that the genes could move within the corn genome from one generation to the next. In 1983, she was awarded the Nobel Prize for her work.

Later generations of scientists discovered a vast menagerie of DNA elements that can move through the genome. About half of the human genome is made up of these mobile genetic elements, which number in the millions. Although most mobile elements are "dead"—that is, they cannot replicate themselves—a few of them still do

Retrovirus is an RNA virus that uses an enzyme called reverse transcriptase to become part of the host cell's DNA. The virus that causes AIDS, the human immunodeficiency virus (HIV), is a type of retrovirus.

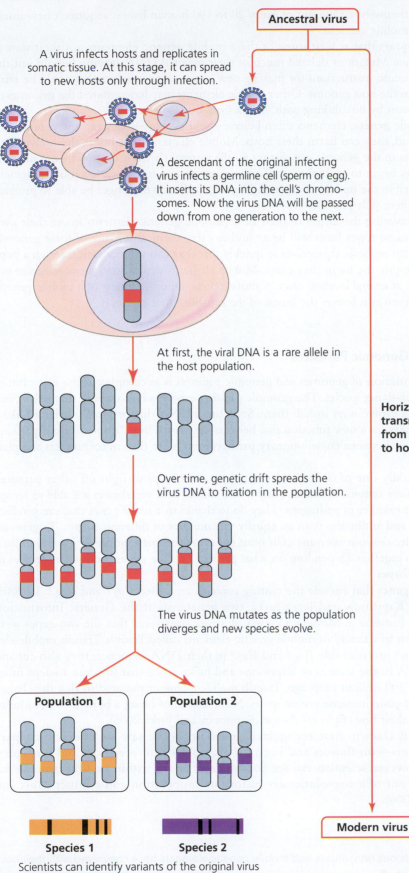

Ancestral virus

A virus infects hosts and replicates in somatic tissue. At this stage, it can spread to new hosts only through infection.

A descendant of the original infecting virus infects a germline cell (sperm or egg). It inserts its DNA into the cell's chromosomes. Now the virus DNA will be passed down from one generation to the next.

At first, the viral DNA is a rare allele in the host population.

Horizontal transmission from host to host

Over time, genetic drift spreads the virus DNA to fixation in the population.

The virus DNA mutates as the population diverges and new species evolve.

Population 1 **Population 2**

Species 1 **Species 2**

Scientists can identify variants of the original virus in the chromosomal DNA of different host species.

Modern virus

Figure 15.23 Endogenous retroviruses sometimes become incorporated into host genomes. They can be recognized in host genomes by comparing them to related viruses that still infect hosts. (Data from Johnson 2010)

replicate themselves. One out of every 20 to 100 human babies acquires a new insertion of a mobile element.

It appears that at least some of these mobile genetic elements got their start as retroviruses. Mutations deleted much of their original DNA, leaving behind just the bare minimum instructions for making new copies of themselves that could be reinserted into the host genome. Other mobile elements may have entered the genomes of our ancestors by hitchhiking with viruses from other species.

Mobile genetic elements often behave—and evolve—like genomic parasites. As they spread, they can harm their hosts. Mobile elements that insert themselves into new places in the genome can disrupt a cell's normal rhythms of growth and division. A cell may begin to multiply out of control, giving rise to cancer. If a mobile element inserts itself in the middle of an essential gene, a cell may no longer be able to produce a vital protein (Chapter 5).

By lowering the fitness of their hosts, mobile genetic elements lower their own fitness because fewer hosts will be around to carry them. But many of these genomic parasites can replicate themselves so quickly that they can still transmit through a population despite the harm they cause. Mobile elements demonstrate how selection can take place at several levels at once. A mutation can raise the fitness of a mobile genetic element, even as it lowers the fitness of the organism that carries it.

Taming Genomic Parasites

The coevolution of genomes and genomic parasites is as complex as the coevolution of two free-living species. The genomic parasites evolve ways to spread themselves, and host genes evolve ways to halt them. Sometimes DNA that originally behaved like a parasite takes on a new function that benefits the host. These "domesticated" parasites blur the line between coevolutionary partners even more than mitochondria and plastids do.

Ironically, one of these domesticated parasites helps us fight off other parasites. The immune systems of sharks, bony fishes, and all land vertebrates are able to recognize a vast number of pathogens. They do so thanks to a set of genes that can produce receptors and antibodies with an equally vast number of different shapes. To generate these molecules, our immune cells must cut apart the corresponding genes and then join them together. Depending on what gets cut out, the genes produce molecules of different shapes.

The genes that encode the cutting proteins are called *Rag1* and *Rag2*. In 2005, Vladimir Kapitonov and Jerzy Jurka, two geneticists at the Genetic Information Research Institute in Mountain View, California, discovered that the two genes were very similar to a family of mobile genetic elements called Transib. Transib mobile elements don't just resemble *Rag1* and *Rag2* in their DNA sequence; they also cut and paste DNA in the same way. Kapitonov and Jurka argue that in some ancient fishes that lived 500 million years ago, Transib mobile elements mutated so that they began to cut and paste immune system genes. Now, instead of being a burden to their hosts, they help their hosts fight off disease (Kapitonov and Jurka 2005).

When Darwin first recognized coevolution, he saw its effects on separate partners—on flowers and bees, for example, or on predators and their prey. Today, however, scientists can see coevolution's marks within us. Our genome has emerged out of a coevolutionary history of cooperation and conflict (Burt and Trivers 2006).

Key Concept
- Endogenous retroviruses and mobile genetic elements have coevolved with their host genomes. ●

TO SUM UP . . .

- Species in ecological relationships can adapt to one another in a process known as coevolution.

- Species can have a range of relationships with each other, from antagonistic to mutualistic.

- Coevolution takes place in a geographic mosaic, where the structure and intensity of selection varies from population to population. The mosaic is defined as coevolved genes from various populations mix together as a result of gene flow.

- Negative frequency-dependent selection can maintain genetic variation within populations of interacting species.

- Parasites and their hosts can evolve in arms races. Predators and their prey can do so as well.

- Mutualists can exert positive frequency-dependent selection on each other, ultimately leading to fixation of alleles that facilitate the interaction in each population.

- Coevolution can promote biodiversity, and at the same time the extinction of a species can endanger its coevolutionary partners.

- During cospeciation, coevolving partners branch into new lineages together.

- Some organisms have become permanent residents inside other organisms, losing their ability to survive on their own. Mitochondria, for example, began as free-living bacteria.

- Some viruses can become established permanently in the genome of their host. The human genome contains 100,000 segments of DNA from viruses.

MULTIPLE CHOICE QUESTIONS   Answers can be found at the end of the book.

1. Which of the following is *not* an example of coevolved adaptation?

    a. Hummingbirds insert their bills into long, slender flower tubes.

    b. Parasitoid wasps lay eggs inside host insects, which die as a result.

    c. Flowers attract pollinators by growing structures that look like female insects; male insects learn to avoid the flowers.

    d. Rough-skinned newts produce toxins to avoid predation; garter snakes evolve defenses against these toxins.

2. Which of these statements about the virulence of rabbit myxoma virus is *true*?

    a. Rabbit myxoma virus needed to become less virulent so it could coexist with its rabbit hosts.

    b. Directional selection favored a coevolutionary escalation where resistance evolved in rabbits and less virulence evolved in the virus.

    c. Rabbit myxoma virus became less virulent over time because natural selection favored strains that did not immediately kill the rabbit hosts, enhancing the likelihood of spreading and infecting other rabbits.

    d. None of the above.

3. What condition(s) is/are necessary for the development of both antagonistic and mutualistic coevolutionary relationships?

    a. Intimate pairwise interactions with one or more species.

    b. The agents of selection that are relevant to the populations must evolve.

    c. Frequency-dependent selection.

    d. Both a and b.

4. Which of the following is *not* an example of diversifying coevolution?

   a. Rough-skinned newts with toxins that only garter snakes can tolerate.

   b. Ants that affect the survival and dispersal of flowering plant seeds that have elaiosomes.

   c. Different populations of a flowering plant that favor different tongue lengths of their moth pollinator.

   d. Pinecones that vary across regions in the thickness of their scales depending on whether squirrel predators are present or not.

5. If an evolutionary biologist hypothesized that a lineage of bacteria experienced a coevolutionary arms race with its hosts, what prediction might she or he make about its evolutionary history?

   a. That the lineage of bacteria was older than the lineage of the host.

   b. That the lineage of the host was older than the lineage of bacteria.

   c. That natural selection was negative frequency dependent.

   d. That the patterns of speciation events of the bacteria lineage would match closely with the patterns of speciation events of the host lineage.

6. What evidence did Gordon Bennett and Nancy Moran use to support their hypothesis about the evolution of the endosymbionts inside the aster leafhopper?

   a. They compared phylogenies of the endosymbionts and their insect hosts and found the patterns of lineages were closely matched.

   b. They used fossils of sap-feeding insects with known ages to establish time frames for the endosymbiotic relationships.

   c. They compared the molecular phylogenies of two endosymbionts, *Nasuia* and *Sulcia,* and found similar patterns in their lineages.

   d. They used all of this evidence to support their hypothesis.

7. Why do different people carry slightly different versions of the same retrovirus segment in their DNA?

   a. Because after the virus invaded the human common ancestor, different versions evolved in different lineages of humans.

   b. Because the genome for every human cell is different.

   c. Because there are almost 100,000 fragments of endogenous retroviral DNA in the human genome, and scientists can't tease them apart very easily.

   d. Because retroviruses became mobile genetic elements in the human genome that can insert themselves anywhere in the genome.

8. We can describe situations in which host and parasite populations generate negative frequency-dependent selection on each other as

   a. reciprocal interactions that can maintain genetic variation within both populations.

   b. reciprocal interactions that can deplete genetic variation from both populations.

   c. nonreciprocal interactions that will harm the host population.

   d. nonreciprocal interactions that will harm the parasite population.

9. When predator populations switch between several different prey populations, this can result in

   a. alternating bouts of pairwise evolution between multiple species.

   b. sequential bouts of pairwise coevolution between species.

   c. Both a and b.

   d. None of the above.

10. Which of the following is true of coevolution?

   a. Coevolution can promote biodiversity.

   b. The extinction of a species can endanger its coevolutionary partners.

   c. Coevolution is a process of reciprocal evolutionary change between ecologically intimate species driven by natural selection.

   d. All of the above.

11. How might the process of coevolutionary alternation depicted in the figure below contribute to a geographic mosaic pattern of coevolutionary interactions across a landscape?

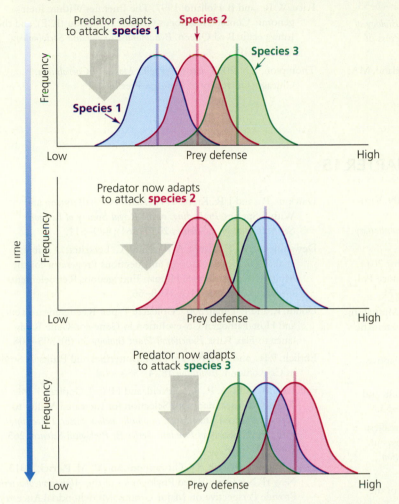

a. At any point in time, different populations/locations may be at different stages of this alternation cycle.

b. Prey species 1 might experience strong selection from this predator in some locations but hardly be preyed on at all by that same predator in other locations.

c. The predator may interact strongly with one species of prey in some locations and another species of prey at other locations.

d. All of the above.

SHORT ANSWER QUESTIONS   Answers can be found at the end of the book.

12. Based on your understanding of geographic mosaic theory, how does this apply to the interaction between long-tongued flies and *Zaluzianskya* flowers?

13. Why is reciprocal coevolution important to the development of evolutionary arms races?

14. Explain why coevolution through Batesian mimicry is most likely to be stable when the mimic species is rarer than the model species.

15. How can diversifying coevolution lead to speciation?

16. What do scientists consider to be the origin of the relationship between eukaryotes and mitochondria?

17. How do retroviruses alter host genomes and affect host fitness?

ADDITIONAL READING

Archibald, J. M. 2015. Endosymbiosis and Eukaryotic Cell Evolution. *Current Biology* 25:R911–21.

Benkman, C. 2010. Diversifying Coevolution Between Crossbills and Conifers. *Evolution: Education and Outreach* 3 (1): 47–53.

Burt, A., and R. L. Trivers. 2006. *Genes in Conflict: The Biology of Selfish Genetic Elements.* Cambridge, MA: Belknap Press of Harvard University Press.

Futuyma, D. J., and M. Slatkin. 1983. *Coevolution.* Sunderland, MA: Sinauer Associates.

Johnson, S., and B. Anderson. 2010. Coevolution Between Food-Rewarding Flowers and Their Pollinators. *Evolution: Education and Outreach* 3 (1): 32–9.

Rice, W. R., and B. Holland. 1997. The Enemies Within: Intergenomic Conflict, Interlocus Contest Evolution (ICE), and the Intraspecific Red Queen. *Behavioral Ecology and Sociobiology* 41 (1): 1–10.

Thompson, J. N. 2005. *The Geographic Mosaic of Coevolution.* Chicago: University of Chicago Press.

PRIMARY LITERATURE CITED IN CHAPTER 15

Alizon, S., A. Hurford, N. Mideo, and M. Van Baalen. 2009. Virulence Evolution and the Trade-Off Hypothesis: History, Current State of Affairs and the Future. *Journal of Evolutionary Biology* 22:245–59.

Anderson, S. H., D. Kelly, J. J. Ladley, S. Molloy, and J. Terry. 2011. Cascading Effects of Bird Functional Extinction Reduce Pollination and Plant Density. *Science* 331 (6020): 1068–71.

Aslan, C. E., E. S. Zavaleta, B. Tershy, and D. Croll. 2013. Mutualism Disruption Threatens Global Plant Biodiversity: A Systematic Review. *PLoS ONE* 8 (6): e66993.

Benkman, C. W. 2003. Divergent Selection Drives the Adaptive Radiation of Crossbills. *Evolution* 57 (5): 1176–81.

———. 2010. Diversifying Coevolution Between Crossbills and Conifers. *Evolution: Education and Outreach* 3 (1): 47–53.

Bennett, G. M., and N. A. Moran. 2013. Small, Smaller, Smallest: The Origins and Evolution of Ancient Dual Symbioses in a Phloem-Feeding Insect. *Genome Biology and Evolution* 5:1675–88.

Brodie, E. D., III. 2011. Patterns, Process, and the Parable of the Coffeepot Incident: Arms Races Between Newts and Snakes from Landscapes to Molecules. In J. B. Losos (ed.), *In the Light of Evolution: Essays from the Laboratory and Field* (pp. 93–120). Greenwood Village, CO: Roberts and Company.

Bull, J. J., and A. S. Lauring. 2014. Theory and Empiricism in Virulence Evolution. *PLoS Pathogens* 10 (10): e1004387.

Burt, A., and R. L. Trivers. 2006. *Genes in Conflict: The Biology of Selfish Genetic Elements.* Cambridge, MA: Belknap Press of Harvard University Press.

Carroll, S. P., J. E. Loye, H. Dingle, M. Mathieson, T. R. Famula, et al. 2005. And the Beak Shall Inherit—Evolution in Response to Invasion. *Ecology Letters* 8 (9): 944–51.

Cuthill, J. H., and M. Charleston. 2012. Phylogenetic Codivergence Supports Coevolution of Mimetic *Heliconius* Butterflies. *PLoS ONE* 7 (5): e36464.

Daltry, J. C., W. Wuster, and R. S. Thorpe. 1996. Diet and Snake Venom Evolution. *Nature* 379 (6565): 537–40.

Davies, N. B., and M. De L. Brooke. 1989. An Experimental Study of Co-evolution Between the Cuckoo, *Cuculus canorus,* and Its Hosts. I. Host Egg Discrimination. *Journal of Animal Ecology* 58 (1): 207–24.

Dawkins, R., and J. R. Krebs. 1979. Arms Races Between and Within Species. *Proceedings of the Royal Society of London. Series B: Biological Sciences* 205 (1161): 489–511.

Dewannieux, M., F. Harper, A. Richaud, C. Letzelter, D. Ribet, et al. 2006. Identification of an Infectious Progenitor for the Multiple-Copy Herv-K Human Endogenous Retroelements. *Genome Research* 16 (12): 1548–56.

Dodds, P., and P. Thrall. 2009. Goldacre Paper: Recognition Events and Host-Pathogen Co-evolution in Gene-for-Gene Resistance to Flax Rust. *Functional Plant Biology* 36 (5): 395–408.

Ehrlich, P. R., and P. H. Raven. 1964. Butterflies and Plants: A Study in Coevolution. *Evolution* 18 (4): 586–608.

Fellowes, M. D. E., A. R. Kraaijeveld, and H. C. J. Godfray. 1998. Trade-Off Associated with Selection for Increased Ability to Resist Parasitoid Attack in *Drosophila melanogaster. Proceedings of the Royal Society of London. Series B: Biological Sciences* 265 (1405): 1553–8.

Ferla M. P., J. C. Thrash, S. J. Giovannoni, and W. M. Patrick. 2013. New rRNA Gene-Based Phylogenies of the *Alphaproteobacteria* Provide Perspective on Major Groups, Mitochondrial Ancestry and Phylogenetic Instability. *PLoS ONE* 8 (12): e83383.

Geffeney, S. L., E. Fujimoto, E. D. Brodie III, E. D. Brodie Jr., and P. C. Ruben. 2005. Evolutionary Diversification of TTX-Resistant Sodium Channels in a Predator–Prey Interaction. *Nature* 434 (7034): 759.

Gilbert, L. E. 1975. Ecological Consequences of a Coevolved Mutualism Between Butterflies and Plants. In L. E. Gilbert and P. H. Raven (eds.), *Coevolution of Animal and Plants* (pp. 210–40). Austin and London: University of Texas Press.

Godsoe, W., J. B. Yoder, C. I. Smith, and O. Pellmyr. 2008. Coevolution and Divergence in the Joshua Tree/Yucca Moth Mutualism. *American Naturalist* 171 (6): 816–23.

Haber, W. A., and G. W. Frankie. 1989. A Tropical Hawkmoth Community: Costa Rican Dry Forest Sphingidae. *Biotropica* 21 (2): 155–72.

Hafner, M. S., and R. D. Page. 1995. Molecular Phylogenies and Host-Parasite Cospeciation: Gophers and Lice as a Model System. *Philosophical Transactions of the Royal Society of London. Series B: Biological Sciences* 349 (1327): 77–83.

Hague, M. T. J., C. R. Feldman, E. D. Brodie Jr., and E. D. Brodie III. 2017. Convergent Adaptation to Dangerous Prey Proceeds

Through the Same First-Step Mutation in the Garter Snake *Thamnophis sirtalis*. *Evolution* 71 (6): 1504–18.

Hague, M. T., G. Toledo, S. L. Geffeney, C. T. Hanifin, E. D. Brodie Jr., et al. 2018. Large-Effect Mutations Generate Trade-Off Between Predatory and Locomotor Ability During Arms Race Coevolution with Deadly Prey. *Evolution Letters* 2 (4): 4060–16.

Hanifin, C. T., E. D. Brodie Jr., and E. D. Brodie III. 2008. Phenotypic Mismatches Reveal Escape from Arms-Race Coevolution. *PLoS Biology* 6 (3): e60.

Heliconius Genome Consortium. 2012. Butterfly Genome Reveals Promiscuous Exchange of Mimicry Adaptations Among Species. *Nature* 487:94–8.

Henter, H. J. 1995. The Potential for Coevolution in a Host-Parasitoid System. II. Genetic Variation Within a Population of Wasps in the Ability to Parasitize an Aphid Host. *Evolution* 49 (3): 439–45.

Henter, H. J., and S. Via. 1995. The Potential for Coevolution in a Host-Parasitoid System. I. Genetic Variation Within an Aphid Population in Susceptibility to a Parasitic Wasp. *Evolution* 49 (3): 427–38.

Hohmann-Marriott, M. F., and R. E. Blankenship. 2011. Evolution of Photosynthesis. *Annual Review of Plant Biology* 62:515–48.

Johnson, S., and B. Anderson. 2010. Coevolution Between Food-Rewarding Flowers and Their Pollinators. *Evolution: Education and Outreach* 3 (1): 32–9.

Johnson, W. E. 2010. Endless Forms Most Viral. *PLoS Genetics* 6 (11): e1001210.

Kapitonov, V. V., and J. Jurka. 2005. *Rag1* Core and V(D)J Recombination Signal Sequences Were Derived from Transib Transposons. *PLoS Biology* 3 (6): e181.

Kerr, P. J., E. Ghedin, J. V. DePasse, A. Fitch, I. M. Cattadori, et al. 2012. Evolutionary History and Attenuation of Myxoma Virus on Two Continents. *PLoS Pathogens* 8 (10): e1002950.

Kiessling, W., and R. C. Baron-Szabo. 2004. Extinction and Recovery Patterns of Scleractinian Corals at the Cretaceous-Tertiary Boundary. *Palaeogeography, Palaeoclimatology, Palaeoecology* 214 (3): 195–223.

Kraaijeveld, A. R., and H. C. J. Godfray. 1999. Geographic Patterns in the Evolution of Resistance and Virulence in *Drosophila* and Its Parasitoids. *American Naturalist* 153 (S5): S61–74.

Kronforst, M. R. 2012. Mimetic Butterflies Introgress to Impress. *PLoS Genetics* 8 (6): e1002802.

Lengyel, S., A. D. Gove, A. M. Latimer, J. D. Majer, and R. R. Dunn. 2009. Ants Sow the Seeds of Global Diversification in Flowering Plants. *PLoS ONE* 4 (5): e5480.

Lively, C. M. 2010. A Review of Red Queen Models for the Persistence of Obligate Sexual Reproduction. *Journal of Heredity* 101 (Suppl. 1): S13–20.

Martin, A., K. J. McCulloch, N. H. Patel, A. D. Briscoe, L. E. Gilbert, et al. 2014. Multiple Recent Co-options of Optix Associated with Novel Traits in Adaptive Butterfly Wing Radiations. *EvoDevo* 5:7.

Maynard Smith, J. 1982. *Evolution and the Theory of Games*. Cambridge: Cambridge University Press.

McGlothlin, J. W., J. P. Chuckalovcak, D. E. Janes, S. V. Edwards, C. R. Feldman, et al. 2014. Parallel Evolution of Tetrodotoxin Resistance in Three Voltage-Gated Sodium Channel Genes in the Garter Snake *Thamnophis sirtalis*. *Molecular Biology and Evolution* 31:2836–46.

Nunn, C. L., V. O. Ezenwa, C. Arnold, and W. D. Koenig. 2011. Mutualism or Parasitism? Using a Phylogenetic Approach to Characterize the Oxpecker-Ungulate Relationship. *Evolution* 65 (5): 1297–1304.

Pardo-Diaz, C., C. Salazar, S. W. Baxter, C. Merot, W. Figuerido-Ready, et al. 2012. Adaptive Introgression Across Species Boundaries in *Heliconius* Butterflies. *PLoS Genetics* 8 (6): e1002752.

Parker, G. A. 1983. Arms Races in Evolution—An ESS to the Opponent-Independent Costs Game. *Journal of Theoretical Biology* 101 (4): 619–48.

Pauw, A., J. Stofberg, and R. J. Waterman. 2009. Flies and Flowers in Darwin's Race. *Evolution* 63 (1): 268–79.

Perry, J. C., and L. Rowe. 2012. Sexual Conflict and Antagonistic Coevolution Across Water Strider Populations. *Evolution: International Journal of Organic Evolution* 66 (2): 544–57.

Reed, R. D., R. Papa, A. Martin, H. M. Hines, B. A. Counterman, et al. 2011. *Optix* Drives the Repeated Convergent Evolution of Butterfly Wing Pattern Mimicry. *Science* 333 (6046): 1137–41.

Rice, W. R., and B. Holland. 1997. The Enemies Within: Intergenomic Conflict, Interlocus Contest Evolution (ICE), and the Intraspecific Red Queen. *Behavioral Ecology and Sociobiology* 41 (1): 1–10.

Roger, A. J., S. A. Muñoz-Gómez, and R. Kamikawa. 2017. The Origin and Diversification of Mitochondria. *Current Biology* 27:R1177–92.

Sapp, J. 1994. *Evolution by Association: A History of Symbiosis*. New York: Oxford University Press.

Thompson, J. N. 2005. *The Geographic Mosaic of Coevolution*. Chicago: University of Chicago Press.

———. 2009. The Coevolving Web of Life (American Society of Naturalists Presidential Address). *American Naturalist* 173 (2): 125–40.

———. 2010. Four Central Points About Coevolution. *Evolution: Education and Outreach* 3 (1): 7–13.

Toby Kiers, E., T. M. Palmer, A. R. Ives, J. F. Bruno, and J. L. Bronstein. 2010. Mutualisms in a Changing World: An Evolutionary Perspective. *Ecology Letters* 13 (12): 1459–74.

Wang, Z., and M. Wu. 2014. Phylogenomic Reconstruction Indicates Mitochondrial Ancestor Was an Energy Parasite. *PLoS ONE* 9 (10): e110685.

West-Eberhard, M. J. 1983. Sexual Selection, Social Competition, and Speciation. *Quarterly Review of Biology* 58 (2): 155–83.

Yoder, J. B., and S. L. Nuismer. 2010. When Does Coevolution Promote Diversification? *American Naturalist* 176:802–17.

CHAPTER 16

Brains and Behavior

Learning Objectives

- Discuss how evolution affects behavioral phenotypes.
- Compare and contrast proximate and ultimate questions.
- Discuss how proximate and ultimate questions influence studies of behavior.
- Identify behavior in organisms without brains.
- Discuss one ultimate cause for the evolution of a behavior and discuss ways to test the hypothesis.
- Explain the different levels of selection at which behavior can evolve.
- Compare the benefits and costs of living in a group.
- Explain how inclusive fitness can lead to kin selection.
- Explain how kin selection can lead to differences in male and female behavior.
- Discuss the trade-offs imposed by the evolution of complex cognition.

New Caledonia, a group of islands 1200 kilometers east of Australia, is one of the world's great havens of biodiversity. Of the 3270 species of plants that grow on New Caledonia, 2432 of them—three-quarters—are found nowhere else on Earth (Conservation International 2007). Evolutionary biologists from all over the world come to New Caledonia to study its extraordinary inhabitants. One of these scientists is Alex Taylor, a professor at the University of Auckland in New Zealand (**Figure 16.1**). Taylor travels to New Caledonia not to study an extraordinary species so much as to study an extraordinary kind of behavior. On New Caledonia, there are crows that can make tools and use them with a sophistication rarely displayed in nonhumans.

For a long time, the scientific community thought that humans were the only animals that could fashion tools. This exceptional skill appeared to explain our remarkable success as a species. In the 1960s, however, primatologists began to observe chimpanzees smashing nuts with rocks and fashioning sticks for fishing termites out of nests. Perhaps, scientists suggested, tool use had evolved in the common ancestors of the great apes and humans, only to reach full flower in our species. And so it came as a great surprise when Gavin Hunt, another scientist at the University of Auckland, discovered in 1992 that New Caledonian crows can fashion tools every bit as impressive as those of chimpanzees.

New Caledonian crows (Corvus moneduloides) can fashion sticks into tools for fishing for insect larvae in tree crevices.

Gavin Hunt

Figure 16.1 Alex Taylor has discovered that New Caledonian crows can use one tool to access another tool that will help them reach food, suggesting these birds have a capacity for abstract thought. (Brenna Knaebe)

New Caledonian crows mostly eat insect larvae nestled away in the crevices of trees. This kind of diet poses a quandary for them, however. Their beaks are too short to reach the grubs, and their skulls are too weak for them to hammer and chisel their way through the trees like woodpeckers. Hunt discovered that the crows solve this problem by making a tool to grab the larvae. They search for a plant with a suitable branch, which they then pry off. Next, they use their beaks to strip away any small branches and curl the end to fashion a hook. The crows can then dip the hook into the crevice and draw out the larvae (Hunt 2000).

But twig hooks are not the only tools in the crows' toolkit. They also pluck leaves from pandan plants and snip off the edges until they look like locksmith picks. The crows can then dip these leaf picks into crevices in trees. When they draw the leaves back out, they also bring out insects snagged on the jagged edges. Hunt and his colleagues have observed that the crows take good care of their tools, holding onto them with their feet as they eat larvae and then storing them on tree perches when they fly to distant hunting grounds.

Alex Taylor wondered what mental activity was going on as the birds made their tools. Did they have an abstract representation of them in their brains, or did they simply tear up branches and leaves without giving much thought to what they were for? To find out, Taylor designed an experiment in which New Caledonian crows could get food only if they could figure out how to use one tool to get another tool.

Taylor put a piece of meat at the far end of a narrow, transparent box. The crows could see the food but not reach it. Near the food box was a second box with a long stick inside. Taylor put bars on the box so that the crow couldn't reach the long stick with its beak. Finally, Taylor tied a short stick to a piece of string, so that it dangled nearby (**Figure 16.2**).

The crows inspected the boxes and the sticks and came up with a solution. They hopped to the string and used it to pull up the short stick. They pulled the short stick free and then used it to fish out the long stick. Then they used the long stick to reach the food. In other words, the birds did not simply see sticks as food hooks. Instead, they could invent new solutions to new problems with the tools they had at hand—or at beak (Taylor et al. 2010).

Behaviors—such as making tools—are part of an organism's phenotype. And, like other phenotypic traits, they can evolve. In this chapter, we will investigate the evolution of behavior. We'll start off with a consideration of behavior itself—what it is, and how scientists study the biological basis of behavior. We'll then examine how behavior evolves through natural selection as the environment favors certain behavioral strategies over others. Behavior is not limited to animals, as we'll see when we consider behavior in plants and microbes. But the evolution of a nervous system some 700 million years ago opened up the opportunity for new types of behavior in animals. We'll explore

Figure 16.2 In Taylor's experiment, food was placed at the end of a transparent box, too far for the birds to reach with their bills. A stick of sufficient length was placed in a cage. The birds figured out they would first have to untie a short stick and then use it to push out the long one. (Data from Taylor et al. 2010)

how brains mediate behavior through neuronal pathways and hormones and how they have evolved as animals have adapted to different niches.

Many behaviors that organisms evolve allow them to find food or escape predation. But *social* behaviors—interactions with other members of the same species—are also important. In this chapter, we give special attention to how social behaviors such as cooperation and cheating have evolved. Finally, we'll survey the animals that are capable of highly sophisticated behaviors, such as tool making. By comparing tool use in different species, scientists can uncover some underlying factors in their evolution. Many of these insights are essential for understanding the evolution of our own species, which displays the most sophisticated behaviors in the whole animal kingdom. We'll consider human evolution in the next chapter. ●

16.1 Behavior Evolves

A behavior is an internally generated response to an external stimulus. The stimulus may come from the abiotic environment (such as a falling rock that triggers a coyote to leap out of the way) or from other organisms (such as the presence of a female fly, which triggers a male fly to flap his wings in a courtship ritual). Like other phenotypic traits, behavior can evolve through natural selection because it fulfills the three criteria we introduced in Chapter 2: it can influence the fitness of individuals, it can vary from individual to individual, and that variation is at least partly due to genes.

Aggression in Foxes

An experiment that's been running in Siberia since 1959 offers a vivid demonstration of how behavior meets these criteria (Dugatkin and Trut 2017). The Russian biologist Dmitry Belyaev launched the experiment because he was curious about how wolves had evolved into dogs. Their domestication brought about a drastic change in their behavior, turning them from wild, aggressive animals into docile, obedient ones.

Figure 16.3 Russian biologist Dmitry K. Belyaev (1917–1985), shown here with foxes he studied to understand the evolution of behavior. His long-running experiment uncovered the genetic underpinning of aggression and tameness in the animals. By selecting tame foxes to reproduce, he and his colleagues produced foxes that were as docile as dogs. (RIA Novosti/Science Source)

Belyaev wondered if he could design an experiment to create a similar change in foxes (**Figure 16.3**).

The experiment revealed that foxes vary in their level of aggression. Starting when the foxes were a month old, Belyaev and his colleagues would approach the cages with food in hand. A fox might run away, snarl, or approach. The scientists then tried to stroke the fox, which might respond by biting, backing away, or tolerating the contact. Belyaev and his colleagues repeated these tests over several months and then gave each animal a tameness score. The foxes ended up with a range of scores, with some individuals displaying more aggression than others. These differences endured throughout each animal's life.

By breeding the foxes, Belyaev and his colleagues were able to demonstrate that the differences in fox behavior were partly due to genetic variation. They performed the same tests on each new generation of fox kits. When they compared the scores of the kits on the behavioral test to those of their parents, they discovered a correlation (**Figure 16.4**). Aggressive foxes tended to have aggressive kits, whereas tame foxes had tame kits (Kukekova et al. 2006). As we saw in Chapter 7, this pattern arises when the variation in a phenotypic trait is at least partly attributable to heritable genetic variation. (Many studies on other animals have also revealed numerous behavioral traits that have a genetic basis; see Bell and Dochtermann 2015.)

Belyaev then imposed selection on the behavior of the foxes. He and his colleagues picked out the individual animals with high tameness scores, and then bred them together. Because tameness is heritable, the foxes responded to this selection: over the generations, they became extraordinarily friendly. Today, after more than 50 generations of selection, the foxes run up to humans even when no food is offered.

This selection then created an opportunity to identify some of the genes that are potentially involved in behavior. In Chapter 7, we introduced the method of quantitative trait locus (QTL) analysis, showing how it allowed researchers to pinpoint genes involved in coat color. In a similar study, Anna Kukekova, then at Cornell University, and her colleagues used QTL analysis on Belyaev's foxes. They bred aggressive and tame foxes together, and then gave the second generation of kits tameness scores (Kukekova et al. 2012). They found certain regions of DNA that were associated with

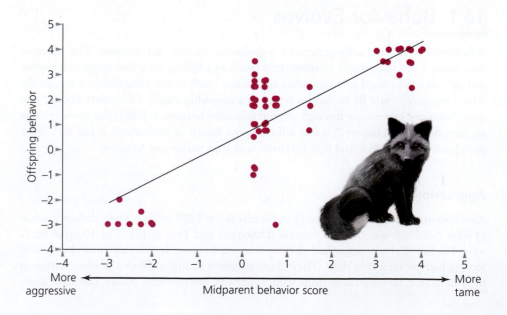

Figure 16.4 Researchers compared the tameness of Belyaev's foxes to that of their offspring. As this graph shows, there was a correlation, indicating that the foxes' behavior was influenced by genes. (Data from Kukekova et al. 2006)

high or low scores, suggesting that genes involved with the evolution of tame behavior might lie nearby (Nelson et al. 2017).

Kukekova and her colleagues then sequenced the entire genome for a sample of tame and aggressive foxes. Lining up the sequences side by side, they searched for regions that differed consistently between the tame and aggressive animals. They also looked at whether any of these regions showed "footprints" of the recent action of selection (F_{ST} outliers; see Chapter 8). Several of them did. Even more helpfully, a few of these regions coincided with the QTL they'd already identified for tame behavior (Kukekova et al. 2018).

Finally, the scientists searched for candidate genes by measuring the levels of expression of thousands of genes in the brains of the foxes (Wang et al. 2018). They've identified 177 genes whose products vary twofold or more between tame and aggressive foxes, suggesting that the regulation of these genes may help produce this behavioral phenotype. Thirty of these genes even sit in the same regions of the genome as the footprints of selection and behavior QTL.

Collectively, these studies point to dozens of genes that may influence tameness in foxes. Studies on other species have also uncovered genes associated with other types of behavior, ranging from courtship to foraging (**Figure 16.5**; see Fitzpatrick et al. 2005; Robinson et al. 2008; and Kent et al. 2009). Likewise, researchers have also run selection experiments that have produced marked changes in a variety of behaviors (**Figure 16.6**; see also Dobzhansky and Spassky 1969; Lynch 1980; and Swallow et al. 1998).

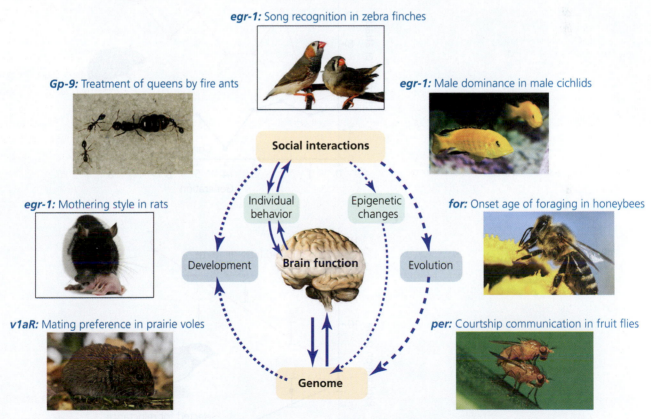

Figure 16.5 Genes influence behavior at three different time-scales. On a moment-by-moment basis (solid arrows), gene expression changes brain function, and vice versa. Social interactions also interact with brain function to influence an individual's behavior. Genes also influence behavior over the course of development (dotted arrows). Finally, genes undergo selection (dashed arrows) in response to the social and physical environment. This complex network of influences makes it challenging to pinpoint the effects of individual genes on behavior. However, scientists have now discovered a small but growing number of these genes such as *Gp-9*, *egr-1*, *for*, *v1aR*, and *per*. (Clockwise from top: Kuttelvaserova Stuchelova/Shutterstock; Andreas Gradin/Shutterstock; Daniel Prudek/Shutterstock; Joppy Mudeng/Pacific Press/Getty Images; Rudmer Zwerver/Shutterstock; Eric Isselee/Shutterstock; SCOTT CAMAZINE/Getty Images; data from Robinson et al. 2008)

Figure 16.6 Selection experiments reveal how behavior can evolve. A: In this experiment, mice that spontaneously ran on a wheel for long periods of time were selected. After 30 generations, they ran about three times more than control mice each day. (Data from Garland 2003) B: Mice selected for aggression made more attacks. (Data from Gammie et al. 2006) C: Great tits (*Parus major*) were selected for either willingness or reluctance to explore. Their behavior shifted in both populations. (Data from Drent et al. 2003)

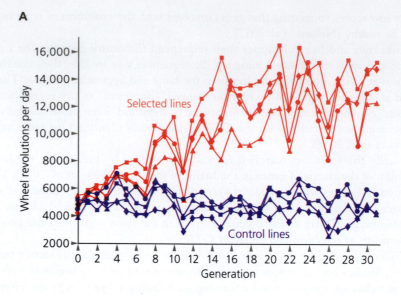

A

Selected lines

Control lines

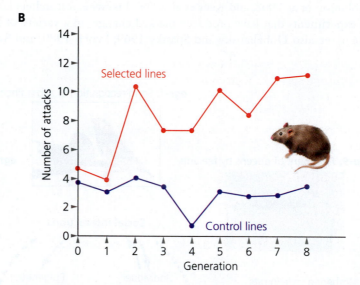

B

Selected lines

Control lines

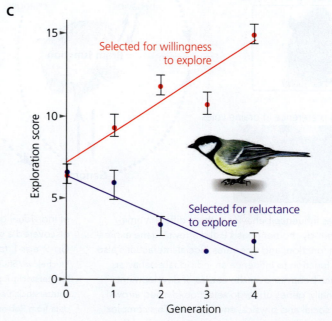

C

Selected for willingness to explore

Selected for reluctance to explore

Documenting Selection in Wild Populations

In artificial selection experiments like Belyaev's fox study, scientists establish the fitness of a particular behavior. In the wild, on the other hand, nature takes charge. Different behaviors can affect fitness in the wild in many ways. They can influence how much food animals find, how much success they have in mating, how likely they are to avoid accidents, and so on.

In Chapter 10, we saw how scientists are documenting selection in wild populations. They are using the same methods to study the evolution of behavior in the wild. Sticklebacks, for example, have evolved new behavior since they began to colonize freshwater lakes, starting about 12,000 years ago (Section 10.5). Carole Di-Poi, of the Université Laval in Quebec, and her colleagues documented this change by comparing the behavior of freshwater sticklebacks in British Columbia to that of marine sticklebacks (Di-Poi et al. 2014). They reared fishes from the two habitats in the same tanks in their lab so that the fishes' behavior developed under identical conditions.

The scientists found that juvenile threespine sticklebacks from freshwater lakes were less social and more aggressive toward conspecifics than their marine counterparts were (**Figure 16.7**). They also were less likely to align their movements with other fishes, a behavior that in marine environments causes fishes to swim together in tight schools (Wark et al. 2011).

Di-Poi and her colleagues hypothesize that the freshwater fishes have evolved different behavior because they've escaped marine predators. As we'll discuss later in this chapter, forming groups can be an effective defense against predators. Once freshwater sticklebacks moved into lakes without predators, schooling no longer offered a selective advantage. Fishes may have been able to get more food on their own.

Pleiotropy may be at work here, too. The same allele—a variant of the *Eda* gene—that's responsible for reductions in armor plating along the sides of the fish also appears to cause fishes to stop schooling (Mills et al. 2014; Greenwood et al. 2016).

Jesse Weber and Hopi Hoekstra have documented an even more dramatic evolutionary shift in the behavior of mice. In Chapter 7, we learned how Hoekstra and her colleagues study the evolution of coat color in oldfield mice (*Peromyscus polionotus*). Hoekstra and her colleagues have also studied how they dig burrows (Weber et al. 2013).

Most mice in the genus *Peromyscus* dig only one short tunnel to their nest. Oldfield mice, on the other hand, dig more elaborate shelters that have a long tunnel to the surface and a second one that goes in the other direction, reaching a sealed end located

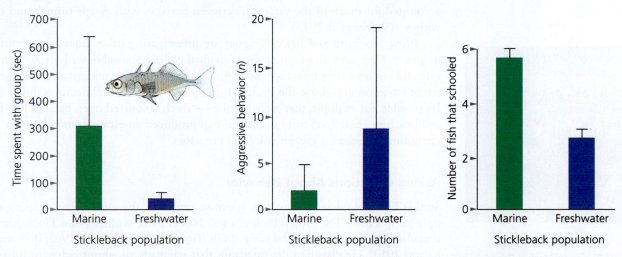

Figure 16.7 Marine sticklebacks have invaded Canadian lakes over the past 12,000 years. Tests reveal differences in behavioral traits such as how much time the fishes spend together, how aggressive they are to conspecifics, and how likely they are to swim together in schools. (Data from Wark et al. 2011; Di-Poi et al. 2014)

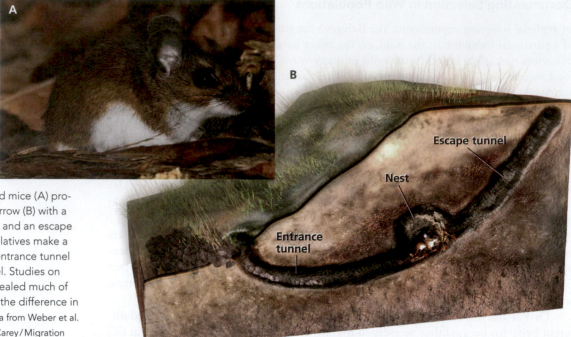

Figure 16.8 Oldfield mice (A) produce a distinctive burrow (B) with a long entrance tunnel and an escape tunnel. Their close relatives make a burrow with a short entrance tunnel and no escape tunnel. Studies on hybrid mice have revealed much of the genetic basis for the difference in these behaviors. (Data from Weber et al. 2013; photo: Shawn P. Carey/Migration Productions)

Escape tunnel

Nest

Entrance tunnel

Figure 16.9 Nikolaas Tinbergen (1907–1988) won the Nobel Prize for his studies on the behavior of birds. Tinbergen developed the conceptual foundations for the modern study of the evolution of behavior. (Lary Shaffer)

just below the surface (**Figure 16.8**). The mice use this second tunnel as an escape route, pushing their way to the surface to get away from predators.

Weber, Hoekstra, and their colleagues hypothesized that digging more elaborate burrows evolved after oldfield mice split off from their closest relatives, deer mice, about 100,000 years ago. To investigate this innovation, Hoekstra and her colleagues took advantage of the fact that deer mice and oldfield mice can still produce viable hybrid offspring. The scientists bred the mice together and observed the burrows they made.

In the second generation, the scientists found, some mice made elaborate burrows, others made simple ones, and still others made intermediate ones. The scientists then used QTL analysis to find loci associated with the differences in burrows. They found that three loci had strong effects on the length of the escape tunnel. These loci acted additively, each one lengthening the tunnel by 3 centimeters, and together they accounted for much of the variation between burrows with escape tunnels and those without (Weber et al. 2013).

Now Hoekstra and her colleagues are investigating other questions about this behavior. They want to pinpoint the individual genes responsible for burrow architecture and see how they influence mice. And they also want to understand how the different environments drove the evolution of different burrow architectures in the mice. It's possible, for example, that oldfield mice—which colonized open habitats like fields and beaches—benefited from mutations that produced an escape tunnel because their environment created a greater risk from predators.

Asking Questions About Behavior

Studies like these demonstrate how scientists can study the same behavior using multiple perspectives. The Dutch biologist and Nobel Prize winner Niko Tinbergen first articulated this approach in the early 1960s (**Figure 16.9**; Tinbergen 1963; Bateson and Laland 2013). He classified the questions that scientists ask about behavior into four types: (1) how does it work, (2) how does it develop, (3) what is its adaptive value, and (4) how did it evolve?

These questions address behavior at two different levels (**Figure 16.10**). The first two questions address the proximate level of a behavior. Scientists studying bird song,

Proximate ("How") Questions about Behavior

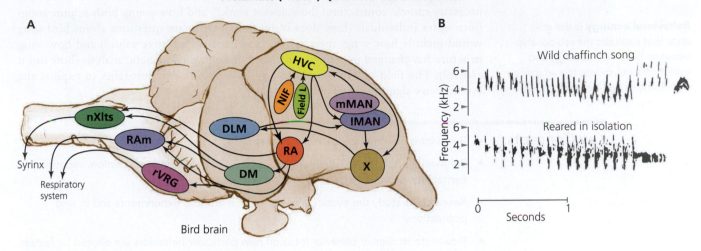

A

Bird brain

Syrinx

Respiratory system

nXIts

RAm

rVRG

DLM

DM

RA

NIF

Field L

HVC

mMAN

lMAN

X

B

Wild chaffinch song

Reared in isolation

Frequency (kHz)

0   Seconds   1

Ultimate ("Why") Questions about Behavior

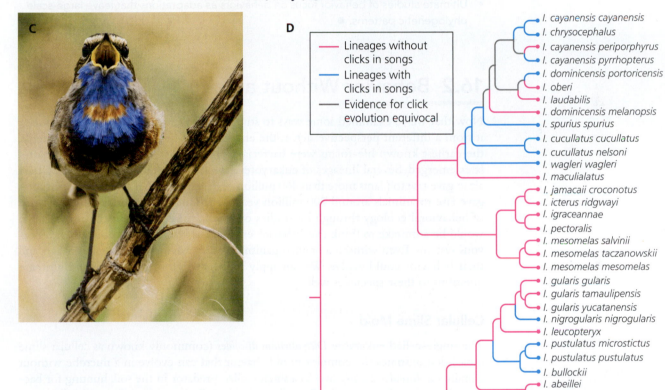

C

D

— Lineages without clicks in songs
— Lineages with clicks in songs
— Evidence for click evolution equivocal

I. cayanensis cayanensis
I. chrysocephalus
I. cayanensis periporphyrus
I. cayanensis pyrrhopterus
I. dominicensis portoricensis
I. oberi
I. laudabilis
I. dominicensis melanopsis
I. spurius spurius
I. cucullatus cucullatus
I. cucullatus nelsoni
I. wagleri wagleri
I. maculialatus
I. jamacaii croconotus
I. icterus ridgwayi
I. igraceannae
I. pectoralis
I. mesomelas salvinii
I. mesomelas taczanowskii
I. mesomelas mesomelas
I. gularis gularis
I. gularis tamaulipensis
I. gularis yucatanensis
I. nigrogularis nigrogularis
I. leucopteryx
I. pustulatus microstictus
I. pustulatus pustulatus
I. bullockii
I. abeillei
I. galbula
I. parisorum
I. chrysater chrysater
I. chrysater giraudii
I. graduacauda audobonii
I. graduacauda graduacauda

Figure 16.10 Tinbergen argued that the study of behavior could be broken down into four main questions. Here are the four questions as applied to the behavior of bird song. A: How does it work? To answer this question, scientists identify neural circuits and genes that control the production of songs. B: How does it develop? By rearing nestlings in captivity isolated from adults, scientists have found that songbirds typically learn songs early in life. If they don't hear songs during a sensitive time window, then they will not acquire an accurate version of the song. This image represents recordings of a normal wild chaffinch song and the song of a chaffinch reared in isolation. C: What's it for? Birds sing to advertise their quality to mates and competitors. Studies of performance in natural populations reveal why males singing a particular song have higher reproductive success than other males singing different songs, or not singing at all. D: How did it evolve? By comparing songs in different species in a phylogenetic context, scientists can reconstruct their evolutionary history, gaining insight into why the behavior arose in the particular lineages that it did. This diagram shows the evolution of clicks in the songs of Icteridae (orioles). (Data from Bateson and Laland 2013; C: Gregory A. Pozhvanov/Shutterstock)

for example, might investigate proximate questions about how neurons produce the necessary muscle contractions (how does it work?), and how young birds acquire songs from older individuals (how does it develop?). Ultimate questions about bird song would include how songs increase fitness (what is its adaptive value?) and how song structure has changed over time as revealed through phylogenetic analysis (how did it evolve?). The field of **behavioral ecology** integrates these approaches to explore the evolutionary significance of behavior.

Behavioral ecology is the science that explores the relationship between behavior, ecology, and evolution to elucidate the adaptive significance of animal actions.

Key Concepts

- Behavior is an internally coordinated response to external stimuli.

- Behavioral phenotypes have the three ingredients required for evolution: variation, heritability, and influence on fitness.

- Researchers study the evolution of behavior in artificial experiments and in wild populations.

- Proximate studies of behavior focus on how particular behaviors are elicited by factors such as hormones, neural signaling, and anatomical structures.

- Ultimate studies of behavior focus on behaviors as adaptations that leave large-scale phylogenetic patterns. ●

16.2 Behavior Without a Brain

Now that we've examined some ways to study the evolution of behavior, let's look at it from a different perspective: across the entire tree of life. As we saw in Chapter 3, the earliest known life-forms were bacteria-like microbes. Single-celled eukaryotes later emerged. Several lineages of eukaryotes gave rise to multicellular species. Green algae gave rise to plants more than 460 million years ago. Another lineage of eukaryotes gave rise to animals around 700 million years ago. Tinbergen pioneered the science of behavioral ecology through his studies on animals (his specialty was birds). But it would be a mistake to think that behavior emerged only in the first animals with nervous systems. Even without a brain, organisms were behaving billions of years ago, and their behavior could evolve. We can apply Tinbergen's four proximate and ultimate questions to these species as well.

Cellular Slime Mold

The single-celled eukaryote *Dictyostelium discoides* (commonly known as cellular slime mold) demonstrates the complexity of behavior that can evolve in a microbe without a brain. *D. discoides* usually lives as a single-celled predator in the soil, hunting for bacteria. But when food supplies dwindle, individual *D. discoides* send out signals to one another. The cells then swarm together to form a slug-shaped mass (**Figure 16.11**). The slug, made up of thousands of cells, then travels through the soil like an animal, moving faster than the individual cells could travel on their own.

A subset of the slug cells, called sentinel cells, sweep through the interior of the slug's "body" as it migrates, engulfing bacteria that may infect the remaining cells. These sentinel cells drop away from the slug and likely die of starvation. The slug eventually comes to a halt, and the cells at the leading tip form an erect stalk. To do so, they produce cellulose and other stiffening molecules, and in the process, they die too. Other cells then migrate up the stalks, forming a bulb at the top of spores that can disperse (Bonner 1967; Strassmann 2010).

This complex repertoire has a genetic basis, which scientists can probe by introducing mutations into *D. discoides* genes. In 2000, Richard Kessin, of Columbia University, and his colleagues discovered one mutation that had a particularly dramatic effect on the slime mold: it turned individual cells into cheaters.

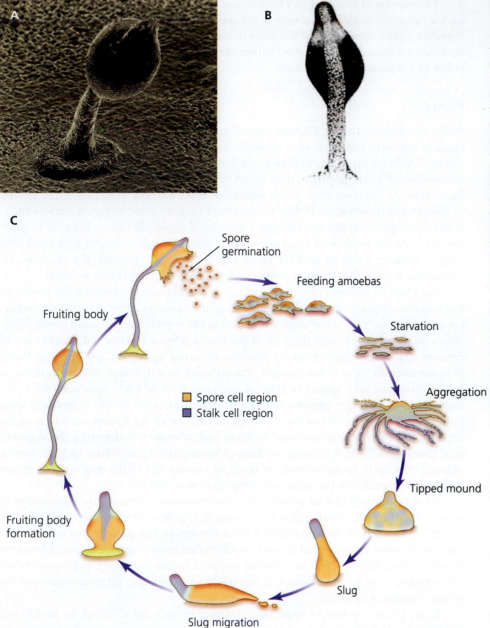

A

B

C

Spore
germination

Feeding amoebas

Fruiting body

Starvation

■ Spore cell region

■ Stalk cell region

Aggregation

Fruiting body
formation

Tipped mound

Slug

Slug migration

Figure 16.11 A: *Dictyostelium discoides*, also known as cellular slime mold, is a single-celled eukaryote that lives in the soil. To cope with starvation, a strategy has evolved in *D. discoides*. The cells join to form a slug, which then grows into a stalk with a patch of spores on top during reproduction. B: The stalk cells support the spores, but die in the process. Kin recognition has evolved in *Dictyostelium* as a way of overcoming cheating, so that relatives are preferentially included in slug formation. C: The life cycle of *Dictyostelium discoides*. (A: David Scharf / Science Source; data from Schaap 2007)

The mutant strain, which they dubbed *chtA*, was incapable of forming a fruiting body of spores on its own (Dao et al. 2000). But when they mixed the mutant strain with wild-type *D. discoides*, the regular development of the slug, stalk, and fruiting body of spores was restored. When Kessin and his colleagues examined the spores in these hybrid fruiting bodies, they found *chtA* mutants there, meaning that the *chtA* cells had managed to crawl up the stalk and initiate their spore-forming pathway. Remarkably, the proportion of mutants in the spores was higher than in the original population. That meant that *chtA* mutants were better at getting into the fruiting body than wild-type cells.

This kind of behavior has important evolutionary consequences because a cell's ability to form a spore can affect its fitness. Kessin and his colleagues demonstrated this importance by combining a single *chtA* mutant with 1000 wild-type cells and allowing them to form a fruiting body. The scientists then used the spores from the fruiting body to create a new population of cells, which then produced a fruiting body of its own. Each time, the cheating *chtA* mutant strain increased in population frequency. After 12 generations, 100% of the spores were *chtA*.

The success of cheating in *Dictyostelium* raises a fascinating question: if it can be such a successful strategy, then why is the species naturally so cooperative? It's a question that applies not just to the behavior of single-celled microbes but also to complex, multicellular animals. Later in this chapter, we'll look at the factors that affect the evolution of cooperation and cheating.

Plants

Plants also behave. In fact, Darwin himself made many pioneering discoveries about plant behavior in the years after he published *The Origin of Species* (Kutschera and Niklas 2009). As plants grow, they reach for the sky, and if they sense they've been overshadowed by other plants, they speed up the growth of their shoots to outgrow their rivals. At the same time, plants send their roots into the ground, sensing gravity to direct them downward. If they encounter a rock or some other obstacle, touch-sensitive receptors on the roots trigger a cascade of signals that cause the root to grow around it. As the tendrils of vines grow, they swing through a circular path until they strike a stem or trunk of another plant. Once they make contact, the tendrils coil around the stem within minutes, allowing the vine to take hold (Braam 2005).

For Darwin, however, the most spectacular of all plant behaviors were produced by carnivorous species such as the Venus flytrap and sundews. Darwin called the behavior of the flytrap "one of the most wonderful in the world" (Ellison and Gotelli 2009). When an insect crawls onto the pads of the flytrap (Figure 16.12A) and touches any of three or four small trigger hairs, intercellular electrical signals collapse the trap. Plants have no neurons or muscles, so their actions work in a different way from those of animals (though they appear to rely on the same flow of Ca^{2+} ions into the cells). Rapid changes in turgor pressure can enlarge or collapse rows of cells in strategic ways, causing plant tissue to fold or bend quickly. In the case of the flytrap, the halves of the trap snap shut in less than a second, holding and subsequently digesting the captured prey (Braam 2005). In Australian sundews (*Drosera glanduligera*; Figure 16.12B), touch-induced changes in turgor pressure of tentacles on one side of the trap catapult insects onto sticky tentacles on the other side (Poppinga et al. 2012).

It's now thought that all plants—not just carnivorous ones—may be sensitive to some form of touch, whether that is the growth responses of shoots to wind, of roots to objects in the soil, or of stamens within flowers to the presence of a pollinator (stamens in some flowers bend inward to dab pollen onto a visiting insect). Mountain laurels (*Kalmia latifolia*) have an extreme response to the touch of a pollinator: when a bumblebee crawls into a flower, it can trigger an explosive catapult release of the pollen (Switzer et al. 2018).

Many plants can mount impressive defenses in response to attack by herbivores. These induced defenses range from production of toxins to release of pheromones that attract natural enemies of the herbivores. Plants even communicate with each other using these volatile signals. If one plant is attacked and begins releasing chemical attractants for predators of the pest, neighboring plants will also begin to produce and emit these attractants—even *before* they've been attacked (Karban 2008). Tobacco plants begin producing the toxin nicotine and shunting it to their leaves when neighboring sagebrush plants are attacked by herbivores. Simply clipping the leaves of sagebrush is enough to trigger this response in other species: the sage plants begin emitting volatile defenses, and the tobacco plants detect and react to these chemicals (Karban et al. 2000).

Each fascinating example of plant behavior raises a host of evolutionary questions. How, for example, did Venus flytraps evolve their insect-snatching behavior? Thomas Gibson and Donald Waller, of the University of Wisconsin, have examined the phylogeny of the plants and found evidence that the ancestors of Venus flytraps had simpler ways to catch prey: by secreting sticky compounds that would snag any insect that happened to land on the plant. This "fly paper" strategy let plants catch small insects but wasn't strong enough for larger ones. Gibson and Waller propose that natural selection favored the more elaborate trap system to let the plants catch bigger prey, which would provide more nutrition (Gibson and Waller 2009).

Figure 16.12 Carnivorous plants catch prey with behaviors that are remarkably similar to those of animals. A: Venus flytraps snap leaves shut to capture insects. B: Australian sundews use a catapult mechanism to trap insects on their sticky tentacles. (A: Marty Pitcairn/Shutterstock; B: Danita Delimont/Getty Images)

- Behavior is not limited to animals with nervous systems.

- Microbes display a range of behaviors, including aggregating to form spores.

- Plants can be sensitive to light and to touch, and they can communicate by sending and receiving chemical signals. Like animals, these behaviors have a genetic basis that can evolve over time. ●

16.3 Behavior and the Origin of Nervous Systems

The behavior of *Dictyostelium* or a Venus flytrap occurs through communication between cells and gene expression changes within cells. That is also true for many other clades, from bacteria to fungi. But when animals emerged, new kinds of behavior became possible thanks to the evolution of a new type of cell: the neuron.

In every animal with a nervous system, from jellyfishes to humans, neurons have the same fundamental morphology and function. Each neuron generates signals with pulses of electric charge that move from one end of the cell to the other. The signals can move from one neuron to the next at a special junction of the two cells called a synapse (**Figure 16.13**). The neuron sending the signal pumps chemicals, known as

Figure 16.13 Almost all animals use neurons to control their behavior. Spikes of voltage travel down the length of a neuron and then trigger the release of neurotransmitters at synapses, which in turn trigger activity in neighboring neurons.

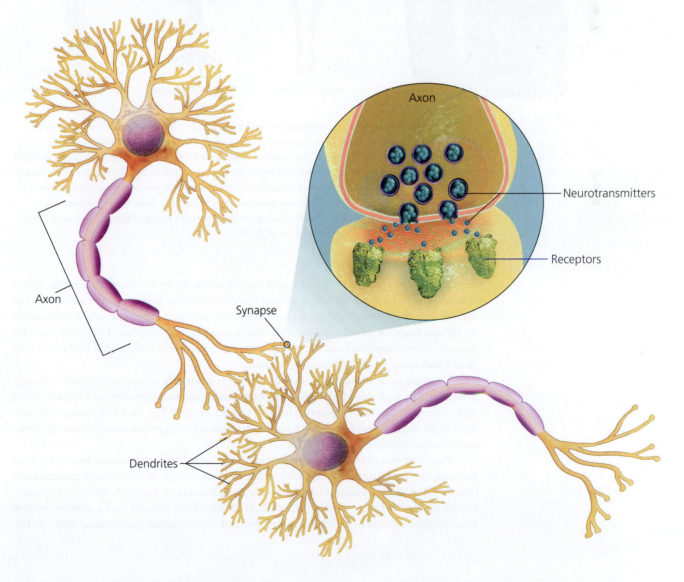

Axon

Neurotransmitters

Receptors

Axon

Synapse

Dendrites

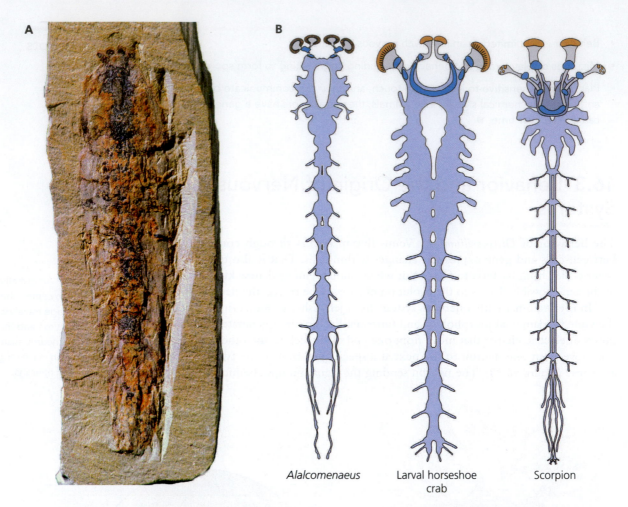

A

B

Alalcomenaeus   Larval horseshoe   Scorpion
crab

Figure 16.14 Fossils help scientists reconstruct the evolution of the animal nervous system. A: In 2013, scientists reported the discovery in China of the fossil of a 520-million-year-old arthropod called *Alalcomenaeus*. The fossil revealed a well-preserved impression of the animal's nervous system. B: Its nervous system was strikingly similar to that of its living relatives. (A: Republished with permission of Nature Publishing Group from "Chelicerate neural ground pattern in a Cambrian great appendage arthropod" by Gengo Tanaka, Xianguang Hou, Xiaoya Ma, Gregory D. Edgecombe & Nicholas J. Strausfeld. Nature 502, 364–367; 2013; permission conveyed through Copyright Clearance Center, Inc.; B: Nicholas Strausfeld; data from Tanaka et al. 2013)

neurotransmitters, into the synapse, and they're taken up by the receiver neuron. If you touch a hot pot lid, the sensation of heat travels from your finger along a series of neurons to the brain—in humans, a fantastically dense, complex organ made of roughly 86 billion neurons interlinked by 100 trillion synaptic connections. The brain uses this sensory information to make decisions and send out commands to the body—to pull your hand away from the heat, for example.

The origin of the animal nervous system is currently an area of intense research. Scientists can find clues to its origin in both living animals (Tosches and Arendt 2013) and exceptionally well-preserved fossils (Tanaka et al. 2013). In basal lineages of animals from the Cambrian period, scientists find traces of nervous systems that are strikingly similar to those of living clades (**Figure 16.14**).

Based on recent evidence, researchers think that the animal nervous system evolved through the co-opting of genes with other functions, such as cell-to-cell communication in single-celled eukaryotes (**Figure 16.15**; Srivastava et al. 2010). But currently, scientists are divided over whether the earliest multicellular animals had a nervous system or a nervous system evolved in a more derived animal clade (Liebeskind et al. 2018).

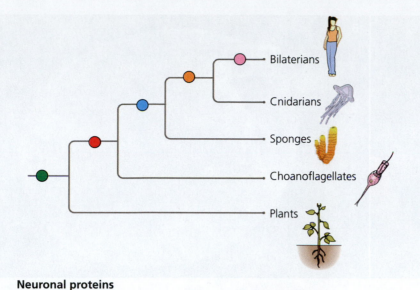

Figure 16.15 A: The network of proteins in neurons evolved in a stepwise fashion in the ancestors of animals. B: Some of the proteins in the synapse are colored in this diagram to indicate when they evolved. (Data from Srivastava et al. 2010)

Neuronal proteins

- Bilaterian/vertebrate origin
- Eumetazoan origin
- Animal origin
- Holozoan origin
- Ancient eukaryotic origin

The Case for Sponges

Sponges are animals. We know that because they share many animal synapomorphies. Some phylogenetic studies identify them as a sister group to all other living animals, giving sponges an important place in our understanding of the evolution of the nervous system. That's because they have no nervous system. If they are indeed the sister group to other animals, then full-blown nervous systems must have evolved *after* sponges diverged from other animals.

There are even intriguing clues suggesting that sponges might represent a transitional stage in the evolution of nervous systems, having evolved some of its building blocks. Most people are familiar with the adult stage of sponges, when they are rooted to the ocean floor, feeding by filtering water through narrow channels. But sponges start out in life as tiny swimming larvae (**Figure 16.16**). These grape-shaped larvae have a crown of hairs at one end, which beat back and forth to propel the animal. Recently, scientists have discovered that cells in these hairs express genes that are also expressed in neurons. This finding has led some scientists to dub these sponge larva hair cells "proto-neurons" (Jékely 2011). Researchers have proposed that these hairs can sense the environment and respond with behavior—the two key functions of neurons. In later animal evolution, these two tasks were divided between specialized neurons. Some neurons only detected sensory information, and others only sent out commands.

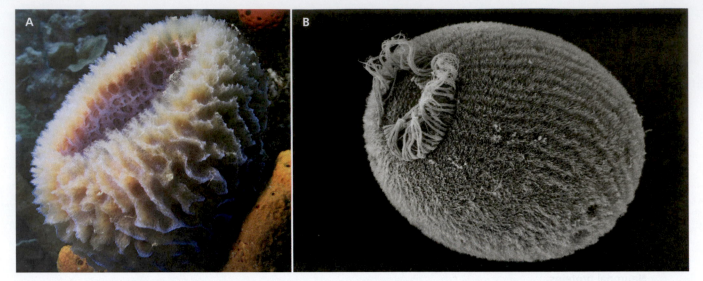

Figure 16.16 A: Sponges lack nervous systems but make a number of proteins that are homologous to proteins found in neurons. B: It's possible that sponge larvae use these proteins to build sensory cells that help them detect good places for them to settle to the seafloor and develop. (A: V Devolder/Shutterstock; B: Photo by Sally Leys)

The Case for Comb Jellies

Although some studies find that sponges are the sister group to all other animals, other studies have come to a different conclusion. They identify ctenophores—also known as comb jellies—as the sister group (**Figure 16.17**). If this is the true shape of the animal evolutionary tree, then the nervous system must have had a different history. That's because ctenophores have a nervous system (Moroz et al. 2014).

Two hypotheses are consistent with this phylogeny: either (1) the common ancestor of all living animals had a nervous system, and it was lost in sponges, or (2) a nervous system evolved in ctenophores independently. Evidence in support of the second hypothesis comes from close studies of ctenophore neurons. They do not express many of the proteins expressed in all other species with nervous systems, for example. Moreover, they use many molecules as neurotransmitters that are not found in other animals.

The origin of the nervous system in animals significantly affected the evolution of behavior. Animals with nervous systems had specialized cells in their body that gathered information about the external world. They could process this information in a dense network of interconnected neurons and then generate motor commands. The nervous system relayed these commands throughout an animal's body in a fraction of a second and produced rapid, coordinated responses.

The impacts of these new behaviors are preserved in trace fossils left behind by the earliest animals during the Ediacaran and Cambrian periods (Carbone and Narbonne 2014). Early trace fossils show evidence of animals grazing on microbial mats, taking undirected paths across the surface. Later, the trace fossils show more complex behaviors, such as spiraling, three-dimensional tunnels through the mats. And later still, trace fossils show evidence of crawling predators tracking their prey with rapid changes in direction. It's amazing that after more than half a billion years, we can still see the vestiges of some of the earliest nervous systems on Earth.

Figure 16.17 Ctenophores, also known as comb jellies, have a pivotal role in current debates about the evolution of nervous systems. Many researchers argue that they belong to a clade that also includes cnidarians and bilaterians, in which the nervous system evolved once. But some studies indicate that ctenophores are the sister group to all other animals, including sponges, that lack nerves. (Reinhard Dirscherl/ullstein bild/Getty Images)

- Current research suggests that the animal nervous system evolved either once or twice.

- The animal nervous system enabled complex behaviors such as three-dimensional burrowing paths. The fossil record preserves evidence of the emergence of this complex behavior. ●

16.4 Innate and Learned Behaviors

Animals with nervous systems display two kinds of behaviors. Some behaviors are innate. In other words, animals develop the same response to the same stimulus, regardless of their experiences. Other behaviors are learned. In these cases, animals update their responses to the environment based on new experiences.

Tinbergen's Herring Gulls

In the 1940s, Tinbergen did pioneering research on innate behaviors in his research on herring gulls (Kruuk 2003). Tinbergen observed that when herring gulls approached their nests, their chicks would beg and peck at their parents' bills, whereupon the adult birds would regurgitate a meal. Tinbergen wondered what information the chicks used to generate this response. And so he ran a simple experiment—but one that would prove to be a classic in the history of behavioral ecology (Tinbergen and Perdeck 1950; ten Cate 2009).

Tinbergen presented the birds with several different stimuli. In one case, he showed the chicks a detailed cutout of an adult bird head, complete with the distinctive red spot that herring gulls have on their beaks. In another, he showed the chicks a cutout head without the red spot. In still other cases, he showed them heads with dots of other colors, including black, blue, and white.

The red-colored dot proved to be the most important feature for triggering the begging in gull chicks (**Figure 16.18**). In other words, the chicks did not need to do

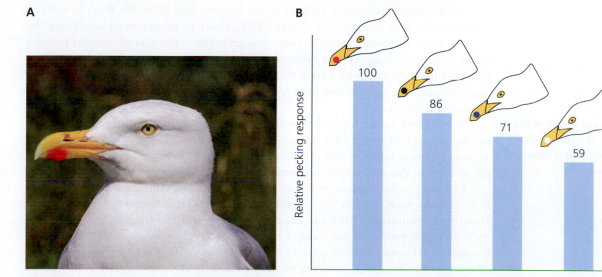

Figure 16.18 Tinbergen ran groundbreaking experiments in herring gulls (A) that revealed the nature of innate behaviors in animals. Herring gull chicks respond to the sight of their parents by begging for food. To determine what the chicks respond to, Tinbergen built cardboard cutouts of adult gull heads. He found that cutouts with a red beak spot triggered the strongest pecking response (B). Cutouts with dots of colors other than red triggered a weaker response, and a head without any dot at all triggered an even weaker one. These experiments show that the red beak spot is mainly responsible for a chick's response. (A: pastie/Getty Images; data from Tinbergen and Perdeck 1950)

any sophisticated three-dimensional visual processing to obtain food. A bias in their sensory systems, making them highly sensitive to a red dot, was sufficient.

Gull chicks peck in response to a red dot the very first time they see one. This reliable response does not mean there's a special gene "for" bill pecking in the gull genome. The response emerges as the bird's nervous system develops, as its eyes begin to recognize shapes and colors, and as it gains control of the muscles in its neck and head. Nevertheless, gulls reared in a normal environment will almost always begin to peck at bill-shaped objects with red dots the first time they see them. The many genes that work together to produce this behavior are passed down from gulls to their chicks.

Aplysia

Along with innate behaviors, animals can also behave in flexible ways. They can learn new responses to their environment and record memories to guide their behavior in the future. Even animals with very simple nervous systems can display this flexibility. Eric Kandel, a Columbia University neurobiologist, won the Nobel Prize in 2000 for his research on learning in a sea slug called *Aplysia*, an animal that has only 20,000 neurons (Kandel 2007).

To breathe underwater, *Aplysia* uses a delicate gill and draws water over it with a siphon. It has evolved behaviors to protect this vital organ from damage. If it senses imminent danger, it draws in its gill and siphon and covers them with a protective flap.

Kandel and his colleagues found that if they brushed a sea slug's siphon with a single nylon bristle, it would respond by only weakly retracting its gill and siphon. If, on the other hand, they gave the sea slug a shock on its tail, the animal had a strong response: sensing a serious threat, it swiftly shut its flap over its gill and siphon.

Then the researchers combined the two stimuli. First, they brushed the sea slugs, and then they applied the shock. The sea slugs learned to associate the two sensations. Eventually, the animals swiftly retracted their gill and siphon if the scientists only brushed the animals and didn't give them a shock. This association lasted for a few minutes, after which the sea slugs returned to their normal, weak response to the brush. But if Kandel and his colleagues repeatedly shocked the animal, the learned association lasted much longer. For weeks after the experiment the slugs would rapidly retract their siphon after a light touch.

By dissecting the sea slug's nervous system, the scientists showed that it formed these new memories through changes in the synapses that allow the slug's neurons to communicate with one another (**Figure 16.19**). Short-term memories associating the shock with being touched resulted from changes in the expression of receptor proteins in the synapses, making certain synaptic connections stronger. Repeated shocking, on the other hand, created longer term memories by adding extra synapses to the connections between the neurons involved in this response, strengthening their linkage and increasing their activity.

Synaptic plasticity occurs when the number or strength of synaptic connections between neurons is altered in response to stimuli.

Subsequent research has shown that other animals, including humans, depend on the same biochemistry for learning. Some synapses become stronger and others weaker, and new synapses join together neurons that were not previously linked. Scientists refer to this malleability as **synaptic plasticity**.

Learning made possible by this synaptic plasticity can enhance the fitness of animals in many ways. *Aplysia*, for example, can adjust its retraction response as it learns about its environment. Withdrawing its gill and siphon rapidly is a good strategy if crabs or other predators are trying to grab it. But responding strongly to harmless brushes of seaweed or other objects would interfere with a sea slug's consumption of oxygen and foraging. Learning enables animals to optimize innate behaviors to a changing environment.

How Learning Evolves

Researchers have found that learning, like other aspects of behavior, has the capacity to evolve. Tadeusz Kawecki and his colleagues at the University of Fribourg in

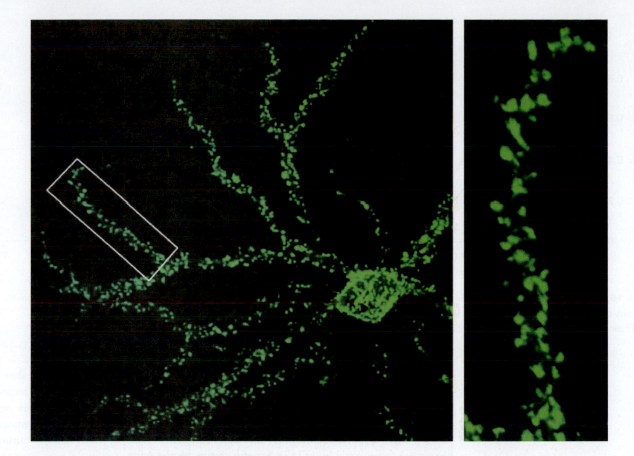

Switzerland, for example, have observed *Drosophila* flies evolve into faster learners in their laboratory (Mery and Kawecki 2003; Burger et al. 2008).

Kawecki and his colleagues trained the flies using orange and pineapple jelly. In each trial, they would spike one of the jellies with quinine—a chemical that is odorless but has a bitter taste. Offered the two flavors of jelly, the flies learned over the course of a few hours to avoid the jelly with quinine and to favor the one without.

Some of the flies needed less time than others to learn to associate the quinine with a particular flavor of jelly. When Kawecki and his colleagues bred these fast-learning flies, they tended to produce fast-learning offspring, demonstrating the heritability of the trait. The scientists then took advantage of this heritability by running an experiment to select for faster learning.

They gave flies 3 hours to learn which jelly was laced with quinine. At the end of this learning period, the flies were allowed to mate, and the females could choose to lay their eggs on either flavor of jelly. Kawecki and his colleagues only collected the eggs from the quinine-free dish. They then reared the eggs and repeated the experiment on the next generation.

This new generation now had to learn for itself which jelly had the quinine; after this learning period, the scientists collected the eggs of that generation and then repeated the experiment. Between each generation, the scientists switched the quinine between the flavors of jelly to make sure they didn't accidentally select for a preference of one flavor instead of learning. After 15 generations of selection, the learning speed of the flies had evolved. Now the selected lineage needed less than an hour to learn to avoid a quinine-laced dish of jelly.

Why aren't all flies naturally smarter? It's possible that learning imposes an evolutionary trade-off between its benefits and its costs. Kawecki and his colleagues found that the evolved, fast-learning flies had shorter life spans than the control flies (**Figure 16.20A**). The scientists got a similar result when they reversed the experiment and selected flies for longevity. Kawecki and his colleagues found that long-lived flies were slower at learning than wild-type flies (**Figure 16.20B**).

Figure 16.19 Synapses change over the lifetime of an animal. New synapses grow, some old ones die back, and other synapses grow stronger or weaker. This photomicrograph shows tiny protrusions, known as dendritic spines, that grow from dendrites during learning. (The right image is an enlarged picture of the dendritic spines in the highlighted box in the left image.) (Jonathan Hanley)

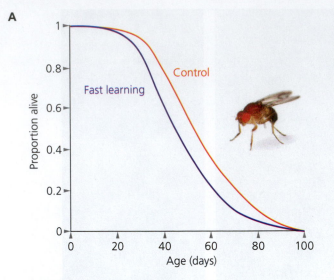

A

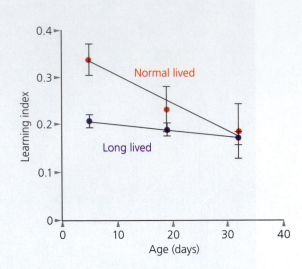

B

Figure 16.20 A: Tadeusz Kawecki and his colleagues at the University of Fribourg in Switzerland selected for high-learning flies. These flies died sooner than control flies. B: Flies selected for longevity were poor learners throughout their entire lives, unlike normal flies. (Data from Burger et al. 2008)

The biological basis for this cost of learning is not yet clear. It may be due to toxic chemicals produced as a side effect of synapse formation, for example. Whatever the reason, natural selection can favor increases in learning only if the costs are outweighed by their benefits. That balance is different for each species. Learning may be favored when a species cannot rely on innate responses—that is, when its environment becomes less predictable, and the benefits outweigh the costs.

Key Concepts

- Animals use innate behaviors to produce adaptive responses to predictable conditions.

- Learning allows animals to modify their behavior to adapt to changes in the environment.

- Synaptic plasticity is thought to be a neurochemical mechanism responsible for learning and memory via interconnected networks of synapses in the brain.

- Like other aspects of behavior, learning can evolve. An evolutionary trade-off between the costs and benefits of learning influences the level of learning and memory in a given population.

- The ability to learn quickly comes at a cost, which can offset the benefits of learning in some situations. ●

16.5 The Vertebrate Brain

Compared to other animals, vertebrates display an extraordinarily rich, complex repertoire of innate and learned behaviors. Part of the reason for this is that they evolved a distinctive nervous system, organized around large, complex brains. Here we will consider how the structure and function of vertebrate brains (including our own) have influenced the evolution of vertebrate behavior.

The oldest signs of the vertebrate nervous system can be found in 530-million-year-old rocks in China. Those rocks contain hundreds of fossilized impressions of bilaterians inferred to be early vertebrates, such as a tiny creature called *Haikouichthys* (**Figure 16.21**; Shu et al. 2003). Measuring about 3 centimeters long, it has many (but

Figure 16.21 A key step in the evolution of vertebrate behavior was the evolution of a more complex brain. Fossils of Cambrian animals such as *Haikouichthys* show that the vertebrate brain originated at least 530 million years ago. (Carl Buell)

not all) of the hallmarks of living vertebrates. Its spinal canal is surrounded by vertebrae, which are supported by a notochord. It has a series of pouches and arches to support gills. Two dark spots at the front of its body appear to be simple eyes. It has holes on the side of its head where sound-sensitive nerves probably grew, and it has another cavity up front that paleontologists suspect was a nostril for smelling. Its head contains a mass of cartilage that appears to have surrounded a primitive brain.

Over the next 100 million years, fish lineages evolved from such humble creatures into the biggest animals in the world. They became predators that could search for prey, in many cases chasing other animals down. As fishes evolved longer bodies, their neurons evolved to great lengths as well. This transformation could not have occurred without the evolution of myelin. This tissue forms an oily sleeve around neurons that acts like the insulation around a wire, preventing the loss of electrical signals over long distances. The resulting lengthened neurons began to supply the vertebrate brain with information from larger sense organs, and new motor neurons allowed fishes to steer their bodies in complex ways.

Early in vertebrate evolution, the brain evolved a divided structure, made up of regions with specialized functions (**Figure 16.22**). These divisions have largely endured for hundreds of millions of years, making it possible for us to recognize some

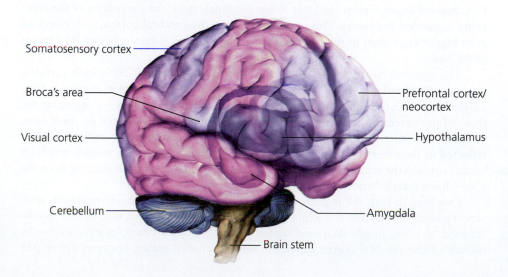

Somatosensory cortex

Broca's area

Visual cortex

Prefrontal cortex/ neocortex

Hypothalamus

Cerebellum

Amygdala

Brain stem

Figure 16.22 The human brain, like other vertebrate brains, is divided into many specialized regions, each helping to carry out certain functions. The cerebellum, for example, is important for balance. The somatosensory cortex organizes sensory information from the skin. Broca's area is involved in language, among other things. Neuroscientists can find homologues to many of these areas in other animals.

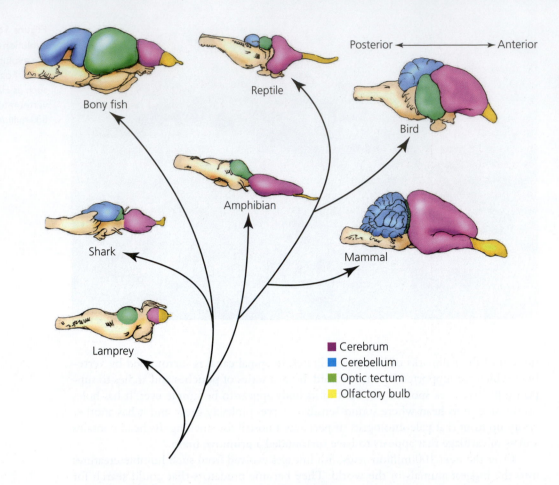

Figure 16.23 Vertebrate brains come in many shapes and sizes, yet they share the same basic organization. (Data from Kardong 2002)

Bony fish

Reptile

Posterior ← → Anterior

Bird

Amphibian

Shark

Mammal

Lamprey

■ Cerebrum
■ Cerebellum
■ Optic tectum
■ Olfactory bulb

homologous brain structures in vertebrate species as different as fishes, birds, and mammals (**Figure 16.23**; Kaas 2013). Some of the functions of these regions have remained the same since these vertebrates diverged from their common ancestor. The cerebellum, for example, plays a vital role in processing information about an animal's balance and sending out motor commands to maintain it. Fishes use the cerebellum to stabilize their body in water so that they don't roll from side to side. For humans, living on land, maintaining our balance keeps us from falling over. People with damage to the cerebellum often have trouble walking.

But the fact that vertebrate brains have these recognizable homologies doesn't mean that they are identical. Evolution has modified the structure of the vertebrate brain in different lineages. When the tetrapod lineage moved on land, for example, they evolved larger cerebrums. Later, when mammals arose, the outer layer of the cerebrum expanded drastically. This region, known as the cerebral cortex, underwent an even bigger expansion in the primate lineage. In humans, it now takes up 90 percent of the brain.

The cerebral cortex has itself become divided into distinct areas, each carrying out specialized tasks, and these areas have evolved to different sizes as mammals have adapted to different niches. Mice have a very large olfactory cortex—a region at the front of the cerebral cortex that receives signals about odors detected by their nose. Humans and other primates rely much more on their eyes than their noses, and this is reflected in their cerebral cortex. Their olfactory cortex has become smaller, and their visual cortex—the region at the back of the brain that gathers signals coming from the eyes—has expanded enormously.

One way to map the regions of the cerebral cortex is to provide mammals with sensory stimuli and then see which parts of the brain become active. When touch-sensitive neurons in our skin send signals to the brain, for example, a region in the middle of the cerebral cortex called the somatosensory cortex becomes active (see

Figure 16.22). It turns out that neurons in different parts of the somatosensory cortex receive signals from different parts of the body—the hands, face, legs, and so on. Remarkably, scientists can draw a "touch map" of the somatosensory cortex that roughly corresponds to the entire body. Some parts of the body send more signals to this region than others. To chart these patterns, neuroscientists can draw the body of a mammal, making highly sensitive parts bigger and those with fewer sensory inputs smaller. These sensory maps can be strikingly different among mammals. As different mammal lineages adapted to different niches, it appears, their somatosensory cortex evolved new topographies.

The sensory maps of some burrowing animals are shown in **Figure 16.24A**. An eastern mole has extremely sensitive front feet, nose, and whiskers. It uses those parts of its body to dig through the dirt and to sense the presence of insects that it can

A

ANATOMICAL PROPORTIONS

Eastern mole   Naked mole rat   Star-nosed mole

CORTICAL MAGNIFICATION

Eastern mole   Naked mole rat   Star-nosed mole

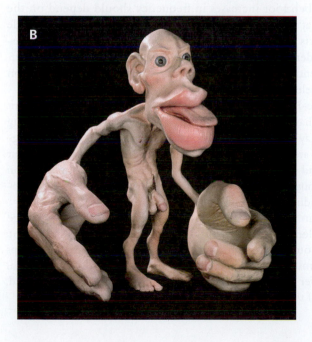

Figure 16.24 A: The evolution of mammalian brains is influenced by the ecological niches in which the animals live. For example, each mammal has some regions of the body that are very sensitive and others that are not. These images show the relative concentration of neurons dedicated to each part of the body in the somatosensory cortex of the brain. Left: Eastern moles use their forelegs to burrow and their noses and whiskers to sense prey. These regions are the biggest in their body map. Middle: The naked mole rat, which spends its life digging tunnels through dry earth, has a body map dominated by its teeth, mouth, and feet. Right: Star-nosed moles, on the other hand, have fleshy appendages on their noses, and these "stars" dominate their body maps. (Data from Alcock 2009; B: This sculpture represents the human body map in the somatosensory cortex. Unlike rodents, our hands and mouths are most sensitive—reflecting our adaptations for using tools and language. (B: Sensory Homunculus (plaster)/English School, (20th century)/NATURAL HISTORY MUSEUM, LONDON/Natural History Museum, London, UK/Bridgeman Images)

eat. A star-nosed mole, by contrast, has long, fingerlike extensions around its nose that it uses to probe the mud around streams. That tiny patch of skin is more sensitive than the rest of its entire body. Naked mole rats also burrow through the ground, but they live in arid regions of Africa, where they use their teeth to dig tunnels. For them, the teeth rather than the nose dominate their somatosensory cortex.

We humans are also biased about the information we receive, but in a different way. As **Figure 16.24B** shows, our hands, lips, and tongues are strongly represented in the somatosensory cortex. This pattern is a result of our own ecology—particularly our adaptations for using tools with our hands (see Chapter 17).

Key Concept

- Vertebrates have large, centralized brains that are divided into specialized regions. The relative size of these regions in different species reflects their evolutionary history and ecological specializations. ●

16.6 Individuals, Groups, and the Evolution of Social Behavior

As the study of animal behavior began to flourish in the mid-1900s, it triggered a profound evolutionary debate. Some researchers argued that selection acts almost entirely on the behavior of *individuals*, whereas others maintained that a significant amount of behavior was the result of selection acting on entire *groups*.

Individual selection occurs when behaviors that cause particular individuals within a population to perform well relative to other individuals are favored and spread. When males compete over access to females, for example, the successful males can transmit more copies of their alleles to subsequent generations.

> **Individual selection** describes selection arising from variation in fitness among individuals.

But individuals often belong to groups—from ant colonies to bison herds to human societies. Groups that perform well may persist and grow in size, while other groups diminish. **Group selection** favors behaviors that contribute to the performance of the group, so that groups performing these behaviors fare better than groups lacking them.

> **Group selection** is selection arising from variation in fitness among groups.

In principle, all behaviors can influence the success of both individuals and groups. Whether or not a particular behavior increases in frequency should depend on the net effects of selection acting at these two levels. Things get interesting when a single behavior has conflicting effects on individuals and groups. A selfish behavior benefiting an individual might undermine the success of an otherwise cooperative group, for example. In such cases, scientists realized, the fate of the behavior should depend on whether individual or group selection is stronger.

The biologist V. C. Wynne-Edwards argued that group selection was an important factor in the evolution of behavior (Wynne-Edwards 1962). Consider the foraging behavior of a bird. A bird may forage at a high or low rate. Wynne-Edwards suggested that if a group of birds ate food at a low rate, they could avoid destroying their food supply and therefore avoid starvation. Wynne-Edwards and other group selection advocates argued that such groups would be more likely to survive, whereas less cooperative groups in which individuals ate at a high rate would perish.

The British biologist John Maynard Smith and the American biologist George Williams pointed out a problem with Wynne-Edwards's group selection argument: selection on individuals can counteract—and even overwhelm—selection on groups (Maynard Smith 1964; Williams 1966). In a group of cooperators, a selfish individual can gain a fitness advantage over other individuals. As a result, selfishness should spread, increasing in frequency even if this new behavior is harmful to the overall group.

To appreciate this argument, it helps to imagine a pair of rocky islands, each with its own colony of a particular species of birds. Let's say that on the southern island, the birds lay six eggs in each clutch. On the northern island, the birds lay only three eggs. The greater reproduction rate on the southern island leads the bird population to become so dense that they devour all the plants on the island, wipe out their food supply, and starve. Meanwhile, on the northern island, the birds have a reproductive rate low enough to allow the plants to regenerate, providing a stable food supply for the birds.

In this scenario, groups with lower reproductive rates are more likely to persist than those that deplete their available resources. They may even spread to neighboring islands, colonizing them after the selfish birds have died and the plants have had a chance to rebound. Because populations with low reproductive rates persist and spread better than populations with higher reproduction, group selection favors a low reproductive rate.

But what if an individual bird on the northern island gains a mutation that causes it to lay six eggs in a clutch instead of three? (For the sake of this thought experiment, let's assume that the birds can rear six offspring as easily as three.) Three extra offspring, on their own, aren't enough to wipe out the island's food supply. Instead, those three extra offspring may mature and then have six offspring of their own. It will take only a few generations for birds laying six eggs per clutch to predominate in the population. The population will thus evolve away from a group-benefiting behavior toward a selfish, individual-benefiting behavior—despite the devastation that these individuals may cause to the group as a whole.

This is exactly what occurs in the experiments on slime molds that we discussed earlier. The *chtA* mutants become spores more often than wild-type slime molds. Slime molds with the *chtA* allele increase in frequency, rapidly replacing other genotypes within the group. These selfish individuals manage to spread even though the net effect on the group is disastrous. Once all of the other genotypes are gone, the group can no longer form a cooperative slug—and, as a result, the group can no longer escape harsh environmental conditions.

The group selection arguments put forward by scientists like Wynne-Edwards were effectively destroyed by Maynard Smith and Williams. They showed that under a range of natural conditions, selection arising from differences in the relative fitness of individuals within a group is stronger than selection acting on those same individuals arising from the success of the group relative to other groups. In practice, this means that most organisms are reproductively selfish. Behaviors that we observe in natural populations today are there because individuals performing those actions did better—they were more likely to survive, and more successful at reproducing—than individuals exhibiting other behavioral phenotypes.

Still, as we'll see later in this chapter, in a few situations this balance can tip the other way, so that selection acting at the level of groups predominates. Cooperative behaviors fare much better, for example, when members of the group are kin (Marshall 2011; McGlothlin et al. 2014).

Playing the Evolution Game

In Chapter 6, we saw how the frequency of different alleles can affect their fitness in a process known as frequency-dependent selection. This phenomenon is especially important to the evolution of social behavior. The behaviors of individuals in a group can affect the fitness of the group's other members. If a group of slime molds is made up of cooperators except for a single cheater, for example, that cheater enjoys a higher fitness because it's more likely to become a spore, while other slime molds contribute to the stalk and die. But if the population is overwhelmingly made of cheaters, their fitness plummets, because there aren't enough cooperators to produce a stalk.

Game theory is a mathematical approach to studying behavior that solves for the optimal decision in strategic situations (games) where the payoff to a particular choice depends on the choices of others.

An **evolutionarily stable strategy** is a behavior that, if adopted by a population in a given environment, cannot be invaded by any alternative behavioral strategy.

Evolutionary biologists can gain insight into these complex situations by borrowing mathematical tools from economists. Economists study how investors and consumers influence each other's behavior in order to find optimal strategies under certain economic conditions. They use an approach called **game theory** that treats economic actors as players in a game, each of which can choose from a variety of strategies. By simulating many rounds of these games, they can then determine which strategies lead to long-term success or failure.

Maynard Smith and other evolutionary biologists realized that economic game theory was well suited for modeling the evolution of behavior in populations (Cowden 2012). Although economists may be interested in the strategy that brings a player the most money, evolutionary biologists can use game theory to see which behaviors lead to the greatest long-term reproductive success.

In some games, the researchers found, a new behavior can "invade" a population and spread to fixation, eliminating the earlier behavior. But in other games, the earlier behavior can withstand these invasions and remain successful. A behavior that resists all possible invasions is known as an **evolutionarily stable strategy**. Sometimes, when the fitness of particular strategies changes depending on the relative frequencies of other strategies, situations can arise where no one strategy is stable.

Game theory has helped scientists to understand how frequency-dependent selection can lead to remarkable diversity in behavior. One well-studied example is displayed by side-blotched lizards (*Uta stansburiana*), which live in the western United States. They produce three strikingly different kinds of males (**Figure 16.25**). The males not only look different from one another; they behave differently, too.

Orange-throated males are big and aggressive and guard large territories containing multiple females. Blue-throated males are smaller and less aggressive than orange-throated males. They defend territories as well, but their territories are only big enough to contain a single female. Yellow-throated males don't defend any territory at all. Instead, they mimic female lizards (which also have yellow throats) to sneak into the territories of the other males. They can then surreptitiously mate with females, passing on their alleles for yellow throats and sneaky behavior.

Barry Sinervo, now at the University of California, Santa Cruz, and his colleagues have found that the lizards are experiencing frequency-dependent selection, with the result that none of the three male strategies are evolutionarily stable. Instead, the lizard populations oscillate in the relative frequencies of the three male forms. When orange males are relatively rare, for example, they fare very well. Orange males can oust blue males from their territories and mate with more females than the other male types.

Figure 16.25 Male side-blotched lizards have three types of behavior. The prevalence of each behavior type raises or lowers the fitness of the others. As a result, the types oscillate in frequency over time. ("Selective loss of polymorphic mating types is associated with rapid phenotypic evolution during morphic speciation" by Corl, Davis, Kuchta and Sinervo. PNAS March 2, 2010 107 (9) 4254–4259)

But as orange males become more common, they become an easier target for the sneaky yellow-throated males. The yellow-throated males sneak into their territory and mate with more females, lowering the relative fitness of the orange males.

The yellow-throated sneaky males eventually become the most successful lizards of all. As they become more common, the orange lizards, struggling to protect several females at once, become increasingly rare. Meanwhile, the blue males experience a boost in their relative reproductive success because they can guard their single female much more effectively. As the blue males spread, they drive the number of yellow males down because there are fewer and fewer opportunities for the yellow males to sneak copulations. And once the blue males predominate, the cycle repeats itself. The orange males, now relatively rare, move back in and begin displacing the blue males from their territories.

Sinervo and his colleagues liken this situation to the children's game of "rock-paper-scissors" (Sinervo and Lively 1996). In the children's game, rock beats scissors, and scissors beats paper, and paper, in turn, beats rock. To understand the side-blotched lizards, we can replace paper, scissors, and rocks with different phenotypes that have changing levels of fitness. The lizard behaviors cycle from common to rare and back.

Key Concepts

- The crucial issue distinguishing the relative importance of individual versus group selection is the fate of alleles that are detrimental to an individual but beneficial to the group in which that individual lives. If individual selection predominates, then these alleles will disappear; if group selection predominates, then these alleles will spread.

- Selection is shortsighted. Immediate fitness consequences determine the success or failure of an allele, irrespective of the ultimate outcome of this process.

- Organisms are reproductively selfish; they behave in ways that enhance the spread of their own genetic material, even if such behavior is detrimental to their population or species.

- An evolutionarily stable strategy is a behavior that, when adopted by a population of players, cannot be invaded by any alternative strategy. ●

16.7 Why Be Social?

If individual selection can undermine group selection so easily, it raises an important question: why are there any stable groups at all? Evolutionary biologists have found that individual selection can indeed favor group living if the gains it brings to an individual outweigh the costs (Alexander 1974; Table 16.1).

Benefits of Living in a Group

One benefit an individual can get from living in a group is increased vigilance. For example, antelopes can reduce their chances of being killed by a leopard if they stick together. A leopard depends on the element of surprise. If the antelope spots the leopard quickly enough, it can run away. A single antelope doesn't have an infinite supply of vigilance, however; for one thing, it has to spend much of its time with its head to the ground to graze. But if antelopes come together in a group, they in effect acquire more eyes and ears. In addition to being less likely to be surprised, the antelopes can also spend more time eating—which can raise their individual fitness even more relative to solitary antelopes. And even if a predator does strike a group

Figure 16.26 Many animals reduce their chances of being attacked by forming groups. A: By jumping off of ice shelves together, Adélie penguins reduce their chances of being eaten by leopard seals waiting in the water. B: Fishes avoid being eaten by dolphins and other predators by forming schools. C: Ostrich chicks are less likely to be captured by predators if they surround themselves with other chicks. D: Impala can increase their vigilance for predator ambushes by living in herds. (A: cascoly/Getty Images; B: Reinhard Dirscherl/WaterFrame/Getty Images; C: MartinMaritz/Shutterstock; D: WLDavies/Getty Images)

The **dilution effect** refers to the safety in numbers that arises through swamping the foraging capacity of local predators.

of antelopes, each group member faces a smaller risk of being the target of the attack (a phenomenon known as the **dilution effect**). Animal groups produced by dilution effects are often called "selfish herds," because their herding behavior is not based on any coordination or cooperation. It's instead the result of a selfish scramble of individuals jockeying for position near the center of the group (**Figure 16.26**).

Adélie penguins, which live on the coast of Antarctica, vividly illustrate how group living allows individuals to escape predators. They dive off ice sheets into the ocean to search for fishes; but when they take the plunge, they sometimes are eaten by leopard seals cruising just offshore. Rather than jump in alone, the penguins crowd together by the hundreds and then leap en masse. Together, they can overwhelm the leopard seals, which can't focus on any single individual. The huge numbers of jumping penguins create both a confusion effect that increases the predator's difficulty in targeting an individual and a dilution effect that reduces the likelihood of any individual being caught (Olson et al. 2013).

Predators can benefit from group living as much as prey (**Figure 16.27**). Lions, dolphins, wolves, and hyenas are more effective at catching prey in packs than alone. American white pelicans practice "fish herding," paddling on the water together in a tactical formation to drive fishes into shallows or into dense concentrations; the birds can then easily scoop up their prey. Even though a pelican has to share the fishes with

A

Figure 16.27 A: Wild dogs benefit from hunting in packs instead of on their own. B: Scott Creel found that when prey is abundant, each dog in a pack gets more energy each day if its pack is bigger. The points in this graph are the mean energy intake. The size of the circles around the point is proportional to the number of observations for each pack size. The dashed line shows the linear regression. (A: imageBROKER/Superstock, Inc.; data from Creel 1997)

B

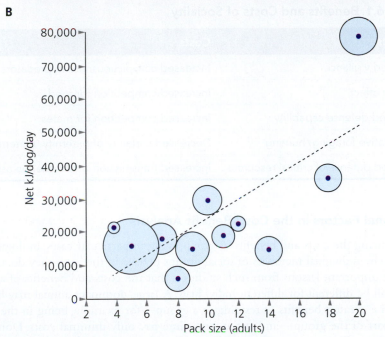

the rest of the group, it still comes out ahead, getting more prey than if it foraged alone (Anderson 1991).

Costs of Living in a Group

Living in groups can bring many benefits, but it can impose costs as well. Individuals living in big groups are at greater risk of getting sick, for instance. Parasites and pathogens can spread more effectively in dense groups of animals than among solitary animals (**Figure 16.28**). Large groups also present more opportunities for extra-pair copulation (Section 11.4). Females may benefit from these opportunities because they can obtain sperm from higher-quality males. Males that can mate with many females can also enjoy higher reproductive success. But in species where males provide parental care, group living may also mean that many males are not actually the fathers of the offspring they are caring for. Finally, living in a group often means more competition for limited food, space, or other resources. See **Table 16.1** for some benefits and costs of living in a group.

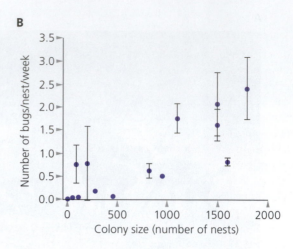

Figure 16.28 A: Cliff swallows build nests in dense colonies. B: This group living comes at a cost: in large colonies, each nest becomes infested with more parasites. (A: Janet Horton/Alamy Stock Photo; data from Brown and Brown 2004)

Table 16.1 Benefits and Costs of Sociality

Benefits	Costs
Increased vigilance	Increased conspicuousness to predators
Dilution effect	Increased competition for food
Enhanced defense capability	Increased competition for mates
Cooperative foraging/hunting	Decreased certainty of paternity/maternity
Improved defense of critical resources	Increased transmission of disease/parasites

Additional Factors in the Cost–Benefit Analysis

By measuring the costs and benefits of group living in particular cases, biologists can explain why individual members of social groups behave in the ways they do. One of the most important lessons from such studies is that the costs and benefits of a given action will be different for different individuals. A large, dominant animal may be able to reap the greatest benefits of residing in a group—for example, being in the safest, central part of the group—and at the same time pay only minimal costs. Dominant individuals are likely to prevail in any competition for food or other resources, so they suffer little loss from being in the group. In contrast, individuals in poor condition or subordinate individuals experience the smallest benefits (they get pushed to the periphery of the group where predation is most likely), and these same individuals pay the highest price (they lose in contests over food or other resources). Consequently, different individuals within the same group can experience very different consequences for choosing to reside in the group. And although sociality may be highly cost-effective for individuals in the best condition, others may actually benefit more if they disperse and live in smaller groups or by themselves. The optimal behavior for one individual may not be the same as the optimal behavior for another.

Key Concepts

- There are costs and benefits to group living.

- The costs and benefits of a behavior may not be the same for all individuals. Factors such as sex or dominance status may shift the relative value of specific behaviors. ●

16.8 The Importance of Kin

In many animal groups, the help that individuals provide each other extends beyond catching prey and escaping predators. They also help each other raise their young. During the debates over selection for behavior, such **altruistic behaviors** seemed to be powerful evidence in favor of group selection. Helping other members of the group benefited the group as a whole. Altruism seemed to pose a contradiction with our understanding of individual selection. Shortsighted natural selection seemed incapable of favoring individuals that lowered their own fitness—in terms of the time spent foraging or rearing young—for the benefit of other individuals.

Since the 1960s, however, a great many studies have documented a number of situations in which helping raises the fitness of the helper, compared to other behaviors. Consider, for example, a young meerkat (**Figure 16.29**). If it lives on its own, it faces a higher risk of being killed by a predator than if it stays with a group. The cost of living in the group means helping to rear the offspring of the dominant meerkats; it's a small price to pay for group membership if it improves the helper's odds of surviving to another season.

Helping may also allow individuals to gain valuable experience that will serve them well when they eventually breed, leading to more successful reproductive efforts and enabling more of their own young to survive. Helping may even give individuals opportunities to acquaint themselves with future mates. In some species of birds, helper males assist other males by bringing food for their mates and their young. If the male getting the help dies, the helper male has a better chance of mating with that female (Clutton-Brock 2002). Helping may even be the best way to acquire a nesting territory, if the helper can inherit it from the breeders when they die.

Altruism can also evolve between relatives under certain circumstances. This happens because relatives share many of the same alleles. A diploid animal shares 50% of its alleles with an offspring. It also shares 50% of its alleles with its siblings and 25% of its alleles with its nephews and nieces (**Table 16.2**). If an animal helps its close relatives to reproduce, some of its own alleles will be carried down to the next generation.

Altruist behavior occurs whenever a helping individual behaves in a way that benefits another individual at a cost to its own fitness.

Figure 16.29 Meerkats live in large groups in southern Africa, where subordinates may spend years helping dominant members rear their young rather than having offspring of their own. The foraging success, growth, and survival of all group members increase with group size. This strategy provides direct fitness benefits to subordinates for helping the group. (namibelephant/Getty Images)

Table 16.2 Genetic Equivalence

Descendant Kin	r	Nondescendant Kin
Offspring	0.5	Full siblings
Grandchildren	0.25	Half siblings, nephews and nieces
Great-grandchildren	0.125	Cousins

The coefficient of relatedness (r) between two individuals is the probability that they share identical copies of a particular allele. It can also be thought of as the proportion of alleles that are likely to be shared between the two individuals. A child inherits half its genetic material from each parent. On average, any allele in the child's genome has a 50% chance of being identical to the allele carried by its mother (or father), so their coefficient of relatedness is 0.5. Brothers and sisters also have a 50% chance of sharing any particular allele, so their coefficient of relatedness is also 0.5. This table shows the coefficients of relatedness for various types of relatives. Because identical alleles can be shared between nondescendant kin (for example, siblings, cousins, nieces, and nephews) as well as between parents and their offspring (descendant kin), helping relatives can raise an individual's inclusive fitness.

Inclusive Fitness

Inclusive fitness describes an individual's combined fitness, including its own reproduction as well as any increase in the reproduction of its relatives due specifically to its own actions.

In the early 1960s, the British biologist William Hamilton presented a mathematical argument for the evolution of altruism among relatives (Hamilton 1964, 1996). He looked at fitness from a new perspective, splitting it into two components known as direct fitness and indirect fitness. Direct fitness results from an organism's own success in transmitting alleles to future generations. Indirect fitness results from the reproductive success of *other* individuals that carry the same alleles. An organism's **inclusive fitness**, Hamilton argued, includes both kinds of fitness.

Inclusive fitness could account for how selection on individuals could lead to altruistic behavior, Hamilton suggested. Under certain conditions, helping a relative rear its offspring would lead to more of an individual's own alleles being passed down to the next generation than if that individual tried to raise its own offspring. Altruism could thus evolve if the cost of helping others were paid back by benefiting genetically similar individuals.

In more precise terms, Hamilton argued that a helping behavior could spread in a population if the benefit to the recipient, weighted by the chance that the recipient shares the same alleles as the donor, exceeded the cost to the donor. Hamilton's rule, as this statement is known, can be expressed formally as

$$rB > C$$

where r is the coefficient of relatedness between donor and recipient, B is the benefit to the recipient arising from help, and C is the cost to the donor from helping.

The coefficient of relatedness can range from 0, for unrelated strangers, to 1, for a pair of identical twins or clones (an extreme case). Helping a clone is genetically equivalent to helping oneself (that is, the behavior is beneficial whenever $B > C$). In most situations, however, the recipient and the donor are not clones. Because the helper and the recipient share a lower proportion of alleles, the net benefits of altruism are lower. The benefits decline in proportion to how distantly related the individuals are. Helping an unrelated individual is not advantageous at all, unless the behavior comes at no cost to the donor ($C = 0$).

Hamilton's rule describes many behaviors in animal groups (**Figure 16.30**; **Box 16.1**). It helps explain, for example, why birds in some species will help their parents raise a clutch of siblings (**Figure 16.31**). This helping typically happens when conditions are poor and the odds are low that a young bird will successfully rear chicks of its own. In such cases, the benefit of rearing relatives outweighs the cost to the helper bird of giving up the chance of rearing chicks of its own. But when conditions improve, the helper birds often abandon their parents to start nests of their own.

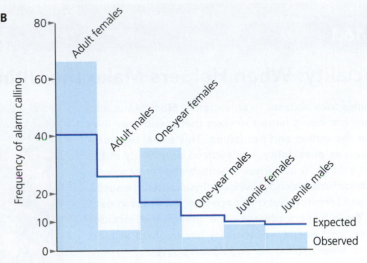

Figure 16.30 A: In colonies of Belding's ground squirrels, individuals sometimes give an alarm call when they detect the approach of a predator. Alarm calling is risky because it draws the attention of the predator toward the caller, but these calls give other individuals in the colony time to escape. B: Unlike males who disperse to unrelated colonies when they reach sexual maturity, female Belding's ground squirrels are often surrounded by close relatives. As a result, females are more likely (and males less likely) to give alarm calls than would be expected based on their relative abundance. The dark blue line shows the number of alarm calls that would be expected for each category of individuals if they were all equally likely to give calls. (A: Richard R. Hansen / Science Source; data from Sherman 1977, 1985)

The Problem of Kin Recognition

Kin selection makes the explicit prediction that animals should proffer their help to relatives. But to do so, they must be able to distinguish their kin from unrelated individuals. In some cases, physical proximity is a reliable indicator that another individual is a relative. If a mother keeps her young inside her body as they develop, then it's a near certainty that the offspring she produces will be hers. In other words, her certainty of maternity will be high.

It can be harder for males to recognize their offspring. As we saw in Chapter 11, females that form long-term bonds with males will sometimes mate surreptitiously

Kin selection is selection arising from the indirect fitness benefits of helping relatives.

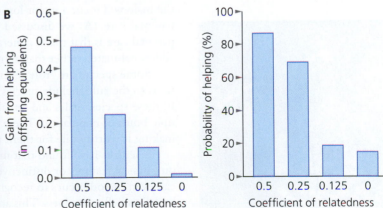

Figure 16.31 A: Instead of attempting to breed, colonial white-fronted bee-eaters sometimes help their parents during their first 2 years of life. Cooperative behavior in this species occurs most frequently during harsh years, when attempts by naive birds to breed would likely have been unsuccessful, and the helping behavior is almost always given to close relatives. B: The more related bee-eaters are to the offspring, the more individual helpers gain; and the more related they are to the offspring, the more likely they are to help. (A: Steffen Foerster / Shutterstock; data from Emlen and Wrege 1988, 1989)

BOX 16.1

Eusociality: When Helpers Make the Ultimate Sacrifice

Ant colonies take altruism to an extreme. Most of the ants in a colony are sterile female workers that spend their lives caring for the queen and her larvae. This social organization, known as **eusociality**, has evolved only rarely in the history of animals. It has arisen in vertebrates, crustaceans, aphids, thrips (small, slender, winged thysanopteran insects), beetles, and termites, but only once or a few times in each lineage. Yet in hymenopterans—the lineage that includes ants, bees, and wasps—eusociality has evolved at least 11 separate times (Hölldobler and Wilson 2009).

Scientists have identified certain factors that favor the evolution of eusociality (**Box Figure 16.1.1**). Eusociality tends to arise in extreme environments, where groups of individuals fare much better than solitary ones. It's also fostered by ecological circumstances that cause the average relatedness of colony members to be unusually high. High coefficients of relatedness ensure that the inclusive fitness benefits to sterile helpers are large.

These environmental factors don't explain why hymenopterans in particular have become eusocial so often. The cause may lie within—in the peculiar mechanism that determines their sex. Hymenopteran insects are haplodiploid, meaning that males develop from unfertilized eggs (they

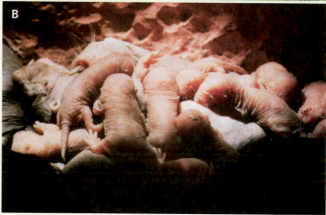

Box Figure 16.1.1 A: Ants live in colonies where sterile workers care for the queens. B: Naked mole rats live in similar eusocial societies. (A: SweetCrisis / Shutterstock; B: Neil Bromhall / Science Source)

with other males. When the females bear their young, their male partner cannot be certain of paternity. If a male invests its resources in rearing the young of another male, the male will incur a major loss of fitness. As a result, evolution has rarely favored male parental care. (As we discussed in Chapter 12, another reason for the rarity of male parental care is that males generally do best by investing time and resources in seeking additional mates, rather than remaining with a single female and her brood.)

Some species are "role reversed," with extensive male parental care. These exceptions to the rule include species of seahorses, pipefishes, giant water bugs, and jacanas. In these species, males and females engage in elaborate courtship behaviors with multiple bouts of copulation. These behaviors give the males a high certainty of paternity, making it worthwhile for them to invest heavily in their offspring (**Figure 16.32**).

Scientists have found that many animals have evolved ways to recognize kin based on physical features or distinctive odors (Hauber and Sherman 2001). Sometimes they can even use odor cues to recognize relatives they have never encountered before: they literally smell relatedness. This ability is called direct recognition because the individual

Figure 16.32 Male giant water bugs carry their brood on their backs, and they aerate the eggs by rocking gently at the water surface. This behavior is dangerous because the ripples attract predatory fishes. Males would not benefit if they incurred these risks for unrelated individuals. But each male accepts eggs only from females that he has copulated with multiple times. As a result, his certainty of paternity is unusually high. (John Cancalosi / Getty Images; data from Smith 1979)

are haploid) and females develop from fertilized eggs (they are diploid). In many hymenopteran species, the queen mates with only one male to produce all her offspring. All the daughters inherit an identical chromosome set from their father. As a result, the daughters have a remarkably high coefficient of relatedness ($r = 0.75$). (The coefficient of relatedness between a daughter and her mother in a diploid species is only 0.5; see Table 16.2.) Thus, daughters are more closely related to each other than they are to their mother. And if the queen's daughters were to have daughters of their own, they too would remain more closely related to their sisters than to their own offspring.

William Hamilton recognized that this unusual situation would cause females in a colony to have higher inclusive fitness from helping their mothers produce more sisters than they would if they reproduced themselves. For this reason, **haplodiploidy** is considered to be a major genetic factor that predisposes hymenopteran lineages toward evolving eusocial behavior. It also helps explain why sterile helpers in the hymenoptera are always female (the unusually high relatedness of siblings applies only to daughters). In contrast, termites, which are all eusocial but are not haplodiploid, can have sterile helpers that are either male or female.

Thrips are haplodiploid, too. And perhaps not surprisingly, they've also been found to contain species with sterile soldier castes—that is, eusocial taxa. In fact, both haplodiploidy and high levels of inbreeding appear to contribute to high coefficients of relatedness among colony mates in thrips ($r = 0.64$ to 0.92). Thus, helping generates strong kin-selected benefits, even if that help requires an individual to completely sacrifice its own direct fitness (Chapman et al. 2000).

Exceptionally high relatedness also occurs in species that reproduce parthenogenetically (females reproducing asexually). In all aphid species, females undergo asexual reproduction for part of the year. For these generations, all of the young are clones ($r = 1.0$), and in some species, where they reside inside defendable woody galls, some of the young become soldiers. These soldiers have thicker cuticles, enlarged forelegs, and sharp spikes protruding from their heads. When they encounter other insects, they grab them and impale them on their spikes. Soldier aphids are sterile, but they derive indirect fitness from their clone siblings (Stern and Foster 1996).

Eusociality is a type of social organization in which species have complete reproductive division of labor. In a eusocial group, many individuals never reproduce, instead helping to rear the offspring of a limited number of dominant individuals.

Haplodiploidy is a mechanism through which sex is determined by the number of copies of each chromosome that an individual receives. Offspring formed from the fertilization of an egg by a sperm (that is, diploids) are female; those formed from unfertilized eggs (that is, haploids) are male.

matches the phenotype of the stranger to itself. In other species, it's not clear exactly how individuals recognize kin, and yet they clearly do.

Key Concepts

- The results of personally reproducing and rearing descendants (direct fitness) and assisting in the rearing of nondescendant kin (indirect fitness) are genetically equivalent. Shared alleles can favor the evolution of behaviors that enhance the reproductive success of close relatives, behaviors often viewed as altruistic.

- Kin selection theory predicts that individuals should help relatives more than non-relatives, and that they should help close relatives more than distant relatives.

- Some animal groups contain helper individuals who forfeit all opportunities to reproduce themselves. Such extreme self-sacrifice can be favored by selection when the relatedness among individuals is exceptionally high because of the indirect fitness benefits the helpers derive from their actions.

- Parental care is the most widespread form of helping behavior, in part because the physical proximity of parents and offspring keeps the certainty of relatedness high.

- One reason male parental care is rare is that certainty of paternity is generally low. Without a high probability of relatedness, parental effort is not evolutionarily cost-effective. ●

16.9 Intelligence and Other Forms of Complex Cognition

One of the big challenges in interpreting behavior is that we are very familiar with how *we* behave, but the behavior of other species is mysterious. This disconnect can lead us to mistakenly ascribe human behaviors to our pets or other animals (**Box 16.2**). On the other hand, this disconnect can also make it hard to appreciate that our own behaviors emerge from the same evolutionary history that produced the behavior in related species.

These challenges are especially daunting when we turn to the subject of intelligence. Although intelligence can have different meanings for different people, scientists define it as the collection of sophisticated cognitive abilities, such as problem solving, complex social cognition, and future planning (Amodio et al. 2018).

By this definition, we humans are unquestionably intelligent. Our social cognition includes language, for example, as well as the ability to organize into large, complex social groups that endure for centuries. Our problem-solving skills are so powerful that they've enabled us to invent boats, computers, and space probes. And our ability to plan for the future has enabled us to establish stable, long-lived communities across the entire world.

No other animals can compare to us in these traits. But does that mean that these other animals lack intelligence? Not necessarily. As scientists spend more time observing other species, they discover more cases of strikingly sophisticated cognition.

Figure 16.33 Chimpanzees use many different gestures to communicate with each other. Two populations of chimpanzees may give two different meanings to the same gesture. Some researchers have suggested that gestures might have played a critical part in the evolution of human language. (Mary Beth Angelo/Science Source)

Some animals can communicate remarkable amounts of information to each other. Bottlenose dolphins, for example, can recognize individual members of their group by their distinctive whistles (Janik et al. 2006). Vervet monkeys produce certain kinds of screams for certain kinds of predators (Seyfarth et al. 1980). This repertoire of alarm calls is adaptive because their fellow monkeys (who typically are close relatives) can make a suitable escape. If an eagle is swooping down at them, they'll want to scramble out of the trees, but if a leopard is running toward them, a tree is exactly where they want to be.

Chimpanzees, our closest living relatives, make an even wider range of sounds, and they can also communicate with gestures—something not observed in other nonhuman primates (**Figure 16.33**). They may wave their arms, reach out their hands, or slap the ground. Chimpanzee gestures generally convey some kind of request—to play, for example, or to share some food—but there's no deep biological impulse linking one gesture to one meaning. In fact, much like human cultural rituals, chimpanzee gestures vary from population to population (Pollick and de Waal 2007; Hobaiter and Byrne 2014).

Making and Using Tools

Some animals are surprisingly adept at making tools. We've seen what New Caledonian crows can do, but they're not alone. Chimpanzees can crack nuts by placing them on one rock and then smashing them with another, and they can also fashion sharp spears to stab bush babies (McGrew 2010).

Animals make tools not just to get food but also to protect themselves. Gorillas will poke a stick into water to test its depth. Orangutans will fashion twigs into probes they can use to pull out irritating hairs from the fruits they eat.

BOX 16.2

The Dangers of Anthropomorphism

Animals make decisions all the time, and the choices they make have consequences. Animals that make good choices and respond appropriately when presented with a challenge tend to fare well; those that make poor choices do not. Over time, effective response mechanisms can evolve—even to the point where animals make astonishingly complex decisions and display a multitude of situation-appropriate behavioral responses. This does not mean that these animals know what they are doing or why. It merely means that they have inherited a nervous system and brain that filter signals effectively and generate behavioral responses that are accurately matched with circumstance. These are genuine decisions, which animals make by actively choosing among suites of possible behavioral responses. But all too often, this behavior leads people to assume that the animals are cognizant of their actions. **Anthropomorphism** is the tendency to attribute human characteristics, such as motives, to other animal species (**Box Figure 16.2.1**).

Anthropomorphism is a dangerous habit because it can give rise to a misleading view of animal actions. It detracts from the objectivity that is necessary when examining the evolutionary significance of behavior. Motive is neither likely nor necessary for animal actions to be adaptive (Wynne 2004).

Ironically, many of our own decisions are made without our knowledge of what we are doing or why. Adaptive behavioral responses may lurk in our genomes, causing us to elicit behavioral preferences that we are utterly unaware of. Yet just because we don't *know* what we are doing does not mean that these behavioral predispositions haven't arisen through fitness advantages accrued in generations past.

Box Figure 16.2.1 We grow up reading stories about animals that act like humans. But just because animal behavior resembles our own doesn't mean we can conclude that they think the way we do. (The Tale of Benjamin Bunny by Beatrix Potter)

Anthropomorphism is the tendency to attribute human characteristics, such as motives, to other animal species.

Dolphins in Shark Bay, Australia, stick sponges on their rostrums to protect their sensitive skin as they probe the rough floor of the bay for food (**Figure 16.34**; Seed and Byrne 2010).

When behavioral ecologists encounter an example of complex cognition in animals, they follow Tinbergen's lead and consider both its proximate and ultimate causes. The ability to make a tool and use it requires more than just innate habits, such as pecking at a red dot. It demands that an animal recognize its long-term goal and find a way to reach it. New Caledonian crows, for example, don't simply start prying at sticks at the sight of food. In Taylor's experiment, the birds recognized a long-term goal—getting food out of a box—and then recognized that a short stick would not do the job. Reaching a goal can also demand the ability to plan ahead. New Caledonian crows transport their tools long distances before using them.

Toolmaking also demands a capacity for innovations—for coming up with new solutions that other animals haven't discovered before. New Caledonian crows can bend pieces of wire to make larva-fishing hooks, for example. Innovations require not just the ability to picture a goal, but a sense about the physical world—how different objects will function in response to gravity, friction, and other forces.

Figure 16.34 A dolphin in Shark Bay, Australia, wears a sponge to protect its rostrum as it hunts for prey on the seafloor. Dolphins are among the few lineages of animals known to make and use tools. (Ewa Krzyszczk / Shark Bay Dolphin Project)

Ecological Intelligence and Social Intelligence

Animals vary in the complexity of their cognition. Some species have a richer communication system than others. Some species show more capacity for innovation. Using tools is an especially unusual trait, found in relatively few species. It's likely that each tool-using lineage independently evolved this capacity. This variation provides empirical data against which scientists can test hypotheses about the evolution of intelligence.

Species with complex cognition tend to have large brains for their size. It's possible that larger brains can carry out more computations because they have more neurons and synapses. The link between intelligence and brain size may offer a clue to why only some animals have evolved to high levels of complex cognition. Just as there is a cost to learning, there are costs to large brains. They demand many more calories per gram than other tissues such as muscle. Only under certain conditions can natural selection readily drive the evolution of intelligence.

There are currently two categories of hypotheses that scientists generally favor for those conditions. According to the ecological intelligence hypothesis, what's most important is the challenge of finding and processing food. Some foods create a bigger cognitive challenge than others. Tropical animals that eat leaves don't have to look very hard for their next meal. For them, the big challenge is breaking down the tough plant material in large guts. But for animals that eat fruit, the next meal can be harder to find. Fruits on different trees may ripen at different times of the year. They may have to distinguish between ripe and unripe fruit. And if there's no fruit to be had, animals may have to be able to switch to other foods.

The social intelligence hypothesis, on the other hand, focuses on the demands of living in a group. Highly social animals maintain long-term social bonds, form alliances, and even deceive rivals. Keeping track of all this social information can be cognitively demanding, and the more individuals an animal interacts with, the bigger those demands may get.

But these hypotheses are not necessarily mutually exclusive. As we discussed earlier, as animals seek out food they can wind up in large social groups. And in those groups, social intelligence can become important. On the other hand, if natural selection favors social intelligence, that trait can enable animals to become better toolmakers because they can learn from other members of their groups. And as we'll see in Chapter 17, both forms of intelligence appear to have been crucial over the past few million years of evolution in our own lineage.

Key Concept   • The evolution of intelligence may be favored by ecological or social factors. •

TO SUM UP . . .

- Behavior is shaped by evolution. Scientists can observe the evolution of behavior in laboratory experiments and in the wild.

- The precursors of neurons evolved millions of years before neurons did.

- The evolution of nervous systems allowed animals to develop more sophisticated behaviors than other taxa.

- Nervous systems allow animals to learn, and learning can evolve. But just like other traits, the evolution of learning can involve trade-offs between costs and benefits.

- Organisms are reproductively selfish; they behave in ways that enhance the spread of their own genetic material, even if this behavior is detrimental to the population.

- Organisms sometimes recognize kin directly by using phenotypic markers that indicate the presence of shared alleles. They share alleles with siblings and cousins as well as with parents and offspring. For this reason, they tend to help relatives more than nonrelatives, and close relatives more than distant relatives.

- The fitness consequences of a behavior often depend on what everyone else is doing.

- Toolmaking is one example of complex cognition in animals.

- Ecological and social selection pressures may drive the evolution of intelligence.

MULTIPLE CHOICE QUESTIONS   Answers can be found at the end of the book.

1. Which of the following is considered a proximate study of behavior?

   a. A phylogenetic examination of the development of complex neural networks in hominids.

   b. Examining a hypothesis about how complex neural networks affect learning behavior in hominids.

   c. Examining a hypothesis about the origin of learning in hominids.

   d. Developing an experiment about the adaptive significance of complex neural networks in hominids.

2. If cheating can be such a successful behavioral strategy, how might slime molds remain cooperative?

   a. Group selection.

   b. Kin recognition.

   c. Inclusive fitness.

   d. Both b and c.

3. A behavioral strategy, once adopted by most of the members of a population, will sometimes be resistant to replacement by another behavioral strategy. This is known as

   a. stasis.

   b. convergent evolution.

   c. an evolutionary arms race.

   d. an evolutionarily stable strategy.

4. Which of the following is *not* a cost of living in a group?

   a. Increased competition for food.

   b. Increased probability of catching a disease.

   c. Increased probability of being singled out by a predator.

   d. Increased competition for mates.

5. If the benefits of helping to raise a sibling's offspring are relatively low, according to Hamilton's rule, what do you predict would happen to helping behavior if predators began using adult activity as a cue to where a nest could be found?

   a. Helping behavior would cease because the costs of helping would increase relative to the benefits of helping.

   b. Helping behavior would become more common because helping would increase the survival of offspring.

   c. Helping behavior would stay the same because adult survival should not affect the benefits of helping.

   d. All of the above.

6. Why is synaptic plasticity an important concept in behavioral ecology?

   a. Learning cannot be mapped onto phylogenies of closely related organisms.

   b. Behavioral ecologists don't understand how learned behaviors evolve.

c. Individuals have to have a variety of behavioral responses to situations.

d. Synaptic plasticity suggests a mechanism for the learning and memory leading to the responses that behavioral ecologists measure.

7. Are all amniotes capable of the same kinds of complex cognition?

   a. Yes. All amniotes can learn, so natural selection should shape their brains similarly when complex cognition is beneficial.

   b. Yes. Because amniotes share a common ancestor, the structure of their brains is similar, so they should all be capable of similar kinds of cognition.

   c. No. All amniotes can learn, but natural selection will shape brains for complex cognition only when it is beneficial to the species.

   d. No. Even though amniotes share a common ancestor, different types of complex cognition have evolved independently in distinct amniotic lineages.

8. An organism using phenotypic markers that indicate the presence of shared alleles to recognize kin is more likely to

   a. cooperate with relatives more than nonrelatives.

   b. favor siblings over cousins.

c. reproduce selfishly as a way of enhancing the spread of its genetic material.

d. All of the above.

9. Proximate studies of behavior focus on all of the following except

   a. how context-specific patterns of gene expression interact with physiology to elicit particular responses to stimuli.

   b. how genetic variation contributes to population differences in the expression of anatomical structures and behaviors.

   c. how genetic variation contributes to individual differences in the expression of anatomical structures and behaviors.

   d. how the relationship between variation in the expression of a behavior and fitness in a particular setting has evolved.

10. Which of the following is true about plant behavior?

   a. Plants can be sensitive to light and touch.

   b. Plants can communicate by sending and receiving signals.

   c. Plant behavior has a genetic basis that can evolve over time.

   d. All of the above.

INTERPRET THE DATA Answer can be found at the end of the book.

11. According to the figure below, males gave fewer alarm calls than females, and fewer than expected overall, simply given their respective numbers in the group. Why might the sexes differ in this fashion?

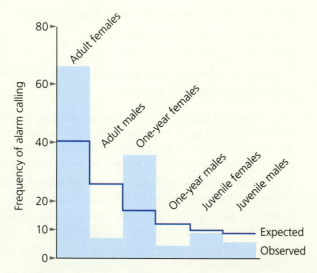

a. Alarm calling is dangerous, and males in many species take more risks than females do.

b. Females are more vulnerable to predators than males in this species.

c. Alarm calling is a form of parental care and thus more likely to have evolved in females.

d. Males tend to be unrelated to other members in their groups and thus derive no indirect fitness benefits from giving alarm calls.

SHORT ANSWER QUESTIONS Answers can be found at the end of the book.

12. As a behavioral ecologist, would you expect variation among individuals in an innate behavior, like pecking at a red dot? Based on your understanding, why might natural selection lead to the evolution of an innate behavior?

13. Does Hamilton's rule support the evolution of altruistic behavior through group selection or individual selection? Why?

14. In which system would you predict finding stronger kin recognition behaviors in males: a role-reversed breeding system where males assume all of the parental care or a breeding system where males and females cooperate in parental care?

15. How are the social behaviors of eusocial organisms and the slime mold *Dictyostelium* similar? How are they different?

16. Do individual side-blotched lizard males make choices about the optimal reproductive strategy to employ based on what other males are doing? Why do all three strategies tend to persist in natural populations (that is, why doesn't one strategy win)?

17. Why is understanding costs and benefits important for understanding the evolution of behavioral traits? How do these trade-offs affect tool use specifically?

ADDITIONAL READING

Alcock, J. 2009. *Animal Behavior: An Evolutionary Approach*. 9th ed. Sunderland, MA: Sinauer Associates.

Braam, J. 2005. In Touch: Plant Responses to Mechanical Stimuli. *New Phytologist* 165 (2): 373–89.

Clutton-Brock, T. 2002. Breeding Together: Kin Selection and Mutualism in Cooperative Vertebrates. *Science* 296 (5565): 69–72.

Emlen, S. T. 1995. An Evolutionary Theory of the Family. *Proceedings of the National Academy of Sciences USA* 92 (18): 8092–9.

Fitzpatrick, M. J., Y. Ben-Shahar, H. M. Smid, L. E. M. Vet, G. E. Robinson, et al. 2005. Candidate Genes for Behavioural Ecology. *Trends in Ecology and Evolution* 20 (2): 96–104.

Krebs, J. R., and N. Davies. 1997. *Behavioural Ecology: An Evolutionary Approach*. 4th ed. Hoboken, NJ: Wiley-Blackwell.

Robinson, G. E., R. D. Fernald, and D. F. Clayton. 2008. Genes and Social Behavior. *Science* 322 (5903): 896–900.

Seed, A., and R. Byrne. 2010. Animal Tool Use. *Current Biology* 20 (23): R1032–9.

Sherman, P., and J. Alcock. 2010. *Exploring Animal Behavior: Readings from American Scientist*. 5th ed. Sunderland, MA: Sinauer Associates.

PRIMARY LITERATURE CITED IN CHAPTER 16

Alcock, J. 2009. *Animal Behavior: An Evolutionary Approach*. 9th ed. Sunderland, MA: Sinauer Associates.

Alexander, R. D. 1974. The Evolution of Social Behavior. *Annual Review of Ecology and Systematics* 5:325–83.

Amodio, P., M. Boeckle, A. K. Schnell, L. Ostojić, G. Fiorito, et al. 2018. Grow Smart and Die Young: Why Did Cephalopods Evolve Intelligence? *Trends in Ecology and Evolution*. https://doi.org/10.1016/j.tree.2018.10.010.

Anderson, J. G. T. 1991. Foraging Behavior of the American White Pelican (*Pelecanus erythrorhyncos*) in Western Nevada. *Colonial Waterbirds* 14 (2): 166–72.

Bateson, P., and K. N. Laland. 2013. Tinbergen's Four Questions: An Appreciation and an Update. *Trends in Ecology and Evolution* 28:712–8.

Bell, A. M., and N. A. Dochtermann. 2015. Integrating Molecular Mechanisms into Quantitative Genetics to Understand Consistent Individual Differences in Behavior. *Current Opinion in Behavioral Sciences* 6:111–4.

Bonner, J. T. 1967. *The Cellular Slime Molds*. Princeton, NJ: Princeton University Press.

Braam, J. 2005. In Touch: Plant Responses to Mechanical Stimuli. *New Phytologist* 165 (2): 373–89.

Brown, C. R., and M. B. Brown. 2004. Empirical Measurement of Parasite Transmission Between Groups in a Colonial Bird. *Ecology* 85 (6): 1619–26.

Burger, J. M. S., M. Kolss, J. Pont, and T. J. Kawecki. 2008. Learning Ability and Longevity: A Symmetrical Evolutionary Trade-Off in *Drosophila*. *Evolution* 62 (6): 1294–1304.

Carbone, C., and G. M. Narbonne. 2014. When Life Got Smart: The Evolution of Behavioral Complexity Through the Ediacaran and Early Cambrian of NW Canada. *Journal of Paleontology* 88:309–30.

Chapman, T. W., B. J. Crespi, B. D. Kranz, and M. P. Schwarz. 2000. High Relatedness and Inbreeding at the Origin of Eusociality in Gall-Inducing Thrips. *Proceedings of the National Academy of Sciences USA* 97 (4): 1648–50.

Clutton-Brock, T. 2002. Breeding Together: Kin Selection and Mutualism in Cooperative Vertebrates. *Science* 296 (5565): 69–72.

Conservation International. 2007. Biodiversity Hotspots. http://www.conservation.org/How/Pages/Hotspots.aspx (accessed January 11, 2018).

Cowden, C. C. 2012. Game Theory, Evolutionary Stable Strategies and the Evolution of Biological Interactions. *Nature Education Knowledge* 3:6.

Creel, S. 1997. Cooperative Hunting and Group Size: Assumptions and Currencies. *Animal Behaviour* 54 (5): 1319–24.

Dao, D. N., R. H. Kessin, and H. L. Ennis. 2000. Developmental Cheating and the Evolutionary Biology of *Dictyostelium* and *Myxococcus. Microbiology* 146 (7): 1505–12.

Di-Poi, C., J. Lacasse, S. M. Rogers, and N. Aubin-Horth. 2014. Extensive Behavioural Divergence Following Colonisation of the Freshwater Environment in Threespine Sticklebacks. *PloS ONE* 9:e98980.

Dobzhansky, T., and B. Spassky. 1969. Artificial and Natural Selection for Two Behavioral Traits in *Drosophila pseudoobscura. Proceedings of the National Academy of Sciences* 62 (1): 75–80.

Drent, P. J., K. van Oers, and A. J. van Noordwijk. 2003. Realized Heritability of Personalities in the Great Tit (*Parus major*). *Proceedings of the Royal Society of London Series B: Biological Sciences* 270 (1510): 45–51.

Dugatkin, L. A., and L. Trut. 2017. *How to Tame a Fox (and Build a Dog): Visionary Scientists and a Siberian Tale of Jump-Started Evolution.* Chicago: University of Chicago Press.

Ellison, A. M., and N. J. Gotelli. 2009. Energetics and the Evolution of Carnivorous Plants—Darwin's "Most Wonderful Plants in the World." *Journal of Experimental Botany* 60 (1): 19–42.

Emlen, S. T., and P. H. Wrege. 1988. The Role of Kinship in Helping Decisions Among White-Fronted Bee-Eaters. *Behavioral Ecology and Sociobiology* 23 (5): 305–15.

Emlen, S. T., and P. H. Wrege. 1989. A Test of Alternate Hypotheses for Helping Behavior in White-Fronted Bee-Eaters of Kenya. *Behavioral Ecology and Sociobiology* 25 (5): 303–19.

Fitzpatrick, M. J., Y. Ben-Shahar, H. M. Smid, L. E. M. Vet, G. E. Robinson, et al. 2005. Candidate Genes for Behavioural Ecology. *Trends in Ecology and Evolution* 20 (2): 96–104.

Gammie, S., T. Garland, and S. Stevenson. 2006. Artificial Selection for Increased Maternal Defense Behavior in Mice. *Behavior Genetics* 36 (5): 713–22.

Garland, J. T. 2003. Selection Experiments: An Under-utilized Tool in Biomechanics and Organismal Biology. In V. L. Bels et al. (eds.), *Vertebrate Biomechanics and Evolution* (pp. 23–56). Oxford: BIOS Scientific Publishers Ltd.

Gibson, T. C., and D. M. Waller. 2009. Evolving Darwin's "Most Wonderful" Plant: Ecological Steps to a Snap-Trap. *New Phytologist* 183:575–87.

Greenwood, A. K., M. G. Mills, A. R. Wark, S. L. Archambeault, and C. L. Peichel. 2016. Evolution of Schooling Behavior in Threespine Sticklebacks Is Shaped by the *Eda* Gene. *Genetics* 203:677–81.

Hamilton, W. D. 1964. The Genetical Evolution of Social Behaviour. I. *Journal of Theoretical Biology* 7 (1): 1–16.

———. 1996. *Narrow Roads of Gene Land: The Collected Papers of W. D. Hamilton.* Oxford: W. H. Freeman/Spektrum.

Hauber, M. E., and P. W. Sherman. 2001. Self-Referent Phenotype Matching: Theoretical Considerations and Empirical Evidence. *Trends in Neurosciences* 24 (10): 609–16.

Hobaiter, C., and R. W. Byrne. 2014. The Meanings of Chimpanzee Gestures. *Current Biology* 24:1596–1600.

Hölldobler, B., and E. O. Wilson. 2009. *The Superorganism: The Beauty, Elegance, and Strangeness of Insect Societies.* New York: W. W. Norton.

Hunt, G. R. 2000. Human-like, Population-Level Specialization in the Manufacture of Pandanus Tools by New Caledonian Crows *Corvus moneduloides. Proceedings of the Royal Society B: Biological Sciences* 267 (1441): 403–13.

Janik, V. M., L. S. Sayigh, and R. S. Wells. 2006. Signature Whistle Shape Conveys Identity Information to Bottlenose Dolphins. *Proceedings of the National Academy of Sciences USA* 103 (21): 8293–7.

Jékely, G. 2011. Origin and Early Evolution of Neural Circuits for the Control of Ciliary Locomotion. *Proceedings of the Royal Society Series B: Biological Sciences* 278:914–22.

Kaas, J. H. 2013. The Evolution of Brains from Early Mammals to Humans. *Wiley Interdisciplinary Reviews: Cognitive Science* 4:33–45.

Kandel, E. R. 2007. *In Search of Memory: The Emergence of a New Science of Mind.* New York: W. W. Norton.

Karban, R. 2008. Plant Behaviour and Communication. *Ecology Letters* 11 (7): 727–39.

Karban, R., I. T. Baldwin, K. J. Baxter, G. Laue, and G. W. Felton. 2000. Communication Between Plants: Induced Resistance in Wild Tobacco Plants Following Clipping of Neighboring Sagebrush. *Oecologia* 125 (1): 66–71.

Kardong, K. V. 2002. *Vertebrates: Comparative Anatomy, Function, Evolution.* New York: McGraw-Hill.

Kent, C. F., T. Daskalchuk, L. Cook, M. B. Sokolowski, and R. J. Greenspan. 2009. The *Drosophila* Foraging Gene Mediates Adult Plasticity and Gene-Environment Interactions in Behaviour, Metabolites, and Gene Expression in Response to Food Deprivation. *PLoS Genetics* 5 (8):e1000609.

Kruuk, H. 2003. *Niko's Nature: The Life of Niko Tinbergen and His Science of Animal Behaviour.* Oxford: Oxford University Press.

Kukekova, A. V., G. M. Acland, I. N. Oskina, A. V. Kharlamova, L. N. Trut, et al. 2006. The Genetics of Domesticated Behavior in Canids: What Can Dogs and Silver Foxes Tell Us About Each Other? *Cold Spring Harbor Monograph Archive* 44:515–37.

Kukekova, A. V., J. L. Johnson, X. Xiang, S. Feng, S. Liu, et al. 2018. Red Fox Genome Assembly Identifies Genomic Regions Associated with Tame and Aggressive Behaviours. *Nature Ecology and Evolution* 2 (9):1479.

Kukekova, A. V., S. V. Temnykh, J. L. Johnson, L. N. Trut, and G. M. Acland. 2012. Genetics of Behavior in the Silver Fox. *Mammalian Genome* 23:164–77.

Kutschera, U., and K. J. Niklas. 2009. Evolutionary Plant Physiology: Charles Darwin's Forgotten Synthesis. *Naturwissenschaften* 96:1339–54.

Liebeskind, B. J., H. A. Hofmann, D. M. Hillis, and H. H. Zakon. 2018. Evolution of Animal Neural Systems. *Annual Review of Ecology, Evolution, and Systematics* 48:377–98.

Lynch, C. B. 1980. Response to Divergent Selection for Nesting Behavior in *Mus musculus. Genetics* 96 (3): 757–65.

Marshall, J. A. 2011. Group Selection and Kin Selection: Formally Equivalent Approaches. *Trends in Ecology and Evolution* 26:325–32.

Maynard Smith, J. 1964. Group Selection and Kin Selection. *Nature* 201 (4924): 1145–7.

McGlothlin, J. W., J. B. Wolf, E. D. Brodie III, and A. J. Moore. 2014. Quantitative Genetic Versions of Hamilton's Rule with Empirical Applications. *Philosophical Transactions of the Royal Society B* 369:20130358.

McGrew, W. C. 2010. Evolution. Chimpanzee Technology. *Science* 328 (5978): 579–80.

Mery, F., and T. J. Kawecki. 2003. A Fitness Cost of Learning Ability in *Drosophila melanogaster*. *Proceedings of the Royal Society B: Biological Sciences* 270:2465–9.

Mills, M. G., A. K. Greenwood, and C. L. Peichel. 2014. Pleiotropic Effects of a Single Gene on Skeletal Development and Sensory System Patterning in Sticklebacks. *EvoDevo* 5:5.

Moroz, L. L., K. M. Kocot, M. R. Citarella, S. Dosung, T. P. Norekian, et al. 2014. The Ctenophore Genome and the Evolutionary Origins of Neural Systems. *Nature* 510:109–14.

Nelson, R. M., S. V. Temnykh, J. L. Johnson, A. V. Kharlamova, A. V. Vladimirova, et al. 2017. Genetics of Interactive Behavior in Silver Foxes (*Vulpes vulpes*). *Behavior Genetics* 47 (1): 88–101.

Olson, R. S., A. Hintze, F. C. Dyer, D. B. Knoester, and C. Adami. 2013. Predator Confusion Is Sufficient to Evolve Swarming Behaviour. *Journal of the Royal Society Interface* 10:20130305.

Pollick, A. S., and F. B. M. de Waal. 2007. Ape Gestures and Language Evolution. *Proceedings of the National Academy of Sciences USA* 104 (19): 8184–9.

Poppinga, S., S. R. H. Hartmeyer, R. Seidel, T. Masselter, I. Hartmeyer, et al. 2012. Catapulting Tentacles in a Sticky Carnivorous Plant. *PLoS ONE* 7 (9): e45735.

Robinson, G. E., R. D. Fernald, and D. F. Clayton. 2008. Genes and Social Behavior. *Science* 322 (5903): 896–900.

Schaap, P. 2007. Evolution of Size and Pattern in the Social Amoebas. *BioEssays* 29 (7): 635–44.

Seed, A., and R. Byrne. 2010. Animal Tool Use. *Current Biology* 20 (23): R1032–9.

Seyfarth, R. M., D. L. Cheney, and P. Marler. 1980. Vervet Monkey Alarm Calls: Semantic Communication in a Free-Ranging Primate. *Animal Behaviour* 28 (4): 1070–94.

Sherman, P. W. 1977. Nepotism and the Evolution of Alarm Calls. *Science* 197 (4310): 1246–53.

———. 1985. Alarm Calls of Belding's Ground Squirrels to Aerial Predators: Nepotism or Self-Preservation? *Behavioral Ecology and Sociobiology* 17 (4): 313–23.

Shu, D. G., S. C. Morris, J. Han, Z. F. Zhang, K. Yasui, et al. 2003. Head and Backbone of the Early Cambrian Vertebrate *Haikouichthys*. *Nature* 421 (6922): 526–9.

Sinervo, B., and C. M. Lively. 1996. The Rock-Paper-Scissors Game and the Evolution of Alternative Male Strategies. *Nature* 380 (6571): 240–3.

Smith, R. L. 1979. Paternity Assurance and Altered Roles in the Mating Behaviour of a Giant Water Bug, *Abedus herberti* (Heteroptera: Belostomatidae). *Animal Behaviour* 27 (3): 716–25.

Srivastava, M., O. Simakov, J. Chapman, B. Fahey, M. E. A. Gauthier, et al. 2010. The *Amphimedon queenslandica* Genome and the Evolution of Animal Complexity. *Nature* 466 (7307): 720–6.

Stern, D. L., and W. A. Foster. 1996. The Evolution of Soldiers in Aphids. *Biological Reviews* 71 (1): 27–79.

Strassmann, J. E. 2010. *Dictyostelium,* the Social Amoeba. In M. D. Breed and J. Moore (eds.), *Encyclopedia of Animal Behavior* (pp. 513–9). Oxford: Academic Press.

Swallow, J. G., P. A. Carter, and T. Garland. 1998. Artificial Selection for Increased Wheel-Running Behavior in House Mice. *Behavior Genetics* 28 (3): 227–37.

Switzer, C. M., S. A. Combes, and R. Hopkins. 2018. Dispensing Pollen via Catapult: Explosive Pollen Release in Mountain Laurel (*Kalmia latifolia*). *The American Naturalist* 191:767–76.

Tanaka, G., X. Hou, X. Ma, G. D. Edgecombe, and N. J. Strausfeld. 2013. Chelicerate Neural Ground Pattern in a Cambrian Great Appendage Arthropod. *Nature* 502:364–7.

Taylor, A. H., D. Elliffe, G. R. Hunt, and R. D. Gray. 2010. Complex Cognition and Behavioural Innovation in New Caledonian Crows. *Proceedings of the Royal Society B: Biological Sciences* 277 (1694): 2637–43.

ten Cate, C. 2009. Niko Tinbergen and the Red Patch on the Herring Gull's Beak. *Animal Behaviour* 77:785–94.

Tinbergen, N. 1963. On Aims and Methods of Ethology. *Zeitschrift für Tierpsychologie* 20:410–33.

Tinbergen, N., and A. C. Perdeck. 1950. On the Stimulus Situation Releasing the Begging Response in the Newly Hatched Herring Gull Chick (*Larus argentatus argentatus* Pont.). *Behaviour* 3:1–39.

Tosches, M. A., and D. Arendt. 2013. The Bilaterian Forebrain: An Evolutionary Chimaera. *Current Opinion in Neurobiology* 23:1080–9.

Wang, X., L. Pipes, L. N. Trut, Y. Herbeck, A. V. Vladimirova, et al. 2018. Genomic Responses to Selection for Tame/Aggressive Behaviors in the Silver Fox (*Vulpes vulpes*). *Proceedings of the National Academy of Sciences USA* 115 (41): 10398–403.

Wark, A. R., A. K. Greenwood, E. M. Taylor, K. Yoshida, and C. L. Peichel. 2011. Heritable Differences in Schooling Behavior Among Threespine Stickleback Populations Revealed by a Novel Assay. *PLoS ONE* 6 (3): e18316.

Weber, J. N., B. K. Peterson, and H. E. Hoekstra. 2013. Discrete Genetic Modules Are Responsible for Complex Burrow Evolution in *Peromyscus* Mice. *Nature* 493:402–5.

Williams, G. C. 1966. *Adaptation and Natural Selection: A Critique of Some Current Evolutionary Thought*. Princeton, NJ: Princeton University Press.

Wynne, C. D. L. 2004. The Perils of Anthropomorphism. *Nature* 428 (6983): 606.

Wynne-Edwards, V. C. 1962. *Animal Dispersion in Relation to Social Behaviour*. Edinburgh: Oliver and Boyd.

CHAPTER 17

Human Evolution

A New Kind of Ape

Learning Objectives

- Describe Darwin's evidence for classifying humans as primates, and contrast this evidence with the more recent data that scientists now use.
- Explain the major splits in the primate lineage.
- Discuss two hypotheses for the evolution of bipedalism.
- Explain how living primates can help us understand fossil hominins.
- Discuss the importance of toolmaking in the evolution of humans.
- Compare and contrast the anatomy of australopithecines and *Homo*.
- Explain the scientific debate about the placement of *Homo naledi* in the human lineage.
- Compare and contrast *Homo neanderthalensis* and *Homo sapiens*.
- Explain how DNA can be used to examine the relationship between Neanderthals and modern humans.
- Describe some of the selective pressures that led to the evolution of the human brain.
- Explain the different ways in which scientists can study the evolution of language.
- Explain how selection is currently acting on human maternity.
- Discuss the homologies between humans and other mammals that underlie our emotions.

Homo naledi, an extinct relative of humans, was discovered in 2013 in a South African cave. It had a brain about a third the size of ours and may have lived as recently as 236,000 years ago, when our own species had already evolved.

I n November 2013, Marina Elliott made her way deep inside a South African cave called Rising Star. Her ultimate destination was a chamber deep inside the cave, which was accessible only by a vertical chute measuring 12 meters long but just 18 centimeters wide. Elliott gazed down the steep narrow chute, lined with stony spikes, and wondered if she had made a wise life choice (Figure 17.1).

Elliott was no stranger to inhospitable places. As part of her work in graduate school at Simon Fraser University, she had already traveled to Siberia and the northern coast of Alaska to

Figure 17.1 Archaeologist Marina Elliott navigated dangerous tunnels to reach a chamber in a South African cave containing extraordinary fossils of *Homo naledi*. (Photo: S. Tucker, image courtesy of M. Elliott and The University of the Witwatersrand)

dig up human remains. Then she heard about a new expedition in South Africa to search for fossils of a mysterious extinct species related to humans. That's how she wound up in such a tight spot.

The expedition was organized by Lee Berger, a paleoanthropologist at the University of Witwatersrand. He had been surveying caves across South Africa for fossils, and in October 2013 two of his team members visited Rising Star. One of the cavers, Steve Tucker, wriggled down the chute into a chamber. When Tucker cast a light around the chamber, he saw bones strewn on the cave floor, which he quickly photographed.

As soon as Berger saw Tucker's photographs, he suspected the bones in the chamber belonged to apelike human relatives. To find out for sure, a team of skilled archaeologists and paleoanthropologists would have to wriggle their way into the chamber and carefully excavate the remains.

Berger put together a team by posting on Facebook. He was looking for people not just with scientific skills but with extensive experience in caving or climbing. And they had to be small and skinny so that they would be less likely to get stuck in the narrow chute.

In very little time, Berger was flooded with dozens of applicants. He selected six people for the team—Elliott and five other women. In November 2013, this remarkable crew clambered through the Rising Star cave and made their way into the fossil chamber.

Working 6-hour shifts day after day, Elliott and her colleagues continued to bring bones out of the cave for other scientists to study. It would have been remarkable enough to find just one skeleton of an ancient human relative. On their first sojourn, Elliott and her colleagues discovered 1550 bones from at least 15 individuals (**Figure 17.2**).

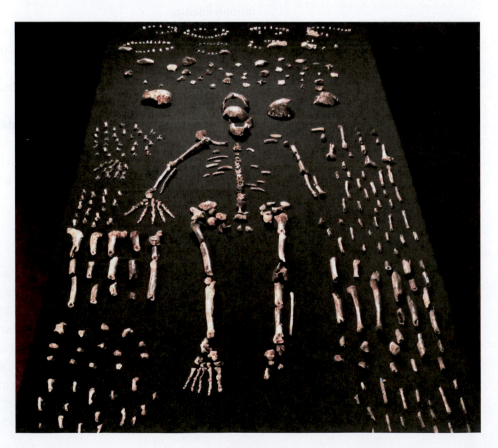

Figure 17.2 Elliott and her colleagues found more than a thousand bones of *Homo naledi*. They discovered it had a mixture of primitive and derived traits. (John Hawks)

In subsequent expeditions, the researchers found more remains in other parts of the Rising Star cave. Elliott, now a post-doctoral researcher at Witwatersrand, and her colleagues have analyzed the material in laboratories, documenting the shapes of the bones and determining how they fit together in skeletons. And the researchers have compared the bones to corresponding bones in our own skeletons, as well as to the skeletons of extinct human relatives—collectively known as **hominins**.

The results of these analyses have been nothing less than astonishing (Yong 2015; Berger et al. 2017). The fossils belong to a new species, which the researchers have named *Homo naledi*. It stood only 1.5 meters tall, about the height of a 12-year-old human child. *H. naledi* had legs adapted for an upright stance and toes and arches to allow it to walk in a humanlike way. But its brain was tiny, and its long fingers show signs of possibly being adapted to climbing in trees. Many other features of *Homo naledi's* anatomy were also primitive hominin traits, previously observed only in species that lived millions of years ago.

But *Homo naledi* was not anywhere near as old. The researchers determined that the *Homo naledi* individuals lived sometime between 236,000 and 335,000 years ago. Our own species, *Homo sapiens,* evolved around then. In other words, humans may have lived side by side with the remarkable *H. naledi*.

Our own species is but one of the millions of species that exist on Earth today. Each species—whether it be the duck-billed platypus, the Venus flytrap, or the death cap mushroom—has its own intriguing evolutionary history. But when we humans look at the history of life, we can't help but wonder how our own species came to be. In this chapter, we will explore what scientists have learned about our evolutionary history. We'll start by looking at our primate relatives and then explore the hominin fossil record—including new discoveries like *Homo naledi*. Additional insights are coming from DNA analysis—from both the genes in the cells of living people and those extracted from ancient fossils. Finally, we'll consider the insights this research gives us about the human mind, certainly one of evolution's most remarkable inventions. ●

A **hominin** is a member of the clade that includes humans as well as all species more closely related to humans than chimpanzees. Humans are the only surviving members of the hominin clade.

17.1 Primate Evolution

When Carolus Linnaeus organized all living things into a single classification system in the mid-1700s, he put humans in the order Primates, along with other species such as apes, monkeys, and lemurs. His reasoning was that humans and other primates shared a number of anatomical traits, from their forward-facing eyes to their gripping thumbs.

That placement was, Linnaeus knew, a controversial thing to do. "It is not pleasing to me that I must place humans among the primates," he wrote in a letter to a fellow naturalist in 1747, "but I desperately seek from you and from the whole world a general difference between men and simians from the principles of Natural History. I certainly know of none. If only someone might tell me one!" (Linnaeus 1747).

Discovering Our Primate Kinship

A century later, Charles Darwin figured out why humans belonged among the primates: they shared a common ancestor with apes, monkeys, and lemurs. But in

The Origin of Species, he said next to nothing about this kinship. "Light will be thrown on the origin of man and his history" was about all he wrote (Darwin 1859).

Historians would later discover why. In a letter to a friend, Darwin wrote that he feared delving deeply into human evolution would prejudice readers against his theory. It was one thing to say that whales had evolved from bearlike terrestrial mammals. It was quite another to say that humans were related to apes.

In 1871 Darwin finally felt the time was right to share his thoughts on human evolution, which he did in a book called *The Descent of Man*. Darwin pointed out many similarities between humans and primates, which he said were best explained as homologies. Some homologous structures, such as our feet and skulls, had obvious similarities. Other homologies were subtler—but just as revealing. Darwin drew his readers' attention, for example, to the base of the human spine, known as the coccyx (**Figure 17.3**). In humans, the coccyx anchors some of the muscles we use to maintain our upright posture. But a comparison to other primates shows a clear homology between the coccyx and primate tails. Darwin argued that the external portion of the primate tail disappeared in apes. This transformation explains why some babies are still born with a stump of a tail emerging from the base of their backs.

Molecular and Fossil Evidence

Today, scientists can look at another line of evidence that Darwin could not. They can compare the genes of humans to those of other species to reconstruct our evolution. And their research has overwhelmingly supported Darwin's hypothesis. **Figure 17.4** shows a phylogeny of primates produced by Mario dos Reis, of Queen Mary University of London, and his colleagues in 2018 (Reis et al. 2018). To generate it, they analyzed the complete genomes of nine primate species, along with smaller samplings of DNA from 358 other primate species. Primates form a monophyletic clade, which branched off from other mammal lineages. Humans share many genetic similarities with primates, revealing that they are deeply nested within the primate clade.

Fossils from the main primate lineages allowed dos Reis and his colleagues to calibrate a molecular clock and estimate when those lineages diverged. They estimated that the most recent common ancestor of all living primates lived at the end of the Cretaceous period, between 70 and 79.2 million years ago. The oldest primate fossils date back 55 million years, leaving a substantial gap for paleontologists to fill in future fossil-hunting expeditions. As of now, the fossil record indicates that the earliest primates were small, long-tailed creatures whose hands and feet were well adapted to

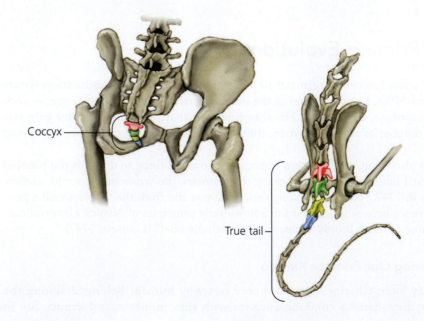

Figure 17.3 The base of the human spine (top) corresponds, bone for bone, to the base of a monkey's tail (bottom). The best explanation for this homology is that we descend from a common ancestor that had a tail. The ancestor of humans and other apes lost a full-blown tail but retained its vestiges. (Data from Coyne 2009)

Coccyx

True tail

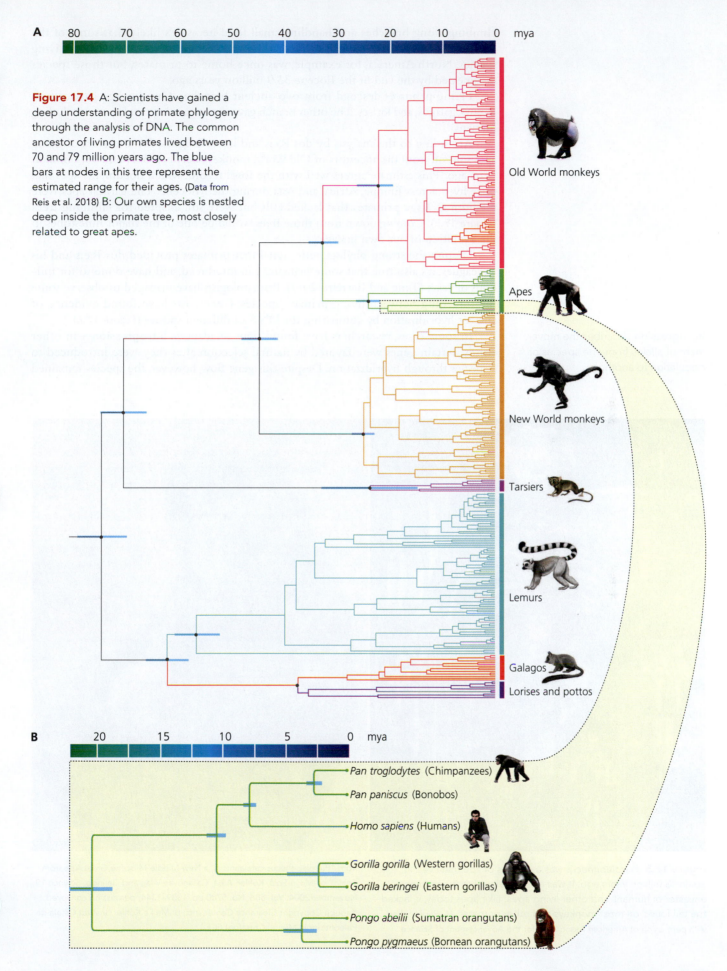

Figure 17.4 A: Scientists have gained a deep understanding of primate phylogeny through the analysis of DNA. The common ancestor of living primates lived between 70 and 79 million years ago. The blue bars at nodes in this tree represent the estimated range for their ages. (Data from Reis et al. 2018) B: Our own species is nestled deep inside the primate tree, most closely related to great apes.

A

80 70 60 50 40 30 20 10 0 mya

Old World monkeys

Apes

New World monkeys

Tarsiers

Lemurs

Galagos

Lorises and pottos

B

20 15 10 5 0 mya

Pan troglodytes (Chimpanzees)

Pan paniscus (Bonobos)

Homo sapiens (Humans)

Gorilla gorilla (Western gorillas)

Gorilla beringei (Eastern gorillas)

Pongo abeilii (Sumatran orangutans)

Pongo pygmaeus (Bornean orangutans)

climbing along branches and handling small food items. It's likely that some of the earliest primate fossils belonged to extinct lineages that aren't closely related to living species. North America, for example, was once home to primates, but those species disappeared by the end of the Eocene 33.9 million years ago.

Living primates descend from two ancient branches. One lineage gave rise to lemurs, galagos, and lorises. The other branch gave rise to monkeys and apes—including humans.

According to the analysis by dos Reis and his colleagues, the ancestors of living apes split off from the ancestors of Old World monkeys between 30.4 and 35.1 million years ago. This estimate agrees well with the fossil record, which documents apes that once lived across Europe, Africa, and Asia during the Miocene. The earliest apes were medium to large primates that lacked tails but were still adapted for climbing trees (**Figure 17.5**). Coming down from those trees would be one of the major milestones in the evolution of our own lineage.

Despite the strong phylogenetic signal that primates provided dos Reis and his colleagues, it's also true that some primates can interbreed, and have done so for millions of years (Tung and Barreiro 2017). Primatologists have managed to observe some interbreeding between living primate species. Geneticists have found evidence of ancient hybridization by comparing the DNA of different species (**Figure 17.6**).

In some cases, researchers have found signs of adaptive **introgression**—in other words, certain genes were favored by natural selection after they were introduced to a species through hybridization. Despite this gene flow, however, the species remained

Introgression describes the movement of alleles from one species or population to another.

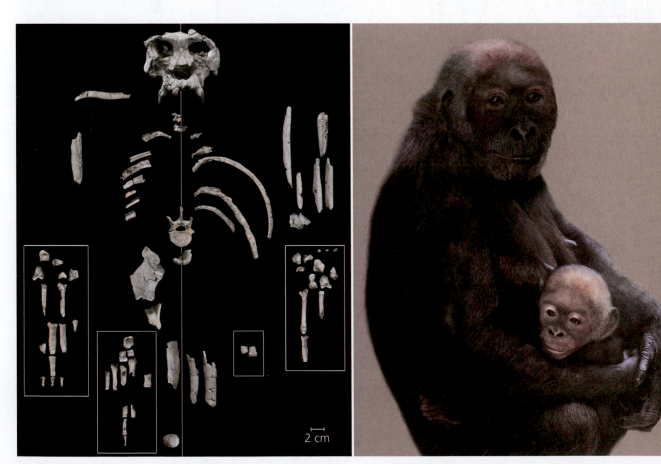

2 cm

Figure 17.5 *Pierolapithecus* was a species of ape that lived in Spain 13 million years ago. It was closely related to the common ancestor of humans and other living apes. Like apes today, it lacked the tail found on most monkeys and other primates. (A: Republished with permission of American Association for the Advancement of Science

from "Pierolapithecus catalaunicus, a New Middle Miocene Great Ape from Spain" by Moyà-Solà, Köhler, Alba, Casanovas-Vilar, and Galindo. Science 19 November 2004: Vol. 306, No. 5700 pp. 1339–1344; permission conveyed through Copyright Clearance Center, Inc.; B: Meike Köhler/Institut Català de Paleontologia Miquel Crusafont (ICP))

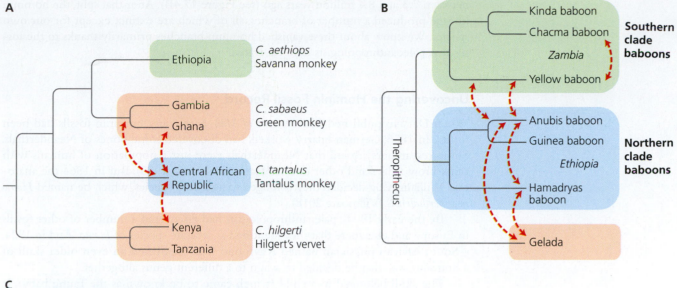

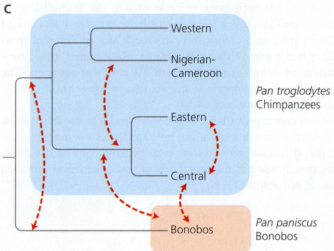

Figure 17.6 While primates have evolved into distinct lineages, some species have interbred, exchanging genes. (Data from Tung and Barreiro 2017)

distinct. As we'll see later in this chapter, hybridization and adaptive introgression have also been important factors in hominin evolution.

Key Concepts

- Independent lines of evidence from morphology and DNA confirm that humans are primates.
- The primate clade originated at the end of the Cretaceous period.
- Within the primate clade, the closest living relatives of humans are chimpanzees and other apes. ●

17.2 The Hominin Lineage

The study by dos Reis and colleagues supports Darwin's theory that humans share a common ancestry with other primates, and it also supports Darwin's proposal that the closest relatives to humans are the great apes of Africa. Darwin based this proposal on the morphological similarities humans shared with apes, to the exclusion of other primates, such as the lack of an external tail. It turns out that apes share genetic markers of a close ancestry as well. In their study, dos Reis and his colleagues estimate that the ancestors of humans diverged from the ancestors of chimpanzees and bonobos between 7.3 and 8.4 million years ago (see Figure 17.4B). After that split, the hominin lineage produced a number of branches, all of which are extinct except for our own species. We know about these vanished hominin branches primarily thanks to the fossils that paleoanthropologists have excavated.

Uncovering the Hominin Fossil Record

When Darwin published *The Descent of Man*, barely any hominin fossils had been found. In 1856, German quarry workers had discovered some bones of Neanderthals. Some researchers argued that Neanderthals were just a population of humans with thick brow ridges and other unusual features (Shreeve 1995). But in 1864 the anatomist William King decided they belonged to a distinct species, which he named *Homo neanderthalensis* (Villmoare 2018).

By the early 1900s, paleoanthropologists had discovered a number of other fossils in Europe and elsewhere that they identified as separate species of *Homo*. And in 1924, a South African physician named Raymond Dart identified an even older skull of a hominin, one that he decided to assign to a different genus altogether.

The skull belonged to a child (which came to be known as the Taung baby). It had forward-facing eyes and a small jaw, much like humans have. But the skull also had many traits that linked it with apes, such as a small braincase. It was so different from us, Dart concluded, that it was not part of the genus *Homo*. He assigned it to a new genus he called *Australopithecus*, meaning "ape of the south." Within that new genus, he dubbed the Taung baby *Australopithecus africanus*, a species that we now know existed from 3 to 2.1 million years ago (Pickering et al. 2018).

For most of the twentieth century, this basic taxonomy served as the framework for interpreting hominin evolution: the first hominins belonged to *Australopithecus*, the thinking went, and from that primitive genus *Homo* later evolved. But by the dawn of the twenty-first century, new fossils and new methods to study them had provided a far more complex picture of our origins.

Hominin Diversity

When paleoanthropologists find a new hominin fossil, they examine its proportions to determine its phylogenetic relationships. Sometimes it belongs to a species that's already known to science. In 2017, for example, scientists announced they had found a 300,000-year-old fossil of our own species, *Homo sapiens*, in Morocco (Hublin et al. 2017). That discovery pushed back the history of our species by about 100,000 years (Chapter 3).

In other cases, researchers recognize that they've found a new hominin species. In 1974, researchers digging in Ethiopia found a remarkably complete hominin they nicknamed Lucy (in honor of the song "Lucy in the Sky with Diamonds," which was playing around the time they found it). Close inspection of Lucy showed she had some similarities to the Taung child—but some distinctive traits, too. The researchers assigned Lucy to a new species, *Australopithecus afarensis*. And since then, other researchers have found much older hominins that are different enough from *Australopithecus* to assign them to genera of their own. So far, the oldest hominin is *Sahelanthropus*, estimated to be about seven million years old.

Quite often, a newly discovered hominin fossil will leave researchers struggling to determine if it belongs to a known species or to a new one (White 2009; Haile-Selassie et al. 2016; de Ruiter et al. 2017). The best way to resolve such a question is to look at a range of fossils within each species to determine the extent of variation for that species. But hominin fossils are scarce compared to other taxa. If a species is known from only a single fossil, it can be hard to know if a new fossil belongs to that species as well. Making matters worse, most hominin fossils are fragments rather than complete skeletons. Frequently, researchers have to diagnose hominin species based on just a skull—or perhaps even a just a fragment of a jaw.

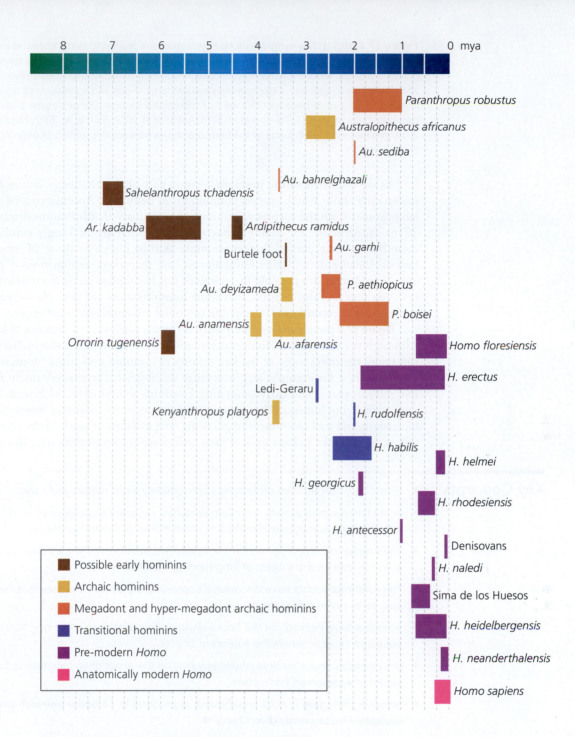

Figure 17.7 Paleoanthropologists have discovered about 30 hominin taxa. Although the phylogeny of hominins remains challenging, these taxa can be grouped according to broad similarities. Italicized names refer to recognized species. Names in regular type, such as Burtele foot, refer to fossils that have not yet been assigned to existing species or formally described as a new species. (Data from Wood and Boyle 2016)

These caveats must be kept in mind when we look at **Figure 17.7**, which shows a synthesis of our current understanding of the hominin fossil record (modified from Wood and Boyle 2016). There are 31 taxa in this figure, many of which coexisted. This figure demonstrates that hominin evolution was not a simple, linear march of progress with one species giving rise to the next. Instead, it has a bushy structure, with several branches simultaneously diversifying.

Hominin Phylogeny

Figure 17.7 does not offer a hypothesis about hominin phylogeny. Although paleo-anthropologists have been able to chart the broad outlines of hominin evolution, a completely resolved evolutionary tree remains elusive.

Figure 17.8 (right) This evogram shows how humans are related to living apes and extinct hominins. The earliest fossils of hominins suggest that our ancestors were partially bipedal seven million years ago. More evidence for bipedality emerges in hominins that lived between four and three million years ago. Later hominins evolved large brains and sophisticated tools. The bonobo arrow is white because there are no fossils of this lineage. The phylogeny shown here is based on Strait et al. 2015; Dembo et al. 2016; Argue et al. 2017; Slon et al. 2017; Hajdinjak et al. 2018; and Reis et al. 2018. Some of the branches shown here are the subject of ongoing scientific debate.

As we discussed in the previous section, part of the challenge comes from the uncertainty about which species hominin fossils belong to. Another part of the challenge comes from the surprising amount of homoplasy that hominins display (Foley 2016). *Homo naledi*, for example, has wrist bones that are strikingly similar to our own. But its long, curved fingers are more akin to Lucy's or those of other members of *Australopithecus*. Other hominin taxa have their own combinations of primitive and derived traits. It's possible that some of this homoplasy arose from interbreeding between taxa. It's also possible that some hominin traits are very plastic, repeatedly evolving into certain forms in response to similar environmental challenges.

Figure 17.8 presents an evogram based on recent phylogenetic analyses of hominins. Although the full story of hominin evolution is more complex than what's shown, this figure illustrates some of the key features of hominin evolution. Hominins originated in Africa, for example, and their range was restricted to that continent for about five million years. The early branching hominins shared only some traits in common with living humans. Three major innovations in the hominin line—big brains, bipedalism, and tool use—evolved gradually over millions of years. In the sections that follow, we will take a closer look at these milestones in our evolutionary history.

Key Concepts

- Hominins diverged from other apes approximately eight million years ago.

- In the nineteenth and early twentieth centuries, paleoanthropologists found fossils of hominin species that they classified as either *Homo* or *Australopithecus*.

- The precise identity of some hominin fossils—as members of known species or new species—can be the subject of long-running debates.

- Paleoanthropologists have discovered approximately 30 hominin taxa that lived in the past six to seven million years.

- Scientists have worked out the broad phylogenetic relationships among hominin taxa, but some aspects remain the subject of debate.

- Uncertainties about hominin phylogeny may be the result of an incomplete fossil record or widespread homoplasy.

- Hominin phylogeny is not a linear march of progress but a bushy tree with many lineages simultaneously diversifying. ●

17.3 Walking into a New Kind of Life

Our closest living relatives can offer some clues to how the first hominins moved around. They may not be our direct ancestors, but a broad comparison of primate locomotor strategies can let us infer some of the ancestral features in hominins (Lieberman 2011). Chimpanzees, for example, do not walk bipedally for long distances. Their anatomy—which is better adapted for climbing in trees—makes bipedalism inefficient. When chimpanzees need to travel on the ground, they lean forward and balance on their knuckles. The ancestors of hominins likely were also adapted for climbing in trees, making it difficult for them to walk bipedally for long distances. But the earliest known fossils of hominins suggest that they were already shifting to a more upright stance.

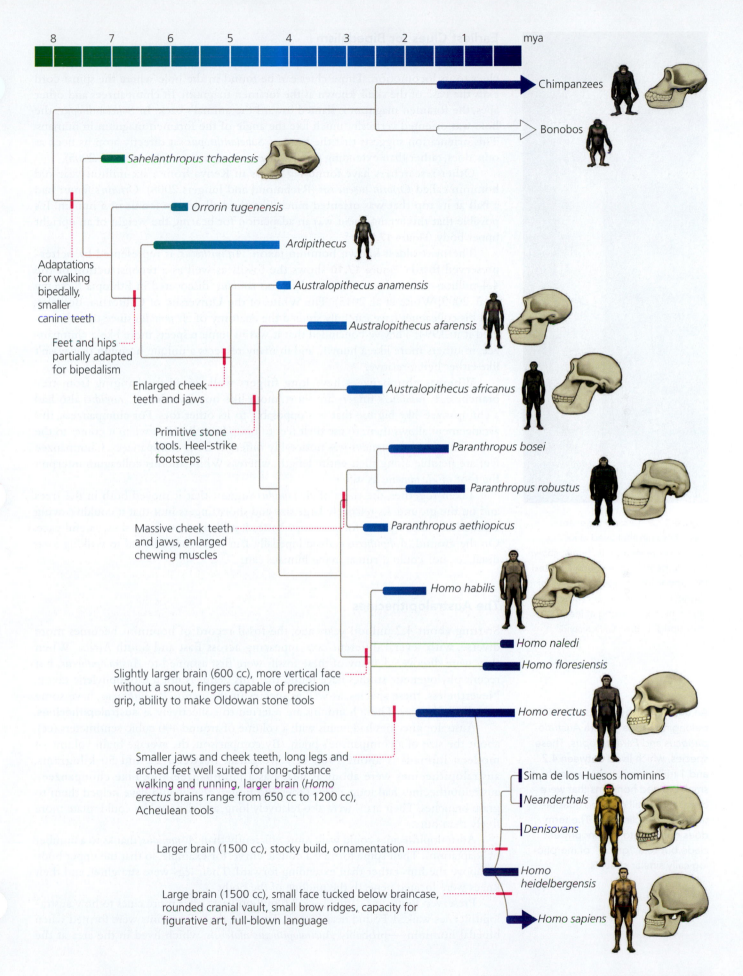

8　7　6　5　4　3　2　1　mya

Chimpanzees

Bonobos

Sahelanthropus tchadensis

Orrorin tugenensis

Ardipithecus

Adaptations
for walking
bipedally,
smaller
canine teeth

Australopithecus anamensis

Australopithecus afarensis

Feet and hips
partially adapted
for bipedalism

Australopithecus africanus

Enlarged cheek
teeth and jaws

Paranthropus bosei

Paranthropus robustus

Primitive stone
tools. Heel-strike
footsteps

Paranthropus aethiopicus

Massive cheek teeth
and jaws, enlarged
chewing muscles

Homo habilis

Homo naledi

Homo floresiensis

Slightly larger brain (600 cc), more vertical face
without a snout, fingers capable of precision
grip, ability to make Oldowan stone tools

Homo erectus

Smaller jaws and cheek teeth, long legs and
arched feet well suited for long-distance
walking and running, larger brain (*Homo
erectus* brains range from 650 cc to 1200 cc),
Acheulean tools

Sima de los Huesos hominins

Neanderthals

Denisovans

Larger brain (1500 cc), stocky build, ornamentation

*Homo
heidelbergensis*

Large brain (1500 cc), small face tucked below braincase,
rounded cranial vault, small brow ridges, capacity for
figurative art, full-blown language

Homo sapiens

Figure 17.9 *Orrorin tugenensis* was a hominin that lived in Kenya six million years ago. Its femur, shown here, appears to have been adapted for supporting an upright torso. This suggests that *Orrorin* and other early hominins were walking at least partially upright. (Brian G. Richmond)

An **australopithecine** is a hominin belonging to the genera *Australopithecus* and *Paranthropus*. These species, which lived between 4.2 and 1 million years ago, were short, small-brained hominins that were bipedal but still retained adaptations for tree climbing. The term does not refer to a monophyletic clade but to a "grade" of morphologically similar species.

Earliest Clues for Bipedalism

Sahelanthropus is only known from a partial skull, but that skull offers some important clues to its locomotion. Those clues can be found in the hole where the spinal cord exits the base of the skull, known as the foramen magnum. In chimpanzees and other apes, the foramen magnum is slanted toward the animal's back. In *Sahelanthropus*, the hole was oriented vertically, much like the angle of the foramen magnum in humans. This orientation suggests that the head of *Sahelanthropus* sat directly atop its neck as ours does, rather than extending forward like a chimpanzee's (Brunet et al. 2005).

Other researchers have found leg bones in Kenya from a six-million-year-old hominin called *Orrorin tugenensis* (Richmond and Jungers 2008). *Orrorin*'s femur had a ball at its top that was oriented much like the ball on the femur of a human. It's possible that this arrangement was an adaptation for bearing the weight of an upright upper body (**Figure 17.9**).

The next oldest known hominin taxon, *Ardipithecus*, is represented by better-preserved fossils. **Figure 17.10** shows the fossils as well as a reconstruction of the 4.4-million-year-old remains of *Ardipithecus ramidus*, discovered in Ethiopia (Lovejoy et al. 2009; White et al. 2015). Tim White, of the University of California, Berkeley, and his colleagues have carefully studied the anatomy of *A. ramidus* since they discovered it in 1994. They've concluded that it was in some respects more like a chimpanzee, in others more like a human, and in many respects a unique hominin that wasn't like either living relative.

Whereas chimpanzees have long fingers well suited for hanging from tree branches, *A. ramidus*'s fingers are short, more like our own. Yet *A. ramidus* also had a chimpanzee-like big toe that was opposable to its other toes. For chimpanzees, this arrangement allows them to use their feet to grasp branches. But when it comes to the rest of the foot, *A. ramidus* was noticeably different from chimpanzees. Chimpanzee feet are flexible along their entire length, whereas White and his colleagues interpret the foot of *A. ramidus* as stiffer.

Taken together, the traits of *A. ramidus* suggest that it moved both in the trees and on the ground. Its relatively large size and short fingers hint that it couldn't swing acrobatically through the forest canopy. Instead, it likely moved at a slow, careful pace. On the ground, *Ardipithecus* walked bipedally. But it was not adapted to walking long distances, nor could it run as living humans can.

The Australopithecines

Starting about 4.2 million years ago, the fossil record of hominins becomes more diverse, with several different taxa appearing across East and South Africa. When they were discovered, many of these fossils were first assigned to *Australopithecus*, but recent phylogenetic studies have found that this genus is not a monophyletic taxon. Nevertheless, these species, as well as species in the genus *Paranthropus*, have some overall similarities. These hominins are referred to collectively as **australopithecines**.

Australopithecines had brains with a volume of around 400 cubic centimeters (cc), about the size of a chimpanzee's brain. (By comparison, the average brain volume of modern humans is about 1400 cc.) Weighing between 25 and 50 kilograms, australopithecines were about as heavy as a chimpanzee, too. Like chimpanzees, australopithecines had long, curved toes and fingers that would have helped them to grasp branches. Their arms were also relatively long, and their ankles could rotate more freely than ours.

Australopithecines could walk more efficiently than *A. ramidus*, thanks to a number of adaptations. Their spine formed a double curve, for example, so that the upper body sat above the hips rather than extending forward. Their legs were straighter, and their knees were located beneath the midline of the body.

Preserved footprints dating back 3.6 million years offer more clues to how australopithecines walked. Found near Laetoli, Tanzania, these footprints were formed when bipedal hominins—probably *Australopithecus afarensis,* which lived in the area at the

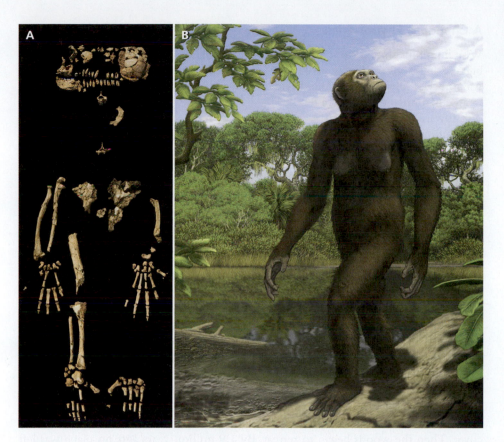

Figure 17.10 A: A relatively complete skeleton of *Ardipithecus*, which lived 4.4 million years ago. B: *Ardipithecus* shows signs of being adapted for both bipedal walking and arboreal life. (A: David L. Brill; B: Carl Buell)

time—walked across wet volcanic ash (Figure 17.11). The impressions show that the hominins landed heel-first with each step and pushed off with the toes—the same way we walk today. But we keep our legs stiff as we walk, whereas the footprints suggest that australopithecines kept theirs flexed. Although the hominins of Laetoli were far more bipedal than chimpanzees, they did not walk like we do (Hatala et al. 2016).

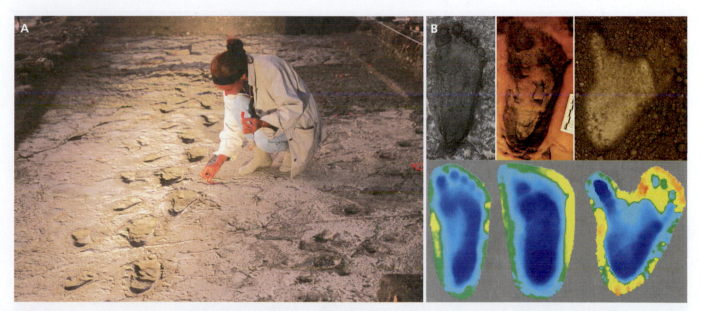

Figure 17.11 A: A volcanic eruption 3.6 million years ago in Tanzania preserved the footprints of bipedal hominins—probably *Australopithecus afarensis*. B: The top row shows, left to right, a human footprint, a Laetoli footprint, and a chimpanzee footprint. The bottom row shows the depth of the footprints. Dark blue corresponds to areas of greatest depth. Shallow depths are shown on a gradient from blue to green to yellow to orange. This analysis shows that the Laetoli hominins had a walking gait strikingly like our own. (A: Kenneth Garrett/Getty Images; B: Kevin G. Hatala)

A Changing Environment

Why hominins shifted from moving in trees to walking on the ground is one of the biggest questions in human evolution—and remains the subject of intense research. A number of scientists have found evidence suggesting that this shift was an adaptation to an environment that was in flux.

Toward the end of the Miocene, the average temperature of the planet dropped. The shifting climate altered weather patterns around the world. The effects were devastating for apes. Most species outside of Africa became extinct. Only gibbons and orangutans still survive, and today they are under serious threat of extinction from deforestation. In Africa, the climate change led to a reduction in rainfall. What was once an unbroken blanket of tropical forests across much of Africa began to fragment into smaller habitats. Meanwhile, woodlands and grasslands began to expand (Figure 17.12). Instead of steady weather patterns throughout the year, these new habitats experienced seasons of rain and drought, making supplies of food less predictable (Klein 2009).

For the most part, African apes stayed in their ecological niches, retreating into the remaining fragmented rain forests. That's one reason gorillas, chimpanzees, and bonobos are at risk of extinction today. Humans are disturbing their forest habitat far faster now than natural global climate changes occurred in the past, and we are reducing their range and preferred habitats. They have nowhere else to hide.

Hominins, on the other hand, adapted to the newly emerging woodlands and grasslands (Cerling et al. 2011; Robinson et al. 2017). Isotopic analysis of hominin fossil sites has revealed that seven million years ago, hominins lived in grasslands with sparse trees. The woody cover increased over the next few million years and reached its greatest extent about 3.6 million years ago, when the sites were 40%–60% woody cover. Then the woods began to retreat. By 1.9 million years ago, no place was left with more than half woody cover. The environments continued to open up, and the trend continues today.

A number of researchers have hypothesized that these environmental changes drove the shift in hominin locomotion recorded in the fossil record. An upright stance may have evolved in early hominins because it allowed them to collect food from trees

Figure 17.12 The emergence of savannas and woodlands in Africa set the stage for the evolution of hominins. (Gil.K/Shutterstock)

A

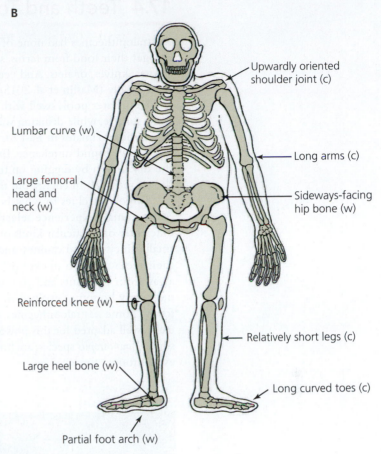

B

Upwardly oriented
shoulder joint (c)

Lumbar curve (w)

Long arms (c)

Large femoral
head and
neck (w)

Sideways-facing
hip bone (w)

Reinforced knee (w)

Relatively short legs (c)

Large heel bone (w)

Long curved toes (c)

Partial foot arch (w)

more efficiently (Hunt 1994). Early hominins were still well adapted for climbing in trees. But as the East African woodlands opened up, traveling from one tree to the next often meant getting down on the ground (**Figure 17.13**). An increasingly upright gait would have made walking more efficient, allowing hominins to save energy (Harcourt–Smith and Aiello 2004).

Peter Wheeler, of Liverpool John Moores University, has championed another hypothesis for the evolution of bipedalism: he argues that walking upright would have helped hominins to stay cool (Wheeler 1991; Lieberman 2015). Moving in open habitats instead of dense forests, hominins would have been pummeled by direct, equatorial sunlight. By standing upright, they exposed less of their skin to the sun, and they could hold their heads up in cooler, breezier layers of air.

Figure 17.13 A: *Australopithecus afarensis*, which lived from 3.7 to 3 million years ago, likely could walk on the ground but also climbed trees. B: The anatomy of *A. afarensis* was an amalgam of adaptations for both types of locomotion (c: inferred adaptations for climbing; w: inferred adaptations for walking). (A: Carl Buell; data from Lieberman 2011)

Key Concepts

- Scientists can gain important clues about the evolution of bipedalism in the human lineage from changes in characteristics of fossils, such as size, shape, orientation of bones, and fossil trackways. They also can examine modes of transportation of living primates and compare their bones to fossils to see how closely they match.

- Bipedalism and an upright stature evolved in hominins while they were still small-brained.

- Shifts in climate brought significant changes to the environment in which hominins evolved.

- The emerging mosaic of open-habitat grasslands and forested areas may have driven the evolution of bipedalism in hominins. ●

17.4 Teeth and Tools

Australopithecines had none of the technology that we depend on today. They could not get their food from farms, supermarkets, or refrigerators. They did not even have arrows, knives, or nets. And yet they survived in an environment with an unreliable food supply (Maslin et al. 2015). In the wet season in East Africa, plants are abundant and freshwater pools swell with rain. In the dry season, plants wither and only tough seeds survive, while drinking water becomes scarce.

Living primates in East Africa offer scientists some clues to how early hominins could have found sustenance. Baboons in Kenya, for example, eke out a living during the dry season by digging up tubers and other buried plant structures. Not only are these foods rich in energy, but they also store water.

Important clues to what hominins were eating come from their jaws and teeth. All mammals experience selection for jaws and teeth that can help them feed most efficiently on particular kinds of food. Compared to chimpanzees, early hominins had smaller incisors and canines, and their molars were bigger and flatter. The enamel on their teeth was also thicker (**Figure 17.14**). These changes may have reflected a shift away from the fruits and young leaves that apes in rain forests eat. To eat seeds and nuts, early hominins evolved an anatomy that generated strong crushing and grinding forces. Some australopithecines, such as *Paranthropus robustus* and *P. bosei*, became especially well adapted for this powerful style of eating. A large ridge ran atop the skull in some *Paranthropus* species, anchoring massive jaw muscles that allowed them to deliver even stronger forces.

Figure 17.14 A: Chimpanzees have large teeth in massive jaws, similar to those of early hominins. B: Human teeth (left) are proportionally much smaller than those of chimpanzees (right), as are human jaws. This transition was driven partly by a shift in diet. It's also possible that a reduction in male–male competition led to a reduction in the canine teeth, which other primates use in displays and fights. (A: kristianq/Getty Images; B: David L. Brill; data from Suwa et al. 2009)

Paranthropus specialized in crushing and grinding with massive jaws; however, other australopithecines retained a more slender build. Rather than gain food by brute force, they appear to have specialized on a different strategy: using tools.

The first major discoveries of early hominin stone tools were made in the 1960s by Louis Leakey and his colleagues in Kenya. They found stones that had been chipped with the clear intention of making tools for cutting, pounding, and other uses. These tools dated back about 1.75 million years ago.

At the same sites, Leakey's team also found fossil jaws and skull fragments from a previously unknown species of *Homo*, the oldest fossils of the genus yet found. Leakey dubbed the fossils *Homo habilis*, which he nicknamed "Handy Man." Finding the oldest known *Homo* fossils and the oldest known stone tools at the same sites led Leakey and his colleagues to argue that the evolution of the genus was linked to the origin of tool use.

This was a profoundly influential hypothesis. And, in the light of recent research, it turns out to be wrong.

Researchers now know that *Homo habilis* was far from the first toolmaker. Working in Ethiopia, Sonia Harmand, a paleoanthropologist at Stony Brook University, and her colleagues have found stone tools dating back 3.3 million years, almost twice as old as Leakey's original discovery (**Figure 17.15**). No *Homo* fossils have been found from that early stage of hominin history (Lewis and Harmand 2016).

When hominins first branched off from other apes about eight million years ago, they likely were already making simple tools. After all, living apes such as chimpanzees craft tools from sticks, leaves, and stones, as we discussed in Chapter 16. But the tools Harmand and her colleagues found in Ethiopia have a workmanship beyond what chimpanzees produce. The hominins that made them started off by carefully selecting certain stones as raw material. They then knocked bits of rock from them using other stones. Harmand and her colleagues suspect that the hominins knocked off pieces of different shapes, with the goal of making tools suited for different tasks.

It's possible that australopithecines initially used stone tools to collect plants—digging up tubers, for example. But at some point hominins started using them to gather meat. The early evidence for hominin meat-eating takes the form of cut marks on the fossil bones of large mammals (**Figure 17.16**). These tool-wielding australopithecines may have functioned ecologically as scavengers, akin to vultures and hyenas. They may have waited for lions and other big African predators to finish feeding on their prey and then approached the carcasses to gather remaining food. The hominins used their sharp stones to slice meat from the bones and then hack the bones open to get the marrow inside.

Figure 17.15 Sonia Harmand and her colleagues have found 3.3-million-year-old stone tools in Ethiopia. In some cases, the tools even offer hints about their production. Top row: A sharp-edged tool (called a flake) still fits snugly against the rock off of which it was broken. Bottom left: Three views of the flake alone. Bottom right: The remaining portion of the rock. (MPK-WTAP; data from Lewis and Harmand 2016)

This innovation in tools led to a profound transformation in hominin biology. Now hominins had access to a new supply of nutrients, proteins, and calories—one that would eventually fuel bigger bodies and bigger brains.

Key Concepts

- The availability of food and water may have been a strong selective factor shaping the evolution of our skulls, jaws, and teeth.

- The oldest evidence of stone tools is from 3.3 million years ago. Australopithecines may have used these early stone tools to gather plants or scavenge meat. ●

17.5 The Emergence of *Homo*

In the decades following Leakey's discovery of *Homo habilis*, other researchers have found other fossils that belonged to early *Homo* species. These fossils support the hypothesis that *Homo* evolved from australopithecines in Africa. But they also display a striking diversity, even at the earliest stages of this lineage.

Early African *Homo* Diversity

In 1984, a paleoanthropologist on Leakey's team named Kimoya Kimeu discovered a 1.5-million-year-old fossil of a 12-year-old boy (Walker and Leakey 1993). Many scientists now assign this nearly complete skeleton, found near Lake Turkana in Kenya, to a species called *Homo erectus*.

At age 12, he was already 160 centimeters (cm) tall, about the same height that 12-year-old boys reach today. His hips were narrow, and his legs were long and slender. His rib cage was cylindrical, and his brain was large. Subtler features also put Turkana Boy in our genus, such as his relatively small teeth and jaws. Taken together, these traits demonstrate that *Homo erectus* was much more like us than like australopithecines (**Figure 17.17**).

But the transition from australopithecines to *Homo* remains murky, and paleoanthropologists are deeply divided about how to interpret the evidence gathered so far (de Ruiter et al. 2017; Villmoore 2018). It is now clear, for example, that *Homo* is a lot older than once thought and may have been remarkably diverse. The fossil record of *Homo habilis* has been pushed back to at least 2.4 million years ago, and researchers

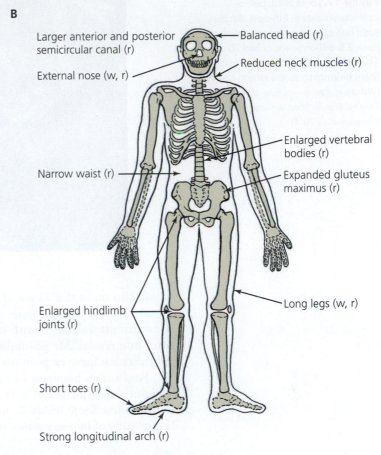

Larger anterior and posterior semicircular canal (r)

Balanced head (r)

External nose (w, r)

Reduced neck muscles (r)

Enlarged vertebral bodies (r)

Narrow waist (r)

Expanded gluteus maximus (r)

Enlarged hindlimb joints (r)

Long legs (w, r)

Short toes (r)

Strong longitudinal arch (r)

Figure 17.17 A: Reconstruction of *Homo erectus*. B: Early members of the genus *Homo* lacked tree-climbing adaptations. They evolved long legs and other traits efficient for walking and running (w: inferred adaptations for walking; r: inferred adaptations for running). (A: Carl Buell; data from Lieberman 2011)

have found fossils of other early *Homo* taxa in Africa. The oldest found so far, dating back 2.8 million years, was discovered in 2013 by the Ethiopian paleoanthropologist Chalachew Seyoum and his colleagues at a site called Ledi-Geraru (Villmoare et al. 2015).

The fossil, known as LD 350-1, consists only of the left side of a lower jaw (**Figure 17.18**). Seyoum and his colleagues found a number of traits in this fragment, such as small teeth, that indicate that LD 350-1 belongs to the *Homo* clade. With only part of a single jaw to study, the researchers refrained from assigning LD350-1 to a species. But they did suggest that it belonged to a lineage that produced all later species of *Homo*.

Homo naledi

Meanwhile, in South Africa, Marina Elliott and the rest of the Rising Star team were uncovering the abundant bones of *Homo naledi*. The remarkable mosaic nature of *H. naledi* made it difficult to determine where it belongs in the hominin evolutionary tree. But in a 2017 review of the evidence, Lee Berger and his colleagues suggested that its lineage might have split off from other hominins at the base of the *Homo* branch (Berger et al. 2017).

Its relatively young age—*H. naledi* may have lived as recently as 236,000 years ago—make this possibility especially intriguing. It would imply that *H. naledi* belonged to a deep lineage of *Homo* reaching back more than two million years. That would mean it had many forerunners yet to be discovered. Such a phylogenetic placement

Figure 17.18 In 2013, paleo-anthropologists in Ethiopia discovered half of a hominin jaw, dating back 2.8 million years, called LD 350-1. They propose that it is the oldest fossil belonging to the *Homo* clade. (William H. Kimbel, Institute of Human Origins, Arizona State University; data from Villmoare et al. 2015)

would also mean that *Homo naledi* independently acquired some anatomical features found only in our own species and close relatives—such as the bones in the wrist and hand that are associated with making sophisticated tools. Berger and his colleagues offer one remarkable possibility: the mosaic anatomy of *H. naledi* is the result of hybridization between primitive and derived hominins.

Fossils from hominins like LD 350-1 and *Homo naledi* leave paleoanthropologists without a clear picture of how *Homo* first evolved. But the intriguing distribution of traits in these fossils makes them among the most exciting avenues for future research in the study of human evolution.

Key Concepts

- Early *Homo* evolved from australopithecines, but the origins of the genus remain the subject of debate.

- *Homo naledi* appears to have branched off early from the *Homo* clade, suggesting that the many traits it shares with modern humans may have evolved independently. ●

17.6 Hominins Expand to New Continents

Starting about 2.1 million years ago, evidence of hominins begins to turn up outside of Africa (**Figure 17.19**). The earliest evidence, not surprisingly, is also the most controversial. Researchers working in China have found what they claim to be 2.1-million-year-old stone tools made by hominins (Zhu et al. 2018). They base this interpretation on a number of lines of evidence, such as the similarity of the shapes of the rocks to well-accepted stone tools of about the same age found in Africa. But a number of experts remain skeptical of this interpretation. Until researchers find fossils of hominins from the same rock layers, these skeptics argue, the identity of the rocks remains uncertain.

The earliest fossils of hominins outside of Africa are about 300,000 years younger than these intriguing rocks. In Indonesia, scientists have found 1.8-million-year-old fossils of *H. erectus*. These fossils belonged to hominins that were similar in many ways to *H. erectus* in Africa from around the same time, such as Turkana Boy. They shared traits such as a tall stature and a large brain. But they also had other traits that set them apart, suggesting that at the very least they belonged to a genetically distinct population.

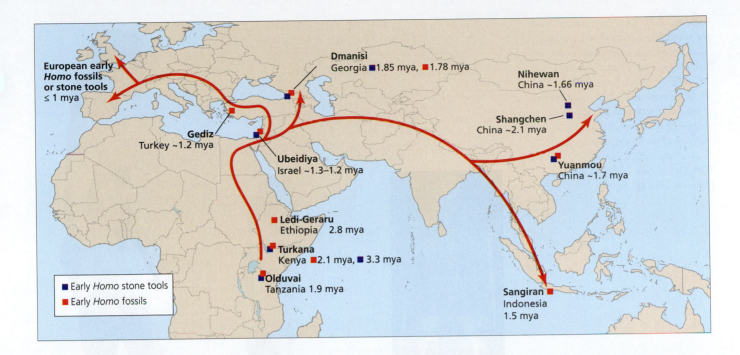

Figure 17.19 The oldest evidence of hominins outside of Africa is 2.1 million years old. Some populations in Europe and Asia probably belonged to *Homo erectus*, but it's possible that other early *Homo* lineages in Africa expanded to other continents as well. (Data from Kappelman 2018)

In the Republic of Georgia, researchers have discovered additional fossils of *Homo* hominins that also date back 1.8 million years (Lordkipanidze et al. 2013). The Georgian hominins have many hallmarks of *Homo*, but they were also very short and had a small braincase, measuring just 600 cc in volume—less than half the size of our own. Some researchers have proposed they belong to a separate species—*Homo georgicus*—whereas others have argued they are another population of *Homo erectus* that developed their dramatically different anatomy in response to their local environmental conditions (Antón et al. 2016).

It's not yet clear why such a wide range of hominins appeared at about the same time outside of Africa. Did *Homo* evolve a capacity to survive in a broader range of habitats? Or did environmental conditions favor their spread into Asia? As scientists discover a greater variety of hominins outside of Africa, they're considering new scenarios to explain the new evidence.

One of the most intriguing species of *Homo* was discovered in 2004 on the Indonesian island of Flores (**Figure 17.20**). Paleoanthropologists digging in a cave there found fossils of tiny hominins who stood just 100 cm tall, with tiny skulls that had a cranial capacity of just 417 cc. Despite their miniature proportions, these hominins share a number of features in common with the *Homo* clade, and the researchers dubbed the fossils *Homo floresiensis* (Aiello 2010). Subsequent research on Flores now indicates that *H. floresiensis* may have been present on Flores at least 700,000 years ago (van den Bergh et al. 2017). They may have endured on the island until as recently as 60,000 years ago.

Some paleoanthropologists initially proposed that *H. floresiensis* evolved from *H. erectus*. The tall *H. erectus* population known from Indonesian fossils might have gotten to Flores, perhaps swept there on vegetation carried out to sea in a storm. Once on the island, they could have evolved to a much smaller size.

One stumbling block with this scenario is that it would also mean that the relatively large *H. erectus* brain shrank down to the size of a chimpanzee's. Our own species has repeatedly evolved to small stature—but the brains of so-called pygmies remain in the normal range for *Homo sapiens*. Another problem with an *H. erectus* ancestry is that *H. floresiensis* lacks some traits that unite *Homo erectus* and closely related species.

In a 2017 phylogenetic analysis, Debbie Argue, of Australia National University, and her colleagues proposed an alternative scenario: *H. floresiensis* shared a close ancestry not with *H. erectus* but with *H. habilis*—Louis Leakey's small Handy Man, which lived in Africa 2.4 million years ago (Argue et al. 2017).

Figure 17.20 Left: *Homo floresiensis*, which lived on the Indonesian island of Flores, stood only 1 meter tall and had a chimpanzee-sized brain. Right: A reconstruction of *H. floresiensis* is shown here next to a modern human for comparison. Although it shows kinship to early species of hominins, it existed on Flores until 60,000 years ago. (Left: Carl Jungers; Right: Mauricio Anton/Science Source)

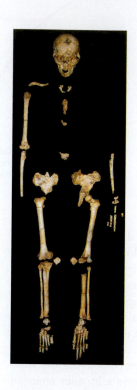

Homo floresiensis   Modern human

This conclusion has a remarkable implication. If true, it would mean that *Homo* expanded out of Africa not as a single wave but at least two times. Small-statured relatives of the *H. habilis* lineage would have migrated from Africa well over two million years ago and reached a remote Indonesian island. In Argue's scenario, the taller *H. erectus* spread out of Africa and into Asia in a separate expansion, perhaps much later.

Key Concepts

- *Homo* appeared in Asia at least 1.8 million years ago, and possibly at least 2.1 million years ago.

- Asian *Homo* fossils display a striking variation of morphology, which may be explained by the expansion of different lineages of *Homo* from Africa.

- *Homo floresiensis*, an enigmatic, small hominin, lived in Indonesia as recently as 60,000 years ago. ●

17.7 The Emergence of Modern Traits in *Homo*

In the previous section, we learned how the origin of *Homo* featured the emergence of a wide range of morphological variation. But within this diversity, we can also trace the emergence of a number of modern traits that our species has inherited.

Bifacial chopper   Discoid   Polyhedron

0   5 cm

Hammer stone   Flake scraper   Flake   Heavy-duty (core) scraper

Figure 17.21 Hominins developed a style of toolmaking called Oldowan. The shapes of the tools show how they were made, by striking off flakes from large, round rocks. Hominins used tools to scavenge meat and possibly to fashion wooden tools. (Data from Boyd and Silk 2009)

Abstraction Revealed in Tools

As we discussed earlier, australopithecines were using stone tools as early as 3.3 million years ago. But the record of hominin stone tools is very sparse until 2.6 million years ago, after which paleoanthropologists find a continuous series of sites with stone tools. The tools at these sites all display the same range of shapes, suggesting they form a single "toolkit." Scientists refer to these tools as **Oldowan technology**, named for an archaeological site in Tanzania (**Figure 17.21**).

Around 1.8 million years ago, a new toolkit appears in the fossil record (**Figure 17.22**). The most spectacular of these new tools were so–called "hand axes," produced by carefully chipping away at large rocks to give them teardrop shapes. The oldest examples of this new toolkit come from Africa, but they appear soon afterward in

Oldowan technology refers to the earliest stone tools continuously associated with hominins, starting 2.6 million years ago. Hominins made these tools for crushing, cutting, and other functions.

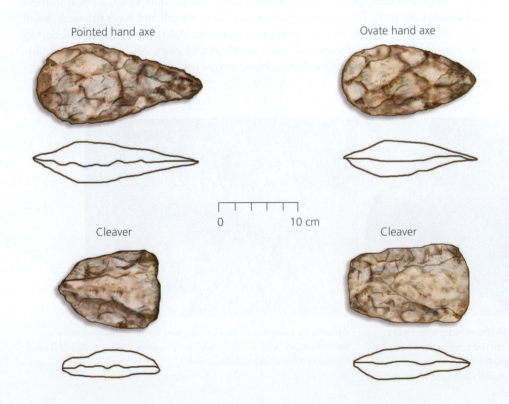

Pointed hand axe   Ovate hand axe

0   10 cm

Cleaver   Cleaver

Figure 17.22 Starting about 1.8 million years ago, early *Homo* began making a more sophisticated technology, called the Acheulean toolkit. Among the Acheulean tools made by hominins were large teardrop-shaped "hand axes." They probably used these hand axes to butcher animals, cut wood, and gather roots and other plant foods. (Data from Toth and Schick 2015)

Acheulean technology refers to tools associated with hominins between 1.8 million and 100,000 years ago. These tools are found across Africa, much of western Asia, and Europe. They are often found in association with *Homo erectus* remains. Acheulean tools, which include hand axes, display more sophistication in construction than Oldowan tools.

Asia and later in Europe. These new tools are known as **Acheulean technology**, named after the site in France where they were first discovered.

Scientists have hypothesized that Acheulean tools were superior to Oldowan technology, allowing *Homo erectus* to cut through the tough skin of elephants and mammoths and obtain a new supply of calories and nutrients (Finkel and Barkai 2018). But some Acheulean hand axes were so big and so elongated that some researchers have questioned whether they had any practical use. Instead, they may have had some kind of symbolic importance for early *Homo*—perhaps as a part of long-lost rituals.

Another fascinating clue to the cognition of *Homo erectus* emerged in 2014 (Joordens et al. 2014). Josephine Joordens, a biologist at Leiden University, was examining freshwater mussel shells that had been found in Indonesia along with the first *Homo erectus* fossils ever discovered in the 1890s. She started out studying them to see how *Homo erectus* collected shellfish for food. But then she noticed signs that *Homo erectus* had modified some of the shells. The most striking of these was a 500,000-year-old shell with geometrical carvings on it (**Figure 17.23**). Although it might be going too far to call this art, it certainly looks like abstract modifications to objects, rather than modifications intended to create a functional tool.

Although early *Homo* was capable of technological innovation, the pace of that innovation was nothing like the swift rush of updates that we expect today from mobile phones or televisions. Even after the development of Acheulean tools, *Homo* continued to use Oldowan tools for half a million years, until about 1.2 million years ago. As for Acheulean technology, hominins used it from 1.8 million years ago to as recently as 100,000 years ago.

During that vast span of time, researchers have found, hominins made only small changes to the design of Acheulean tools. They made hand axes more symmetrical and gave them a thinner tip, for example. But this transformation took hundreds of thousands of years—suggesting that *Homo erectus* lacked language and other cognitive innovations that allow living humans to carry out rapid technological change (**Figure 17.24**; Beyene et al. 2013).

A Hunter's Anatomy?

Homo erectus and later species, such as *Homo sapiens*, share a suite of anatomical features that distinguish them from australopithecines. *Homo* are tall and have flat faces, small teeth, and fingers capable of more precise movements. Australopithecines had flaring rib cages, whereas *Homo erectus* had more barrel-shaped ones. And although australopithecines have chimpanzee-like brains, *Homo erectus* began a trend toward much bigger brain sizes.

Figure 17.23 A 500,000-year-old shell (A) from Indonesia bears geometrical engravings (B), suggesting abstract thought in *Homo erectus*. (Republished with permisson of Nature Publishing Group from "Homo erectus made world's oldest doodle 500,000 years ago" Ewen Callaway, 03 December 2014, Nature; photo by Henk Caspers/Naturalis; permission conveyed through Copyright Clearance Center; B: W. J. M. Lusterhouwer/VU/University of Amsterdam)

Paleoanthropologists have proposed a number of hypotheses to explain this combination of traits that are so important to our own anatomy. Natural selection may have favored hands with finer motor control because hominins could make more sophisticated tools. Those tools, in turn, allowed hominins to scavenge more food from animals. Other shifts in the anatomy of *Homo* may have led to new ways of moving across the African landscape (Lieberman 2015). Their long legs and narrow hips would have enabled efficient, long-distance walking. Daniel Lieberman, of Harvard University, has argued that other adaptations, such as head-stabilizing neck muscles, allowed *Homo* to also run.

Running on the African savanna would have generated large amounts of heat in the hominin body. This may have created a selective pressure that caused *Homo* to evolve a mostly hairless body. Humans also have a greater density of sweat glands than other apes; this evolutionary transition could have also helped keep running hominins cool. By running, *Homo* could cover more ground and find more carcasses to scavenge. Some researchers have argued that as early as 1.5 million years ago, these running hominins started to hunt for themselves, rather than scavenge (Roach et al. 2018).

A shift to energy-rich meat may also explain the change in the hominin rib cage. The flaring shape of australopithecine rib cages suggests that they had a massive digestive system necessary to break down tough plant matter. *Homo erectus*'s narrow rib cage suggests a shorter digestive system. The anthropologist Leslie Aiello, professor emeritus of the University College London, has proposed that this shift was crucial to the expansion of the hominin brain. Brains require huge amounts of energy; one out of every five calories we eat goes to fueling our brains. Aiello argues that a shift in diet led to a smaller digestive system, freeing up developmental resources and energy that could be directed away from maintaining intestinal tissues and toward an expanding brain (Aiello and Wheeler 1995). We'll discuss the evolution of the human brain in more detail later in this chapter.

Figure 17.24 Early hominins demonstrated technological innovations, but the pace was inconceivably slow. These Acheulean hand axes were produced over a 900,000-year span at a single site in Ethiopia. Top and bottom rows show opposite sides of the same tools. (Fig 4 from "The characteristics and chronology of the earliest Acheulean at Konso, Ethiopia" by Yonas Beyene et al. PNAS January 29, 2013 110 (5) 1584–1591; data from Beyene et al. 2013)

Key Concepts

- Hominin toolkits changed from simple Oldowan technology to more complex Acheulean tools such as hand axes.

- *Homo erectus* displayed some evidence of abstract thought by making geometrical markings on shells.

- Hominin brain evolution may have been coupled with changes in the digestive tract and adaptations for running. ●

17.8 Parallel Humans

Starting around 800,000 years ago, some striking new species of *Homo* arose in Africa, Asia, and Europe. It is not yet clear exactly how they are related to each other, or to us. But it is clear that these lineages bore significant similarities with our own species that were not present in earlier species of *Homo*.

Homo heidelbergensis

In Africa and Europe, paleoanthropologists have found fossils of a species called *Homo heidelbergensis*. It had a larger brain than *Homo erectus,* measuring about 1200 cc—just 200 cc shy of the typical size of brains in living humans (Buck and Stringer 2014). It appears that *H. heidelbergensis* evolved a new toolkit. In Europe 400,000 years ago, *Homo heidelbergensis* made wooden spears up to 3 meters long (Thieme 1997). They also produced new stone tools known as **Levallois tools**. To make them, *H. heidelbergensis* patiently knocked off small bits from a rock until they achieved a broad, flat shape. Then they delivered a sharp blow across the top of the rock, shearing off a large flake. They used the tools to cut up food and plant material, and they even may have attached some of the stone blades to the ends of spears.

H. *heidelbergensis* used these tools to carry out sophisticated hunts. Near the site where the wooden spears were found in Germany, archaeologists have found the butchered bones of wild horses. On the island of Jersey, in the English Channel, the fossils of rhinos and other big mammals have been found at the bottom of cliffs, where they show signs of having been butchered. It's likely that they were killed by hunters who chased them off a precipice and then finished them off at close range. These sites hint that hominins could cooperate to carry out complex strategies to take down big game. Charred wood and soil in the fossil record indicate that these hominins were probably regularly cooking the meat from these hunts at campfires (Gowlett 2016).

Neanderthals

A number of scientists have argued that *Homo heidelbergensis* was the direct ancestor of Neanderthals, or at least a close relative (Buck and Stringer 2014). Neanderthal-like hominins first appear in the fossil record about 400,000 years ago (De Groote et al. 2017). Over the following 360,000 years, the Neanderthal lineage left behind a rich record of fossils and artifacts that scientists have used to reconstruct their existence.

Compared to living humans, Neanderthals had wide shoulders and hips, along with muscle-bound bodies (**Figure 17.25**). An adult Neanderthal could grow as tall as 1.75 meters and weigh up to 82 kilograms. Their skulls were relatively long and low, with a distinctive brow ridge over their eyes. They had wide noses and almost no chin at all.

But it would be wrong to dismiss Neanderthals as brutish cavemen. Neanderthals had brains as big as our own—even somewhat larger in some cases. Neanderthals gathered an impressive range of food. There's evidence that they hunted rhinoceroses, caught dolphins in the ocean, and supplemented this meat with a number of plant species. Their range spread across much of Eurasia, from the edges of glaciers in northern Europe to the arid landscape of the Near East to the mountains of Siberia.

Neanderthals even showed some capacity for self-expression. They colored shells with pigment and drilled holes into them, perhaps to string them on necklaces. Eagle talons discovered in Croatia also show signs of having been used by Neanderthals as

A **Levallois tool** is one formed by a distinctive method of stone knapping involving the striking of flakes from a prepared core. This technique was much more sophisticated than earlier toolmaking styles, and flakes could be shaped into sharp scrapers, knives, and projectile points.

Figure 17.25 Top: Neanderthals belong to a lineage that diverged from our own lineage about 500,000 years ago. They lived across Europe, the Near East, and Western Asia before becoming extinct 40,000 years ago. Bottom: Neanderthals had brains as large as ours, and they had a stockier build. (Top: Carl Buell; bottom: Blaine Maley)

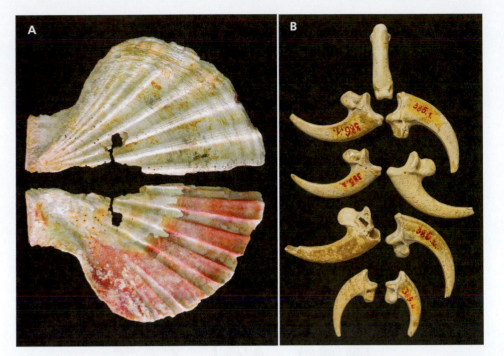

Figure 17.26 Neanderthals left behind evidence of sophisticated self-awareness. A: Shells found at a Neanderthal site show signs of being manipulated. This shell has a drill hole and is painted white on one side. Scientists suspect that these clues indicate Neanderthals wore jewelry. (Jaao Zihão; data from Zilhão et al. 2010) B: Eagle talons bearing notches, polished facets, and smoothed cut marks suggest they were used as jewelry. (Luka Mjeda; data from Radovčić et al. 2015)

jewelry (**Figure 17.26**). And deep inside a French cave, researchers have found broken pieces of stalagmites arranged into a ceremonial circle, dating back 47,600 years ago when Neanderthals occupied the site.

Homo sapiens

Over the same period of time that Neanderthals were thriving, our own lineage was evolving in Africa. Fossils dating back 300,000 years in Morocco constitute the oldest evidence of *Homo sapiens*. They were distinctive in their slender build, their flat face, and other skeletal traits. Over the next 200,000 years or so, *Homo sapiens* remained limited to Africa but spread across much of that vast continent (**Figure 17.27**). The artifacts that *H. sapiens* left behind document an accelerating pace of change. Across Africa, tools acquired a local flavor (**Figure 17.28**). In addition, *H. sapiens* also began trading their tools, and some of them ended up hundreds of kilometers from where they had

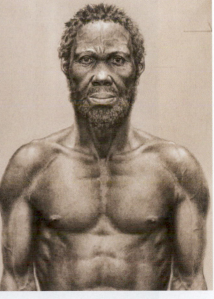

Figure 17.27 After the origin of *Homo sapiens* 300,000 years ago, the fossil record of our species remained limited to Africa for nearly 200,000 years. This skull, found in Ethiopia, dates back 160,000 years. (Left: David L. Brill; right: © 2003 Jay H. Matternes)

Figure 17.28 Early *Homo sapiens* in Africa made stone tools in a wide range of styles, reflecting a growing cultural diversity. (Data from Boyd and Silk 2009)

Pre-Aurignacian
Arterian
Nubian complex
Ethiopian
Kenya Rift
Lupemban
Mumba industry
Katanda
Bambatan/
Pietersburg
MSA I-IV
Howleson's Poort
Stillbay

been made. By 70,000 years ago, *H. sapiens* began to show signs of self-expression, such as pierced snail shells that might have gone on a necklace and abstract decorations on egg shells (**Figure 17.29**).

The descendants of early *H. sapiens* would spread throughout Africa and then the world (Bae et al. 2017). And during their expansion, all other species of hominins would become extinct, leaving our species the last survivor of our lineage (Stringer 2012).

Figure 17.29 *Homo sapiens* in Africa displayed an increasing level of creativity and self-expression. In South Africa 60,000 years ago, humans made water containers out of ostrich eggs and decorated them with geometric patterns. (Pierre-Jean Texier)

- Larger-brained hominins emerged within the last 800,000 years.

- Multiple lineages of these large-brained hominins lived as recently as 40,000 years ago.

- Anatomically modern humans expanded out of Africa and settled across the entire world. ●

17.9 New Discoveries from Ancient Genes

In the 1990s, scientists began learning how to extract fragments of DNA from fossils. Although DNA degrades over time, it is possible under certain conditions for it to survive in a fossil for vast amounts of time (Pääbo 2014). The oldest hominin DNA scientists have discovered so far dates back 430,000 years, coming from remains found in a cave in Spain called Sima de los Huesos (Meyer et al. 2016).

The growing number of ancient DNA sequences from fossils allows scientists to use molecular phylogeny to explore hominin evolution. In Chapter 8, we saw how Sarah Tishkoff and her colleagues have used molecular phylogeny to compare living humans. Thanks to ancient DNA, researchers can add branches of extinct hominin taxa to the same evolutionary tree.

For the most part, the results from ancient DNA align well with earlier studies on the morphology of hominin fossils. Researchers have extracted DNA from a number of fossils that researchers had previously identified as Neanderthals, for example. The genetic material from these fossils confirms that they belong to a monophyletic clade—one that's much closer to living humans than to chimpanzees and other apes (Hajdinjak et al. 2018).

When paleoanthropologists examined the fossils at Simo de los Huesos, they speculated that the hominins there belonged to the lineage that gave rise to Neanderthals. Their DNA supports this hypothesis, showing that the Spanish hominins are closely related to Neanderthals.

But ancient DNA has also delivered some profound surprises. Svante Pääbo, of the Max Planck Institute for Evolutionary Anthropology, and his colleagues retrieved DNA from a nondescript finger bone found in a cave called Denisova in the Altai Mountains of Siberia (**Figure 17.30**). At first, Pääbo and his colleagues speculated that the bone belonged to either a modern human or a Neanderthal. But its genome revealed that it belonged to a previously unknown lineage of hominins. Since then, the same cave has yielded DNA-bearing teeth from four individuals of the same lineage. In honor of the cave, these hominins were dubbed Denisovans (Slon et al. 2018).

Using molecular clock methods, researchers have found that humans, Neanderthals, and Denisovans all descend from a common ancestor. The lineage leading to modern humans split from that of Neanderthals and Denisovans about 530,000 years ago (Hajdinjak et al. 2018). Neanderthals and Denisovans then split from each other about 400,000 years ago.

It's possible that geography caused Neanderthals and Denisovans to diverge. Neanderthals may have been spread across western Eurasia while Denisovans lived in the east. But until scientists start finding more fossils of Denisovans beyond the Denisova cave, this will remain a hypothesis. One way to test it will be to search for ancient DNA in hominin fossils less than 500,000 years old that have already been found elsewhere in East Asia. This hypothesis predicts they will turn out to belong to the Denisovan lineage.

Studies of ancient DNA have also helped clarify the evolution of modern humans. Living populations of Africans have genetic differences that arose more than 200,000 years ago (Schlebusch and Jakobsson 2018). The oldest fossil evidence of *Homo sapiens* beyond Africa, found in the Near East, dates back about 120,000 years

A

B

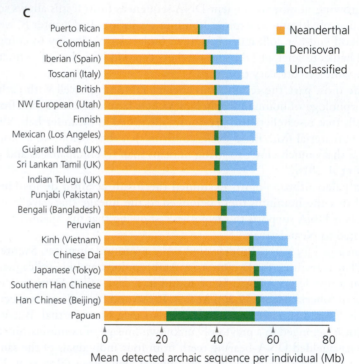

C

Figure 17.30 A Siberian cave called Denisova (A) has yielded 50,000-year-old hominin fossils including a tooth (B). A few teeth and bones in the cave preserved enough DNA to reconstruct an entire genome, which turned out to be different from both humans and Neanderthals. Scientists have dubbed this hominin lineage the Denisovans. C: Scientists have discovered segments of Denisovan and Neanderthal DNA in living humans. Some individuals have a few percent DNA from these extinct humans. This pattern indicates that both lineages interbred with the ancestors of living humans. (A: Ann Gibbons; B: Max Planck-Institute of Evolutionary Anthropology; data from Browning et al. 2018)

Puerto Rican
Colombian
Iberian (Spain)
Toscani (Italy)
British
NW European (Utah)
Finnish
Mexican (Los Angeles)
Gujarati Indian (UK)
Sri Lankan Tamil (UK)
Indian Telugu (UK)
Punjabi (Pakistan)
Bengali (Bangladesh)
Peruvian
Kinh (Vietnam)
Chinese Dai
Japanese (Tokyo)
Southern Han Chinese
Han Chinese (Beijing)
Papuan

- Neanderthal
- Denisovan
- Unclassified

Mean detected archaic sequence per individual (Mb)

ago. But it's not clear if these early wanderers were the ancestors of living humans. It's possible that they died out (**Figure 17.31**).

The genes of living people outside of Africa suggest that they belong to one main lineage, which arose from a small population of East Africans who expanded into the Near East. This history is reflected in the genetic diversity of living people: Africans today have much more genetic diversity than all other people on Earth combined because their populations have accumulated variation in their DNA for much longer.

When humans expanded across Eurasia, they encountered Neanderthals and Denisovans—and the different lineages of humans interbred (**Figure 17.32**). We know this because living humans carry a few percent DNA from both lineages (Wolf and Akey 2018). Ancient DNA thus gives us a new way of thinking about the most recent chapter of hominin evolution. During the 300,000–year-old history of our own species, we know that at least five other lineages of hominins existed (*H. erectus*, *H. floresiensis*, *H. naledi*, Neanderthals, and Denisovans). The youngest material evidence of these other hominin taxa are Neanderthal fossils and tools—dating back 40,000 years ago (Higham et al. 2014).

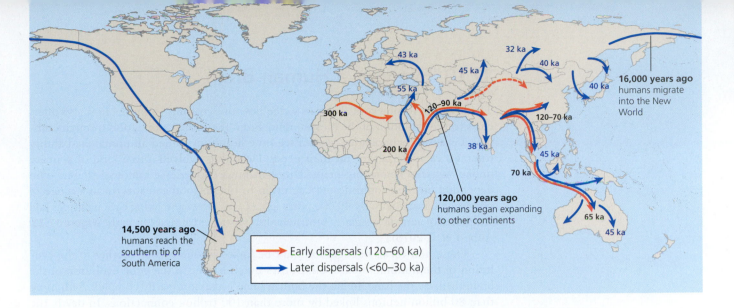

Figure 17.31 Studies on fossils and DNA indicate that humans originated in Africa and expanded throughout the continent. Starting about 120,000 years ago, humans began expanding to other continents. (Data from Bae et al. 2017)

Scientists have proposed a number of different hypotheses for why other hominins became extinct, but we survived (Kolodny and Feldman 2017). Some argue that changes in the climate such as ice age cycles pushed down some population of hominins. It's also possible that modern humans helped drive other hominins extinct through competitive replacement.

But as indicated earlier, at least two of these hominin lineages are not exactly extinct after all. The genes of Neanderthals and Denisovans have been passed down through human generations for tens of thousands of years, and today they endure in billions of people. It's even possible that natural selection has favored some of their genes in modern humans, helping us to fight diseases or otherwise increasing our fitness (Danneman and Racimo 2018).

Key Concepts

- Both molecular and fossil evidence suggests that humans, Neanderthals, and Denisovans represent separate, parallel lineages descended from a common ancestor living in Africa several hundred thousand years ago.

- Molecular studies indicate that modern humans interbred with Neanderthals and Denisovans. As a result, most living people carry a small amount of Neanderthal or Denisovan DNA (or both). ●

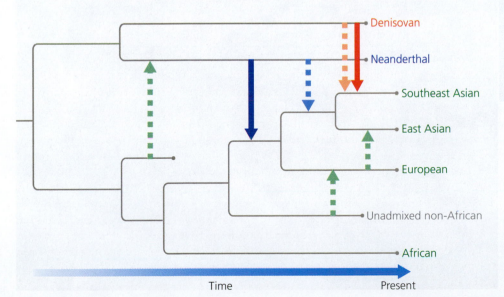

Figure 17.32 The recovery of DNA from Denisovans and Neanderthals has revealed interbreeding between different lineages of hominins. The dashed arrows represent preliminary evidence of interbreeding between deeply diverged human populations. (Data from Wolf and Akey 2018)

17.10 Evolving a Human Brain

The hominin fossil record allows us to see how our anatomy evolved, but it also offers clues to the evolution of the human mind. From the earliest stone tools through complex Levallois technologies, we can see how hominins gained an unmatched ability to craft raw materials to suit their needs. Hominins evolved into a kind of primate unlike any seen before: big-brained, tool-making, cooperative hunter-gatherers who depended on their sophisticated cognition to obtain food and survive in a range of physical environments. By 40,000 years ago, *H. sapiens* were making cave paintings and sculptures of animals—signs of unmatched sophisticated abstract thought (**Figure 17.33**; Aubert et al. 2018).

These new forms of cognition must be linked in one way or another to the evolution of the hominin brain itself. But it's much harder to track the evolution of the brain than, say, the hand. In life, the human brain is extraordinarily complex, with more than 80 billion neurons linked by more than 100 trillion connections. In death, the brain typically decays quickly, leaving nothing to form a fossil.

Although brains themselves may be almost entirely absent from the hominin fossil record, braincases are not. By studying hominin skulls, paleoanthropologists can estimate the size of the brains they once contained. **Figure 17.34** plots a sample of hominin brain sizes over time (Montgomery 2018). Overall, these measurements display a remarkable trend. Hominin brains have tripled in size over the past four million years. Today, humans have the biggest brains of any animal relative to their body. This expansion is all the more remarkable when we consider just how much energy that brain tissue demands compared to other cell types.

When scientists first saw the outlines of this trend, some proposed that long-term directional selection was responsible. Today, with more fossils and more careful measurements, the picture has gotten more complicated (Du et al. 2018). Andrew Du, of George Washington University, and his colleagues found that only some hominin lineages experienced patterns of brain size evolution consistent with the action of directional selection. Brains in other hominin lineages experienced stasis or fluctuated in size due to drift.

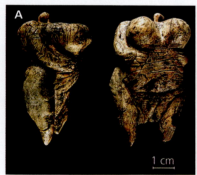

Figure 17.33 Figurative art begins to appear in the fossil record roughly 40,000 years ago. A: A female sculpture dating back 35,000 years, found in Germany. B: Images of hands and animals on Indonesian cave walls date as far back as 40,000 years. (A: Hilde Jensen/University of Tübingen; B: Photograph by Maxime Aubert with permission of Balai Pelestarian Cagar Budaya (Preservation for Archaeological Office) Makassar)

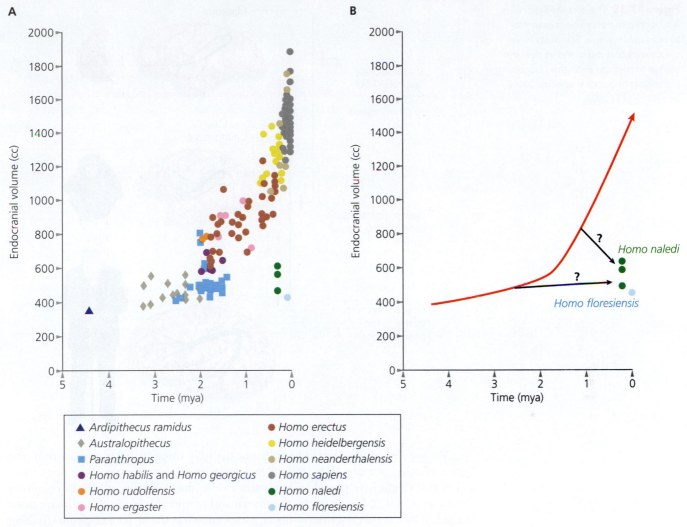

Figure 17.34 Early hominins had brains about the size of a chimpanzee's. A: About two million years ago, hominin brains began to increase; after one million years, their growth accelerated. Today, humans have brains more than three times the size of those of the earliest hominins. B: Two species, *H. floresiensis* and *H. naledi*, complicate the trend toward greater brain size. Depending on their precise location on the hominin evolutionary tree, their brains remained small or dramatically shrank. (Data from Montgomery 2018)

Some of the overall trend may also be the result of different rates of speciation and extinction: small-brained hominins appear to have been more prone to extinction than bigger-brained ones. But the recent discoveries of tiny brains in *Homo naledi* and *Homo floresiensis* complicate hominin brain evolution even further. Depending on their phylogenetic position, their small brains may be the result of long-term stasis over two million years or a dramatic reduction from larger-brained ancestors (Montgomery 2018).

As we saw in Chapter 16, researchers have investigated different selective forces that can explain the variation in brain size in vertebrates. They've extended these studies to hominin evolution. In a 2018 study, Mauricio González-Forero and Andy Gardner, of the University of St Andrews, tested the ecological and social hypotheses for intelligence to see which combination of factors might best explain the evolution of brain size documented in the hominin fossil record. They concluded that ecological intelligence accounted for most of the increase. Their study suggests that making tools and using them to process food was a key driving force in the evolution of our big brains (González-Forero and Gardner 2018).

Figure 17.35 The arcuate fasciculus is a bundle of nerve fibers essential to human language. Scientists have discovered that it expanded over the course of primate evolution, becoming largest in our own species. (Data from Rilling et al. 2008)

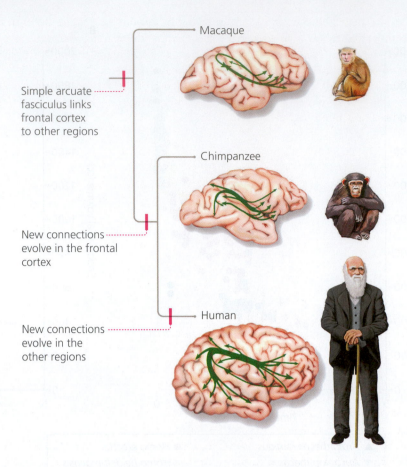

Macaque

Simple arcuate fasciculus links frontal cortex to other regions

Chimpanzee

New connections evolve in the frontal cortex

Human

New connections evolve in the other regions

Yet the size of the human brain is not the only thing that sets it apart from the brains of other primates. Researchers have pinpointed certain changes in neural circuitry that might have played an important role in the evolution of human cognition (Sousa et al. 2017). For example, the frontal and temporal lobes in the primate brain are linked by long-distance nerve fibers, a tract known as the arcuate fasciculus (Rilling et al. 2008). The common ancestor of chimpanzees and humans evolved a denser bundle of fibers than did other primates. And humans evolved even more connections (**Figure 17.35**).

From this kind of evidence from living species, we can't know when the arcuate fasciculus became denser in the human lineage. It's conceivable that australopithecines experienced this change; on the other hand, it's possible that it didn't occur until after our lineage split from Neanderthals. But scientists are now starting to gain some important clues about the brain structure of extinct hominins. Although Neanderthals had brains of the same volume as ours, their brains had a different shape. Living humans have a more globe-shaped brain, whereas Neanderthal brains are more elongated (**Figure 17.36**). The oldest *Homo sapiens* fossils started out with a similarly elongated shape,

Figure 17.36 A: The average Neanderthal brain (left) was as large as a modern human one (right). But it was more elongate than our more globular brain. B: Some evidence suggests that the enlargement of the putamen (red) and cerebellum (blue)—both involved in speech—were the cause of this change in shape. (A, B: Philipp Gunz)

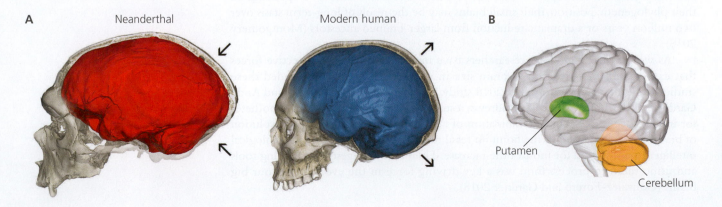

A   Neanderthal   Modern human   **B**

Putamen

Cerebellum

and over the next 200,000 years or so, they became rounded. Studies on both fossils and Neanderthal DNA suggest that this change was the result of changes in certain regions of the brain. Two regions, known as the cerebellum and the putamen, seem to have become enlarged in the modern human lineage (Gunz et al. 2018).

Scientists are still a long way from making sense of these evolutionary changes in the brain. But what makes these findings especially exciting is that all of these features are important for one uniquely human trait: language.

- Hominin brains have evolved dramatically over the past four million years. Hypotheses to account for this expansion are based on ecological, social, and cultural factors. ●

Key Concept

17.11 The Language Instinct

As far as we know, no other species can communicate with full-blown language—a system made up of sounds, gestures, or written symbols that convey information not just about what's immediately in front of us but about what lies in the distant past, in the far-off future, or in a world that never will be.

Language, many scientists argue, is at the core of human nature. To use it, we have to be able to understand abstract concepts instead of just using labels for obvious things like snakes or birds. Language lets us do things other animals cannot do, such as make complex plans together, gain a deep understanding of the inner lives of other humans, and teach each other new concepts and technologies (Pinker 1995).

There are more than six thousand languages on Earth, each the product of a particular culture with a particular history. But underlying this staggering diversity, languages have a lot in common. All spoken languages are based on sets of sounds, which can be combined into thousands of words, which in turn can be combined into a countless number of sentences. Words cannot be tossed randomly into a sentence in any language; they all share some basic rules of grammar and depend on **syntax** for their meaning.

Syntax is the arrangement of words and phrases to create well-formed sentences in a language.

Despite language's complexity, children don't need to attend a linguistics class to learn how to speak. They quickly pick up the basic rules of grammar for themselves in the first few years of life. Along with having a capacity for learning the rules of language, our brains are well adapted to hearing speech. When we listen to someone speaking, the network of brain regions that becomes active is not the same as the one we use to listen to ordinary sounds. Certain kinds of brain damage cause "word deafness," leaving people unable to understand speech but still able to hear other sounds.

All of this evidence suggests that language in humans is an adaptation that was shaped by natural selection. Scientists are investigating this evolutionary transformation at several different levels. Some researchers search for language-related behaviors in our nonhuman relatives to isolate the novel features that evolved in the hominin lineage (Fitch 2017; Dediu and Levinson 2018). We have an unusually powerful, precise control over the larynx, for example, and are unique among primates in quickly learning new vocal patterns. We can give these vocal patterns a practically infinite variety of meanings, thanks to syntax. We can organize words into a vast number of combinations, and we use the rules of syntax to understand the meaning of those combinations.

Other researchers are searching for the genes underlying human language in the hope of finding clues to its molecular evolution. Currently, they are only beginning to work out the genetic basis of language (Deriziotis and Fisher 2017; Fisher 2017). Some of these genes have come to light in the study of hereditary language disorders such as stuttering and dyslexia. Some genes appear to lead to different growth patterns of certain populations of neurons.

A third line of research, which we touched on earlier, is to track the evolution of the brain itself. The expansion of the arcuate fasciculus, for example, may have created stronger links between different parts of the brain that have to cooperate to make language possible. The cerebellum and putamen are both important for muscles to learn new patterns—including the larynx. Such studies have raised a fascinating possibility: maybe full-blown language only evolved in our own species.

Taken together, these three lines of research suggest that language emerged gradually (Fitch 2017). Parents may have begun to use a singsong pattern of sounds to transmit information to their young. Children may have evolved the capacity to extract meaning from their parents' sounds, and eventually adults organized their sounds with syntax, allowing them to develop more complex social networks. It's possible—although far from established—that once full-blown language emerged in our own species, we gained a crucial advantage over other hominins.

Key Concept
- Many of the adaptations for understanding language likely evolved early in the hominin lineage, but some specializations are uniquely human. ●

17.12 The Continuing Evolution Of *Homo sapiens*

By 40,000 years ago, modern humans were pretty much as living humans are today. Their anatomy was the same, they could make figurative art, and they likely spoke full-blown language. If you could reconstruct the genome of a 40,000-year-old fossil and place her DNA in a donor egg, the resulting offspring would grow up into a person who would fit comfortably in the range of living humans. Not only would she be indistinguishable physically, but she would have the capacity to learn any language and adapt to life in the twenty-first century.

But none of this means that humans have stopped evolving in the past 40,000 years. Allele frequencies have changed in modern populations due to the mechanisms we discussed in Chapter 6, such as drift and selection.

Bottlenecks and Genetic Drift

Genetic drift causes certain alleles to change frequency not due to their effects on fitness but due to random events. Drift is stronger in smaller populations than in larger ones, as well as in populations that go through bottlenecks. And it is pervasive whenever small numbers of individuals found new populations. Human history has been punctuated by such conditions as our species has migrated around the world. Small groups of Africans migrated into the Near East, for example, before spreading into Europe to the west and Asia and Australia to the east. About 16,000 years ago, a small population of Siberians migrated over the Bering Land Bridge, giving rise to all the indigenous people of the Americas. Long after these small founding populations expanded to much larger sizes, they retained the legacy of their bottleneck events and genetic drift in their distinctive patterns of genetic variation.

Scientists have been able to uncover these patterns by carrying out large-scale genetic studies on human populations. In one of these studies, Jun Li and colleagues sampled the genotypes of 938 individuals spanning 51 populations worldwide, and they examined variation at more than 630,000 loci. They found that the farther from Africa a population was, the smaller its level of genetic diversity. That's what you'd expect if our species originated in Africa and then populated the rest of the planet in a series of successive founding events (**Figure 17.37**; Li et al. 2008).

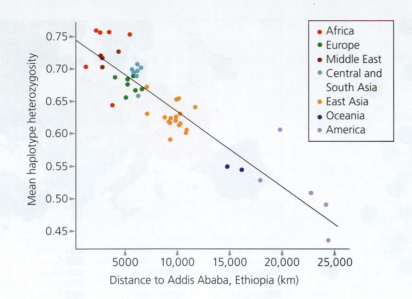

Figure 17.37 All non-Africans descend from a small population of humans who lived in East Africa. The farther populations are from Ethiopia, the lower their allelic diversity. This decline reflects founder effects caused when small populations settled new regions of the world. (Data from Li et al. 2008)

Once people established themselves in different parts of the world, they did not stay put (Reich 2018). From time to time, certain populations have expanded into places occupied by other populations, where they have either interbred or pushed them out. We can look at written history for stories of recent migrations, such as the European colonization of the Americas starting in the fifteenth century. But ancient DNA has also revealed its own stories of populations on the move. In Africa, for example, a group of farmers in Cameroon known as the Bantu expanded east and south starting two thousand years ago, moving into territories previously occupied by hunter-gatherers (Skoglund et al. 2017). Around the world, populations today carry genetic legacies of such mixtures.

Although these movements had a major impact on the world's populations, they did not happen frequently. Most of the time, human populations remained fairly isolated. Hunter-gatherers lived in small bands that moved around a relatively small territory, making contact with relatively few other groups. Even after the agricultural revolution, most people lived around the same farming villages where their forebears had lived. Isolated, geographically separated populations began to diverge from each other due to genetic drift.

John Novembre, then at the University of California, Los Angeles, and colleagues examined the genetic structure that resulted from this drift by studying people from across Europe. They looked at allelic variation in a sample of three thousand individuals genotyped at more than 500,000 variable loci. They then sorted individuals based on overall genetic similarity and plotted these points in a two-dimensional graphical summary.

The genetic pattern had striking parallels with geography: individuals who were close to each other genetically also tended to be close geographically, and vice versa (Figure 17.38). In fact, the spatial resolution of this genetic pattern was so fine that Novembre and his colleagues were able to distinguish between French-, German-, and Italian-speaking groups within the small country of Switzerland. They suggest that the genetic structure of human populations is sufficiently precise that forensic anthropologists may someday be able to use DNA samples to localize the geographic origin of individuals to within a few hundred kilometers (Novembre et al. 2008).

Such insights into genetic drift across our species are the foundation of direct-to-consumer ancestry tests from companies such as 23andMe and Ancestry.com. When customers submit a saliva sample, the companies extract DNA and genotype it. They compare the genetic markers in each customer's genome to a database of tens of thousands of people from populations around the world. They can then estimate the percentage of DNA that each customer inherits from each ancestral population (Zimmer 2018).

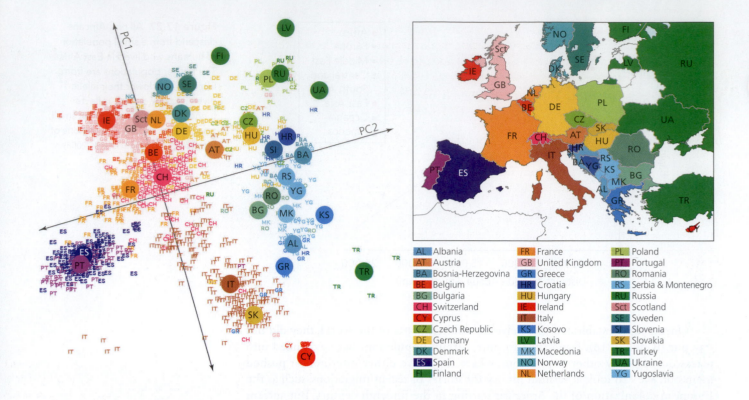

AL	Albania	FR	France	PL	Poland
AT	Austria	GB	United Kingdom	PT	Portugal
BA	Bosnia-Herzegovina	GR	Greece	RO	Romania
BE	Belgium	HR	Croatia	RS	Serbia & Montenegro
BG	Bulgaria	HU	Hungary	RU	Russia
CH	Switzerland	IE	Ireland	Sct	Scotland
CY	Cyprus	IT	Italy	SE	Sweden
CZ	Czech Republic	KS	Kosovo	SI	Slovenia
DE	Germany	LV	Latvia	SK	Slovakia
DK	Denmark	MK	Macedonia	TR	Turkey
ES	Spain	NO	Norway	UA	Ukraine
FI	Finland	NL	Netherlands	YG	Yugoslavia

Figure 17.38 The genetic structure of European populations reflects their geography. Plotting the variation among genotypes reveals significant clumping. Slightly tilting the graph reveals its alignment with the geography of Europe. (Data from Novembre et al. 2008)

Recent Human Selection

Natural selection has been an important mechanism throughout hominin evolution. There's no other known mechanism that can give rise to adaptations such as efficient bipedalism or language. We can only study how natural selection shaped hominins millions of years ago indirectly, by studying fossils. But in more recent times, the DNA of living humans and ancient DNA allow scientists to inspect natural selection on a molecular level.

As modern humans have spread over the planet and developed new ways of finding food and shelter, they have experienced natural selection (Marciniak and Perry 2017). In some parts of the world, researchers have obtained ancient DNA from many skeletons across thousands of years. They can then compare the alleles over time, searching for changes in frequency that may be the result of selection.

In Chapter 8, for example, we considered the case of lactase persistence—the ability of some human populations to produce lactase as adults, enabling them to digest milk. The first evidence for this adaptation emerged when scientists compared levels of lactose tolerance in different countries, discovering that it was high in places where cattle herding had been common for thousands of years. Then researchers compared the DNA of these populations and discovered that markers adjacent to the *LCT* allele for lactase persistence were in a state of linkage disequilibrium—a sign of recent, strong natural selection.

Researchers have combined these findings with archaeological evidence about the history of cattle herding to come up with hypotheses for how this natural selection unfolded. In Europe, for example, the oldest evidence for farming—both raising crops and herding livestock—reaches back about 8500 years. Once agriculture made it possible for people to drink milk, the *LCT* allele started to be favored by natural selection—presumably because people who could digest milk after infancy were more likely to survive.

Now DNA recovered from ancient skeletons is allowing scientists to track the rise of lactase persistence in real time. The new evidence does not support the earlier hypotheses (Mathieson and Mathieson 2018).

European hunter-gatherers, it turns out, did not invent agriculture, nor did they borrow it from neighboring societies. Instead, farmers brought their crops, animals, and agricultural technology to Europe. After crops were first domesticated in the Near East about 10,500 years ago, farmers from that region expanded northwest through Anatolia (where modern Turkey is now located) and into Europe.

European hunter–gatherer populations either collapsed or intermixed with the farmers. And then, about 4500 years ago, another wave of people arrived in Europe from the steppes of what is now Russia. This new population—originally horse-riding nomads who herded cattle and sheep—spread across the entire continent, reaching as far west as Ireland and Spain. The ancestry of living Europeans is largely a mixture of these three populations, dominated by the DNA of the steppe nomads.

Ancient DNA recovered from Europe's first farmers reveals that they did not carry the *LCT* allele for lactase persistence. This is fascinating when you consider the archaeological evidence left behind by these early farmers. Archaeologists have found pots, for example, with traces of milk still embedded in them.

It turns out that the *LCT* allele came to Europe with the arrival of nomads from the Russian steppes. It was rare in that population, and it remained rare in Europe even as the nomads expanded rapidly across the continent. Only centuries later did the frequency of the *LCT* allele begin a steady increase (**Figure 17.39**).

A

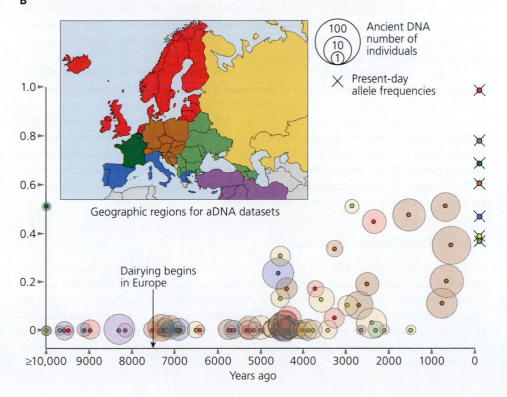

B

Figure 17.39 A: The origin of milking in regions such as Europe led to the selection for lactase persistence. B: Ancient DNA from European skeletons shows that this trait did not become widespread for more than two thousand years after the rise of dairying in Europe. (A: JT Vintage/Glasshouse Images/Alamy Stock Photo; B: Republished with permission of Nature Publishing Group, from "Harnessing ancient genomes to study the history of human adaptation" by Stephanie Marciniak & George H. Perry in Nature Reviews Genetics volume 18, pages 659–674 (2017), permission conveyed via Copyright Clearance Center, Inc.; data from Mathieson and Mathieson 2018)

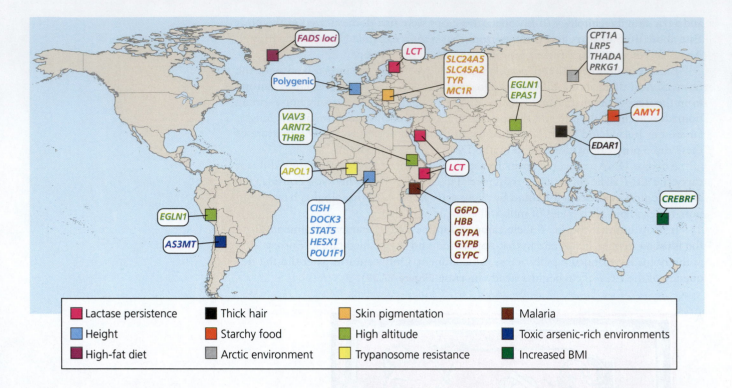

Figure 17.40 Studies on DNA from living people and ancient skeletons have documented natural selection in response to local environmental conditions around the world. (Data from Fan et al. 2016)

It's possible that early farmers turned milk into low-lactose dairy products like cheese and yogurt and only ate them in small amounts. Even after nomads introduced the *LCT* allele to Europe, it may have remained rare because Europeans continued to supplement their diets with wild plants and game. Alternatively, the selective advantage of agriculture-related alleles may only have emerged once the population of Europe became dense (Mathieson and Mathieson 2018). These new hypotheses, like the ones they replace, remain to be tested against new evidence.

Research on other continents—on DNA from fossils and from living people—has revealed many other examples of recent human selection. **Figure 17.40** shows some examples. The domestication of plants and animals has driven the spread of many alleles enabling the digestion of foods like milk. In South America, for example, the domestication of potatoes five thousand years ago in the Andes led to natural selection favoring an allele for better digestion of starch. Diseases such as malaria have driven natural selection on a variety of alleles for resistance.

In just the past two centuries, human cultural evolution has caused profound changes in our environment. Much of the world has gained access to clean water, new kinds of medicine and vaccines, and a more reliable supply of food. As a result, rates of mortality have dropped, especially among children. In Germany, for example, 270 in 1000 children died in infancy in 1880. Today, fewer than 4 in 1000 die. Average human height has increased in many countries, but not due to natural selection on certain height alleles. Instead, the increase is an example of human phenotypic plasticity (Chapter 7). An increased supply of food has reprogrammed human growth rates to higher and higher levels with each generation.

Twenty-First-Century Evolution

In some ways, modern life has reduced the opportunities for natural selection by reducing the relative fitness consequences of certain genotypes. An allele that made a child more likely to die from an infection a hundred years ago has much less effect on fitness today, when a child can be cured with antibiotics. Even seemingly minor inventions like eyeglasses may be affecting human evolution. People who would have been practically blind a thousand years ago can (with help) see about as well as people with perfect eyesight.

Nevertheless, natural selection continues to act on certain traits, even in the most affluent societies on Earth. Stephen Stearns, an evolutionary biologist at Yale University, and his colleagues studied natural selection in 2238 U.S. women. The women were participants in a major medical study tracking health in the town of Framingham, Massachusetts, since 1948. The scientists searched for traits that were correlated with having a higher number of children. Then they checked to see whether those traits tended to be passed down from mother to child—in other words, whether they were heritable (Byars et al. 2010).

The scientists discovered that a handful of traits are indeed being favored by natural selection. Women with a genetic tendency for low cholesterol, for example, had more children on average than women with high cholesterol. A greater body weight also led to greater reproductive success, shorter height, lower blood pressure, an older age at menopause, and having one's first child at an earlier age.

Stearns and his colleagues have yet to determine exactly what advantage each of these traits confers—a situation that evolutionary biologists often face when documenting natural selection in action. Still, based on the strength of this natural selection and with all else being equal, the scientists made these predictions: ten generations from now, the women of Framingham will first give birth, on average, a few months younger than today; they will have 3.6% lower cholesterol, and they will be 1.3% shorter.

Key Concepts

- Geographic isolation and genetic drift have resulted in the accumulation of local differences in allele frequencies across populations. This genetic structure forms the basis of ancestry tests like 23andMe and Ancestry.com.

- Humans have adapted genetically to recent changes in their selection environment, such as changes in our diet due to the advent of agriculture and domestication of animals. ●

17.13 Emotions as Evolutionary Legacies

There are two sides to the study of human evolution. On one side, we can learn a lot about ourselves by tracing the origin of innovations in the hominin lineage. On the other, we can also gain insights from the deep connections we still have with other animals. Our distant ancestors evolved a left–right symmetry to their bodies, for example, when they were still wormlike animals crawling on the seafloor 600 million years ago. Even today, our bodies retain that same symmetry. As Darwin himself wrote, "Man still bears in his body the indelible stamp of his lowly origin" (Darwin 1871).

Our evolutionary legacy also extends into our minds. Although we may be unique in the animal kingdom for our abilities to reason, plan ahead, and use language, our behavior still shares profound similarities with that of other animals (LeDoux 2012).

The Emotional Brain

By 100 million years ago, the basic systems that are essential for many of our feelings had already evolved. Fear, for example, triggers responses from an almond-shaped region on the underside of the brain called the amygdala. The amygdala becomes active at the sight of fearful things, such as a picture of an angry face. In fact, neuroscientists can observe activity in the amygdala when people see these pictures for just a tenth of a second, which is not enough time for them to become aware they've even seen anything. Our brains have shortcuts that can relay information from our eyes and ears to the amygdala without passing through the cerebral cortex. The amygdala, in turn, sends signals to other parts of the brain that produce changes in the body, such as a rapid heartbeat and heightened attention (LeDoux 2007).

Some of the earliest insights about the human amygdala came from studies on the brains of mice and rats. In the mid-1900s, a number of scientists began to experiment on rodents to understand the biology of fear. Like all other mammals, rats and mice have amygdalas, which are linked to other parts of the brain in much the same arrangement as in the human brain. By implanting electrodes in the brains of the animals, scientists can observe how rodents learned fear as synapses in the amygdala became stronger.

The fear we feel is similar to the fear experienced by a rat, but not identical. Reading threatening words, like *poison* or *danger*, is enough for the human amygdala to become active. Obviously, that's an experience a rat will never have. But our reactions to those words evolved from the same underlying circuitry we share with our rodent cousins—reactions that likely existed in our common mammalian ancestors. This capacity to feel fear allowed them to respond quickly to threats. If they saw a predator about to attack suddenly, for example, those early mammals could freeze, flee, or retaliate. Some of the dangers that mammals faced were reliable enough that natural selection could produce instinctive fears. Some studies suggest, for example, that we are born with an innate fear of snakes, which would have threatened our primate ancestors for millions of years (Van Le et al. 2013). But mammals did not have to rely only on hardwired fears. They could also learn a healthy fear for any new dangers they might encounter.

The Chemistry of Reward and Addiction

There's more to life than being scared. Motivations help mammals reach important goals such as finding food or mates. The most important region for generating these motivations is a small cluster of neurons in the brain stem. If a rat, for example, is searching for food and unexpectedly gets a whiff of something delicious, those neurons will release a tiny surge of a neurotransmitter called dopamine. The dopamine-producing neurons have a vast number of connections to many networks in the brain, so they can quickly alter how the entire brain functions. Dopamine arouses an animal's attention and also makes it easier for neurons to form stronger connections with other neurons. A rat's brain can begin to associate cues like odors with its long-term goals, such as finding food.

The power of dopamine over the mammalian brain is astonishing. One way to demonstrate its importance is by genetically engineering mice so that they can't produce it. These dopamine-free mice are in many ways perfectly normal. They still prefer the taste of sucrose to other foods, and they can learn where food is located. But they lose the motivation to pursue any goals. They will simply starve from that lack of motivation to eat less than a month after they're born. If scientists give these mice injections of dopamine, however, they will feed for about 10 hours, until the motivation disappears again (Palmiter 2008).

Too much dopamine can be just as dangerous as too little. Scientists often reward rats by giving them food if they press a lever in response to the right signal—in response to a green light but not a red one, for example. The rat's brain produces surges of dopamine as it learns the rule. If the scientists give the rats an injection of dopamine each time they press the lever, however, something else happens. The rats will keep pressing the lever again and again. They will do nothing else, not even eat. Ultimately, they may die of starvation.

Humans have inherited the same dopamine delivery system. It doesn't make us feel happy so much as eager with anticipation. The rewards that can trigger a release of dopamine are, like our fears, more sophisticated than those that occur in a rat's brain. Winning at a slot machine, feeling the lunge of a fish on a lure, the sight of an attractive face, or even hearing a joke can trigger a dopamine release. Unfortunately, the reward system can also be hijacked by substances that cause the brain to release unnaturally large amounts of dopamine. Cocaine and other drugs do just this, and it's why they can become so addictive (Volkow et al. 2010).

Deep Bonds

Our relationships with other people lead to many of our most intense emotions—parental love for children, for example, or the romantic love between adult partners. Psychologists have found that certain hormones in the brain play crucial roles in forming these bonds (Walum and Young 2018). It also turns out that these hormones, such as oxytocin and vasopressin, have an evolutionary history reaching back 600 million years (**Figure 17.41**). By comparing their effects in other species, we can discern the evolutionary roots of these emotion-controlling hormones.

A hormone called oxytocin, for example, is essential for the bond between mothers and their offspring in all living mammals—humans included (Donaldson and Young 2008; Saltzman and Maestripieri 2010; Carter 2014). A region of the brain known as the hypothalamus produces oxytocin late in pregnancy. After being released into the bloodstream, some of these oxytocin molecules latch onto receptors in the mammary glands, causing them to begin producing milk. Some oxytocin molecules latch onto neurons in the brain, altering a mother's behavior.

In sheep, for example, oxytocin causes ewes to bond with their lambs just after birth. They will be able to recognize the smell and bleat of their own lambs for the weeks that they spend nursing. If scientists block the uptake of oxytocin, however, ewes reject their lambs. If ewes that aren't even pregnant get an injection of oxytocin, they start behaving like a mother to an unrelated lamb.

Oxytocin is also produced by the brains of human mothers during pregnancy and after a baby's birth. The touch of a baby during nursing is enough to trigger an increase of the hormone. Oxytocin tends to cause women to bond more with their babies, as measured by the sounds they make, the number of times they check in on the children, and how much they gaze at them.

Figure 17.41 Vasopressin and oxytocin are important hormones for human emotions and social behavior. The genes for these two hormones belong to a family of homologous hormone genes found in other animals. These other hormones play a similar role in behavior. Vasopressin has close homologs in mammals, including lysipressin and phenypressin, and vasotocin in other vertebrates. Likewise, oxytocin in mammals is homologous to mesotocin and isotocin. Invertebrates have more distantly related hormones, shown in the bottom box. Amino acids in bold are found at the same positions in all these species. (Data from Donaldson and Young 2008)

Vasopressin (mammals)
Cys · Tyr · Phe · Gln · **Asn · Cys · Pro** · Arg · **Gly** · NH₂

Lysipressin (pigs, marsupials)
Cys · Tyr · Phe · Gln · **Asn · Cys · Pro** · Lys · **Gly** · NH₂

Phenypressin (marsupials)
Cys · Phe · Phe · Gln · **Asn · Cys · Pro** · Arg · **Gly** · NH₂

Vasotocin
Cys · Tyr · Ile · Gln · **Asn · Cys · Pro** · Arg · **Gly** · NH₂

Oxytocin
Cys · Tyr · Ile · Gln · **Asn · Cys · Pro** · Leu · **Gly** · NH₂

Mesotocin
Cys · Tyr · Ile · Gln · **Asn · Cys · Pro** · Ile · **Gly** · NH₂

Isotocin
Cys · Tyr · Ile · Ser · **Asn · Cys · Pro** · Leu · **Gly** · NH₂

Annepressin (annelid worms)
Cys · Phe · Val · Arg · **Asn · Cys · Pro** · Thr · **Gly** · NH₂

Conopressin (snails, cones, sea hare, leeches)
Cys · Phe/Ile · Ile · Arg · **Asn · Cys · Pro** · Lys/Arg · **Gly** · NH₂

Inotocin (some insects)
Cys · Leu · Ile · Thr · **Asn · Cys · Pro** · Arg · **Gly** · NH₂

Hormones are also important for romantic love. Although romantic love may be one of the highlights of human existence, it's a peculiar feature of humanity from a biologist's point of view. It leads men and women to form long-term bonds, having sex with their partner to the exclusion of other partners. In most mammalian species, males and females have sex only when the female is in estrus, and males and females often mate with multiple partners in a single mating season.

The unusual pair-bonding in our species may be the result of our unusual brains. Human babies are born with large brains, which continue to grow and develop mature connections throughout childhood. This life history strategy is different from that of other primates, and it leaves human children exquisitely vulnerable. They are born in need of lots of calories to nourish their big developing brains. But they're unable to get that food on their own for many years. Romantic love may have its origins in natural selection for long-term bonds between human parents. Emotional connections that keep mothers and fathers together may lead to their children getting a better supply of food.

To better understand the evolution of long-term pair bonds, Larry Young, of Emory University, studies rodents. Two closely related species, prairie voles and meadow voles, have dramatically different mating systems. Male meadow voles spend very little time with females beyond that needed to mate, and they mate with several different females in a season (a system known as polygyny; Chapter 11). Prairie voles, on the other hand, are more like us. A male prairie vole will mate with a single female and live with her (a system known as monogamy). Young wondered if the two species differed in their hormones and brains in ways that might account for their different mating systems.

Young and his colleagues have been able to make precise measurements— and manipulations—of hormones such as oxytocin and vasopressin in these species. Prairie voles have a higher density of oxytocin-producing neurons in their brains than meadow voles, for example. What's more, prairie vole males have higher levels of expression of the vasopressin receptor gene *AVPR1a* in their forebrains than do the meadow voles. Young and his colleagues can make the polygynous meadow vole males behave like monogamous prairie voles by increasing the expression of this receptor in their brains (**Figure 17.42**; Lim et al. 2004; Donaldson and Young 2008).

Oxytocin and vasopressin may have also been co-opted in humans as we evolved romantic bonds. Researchers have found, for example, that new romantic partners have higher levels of oxytocin in their blood and have greater activity in the regions of the brain that contain oxytocin-producing neurons (Feldman 2017).

To probe the effects of these hormones, researchers have fashioned sprays that can deliver the molecules into the nose, from where they can enter the bloodstream and eventually reach the brain. Studies on rodents and humans alike suggest that oxytocin and other hormones can influence behaviors toward other individuals, even if they're not relatives or partners. People given oxytocin can become more trusting of others and more willing to forgive. They even do a better job of empathizing with other people simply by looking at their facial expressions.

These results have triggered a wave of research into using these hormone sprays as psychiatric treatments (Andari et al. 2017). Extra oxytocin, some researchers have hypothesized, can help alleviate the symptoms of conditions ranging from anxiety disorders to the autism spectrum. But Young and other researchers have cautioned that this line of research will only have a chance of translating into clinical treatments if scientists gain a firm understanding of the evolution of our emotions (Putnam et al. 2018).

Researchers have learned much more about how oxytocin and vasopressin work in rodents than in humans. That's because they can directly manipulate the brains of the animals in their experiments. Although the parallels between rodents and humans are indeed striking, the fact remains that the two lineages of mammals are separated by 100 million years of evolution. Over the course of primate evolution, the pathways of oxytocin-producing neurons in the brain have been rerouted, responding more to visual stimuli than to smell. It remains to be discovered how this hormone affects

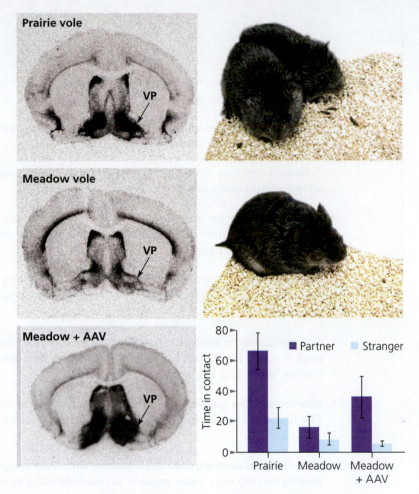

Figure 17.42 Prairie voles form strong bonds with their mates, whereas their relatives, the meadow voles, mate with many females. Vasopressin plays a critical role in driving the social behavior of prairie voles, and they have higher levels of expression of a vasopressin receptor, *AVPR1a*, in their ventral forebrains than meadow voles do. Artificially increasing expression of this gene in the forebrains of meadow voles (AAV) is sufficient to induce pair-bonding behavior, effectively switching the mating system of this species from polygyny to monogamy. (Republished with permission of American Association for the Advancement of Science from "Oxytocin, Vasopressin, and the Neurogenetics of Sociality" by Donaldson and Young, Science, 7 November 1008: Vol. 322 No. 5903, pp. 900–904; permission conveyed through Copyright Clearance Center; data from Donaldson and Young 2008)

the complex social networks in the primate brain—and how those connections have changed in the human brain. This evolution-based understanding of our emotions, Young and his colleagues argue, will allow for the precise manipulations that can offer safe and effective psychiatric therapies.

This research demonstrates how human evolution is about much more than fossil fragments. Human evolution has shaped our lives up to the present day, and by exploring our evolution, researchers can gain new insights that can advance medical research. Scientists are gaining insights into conditions ranging from infectious diseases to cancer to obesity by appreciating how our bodies are the product of millions of years of evolution. In Chapter 18, we will end our exploration of evolution by considering the emerging field of evolutionary medicine.

Key Concepts

- The chemistry of human emotions such as fear, anticipation, nurturing, and love is the result of a history of selection and adaptation. Many homologies are shared between humans and other mammals.

- Studying animal models of emotion can help us understand psychiatric disorders, but only if we take evolution seriously into account. ●

TO SUM UP . . .

- Hominins include humans and all other species more closely related to us than to chimpanzees and bonobos.

- The hominin fossil record extends back seven million years, agreeing roughly with molecular clock studies.

- Scientists continue to clarify the relationships among taxa in the human lineage as more fossil evidence is discovered and interpreted.

- Hominin bipedalism gradually became an efficient mode of transport. Hypotheses for the evolution of bipedalism include benefits from more efficient feeding and heat dissipation.

- The ability to make and use stone tools evolved perhaps as long ago as 3.3 million years. Hominins used these tools to scavenge meat, among other uses.

- The genus *Homo* evolved more than two million years ago. Mosaicism is common in early *Homo* taxa, but later species were characterized by larger brains, smaller teeth, and longer legs for efficient walking.

- Members of the genus *Homo* were the first known hominins to leave Africa. The first wave left Africa about 1.8 million years ago and spread across Eurasia.

- Neanderthals were sophisticated hunters who were adept at making tools, and their brain may have been capable of understanding symbols.

- *Homo sapiens* evolved in Africa 300,000 years ago and expanded out of Africa about 70,000 to 50,000 years ago. Our species interbred with at least two other hominin lineages after leaving Africa.

- The expansion of the hominin brain may have been driven by the demands of ecological, social, and cultural intelligence.

- Language evolved as a tool for communication, through changes in connections in the brain.

- The spread of humans across the planet and the rise of civilization allowed for new opportunities for drift and natural selection to reshape the human genome and change the trajectory of human evolution.

- Emotions such as love and fear have evolutionary roots in distant mammalian ancestors.

1. All of the following are true of primates *except*

   a. independent lines of evidence from morphology and DNA confirm that humans are primates.

   b. Charles Darwin was the first to put humans in the order Primates, along with species such as apes, monkeys, and lemurs.

   c. the primate clade originated at the end of the Cretaceous period.

   d. within the primate clade, the closest living relatives of humans are chimpanzees and other apes.

2. All of the following are features related to bipedalism that modern humans share with early hominin fossils *except*

   a. similar anchors for muscles of the pelvis.

   b. vertically oriented foramen magnum.

   c. stiffened foot arch.

   d. lighter, elongated pelvis.

3. Which of these statements is *true*?

   a. Efficient bipedal walking evolved before the species *Homo floresiensis* diverged from the closely related *Homo erectus*.

   b. Efficient bipedal walking evolved before the species *Homo floresiensis* diverged from the closely related *Homo sapiens*.

   c. Efficient bipedal walking evolved in *Homo floresiensis* after the species diverged from a common ancestor shared with *Homo erectus* and *Homo sapiens*.

   d. All of the above could be true depending on the phylogenetic hypothesis used.

4. What might additional fossils of early hominins add to our understanding of the origin of humans?

   a. Additional fossils could provide evidence to support the hypothesis that bipedal walking evolved before human ancestors left Africa.

   b. Additional fossils could provide evidence about the timing of speciation within the genus *Homo*.

   c. Additional fossils could provide evidence to support the hypothesis that *Homo erectus* shares a more recent common ancestor with *Homo floresiensis* and/or *Homo georgicus* than with *Homo sapiens*.

   d. All of the above.

5. The hypothesis that contends that the evolution of large brain size in modern humans is a result of our ability to learn from others is called

   a. the cultural evolution hypothesis.

   b. the ecological intelligence hypothesis.

   c. the foramen magnum hypothesis.

   d. the social intelligence hypothesis.

6. Which pattern in allelic diversity would you expect to find if you hypothesized that humans first left Africa and then spread across all of the continents?

   a. Allelic diversity would increase continuously as populations moved farther and farther from Africa.

   b. Allelic diversity would be similar between *Homo sapiens* and *Homo neanderthalensis* but different from *Homo heidelbergensis*.

   c. Allelic diversity for populations outside of Africa would form a monophyletic group that nests within the diversity present in Africa.

   d. Allelic diversity is far too high to detect any pattern because the human population is so large.

7. Why do *Homo sapiens* and *Homo neanderthalensis* share DNA if not from common recent descent?

   a. Because *Homo sapiens* and *Homo neanderthalensis* hybridized during their coexistence.

   b. Because techniques for examining DNA are imperfect.

   c. Because *Homo sapiens* and *Homo neanderthalensis* are not really distinct species.

   d. Because of sampling error in the methodology.

8. Is natural selection currently favoring lower cholesterol levels in women?

   a. No. Cholesterol is necessary for body maintenance and growth and generally causes heart disease only later in life; it is likely a result of senescence with very little effect on reproduction in women.

   b. No. Individual women vary in the tendency to metabolize cholesterol, but that variation is largely due to diet, and diet is more related to culture—not genetically transmitted.

   c. Yes. Women with greater body weight have greater reproductive success and lower cholesterol.

   d. Yes. Individual women vary in the tendency to metabolize cholesterol, this ability has a genetic component that can be transmitted to offspring, and individuals with a genetic tendency for low cholesterol have greater reproductive success than women with a genetic tendency for high cholesterol.

9. Which molecule(s) related to our emotions likely evolved early in our mammalian history?

   a. Oxytocin.

   b. Dopamine.

   c. Vasopressin.

   d. All of the above.

10. How has human culture altered the selection environment experienced by humans?

   a. Technology has relaxed the intensity of selection by minimizing the fitness costs associated with previously deleterious traits (for example, poor eyesight).

   b. Technology has altered the nature of selection experienced by people, such as by changing the types of foods we eat (for example. domesticated, processed foods).

   c. Increased group sizes (for example, towns, cities) have resulted in selection for heightened social awareness and social intelligence.

   d. All of the above.

11. Which evolutionary scenario best explains the relatively small brain size of *H. floresiensis*?

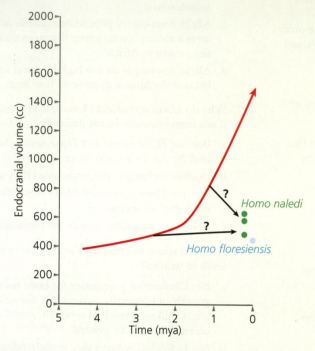

a. Gene flow resulting from interbreeding with *Australopithecus* populations on the island of Flores prevented evolution of increased brain size in *H. floresiensis*.

b. *H. floresiensis* actually represents a relict population of an older lineage of *Homo* that never evolved a large brain size.

c. Social/ecological environments resulted in selection for a secondary reduction in brain size after these hominins colonized the island of Flores.

d. Both b and c.

SHORT ANSWER QUESTIONS   Answers can be found at the end of the book.

12. How can studying living primates provide clues to our hominin ancestry?

13. Would you expect the common ancestor of chimpanzees and early hominins to have used tools? Why or why not?

14. What evidence indicates that our species, *Homo sapiens*, split from *Homo heidelbergensis* and not *Homo neanderthalensis*?

15. Is the capacity for complex language unique to humans? Explain your reasoning.

16. How has genetic drift influenced the evolution of humans?

17. How do our genomic legacies shape human evolution today? Explain.

ADDITIONAL READING

Boyd, R., and J. Silk. 2009. *How Humans Evolved*. 5th ed. New York: W. W. Norton.

Gibbons, A. 2006. *The First Human: The Race to Discover Our Earliest Ancestors*. New York: Doubleday.

Maslin, M. A., S. Shultz, and M. H. Trauth. 2015. A Synthesis of the Theories and Concepts of Early Human Evolution. *Philosophical Transactions of the Royal Society B* 370:20140064.

Pinker, S. 1995. *The Language Instinct*. New York: Harper Perennial.

Reich, D. 2018. *Who We Are and How We Got Here*. New York: Oxford University Press.

Stringer, C. 2012. *Lone Survivors: How We Became the Only Humans on Earth*. New York: Times Books.

PRIMARY LITERATURE CITED IN CHAPTER 17

Aiello, L. C. 2010. Five Years of *Homo floresiensis*. *American Journal of Physical Anthropology* 142:167–79.

Aiello, L. C., and P. Wheeler. 1995. The Expensive-Tissue Hypothesis: The Brain and the Digestive System in Human and Primate Evolution. *Current Anthropology* 36 (2): 199–221.

Andari, E., R. Hurlemann, and L. J. Young. 2017. A Precision Medicine Approach to Oxytocin Trials. *Current Topics in Behavioral Neurosciences* 35:559–90.

Antón, S. C., H. G. Taboada, E. R. Middleton, C. W. Rainwater, A. B. Taylor, et al. 2016. Morphological Variation in *Homo erectus* and the Origins of Developmental Plasticity. *Philosophical Transactions of the Royal Society B* 371:20150236.

Argue, D., C. P. Groves, M. S. Y. Lee, and W. L. Junger. 2017. The Affinities of *Homo floresiensis* Based on Phylogenetic Analyses of Cranial, Dental, and Postcranial Characters. *Journal of Human Evolution* 107:107–33.

Aubert, M., P. Setiawan, A. A. Oktaviana, A. Brumm, P. H. Sulistyarto, et al. 2018. Palaeolithic Cave Art in Borneo. *Nature* 564:254–7.

Bae, C. J., K. Douka, and M. D. Petraglia. 2017. On the Origin of Modern Humans: Asian Perspectives. *Science* 358:eaai9067.

Berger, L. R., J. Hawks, P. H. G. M. Dirks, M. Elliott, and E. M. Roberts. 2017. *Homo naledi* and Pleistocene Hominin Evolution in Subequatorial Africa. *eLIFE* 6:e24234.

Beyene, Y., S. Katoh, G. WoldeGabriel, W. K. Hart, K. Uto, et al. 2013. The Characteristics and Chronology of the Earliest Acheulean at Konso, Ethiopia. *Proceedings of the National Academy of Sciences USA* 110:1584–91.

Boyd, R., and J. B. Silk. 2009. *How Humans Evolved*. 5th ed. New York: W. W. Norton.

Browning, S. R., B. L. Browning, Y. Zhou, S. Tucci, and J. M. Akey. 2018. Analysis of Human Sequence Data Reveals Two Pulses of Archaic Denisovan Admixture. *Cell* 173:1–9.

Brunet, M., F. Guy, D. Pilbeam, D. E. Lieberman, A. Likius, et al. 2005. New Material of the Earliest Hominid from the Upper Miocene of Chad. *Nature* 434 (7034): 752–5.

Buck, L. T., and C. B. Stringer. 2014. Quick Guide: *Homo heidelbergensis*. *Current Biology* 24:R214.

Byars, S. G., D. Ewbank, D. R. Govindaraju, and S. C. Stearns. 2010. Colloquium Papers: Natural Selection in a Contemporary Human Population. *Proceedings of the National Academy of Sciences USA* 107 (Suppl. 1): 1787–92.

Carter, C. S. 2014. Oxytocin Pathways and the Evolution of Human Behavior. *Annual Review of Psychology* 65:17–39.

Cerling, T. E., J. G. Wynn, S. A. Andanje, M. I. Bird, D. K. Korir, et al. 2011. Woody Cover and Hominin Environments in the Past 6 Million Years. *Nature* 476 (7358): 51–6.

Coyne, J. A. 2009. *Why Evolution Is True*. New York: Viking.

Danneman, M., and F. Racimo. 2018. Something Old, Something Borrowed: Admixture and Adaptation in Human Evolution. *Current Opinion in Genetics and Development* 53:1–8.

Darwin, C. 1859. *On the Origin of Species*. London: Murray.

Darwin, C. 1871. *The Descent of Man, and Selection in Relation to Sex*. New York: D. Appleton and Company.

Dediu, D., and S. C. Levinson. 2018. Neanderthal Language Revisited: Not Only Us. *Current Opinion in Behavioral Sciences* 21:49–55.

De Groote, I., M. Lewis, and C. Stringer. 2018. Prehistory of the British Isles: A Tale of Coming and Going. *Bulletins et Mémoires de la Société d'Anthropologie de Paris* 30 (1–2): 1–13.

Dembo, M., D. Radovčić, H. M. Garvin, M. F. Laird, L. Schroeder, et al. 2016. The Evolutionary Relationships and Age of *H. naledi*: An Assessment Using Dated Bayesian Phylogenetic Methods. *Journal of Human Evolution* 97:17–26.

Deriziotis, P., and S. E. Fisher. 2017. Speech and Language: Translating the Genome. *Trends in Genetics* 33:642–56.

de Ruiter, D. J., S. E. Churchill, J. Hawks, and L. R. Berger. 2017. Late Australopiths and the Emergence of *Homo*. *Annual Review of Anthropology* 46:99–115.

Donaldson, Z. R., and L. J. Young. 2008. Oxytocin, Vasopressin, and the Neurogenetics of Sociality. *Science* 322 (5903): 900–904.

Du, A., A. M. Zipkin, K. G. Hatala, E. Renner, J. L. Baker, et al. 2018. Pattern and Process in Hominin Brain Size Evolution Are Scale-Dependent. *Proceedings of the Royal Society B* 285:20172738.

Fan, S., M. E. B. Hansen, Y. Lo, and S. A. Tishkoff. 2016. Going Global by Adapting Local: A Review of Recent Human Adaptation. *Science* 354 (6308): 54–9.

Feldman, R. 2017. The Neurobiology of Human Attachments. *Trends in Cognitive Sciences* 21:80–99.

Finkel, M., and R. Barkai. 2018. The Acheulean Handaxe Technological Persistence: A Case of Preferred Cultural Conservatism? *Proceedings of the Prehistoric Society* 84:1–19.

Fisher, S. E. 2017. Evolution of Language: Lessons from the Genome. *Psychonomic Bulletin and Review* 24:34–40.

Fitch, W. T. 2017. Empirical Approaches to the Study of Language Evolution. *Psychonomic Bulletin and Review* 24:3–33.

Foley, R. A. 2016. Mosaic Evolution and the Pattern of Transitions in the Hominin Lineage. *Philosophical Transactions of the Royal Society B* 371:20150244.

González-Forero, M., and A. Gardner. 2018. Inference of Ecological and Social Drivers of Human Brain-Size Evolution. *Nature* 557:554–7.

Gowlett, J. A. J. 2016. The Discovery of Fire by Humans: A Long and Convoluted Process. *Philosophical Transactions of the Royal Society B* 371:20150164.

Gunz, P., A. K. Tilot, K. Wittfeld, A. Teumer, C. Y. Shapland, T. G. M. van Erp, et al. 2018. Neandertal Introgression Sheds Light on Modern Human Endocranial Globularity. *Current Biology* 29:120–7.

Haile-Selassie, Y., S. M. Melillo, and D. F. Su. 2016. The Pliocene Hominin Diversity Conundrum: Do More Fossils Mean Less Clarity? *Proceedings of the National Academy of Sciences USA* 113:6364–71.

Hajdinjak, M., Q. Fu, A. Hübner, M. Petr, F. Mafessoni, et al. 2018. Reconstructing the Genetic History of Late Neanderthals. *Nature* 555:652–6.

Harcourt-Smith, W. H., and L. C. Aiello. 2004. Fossils, Feet and the Evolution of Human Bipedal Locomotion. *Journal of Anatomy* 204:403–16.

Hatala, K. G., B. Demes, and B. G. Richmond. 2016. Laetoli Footprints Reveal Bipedal Gait Biomechanics Different from Those of Modern Humans and Chimpanzees. *Proceedings of the Royal Society B* 283:20160235.

Higham, T., K. Douka, R. Wood, C. B. Ramsey, F. Brock, et al. 2014. The Timing and Spatiotemporal Patterning of Neanderthal Disappearance. *Nature* 512:306–9.

Hublin, J-J., A. Ben-Ncer, S. E. Bailey, S. E. Freidline, S. Neubauer, et al. 2017. New Fossils from Jebel Irhoud, Morocco and the Pan-African Origin of *Homo sapiens*. *Nature* 546:289–92.

Hunt, K. D. 1994. The Evolution of Human Bipedality: Ecology and Functional Morphology. *Journal of Human Evolution* 26:183–202.

Joordens, J. C., F. d'Errico, F. P. Wesselingh, S. Munro, J. de Vos, et al. 2014. *Homo erectus* at Trinil on Java Used Shells for Tool Production and Engraving. *Nature* 518:228–31.

Kappelman, J. 2018. An Early Hominin Arrival in Asia. *Nature* 559 (7715): 480–1.

Klein, R. G. 2009. *The Human Career: Human Biological and Cultural Origins.* Chicago: University of Chicago Press.

Kolodny, O., and M. W. Feldman. 2017. A Parsimonious Neutral Model Suggests Neanderthal Replacement Was Determined by Migration and Random Species Drift. *Nature Communications* 8:1040.

LeDoux, J. 2007. The Amygdala. *Current Biology* 17 (20): R868–74.

———. 2012. Rethinking the Emotional Brain. *Neuron* 73: 653–76.

Lewis, J. E., and S. Harmand. 2016. An Earlier Origin for Stone Tool Making: Implications for Cognitive Evolution and the Transition to *Homo*. *Philosophical Transactions of the Royal Society B* 371:20150233.

Li, J. Z., D. M. Absher, H. Tang, A. M. Southwick, A. M. Casto, et al. 2008. Worldwide Human Relationships Inferred from Genome-wide Patterns of Variation. *Science* 319:1100–1104.

Lieberman, D. 2011. Four Legs Good, Two Legs Fortuitous: Brains, Brawn, and the Evolution of Human Bipedalism. In J. B. Losos (ed.), *In the Light of Evolution: Essays from the Laboratory and Field* (pp. 55–72). Greenwood Village, CO: Roberts and Company.

Lieberman, D. E. 2015. Human Locomotion and Heat Loss: An Evolutionary Perspective. *Comprehensive Physiology* 5 (1): 99–117.

Lim, M. M., Z. Wang, D. E. Olazabal, X. Ren, E. F. Terwilliger, et al. 2004. Enhanced Partner Preference in a Promiscuous Species by Manipulating the Expression of a Single Gene. *Nature* 429 (6993): 754–7.

Linnaeus, C. 1747. Carl Linnaeus to Johann Georg Gmelin, 25 February 1747. In *The Linnaean Correspondence.* http://linnaeus.c18.net, letter L0783 (accessed January 24, 2019).

Lordkipanidze, D., M. S. Ponce de León, A. Margvelashvili, Y. Rak, G. P. Rightmire, et al. 2013. A Complete Skull from Dmanisi, Georgia, and the Evolutionary Biology of Early *Homo*. *Science* 342:326–31.

Lovejoy, C. O., G. Suwa, L. Spurlock, B. Asfaw, and T. D. White. 2009. The Pelvis and Femur of *Ardipithecus ramidus*: The Emergence of Upright Walking. *Science* 326:71–6.

Marciniak, S., and G. H. Perry. 2017. Harnessing Ancient Genomes to Study the History of Human Adaptation. *Nature Reviews Genetics* 18:659–74.

Maslin, M. A., S. Shultz, and M. H. Trauth. 2015. A Synthesis of the Theories and Concepts of Early Human Evolution. *Philosophical Transactions of the Royal Society B* 370:20140064.

Mathieson, S., and I. Mathieson. 2018. *FADS1* and the Timing of Human Adaptation to Agriculture. *Molecular Biology and Evolution* 35 (12): 2957–70.

McPherron, S. P., Z. Alemseged, C. W. Marean, J. G. Wynn, D. Reed, et al. 2010. Evidence for Stone-Tool-Assisted Consumption of Animal Tissues Before 3.39 Million Years Ago at Dikika, Ethiopia. *Nature* 466:857–60.

Meyer, M., J-L. Arsuaga, C. de Filippo, S. Nagel, A. Aximu-Petri, et al. 2016. Nuclear DNA Sequences from the Middle Pleistocene Sima de los Huesos Hominins. *Nature* 531:504–7.

Montgomery, S. 2018. Hominin Brain Evolution: The Only Way Is Up? *Current Biology* 28:R788–90.

Novembre, J., T. Johnson, K. Bryc, Z. Kutalik, A. R. Boyko, et al. 2008. Genes Mirror Geography Within Europe. *Nature* 456 (7218): 98–101.

Pääbo, S. 2014. *Neanderthal Man: In Search of Lost Genomes.* New York: Perseus Books.

Palmiter, R. D. 2008. Dopamine Signaling in the Dorsal Striatum Is Essential for Motivated Behaviors: Lessons from Dopamine-Deficient Mice. *Annals of the New York Academy of Sciences* 1129:35–46.

Pickering, R., A. I. R. Herries, J. D. Woodhead, J. C. Hellstrom, H. E. Green, et al. 2018. U–Pb-Dated Flowstones Restrict South African Early Hominin Record to Dry Climate Phases. *Nature* 565:226–9.

Pinker, S. 1995. *The Language Instinct.* New York: Harper Perennial.

Putnam, P. T., L. J. Young, and K. M. Gothard. 2018. Bridging the Gap Between Rodents and Humans: The Role of Non-Human Primates in Oxytocin Research. *American Journal of Primatology* 80 (10): e22756.

Radovčić, D., A. O. Sršen, J. Radovčić, and D. W. Frayer. 2015. Evidence for Neanderthal Jewelry: Modified White-Tailed Eagle Claws at Krapina. *PLoS ONE* 10 (3): e0119802.

Reich, D. 2018. *Who We Are and How We Got Here.* New York: Oxford University Press.

Reis, M. D., G. F. Gunnell, J. Barba-Montoya, A. Wilkins, Z. Yang, et al. 2018. Using Phylogenomic Data to Explore the Effects of Relaxed Clocks and Calibration Strategies on Divergence Time Estimation: Primates as a Test Case. *Systemic Biology* 67 (4): 594–615.

Richmond, B. G., and W. L. Jungers. 2008. *Orrorin tugenensis* Femoral Morphology and the Evolution of Hominin Bipedalism. *Science* 319:1662–65.

Rilling, J. K., M. F. Glasser, T. M. Preuss, X. Ma, T. Zhao, et al. 2008. The Evolution of the Arcuate Fasciculus Revealed with Comparative DTI. *Nature Neuroscience* 11 (4): 426–8.

Roach, N. T., A. Du, K. G. Hatala, K. R. Ostrofsky, J. S. Reeves, et al. 2018. Pleistocene Animal Communities of a 1.5 Million-Year-Old Lake Margin Grassland and Their Relationship to *Homo erectus* Paleoecology. *Journal of Human Evolution* 122:70–83.

Robinson, J. R., J. Rowan, C. J. Campisano, J. G. Wynn, and K. E. Reed. 2017. Late Pliocene Environmental Change During the Transition from *Australopithecus* to *Homo*. *Nature Ecology and Evolution* 1 (6): 0159.

Saltzman, W., and D. Maestripieri. 2010. The Neuroendocrinology of Primate Maternal Behavior. *Progress in Neuropsychopharmacology and Biological Psychiatry* 35 (5): 1192–1204.

Schlebusch, C. M., and M. Jakobsson. 2018. Tales of Human Migration, Admixture, and Selection in Africa. *Annual Review of Genomics and Human Genetics* 19:405–28.

Shreeve, J. 1995. *The Neanderthal Enigma: Solving the Mystery of Modern Human Origins*. New York: William Morrow.

Skoglund, P., J. C. Thompson, M. E. Prendergast, A. Mittnik, K. Sirak, et al. 2017. Reconstructing Prehistoric African Population Structure. *Cell* 171 (1): 59–71.

Slon, V., B. Viola, G. Renaud, M-T. Gansuage, S. Benazzi, et al. 2017. A Fourth Denisovan Individual. *Science Advances* 3 (7): e1700186.

Sousa, A. M., K. A. Meyer, G. Santpere, F. O. Gulden, and N. Sestan. 2017. Evolution of the Human Nervous System Function, Structure, and Development. *Cell* 170 (2): 226–47.

Strait, D., F. E. Grine, and J. G. Fleagle. 2015. Analyzing Hominin Phylogeny: Cladistic Approach. In W. Henke and I. Tattersall (eds.), *Handbook of Paleoanthropology*, 2nd ed. (pp. 1989–2014). Berlin: Springer-Verlag.

Stringer, C. 2012. *Lone Survivors: How We Became the Only Humans on Earth*. New York: Times Books.

Suwa, G., R. T. Kono, S. W. Simpson, B. Asfaw, C. O. Lovejoy, et al. 2009. Paleobiological Implications of the *Ardipithecus ramidus* Dentition. *Science* 326 (5949): 69, 94–99.

Thieme, H. 1997. Lower Palaeolithic Hunting Spears from Germany. *Nature* 385 (6619): 807–10.

Toth, N., and K. Schick. 2015. Overview of Paleolithic Archaeology. In W. Henke and I. Tattersall (eds.), *Handbook of Paleoanthropology* (pp. 2441–64). Berlin: Springer-Verlag.

Tung, J., and L. B. Barreiro. 2017. The Contribution of Admixture to Primate Evolution. *Current Opinion in Genetics and Development* 47:61–8.

van den Bergh, G. D., Y. Kaifu, I. Kurniawan, R. T. Kono, A. Brumm, et al. 2017. *Homo floresiensis*-like Fossils from the Early Middle Pleistocene of Flores. *Nature* 534:245–8.

Van Le, Q., L. A. Isbell, J. Matsumoto, M. Nguyen, E. Hori, et al. 2013. Pulvinar Neurons Reveal Neurobiological Evidence of Past Selection for Rapid Detection of Snakes. *Proceedings of the National Academy of Sciences USA* 110 (47): 19000–19005.

Villmoare, B. 2018. Early *Homo* and the Role of the Genus in Paleoanthropology. *American Journal of Physical Anthropology* 165 (S65): 72–89.

Villmoare, B., W. H. Kimbel, C. Seyoum, C. J. Campisano, E. DiMaggio, et al. 2015. Early *Homo* at 2.8 Ma from Ledi-Geraru, Afar, Ethiopia. *Science* 347 (6228): 1352–5.

Volkow, N. D., G-J. Wang, J. S. Fowler, D. Tomasi, F. Telang, et al. 2010. Addiction: Decreased Reward Sensitivity and Increased Expectation Sensitivity Conspire to Overwhelm the Brain's Control Circuit. *BioEssays* 32 (9): 748–55.

Walker, A., and R. E. Leakey. 1993. *The Nariokotome Homo erectus skeleton*. Cambridge, MA: Harvard University Press.

Walum, H., and L. J. Young. 2018. The Neural Mechanisms and Circuitry of the Pair Bond. *Nature Reviews Neuroscience* 19:643–54.

Wheeler, P. 1991. The Thermoregulatory Advantages of Hominid Bipedalism in Open Equatorial Environments: The Contribution of Increased Convective Heat Loss and Cutaneous Evaporative Cooling. *Journal of Human Evolution* 21 (2): 107–15.

White, T. D. 2009. Ladders, Bushes, Punctuations, and Clades: Hominid Paleobiology in the Late Twentieth Century. In D. Sepkoski and M. Ruse (eds.), *The Paleobiological Revolution: Essays on the Growth of Modern Paleontology* (pp. 122–48). Chicago: University of Chicago Press.

White, T. D., C. O. Lovejoy, B. Asfaw, J. P. Carlson, and G. Suwa. 2015. Neither Chimpanzee nor Human, *Ardipithecus* Reveals the Surprising Ancestry of Both. *Proceedings of the National Academy of Sciences USA* 112 (16): 4877–84.

Wolf, A. B., and J. M. Akey. 2018. Outstanding Questions in the Study of Archaic Hominin Admixture. *PLoS Genetics* 14 (5): e1007349.

Wood, B., and E. K. Boyle. 2016. Hominin Taxic Diversity: Fact or Fantasy? *Yearbook of Physical Anthropology* 159:S37–78.

Yong, E. 2015. 6 Tiny Cavers, 15 Odd Skeletons, and 1 Amazing New Species of Ancient Human. *The Atlantic*. https://www.theatlantic.com/science/archive/2015/09/homo-naledi-rising-star-cave-hominin/404362/ (accessed January 24, 2019).

Zhu, Z., R. Dennell, W. Huang, Y. Wu, S. Qiu, et al. 2018. Hominin Occupation of the Chinese Loess Plateau Since About 2.1 Million Years Ago. *Nature* 559:608–12.

Zilhão, J., D. E. Angelucci, E. Badal-García, F. d'Errico, F. Daniel, et al. 2010. Symbolic Use of Marine Shells and Mineral Pigments by Iberian Neandertals. *Proceedings of the National Academy of Sciences USA* 107 (3): 1023–8.

Zimmer, C. 2018. *She Has Her Mother's Laugh: The Powers, Perversions, and Potential of Heredity*. New York: Dutton.

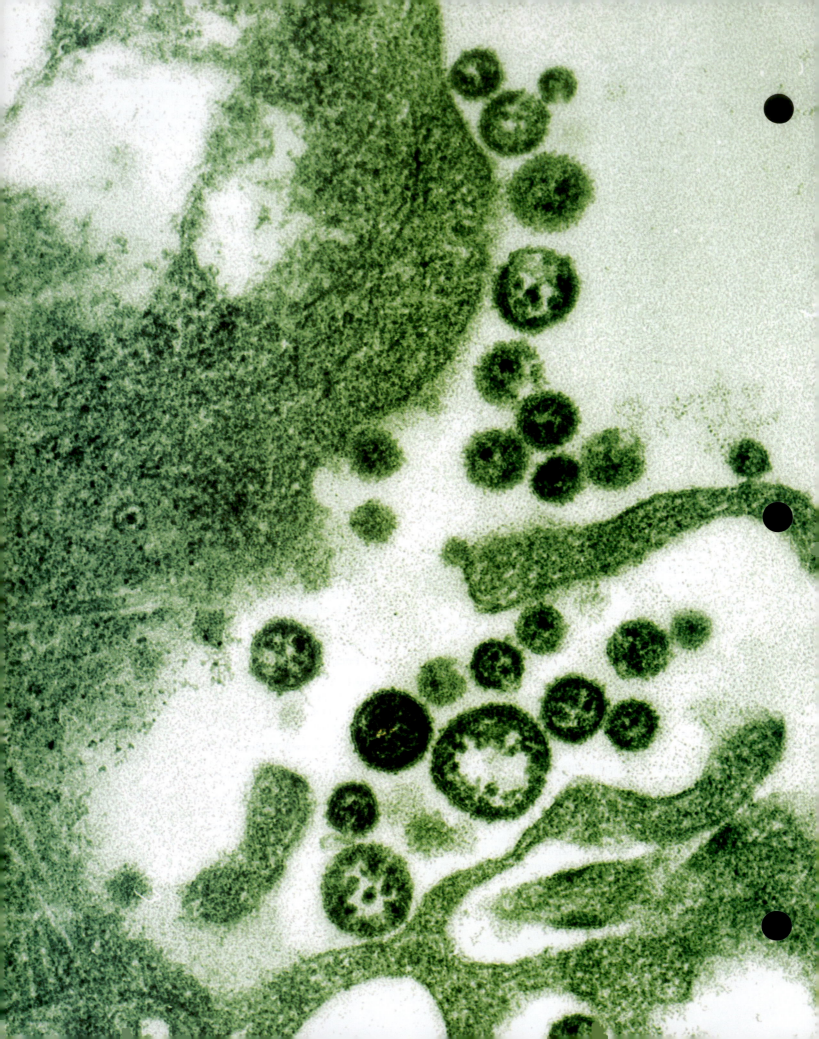

CHAPTER 18

Evolutionary Medicine

Learning Objectives

- Explain how maladaptation and pathogen evolution are relevant to medicine.
- Describe the similarities between the evolution of pathogens and cancer cells within a patient.
- Describe the history of antibiotic resistance since the invention of these drugs.
- Explain why phylogenies are effective tools for understanding the origins of infectious diseases.
- Discuss how natural selection and the founder effect can increase the frequency of disease-related genetic variants.
- Explain why an understanding of human genetic variation is important when considering drug treatment options.
- Review the pleiotropic effects of genes and their role in senescence.
- Apply models used to understand the rise of antibiotic-resistant bacteria to anticancer drug resistance.
- Explain the "Old Friends" hypothesis.
- Describe the thrifty genotype hypothesis and its recent critiques.
- Explain how understanding evolutionary biology leads to new avenues of treatment and prevention.

Throughout her career, Pardis Sabeti has battled against some of the deadliest viruses on Earth (**Figure 18.1**). But her weapon of choice is not a vaccine or an antiviral drug. Instead, it's a cluster of powerful computers that crunch data in Cambridge, Massachusetts.

Sabeti joined the fight while she was in graduate school at Oxford University and Harvard University, where she pioneered statistical methods for detecting signals of natural selection in DNA sequences. To do so, she had to analyze billions of base pairs. And so she needed powerful computers. Fortunately, she had access to the world-class facilities at the Broad Institute, a Cambridge-based research center for the study of genomes.

Sabeti's analysis revealed a number of genes undergoing natural selection in humans. One gene in particular intrigued her. She found that 34% of people in Nigeria shared the same allele of a gene called *LARGE*. Outside of West Africa, almost

Lassa virus is a highly deadly pathogen widespread in West Africa. It arose at least 1000 years ago and circulates between rats and humans.

Figure 18.1 Pardis Sabeti studies the evolution of Lassa virus. Her insights help guide efforts to contain and treat the disease. (David L Ryan/Boston Globe/Getty Images)

no one carries the variant. When she investigated the scientific literature on the allele, she learned that it plays a role in a disease called Lassa fever.

Discovered only in 1969, the Lassa virus is one of the most dangerous pathogens known to medicine (Andersen et al. 2015; Sabeti and Salahi 2018; Siddle et al. 2018). Only about half of those who contract Lassa fever survive. Researchers had found that if they knocked out *LARGE* in mice, the Lassa virus could not infect their cells. Sabeti hypothesized that the variant favored by natural selection provided resistance to the disease.

If her hypothesis was correct, the true impact of Lassa on West Africa might be drastically underestimated. If she could better understand the evolution of the virus, she reasoned, she might be able to help fight it.

When Sabeti got a job at Harvard University in 2008, she launched a major research project on Lassa fever. She traveled to Nigeria and Sierra Leone, where she set up partnerships with hospitals in both countries. Together, Sabeti and her network of coworkers implemented a quick, accurate test for Lassa fever. They began systematically identifying people who had been exposed to the virus and then started taking blood samples from them. Isolating the Lassa virus from the samples, the researchers could analyze the viral genes. Some of this research took place at the hospitals themselves, while Sabeti oversaw far more intensive genome sequencing back at the Broad Institute.

As Sabeti and her colleagues gathered more Lassa virus genomes, they could begin to uncover its evolutionary history. They determined that it originated in Nigeria at least 1000 years ago and subsequently spread into other West African countries. Scientists have long known that Lassa virus can replicate not just in people but also in the African soft-furred rat (*Mastomys natalensis*). Sabeti and her colleagues constructed phylogenies of the virus during Lassa fever outbreaks to understand its transmission patterns. They found that the viruses isolated from people in different regions were not closely related to each other. This pattern suggests that Lassa virus tends to spread repeatedly from rats to people, rather than from one person to another.

In 2018, a massive new outbreak of Lassa fever struck Nigeria. Its scale worried Nigerian doctors, raising the possibility that it had evolved into a new form that could easily spread between humans. If that were true, then the outbreak might explode unless doctors established quarantines and other extensive public health controls.

Fortunately, the network Sabeti helped build was ready to test the hypothesis. They sequenced genomes of Lassa viruses from 220 patients and tested the human transmission hypothesis, even as the outbreak was still raging. The phylogeny of the viruses revealed that fewer than 10% of the viruses had spread from one person to another. Somehow, Lassa was jumping repeatedly from rats to people. It was possible that the population of rats had exploded or that people were coming into closer contact with the animals (Siddle et al.

2018). It's even possible that a raised awareness of Lassa led more people to get tested.

Nigerian doctors used this discovery to guide their management of the outbreak. And Sabeti is also using these insights to develop even better diagnostic tests and help in the search for a vaccine for Lassa.

Sabeti's research is part of a growing trend in medical studies: exploring evolution in search of clues to better health. Pathogens undergo rapid evolution, spreading from species to species and person to person. They can even evolve within an individual host. But the evolution of our own species also offers insights about why we get sick. These insights offer deep lessons about what it means to be human. Natural selection may be able to shape complex adaptations, but it has not made our bodies perfect. We are still left vulnerable to many disorders. In some cases, evolution has actually made us more likely to get sick, not less. In other words, **evolutionary medicine** helps us understand our maladaptations as well as our adaptations (Stearns 2018). In this chapter, we'll explore how evolution shapes our health; we'll close out the chapter, and the book, by looking at the new ideas for medical treatments that evolution can inspire. ●

> **Evolutionary medicine** is the integrated study of evolution and medicine to improve scientific understanding of the reasons for disease and actions that can be taken to improve health.

18.1 Medicine and the Limits of Natural Selection

Throughout this book, we've explored case after case in which natural selection has acted as a potent force for reshaping life. It can eliminate deleterious mutations from populations while spreading beneficial ones. Over long periods of time, it can also drive the evolution of complex adaptations. But natural selection has not made us perfect. It has left us vulnerable to a host of ailments (Nesse and Williams 1994; Lents 2018).

Take the human eye. As we discussed in Chapter 9, the eye is a remarkable and complex adaptation. Although there's much to admire about the eye, it's also flawed. Nearly a third of people have hereditary nearsightedness. Most people beyond the age of 55 can't read fine print without help from glasses. Likewise, lens proteins cannot be repaired; as a result, by age 80, more than half of all adults have developed cataracts.

The human eye also has a blind spot where a bundle of nerves and blood vessels penetrates the eyeball and spreads out along the interior surface of the retina. We have to compensate for this defect by constantly darting our eyes around to fill in the obscured part of our field of view. Comparing our own eyes to the eyes of other animals, we can see that this arrangement is not a biological imperative. Octopuses have camera-vision eyes, too, but they don't have blind spots. Their retinal blood vessels and nerve endings rise up from underneath the eye. The structure of the human eye also makes it vulnerable to damage, making the retina easily detached. All these defects make it clear that to explain the eye, we must understand it not only as an adaptation but also as a maladaptation.

Understanding our maladaptations provides us with a new way to think about diseases and why we are vulnerable to them. We can categorize this vulnerability into six evolutionary explanations (Nesse 2005):

1. Pathogens evolve faster than their hosts.

2. Natural selection often lags behind environmental change.

3. Trade-offs make it nearly impossible for natural selection to "solve" certain biological problems.

4. A species' evolutionary history puts constraints on the potential changes natural selection can bring about.

5. Some traits increase reproductive fitness at the cost of increasing vulnerability to disease.

6. What may appear to be simply disease may actually be an adaptation.

We will explore these explanations in the rest of the chapter, using them to better understand diseases ranging from malaria to cancer to obesity.

Key Concept • Understanding maladaptations provides us with new ways to think about diseases. •

18.2 Evolving Pathogens

We have natural selection to thank for a remarkably sophisticated immune system. The human immune system can learn how to defeat new pathogens with antibodies and other chemical weapons, marshaling this arsenal if the pathogens ever return. It can defeat viruses, bacteria, protozoans, fungi, and even parasitic worms. And yet our immune systems provide far from perfect protection. Even with the help of modern medicine, people become infected in vast numbers every day. Tuberculosis alone kills 1.2 million people worldwide each year; HIV claims another one million (World Health Organization 2018).

One reason that pathogens still make us sick is that they evolve, too. By understanding this evolution, we can understand how old pathogens continue to evade our immune systems and how new diseases emerge.

The Diversification of HIV Within the Body

For human pathogens, our body is the environment in which their fitness is determined. In this environment, certain mutations will experience natural selection because they allow pathogens to make better use of their host's resources. A virus, for example, may acquire a mutation that allows it to do a better job of hijacking an infected cell. The virus makes more copies of itself than the other viruses in the body, with the result that the frequency of the mutation increases in the viral population inside the host. The immune system presents another selective pressure for pathogens. As the immune system wipes out a population of viruses, natural selection will favor a mutation that allows viruses to evade destruction.

Pathogen populations can respond to selection with remarkable speed. One reason they can evolve so quickly is that many reproduce on a staggering scale. In a single day, an HIV virus may give rise to a billion descendants. The other reason for the rapid response to selection is that many pathogens have very high mutation rates. Some viruses, for example, lack genes for repair enzymes that would normally fix errors that arise during the replication of their genomes. Instead, the errors go unfixed, introducing new genetic diversity into the viral population. Viruses and bacteria can also generate additional genetic variation thanks to horizontal gene transfer (see Box 4.2).

Scientists can track this evolution within individual patients. In 2015, a team of researchers at the Max Planck Institute for Developmental Biology in Tübingen, Germany, gathered serum samples collected from nine HIV patients who had been diagnosed as early as the 1990s (Zanini et al. 2015). They sequenced the genomes of some of the viruses present in the serum at 4 months after infection, 5 years, and 8 years. The researchers then looked for single nucleotide polymorphisms (SNPs) in

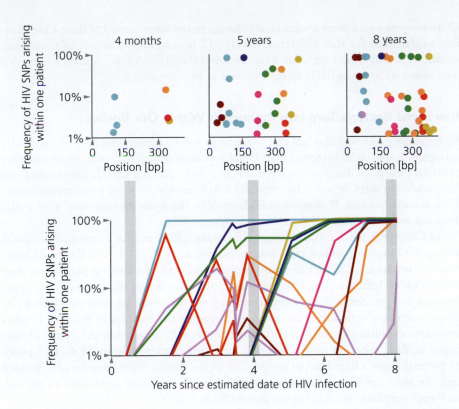

Figure 18.2 A: Scientists sampled HIV at three points during a single patient's infection. Each colored dot represents a SNP that arose in a region called p17 after the patient became infected. B: This chart shows how some SNPs became fixed in the patient's viral population, whereas others rose and then declined in frequency. (Data from Zanini et al. 2015)

the viruses at each time point. **Figure 18.2** illustrates the SNPs found in just one patient and in just one small region of the HIV genome known as p17. New SNPs arose, some of which spread to fixation, whereas others later disappeared.

We can visualize the evolutionary change that occurred in this one patient with an evolutionary tree (**Figure 18.3**). This phylogeny shows how p17 gained new mutations and how these haplotypes became more common in the population of viruses. The length of the branches shows the overall genetic difference that accumulated in this region as the viruses replicated. The length of these branches is especially intriguing when we compare the p17 phylogeny to that of another region, called V3.

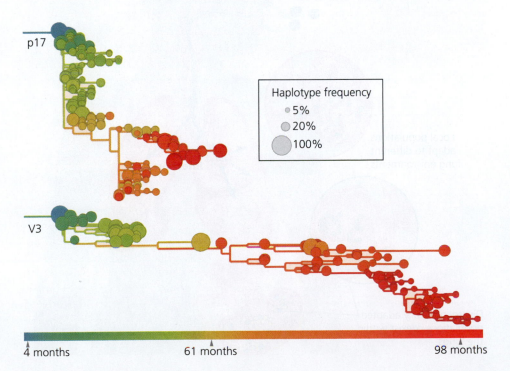

Figure 18.3 This figure shows the diversity of variants in the same HIV patient, represented as two evolutionary trees. Each one represents the evolution of a different gene in the HIV population inside the patient. V3 experienced much more evolutionary change than p17, suggesting that p17 is less tolerant of mutations (From © 2015, Zanini et al. Population genomics of intrapatient HIV-1 evolution by Fabio Zanini et al. Dec 10, 2015 eLife 2015;4:e11282; CC Attribution License)

V3 underwent much more evolutionary change in the same period of time. One plausible explanation for this difference is that p17 is much less tolerant of mutations. Mutations to that region are more likely to lower the fitness of the viruses, whereas V3 mutations may be more likely to have neutral or positive effects.

How Some Bacteria Turn into Pathogens Within Our Bodies

Within-host evolution also has the capacity to turn harmless microbes into deadly pathogens. A species of bacteria called *Pseudomonas aeruginosa* thrives in many environments, ranging from damp soil to rivers, kitchen sinks, and hot tubs. It sometimes floats into people's mouths or noses, but typically fails to survive in these new environments. If, on the other hand, *P. aeruginosa* finds itself in the body of someone with cystic fibrosis, it may establish itself there for its host's entire life.

In Chapter 6, we discussed how cystic fibrosis is the result of a recessive mutation, causing the lungs to produce extra mucus (Ratjen et al. 2015). In this disturbed environment, immune cells can't function properly. Instead of targeting pathogens, they produce excess inflammation. If *P. aeruginosa* enters the airway of someone with cystic fibrosis, it can proliferate in the lungs. But along the way, these bacteria experience significant natural selection. You can think of them like the bacteria in Richard Lenski's long-term evolution experiments (Chapter 6). For *P. aeruginosa,* the cystic fibrosis lung is its flask, where it must adapt to a new environment (**Figure 18.4**). In a pond, the bacteria enjoy a high level of oxygen, but in the mucus-filled crannies of a diseased lung, the levels are very low. Along with oxygen, *P. aeruginosa* needs iron to survive, but there's very little free iron in the human body.

Studies on bacterial samples from people with cystic fibrosis show that each time the microbes invade a new host, they can re-evolve adaptations for challenges such as these (Didelot et al. 2016). Bacteria gain mutations that change them from

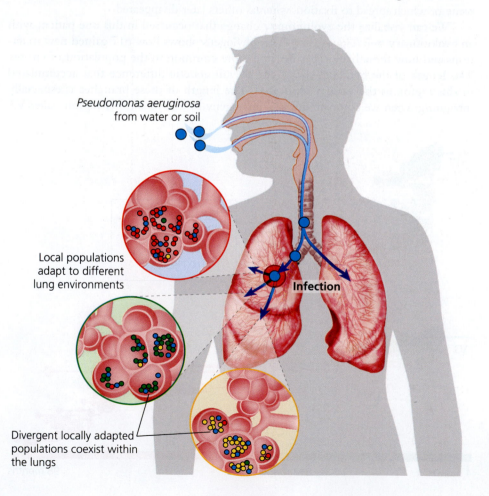

Pseudomonas aeruginosa
from water or soil

Local populations
adapt to different
lung environments

Infection

Divergent locally adapted
populations coexist within
the lungs

Figure 18.4 A species of bacteria called *Pseudomonas aeruginosa* lives in soil and water. It can also colonize people with cystic fibrosis. As it moves into the lungs, it undergoes adaptive evolution for this new niche.

free-floating individual cells into clusters that stick to the lining of the lung (Winstanley et al. 2016). They also evolve proteins that can wrest iron atoms from their hosts. *P. aeruginosa* doesn't simply evolve into a lung-adapted form, however. Different regions of the lung pose different challenges to the bacteria, and in response they acquire mutations that are suited to their particular micro-environments (Schick and Kassen 2018). Intriguingly, some bacteria become "hypermutators," evolving an accelerated mutation rate. Although hypermutation may be risky in a less challenging environment, it may allow bacteria inside human lungs to adapt quickly to new, extreme challenges.

18.3 The Evolution of Virulence

When *P. aeruginosa* bacteria first enter a cystic fibrosis lung, they are aggressive invaders. They swim deep into the lungs by spinning their tails. They secrete cyanide and other toxins, killing host cells to feed on the cells' contents. But as the bacteria adapt to the lung, something remarkable happens. In host after host, they consistently lose many of those invasive traits. They acquire mutations that disable the genes required for swimming, releasing toxins, and so on.

The level of harm that a pathogen causes its hosts as it uses their resources is known as **virulence** (Bull and Lauring 2014). The level of virulence varies from species to species—a fact that's familiar to anyone who has gotten sick with the flu and the cold. A cold may slow you down for a day or two, whereas the flu may keep in you in bed for a week. But the virulence of a pathogenic species is not fixed. Over time, it can evolve to different levels.

Figure 18.5 shows an experiment to determine the virulence of *Salmonella typhimurium*, a species of bacteria that causes severe food poisoning (Ebert 1998). Researchers infected mice with a mild strain of *S. typhimurium*. They allowed the bacteria to replicate and passed their descendants on to healthy mice. To measure the virulence of the bacteria, they infected more mice and observed how sick they got. This procedure, known as serial passage, had a striking effect on the microbes. The bacteria started out relatively mild, with the vast majority of mice recovering. But in just ten passages, the bacteria became so virulent that every infected mouse died.

Virulence describes the ability of a pathogen to cause disease.

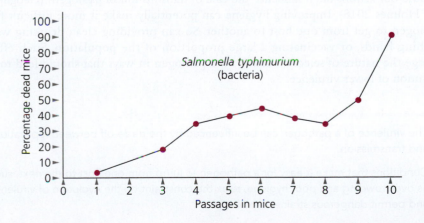

Figure 18.5 If bacteria are passed from one mouse to another, they evolve from low to high virulence. (Data from Ebert 1998)

Figure 18.6 A: The virulence of a pathogen is influenced by two opposing agents of selection: selection for within-host replication and selection for between-host transmission. B: Stronger selection within hosts favors rapid replication and increased virulence. C: Stronger selection across hosts favors reduced virulence.

A

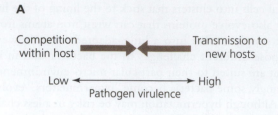

Competition within host ⟶⟵ Transmission to new hosts

Low ⟵⟶ High
Pathogen virulence

B

Stronger selection for competition **within hosts** favors rapid replication and increased virulence

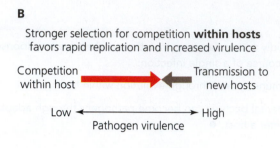

Competition within host ⟶⟵ Transmission to new hosts

Low ⟵⟶ High
Pathogen virulence

C

Stronger selection for movement **across hosts** favors reduced virulence

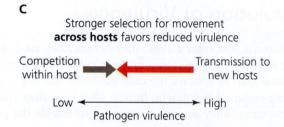

Competition within host ⟶⟵ Transmission to new hosts

Low ⟵⟶ High
Pathogen virulence

Virulence can evolve because it is determined by two factors, each of which can experience natural selection. Within a host, a pathogen reproduces, and traits that allow it to compete for host resources raise its fitness. But the long-term fitness of a pathogen also depends on how likely it is to spread from one host to another. The faster a pathogen replicates, the more harm it may do to its host. If its hosts end up dying quickly, then it will have a lower probability of reaching a new host. Such evolutionary trade-offs can keep pathogens from becoming more virulent (**Figure 18.6**).

Normally, *S. typhimurium* spreads by being shed in feces. The longer an infected host survives, the more feces it can release into the environment. In a serial passage experiment, this requirement disappears. The bacteria always get transmitted no matter how deadly they become. The only thing that matters to their fitness is how fast each strain replicates within the host mice.

Understanding this balance of selection is critically important for public health because our actions often influence the ease of transmission of pathogens (Geoghegan and Holmes 2018). Improving hygiene can potentially make it more difficult for a pathogen to get from one host to another. So can providing clean drinking water, washing hands, or vaccinating a large proportion of the population. Such efforts change the nature of selection acting on a pathogen in ways that should lead to the evolution of lower virulence.

Key Concepts

• The virulence of a pathogen can be influenced by the trade-off between replication and transmission.

• Conditions that make it easy for a pathogen to jump from one host to the next, such as overcrowding and poor hygiene, relax the constraint on the evolution of virulence and permit dangerous strains to evolve. ●

18.4 Defeating Antibiotics

In the mid-1900s, scientists discovered that fungi and bacteria make compounds that can kill off pathogens. Known as antibiotics, these drugs revolutionized medicine (Figure 18.7). Some optimists declared that infectious diseases would soon be a thing of the past.

But not long after antibiotics first became available, doctors began reporting that they sometimes failed. In the 1950s, Japanese doctors used antibiotics to battle outbreaks of dysentery caused by *Escherichia coli*, only to watch the bacteria gain **resistance** to one drug after another. Bacterial resistance to drugs has become more widespread over the years. Today, some bacteria strains are resistant to every antibiotic in use (McKenna 2017).

These bacteria gained resistance to antibiotics through evolution, in much the same way that insects have evolved resistance to pesticides and weeds to herbicides (Chapter 10). In all three cases, we humans have drastically altered the environment with a deadly chemical. Organisms with mutations that enable them to defend against the threat have been able to survive and reproduce, and their mutations have become more frequent in the population.

Antibiotic resistance can take many forms. Some microbes close off the passages through which antibiotics slip inside their cell walls. Some make enzymes that can break down the drugs into harmless components. Some push the antibiotics out before they can cause harm (Munita and Arias 2016).

Evolution of Resistance Within a Host

Bacteria have such large populations and reproduce so quickly that they can evolve significant antibiotic resistance even within a single host. One of the most striking cases of this rapid evolution was documented by Vegard Eldholm and his colleagues at the Norwegian Institute of Public Health in a patient with tuberculosis (Eldholm et al. 2014).

When the patient was initially diagnosed at a clinic in Norway, tests on the bacteria showed that they were susceptible to the standard antibiotics used against tuberculosis. The patient's doctors prescribed three of those drugs: isoniazid, pyrazinamide, and rifampicin. Despite receiving these antibiotics, the patient's condition gradually worsened, and after 8 months a cavity formed in his lung. Tests revealed that only pyrazinamide still had an effect on the bacteria: they had evolved resistance to isoniazid and rifampicin. Sequencing the DNA of these bacteria, the researchers pinpointed the new mutations that now shielded the bacteria to these drugs.

The patient was admitted to Oslo University Hospital, where his doctors continued giving him pyrazinamide and added four new antibiotics. For a year, they could not isolate any bacteria from him, but then the microbes returned with a vengeance. They were now resistant to three of the new antibiotics. The doctors swapped those out for two other antibiotics, but those soon failed as well. Eventually, the bacteria became resistant to almost every drug available to treat tuberculous—a type of infection known as extensively drug-resistant tuberculosis, or XDR-TB. Finally, the doctors found that a drug called linezolid was able to clear the infection. But after 4 years of treatment, the patient's lung was so damaged that surgeons had to remove it.

Figure 18.8 summarizes the course of the disease. Eldholm and his colleagues found that after new antibiotics were added to the treatment, more than one lineage of bacteria would gain a mutation conferring resistance. Only some of those mutations would become fixed in the population living inside the patient's lung, however. When Eldholm and his colleagues grew these resistant lineages together in a petri dish, exposed to antibiotics, they found that the lineage that became fixed in the patient also outcompeted the other strains in the petri dish.

Studies such as these show how natural selection and other mechanisms of evolution can play out within a single patient. In effect, a patient's body becomes an

Resistance refers to the capacity of pathogens to defend against antibiotics or other drugs.

Figure 18.7 When antibiotics were discovered in the mid-1900s, they were hailed as a miracle drug. Since then, bacteria have evolved increasing resistance to them. (National Institutes of Health)

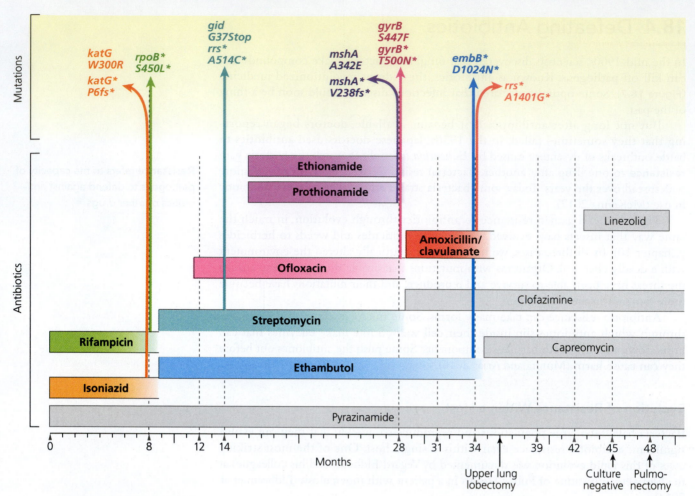

Figure 18.8 A patient with tuberculosis was initially treated with three antibiotics (shown as three horizontal bars). This chart shows how new mutations (top) arose that made the bacteria in his body resistant to two of these drugs, requiring several new antibiotics over the course of 4 years, before the patient's lung had to be surgically removed. The mutations marked by asterisks spread to fixation. (Data from Eldholm et al. 2014)

experimental flask into which susceptible bacteria go, and out of which resistant bacteria come.

The evolution that Eldholm and his colleagues observed arose through a series of mutations accumulating along lineages of bacteria. The danger of these mutations is amplified by horizontal gene transfer (Chapter 5). Once a new resistance gene arises in one lineage of bacteria, it can jump to another lineage—even a different bacteria species. Bacteria can accumulate these resistance genes, and mutations can move them close together. This bundle of resistance genes, known as a cassette, can then be transferred all at once from one microbe to another. Eventually, such a cassette can endow a microbe with resistance to a host of antibiotics in one fell swoop.

Combating Antibiotic Resistance

As we have increased our use of antibiotics, resistance has increased (**Figure 18.9**). One by one, drugs that were once effective against a wide swath of bacteria have grown weak (**Figure 18.10**). If bacteria continue to evolve along the same trajectory, by the year 2050, antibiotic resistance may cause a staggering ten million deaths per year (Baker et al. 2018).

Around the world, researchers and doctors have called for a concerted campaign to halt this dangerous form of evolution (Martens and Demain 2017). Searching for new antibiotics will be essential, but these drugs will probably fail as well if we keep

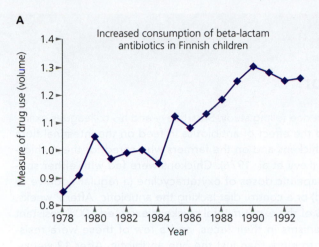

A

Increased consumption of beta-lactam antibiotics in Finnish children

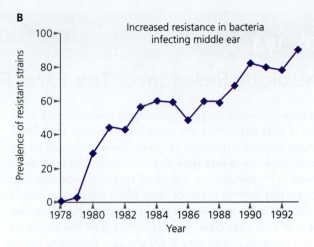

B

Increased resistance in bacteria infecting middle ear

driving the evolution of resistance. One way to put a brake on this evolution is to find ways to reduce our use of antibiotics. The vast majority of antibiotics are given not to sick people but to healthy farm animals (**Box 18.1**). Reducing agricultural use of antibiotics could reduce the selection pressure for resistance while not putting people's lives at risk.

Doctors can also help improve their stewardship of antibiotics. When patients complain of fevers or scratchy throats, it can be hard to determine at first whether viruses or bacteria are responsible. Antibiotics are useless against viral infections. But they can still drive the evolution of resistance in the harmless bacteria in our body, which can then pass on resistance genes to pathogens. The Centers for Disease Control estimates that about half of all antibiotics prescribed by doctors on an outpatient basis are inappropriate (Imanpour et al. 2017). As we'll discuss later, reckless prescription of antibiotics to children is dangerous not just for the resistance it fosters but for the potential impact it can have on the development of children's immune systems.

Figure 18.9 A: In Finland, children were increasingly prescribed antibiotics in the 1980s for ear infections. B: That increase was accompanied by a rise in resistance to those antibiotics in the bacteria causing the ear infections. (Data from Nissinen et al. 1995)

- Because bacteria are capable of acquiring new mutations quickly, antibiotic resistance can evolve especially quickly.

- Understanding the evolutionary biology of these important pathogens could influence treatment options and slow the pace of the evolution of resistance. ●

Key Concepts

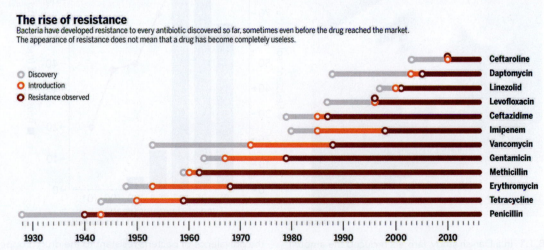

The rise of resistance
Bacteria have developed resistance to every antibiotic discovered so far, sometimes even before the drug reached the market. The appearance of resistance does not mean that a drug has become completely useless.

○ Discovery
○ Introduction
● Resistance observed

Ceftaroline
Daptomycin
Linezolid
Levofloxacin
Ceftazidime
Imipenem
Vancomycin
Gentamicin
Methicillin
Erythromycin
Tetracycline
Penicillin

Figure 18.10 The history of antibiotic drug development is one of rapid failure. When antibiotics are first introduced, they are overwhelmingly effective. But within a matter of years, strains of resistant bacteria evolve. (Reprinted with permission of American Association for the Advancement of Science from "Resistance fighters", Kai Kupferschmidt, Science 13 May 2016: Vol. 352, Issue 6287, pp. 758–761. Permission conveyed via Copyright Clearance Center, Inc.)

BOX 18.1

Antibiotic Resistance: The Farm Factor

One reason for the continuing evolution of antibiotic resistance is that antibiotics are used for purposes other than treating human infections: farmers spray them on crops, aquaculture managers give them to their fishes, they're present in household and industrial cleaning products, and ranchers put them in livestock feed. Most antibiotics used in the United States are given to healthy farm animals. For reasons that aren't yet clear, the antibiotics give the animals a 3%–5% increase in their rate of weight gain (McKenna 2017).

These low levels of antibiotics foster the evolution of resistance. Although they can slow the growth of bacteria populations by killing some of the bacteria, the doses aren't strong enough to kill all the bacteria outright. Microbes with mutations that confer resistance to antibiotics perform better than those that are susceptible. As the resistance alleles accumulate in the bacteria in healthy livestock, they become reservoirs for the evolution of bacteria resistant to modern antibiotics (Box Figure 18.1.1).

Resistant bacteria escape the animals via their feces, and they can then spread through the environment, even infecting humans. The drugs they are resistant to are the same ones we rely on to treat illnesses (Khachatourians 1998; Teuber 2001; Shea 2003).

In one telling study, Stuart Levy and his colleagues examined the effect of antibiotics in feed on the intestinal flora of chickens and on the farmers who cared for these chickens (Levy et al. 1976). Chickens were fed with either subtherapeutic doses of oxytetracycline (a regular additive to feed) or a control diet lacking the antibiotic. After 2 weeks, 90% of the experimental chickens excreted 100% resistant organisms in their feces, and a few of these were resistant to more than just the one antibiotic. After 12 weeks, almost two-thirds of the experimental chickens excreted microbes resistant to all four of the tested antibiotics (that is, multidrug resistance). By 4 months, these resistance genes had transferred to the control chickens, which now began to excrete resistant microbes, and by 6 months, fecal samples of the farmers and other people on the farm now contained multidrug-resistant bacteria. The same pattern of four-drug resistance was found in the experimental chickens and farm dwellers, but not in control families tested from neighboring (and antibiotic-free) farms. Antibiotics added to chicken feed had selected for the evolution of antibiotic resistance, and these resistant microbes had contaminated other animals on the farm as well as the farmers themselves.

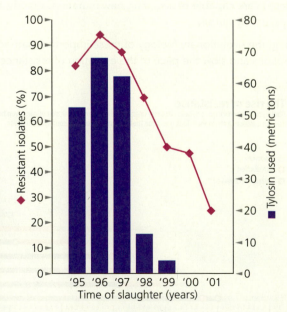

Box Figure 18.1.1 In a Danish study, farmers reduced the amount of tylosin, an antibiotic, they supplied to healthy pigs. As a result, the prevalence of bacteria resistant to the drug dropped. (Photo: Adriano Castelli/Shutterstock; data from Aarestrup et al. 2000)

18.5 The Origin of New Diseases

The evolution of pathogens helps explain why we are vulnerable to infectious diseases. It also helps explain how new diseases emerge.

To trace the long-term history of diseases, evolutionary biologists study their phylogenies as well as the phylogenies of their hosts (see Box 15.2). When scientists reconstruct the evolutionary history of pathogens, they sometimes find that pathogens have been coevolving with our ancestors for many millions of years. Others have shifted from other hosts to humans more recently—sometimes in just the past few years.

Loyal Pathogens and Jumping Viruses

By the time Americans reach the age of 40, more than half of them will be infected with a virus called cytomegalovirus (it also goes by the name human herpesvirus 5). In most people, the virus causes no symptoms at all; in others, it may cause swollen glands, a fever, and a lingering fatigue. But cytomegalovirus is sometimes far less benign. If it infects babies before birth, for example, they may become deaf, blind, or mentally disabled.

Virologists have discovered cytomegalovirus strains in other mammals as well. Each strain is adapted only for infecting a particular species. Human cytomegalovirus cannot infect tree shrews, for example, nor can tree shrew cytomegalovirus infect us.

In 2009, Fabian H. Leendertz and his colleagues at the Robert Koch Institute in Berlin constructed a phylogeny of human cytomegalovirus and its closest relatives (**Figure 18.11**). The scientists found that the branching pattern of the viral phylogeny mirrored the phylogeny of their host species (Leendertz et al. 2009; see also Anoh et al. 2018). The closest relative of human cytomegalovirus, a lineage of viruses called CG2, infects our closest relatives, the chimpanzees.

This mirrorlike phylogeny suggests that cytomegalovirus has been tracking the evolution of its hosts for a very long time. The common ancestor of humans, apes, and monkeys, which lived some 40 million years ago, was host to a cytomegalovirus. As that ancestor's descendants diverged into new lineages, the virus diverged as well. It did not leap to distantly related animals, like turtles or sharks. Instead, it continued to adapt to its evolving hosts.

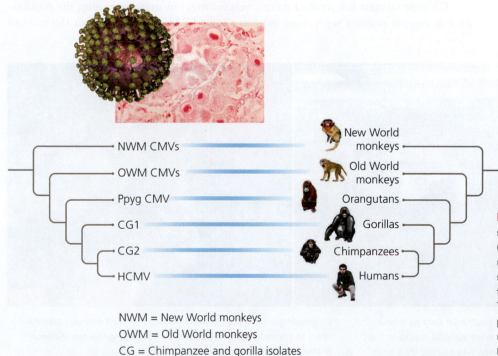

NWM = New World monkeys
OWM = Old World monkeys
CG = Chimpanzee and gorilla isolates

Figure 18.11 This phylogenetic tree shows how human cytomegalovirus (HCMV, or human herpesvirus 5) is related to other viruses. The relationship of the viruses (left) is a mirror of the relationship of their hosts (right). This pattern suggests that hosts and parasites have been cospeciating for more than 30 million years. (CDC/Dr. Haraszti; data from Leendertz et al. 2009)

In other cases, however, the phylogeny of a pathogen does not perfectly mirror that of its host. Their discordance can reveal instances where a pathogen moved from one species to another, giving rise to a new disease. HIV-1 and related viruses known as simian immunodeficiency virus (SIV) do not have a phylogeny like that of cytomegalovirus. As we discussed in Chapter 8, the phylogeny of HIV-1 and its closest relatives does not mirror the phylogeny of humans and their closest relatives. Different lineages of HIV-1 jumped to humans from other apes in the early 1900s—twice from chimpanzees and twice from gorillas.

It's not uncommon for pathogens to come into contact with new hosts. Once in a new host, pathogens may be unable to replicate, or they may only be able to replicate slowly. But natural selection can drive the spread of mutations that make them better adapted to their new host. In many cases, pathogens face a trade-off: adaptations that make them better adapted to spread in a new species will make them less fit in their original host (Longdon et al. 2014).

Where New Diseases Come From

In the century since HIV-1 emerged, other diseases have jumped into our species. Thanks to advances in gene-sequencing technology and phylogenetic analysis, scientists can reconstruct their evolution soon after they strike.

In November 2002, a deadly new virus came to light. A Chinese farmer with a high fever came to a hospital and died within a matter of days. Other people from the same region began to develop the disease as well. About 10% of them died. But the new fever didn't reach the world's attention until an American businessman flying back from China developed a fever. The flight stopped in Hanoi, where the businessman died. Soon, people were falling ill in countries around the world, although most of the cases turned up in China and Hong Kong (**Figure 18.12A**). The disease was not the flu, not pneumonia, nor any other known disease. It was given a new name: severe acute respiratory syndrome, or SARS (Drexler et al. 2014; Cui et al. 2018).

In March 2003, the World Health Organization established a network of labs around the globe to pin down the cause of SARS. Just a month later, they had isolated the virus responsible for the outbreak. Soon after that, they got a clue to the origin of the SARS virus. They found a related virus in a catlike mammal called the masked palm civet (*Paguma larvata*).

Chinese farmers sell masked palm civets for meat in markets, raising the possibility that regular contact with these animals allowed SARS to jump from the masked

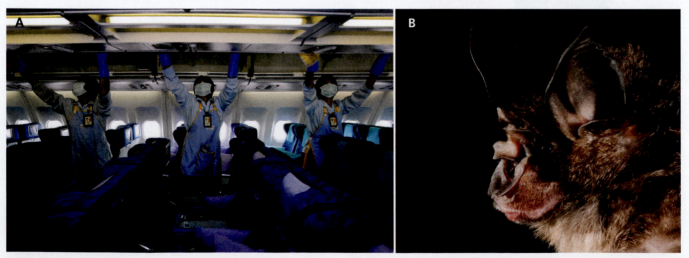

Figure 18.12 A: In 2003, panic swept Southeast Asia as a new disease called SARS emerged. People wore surgical masks in the hope of avoiding infection. B: The disease was caused by a virus originating in Chinese horseshoe bats that had evolved adaptations to spread between people. (A: Kin Cheung/Reuters/Newscom; B: Merlin D. Tuttle/Science Source)

palm civets to a new host—us. But the story of SARS proved to be more complex. Researchers studying viruses in bats discovered a large number of SARS-like strains in the horseshoe bats of China (**Figure 18.12B**). This research shows that the virus likely first evolved in bats, moved to masked palm civets, and finally reached humans. There was just one problem with this hypothesis—some genes in the human version of the virus didn't match any of the versions present in the horseshoe bats.

For years, Zhengli Shi and Jie Cui, of the Wuhan Institute of Virology in China, kept investigating this hypothesis by searching for SARS-like viruses in Chinese bats (Cyranoski 2017). In one cave in southwestern China they hit the viral jackpot. The bats there had a remarkably wide range of viruses. For 5 years they collected fresh guano and took anal swabs to survey the viruses' genetic diversities. Every genetic component of the human SARS virus was carried by at least one strain in the cave's bats.

Shi and Cui propose that the human SARS virus came into being in this cave, or a similar one nearby. It arose through a merging of the genes of other viruses. As the bats in the cave swapped viruses, one animal became simultaneously infected with two strains. One of its cells was invaded by the two strains at once, and both kinds of viruses started replicating inside of it. But as the cell made new copies of their genes, it scrambled their RNA. A viral polymerase began making a gene using the template from one of the viruses, and then switched to the other. Through a process of recombination, the bat cell produced new viruses that had mixtures of the genetic material from both ancestors. One such hybrid combined the right genes together to efficiently infect other species—in particular, masked palm civets and humans.

Thankfully, the SARS outbreak did not last long. Doctors found that they could stop the transmission of the virus by identifying infected patients and treating them in isolation. Eight months after the first case, the outbreak came to a halt in July 2003. Over the course of its spread, the SARS virus infected 8096 people worldwide and killed 774 of them. Since then, not a single person has been infected by SARS. We may very well have eradicated it from the planet.

Our victory over SARS should not make us complacent. More viruses have since leaped from obscurity into international epidemics—MERS in the Middle East, for example, along with Zika and chikungunya in the Americas. And we should expect that even more emerging pathogens will surface as humans disrupt wild regions and come into more contact with pathogen-carrying animals such as primates and bats (Morens and Fauci 2013; Plowright et al. 2017). The Chinese cave where SARS may have been born remains a cauldron of viral evolution, and there's a village just a kilometer away where a new version of SARS may spill over.

Key Concept

- Phylogenetic studies of pathogen evolution can provide important clues to the native hosts—and therefore origins—of emerging infectious diseases. •

18.6 Ever-Evolving Flu

In Chapter 1, we introduced some of the themes of this book by briefly discussing the influenza virus. Now, as we near the conclusion of the book, we return to the flu for a deeper exploration. As we'll see, this virus puts on a spectacular display of evolutionary mechanisms—one that leads to about a billion infections worldwide each year and ultimately causes as many as half a million deaths annually (Krammer et al. 2018).

The Mutant Swarm

The global devastation caused by influenza virus is all the more remarkable when you consider how simple it is. The flu virus consists of a tiny membrane envelope

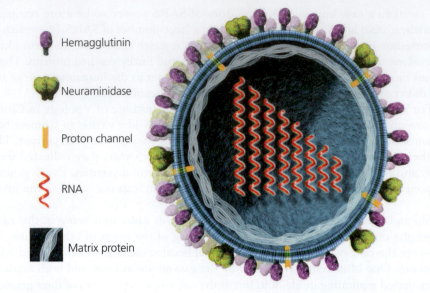

Hemagglutinin

Neuraminidase

Proton channel

RNA

Matrix protein

that encloses eight short pieces of RNA (**Figure 18.13**). Yet this is enough genetic information to allow the influenza virus to spread from one person to another, reach the respiratory tract, and invade epithelial cells (**Figure 18.14**). It gains entry to these cells thanks to one of its 13 genes, which encodes a surface protein called hemagglutinin. The hemagglutinin protein latches onto sialic acids on the host cell, and then allows the virus to slip inside. Once inside, the influenza virus recruits the cell's genetic machinery to mass-produce copies of itself. Each infected cell can make anywhere from 100,000 to one million new copies of the viral genome, which are then packaged into new membrane envelopes and escape the cell to infect a new host.

During this replication explosion, a vast number of mistakes are made. The influenza virus accumulates mutations a million times faster than our genomes do. In fact, these viruses accumulate mutations so fast that they are often called mutant swarms.

Figure 18.14 People with the flu cough and sneeze, projecting virus-laden droplets into the air. Natural selection within our body favors viral genotypes that are the most efficient at replicating themselves. These genotypes will be represented disproportionately in the infectious spray, so they will most likely be transmitted to subsequent hosts. (CDC/James Gathany)

Many mutations lower fitness, but some are beneficial. They can speed up the replication of the viruses or allow them to evade the immune system. Rival viral genotypes with fitness-enhancing mutations compete with each other.

As these viruses replicate inside our bodies, they bring on a powerful immune response that causes aches, high fever, and a painful cough. For most of us, this is as far as the flu gets. We produce antibodies against hemagglutinin and other surface proteins, which stop them from replicating. Other immune responses cause infected cells to commit suicide. But for some, the toll is far worse. The immune system may create an overwhelming response, creating so much inflammation that it damages the patient's body. In the battle between host and pathogen, cellular debris can build up in the airway, allowing bacterial infections to take hold. These complications can be fatal—especially for children and the elderly.

Influenza is a worldwide phenomenon, moving across the entire planet every year. In the tropics, the rate of flu infections stays fairly steady year-round, whereas it surges in temperate latitudes each winter. As they spread from person to person, flu viruses experience ongoing natural selection (Xue et al. 2017). This selection leads to, among other things, the ability to evade antibodies. Flu viruses can escape the immune system by accumulating mutations that alter the tip of the hemagglutinin stalk. Antibodies are less likely to bind to the new shape, enabling the viruses to go on with their life cycle. From one flu season to the next, the viruses may change so much that people's antibodies from the previous year do a poor job of stopping infections (**Figure 18.15**).

Figure 18.15 A phylogeny for sequence evolution of the hemagglutinin gene in influenza taken from sick people sampled over two winters in France. The strain had evolved (antigenic drift) so drastically from the winter of 2002–03 (blue) and 2003–04 (red) that new vaccines had to be developed. (Data from Al Faress et al. 2005)

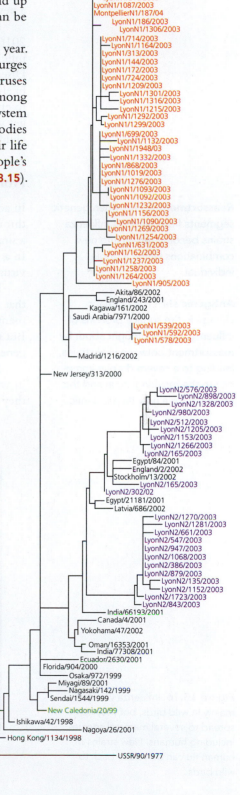

0.01

Antigenic drift describes natural selection on seasonal influenza viruses that alters the structure of surface proteins, allowing them to evade the antibody-mediated immunity induced during previous infections or vaccinations.

Escaping from Vaccines

This seasonal evolution, called **antigenic drift**, is the reason why we have to get flu vaccines every year. Standard flu vaccines, made of dead viruses or protein fragments, stimulate the immune system to make antibodies against hemagglutinin. It takes months to create a new batch of the vaccines, a time during which the virus population rapidly evolves. As a result, researchers have to guess in advance about each upcoming flu season, deciding which antibodies will be most likely to be effective. The effectiveness of flu vaccines typically fluctuates between 40% and 60%—and that's only for a single flu season. By the next year, antigenic drift may produce a new type of flu that requires a new vaccine.

Even in their imperfect state, influenza vaccines are well worth getting. In the 2016–17 flu season, vaccines prevented 5.3 million influenza illnesses and 85,000 influenza-associated hospitalizations in the United States, according to the Centers for Disease Control. And vaccinated people who end up getting the flu anyway are less likely to suffer severe symptoms. People who get sick with the flu are 59% less likely to end up in an intensive care unit if they have been vaccinated. Healthy children who get vaccinated are two-thirds less likely to die from the flu compared to unvaccinated children.

From Antigenic Drift to Antigenic Shift

In addition to vertically inherited mutations, influenza can gain new genetic variation through a process that's similar to recombination. When two influenza viruses infect a single cell, its host produces the eight segments of RNA contained in each of them. In a process called **reassortment**, the segments may wind up mixed together in new viruses.

Reassortment occurs when genetic segments from different influenza strains become mixed into new combinations within a single individual.

Antigenic shift is a dramatic change in the surface proteins of influenza viruses brought about by reassortment between viral strains, leading to a severe drop in immunity to the resulting strain and the potential for a new flu pandemic.

Reassortment can occur between flu viruses of the same strain, or of two strains that are both circulating from person to person. Like accumulating mutations, reassortment can produce viruses that can evade the antibodies produced through vaccination. But reassortment can lead to a much more dramatic change when human viruses swap genetic segments with viruses from other species—a process called **antigenic shift**.

As shown in **Figure 18.16**, humans are just one node in a huge network of animals in which influenza viruses circulate. Wild birds are the primary source of influenza; they harbor an enormous diversity of flu viruses that are adapted to living harmlessly

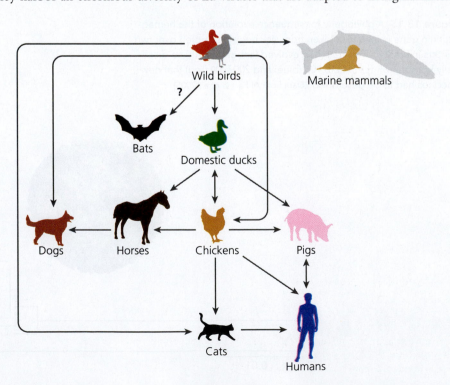

Figure 18.16 Influenza viruses live mainly in wild birds, but they can spread to several mammal species, including humans. New strains of human flu can be traced back to wild birds.

in their guts. Those viruses sometimes shift to other host species, which, in turn, can pass them on to humans. In recent years, several outbreaks of so-called "bird flu"—spread mainly from domesticated birds to people—have claimed hundreds of lives. Despite being highly virulent, avian influenza viruses are poorly adapted to infecting human airways. As a result, they almost never spread from one person to another.

Reassortment can allow bird flu to complete the jump to humans (**Figure 18.17**). On rare occasions, a bird flu and human flu virus may end up in the same cell. In some cases, the cell may be inside the body of an unfortunate human; in other cases, pigs may serve as a so-called "mixing vessel," because the receptors on their cells allow them to harbor both human and bird flu viruses. The hybrid virus produced through reassortment can then spread swiftly through the human population because it contains human-adapted genes. But if its surface proteins come from birds, it can go almost entirely unnoticed by antibodies that target human strains in circulation. Once a single virus is produced through reassortment, it can spread across the planet, driving older strains out of existence. This massive spread is known as a pandemic, and it can claim millions of lives on its march across Earth (**Figure 18.18**).

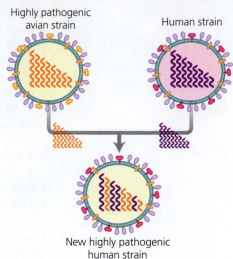

Figure 18.17 Reassortment occurs when two strains of influenza co-occur in the same cell. Their genetic segments get shuffled in new viruses, producing new combinations of genes.

The 1918 Flu

It's not clear when flu pandemics began, but over the course of the past century, we have been hit by four. The biggest got its start in spring 1918, and more than a century later, scientists are still trying to figure out exactly what happened (Barry 2005; Humphreys 2018; Medina 2018).

Here's what we know. In rural Kansas, a traveling doctor named Loring Miner observed that his patients were coming down with a particularly severe form of the flu.

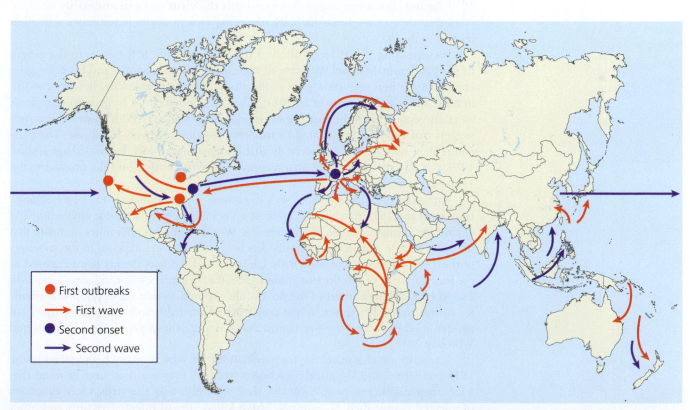

- 🔴 First outbreaks
- → First wave
- 🟣 Second onset
- → Second wave

Figure 18.18 In 1918, antigenic shift gave rise to a flu pandemic that spread quickly around the world and killed an estimated 50 million people. Three pandemics have arisen since then. Each time a new pandemic strikes, it eventually gives rise to less severe seasonal flu strains. (Data from Krammer et al. 2018)

Some of them died; the disease was so lethal that it could claim some victims within a single day. The same week that Loring Miner's patients fell ill, two Haskell County recruits named Ernest Elliot and Dean Nilson took the train to Camp Funston, the second-largest training camp in the country. Within 3 weeks, more than 1100 soldiers at the camp were hospitalized with the flu. Thousands more were treated at infirmaries across the base.

All the while, troops were being shipped from Camp Funston to other camps. Two weeks after soldiers at Camp Funston got sick, influenza outbreaks hit 24 additional training camps and 30 of the largest cities in the United States. American soldiers were also being shipped from stateside camps to the European front, and very soon the virus was raging through the trenches. From there, it swept across Europe and then into China, New Zealand, and Australia, before circling back again to the United States. By the time it reestablished itself in the United States, influenza was killing almost one in every ten people it infected. It crippled the country, and the rest of the world, during October and November of that year. Incredibly, a third of the world's entire population got sick, and an estimated 50 million people died before the pandemic came to an end.

Resurrecting a Killer

The 1918 pandemic struck at a time when scientists understood almost nothing about influenza. In 1951, a team of University of Iowa scientists traveled to Alaska and dug up the frozen bodies of people who died in the 1918 pandemic, hoping to isolate the flu viruses that killed them. They failed. Five decades later, another team of scientists from the National Institutes of Health tried again. This time, they used a new strategy. Rather than trying to grow frozen viruses, they extracted viral DNA. And this time they succeeded. In 2005, they were able to synthesize the virus's genetic segments and generate full-blown viruses from them (Belser et al. 2018).

The researchers resurrected the viruses in laboratories designed for Lassa and other dangerous pathogens. They then infected mice, ferrets, and monkeys with the 1918 flu and (not surprisingly) discovered that the virus was extraordinarily virulent, often spreading throughout the body, triggering violent immune responses, and often leading to death.

Resurrecting the 1918 flu allowed scientists to study it for clues to the origin of the pandemic. Michael Worobey, of the University of Arizona, and his colleagues compared the hemagglutinin gene in this strain of flu to that in closely related strains in birds and pigs (Worobey et al. 2014). They concluded that a human flu virus with an avian version of the hemagglutinin gene arose in the late 1800s. Historical records suggest that this new virus caused a pandemic—but a relatively small one. The pandemic then switched to a seasonal flu that returned every year. Later, perhaps around 1917, this flu virus went through a reassortment with another bird flu, producing the 1918 pandemic.

Although preliminary, Worobey's hypothesis might explain some of the strange features of the 1918 pandemic. The most vulnerable people were not children or the elderly, for example, but people in their 20s. Worobey and his colleagues speculate that the older people had protective antibodies because they were exposed to the ancestors of the 1918 viruses in the late 1800s. Children in 1918 had already been exposed to the seasonal form of this flu. But around 1900, there's evidence of a different pandemic, caused by a different strain of the flu. People exposed to that flu developed antibodies against that other strain. When they reached their 20s and were hit by the 1918 pandemic, they lacked the protection that old and young people had (Nelson and Worobey 2018).

Scientists are also searching for explanations for why so many people died in the 1918 pandemic, and so quickly. To begin with, it's important to bear in mind that a far larger number of people got sick than were killed by the virus. Only about 2% of infected people died. By contrast, SARS killed 10% of infected victims, and other pathogens have far higher mortality rates. Even so, the 1918 flu remains unmatched in modern history in its combination of massive death toll and explosive spread. No other flu pandemic since has come close to its devastation.

It's possible that the 1918 flu was so bad thanks to an unlucky combination of historical turns of events. When World War I broke out, the modern network of trains could quickly bring millions of young men together from across the country into military camps, where a virus could find many susceptible young people to infect. On the battlefield, the unsanitary conditions of trench warfare made it even easier for the virus to spread. Wounded soldiers then delivered the virus to military hospitals and troop ships, and ultimately to their homes, where the viruses could find new hosts.

Yet it's also clear from experiments on the resurrected virus that it was intrinsically more harmful than other influenza strains. Paul Ewald, an evolutionary biologist at the University of Louisville, has argued that the conditions at the end of World War I drove the virus to evolve into a more virulent pathogen (Ewald 1996). The viruses could replicate faster, despite the attendant virulence, because they didn't have to travel far to infect someone else. Even a soldier in a coma could come into contact with many people as he was moved from place to place. Ewald argues that the trade-off between virulence and transmission became relaxed. As a result, its virulence evolved to devastating levels.

Three more pandemics have swept the planet since 1918, most recently in 2009. Since then, bird flu outbreaks have flared up from time to time, mainly in Southeast Asia. Reassortment continues to shuffle flu genes into new combinations. Scientists have no way to predict exactly when the next pandemic will strike. But it is only a matter of time before it does. And a century after the 1918 flu, we have created many new opportunities for the virus to spread (**Figure 18.19**). Today, the world population lives mostly in cities, wwhere viruses can move short distances to new hosts. There are more than 68 million forcibly displaced people around the world, many living in unsanitary refugee camps where diseases frequently flare up. Airplanes, meanwhile, shuttle people quickly around the planet. A virus can thus infect a person in one part of the world and end up on another continent in a few hours, even before it causes any symptoms (Colizza et al. 2007; Epstein et al. 2007).

How will humanity fare against the next flu pandemic? In some respects, the world is better prepared than ever before. We can track the spread of the flu, design new vaccines, treat the sick with antiviral drugs, and use antibiotics against secondary bacterial infections. But it takes a serious dedication of resources to make sure these tools get put to good use. The sooner we know that a new pandemic has arisen, the faster we can act to stop its transmission—but this knowledge requires continual

Figure 18.19 A: Crowded troop ships and unprecedented world travel facilitated the spread of a deadly influenza strain in 1918. B: Scientists worry that overpopulated cities, refugee camps, and rapid air travel (C) could have a similar effect. As in 1918, these human conditions could permit the spread of another deadly strain. (A: Interim Archives/Getty Images; B: Alison Wright/Getty Images; C: CARL DE SOUZA/AFP/Getty Images)

worldwide monitoring. Antiviral drugs can cut infections short, but they need to be stored up in vast quantities in advance—in both wealthy countries and poor ones—to be effective against a global pandemic. And for antibiotics to remain effective against secondary bacterial infections, we have to stop the rise of resistance.

Key Concepts

- The influenza virus can evolve rapidly to evade detection by our immune system and our vaccines.

- Antigenic drift leads to changes in seasonal flu from one year to the next. Antigenic shift produces new strains and pandemics.

- Wartime conditions changed the nature of selection acting on the influenza virus, permitting the evolution and spread of an extraordinarily deadly strain in 1918.

- Modern world travel may be creating a selection environment similar to the one that facilitated the 1918 pandemic: mass movement and crowding relax the selective constraint on virulence. ●

18.7 Molded by Pathogens

The constant threat of pathogens over billions of years has shaped the genomes of their hosts. Single-celled bacteria, archaea, and eukaryotes all produce proteins and RNA molecules that enable them to attack viruses that enter their interiors. When animals evolved multicellular bodies, they also evolved a symbiotic relationship with bacteria that began to grow inside them. They evolved pattern-recognizing proteins to distinguish these beneficial microbes from pathogens (**Figure 18.20**; Degnan 2015). Later, vertebrates evolved the ability to generate antibodies (Boehm et al. 2012).

But even after these defenses emerged, pathogens continued to shape our genome. That's because we are engaged in an epic arms race with our enemies. For example, viruses can only replicate if they can interact with our cellular proteins. They use some proteins to enter a cell, others to read their genes, to assemble their new shells, and so on. Certain mutations to our proteins can change their shape in a way that makes it harder for viruses to interact with them—and thus to make us sick. Once our proteins have changed shape, however, the viruses have a selective pressure to change their shape as well so that they can get better at manipulating us again.

This back-and-forth evolution has altered thousands of our proteins for millions of years. In 2016, David Enard, then at Stanford University, and his colleagues compared human genes to those of other apes to estimate how extensive this particular form of adaptive evolution has been in the hominin lineage. Astonishingly, they concluded that in that time one-third of all our proteins have evolved new amino acids under the pressure of viruses (Enard et al. 2016).

Even within the last few thousand years, humans have evolved new defenses against certain diseases (Karlsson et al. 2014). Malaria, for example, has had an especially powerful effect on the

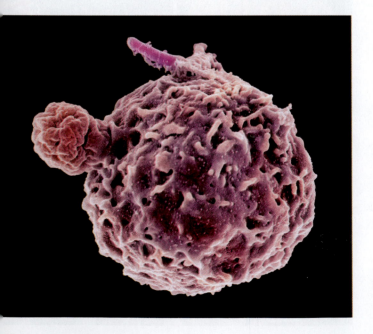

Figure 18.20 A white blood cell (right) attacks an invading bacterium. Mutations that improve the defenses of hosts are strongly favored by natural selection. (SPL/Science Source)

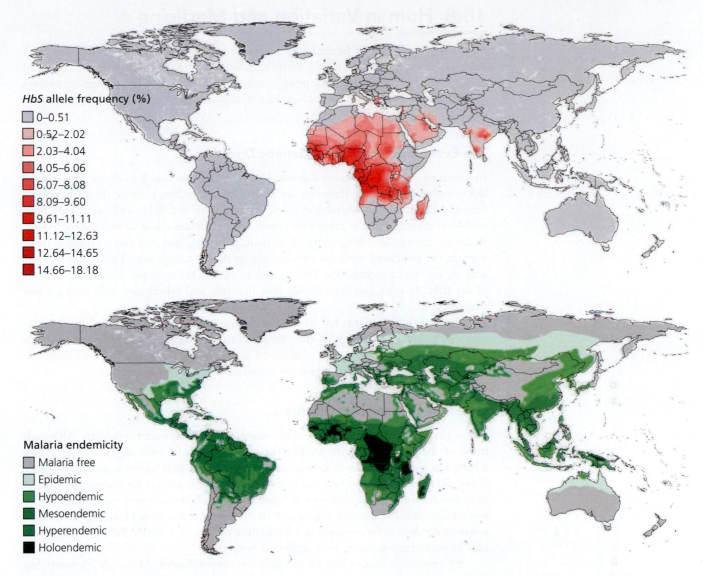

HbS allele frequency (%)
- 0–0.51
- 0.52–2.02
- 2.03–4.04
- 4.05–6.06
- 6.07–8.08
- 8.09–9.60
- 9.61–11.11
- 11.12–12.63
- 12.64–14.65
- 14.66–18.18

Malaria endemicity
- Malaria free
- Epidemic
- Hypoendemic
- Mesoendemic
- Hyperendemic
- Holoendemic

Figure 18.21 Scientists developed a statistical approach to mapping the frequency of the *S* allele of the hemoglobin gene. In Africa, the prevalence of malaria matches up strongly with the *S* allele frequencies. This alignment shows how balancing selection can result in high frequencies of alleles that can cause disease, thanks to heterozygote advantage. In this case, a single *S* allele protects against malaria, whereas two copies cause sickle-cell anemia. (Data from Piel et al. 2010)

human genome. The disease appears to have emerged as a serious threat as agriculture spread through Africa. As farmers cleared forests, malaria-spreading mosquitoes could breed in the standing water in their fields and then infect the farmers as they slept in nearby villages. Numerous mutations that confer resistance to malaria spread quickly in regions where malaria is common. In Chapter 6, we introduced one of these mutations—the *S* allele for hemoglobin. The geographic distribution of the allele in people today reflects its advantage against malaria. Human populations where malaria is prevalent show much higher frequencies of the allele than other areas where malaria is less common or absent (**Figure 18.21**; Piel et al. 2010).

Key Concept

- The human genome has been profoundly shaped by coevolution with pathogens. Some defensive variants became fixed in our species. Others show geographic variation that matches the ranges of diseases they protect against. ●

18.8 Human Variation and Medicine

Medicine is increasingly becoming personalized. Instead of treating all patients alike, doctors are starting to consider genetic variations that make them different. Some of these genetic variants can make people vulnerable to diseases. Others render certain medications useless. In all these cases, an evolutionary perspective can help us understand the nature of these conditions.

The Evolutionary History of Genetic Disorders

In previous chapters, we have discussed genetic disorders such as cystic fibrosis and sickle-cell anemia. These disorders are known as Mendelian, because their phenotype typically follows simple Mendelian rules. Cystic fibrosis and sickle-cell anemia are both recessive, meaning that it takes two copies of the disease allele to make people sick. A disease such as Huntington's is dominant, requiring just one copy. Other genetic variants are associated with diseases, but not in such a simple way. Thousands of variants are known to increase the lifelong risk of cancer, for example. Some, like variants of the *BRCA1* gene, can dramatically raise that risk, and others may only raise it a few percentages.

Each of these variants has its own evolutionary history. Understanding this history can help us to understand the characteristics of Mendelian diseases and genetic risk factors—why some are rare and others are relatively common, and why some are found around the world and others are found only in tiny populations.

Selection on Mendelian Disorders

As we discussed in Chapter 5, random mutations arise in every newborn. Although most of these mutations are predicted to be harmless or only slightly deleterious, a baby has a slight chance of acquiring a new mutation that causes a genetic disorder. If the disorder is fatal in childhood, the new mutation cannot be transmitted to the next generation. Its frequency in the population will remain very rare, equal to the probability that the corresponding site in the genome mutates. Hutchinson-Gilford progeria syndrome, for example, is a devastating disease that causes children to age rapidly. Its prevalence is one in four million newborns.

By contrast, a disease like Huntington's is very different. Although it's invariably fatal, its symptoms only emerge when people reach their 40s. That is enough time for people with the disease to have children who can inherit the allele and increase its frequency in a population. In people with European ancestry, the disorder has a prevalence as high as five in 100,000.

Paradoxically, natural selection makes certain genetic disorders even more common. Sickle-cell anemia is diagnosed in one in 365 African American births. That's a rate 10,000 times higher than Hutchinson-Gilford progeria syndrome. It's so much more common because, as we discussed earlier, the *S* allele provides a heterozygous advantage in places with high levels of malaria. Other recessive diseases are much rarer than sickle-cell diseases because natural selection does not favor the allele in heterozygotes.

The Medical Impacts of the Founder Effect

Genetic drift can also drive disease alleles to higher frequencies, despite lowering fitness. As we saw in Chapter 6, founder effects can produce high frequencies of deleterious alleles in later generations. In addition, when these deleterious alleles are recessive, inbreeding can make the genetic disorder they underlie much more common. As a result, many isolated populations descended from small groups of founders have unusually high rates of genetic disorders.

On the remote Pacific island of Pingelap, for example, 5% of the population is completely color-blind (Hussels and Morton 1972). By comparison, only 0.003%

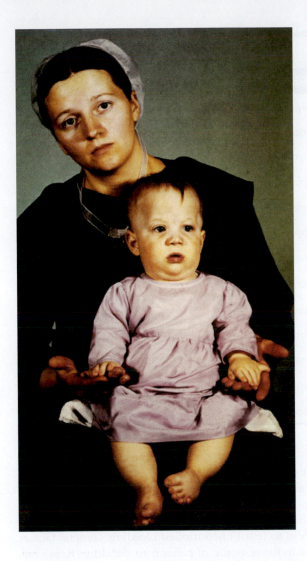

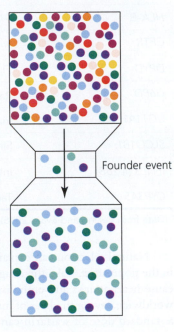

Founder event

Figure 18.22 Some human populations suffer from high rates of certain genetic disorders. The Amish, for example, are unusually likely to suffer from Ellis-van Creveld syndrome, which can cause heart defects, along with deformities to the skeleton, such as additional fingers and toes, and limb dwarfism. The Amish descend from a small group of immigrants from Europe who settled in the United States. Their genotypes are not a representative sample of the genotypes in their homeland because the initial population was so small. Due to these types of founder effects, even one person carrying genes for a rare genetic disorder can make it much more common in the new population. (Photo: Alan Mason Chesney Medical Archives of The Johns Hopkins Medical Institutions)

of the U.S. population suffers from this condition, called achromatopsia. Historical research indicates that around 1775, a typhoon reduced the island's already small population to just 20 survivors. One of those survivors carried one copy of the achromatopsia allele. Thanks to that fluke of history, 5% of Pingelap's current residents now suffer from the condition because they carry two copies of the original allele (30% carry a single copy).

Islands are not the only places where genetic drift drives up genetic disorders. In the late 1600s, a small group of closely related farmers traveled from Switzerland to Germany and finally to the United States, where they became known as the Amish. Shunning intermarriage with other groups, they kept to themselves, and they still do today. In effect, they've created a "genetic island" in the middle of a continent. Not surprisingly, the Amish suffer high rates of certain genetic disorders, such as Ellis–van Creveld syndrome, which leads to extra fingers and dwarfism (Figure 18.22). Scientists have been able to trace that disorder to a single Amish couple that immigrated to Pennsylvania in 1744 (McKusick 1978).

How Genes Affect Our Response to Drugs

Certain alleles can lead to harmful side effects from drugs that are safe in most people. Other alleles can lead some people to require a larger dose of a drug than others to get an effective treatment. Scientists still know relatively little about how genes determine drug responses; the more they learn, the better they can personalize the medicine doctors give to their patients (Table 18.1; Relling and Evans 2015; Maron 2016).

Table 18.1 Examples of Human Genetic Polymorphisms and the Drugs They Influence

Genetic Variation	Medications
TPMT	Mercaptopurine, thioguanine, azathioprine
CYP2D6	Codeine, tramadol, tricyclic antidepressants
CYP2C19	Tricyclic antidepressants, clopidogrel, voriconazole
VKORC1	Warfarin
CYP2C9	Warfarin, phenytoin
HLA-B	Allopurinol, carbamazepine, abacavir, phenytoin
CFTR	Ivacaftor
DPYD	Fluorouracil, capecitabine, tegafur
G6PD	Rasburicase
UGT1A1	Irinotecan, atazanavir
SLCO1B1	Simvastatin
IFNL3 (IL28B)	Interferon
CYP3A5	Tacrolimus

(Data from Relling and Evans 2015)

Natural selection can explain why some of these variants exist at high frequencies in the population. Take the drug warfarin, which is used to destroy blood clots that can cause heart attacks and embolisms. It is the most commonly prescribed anticoagulant worldwide, saving the lives of millions of people. In a small fraction of people, however, a standard dose of warfarin can lead to lethal uncontrolled bleeding. Genetic factors explain about half of the variance in the response of patients to the drug (Ross et al. 2010).

Kendra Ross, of the University of Toronto, and her colleagues surveyed alleles of warfarin-sensitive genes in 1279 people representing a number of major human populations worldwide. They found that one of these genes, *VKORC1*, has a striking pattern of geographic variation. People with a particular allele of the gene, known as the *rs9923231 T* allele, require only a low dose of warfarin and are at greater risk of bleeding. Ross and her colleagues found this allele in 100% of Han Chinese but in only 10% of Africans. Using several independent tests, the scientists found strong evidence for natural selection as the cause of the fixation of the *rs9923231 T* allele in the Han population.

It will be intriguing to learn the cause of selection for this allele. The protein encoded by *VKORC1* is involved in blood clotting as well as other functions such as bone mineralization. Whatever the cause, natural selection has also made many people vulnerable to warfarin. Just as drugs have side effects, natural selection can have side effects of its own.

Key Concepts

- Many genetic disorders are caused by mutations resulting in defective copies of genes. The relative frequencies of these alleles—and the diseases they cause—are determined by genetic drift and natural selection, which vary geographically and

- Understanding the role of genetic variation in human diseases can be critical not just for treatment of genetic disorders but also for diagnosing the drug reactions that can accompany treatment. ●

18.9 Getting Old, Getting Sick

Even if we have the good fortune to avoid fatal genetic disorders and infections, we don't stay young forever. Getting old is an inescapable part of life. Despite the advances of modern medicine, people today still age in much the same way that people did a hundred years ago. Our bones get brittle, we lose our stamina, and our defenses against infections weaken. In addition, the rate of certain diseases, such as cancer, heart disease, and Alzheimer's disease, are very low in young people but climb as people get older. Even our athletic performance shows a steady decline throughout adulthood (**Figure 18.23**).

As we saw in Chapter 12, aging arises through natural selection on life history traits that influence survival and reproduction. Mutations that cause death at an early age experience strong negative selection because they prevent children from growing up and having children of their own. Mutations that cause diseases only after individuals have reproduced, on the other hand, can be shielded from selection. And natural selection can also increase the frequency of mutations contributing to aging if those same mutations also promote fitness earlier in life—a case of antagonistic pleiotropy (Austad and Hoffman 2018).

Evolutionary biologist George Williams first proposed this trade-off in the 1960s, and many of the predictions from his antagonist pleiotropy hypothesis have been borne out. Williams developed his hypothesis decades before scientists could read the sequence of genes or compare the fitness of their alleles. But today, geneticists are finding that antagonistic pleiotropy is widespread in the human genome (Corbett et al. 2018).

Sean Byars, of the University of Melbourne, and his colleagues studied 76 genes known to have a role in the risk of cardiovascular disease—a disease that primarily affects people in old age (Byars et al. 2017). Forty of those genes showed signs of positive selection. And when the researchers examined these 40 genes in detail, they found that they were associated with early-life reproductive traits, such as the number of children women gave birth to and how much milk they produced.

Cancer-linked genes also show signs of antagonistic pleiotropy. Mutations to the *BRCA1* gene can drastically increase the risk of breast and ovarian cancer in women. But studies on women carrying these cancer-promoting mutations show that they have more children over their lifetimes than women with harmless *BRCA1* genes (Smith et al. 2012).

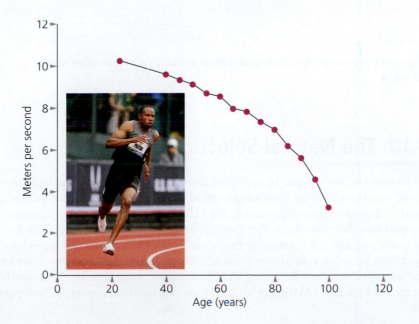

Figure 18.23 The performance of the human body declines over time due to aging. Using world records for the men's 100-meter run by age groups, scientists can calculate the average speed of each champion runner. The maximum speed starts dropping in the third decade of life. Evolutionary biologists seek to understand why this decline evolved. (Photo: Andy Lyons/Getty Images; data from Stearns and Koella 2008)

BOX 18.2

Genes, Environments, and Cancer Risk

Cancer is a genetic disease. For cells to start dividing uncontrollably, genes that regulate cell growth must be damaged. The likelihood of this occurring depends on the particular alleles that each person inherits as well as on the DNA-damaging (mutagenic) conditions they are exposed to.

In 1971 the American geneticist Alfred Knudson proposed a mechanism to determine why cancer risk clearly ran in families. His "two-hit hypothesis" proposed that an inherited germline mutation in a **tumor-suppressor gene** would cause cancer only if another mutation event occurred later in life that inactivated the remaining allele of that gene. Individuals could inherit a predisposition toward developing cancer if either of their maternal or paternal alleles for cancer-causing genes were defective. The risk was greater for these individuals because they needed only one further mutation to these genes to begin the process of uncontrolled cell division. People inheriting two viable alleles for these genes, on the other hand, have less risk because it would take "two hits" to the same gene in the same cell to begin the process (Knudson 1971).

Scientists now recognize that the process is more complex. It often involves multiple mutations to both **oncogenes** and tumor-suppressor genes as well as epigenetic events occurring during early development (silencing a copy of one of these genes can also predispose an individual to cancer). But the basic idea holds: defective or silent alleles inherited from parents predispose individuals to greater susceptibility because these individuals require fewer spontaneous somatic mutations to initiate cancerous growth.

The other half of the equation is the environment. Scientists have identified many chemical and other factors (for example, radiation) that can lead to mutations in both somatic and germline cells, and all these factors increase the risk of cancer if the mutations affect the expression of oncogenes or tumor-suppressor genes. Mutations arise as DNA is replicated (that is, when cells are dividing). This means that all else being equal, developing fetuses and young children are especially sensitive to exposure to environmental sources of mutation. For this same reason, tissues that routinely undergo cell division are more vulnerable than other tissues. Skin, colon, lung, breast, prostate, and bone marrow cells all are at unusually high risk for developing cancer because these tissues require some degree of continuous cell division to function. Increased rates of cell proliferation bring with them greater vulnerability to mutagenic effects of environmental toxins or radiation (**Box Figure 18.2.1**).

Cells in our bodies accumulate mutations gradually over time, and these events occur randomly and uniquely from cell to cell. For cells to become cancerous, several different genes must be damaged in the same cell. All of the redundant mechanisms of proliferation regulation must be inactivated together before any specific cells can begin to divide uncontrollably. This is one reason that cancers primarily strike the elderly. It takes a lifetime of accumulation of mutations before the requisite hits occur together in the same cells. But cancer can strike earlier in life if we expose particular tissues to unusually large doses of mutation-causing environmental stimuli. In these instances it may not take long at all for the required mutations to strike the same cells—even less time if

---

Key Concept • Aging is a pleiotropic by-product of natural selection acting at other stages of life. ●

18.10 The Natural Selection of Cancer

We humans are not unique in suffering from cancer. Roughly 750 million years ago, our ancestors evolved from single-celled protozoans into multicellular animals. Instead of reproducing independently, the cells in animals began to cooperate. They developed into different tissues that together produced a working body. The fitness of their genes rose and fell together as they enabled animals to feed and reproduce. The cooperation between cells allowed animals to occupy new niches that their single-celled ancestors could not. But along with these evolutionary benefits, these new bodies presented a new risk (Aktipis et al. 2015). As we saw in Chapter 16, cooperation

we start the process with defective alleles inherited from our parents.

It's no accident that types of cancer correlate with exposure to specific sources of mutagenic environmental stimuli. Inhalation of asbestos or tobacco smoke is most likely to affect lung cells, and ultraviolet radiation from sunlight or from tanning machines is most likely to induce mutations in skin cells. Our genotypes and environments interact to determine the types and amounts of cancer risks that we face.

A **tumor-suppressor gene** is a gene that suppresses cell growth and proliferation. Many tumor-suppressor genes are transcription factors activated by stress or DNA damage that arrest mitosis until DNA can be repaired. Mutations interfering with their expression can lead to excessive proliferation and cancer.

An **oncogene** is a normal gene whose function, when altered by mutation, has the potential to cause cancer.

A

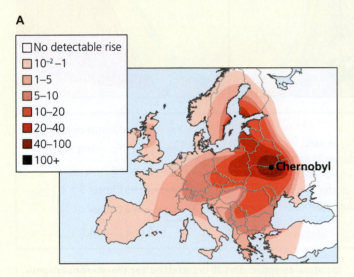

B

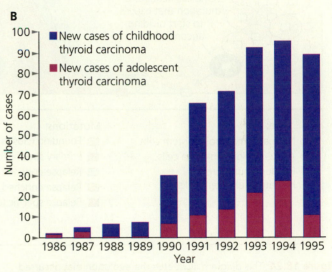

Box Figure 18.2.1 Cancer risk is often localized around geographic sources of mutagens. A: After the nuclear reactor meltdown in Chernobyl in 1986, thyroid cancer rates increased dramatically in areas exposed to the most radioactive fallout. B: Young children were more likely to develop cancer than adults were. (Data from Pacini et al. 1997)

provides the opportunity for cheaters to thrive. In our own bodies, this cheating is known as cancer.

Every time a cell divides, there's a tiny chance that a mutation will occur. Smoking and other environmental influences can increase the chances of mutations in certain tissues (**Box 18.2**). Although these mutations arise across the genome at random, on rare occasion they strike genes that control the rate at which cells divide. These cells start to grow faster than their neighbors. In some cases, these runaway mutant cells can form a tumor.

Within a developing tumor, cells continue to acquire mutations as they divide, and cells with mutations that speed up their growth come to dominate the population of cancer cells. Some genes, for example, normally become active only in sperm cells, helping them to grow rapidly throughout a man's adult life. Normally these sperm-growth genes are kept silent in other parts of the body. But mutations can switch them on in cancer cells, making them divide faster in both men and women.

Once cells begin to divide uncontrollably, selection begins to act at the level of individual cell lines within the tumor as they compete for space and resources

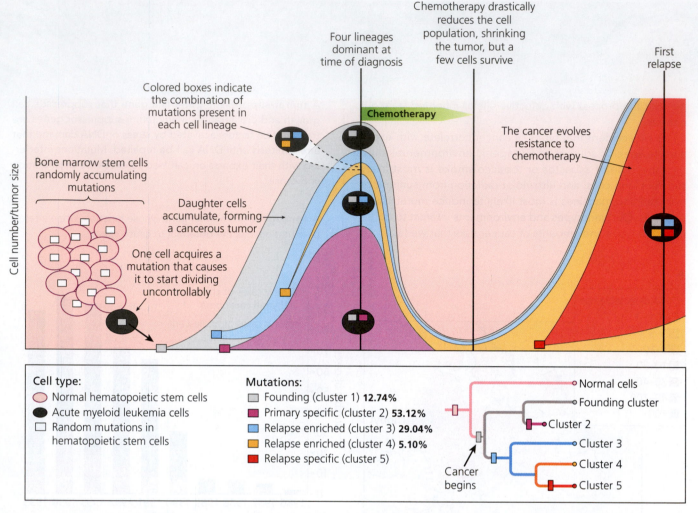

Figure 18.24 This diagram illustrates the evolution that occurred to cancer cells in a woman who died of acute myeloid leukemia. A lineage of hematopoietic stem cells (progenitors of immune cells) acquired a mutation that caused it to begin dividing uncontrollably, creating a lineage of rapidly dividing cells (gray box). The cancerous cells became more common, as indicated by the increasing width of the gray curve. Within this lineage, some cancer cells acquired new mutations, three of which are represented here by the blue, purple, and yellow boxes. The descendants of the cells with these mutations rapidly became more common, thanks to natural selection. This all occurred before the woman's diagnosis. Chemotherapy drastically reduced the population of all four lineages present at diagnosis, but it did not cure her. Instead, the chemotherapy acted as a powerful agent of selection. A resistant lineage of cancer cells, marked here in red, became dominant as other cells died. New mutations within this lineage then provided even stronger resistance, and it came to dominate the population. The cancer thus evolved complete resistance to chemotherapy. (Macmillan Learning; data from Ding et al. 2012)

(Shpak and Lu 2016). As cells in the tumor accumulate new mutations, they diverge genetically from each other. Just like populations of pathogens, the rising disparity among the cell lines makes the competition between them more intense, speeding up evolution.

Tumors are notoriously prone to become resistant to drugs; here, too, natural selection is responsible. Much like bacteria, cancer cells that acquire mutations that allow them to resist chemotherapy will be able to grow faster than susceptible cancer cells during treatment (but not necessarily in the absence of the drugs). Scientists can document this evolution by surveying cells at different stages of a cancer. Rare resistant cells become more common, and new mutations arise and spread. **Figure 18.24** shows the evolution that occurred in a single person before and after chemotherapy (Ding et al. 2012).

- Under normal conditions, the balance between proliferation and programmed cell death is tightly regulated. Somatic mutations that break down these regulatory processes can lead to uncontrolled cell proliferation and cancer.

- Once cells start to divide uncontrollably, they begin to evolve by natural selection inside our body. Their fitness is no longer aligned with ours, and selection acts at the level of individual cell lines within the tumor.

- Somatic cell-line evolution within a cancerous tumor exemplifies how natural selection can be "shortsighted." Immediate fitness gains drive the evolution of increased cell division and metastasis even though this activity ultimately kills the host and, with it, all of the cancer cells in the tumor. ●

18.11 Mismatched with Modern Life

The medical picture of humanity is profoundly different in the twenty-first century than it was in 1900 (**Figure 18.25**). Across the world, childhood mortality has dropped dramatically thanks to a number of advances, from cleaner water to vaccination to safe milk supplies. As children started surviving more often to adulthood, the average life span increased, even though the maximum life span barely budged. In recent decades, the average life span has continued to increase, but most of the gains have come from

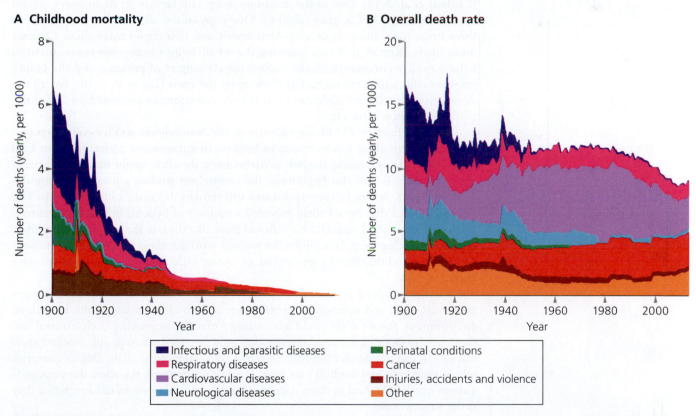

Figure 18.25 A: Childhood mortality declined dramatically in England and Wales in the twentieth century, thanks to improved hygiene, clean drinking water, better food, and improvements in medicine. This trend has been reproduced around the world. B: The overall death rate in England and Wales has dropped both as a result of the drop in childhood mortality, and, more recently, due to improvements in treating diseases in adults. This trend is likewise reflected in other countries, leading to a remarkable shift: infectious diseases are becoming responsible for fewer deaths than noncommunicable diseases such as cancer and heart disease. (Data from Corbett et al. 2018)

improved survival of people over 45 (Corbett et al. 2018). An American born in 1880 had a life expectancy of 39 years. People in the United States born in 2015 can expect, on average, to live longer than 79 years.

The modernization of human health has meant that today fewer people die of infectious diseases. Instead, the major causes of mortality have shifted from communicable diseases to noncommunicable ones, such as cancer, Alzheimer's disease, and heart disease. Obesity has reached epidemic proportions, and affluent countries have seen an explosion of disorders of the immune system. Crohn's disease, for example, occurs when the immune system attacks the lining of the intestines. Asthma is caused by inflammation in the lungs. Type 1 diabetes occurs when immune cells attack insulin-producing cells in the pancreas. All these diseases were once rare in the United States and other developed countries but now are increasingly common (**Figure 18.26**). Evolution provides us with a conceptual foundation to make sense of this radical shift in human health.

Unmasking Antagonistic Pleiotropy

One simple reason for the shift in diseases is that, until recently, most people died long before old age. Because they were already dead—killed by a pathogen, starvation, an accident, or some other cause—they never had a chance to experience the cost of antagonistic pleiotropy. Only after a large number of people began to live long lives was the true scope of antagonistic pleiotropy unmasked. Diseases such as cancer that can take decades to manifest began to make up a bigger burden on public health.

Alzheimer's disease provides a particularly striking example of this unmasking (Corbett et al. 2018). One of the most important risk factors for Alzheimer's disease is an allele of the APOE gene called *ε4*. One copy of the allele makes people two to three times more likely to develop Alzheimer's, and two copies make them 12 times more likely (Kim et al. 2009). Inserting the *ε4* allele into brain cells causes them to behave oddly—neurons with the variant release tangles of proteins, and the brain's immune cells, called microglia, fail to clean up the mess (Lin et al. 2018). For all the devastation that the *ε4* allele causes, it is very common: an estimated 14.5% of the world population carries it.

Eric van Exel, of VU Medical Center in the Netherlands, and his colleagues suspected that this allele was so common because of antagonistic pleiotropy (van Exel et al. 2017). Despite raising the risk of Alzheimer's, the allele might somehow increase fitness overall. To test this hypothesis, the researchers studied a poor, rural population in Ghana, where infectious diseases still remain the main cause of death. The scientists found that the *ε4* allele provided a number of benefits in this environment. In poorly nourished infants who suffered from diarrhea, it appears to protect their cognitive development. In addition, in women who got their water from open wells and rivers—and thus faced a greater risk of disease—the *ε4* allele was associated with higher fertility.

For thousands of years, most people lived as subsistence farmers, drinking contaminated water and suffering from infections as both children and adults. In such an environment, the *ε4* allele could have raised fitness by improving both survival and fertility. The fact that it caused diseases that only emerged when people reached their 60s or 70s did not detract from the selective advantage of the allele. But in countries where sanitation and medical care have drastically improved, the allele does not provide so much help. And in these countries, more people survive to old age, when they suffer its downsides.

Maladapted to the Modern Diet

As we saw in Chapter 17, food has played a pivotal role in human evolution. Starting perhaps 3.3 million years ago, our hominin ancestors evolved a diet mixing together plants and meat. Over millions of years, their bodies were reshaped by their food, from their teeth to their feet to their guts. Our own species, which emerged some

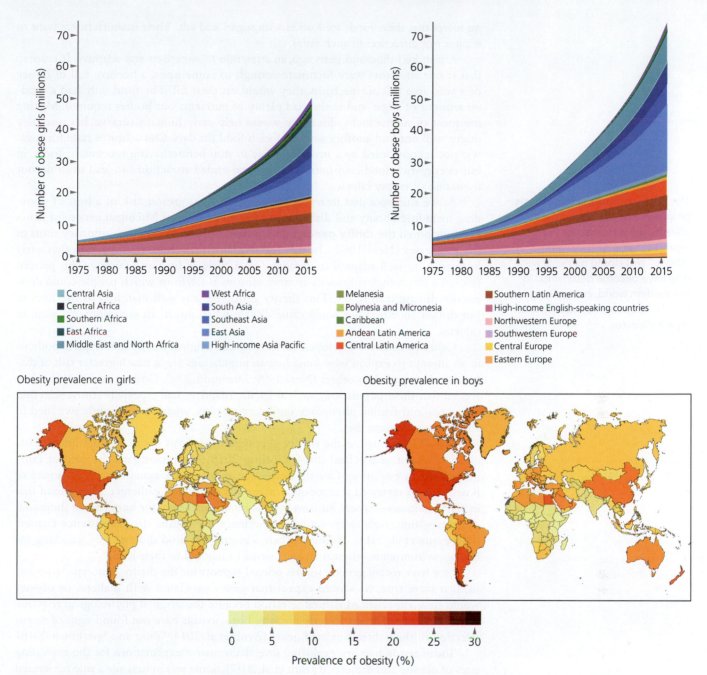

Central Asia
Central Africa
Southern Africa
East Africa
Middle East and North Africa

West Africa
South Asia
Southeast Asia
East Asia
High-income Asia Pacific

Melanesia
Polynesia and Micronesia
Caribbean
Andean Latin America
Central Latin America

Southern Latin America
High-income English-speaking countries
Northwestern Europe
Southwestern Europe
Central Europe
Eastern Europe

Obesity prevalence in girls

Obesity prevalence in boys

Prevalence of obesity (%)

Figure 18.26 A: The number of obese children has increased dramatically around the world since the 1970s. Initially, most obese children lived in Western affluent countries. But in recent decades, other regions of the world have been catching up. In 2016, 124 million children were obese. B: Globally, 5.6% of girls and 7.8% of boys are now obese, but some countries are experiencing far higher rates. (Data from: Worldwide trends in body-mass index, underweight, overweight, and obesity from 1975 to 2016: a pooled analysis of 2416 population-based measurement studies in 128.9 million children, adolescents, and adults Lancet 2017; 390: 2627–42 Published Online, October 10, 2017. http://dx.doi.org/10.1016/S0140-6736(17)32129-3)

300,000 years ago, has for most of its history survived on a hunter-gatherer lifestyle. Less than 10,000 years ago agriculture arose, shifting an increasing number of people to a diet based on grains and other crops. The rise of agriculture created selective pressure on a number of alleles that helped people digest their new sources of food, from milk to potatoes.

But in the twentieth century, the world's diet suddenly changed. Modern farming, industrial food processing, and global trade radically altered our eating habits. Walk into an American convenience store and spread out before you are all manner of energy-dense foods—soda, candy, donuts, potato chips, ice cream, and on and on. It is no secret that these foods are loaded with sugars and salt. Their manufacturers want to exploit our attraction to such tastes.

A hundred thousand years ago, an attraction to sweetness was adaptive. It ensured that if our ancestors were fortunate enough to come upon a beehive full of honey or a wild tree full of ripe fruits, they would eat their fill. The food only had a modest amount of sugar, and it also had plenty of nutrients our bodies required. Making the most of such a lucky discovery would help early humans survive, because they might well not find another source of such food for days. Our adaptive taste for sugar was not accompanied by a strong impulse to stop before having too much. Today, in our energy-rich food environment, that open-ended attraction can lead us to devour thousands of empty calories.

As we shift to a diet heavy in sugar, it puts us at greater risk of a host of disorders, including obesity and diabetes. In 1962, University of Michigan geneticist James Neel proposed the **thrifty genotype hypothesis** to explain the evolutionary roots of these diseases (Neel 1962). In this influential explanation, he proposed that early humans were well adapted to surviving on a cycle of feast and famine. This pattern favored a physiological capacity to store calories as fat, from which people could draw energy during lean times. This thrifty genotype was well matched to the lives of our distant ancestors but could cause disease when put in an environment awash in calories.

Other researchers developed variants on the original thrifty genotype hypothesis in an attempt to explain why some human populations are at much greater risk of diabetes and obesity than others (Neel 1999; Diamond 2003; Gosling et al. 2015). They argued that these populations, such as Pacific Islanders, had especially strong selection for a feast-and-famine physiology until very recently, when they abruptly switched to a high-sugar Western diet.

For all its influence, the thrifty genotype hypothesis has received a lot of criticism in recent years (Beil 2014; Gosling 2015). To investigate the role of famine in human evolution, Colette Berbesque and her colleagues at the University of Roehampton reviewed anthropological reports on hunter-gatherers. They found that in warm climates—where humans lived as hunter-gatherers for hundreds of thousands of years—hunter-gatherers actually experience *less* famine than subsistence farmers (Berbesque et al. 2014). Hunter-gatherers respond to food shortages by searching for better environments, whereas farmers remain anchored to their land.

Nor have recent genetic studies offered support for the thrifty genotype hypothesis. If it were true, we would expect that genes associated with diabetes or obesity would have experienced natural selection because the original physiological response would have raised fitness in early humans. But scientists have not found signs of recent selection in these disease-related genes (Ayub et al. 2014; Wang and Speakman 2016).

Today, researchers are exploring several alternative explanations for the exploding rates of obesity and diabetes (Qasim et al. 2017). Some researchers see a role for natural selection favoring certain traits in the past, which then left people vulnerable to the modern Western diet. For example, Anna Gosling, of the University of Otago in New Zealand, and her colleagues observe that Pacific Islanders were devastated by diseases introduced by European colonists in the 1800s. Natural selection may have favored alleles providing resistance to diseases such as measles—but these alleles might have also left them vulnerable to an energy-dense Western diet (Gosling et al. 2015).

Dyan Sellayah, of the University of Reading, and his colleagues have argued that the difference in susceptibility to diabetes and obesity arose as populations adapted to local climates (Sellayah et al. 2014). People living in cold climates evolved a higher metabolic rate to keep warm, which protected them from Western diets. People in warm climates, with a low metabolic rate, lacked that protection.

John Speakman, of the University of Aberdeen, meanwhile, has argued that our ancestors experienced negative selection for high levels of fat storage because they could be more easily caught by predators (Speakman 2018). Once humans became weapon-carrying hunters, that risk declined and heavier people were more likely to survive and pass on their genes. The selective pressure against fat storage disappeared, allowing obesity-related genes to rise in frequency due to drift.

The **thrifty genotype hypothesis** proposes that alleles that were advantageous in the past (for example, because they were "thrifty" and stored nutrients well) may have become detrimental in the modern world, contributing to metabolic syndrome, obesity, and type 2 diabetes.

Other scientists have considered human development in their search for an explanation. In 1992, David Barkman, a British epidemiologist, found that low birth weight was a strong predictor of obesity and diabetes later in life. One cause of low birth weight is an unpredictable food supply during pregnancy. Barker proposed that fetuses could be "programmed" for different patterns after birth, depending on their food supply. Storing more fat would be adaptive if a child had to develop in an environment where there was no guarantee of where the next meal might come (Hales and Barker 1992; Edwards 2017).

In a 2017 meta-analysis, Daniel Nettle, of Newcastle University, and his colleagues found that food insecurity is a predictor of high body weight in humans. Nettle and his colleagues also suggest that people remain sensitive to cues of food insecurity as they grow up. It may give them a greater motivation to eat—and thus to put on more weight (Nettle et al. 2017).

Losing Our Old Friends

Many of the current hypotheses for the modern rise in obesity and diabetes are based on the concept of an evolutionary mismatch (Nesse and Williams 1995; Gluckman and Hanson 2006). Our ancestors evolved a set of adaptations for an ancestral environment. These traits are mismatched with our new environment and produce negative consequences.

Evolutionary mismatches may also be the cause of other afflictions that are on the rise in the modern era. According to the "Old Friends" hypothesis, the rise in immune disorders such as allergies and asthma is due to a mismatch between our immune systems and the modern environment—one in which certain species of microbes and parasites are missing (Bloomfield et al. 2016; Scudellari 2017).

The rise of immune disorders in recent decades has had a striking geographic pattern. These disorders first emerged in countries such as England and the United States, which were among the first nations to go through an industrial revolution and then improve their public health (**Figure 18.27**). Later, when other countries went through the same transition, they also saw a rise in autoimmune diseases. Even within countries, a similar pattern can be found. In Venezuela, for example, the population is split mainly between cities and farms, but some indigenous tribes still live in isolated rain forest villages. Venezuelan city dwellers have higher rates of allergies than Venezuelan farmers do, and inhabitants of the forests have no allergies to speak of.

In 1989, David Strachan, an epidemiologist at the London School of Hygiene and Tropical Medicine, suggested that these autoimmune diseases were breaking out

Figure 18.27 The decline of infectious diseases in the United States has been mirrored by a rise in immune disorders. (Data from Bach 2002)

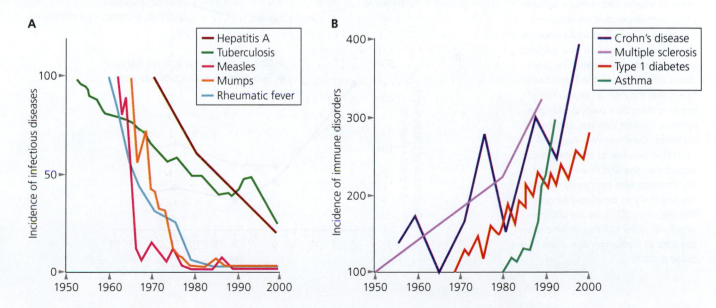

18.11 Mismatched with Modern Life **719**

because children living in urbanized homes were no longer exposed to dirt and dust, and they were not being infected by bacteria and parasitic worms. As a result, their immune systems were attacking themselves, rather than pathogens.

Initially, Strachan's proposal came to be known as the "hygiene hypothesis," but many researchers now consider it a misleading and potentially dangerous term (Bloomfield et al. 2016). It would be a mistake for people to give up sensible hygiene in the hopes of promoting a balanced immune system, only to get dangerously ill from food poisoning or influenza. Graham Rook coined the term "Old Friends" to better describe the relationship that humans have evolved with commensal species—a relationship that is unquestionably disturbed by modern life (Rook 2010).

In its current formulation, the Old Friends hypothesis holds that the immune system evolved a fine-tuned response to organisms it encountered in our bodies. It could respond quickly to potentially dangerous infections, but also was subject to constraints. Rather than eliminate some enemies, the immune system simply held them in check. This restraint allowed the immune system to spare beneficial microbes in our bodies and to minimize the risk of erroneously learning to treat our own tissues as the enemy.

This delicate balance is not hard-wired into our genes, however. It emerges out of a genetically programmed course of development—one in which microbes are present in the body from birth. The immune system learns tolerance, in effect. This view of the immune system is borne out by experiments in which mice are experimentally reared with no bacteria in their bodies. As they grow, they frequently develop auto-immune disorders.

According to the Old Friends hypothesis, modern life has disrupted the close relationship children have with microbes (**Figure 18.28**). Children who are fed formula

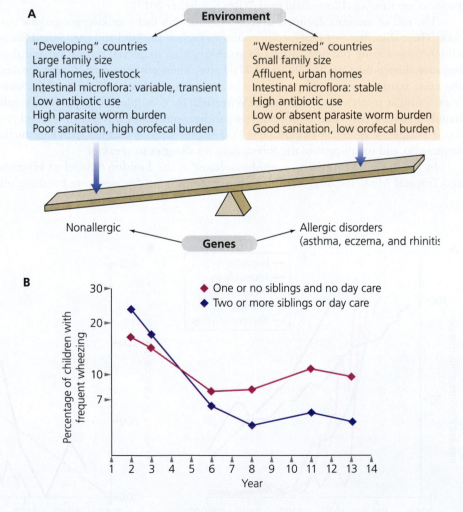

Figure 18.28 A: The Old Friends hypothesis predicts that the development of a balanced immune system previously depended on children's exposure to microbes and parasites. When children grow up in urbanized environments and receive frequent doses of antibiotics, the hypothesis holds that they run a greater risk of immune disorders such as allergies, asthma, Crohn's disease, and type 1 diabetes. (Data from Wills-Karp et al. 2001) B: A number of studies are consistent with the Old Friends hypothesis. Children who are raised on farms are less likely to develop allergies (Riedler et al. 2001), as are children who have older siblings or who enter day care at a younger age. (Data from Ball et al. 2000)

don't get a supply of commensal bacteria that nursing delivers, for example. Playing predominantly inside houses means they aren't exposed to microbes in the outdoor environment or carried by livestock. Many children get abundant doses of antibiotics—often for viral infections for which antibiotics do no good anyway. These chemical blasts don't just kill pathogenic bacteria, but also wipe out beneficial species. If these bacteria go missing from a child's body, they cannot help teach the immune system the proper restraint (Blaser 2014).

The Old Friends hypothesis does not imply that we must abandon antibiotics to get back to a Stone Age harmony with nature. But we do need to recognize that for all the benefits that antibiotics offer, they can impose a cost. In the future, scientists may invent precise antibacterial drugs that kill only certain pathogens while sparing commensal species. Until then, doctors can exercise more careful stewardship of antibiotics, not wasting them on young patients who don't need them. That restraint will not just benefit the children but will also help fight the rise of antibiotic resistance.

Key Concepts

- Modern life has improved human health in many ways. But it has also uncovered vulnerabilities shaped by natural selection.

- Genetic variants that promoted fertility in the past may now be risk factors for Alzheimer's disease, heart disease, and other major noncommunicable diseases.

- The modern diet has triggered an explosion of obesity and diabetes. Our vulnerability to these disorders may arise from ancient adaptations to hunter-gatherer diets.

- The Old Friends hypothesis holds that immune disorders such as asthma have risen in recent decades due to disruptions to the community of commensal organisms inside our bodies. ●

18.12 Harnessing Evolution for Medicine

Evolutionary medicine is a sobering science. It reveals to us the relentless origin of new diseases, the rapid adaptation of bacteria, and the deep vulnerabilities that make us susceptible to illnesses. Medical treatments that seemed to triumph over diseases, such as antibiotics, have faltered and failed because physicians did not take evolution seriously.

But evolutionary medicine also offers a lot of hope. Scientists can harness evolution to treat diseases in new ways. It can lead them to ask new questions and discover new ideas. Here we will explore a few case studies.

Evolving Vaccines

The goal of vaccines is to let people develop a long-lasting immune response to a pathogen without experiencing an actual infection. Some vaccines achieve this goal by delivering live viruses that are so maladapted that they cannot make us sick.

To produce these vaccines, researchers create the conditions in which viruses evolve reduced virulence. This process is known as attenuation. To make a vaccine against polio, for example, scientists mix the viruses with the cells of monkeys or chimpanzees—species in which the polio viruses don't replicate in nature.

The viruses still manage to replicate, albeit slowly. And as they do, they mutate. Selection favors the strains best adapted to these nonhuman host cells. As they adapt better to those cells, the viruses become less suited to infecting humans. Once the viruses have become weak enough, they can act as vaccines. When they're injected into people's bodies, they cannot cause an infection, but they can still trigger a strong immune response, which leads to lifelong protection (**Figure 18.29**).

Better Vaccines

As we saw earlier, evolution undermines the effectiveness of current influenza vaccines. As a result, we have to get a new vaccine each year, and it offers only partial protection. Scientists recognize that many more infections could be avoided—and lives saved—if they can find a more effective vaccine. A number of them are searching for a so-called universal flu vaccine (Cohen 2018). Such a vaccine would cause our immune systems to make a different kind of antibody against the influenza virus. Rather than bind to the tip of the hemagglutinin stalk, as conventional vaccine antibodies do, these would bind to its base.

This strategy is based on the study of flu evolution. Scientists have found that the base of the hemagglutinin stalk is nearly identical from one strain of flu to another. Mutations affecting this portion of the protein are overwhelmingly harmful, these studies suggest, because the base of the stalk can't function properly if its shape changes even a little. In theory, a vaccine that attacked this new target would work effectively against any strain of the flu and would provide protection year after year.

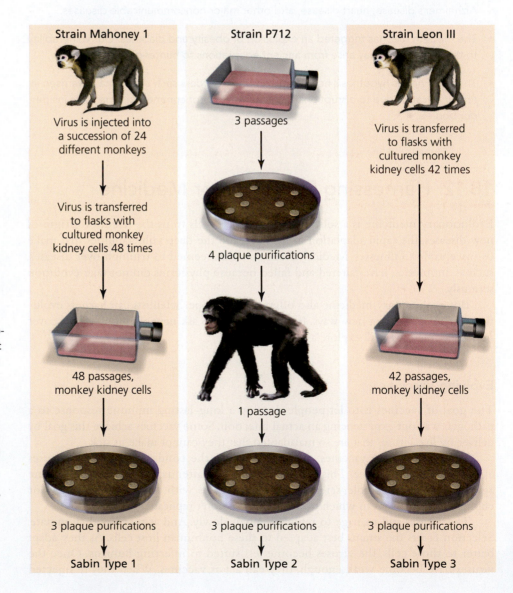

Figure 18.29 Evolution can be harnessed to make vaccines. One type of vaccine against polio contains weakened polio viruses. They can provoke an immune response but cannot cause the devastating symptoms that normal polio virus causes, such as paralysis. To create these live-virus vaccines, researchers present the viruses with cells of nonhuman primates. They collect the viruses from one host individual and transfer them to another (each transfer is called a passage). As selection acts on the viruses for this new host species, they become less adapted for infecting humans. Once the virus has become sufficiently weak, it can be safely injected into people. (Data from Minor 2004)

Strain Mahoney 1

Virus is injected into a succession of 24 different monkeys

Virus is transferred to flasks with cultured monkey kidney cells 48 times

48 passages, monkey kidney cells

3 plaque purifications

Sabin Type 1

Strain P712

3 passages

4 plaque purifications

1 passage

3 plaque purifications

Sabin Type 2

Strain Leon III

Virus is transferred to flasks with cultured monkey kidney cells 42 times

42 passages, monkey kidney cells

3 plaque purifications

Sabin Type 3

New Prescriptions for Old Medicines

Cancer can evolve resistance to chemotherapy in a matter of months through natural selection. But the strength of natural selection depends on many factors that can potentially be manipulated. Scientists have been developing mathematical models of cancer evolution to find ways to stop the rise of resistant tumors (Staňková et al. 2018).

Robert Gatenby, of the Moffitt Cancer Center in Florida, and his colleagues used these models to devise a new course of treatment for patients with aggressive prostate cancer. Rather than continuously giving the drugs to cancer patients, their new regimen called for only giving them in pulses. Each pulse drastically reduced the overall population of the tumors in the patients. During the pauses between pulses, the tumors partially grew back. But in the absence of the drug, the susceptible cancer cells had higher fitness and thus increased in frequency. When Gatenby's team applied the next pulse of chemotherapy, the tumors were just as vulnerable as before.

In the first 27 months of treatment, the researchers found that the cancers in 10 of 11 patients stopped growing. What's more, thanks to the precision of their evolution-based models, they were able to reach these impressive results with only 47% of the standard dose for metastatic prostate cancer, reducing the side effects of the drugs.

Mathematical models of evolution may help fight more than just cancer. Pulses of antibiotics may also slow the evolution of resistance in bacteria, for example (Kupferschmidt 2016).

Evolution-Proof Drugs

Antibiotics work by directly killing bacteria. Some of them lock onto proteins that help replicate DNA. Some tear open cell walls. These deadly attacks create intense selection on mutants with even a little resistance and favor subsequent mutations that make bacteria even more resistant. Evolutionary biologists and drug designers have collaborated in recent years to search for drugs that can stop bacterial infections without creating selection pressure against the drugs (Allen et al. 2014; Dickey et al. 2017; Whiteley et al. 2017).

Consider *Pseudomonas aeruginosa*, the soil microbe that infects the lungs of people with cystic fibrosis. To obtain the iron these microbes need to grow, they release molecules called siderophores. After the siderophores bind iron from the host, the bacteria can absorb them, consume the iron, and release the siderophore back into the environment. Remarkably, every member of a *P. aeruginosa* colony can take in an iron-binding siderophore produced by any other member of the colony.

A drug that targets siderophores might be able to starve the bacteria without promoting the evolution of resistant mutants. Researchers at the University of Zurich in Switzerland have tested this idea on animals. They gave infected animals gallium, an ironlike metal. The bacterial siderophores took up the gallium instead, causing the bacteria to become iron-starved. But no mutants arose that could make siderophores that could avoid the gallium (Ross-Gillespie et al. 2014).

The researchers hypothesize that resistance may be difficult or even impossible to evolve. If a mutant microbe does arise that can make resistant siderophores, it gains no fitness increase. That's because all the bacteria in the colony may take up the iron that the mutant captures with its siderophores. The individual mutant has no relative growth advantage, and so selection cannot increase the frequency of the mutation.

How Elephants Fight Cancer

Once animals evolved multicellularity, cancer became an inescapable threat. When some animals evolved to large sizes, they faced a bigger threat because their bodies contained more cells that could potentially gain cancerous mutations. Yet large species such as whales and elephants don't seem to get cancer any more often than smaller ones (Caulin and Maley 2011).

Figure 18.30 Elephants should have high rates of cancer because their bodies are so large. Instead, they have low rates. Scientists are uncovering the unique defenses they have evolved against cancer—defenses that may inspire cancer treatments for humans. (Tier Und Naturfotografie J und C Sohns/Superstock, Inc.)

This paradox led scientists to wonder if big animals have evolved unique weapons against cancer. We have one copy of the *p53* gene, which is vital for stopping mutant cells from becoming cancerous. Elephants, scientists have now found, have evolved 20 copies (**Figure 18.30**; Abegglen et al. 2015). Scientists know that *p53* works by sensing aberrant cell division and switching on a lot of genes to kill the cell. Scientists have discovered a unique gene in elephants that is switched on by *p53*, called *LIF6*. Its proteins seek out the mitochondria, which they destroy (Vasquez et al. 2018).

The elephant's strategies, revealed through an evolutionary perspective on cancer, may offer a new model for how to design anticancer drugs.

Forecasting Diseases

In our personal lives, weather forecasting may help us plan whether to bring an umbrella to work or take in the patio furniture before a storm. But it has had much more profound benefits to society as a whole. It allows governments to make plans to cope with large-scale disasters. As meteorologists get better at predicting hurricanes, for example, people can be evacuated in advance, rather than face the wrath of a storm. If a drought is forecast, governments can take steps to ensure that people have access to water. A number of scientists are exploring ways to forecast the future of diseases. In essence, what they're trying to do is predict evolution.

Seasonal flu poses one of the most pressing needs for disease forecasting. Until someone invents a universal flu vaccine (if they ever do), vaccine designers will still need to determine which flu strain to target months before the next flu season arrives. Scientists are developing predictive computer models that can use observations about existing strains to determine which ones will come to dominate the influenza viral population (Morris et al. 2018).

Scientists are also trying to monitor the ongoing evolution of other diseases. In West Africa, Pardis Sabeti is helping to track Lassa viruses, for example, as well as other animal-borne pathogens such as Ebola virus. For the time being, these viruses are not well adapted to spreading in humans. After periodic outbreaks, they retreat back to their nonhuman animal hosts. By tracking the genetic diversity of these viruses, Sabeti and other researchers hope to catch the early warning signals of evolution that could make the viruses even more dangerous than they already are.

Flu researchers are also anticipating a jump of influenza viruses from wild birds to humans. Exactly when antigenic shift will produce a new pandemic flu no one can say. And even when it does arise, it will be hard to predict how bad it will be. When the latest pandemic arose in 2009, researchers could not say whether the world was facing a disaster on the scale of the 1918 pandemic or something of much less impact (Fineberg 2014). The 2009 flu ended up claiming far fewer deaths—somewhere in the neighborhood of 200,000 or 300,000 worldwide. Although that's a tremendous loss of life, it's not exceptional when compared to the toll that seasonal flu takes each year.

Some researchers have wondered if they can get an earlier preview of the emergence of future pandemics. They have been studying some of the strains of bird flu that have been infecting people in recent years. These flu strains appear to have the capacity to establish themselves in the human airway, but they haven't yet evolved the capacity to spread readily from one person to another. Precisely what mutations would provide them with that capacity is an open question.

Ron Fouchier, of the Erasmus Medical Center in the Netherlands, and his colleagues have investigated what it would take for a bird flu strain called H5N1 to adapt

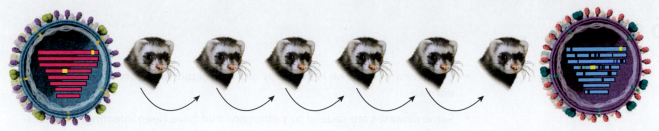

H5N1 bird flu

Evolved flu virus capable of
transmission from ferret to ferret

to mammals. They randomly introduced mutations into the viruses and then sprayed virus-laden droplets on a ferret (Herfst et al. 2012). Some mutations enabled viruses to replicate faster than others. They came to dominate the virus population inside the ferret.

Fouchier and his colleagues then used the viruses from that initial ferret to inoculate another one. Then they repeated the procedure, through a succession of ten ferrets in total. They found that after these passages, the virus could grow faster in ferrets than it had at the start of the experiment. And when Fouchier and his colleagues placed a sick ferret next to a healthy one, the virus was able to travel on its own from one mammal host to the next (**Figure 18.31**).

The scientists then examined the genetic makeup of the evolved viruses. Only a handful of mutations were necessary to allow the pathogen to make the jump from birds to mammals. Some of the mutations altered the proteins that the flu viruses use to latch onto host cells. Others altered how it hijacks the cell to make new viruses (Herfst et al. 2012).

Fouchier and his colleagues urge that public health workers monitor H5N1 strains for these mutations so that we can know in advance if the virus is evolving into a human pathogen. But how likely are those mutations to occur outside of a laboratory? That's still an open question.

Basing their study on Fouchier's work, Derek Smith, an evolutionary biologist at the University of Cambridge, and his colleagues created an evolutionary model for the emergence of H5N1 as a human disease. They concluded that it is possible for the virus to acquire all the mutations it needs to make the jump while in a single person (Russell et al. 2012). Although the likelihood of this actually happening remains unclear, results like the ones from Smith's group highlight just how dangerous it can be when people become infected with multiple strains of the flu simultaneously.

These experiments, it must be noted, are intensely controversial. To predict the future of flu evolution, critics argue, we should not actually carry it out (Lipsitch 2018). If scientists use serial passage experiments to produce an airborne influenza that can spread between mammals, there's a chance it could escape and cause a global catastrophe. These critics urge that the scientific community must find ways to prepare for future flu pandemics that don't carry such huge risks.

It may very well be that the next major medical emergency we face is not a new influenza virus but a pathogen lurking in some wild animal that we don't even know about yet. Some researchers are trying to forecast the future of diseases by cataloging the diversity of viruses and other disease agents that are most likely to spill over into our species. Some of these pathogens may be especially well suited to crossing the species barrier. Researchers are also stepping up monitoring of hunters, farmers, and others who have frequent contact with animals and might become the first stepping stones for a new human pathogen (Epstein and Anthony 2017; Bird and Mazet 2018). They hope these efforts will help the world become better prepared for evolution's relentless creativity.

Figure 18.31 Scientists inserted mutations into the H5N1 bird flu virus and then passed it from ferret to ferret. The virus evolved along the way. When the experiment was finished, the virus had gained the ability to spread between the ferrets by air. (Data from Herfst et al. 2012)

- Evolution-based research is providing new avenues for predicting, preventing, and treating diseases. ●

Key Concept

TO SUM UP . . .

- Evolutionary medicine is the study of the evolutionary roots of health and sickness.

- Some diseases are caused by pathogens that have been infecting our ancestors for millions of years. Some have begun infecting humans in just the past few decades.

- Human pathogens are always rapidly evolving, adapting to their human hosts.

- The virulence of pathogens emerges from an evolutionary trade-off between replicating quickly and moving easily from host to host.

- Reconstructing the evolutionary trees of emerging pathogens allows scientists to determine their origins.

- Pathogens evolve resistance to antibiotics.

- Pathogens have driven the evolution of defenses in human populations.

- Genetic drift can lead to harmful mutations becoming common in a population.

- Aging is an evolutionary by-product of natural selection for traits that help animals to survive during their reproductive years.

- Cancer develops in an evolutionary process, during which cells compete within a tumor in the same way that organisms compete in a population.

- Our bodies are adapted to a preindustrial life. Some diseases are the result of the mismatch between our bodies and modern life.

- Evolutionary biology reveals the great obstacles to treating some diseases, but it also opens new avenues of research for potential new treatments.

MULTIPLE CHOICE QUESTIONS Answers can be found at the end of the book.

1. Which cytomegalovirus is the Ppyg cytomegalovirus (CMV) more closely related to?

   a. Human cytomegalovirus.

   b. New World monkey (NWM) cytomegalovirus.

   c. Old World monkey (OWM) cytomegalovirus.

   d. It is equally related to all the other cytomegaloviruses.

2. How important are mutations in the evolution of HIV once the virus has infected a host, and why?

   a. Very important. Mutations will weaken the virus, allow-ing the immune system to mount a response.

   b. Very important. Mutations give rise to variation that can generate fitness advantages within a host and lead to evolution of HIV resistance to immune system responses.

   c. Not important. Mutations are always deleterious; a virus experiencing mutations would eventually be destroyed by the host's immune system.

   d. Not important. HIV cannot evolve once it has entered a host because it does not reproduce sexually.

3. How did conditions during World War I operate to maintain exceptionally high virulence in the influenza virus?

   a. Crowded conditions increased the relative importance of selection for efficient transmission across hosts, leading to increased spread of the virus and evolution of increased virulence.

   b. The high mobility of infected patients reduced the rel-ative importance of selection for efficient transmission across hosts, favoring selection acting within hosts and driving the evolution of increased virulence.

c. Vaccines caused the evolution of increased virulence in the virus.

d. All of the above.

4. What factors associated with the evolutionary biology of bacteria facilitate the evolution of resistance to antibiotics?

a. High mutation rates.

b. Horizontal gene transfer.

c. Sublethal doses of antibiotics.

d. All of the above.

5. Which of the following statements about viral reassortment is *false*?

a. Viral reassortment is a type of horizontal gene transfer.

b. Viral reassortment can add to the genetic variation of virus populations, speeding their evolution as they adapt to different hosts.

c. Viral reassortment can add to the genetic variation of a virus within a host by producing new combinations of genes not seen before.

d. None of the above.

6. Why is a child born on the island of Pingelap more than 1500 times more likely to be completely color-blind than a child born in the United States?

a. Because of a founder event several generations ago that reduced the Pingelap population to 20 individuals.

b. Because achromatopsia is caused by excessive exposure to sunlight.

c. Because 100% of Han Chinese carry the *rs9923231 T* allele.

d. Because sufferers of this disorder are only able to perceive black, white, and shades of gray due to the absence of functioning photoreceptors in the retina.

7. What types of genetic variation might contribute to differences in how individuals metabolize drugs?

a. Variation among individuals in alleles for protein-coding genes.

b. Variation among individuals in the pleiotropic effects of genes.

c. Variation among individuals in segments of DNA that affect gene expression.

d. All of the above.

8. An example of antagonistic pleiotropy is

a. a gene that produces a protein that can stop cells from turning cancerous but damages tissues that may be more likely to become cancerous later in life.

b. a gene with two alleles: one that increases susceptibility to cancer and one that decreases it.

c. a developmental gene that is influenced by the environment.

d. a mutation that produces a protein that can enhance development both early and later in life.

9. Which statement is *not* a prediction of the "Old Friends" hypothesis?

a. Children who frequently play outside should develop allergic diseases less often than children who don't play outside.

b. Children who never wash their hands or bathe will be less likely to develop immune disorders than children who wash regularly.

c. The more infectious diseases an individual is exposed to as a child, the less likely that person will be to develop immune disorders as an adult.

d. Extensive use of antibiotics as a child may affect how the body responds to future bacterial invasions.

10. Why is understanding evolutionary biology important to the future of medicine?

a. Because phylogenies can be used to predict which influenza viruses may be particularly likely to cause future epidemics.

b. Because studying how other species fight diseases like cancer may lead to innovative strategies or drugs.

c. Because evolutionary biology can guide the development of antibiotic treatments, antiviral drugs, and chemotherapy.

d. All of the above.

11. In the figure below, what do the different colored wedges depict?

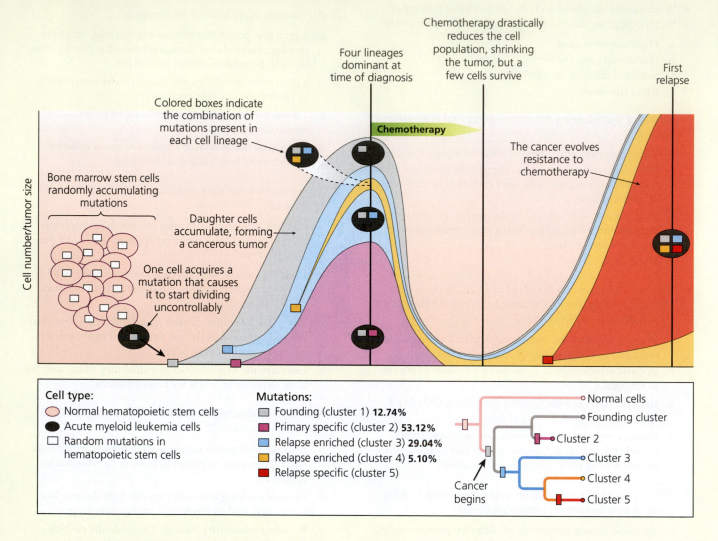

a. Different cancerous tumors within the patient's body.

b. Competing cell lineages that co-occurred within a single tumor at successive points in time.

c. A nested phylogeny of cell lineages defined by a series of sequential mutations.

d. Both b and c.

SHORT ANSWER QUESTIONS Answers can be found at the end of the book.

12. In 2014, the Ebola virus went from highly localized outbreaks to a problem of international concern. Why do you think outbreaks of the Ebola virus tended to be localized before 2014, and what ecological conditions may have changed in 2014?

13. Would you consider the *S* allele to be an adaptation to malaria? Why or why not?

14. How does natural selection operate within a cancerous tumor?

15. Why might a mismatch between our inherited physiology and modern, calorie-rich diets be contributing to the modern epidemic in obesity and type 2 diabetes?

16. How have evolutionary theory and population genetics (for example, $R = h^2 \times s$) contributed to current approaches for treating cancerous tumors?

17. Give an example of how genetic drift can lead to harmful mutations becoming common in a population.

ADDITIONAL READING

Despommier, D. D. 2013. *People, Parasites, and Plowshares: Learning from Our Body's Most Terrifying Invaders.* New York: Columbia University Press.

Ewald, P. 1996. *Evolution of Infectious Disease.* London: Oxford University Press.

Gluckman, P. D., A. Beedle, and M. Hanson (eds.) 2009. *Principles of Evolutionary Medicine.* Oxford: Oxford University Press.

Lents, N. 2018. *Human Errors: Pointless Bones, Runaway Nerves and Other Human Defects.* New York: Orion.

Levy, S. 2002. *The Antibiotic Paradox: How the Misuse of Antibiotics Destroys Their Curative Powers.* New York: Perseus Publishing.

Longdon, B., M. A. Brockhurst, C. A. Russell, J. J. Welch, and F. M. Jiggins. 2014. The Evolution and Genetics of Virus Host Shifts. *PLoS Pathogens* 10 (11):e1004395.

McKenna, M. 2017. *Big Chicken: The Incredible Story of How Antibiotics Created Modern Agriculture and Changed the Way the World Eats.* Washington DC: National Geographic Books.

Nesse, R. M., and G. C. Williams. 1994. *Why We Get Sick: The New Science of Darwinian Medicine.* New York: Times Books.

Quammen, D. 2012. *Spillover: Animal Infections and the Next Human Pandemic.* New York: W. W. Norton.

Stearns, S. C., and J. C. Koella. 2008. *Evolution in Health and Disease.* 2nd ed. Oxford: Oxford University Press.

Zimmer, C. 2000. *Parasite Rex: Inside the Bizarre World of Nature's Most Dangerous Creatures.* New York: Free Press.

PRIMARY LITERATURE CITED IN CHAPTER 18

Aarestrup, F. M., H. Kruse, E. Tast, A. M. Hammerum, and L. B. Jensen. 2000. Associations Between the Use of Antimicrobial Agents for Growth Promotion and the Occurrence of Resistance Among *Enterococcus faecium* from Broilers and Pigs in Denmark, Finland, and Norway. *Microbial Drug Resistance* 6 (1): 63–70.

Abarca-Gómez, L., Z. A. Abdeen, Z. A. Hamid, N. M. Abu-Rmeileh, B. Acosta-Cazares, et al. 2017. Worldwide Trends in Body-Mass Index, Underweight, Overweight, and Obesity from 1975 to 2016: A Pooled Analysis of 2416 Population-Based Measurement Studies in 128.9 Million Children, Adolescents, and Adults. *The Lancet* 390:2627–42.

Abegglen, L. M., A. F. Caulin, A. Chan, K. Lee, R. Robinson, et al. 2015. Potential Mechanisms for Cancer Resistance in Elephants and Comparative Cellular Response to DNA Damage in Humans. *JAMA* 314 (17): 1850–60.

Al Faress, S., G. Cartet, O. Ferraris, H. Norder, M. Valette, et al. 2005. Divergent Genetic Evolution of Hemagglutinin in Influenza A H1N1 and H1N2 Subtypes Isolated in the South-France Since the Winter of 2001–2002. *Journal of Clinical Virology* 33 (3): 230–6.

Allen, R. C., R. Popat, S. P. Diggle, and S. P. Brown. 2014. Targeting Virulence: Can We Make Evolution-Proof Drugs? *Nature Reviews Microbiology* 12:300–308.

Andersen, K. G., B. J. Shapiro, C. B. Matranga, R. Sealfon, A. E. Lin, et al. 2015. Clinical Sequencing Uncovers Origins and Evolution of Lassa Virus. *Cell* 162:738–50.

Anoh, A. E., S. Murthy, C. Akoua-Koffi, E. Couacy-Hymann, F. H. Leendertz, et al. 2018. Cytomegaloviruses in a Community of Wild Nonhuman Primates in Taï National Park, Côte D'Ivoire. *Viruses* 10:E11.

Aktipis, C. A., A. M. Boddy, G. Jansen, U. Hibner, M. E. Hochberg, et al. 2015. Cancer Across the Tree of Life: Cooperation and Cheating in Multicellularity. *Philosophical Transactions of the Royal Society B* 370:20140219.

Austad, S. N., and J. M. Hoffman. 2018. Is Antagonistic Pleiotropy Ubiquitous in Aging Biology? *Evolution, Medicine, and Public Health* 2018 (1): 287–94.

Ayub, Q., L. Moutsianas, Y. Chen, K. Panoutsopoulou, V. Colonna, et al. 2014. Revisiting the Thrifty Gene Hypothesis via 65 Loci Associated with Susceptibility to Type 2 Diabetes. *The American Journal of Human Genetics* 94:176–85.

Bach, J-F. 2002. The Effect of Infections on Susceptibility to Autoimmune and Allergic Diseases. *New England Journal of Medicine* 347:911–20.

Baker, S., N. Thomson, F-X. Weill, and K. E. Holt. 2018. Genomic Insights into the Emergence and Spread of Antimicrobial-Resistant Bacterial Pathogens. *Science* 360 (6390): 733–8.

Ball, T. M., J. A. Castro-Rodriguez, K. A. Griffith, C. J. Holberg, F. D. Martinez, et al. 2000. Siblings, Day-Care Attendance, and the Risk of Asthma and Wheezing During Childhood. *New England Journal of Medicine* 343 (8): 538–43.

Barry, J. 2005. *The Great Influenza: The Story of the Deadliest Pandemic in History.* New York: Penguin Publishing.

Beil, L. 2014. Ancient Genes, Modern Meals: Poking Holes in the Thrifty Gene Hypothesis. *Science News* 186 (6): 18–22.

Belser, J. A., T. R. Maines, and T. M. Tumpey. 2018. Importance of 1918 Virus Reconstruction to Current Assessments of Pandemic Risk. *Virology* 524:45–55.

Berbesque, J. C., F. W. Marlowe, P. Shaw, and P. Thompson. 2014. Hunter-Gatherers Have Less Famine Than Agriculturalists. *Biology Letters* 10:20130853.

Bird, B. H., and J. A. K. Mazet. 2018. Detection of Emerging Zoonotic Pathogens: An Integrated One Health Approach. *Annual Review of Animal Biosciences* 6:121–39.

Blaser, M. J. 2014. *Missing Microbes: How the Overuse of Antibiotics Is Fueling Our Modern Plagues.* New York: Henry Holt and Company.

Bloomfield, S. F., G. A. W. Rook, E. A. Scott, F. Shanahan, R. Stanwell-Smith, et al. 2016. Time to Abandon the Hygiene Hypothesis: New Perspectives on Allergic Disease, the Human Microbiome, Infectious Disease Prevention and the Role of Targeted Hygiene. *Perspectives in Public Health* 136:213–24.

Boehm, T., N. Iwanami, and I. Hess. 2012. Evolution of the Immune System in the Lower Vertebrates. *Annual Review of Genomics and Human Genetics* 13:127–49.

Bull, J. J., and A. S. Lauring. 2014. Theory and Empiricism in Virulence Evolution. *PLoS Pathogens* 10 (10): e1004387.

Byars, S. G., Q. Q. Huang, L-A. Gray, A. Bakshi, S. Ripatti, et al. 2017. Genetic Loci Associated with Coronary Artery Disease Harbor Evidence of Selection and Antagonistic Pleiotropy. *PLoS Genetics* 13 (6): e1006328.

Caulin, A. F., and C. C. Maley. 2011. Peto's Paradox: Evolution's Prescription for Cancer Prevention. *Trends in Ecology and Evolution* 26 (4): 175–82.

Cohen, J. 2018. Universal Flu Vaccine Is an 'Alchemist's Dream.' *Science* 362 (6419): 1094.

Colizza, V., A. Barrat, M. Barthelemy, A. J. Valleron, and A. Vespignani. 2007. Modeling the Worldwide Spread of Pandemic Influenza: Baseline Case and Containment Interventions. *PLoS Medicine* 4 (1): e13.

Corbett, S., A. Courtiol, V. Lummaa, J. Moorad, and S. Stearns. 2018. The Transition to Modernity and Chronic Disease: Mismatch and Natural Selection. *Nature Reviews Genetics* 19:419–30.

Cui, J., F. Li, and Z-L. Shi. 2018. Origin and Evolution of Pathogenic Coronaviruses. *Nature Reviews Microbiology* 17:181–92.

Cyranoski, D. 2017. SARS Outbreak Linked to Chinese Bat Cave. *Nature* 552:15–6.

Degnan, B. 2015. The Surprisingly Complex Immune Gene Repertoire of a Simple Sponge, Exemplified by the NLR Genes: A Capacity for Specificity? *Developmental and Comparative Immunology* 48 (2): 269–74.

Diamond, J. 2003. The Double Puzzle of Diabetes. *Nature* 423 (6940): 599–602.

Dickey, S. W., G. Y. C. Cheung, and M. Otto. 2017. Different Drugs for Bad Bugs: Antivirulence Strategies in the Age of Antibiotic Resistance. *Nature Reviews Drug Discovery* 16:457–71.

Didelot, X., A. S. Walker, T. E. Peto, D. W. Crook, and D. J. Wilson. 2016. Within-Host Evolution of Bacterial Pathogens. *Nature Reviews Microbiology* 14:150–62.

Ding, L., T. J. Ley, D. E. Larson, C. A. Miller, D. C. Koboldt, et al. 2012. Clonal Evolution in Relapsed Acute Myeloid Leukaemia Revealed by Whole-Genome Sequencing. *Nature* 481 (7382): 506–10.

Drexler, J. F., V. M. Corman, and C. Drosten. 2014. Ecology, Evolution and Classification of Bat Coronaviruses in the Aftermath of SARS. *Antiviral Research* 101:45–56.

Ebert, D. 1998. Experimental Evolution of Parasites. *Science* 282:1432–6.

Edwards, M. 2017. The Barker Hypothesis. In V. Preedy and V. Patel (eds.), *Handbook of Famine, Starvation, and Nutrient Deprivation.* Cham, CH: Springer.

Eldholm, V., G. Norheim, B. von der Lippe, W. Kinander, U. R. Dahle, et al. 2014. Evolution of Extensively Drug-Resistant *Mycobacterium tuberculosis* from a Susceptible Ancestor in a Single Patient. *Genome Biology* 15:490.

Enard, D., L. Cai, C. Gwennap, and D. A. Petrov. 2016. Viruses are a Dominant Driver of Protein Adaptations in Mammals. *eLife* 5:e12469.

Epstein, J. H., and S. J. Anthony. 2017. Viral Discovery as a Tool for Pandemic Preparedness. *Scientific and Technical Review of the Office International des Epizooties* 36 (2): 499–512.

Epstein, J. M., D. M. Goedecke, F. Yu, R. J. Morris, D. K. Wagener, et al. 2007. Controlling Pandemic Flu: The Value of International Air Travel Restrictions. *PLoS ONE* 2 (5): e401.

Ewald, P. 1996. *Evolution of Infectious Disease.* London: Oxford University Press.

Fineberg, H. V. 2014. Pandemic Preparedness and Response — Lessons from the H1N1 Influenza of 2009. *The New England Journal of Medicine* 370:1335–42.

Geoghegan, J. L., and E. C. Holmes. 2018. The Phylogenomics of Evolving Virus Virulence. *Nature Reviews Genetics* 19:756–69.

Gluckman P., and M. Hanson. 2006. *Mismatch: Why Our World No Longer Fits Our Bodies.* Oxford: Oxford University Press.

Gosling, A. L., H. R. Buckley, E. Matisoo-Smith, and T. R. Merriman. 2015. Pacific Populations, Metabolic Disease and 'Just-So Stories': A Critique of the 'Thrifty Genotype' Hypothesis in Oceania. *Annals of Human Genetics* 79:470–80.

Hales, C. N., and D. J. P. Barker. 1992. Type 2 (Non-Insulin-Dependent) Diabetes Mellitus: The Thrifty Phenotype Hypothesis. *Diabetologia* 35:595–601.

Herfst, S., E. J. Schrauwen, M. Linster, S. Chutinimitkul, E. de Wit, et al. 2012. Airborne Transmission of Influenza A/H5N1 Virus Between Ferrets. *Science* 336:1534–41.

Humphreys, M. 2018. The Influenza of 1918: Evolutionary Perspectives in a Historical Context. *Evolution, Medicine, and Public Health* 2018 (1): 219–29.

Hussels, I. E., and N. E. Morton. 1972. Pingelap and Mokil Atolls: Achromatopsia. *American Journal of Human Genetics* 24 (3): 304–9.

Imanpour, S., O. Nwaiwu, D. K. McMaughan, B. DeSalvo, and A. Bashir. 2017. Factors Associated with Antibiotic Prescriptions for the Viral Origin Diseases in Office-Based Practices, 2006–2012. *Journal of the Royal Society of Medicine Open* 8 (8): 1–8.

Karlsson, E. K., D. P. Kwiatkowski, and P. C. Sabeti. 2014. Natural Selection and Infectious Disease in Human Populations. *Nature Reviews Genetics* 15:379–93.

Khachatourians, G. G. 1998. Agricultural Use of Antibiotics and the Evolution and Transfer of Antibiotic-Resistant Bacteria. *Canadian Medical Association Journal* 159 (9): 1129–36.

Kim, J., J. M. Basak, and D. M. Holtzman. 2009. The Role of Apolipoprotein E in Alzheimer's Disease. *Neuron* 63:287–303.

Knudson, A. G. 1971. Mutation and Cancer: Statistical Study of Retinoblastoma. *Proceedings of the National Academy of Sciences* 68:820–23.

Krammer, F., G. J. D. Smith, R. A. M. Fouchier, M. Peiris, K. Kedzierska, et al. 2018. Influenza. *Nature Reviews Disease Primers* 4:3.

Kupferschmidt, K. 2016. Resistance Fighters. *Science* 352 (6287): 758–61.

Leendertz, F. H., M. Deckers, W. Schempp, F. Lankester, C. Boesch, et al. 2009. Novel Cytomegaloviruses in Free-Ranging and Captive Great Apes: Phylogenetic Evidence for Bidirectional Horizontal Transmission. *Journal of General Virology* 90 (10): 2386–94.

Lents, N. 2018. *Human Errors: Pointless Bones, Runaway Nerves and Other Human Defects.* New York: Orion.

Levy, S. B., G. B. FitzGerald, and A. B. Macone. 1976. Changes in Intestinal Flora of Farm Personnel After Introduction of a Tetracycline-Supplemented Feed on a Farm. *New England Journal of Medicine* 295:583–8.

Lin, Y-T., J. Seo, F. Gao, H. M. Feldman, H-L. Wen, et al. 2018. *APOE4* Causes Widespread Molecular and Cellular Alterations Associated with Alzheimer's Disease Phenotypes in Human iPSC-Derived Brain Cell Types. *Neuron* 98:1–14.

Lipsitch, M. 2018. Why Do Exceptionally Dangerous Gain-of-Function Experiments in Influenza? In Y. Yamauchi (ed.), *Influenza Virus: Methods and Protocols*, pp. 589–608. New York: Humana Press.

Longdon, B., M. A. Brockhurst, C. A. Russell, J. J. Welch, and F. M. Jiggins. 2014. The Evolution and Genetics of Virus Host Shifts. *PLoS Pathogens* 10 (11): e1004395.

Maron, D. F. 2016. The Right Pill for You. *Scientific American* 315 (4): 46–51.

Martens, E., and A. I. Demain. 2017. The Antibiotic Resistance Crisis, with a Focus on the United States. *The Journal of Antibiotics* 70:520–6.

McKenna, M. 2017. *Big Chicken: The Incredible Story of How Antibiotics Created Modern Agriculture and Changed the Way the World Eats.* Washington DC: National Geographic Books.

McKusick, V. A. 1978. *Medical Genetic Studies of the Amish: Selected Papers.* Baltimore: Johns Hopkins University Press.

Medina, R. A. 2018. 1918 Influenza Virus: 100 Years On, Are We Prepared Against the Next Influenza Pandemic? *Nature Reviews Microbiology* 16:61–2.

Minor, P. D. 2004. Polio Eradication, Cessation of Vaccination and Re-Emergence of Disease. *Nature Reviews Microbiology* 2 (6): 473–82.

Morens, D. M., and A. S. Fauci. 2013. Emerging Infectious Diseases: Threats to Human Health and Global Stability. *PLoS Pathogens* 9 (7): e1003467.

Morris, D. H., K. M. Gostic, S. Pompei, T. Bedford, M. Łuksza, et al. 2018. Predictive Modeling of Influenza Shows the Promise of Applied Evolutionary Biology. *Trends in Microbiology* 26 (2): 102–18.

Munita, J. M., and C. A. Arias. 2016. Mechanisms of Antibiotic Resistance. *Microbiology Spectrum* 4 (2). dx.doi.org/10.1128/microbiolspec.VMBF-0016-2015.

Neel, J. V. 1962. Diabetes Mellitus: A "Thrifty" Genotype Rendered Detrimental by "Progress"? *American Journal of Human Genetics* 14 (4): 353–62.

Neel, J. V. 1999. The "Thrifty Genotype" in 1998. *Nutrition Reviews* 57:(II)S2–9.

Nelson, M. I., and E. C. Holmes. 2017. The Evolution of Epidemic Influenza. *Nature Reviews Genetics* 8:196–205.

Nelson, M. I., and M. Worobey. Origins of the 1918 Pandemic: Revisiting the Swine "Mixing Vessel" Hypothesis. *American Journal of Epidemiology* 187 (12): 2498–2502.

Nesse, R. M. 2005. Maladaptation and Natural Selection. *The Quarterly Review of Biology* 80:62–70.

Nesse, R. M., and G. C. Williams. 1994. *Why We Get Sick: The New Science of Darwinian Medicine.* New York: Times Books.

Nettle, D., C. Andrews, and M. Bateson. 2017. Food Insecurity as a Driver of Obesity in Humans: The Insurance Hypothesis. *Behavioral and Brain Sciences* 40:e105.

Nissinen, A., E. Herva, M-L. Katila, S. Kontiainen, O. Liimatainen, et al. 1995. Antimicrobial Resistance in *Haemophilus influenzae* Isolated from Blood, Cerebrospinal Fluid, Middle Ear Fluid and Throat Samples of Children. A Nationwide Study in Finland in 1988–1990. *Scandinavian Journal of Infectious Diseases* 27 (1): 57–61.

Pacini, F., T. Vorontsova, E. P. Demidchik, E. Molinaro, L. Agate, et al. 1997. Post-Chernobyl Thyroid Carcinoma in Belarus Children and Adolescents: Comparison with Naturally Occurring Thyroid Carcinoma in Italy and France. *Journal of Clinical Endocrinology and Metabolism* 82 (11): 3563–9.

Piel, F. B., A. P. Patil, R. E. Howes, O. A. Nyangiri, P. W. Gething, et al. 2010. Global Distribution of the Sickle Cell Gene and Geographical Confirmation of the Malaria Hypothesis. *Nature Communications* 1:104.

Plowright, R. K., C. R. Parrish, H. McCallum, P. J. Hudson, A. I. Ko, et al. 2017. Pathways to Zoonotic Spillover. *Nature Reviews Microbiology* 15 (8): 502–10.

Qasim, A., M. Turcotte, R. J. de Souza, M. C. Samaan, D. Champredon, et al. 2018. On the Origin of Obesity: Identifying the Biological Environmental and Cultural Drivers of Genetic Risk Among Human Populations. *Obesity Reviews* 19:121–49.

Ratjen, F., S. C. Bell, S. M. Rowe, C. H. Goss, A. L. Quittner, et al. 2015. Cystic Fibrosis. *Nature Reviews Disease Primers* 1:15010.

Relling, M. V., and W. E. Evans. 2015. Pharmacogenomics in the Clinic. *Nature* 526:343–50.

Riedler, J., C. Braun-Fahrländer, W. Eder, M. Schreuer, M. Waser, et al. 2001. Exposure to Farming in Early Life and Development of Asthma and Allergy: A Cross-Sectional Survey. *Lancet* 358 (9288): 1129–33.

Rook, G. A. W. 2010. 99th Dahlem Conference on Infection, Inflammation and Chronic Inflammatory Disorders: Darwinian Medicine and the "Hygiene" or "Old Friends" Hypothesis. *Clinical and Experimental Immunology* 160 (1): 70–9.

Ross, K. A., A. W. Bigham, M. Edwards, A. Gozdzik, G. Suarez-Kurtz, et al. 2010. Worldwide Allele Frequency Distribution of Four Polymorphisms Associated with Warfarin Dose Requirements. *Journal of Human Genetics* 55 (9): 582–9.

Ross-Gillespie, A., M. Weigert, S. P. Brown, and R. Kümmerli. 2014. Gallium-Mediated Siderophore Quenching as an Evolutionarily Robust Antibacterial Treatment. *Evolution, Medicine, and Public Health* 2014:18–29.

Russell, C. A., J. M. Fonville, A. E. Brown, D. F. Burke, D. L. Smith, et al. 2012. The Potential for Respiratory Droplet–Transmissible A/H5N1 Influenza Virus to Evolve in a Mammalian Host. *Science* 336:1541–7.

Sabeti, P., and L. Salahi. 2018. *Outbreak Culture: The Ebola Crisis and the Next Epidemic.* Cambridge, MA: Harvard University Press.

Schick, A., and R. Kassen. 2018. Rapid Diversification of *Pseudomonas aeruginosa* in Cystic Fibrosis Lung-like Conditions. *Proceedings of the National Academy of Sciences USA* 115: 10714–9.

Scudellari, M. 2017. Cleaning Up the Hygiene Hypothesis. *Proceedings of the National Academy of Sciences USA* 114: 1433–6.

Sellayah, D., F. R. Cagampang, and R. D. Cox. 2014. On the Evolutionary Origins of Obesity: A New Hypothesis. *Endocrinology* 155 (5): 1573–88.

Shea, K. M. 2003. Antibiotic Resistance: What Is the Impact of Agricultural Uses of Antibiotics on Children's Health? *Pediatrics* 112 (Suppl. 1): 253–8.

Shpak, M., and J. Lu. 2016. An Evolutionary Genetic Perspective on Cancer Biology. *Annual Review of Ecology, Evolution, and Systematics* 47:25–49.

Siddle, K. J., P. Eromon, K. G. Barnes, S. Mehta, J. U. Oguzie, et al. 2018. Genomic Analysis of Lassa Virus During an Increase in Cases in Nigeria in 2018. *New England Journal of Medicine* 379:1745–53.

Smith, K. R., H. A. Hanson, G. P. Mineau, and S. S. Buys. 2012. Effects of *BRCA1* and *BRCA2* Mutations on Female Fertility. *Proceedings of the Royal Society B* 279:1389–95.

Speakman, J. R. 2018. The Evolution of Body Fatness: Trading Off Disease and Predation Risk. *Journal of Experimental Biology* 221:jeb167254.

Staňková, K., J. S. Brown, W. S. Dalton, and R. A. Gatenby. 2018. Optimizing Cancer Treatment Using Game Theory: A Review. *JAMA Oncology.* 5:96–103.

Stearns, S. C. 2018. Outstanding Research Opportunities at the Interface of Evolution and Medicine. *Nature Ecology and Evolution* 2:3–4.

Stearns, S. C., and J. C. Koella. 2008. *Evolution in Health and Disease.* 2nd ed. Oxford: Oxford University Press.

Teuber, M. 2001. Veterinary Use and Antibiotic Resistance. *Current Opinion in Microbiology* 4 (5): 493–9.

van Exel, E., J. J. E. Koopman, D. van Bodegom, J. J. Meij, P. de Knijff, et al. 2017. Effect of APOE ε4 Allele on Survival and Fertility in an Adverse Environment. *PLoS ONE* 12 (7): e0179497.

Vasquez, J. M., M. Sulak, S. Chigurupati, and V. J. Lynch. 2018. A Zombie LIF Gene in Elephants Is Upregulated by TP53 to Induce Apoptosis in Response to DNA Damage. *Cell Reports* 24:1765–76.

Wang, G., and J. R. Speakman. 2016. Analysis of Positive Selection at Single Nucleotide Polymorphisms Associated with Body Mass Index Does Not Support the "Thrifty Gene" Hypothesis. *Cell Metabolism* 24:531–41.

Whiteley, M., S. P. Diggle, and E. P. Greenberg. 2017. Progress in and Promise of Bacterial Quorum Sensing Research. *Nature* 551:313–20.

Wills-Karp, M., J. Santeliz, and C. L. Karp. 2001. The Germless Theory of Allergic Disease: Revisiting the Hygiene Hypothesis. *Nature Reviews Immunology* 1 (1): 69–75.

Winstanley, C., S. O'Brien, and M. A. Brockhurst. 2016. *Pseudomonas aeruginosa* Evolutionary Adaptation and Diversification in Cystic Fibrosis Chronic Lung Infections. *Trends in Microbiology* 24:327–37.

World Health Organization. 2018. *Global Health Estimates 2016: Deaths by Cause, Age, Sex, by Country and by Region, 2000–2016.* Geneva: World Health Organization.

Worobey, M., G. Han, and A. Rambaut. 2014. A Synchronized Global Sweep of the Internal Genes of Modern Avian Influenza Virus. *Nature* 508:254.

Xue, K. S., T. Stevens-Ayers, A. P. Campbell, J. A. Englund, S. A. Pergam, et al. 2017. Parallel Evolution of Influenza Across Multiple Spatiotemporal Scales. *eLife* 6:e26875.

Zanini, F., J. Brodin, L. Thebo, C. Lanz, G. Bratt, et al. 2015. Population Genomics of Intrapatient HIV-1 Evolution. *eLife* 4:e11282.

Timeline

4.57 bya

Earth forms

4.51 bya

Origin of Moon

3.9 bya

Intense bombardments melt parts of Earth's crust

3.5 bya

Possible biomarkers of methane-producing archaea and oldest fossils of bacteria

2.5 bya

Earliest known fossils of bacteria on land

2.4 bya

Previously oxygen-free atmosphere acquires 1% of today's oxygen level

2.1 bya

Possible fossils of multicellular organisms

HADEAN EON

ARCHEAN EON

4.6 billion years ago (bya)   4 bya   3.5 bya   3 bya   2.5 bya

4.1 bya

Oldest known possible markers of life (carbon isotope ratio)

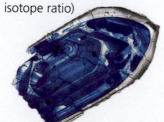

Page 734: top row: Earth: NASA; moon: NASA/JPL//USGS; fossil: © Abderrazak El Albani/Arnaud Mazurier/CNRS Photo Library; bottom: John Valley, University of Wisconsin–Madison)

Page 735: top row: Bengtson S, Sallstedt T, Belivanova V, Whitehouse M (2017) Three-dimensional preservation of cellular and subcellular structures suggests 1.6 billion-year-old crowngroup red algae. PLoS Biol 15(3); bottom: Yin, Z., M. Zhu, E. H. Davidson, D. J. Bottjer, F. Zhao, and P. Tafforeau. 2015. Sponge Grade Body Fossil with Cellular Resolution Dating 60 Myr Before the Cambrian. Proceedings of the National Academy of Sciences 112: E1453–E1460.

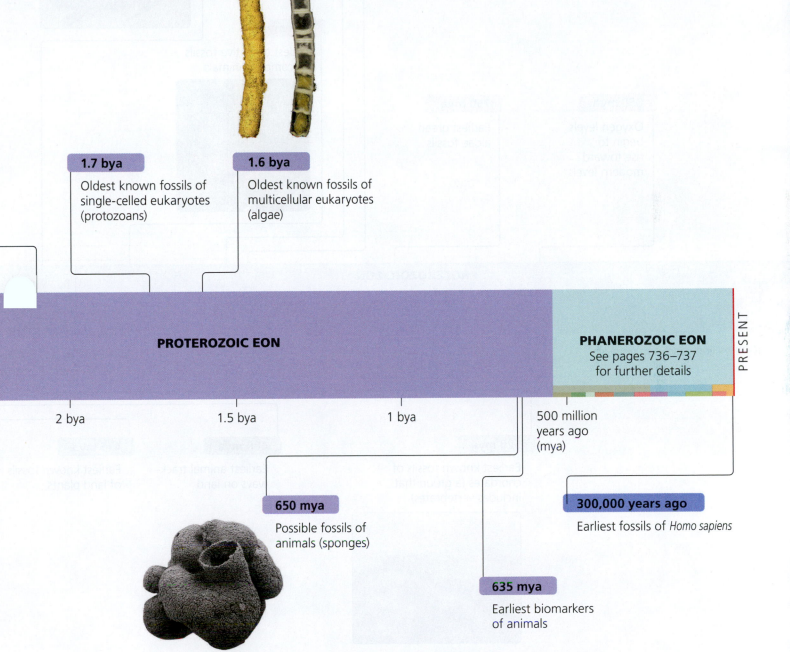

1.7 bya

Oldest known fossils of
single-celled eukaryotes
(protozoans)

1.6 bya

Oldest known fossils of
multicellular eukaryotes
(algae)

PROTEROZOIC EON

PHANEROZOIC EON
See pages 736–737
for further details

PRESENT

2 bya   1.5 bya   1 bya   500 million
years ago
(mya)

650 mya

Possible fossils of
animals (sponges)

300,000 years ago

Earliest fossils of *Homo sapiens*

635 mya

Earliest biomarkers
of animals

For a detailed geological timescale, visit http://www.stratigraphy.org/index.php/ics-chart-timescale.

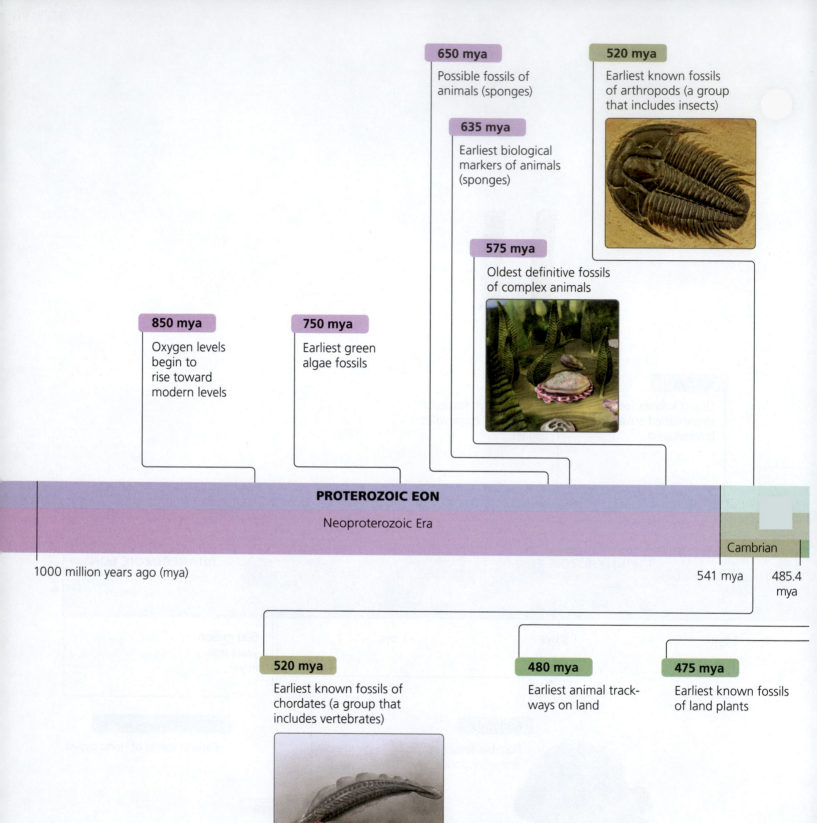

650 mya

Possible fossils of animals (sponges)

520 mya

Earliest known fossils of arthropods (a group that includes insects)

635 mya

Earliest biological markers of animals (sponges)

575 mya

Oldest definitive fossils of complex animals

850 mya

Oxygen levels begin to rise toward modern levels

750 mya

Earliest green algae fossils

PROTEROZOIC EON

Neoproterozoic Era

Cambrian

1000 million years ago (mya)

541 mya

485.4 mya

520 mya

Earliest known fossils of chordates (a group that includes vertebrates)

480 mya

Earliest animal track-ways on land

475 mya

Earliest known fossils of land plants

443.8 mya

Ordovician-Silurian mass extinctions

428 mya

Earliest known animal fossil on land (a millipede)

400 mya

Earliest known fossils of fungi on land

360 mya

Earliest known fossils of land vertebrates

252 mya

Permian-Triassic mass extinctions (largest extinction event in the past 500 million years)

230 mya

Earliest known fossils of dinosaurs

201 mya

Triassic-Jurassic mass extinctions

180 mya

Common ancestor of living mammals

150 mya

Earliest known fossils of birds

100 mya

Rapid rise of forests dominated by flowering plants

66 mya

Cretaceous-Tertiary mass extinctions (large dinosaurs and other species disappear)

55 mya

Earliest known primate fossils
Earliest known semi-aquatic whale fossils

20 mya

Earliest known fossils of apes
First large grasslands expand

7 mya

Earliest known hominin fossils

3.3 mya

Earliest known hominin stone tools

PHANEROZOIC EON											PRESENT

	Paleozoic Era					Mesozoic Era				Cenozoic Era	

Ordovician	Silurian	Devonian	Carboniferous	Permian	Triassic	Jurassic	Cretaceous	Paleogene	Neogene

443.8 mya | 419.2 mya | 358.9 mya | 298.9 mya | 251.9 mya | 201.3 mya | 145mya | 66 mya | 23.03 mya | 2.58 mya

Quaternary

Quaternary

2.6 mya

1.8 mya

Oldest hominins out of Africa

400,000 yrs ago

Neanderthals evolve

300,000 yrs ago

Oldest known fossils of *Homo sapiens*

40,000 yrs ago

Earliest figurative art

Neanderthals become extinct

Glossary

An **adaptation** is an inherited aspect of an individual that allows it to outcompete other members of the same population that lack the trait (or that have a different version of the trait). Adaptations are traits that have evolved through the mechanism of natural selection.

An **adaptive radiation** is an evolutionary lineage that has undergone exceptionally rapid diversification into a variety of lifestyles or ecological niches.

An **allele** refers to one of any number of alternative forms of the DNA sequence of the same locus.

Allopatric speciation is the evolution of new species after populations have been separated geographically, preventing gene flow between them.

Allopatry occurs when populations are in separate, nonoverlapping geographic areas (that is, they are separated by geographic barriers to gene flow).

Allopolyploidy refers to polyploidy (more than two paired chromosomes) resulting from interspecific hybridization. (If polyploidy arises within a species, it's called autopolyploidy.)

Alternative splicing is the process of combining different subsets of exons together, yielding different mRNA transcripts from a single gene.

Altruistic behavior occurs whenever a helping individual behaves in a way that benefits another individual at a cost to its own fitness.

An **amino acid** is the structural unit that, among other functions, links together to form proteins.

Anagenesis refers to wholesale transformation of a lineage from one form to another. In macroevolutionary studies, anagenesis is considered to be an alternative to lineage splitting or speciation.

Anisogamy refers to sexual reproduction involving the fusion of two dissimilar gametes; individuals producing the larger gamete (eggs) are defined as female, and individuals producing the smaller gamete (sperm) are defined as male.

Antagonistic pleiotropy is the condition that occurs when a mutation with beneficial effects for one trait also causes detrimental effects on other traits.

Anthropomorphism is the tendency to attribute human characteristics, such as motives, to other animal species.

Antigenic drift describes natural selection on seasonal influenza viruses that alters the structure of surface proteins, allowing them to evade the antibody-mediated immunity induced during previous infections or vaccinations.

Antigenic shift is a dramatic change in the surface proteins of influenza viruses brought about by reassortment between viral strains, leading to a severe drop in immunity to the resulting strain and the potential for a new flu pandemic.

Archaea are one of the two prokaryotic domains of life. Archaea superficially resemble bacteria, but they are distinguished by a number of unique biochemical features.

Artificial selection is the selective breeding of animals and plants to encourage the occurrence of desirable traits. Individuals with preferred characteristics are mated or cross-pollinated with other individuals having similar traits.

An **autosome** is a chromosome that does not differ between sexes.

Background extinction refers to the normal rate of extinction for a taxon or biota.

Bacteria are one of the two prokaryotic domains of life. Domain Bacteria includes organisms such as *Escherichia coli* and other familiar microbes.

Balancing selection describes the type of selection that favors more than one allele. This process acts to maintain genetic diversity in a population by keeping alleles at frequencies higher than would be expected by chance or mutation alone.

Base refers to one of four nitrogen-containing molecules in DNA: adenine (A), cytosine (C), guanine (G), and thymine (T). (In RNA, uracil (U) replaces T.)

Batesian mimicry occurs when harmless species resemble harmful or distasteful species, deriving protection from predators in the process.

A **Bateson–Dobzhansky–Muller incompatibility** is a genetic incompatibility in hybrid offspring arising from epistatic interactions between two or more loci.

Bayesian methods refer to tests that are similar to maximum likelihood; they use statistical models to determine the probability of data given an evolutionary model and a phylogenetic tree.

Behavioral ecology is the science that explores the relationship between behavior, ecology, and evolution to elucidate the adaptive significance of animal actions.

Biogeography is the study of the distribution of species across space (geography) and time.

Biological evolution is any change in the inherited traits of a population that occurs from one generation to the next (that is, over a time period longer than the lifetime of an individual in the population).

The **biological species concept** holds that species are groups of actually (or potentially) interbreeding natural populations that are reproductively isolated from other such groups.

A **biomarker** is molecular evidence of life in the fossil record. Biomarkers can include fragments of DNA, molecules such as amino acids, or isotopic ratios.

Bootstrapping is a statistical method that allows for assigning measures of accuracy to sample estimates; scientists use it for estimating the strength of evidence that a particular branch in a phylogeny exists.

A **branch** is a lineage evolving through time that connects successive speciation or other branching events.

Broad sense heritability (H^2) is the proportion of the total phenotypic variance of a trait that is attributable to genetic variance, where genetic variance is represented in its entirety as a single value (that is, genetic variance is not broken down into different components).

Burgess Shale is a Lagerstätte in Canada in which there is a wealth of preserved fossils from the Cambrian period.

Certainty of paternity is the probability that a male is the genetic sire of the offspring his mate produces.

A **character** is a heritable aspect of organisms that can be compared across taxa.

A **chordate** is a member of a diverse phylum of animals that includes the vertebrates, lancelets, and tunicates. As embryos, chordates all have a notochord (a hollow nerve cord), pharyngeal gill slits, and a post-anal tail. Many present-day chordates lose or modify these structures as they develop into adults.

A **cis-acting element** is a stretch of DNA located near a gene—immediately upstream (adjacent to the promoter region), downstream, or inside an intron—that influences the expression of that gene. Cis regions often code for binding sites for one or more transposable factors.

A **clade** is a single "branch" in the tree of life; each clade represents an organism and all of its descendants.

Coalescence is the process in which the genealogy of any pair of homologous alleles merges in a common ancestor.

Coevolution is reciprocal evolutionary change between interacting species, driven by natural selection.

Coevolutionary alternation occurs when one species coevolves with several other species by shifting among the species with which it interacts over many generations.

A **coevolutionary arms race** occurs when species interact antagonistically in a way that results in each species exerting reciprocal directional selection on the other. As one species evolves to overcome the weapons of the other, it, in turn, selects for new weaponry in its opponent.

Commensalism describes a relationship between species in which one species benefits and the other is neither helped nor harmed.

A **complex adaptation** is a suite of coexpressed traits that together experience selection for a common function. Phenotypes are considered complex when they are influenced by many environmental and genetic factors and when multiple components must be expressed together for the trait to function.

Convergent evolution is the independent origin of similar traits in separate evolutionary lineages.

Cospeciation occurs when a population speciates in response to, and in concert with, another species.

Cryptic female choice refers to a form of sexual selection that arises after mating, when females store and separate sperm from different males and thus bias which sperm they use to fertilize their eggs.

Cryptic species are lineages that historically have been treated as one species because they are morphologically similar but that are later revealed to be genetically distinct.

Deep homology is a condition that occurs when the growth and development of traits in different lineages result from underlying genetic mechanisms (for example, regulatory networks) that are inherited from a common ancestor.

Deoxyribonucleic acid (**DNA**) is an essential macromolecule containing genetic information for development, life, and reproduction. DNA is a long, double-stranded, helical nucleic acid molecule capable of replicating and determining the inherited structure of a cell's proteins.

Descent with modification refers to the passing of traits from parents to offspring. Darwin recognized that, over time, this process could account for gradual change in species' traits and homology.

The **dilution effect** refers to the safety in numbers that arises through swamping the foraging capacity of local predators.

A **direct benefit** is a benefit that affects a particular female directly, such as food, nest sites, or protection.

Directional selection favors individuals on one end of the distribution of phenotypes present in a population.

Dispersal describes the movement of populations from one geographic region to another with very limited, or no, return exchange.

Disruptive selection favors individuals at the tails of the distribution of phenotypes present in a population (for example, by acting against individuals with intermediate trait values).

A **distance-matrix method** is a procedure for constructing phylogenetic trees by clustering taxa based on the proximity (or distance) between their DNA or protein sequences. These methods place closely related sequences under the same internal branch, and they estimate branch lengths from the observed distances between sequences.

Diversifying coevolution refers to an increase in genetic diversity caused by the heterogeneity of coevolutionary processes across the range of ecological partners.

Dominant allele describes an allele that produces the same phenotype whether it is paired with an identical allele or a different allele (that is, a heterozygotic state).

Ecological character displacement refers to evolution driven by competition between species for a shared resource (for example, food). Traits evolve in opposing directions, minimizing overlap between the species.

Ecological speciation is the evolution of reproductive barriers between populations by adaptation to different environments or ecological niches.

Ediacaran fauna is a group of animal species that existed during the Ediacaran period, just before the Cambrian. The oldest Ediacaran fossils are about 570 million years old. Ediacarans included diverse species that looked like fronds, geometrical disks, and blobs covered with tire tracks. They appear to have become extinct 540 million years ago, although some researchers argue that certain Ediacaran taxa belong to extant animal clades.

An **endosymbiont** is a mutualistic organism that lives within the body or cells of another organism.

An **enhancer** is a short sequence of DNA within the gene control region where activator proteins bind to initiate gene expression.

Epigenetic refers to the functional modifications to DNA that don't involve changes to the sequences of nucleotides. Epigenetics is the study of the heritability of these modifications.

Eukarya is the third domain of life, characterized by traits that include membrane-enclosed cell nuclei and mitochondria. Domain Eukarya includes animals, plants, fungi, and protists (a general term for single-celled eukaryotes).

Eusociality is a type of social organization in which species have complete reproductive division of labor. In a eusocial group, many individuals never reproduce, instead helping to rear the offspring of a limited number of dominant individuals.

An **evolutionarily stable strategy** is a behavior that, if adopted by a population in a given environment, cannot be invaded by any alternative behavioral strategy.

Evolutionary medicine is the integrated study of evolution and medicine to improve scientific understanding of the reasons for disease and actions that can be taken to improve health.

Evolutionary reversal describes the reversion of a derived character state to a form resembling its ancestral state.

An **exaptation** is a trait that initially carries out one function and is later co-opted for a new function. The original function may or may not be retained.

An **extended phenotype** is a structure constructed by an organism that can influence its performance or success. Although it is not part of the organism itself, its properties nevertheless reflect the genotype of each individual. Animal examples include the nests constructed by birds and the galls of flies.

Extinction refers to the permanent loss of a species. It is marked by the death or failure to breed of the last individual.

A **fauna** is an assemblage of many different species that live together. A fauna may comprise the species in a single ecosystem, in a region, or across the entire planet.

Fecundity describes the reproductive capacity of an individual, such as the number and quality of eggs or sperm. As a measure of relative fitness, fecundity refers to the number of offspring produced by an organism.

A **fossil** is preserved evidence of life from a past geological age, including impressions and mineralized remains of organisms embedded in rocks.

F_{ST} is a measure of genetic distance between subpopulations.

The F_{ST} **outlier** method detects loci with allele frequencies that are more different than expected between populations. These outlier loci are likely to be near to regions of the genome experiencing strong selection.

Game theory is a mathematical approach to studying behavior that solves for the optimal decision in strategic situations (games) where the payoff to a particular choice depends on the choices of others.

Gametic incompatibility occurs when sperm or pollen from one species fails to penetrate and fertilize the egg or ovule of another species.

A **gene** is a segment of DNA whose nucleotide sequence codes for proteins, codes for RNA, or regulates the expression of other genes.

Gene control region refers to an upstream section of DNA that includes the promoter region as well as other regulatory sequences that influence the transcription of DNA.

Gene expression is the process by which information from a gene is transformed into a product.

Gene flow describes the movement, or migration, of alleles from one population to another.

Gene recruitment refers to the co-option of a particular gene or network for a totally different function as a result of a mutation. The reorganization of a preexisting regulatory network can be a major evolutionary event.

Gene tree refers to the branched genealogical lineage of homologous alleles that traces their evolution back to an ancestral allele.

The **general lineage species concept** is the idea that species are metapopulations of organisms that exchange alleles frequently enough that they constitute the same gene pool and therefore the same evolutionary lineage.

Genetic distance is a measure of how different populations are from each other genetically. Genetic distance can inform population geneticists about levels of inbreeding within a population or about the historic relationships between populations or species.

Genetic drift is evolution arising from random changes in the genetic composition of a population from one generation to the next.

Genetic hitchhiking occurs when an allele increases in frequency because it is physically linked to a positively selected allele at a nearby locus.

Genetic load is the burden imposed by the accumulation of deleterious mutations.

Genetic polymorphism is the simultaneous occurrence of two or more discrete phenotypes within a population. In the simplest case, each phenotype results from a different allele or combination of alleles of a single gene. In more complex cases, the phenotypes result from complex interactions between many different genes and the environment.

Genetic recombination is the exchange of genetic material between paired chromosomes during meiosis. Recombination can form new combinations of alleles and is a source of heritable variation.

Genetics is the study of heredity, or how characteristics of organisms are transmitted from one generation to the next.

A **genome** includes all the hereditary information of an organism. The genome comprises the totality of the DNA, including the coding and noncoding regions.

Genome-wide association (**GWA**) mapping involves scanning through the genomes of many different individuals, some with, and others without, a focal trait of interest, to search for markers associated with expression of the trait.

Genomic imprinting occurs when genes inherited from one or the other parent are silenced due to methylation. Imprinting can result in offspring who express either the maternal or paternal copy of the gene, but not both.

Genomics is the study of the structure and function of genomes, including mapping genes and DNA sequencing. The discipline unites molecular and cell biology, classical genetics, and computational science.

Genotype describes the genetic makeup of an individual. Although a genotype includes all the alleles of all the genes in that individual, the term is often used to refer to the specific alleles carried by an individual for any particular gene.

Genus (plural, **genera**) is a taxonomic group that includes species. (*Canis*, for example, is the genus that includes dogs, wolves, and coyotes.)

The **geographic mosaic theory of coevolution** proposes that the geographic structure of populations is central to the dynamics of coevolution. The direction and intensity of coevolution vary from population to population, and coevolved genes from these populations mix together as a result of gene flow.

A **germline mutation** is a mutation that affects the gametes (eggs, sperm) of an individual and can be transmitted from parents to offspring. Germline mutations create the heritable genetic variation that is relevant to evolution.

Group selection is selection arising from variation in fitness among groups.

Haplodiploidy is a mechanism through which sex is determined by the number of copies of each chromosome that an individual receives. Offspring formed from the fertilization of an egg by a sperm (that is, diploids) are female; those formed from unfertilized eggs (that is, haploids) are male.

Heredity is the transmission of characteristics from parent to offspring.

A **hermaphrodite** is an individual that produces both female and male gametes.

Heterozygote advantage occurs when selection favors heterozygote individuals over either the dominant homozygote or the recessive homozygote.

A **hominin** is a member of the clade that includes humans as well as all species more closely related to humans than chimpanzees. Humans are the only surviving member of the hominin clade.

A **homologous characteristic** is similar in two or more species because it is inherited from a common ancestor.

Homology refers to the similarity of characteristics resulting from shared ancestry.

Homoplasy describes a character state similarity *not* due to shared descent (for example, produced by convergent evolution or evolutionary reversal).

Horizontal gene transfer is the transfer of genetic material—other than from parent to offspring—to another organism, sometimes a distantly related one, without reproduction. Once this material is added to the recipient's genome, it can be inherited by descent.

A **hormone** is a molecular signal that flows from cells in one part of the body to cells in other parts of the body. Hormones act directly or indirectly to alter the expression of target genes.

A **hypothesis** is a tentative explanation for an observation, phenomenon, or scientific problem that can be tested by further investigation or experimentation.

Inbreeding coefficient (**F**) refers to the probability that the two alleles at any locus in an individual will be identical because of common descent. F can be estimated for an individual, $F_{pedigree}$, by measuring the reduction in heterozygosity across loci within the genome of that individual attributable to inbreeding, or it can be estimated for a population, by measuring the reduction in heterozygosity at one or a few loci sampled for many different individuals within the population.

Inbreeding depression is a reduction in the average fitness of inbred individuals relative to that of outbred individuals. It arises because rare recessive alleles become expressed in a homozygous state where they can affect detrimentally the performance of individuals.

Inclusive fitness describes an individual's combined fitness, including its own reproduction as well as any increase in the reproduction of its relatives due specifically to its own actions.

Incomplete lineage sorting occurs when a genetic polymorphism persists through several speciation events. When fixation of alternative alleles eventually occurs in the descendent species, the pattern of retention of alleles may yield a gene tree that differs from the true phylogeny of the species.

Independent assortment is the random mixing of maternal and paternal copies of each chromosome during meiosis, resulting in the production of genetically unique gametes.

An **indirect benefit** is a benefit that affects the genetic quality of a particular female's offspring, such as male offspring that are more desirable to females.

Individual selection describes selection arising from variation in fitness among individuals.

An **internal node** is a node that occurs within a phylogeny and represents ancestral populations or species.

Intersexual selection occurs when members of the limiting sex (generally females) actively discriminate among suitors of the less limited sex (generally males). Often called female choice.

Intralocus sexual conflict is a conflict between the fitness effects of alleles of a given locus on males and females.

Intrasexual selection occurs when members of the less limiting sex (generally males) compete with each other over reproductive access to the limiting sex (generally females). Often called male–male competition.

Introgression describes the movement of alleles from one species or population to another.

An **isochron** is a line on a graph, connecting points at which an event occurs simultaneously. Isochron dating is a technique to measure the age of rocks by determining how ratios of isotopes change in them over time.

An **isolating barrier** refers to an aspect of the environment, genetics, behavior, physiology, or ecology of a species that reduces or impedes gene flow from individuals of other species. Isolating barriers can be geographic or reproductive.

Isolation by distance is a pattern in which populations that live in close proximity are genetically more similar to each other than populations that live farther apart.

A **key innovation** is a novel trait that allows the subsequent radiation and success of a clade.

Kin selection is selection arising from the indirect fitness benefits of helping relatives.

Lagerstätte (plural, **Lagerstätten**) is a site with an abundant supply of unusually well-preserved fossils—often including soft tissues—from the same period of time.

A **lek** is an assemblage of rival males who cluster together to perform courtship displays in close proximity.

Life history refers to the pattern of investment an organism makes in growth and reproduction. Life history traits include an organism's age at first reproduction, the duration and schedule of reproduction, the number and size of offspring produced, and life span.

Lineage refers to a chain of ancestors and their descendants. A lineage may be the successive generations of organisms in a single population, the members of an entire species during an interval of geological time, or a group of related species descending from a common ancestor.

Linkage disequilibrium exists for any two loci if the occurrence of an allele at one of the loci is nonrandomly associated with the presence or absence of an allele at the other locus.

Linkage equilibrium exists for any two loci if the occurrence of an allele at one of the loci is independent of the presence or absence of an allele at the other locus.

Macroevolution is evolution occurring above the species level, including the origination, diversification, and extinction of species over long periods of evolutionary time.

A **magic trait** is a trait that simultaneously confers ecological divergence and reproductive isolation.

Mass extinction describes a statistically significant departure from background extinction rates that results in a substantial loss of taxonomic diversity.

The **maximum likelihood** is an approach used to estimate parameter values for a statistical model. Maximum likelihood and the similar Bayesian method are used in phylogeny reconstruction to find the tree topologies that are most likely, given a precise model for molecular evolution and a particular data set.

Maximum parsimony is a statistical method for reconstructing phylogenies that identifies the tree topology that minimizes the total amount of change, or the number of steps, required to fit the data to the tree.

Meiosis, which occurs only in eukaryotes, is a form of cell division that cuts the number of chromosomes in half.

Messenger RNA (mRNA) consists of molecules of RNA that carry genetic information from DNA to the ribosome, where it can be translated into protein.

A **metapopulation** is a group of spatially separated populations of the same species that interact at some level (for example, exchange alleles).

Methylation is the process by which methyl groups are added to certain nucleotides. Methylation is associated with altered gene expression, thereby reducing or eliminating the production of proteins or RNA molecules.

Microevolution is evolution occurring within populations, including adaptive and neutral changes in allele frequencies from one generation to the next.

MicroRNA describes one group of RNAs that act as post-transcriptional regulators of gene expression. MicroRNAs bind to complementary sequences on specific mRNAs and can enhance or silence the translation of genes. The human genome encodes more than 1000 of these tiny RNAs.

A **microsatellite** is a noncoding stretch of DNA containing a string of short (one to six base pairs), repeated segments. The number of repetitive segments can be highly polymorphic, and for this reason microsatellites are valuable genetic characters for comparing populations and for assigning relatedness among individuals (DNA fingerprinting).

A **mobile genetic element** is a type of DNA that can move around in the genome. Common examples include transposons ("jumping genes") and plasmids.

A **molecular clock** is a method used to determine time based on base pair substitutions. Molecular clocks use the rates of molecular change to deduce the divergence time between lineages in a phylogeny, for example. They work best when they can be "calibrated" with other markers of time, such as fossils with known ages and placements.

Monogamy is a mating system in which one male pairs with one female.

Monophyletic describes a group of organisms that form a clade.

A **morphogen** is a signaling molecule that flows between nearby cells and acts directly to alter the expression of target genes.

Morphology refers to the form and structure of organisms.

Müllerian mimicry occurs when several harmful or distasteful species resemble each other in appearance, facilitating the learned avoidance of predators.

Muller's ratchet describes the process by which the genomes of an asexual population accumulate deleterious mutations in an irreversible manner.

A **mutation** is any change to the genomic sequence of an organism.

Mutualism describes a relationship between species in which both species benefit.

Narrow sense heritability (h^2) is the proportion of the total phenotypic variance of a trait attributable to the *additive effects of alleles* (the additive genetic variance). This is the component of variance that causes offspring to resemble their parents, and it causes populations to evolve predictably in response to selection.

Natural selection is a mechanism that can lead to adaptive evolution, whereby differences in the phenotypes of individuals cause some of them to survive and reproduce more effectively than others and therefore outcompete them.

Neighbor joining is a distance method for reconstructing phylogenies. Neighbor joining identifies the tree topology with the shortest possible branch lengths given the data.

A **node** is a point in a phylogeny where a lineage splits (a speciation event or other branching event, such as the formation of subspecies).

A **nonsynonymous substitution** is a substitution that alters the amino acid sequence of a protein. Nonsynonymous substitutions can affect the phenotype and are therefore more subject to selection.

A **notochord** is a flexible, rod-shaped structure found in the embryos of all chordates. Notochords served as the first "backbones" in early chordates, and

in extant vertebrates the embryonic notochord becomes part of the vertebral column.

A **nucleotide** is the structural unit that links together to form DNA (and RNA). Each nucleotide includes a sugar (like deoxyribose or ribose) and a base.

Offspring–offspring conflict (sibling rivalry) occurs when siblings compete for parental care or limited resources.

An **oncogene** is a normal gene whose function, when altered by mutation, has the potential to cause cancer.

Operational sex ratio (OSR) is the ratio of male to female individuals who are available for reproducing at any given time.

Opportunity for selection refers to the variance in fitness within a population. When there is no variance in fitness, there can be no selection; when there is large variance in fitness, there is a great opportunity for selection. In this sense, the opportunity for selection constrains the intensity of selection that is possible.

An **ortholog** is one of two or more homologous genes separated by a speciation event (as opposed to paralogs—homologous genes, produced by gene duplication, that are both possessed by the same species).

An **outgroup** is a group of organisms (for example, a species) that is outside of the monophyletic group being considered. In phylogenetic studies, outgroups can be used to infer the ancestral states of characters.

Paleontology is the study of prehistoric life.

Parallel evolution refers to independent evolution of similar traits, starting from a similar ancestral condition.

A **paralog** is a homologous gene that arises by gene duplication. Paralogs together form a gene family.

Parapatric speciation is the evolution of new species within a spatially extended population that still has some gene flow.

Paraphyletic describes a group of organisms that share a common ancestor, although the group does not include all the descendants of that common ancestor.

Parental conflict occurs when parents have an evolutionary conflict of interest over the optimal strategy for parental care.

Parent–offspring conflict occurs when parents benefit from withholding parental care or resources from some offspring (for example, a current brood) and invest in other offspring (for example, a later brood). Conflict arises because the deprived offspring would benefit more if they received the withheld care or resources.

Parent-of-origin effect describes an effect on the phenotype of an offspring caused by an allele inherited from a particular parent.

Parsimony is a principle that guides the selection of the most compelling hypothesis among several choices. The hypothesis requiring the fewest assumptions or steps is usually (but not always) best. In cladistics, scientists search for the tree topology with the least number of character-state changes—the most parsimonious.

Parthenogenesis is a mode of reproduction in which female sex cells undergo meiosis but are not fertilized by sperm; in parthenogenetic species, females produce only daughters.

Phenotype is an observable, measurable characteristic of an organism. A phenotype may be a morphological structure (for example, antlers, muscles), a developmental process (for example, learning), a physiological process or performance trait (for example, running speed), or a behavior (for example, mating display). Phenotypes can even be the molecules produced by genes (for example, hemoglobin). Genes interact with other genes and with the environment during the development of the phenotype.

Phenotypic plasticity refers to changes in the phenotype produced by a single genotype in different environments.

The **phylogenetic species concept** is the idea that species are the smallest possible groups whose members are descended from a common ancestor and who all possess defining or derived characteristics that distinguish them from other such groups.

Phylogeny is a visual representation of the evolutionary history of populations, genes, or species.

Physical linkage is the adjacency of two or more loci on the same chromosome.

A **plasmid** is a molecule of DNA, found most often in bacteria, that can replicate independently of chromosomal DNA.

Ploidy refers to the number of copies of unique chromosomes in a cell (*n*). Normal human somatic cells are diploid (2*n*); they have two copies of 23 chromosomes.

Polyandry is a mating system where females mate (or attempt to mate) with multiple males.

Polygyny is a mating system where males mate (or attempt to mate) with multiple females.

Polyphenism is a trait for which multiple, discrete phenotypes can arise from a single genotype, depending on environmental circumstances.

Polyphyletic describes a taxonomic group that does not share an immediate common ancestor and therefore does not form a clade.

Polytomy describes an internal node of a phylogeny with more than two branches (that is, the order in which the branchings occurred is not resolved).

Population structure (or **population subdivision**) refers to the occurrence of populations that are subdivided by geography, behavior, or other influences that prevent individuals from mixing completely. Population subdivision leads to deviations from Hardy-Weinberg predictions.

A **postzygotic reproductive barrier** is an aspect of the genetics, behavior, physiology, or ecology of a species that prevents hybrid zygotes from successfully developing and reproducing themselves.

A **prezygotic reproductive barrier** is an aspect of the genetics, behavior, physiology, or ecology of a species that prevents sperm from one species from fertilizing eggs of another species. Prezygotic barriers reduce the likelihood that a zygote will form.

A **prokaryote** is a microorganism lacking a cell nucleus or any other membrane-bound organelles. Prokaryotes comprise two evolutionarily distinct groups, the Bacteria and the Archaea.

A **promiscuous protein** is a protein capable of carrying out more than one function, such as catalyzing reactions of different substrates.

A **protein** is an essential macromolecule for all known forms of life. Proteins are three-dimensional biological polymers constructed from a set of 20 different monomers called amino acids.

A **pseudogene** is a DNA sequence that resembles a functional gene but has lost its protein-coding ability or is no longer expressed. Pseudogenes often form after a gene has been duplicated, when one or more of the redundant copies subsequently lose their function.

Punctuated equilibria is a model of evolution that proposes that most species undergo relatively little change for most of their geological history. These periods of stasis are punctuated by brief periods of rapid morphological change, often associated with speciation.

Purifying selection (also called negative selection) removes deleterious alleles from a population. It is a common form of stabilizing selection.

Quantitative genetics is the study of continuous phenotypic traits and their underlying evolutionary mechanisms.

A **quantitative trait** is a measurable phenotype that varies among individuals over a given range to produce a continuous distribution of a phenotype. Quantitative traits are sometimes called complex traits; they're also sometimes called polygenic traits because their variation can be attributed to polygenic effects (that is, the cumulative action of many genes).

Quantitative trait locus (QTL) is a stretch of DNA that is correlated with variation in a phenotypic trait. These regions contain genes, or are linked to genes, that contribute to population differences in a phenotype.

Radiometric dating is a technique that allows geologists to estimate the precise ages at which one geological formation ends and another begins.

Reaction norm refers to the pattern of phenotypic expression of a single genotype across a range of environments. In a sense, reaction norms depict how development maps the genotype into the phenotype as a function of the environment.

Reassortment occurs when genetic segments from different influenza strains become mixed into new combinations within a single individual.

Recessive allele refers to an allele that produces its characteristic phenotype only when it is paired with an identical allele (that is, in homozygous states).

Reciprocal selection describes selection that occurs in two species due to their interactions with one another. Reciprocal selection is the critical prerequisite of coevolution.

The **Red Queen effect** describes a phenomenon seen in coevolving populations—to maintain relative fitness, each population must constantly adapt to the other.

A **regulatory network** is a system of interacting genes, transcription factors, promoters, RNA, and other molecules. It functions like a biological circuit, responding to signals with output that controls the activation of genes during development, the cell cycle, and the activation of metabolic pathways.

Reinforcement refers to the increase of reproductive isolation between populations through selection against hybrid offspring.

A **repressor** is a protein that binds to a sequence of DNA or RNA and inhibits the expression of one or more genes.

Reproductive isolation occurs when reproductive barriers prevent or strongly limit reproduction between populations. The result is that few or no genes are exchanged between the populations.

Resistance refers to the capacity of pathogens to defend against antibiotics or other drugs.

Retrovirus is an RNA virus that uses an enzyme called reverse transcriptase to become part of the host cell's DNA. The virus that causes AIDS, the human immunodeficiency virus (HIV), is a type of retrovirus.

Ribonucleic acid (RNA) is an essential macromolecule for all known forms of life (along with DNA and proteins). RNA differs structurally from DNA in having the sugar ribose instead of deoxyribose and in having the base uracil (U) instead of thymine (T).

RNA polymerase is the enzyme that builds the single-stranded RNA molecule from the DNA template during transcription.

RNA splicing is the process of modifying RNA after transcription but before translation, during which introns are removed and exons are joined together into a contiguous strand.

Selection differential (S) is a measure of the strength of phenotypic selection. The selection differential describes the difference between the mean of the reproducing members of the population that contribute offspring to the next generation and the mean of all members of a population.

Selective sweep describes the situation in which strong selection can "sweep" a favorable allele to fixation within a population so fast that there is little opportunity for recombination. In the absence of recombination, alleles in large stretches of DNA flanking the favorable allele will also reach high frequency.

Senescence refers to the deterioration in the biological functions of an organism as it ages.

A **sex chromosome** is a chromosome that pairs during meiosis but differs in copy number between males and females. For organisms such as humans with XY sex determination, X and Y are the sex chromosomes. Females are the homogametic sex (XX) and males are the heterogametic sex (XY).

Sexual conflict is the evolution of phenotypic characteristics that confer a fitness benefit to one sex but a fitness cost to the other.

Sexual dimorphism is a difference in form between males and females of a species, including color, body size, and the presence or absence of structures used in courtship displays (elaborate tail plumes, ornaments, pigmented skin patches) or in contests (antlers, tusks, spurs, horns).

Sexual selection refers to differential reproductive success resulting from the competition for fertilization, which can occur through competition among individuals of the same sex (intrasexual selection) or through attraction to the opposite sex (intersexual selection).

A **somatic mutation** is a mutation that affects cells in the body ("soma") of an organism. These mutations affect all the daughter cells produced by the affected cell and can affect the phenotype of the individual. In animals, somatic mutations are not passed down to offspring. In plants, somatic mutations can be passed down during vegetative reproduction.

Speciation is the evolutionary process by which new species arise. Speciation causes one evolutionary lineage to split into two or more lineages (cladogenesis).

Sperm competition is a form of sexual selection that arises after mating, when males compete for fertilization of a female's eggs.

Stabilizing selection favors individuals in the middle of the distribution of phenotypes present in a population (for example, by acting against individuals at either extreme).

Stable ecotype model is a species concept constructed for microbes in particular. In this model, a species is a long-lived population of genetically related individuals that share a stable set of adaptations for the same ecological niche.

Standing diversity is the number of species (or other taxonomic unit) present in a particular area at a given time.

Stratigraphy is the study of layering in rock (stratification).

Stromatolites are layered structures formed by the mineralization of bacteria.

A **supergene** is a group of functionally related genes located close enough together that they segregate as a single unit.

Sympatric speciation is the evolution of new species within a contiguous population that still has extensive gene flow.

Sympatry occurs when populations are in the same geographic area. A synapomorphy is a derived form of a trait that is shared by a group of related species (that is, one that evolved in the immediate common ancestor of the group and was inherited by all its descendants).

Synapsids are a lineage of tetrapods that emerged 300 million years ago and gave rise to mammals. Synapsids can be distinguished from other tetrapods by the presence of a pair of openings in the skull behind the eyes, known as the temporal fenestrae.

Synaptic plasticity occurs when the number or strength of synaptic connections between neurons is altered in response to stimuli.

A **synonymous substitution** is a substitution that does not alter the amino acid sequence of a protein. Because synonymous substitutions do not affect the protein an organism produces, they are less prone to selection and often free from selection completely.

A **taxon** (plural, **taxa**) is a group of organisms that a taxonomist judges to be a cohesive taxonomic unit, such as a species or order.

Taxonomy is the science of describing, naming, and classifying species of living or fossil organisms.

Teleosts are a lineage of bony fishes that constitute most living species of aquatic vertebrates. They can be distinguished from other fishes by unique traits, such as the mobility of an upper jawbone called the premaxilla.

A **tetrapod** is a vertebrate with four limbs (or, like snakes, descended from vertebrates with four limbs). Living tetrapods include mammals, birds, reptiles, and amphibians.

A **theory** is an overarching set of mechanisms or principles that explain major aspects of the natural world. Theories are supported by many different kinds of evidence and experimental results.

The **thrifty genotype hypothesis** proposes that alleles that were advantageous in the past (for example, because they were "thrifty" and stored nutrients well) may have become detrimental in the modern world, contributing to metabolic syndrome, obesity, and type 2 diabetes.

A **tip** is the terminal end of an evolutionary tree, representing species, molecules, or populations being compared.

A **trans-acting element** is a sequence of DNA located away from the focal gene (for example, on another chromosome). These stretches of DNA generally code for a protein, microRNA, or other diffusible molecule that then influences expression of the focal gene.

Transcription is the process that takes place when RNA polymerase reads a coding sequence of DNA and produces a complementary strand of RNA, called messenger RNA (mRNA).

A **transcription factor** is a protein that regulates the expression of a gene by binding to a specific DNA sequence in association with the gene sequence. A single transcription factor can regulate many genes if they share the same regulatory sequence.

Transfer RNA (**tRNA**) is a short piece of RNA that physically transfers a particular amino acid to the ribosome.

Translation is the process that takes place when a strand of mRNA is decoded by a ribosome to produce a protein.

A **trilobite** was a marine arthropod that diversified during the Cambrian period. Trilobites gradually died out during the Devonian period.

A **tumor-suppressor gene** is a gene that suppresses cell growth and proliferation. Many tumor-suppressor genes are transcription factors activated by stress or DNA damage that arrest mitosis until DNA can be repaired. Mutations interfering with their expression can lead to excessive proliferation and cancer.

Turnover refers to the disappearance (extinction) of some species and their replacement by others (origination) in studies of macroevolution. The turnover rate is the number of species eliminated and replaced per unit of time.

Twofold cost of sex refers to the disadvantages of being a sexual rather than an asexual organism. Asexual lineages multiply faster than sexual lineages because all progeny are capable of producing offspring. In sexual lineages, half of the offspring are males who cannot themselves produce offspring. This limitation effectively halves the rate of replication of sexual species.

Uniformitarianism is the idea that the natural laws observable around us now are also responsible for events in the past. One part of this view, for example, is the idea that Earth has been shaped by the cumulative action of gradual processes like sediment deposition and erosion.

Variance is a statistical measure of the dispersion of trait values about their mean.

Vertical gene transfer is the process of receiving genetic material from an ancestor.

Vicariance is the formation of geographic barriers to dispersal and gene flow, resulting in the separation of once continuously distributed populations.

Viral reassortment occurs when genetic material from different strains is mixed into new combinations within a single cell.

Virulence describes the ability of a pathogen to cause disease.

Answers to Multiple Choice and Short Answer Questions

Chapter 1

Multiple Choice

1. d 2. d 3. b 4. c 5. b 6. c 7. d 8. d 9. b 10. d

Interpret the Data

11. d

Short Answer

12. Scientists understand that biological evolution is simply the processes by which populations of organisms change over time, and variation among individuals is the fundamental reason for that change. If some individuals survive better or produce more offspring because they vary genetically from other individuals, and they can pass those traits to their offspring, then offspring with those traits will have a better chance of surviving and reproducing than individuals without them. In time, those traits will be more common in the population. Scientists have observed these phenomena over and over, in nature and in their laboratories. These evolutionary principles explain biological observations as diverse as the streamlined bodies of fishes and whales, the massive horns on beetles and antlers on elk, and why humans see one range of colors whereas other animals see other ranges, for example.

13. Evolutionary biologists consider observations of living species, DNA, and fossils. Scientists make comparisons between living species and fossils of extinct species to study shared anatomical traits. More recently, they have begun comparing DNA. Close relatives will share more traits, both physical and genetic, inherited from their common ancestor. So scientists can develop hypotheses for how these traits changed over time and how they evolved between different lineages and test them with additional evidence.

14. Although whales have fishlike bodies, with the same sleek curves and tails you can find on tunas and sharks, they have a number of distinguishing characters. Whales do not have gills, so they cannot extract dissolved oxygen from the water in which they live. Whales must rise to the surface of the ocean in order to breathe. Whales and dolphins have long muscles that run the length of their bodies — much like the long muscles running down your back — whereas tuna have muscles that form vertical blocks from head to tail. Whales lift and lower their tails to generate thrust. Sharks and tunas move their tails from side to side. And whales give birth to live young that cannot get their own food; instead, the young must drink milk produced by their mothers. Only some species of sharks and fishes give birth to live young, but those offspring can feed themselves; sharks and tuna do not produce milk. So despite their fishlike appearance, whales are very different from sharks and tuna.

15. Mutations may be detrimental (or even lethal); they may be harmless; or they may be beneficial in some way. Detrimental mutations should become less common over the course of generations because individuals with those mutations shouldn't do very well relative to other individuals. If a mutation is beneficial, however — one that helps an organism fight off diseases, thrive in its environment, or improve its ability to find mates, for example — that individual should produce more offspring on average than individuals without the mutation. Mutations can also become more or less common in a population due to chance, a process called genetic drift.

16. When two viral strains infect the same cell, their genetic material can become mixed as copies of their RNA are bundled into new virus particles. The new combinations of genetic material can give rise to new beneficial characteristics of the virus, for example, characteristics that permit bird flu to invade human cells. Humans who have never been exposed to the characteristics of bird flu may have no antibodies to fight the infection, and the virus can reproduce rapidly. A bird flu virus that gains the capacity to spread from human to human through reassortment may spread rapidly over large geographic areas, leading to a serious pandemic.

17. The simplest answer is that phylogenies can relate the entire evolutionary history of a species in a single image. From this image, one can infer the myriad genetic changes that had to occur for related species to branch off from one another and form new populations of species.

Chapter 2

Multiple Choice

1. b 2. a 3. a 4. b 5. d 6. a 7. c 8. a 9. b 10. c

Interpret the Data

11. c

Short Answer

12. The most valuable contribution made by Carolus Linnaeus was the idea that all of life could be organized in a single hierarchy, but he did not think that all organisms were directly related through common descent. Although he did believe that species could sometimes change or bring about new hybrid species, he thought that most species were created in their present form and did not undergo evolution.

13. James Hutton's view of a world slowly changing over vast periods of time led to William Smith's observation that layers of rock could be identified by the kinds of fossils that were found in them. Georges Cuvier studied these fossils to identify the age of the rocks found in different locations and to demonstrate that these geographically distant rock layers were deposited at the same time.

14. Stratigraphy is the study of layering in rocks (from old to young) as a way to understand the geological history of Earth. Recognizing that the planet keeps a sequential record of its own history allowed early biologists to understand that the planet—and the life that has occupied certain regions of it—has changed over time. Later (see Chapters 3 and 4), stratigraphy developed further as a way to date the fossilized remains of life tucked within the planet's strata.

15. Thomas Malthus was referring to human populations in his writings, but they showed Darwin and Wallace that the number of organisms that could be produced generation after generation would quickly surpass the amount of resources that would be needed for all of them to survive. This idea, that reproduction would outstrip availability of resources and lead to competition, provided the key for both Darwin and Wallace in understanding the differential survival of varying forms of organisms and the selection of better-performing variants.

16. Darwin noticed that breeders, by choosing which members of their flocks reproduced, could encourage the occurrence of traits they desired in the next generation of offspring. This artificial selection was analogous to Darwin's natural selection. Yet instead of a breeder's eye judging who would be allowed to reproduce, it would be the adaptations and traits of the organisms themselves that would "choose" who would survive to breed. Thus, the traits passed on wouldn't be for aesthetics of a pigeon breed but rather for the reproductive success of individuals in their environment.

17. Changes to humans within their lifetime were readily observable (such as the blacksmith example used in Box 2.1). To show that these characteristics were heritable, one piece of necessary evidence would be to show that all of a blacksmith's sons and daughters were strong simply because the blacksmith needed to be strong and not because the children performed work similar to their father's work as they grew up. Lamarck would have needed to show that the sons and daughters would be strong even if they became scholars and spent most of their time reading.

Chapter 3

Multiple Choice

1. c   2. d   3. a   4. b   5. d   6. b   7. a   8. a   9. c   10. d

Interpret the Data

11. c

Short Answer

12. Kelvin didn't know that Earth's interior was dynamic—that within it hot rock was rising, cooling, and sinking. He assumed that the planet was a rigid sphere, so his model didn't account for the greater heat flow that results from this movement. More important, scientists figured out a way to measure the absolute age of rocks using decay rates of isotopes. Radiometric dating indicates that Earth formed as part of a dust cloud around the Sun 4.568 billion years ago.

13. Fossils are rare for several reasons. Most important, not all animals' remains are left to be fossilized. Most animals are food for other organisms, so their bodies may not even be available to be fossilized. Moreover, organisms that die are left to the elements, like wind and rain, which can destroy any remaining evidence of their existence. If an organism is buried by sediments and its remains fossilize, paleontologists have to be able to access the rocks—and access them quickly. Over time the fossils themselves can again be exposed to the elements, destroying what is now pretty rare evidence. Soft tissues, like skin and organs, can be especially difficult because it takes an especially rare set of circumstances to mineralize and preserve these tissues.

14. Scientists have found fossils that actually show different behaviors, such as live birth, predation, herding, and even parental care. In addition, scientists can examine the behavior of animals today and combine that knowledge with the physical evidence from fossils to develop hypotheses about how the extinct creatures lived. They can test their predictions with additional lines of evidence. So scientists can examine muscle attachment sites in the legs of *T. rex*, the organization of the melanosomes of fossil feathers, or the structure of nasal passages of hadrosaurs and compare those data to data from living animals. As a result, they can gain insight into how dinosaurs ran, how they may have used feathers in courtship display (a behavior common in modern birds), and how they communicated.

15. Biomarkers are specific kinds of traces left by living organisms; they are molecules that were formed through biological processes. For example, okenane functions as a biomarker because it is produced only by purple sulfur bacteria. Scientists don't know of any nonbiological source of okenane. As a result, the presence of okenane in rocks means that purple sulfur bacteria must have been around producing it before those rocks formed. Radiometric dating

can indicate the age of the rocks, and scientists can gain insight into the history of life.

16. Stromatolites are ancient rocks found in some of the oldest geological formations on Earth. They bear striking microscopic similarities to large mounds built by colonies of bacteria alive today. Although these formations are rare on Earth today, they are abundant in the early fossil record. So stromatolites are not only evidence of early bacterial life, they may also be important to understanding life on Earth 3.7 billion years ago.

17. Scientists believe that plants and fungi may have been integral to each other's colonization of dry land because the oldest fossil fungi are found mingled with the early land plant fossils. Today, similar fungi live in close association with land plants and supply nutrients to their roots in exchange for the organic carbon that the plants create in photosynthesis.

Chapter 4

Multiple Choice

1. a   2. b   3. c   4. d   5. a   6. d   7. d   8. c   9. b   10. a

Interpret the Data

11. d

Short Answer

12. A clade is a common ancestor and all the descendants of that common ancestor, including any new lineages. It is a single "branch" of an evolutionary tree. A phylogenetic tree represents many different clades, and depicting a clade depends on the level of interest. A specific clade starts at a node and includes all the branches and nodes below it, and clades can be nested within clades. Similarly, a big branch of a tree can have many small branches. Either a big or a small branch can be considered a clade as long as the branch includes all of the descendants only.

13. The order of the terminal nodes does not represent any information about the pattern of branching in the phylogeny. It's easy to be fooled into interpreting the terminal nodes to mean some kind of order within the phylogeny, and the danger stems from applying causal mechanisms to explain the illusion of pattern. Like a mobile, the internal nodes rotate freely, each turning on its central axis without affecting the pattern of branching or the topology. Phylogenies do not have order at the tips—only branching order affects topology.

14. Including fossils can change the hypothesis generated by the phylogeny. Fossils can define the timing of branching events, and they can affect understanding of traits in common ancestors. The discovery of new fossils can also generate new questions about clades.

15. *Tiktaalik* definitely resolves some of the questions about the evolution of the tetrapod clade. For example, the common ancestor of tetrapods and their closest living relatives had stout, paddle-shaped fins and a neck. As a result, *Tiktaalik* could be considered a so-called missing link in the evolution of tetrapods. However, the term *missing link* is often used incorrectly to describe a direct ancestor between two species or groups, with the expectation that the missing link will explain all the traits the two have in common. Generally, scientists do *not* go looking for missing links, and they do not expect to find them thanks to the immense diversity of historic life and the extremely low probability of fossilization.

16. There are two predominant techniques that contribute to determining the timing of the branching events for phylogenetic trees. The one discussed in this chapter uses the fossil record: if you can determine that a set of fossils are related to one another, then you can extrapolate which fossils represent either the ancestral or the derived states. With this knowledge, *relative* branching times may be inferred. Finding the *absolute* time, however, would require knowing the age of the strata in which the fossils were found, but even this gives only an estimated range of branching. The second technique, discussed in Chapter 9, uses DNA and "molecular clocks" to estimate the ages of branch points. In the end, a robust analysis of timing makes use of both techniques—the fossil record and DNA.

17. No. The arrows incorrectly indicate that time is going from left to right, rather than from top to bottom. On this tree, the oldest nodes are at the top, so the branching event occurred earlier at those nodes than at nodes further down the tree. So the lizard lineage branched off earlier than the bird lineage, but that doesn't mean that within these taxonomic groupings, *all* lizards are older than *all* birds.

Chapter 5

Multiple Choice

1. c   2. a   3. b   4. b   5. b   6. c   7. c   8. d   9. d   10. a

Interpret the Data

11. b

Short Answer

12. Messenger RNA is made from DNA code in the nucleus and travels outside of the nucleus to act as a template for the construction of a protein. RNA also forms the core of the ribosome, where amino acids are built into proteins. As micro-RNAs, RNA can also act as post-transcriptional regulators of gene expression by binding to complementary sequences on specific messenger RNAs and silencing the translation of these genes.

13. Alternative splicing can combine different sets of exons to yield many alternative mRNA transcripts, which in turn will code for many different proteins.

14. It's possible that nothing would happen to the protein if the point mutation does not alter the sequence of amino acids. If the mutation does cause a different amino acid to be placed in the final protein, then there's a good chance that the protein will not fold in the same way. This could either alter the function of the protein or make it completely nonfunctional.

15. Unlike cis-acting elements, trans-acting elements are not located near the gene being affected, but rather at another location in the genome. Cis-elements can be found directly upstream or downstream of the gene or in an intron inside the gene. A hormone produced by the body in reaction to environmental stress can bind to a cis-regulatory region and alter the expression of the gene.

16. Genetic recombination results in the exchange of DNA segments between pairs of chromosomes. This results in new combinations of alleles along the length of the chromosome. Only one of the pair of chromosomes will be included in each gamete, and thanks to independent assortment, the particular copy (for example, maternal or paternal) of each chromosome included in a gamete is random. A sperm might inherit the maternal copy of one chromosome but the paternal copy of another. Mixed and matched across all of the different chromosomes, this process results in very different combinations of alleles from one sperm (or egg) to the next, which can cause sibling offspring to be genetically different from each other as well as from either parent.

17. Phenotypic traits are rarely determined by single Mendelian loci. Variation in some traits can be attributed to the cumulative action of many genes, but the environment can also influence phenotypic traits. Variation in the environment can lead to variation in the phenotypes that arise from a single genotype, as do the complex interactions between many different genes and the environment.

Chapter 6

Multiple Choice

1. a 2. b 3. c 4. a 5. d 6. c 7. d 8. c 9. d 10. d

Interpret the Data

11. b

Short Answer

12. You must calculate both the allele and genotype frequencies using the data, because you don't know whether this population meets Hardy–Weinberg assumptions. First, you have to figure out the frequency of the alleles. The frequency of the A allele will equal the total number of A alleles in the population relative to all the alleles in the population.

So you have to calculate the number of alleles present in both the homozygotes and the heterozygotes. Because 1027 individuals are heterozygous for the odor detection allele, 1975 individuals must be homozygous (3002 total that can smell the urine—1027 heterozygotes). Because homozygous individuals carry two copies of the A allele and heterozygous individuals carry one copy, the frequency of the A allele is 2 times the number of homozygotes plus 1 times the number of heterozygotes, all divided by the total number of alleles in the population for the locus (the total number of individuals times 2 because each locus has two alleles):

$$f(A) = [(1975 \times 2) + 1027]/(4737 \times 2) = 0.53$$

The frequency of the G allele will equal the total number of G alleles in the population (homozygotes and heterozygotes) divided by the total number of alleles in the population for the locus:

$$f(G) = [(1735 \times 2) + 1027]/(4737 \times 2) = 0.47$$

$$F(G) \text{ also equals } 1 - f(A)$$

Second, to determine genotype frequencies, divide each genotype by the population size:

$$f(AA) = 1975/4737 = 0.42$$
$$f(AG) = 1027/4737 = 0.22$$
$$f(GG) = 1735/4737 = 0.37$$

But if the Hardy–Weinberg assumptions are met, and no evolutionary mechanisms are operating on this locus, then the theorem predicts that the genotype frequencies should equal p^2, $2pq$, and q^2:

$$p^2 = 0.53 \times 0.53 = 0.28$$
$$2pq = 2 \times 0.53 \times 0.47 = 0.50$$
$$q^2 = 0.47 \times 0.47 = 0.22$$

The observed genotype frequencies do not equal those predicted by the Hardy–Weinberg theorem, but you can't necessarily conclude that evolutionary mechanisms operate on this locus in this population. The chi-squared (χ^2) test is a statistical test used by population geneticists to determine whether the difference between observed and expected frequencies is significant (that is, the likelihood that the difference may be due to chance). Pearson's chi-squared test is calculated as the square of the sum of each observed genotype frequency minus its expected frequency, divided by its expected frequency. The sum for all genotypes is compared to a distribution to determine significance. In this case, the frequencies are statistically significantly different from those expected.

13. Many alleles were lost in the bottleneck event due to the death of the individuals that carried those alleles. The small number of survivors had only a subset of the original genetic diversity. Because the recovering populations initially were also very small, and the mutation rate is relatively low, genetic drift had a much stronger effect, which could possibly eliminate any new alleles that arose or make it harder for genetic diversity to increase.

14. The frequency of *AS* individuals would decrease without balancing selection. Without mosquitoes there would be no transmission of malaria, so the *S* allele would lose its advantage of protecting against death from malaria. It would not likely disappear right away because as the *S* allele became rarer and rarer, its probability of occurring in a homozygous genotype (that is, paired with another *S* allele) would be low. Once in this rare state, drift alone would determine whether it persisted in the population. Selection could act on the *S* allele only when in the homozygous *SS* genotype, which causes sickle-cell anemia in the phenotype.

15. Mathematically, Δp is the frequency of the allele, p, times the average excess of fitness for that allele divided by the average fitness of the population. In other words, the change in frequency of an allele is a function of how common that allele is and how it affects fitness relative to other alleles in the population. When p is close to zero, Δp will be close to zero because multiplying any number by a number that is essentially zero yields a number that is essentially zero. On the other hand, the average excess of fitness of an allele is $[\,p \times (w_{11} - \overline{w})\,] + [\,q \times (w_{12} - \overline{w})\,]$ or $[\,p \times (w_{12} - \overline{w})\,] + [\,q \times (w_{22} - \overline{w})\,]$—depending on the allele being described. As a result, a rare allele (even if it is only slightly greater than zero) can eventually become more common if it has a higher fitness relative to other alleles, and it can become less common if it has a lower fitness relative to other alleles.

16. A smaller population size makes it more likely that two closely related individuals will mate, which also makes it more likely that their offspring will get two copies of a rare recessive trait with harmful effects. As for the genes, there will be more harmful homozygous combinations; this will result in a reduction of the fitness of these individuals and a reduction in the average fitness of the population.

17. Pleiotropy is the condition when a mutation in a single gene (like a regulatory gene) affects the expression of more than one phenotypic trait. Therefore, even if a mutation on an allele has a *positive* effect on the fitness of an organism under a specific selective pressure, there may be a *negative* effect on other genes that, overall, reduces the ability of the organism to reproduce and pass on that gene. Thus, selection for that allele would decrease, even if it has a positive effect in isolation. In essence, then, pleiotropy represents the interconnectedness of biology and emphasizes that one needs to consider the *net* effect of an allele on an organism before judging whether or not that allele (again, under specific selective conditions) positively or negatively impacts fitness.

Chapter 7

Multiple Choice

1. d   2. b   3. c   4. d   5. b   6. a   7. c   8. b   9. a   10. a

Interpret the Data

11. b

Short Answer

12. Evolution is any change in allele frequencies over time, but selection does not act on allele frequencies directly. Selection acts on phenotypes, and it can potentially influence allele frequencies if the variation in a phenotypic trait is tied to fitness, and this variation is caused in part by variation in alleles. The selection differential is a measure of how much the mean value of a trait differs between reproducing individuals and the general population. The evolutionary response to selection in a population is the effect on the frequencies of alleles themselves. In quantitative genetics, a population will, or will not, evolve depending on the additive effects of alleles and the strength of selection. So the response to selection equals the proportion of the total phenotypic variance of a trait that can be attributed to the additive effects of alleles times the difference between a mean trait value of reproducing individuals and the total population: $R = h^2 \times S$.

13. Dominance effects and epistatic interactions don't generally contribute to the phenotypic resemblance among relatives, because they are genotype dependent: the effect depends on what the alleles are paired with. That pairing can change in every generation because of meiosis. For example, an allele for a gene influencing mouse coat color (*Mc1r*) may be paired with an allele for a gene (*Agouti*) that shuts down the *Mc1r* allele's signaling capacity. The adult phenotype will have a light coat color. But the offspring of that adult may or may not have a light phenotype. If the *Agouti* allele is not passed down, *Mc1r* may produce dark or light coat colors, depending on the allele passed down to offspring. So offspring coat color depends on which alleles are paired with which after meiosis and fertilization—although this pairing contributes to variation, the effect disappears each generation. Independent assortment and recombination also break down associations between alleles at the same locus, which can cause dominance interactions to disappear. The additive effect of genes, V_A, on the other hand, leads to resemblance among relatives because additive effects influence the phenotype in the same way regardless of which other alleles they end up paired with. When alleles with additive effects are passed from parent to offspring, they cause these individuals to resemble each other.

14. Reaction norms are a way to visualize phenotypic plasticity. Specifically, they show the range of possible phenotypes that would be produced by a single genotype, were it raised in different environments. For example, if genetically identical seeds of a mustard plant were grown along a gradient of soil moistures, then the relationship between soil moisture and plant height would indicate the extent and nature of moisture-induced plasticity. If the reaction norms for many different genotypes are shown together, it's possible to visualize genetic variation for plasticity.

15. Most of the snowshoe hare populations respond to changing weather with changes in the color of their pelage from brown to white. This phenotypic response is plastic, and genotypes

vary in their responses to the environment—some turn white earlier than others, and some no longer turn white at all. Warming caused by climate change will lead to a new selective environment—snow will likely come later and melt sooner. If snowshoe hare fitness is measured in terms of survival, genotypes that turn color too quickly will be selected against—they will produce fewer offspring relative to genotypes that turn later in the season, because their pelage will not match the snowless environment and they should experience greater predation rates. If there is sufficient genetic variation in hare populations, then the timing of the phenotypically plastic response should evolve (shift) in response to selection for later fall (and earlier spring) switching (for example, by favoring genotypes that have slightly different critical sensitivities to seasonal photoperiod or temperature).

16. QTL mapping studies may be considered difficult because of the sheer work involved in competently performing one. Recall that a researcher is looking for a genetic marker for a trait, which requires first having organisms that are purebred to be homozygous at multiple loci for the trait in question. Once two populations of organisms have measurably distinct and heritable traits for the phenotype in question, the researcher has to cross them. After generating hundreds of F_2 organisms, the researcher needs to measure the traits, and then sequence certain loci to identify which genes were inherited from what parent (by first establishing which DNA markers are statistically significant and not just error). When you consider the generation times of some organisms, the labor adds up. Fortunately, DNA sequencing is becoming faster and cheaper, and with the advent of GWA techniques making sense of genetic differences at the genome scale, processing the data is becoming more efficient as well. Nevertheless, such experiments are called "heroic" for a reason.

17. Directional selection occurs when selective pressures favor heritable phenotypes at one end of a distribution of values for a trait. Specifically, the mean characteristic for the surviving/breeding parents differs from the mean of the population prior to selection. The chapter uses the example of fishes in a lake where a drought reduces the amount of food for a population. Small fishes, needing less nutrition to maintain their body mass (a heritable trait), thrive, whereas every fish equal to and greater than the mean dies. Thus, in the next generation, the small fishes, having been able to pass on the genes for their particular size, will shift the mean size of the newborn fishes in the lake to the small extreme.

Chapter 8

Multiple Choice

1. b   2. d   3. d   4. a   5. b   6. d   7. a   8. d   9. b   10. a

Interpret the Data

11. b

Short Answer

12. Different methodologies are important for building phylogenies because scientists are working with information about the largely unknown history of organisms. But the data that they do have are quite valuable, providing evidence that supports or refutes hypotheses about relationships. Developing molecular phylogenies is complicated by the structure of DNA. Unlike other character-state changes (for example, the number of bones in the middle ear or height of an individual), the bases in a segment of DNA can be in only one of four states. As a result, the probability of molecular homoplasy (separate lineages arriving independently at the same character state) can be high, and reversals can lead to false signals. In addition, the differences in rates of change in coding versus noncoding regions, and synonymous versus nonsynonymous substitutions, provide a wealth of information that can be incorporated into some models.

Whether the data are based on physiological character states or molecular character states, some methodologies help scientists build trees (maximum parsimony, distance-matrix methods), whereas other statistical tools help them decide how well hypothetical trees fit the data (bootstrapping, maximum likelihood, Bayesian). Parsimony relies on the principle that the simplest solution is the most reasonable way to account for similarities; the tree with the fewest changes (in total amount or number of steps) is considered the most parsimonious. But a group of trees may be equally parsimonious. Distance-matrix methods, like neighbor joining, do not assume that all lineages evolve at the same rate, and they are fairly easy to compute even with large data sets. Bootstrapping can be used to test how well the data actually support a particular tree. Maximum likelihood and Bayesian methods use models that can incorporate different substitution rates and probabilities of substitution, for example, then find the tree that is most likely. In any case, all of these methods generate hypotheses about the historical relationships among taxa that can be tested as additional evidence becomes available.

13. The theory of neutral evolution addresses variation in nucleotide sequences and makes two predictions: that this variation is largely due to differences that are selectively "neutral" and that most evolutionary change is the result of genetic drift acting on neutral alleles. Variation can be selectively neutral in several ways. Noncoding DNA, such as pseudogenes, is not likely to affect phenotypes, so mutations in those portions of the genome should not experience natural selection. Mutations to protein-coding genes may or may not affect the phenotype—the amino acid coding structure contains a lot of flexibility. For example, sequences of three nucleotides (codons) may differ and yet encode the same amino acid (CUU, CUC, CUA, CUG, UUA, and UUG all encode for leucine). So, potentially, many single-nucleotide mutations are synonymous substitutions that are not expressed.

Neutral theory also predicts that changes in the frequency of these neutral alleles will result from drift.

Because genetic drift occurs when a random, nonrepresentative sample from a population produces the next generation, the frequencies of neutral alleles can change radically, being driven to fixation or extinction. This prediction leads to testable hypotheses about natural selection because scientists can compare the substitutions that occur in replacement sites to the substitutions that occur in synonymous sites.

The theory of evolution by natural selection requires that some aspect of the phenotype be affected by mutation—a single-nucleotide mutation that is a replacement substitution and affects the behavior of a protein, for example—leading to heritable variation among individuals. In this case, changes in allele frequency occur because different alleles have different fitnesses; therefore, some alleles leave more copies of their kind to subsequent generations than others.

Like natural selection, neutral evolution in sexually reproducing organisms requires mutations to affect germlines in order to be heritable (in plants, somatic mutations may be transferred to offspring). Both the theories of neutral evolution and evolution by natural selection explain variation among individuals, and both predict that allele frequencies will change over time.

14. Humans and dogs might use the same genes to locally adapt to high-elevation populations because the genes that control the uptake of oxygen in the blood are likely to be highly conserved for most animals, especially, like in this case, if they are of the same taxonomic class (Mammalia). Therefore, under the same selective pressure (low partial pressure of oxygen at high altitude), it follows that if two different animals have homologous genes, then that is what natural selection would act upon. Barring temporary physiological responses to high altitude that might be different between humans and dogs, there's only so much that can change if a transmittable, genetic adaptation is needed.

15. Sarah Tishkoff and her colleagues examined more than 1300 genetic loci for patterns of variation. Based on the quantity of that variation, they used a neighbor-joining distance matrix to develop a phylogenetic tree. They found that all non–Africans form a monophyletic group, indicating that non–Africans share more derived characters with each other than with Africans. They also found that genetic variation was greater in Africans than in these non–African groups. This pattern is typical of a founder event (a type of genetic drift that accompanies the founding of a new population from a very small number of individuals). These results support the hypothesis that non–Africans are descendants of a common ancestor shared with Africans and that the ancestor likely migrated out of Africa.

16. Conserved sequences are essentially ancestral "character states." Unlike other regions of the genome, which steadily accumulate substitutions over time, conserved regions remain relatively unchanged across multiple speciation events. In fact, the deeper a conserved sequence is found within a phylogeny, the more highly conserved it is considered. The similarity of conserved sequences across taxa generally indicates some kind of important functional value that was maintained over time through natural selection. Mutations arise in these genome regions, but new alleles are purged from the population by purifying selection, causing these sequences to remain largely unchanged across vast expanses of time. Beneficial mutations do occasionally arise, and these can spread, but the evolution of these regions is very slow compared with other regions of the genome. Both protein-coding and noncoding (for example, regulatory) segments can be conserved. For example, physiologically important noncoding elements have been identified, in part because of their conserved nature.

Phylogenies are constructed using shared derived characters, and conserved sequences can help identify the patterns of ancestry. They are especially useful for resolving deeper branches in these trees. Rapidly evolving genes often have accumulated so many changes that comparing them across distantly related taxa is uninformative—the sequences may be so divergent that they are difficult to align in the first place, and, even if they are aligned, so many substitutions have occurred that many base pairs have reverted to earlier states, masking important evolutionary changes, and others have independently converged on the same state. Rapidly evolving sequences thus have the problem of too much "noise" masking the actual signal of ancestry.

Conserved genes, on the other hand, evolve more slowly. These sequences are still recognizable even across distant taxa, making them easier to align and compare, and the few changes that have occurred are more likely to be informative (for example, substitutions occur sufficiently rarely that the chances of back-mutations at any site are extremely low). Interestingly, the same features that make conserved genome regions useful for resolving deep nodes in a tree—relationships among distantly related taxa—also make them largely uninformative for resolving relationships closer to the tips of a tree. Closely related species simply have not been separated from each other long enough to have accumulated very many changes. Other genome regions that evolve more rapidly are the best choices for resolving closely related groups or species.

17. A species tree represents the divergence of populations into reproductively isolated species. Once a branch splits (represented by a node on the tree) the daughter lineages begin evolving independently, steadily accumulating genetic differences as new mutations arise and eventually become fixed within each lineage. Each of these branches may split again, all the while accumulating new genetic variation. Most of the time, these accumulated genetic differences provide a signal biologists can use to infer the topology of the species tree, because species that are closely related should be more similar to each other then they are to more distantly related taxa. Occasionally, however, the history inferred from a particular gene may differ from the rest of the signal, yielding a gene tree that is incongruent with the overall species tree. If multiple alternative alleles for a gene persist for a very long time (as occurs under some forms of

balancing selection, for example), or if speciation happens very fast, then the same alternative alleles may be carried through several speciation events. When, eventually, some of these alleles are lost to drift, they may be lost in a pattern that does not reflect the branching of the species (called incomplete lineage sorting). Gene trees may also be complicated by introgression with separate species—for example, finding Denisovan DNA in a human being.

Chapter 9

Multiple Choice

1. b 2. d 3. a 4. d 5. a 6. c 7. d 8. a 9. b 10. d

Interpret the Data

11. a

Short Answer

12. In bacteria and archaea, gene transfer between species is especially common. This phenomenon can act like a gene duplication event. One of the two genes would then be under less selective pressure and could evolve a different function, or it could be recruited into an existing gene network, where it may give rise to a new adaptation.

13. Gene duplication involves the copying of an existing gene, which may allow one of the copies to take on a new role. Gene recruitment occurs when a gene or gene network is transformed through mutation to perform a totally different function. Both processes can involve genes switching roles, but they differ in the source of the gene and how similar or different the final role will be from the original function. Gene duplication events often facilitate subsequent gene recruitment because the duplicated genes are no longer constrained by selection to perform their original function.

14. *Defensin* genes, used to produce proteins that fight pathogens, were originally expressed in the pancreas of snakes and their ancestors. These genes began to be expressed in the mouth of the snakes as the result of a regulation mutation. Further mutations to the protein product allowed it to break down the muscles of the snake's prey. And finally, additional mutations combined with natural selection increased its toxicity to become potent venom.

15. Both mice and flies have bilateral body symmetry, and bilateral symmetry is an ancestral trait. The basic genetic toolkit for bilateral symmetry was inherited from the ancestor of all bilaterians. The presence or absence of the gene products of *Engrailed* and *Hedgehog* initiate a cascade of genetic signals that map out the major axes of the developing limb.

16. Opsins are proteins that allow cells to detect light and transmit this information to other cells. These molecules were present in the ancestor to both cnidarians and bilaterians. Both groups inherited opsins and used them to detect the presence of light and eventually form images. Transparent crystallins evolved—recruited from stable heat-shock proteins—that could focus light. Ultimately, natural selection favored mutations that improved the function of the opsins and the crystallins for vertebrate eyesight.

17. Two traits that are homologous between sharks and dolphins are their eyes and jaws. They are homologous because both traits had already evolved in the common ancestor of the two groups—evidence that can be seen in their fossil forms. Further evidence includes the common genes that are involved in development of these traits as well as the same molecules used for vision in the eye.

One convergent feature is the streamlined shape of the bodies of both sharks and dolphins. This shape was not inherited from a common ancestor. The ancestors of dolphins were tetrapods, and dolphins begin to develop hindlimb buds as embryos, just like their legged ancestors. The limb buds stop growing during development, however, contributing to the streamlined (hindleg-free) shape of the dolphin body. Sharks lack the genes necessary to produce hindlimbs in the first place. A second convergence is the evolution of fins and flippers. The flat, paddle-like shape of fins and flippers evolved independently in dolphins and were features their direct terrestrial ancestors did not have.

Chapter 10

Multiple Choice

1. c 2. a 3. c 4. d 5. d 6. b 7. a 8. a 9. d 10. c

Interpret the Data

11. b

Short Answer

12. Peter and Rosemary Grant's team needed to measure the beak size of many individuals in multiple generations to determine how this trait is inherited and how it changes from one generation to the next. They also had to measure the types and sizes of seeds that were available, and how well beaks of different sizes performed when cracking these seeds, to understand the selection pressures on beak size.

13. Both directional and stabilizing selection can cause populations to evolve; both require that there be variation in the phenotypes of individuals, and both require that this variation in trait values influences survival and/or reproductive success. They differ because directional selection favors increases, or decreases, in the size or dimensions of a trait, whereas stabilizing selection favors an intermediate value for the trait. Directional selection will always change the average value of a trait in a population (provided there is heritable variation), whereas stabilizing selection will keep it the same.

14. If all individuals had identical coat colors, then there would be no opportunity for selection, and variation in the probability of survival or reproduction would not be explained by this trait. If, on the other hand, individuals differed markedly in the darkness of their coat color, then the raw material for selection would be present. If individuals with lighter (or darker) fur survived better than individuals with other coat colors, then selection could favor lighter (or darker) coat colors, respectively. However, whether or not the population evolved in response to this selection would depend on the extent to which offspring resembled their parents for this trait. That is, at least some of the variation in coat color would have to be heritable.

15. The answer depends on what environment is being considered. Under natural grassland conditions with diverse communities of plant competitors, maize would be at a disadvantage thanks to the domesticated traits that it possesses. In a domestic environment, maize is at an advantage because of human-based artificial selection. Having fewer branches and many large starchy kernels that stay on the cob are adaptive traits only in a human-controlled environment.

16. Herbicides and pesticides can impose extremely strong selection, and weed and pest populations generally contain vast numbers of individuals harboring large amounts of genetic variation. The speed of any evolutionary response to selection will be the product of the strength of selection times the amount of heritable genetic variation, so application of herbicides and pesticides is expected to drive extremely rapid evolution in target populations. Scientists advise farmers to create "refuges" where pests experience no selection (because no pesticide is applied). These populations remain susceptible, and gene flow between these refuges and other fields can slow the rate of evolution of resistance overall.

17. Snake populations had to contain heritable variation for gape width, and this variation had to affect survival or reproduction. In this case, snakes with a wider gape were able to eat the largest cane toads, and they suffered increased mortality as a result. Snakes with smaller gape widths could only eat small cane toads, which were less toxic. As a result, snakes with small gape widths were more likely to survive and reproduce, passing alleles for small gape width to their offspring. Across several successive generations, the result was an evolutionary reduction in snake gape width.

Chapter 11

Multiple Choice

1. a 2. b 3. d 4. b 5. a 6. b 7. b 8. d 9. a 10. b

Interpret the Data

11. d

12. Sexual reproduction is so common because of the genetic diversity advantage it offers. Sexual reproduction allows the combination of genomes, bringing together beneficial adaptations much faster than if they arose separately. Recombination during meiosis also creates new genotypes with various combinations of alleles, and recombination can lead to the exclusion of deleterious mutations from some genotypes. The genetic diversity created by sexual reproduction can lead to rapid evolution because selection has so much variation to act upon. For example, even though diseases and parasites can be transmitted through sexual contact, the enhanced genetic diversity that results from sexual reproduction provides a strong foundation for selection to weed out those more susceptible from those less susceptible.

13. Natural selection is a mechanism that can lead to adaptive evolution when differences in the phenotypes of individuals cause some to survive and reproduce at greater rates than others. Natural selection includes all aspects of organismal performance, ranging from successful establishment of seeds or eggs, successful development, and competition at any stage of the life cycle to the abilities to thrive under particular physical surroundings, forage, avoid predation, and find safe locations to hide or reproduce. Sexual selection is related specifically to differences in the *reproductive* success of individuals, and it arises whenever individuals of one sex must compete for access to the other sex. Sexual selection results from the competition for fertilization. That competition can occur among individuals of the same sex (intrasexual selection, for example, male competition) or through attraction to the opposite sex (intersexual selection, for example, female choice).

Owing to the twofold cost of reproduction, the explanation for the evolution of sex has not been completely resolved. Clearly, though, the combining and mixing of chromosomes in sexual reproduction generate more genetic variation than asexual reproduction, leading to tremendous capacity for populations to respond to fluctuating environments. As soon as sex evolves, however, differences begin to arise between the two sexes (for example, anisogamy evolves as eggs and sperm specialize for quality/size or number, respectively). Once this occurs, natural selection should act differently on the two sexes—males and females begin to maximize their reproductive success in different ways, and, often, one of them becomes relatively limiting compared with the other. The stage is set for competition. Thus, sexual selection is the ultimate outcome of the evolution of sex.

14. Sexual cannibalism is likely to be most beneficial to females when the number of males does not limit female reproductive success. This is more likely to be the case in polygynous mating systems, where an excess of males compete for access to a limited number of females. Sexual cannibalism

is only rarely advantageous to the males. However, the costs of self-sacrifice can be offset to some extent if the offspring that benefit are the sacrificial males. Polyandrous mating systems would undermine these benefits to the males because the certainty of paternity of males is lower when females mate with multiple males. Extreme sacrifice is likely to be cost-effective to the male only in species that incorporate elaborate behaviors ensuring a high probability of paternity for that male.

15. She would benefit if the brightest males also were consistently the best-quality males. This could occur if the pigments necessary for bright red feathers are difficult to acquire because males would have to compete with each other for access to them or because only the best-quality parent birds would be able to feed their sons sufficient amounts of the nutrients that serve as sources of pigment. If bright colors attract predators, then only the fittest males may be able to survive despite the handicap of being bright. For this reason, too, a bright red male may be of unusually high quality. Females may benefit by picking bright males as mates because these males are likely to sire the brightest male offspring. Her sons therefore would also be especially likely to mate.

16. Polyandry occurs when females mate with more than a single male. By mating with multiple males, a female increases her chances of producing offspring that combine her genes with the highest-quality male genes (more males equals more possible combinations of alleles, some of which will be more beneficial than others). She also increases the chances of finding a male that is not too genetically similar to her, thus giving her offspring the ability to resist a wider range of pathogens. Females may also be able to "upgrade" the quality of the sperm they use to fertilize their eggs, by soliciting matings from males other than her mate, if these males are especially attractive or otherwise higher quality than her mate.

17. The longer a female waits after mating before she lays her eggs, the greater the opportunity for another male to intercede and displace sperm from the original male. Chemicals that alter a female's receptivity or induce her to lay her eggs sooner will minimize opportunities for rival males to mate, increasing the likelihood of fertilization for the male that transfers them. These chemicals may hurt the female, shortening her life span overall or diminishing her potential for future reproductive events. However, because the male injecting the chemicals is not likely to sire later clutches of eggs by that female, these costs do not apply to him. For him, the benefits of seminal fluid chemicals outweigh the costs, favoring their evolution.

Chapter 12

Multiple Choice

1. c   2. a   3. c   4. c   5. d   6. d   7. d   8. d   9. a   10. c

Interpret the Data

11. In the chart comparing the number of offspring sired (top two), male elephant seals have a greater variance in reproductive success compared to females. In the chart comparing the number of mates (bottom), on the other hand, female pipefish had a greater variety in reproductive success (at least concerning number of mates). In both cases, the sex with the lower variance in reproductive success (females in elephant seals, males in pipefish) is investing relatively more in each reproductive event. As a result, they require longer to recycle between successive reproductive events/broods, and thus become limiting. Members of the opposite sex are forced to compete for access to them. The resulting competition—sexual selection—can drive the evolution of traits that aid in the battle for access to mates, including costly and showy ornaments or displays, and/or costly weapons. Intense competition creates opportunities for a small number of individuals to monopolize access to a disproportionate number of the limiting sex, creating huge disparities in reproductive success between the winners and losers. For this reason, the variance in reproductive success will be larger in the sex that is forced to compete, and the greater the disparity between winners and losers (that is, the greater the variance), the stronger will be the resulting selection.

Short Answer

12. Individuals should be maximizing their reproductive effort, but investment in reproduction often involves trade-offs between reproduction and growth/body maintenance or trade-offs between the ability to breed early or later in life. Although individuals experience the trade-offs associated with the schedule and duration of key events in their lives, those trade-offs are subject to natural selection only if (1) there is variability among individuals, (2) that variability is heritable, and (3) that variability confers some advantage or disadvantage to the individuals in terms of reproduction or survival. Some mutations may affect an individual's sensitivity to external conditions, making reproduction more or less likely. Others may be more or less sensitive to hormones that influence behaviors, such as the release of ova in females. Still other mutations may lead to higher or lower levels of egg dumping, for example, or egg consumption. Selection should favor the optimal trade-off that maximizes the number of offspring that survive to maturity over the entire course of an organism's life.

13. Gulf pipefish are a sex-role-reversed species; males provide most of the parental care, and the females compete for access to males. Males prefer to mate with large females over small females, and this mate choice affects the reproductive success of both big and small females. Big females end up transferring more eggs to males than small females, and males devote more resources to the eggs of big females. As a result, big females gain more access to males and have

higher brood success than small females. Males are not limited by the number of females they can mate with, but by the space they have available for brooding eggs. They may use a flexible strategy to raise offspring from multiple broods. If a male mates with a small female, he may be able to reduce the amount of his current investment in favor of future reproductive efforts.

14. Female Seychelles warblers tend to stay near their natal site, and males tend to leave. The environmental conditions females experience around their nest site may vary from season to season, and research has already shown that when environmental conditions are good, females produce more daughters. So environmental conditions may act as cues for the optimum ratio of sons and daughters to produce in the current breeding attempt. Drought, for example, may affect the amount of foods available to reproductively active females. If females have to spend more time searching for food, they may produce fewer maternal hormones.

Maternal hormones may be instrumental in determining sex ratios before or after the eggs are fertilized. For example, female birds are the heterogametic sex, and the number of oocytes a female produces with the W chromosome (the smaller chromosome that determines male gender) may be influenced by the amount of circulating hormones. Maternal hormones may affect resources given to the ova before their release, so that ova with the small W chromosomes may not be provisioned as well as ova with Z chromosomes and could be less likely to be successfully fertilized. Similarly, after fertilization, the quantity of maternal hormones could affect egg formation, reducing the likelihood of successful hatching of embryos with both chromosomes.

One possible test to determine whether females were adjusting the sex ratio before fertilization could be to restrict food to females and determine whether the ova were biased toward one sex or the other. The hypothesis could be as follows: if females were adjusting the sex ratio before fertilization, then the proportion of ova that are male should be high; if females were adjusting the sex ratio after fertilization, then the proportion of ova that are male should be 1:1. Natural selection will act to hone the females' sensitivity to environmental conditions as long as the sensitivity varies among individuals and is heritable.

15. During sex between diploid individuals, one parent contributes one set of alleles to the offspring and the other parent contributes the other. Although independent assortment and genetic recombination can mix how these alleles are passed down to different individuals, the parent that contributed those alleles can still have an effect. Methylation can silence the alleles contributed by either the mother or the father. At least in mules and hinnies, these imprinted genes may have the greatest influence on the offspring in the placenta of the mother. So, even though scientists are still trying to understand the exact details of this genetic phenomenon, breeders know that a male donkey mated with a female horse will produce a mule, but a female donkey mated with a male horse will produce a hinny.

16. Senescence is the deterioration in the biological functions of an organism that occurs as it ages—it is a by-product of natural selection for alleles that improve reproduction and growth earlier in life, and it materializes later in life. Theoretically, organisms must expend energy maintaining and repairing the cells in their bodies. Natural selection should favor only the levels of self-repair that keep an organism in sound condition during its reproductive period.

Scientists are beginning to examine some of these effects specifically. For example, p53 helps control cells from developing into a cancer, but only when an animal is young. The protein also causes damage that accumulates over time, affecting individuals as they age. Stress-fighting genes in nematode worms help worms earlier in their lives, perhaps as they cope with stressful environmental conditions; worms with mutations that keep these stress-fighting genes turned on have longer life spans than other worms. The mutations have a cost, however: lower fertility, which is enough of a cost for natural selection to favor worms without the mutations. So worms with shorter life spans produced more offspring than long-lived worms, and eventually the alleles that kept the stress-fighting genes turned on disappeared from an experimental population.

17. No, the statement does not accurately reflect the current understanding of evolutionary biologists. Selection would favor the optimal trade-off that maximized reproductive fitness, but not because of an individual's need. The individual female is making a "choice," but that choice is not governed by a sense of need or an understanding of future breeding opportunities. Different species, even different populations, will respond differently to the selection pressures that shape their environment. Much like behavioral plasticity, natural selection can shape mechanisms that affect how individuals allocate resources to reproduction or growth/survival in response to conditions. Heritable mechanisms that affect the lifetime fitness of individuals will have more or less presence in future generations based on their success.

Chapter 13

Multiple Choice

1. b 2. d 3. b 4. d 5. d 6. a 7. a 8. a 9. c 10. b

Interpret the Data

11. a

Short Answer

12. No. Geographic isolating barriers can interrupt gene flow between populations of individuals. These populations may then evolve independently due to drift, mutation, and natural selection. According to the phylogenetic species concept, as long as the populations have evolved independently, so that there are recognizable sets of individuals with distinct patterns of ancestry and descent, then the populations

can be considered species. The biological species concept, however, would consider the two populations as isolated, though they could still potentially interbreed. Geographic isolation may or may not lead to the evolution of reproductive isolation. Because the general lineage species concept defines species based on the frequency of allele exchange, geographic isolation alone may be enough to consider the populations to be different species, if the barrier restricts exchange and drift affects the composition of alleles.

13. Bacteria and archaea reproduce asexually, dividing into genetically identical copies. So reproductive barriers used in the biological species concept are meaningless for them. Genetic and ecological differences might be more useful in classifying bacteria and archaea.

14. The similarities occur because both processes lead to the splitting of one species into two. Both processes involve genetic diversification and reproductive isolating mechanisms. The differences are in where these events happen and how those circumstances affect the conditions of speciation. Allopatric speciation happens in geographically separate areas, so potentially reproducing individuals are already isolated. Sympatric speciation happens in the same geographic location where individuals can come in contact with each other, so some other mechanism (for example, divergent ecological selection) functions to isolate the populations.

15. Sympatric speciation is most likely when individuals from the same species and same location become separated by preferences for habitat or resources, keeping them from coming in contact with each other, especially during mating. The speciation of *Rhagoletis* flies is considered sympatric because the hawthorn trees and apple trees exist in the same field (they are not geographically isolated).

16. Because the biological species concept identifies species as groups of actually or potentially interbreeding populations, it may not universally apply to plants. Clearly, reproductive isolating barriers break down or do not exist among many easily identifiable species (to humans). Several known cases of speciation have resulted from hybridization and allopolyploidy. The phylogenetic species concept may be more appropriate when studying plant speciation, especially speciation due to allopolyploidy, because groups descended from the common ancestor whose members all possess the defining or derived characteristics can be easily distinguished from other groups. The general lineage species concept is also appropriate, but it doesn't include the history of the lineages.

17. Humans have the tendency to categorize, and we use human senses to make decisions about including and excluding members in those categories. Before the technological development of genetic tools, humans used visual, physical clues to determine membership because we could see those differences. As our understanding advanced, so did our acceptance that we cannot personally experience all the factors that affect all other organisms. With the development of DNA barcoding, for example, scientists have discovered that what looks indistinguishable to humans does not necessarily apply to the organisms themselves. Although scientists clearly recognize that a species is some biological unit, they also recognize that not having clearly defined boundaries is an artifact of our need to classify. The inherent variation within these taxonomic groups that makes them so difficult to classify is exactly the raw material of evolution that has led to such grand diversity of life.

Chapter 14

Multiple Choice

1. a 2. b 3. d 4. d 5. a 6. c 7. b 8. d 9. d 10. a

Interpret the Data

11. b

Short Answer

12. Microevolution and macroevolution are different emphases along the continuum that is the theory of evolution. Microevolution is the study of small-scale changes, such as changes in allele frequencies that are affected by selection and drift and their effects on the phenotypes within a population or species. Macroevolution is the study of the large-scale changes that are outcomes of these small-scale processes within the landscape over long periods of evolutionary time. It refers to any evolutionary change at or above the level of species.

Microevolution and macroevolution have both been observed, but the two emphases focus on different lines of evidence. Evolutionary biologists are homing in on the microevolutionary processes that affect phenotypes as new technologies for studying alleles and developmental pathways, for example, become available. Mutation and natural selection have been observed in experimentally controlled populations of *Drosophila*, in wild populations from weeds such as Palmer amaranth, and in oldfield mice. Advances in the study of biogeography and paleontology, on the other hand, have been instrumental to the understanding of macroevolutionary processes. Speciation has been observed in apple maggot flies, *Rhagoletis pomonella*, for example, and incipient stages of speciation are being studied as they unfold in numerous species. In addition, the plethora of fossil evidence that is constantly being discovered, including a diversity of transitional forms, is providing evidence for the originations and extinctions of species of the historic past.

The evidence for the theory of evolution comes from myriad sources, and all the various lines of evidence serve to support that theory. Scientists may disagree on specific aspects of both microevolution (for example, whether menopause is adaptive) and macroevolution (for example, how to define species) and on how microevolution and macroevolution interact, but neither microevolution nor macroevolution has more or less evidentiary support.

13. Macroevolutionary biologists used three very different lines of evidence to determine the origin of marsupials. The first line of evidence came from an understanding of micro-evolutionary processes. Scientists started by developing a molecular phylogeny of the living marsupials, using mobile genetic elements called retroposons. They found that marsupials form a monophyletic clade, indicating that all living marsupials are more closely related to each other than they are to other mammals — marsupials share a single common ancestor. This marsupial clade is also completely nested within a clade of mammals found only in South America, so the common ancestor of marsupials and their nearest relatives likely originated there.

The second line of evidence came from the fossil record. Fossil marsupials are known from Asia, Europe, North America, South America, and Antarctica — not just Australia. Scientists were able to construct a phylogeny of the major marsupial lineages by using morphological traits of these fossil marsupials. Just like the molecular phylogeny, Australian marsupials formed a monophyletic clade. But unlike the molecular phylogeny, the phylogeny based on fossils indicated that marsupials from Antarctica were more closely related to marsupials from Australia than to marsupials from South America. Fossils from North American marsupials were the most distantly related. The ages of fossils indicate that the animals originated in Asia, dispersed to North America, and then spread to Europe (and even northern Africa) and to South America, Antarctica, and Australia.

The third line of evidence came from geologists and the reconstruction of Earth's continents through plate tectonics. More than 100 million years ago, Asia and North America were linked, as were South America, Antarctica, and Australia. Dispersal from Asia to North America and then to South America, Antarctica, and Australia would have been relatively unimpeded. These continents started to drift apart tens of millions of years ago, which would have isolated populations of marsupials in Antarctica and Australia.

These lines of evidence are distinctly different — molecular biology, paleontology, and geology — but together they provide evidence to support the phylogenetic hypothesis for the origination, diversification, and extinction of marsupials.

14. Abiotic factors are nonliving, chemical, or physical causes, whereas biotic factors are biological. An asteroid that strikes Earth and affects the particulate matter in the atmosphere that blocks out sunlight is an abiotic factor (and extraterrestrial). The collapse of a food web caused by the blocked sunlight is a biotic factor. Collapsing food webs can cause mass extinctions as organisms dependent on each other for sustenance disappear. Climate change is also an abiotic factor that has been linked to species diversity in the fossil record, but both abiotic and biotic factors can contribute to climate change. Volcanic eruptions can emit carbon dioxide, trapping heat near Earth's surface and affecting climate, but photosynthesizing plants can remove carbon dioxide. Over

the past 300 million years, extinction rates increased (leading to low diversity) when Earth was warm and decreased (and led to higher diversity) when Earth cooled.

15. Yes. Although a single extinction may not significantly affect an ecosystem, mass extinctions can have cascading effects that may ultimately affect organisms that humans rely on directly. The indirect costs may be higher than we can even imagine. Even though humans may be able to control some of the organisms necessary for our survival through farming and ranching, if ecosystems begin collapsing at a similar pace to extinctions in Earth's past, humans will definitely be affected. That effect is likely to include drastic political upheaval and sudden human population declines, though when, where, and how, precisely, this would occur are difficult to predict. However, the decision to contribute to the pace of extinctions is under our control.

16. Macroevolution looks at the long-scale, geological-time events of speciation (that is, origination, diversification, and extinction). Fossils, which form over geological time, can therefore offer snapshots of species' prevalent morphologies, their geographic range/global distribution, the biotic and/or abiotic components of their environment, and even some information on the climate of their existence. This is all because fossils don't occur in isolation. For geographic range, paleontologists and geologists can determine where the surrounding strata would have been at the time of the fossil's burial as well as how old that fossil is based on the age of the surrounding rock. For biotic and abiotic components, fossil beds and Lagerstätten can show what other organisms may have shared the environment with the fossil; as for abiotic conditions, certain clues — like animals suddenly forming shells — can indicate high calcium levels in the primordial seas. As for climate data, very skilled scientists can extrapolate global conditions from the relative amounts of biodiversity found in the fossil record. Morphology is self-explanatory. So, even if fossils offer a relatively limited view of the rise and fall of species, there is still a great deal of information that can be had from them.

17. There is no single mechanism that explains all of the mass extinctions — but, like the question implies, there are two common culprits. The primary would be Large Igneous Provinces — large-scale (up to millions of cubic kilometers in volume), thousand- or million-year-long volcanic eruptions that coincided with four of the five mass extinctions. The resulting climate change that these volcanoes created, as geological evidence shows, would have been more than sufficient to wreak havoc on living things (for example, belching toxic gases and metals from the earth, destroying the ozone layer, releasing CO_2 — acidifying oceans — and, of course, acid rain). The other famous culprit for at least one mass extinction comes in the form of the 10-mile-wide asteroid that impacted off the coast of Mexico about 66 million years ago. This would have created the chronic events of a Large Igneous Province in a relatively rapid amount of time. Nevertheless, and regardless of what causes

these major extinction events, history is clear in suggesting that changes across the global ecosystem must happen for an event on the scale of a mass extinction.

Chapter 15

Multiple Choice

1. b 2. c 3. d 4. a 5. d 6. d 7. a 8. a 9. c 10. d

Interpret the Data

11. d

Short Answer

12. The three components of geographic mosaic theory include the geographic variation in (a) the distribution of traits that evolve in response to selection, (b) the types of selection on the interacting species, and (c) the strength of the reciprocal selection. Long-tongued flies and *Zaluzianskya* flowers have a mutualistic relationship; as the flies drink nectar from the flowers, they pick up pollen on their heads and transfer it to other flowers as they continue to feed. The flies' tongues and the flower tube depth match at different sites. They are long where long-tubed *Zaluzianskya* flowers are common and short where short-tubed *Zaluzianskya* flowers are common, so there's tremendous variation in the tongue length of long-tongued flies and in the depth of *Zaluzianskya* flower tubes. The type of selection and the strength of reciprocal selection depend on whether *Zaluzianskya* flowers are the dominant source of nectar for the flies, however. Where *Zaluzianskya* flowers are common, directional selection coupled with strong reciprocal selection leads to long tongues and long tubes. But where other flowers are common, competition for pollinators dilutes the strength of reciprocal selection, and directional selection favors shorter tubes and shorter tongues. A geographic mosaic results, where strong mutualistic relationships are common in some locations and not in others.

13. Reciprocal coevolution is important to the development of evolutionary arms races because if the interacting organisms did not act as agents of selection on each other, then neither would be affected by changes in the other. Individuals would not experience better or worse reproduction or survival due to changes in individuals of other populations, and the frequency of the alleles they carried would not be affected as a result. Evolutionary arms races not only require conflict, but the changes in one of the interacting populations must affect the fitness of the other. Once populations respond to the selection exerted by the other, however, if the agents of selection themselves don't evolve, the interaction will stall. Evolutionary arms races result because the agents of selection did evolve; each population acted on the other, and each population responded to the selective pressure.

14. Batesian mimicry is a coevolutionary relationship where one species (the mimic) benefits from appearing distasteful or harmful (like the model) when in fact it is not. If the mimics become abundant, predators that previously avoided the model may suddenly switch back to the model once they discover that mimics are actually palatable. At this point, the model would evolve away from the patterning of the mimic, and the two species would begin a coevolutionary game of chase. If, however, the mimics don't get too widespread, enough of the models would still be around to enforce the avoidance of predation and all would remain copacetic between predator, model, and mimic.

15. Diversifying coevolution is the increase in genetic diversity caused by the heterogeneity of coevolutionary processes across the range of ecological partners. Speciation is one predicted outcome of geographic mosaic theory. When local populations differ in the details of their coevolutionary partnerships or in the strength or direction of selection resulting from those coevolutionary interactions, they may diverge in ways that reduce the likelihood of exchange of gametes between them. These reproductive barriers to gene flow will further reduce the likelihood of interbreeding between individuals of different populations. Depending on the strength and types of selection, one evolutionary lineage may become two or more lineages, representing a speciation event. Coevolution, because it can drive unusually rapid evolution, may dramatically accelerate how quickly such divergence arises and thus may accelerate the rate of speciation.

16. In the 1960s, Lynn Margulis revived the hypothesis that mitochondria were bacteria that started as endosymbionts in early, free-living cells. Indeed, once scientists were able to test this hypothesis in the 1970s (by sequencing mitochondrial DNA), they discovered it resembled bacteria more than any other domain of life. What we hypothesize now—new discoveries are still being made—is that mitochondria are related to a clade containing both free-living and parasitic oxygen-consuming bacteria. Thus, some two billion years ago, the ancestral mitochondria invaded free-living cells likely to consume their ATP. However, those early host cells exploited the energy provided by the new residents until they could actually drop their own energy-producing mechanisms. Therefore, a relationship that started parasitic ended up mutualistic, and mitochondria became incorporated into their host's physiology and genome and have been ever since.

17. Retroviruses store their genome on RNA, and for them to reproduce in a host once they infect one, they need to turn that genome into DNA. As is often the case, when the virus accomplishes this, its DNA becomes incorporated into the host's genome. Then, in the process of the host carrying out its own day-to-day genetic expression needs, the virus's proteins become expressed and begin to take over the physiological function of the host, turning it, essentially, into a virus factory. For this reason, viruses are

often said to "hijack" a cell, and in the retrovirus case, by physically altering the host's DNA, the virus will also hijack any offspring that the host cell produces—if host cell survives *to* reproduce. Naturally, for a host, this is not good for fitness, whether that host be a single cell or a multicellular organism.

Chapter 16

Multiple Choice

1. b   2. d   3. d   4. c   5. a   6. d   7. d   8. d   9. d   10. d

Interpret the Data

11. d

Short Answer

12. Yes, even though a behavior is innate (that is, a reliable behavioral response found in all members of a population), the behavior may be polygenic, resulting in a continuous phenotypic trait like running speed, rather than a discrete trait like eye color. So, although all chicks peck at a red dot, some chicks might respond more quickly, some might respond more forcefully, and some might respond for a longer period of time than others. Or individuals may vary in their response threshold to the contrast of colors. In either case, behavioral ecologists studying a particular innate behavior might look at the proximate mechanisms—the neurological or genetic framework—that is the underlying basis for the variation.

     The innate behavior may have an ultimate foundation, however, that led to its evolution. For example, pecking at a specific point on an adult's bill may elicit a feeding response—the sooner the chick can get fed, the sooner the adult can get back out hunting for more food, and the faster the chick will grow, leading to heavier offspring, better survival, and increased fitness. (Why that spot specifically would elicit feeding would be a proximate study.)

13. Individual selection is the differential success of alleles that arises when individuals carrying the allele perform better than other individuals in the same population carrying different alleles. Group selection is the differential success of alleles that occurs when individuals are members of groups, and some of these groups survive and spread better than other groups. Often, the directional effects of individual and group selection conflict, as occurs when selfish behaviors benefiting an individual are harmful to group success. Hamilton's rule reconciles this conflict for situations where individuals in groups are closely related. Because individuals may pass on their alleles to subsequent generations both directly, via their own offspring, and indirectly, through aid they give that increases the number of offspring produced by close relatives, group-benefiting behaviors that would otherwise appear to be altruistic may evolve. For example, individuals may forgo

their own reproduction to help other individuals rear offspring. Alleles causing this behavior may still spread within the population, however, if the individuals receiving the aid are closely related to the helper. Specifically, Hamilton's rule states that a helping behavior will spread in a population if the benefit to the recipient, devalued by the coefficient of relatedness between the helper and recipient, exceeds the cost to the donor, or when $rB > C$.

14. In role-reversed mating systems, where males assume all of the parental care, selective pressures to raise one's own offspring would be high. However, if behaviors in role-reversed systems, such as elaborate courtship with multiple bouts of copulation, increase the certainty of paternity, kin recognition behaviors would be redundant. As a result, the selective pressures for kin recognition as a means to raise a male's own offspring would be low.

     In cooperative breeding systems where male parental care increases offspring survival, the selective pressures to raise an individual's own offspring could also be high, especially if there are costs to the helping male (for example, in terms of other opportunities to mate or even his own survival). Without the certainty of paternity that makes investing heavily in offspring cost-effective, selection should favor kin recognition behaviors. Therefore, behavioral ecologists would predict that male kin recognition would be stronger in these systems than in systems with alternative means for assuring paternity.

15. Eusociality is an unusually high level of social organization that involves a division of reproductive labor among members of a group (that is, when some members of the group are reproductively sterile). Eusociality tends to arise in extreme environments where individuals are not likely to survive outside of a group and where the average relatedness of colony members is very high. For example, in haplodiploid organisms such as ants, full-sibling workers within a colony are more closely related to each other than they would be to their own offspring (were they to produce them).

     Dictyostelium shows a high degree of social organization that involves a division of reproductive labor among members of the group. When the environment becomes extreme (for example, food becomes scarce, and the organisms are close to starvation), the cells join to form a slug, which then grows into a stalk with a patch of spores on top. The stalk cells do not reproduce; they support the spores and die in the process.

     Although these systems are similarly organized in the sense that they each include nonreproductive members of their respective groups, one key difference is the length of time they remain as a social group. Most eusocial species can persist only in groups. *Dictyostelium* cells, on the other hand, disperse and act independently when environmental conditions are good. Only when conditions become poor do they aggregate to form a social group.

16. No. Individual side-blotched lizard males do not "choose" their optimal reproductive strategy. That behavioral

strategy depends on male throat color, and male throat color is a genetic polymorphism that responds to negative frequency-dependent selection. A genetic polymorphism can result from a different allele or combination of alleles of a single gene, or from complex interactions between many different genes and the environment. So, in the simplest case, a single allele could determine whether a male has an orange, blue, or yellow throat. Although the individual male has no "choice" in the reproductive strategy he can employ, his fitness outcome is affected by the other members of the population. Negative frequency-dependent selection raises the relative fitness of a genotype for throat color when it is rare and lowers the relative fitness of a genotype when it is common. If the orange-throated side-blotched males' reproductive strategy dominated both blue and yellow male reproductive strategies, then selection would become directional, favoring orange-throated males. But the winning order in the rock-paper-scissors game in lizards goes like this: orange beats blue, blue beats yellow, and yellow beats orange. So, how well orange does depends on how many blues and yellows are in the population. Same with blue and yellow—their reproductive success depends on how many of the other phenotypes are in the game. The frequencies of the orange, blue, and yellow phenotypes that result from different alleles in each generation fluctuate depending on the parents' relative reproductive success.

The optimal reproductive strategy is a mathematical outcome of game theory models—not an individual outcome—based on the relative payoffs of each strategy. In the case of side-blotched lizards, the payoff is fitness, and game theory provides models of fitness outcomes that predict which color will be relatively more successful depending on the frequency of other colors.

17. The evolution of tool use requires both proximate and ultimate mechanisms whose costs and benefits can be evaluated. For example, tool use requires recognition, goal setting, planning, and innovation. This complex cognition can be affected by context-specific patterns of gene expression, neural signaling, and physiology, all of which can come with costs that may affect expression of other genes, signaling pathways, or even the cell repair mechanisms leading to senescence. Research examining these types of proximate mechanisms has shown that there are indeed costs to the evolution of learning; fruit flies selected to learn quickly had shorter life spans than fruit flies that were not selected to learn quickly. Likewise, mammalian brain size relative to body size tends to be higher in species that are more social—where tool use may be more common. Large brains can be quite costly to maintain (for example, brain temperature, oxygen regulation) and reproduce (for example, live birth, parental care).

Understanding why tool use evolved is also related to the costs and benefits of that adaptation. Phylogenetic relationships offer insight into shared ancestry, but the evolution of tool use within lineages is inconsistent. Not all animals with the faculties for complex cognition develop

tools, and not all environments or ecological challenges may favor the evolution of tool use. For example, dogs and chimpanzees are both highly social species, but only dogs have the capacity to understand behaviors in other species (pointing by humans). Learning to use tools may evolve in unpredictable environments where a species cannot rely on innate responses. There are obvious costs to living in unpredictable environments (for example, food supply), but the benefits may be less clear (for example, enhanced nutrition). Similarly, the capacity for living in groups may lead to the ability to use tools, but sociality comes with its own costs and benefits. Natural selection favors learning and tool use only when the benefits outweigh the costs, and costs and benefits differ among species. So understanding costs and benefits may provide useful insights into the course evolution has taken in the development of these proximate and ultimate mechanisms.

Chapter 17

Multiple Choice

1. b   2. d   3. d   4. d   5. d   6. c   7. a   8. d   9. d   10. d

Interpret the Data

11. d

Short Answer

12. Living primates are adapted to diverse habitats and foods, and thanks to the concept of uniformitarianism, that diversity is highly valuable to understanding the historical past. Uniformitarianism is the idea that the natural laws that we can observe are operating the same way they did in the past. Scientists can compare similar features between these living relatives and our ancestors, and they can develop hypotheses about the habitats and foods that our hominin ancestors had to contend with. For example, scientists can examine diets and physiology of baboons and their survival during drought and hypothesize that our shared common ancestors may have faced similar issues in their evolution. They can use the relationships among our common ancestors to determine where and when these adaptations may have evolved, and they can use other lines of evidence, such as biogeography and ecology, to gain valuable insight into how our hominin ancestors adapted and survived.

13. Yes, the common ancestor of chimpanzees and early hominins likely used tools. Chimpanzees learn and use tools, and there is ample evidence that early hominins did so as well. In fact, different types of tools can be associated with different hominin species. Although tool use evolved independently a number of times in a number of different amniote lineages, several factors point to a shared ancestry. The evolution of tool use may be associated with lineages that already have complex social cognition and unusually

large brains. Evidence indicates that early hominins lived in social groups, like chimpanzees, and shared behavioral and physiological adaptations for living in those circumstances. Indirect evidence also indicates that even the cultural evolution of tools across chimpanzee groups is similar to the cultural evolution of tools in modern-day humans. So the most parsimonious explanation for these shared abilities in chimpanzees and early hominins is that the common ancestor of the two lineages could also use tools.

14. Both the fossil record and DNA evidence indicate that *Homo sapiens* split from *H. heidelbergensis* and not *H. neanderthalensis*. Fossil evidence indicates that *H. heidelbergensis* spread from Africa across Europe about 600,000 years ago, before the evolution of *H. sapiens* (before the oldest fossil evidence belonging to *H. sapiens*). The oldest fossils of *H. neanderthalensis* were found in Europe and date to about 300,000 years ago. Neanderthals represented a new lineage distinguishable from *H. heidelbergensis*, and the species spread across the continent until relatively recently. Bones and artifacts have been discovered in a rock shelter in Jordan from Neanderthals who lived between 69,000 and 49,000 years ago and in a cave in Spain from 40,000 years ago. The oldest fossil evidence belonging to *H. sapiens* dates back to 200,000 and 160,000 years ago from Ethiopia, indicating that *H. sapiens* and *H. neanderthalensis* existed at the same time.

   DNA evidence provides additional support for the hypothesis that Neanderthals and humans represent two separate lineages of *Homo* descending from a common ancestor. Comparisons of alleles between the Neanderthal genome and the human genome point to a shared common ancestor that lived about 400,000 years ago. Additional DNA evidence indicates that Neanderthals also share a common ancestor that excludes humans. All of the humans share a recent common ancestor who lived (by molecular clock estimates) roughly 150,000 years ago. And variation in the DNA of living humans shows that Africans have the greatest genetic diversity, suggesting that *H. sapiens* evolved in Africa and then expanded out to other continents.

15. Humans are the only species with a language that can be spoken and written, but the capacity for language likely evolved early in our lineage. Our understanding of our language comes from our legacy, but recent research has shown that Neanderthals may have had language, too. In reality, many of the factors that likely led to the evolution of complex language in our own species influence other species as well. For example, dolphins and some birds live in highly social groups, and they have developed the capacity to communicate (albeit through languages we are just beginning to understand). So the capacity for complex language may not be unique to humans.

16. Genetic drift probably has been very influential in human evolution. It took place when small numbers of people dispersed to colonize new areas (founder events) and when local populations shrank during periods of hardship or disease (bottlenecks). The descendants of the crew of the *Bounty* on Pitcairn and Norfolk Islands, for example, suffer migraines at much higher rates than the rest of us—a disorder linked to an allele on the X chromosome. Although the crew of the *Bounty* represents an obvious case of a small and isolated population, humans traveled in bands across vast expanses of habitat that likely had the same effect on genetic diversity. Genetic drift would have been affecting allelic diversity in our lineage.

17. Although we humans may think we have conquered most of the predators that shaped our evolutionary past, we are still subject to selection—and our lineage is still evolving. Indeed, our genomic legacies are likely influencing our responses to factors that influence our survival and reproduction, such as the characteristics we prefer in our mates, the strength and quality of our social bonds, our risk to particular diseases, our susceptibility to allergens, and even our reactions to specific medicines. Our evolutionary path cannot be predicted, however. Scientists can only look at our past and offer possibilities for our future.

Chapter 18

Multiple Choice

1. a   2. b   3. b   4. d   5. d   6. a   7. d   8. a   9. b   10. d

Interpret the Data

11. d

Short Answer

12. A virus's success depends on both its replication within hosts and its ability to infect new hosts. Within hosts, genotypes that replicate quickly do better than their rival genotypes that replicate slowly. To infect new hosts, however, genotypes with reduced virulence do better than genotypes with high virulence because these latter genotypes tend to kill the host before they can spread.

   The Ebola virus is extremely virulent in humans. Once a human is infected, the virus is transmitted through contact with body fluids—not through the air—so it spreads quickly within families and villages, where people have direct contact with body fluids, but it is not likely to spread much beyond that. Historically, this mode of transmission coupled with high virulence reduced the likelihood that the virus would spread beyond the immediate local area of outbreak. Quarantine would have been a highly effective control measure.

   In 2014, the spread of Ebola quite likely resulted from a combination of factors similar to the 1918 flu pandemic. The crowded and unsanitary conditions in the affected region permitted the virus to infect new hosts—increased population density made it easier than in the past for fluids to come into contact with many different people. And infected people were able to move from place to place (including health-care workers from around the world) much faster and farther than in the past. Thanks to unprecedented population densities and rapid long distance travel, this Ebola strain was able to spread despite being highly virulent. The spread of Ebola was controlled only when local quarantines and travel restrictions were enforced (the development of vaccines also helped).

13. Yes, the *HbS* allele could be considered an adaptation to malaria. Adaptations are heritable traits that allow an individual to outcompete other individuals that lack that trait. Although individuals carrying two copies of the *HbS* allele suffer from sickle-cell anemia, individuals carrying a single *HbS* allele are only one-tenth as likely to get severe malaria and so have a major fitness advantage. This heterozygote advantage is apparent in populations where malaria is prevalent—the *HbS* allele occurs at much higher frequencies than in areas where malaria is less common. In addition, the high level of linkage disequilibrium in the DNA surrounding the *HbS* allele indicates strong recent natural selection. So, although the *HbS* allele can reduce fitness in homozygotes, clearly natural selection is acting to maintain the *HbS* allele in populations regularly exposed to malaria.

14. For an evolutionary response to natural selection, three conditions are necessary: (1) traits must vary among individuals; (2) that variation must have a heritable component; and (3) individuals possessing certain traits must fare better than individuals possessing other traits. This model can be applied to the population of cells within a cancerous tumor and the evolution of resistance to chemotherapy. Once an oncogene mutates to a proto-oncogene within a cell, that cell can grow and divide faster than its neighbors. The mutation gives the cell an immediate fitness gain, driving the evolution of increased cell division and metastasis at the host's expense.

Resistance to chemotherapy and drugs can evolve in a similar way. Some cell lineages may be less susceptible to drugs than others, or mutations may arise that confer resistance. Once this kind of heritable variation arises, natural selection will strongly favor individual cells that can resist drugs, leading to faster growth than in susceptible cancer cells. Ultimately, these resistant cells will dominate the tumorous growth. Paradoxically, the evolution of these traits leads not only to the death of the host but also to the death of all the cancer cells.

15. There are several hypotheses about what is contributing to our modern struggles with nutrition-based diseases. Most of them consider what humans were likely to have been eating some 300,000 years ago when we first came about and then how those habits changed after we developed agriculture (about 10,000 years ago). Speculation is that in our hunter-gatherer days, an attraction to sweet, calorie-dense food would have been adaptive, as gorging oneself beyond reason was a good idea if you didn't know what your next food source was going to be. In other

words, it was an open-ended craving that was good for consuming empty calories every so often, but this same craving later became "mismatched" with our modern world when humans began to have access to high-calorie foods all the time. Nevertheless, such a "thrifty genotype" hypothesis (and its derivatives) hasn't held up as well with recent research. For one, it seems that hunter-gatherers had no need to crave sugary food for the days they wouldn't be able to eat: if they didn't have food, they could just travel until they found some more. Similarly, there is no early-life fitness advantage for the genes behind type 2 diabetes and obesity. In fact, other research indicates that fat storage would have been detrimental to survival; it wouldn't be until after we developed weapons that that selective pressure flattened out. So in the end, understanding our early physiology presents more questions about obesity and type 2 diabetes than it answers.

16. We know that tumors represent a population of cancerous cells that vary in their sensitivity to conventional chemotherapeutic treatments. As such, when a conventional treatment regimen is applied—such as continuous administration of a drug—the tumor's cells evolve under directional selection, pushing the population to become, on average, more resistant with each generation. In 2018, researchers published a different regimen that accounted for this challenge. Instead of giving a continuous dose, doctors applied short pulses of the drug, bottlenecking the tumor population. Then they paused treatment. The tumor initially shrunk in size, but it did grow back slightly during the pause. However, instead of a tumor composed of mostly resistant cells, cells with lesser resistance also grew back. Recall why this might be important: often, the adaptations that give resistance reduce general fitness without the offending selective pressure. And so when the researchers treated the tumor again, it still remained susceptible to the initial chemotherapeutic dose because nonresistant cells were able to come back and maintain higher fitness in the population than the resistant cells. Impressively, the study indicated that this technique halted tumor growth in 10 out of the 11 patients treated. In addition, this regimen used only 47% of the standard chemotherapeutic dose, meaning that side effects to the patients were reduced compared to conventional treatment.

17. The chapter lists two examples: either colorblindness on the island of Pingelap or a prevalence of Ellis–van Creveld syndrome in the Amish community of the United States. In both cases, when the populations lost genetic diversity—through either isolation or a bottleneck effect—any deleterious mutations that resulted from genetic drift became a significant part of the genetic makeup of the population. In the course of breeding from the same isolated pedigrees, the likelihood of inheriting one or two copies of the deleterious mutation increased for each generation.

Index

Page numbers in bold indicate glossary terms. *Page numbers in italics* indicate boxes, figures, and tables.

765

operons, 154
Ophioglossum, 146
opossums, 125, *127*, *434*, 435–6, 437–40, *439*, 513, *514*, 515
opsins, 337–9, *339*
optic tectum, *610*
optix gene, 567, *567*, *569*
orangutans
    foot of, 124, *124*
    great apes evolution and, 274–6, *275*
    sperm competition in, *421*
origin of species, *see* speciation
orioles, 597
Ornithischians, *131*
Ornithomimosaurs, *131*
Orr, H. Allen, 492–3
Orrorin tugenensis, 641, 643, 644, 644
orthologous genes, 272, 299
orthologs, **332**
Orthrozanclus, 85
OSR (operational sex ratio), *see* operational sex ratio (OSR)
ostriches, *616*
Ottoia, 85
OTU (operational taxonomic unit), 497
outgroups, 109, **109**
oviducts, 392
Oviraptosaurs, *131*
Ovis canadensis, 214, 214–15, 217, 220, 383, 384
ovules, 393–4, 483, *483*
ox (*Bos*), astragalus of, *8*
oxpeckers, *551*, 551–2
oxygen
    cyanobacteria and atmospheric increase in, 78
    isotope ratios in fossil and modern cetaceans, 10, *10*, 72
    physiology for low levels of, 295–7, *296*
    Pseudomonas aeruginosa and, 690
oxytetracycline, 696
oxytocin, *675*, 675–7

p17, *689*, 689–90
p53 tumor-suppressor protein, 457, 724
Pääbo, Svante, 661
Padieu, Philippe, *286*, 286–7, *287*
Paguma larvata, 698, 698–9
pair bonding, 676–7, *677*
Pakicetus, 6, 6–8, *8*, *9*, 10, *10*
paleobiology, 518–21, *519*
paleontology
    animal behavior studies, 66, *66*, *67*
    Cambrian fauna, *85*, 86, *86*
    concepts overview, 53–4, 93
    defined, **35**
    Ediacaran fauna, 83, *84*, 86
    fossil age determination, *64*, 64–5, *65*
    fossil record and, 56, 63–4
    history of life on Earth, 73, 74–5, 76, *76*
    hominin evolution, 639–42, *641*, *643*
    life's early signatures, 77–8
    microbial life-forms, 78, *79*, 80
    movement to land, 86–9, *87*, *88*, *89*
    multicellular life, 80–1, *81*, *82*
    physics application to, 67–9, *68*, *70*
    plants and grasses, *87*, 87–8, 90–1
    of Precambrian, 78, *79*, 80
    sponges, 82–3, *83*
    synapsids, 89, *90*
    teleosts, 89
Paleozoic fauna, 519, *519*, 520

Paley, William, 34
Palmer amaranth, 379–80, *380*
pandemic, 703
Panderichthys, 122
Pang, Kevin, 329
Pang, Tin Yau, 315–16
Pan troglodytes troglodytes, 284, 285
Pappa, Riccardo, 567
parallel evolution, **346**
paralogs, 274, **316**, *316*, 316–17
Paranthropus, 73, 73, 644–5, 645, 649, 665
Paranthropus aethiopicus, 641, 643
Paranthropus bosei, 641, 643, 648
Paranthropus robustus, 641, 643, 648
parapatric speciation, 481, **481**
paraphyletic groups, 106, **106**, *107*
parasites
    coevolution and, *553*
    "egg parasites," 558, *559*
    genomic, 580–2, *581*
    Red Queen effect and, 397–9, *398*, *399*
    virulence and trade-off hypothesis, 560–1, 564
parasitoid wasps, 368–9, *369*, *552*, *554*, 554–6, *556*
parental care
    in anolis lizards, 441–2, *443*
    concepts overview, 435, 460
    in Eurasian penduline tits, 450, *450*, 452
    genomic imprinting, *454*, 454–6, *456*
    in guppies, *440*, 440–1, *441*
    intralocus sexual conflict, 456
    life history, *see* life history
    menopause, *458*, 458–9
    miscarriages, 447
    offspring number adjustments, 446–7, *447*
    offspring–offspring conflict, 452, *453*, 453
    in opossums, *434*, 435–6, 437–40, *439*
    parental conflict, 450, *450*, 452
    parent–offspring conflict, *452*, 452–3
    parent-of-origin effect, *454*, 454–5
    in sand gobies, 447, *447*
    sex ratio adjustments, *448*, 448–51, *449*, *451*
    sexual selection and, *402*, 402–3
    in Seychelles warblers, *449*, 449–50
    in syngnathids, *445*, 445–6
    in wattled jacanas, 444, *444*
    see also lifetime fitness
parental conflict, 450, **450**, *450*, 452
parental investment, 443–4
    female-biased operational sex ratios, *444*, 444–6, *445*
    female–female competition, 446
    offspring number adjustment, 446–7, *447*
    sex ratio of offspring adjustment, *448*, 448–51, *449*, *451*
parent–offspring conflict, **452**, *452*, 452–3
parent–offspring regression, 232, *233*, 240–1, *241*
parent-of-origin effect, **454**, *454*, 454–5
parsimony, *111*, 112, **112**, *113*
parthenogenesis, **393**, 395
Parus major, 594
Pasteur, Nicole, 178
paternity, certainty of paternity, 402, *402*
pathogens
    development of, *690*, 690–1
    evolution of, 688–91, *689*, *690*
    host molding by, *706*, 706–7, *707*
    Red Queen effect and, 397–9, *398*, *399*
    virulence of, *691*, 691–2, *692*
    see also disease
Pax-6 gene, 346

peacocks, *404*
Pearson, Karl, 169, *169*
Pecoits, Ernesto, 83
pedigree *F* and population heterozygosity, 216–17
pedipalps, *408*, *409*, 411
pelicans, "fish-herding" and, 616–17
Pellmyr, Olle, 571–2
Peromyscus maniculatus, 421, 421
Peromyscus polionotus, see oldfield mice
pesticide resistance, 177–8, *178*, *179*, 377–81, *378*, *379*, *380*
Peters, Shanan, 520
Phaseolus vulgaris, 243
phenotypes
    defined, **23**, **163**
    environmental effects on, 167–70, *168*, *169*
    extended, 367
    fitness and reproductive success, 192–3
    genetic polymorphism and polyphenism, *166*, 166–7, *167*
    height, 167–70, *168*, *169*
    quantitative traits and, 229, *230*
    selective neutrality, 193
    see also quantitative genetics
phenotypic plasticity
    in *Caenorhabditis elegans*, 256, 258, *259*
    defined, **171**
    described, 229, *230*, 260
    in *Drosophila*, 259–60, *260*
    environmental response, 171–2, 256, 258
    height and, 672
    Lamarck and, 39
    phenotypic variation calculations for, 256, 258
    plants grown in low vs. high light environments, *255*, 255–6, *257*, 258
    reaction norm, *257*, 258, *259*
    in snowshoe hares, 255–6, *256*, 260
phenypressin, *675*
Phillips, Ben, 382
Phipps, Constantine John, 490
Phoenix retrovirus, 580
photoreceptors, 336, 337, *337*
photosynthesis and fossil biomarkers, 72–3, *73*
phylogenetic species concept, 469, **469**
    microbe application of, 497–8
phylogenetic trees
    of cetaceans, 8, *9*
    construction of, 101–3, *102*, *103*
    of domains, 78, *79*
    of H3N2 influenza, *19*
    reading of, 103–5, *104*, *105*
phylogeny
    carnivoran phylogeny reconstruction, *108*, 108–10, *109*, *110*, *111*, *112*, *113*
    of cetaceans, 8, *9*, 12
    characters and, *see* characters, phylogenetic
    consensus trees, 112, *113*
    defined, **8**, 101, **101**
    endosymbiosis and reconstructed phylogenies, 577, *577*
    evolution as tree of life, 101, 123, 499, *499*
    extinction risk and, *114*, 114–15, *115*
    feathered dinosaurs and bird evolution, *130*, 130–4, *131*, *132*, *133*, *134*
    fossils and timing of evolution and, *116*, 116–17
    gene trees and, 272
    of hominins, 641–2, *643*
    homology and, *124*, 124–5, *125*, *126*, *127*, 127–8, *128*
    of Lassa virus, 686–7

Snowshoe hares have evolved adaptations for staying well camouflaged throughout the year, even as their background changes over the seasons. Natural selection has adapted populations of snowshoe hares to the timing of snowfall and spring plant growth in different regions. Unfortunately, climate change in recent decades has raised temperatures worldwide. In parts of the snowshoe hares' range, snow arrives weeks later in the fall. Some hares now turn white weeks before the arrival of snow, and these mismatched animals are easy for predators to spot. In *Evolution: Making Sense of Life*, Third Edition, we tell the story of scientists who study how hares are now evolving to adapt to this changing world.

BUDGET BOOKS This book does not have a traditional binding; its pages are loose and 3-hole punched to provide flexibility and a low price to students.

Cover images:
photos_martYmage/iStock/Getty Images (front);
Art Wolfe/Science Source (back)

AUTHENTIC

macmillan learning

macmillan learning
macmillanlearning.com

ISBN-13: 978-1-319-23522-2
ISBN-10: 1-319-23522-0

90000

9 781319 235222